Women in the Third World

Garland Reference Library of Social Science (Volume 760)

Advisors

Neuma Aguiar
Instituto Universitário de Pesquisas de Rio de Janeiro
Rio de Janeiro, Brazil

Almaz Eshete
Addis Ababa University
Addis Ababa, Ethiopia

Vandra Masemann
Florida State University
Tallahassee, Florida

Filomina Steady
University of Sierra Leone
Freetown, Sierra Leone

Irene Tinker
University of California at Berkeley
Berkeley, California

Virginia Vargas
Centro de la Mujer Peruana Flora Tristan
Lima, Peru

Women in the Third World

An Encyclopedia of Contemporary Issues

Editor
Nelly P. Stromquist

Assistant Editor
Karen Monkman

Garland Publishing, Inc.
A member of the Taylor & Francis Group
New York & London, 1998

Library of Congress Cataloging-in-Publication Data

Women in the Third World : an encyclopedia of contemporary issues /
    edited by Nelly P. Stromquist.
            p.   cm. — (Garland reference library of social science ; v.
760)
        Includes bibliographical references and index.
        ISBN 0-8153-0150-2 (alk. paper)
        1. Women-Developing countries—Social conditions.   2. Women—
Developing countries—Economic conditions.   3. Women—Legal status,
laws, etc.—Developing countries.   I. Stromquist, Nelly P.   II. Series.
HQ1870.9.W6548   1998
305.42'09172'4—dc21   98–14689
                                                                                CIP

Cover art: Woman at the market, Tichit, Shan State, Burma. Photography by
        Jean Léo Dugast/Panos Pictures, London.

Cover design: by parlour, New York.

Printed on acid-free, 250-year-life paper
Manufactured in the United States of America

Contents

IX. Movements for Change

X. Geographical Entries

Acknowledgments

Numerous are the persons and institutions whose support was crucial to the production of this encyclopedia, a journey that took three years from the time I first received a letter of invitation from the publishers.

My first thanks go to Garland Publishing for inviting me to embark on the long, tortuous, but also exhilarating venture of producing an encyclopedia. By the time it ended, it had involved the participation of more than 200 scholars and activists, either as authors and reviewers or as sources of information about pertinent scholars. Marie Ellen Larcada of Garland Publishing also receives my deep gratitude for her unending support. The careful and consistent copy editing by Mary Cooney and the work of managing editor Helga McCue are gratefully acknowledged.

The School of Education at the University of Southern California, Los Angeles (USC), through Professors William Rideout and Robert Ferris and Dean Gilbert Hentschke, manifested enormous generosity in providing me with access to clerical and communication support during most of the project. The Ford Foundation provided critical financial assistance in the last stage, to accelerate the editing and communication work necessary to bring the entries to a polished condition. Ford Foundation officials Janice Petrovich and Mary Ann Burris demonstrated a strong conviction that the encyclopedia project had merit and provided partial financial support for its completion.

To the authors of the encyclopedia entries go my deepest thanks, not only for writing their essays so readily and ably but also for demonstrating the commitment it takes to produce syntheses that are difficult in nature, as was the attempt to cover the various regions of the developing world. In many cases, this called for the contributors' willingness to expand their own knowledge and understanding. Compiling these essays entailed constant searching for key topics and for authors to address them. Since the encyclopedia covers many issues with distinct disciplinary emphases, in the search for authors I was called upon to expand my network of contacts of academics and activists in the social sciences, a network that had emphasized my own field—international development education—and to move into circles of academicians and activists I did not know before.

Beyond the search for authors, another major effort was made to ensure that all entries would be reviewed by peers cognizant of the topics under their responsibility—a quality-control strategy. In most instances, each entry was reviewed by two peers, who generally demonstrated considerable rigor in their expectations for the encyclopedia contributions, not hesitating to identify omissions, weaknesses, and additional aspects to be covered in the entries they read. It was my pleasant surprise to realize the willingness manifested by most of those contacted to serve as peer reviewers. In all, 74 persons served as peer reviewers; their names and institutions grace the first pages of this encyclopedia. Without the contribution of the authors and the complementary support of the peer reviewers, this encyclopedia would not have seen the light of day.

At USC, I had not only the support of colleagues as informal critics but also the continuous support of advanced doctoral students who volunteered as critics of chapters in their area of competence. Indispensable in the effort to complete the encyclopedia has been the work of Karen Monkman, who started as research assistant and ended in the well-deserved position of assistant editor. Her ability to remind authors to be on time, to suggest pertinent reviewers, and to keep the process of manuscript completion and editing moving along was extraordinary; her contribution as an author was equally valuable. Richard Gilbert, another advanced doctoral student, furnished instant problem solving of erratic-disk and word-processing difficulties. (Both Karen and Richard received their doctorate in 1997.) Margaret Cyrus in the Division of Policy and Administration provided constant clerical support in retyping manuscripts with meticulous care. Her patience, precision, and sisterhood were indispensable. Mary

Orduno, Cassandra Davis, and Quiana Jackson from the same division, always found time and energy to lend a hand in the copying and reformatting of chapters when diskettes or language necessitated extensive reworking.

Finally, I wish to thank my husband, Eric Stromquist, who throughout this journey was willing to review entries written by contributors whose primary language was not English. The long hours involved in rephrasing and capturing meaning resulted in crisp and accessible manuscripts. My daughter, Pia, and my mother-in-law, Anne, made sure I always had time free from domestic work to concentrate on the production of the encyclopedia. Karen Monkman, in turn, extends her gratitude to her husband, Jim Kotz, and her daughter, Gaelin for their support and patience. This work, which embodies the love, assistance, and trust of many, will have been worth these efforts if its readers now find they can use it as a comprehensive, timely, and pertinent reference source.

Preface

This encyclopedia owes its intellectual origins to Gail Kelly, professor in comparative education at the State University of New York College at Buffalo, who several years ago—before death claimed her in the most dynamic years of her scholarly work—envisaged this publication. Through her steadfast commitment to expanding the knowledge of gender, as well as her own scholarly works on the subject, she was aware of the need for a comprehensive volume covering the contributions and initiatives in the field of gender and development. Her pioneering notion has proven prophetic. As women have faced constantly shifting economic and political terrains and gone through four (as of 1996) international world conferences on women, there is a growing and inescapable need to understand the increasing complexity of new information and developments, to cross disciplinary boundaries, and to gain knowledge of the many geographic settings. The initial topics Gail identified for the encyclopedia have been kept, and others have been added based on more recent events and circumstances. Were Gail with us today, she would have been surprised at the length of the new list, but, I am certain, she would have agreed about the importance of the added entries.

It was a daunting task to locate scholars in such a wide range of fields, made all the more difficult because my initial desire was to have as authors primarily scholars from the Third World, expressing their own voices and perceptions. I found busy people, involved in many projects, but many of them were willing to become part of the encyclopedia's effort. The enterprise had uneasy moments given its comprehensive focus and the international nature of the collaboration, but it was also a pleasant opportunity to make friends and learn about the work of numerous feminist researchers.

Production of the encyclopedia clearly demonstrated the obstacles that researchers, and feminist researchers in particular, face in developing countries. Scholars in the Third World often reflected an empirical knowledge limited to their country of residence since their access to bibliographic materials from other countries and regions is so restricted; relying on poorly supplied libraries and unable to subscribe to journals, researchers in the Third World do not often encounter the luxurious situation of ample access to books, journals, and Internet resources that many of us in industrialized countries take for granted. Further limiting their knowledge base is the precarious state of research funding. Many Third World researchers, even the best known in their fields, do not have the financial resources to conduct detailed or complex investigations; consequently, their database is modest, and crucial pieces of evidence remain fragmentary. Feminist scholars often develop knowledge limited to their own country or, at best, to their region. Few could claim more than a passing knowledge of global situations. In the end, this problem could not be entirely resolved. I have let the voices speak the truth they know.

Each author was given the basic parameters for her essay. All were given freedom, however, to include any additional topics they considered important. Nonetheless, the reader will find substantial variation in approach to the topics and ways of presentation. Persons in the academic world are more likely to cite the available literature and to produce statements linked to empirically-based conditions. Activists, as might be expected, are more likely to identify future than past actions to describe particular struggles and meetings. They also use a language that is more hortatory and certain than the voices of academics for whom contingency factors and qualified statements are more frequent. Their work tends to be short on theory; on the other hand, their contributions are extremely innovative and identify emergent issues. In some cases, the depth of the discussion was affected by the newness of the problem in the Third World. For such topics as urban space and women, environmental activism, women's control over their bodies, and new labor processes, the body of research in less developed countries is just shaping up. In some areas, such as education, health, and law, there is considerable research. As the editor, I had to make the decision about when to accept significant variation from expected research norms. I also had to decide that entries, in some cases, would not be supported by extensive references or that the writing style would be more that of an essay type than an aggregation of research findings. I decided to accept less-than-orthodox entries when they contributed innovative perspectives and high quality in other respects.

The contributors to the encyclopedia are a combination of established scholars, well-known activists, and new figures in the feminist literature. The reader is invited to peruse their biographical descriptions after the preface of this volume. To ensure that the entries satisfied the content expected from them, the mechanism of peer review was necessary. This meant that, in general, in addition to the encyclopedia's editor and assistant editor, two scholars reviewed each entry. Reviewers were selected not only for their ability to represent the voices and experiences of women in Third World countries but also because of their

known expertise in the area under discussion. Their work was swift and honest—they minced no words in their critiques. For several authors, this meant extensive revisions that in the end produced stronger chapters.

The encyclopedia's voices are different in numerous ways, including the territorial setting they address and the approach they bring. Entries run the gamut from synthesis of the literature, to interpretation of current events, to advocacy of special issues. In several cases, the reader will find new ways of looking at old problems, and efforts to set new conceptual frameworks. In the end, I decided to let these voices maintain their specificity. While it means that the entries do not always offer a uniform approach, they represent valuable experiences and positions concerning the topics treated. Finally, an effort has been made to render the encyclopedia as user-friendly as possible. As a result, the reader will often find parenthetical definitions of words, concepts, and events. So, although writing styles differ, all entries attempt to present accessible perspectives about and from the Third World.

Contributors

Jacob Adetunji is a demographer (Ph.D., Australian National University, Camberra) who is a demographic and health-surveys fellow at Macro International Inc. in Maryland. His publications include papers on infant and child mortality and their determinants in Africa. He is studying the causes of death in Africa as well as the male role in family planning and reproductive health in that region.

Samia Ahmad is a doctoral student in the Department of Sociology at the University of Maryland, College Park. Her areas of concentration are demography and development. She received her M.A. from the same department for her thesis, "Female Labor Force Participation and Its Impact on Contraceptive Use in Rural Bangladesh."

Sara Ahmed (Ph.D., University of Cambridge, 1991) is an assistant professor at the Institute of Rural Management in Anand, India. She has been researching the relationship among the state, development organizations, and local communities in the management of natural resources in India. Since joining the institute in 1992, Ahmed has studied the impact of natural-resource degradation on gender relations and the role of women and men in the management of their resource base.

Nahid Aslanbeigui is an associate professor of economics and finance at Monmouth University in New Jersey. Her main areas of research are economics education, women in development, and history of economic thought. She is co-editor with Steven Pressman and Gale Summerfield of *Women in the Age of Economic Transformation* (Routledge, 1994). She was the co-winner in 1985–1986 of the Henry George Prize awarded by St. John's University and the Robert Shalkenbauck Foundation.

N'Dri Assié-Lumumba, an education specialist with a particular interest in gender, Africa, and the African diaspora, teaches at Cornell University in Ithaca, New York. She has taught and conducted research in several institutions, including the Université Nationale de Côte d'Ivoire, CIRSSED (Togo), the University of Houston, Vassar College, and Bard College, and her publications include *Les Africaines dans la Politique: Femmes Baoulé de Côte d'Ivoire* (Harmattan, 1996).

Marta Benavides, a Salvadorian educator, biologist, and theologian, is committed to social justice. In that context, she has worked in rural development and on soil and water recovery. She works with the Women's Environment and Development Program in rural communities in El Salvador to improve the quality of life there; to create a process of personal, family, and community stability; and to help people work toward a sustainable peace.

Rosa Briceño is a Latin American educator who is assisting the Association for Women in Development (AWID) with the planning and implementation of its program in Washington, D.C. During the years 1994–1995, she was senior program associate with Women, Law, and Development International, Washington, D.C., where she worked on the planning and implementation of a number of projects related to the Fourth World Conference on Women.

Birgit Brock-Utne is a professor at the University of Oslo, Norway. She has written extensively in the fields of women's studies, peace studies, and, more recently, African studies. Among her publications are *Educating for Peace: A Feminist Perspective* (Pergamon, 1985, 1989) and *Feminist Perspectives on Peace and Peace Education* (Pergamon, 1989).

Judith K. Brown is a professor of anthropology at Oakland University in Rochester, Michigan. She has coedited books dealing with middle-age women and with wife beating and has done postdoctoral research as a visiting scholar at Bryn Mawr College and at Stanford University and as a fellow of the Bunting Institute. A recent work is *Sanctions and Sanctuary: Cultural Perspectives on the Beating of Wives* (Westview, 1992).

Charlotte Bunch helped create, and is the director of, the Center for Women's Global Leadership at Rutgers University, New Brunswick, New Jersey. A lawyer by training, she has published numerous articles and books on the issue of human rights for women.

Wha Soon Byun is a fellow, and the chief of welfare and family research, at the Korean Women's Development Institute in Seoul, Korea. Her research interests are family relations, family policy, and women's status in the family. Recent publications include "The Influence of Patriarchal Militarism on the Life of Women" (1995), and "L'Exploitation Sexuelle et la Prostitution en Corée: De la Colonisation Japonais à Nos Jours" (1995).

Sônia Alves Calió, who studied geography and urban geography (M.A., University of Paris VIII; Ph.D., University of Sao Paulo, Brazil), is a researcher in the Unit for Women's Studies and Social Relations of Gender at the University of Sao Paulo and has assisted in mobilization activities of low-income women in the industrial surroundings of Sao Paulo. A recent article, "The Absence of Housing Policies Worsens the Social Crisis," appeared in the *Feminist Perspective Journal*, vol. 3, no. 6, August 1994.

Roxanna Carrillo is a Peruvian active in the women's movement for more than 20 years. She was a founder of Centro de la Mujer Peruana Flora Tristán, one of the strongest and oldest feminist institutions in Latin American. Carrillo received her M.A. in political science from Rutgers University, New Brunswick, New Jersey; she currently works as a human rights expert at United Nations Development Fund for Women (UNIFEM).

Elsa M. Chaney, a political scientist, is visiting research scholar at the Center for International and Comparative Studies, University of Iowa, Iowa City. She has spent the past 25 years researching women in development, politics, and employment in Latin America and the Caribbean. Her research centers on household (domestic) workers and older women in those regions.

Suma Chitnis is a sociologist who, until recently, was vice chancellor of Shrimati Nathibai Damodar Thackersey Women's University, Bombay, India. Professor Chitnis has written extensively on education and on women's issues. She was a Fulbright Senior Scholar in Residence at the State University of New York (SUNY) Buffalo in 1989, Jawaharlal Nehru Fellow in the years 1983–1985, and University Grants' Commission National Lecturer in 1981–1982.

Thais Corral is a journalist and activist on issues related to women, health, and the environment. She is co-chair of WEDO (Women, Environment, and Development Organization). She organized Planeta FEMEA, a women-and-environment event at the Global Forum at the United Nations Conference on Environmental Development, Rio de Janeiro (UNCED). She has coordinated a women's environment-and-development radio campaign in Brazil and is the coordinator and founder of the Brazilian women's health and environment network, Rede de Defesa da Especie Humana (REDEH).

Monica Das Gupta, is a senior fellow at the Center for Population and Development Studies at Harvard University, Cambridge, Massachusetts. Her research, reported in numerous articles, has examined demographic processes and social organization in societies, particularly issues of family, gender, and health. Recent works include *Women's Health in India: Risk and Vulnerability* (Oxford University Press, 1995) and *Health, Poverty, and Development in India* (Oxford University Press, 1996).

Jean Davison, of the International Development and Education Association, is an anthropologist and international educator who specializes in gender impact assessments and project design. She is the author of *Agriculture, Women and Land: The African Experience* (Westview, 1988) and *Voices from Mutira: Lives of Rural Gikuyu Women* (Lynne Rienner, 1989), as well as a second edition of *Voices from Mutira: Change in the Lives of Rural Gikuyu Women, 1910–1995* (Lynne Rienner, 1996). She has completed the manuscript for another book, *Gender, Lineage, and Ethnicity* in Southern Africa (Westview, 1997).

Sonalde Desai is associate professor of sociology and faculty associate at the Center on Population, Gender, and Social Inequality at the University of Maryland, College Park. Her primary research interests are social demography, gender and development, and the politics of health and population policy. Recent publications include *Gender Inequality and Demographic Behavior: India* (Population Council, 1994).

Anita Dighe is a senior fellow at the National Institute of Adult Education, New Delhi, India. With her Ph.D. in communications from Ohio State University, she has conducted her research mainly in the areas of communications and adult education. In recent years, though, she has focused exclusively on women and literacy issues. She has coauthored two books and has written many articles for professional journals in India and abroad.

Graciela di Marco received her *licencia* in sociology from the Catholic University of Argentina, where she is pursuing a doctorate in the same field. Her interests center on issues of power, social action, and democratization. She is a full professor at the Universidad de la Pampa, where she teaches gender theory and citizenship.

Noemí Ehrenfeld Lenkiewicz is an M.D. and researcher at the Hospital General Manuel Gea González at the Universidad Autónoma Metropolitana in Mexico. Her research interests include adolescent pregnancy; induced abortion and sequelae in young women; violence and rape and their impact on reproductive, sexual, and mental health; and early parenthood and social meanings for adolescent boys in Mexico City.

Nagat El-Sanabary, an Egyptian-American scholar, received her Ph.D. in educational planning and administration from the University of California, Berkeley. She has held the positions of university professor, academic administrator, consultant to government and international organizations, and women-in-develoment adviser at the U.S. Agency for International Development (USAID). Her primary area of research is girls' education in Arab and Islamic countries. (Her consulting agency is Gender Ed Consulting Services in Washington, D.C.)

Judith L. Evans, codirector of the Consultative Group on Early Childhood Care and Development (ECCD), has a Ph.D. in developmental psychology and has been involved in the field of international ECCD for the past 30 years. Her primary interest is in the synthesis and dissemination of knowledge, gained through research and program experience, to planners, policymakers, and practitioners. She has worked and/or lived in more than 25 Third World countries.

Shelley Feldman is associate professor of rural sociology and director of the South Asia program at Cornell University in Ithaca, New York. She has written extensively on issues concerning social regulation, the gender division of labor, and household food production systems, as well as on the political economy of development and change. She is coeditor (with Lourdes Benería) of *Unequal Burden: Economic Crises, Persistent Poverty, and Women's Work* (Westview, 1992) and joint author of *Rural Poverty in Bangladesh: A Report to the Like-Minded Group.*

Susana T. Fried is completing an interdisciplinary doctoral program in political science at Rutgers University, New Brunswick, New Jersey. Fried works on reconceptualizing economic and political development, gender and international political economy, and human rights. Her publications include "From the Third World to the First: Women's Experiences in Microenterprises" in *Equal Means* (forthcoming) and *Women, Culture, and Society: A Reader* (Kendall/Hunt, 1994), which she coedited.

Sandra Mara Garcia is a technical assistant at the Commission for Citizenship and Reproduction, Sao Paulo, Brazil. She holds an M.A. in Gender and Development Studies from the University of Sussex and is currently working on a doctorate in demography at the University of Campinas, Brazil. Her areas of interest include gender

relations, sexuality and reproductive rights, populations studies, and family studies. Her publications include "Women and Population Issues" in *Genero e Cidadania* (Universidade Federal de Alagoas, Edufal, 1994) and "Undoing the Natural Links Between Gender and the Environment," *Revista de Estudos Feministas* (1992).

Zelda Groener, a South African, teaches at the University of the Western Cape, South Africa. She has a Ph.D. from the University of California at Los Angeles. Her academic interests include education policy, international and comparative education, social movements and popular organizations, and gender issues.

Mary E. Haas is a professor of curriculum and instruction at West Virginia University, Morgantown. Her research centers on women and war and how social studies is learned and taught. Published works include *Social Studies for the Elementary/Middle School Student* (Harcourt Brace) and *Toward Excellence: Readings in Social Studies K–6* (National Council for Social Studies).

Manar Hasan, a Palestinian who was born in Galilee, is working towards a Ph.D. at Tel Aviv and Paris VII Universities. She is the author of several articles, including "Growing up Female and Palestinian in Israel" and "Fundamentalism in Our Land." In 1991, she cofounded Al-Fanar, the Palestinian feminist organization that combines the feminist struggle with the national democratic one.

Berta Esperanza Hernández Truyol, (J.D., Union University; LL.M., New York University) is a professor of law at St. John's University School of Law. She teaches international law, international human rights, and civil/human rights. Her extensive publications focus on issues concerning the rights of women; racial, ethnic, and cultural minorities; and indigenous populations. She has also written on issues concerning Cuba.

Pierrette Hondagneu-Sotelo (Ph.D. sociology, University of California at Berkeley) is assistant professor of sociology at the University of Southern California, Los Angeles, where she also teaches in the Program for the Study of Women and Men in Society and in Chicano/Latino studies. She is author of *Gendered Transitions: Mexican Experiences of Immigration* (University of California Press, 1995), and coeditor of *Through the Prism of Difference: Readings on Sex and Gender* (Allyn and Bacon, 1997), and *Challenging Fronteras: Structuring Latina and Latino Lives in the U.S.* (Routledge, 1997).

Benson Honig (Ph.D., Stanford University) is an assistant professor at Tel Aviv International School of Management, Israel. He has conducted research in the Caribbean and Africa, examining the role of education in economic development, including nonformal education, macro policy, and the informal sector, about which he has written and published numerous articles. His research interests include the study of organizations, entrepreneurship, technical education, and the links between a country's educational system and the labor market.

Liang Jun, is associate professor of women's education and vice president of the International Women's College, Zhengzhou University, China.

Jae In Kim (Ph.D., from Ewha Women's University, Korea) is a fellow and chief of education and culture research at the Korean Women's Development Institute. She has conducted research on formal as well as nonformal education of women for more than 10 years. She has written numerous articles and books, including "Gender Equity in Education and Training in Korea" (1995) and "The Impact of Women's Education on the Status of Women in Korea" (1995).

Yanghee Kim is a senior researcher at the Korean Women's Development Institute. She is a social psychologist (Ph.D., Loyola University, Chicago) with interests in ecofeminism, women and mass media, and feminist leadership. She has coauthored *An Introduction to Psychology* (1996) and *Program to Empower Women Television Audiences* (1995).

Young Hee Kim (J.D., Universität Bayreuth, Germany) is a senior researcher at the Korean Women's Development Institute. Her research interests are constitutional law and public law. Recent publications include "State and Economy in the German Basic Law" (1993), and "Verfassungsgerichtsbarkeit in der Republik Korea und der Bundesrepublik Deutschland" (1993).

Young Ok Kim is a senior researcher at the Korean Women's Development Institute. She earned an M.A. in women's studies from the Institute of Social Studies in the Netherlands (1985) and an M.A. in economics from Korea University in Korea in 1991 and is a candidate for a Ph.D. in economics from Korea University. She has authored books entitled *Unstable Situation of Women Workers and the Policy Implications* and *Women Workers' Career Patterns.*

Molly Lee is an associate professor of education at Universiti Sains Malaysia, where she teaches courses in the sociology of education and in science teaching methods. She has written a number of articles on education in Malaysia concerning higher education, private education, science education, teacher education, school curriculum, and language policies.

Virginia W. Leonard, who received her Ph.D. in Latin American history from the University of Florida, Gainesville, is a professor of history at Western Illinois University, Macomb. She has published numerous articles, one of which won the Peter Guilday Prize, and a book, *Politicians, Pupils, and Priests: Argentine Education Since 1943* (Peter Lang, 1989).

Lynellyn D. Long (Ph.D., Stanford University) is a research associate at Johns Hopkins University School of Public Health, Baltimore, Maryland, and at the Population Council, currently working in Vietnam. She has worked in areas of health and population, including HIV/AIDS, refugee and displaced persons and emergency aid, as well as in education and gender issues. Recent publications include *Women's Experiences with HIV/AIDS* (Columbia University Press, 1996), which she coedited, and *Ban Vinai: The Refugee Camp Experience* (Columbia University Press, 1993).

Madhuri Mathema received her Ph.D. from Stanford University. She has served as education officer for the United Nations Children's Fund (UNICEF) in Uganda, where she helped develop an alternative to formal education for disadvantaged children. Her main research interest is the construction and maintenance of gender inequality and its alleviation. She also works toward improving access and effectiveness of education for girls.

Golnar Mehran is assistant professor of education at Al-Zahra University, Tehran, Iran. Her publications have addressed such topics as the education and creation of the New Islamic woman in postrevolutionary Iran; the study of Islamic ideology and its impact on education; female literacy in Iran; and the Islamization and politicization of schoolchildren in the Islamic Republic of Iran.

Kathleen M. Merchant is an assistant professor in the College of Health Sciences at the University of Nevada, Las Vegas. She conducts research in international nutrition, investigating the effects of nutritional deprivation and reproduction on women. She has authored several research articles addressing issues of maternal nutritional depletion and the effect of maternal stunting on maternal delivery complications.

Lisa J. Messersmith (Ph.D., anthropology, University of California at Los Angeles), a specialist in gender and HIV/AIDS, is a fellow and adviser at the U.S. Agency for International Development (USAID) and research associate at Johns Hopkins University, Baltimore, Maryland, currently working in Bangladesh. She is the leading author in an article on sexual behavior in Nigeria in *Health Transition Review,* and wrote a publication entitled *Effectiveness of HIV Counseling and Testing in Changing High-Risk Sexual Behavior: An Annotated Bibliography* (AIDSCAP, Family Health International, 1992).

Maria Mies is professor of sociology at Fachhochschüle Cologne, Germany. Among her areas of work and concern are feminist theory and methodology; the relationship between patriarchy, colonialism, and capital accumulation; the women's movement in the international division of labor; women and ecology; women and work; and the feminist critique of science and technology. Her main publications are *Ecofeminism* (with Vandana Shiva) (Zed, 1993),

Women: The Last Colony (with Veronika Bennholdt-Thomsen and Claudia von Werlhof) (Zed, 1988), and *Patriarchy and Accumulation on a World Scale* (Zed, 1986).

Valentine M. Moghadam is a sociologist with interests in development, gender, the labor force, and women's movements. She is the director of Women's Studies at University of Illinois, Normal, and has conducted research on women, work, and social change in the Middle East and North Africa, on gender and economic transition in former socialist countries, and on nongovernmental organizations (NGOs). From 1990 to 1995, she was senior researcher and coordinator of the Research Program on Women and Development at the World Institute for Development Economics Research, United Nations University, Helsinki, Finland.

Karen Monkman is an adjunct assistant professor at the University of Southern California, Los Angeles, with interests in gender, nonformal education, literacy, and social change. Her dissertation focused on migration processes, gender dynamics, and adult learning. She has worked and done research in Burkina Faso and Ecuador in the areas of language, literacy, and community-based cross-generational nonformal education. She is assistant editor of this encyclopedia, wrote/co-wrote two essays for it, and compiled the Annotated Bibliography.

Robert G. Myers (Ph.D., economics of education, University of Chicago) is co-coordinator of the Consultative Group on Early Childhood Care and Development in Mexico and division director of the High/Scope Educational Research foundation. His writing includes books dealing with programs of early childhood care and development (*The Twelve Who Survive* [1992]; *A Fair Start for Children* [1990]), the organization and conduct of research (*Connecting Worlds of Research* [1981]), and migration of high-level talent (*Education and Emigration,* [1972]).

Cynthia Myntti, an anthropologist with training in public health, is codirector of the Center on Women and Public Policy at the Humphrey Institute of Public Affairs at the University of Minnesota. She has published work, much of it based on field research in the Arab world, on the social determinants of health, the changing use of traditional medicine, and the problems of target-setting in family planning programs.

Maria Nzomo is an associate professor of political economy and international studies at the University of Nairobi, Kenya, and chair of one of the leading national women's nongovernmental organizations (NGOs) in Kenya: The National Commission on the Status of Women (NCSW). She has published widely on democratization in Africa, structural-adjustment programs, the role of multinational corporations in Africa, and women in politics and public decision making.

Chloe O'Gara (Ed.D.), is director of research and gender issues at the Academy for Educational Development. Until 1997 she was a senior population fellow at the University of Michigan School of Public Health and served as the adviser for communication in the Office of Population, U.S. Agency for International Development (USAID). She was formerly the deputy director for the Women in Development Program at USAID. Her publications have focused principally on women in development, women's work and childcare, breast-feeding, and AIDS.

Ila Patel, an associate professor at the Institute of Rural Management in Anand, India, is a specialist in adult education, nonformal education and literacy; broadcast media and development; and women's studies. She received a Ph.D. in international development education and an M.A. in communication from Stanford University, Stanford, California.

Lisa Petrides is a postdoctoral fellow at Educational Testing Service, Princeton, N.J. She obtained her Ph.D. in higher education policy analysis at Stanford University, Stanford, California. Her dissertation research explored differences in experiences, beliefs, and expectations among male and female engineering graduate students. She is cofounder of the Bay Area Girls Education Network, which has designed an action guide for educational access and equity for girls.

Elisabeth Prügl is an assistant professor at Florida International University, Miami, where she teaches international relations (IR) theory, gender and IR, and IR of Europe. She is coeditor (with Eileen Boris) of *Homeworkers in Global Perspective: Invisible No More* (Routledge, 1995). Her latest research explores gendered discourse in the International Labor Organization (ILO).

Jean Larson Pyle is a professor of economics at the University of Massachusetts, Lowell. She specializes in the overlapping areas of economic development, labor, and public policy, with particular attention to gender issues. Recent works include chapters in *Women in the Latin American Development Process* (Temple University Press, 1995) and in *Women in the Age of Economic Transformation* (Routledge, 1994), and a special issue of *Development* (1995). Pyle is a member of the International Association for Feminist Economics.

Agnes R. Quisumbing is a research fellow at the International Food Policy Research Institute (IFPRI), Washington, D.C., where she heads a multicountry research program on intrahousehold and gender issues in food and agricultural policy. Her main research interests lie in intrahousehold allocation and property rights. Prior to joining IFPRI, she was a faculty member at the University of the Philippines and an economist at the World Bank.

Eva M. Rathgeber is regional director for eastern and southern Africa for Canada's International Development Research Center (IDRC). She is based in Nairobi, Kenya. Rathgeber is a social scientist with many publications in the areas of development studies, including higher education, science and technology, policy, and gender and development.

Bryant Robey is an associate in the Department of Population Dynamics and senior writer in the Population Information Program at Johns Hopkins University School of Public Health, Baltimore, Maryland. He is principal author of "The Fertility Decline in Developing Countries" in *Scientific American* (December 1993) and "The Reproductive Revolution: New Survey Findings" in *Population Reports* (Johns Hopkins University Press, 1992).

Martha Roldán is a senior research associate at the Facultad Latinoamericana de Ciencias Sociales (FLACSO) in Buenos Aires, Argentina. She is a sociologist with wide experience in the investigation of labor processes and their linkage to the formation of gender identity and the negotiation of household relations. One of her recent books (coauthored with Lourdes Benería,) is *The Crossroads of Class and Gender* (University of Chicago Press, 1987).

Patricia Ruiz Bravo is a sociologist and a professor in the Faculty of Social Sciences at the Pontificia Universidad Católica del Peru. She is a specialist in the areas of development, regionalism, and gender and a consultant on development projects. The books she has written include: *Género, Educación y Dessarrollo* (Gender, Education, and Development) (UNESCO, 1994), and *Con los Zapatos Sucios: Promotores de ONGs* (With Dirty Shoes: Promoters of NGOs), (Escuela para el Desarrollo, 1993).

Graciela Sapriza, with a *licencia* in historic science (Universidad de la República, Uruguay), is a researcher at GRECMU (Grupos de Estudios Sobre la Condición de la Mujer en el Uruguay) and a faculty member at the Universidad de la República. She has worked in areas concerning social history and women, especially the social and political participation of women.

Violeta Sara-Lafosse is a professor of sociology at the Pontificia Universidad Católica del Perú in Lima. She did graduate studies at the Catholic University of Louvain in Belgium. Her research interests and numerous publications address questions of the family and socialization, women's exploitation, women's organizations, co-education, and discriminatory practices in teacher training. She is one of the founders of the first women's studies program in Peru.

Haneen Sayed is a human-resources economist in the East Asia Department of the World Bank in Washington, D.C. She is a graduate of Stanford University, Stanford, Califor-

nia and Columbia University, New York City. Prior to joining the World Bank, she taught gender and development at Barnard College and Middle East economic development at Fordham University.

Beatriz Schmukler, chair of the master program in political sociology, Instituto Mora, Mexico City, has published extensively on the role of women in social democratization and on the connections between women's participation in community organizations and the transformations of their gender identities. She is editing a book on the changes in gender power relations in families in Latin America (Population Council, forthcoming).

Ella Habiba Shohat is professor of cultural studies and women's studies at the City University of New York Graduate Center. She is the author of *Israeli Cinema: East/West and the Politics of Representation* (University of Texas Press, 1989), the coauthor (with Robert Stam) of *Unthinking Eurocentrism: Multiculturalisms and the Media* (Routledge, 1994), and an editor of the journal *Social text*. Her work has appeared in numerous journals and edited books related to film, culture, feminism, and the Middle East.

Rima Shore is a feminist scholar specializing in women's human rights and domestic violence. She works at the Center for Women's Global Leadership at Rutgers University, New Brunswick, New Jersey.

Neera Kuckreja Sohoni has a Ph.D. in economics and an M.A. in history and public administration. An affiliated scholar at Stanford University's Institute for Research on Women and Gender, she is the author of *Women Behind Bars* (Vikas, 1989), a study of women prisoners; *People in Action* (Vikas, 1990); *Sketches from My Past: Encounters with India's Oppressed* (Translation of Mahadevi Varma's *Ateet Ke Chalchitra*) (Northeastern University Press, 1994); and *The Burden of Girlhood: A Global Inquiry into the Status of Girls* (Third Party, 1995).

Kathleen Staudt, professor of political science at the University of Texas at El Paso, has published articles and edited books on women, development, and politics in Africa and Latin America. The author of *Managing Development* (Sage, 1991), she is writing up research on culture, politics, and informal economies at borders.

Nelly P. Stromquist, professor of education at the University of Southern California, Los Angeles, specializes in international development education, which she observes from a sociological perspective. Her work has addressed questions of gender, equity policy, and adult education in developing countries, particularly Latin America and West Africa. Her most recent books are *Literacy for Citizenship: Gender and Grassroots Dynamics in Brazil* (State University of New York Press, 1997) and *Gender Dimensions in Education in Latin America* (Organization of American States,

1996). She is past president of the Comparative and International Education Society.

Gale Summerfield, associate professor of economics at the Monterey Institute of International Studies, received her Ph.D. in economics from the University of Michigan, Ann Arbor. Her research focuses on economic development, particularly gender aspects of economic reform and environmental interactions. She co-edited *Women in the Age of Economic Transformation: Gender Impacts of Economic Reforms in Post-Socialist and Developing Countries* (Routledge, 1994). She won the Helen Potter Award from the Association of Social Economics in 1994.

Margaret Sutton, on the faculty at Indiana University at Bloomington, is a comparative educator and women's studies specialist. Until 1997 she worked for the Academy for Educational Development in Washington, D.C. as the director of research and gender issues in the Department of International Basic Education. Most of her international work has been based in Asia and has focused on improving educational systems, particularly as they benefit girls and women. She has taught and published in the fields of feminist epistemology, education and cultural change, and educational practices.

Barbara Swirski is an Israeli feminist, director of the Adva Center for the study of Israeli Society, cofounder of Israel's first shelter for battered women (1977), coeditor of *Calling the Equality Bluff: Women in Israel* (1993), and author of *Daughters of Eve, Daughters of Lilith: On the Lives of Israeli Women* (1984, in Hebrew) and *Legal Guide for Women in Domestic Matters* (1987, in Hebrew).

Sharon W. Tiffany, professor of anthropology and women's studies at the University of Wisconsin at Whitewater, conducted her initial field work in Western and American Samoa. Since then she has worked elsewhere in Oceania and Latin America. Her research interests in feminism and ethnographic narratives of gender are reflected in a recent article (coauthored with Kathleen J. Adams), "Housewives of the Forest: Representation in Ethnographic Film," published in a special "Women and Film" issue of *Women's Studies*, vol. 25, 1996.

Zafiris Tzannatos is a senior economist at the World Bank in Washington, D.C. His publications in the general area of labor economics include *Current Issues in Labour Economics* (1990) and *Labor Market Economics*, (1993), which he coauthored. On gender issues, he coauthored *Women and Equal Pay: The Effects of Legislation in Britain* (1985) and *Women in Latin America: Overview and Methodology* (1992) and coedited *Women in Latin America: 15 Country Studies* (1992) and *Out of the Margin: Feminist Perspectives on Economics* (1995).

Maria Elena Valenzuela received her *licencia* in sociology from the Catholic University of Chile and is director of the Women' Unit at the National Service for Women in Chile. She has written on gender, authoritarianism, and democracy; more recently, she has addressed the intersection of gender and poverty. Her latest book is *Genero, Pobreza y Políticas Públicas,* a case study of Chile, cowritten with Sylvia Venegas and Carmen Andrade.

Claudia María Vargas is visiting assistant professor of education at the University of Vermont. She received her B.A. at Pitzer College of the Claremont Colleges (California) and her M.A. and Ph.D. at the University of Southern California, Los Angeles. She coedited *Implementing Sustainable Development* (United Nations, 1995). Her research interest is refugee services, with a particular focus on educational issues.

Moema Viezzer is a Brazilian feminist writer and environmentalist. She has an M.A. in the social sciences, specializing in adult education. She is founder and president of Rede Mulher de Educação, from which she has promoted programs of environmental education at local, national, and international institutions. She has been a facilitator of the Environmental Education Treaty for Sustainable Societies and Global Responsibility.

Carol Vlassoff is a Canadian demographer (Ph.D., University of Poona, India, 1978) working in the World Health Organization with the Special Program for Research and Training in Tropical Diseases (TDR). There, she is responsible for the establishment of research priorities, in collaboration with top scientists from all areas of the world, on gender and tropical diseases. Vlassoff has published on the relationship between human behavior and health, including gender differences in tropical diseases, rapid-assessment methods for health planning, and social constraints on family planning in developing countries.

Sima Wali is president and CEO of Refugee Women in Development (RefWID), a non-profit organization in Washington, D.C. devoted to supporting Third World refugee women in their attainment of social and economic independence. Wali has represented refugee women in national and international conferences. In 1985, she was invited by the U.S. Office on Refugee Resettlement, a federal agency, to advise the Task Force on Refugee Women on policy and program recommendations concerning refugee women.

Shirley Walters is founding director of the Center for Adult and Continuing Education (CACE) and professor of adult and continuing education at the University of the Western Cape, South Africa. Her research interests include gender and popular eduation, adult education, and training policy as related to the establishment of new provincial and national systems of education in South Africa.

Li Xiaojian is president of, and professor of literature and women's studies at International Women's College, Zhengzhou University, in China. In addition, she directs the Women's Museum Preparatory Committee and is chief editor of the *Women's Series*. She writes about women's history and philosophy.

Reviewers

Bina Agarwal
Institute of Economic Growth
Delhi, India

Sara Ahmed
Institute of Rural Management
Anand, India

Sajeda Amin
Population Council
New York, New York

Jeanine Anderson
Pontificia Universidad Catolica del Peru
Lima, Peru

Venkatesh Athreya
Bharathidasan University at Tiruchirapalli
Tiruchirapalli, India

Keziah A. Awosika
Women Law and Development Center
Lagos, Nigeria

Carolyn Benson
Ministry of Education
Maputo, Mozambique

Judith K. Brown
Oakland University
Rochester, Michigan

Mary Ann Burris
Ford Foundation
Nairobi, Kenya

Irene Campos Carr
Northeastern Illinois University
Chicago, Illinois

Martin Carnoy
Stanford University
Stanford, California

Ankila S. Chandran
University of Southern California
Los Angeles, California

Kowsar P. Chowdhury
World Bank
Washington, D.C.

Sheela Rani Chunkath
Government of Tamil Nadu
Madras, India

Jane F. Collier
Stanford University
Stanford, California

Dorothy Counts
University of Waterloo
Waterloo, Canada

Doria Daniels
Rand Afrikaans University
Johannesburg, South Africa

Linda Grant De Pauw
George Washington University
Washington, D.C.

Isabela Cabral Felix de Sousa
Fundação Oswaldo Cruz
Rio de Janeiro, Brazil

Carmen Diana Deere
University of Massachusetts at Amherst
Amherst, Massachusetts

Anita Dighe
National Institute of Adult Education
New Delhi, India

Shari Dworkin
University of Southern California
Los Angeles, California

Susan Eckstein
Boston University
Boston, Massachusetts

Jean Ellickson
Western Illinois University
Macomb, Illinois

Erwin H. Epstein
Ohio State University
Columbus, Ohio

Irving Epstein
Illinois Wesleyan University
Bloomington, Illinois

Shelley Feldman
Cornell University
Ithaca, New York

Nancy Flowers
Amnesty International
New York, New York

Susana T. Fried
Rutgers University
New Brunswick, New Jersey

Rita Gallin
Michigan State University
East Lansing, Michigan

Esther Gottlieb
West Virginia University
Morgantown, West Virginia

Mary E. Haas
West Virginia University
Morgantown, West Virginia

Catherine Hakim
London School of Economics and Political Science
London, England

Julia Havelin
University of Southern California
Los Angeles, California

Berta Esperanza Hernández Truyol
St. John's University School of Law
Jamaica, New York

Susan Horton
University of Toronto
Toronto, Canada

Lynn Ilon
State University of New York at Buffalo
Buffalo, New York

Cecile Jackson
University of East Anglia
Norwich, England

Marsha Kinder
University of Southern California
Los Angeles, California

Virginia W. Leonard
Western Illinois University
Macomb, Illinois

Maria Lepowsky
University of Wisconsin at Madison
Madison, Wisconsin

Cynthia B. Lloyd
Population Council
New York, New York

Lynellyn D. Long
Population Council
Hanoi, Vietnam

Martha MacDonald
St. Mary's University
Halifax, Canada

Vandra Masemann
Florida State University
Tallahassee, Florida

Doe Mayer
University of Southern California
Los Angeles, California

Valentine M. Moghadam
University of Illinois at Normal
Normal, Illinois

Caroline Moser
World Bank
Washington, D.C.

Bertha Msora
University of Southern California
Los Angeles, California

Judith Narrowe
University of Stockholm
Stockholm, Sweden

Jill Nash
State University of New York College at Buffalo
Buffalo, New York

Marysa Navarro-Aranguren
Dartmouth College
Hanover, New Hampshire

Judy Norsigian
Boston Women's Health Book Collective
Boston, Massachusetts

Shannon O'Grady
University of Southern California
Los Angeles, California

Jane Parpart
Dalhousie University
Halifax, Canada

Ila Patel
Institute of Rural Management
Anand, India

Diana Pearce
Wider Opportunities for Women
Washington, D.C.

Patricia Poppe
Johns Hopkins University
Baltimore, Maryland

Elisabeth Prügl
Florida International University
Miami, Florida

Laura Pulido
University of Southern California
Los Angeles, California

Joel Samoff
Stanford University
Stanford, California

Eliz Sanasarian
University of Southern California
Los Angeles, California

Marla Shelton
University of Southern California
Los Angeles, California

Awatef Siam
United Nations Relief and Works Agency (UNRWA)
Ramallah, Israel

Cristine Smith
World Education
Boston, Massachusetts

Neera Kuckreja Sohoni
Stanford University
Stanford, California

Kathleen Staudt
University of Texas at El Paso
El Paso, Texas

Judith Stiehm
Florida International University
Miami, Florida

Madeleine Stoner
University of Southern California
Los Angeles, California

Myra H. Strober
Stanford University
Stanford, California

Gale Summerfield
Monterey Institute of International Studies
Monterey, California

Susan Tiano
University of New Mexico
Albuquerque, New Mexico

Maura I. Toro-Morn
Illinois State University
Normal, Illinois

Carol Vlassoff
World Health Organization (WHO)
Geneva, Switzerland

Sally Yudelman
International Center for Research on Women (ICRW)
Washington, D.C.

Introduction

Nelly P. Stromquist

A preoccupation in the making of this encyclopedia revolved around its title. It certainly has to do with women, and with women who live in certain areas of the world. But what to call these areas collectively? Many women believe that conventions of language can serve to perpetuate undesirable situations. Should the nations that are the focus of this encyclopedia constitute the "Third World"? The "South"? The "developing countries"? The "less-developed countries"? The Third World as a concept originated in the 1950s. At that time, the First World represented capitalism or what the United Nations calls "development market economies." The Second World referred to industrializing socialism or "centrally planned economies." The Third World was the remainder; it also meant the nonaligned countries.

Some scholars, consider both "developing" and "Third World" to be misleading, restraining, and demeaning. They would argue that the term "developing country" devalues a country's cultural, esthetic, historical, and humanistic achievement or capability, imposing, instead, Western norms to assess the level of human progress. Parpart (1995) holds that the term "Third World," when applied to women, is usually taken to mean that they are "backward, premodern persons" or "vulnerable 'others.'"

Other scholars, including some contributors to this encyclopedia, maintain that the term "Third World" is outmoded or inaccurate in that it generalizes across a broad spectrum of nation-states and economic conditions. Still others argue that, with the collapse of the Soviet bloc, the "Second World" is gone, most of it inherited by the "Third," which, therefore, no longer applies. There are no easy solutions.

The term "developing countries" recognizes the multiplicity of nation-states, each with varying levels of socioeconomic development, but the term does not avoid a generalization of countries. A typical developing country is a combination of different sectors at different levels of development with different types of problems (Sengupta,

1993). The term "industrializing countries" recognizes current conditions as part of an evolving process, yet the term emphasizes economic productivity over cultural, social, and ideological features. Further, it implies a linear path, by which less-developed countries will eventually become developed.

Other terms could have been used. "North–South" is another dichotomy that fails to distinguish variation in the North and the South and contains distortions resulting from the location of South Asia, two-thirds of Africa, and all of Central America in the Northern Hemisphere (Merriam, 1988). Other terms such as "less-developed countries" (LDCs) and "newly industrialized countries" (NICs) seem no better because they, too, tend to emphasize economic development.

The "Third World" is a term widely accepted and used by economically poor countries themselves, especially in their negotiations with economically rich nations on critical issues relating to trade, energy, natural-resource depletion, and dwindling food supplies (Todaro, 1985). "Third World" seems to convey an overall subordination to central countries against which many of these nations are struggling. Yet, I do not consider that acknowledging one's present subordination condemns one to inferior status. Rather, the perception of inequalities and disadvantage is usually the point of departure for searches for better self-identities and feelings of self-worth. Despite its limitations, no other concept seems to achieve greater clarity and simplicity. As Merriam (1988) observes, "Third World" is a socioeconomic concept and refers as well to a state of mind that addresses the hopes and aspirations of three-fourths of humanity.

Ayoob (1995) argues well for the continuing relevance of "Third World" as concept and category:

The end of the Cold War has not resulted in making the Third World irrelevant as an explanatory category in international politics. By freeing the concept from its original bondage to the concepts of the First and Second Worlds, the demise of the Cold War has provided it greater relevance as an analytical device with the power to explain much of what goes on in the international system, especially in terms of hierarchy and inequality, order and disorder, and conflict and security. The dismemberment of the Soviet Union and Yugoslavia at the end of the Cold War also has led to the geographical expansion of the Third World, which now embraces Central Asia, the Caucasus, and the Balkans. The relevance of the Third World, or the South, has been reinforced by the end of the Cold War because now the dichotomy between the developed, affluent, and powerful North and the underdeveloped, poor, and weak South is visible in even starker terms than earlier. Formerly, the Second World had allied itself on several crucial issues with the Third World and provided some balance to the West/North in the international system. This gave the Third World political and economic leverage vis-à-vis the industrialized democracies. Moreover, the Second World had provided greater complexity to the international political and economic scene during the Cold War, making the North–South division somewhat fuzzy. The end of the Cold War clearly has juxtaposed the vulnerabilities and insecurities of the South—an increasingly popular synonym for the Third World—with the power and affluence of the industrialized North (Ayoob, 1995, p. 61).

In the end, I retained "Third World" in the title. It is not a perfect term, but it conveys two important aspects. First, regardless of political transformations in the former socialist world, there are vast regions that continue to live in great poverty and to receive unfair treatment from industrialized countries. Second, these regions, although diverse, maintain significant ties of collective identification, seeing themselves as searching for ways that are not totally capitalistic and that still retain good features of a socialist order.

The term "Third World" does raise representations of inferiority and superiority, but it also offers a sense of common identity and a growing unit of purpose (Todaro, 1985, p. xxxvi). The term applies to more than 143 countries of Asia, Latin America, Africa, and the Middle East characterized by low living standards, high levels of population growth, low per-capita income, and general economic and technological dependence on industrialized countries (Todaro, 1985, p. 610).

"Third World" highlights the fact that there is a bloc of countries that share unequal relations in the global economy. It also underscores the internal differentiation within these countries and their great antagonisms along social, cultural, and economic lines. Upper classes of all countries have more in common with each other than with other classes within their own societies.

Actually, if we are going to be sensitive to generalities, we have to admit to two: women and Third Word. Women are certainly not a homogeneous class, yet it is undeniable that women of all classes share some crucial interests. Women in industrialized countries have many unmet needs regarding equality and power in economic and political arenas and also in basic human rights. In former socialist countries, the problem now is employment and keeping proper representation in government and management. In the Third World, women's struggles are severe. Their status is much more marginal as they must fight for access to food, health care, education, and housing, in addition to bearing heavier burdens for house maintenance, child care, and family well-being and having to satisfy social claims on fertility and on being responsible for fidelity in marriage. There are variations to be sure, some due to custom, others to law, still others to religion, but many are mainly the result of poverty and lack of social development.

A Feminist Perspective

This encyclopedia seeks to provide a feminist perspective on issues involved in socioeconomic development of Third World countries. There are multiple "feminist perspectives" in the social sciences and among activist groups. The encyclopedia embodies this diversity, while keeping a solid core of feminist principles. These are: the further detection of women's unequal and unfair conditions in contemporary life; the attribution of these inequalities to policies, decisions, and dynamics derived from beliefs that masculinity and femininity constitute distinct natures; and the firm resolve to correct these inequalities, particularly by envisaging alternative social arrangements that do not impose definitions of self upon either men or women.

Various conceptual and analytical frameworks can be employed to analyze the condition of women. Each of these frameworks places the site of oppression in different locations and advocates its own methods of reform. The encyclopedia has not one voice but many. Thus, in some essays, inequality is clearly traced to pervasive ideologies supported by patriarchal institutions. In others, arguments about the importance of economic and class distinctions dominate. Most often, the essays consider various forms of oppression that arise from patriarchal ideology, as well as its intersection with class, race, and ethnicity. The entries do pay attention to laws and practices and, hence, to behaviors by the state and its multiple public institutions; they also consider that the work of civil society, in the form of either nongovernmental organizations (NGOs), individual agency, or, in some cases, personal autonomy, are just as important in the quest for altered and improved social orders.

Another element of the feminist perspective in evidence in some of the articles is the interaction between the subject or self (the researcher) and her object of research (the women of the Third World). Feminism being a social movement, in addition to providing an analytical perspective, compels adoption of advocacy. This necessitates what Mies has called "conscious partiality" (Mies, 1983, p.

122)—that is, empathizing with the oppressed group but also being conscious that one is taking sides and why. This modification of the conventional subject/object separation is reflected in the work of the activists who produced essays for the encyclopedia in such areas as ecology, women in war, refugee and displaced women, and women in the city. Their work not only identifies problematic areas affecting women but also conveys recommendations for action; it clearly is oriented toward the future—a better one, always.

A feminist perspective is manifested in the encyclopedia in yet one additional form: the identification of gender issues in problems and arenas considered less touched by gender principles. That women's issues are human issues becomes evident as repercussions of many of the latter have strong gender components. Examples of these occur in the entries on human rights, transitions to democracy, anticolonial movements, and multinational corporations.

One of the main contributions of this encyclopedia is that of locating the forms and dimensions of women's subordination in the wide array of issues that constitute everyday life and showing that gender does not function merely as one additional marker in social and cultural transactions but as a force that affects context and process simultaneously, as a major organizing schema. In this respect, gender is subject to a multiplicity of discourses, but the diversity of perspectives does not mean that gender as an organizing principle of society is weak. The collection of essays reveals that gender functions as a massive project of social relations involving the state, the economy, culture, kinship, childrearing, sexuality, and communications.

Gender-sensitive approaches require both men and women to (1) recognize and revalue women's experiences, skills, and contributions to social life, the economy, and culture; (2) share domestic and caring responsibilities, thus promoting women's participation in decision making in economic, social, and political institutions; and (3) participate in society as conscious, active, and responsible citizens.

The Conditions of Third World Women

Since the 1970s, significant advancements have been made in women's access to education, legislation on family rights, employment, professional fields, and representation in political life. The international women's movement has made a noticeable impact upon the social fabric and public agenda, showing that women can and will reconstruct society on the principles of equality, justice, and freedom for all.

Yet, women in developing countries have unsatisfied basic needs in education, employment, health, and the law. Nearly 95 percent of 5–17-year-olds not attending school and 90 percent of the world illiterates are in developing countries. According to the *Human Development Report 1993* (UNDP, 1993), fewer than 10 percent of the world's people participate fully in political, economic, social, and cultural life. Changes have occurred with the expansion of market economies, the growth of multiparty democracies,

and the increased recognition of NGOs. Many people in the Third World, however, continue to be disempowered, and the proportion of women among them is substantial.

Women have low employment rates in the formal sector of the economy, and their aggregate salaries are still lower than those of men. In most countries, women have gained access to the vote but have achieved limited political representation. In 1995 women accounted for a little more than 10 percent of all world legislative representatives and fewer than 4 percent of cabinet ministers. On the other hand, women have increased their ability to control their fertility, and access to the labor market has given them a platform from which to raise more demands.

Women are obviously not a homogeneous group; differences emerge in terms of social class, ethnicity, religious affiliation, and age. Consequently, they face a variety of conditions. Women in Lagos as a group have very little in common with women in San Francisco. And within countries themselves, differentiation may be enormous. Wealthy women living in splendid isolation and in the luxury that affords them many servants on most counts cannot be compared to the women working as maids and food vendors in streets of developing countries. At the same time, there is also a commonality of experiences and interests among women. Similar sexuality issues tend to be faced by women regardless of social class and wealth: Men's control over women's bodies, restrictions about abortion, the fear and reality of rape and sexual harassment, restricted social expectations about fields of study and employment are but a few of the commonalities.

This encyclopedia appears at a moment of intense social, political, cultural, and economic transformation in many parts of the world. Changes, however, do not always mean progress. Since the early 1990s, the world has moved toward a freer and more competitive market. Communism as an ideology has been rejected by former Soviet-bloc countries. The horrors of poverty, inequality, unemployment, rural neglect, and population increase continue, and new, brutal realities have emerged: AIDS, ethnic cleansing, environmental deterioration. These changes affect women, certainly. But women are not mere victims of rapid transformations; their ability to cope with change as either researchers or activists is considerable. And so is their capacity to generate change itself.

World Changes: Positive Impact on Women

Conceptually and, to a lesser extent, programmatically, the field of international development has benefited from a wider definition of its key concept. Development is now increasingly understood to include human development, defined as "a process of enlarging choice for all people, not just for one part of society" (UNDP, 1995, p. 1). This approach intends to be different from the previous efforts to concentrate on narrow measures of economic growth that paid little or no attention to the questions of distribution of benefits and costs implied in that growth. It also intends to be different from "human resource development" efforts that treat human beings as an input into the production process rather and as an end in themselves. The "human development" approach, in contrast, sees individuals not merely as beneficiaries, but primarily as agents, of change (UNDP, 1995).

The concept of "human development" is built on four elements: productivity, equity, sustainability, and empowerment. Human development thus becomes impossible without gender equality. Beginning with the United Nations Decade for Women (1976–1985), many writings have centered on the condition of women in developing countries. These works have been both empirical and conceptual in nature, augmenting our knowledge of the variety and range of women's conditions and our comprehension of the many forces that contribute to create, sustain, and, in fewer cases, transform the gendered nature of people's lives. The application of greater resources from international donors to women's development will create possibilities for bringing women's concerns into the development process. It is hoped that with the ending of the Cold War, both industrialized and developing countries will be able to generate new resources, as savings from armaments and large armies materialize. It has been estimated by U. N. agencies that even a portion as small as 3 percent of these savings would go a long way toward addressing basic needs in education in developing countries.

NGOs, which are vital expressions of civil society, are now recognized as essential partners in the process of development, a position further accentuated in the *Platform for Action,* the document derived from the Fourth World Conference on Women (United Nations, 1996) and clearly revealed in the large and enthusiastic participation of women at that conference. It is also a welcome development that the feminist movement, particularly as represented by women's NGOs, has grown less reluctant to face the question of power. Initially dismissed as a product of the patriarchal order, where hierarchies and force were the rule, power is now seen to have positive, exploitable aspects, particularly in the empowerment of women themselves. NGOs are more mature and powerful actors than in previous decades; their members, more knowledgeable in gender issues, more creative in the production of multifaceted solutions. A key question they must address is how to establish effective alliances between NGOs and the state and other expressions of civil society.

The United Nations Convention on the Elimination of All Forms of Discrimination Against Women (CEDAW), which was adopted in 1979 and has been in force since 1981, must be considered one of the most promising legal mechanisms for change in the social and economic relations of gender. By early 1998, it had been ratified by 161 governments, even though many governments have reservations about some of its articles. Nonetheless, CEDAW enshrines the principles that protect women against discrimination as international law, thus making them universally applicable, regardless of culture and religion. Revolutions in communications, with faxes and e-mail (particularly the latter), have made information less expensive and easier to locate and transmit. The internationalization of communications is making it possible for the powerless and poor—including a large number of women's NGOs—to become more capable of collective organization and less vulnerable to political maneuvers by unfriendly groups.

Women, especially in Latin America, have made contributions to an expanding definition of democracy and citizenship. Their political mobilization has been essential to the return of democratic rule in many countries in the region. This transition has benefited from new ways of conceiving political life and relations. It has shown the need to imagine democratic rule beyond representative and electoral procedures to incorporate sustained dialogue among individuals at micro levels, particularly the home, the school, and the community. The participation of women in social mass movements has also produced a larger segment of female leaders, totally committed to emancipatory claims on existing society.

The informal sector of the economy is attracting and capturing increasingly more women. This phenomenon blurs the lines between household and enterprise with the growth of subcontracting, home-based work, and self-employment. Although this employment often brings with it exploitive conditions for women's work, it also is creating spaces of economic autonomy that women are beginning to use to renegotiate conditions within their households.

The conceptual, theoretical, and practical understanding of gender as a force that shapes the social order in decisive ways has matured and grown increasingly complex. While there are diverse voices within the feminist movement, there is also consensus that women's needs are not limited merely to entering the economy and being able to contribute their talents and productive skills but also include the creation of a new social configuration in which inequalities of all kinds—race, class, ethnicity, and gender—are recognized as arbitrary markers and made to disappear.

World Changes: Negative Impact on Women

Despite currents of triumphalism in many Western nations, particularly the United States, the world is more chaotic than before. While the bipolar capitalist/communist world is no longer, large industrial countries still compete for power and influence in the developing world. The disorder within countries has produced violent acts. Under those conditions, issues of gender become relegated to second and even third place.

Many of the positive changes described above seem to have come with negative counterparts. Notable about this other side of the coin is the emergence of old and unresolved ethnic and nationalistic conflicts in the wake of the Cold War. These unexpected conflicts, particularly those in presumed "developed" countries—that is, the former socialist regimes—have been costly and have become a major object of assistance, requiring peace missions that consume large amounts of funds that could have gone into construction in other parts of the world rather than into reconstruction of broken economies and societies.

A parallel development set off by the end of the Cold War has been the dire need for financial and technical resources by the previously communist countries as they embark on a transition toward market economies and more democratic rule. Again, this has entailed the diversion and consumption of important resources that otherwise would have gone to the Third World.

Industrialized countries agreed over two decades ago to give 0.7 percent of their GNP (gross national product) for development purposes in Third World countries, a commitment restated at the World Conference on Sustainable Development in Copenhagen (1993). Further, it was anticipated that additional funds would be created from the "peace dividend"—that is, the financial resources no longer needed for the U. S. arms race with the Soviet Union. What happened to the peace dividend? Instead of acknowledging additional resources, some of the industrialized countries have been torn by internal conflicts about budget reductions, as if financial resources were smaller than before. Times of scarcity seem to prevail rather than the expected times of riches.

Political changes in the European scene have brought with them an agenda that is less gender-sensitive than anticipated. "Emergencies" and "crises" have a way of being defined as if women do not matter, even though the majority of those who suffer, or at least survive the immediate violence, are women.

On the eve of the twenty-first century, we face a changed environment and additional forces affecting the condition of women in society: global economic restructuring, a new global political and cultural order, neoconservative challenges to the legitimacy of the politics of the state, and a revaluing of the nuclear family giving its women responsibilities for social caring.

In many respects, we live in a world that is moving toward a new dichotomy. As Goetz (1995) remarks, there are two current tendencies in international politics: one toward democratization and recognition of human rights, and the other toward nationalism, religious fundamentalism, and ethnic mobilization. This dichotomy is creating opposing camps and consuming energies for a politics of difference that promises little that is constructive. The religious fundamentalism that is emerging has no ideological boundaries; it is affecting all three major religions: Islam, Hinduism, and Christianity. For women, this religious fundamentalism is ominous, for there are signs of a conservative pact between fundamentalists from the West and the East to restrain women's power and even their aspirations. These conservative forces manifested themselves in recent world conferences (on population and development in Cairo in 1994 and on women in Beijing in 1995). The resolutions sought by this coalition were defeated in both conferences after much struggle,

but they continue their work, successfully, within many countries.

The unstoppable trends toward globalization of the economy have resulted in a greater participation of women in the labor force. At the same time, the empirical evidence indicates that their incorporation into remunerated employment is occurring under conditions of inferiority and subordination. A few areas of the world—China, Malaysia, South Korea, and Thailand—have registered strong economic growth, but most others are caught in economic exchanges that do not generate the wealth needed for satisfaction of even basic needs; thus poverty is growing much faster than wealth. Latin America was a promising region a generation ago. By the early 1990s, 200 million people (about 46 percent of its total population) were unable to satisfy their basic needs, and 94 million people (22 percent) were living in extreme poverty (ECLAC, 1994). Poverty is ugly for men and women, but we know that it is emerging increasingly with a woman's face.

Multinational corporations (MNCs) have increased in power. Already by 1970, 51 of the 100 largest entities in the world were MNCs, thus surpassing governments in terms of wealth (Haavio-Manila et al., 1985). With the creation of new economic pacts, such as the European Union, the North American Free Trade Agreement (NAFTA), and others in Asia and Latin America, the economic environment has become less friendly and more rigid for many of the developing countries. There is much greater competition among industrialized powers, with the result that they are less concerned with the problems of the Third World, beyond those that may affect the limited consumer markets they represent.

Given relatively unsupportive states, one of the main sources of assistance for women in the Third World is likely to be the women from the industrialized countries. Their knowledge of gender and their conviction that efforts must be made to improve the social and economic relations of gender should result in an application of pressure upon donor agencies to assign more resources to women in developing countries. But women in the First World also face conflictual interests: Industrialized countries consume 75 percent of the world's resources though they represent 20 percent of the global population. This unbalanced consumption hampers the creation of processes for a just and sustainable livelihood and widens distance between rich and poor countries, making attention to gender issues less urgent than mere survival. Increasing environmental degradation forces women in rural and tribal areas to redouble their efforts to locate diminishing fuel and food, with the consequence that they must spend more time and energy to accomplish regular needs.

The winners in the new economic world order will be men from the First World, and the losers will be women from the Third World, hurt by their growing economic separation from women in industrialized countries.

The impact of structural adjustment in many African and Latin American countries (less so in Asian countries, except for the Philippines) has resulted in substantial regression of many of the gains made by women. Despite the creation of "safety nets," many women have seen their incomes decrease and their possibilities of employment in the formal sector, particularly in government, drastically reduced. With the state less active than in the past, legislation to improve women's conditions will be more difficult to attain. The old states were not friendly to women, but their growing weakness means that massive social changes are becoming less likely.

Objectives of This Encyclopedia

The editor and contributors have pursued four basic goals for this encyclopedia: (1) to present in a single volume a large amount of information on various subjects having definite implications for women in the Third World; (2) to stimulate border crossing of disciplines to enhance our understanding of the limits and possibilities of each field. Broadening our critical gaze is a political and academic obligation; (3) to inform readers of the condition of women across different settings and countries in the belief that an understanding of the myriad conditions and situations will help them grasp the pervasive reach of gender ideology; and (4) to demonstrate the relation between gender and power in the hope that the insights gleaned will help all parties reconceptualize problems—and solutions—affecting women and men.

We also hope that this encyclopedia will offer a much needed connection between women's studies and women-in-development studies. In many industrialized countries, particularly the United States, the active and otherwise innovative programs dealing with women and feminist issues tend to focus on women in American society. Not infrequently, a result is that the humanities and the arts are privileged, and the questions of poverty, oppressive sexual ideologies, and low levels of technological development are not linked to the exploration of women's conditions. We think that the consideration of race, gender, and class in a global context, beyond the parameters of Western and U.S. society, should provide much-needed points of comparison. In essence, the encyclopedia seeks to reduce the invisibility of women to men and the invisibility of women in the Third World to women in industrialized countries.

Because we wanted to provide a concise, yet comprehensive, treatment of the issues, this encyclopedia departs from the norm in the length of its entries. Unlike typical reference books that present brief printed entries, we requested contributions of about 30 double-spaced pages for the topical entries and about 20 such pages for the geographical entries. The result has been a deeper and more exhaustive treatment of the selected topics. Each entry comes with a selected list of references. Keywords in the index of the encyclopedia identify topics within and across entries.

Who is the audience for this encyclopedia? We are aiming at a broad audience, comprising discipline specialists, feminists, educators, practitioners, policymakers, and university students.

The discipline specialist will find articles of importance in other fields. Feminists will develop a global picture of the issues involved affecting women. Educators and practitioners will find current material they can incorporate into their pedagogical activities. Policymakers, especially those not familiar with gender issues, will have access to a complete compendium of contemporary discussion and pertinent sources. University students will have access to comprehensive treatment of issues and to the pertinent literature: Reference sources are cited not only within each chapter but also in the annotated bibliography at the end of the encyclopedia.

While we use a feminist approach in the analysis and discussion of the encyclopedia topics, it is our hope that the audience will also include pre-feminists and "antifeminists." Although we do not subscribe to a world ruled entirely by reason, we do believe that knowledge often brings a new awareness and, with it, a weakening of misconceptions and outright social and economic subordination of women.

Entry Topics

The encyclopedia presents updated information, within a feminist framework, on a number of traditional and well-identified issues. It also brings up emerging issues, some of which have been more acted than reflected upon. Among them: domestic and sexual violence, creation of women-friendly cities, religious beliefs as patriarchal ideologies, older women and their needs, new labor processes in industrial production, AIDS, and the gender consequences of ecological deterioration and of war.

The encyclopedia comprises 10 sections. Its content begins with a discussion of some central conceptual and theoretical issues and then moves to major areas concerning gender and development. These areas can be classified into two categories: areas that document and explain the situation of women, such as political and legal contexts, sex role ideologies, demographics and health, marriage and the family, women and production; and areas that inform the reader of more transformative positions, such as the sections on women and the environment, enabling conditions for change, and movements for change. The final section of the encyclopedia contains geographical entries. The coverage of these areas is not exhaustive but those within the greatest dimensions of gender in social and economic life are highlighted.

Although all entries have some geographical references, separate geographical entries are presented for some regions and countries. The geographical discussion of a particular country was considered appropriate in cases where its large size (India and China) or its peculiar circumstance (Cuba and Israel) made this treatment appropriate. Countries with some commonalities are presented as a group. The purpose of the geographical entries is to give concrete form and context to substantive issues discussed in earlier sections of the encyclopedia: They aim not to present an exhaustive discussion of gender issues in the selected geographical space but to discuss the major issues that are being addressed in that region or country and the major contributions being made to feminist thought and achievement.

An annotated bibliography comprising 60 books and reports is included for those wishing to pursue more in-depth study of any of the selected topics. The books cover all disciplines; they have been selected using such criteria as topic, timeliness, and geographical coverage. That many of them are in English reflects the dominance of the English language in late-twentieth-century scholarship on gender.

On Comparison

The Third World encompasses a wide range of countries, diverse both among and within themselves. Scientific knowledge develops by identifying common patterns among a seemingly disorderly multitude of voices. The examination of issues across developing countries should enhance our awareness of the problems that are unique to developing countries and those that can be found in all parts of the globe.

By bringing together a large number of situations, patterns of diversity and similarity across countries and societies will surface. Some striking similarities may be found not only among developing countries but also between societies in industrialized and developing countries. The implicit comparison contained in the encyclopedia's entries should produce new insights.

It is our hope that the encyclopedia will play a role in reconceptualizing problems and topics to include sufficient appreciation of gender in the configuration of social life and society, and that those who study gender may see the intimate interconnection that exists between women in industrialized countries and those in developing countries. Third World women account for more than three-quarters of the women of the world, yet, as Samarasinghe (1994) observes, "the study and analysis of their reality is relegated to a peripheral segment, an area study, if you will, under the rubric of Third World Development" (Samarasinghe, 1994, p. 220). Samarasinghe makes the point that a reverse flow of ideas about "the South" may stimulate a dialogue of the issues shared by women in the economically advanced West and the economically poor Third World. He identifies as relevant issues the intersection of gender and social class, the various kinds of productive work that result in women's subordination in the labor market, and household-observation and time-allocation methodologies.

Women are diverse in terms of ethnicity, national origin, and religious affiliation. They belong to different classes—rich and poor, middle class and destitute. There are also Black women, indigenous women, caste women. Is it possible to speak about women in general? I argue that we can, that the diversity of women does not preclude their existence as a group subjected to systematic marginalization and inequality.

In the academic world, postmodernist perspectives have highlighted the existence of multiple voices of women, as well as the arbitrary nature of gender ideologies in cre-ating only two categories of human beings: women and men. The refusal of some feminist groups to endorse arbitrary categories might at some point present conflicting arguments regarding the work on gender transformation since "women" as a category stands as artificial and arbitrary. If reality is arbitrary, if it is primarily a social construction, why would one perspective have more validity than others? The issue is not simple. There is a movement away from great and unchangeable truths. However, there is also a recognition that certain principles may assure the benefit of the largest number of people. Some social constructions, therefore, are more fair and just than others. This encyclopedia does not argue for universal and fixed truth; it does highlight contemporary phenomena that create asymmetries in our social world with no other justification than that what is cultural is valid in its own right.

Journey to Solidarity

As you read through the entries, I believe you will find abundant new knowledge, reflective of the significant intellectual production that has occurred in the sphere of women and development during the 1990s. Women's activism and vitality have also produced challenging insights in fields and areas that have begun to take shape. It is evident that we cannot relegate gender only to topics such as family, education, health, and nutrition without engaging in misconceptions of women's roles in developing societies and without forcing preconceived notions of their abilities, achievements, and dreams.

The slow pace of change in women's conditions reflects the complex nature of the problem and the deep-rooted forces that necessitate the maintenance of women's subordination. These forces feed on symbolic efforts for social change. One such force occurs in schooling, where policies generally focus on access to schooling and fail to consider both the need for drastic changes in the content of what is learned and the need to modify authoritarian patterns of governance and interaction that influence gender differentials.

Developing countries remain handicapped in terms of research capabilities and the financial and human resources needed for innovation. Yet, what women in developing countries have been able to accomplish with their meager resources and with the support of feminist scholars and activists in industrialized countries gives room for great hopes and expectation. The solidarity of women on a global scale is crucial—and, as the entries in this encyclopedia demonstrate, quite possible as well.

References

Ayoob, Mohammed. "The New-Old Disorder in the Third World," *Global Governance,* vol. 1, no. 1, 1995, pp. 59–77.

Economic Commission for Latin America and the Caribbean (ECLAC). "The Social Summit: A View from Latin America and the Caribbean. Note by the Secretariat." Santiago: ECLAC, 1994.

Goetz, Anne Marie. *The Politics of Integrating Gender to State Development Processes.* Geneva: United Nations Research Institute for Social Development and United Nations Development Program, 1995.

Haavio-Mannila, Elina, Drude Dahlerup, Maud Edwards, Esther Gudmundsdóttir, Beatrice Halsaa, Helga Hernes, Eva Hänninen-Salmelin, Bergthora Sigmundsdóttir, Sirka Sinkkonen, and Torild Skard (eds.). *Unfinished Democracy: Women in Nordic Politics.* Oxford: Pergamon, 1985.

Merriam, Allen. "What Does 'Third World' Mean?" In Jim Norwine and Alfonso González (eds.), *The Third World: States of Mind and Being.* Boston: Unwin Hyman, 1988, pp. 15–22.

Mies, Maria. "Towards a Methodology for Feminist Research." In Gloria Bowles and Renate Klein (eds.), *Theories of Women's Studies.* London: Routledge and Kegan Paul, 1983, pp. 117–139.

Parpart, Jane. "Deconstructng the Development 'Expert.'" In Marianne Marchand and Jane Parpart (eds.), *Feminism/Postmodernism/Development.* London: Routledge, 1995, pp. 218–231.

Samarasinghe, Vidyamali. "The Place of the WID Discourse in Global Feminist Analysis: The Potential for a 'Reverse Flow.'" In Gay Young and Bette Dickerson (eds.), *Color, Class, and Country: Experiences of Gender.* London: Zed, 1994, pp. 218–231.

Sengupta, Arjun. *Aid and Development Policy in the 1990s.* Helsinki: United Nations University World Institute for Development Economics, 1993.

Todaro, Michael. *Economic Development in the Third World.* 3rd ed. New York: Longman, 1985.

United Nations. *Platform for Action and the Beijing Declaration.* New York: Department of Public Information, United Nations, 1996.

United Nations Development Program (UNDP). *Human Development Report 1993.* New York: UNDP, 1993.

United Nations Development Program (UNDP). *Human Development Report 1995.* New York: UNDP, 1995.

Conceptual and Theoretical Issues

Roles and Statuses of Women

Nelly P. Stromquist

Introduction

According to Western analytical frameworks of society, the interaction—or act—among individuals is the key element of the social system. Roles and statuses are the basic features of each act. Role, here, is the patterned interactive relationship among people, specifically, what an individual does in his/her relations with others; status is the location of an individual in the social system vis-à-vis others (Parsons, 1964). While it is tempting to focus on individuals, Parsons makes it clear that "statuses and roles are not in general attributes of the actors but are units of the social system" (Parsons, 1964, p. 25). In other words, roles and statuses are assumed by individuals but they are not created by them.

There is less agreement on the forces that create roles and statuses. From a functionalist-theoretical perspective, social processes evolve to find the forms and expressions most appropriate to the particular environment of a society. From a conflict perspective, social evolution is the result of constant struggle in which the dominant forces shape reality via the establishment of particular structures and the production of concomitant ideologies.

In most societies, women and men perform different roles and hold different statuses. Are these the extension of natural differences or primarily the result of arbitrarily established dichotomies regarding femininity and masculinity? Verba (1990) notes that women have a "special relationship to human reproduction, even though the extent to which biological differences should determine patterns of social, economic, and political activity is a matter of debate" (Verba, 1990, p. 560). Should differences between men and women regarding pregnancy, delivery, and postpartum periods affect their entire lives? Aside from these periods, should not both parents have equal responsibility—in nurturing and guidance, time and effort—in bringing an infant to adulthood? Setting a boundary between the natural and the real is part of the struggle of interests and ideas (Peattie and Rein, 1983). Ideas of the natural shape

and constrain women's roles; thus, the notion of masculinity is counterpoised to the notion of femininity. The debate continues, and it affects policy and even research frameworks.

This essay examines the various gender roles and statuses across societies, noting the similarities and differences among them. It also shows how these roles and statuses are intimately intertwined and how their existence is firmly embedded in structural forces. Not only have roles been made different between women and men, but the rewards and the prestige attached to those roles—and, thus, their statuses—are also different.

To discuss variance in roles is not to subscribe to sex-role theory. The latter assumes a relatively harmonious socialization process, in which individuals learn in passive ways the messages and expectations transmitted to them. Sex-role theory has been criticized for emphasizing individual agency and "dissolving structure into agency" (Connell, 1987, p. 50). The sex-role literature has also been accused of simplifying reality by promoting binary distinctions based on "feminine" and "masculine" roles and not considering the actual behavioral practices of women and men in everyday life. Role theory tends to be apolitical and assumes that differences between men and women are only a function of proportion; that, as more women fill roles now occupied by men, these differences will be erased (Eisenstein, 1993).

Conceptual Framework

In observing the much more frequent outcomes of maintenance rather than transformation linked to gender roles and statuses, some feminist researchers prefer to talk about "gender-identity formation" rather than "socialization practices" to highlight the power-based nature of the process. A feminist theoretical framework that recognizes the intertwined dynamics of the sexual division of labor and men's control of women's sexuality helps to understand the enduring nature of gender roles and statuses in society and to question taken-for-granted assumptions of the social sciences. The framework used in this essay asserts that the subordination of women is based on the combined material and ideological benefits men derive from the sexual division of labor and control of women's sexuality. This perspective also acknowledges the powerful impact of class and ethnicity on gender but holds that gender distinctions produce dynamics of their own that are pervasive and extremely strong.

Gender Roles

Roles constitute basic scripts for performing and understanding social behaviors. The critical issue is not the social diversity that stems from the variety of roles but rather the disadvantaged position many of these roles create for certain groups in society—in this case, women. Most societies tend to uphold male behaviors as the norm and to evaluate female behavior against that standard. If female behavior matches sex-role expectations, women are considered inferior. If it does not, they are seen as deviant. Gender as a social marker structures our everyday life in multiple ways, although other forces such as social class, ethnicity, and religious beliefs further shape social acts.

Roles can be further defined as expectations regarding the skills, rights, and duties of individuals and thus function as prescriptions for interpersonal behaviors that are associated with particular, socially recognized categories of persons (Scanzoni, 1978). Roles are learned in the process of social interaction and through the imitation of people whose tasks we grow to consider "appropriate" for women and men. When people interact with others, they see themselves and others increasingly as occupants of particular roles.

Roles are transmitted in specific contexts, such as marital relationships, parent–child relations, work, and schooling:

> The context is an arena in which sex-role socialization occurs and sex-role performances are enacted. In a cyclical fashion, subtle and explicit sex-role expectations as well as appropriate sex-role performances reinforce each other to comprise the socialization process (Berryman-Fink et al., 1993, p. 65).

Roles provide the scripts people adopt in their interaction with others. These scripts are not invariable and can be challenged by others. But every social system tends to preserve a core set of roles through mechanisms of social persuasion and social control.

In all societies, feminine roles are linked to motherhood and considered to operate primarily in the domestic/family sphere. While men are given the freedom to select a variety of roles—from the vile to the sublime—women must select from a much narrower range. The family is seen as the pillar of society, and women as the agents responsible for its efficient functioning, which implies taking care of children, spouse, and household management. These beliefs permeated scholarly work, particularly before feminist research. Thus, Parsons and Bales (1954) saw the family as a "solidarity unit," which necessitated that one of its members (read woman) not have an occupational role outside the home in order to avoid conflict and to provide psychological security to all family members. Parsons and Bales considered that men were more task oriented (instrumental rationality) where women were more predisposed toward caring for others (affective rationality). While there have been shifts over time in the ideological construction of femininity and masculinity, and while women have gained greater degrees of freedom, instrumental rationality continues to be associated with masculinity, and expressive and emotional features are constantly linked with femininity. This is particularly true in the literature on national development.

Because motherhood is seen as the most important role women can have, in many developing countries girls marry at a young age and generally to older men, which ensures the wives' subordination to their husbands since these young wives are usually less educated and thus less able to generate an independent or substantial income. In India, 20 percent of the girls in the northern states marry before 17 years of age, and half of them become pregnant before their 17th birthday. According to U. N. data (1985), between 54 and 70 percent of girls are married before 20 years of age in such countries as Burkina Faso, Ethiopia, Cameroon, India, and Bangladesh. In countries in Africa and the Middle East, where polygamy is still common, older men usually take much younger women as second wives.

In most societies, women are expected to be soft, sweet, affective, intuitive, obedient, and dependent. Men must be intellectual, competitive, ambitious, curious, and rational decision makers and planners (Millett, 1970; Bonaparte, 1979). While women are expected to display more feelings (to be "emotional") and be nurturing and caring, men are expected to be abstract, aggressive, and strong. Most societies assign women the roles of home managers, mothers, and, by extension, community organizers. In contrast, males are depicted as providers, drawing economic resources from the world of work. Women are assigned supportive endeavors, as caretakers and producers of goods and services in the nonmonetarized household economy (Peattie and Rein, 1983).

In industrialized countries, clear sexual divisions of labor exist for the home, even though deviations from these roles are increasingly permitted. Women assume most of the tasks concerning food preparation and cleanup, housecleaning, care of clothes/laundering, and physical care of children. Men are responsible for home repairs, auto repairs, lawn care, and outdoor cooking. There are some tasks not as closely related to gender, such as shopping for food, gardening, budgeting of income and expenditures, and the moral upbringing of children (Davidson and Gordon, 1979). With the extraordinary penetration of the mass media, the value of a more or less universal adherence to many of these gendered tasks is being disseminated throughout the world. An extension of the domestic stereotypes associated with women is the belief that women have greater dexterity, attention to details, and patience than men; hence, women in the Third World, particularly in Asia and Latin America, have been incorporated in industries such as textiles, garments, and electronics. But since, at the same time, women's roles are perceived as primarily related to the home, the women workers receive low wages and face unstable employment conditions.

In most contemporary societies, women are socialized to be moderate and reactive (not aggressive) in their sexual expression and response. Men, in contrast, are encouraged to engage in sexual initiatives. On the question of sexuality, women of minority status in some societies, such as Black and indigenous women, experience double social markers. For instance, Black women (and men) are portrayed in North and South American countries as lustful, passionate, and sexually aggressive. In some regions, such as Latin America, sexual socialization is strong; its cultural expression, machismo, encourages the emergence of men as sexually active, capable of procreating several offspring, and in command of social and sexual relations. Machismo results in extramarital relations for men and in decisions by women to participate in and become victims of these relationships. The women who are willing to engage in these subordinate relations with men are persuaded by the widespread social myth that "men are like that" and by the financial support these men are able to provide them. Ideological and material conditions combine to render machismo in Latin America and other regions an enduring social feature.

In Japanese, Chinese, and Korean cultures, gender roles are influenced by Confucianism. "Confucianism made men alone the structurally relevant members of the society and relegated women to social dependence" (Kim, cited in Samovar and Porter, 1995, p. 13). As a result, in large segments in those societies, women tend to accept their roles as complementary to those of men, not always recognizing the subordinate position of their "complementariness." Cultural manifestations and distortions of the Qur'an (Koran) in Middle Eastern countries have also produced very domestic roles for women. In Shiite-dominated Iran, women are defined as dangerous and destructive if not under the control of their husbands.

In practically all cultures, the father is the ultimate authority in the household, while women are in charge of caring for family members in such situations as sickness, injury, or old age. Since childhood, men in India are given more freedom of expression, allowed to undertake more risks in play, and stimulated more to participate in cultural activities such as religious festivals, whereas women are asked to help with the chores. Domestic expectations and obligations shape and constrain women's actions. Molyneux (1985) has observed that women are much more likely to engage in "practical" rather than "strategic" actions. She defines the former as those addressing the satisfaction of basic needs women encounter in the performance of their domestic duties and family responsibilities,

such as housing, food, water, sanitation, and health. The latter, which refer to medium- and long-term strategies to modify the gender order in society through policies and the questioning of major institutions, are much less frequently assumed.

Mechanisms of social control operate jointly with socialization messages. Messages for conventional gender behaviors are accompanied by heterosexism—the constant depiction of the world as based on masculine/feminine arrangements—and by homophobia—the fear of being or becoming homosexual. These mechanisms are especially emphasized in the early years; thus, childhood games are heavily characterized by sexual stereotypes. Moreover, to prevent the devaluation of masculine characteristics, sanctions and ridicule are more frequently exercised in cases of gender crossing by boys than by girls. Domestic violence, especially from husband to wife, is frequent and relatively unaffected by legal strictures. Indeed, in some societies, wife beating is expected and justified as a form of the husband's love and virility. Violence against wives has been found to be associated with violence in other family roles, such as father; hence, domestic violence functions as a transgenerational form of social control.

Another mechanism of social control is the use of sexual stereotypes regarding women's abilities and skills. These stereotypes feed on the dominant gender roles and, at the same time, support them by creating widespread expectations about personality traits, attributes, and characteristics of women and men and about how they are likely to perform various activities. Schools, particularly through textbooks, have been found to reinforce sexual stereotypes.

In modern societies, however, women receive conflicting messages about their possibilities. While it is true that women are expected to marry and form a family, they are also exposed to increasingly higher levels of education that make the selection of professional careers almost unavoidable. Married women who seek professions that demand long hours away from home experience one form of role conflict: They feel guilty about not being good mothers when they work outside the home in such a way that schedules and activities within the household are affected. Women also face role conflict when they select professions or occupations that are not "feminine," such as becoming a politician or a firm manager, both of which require engaging in discussions, conflicts, controversies. Not infrequently, role conflict is solved by avoiding occupations or situations that affect conventional definitions of femininity.

Very few studies have examined empirically the combined impact of gender, race (or ethnicity), and class, despite conceptual advances on this interrelation. One study, albeit exploratory since it was based on a sample of convenience of 500 White and African-American men in the United States, found that, regardless of race, American men shared dominant masculine values (Harris et al., 1994). The study, based on an instrument measuring 24 roles commonly attributed to men found that, by the time the men were age 18, only 13 percent of their responses showed differences between the groups. These differences increased with age, but they never became substantial (reaching, at most, a 20 percent difference). This finding is particularly intriguing because, in the United States, there is a clear separation (de facto, not legal) between Blacks and Whites in terms of residence, economic level, and cultural practices. It is impressive, therefore, that men of both races would have so many masculine traits in common.

Role learning occurs through social imitation and through the development of particular cognitive schemes as the mind organizes and identifies patterns for future reference. These patterns are based on simplified representations and are adopted by both men and women even though the effects may be more favorable to one group than the other. As a result of constant socialization practices, individuals internalize messages and expectations. Feminist and gender researchers have demonstrated that women respond differently than do men to achievement, moral dilemmas, work, and pressures to conform. Gender theorists have detected gender differences in relation to cognition (Belenky et al., 1986), value development (Gilligan, 1982), and language uses (Lakoff and Johnson, 1980; Tannen, 1990). The mapping of these differences has taken place mostly in industrialized countries, but evidence from developing countries is beginning to accumulate.

Women live a simultaneity of roles; the practice of these roles enhances certain skills but hampers the development of others. Women have been found to be reactive when bargaining in public spheres. They do not sort out opponents' behaviors and they respond to conflicts mostly in a narrow sense. Men tend to be more goal-oriented bargainers (Scanzoni, 1978). That women are less-effective bargainers is partly reflected in the low salaries they command in the marketplace, even at higher levels of education than men.

Often, ideological beliefs about women's roles distort our appreciation of women's contributions. In Africa, women play the primary role in subsistence farming, although they are less involved in cash crops. Much of the international assistance and national efforts to improve agriculture in Africa have been notoriously blind to the contributions of women and have not provided them with the technical support or the tools that would enable them to increase their productivity. In Asia, men produce the major cash crops, while women contribute to the production of home crops; nonetheless, women's participation in agricultural work is far from negligible. Similarly, a stereotypic view of women as housewives detracts from understanding in full women's roles in agricultural production. This continues to be a persistent problem in the Third World.

Socialization messages create a number of interesting contradictions. On the one hand, men and women are depicted as having different sets of interests and abilities. On the other hand, they are expected to live in harmony within the family. One such expression of harmony is the assumed "pooling" of economic resources within the family. An increasing number of studies show that decisions at this level are far from conflict-free or apolitical since they favor some family members over others. Husbands, as the real or assumed income earners, have authority over wives; sons, as future income earners, have privileges and, in some cases, authority over daughters (who are often perceived as lost to their husband's family).

Education has been proposed as a crucial means to alter gender roles. Unquestionably, access to schooling facilitates the acquisition of skills to observe and analyze, but the content and experience of schooling may not always be conducive to the questioning of traditional gender roles. Conventional roles for men and women are learned through knowledge and practices at school: who talks, who listens, who is rewarded. The messages of textbooks have been found remarkably consistent across countries. A study of primary textbooks in Peru found that men outnumbered women 5 to 1 in giving orders, providing a strong message about masculine authority and leadership (Anderson, in Fernández, 1993). Dominant figures in textbook illustrations, stories, examples, and historical accounts construct representations of what is "suitable" for men and women. Schools themselves evince a sexual division of labor as women usually teach while men teach and administer. While schooling may be used to transform gender ideologies, most often it is not. Policies to promote girls' enrollment in primary education, especially in countries in which gender-segregated schools are the norm, have a weak impact on women's participation in the workforce if not accompanied by measures in other sectors of society, particularly the economy. In some Muslim countries, for instance, women's school attendance and completion rates far exceed their employment rates (e.g., Tunisia, Algeria, Saudi Arabia).

The Life Cycle and Gender Roles

Given the intimate connection of women's lives to their families, where women are at in the life cycle is a crucial factor affecting their roles. At younger ages, including those at which women have babies and young children, domestic concerns are paramount in women's decisions. This can be clearly observed in the kinds of constraints low-income women face in attending literacy and other education and training programs. While the role of the young wife is associated with restricted physical mobility—to allay fears concerning her sexual behavior, such as having lovers or other men's children—these restrictions are relaxed with age. For instance, older, upper-class, urban Bengali women can make one or two annual religious pilgrimages to relatively distant places (Brown, 1982). As a married woman becomes a mother-in-law, she gains authority and power over the young wife of her son. The daughter-in-law often assumes much of the housework, releasing her mother-in-law from other tasks, except from criticism of her performance. Across many cultures, older women have an important role in the distribution of food and in the supervision of its preparation. Older women sometimes become the enforcers of male domination, as in Sudan, where older women insist that infibulation (sewing shut the labia) is necessary to "insure the moral character of women and the honor of the lineage" (Brown, 1982, p. 144). In some Middle Eastern countries, older women may also play key roles in the selection of their sons' wives. Strong support has been found across cultures for the mother-in-law's supervision and punitive behavior in cases in which the younger woman's contribution to subsistence is minimal, marital residence is patrilocal, and descent is patrilineal (Brown et al., 1994).

As women's domestic roles change over their life cycle, so does their status. Older wives tend to have more egalitarian relations with their husbands, and, as mothers-in-law, they gain authority over daughters-in-law. In some Muslim countries, an older woman has the power to persuade her son to send a wife away or may overwork and punish a daughter-in-law she does not like. In Egypt, an elderly male household head may relinquish to his wife authority over the division of labor and the distribution of resources within the extended-family household. Older women may become matchmakers, medicine women, mistresses of ceremonies at initiation rites (Brown, 1982). Older women as mothers-in-law can also assume negative roles, such as becoming active participants in dowry deaths (killing of wives and brides because of disputes over dowry amounts) in India. Middle age brings women increased authority over kinsmen, fewer restrictions, and greater physical mobility because the end of fertility and menstruation makes a woman simultaneously less vulnerable and less dangerous. While this may be seen as a manifestation of women's power, it is a form of residual and delegated power by which women assume the enforcement of patriarchal ideologies and thus engage in the reproduction of asymmetrical gender relations. With modern-day shifts in migration and technology, older women in more transitional societies are being displaced from these roles.

Roles and Language

An obvious and clear manifestation of role differences occurs through the way language is used. Since knowledge is situated and conditioned by one's experience, women and men tend to do things differently, particularly in the way they speak. In both Western and Eastern languages, women's speech is characteristically less assertive and more deferential than that of men. Even in societies in which there is parity along several social indicators between men and women, such as the United States, women are more conscious than men of "proper speech" and speech strategies. They avoid taking a strong stand; they make assertions with a tag line to make them milder; and they use more adverbs and adjectives—"That was a great show, wasn't it?" "Maybe you should go." "Perhaps you are driving too fast." "I guess we can do it" (Lakoff, 1975). In mixed groups, women generally talk less frequently than men and get everyone's attention less often, and, when they do get it, it is for shorter periods. Women interrupt less than men and have less control over the conversation; men tend to make

more statements and to express ideas in a less qualified manner; and women value more highly the role of listening (Coates, cited in Samovar and Porter, 1995). Some cultures have gender differences built into the structure of the language itself, such as Korean and Japanese, in which there are specific forms of addressing men and women, young and old.

Gender Statuses

Ruth Dixon, a pioneer author on women and development, defines women's status as the degree to which they can have access to, and control over, material resources (food, land, income, and other forms of wealth) and social resources (knowledge, power, and prestige) within the family, the community, and society (Dixon 1978). Increasingly, social status is derived from one's ability to command economic resources. Women's status, however, continues to be derived from their family, while men's status is linked to their jobs. Women cannot ignore this connection because their status derives strongly from motherhood, wifehood, and being feminine (Afshar, 1992). Despite women's significant status as mothers and wives, men often have higher authority than women within the household. Keeping house has been termed the "most singular of occupations," since,

> by its informality, its irrationality, and its cultural importance, the whole situation of the housewife stands in violent contrast to the rest of the occupations system. . . . The system of motivation attached to the work of the housewife (who cannot choose freely among "employers") bears no resemblance to any other (Caplow, cited in Peattie and Rein, 1983, p. 41).

It is clear that the domestic work of the housewife, although productive in nature, cannot be explained in economic terms but must be understood in the context of powerful social norms that assign women responsibility for the domestic domain.

Cross-cultural research shows that women tend to have the most egalitarian status in those societies where they directly contribute to the production of goods, as in hunting-and-gathering societies, where women produce most of the food supply. In agricultural societies, women's status deteriorates as land and economic surplus become concentrated in the hands of males.

Westernization has contributed to making status much less ascriptive than in the past, but it has also worsened women's status. The colonial experience in Africa greatly diminished the position of women through the privatization and ownership of land, which eliminated their right to land and relegated them to the private sphere (Boserup, 1970). On the other hand, women's access to employment, however small their wages, constitutes a crucial means of becoming less dependent on their husbands and, thus, of exercising more assertiveness in domestic decision making, as studies of female workers in Cuba,

Puerto Rico, and the Dominican Republic demonstrate (Safa, 1995).

Women are less likely to be beaten in societies where they are regarded as autonomous adults (Brown, 1992). Wife-beating is more likely in societies in which men control family wealth, in which conflicts are resolved by physical violence, and in which women do not have equal rights to obtain a divorce (Brown, 1992). In Latin America, women who have been able to secure employment often escape violent situations in the home by leaving their husbands.

Modernizing forces, such as work in urban industrial sectors, are fostering the abandonment of families by men. In Kenya, women-led rural households are estimated to represent 27 percent of the families, but women-managed households (with migrant husbands) account for another 47 percent (Thomas-Slayter and Rocheleau, 1995). The socioeconomic development of women is reduced in a predominantly agricultural economy (Stewart and Winter, 1977); however, the move away from communal lands and into capitalist farm production is further undermining the status of African women.

The status of women includes issues of divorce, abortion, and legal age of marriage. In many countries, they are not treated as equal to men in property rights, in rights regarding inheritance, marriage, and divorce, or in rights to acquire nationality, manage property, or seek employment (UNDP, 1995).

On average, women live seven years longer than men despite an existence that makes serious demands on women's time and energy, particularly through childbearing. Not included in the statistics about women's longer life span is the fact that many women are not even allowed to be born: The devaluation of women affects the possibility of female fetuses coming to full term in countries such as China and India. Because of patterns of care and feeding that prefer sons to daughters, infant mortality is higher among girls in several Asian and North African countries. Considering the natural birthrate of male and female children, it has been estimated that about 100 million women are "missing" in demographic comparisons in Asia and North Africa (UNDP, 1995, p. 35).

Since status in most modernizing societies is equated with earning power, women hold subordinate economic status despite the fact that women perform a substantial amount of work. Status is often measured in terms of prestige of occupation. Women are more likely to be employees than employers and more likely to be in clerical and administrative positions than in professional, technical, and industrial-production occupations. There is a universal concentration of women in a few occupations, although occasionally concentrations of women are found in more diverse occupations and nontraditional fields such as medicine, dentistry, architecture, and psychology.

Gender Inequality: Cross-National Comparisons

The influential *Human Development Report,* produced annually by the United Nations Development Program since 1990, developed in 1995 two indices to measure women's status. The first, termed the gender-related development index (GDI), measures achievements in life expectancy (a proxy for a long and healthy life), educational attainment (an approximate indicator of knowledge), and gross national product (GNP) per capita in terms of equality between women and men. The formula used produces a decrease from a perfect score of 1 as the disparity increases. The second measure, the gender empowerment measure (GEM), is an attempt to assess women's ability to participate actively in economic and political life—for example, women's representation in parliaments, share of managerial and professional positions, participation in the active labor force, and share of the national income. As with the GDI, the perfect GEM score is 1. As noted in the *Human Development Report* 1995: "while the GDI focuses on expansion of capabilities, the GEM is concerned with the use of those capabilities to take advantage of the opportunities of life" (UNDP, 1995, p. 73). In all cases, GDI values are smaller than overall human (that is, including male) development standards. GDI values are 0.67 for Latin America and the Caribbean, 0.56 for Asia, 0.52 for Arab states, and 0.38 for sub-Saharan Africa. GEM values are much lower, with 0.42 for Latin America and the Caribbean, 0.30 for Asia, 0.28 for sub-Saharan Africa, and 0.25 for the Arab states. According to these rankings, it would appear that women in Latin America and the Caribbean enjoy higher status than women in other regions of the Third World and that the Arab states offer women the worst status in terms of access to economic and political power. The measures reported in the *Human Development Report* 1995 (UNDP, 1995) indicated that only nine countries (all industrialized) have GEM values above 0.6, compared with 66 countries with a GDI value above 0.6. Although, registration of statistics is still incomplete and faulty, the application of similar indicators and definitions across countries represents a significant improvement in our understanding of women's conditions. The existence of GDI and GEM indicators—systematically defined and collected—will facilitate tremendously the assessment of country performance in moving toward equality between women and men.

One of the rare studies of cross-national data using multiple-regression techniques found that socioeducational status and economic status are two completely independent aspects of women's status (Stewart and Winter, 1977). Both conditions have been found to be improved by higher levels of education and by the adoption of socialist policies by the state. Education helped directly in terms of participation in economic life and indirectly through fostering norms of universalism, merit, and rationality. Socialism helped by shifting control over wealth, capital, and production from particularistic individuals to those who, in principle, sought to speak for the public as a whole. The authors hypothesized that, as education and the economy enter the political nexus, universalistic procedures such as nondiscrimination, affirmative action, and direct employment of women by the government itself become more important and effective.

Women suffer from an interlocking complex of lower status and limited opportunities that hinders any thought or efforts they might direct toward access to, or acquisition of, economic, political, and social power. Since women work in less prestigious occupations, they get lower wages than men. The salary gap varies by occupation, but even where the gap is the smallest—usually in the professional-technical and sales categories—women earn about two-thirds of the salary of men. Some economists see women's lower earnings as a condition of "comparative advantage" for women to assume domestic roles and for men to enter the regular labor force (Becker, 1971). This argument turns an unfair condition on its head by seeing economic rationality in women's "decisions" and "choices" to stay at home. At the same time, this notion of women's comparative advantage ignores the fact that there is an increasing number of female-headed households in most developing countries as well as a progressively larger proportion of women in the informal sector of the economy. Moreover, in both industrialized and developing regions, particularly in Latin America, women and non-White males are often engaged in work far below their abilities and educational qualifications.

Change Processes

Socialization mechanisms account for maintenance, but they do not explain the origins of the social structure of gender inequality nor account for changes in norms, beliefs, and practices over time. Many factors underlie ongoing changes in the roles and statuses of women. Many damaging practices associated with gender roles have been greatly reduced in the twentieth century, including foot binding in China, widow burning in India, and genital mutilation of girls and women in Africa and Asia. The first two were brought about by their clear deviation from dominant notions of human rights, which prompted men in power to repudiate them. A number of rights pertaining to citizenship, such as voting and the right to education and work, have been given to women as extensions of universal democratic features. Some changes (the banning of genital mutilation in some countries and of dowry deaths in India) have been brought about by participants in the feminist movement. Other rights exist under great contestation, such as the right to abortion, to protection from rape and sexual harassment, and to the free expression of sexuality; conversely, some practices persist despite heavy feminist attacks and even legal prohibitions, such as dowry deaths in India and infanticide in China and India. In many other countries, oppressive practices (genital mutilation) exist under the name of culture and religion.

Since the 1960s, substantial transformations in the social relations of gender have taken place. While the pace relative to the need is slow, these changes can be traced to multiple sources. One is the power of new ideas that resonate with women's experience despite their location in different cultures and levels of national development. Prior to the women's movement, the roles of women and men were seen as complementary and unproblematic, different roles but of equal value. The ideas emanating from the women's movement, from the popular writing in Betty Friedan's *Feminine Mystique* (Friedan, 1963) to the more intellectual discussion in Kate Millett's *Sexual Politics* (Millett, 1970), have been disseminated through the feminist literature in both academic and popular journals. A second source of change is the international conferences sponsored by the United Nations, which have also provided forums for dialogue and the identification of gender problems and issues in social, economic, and political dimensions of life. These mechanisms have fostered a process of questioning and unlearning conventional gender roles.

Perhaps some of the strongest forces of change in the roles and statuses of women can be attributed to women's access to contraceptives. As Bonaparte (1979) remarks, contraceptives amount to a Copernican revolution in the sexual life of women. Many married women in developing countries have access to, and use, contraceptives, but, for a full sexual revolution, it is necessary that women become liberated from the feelings of guilt associated with the use of contraceptives—a concern prevalent among many young women. In the period from 1970–75 to 1990–95, fertility rates declined from 4.7 to 3.0 live births per woman; maternal mortality rates were halved in the same period (UNDP, 1995, p. 3).

A strong drive for gender equality also comes from working women. The increasing percentage of wives and mothers who work is introducing concrete problems into the search for immediate solutions. The presence of women with higher levels of education, which increases their tendency to be engaged in paid employment regardless of marital status, is also introducing new concerns and voices. Feminist activism in Latin America can be traced to women's increased levels of education and participation in the labor force.

The condition of women has also changed because of economic crises and innovations. The spread of domestic technologies—potable water, electricity, electric or gas stoves, processed foods—has given women more time to allocate to other functions and concerns. In industrialized countries, the two-earner family has become increasingly common since the 1960s. The impact of structural-adjustment policies in African and Latin American countries has increased the number of women who have joined the labor market in order to be able to support their families. Women who have entered the labor force are gaining independent income and thus seek new possibilities along political, economic, and social lines. An important lesson deriving from research in Nordic countries, where women's status is one of the highest in the world, is that much of the impetus for change is largely attributable to labor market–related measures rather than equality policies (Eduards et al., 1985)

By the end of 1995, international laws to protect

women embodied in the United Nations Convention on the Elimination of All Forms of Discrimination Against Women (CEDAW), which the UNDP terms "a path breaking charter of the legal and human rights of women" (UNDP, 1995, p. 7), had been ratified by 148 countries, even though many countries have expressed reservations to some of the provisions. On the other side, it must be noted, implementation of CEDAW has remained weak as a result of inadequate funding.

Women and the State

Gender roles are affected by state laws and policies, which establish and maintain the structure of women's subordinate status in families by allowing differentiation in legal definitions of adulthood or in inheritance rights between women and men and by failing to punish domestic and sexual violence. State behaviors also affect women's status in the labor market by not questioning differential wages for equal or comparable work (Goetz, 1995). These state policies generally make women dependent on men. Often, welfare policies, designed to help women, assume women's dependency on men or further crystallize it. Women's claims impinge on the state in its capacity as provider, regulator, and employer. It has been observed by feminist scholars that states respond to women as if they were a small and narrow social group. Thus, states often tend to consider welfare policies that recognize only low-income women.

Many feminist demands, such as equal status, antidiscrimination laws, quota systems, and regulations against sexual harassment, are directed toward the regulatory functions of the state and seek to modify social expectations about women's roles. Government bureaucracies are major employers of women, at both professional and clerical levels. Women's wages within governments tend to be higher than in private firms. Thus women experience an ambivalent situation toward the state: In terms of salaries, it is more fair than the market; at the same time, it usually operates to maintain the social and economic conditions that favor the gender status quo. Moreover, the socialization of women to distance themselves from political issues in practically all countries discourages them from seeking positions of political leadership, thus delaying the possibility of gender transformation through state policies.

Fortunately, the women's movement has grown more predisposed to accept the notion of power—a concept considered in the 1970s and 1980s as evil or too reflective of a patriarchal order. As manifested in the feminist input that shaped the strategic actions of the *Platform for Action,* the document produced at the Fourth World Conference on Women (Beijing, 1995), many women in both industrial and developing countries understand that, insofar as equality for women presupposes social reforms, there will be conflict (Eduards et al., 1985) and that power will be needed to bring this conflict to a successful resolution.

In the journey to equality, three stages are discernable. The first, the moment of denunciation, during which the many inequalities of women are documented, seems to

have been substantially completed. Research on these issues is not final, but it has been sufficient to detect the existence of inferior situations and conditions of women in many societies. The second stage, the moment of legitimation, is being achieved in the late twentieth century in an increasing number of countries. Declarations at the end of the four (as of 1995) world conferences on women sponsored by the United Nations openly acknowledge the need to address women's problems and needs. The growing acceptance of CEDAW is one of the most significant recognitions of the legitimacy of women's issues.

It appears that we are right in the middle of the third stage, concrete actions to address the problems of equality and equity that women encounter, and facing two paramount obstacles in this struggle. The first is the urban-industrial bias of development policies, which has led to a widening gap in income for men and women in the Third World through the incorporation of women into the labor force but in occupations that provide lower remuneration, authority, and mobility. Without economic autonomy for women, which comes from both better wages and more control over those wages, attempts at changes in gender roles will have limited success. Policies to eliminate poverty will improve the status of poor women through such mechanisms as education, employment, and access to the means of production. But gender policies will have to move beyond attention to the "most vulnerable" group if the condition of women is to be modified.

The second big obstacle to equality is the fact that ideological control through institutions such as the school and the mass media continues relatively unchallenged. Demands for greater schooling provide women with more skills and some access to new fields; most often, though, the received knowledge does not question gender roles but, on the contrary, reinforces them along conventional lines of femininity and masculinity. When women receive feminist messages—which are much less frequent than the institutional ones and, thus, have a weaker impact—many develop ambivalence about their roles. They accept marriage and motherhood; they also want professions and often believe they can assume private and public roles equally well. In the end, many women end up sacrificing professional ambitions in favor of a happy family and children. They console themselves with the dream deferred: It is the daughters who will be able to attain what their mothers could not.

Conclusion

Women have made progress in redefining roles and statuses. Yet, their continued subordinate condition is evident in multiple facets of social and economic life. Class and ethnic differences further affect the condition of women, but neither of these can claim a stronger impact than the other.

Women's roles and statuses are intertwined and embedded in a dense network of material and ideological forces. The transformation of sex roles will necessitate strong questioning of the dynamics and messages now promoted by major institutions such as schools, families, and the mass media. At the same time, the promotion of new ideas and visions will not take root unless women are in a position to develop greater economic independence, a situation likely to emerge only if they are eager to fight for better access to the marketplace.

Poor women suffer the most, yet they do not represent the entire category of gender. State policies must address economic issues to modify the status of women, but additional policies must address social and ideological issues to transform the gender roles that later shape the status of women. Since most governments fail to address women's roles in their totality, feminist advocacy will be essential to redefine gender roles and to demand new statuses for women. Appropriate roles are those that recognize the diversity of human actions. The goal here is not uniformity but the elimination of the asymmetric assignment of roles that gives women reduced possibilities and men a much wider range of behavioral and occupational options. The envisaged widening of roles for all will enable men and women to have more equitable power balances in marriage, the home, the marketplace, and other social institutions.

Since each culture creates its own mix of statuses and roles, it is easy to justify these differences in terms of cultural diversity and preference. But when such social creations function to the chronic disadvantage of some groups in society, invoking culture is nothing more than making a claim for exemption from the increasingly accepted standards of justice and democracy. Cultures over time create a bond between ideology and power. We must be careful, therefore, not to accept too readily cultural autonomy which is often a mantle for inequality.

References

Afshar, Haleh. "Iran, Women, and Work: Ideology Not Adjustment at Work in Iran." In Haleh Afshar and Carolyne Dennis (eds.), *Women and Adjustment Policies in the Third World.* New York: St. Martin's, 1992.

Becker, Gary. *The Economics of Discrimination.* 2nd ed. Chicago: University of Chicago Press, 1971.

Belenky, M., B. Clinchy, N. Goldberger, and J. Tarule. *Women's Ways of Knowing: The Development of Self, Voice, and Mind.* New York: Basic Books, 1986.

Berryman-Fink, Cynthia, Deborah Ballard-Reisch, and Lisa Newman (eds.). *Communication and Sex-Role Socialization.* New York: Garland, 1993.

Bonaparte, Hilda de. "Los roles masculino-feminino y la educación sexual," *Temas: Infancia, Adolescencia, Familia,* vol. 2, no. 9, 1979, pp. 15–21.

Boserup, Ester. *Woman's Role in Economic Development.* New York: St. Martin's, 1970.

Brown, Judith. "Cross-Cultural Perspectives on Middle-Aged Women," *Current Anthropology,* vol. 23, no. 2, April 1982, pp. 143–156.

———. "Introduction: Definitions, Assumptions, Themes, and Issues." In Dorothy Counts, Judith Brown, and Jacquelyn Campbell (eds.), *Sanctions and Sanctuary: Cultural Perspectives on the Beating of Wives.* Boulder: Westview, 1992, pp. 1–18.

Brown, Judith, Perla Subbaiah, and Therese Sarah. "Being in Charge: Older Women and Their Younger Female Kin," *Journal of Cross-Cultural Gerontology,* vol. 9, no. 2, April 1994, pp. 231–254.

Connell, Robert. *Gender and Power.* Stanford: Stanford University Press, 1987.

Davidson, Laurie, and Laura Gordon. *The Sociology of Gender.* Chicago: Rand McNally, 1979.

Dixon, Ruth. *Rural Women at Work.* Baltimore: Johns Hopkins University Press, 1978.

Eduards, Maud, Beatrice Halsaa, and Hege Skjeie. "Equality: How Equal? Public Equality Policies in the Nordic Countries." In Elina Haavio-Mannila, Drude Dahlerup, Esther Gudmundsdöttir, Beatrice Halsaa, Helga Hernes, Eva Hänninen-Salmelin, Bergthora Sigmundsdöttir, Sirka Sinkkonen, and Torild Skard (eds.), *Unfinished Democracy: Women in Nordic Politics.* Oxford: Pergamon, 1985, pp. 134–159.

Eisenstein, Hester. "A Telling Tale from the Field." In Jill Blackmore and Jane Kenway (eds.), *Gender Matters in Educational Administration and Policy: A Feminist Introduction.* London: Falmer, 1993, pp. 1–8.

Fernández, Hernán. "Persistent Inequalities in Women's Education in Peru." In Jill Ker Conway and Susan C. Bourque (eds.), *The Politics of Women's Education: Perspectives from Asia, Africa, and Latin America.* Ann Arbor: University of Michigan Press, 1993, pp. 207–216.

Friedan, Betty. *The Feminine Mystique.* New York: Norton, 1963.

Gilligan, Carol. *In A Different Voice: Psychological Theory and Women's Development.* Cambridge: Harvard University Press, 1982.

Goetz, Anne Marie. *The Politics of Integrating Gender to State Development Processes.* Geneva: United Nations Research Institute for Social Development, 1995.

Haavio-Mannila, Elina, Drude Dahlerup, Maud Eduards, Esther Gudmundsdöttir, Beatrice Halsaa, Helga Hernes, Eva Hänninen-Salmelin, Bergthora Sigmundsdöttir, Sirka Sinkkonen, and Torild Skard, (eds.). *Unfinished Democracy: Women in Nordic Politics.* Oxford: Pergamon, 1985.

Harris, Ian, Jose Torres, and Dale Allende. "The Responses of African American Men to Dominant Norms of Masculinity Within the United States," *Sex Roles,* vol. 31, nos. 11–12, 1994, pp. 703–719.

Lakoff, Robin. *Language and Woman's Place.* New York: Harper and Row, 1975.

Lakoff, George, and Mark Johnson. *Metaphors We Live By.* Chicago: University of Chicago Press, 1980.

Millett, Kate. *Sexual Politics.* Garden City: Doubleday, 1970.

Molyneux, Maxine. "Mobilization Without Emancipation: Women's Interests, State, and Revolution in Nicaragua," *Feminist Studies,* vol. 11, no. 2, 1985, pp. 227–254.

Parsons, Talcott. *The Social System.* Toronto: Collier-Macmillan, 1964.

Parsons, Talcott, and Robert Bales. *Family Socialization and Interaction Process.* Glencoe: Free Press, 1954.

Peattie, Lisa, and Martin Rein. *Women's Claims: A Study in Political Economy.* Oxford: Oxford University Press, 1983.

Safa, Helen. "Restructuración Económica y Subordinación de Género." In Rosalba Todaro and Regina Rodríguez (eds.), *El Trabajo de las Mujeres en el Tiempo Global.* Santiago: ISIS Internacional, 1995, pp. 161–179.

Samovar, Larry, and Richard Porter. *Communication Between Cultures.* Belmont: Wadsworth, 1995.

Scanzoni, John. *Sex Roles, Women's Work, and Marital Conflict.* Lexington: Lexington Books, 1978.

Stewart, Abigail, and David Winter. "The Nature and Causes of Female Suppression," *Signs,* vol. 2, no. 3, 1977, pp. 531–553.

Tannen, Deborah. *You Just Don't Understand: Men and Women in Conversation.* New York: Morrow, 1990.

Thomas-Slayter, Barbara, and Diane Rocheleau. *Gender, Environment, and Development in Kenya.* Boulder: Lynne Rienner, 1995.

Todaro, Michael. *Economic Development and the Third World.* 3rd ed. New York: Longman, 1985.

United Nations Development Program (UNDP). *Human Development Report 1995.* New York: UNDP, 1995.

Verba, Sidney. "Women in American Politics." In Louise Tilly and Patricia Gurin (eds.), *Women, Politics, and Change.* New York: Russell Sage Foundation, 1990, pp. 555–572.

See also Judith K. Brown, "Lives of Middle-Age Women"; Noemí Ehrenfeld Lenkiewicz, "Women's Control Over Their Bodies"; Agnes R. Quisumbing, "Women in Agricultural Systems"; Beatriz Schmukler and Maria Elena Valenzuela, "Women in Transition to Democracy"; Beatriz Schmukler, Maria Elena Valenzuela, Sandra Maria Garcia, Graciela Sapriza, and Graciela di Marco, "Women in the Southern Cone and Brazil"; Nelly P. Stromquist, Molly Lee, and Birgit Brock-Utne, "The Explicit and the Hidden School Curriculum"; Zafiris Tzannatos, "Women's Labor Incomes"

Feminist Epistemology and Research Methods

Margaret Sutton

Introduction

Feminist epistemology and research methods have become subjects of vital debate over the last 15 years. There are now numerous books and articles with titles that juxtapose women, gender, or feminism, with science, research methods, or knowledge. The debate encompasses a wide range of concerns, from how science defines gender to what methods are most responsive to women's interests. This essay summarizes discussions that the author believes have particular relevance to international women's studies. Although there are many interesting pieces on the masculinity of the physical sciences, for example, the focus here is on discussions of the social sciences, including psychology, sociology, economics, anthropology, history, political science, and related applied sciences. These sciences have a direct influence on the ways in which women live, all over the world, and on the policies that are designed, purportedly, to improve their lives.

As an initial definition, feminist epistemology is a field of inquiry devoted to understanding what it has meant for knowledge and what it has meant for women that the production of scientific and other formal knowledge historically has been the preserve of men. "Epistemology" is the name of the branch of Western philosophy that questions the nature of knowledge. It asks, What is true? and How do we know that it is true? Feminist epistemology grows from the consciousness that science is a profoundly political activity and that those politics have had negative consequences for women.

Feminist epistemology begins by understanding that scholarly practices are located in time and place and are carried out by people with social and political identities. This perspective inevitably confronts us with the limitations of scientific and other scholarly knowledge. In contrast to the historically dominant ideology of science as an activity that produces universal and unchanging Truths, feminists recognize that truths are partial and situated. Like life as a whole, truth grows and changes. To practice science with this belief requires a fresh understanding of what it means to create valid knowledge. Thus, the field of feminist epistemology encompasses several sets of issues, from research methods to the possibility of objectivity.

The claim that science somehow supports unequal relations between men and women can, at first blush, seem audacious, particularly to those who view themselves as scientists working for the good of humanity. This claim derives first from analysis of the ways in which the absence of women from science has influenced the content of scientific knowledge. This analysis is discussed in the first section of this essay. The critique of content leads feminist scholars to advocate certain principles and approaches to research methods, which are elaborated in the second section.

Considerations of how knowledge is made run throughout the discussions on feminist research methods. Thus, in practice, it is difficult to draw clear lines between feminist epistemology and research methods per se. Research methods embody principles of knowledge; these principles, in turn, determine the types of methods that are considered acceptable within scientific practice. Recognizing how deeply sexual politics are insinuated into the fabric of scientific practices, feminists confront a central problem: How do we conduct research that is consonant with our respective politics and ethics? To meet this challenge, feminist epistemologists must go beyond "the elimination of bias [in science] . . . to include the detection of limiting and interpretive frameworks and the finding or construction of more appropriate frameworks" (Longino, 1989, p. 54). The discussions within feminist epistemology are many and diverse. Those summarized in the final section speak to three major and interconnected issues. First, feminist epistemology is concerned with reforming or reconstructing scientific practice in a manner that equalizes power between social scientists and the people they study. Second, feminist epistemology, like other contemporary modes of critical thought, seeks to construct scientific analysis in a manner that is responsible about the specificity of persons and contexts. Finally, feminist epistemology considers not only the construction, but also the validation of knowledge as a social process and proposes means for making such validation into a more democratic process.

Although a large number of the authors who refer to their work as "feminist epistemology" live and work in the context of European and North American universities, the subject matter of the field has evolved in dialogue with Third World scholars, as well as other critical scholars in the West. Whether as colonial subjects or as citizens of independent nations, Third World scholars and intellectuals have long noted the relationship of scientific practices to the exercise of power. The field of anthropology has come under particular scrutiny because of its connections to colonial rule. Scholars and activists from disenfranchised ethnic groups in the United States have for more than a century challenged the racially restrictive character of science. Marxist intellectuals have critiqued the service of science to capitalism and speculated on how to bend the will of science to egalitarian purposes. Analyses of national, racial, and class-based characteristics of scientific practices intersect at many points with feminist assessments of the "maleness" of science. As Fee (1986) observes:

> . . . the conception of science defined as male in much of the feminist literature belongs to a specific period of the last 300–400 years and is characteristic of the period of capitalist development; it *is* European and also male; and white, and bourgeois . . . (Fee, 1986, pp. 52–53).

It is difficult in the course of one essay to do justice to the specificity and richness in the critiques of science written by feminists, both female and male, addressing the politics of science in Asia, Africa, South and North America, Europe, and Australia, from the perspective of many disciplines and through the lived experiences of all ethnicities and classes. The juxtaposition of analyses by, for example, African-American and Indian feminists should not lead the reader to conclude that any two authors are identical in their thought or, even less, that science has the same impact on women's lives everywhere. The strength of feminist epistemology, like other fields of feminist work, resides in the interplay of convergent and divergent beliefs, values, experiences, and interests of its practitioners.

Feminist Critiques of the Scientific Canon

The English writer Aphra Behn noted that, in her seventeenth-century world, women were caught in a curious bind when it came to higher learning. On the one hand, girls and women were judged unfit for intellectual labor. It was said, in fact, that serious study could harm females both physically and spiritually. Thus very few girls and women were granted the educational resources that would provide them such learning. At the same time, women's socially inferior position to men was justified precisely on

the ground that women had not acquired that same knowledge to which they were systematically denied access. In the twentieth century, women have entered scientific professions in unprecedented numbers. In 1960, women received fewer than 7 percent of the doctorates granted in the United States in the physical, social, and life sciences and engineering. In 1990, 34 percent of the new U. S. doctorates in these fields were conferred to women. The growth of universities throughout the Third World has resulted in a similar influx into scientific practice of both men and women who historically had been excluded from the educational systems that prepare people to become producers of modern scientific knowledge.

The entry of women, ethnic minorities, and citizens of the Third World into scientific professions does not, in and of itself, change the ways in which scientific knowledge is created. Rather, women and members of other historically marginalized groups have found themselves faced with a scientific canon that has, many argue, been biased by the interests and perspectives of male elites. The rules, principles, and standards that govern the practice of science, feminists argue, have reflected and reinforced male domination over women. Feminist epistemologists identify three ways in which the content of social science reflects the historical absence of women from its practice. Scientific views of society, feminists argue, have overlooked and silenced women. From the silence and invisibility of real women have grown various "scientific" renderings of women as somehow less than fully human. With the weight of scientific authority behind them, these representations manifest the power to define the lives of people who are not scientists.

Overlooking Women

Schoolchildren around the world read textbooks that are filled with the stories of men, most likely with an emphasis on Western men. Social-science accounts of the world, though more nuanced and complex, also direct many more questions and devote many more words to men than to women. One field in which the near-absence of women has been most striking is history:

> In 1238, only one maidservant, "awake by night and singing psalms" saw the assassin who gained entry to the bedchamber of the king of England, knife in hand. She changed the course of history—and the chronicler, Matthew de Paris, did not even get her name (Miles, 1990, p. xi).

The records of history name, describe, and biographize men whose actions have influenced their societies. Consequently, children and adults learn a great deal about princes, generals, and popes and rather little about the anonymous maidservant and countless other women whose actions have transformed their societies. The relative invisibility of women in the historical record is hard to dispute. Opinions vary, however, on the issues of why this is so and what difference it makes.

As in other scientific and scholarly fields, history in Europe and the Middle East has long been the domain of men: "Males gained entry to the business of recording, defining, and interpreting events in the third millennium BC; for women, this process did not even begin until the nineteenth century" (Miles, 1990, p. xii). When people become scholars or scientists, they do not automatically shed the biases, beliefs, and norms with which they were raised. Feminist researchers believe that scientists, like other people, "are not divorced from time and place, housed in some conflict-free world populated only by themselves" (Farganis, 1989, p. 208). Rather, scientists bring to their professional lives both the common sense of their time and the interests of their place in society. When these ideas cast women as domestic actors only, or as incapable of leadership, male historians are unlikely to focus their scholarly efforts on understanding how the actions of women have shaped history. In the study of history, women have been largely assumed away. To the extent that scholars believe that women don't or shouldn't act in the public sphere, they are unlikely to see the ways in which they do, let alone to record or analyze them.

A similar blindness toward the actions of women marks other social science disciplines. Millman and Kantor (1987) take sociologists to task for documenting and analyzing a limited "field of social action":

> When focusing only on "official" actors and actions, sociology has set aside the equally important locations of private, supportive, informal, local social structures in which women participate most frequently. In consequence, not only do we underexamine and distort women's activities in social science, but we also fail to understand how social systems actually function . . . (Millman and Kantor, 1987, p. 32).

The authors provide examples of sociological inquiries that are hampered by neglecting to study the diffuse effects of informal social organizations. One of these is the analysis of the attainment of professional status. In fields as diverse as medicine and business, access to powerful or prestigious positions is determined not only by achievement but also through connections to people who are already in positions of power and prestige. To overlook the informal socialization and selection processes within "old boys networks" is to fall short of accounting for the social organization of professions.

Scholars in international women's studies have identified several blind spots in the social-science canon on women in the Third World. One of these is in the assumptions that underlie the study of households as basic economic units. Household economic surveys in most countries presume that the head of the household is male, barring the legal absence of a husband because of divorce, death, disability, or other causes. This assumption is reinforced by "standard survey methodologies [that] do not assist us in accurately viewing the household and locating economic authority within it when adult males are mechanically identified as heads of household" (Bruce, 1991, p. 56).

Within this framework, the monetary rewards to a woman's agricultural, domestic, or informal-sector activities are often considered incidental to men's income:

> In farm families, an end result [of stereotyping heads of household as males] is the computation of a woman's nonpaid labor through the male head of household and the counting of her monetary earnings, if any, as supplementary to his (Karim, 1991, p. 153).

The historical pattern of overlooking women in research on domestic production, consumption, and investment is of great relevance to policy-related research. Examining the organization and distribution of food resources in an aboriginal Australian community, Povinelli (1991) characterizes government support available to women, who head 60 percent of the households:

> Widow, child-support, and unemployment benefits are designed to provide for senior women deprived of their major income earner, single mothers without extensive kin-based support, and the unemployed who are deprived access to any means of production. Such conditions rarely occur in rural aboriginal groups (Povinelli, 1991 p. 243).

What is at stake in such oversight is, in part, the quality of scientific knowledge that is based on a bad sample. Both the problems and the consequences of excluding women from the vision of science, in addition, spill over into the daily lives of women, men, and children everywhere. Misperceptions of the sources and uses of family resources can lead, and have led, to ill-advised policy interventions in the areas of agriculture, health, nutrition, and skills training.

Silencing Women

Exclusion of women from the ranks of scientists has created a knowledge base in which women are only faintly seen. It has also denied women a voice in scientific matters, over what is to be studied, how, and to what purpose. Gilligan's (1982) landmark work on girls' moral development in the United States, *In a Different Voice,* showed that girls did not formulate and resolve moral dilemmas in the way that her professor, Lawrence Kohlberg, had found was the case for boys (Kohlberg, 1981). Rather than appeal to universal principles as the basis for moral actions, girls were more likely to consider the human context, asking who would be helped or harmed. Kohlberg's claim to have discerned a universal pattern of moral development, Gilligan found, was flawed by the absence of girls' voices. So long as social theories are built on the experiences of boys and men, the voices of women are muted.

In addition to their experiences, women have been

restricted from voicing either their interests or their knowledge within the institutions of scientific practice. As Harding (1987) observes, scientific research is silent on many issues with a pressing relevance for women:

> Why do men find child care and housework so distasteful? Why do women's life opportunities tend to be most constricted exactly at the moments traditional history marks as most progressive? Why is it hard to detect black women's ideals of womanhood in studies of black families? Why is risking death said to represent the distinctively human act but giving birth regarded as merely natural? (Harding, 1987, p. 6).

Tautologically, scientific problems are those that are defined by scientists. Perhaps less obvious, what appears problematic to those people who are able to become scientists may seem inconsequential to socially different groups of people, while questions with burning importance to nonscientists do not make it onto the scientific agenda. As increasing numbers of women have become scientists, however, many have given voice to questions rarely raised by male scientists.

The field of history again provides good examples. Social relations between men and women vary across time and space, with some societies supporting relatively more egalitarianism and others less. Within the period of recorded history, however, there are few societies in which women enjoyed legal and economic rights that are fully equivalent to those of men. So long as such a state of affairs is considered normal or natural by those who write about history, it is unlikely that this inequality will be analyzed or even well documented. For feminist historians, however, the widespread asymmetry of power between women and men raises fundamental questions. Miles (1990) maintains that, whatever else it considers, a women's history must address two questions: "How did men succeed in enforcing the subordination of women? And why did women let them get away with it?" (Miles, 1990, p. xiii). The question can be reformulated in less accusatory terms: How have power relations between genders been constructed in different places and transmitted over time?

Scholars in U. S. ethnic studies have made similarly pointed observations about how the questions asked by scientists about minorities have been shaped by dominant-group interests. Summarizing the history of Chicano studies, Zavella (1991) identified a common thread that wove through this field in the 1970s. This was the realization that "Chicanos had been studied as if they were 'traditional' Mexican peasants, and whether they would assimilate into American life or retain their alien, backward status was the question usually raised by white researchers" (Zavella, 1991, p. 316). For Chicano scholars concerned with improving their own communities, however, the compelling questions were less about value differences between Chicano and Anglo cultures, and more about structural

inequalities in the labor market and the civil society. As with women, the silence of ethnic minorities and formerly colonized people in the formulation of social sciences has thus restricted the scope of questions that are raised by scientists.

Science is also the poorer for neglecting the contributions that women have made, outside of the institutions of science, to our collective knowledge of nature and society. The socially defined and culturally specific environments in which women live and work provide rich contexts for innovation and experimentation. These settings have included

> . . . that place of discovery and fact making, the household, where women explore and use our brands of botany, chemistry, and hygiene in our gardens, nurseries, and sick rooms. The fact is that much of the knowledge women have acquired in those places is systematic, communicated, and effective (Hubbard, 1989, p. 128).

Along with the botanical, medical, and cosmological knowledge sustained in many Asian, African, and indigenous American cultures, the "folk wisdom" of women of most cultures has long remained unspoken in the halls of science.

The knowledge produced by scientific practices that neither hear nor respond to the experiences, interests, and knowledge of any group of people is unlikely to provide great utility to them. Gran (1986) makes this point well in his study of the agricultural practices promoted by international assistance agencies, "Beyond African Famine: Whose Knowledge Matters?" Gran argues that the wholesale promotion of agricultural technologies inappropriate to the local environment has been as much responsible for famine in the Horn of Africa as meteorological factors have. Gran urges the researchers and policymakers who formulate agricultural strategies to listen to the people with the deepest knowledge of, and interest in, the land—the farmers who have learned from their elders the patterns and habits of their ecosystem.

It is not clear whether Gran is thinking of women, who conduct the majority of agricultural labor in Africa, but his point applies well to policy-related research that is connected to women. Many income-generation programs for women have failed because they focused on occupations that women were unable to pursue within their specific sociocultural context. Other programs have wrongly defined their purpose and starting point. Some programs to train midwives in modern birthing techniques, for example, have been insufficiently cognizant of the knowledge that midwives already possessed. The implications are clear. The outcomes of a program are unlikely to be beneficial unless they are informed by the knowledge and interests of the intended beneficiaries.

Defining Women

The practices of silencing and overlooking women in the

construction of science have enabled scientists to arrogate to themselves the power to define women's characters and needs. Feminist epistemologists have explored a number of mechanisms and means by which mainstream science has perpetuated and reinforced prevailing social biases that define women as somehow less than men. For women, as for colonized people and minorities, exclusion from scientific practices has enabled social scientists to define male (and White and Western) experiences as the primary reality and other sociocultural spheres as "deviant." In connection with her study of Black adolescents entering womanhood in the United States, Ladner (1987) argues that the scientific construction of deviance arises from social inequalities between researchers and their subjects:

> Placing Black people in the context of the deviant perspective has been possible because Blacks have not had the necessary power to resist the labels. This power could have come only from the ability to provide *definitions* of one's past, present, and future (Ladner, 1987, p. 75).

In forms both subtle and blatant, female behaviors and experiences have been defined within the social sciences as deviating from the presumed male norm. In this way, the social sciences treat "male difference as male superiority" (Longino, 1993, p. 105). In the process, existing political asymmetries between women and men are "naturalized" into biological, psychological, and other scientific accounts.

Within each social-science discipline, feminists have critiqued the arbitrary divisions posited between women and men. Within psychology, "sex difference" studies have been strongly challenged for their "near-obsessive focus on discovering a biological basis for small average differences between the sexes in behavior, or in the scores achieved on some kind of cognitive test" (Namenwirth, 1986, p. 25). Primatology and sociobiology, two other fields that are centrally concerned with links between biological and social natures, have also been found guilty of distorting evidence through an ideological lens that takes male superiority as given (see Haraway, 1986).

These feminist critiques of dualistic definitions have inspired critical scholars to examine oppositional definitions in areas other than gender. Said (1978) argues that the ability to define a less powerful group of people as diametrically distinct from the dominant group derives from logical fallacy and political privilege. The logical fallacy is to define a set of characteristics as intrinsic to one group—Europeans, males, White people—and to ascribe opposite characteristics to a subordinated group—historically colonized people, women, non-Whites. Concerning the scientific canon on women, Bleier (1986) observes: "Woman as reproductive being embodie[s] the natural, the disordered, the emotional, the irrational; man as thinker epitomize[s] objectivity, rationality, culture, and control" (Bleier, 1986, p. 6). Indeed, the categories employed, and their assignment to women, echo those identified by Said as central

to the construct of Orientalism. Reason and culture are allocated to men and to Europeans. Emotion and nature are assigned by the scientific canon to women everywhere and to colonized people of both genders.

In her critical analysis of Western feminist scholarship that focuses on Third World women, Mohanty (1991) identifies the source of structural similarity in the scientific canon on colonized people and on women. Her own argument, she states, "holds for any discourse that sets up its own authorial subjects as the implicit referent, i.e., the yardstick by which to encode and represent Others. It is in this move that power is exercised in discourse" (Mohanty, 1991, p. 55). Like Said, Mohanty argues that scientists and scholars in the West have assumed that their own reality defines the norm of human behavior. In representing socially different people as essentially different, scientists exercise the power to define the reality of others.

The critique of science that is constructed by encompassing dichotomies is thus particularly salient to the field of women in development, in which so much of the work is conducted by Euro-American women about women in diverse societies throughout the Third World. As Mohanty observes, far too much of this work constructs an "average third world woman," who "leads an essentially truncated life based on her feminine gender . . . and her being 'third world'" (Mohanty, 1991, p. 56). Colonialism and class structure have created commonalities of economic and political disempowerment of women across diverse cultures. At the same time, each culture and local economy carries its own opportunities for exercising power and potential for change, which outsiders may often have difficulty seeing. The implications of Mohanty's critique are challenging. In the context of feminist epistemology, the most immediate is that Western feminists must attend not only to the power relations that restrict women in science but also those that afford relative privilege to Western women in its practice.

Feminist Research Methods

Feminists in growing numbers have worked to rectify male bias in the scientific canon by bringing women into the practices and purview of scholarship. Perhaps the earliest fields to benefit significantly from these efforts have been the disciplines that straddle the humanities and the social sciences. History, anthropology, and literary criticism are notable examples. Histories of women, both exceptional and ordinary, are available in English and many other languages for children, adults, specialists, and the general reader. By documenting that which has been excluded from standard historical texts, women's history provides "the body of experience against which new theory can be tested and the ground on which women of vision can stand" (Lerner, 1986, p. 229). Feminist literary criticism of the 1980s and 1990s has extended the centuries-old practice of learning about women's experiences through the works of women writers: "Nineteenth-century female writers avidly read the work of eighteenth-century female novelists; over and over again they read the "lives" of queens, abbesses, poets, learned women" (Lerner, 1986, pp. 225–226). Feminist anthropologists have reinterpreted kinship and cosmology and have attended to patterns of oratory in public life in order to understand the political channels and strategies available to women in different societies (see Rosaldo and Lamphere, 1974; Atkinson and Errington, 1990).

In these fields and others, including sociology, psychology, economics, and policy analysis, feminist social scientists have had to exercise considerable creativity in adapting or designing techniques that enable them to research women's lives. In some cases, including women as research subjects has been a relatively straightforward matter of selecting gender-balanced samples or analyzing data separately for women and men. However, the long-standing focus of social sciences on men, on formal institutions, and on political leaders has fostered development of a specific set of research techniques. Historical research, for example, relies heavily on written records. If women were less literate and, therefore, less likely to write than men, their voices must be sought elsewhere, such as in oral history and the analysis of popular culture. Feminist researchers have expanded the scientific canon by introducing new techniques of information gathering that enable us to address questions framed by a feminist perspective.

Counting Women In

Among the earliest and most enduring techniques for redressing gender imbalance in the content of science has been to focus analytic attention specifically on women. This can be accomplished either by conducting studies of exclusively female samples or institutions or by analyzing both male and female experiences and comparing them. Helen Thompson Wooley, a pioneer in the study of sex differences in cognition, chose the latter approach for her 1903 dissertation on the mental characteristics of women and men. Applying the experimental techniques of the time, Wooley "demonstrated that the distribution curves of numerous types of behavior performed by men and women actually overlap," leading her to conclude that commonly held beliefs concerning sex differences in thought were exaggerated (Reinharz, 1992, p. 96).

Like Wooley's work nearly a century ago, studies by contemporary feminist researchers often challenge prevalent beliefs about women with the authority of scientific method. Such work also serves to underscore the reality of women's problems and perceptions. Reinharz makes this point in her analysis of feminist survey research: "Survey research can put a problem on the map by showing that it is more widespread than previously thought. Feminists have used survey research for precisely this purpose, dispelling the common argument that the complaint of a particular woman is idiosyncratic" (Reinharz, 1992, p. 78). Among the examples Reinharz cites in support of this point are surveys of sexual harassment in U. S. schools and workplaces, which have contributed to greater social awareness

of sexual harassment and have supported legal and institutional efforts to reduce it.

By increasing social demands to understand the contours of women's lives, feminist researchers support activism by and for women. In the process of such work, feminist research also defines baselines for assessing change in the conditions of women, an important component of effective policy research. Feminist researchers have both influenced and benefited from the growing practice of collecting gender-specific data within the international system of development assistance, such as the U. N. agencies that compile cross-national data on disease and morbidity (World Health Organization), employment (International Labor Organization), education (United Nations Educational, Scientific, and Cultural Organization), and other fields. These data enable concerned researchers to locate and verify gender inequities and to analyze some of the impacts of policy interventions. In the process, these researchers have pushed for more comprehensive information to define and monitor policy processes. Maternal mortality is a case in point. It is arguably the single most important quantitative indicator of women's well-being, but the majority of national statistical offices are still hard-pressed to estimate reliably the number of maternal deaths per 100,000 births.

These examples of feminist research methods illustrate what Harding (1987) has called "feminist empiricism," the modification and application of standard scientific methods by scholars with feminist concerns. Underlying feminist empiricism is the belief that "sexist and androcentric biases [in science] are eliminable by stricter adherence to the existing methodological norms of scientific inquiry" (Harding, 1987, p. 182). This belief is manifested through careful research devoted to understanding and changing women's lives. Feminist empiricism remains the most visible manifestation of scholarship dealing with women today, no doubt because it conforms to widely accepted norms of scientific practice.

Although feminist researchers in increasing numbers have advocated alternatives to standard scientific practices such as laboratory and other experimental research, it remains the case that feminist empiricism has, in Harding's words, "a radical future" (Harding, 1987, p. 183). Feminist empiricism contributes to the overall dynamic of feminist research by challenging unexamined beliefs about women and by documenting the prevalence of particular social problems that affect women. The field of women and development is strongly indebted to feminist empiricism. Boserup's (1970) pioneering work, *Woman's Role in Economic Development,* extends such staples of economic analysis as agricultural productivity measures to women's work in developing countries. Boserup's work contributed greatly to putting "women and development" onto the scholarly and policymaking maps. Reflections on this work have led feminists concerned with development issues into several fruitful enquiries. Independent researchers have greatly expanded our understanding of the various ways in which women live and work throughout the Third World, by studying and reporting on household economics, social organizations of women, and women in the labor market. It has also become increasingly common for international-assistance agencies to commission research focused on women, in support of specific policy measures. This research may employ surveys, individual and group interviews, or short ethnographic studies of the women who are expected to be affected by changes in the delivery of health, education, credit, water supplies, and other social interventions.

Documenting Women's Lives

The recognition that women's lives are largely excluded from the standard records has led many feminist social scientists to expand on qualitative research techniques in an effort to understand women's historical and current circumstances. One broad source of information for all social scientists is "cultural artifacts." The term may evoke images of potsherds and paintings, but material productions of culture are far more encompassing than this. Reinharz defines cultural artifacts as "the products of individual activity, social organization, technology, and cultural patterns" (Reinharz, 1992, p. 147). She defines four major types of artifacts studied by feminist researchers, providing examples of each:

> They are *written records* (e.g., diaries, scientific journals, science fiction, and graffiti); *narratives and visual texts* (e.g., movies, television shows, advertisements, and greeting cards); *material culture* (e.g., music, technology, contents of children's rooms, and ownership of books); and *behavioral residues* (e.g., patterns of wear in pavement) (Reinharz, 1992, p. 147).

These artifacts are, in principle, available to all researchers, whether feminist or not. For feminist researchers, however, the less-official records of culture take on special weight. Feminist critiques of the scientific canon have argued that the exclusion of women from the content of social sciences is one dimension of social systems that are organized to uphold male privilege. To the extent that scholars ignore, downplay, or deny opposition to gender inequity, science maintains the credibility of such organizations. The artifacts that are neglected in the process are likely to be those that convey opposition to prevailing norms and practices.

The challenges of locating and listening to women's voices increases when we seek to understand the perspectives and experiences of women who are not linked to male elites through race, class, or nationality. Collins (1990) observes that

> The suppression of Black women's efforts for self-definition in traditional sites of knowledge production has led African-American women to use alternative sites such as music, literature, daily conversations, and everyday behavior as important locations for articulating the core themes of a Black feminist consciousness (Collins, 1990, p. 202).

For this reason, a social scientist cannot begin to understand the perceptions and actions of African-American women without identifying cultural artifacts and other forms of expression created by the women themselves.

Expanding the range of cultural artifacts that are considered legitimate sources of scientific evidence is one dimension of feminist research. In addition, many feminist researchers confer great weight on the direct testimony of women concerning their own lives and beliefs. Reinharz begins her treatise on research methods employed by self-identified feminists with an analysis of interview research. Feminist researchers, she observes, make extensive use of interviewing, because

> . . . interviewing offers researchers access to people's ideas, thoughts, and memories in their own words, rather than in the words of the researcher. This asset is particularly important for the study of women because in this way learning from women is an antidote to centuries of ignoring women's ideas altogether or having men speak for women (Reinharz, 1992, p. 19).

When thoughtfully conducted, interviewing provides a means for participants to define problems as they see them, as well as the terms in which they are considered.

Ethnographic research methods, which are likely to include interviewing and the analysis of cultural artifacts, are favored by many feminist researchers for their potential to illuminate the context in which women live, work, and create meanings. The long periods of "participant observation," which distinguish ethnographic research from other modes of information gathering, can also include quantitative data gathering and analysis. Peacock's (1991) work with the Efe of Zaire exemplifies such a combination of qualitative and quantitative methods in ethnographic research. During her two years of residence with the Efe in the 1980s, Peacock interviewed and spent time with 27 women in three "residence groups." She also developed an instrument for recording how each of these women spent their time, specifying the nature and duration of work and leisure. One of her findings is that the physically strenuous work of women in this foraging society is supported by a distribution of labor among women: "The Efe case demonstrates that an intricate and varied pattern of cooperative work and mutual caretaking among women permits combinations of subsistence work and childcare that would at first glance seem unworkable" (Peacock, 1991, p. 358). The careful analysis of women's labor that formed the core of this study enabled the author to "illustrate the importance of looking at behavior from a collectivist as well as individualist perspective" (Peacock, 1991, p. 358). Other feminist ethnographers have come to similar conclusions concerning the importance of women's social organizations. Wolf (1974) observed that in a Taiwanese village where she conducted research, women "who worked through the women's community" were successful at swaying public opinion on "matters as apparently disparate as domestic conflicts and temple organization" (Wolf, 1974, p. 162).

Feminist ethnographers have also generated a rich literature that documents the life histories of women in diverse cultures. Shostak's (1983) presentation of the life of Nisa, a !Kung woman, is one of the best known of these. Like Shostak, Barnes and Boddy (1994) cast the story of Aman, a Somali girl, in her own words. These and many other life histories of individual women show how women take active control of their lives, make decisions, and affect the lives of their communities in ways that are outside the grasp of community- or organizational-level analysis.

Ethnographic research focused on women thus can begin to uncover the specific circumstances in which women construct their lives. Such understanding would appear invaluable as an input to policy processes. However, ethnographic research, in particular, and qualitative research, in general, require that researchers be granted ample time and sufficient material resources for their implementation. These conditions are all too rare for research that is commissioned in support of policymaking. Two additional obstacles limit the use of qualitative methods for feminist research in the policy context. First, it is difficult to distill the findings of qualitative research into the succinct packages of information on which policymakers have come to depend. In addition, policymakers who have themselves been educated in the norms of quantitative research, like their counterparts in universities, often doubt the validity of conclusions drawn from in-depth, rather than "in-breadth," research. In both contexts, feminist researchers who employ qualitative methods may be met with a double suspicion. The questions explored by feminists researchers may be held suspect, focusing as they do on the interests of women. In addition, the use of qualitative methods, which is so common to feminist research, may be received with doubts as to its "scienticity."

Pushing Back the Boundaries

Feminist social scientists face their own sets of doubts and suspicions, which center on the politics and ethics of the research process. Many feminist researchers have commented on the "dual consciousness" that arises from being aware of one's own social position as a woman, on the one hand, and from working within a profession that is intrinsically biased against women, on the other. Smith (1987) speaks of the "bifurcation of consciousness" between the "special activity of thought" that defines scientific work and "the world of concrete practical activities" that most women inhabit regardless of professional role. This dual consciousness extends into professional practice, as the "frames of inquiry which order the terms upon which inquiry and discussion are conducted originate with men" (Smith, 1987, pp. 90–91). It is the driving force behind the construction of feminist research methods that raise the visibility and increase the voice of women in the content of scientific analysis. At the same

time, putting these principles into practice raises deeper questions about the processes by which scientific knowledge is constructed and validated. These are the concerns that are at the center of the discussion called feminist epistemology.

Equalizing Relations

The critiques of the scientific canon on which feminist epistemology and research methods are based suggest several ways in which feminist scientists can challenge the gendered power system intrinsic to scientific practice. Foremost among these are the efforts by feminists to conduct research in ways that shift the center of authority from scientists to the people they study. Employing subject-centered methods like open-ended interviews and oral history creates the potential for participants in the research process to define their own perceptions, circumstances, and needs, rather than have these defined by the authorial voice of the scientist. For such methods to be used to this end, social scientists must attend not only to how information is gathered but also to how it is presented.

Many feminists go a step further in the move to equalize power by employing research methods that enable the people being studied to define the questions addressed and the methods used to answer them. Such research approaches often come under the name of participatory, or collaborative, research, in which "the researcher abandons control and adopts an approach of openness, reciprocity, mutual disclosure, and shared risk" (Reinharz, 1992, p. 181). Participatory research can make valuable contributions to the policymaking process. In China from 1992 to 1993, the Ford Foundation supported an effort in which rural women took photographs that reflected their lives and circumstances, then met on a monthly basis to discuss their meanings (see Yunnan Women's Federation, 1995). These discussions culminated in meetings with provincial health authorities to design local programs in reproductive health that would serve the specific needs of the communities involved.

In one innovative project, Mies (1991) created the conditions for a group of Third World women in the Netherlands and a group of Dutch feminists to study each other's lives. Mies reported that the activity resulted in enriched understanding among women of both groups:

> The Dutch women saw that not all women in the Third World are "poor" and that despite "underdevelopment," Third World women were freer in some respects than they themselves. Conversely, the women from the Third World saw that capitalist development of material riches and the supposedly progressive small family do not free women but instead have made them profoundly more dependent (Mies, 1991, p. 78).

This collaborative, or interactive, research process illustrates an often-noted phenomenon: The more that researchers "give over" the power to name and define, the more likely it is that they themselves will be changed in the process.

Specificity and Diversity

The research process that Mies helped structure illustrates a second major principle of feminist epistemology: the need to attend to the specificity of people and their circumstances. In part, this principle reflects the importance that feminists place on understanding contexts, rather than only discrete facts. The natural-science model on which much social science is built favors looking at social phenomena "in small chunks and as isolated objects" (Hubbard, 1989, p. 125). This analytic act of "context stripping" enables social scientists to obscure problematic assumptions behind ostensibly neutral terms. Hubbard points to the acceptance by economists of some rate of unemployment as "normal," which obscures "a wealth of political and economic relations which are subject to social action and change" (Hubbard, 1989, p. 126). Paying attention to the contexts in which people live and in which social phenomena occur is critical for attaining a full understanding of their meaning. An enriched understanding, in turn, provides the groundwork for strategies of social and political change.

Increasingly, feminists have also become critically aware of the many dimensions of power relations. Not only gender, but class, race, nationality, age, and physical and sexual difference are vectors along which power relations are ordered. Feminists who are neither White nor Western have understandably led White Western feminists to recognize and take seriously the diversity of women's experiences and, so, the impossibility of assuming universal definitions of what it means to be a woman. Mohanty's trenchant analysis of the "totalizing" tendencies of Western feminist thought challenges all feminists to come to terms with "the pluralities of the simultaneous location of women in social class and ethnic frameworks" (Mohanty, 1991, p. 72).

The inherent potential for social sciences to misrepresent the meaning of gender in non-European cultures leads Narayan (1989) to advocate other means for communicating gendered experiences across political and cultural boundaries: "'Nonanalytical' and 'nonrational' forms of discourse, like fiction or poetry, may be better able than other forms to convey the complex life experiences of one group to members of another" (Narayan, 1989, p. 264). At the same time, Narayan recognizes the potential of social research for clarifying the problems faced by particular groups of women and for suggesting means of addressing them. The more that this work is carried out by women who are members of the society they are studying, the more likely it is that the results will reflect the context and specificity of women's lives. The implication for "transnational" feminists, Western or otherwise, is that one of the greatest contributions we can make to feminist research in other countries is to support the development of local capacities and practices of woman-centered research.

Building Knowledge Through Dialogue

Feminist epistemologists, like other critics of the politics of science, are struggling with the implications of their own critique. Recognizing that science reinforces relations of inequality, what can researchers do to overturn them? Understanding that generalizations about groups of people obscure important differences within groups, how can we represent social life without distortions? Some feminist epistemologists maintain that these practical questions are far from being resolved. Farganis (1989), for example, finds in feminist epistemology "a feminist critique of (social) science but not a feminist (social) science" (Farganis, 1989, p. 218). Although there are substantial issues that are very much open to debate and reformulation, reflections of feminists working within the social sciences have begun to create models for feminist research.

Feminist research methods are diverse in technique, perspective, and underlying theory. The "context-stripping" quantitative methods of traditional sociological and economic research have their place in research, by, about, and for women, in defining and verifying the problems women face in particular societies. The qualitative approaches to research, including ethnography, interviewing, and life histories, provide the context and details of how women actively construct their lives and societies, in all varieties of social orders. Because feminist epistemology starts from the premise that all research is conducted by individuals who occupy specific social, political, and economic roles, all research products and findings are viewed as partial and provisional. Recognizing that scientific truths are never complete or final does not undermine either the validity or the relevance of scientific work. Rather, it demands of feminists an openness to challenge and growth.

Collins, among others, argues that the process of dialogue provides the strongest grounds for validating knowledge. She provides a model for sharing knowledge across communities of difference:

> Each group speaks from its own standpoint and shares its own partial, situated knowledge. But because each group perceives its own truth as partial, its knowledge is unfinished . . . dialogue is critical to the success of this epistemological approach, the type of dialogue long extant in the Afrocentric call-and-response tradition whereby power dynamics are fluid, everyone has a voice, but everyone must listen and respond to other voices in order to be allowed to remain in the community (Collins, 1990, p. 236–237).

Collins's formulation provides valuable guidance for the practice of research in the field of women and development. Each scientist, whether conducting research in her native land or on foreign soil, brings a singular and necessarily limiting perspective to the task. For the growth of valid knowledge, all relevant voices must be heard with an openness to learning.

References

Atkinson, Jane Monnig, and Shelly Errington (eds.). *Power and Difference: Gender in Southeast Asia.*

Stanford: Stanford University Press, 1990.

Barnes, Virginia Lee, and Janice Boddy. *Aman: The Story of a Somali Girl.* New York: Pantheon, 1994.

Bleier, Ruth (ed.). *Feminist Approaches to Science.* New York: Pergamon, 1986.

Boserup, Ester. *Woman's Role in Economic Development.* New York: Saint Martin's, 1970.

Bruce, Judith. "A Home Divided: New Models for Research and Policy on Women and Income in the Third World." In Aruna Rao (ed.), *Women's Studies International: Nairobi and Beyond.* New York: Feminist Press at the City University of New York, 1991, pp. 55–63.

Collins, Patricia Hill. *Black Feminist Thought: Knowledge, Consciousness, and the Politics of Empowerment.* London: Harper Collins Academic, 1990.

Farganis, Sondra. "Feminism and the Reconstruction of Social Science." In Alison Jaggar and Susan Bordo (eds.), *Gender/Body/Knowledge: Feminist Reconstructions of Being and Knowing.* New Brunswick: Rutgers University Press, 1989, pp. 207–223.

Fee, Elizabeth. "Critiques of Modern Science: The Relationship of Feminism to Other Radical Epistemologies." In Ruth Bleier (ed.), *Feminist Approaches to Science.* New York: Pergamon, 1986, pp. 42–56.

Gilligan, Carol. *In a Different Voice: Psychological Theory and Women's Development.* Cambridge: Harvard University Press, 1982.

Gran, Guy. "Beyond African Famine: Whose Knowledge Matters?" *Alternatives,* vol. 11, no. 2, 1986, pp. 275–296.

Haraway, Donna. "Primatology Is Politics by Other Means." In Ruth Bleier (ed.), *Feminist Approaches to Science.* New York: Pergamon, 1986, pp. 77–118.

Harding, Sandra (ed.). *Feminism and Methodology.* Bloomington: Indiana University Press, 1987.

Hubbard, Ruth. "Science, Facts, and Feminism." In Nancy Tuana (ed.), *Feminism and Science.* Bloomington: Indiana University Press, 1989, pp. 119–131.

Karim, Wazir-Jahan. "Research on Women in Southeast Asia: Current and Future Directions." In Aruna Rao (ed.), *Women's Studies International: Nairobi and Beyond.* New York: Feminist Press at the City University of New York, 1991, pp. 142–155.

Kohlberg, Lawrence. *The Philosophy of Moral Development: Moral Stages and the Idea of Justice.* San Francisco: Harper and Row, 1981.

Ladner, Joyce A. "Introduction to Tomorrow's Tomorrow: The Black Woman." In Sandra Harding (ed.), *Feminism and Methodology.* Bloomington: Indiana University Press, 1987, pp. 74–83.

Lerner, Gerda. *The Creation of Patriarchy.* New York: Oxford University Press, 1986.

Longino, Helen. "Can There Be a Feminist Science?" In Nancy Tuana (ed.), *Feminism and Science.* Bloomington: Indiana University Press, 1989, pp. 45–57.

———. "The Weaker Seed: The Sexist Bias of Reproductive Theory." In Nancy Tuana (ed.), *Feminism and Science.* Bloomington: Indiana University Press, 1989, pp. 147–171.

———. "Subjects, Power and Knowledge: Description and Prescription in Feminist Philosophies of Science." In Linda Alcoff and Elizabeth Potter (eds.), *Feminist Epistemologies.* New York: Routledge, Chapman and Hall, 1993, pp. 101–120.

Mies, Maria. "Women's Research or Feminist Research? The Debate Surrounding Feminist Science and Methodology." In Mary Margaret Fonow and Judith A. Cook (eds.), *Beyond Methodology: Feminist Scholarship As Lived Research.* Bloomington: Indiana University Press, 1991, pp. 60–84.

Miles, Rosalind. *The Women's History of the World.* New York: Harper and Row, 1990.

Millman, Marcia, and Rosabeth Moss Kantor. "Introduction to Another Voice: Feminist Perspectives on Social Life and Social Science." In Sandra Harding (ed.), *Feminism and Methodology.* Bloomington: Indiana University Press, 1987, pp. 29–36.

Mohanty, Chandra. "Under Western Eyes: Feminist Scholarship and Colonial Discourses." In Chandra Mohanty, Ann Russo, and Lourdes Torres (eds.), *Third World Women and the Politics of Feminism.* Bloomington: Indiana University Press, 1991, pp. 51–80.

Namenwirth, Marion. "Science Seen Through a Feminist Lens." In Ruth Bleier (ed.), *Feminist Approaches to Science.* New York: Pergamon, 1986, pp. 18–41.

Narayan, Uma. "The Project of Feminist Epistemology: Perspectives from a Non-Western Feminist." In Alison M. Jaggar and Susan R. Bordo (eds.), *Gender/Body/Knowledge: Feminist Reconstructions of Being and Knowing.* New Brunswick: Rutgers University Press, 1989, pp. 256–269.

Peacock, Nadine R. "Rethinking the Sexual Division of Labor: Reproduction and Women's Work Among the Efe." In Micaela di Leonardo (ed.), *Gender at the Crossroads of Knowledge: Feminist Anthropology in the Postmodern Era.* Berkeley: University of California Press, 1991, pp. 339–360.

Povinelli, Elizabeth A. "Organizing Women: Rhetoric, Economy, and Politics in Process." In Micaela di Leonardo (ed.), *Gender at the Crossroads of Knowledge: Feminist Anthropology in the Postmodern Era.* Berkeley: University of California Press, 1991, pp. 235–254.

Reinharz, Shulamit. *Feminist Methods in Social Research.* New York: Oxford University Press, 1992.

Rosaldo, Michelle Zimbalist, and Louise Lamphere (eds.). *Woman, Culture, and Society.* Stanford: Stanford University Press, 1974.

Said, Edward. *Orientalism.* New York: Pantheon Books, 1978.

Shostak, Marjorie. *Nisa: The Life and Words of a !Kung*

Woman. New York: Vintage, 1983.

Smith, Dorothy. "Women's Perspective As a Radical Critique of Sociology." In Sandra Harding (ed.), *Feminism and Methodology.* Bloomington: Indiana University Press, 1987, pp. 84–96.

Wolf, Margery. "Chinese Women: Old Skills in a New Context." In Michelle Zimbalist Rosaldo and Louise Lamphere (eds.), *Woman, Culture, and Society.* Stanford: Stanford University Press, 1974, pp. 157–172.

Yunnan Women's Federation. *Visual Voices: One Hundred Photographs of Village China by the Women of Yunnan Province.* Kunming: Yunnan People's Publishing House, 1995.

Zavella, Patricia. "Mujeres in Factories: Race and Class Perspectives on Women, Work, and Family." In Micaela di Leonardo (ed.), *Gender at the Crossroads of Knowledge: Feminist Anthropology in the Postmodern Era.* Berkeley: University of California Press, 1991, pp. 312–336.

Conceptualizing Change and Equality in the "Third World" Contexts

Shelley Feldman

Introduction

Women constitute a growing proportion of labor forces throughout the world (ILO/INSTRAW, 1985; Nuss, 1989). Their integral role in global manufacturing networks and in household production systems that contribute to market and home consumption, and their access to educational and credit resources, have been well documented (World Bank, 1990). Noting their prior exclusion from national labor-force surveys, scholars have used women's participation rates as agricultural laborers, unpaid household workers, and urban workers in the informal sector of the economy to reconstitute the category of employment (Anker, 1983; Leon, 1984; Tomoda, 1985; Wainerman, 1992).

Why, then, is it important to revisit changes in women's economic and social participation in the 1990s? What more can we learn from retelling the story of the changing configuration of women's participation in the global economy? If we seek merely to acknowledge and document women's participation in economic and social production, revisiting these questions will add little to the discussion. However, if we want to probe more deeply into questions about how the goal of equal opportunity has been promoted and implemented in development practice, we must explore the bases upon which gender equity is assessed and reexamine how changes in gender relations are conceptualized and measured. This essay takes such reexamination as its point of departure.

The dramatic changes in power relations brought about by the globalization of financial and labor markets and by critical reinterpretations of modernity provide fertile ground for rethinking changing gender relations and development practice in the Third World. The term "Third World" is used here as a way to specify postcolonial settings in which aid dependence and structural adjustment characterize the political and economic context of resource distribution and policy initiatives. The term is not meant to homogenize a particular group of countries nor under-estimate or ignore their historical and contextual specificities and national capacities. Understanding how colonial discourses shape interpretations of Third World change and transformation invites attention to the following questions: How does the modernization narrative mask the specific and heterogeneous conditions and relations of Third World women? (By modernization, I refer to particular objectives that are assumed to support a country's development: the acceptance of modern agricultural practices, such as those embodied in the Green Revolution, and the use of modern machinery and technology; a linear understanding of change that is characterized by increasing industrialization and urbanization; an increase in capitalist forms of accumulation; and a trickle-down notion of the adoption, diffusion, and benefits of these changes.) How do normative interpretations of changing gender relations lead to (mis)understanding the achievements won by, and for, women during this period? What markers have been employed to assess alterations in these patterns, and how well do they account for women's varied opportunities and diverse needs in the contemporary global economy?

These questions pose fundamental challenges to previous thinking that holds that individualism and enlightenment rationality are the core assumptions for interpreting the cultures and histories of non-Western societies. For many policymakers, analysts, and practitioners working in postcolonial contexts, the unproblematic acceptance of these core assumptions has meant that differences between Western and non-Western societies are understood to consist primarily of whether modernization serves as a vehicle for women's emancipation or simply reinforces systems of gender inequality. Both interpretations depend upon and share assumptions about measures of progress and change and about the objectives and goals of the development model. In assessing changes in women's status or control of resources, both interpretations also tend to compare nonfree, often collective, labor forms to modern, albeit limited, expressions of possessive individualism.

I begin with the assumption that knowledge is socially produced and embodies specific interests and historical circumstances. In this view, "knowledge is situated and limited by its positioning . . . (and) reflects our social experience, our understanding of our interests and our values" (Strickland, 1994, p. 265). Different strategies of description and understanding are products of values and objectives created in a specific "transnational network of power relations" (Ong, 1994, p. 372–373) within an interstate system of national interdependence. Historical circumstances embody different relations of power among and between nations and peoples characterized by forms of class, ethnic, and gender inequality. I use this materialist understanding of power to explore here how change and equality have been conceptualized and measured in different development contexts. Throughout, my emphasis is on how new interpretations of gender inequality are incorporated into programs and policies for national development—rather than on reinterpretations of gender and inequality that have been central to the scholarly community.

Probing further the idea that paradigms, discourses, and practices are socially constructed, I critically examine how concepts of change and equality have been formalized into measures or categories for comparing gender differences. Here I focus on the way established measures of change are employed to justify programmatic intervention. I next show how initiatives to enhance gender equity have tended to homogenize the needs of women by embodying meanings that are constructed in the narrative of the generic (wo)man. I argue that simple comparisons of change in work relations and labor-market participation across place and social context can be misleading, since they often depend upon Western values and assumptions. I premise this argument on what Ong identifies as the need to jettison our conceptual baggage "to produce an understanding of gender as constructed by, and contingent upon, the play of power relations in (historically specific social and) cultural contexts" (Ong, 1994, p. 378). This "requires moving beyond a view of the non-Western woman . . . as an unproblematic universal category" (Ong, 1994, p. 374) or as a victim of development (Mohanty, 1988). It requires, as well, an examination of the assumptions framing development—that is, purposive and planned change or transformation with the objectives of enhancing individual choice and equalizing opportunity—assistance, since these assumptions provide the backdrop for the various initiatives taken to alter women's position and to assess changes in gender relations in selected peripheral contexts.

Before examining the consequences of varied interpretations of changing gender relations, I briefly show how modernist discourses have shaped both policies and programs that characterize the period since the United Nations Decade for Women (1975–1985), as well as gender-and-development scholarship. I also indicate the prominence of these discourses in recent discussions about economic restructuring.

The "Women in Development" Debates and Struggles for Equity

The women-in-development (WID) nomenclature embraces analyses of women *in* development to indicate efforts to add women to those who are affected by changes in the political and social economy. Also, the shift from the term "women" to that of either "gender" or "gender relations" often has been nominal at best. Despite this, a number of scholars and a still smaller proportion of practitioners have used the shift from woman to gender to recognize the power dynamics shaping relations between women and men and to incorporate these dynamics in their analyses and practice. Initiatives, ranging from multilateral and bilateral lending agreements to national policy prescriptions, to alter the conditions of Third World women's lives were premised upon two critical themes: first, the fact that states must retain "primary responsibility to safeguard domestic welfare and levels of employment and economic activity . . . aimed at the harmonisation of different national economic policies"; and, second, fear about the consequences of increasing social instability and rising poverty and fertility (Cox, 1993, p. 260). Such themes promote the extensive reorganization of social relations into commodity forms, where states "become the instruments for adjusting national economic activities to the exigencies of the global economy" (Cox, 1993, p. 260). These changes proved to be costly both for women and for other dispossessed groups, resulting in a series of challenges to the state that began in the 1960s and 1970s (Maguire, 1984). In the South, these challenges included struggles for national independence and against the social and economic injustices brought about by the increased integration of the world economic order. In Europe and North America, antiwar and anticolonialist struggles were organized around movements for class, race, and gender justice. Challenges were also levied against the trickle-down development models of the post–World War II period, which, by the early 1980s, were accompanied by increasing unemployment, inequality, and absolute poverty. Common to each of these challenges were moves toward decentralization, local development, and grass-roots initiatives that focused attention on the plight of the excluded—the rural landless, the urban poor, and women—and a shift in development discourse to "meeting basic needs."

Efforts to meet basic needs were accompanied by a growing urgency to incorporate the dispossessed more explicitly in the rewards of development assistance. Women were targeted as one group whose needs could be met by adding programs to increase their access to resources. The implicit objective of many of these initiatives was to transform agricultural commodity production, heretofore focused primarily on middle- and large-scale producers, and to mediate the costs of this transformation for small-scale agricultural producers and the increasing numbers of landless who were dependent on the rural nonfarm sector for employment. For example, the integrated-rural-development model, supported by the World Bank and popular in places across the globe, provided women with training and credit in sewing, literacy, and selected home-based activities. These resources tended, however, to reconfirm women as those who supplement the family income and who, with modest additional resources, can increase their productive capacity and efficiency. Thus, while modifying the language of development, these "women's programs" did not significantly alter extant interests; challenge the commitment to increase efficiency and productivity; question the household as an integrated, homogeneous unit; or move away from macroeconomic adjustment policies and politics.

Scholars during this period also focused on women's roles in economic development and contributed to realizing these development objectives. Boserup's (1970) landmark study demonstrated the costs of women's exclusion from access to agricultural resources, challenged generalized and naturalized notions of women's exclusion from field production, and emphasized the removal of institutional barriers that limit women's productive capacity and competitive position in the market. The premise that productive work is of greater value than unpaid household activity provides the key to her argument that women's economic activities underlie their differential status. Rogers (1980) shared Boserup's commitment to market competition as the means for increasing gender equity: "Ways [must] be found to . . . improve [women's] productivity and, most important, to ensure that 'incentives for increased production are channelled to the women as well as the men, in proportion to the contributions made by each'" (Rogers, cited in Kabeer, 1994, p. 23). In arguing that gender inequality could be reduced by removing the bureaucratic restrictions inhibiting women's participation in remunerative work, both sought technical solutions for altering the institutional constraints of the capitalist market. But while Boserup and Rogers sought to improve women's status through participation in market-based exchange relations, they underestimated women's reproductive responsibilities in status production.

Thus, while these authors acknowledge women's responsibility for productive agriculture, their goal is to increase productive capacity by making economic relations more competitive and commoditylike. Rather than challenge the existing development paradigm, these scholars, like their practitioner counterparts who seek increased gender equity, merely identify a new constituency to increase Third World output and raise household incomes. Given these objectives, it is not surprising that women became *targets* for meeting productivity and efficiency goals. But it also is not surprising that this framework provided no challenge to the idea that women were solely responsible for the tasks associated with reproduction, including child care, domestic work, and other unpaid household tasks. Although issues of nutrition and health care were addressed, and contraception and sterilization campaigns flourished throughout this period, not until the later 1980s did discussions include reproductive rights and issues of sexuality. Even among what are defined as women's issues, there is a particular gendering of how and when these issues are discussed and acted upon. Despite its limited challenge, recognizing women as producers as well as those responsible for household reproduction contributed to shifting development assistance and medium-term development objectives toward organizing special programs to provide women access to resources that would enhance their financial contribution to the household.

Indeed, as targets of development assistance, women were now charged with the additional responsibility of securing an independent income. What is important to note here is that these demands rested on a logic of individualism, a logic that presumes an oppositional relationship between women and the institutional settings that frame their daily lives. For example, targeted as captive recipients, women were often presumed to reside in extended or nucleated households, bastions of patriarchal control, from which they must be helped to escape in order to realize the value of wage employment. As such, women were cast simply as victims who are unable to compete for skills and resources because of traditional values associated with subsistence-based production. It must be noted that while studies of women's resistance and intrahousehold bargaining have more recently been elaborated (see Sen, 1990), earlier conceptions of the relationship between women and their households were framed by patriarchal familial relations limiting women's voice and opportunity. The policy goal was to free women from backward and oppressive households and patriarchal relations that prohibited their participation in the labor market. Increasing women's wage employment, through employment-generation schemes sustained by training and primary-education projects and credit for microenterprise development, was central to realizing this goal. Becoming wage laborers and self-reliant individuals then became the criteria for measuring women's status, mobility, and achievement.

Technical training projects and credit schemes were initially limited to capitalizing agricultural activities for which rural women often held responsibility—cow and goat fattening, fish and vegetable production, and the preparation of food (beer, tortillas, and small snacks) for street sale. The burgeoning informal economy in both rural and urban communities also garnered the support of special credit schemes to capitalize craft production, petty trade, and grain and other food processing (rice, wheat, pickles). Almost all of these projects distribute credit on an individual basis, although they often rely upon village groups to reduce the institutional costs of individual lending and to provide peer pressure, which is assumed to ensure high repayment rates. The Grameen Bank, an NGO in Bangladesh founded in 1976, is perhaps the most well-known decentralized rural banking scheme for women and poor men. It has since begun to organize women to pool resources for joint projects.

With the more complete reorganization of agricultural

production by the mid-1980s, research and policy attention turned from investments in agricultural and agroindustrial projects to providing women with skills that would enable them to better compete in the broader national, as well as global, marketplace. This included a focus on the nonfarm rural economy and, increasingly, on the growing numbers of urban migrants and urban poor who sought employment in both the nation's capital and the provincial cities. Participation in "putting out" networks (subcontractors to large, often multinational, firms) and in home-based family enterprises expanded, as did recognition that "informal" labor relations were already part of household production (Mies, 1982; Roldán, 1985; Benton, 1989). Thus, the availability of credit, in combination with new skills and training, transformed subsistence and nonwage relations into marketable skills, enabling women to engage in enterprises in ways that were increasingly similar to those of men.

At this same time, garment and electronic manufacturers relocated their production from Taiwan, South Korea, Singapore, and Hong Kong (where off-shore production initially had been established) to countries that had no quota restrictions on exports to the United States and significantly lower labor costs. Moreover, in efforts to lure investment and expand employment opportunities, governments in Thailand, Malaysia, and Bangladesh subsidized the development of special zones or enclaves by providing water, power, and other infrastructural investments and supported tax holidays, low wages, and limits on worker mobilization.

These changes were consistent with World Bank and International Monetary Fund (IMF) demands that countries renegotiate their approach to development, a process that shifted national economic planning from import substitution and agricultural self-sufficiency to strategies of comparative advantage and export-led growth. Moreover, the need to service the national debt and retain favor with the donor community encouraged governments to expand wage employment and to modify their labor policies toward increased support for private entrepreneurship. Export-processing zones (EPZs) in these countries and elsewhere built upon the training and skills acquired by women's participation in both government and nongovernmental-organization (NGO) projects and increased nonfarm employment opportunities at a time of relative stagnation or decline in the demand for agricultural workers. The demand for labor in the EPZs also complemented declines in rural investment and the consolidation and capitalization of agriculture, which reduced rural labor demand as well as the family wage. These shifts led to the definition of women as economic actors, disconnecting them from their work as unpaid family laborers, wives, and mothers. Deeply insinuated in this strategy was a vision of women as a source of cheap labor.

By the later 1980s, then, the gender agenda increasingly included the concerns of women within the burgeoning urban informal economy. Programs for urban women focused on capitalizing and extending unpaid household work and cottage-industry production in processes of economic reorganization. In some cases, income from these enterprises was greater than that earned from government service or from government-registered, private-sector firms (Tripp, 1992), and women's income was acknowledged as part of a household's survival strategy. As families were no longer able to manage on the earnings of a single household member, the ideology of the family wage was discredited.

However, rather than view women's work as integral to processes of economic reorganization and central to household survival, policymakers and development practitioners alike continued to focus on the short-term and adverse effects of structural-adjustment policy reform on women. They felt that these negative effects could be mediated by special programs for women and children until economic equilibrium could once again be realized (Cornia et al., 1987). Despite the problems entailed in this analysis, the discussions nonetheless linked the costs of economic reorganization directly to poverty, motherhood, and the privatization of welfare institutions (Cornia et al., 1987; Bakker, 1994). This represents a significant departure from prior research, which attributed women's unequal status to their lack of access to resources, their limited skills, traditional cultural practices, and their relationship to patriarchal household structures. In the 1990s, scholars acknowledging the long-term consequences of reductions in social expenditures have recognized that women bear the unequal burden introduced by economic restructuring and social dislocation and have concluded that these costs are not temporary (Elson, 1990, 1992; Afshar and Dennis, 1992; Benería and Feldman, 1992; Feldman, 1992; Bakker, 1994). At the same time, more than 10 years of targeting resources to women has done little to successfully mediate various forms of gender inequity.

Autonomy and Empowerment: The New Challenge?

Discourses of empowerment in development research and practice, popularized during the early 1990s, reflect a significant shift away from a WID approach to a gender-and-development (GAD) perspective. Whereas WID initiatives were deployed to put women back into a social science and a development practice that had excluded them, GAD sought to "... look for the potential in development initiatives to transform unequal social/gender relations and to empower women" (Canadian Council for International Cooperation, in Braidotti et al., 1994, p. 82). Drawing upon poststructuralist and postmodernist critiques of determinism, the politics of empowerment stresses the salience of power, personal experience, and agency as sources of change: People act as subjects who constitute their worlds, rather than as objects of structural demands placed upon their time and interests.

In contrast to approaches to empowerment that alter women's ascribed status, the GAD approach challenges the institutional bases of gender inequality by emphasizing the unequal costs of economic reorganization for women and men. By so doing, they call into question the social costs of processes of privatization and global integration. Particularly noteworthy are the collective efforts of Latin American women who organized *comedores populares* (community kitchens [in fact]) as alternative forms of subsistence in Peru and Chile during the 1980s (Moser, 1989; Lind, 1992; Sternbach et al., 1992). These efforts illustrate the transformative potential of alternative survival strategies constructed around a politics of consumption, strategies that, as Elson argues, "transform gender relations by meeting both women's practical needs and their strategic gender needs" (Elson, 1992, pp. 41–42). Her argument is based on Molyneux's (1985) important essay that distinguishes between women's everyday condition and women's social position, between practical and strategic gender interests. These examples show that, by the mid-1980s, strategies to alter and reconstitute gender relations were framed not only in terms of achieving equity for women but, more fundamentally, as challenges to the institutional bases of gender inequality.

Despite this understanding, many development programs continue in the 1990s to assume that women can be empowered through external initiatives. Some programs, for instance, assume that offering women opportunities to engage in cottage production will increase gender equity and empower women. In this view, women's contribution to family income through wage or self-employment serves as a key indicator of women's autonomy and empowerment, because this contribution is assumed to strengthen women's bargaining position within the household. The increased independence, autonomy, and self-reliance that come from contributing to household income also are assumed to make women independent decision makers with regard to family planning. Literacy and educational programs likewise provide women with opportunities for income earning, help them build self-esteem, and contribute to the resources women can negotiate in marriage arrangements. The aim of these activities is to ensure that women participate in, and benefit from, the resources provided by development assistance, including the goal of overall fertility decline.

But, while many development programs incorporate the language of autonomy and empowerment, support for women's autonomy and empowerment too often has been dissociated from long-term strategic demands for institutional transformation. This is because programs retain a focus on the self-reliant individual, even in activities that require group formation, and organize and distribute credit only to these local groups. For instance, rather than encourage group lending as a way to reorganize production, these groups merely enhance institutional viability and enable better monitoring and credit repayment. Such efforts may even subvert collective production opportunities and indigenous forms of labor organizing by holding group members responsible for ensuring each individual's repayment. Under pressure to provide surveillance for one another,

these groups are as likely to foster inequality and tension as they are to build community and solidarity.

Fostering self-reliance, an implicit goal of many versions of empowerment politics, may thus actually diffuse potential efforts to transform social and economic inequality. It also may reproduce liberal notions of individual responsibility and leave unquestioned structural inequities and particular articulations of gender difference, as well as help consolidate state power by incorporating indigenous forms of organizing and local expressions of authority under the umbrella of nongovernmental development organizations. Although nongovernmental organizations vary, the successful efforts of individual states and the donor community to incorporate many of their challenges generate contradictory results for participants. NGO demands have improved conditions of resource access for some women. However, the development of NGO programs has also diffused claims that challenge the deep roots of structural inequality. The goal of increasing women's autonomy may thus be a way to realize the objectives of development—greater economic productivity and efficiency—by giving women access to more resources, while mediating deeper challenges to forms of structural inequity and the materiality of women's subordination. These limitations notwithstanding, challenges to gender inequality may generate outcomes that embody transformative possibilities able to create relations of gender equity, autonomy, and empowerment.

Measurement, Assessment, and the Development Model

Assessments of changes in Third World women's status have depended on critical evaluations of programs that have provided women access to an increasing array of productive resources; most research has employed what are presumed to be objective categories to measure changes in literacy and educational levels, employment status, age at marriage, household and/or family size, and fertility levels (World Bank, 1990). The most critical contribution made by gender researchers using these measures during the 1970s and early 1980s was to document the range of activities and descriptions of the social context in which women's labor has been deployed in everyday life. These efforts identified women's direct contribution to the gross national product (GNP) and acknowledged as well the subsidy women provide to expanding capitalist forms of production. Researchers documented the transition from capital investments to the development of human capital—improvements in education, health, and social security, and the more equitable distribution of income and social justice—and argued that the latter would increase the productivity of the poor and support a politics of growth with redistribution. While some studies did examine how middle-class, rural, and urban educated women met the responsibilities of wives, mothers, and workers, attention focused primarily on poor rural and urban populations and female-headed households. Research that highlighted the complexity of women's work in agriculture; unpaid, home-based activity; and informal enterprises has been particularly important since, in some cases, it provides the only benchmark data for assessing further changes in women's work relations and status as well as control and access to resources (Safa, 1990; Scott, 1990; Safa and Antrobus, 1992; Mohiuddin, 1993).

To reiterate a well-documented but critical point, prior to the 1980s, employment and agricultural surveys usually excluded women from the categories of producers and workers because women's economic participation, being more temporary and flexible than that of men, was difficult to document accurately. Additionally, work was defined primarily by male activities, and the incomes earned or subsistence produced by female household members were masked by constructions of the male family wage. Also, surveys of institutional access to resources often were incomplete since data on public and private education, health-care utilization, and disease monitoring rarely were systematically collected. In some contexts, moreover, women's access to services, particularly their use of health facilities, was negotiated through male kin who brought women's symptoms to the doctor, secured a prescription or treatment plan, and decided whether to provide her with medicine. These practices contribute to the absence, underreporting, or misrepresentation of data on women's work and resource use that are the bases of measures of gender equity and change.

While these measures were critical in documenting women's differential access to economic and social resources, and while they identified women's hidden contributions to household and subsistence production, agriculture, and other forms of unpaid family labor, they remain problematic. First, seen as value neutral, these categories are wrongly assumed to measure the same things and to have the same meaning in different social contexts. Information collected under assumptions of value neutrality and standardization is used to measure improvement and progress toward goals of economic independence and the increasing productive capacity of women across time and place.

Second, in addition to the technical reasons constraining data accuracy are the problems incurred by ignoring women's labor in nonmarket exchanges and by subsuming within the family wage women's unpaid work in food production and in home-based enterprises. The lack of critical attention to these problems has resulted in a limited basis for imagining alternative conceptions of labor relations, production processes, and ways to negotiate resource use. Without the categories to help identify the range of activities that women undertake in subsistence households, neither researchers nor women engaged in these activities forced their inclusion in assessments of the work that constitutes subsistence production. For instance, rice production was understood to include sowing, transplanting, irrigating, weeding, harvesting, and threshing, thus ignoring the labor embodied in women's contributing to production—rice drying, husking, winnowing, and parboiling.

Using a definition of rice production that is external to the logic engaged by women makes those activities that "have no name" invisible. The logic imposed on such settings, often constructed in the West, tends to preclude understanding different organizational strategies, such as those that frame domestic economies and extended-kin exchanges. As a consequence, various forms of labor obligation, barter, and reciprocity may be occluded from a view that imagines social organization and exchange as carried out among individuals and free-wage laborers in more contractual terms.

Because of the assumptions imbedded in the categories and indicators that measure women's participation in wage employment, it is not surprising that women's increasing responsibility for individual subsistence is readily accommodated in the reorganization of social and economic reproduction. Increases in informal, small-scale commodity production, the decentralization of production, and the delivery of credit and other resources, as well as reductions in bureaucratic support—each contributes to the likelihood that women secure wage or self-employment. The growing number of women employed in government service and the incentives offered to employers to hire women also increase women's wage employment. Likewise, women's increased access to agricultural resources and their participation as extension workers and health-care and family-planning workers all become factors that imply an increased gender equity, as do campaigns to encourage the enrollment of girls in primary school and efforts to bring medical services to local communities. All of these indicators gain credibility under the political impetus of neoliberal efforts to transfer accountability for social subsistence from states to individuals and to expand women's participation in commodity production and exchange (Feldman, 1992).

Similarly, gender research that has provided information on the ways bureaucratic arrangements have restricted women's access to development resources has been used to support policies promoting women's need for work, income, and services. This information was deployed to mobilize international and national resources to expand women's access to primary school, maternal and child health care, and sterilization and family-planning services. But, here, too, one can see how data were collected for the purpose of enhancing the ways in which women could be incorporated into processes that capitalize extant relations and further the objectives of the modernization project.

Measures that examine women's increased control of particular resources while ignoring how these resources are integrated into women's lives are likely to underestimate or conceal demands within the household for which women generally assume sole responsibility. Such measures discount both the different meanings women and men give to their control of resources and their different assessments of gender parity. In presenting what are presumed to be neutral categories of comparison, such measures mask the different ways that activities combine to shape the varied needs and obligations of women and men. What is called for, instead, are categories that illuminate the ways that social activities are situated in everyday practices. This situatedness, usually unaccounted for in existing categories for measuring gender equity and change, is nonetheless integral to notions of comparative advantage and opportunity costs and in interpretations of gender differences in pay, seniority, and mobility. Constructing comparative categories is usually premised on ignoring contextual specificities, since the categories themselves are offered as objective, external criteria not muddled by local contingencies. "Social research depends on standardization. . . . [M]easures must be made interchangeable; land and commodities, labor and its products, must be conceived as represented by an equivalent in money." This entails much of what Max Weber called rationalization and a good deal of centralization (Porter, 1994, p. 201).

Current measures also mask the meaning and impact of rapidly changing technologies and new social practices on gender relations among Third World countries. For example, innovations in production generally undermine previous conditions of economic security. While this cost is often justified because such changes are said to reduce the arduousness of women's work and to promote greater gender equity, they nonetheless erode women's control of selected production processes.

A well-documented case concerns the mechanization of rice processing in many parts of Asia (McCarthy, 1981). With mechanization, the centralization of rice mills, and the simultaneous transformation of home-based processing, women lost control of, and eventually the knowledge and skills required to engage in this economic activity. This transformation also reshaped intrahousehold and community networks and forms of labor organization. For poor women able to secure wage or in-kind work as processors for surplus producers, it resulted in significant declines in employment. Women also lost control of the product that, when integrated in the household economy, provided animal feed and fuelwood for a relatively sustaining reproduction cycle. With mechanization, these by-products were removed from the control of producers, an innovation that also altered the demand for purchased goods, women's dependence on the wage economy, and intrahousehold gender divisions of labor.

This example suggests that freeing women from the unpaid labor of rice processing and applauding their increased participation in the labor market must be examined against their vulnerability in the labor market, their increasing dependence on market goods to secure household sustenance, and their need to renegotiate waged and unpaid claims on their time, energy, and resources. For those who lose informal employment opportunities as rice processors, the result may be devastating, since poor, but specially skilled, laborers are among those least likely to find secure employment in a rapidly changing rural economy. In these cases, employment in the interstices of the economy is likely to be unaccounted for in existing measures of labor-force participation, resulting in inflated estimates of women's increasing employment.

Such consequences suggest that caution is called for in reassessing changes in family status as women move from resource controllers to those dependent on insecure forms of employment. While initial evidence indicates that labor-force participation increases women's status and decision making in the household, it is important to keep in mind that this assessment is contingent on specific historical and social circumstances. When values associated with what it means to be modern become dominant, women's status may be enhanced through employment, particularly in a differentiated economy in which employment can be secured in new, hi-tech, factory production, or women may lose family status and authority if they are able to secure work only in craft or home-based activity.

Alternatively, in situations where the political discourse is framed as a challenge to Western, capitalist intervention, women's paid work may be viewed negatively, and employed women may have lower family status than those engaged in unpaid domestic activity. In these circumstances, employed women may face harassment, violence, and other attacks on their personal security, particularly during periods of economic and social transition and dislocation. These examples emphasize that women's status is contingent upon particular social circumstances and embodies different values and meanings across time and place; social meanings, in other words, are neither self-evident and unchanging nor readily transferred with new employment opportunities or alternative social practices. Also evident: Measures of apparent similarity in the opportunities or behaviors of women and men may, in fact, reflect quite different meanings and have quite different gender consequences across time and place.

Training programs for women also are not value neutral and may have contradictory consequences, increasing women's vulnerability, mobility, and independence: Specialized training projects for women often help (re)produce divisions of labor whereby men become more competitive for selected, usually higher status and higher salaried, posts, while women compete for special "women's positions." With demands for fiscal austerity, denationalization, and privatization, special women's positions have been among those more likely to be eliminated. These differences must be included in constructing measures of program success and in assessing the adequacy of such measures in accounting for changes in gender parity.

Programs also differentiate the kinds of resources made available to women and men based on stereotypic and homogenized understandings of women's experience. The Integrated Rural Development Programs (IRDP) supported in many countries by the World Bank during the mid-1970s differentiated between the resources provided to women and to men. In Bangladesh, for example, men's programs focused on increasing agricultural productivity through credit, training, and infrastructure-development schemes, whereas women's programs were inaugurated as population-control initiatives based on the assumption that increasing women's income would enhance their status, reduce their desired family size, and thus lower the national reproduction rate. Credit was a way to increase women's incomes but was limited to particular women's activities and was offered in amounts that were insufficient to undertake projects that would enable accumulation and expansion (Feldman et al., 1980). Among nongovernmental organizations at the time, differences between women's and men's credit access also prevailed, even when only implicitly acknowledged. Women were given credit for activities perceived to be in areas of their competence, such as in poultry rearing and animal fattening, but in such small amounts and under repayment schemes that inhibited its productive use. In the CARE program, for example, women had to begin repayment a week after receiving a loan and continue repaying 5 percent of their loan each week for the next 19 weeks. Not surprisingly, the women wondered, how they could sell a fattened goat after one week and how they could meet their repayment requirement if they undertook this activity (Feldman, 1978–82).

Gender differences in credit amounts are commonly justified by institutional, rather than individual, criteria. For instance, concerned with averting risk, credit programs presuppose that the experiences of women limit their productive use of credit, and their marked exclusion from credit institutions constrains their ability to meet repayment schedules. Programs thus justify offering women less credit than they do men using the discourse of the market; insufficient documentation of women's productive capacity, efficiency, and repayment history and their general lack of collateral, particularly landed property, make women poor credit risks. (For poor men, most programs ignore their lack of experience with banking institutions and the absence of a credit history.) Such differences, in concert with those resulting from dissimilar training opportunities, limit women's remunerative choices and reconstitute, if differently, a segregated and segmented labor market.

A growing number of program assessments recognize the achievements women have made in increasing their access to selected resources. These achievements usually are measured by comparing women's past and more recent access to credit, training, and other opportunities. The above-mentioned internationally recognized Grameen Bank in Bangladesh, for example, has placed women at the center of its campaign for credit decentralization worldwide and highlights women's successful use of credit in formal activity. The Self Employed Women's Association (SEWA) and the Working Women's Forum (WWF), both begun in India in the 1970s, highlight how women working in the informal economy as petty traders, hawkers, and laborers formed, registered, and organized as an independent trade union. Both SEWA and WWF are organizational achievements that exemplify the potential of women to organize, demand recognition and rights within the international labor movement, and extend the principles of democracy to poor women (Jhabvala, 1994; Mitter, 1994).

This new focus on women as active agents of change epitomizes the shift in recent scholarship from analyses of women as victims of economic crises to the centrality of women's work as an organizing premise of shifts in the broader political economy. These recent efforts also serve to identify the meaning of these shifts for understanding women's changing status and economic security, as well as their autonomy and empowerment.

Excluding the examples from India, most credit programs for women are directly connected to questions of inheritance, private property, and alternative forms of labor organization, all of which characterize institutionalized inequalities that rarely become the focus of organized women's programs or challenges to economic insecurity or gender inequity (Stivens, 1994). When programs organized to enhance women's autonomy and independence do not challenge these institutionalized inequalities, they reproduce, rather than call into question, existing social arrangements. Importantly, the failure to question such inequalities reinforces neoliberal assumptions—a commitment to private property, individualism, and self-reliance—that underlie current development discourse.

Finally, it is important to underscore the failure to question how women's programs generate changes that obscure the role of measurement in constructing both new social practices and particular interpretations of change. Embodied in the failure to question initiatives to alter gender inequities is a presumption that such initiatives, like the measures used to assess them, are value neutral. As at least one researcher has argued, "society must be remade *before* it can be the object of quantification [and measurement]" (Porter, 1994, p. 201, italics mine). Instead, I argue that efforts to bring about change and to measure it are not value-neutral endeavors in the search to understand the reorganization of gender relations. Rather, in contributing to assessments of the changing status of women, measurement categories, drawn from Western, positivist science, themselves constitute the very processes by which social contexts in the Third World are transformed.

Some of these issues have been explored in research on Third World women by social scientists and humanists engaged in postcolonial studies. Postmodernist challenges to development theory highlight the imperial bases of Western analyses. Feminist challenges to positivist notions of objectivity and the truth claims of normative science also have made important contributions to the practices of microlevel analyses, particularly those that interrogate the researcher's location and positionality (i.e., her perspective in addition to the researched–researcher tension) in interpretations of social change and difference. Additionally, efforts to integrate the ways that social values help frame research questions, characterize objects of inquiry, justify forms of explanation and argument, and embody interpretations of social change have been of critical importance in constructing emergent discourses on women and in framing our understanding of social analysis. However, through the mid-1990s, these breakthroughs in conception and interpretation have yet to be fully incorporated into an understanding of how macro-level changes in gender relations are examined, measured, and interpreted.

Envisioning It Differently

The above overview argues that reforms that create new opportunities for women by channeling or mainstreaming them into the institutions and programs that support economic reorganization fail to challenge the fundamental assumptions of the modernity narrative. This narrative (or discourse) is premised on a view of change as progress measured in terms of the increased capitalization of commodity production, greater participation in remunerated labor, and the increasing individuation of social life. The measures used within this narrative to assess women's status employ categories that assess changes in gender relations through women's acquisition of resources that were previously available only to men.

Moreover, this narrative presumes a set of social categories in which selfhood is constructed in industrial contexts and in which participation in wage employment, market consumption, autonomy, and independence make up the essence of what it means to be a modern, rational individual. These categories draw on what Spivak (1983) refers to as "an ideology of free enterprise" founded upon the sovereignty of individual subjects and their right and ability to choose from among a number of different opportunities. Shifts toward an ideology of free enterprise are contrasted with practices considered either backward or representative of the transformation of a romanticized "natural economy" (Ong, 1994).

Imposing such categories on the activities of Third World women may transform the meaning of these activities, reducing the comparison to an apparent, rather than a real, measure of similarity or difference. Since categories are already and always embodied in particular meanings, imposing them on activities grounded in other social contexts risks transforming their meanings into paradigmatic frameworks or naturalized concepts that erase the possibility for any other understandings. This imposition also leaves aspects of life unexamined either because they are invisible to those for whom there are no concepts to capture what is going on or because such activities have no salience for those unable to recognize the complexity of women's everyday lives.

When researchers or program evaluators transform, ignore, or cannot even imagine an explanation for what they view as different, the measures of change they employ merely embody the values of outsiders who wish to implement programs that endeavor to alter the very diversity that constitutes the lives of women worldwide. Given such interests, these efforts to understand patterns of equity tend to conform to an imperial logic of Western domination realized in the expanding capitalization of production and

exchange. Moreover, like other hegemonic constructs, there is a conflation, or confusion, of a *particular* social order with one that is asserted to be natural, neutral, and value free. And, in its assumptions about objectivity, there is a denial of temporal, spatial, and cultural distinctions that would enable self-conscious examination of these processes.

Two significant consequences follow from this way of examining gender equity and patterns of change. First, this construction naturalizes economic and social dualisms and normalizes that which is male and part of the public domain—wage employment, political participation, and individual mobility. In this view, the individual is posed in opposition to society, always forced to challenge the cultural and structural constraints that impede her/his freedom and autonomy. Also, values associated with the market normalize the public domain as the site of status construction against which women's worlds are assumed to be separate and distinct. This separation has been used to explain why the domain of domestic responsibility, and the complex intrahousehold division of labor that makes possible particular production and exchange relations, have only recently been theorized (Folbre, 1986; Sen, 1990). Moreover, distinctions between the private domain of the household and the public domain of the labor market and politics, or between subsistence production and remunerative labor, are conceived as oppositional—or at best autonomous—rather than embedded social relations. While such presuppositions have been used to explore the complementarity of tasks between women and men, they cannot be the basis for apprehending how social processes are constitutive of one another, making possible the range of activities that comprise daily life.

Research that presupposes the independence of these domains can, at its best, draw attention to the relationship between them, but it cannot move beyond the notion of separate spheres. Reifying (objectifying, or accepting as real) the socially constructed categories embodied in public and private space, work and nonwork, and productive and reproductive labor makes it impossible to account for the ways in which these spatial domains and relations are determinant of each other. This dualist construction limits our ability to comprehend different conditions of situatedness, different meanings, and different measures of social reproduction, change, and identity formation; it also hinders an appreciation of how the category "women" is itself a social construction, since it does not call into question how women's and men's identities are constructed by, embedded in, and integral to specific (re)productive domains, including those that are characterized by reciprocal and redistributive social exchanges.

The second consequence: Measures of change premised on Western notions of progress and modernity, when not historicized, fail to account for actual transformations in social forms and relations. This means that changes in gender relations and assessments of gender equity are compared on the basis of nonequivalents. For instance, the shift from feudal and patron–client relations to capitalist social relations and commodity exchanges cannot be captured by measures of women's increasing wage employment, since this leaves unexplained the meanings of change in other forms of labor deployment and other constructions of the self and the social. Also ignored are the ways in which nonmarket relations structure the meanings and configuration of the activities being compared. In this context, reciprocal and in-kind exchanges among community members, grounded in meanings and understandings quite different from commodified forms of hired labor, are likely to be excluded from the construction of measures of gender equity. Or, when these exchanges are included, they are interpreted as insignificant, backward, or patriarchal and, therefore, inconsequential for interpreting changing gender relations. Such normative assessments of participation in wage work and publicly recognized activity also fail to account for the different social costs resulting from resources that have been lost or transformed.

Research acknowledging interactions between household labor and employment has unmasked unpaid and nonmarket labor forms and has credited women for the kind and duration of their varied labor investments (Benería and Roldán, 1987; Stichter and Parpart, 1990; Scott, 1994). Time-allocation studies are good examples of these efforts. However, the logic of these studies is often additive, rather than constitutive (to reconstitute social relations). Other studies show that women's increasing expectations and their actual participation in the labor market arise simultaneously with their increasing unpaid labor demands; both are consequences of declining economic security. A rise in subsistence production is one consequence of reduced household incomes especially when households come to depend on the purchase of poorer-quality food and other provisions. This, in turn, increases the amount of time women are likely to expend to secure and prepare food, purchase or make clothes, and maintain the household. Also, when commodities are purchased on credit, in small amounts, and from places proximate to residential communities, there often is a rise in the cost of commodities that places an increased burden on women to secure wage employment. These changes have been a particular focus of research examining the effects of structural-adjustment programs and declines in social expenditures on women's time and resources (Benería and Feldman, 1992). Such analyses appreciate the effects of particular state practices or interventions *on* women, but they ignore how women themselves, and particular gender relations, construct and constitute new social relations.

Conclusion

To imagine an alternative interpretation, we need to reject ideal-typical frameworks within which we fit assumed similarities in conditions, patterns, and relations between women and men. We need to move away from the tendency to impose an a priori logic on situations and reject, as well, nonreflexive understandings of social dynamics. We also need to depart from interpreting individuals as independent from, and in an antagonistic relationship to, family, kin, and community, as well as government, or state, practice. Instead, we must read, see, and understand situations from within and interpret different situations through interrogations of how they may depart from our expectations.

Finally, we must challenge the interpretations of events and practices as they are presented to us, rather than simply appropriate and reconstitute them through already existing categories and frameworks. This approach does not mean that there is a "way of knowing" already existing completely outside of our experience, nor does it mean that we cannot question our experience and find comparisons and equivalencies across space and time. Rather, it suggests that frameworks and systems of logic need to be understood in their own terms if we are to respect and understand their meanings and practical manifestations. This will help us comprehend how people make sense of social relations and practices in contexts that differ from those with which we are familiar. (While my reference here is to cross-cultural differences, it is important to emphasize how ethnic, class, and spatial experiences are also stylized and homogenized, how they also structure knowledge production, and how we interpret different social contexts.)

It is important to recognize not just that experiences differ, but that "my understanding of my own experience may not be adequate or sufficient" to enable an understanding of another's experience. "I have to accept that others are offering their different experience at least partly in critique of my understanding of the world, and that I might need to change or revise my categories and understandings in light of it" (Strickland, 1994, p. 268). Such an approach is premised on interpreting difference as partial and multiple (Haraway, 1991), and recognizes that different logical constructs are the means by which different people make sense of their lives, derive meaning from their activities, and produce social situations that create their communities.

Processes of economic or global integration incorporate diverse communities and activities; but they do not do so in a determinate fashion. Throughout these processes of incorporation, people may contest and challenge efforts to undermine their self-determination, but their efforts are not solely reactions to processes of broad structural reorganization. On the contrary, processes of global integration are constituted by various forms of difference that are themselves implicated in subjective identities and various expressions of national identity, as well as in social relations that create particular expressions of equity and particular forms for social change. Such differences demand that, even when efforts are made to implement similar programs or to secure similar outcomes from various initiatives, our understanding of the heterogeneity of different communities and interest will make visible the divergence as well as unevenness of processes of change.

We need to construct new ways to view and interpret changes in gender relations and patterns of gender equity that can conceptualize knowledge construction as situated within relations of interlocking membership in systems of kin and nonkin networks of support, solidarity, and exchange. These conceptualizations, moreover, need to be constructed internally rather than derived from the reified categories of neoliberalism. An important contribution in directing our attention toward this goal is found in interrogations framed by methodological challenges to normative science (Farganis, 1989; Smith, 1990; Haraway, 1991). These challenges include the resurgence of interest in ethnographic practice and its emphasis on the importance of meanings and how they are constructed by those we study. They include as well efforts to rethink notions of objectivity, the dualisms characterized by subject–object and research–researcher, and the complexities of interpreting and accounting for the position of the researcher in the research process (Narayan, 1989; Abu-Lughod, 1991; Scott, 1992; Bar On, 1993). These challenges also include historical reconceptualizations of patriarchy, the engendered nation, and the embodiment of gender in constructions of nationalism and Islam (Sangari and Vaid, 1989; Ahmed, 1992; Chatterjee, 1993). Innovative feminist approaches must be able to imagine identity formation within an entirely different cultural politic; they need as well to envision different principles of social organization that are not hierarchically located on the linear continuum offered by the Western development paradigm.

This reconceptualization challenges us to move beyond productionist arguments that examine women's position, place, and situation as automatically improving with capitalist transformation. We also must move beyond examining the effects on women of development processes and forms of economic and cultural change to explore how gender constructions operate to make history. We need to know how "gender relations have been part of . . . [historical processes] and to detail the linkages between local level and larger political and economic forces" (Stivens, 1994, p. 373). Such an effort will draw attention to discontinuities in, and challenges to, predefined interpretations of women and notions of equity and change. It will also enable us to employ a dialectical and dialogical construction of the various patterns of gender relations that characterize the heterogeneous global economy. This exploration jettisons our fixed conceptual baggage, enabling us to engage possibilities for understanding relations in their more dynamic and varied expressions. Finally, it will help us to recognize how imperial constructs used to alter or interpret gender relations in Third World contexts have characterized a significant proportion of the research, policy analysis, and program development that has provided the bases for gender analyses, meanings of gender equity, and contemporary articulations of capitalist development.

References

Abu-Lughod, Lila. "Writing Against Culture." In Richard Fox (ed.), *Recapturing Anthropology: Working in the Present*. Santa Fe: School of American Research Press, 1991, pp. 137–162.

Afshar, Haleh, and Carolyne Dennis. *Women and Adjustment Policies in the Third World*. New York: St. Martin's, 1992.

Ahmed, Leila. *Women and Gender in Islam: Historical Roots of a Modern Debate*. New Haven: Yale University Press, 1992.

Anker, Richard. "Female Labour Force Participation in Developing Countries: A Critique of Current Definitions and Data Collection Methods," *International Labour Review*, vol. 122, no. 6, 1983, pp. 709–723.

Bakker, Isabella (ed.). *The Strategic Silence: Gender and Economic Policy*. London: Zed, 1994.

Bar On, Bat-Ami. "Marginality and Epistemic Privilege." In Linda Alcoff and Elizabeth Potter (eds.), *Feminist Epistemologies*. New York: Routledge, 1993, pp. 83–100.

Benería, Lourdes, and Shelley Feldman (eds.). *Unequal Burden: Economic Crises, Persistent Poverty, and Women's Work*. Boulder: Westview, 1992.

Benería, Lourdes, and Martha Roldán. *The Crossroads of Class and Gender: Industrial Homework, Subcontracting, and Household Dynamics in Mexico City*. Chicago: University of Chicago Press, 1987.

Benton, Laura. "Homework and Industrial Development: Gender Roles and Restructuring in the Spanish Shoe Industry," *World Development*, vol. 17, no. 2, 1989, pp. 255–266.

Boserup, Ester. *Woman's Role in Economic Development*. New York: St. Martin's, 1970.

Braidotti, Rosi, Ewa Charkiewicz, Sabine Hausler, and Saskia Wieringa. *Women, the Environment, and Sustainable Development: Towards a Theoretical Synthesis*. London: Zed, in association with United Nations International Research and Training Institute for the Advancement of Women (INSTRAW), 1994.

Chatterjee, Partha. *The Nation and Its Fragments: Colonial and Postcolonial Histories*. Princeton: Princeton University Press, 1993.

Cornia, Giovanni A., Richard Jolly, and Francis Stewart (eds.). *Adjustment with a Human Face*, vol. 1. New York: UNICEF/Clarendon, 1987.

Cox, Robert W. "Structural Issues of Global Governance: Implications for Europe." In Stephen Gill (ed.), *Gramsci, Historical Materialism , and International Relations*. Cambridge: Cambridge University Press, 1993, pp. 259–289.

Dey, Jennie. *Women in Rice Farming Systems*. Rome: Food and Agriculture Organization, 1984.

Dixon, Ruth. "Women in Agriculture: Counting the Labor Force in Developing Countries," *Population and Development Review*, vol. 8, no. 3, 1982, pp. 539–565.

Elson, Diane. "Male Bias in Macro-economics: The Case of Structural Adjustment." In Diana Elson (ed.),

Male Bias in the Development Process. Manchester: Manchester University Press, 1990, pp. 164–190.

———. "From Survival Strategies to Transformation Strategies: Women's Needs and Structural Adjustment." In Lourdes Benería and Shelley Feldman (eds.), *Unequal Burden: Economic Crises, Persistent Poverty, and Women's Work.* Boulder: Westview, 1992, pp. 26–48.

Farganis, Sondra. "Feminism and the Reconstruction of Social Science." In Alison Jaggar and Susan Bordo (eds.), *Gender/Body/Knowledge: Feminist Reconstructions of Being and Knowing.* New Brunswick: Rutgers University Press, 1989, pp. 207–223.

Feldman, Shelley. "Fieldnotes." Bangladesh, 1978–82.

———. "Crises, Poverty, and Gender Inequality: Current Themes and Issues." In Lourdes Benería and Shelley Feldman (eds.), *Unequal Burden: Economic Crises, Persistent Poverty, and Women's Work.* Boulder: Westview, 1992, pp. 1–25.

Feldman, Shelley, Fazila Banu Lily, and Farida Akhter. "The IRDP Women's Programme in Population Planning and Rural Women's Cooperatives," working paper, Ministry of Local Government and Rural Development, Dacca, Bangladesh, 1980.

Folbre, Nancy. "Hearts and Spades: Paradigms of Household Economics," *World Development,* vol. 17, no. 7, 1986, pp. 979–991.

Haraway, Donna. "Situated Knowledges: The Science Question in Feminism and the Privilege of Partial Perspective." In Donna Haraway (ed.), *Simians, Cyborgs, and Women: The Reinvention of Nature.* New York: Routledge, 1991, pp. 183–204.

International Labour Organization (ILO) and United Nations International Research and Training Institute for the Advancement of Women (INSTRAW). *Women in Economic Activity: A Global Statistical Survey (1950–2000).* Dominican Republic: ILO/INSTRAW, 1985.

Jhabvala, Renana. "Self-Employed Women's Associations: Organising Women by Struggle and Development." In Sheila Rowbotham and Swasti Mitter (eds.), *Dignity and Daily Bread: New Norms or Economic Organising Among Poor Women in the Third World and the First.* London: Routledge, 1994, pp. 114–138.

Kabeer, Naila. *Reversed Realities: Gender Hierarchies in Development Thought.* London: Verso, 1994.

Leon, Magdalena de. "Measuring Women's Work: Methodological and Conceptual Issues in Latin America," *IDS Bulletin* (Institute of Development Studies, Sussex), vol. 15, no. 1, 1984, pp. 12–17.

Lind, Amy Conger. "Power, Gender, and Development: Popular Women's Organizations and the Politics of Needs in Ecuador." In Arturo Escobar and Sonia Alvarez (eds.), *The Making of Social Movements in Latin America: Ideology, Strategy, and Democracy.* Boulder: Westview, 1992, pp. 134–149.

McCarthy, Florence. "Patterns of Involvement and Participation of Rural Women in Post Harvest Processing Operation." Dacca: Ministry of Agriculture of Bangladesh, 1981.

Maguire, Patricia. *Women in Development: An Alternative Analysis.* Amherst: Center for International Education, 1984.

Mies, Maria. *The Lacemakers of Narsapur: Indian Housewives Produce for the World Market.* London: Zed, 1982.

Mitter, Swasti. "On Organising Women in Casualised Work: A Global Overview." In Sheila Rowbotham and Swasti Mitter (eds.), *Dignity and Daily Bread: New Forms of Economic Organising Among Poor Women in the Third World and the First.* London: Routledge, 1994, pp. 14–52.

Mohanty, Chandra. "Under Western Eyes: Feminist Scholarship and Colonial Discourses," *Feminist Review,* vol. 30, no. 3, 1988, pp. 61–89.

Mohiuddin, Yasmeen. "Female-Headed Households and Urban Poverty in Pakistan." In Nancy Folbre, Barbara Bergmann, Bina Agarwal, and Maria Floro (eds.), *Women's Work in the World Economy.* New York: New York University Press, 1993.

Molyneux, Maxine. "Mobilization Without Emancipation: Women's Interests, State, and Revolution in Nicaragua," *Feminist Studies,* vol. 11, no. 2, 1985, pp. 227–254.

Moser, Caroline. "Gender Planning in the Third World: Meeting Practical and Strategic Gender Needs," *World Development,* vol. 17, no. 11, 1989, pp. 1799–1825.

Narayan, Uma. "The Project of Feminist Epistemology: Perspectives From a Nonwestern Feminist." In Alison Jaggar and Susan Bordo (eds.), *Gender/Body/Knowledge: Feminist Reconstructions of Being and Knowing.* New Brunswick: Rutgers University Press, 1989, pp. 256–269.

Nuss, Shirley (with Ettore Denti and David Viry). *Women in the World of Work.* Geneva: International Labour Organization (ILO), 1989.

Ong, Aihwa. "Colonialism and Modernity: Feminist Representations of Women in Non-western Societies." In Anne C. Herrmann and Abigail Stewart (eds.), *Theorizing Feminism: Parallel Trends in the Humanities and Social Sciences.* Boulder: Westview, 1994, pp. 372–381.

Porter, Theodore M. "Objectivity As Standardization: The Rhetoric of Impersonality in Measurement, Statistics, and Cost-benefit Analysis." In Allan Megill (ed.), *Rethinking Objectivity.* Durham: Duke University Press, 1994, pp. 197–237.

Rogers, Barbara. *The Domestication of Women: Discrimination in Developing Societies.* New York: St. Martin's Press, 1980.

Roldán, Martha. "Industrial Outworking, Struggles for the Reproduction of Working Class Families, and

Gender Subordination." In Naneke Redclift and Enzo Mingione (eds.), *Beyond Employment: Household, Gender, and Subsistence.* Oxford: Blackwell, 1985, pp. 248–285.

Safa, Helen. "Women and Industrialisation in the Caribbean." In Sharon Stichter and Jane Parpart (eds.), *Women, Employment, and the Family in the International Division of Labor.* Philadelphia: Temple University, 1990, pp. 72–97.

Safa, Helen, and Peggy Antrobus. "Women and the Economic Crisis in the Caribbean." In Lourdes Benería and Shelley Feldman (eds.), *Unequal Burden: Economic Crises, Persistent Poverty, and Women's Work.* Boulder: Westview, 1992, pp. 49–82.

Sangari, KumKum, and Sudesh Vaid. "Recasting Women: An Introduction." In KumKum Sangari and Sudesh Vaid (eds.), *Recasting Women: Essays in Colonial History.* New Delhi: Kali for Women, 1989, pp. 1–26.

Scott, Alison. "Patterns of Patriarchy in the Peruvian Working Class." In Sharon Stichter and Jane Parpart (eds.), *Women, Employment, and the Family in the International Division of Labor.* Philadelphia: Temple University, 1990, pp. 198–220.

Scott, Alison. *Divisions and Solidarities: Gender, Class, and Employment in Latin America.* London: Routledge, 1994.

Scott, Joan. "Experience." In Judith Butler and Joan Scott (eds.), *Feminists Theorize the Political.* New York: Routledge, 1992.

Sen, Amartya K. "Gender and Cooperative Conflicts." In Irene Tinker (ed.), *Persistent Inequalities: Women and World Development.* New York: Oxford University Press, 1990, pp. 123–149.

Smith, Dorothy. *The Conceptual Practices of Power: A Feminist Sociology of Knowledge.* Boston: Northeastern University Press, 1990.

Spivak, Gayatri Chakravorty. "The Politics in Interpretation." In W.J.T. Mitchell (ed.), *The Politics of Inter-*

pretation. Chicago: University of Chicago Press, 1983.

Sternbach, Nancy Saporta, Marysa Navarro-Aranguren, Patricia Chuchryk, and Sonia Alvarez. "Feminism in Latin America: From Bogota to San Bernardo." In Arturo Escobar and Socia Alvarez (eds.), *The Making of Social Movements in Latin America: Ideology, Strategy, and Democracy.* Boulder: Westview, 1992, pp. 207–239.

Stichter, Sharon, and Jane Parpart (eds.). *Women, Employment, and the Family in the International Division of Labor.* Philadelphia: Temple University Press, 1990.

Stivens, Maila. "Gender at the Margins: Paradigms and Peasantries in Rural Malaysia," *Women's Studies International Forum,* vol. 17, no. 4, 1994, pp. 373–390.

Strickland, Susan. "Feminism, Postmodernism, and Difference." In Kathleen Lennon and Margaret Whitford (eds.), *Knowing the Difference: Feminist Perspectives in Epistemology.* London: Routledge, 1994, pp. 265–274.

Tomoda, S. "Measuring Female Labour Force Activities in Asian Developing Countries: A Time Allocation Approach," *International Labour Review,* vol. 124, no. 6, 1985, pp. 661–676.

Tripp, Aili Mari. "The Impact of Crisis and Economic Reform on Women in Urban Tanzania." In Lourdes Benería and Shelley Feldman (eds.), *Unequal Burden: Economic Crises, Persistent Poverty, and Women's Work.* Boulder: Westview, 1992, pp. 159–180.

Wainerman, Catalina. "Improving the Accounting of Women Workers in Population Censuses: Lessons from Latin America," Population and Labor Policies Program working paper no. 178, World Employment Program Research, International Labour Organization, Geneva, 1992.

World Bank. *World Development Report 1990.* Washington, D.C.: World Bank, 1990.

World Economy, Patriarchy, and Accumulation

Maria Mies

Introduction

The very fact that this encyclopedia deals with Third World women as a separate category and not with women generally cannot be explained merely by pointing to the differences that exist among women worldwide; it must be seen as a manifestation of a political and economic global system that emerged historically in the fifteenth and sixteenth centuries with European colonialism, capitalism, and, later, industrialism. It is due to the divide-and-rule strategies of this system that the world is dichotomized in First and Third, "developed" and "underdeveloped," "industrialized" and "nonindustrialized," "central" and "peripheral" countries or regions. The efforts to find nicer labels for the exploitive, colonial, hierarchical relationship between the two—or three—parts of the world, the last one being the "North" and "South" divide, have not changed this reality. I am not going to quarrel with the concept "Third World"; I use it here pragmatically, always aware of the system it represents. What is more important is to understand this modern, global world system and its impact on women, particularly on Third World women. It is impossible to understand the situation and the problems of Third World women unless one's analysis is grounded in this world system and its main driving force—capital accumulation, or unlimited economic growth. Nor is it possible to understand the relationship between sexism, classism, and racism without first looking at colonialism.

I shall, first, show that, on the basis of this approach, a number of common assumptions on the workings of this system of capital accumulation have to be demystified and that we have to arrive at a different understanding of its functioning. Among these—in my view, false—assumptions are that progress is a unilinear, evolutionary process; that the process of capital accumulation is driven only by the development of productive forces—science and technology and the exploitation of wage labor; that colonialism is a thing of the past; that violence is a thing of premodern, traditional societies; that patriarchy has disap-

peared with modernization; and that, therefore, the strategy for women's liberation could be summarized as overcoming the shackles of traditional, patriarchal societies, or as catching up with the men, or as a politics of equalization only.

As the women's movement has amply demonstrated, men's dominance over women, or patriarchy, has not disappeared even in modern, industrialized societies. A question therefore arises as to what this historical fact has to do with the functioning of the dominant global world economy. Why does patriarchy not disappear with capitalist development? In this essay, I present a discussion of this question, followed by an analysis of the historical and social roots of patriarchal male–female relations, particularly the hierarchical division of labor and the role of violence in this process. I then extend this analysis to the processes happening now, both in the Third World and in the First, showing that, and how, these processes are interrelated. Next is a discussion of how and why the processes of labor control in the Third World, particularly the housewifization of labor, are being reimported into the core countries of industrial society to counter economic and ecological crises. From this overall analysis follows an account of Third World women's resistance to these processes and some of their visions that go beyond the well-known paradigm of "catching-up development."

Women, Colonialism, and the Capitalist World Economy

Several theses can be offered on the functioning of an economic system whose very motor is capital accumulation or permanent unlimited economic growth.

It is usually assumed that progress is an evolutionary process starting from a primitive, backward stage and, driven by the development of science and technology (in Marxist terms, "productive forces") moving upward in unlimited progression. This Promethean project, however, does not respect the limits of this globe, of time, of space, of our human existence. Within a limited world, aims like unlimited growth can be realized only at the expense of others. There cannot be progress of one part without regression of another part; there cannot be development of some without underdevelopment of others; some cannot be wealthy without impoverishing others. Thus, concepts like "unlimited growth" or "capital accumulation," necessarily imply that this growth (progress, development, wealth) is at the expense of some "others," given the limits of our world. This means progress or development can no longer be conceived of as an evolutionary, upward, linear movement but must be seen as a *polarizing process,* following a dualistic worldview (Plumwood, 1993).

This means the process of capital accumulation cannot function solely on the basis of the "normal" exploitation of the wage laborers in the capitalist centers, as Karl Marx assumed. According to Marx, the secret of capital accumulation, of "money that ever breeds more money," is the fact that the capitalist does not pay for the whole labor time a wage worker works, but only for the "necessary" labor time—time that, according to Marx, is expended to earn the money to buy the goods that are necessary for the daily production and intergenerational reproduction, of the laborers. The time and the value produced over and above this necessary labor are appropriated and reinvested by the capitalist. According to this analysis, the exploitation of the workers can be scientifically explained; it does not need extra-economic violence (Marx, 1974).

Luxemburg (1913) has shown that capital accumulation presupposes the exploitation of ever more "noncapitalist" milieux and areas for the appropriation of more labor, more raw materials, and more markets. I call these milieux and areas "colonies." Colonies were necessary to initiate the process of capital accumulation in what Marx called the period of "primitive accumulation" in the beginning of capitalism; they continue to be necessary even today to keep the growth machine going. Therefore, we talk of the need for "ongoing primitive accumulation" and ongoing colonization (Mies, 1988). The main colonies of capital in this process are countries in the Third World, women, and nature. These colonies constitute the hidden underground of the economic system, which can be compared to an iceberg. The visible part above the water consists of capital and wage labor, the invisible part under the water represents nonwage labor—particularly by women labor, nature, and the Third World.

There is no colonization without violence. Whereas the relationship between the capitalist and the wage laborer is legally one of owners (the one of capital, the other of labor power) who enter a contract of exchange of equivalents, the relationship between colonizers and colonies is never based on a contract or an exchange of equivalents. It is enforced and stabilized by direct and structural violence. Hence, violence is still necessary to uphold a system of dominance oriented toward capital accumulation.

This violence is not gender neutral; it is basically directed against women. It is usually assumed that with modernization, industrialization, and urbanization, patriarchy as a system of male dominance would disappear and make way for equality between the sexes. Contrary to this assumption, patriarchy did not disappear in this process, which is identical with the spreading of the modern capitalist world economy; rather, the ever-expanding process of growth or capital accumulation is based on the maintenance or even recreation of patriarchal, or sexist, male–female relations, an asymmetric sexual division of labor within and outside the family and the definition of all women as dependent housewives and of all men as breadwinners. This sexual division of labor is integrated with an international division of labor in which women are manipulated both as producer-housewives and as consumer-housewives.

With crises in the global world system increasing, we can observe an increase of violence, particularly against women, not only in the Third but also in the First World, which is supposed to represent "civil society." As this violence is part and parcel of a political-economic system based on colonization and limitless growth, it cannot be overcome by a strategy aiming only at gender equality. Within a colonial context, "equality" means catching up with the colonial masters (Memmi, 1965), not doing away with colonialism. This is the reason feminists cannot be satisfied with a policy of equal opportunities but must strive to overcome all relationships of exploitation, oppression, and colonization necessary for the maintenance of global capitalist patriarchy (Plumwood, 1993).

What Is Patriarchy?

Before elaborating on this thesis, it is necessary first to clarify the concept "patriarchy." Patriarchy is a historical system of dominance of men over women. It is much older than capitalism, having emerged 5,000–6,000 years ago with tribes who made warfare and conquest, subjugating other tribes and their territories as their main source of wealth. In this process, first the foreign women, then their own women were forcibly domesticated and put under the control of the dominant male. The secret of the success of such tribes, or peoples, or classes was not their superior brainpower or culture or the superiority of their means of production, but rather the monopoly of arms in the hands of some men. Arms gave these men power over women, other people, and nature, a power that did not stem from their own productive work. This power changed the relationship between man and nature, between man and woman, and between different peoples, from one of reciprocity to one of exploitation. At the core of this system, we can identify violence and coercion as the main mechanisms of control (Mies, 1986).

Having said this, it should be clear that (1) patriarchy is not a universal system that existed everywhere and always but a system that had a distinct historical beginning in particular geographical areas and tribes, and (2) patriarchy is not rooted in the biological differences between females and males but is a historical human creation and can, therefore, be changed by us.

The great patriarchal civilizations that we know—the European, the Arab, the Indian, the Chinese, and the Japanese—all have their origins in the rise of some martial tribes or classes, sometimes pastoral tribes, as was the case with the Aryans, who invaded Europe and India, or with the Hebrews and the Arabs, who won supremacy over other tribes by way of conquest. Everywhere in these civilizations, this supremacy included not only man's lordship over the earth, but also over women. These social relationships, based on exploitation and oppression, were then legitimized by priests' castes or philosophers as nature- or God-ordained. In India, patriarchy was introduced by the vedic Aryans and upheld by the Kshatria and Brahmin castes—the warriors and the priests. The Roman and Greek patriarchies were based on warfare and conquest and were legitimized by philosophers. In China, the dominant patriarchal philosophy, Confucianism, arose only after feudal lords had gained power and prestige through warfare, mainly against border peoples, and had established a hierarchical social and sexual division of labor.

Chinese and Japanese examples may serve to illustrate this point: Granet (1985) notes that the Chinese usually say that the hierarchical social relationships between subjects and the ruler, today between people in authority and those dependent on them, have their roots in patriarchal family relations, particularly the filial piety of the son toward the father, which they consider as *the natural* relationship. But by studying the historical evidence, Granet comes to the conclusion that one has to turn this view of things upside down to reach the truth. In earlier matrilineal clans, not even a kinship relationship existed between father and son. Such a relationship was established only by the feudal aristocratic class, and it was a feudal relationship, identical to the relationship between a feudatory lord and his vassal. There was nothing natural about this relationship; it had to be artificially constructed and stabilized by cult and ritual. In one such ritual, the connection between warfare and fatherhood becomes clear: A victorious warlord would bury the head of his slain enemy under the threshold of his house. In so doing, he won a name for his son. By granting the name of the killed enemy to his son, he also granted the enemy's soul to him, a soul that he had conquered and appropriated. From then on, the father's rights over the son were stronger than the mother's. Patriarchal father–son and male–female relationships have their origin in this military feudal relationship. As Granet says: "Civil morality is not a projection of domestic morality; on the contrary, the rights of the feudal town permeate domestic life" (Granet, 1985, p. 173).

A similar development took place in Japan, as Aoki (n.d.) has shown. According to her, patriarchal male–female relationships did not exist in earlier Japan among the majority of people, the peasants and the artisans. It was only the warrior class, the samurai, who made up 6 to 8 percent of the population, who had, prior to the introduction of Confucianism, domesticated their women, put them under a strict patriarchal discipline, and established

the male line of succession. It was also this class that adopted and spread Confucian ideology and morality. What is more important for our question, however, is Aoki's observation that the Meiji reforms of the nineteenth century, which ended Japan's feudalism and set the nation on the way to modernity, were promoted by the same warrior class. These samurai matched this feudal-patriarchal ideology with the Western values of "civilization and enlightenment" and the work discipline of industrial society. The "fit" between a feudal-patriarchal sexual division of labor based on militarism and the capitalist-patriarchal sexual division of labor was ideal. Aoki writes:

> Certainly, the sexual division of labor is inherent in the economic system brought about by the necessities of the industrial society. But in Japan an analogous model already existed in the form of the families of the warrior class. Thus the families of the so-called "company men" may appear to have a modern character, but in fact are supported by a web of traditional culture, reworked by Confucian ideology. In this way, the men who work as "salarymen," although they appear to have a modern character, form a group with a strong sense of class and a diligence bordering on self-sacrifice. They have a deep sense of loyalty to the company, much stronger than in other countries where workers simply see work as a means to earn their daily bread. This is the modern version of the mentality which saw Samurai serving their local fief. The morals of the warrior class are also preserved in the lives of the wives who see their jobs as looking after the home, who do not bother their husbands with details of the education of their children, and see it as their duty to devote themselves to the husband who comes home tired from work (Aoki, n.d., p. 13).

Hence, there was not, as functionalist sociologists believe, a break between tradition and modernity but rather a continuity, from early patriarchal tribes and classes, to feudalism, right into capitalism, and even into socialism (Mies, 1986). This continuity was closely interwoven with the establishment of capitalist class relations.

Not only is there historical continuity with regard to patriarchy in these civilizations, there is also cross-cultural continuity. In spite of all cultural differences that exist between the great patriarchal civilizations—which, not accidentally, are identical with the great religions and philosophies: Hinduism, Buddhism, Confucianism, Islam, Judaism, Greek philosophy, and Christianity—when it comes to the relationship between man and woman they have more in common than is usually thought. We find virtually identical statements in the old texts about women: that they are only vessels for the procreation of sons and, hence, necessary instruments for the maintenance of the male lineage (most religions); that they should always be dependent on a man and, thus, obedient to him (Hinduism, Confucianism); that they are the embodiment of sin and chaos (Christianity, Judaism, Islam, Hinduism). It is perhaps this commonality of a basic devaluation of women, based on an exploitive male–female relationship, that has its direct parallel in the exploitive relationship between such patriarchal men and Mother Earth, which constitutes the structural and ideological fit between such "traditional" patriarchies and the modern capitalist patriarchy. The rise of the "White Man," of his model of economy and society, of his rationality and his scientific progress is based on the colonization of women, of Mother Earth, and of foreign peoples. In this process of colonization, he could count on the loyalties of local patriarchal men and the continuity of their patriarchal values and institutions as a base for modernization and capital accumulation.

In addition to continuities between the old patriarchies and the new capitalist one, there are also discontinuities, because only capitalism has been able to create a worldview in which man appears as totally independent of nature (or God, as one would have said earlier). The new power by which man was able to set himself over and above nature—and women—is, together with modern science and technology, the power of *money,* or capital. Capital here means money that is capable of "giving birth" to ever more money, the "money-breeding-money." Money is the real god of capitalism; it appears as the creator and preserver of all life. Those who have no access to money cannot live. This new life-giving power of money seems to have replaced the life-giving capacity of women and of the earth. The new sexual division of labor under capitalism is one in which a male "breadwinner" sells his labor power for a wage or a salary to an employer, then brings home the money by which he can maintain a housewife and his children. The institution within which the housewife is responsible for the production of life is the nuclear family.

Patriarchy, Housewifization, and Colonization

The propagation and universalization of the housewife as *the* image of modern woman paralleled the propagation of the nuclear family as the universal family. The housewife is an ideal type; however, she is not the strong, self-sufficient, independent woman of prepatriarchal times but the weak, dependent, domesticated woman who is isolated from public life, and whose only concern is her husband and her children. An important feature of this image is the ideology of sex appeal and romantic love as the main mechanisms whereby modern woman can hope to attract a breadwinner with a lot of money. In the Third World, this family model is propagated above all through the discourse on the population explosion. The small family is considered to be a happy family.

The creation of the modern housewife is not the result of some inborn male sadism but is a structural necessity of the process of capital accumulation. Feminist analysis has shown that the housewife, who reproduces the labor power of the salaryman, or wage worker, contributes to the production of surplus value, particularly because her own labor is not attributed any value, is not paid and, therefore,

is not included in the calculations of the gross national product (GNP)/gross domestic product (GDP). It is not even called labor but love, is not unionized, has no time limit, and seems to be freely available, like sun and air, like a natural resource. Economists call it a "free good." It is precisely the exploitation of this invisible housewife-labor that feminists have identified as one of the secrets of capital accumulation (Dalla Costa, 1972; Bock and Duden, 1984; Mies, 1986; Bennholdt-Thomsen, 1988; Waring, 1988; Werlhof, 1988).

This modern image of the woman as housewife is not even challenged when women are being gainfully employed, or, as is increasingly the case in many countries, particularly in the Third World, even when they are the de facto heads of household and breadwinners of the family. Women's salaries or wages are invariably lower than men's on the assumption that their income is only supplementary to that of the male head of the family. Housewives are not counted as workers in national statistics; their economic contribution is not counted in the GNP or in the UNSNA (United Nations System of National Accounts) (Waring, 1989). Trade unions have never bothered to include housework in the definition of "work" nor have they fought for the interests of housewives. Women are often overlooked for secure jobs because the employer expects their ultimate return to their "real" vocation: of housewife or mother. In times of economic recession, they are the first to be fired. Their chances to move up to the higher echelons in their professions are, therefore, seriously limited. The findings of the United Nations are well known: that women constitute half of the world's population, perform almost two-thirds of the world's working hours, receive only one-tenth of the world's income, and possess less than one-hundredth of the world's property.

But what is perhaps more alarming than these stark facts is the continued fascination with this image of the domesticated, dependent woman. It has become the symbol of progress, the image of the future for many other women: peasant women, working-class women, women who are still capable of maintaining themselves by their own subsistence production, women in the Third World.

This does not mean that the majority of Third World women are de facto housewives in the above-mentioned sense. But the ideal type of man—the breadwinner—and woman—the "nonworking" housewife—is at the base of most policies regarding such women. It is one of the reasons why working women there are concentrated in the informal sector.

The Making of the Housewife

It is not enough merely to describe the present sexual division of labor in which the definition of women as housewives plays the central role; we must also explain how this particular image of woman could emerge in history. The domestication of European women had by no means a peaceful history but was accompanied by at least three centuries of brutal violence against millions of women in the witch hunt. This witch hunt, which raged through Europe from the fourteenth to the seventeenth century, is the largest mass killing of women in modern history. We have no exact figures of how many people were killed as witches, but estimates range from 500,000 to several million (Heinsohn and Steiger, 1985, pp. 136–143). At least 80 percent of these were women. It is important to note that this massive holocaust against women did not occur, as is commonly believed, in the Dark or Middle Ages as a result of superstition but in those centuries at the beginning of the modern era, the beginning of modern science and technology, of modern medicine, of modern rationality, of the modern economy, and of the modern state. Feminist research has shown the interconnection between this brutal attack on women, particularly on healers and midwives, the doctors of the poor, and the rise of modern school medicine, which subsequently became a monopoly of men (Becker et al., 1977; Ehrenreich and English, 1979; Merchant, 1983). Merchant (1983), particularly, has demonstrated that not only modern medicine, but also modern physics and mechanics and the new "scientific" relation to nature, rose up from the fires of the witches. In these fires, the old organic relation of women to their own bodies, their knowledge about health and disease and about the healing powers of nature—as well as men's knowledge that they are children of Mother Earth—were all burnt to ashes (Merchant, 1983). One aspect is often overlooked in the analyses of the witch pogroms; namely, the accumulation of wealth through such witch trials. Wars and modern projects were being financed out of the property confiscated from the witches and their families (Mies, 1986).

After these centuries of brutal destruction and subordination of the "bad woman," we find by the eighteenth century the domesticated, tamed, weak, dependent "good woman," the housewife in the households of the rising bourgeois class. This woman became the symbol of progress and the model for all other lower-class women. Her image was carried by missionaries and colonialists all over the globe as a model of "civilization and enlightenment" (Aoki, n.d., p. 19). What is worse, the European working classes, which fought against capitalism, now adopted this image of woman and the bourgeois family as models of progress. It was an important aim of trade-union struggles that the proletarian man should also earn a "family wage" so he could keep a "nonworking" housewife at home (Mies, 1986, pp. 107–108).

Occurring alongside the process of housewifization was the process of colonization of far-away lands in Asia, Africa, and South America. These two processes were directly and causally linked. Without the conquest of colonies and the robbing of precious metals, and later the exploitation of the land, the resources, and the people of these colonies for luxury items like coffee, sugar, tea, cotton, and the like, the European bourgeois class would not have been able to start its Industrial Revolution, the European scientists would not have found capitalists with an interest in their inventions, the European salaried classes would not have been rich enough to afford a "nonworking" housewife,

and the European working class would have continued to live a miserable proletarian life. Colonialism provided the material source for the increase of productivity of human labor, which then gave a boost to industrial expansion. Colonialism—today we call it the international division of labor—brought, and still brings, such an influx of wealth to the metropolises that now even proletarian men can afford to have a "nonworking" housewife.

Divide and Rule: Housewifization International

Due to the polarized structure of the modern world economy, however, the double process of housewifization and colonization did not affect women in the metropolises and in the colonies or underdeveloped countries in the same way. In the classes and nations that profited most from colonial plunder, the majority of the women could de facto become housewives. But this was, and is, not the case for the vast majority of women in the colonies or underdeveloped countries. They cannot hope to rise to the status of a woman who is being fed and maintained by a wage-working or salaried husband only; instead, they have to work in the fields, factories, sweat-shops, and household industries as petty producers, petty vendors, or even prostitutes to make a living. In India, for example, about 80 percent of the adult female workers are employed in agriculture (Mies, 1988).

Yet, international development organizations, national governments, national census authorities, and international organizations like the United Nations or the World Bank also project the image of the housewife onto these women as the symbol of *the* modern woman. In the so-called income-generating activities, which are proposed to alleviate poor women's lot, women are not defined as workers but as housewives, who can supplement the insufficient income of their husbands by some small side activity, like making handicrafts or raising chickens for an external urban or foreign market. I discovered the prototype of such income-generating activities in a home-based industry in southern India, among the lace makers of Narsapur (Mies, 1982). In this industry, poor rural women make crochet lace, which is all exported to Europe, Australia, and the United States. The lace making had been introduced in the nineteenth century by Scottish missionaries, who also defined these women as housewives. In this case, the women used a pattern that already existed in the area. The Kshatria (warrior) caste of the Kapus had always kept their women as "gosha women," or "women sitting in the house." As we saw in the case of the Japanese samurai, there was a direct fit between the Kapu ideology on women and the ideology of the modern, Western missionaries. Today the exporters of lace have become millionaires. Their fortunes are based directly on the definition of these lace makers as housewives, who, according to them, use their idle time productively. Due to this definition, their wages were appallingly low, only 0.56 rupees per day, whereas the minimum wage of an agricultural laborer was between 2.00 and 4.00 rupees.

This case study taught me that housewifization of women is not only the cheapest method for capital to reproduce the labor power but also the cheapest way of producing commodities for the world market. This is the reason we find housewifized producers not only in the handicrafts sector in Third World countries but also in the most modern factories in the free-trade zones in South and Southeast Asia, Latin America, and Africa. It is a common characteristic of all world-market factories and relocated industries, such as electronics industries and garment and toy factories, that the vast majority of their workers (up to 80 percent) are women, mostly young and unmarried (Fröbel et al., 1986). The firms are mainly relocated American, Japanese, and European concerns that, as Fröbel and his associates and several feminist scholars have shown, can thus reduce the labor costs considerably. On average, the wages in such free-trade or export-production zones are one-fifth or one-tenth, sometimes even one-twentieth of the wages in similar factories in the old industrialized countries. Moreover women get much lower wages, even for the same work, than men. In South Korea, in the early 1980s, women's wages in the textile industry were only half of men's wages (Fröbel et al., 1986, pp. 468–469). Working hours generally were much longer than in the West, with a work week of 50 to 60 hours the norm. In the Philippines, in 1982–84 the average weekly labor time was 54 hours. In spite of these long working hours, which also often include night shifts for women, the wages are too low for the women to maintain themselves and their families. Therefore, they often have to supplement their income by some other subsistence production in the informal sector (Fröbel et al., 1986, pp. 468–469).

The mechanisms by which multinational and national firms keep up this system of production, which has led to an enormous growth of industrially produced consumer goods for foreign markets, are not only the prohibition of trade-union activities but also a particularly sexist system of labor control. This system was first analyzed by Grossman (1979), who found that U.S. and Japanese managers in the electronics industries in Malaysia and the Philippines not only used quotas that women had to fulfill but also held beauty contests and sold cosmetics in the factories to promote a sexist image of the modern woman as a consumer and to cultivate the competitiveness among the women, not only as workers but as women competing for the favors of their male managers.

It is not a stretch to conclude that it is the images of woman as sex symbol and as housewife, not as worker, that are used to keep women's wages low, to keep women disorganized, to force them into long working hours, and to fix their attention on their sex appeal to attract a man—and that exposes them increasingly to sexual abuse even at their workplace. Also in Third World agriculture and in the informal sector (i.e., the unregulated labor market), the image of woman as housewife plays a strategic role in keeping production costs low (Mies, 1986). Had these firms employed men, with their claim of being the "breadwinners" of their families and their demands for unionization, their profits and growth rates would have been much lower. According

to Werlhof, the housewife, not the male proletarian, is the optimal labor force for capital. The male wage worker is too expensive, too unproductive, and too unreliable (Werlhof 1988, pp. 177–179).

This system of labor control for women was not basically changed when, after the First United Nations World Conference on Women in Mexico in 1975, the discourse started on "integrating women into development." The strategy of integrating Third World women's labor into the global market system was also not basically changed when the World Bank and other international organizations stopped talking about integrating women into development and around 1988 began a discourse on "investment in women." In this new discourse, poor Third World women who produce for the world market are called entrepreneurs, not housewives. At the 1988 annual conference of the World Bank/IMF in Berlin, Barbara Herz, head of the Women in Development Division of the World Bank, stated in a briefing paper that "investing in women" was necessary not only to enhance their labor productivity through education and training but also to draw them away from subsistence production into production for an external market.

> As a general proposition, it makes sense to allow women, like other entrepreneurs, an expanded range of economic opportunities and let them weigh market potential and family concerns rather then assuming they "should" stay in certain lines of activity. Culture may limit the scope and pace of such expansion, but the economic virtue of deregulation ought to be clear (Herz, 1988, p. 2).

The term "entrepreneur" does not relieve Third World women of their responsibilities for their families. Male entrepreneurs never have to "weigh" housework and market production. This means, in spite of the new terminology, that the basic strategy of treating women's labor as a more or less free resource is not fundamentally changed (Mies, 1994).

Another factor in the process of the internationalization of the housewife model is the differences between the two sets of women: the housewife-consumers in the First World and in the middle classes in the Third World, and the housewife-producers, mainly in the Asian, African, and Latin American countries. In the old industrialized countries in Europe and in the United States, women are constantly mobilized as consumers by the advertising industries and by an overflowing commodity market. The consumer-housewife and the producer-housewife are, however, connected by the world market. Without the superexploitation of the latter, the former would not be able to buy so many commodities at such a low price. Moreover, these commodities would not even be in the supermarkets in such enormous quantities without such exploitation.

Having said this, we must also recognize that, as far as these material interests are concerned, international capital has created a contradictory, if not antagonistic, relationship between these two sets of women. If the producer-housewife demands higher wages and better work conditions similar to those of the male workers in the industrial countries of the West, the price of the commodities will be pushed up for the consumer-housewives in these countries. They will no longer be able to buy a T-shirt for 5 deutsche marks (DM) and a pair of trousers for DM 15. Moreover, if conditions in these export-production zones and world-market factories were equal to those in the United States or Europe, the multinational firms would most probably lose their interest in these countries as locations for their industrial production and move on to other countries of the Third World for still cheaper labor. This has been the strategy of some South Korean garment industries, which have moved in the early 1990s to Bangladesh. With the new free trade policy and the globalization of the economy promoted by institutions such as the General Agreement on Tariffs and Trade (GATT), the World Bank, and the International Monetary Fund (IMF), the tendencies toward flexibility, deregulation, or housewifization of women's labor in the Third World are being strengthened.

The Universalization of the Modern Standard of Living and the Ecology Crisis

We have seen that, since the nineteenth century, the Western model of development based on industrialism, permanent growth, and the never-ending progress of science and technology held a great fascination for people in Asia, Latin America, and Africa. We have also seen that the image of woman as housewife is closely bound up with this model as a symbol of progress. The goal of most societies that have followed this model was, and is, to reach a standard of living and a level of consumerism comparable to the United States and Western Europe. This is what "development" means.

However, this model has always been based on a sexual and an international division of labor through which the costs of development are systematically pushed outside their own borders to the three colonies of White Man: women, nature, and other peoples. The ecological crises and industrial catastrophes that, since the 1970s, have haunted the developed and developing countries alike are strong evidence that this model, based on constant growth and unlimited use of nonrenewable energy, has reached its natural limits. It is well-known that a person in the industrialized countries uses at least ten times more energy than a person in a poor country (Mies and Shiva, 1993). The production of industrial and household waste is so high that it threatens the ecosystem everywhere. If one tried to universalize this model to project the average standard of living of all people in the world—one would quickly come to the conclusion that the world's energy resources wouldn't last more than a few years. The whole world, including the seas, would be buried under the waste produced by the world's five billion consumers. Thus, although modernization and progress along the lines of this growth model are preached everywhere and retain a great attraction for those countries defined as "backward," a generalization of this model to *all* people is ecologically impossible.

It is also undesirable, unless we want to destroy the very foundations of our existence: water, air, nutritious food, forests, and animal life. Already drinking water and even breast milk in Germany are so poisoned that experts warn people about their safety. Universalization of this growth model is also undesirable from a feminist point of view—not only because the degradation and domestication of the Western housewife have been paid for in part by an even greater exploitation of women in the Third World, but also because the development model has not liberated women, even in the West. This is the reason why the new women's movement (i.e., a vindication going beyond voting rights) started there at the end of the 1960s.

The universalization of the Western model of woman and her standard of living, based on the capitalist growth model, cannot be in the interest of women's liberation, since it presupposes the dominance of man over nature, women, and other peoples. A different perspective of a new society cannot be based on capital accumulation or high tech or be bound up with modern consumerism; rather, it will emerge in the struggles for survival of Third World women.

Resistance

The processes of colonization and capital accumulation have affected women negatively; they have also given rise to different forms and strategies of resistance. Whereas the new women's movement in both the developed and the developing countries of the world took up the issue of violence against women as one of the first and most crucial struggles, Third World women, particularly those affected by the new globalization strategy of capital, fashioned a whole range of creative actions, new strategies, movements, and organizations against the destruction of their survival base and their human dignity. Such resistance movements were not always explicitly feminist in their orientation, but the patriarchal character of modern work relations was often difficult to ignore in the course of the struggle, so new forms and methods of resistance had to be devised that did (and do) not always fit the classical type of male working-class resistance. Brief accounts of some such struggles in the Third World are presented below. They show that women are not only victims in these processes but also active subjects, who are able to develop a correct analysis of their situation and a vision of a society that does not colonize women, nature, and other countries in the name of profit and accumulation.

Such resistance movements against the deterioration of the basis of everyday life as a result of modern "development" can be observed in most sectors of the economy in the Third World from agriculture, to industry, to the informal (cheap-labor) sector, and among urban as well as rural women. Here are a few examples:

1. In many Third World countries, women are demanding not only better wages as agricultural

laborers but also control over land, and land titles in their own name. Omvedt describes the success of such a movement in Maharashtra, India (Omvedt, 1993).

2. Sometimes such rural movements for women's control over their own resources are combined with ecological concerns. This has been the case with the Chipko movement in India (Shiva, 1988), the Green Belt movement in Kenya, the movement to save the mangrove forests in Bolivia, and many other such ecofeminist struggles (Dankelman and Davidson, 1988; Mies and Shiva, 1993).

3. The peasant movement in India against the GATT treaty and trade-related intellectual-property rights (TRIPs), and the Seed-Satyagraha and Navdanya movements for the autonomy of peasants over their own seed production and genetic resources, show that Third World peasants, particularly women, not only understand the onslaught of the new globalization strategy of international capital on their autonomous subsistence base but also how to counter this strategy (Shiva et al., 1994).

4. The exploitive conditions in relocated industries, free-production zones (FPZs), and the world-market factories were also answered by women's resistance, right from the beginning. These struggles ranged from cases of mass hysteria in the microchip industries in Malaysia (Grossman, 1979) to the struggles for legal trade unions and better working conditions in South Korea and Mexico. The women workers' resistance played a major role in the democratic movement in South Korea in the late 1980s (Ching yoon Louie, 1995). However, when this movement began to show its first fruits in the form of better labor laws, of allowing trade unions, of better pay and better labor conditions, the neoliberal strategy of globalization and industrial restructuring set in at the beginning of the 1990s. Many foreign and Korean companies in the garment and textile industries left Korea and reopened their factories elsewhere. The women workers were the losers again (O'Sullivan, 1995).

Similar processes are happening in most countries of the Third World under the impact of globalization of the economy. Apart from this, the Structural Adjustment Programs (SAPs) imposed on the indebted Third World nations by the IMF have eroded whatever these nations have done to fight poverty and inequality through state subsidies for health, education, food for the poor, and so forth. The policy of deregulation also undermines whatever labor legislation had been achieved. All of these measures, particularly the SAPs, affect poor, Third World women most. These women are also the ones who suffer most from environmental degradation. Hence, their very survival base is being threatened by this neoliberal strategy of the "free" world market. But, unlike middle class women, these

women cannot expect that they will ever benefit from "catching-up development." Therefore, fewer and fewer women in grass-roots movements in the Third World harbor any illusions about the global economic system.

They do not simply demand a return to the previous state of affairs and the continuation of some state welfare measures. They question the whole paradigm of development, with its intrinsic tendency toward social injustice, sexism, growth mania, and ecological destruction. Instead, they reclaim access to, and control over, their local resources, land, water, forests, production processes, and culture. They do not want to be integrated into the world market but have already started to establish direct links between different categories of local and regional producers. Thus, the participants in a workshop in Brazil on "Women and a Sustainable Economy," which took place during the United Nations Conference on Environment and Development (1992), declared that they—the fisherwomen, rubber tappers, coconut gatherers, small peasants, and small urban producers—were fed up with the economic model and that they wanted to promote a really sustainable economy and ecology based on direct exchange between different producers and between urban and rural producers in a regional economy. In this strategy, they claimed access to land, to the forests, to the rivers. These common resources should no longer be used for the extraction of profit but for the livelihood of the people. And the people, particularly the women, would see to it that these resources would be sustained for future generations (Viezzer et al., 1992).

Poor women's resistance in the Third World against the destruction of the basis of their livelihood is not motivated only by narrow self-interest and provincialism. The new visions of a better society and economy emerge from such grass-roots movements.

References

Aoki, Yayoi. "Thoughts on Japan's Modernization and Confucian Ideology." In T. Yamamoto (ed.), *Puragu o nuku: I Keizai sekkusu tojendsa.* Translated by Vera C. Mackie. Tokyo: Shinhyoron, n.d.

Becker, B., S.H. Bovenschen, and N. Brackert (eds). *Aus der Zeit der Verzweiflung: Zur Genese und Aktualität des Hexenbildes.* Frankfurt am Main: Suhrkamp, 1977.

Bennholdt-Thomsen, V. "The Future of Women's Work and Violence Against Women: Why Do Housewives Continue to Be Created in the Third World, Too?" In M. Mies, V. Bennholdt-Thomsen, and C. von Werlhof, *Women: The Last Colony.* London: Zed, 1988, pp. 113–129.

Bock, G., and B. Duden. "Labor of Love—Love As Labor," *Development,* no. 4, 1984, pp. 6–14. Special Issue: "Women, Protagonists of Change."

Ching yoon Louie, Miriam. "Minjung Feminism: Korean Women's Movement for Gender and Class Liberation," *Women's Studies International Forum,* vol. 18, no. 4, 1995, pp. 417–430.

Dalla Costa, M. *The Power of Women and the Subversion*

of the Community. Bristol: Falling Wall, 1972.

Dankelman, Irene, and Joan Davidson. *Women and Environment in the Third World: Alliance for the Future.* London: Earthscan, 1988.

Ehrenreich, B., and D. English. *For Her Own Good: 150 Years of Experts' Advice to Women.* London: Pluto, 1979.

Fröbel, F., J. Heinrichs, and O. Kreye. *Umbruch in der Weltwirtschaft.* Reinbek: Rororo, 1986.

Granet, M. *Die chinesische Zivilisation.* Frankfurt am Main:: Suhrkamp, 1985.

Grossman, R. "Women's Place in the Integrated Circuit," *South East Asian Chronicle,* no. 66, vol. 9, January/February 1979, pp. 2–17.

Heinsohn, G., and O. Steiger. *Die Vernichtung der Weisen Frauen.* Herbstein: März-Verlag, 1985.

Herz, Barbara. "Briefing on Women in Development," paper presented at the annual meeting of the World Bank/International Monetary Fund (IMF), Berlin, 1988.

Luxemburg, Rosa. *Die Akkumulation des Kapitals, Ein Beitrag zur ökonomischen Erklärung des Kapitalismus.* Berlin: Buchhandlung Vorwärts, Paul Singer, 1913.

Marx, Karl. *Capital: A Critique of Political Economy.* London: Lawrence and Wishart, 1974.

Memmi, Albert. *The Colonizer and the Colonized.* New York: Orion, 1965.

Merchant, Carolyn. *The Death of Nature: Women, Ecology, and the Scientific Revolution.* San Francisco: Harper and Row, 1983.

Mies, Maria. *The Lace Makers of Narsapur: Indian Housewives Produce for the World Market.* London: Zed, 1982.

———. *Patriarchy and Accumulation on a World Scale: Women in the International Division of Labour.* London: Zed, 1986.

———. "Capitalist Development and Subsistence Production: Rural Women in India." In M. Mies, V. Bennholdt-Thomsen, and C. von Werlhof, *Women: The Last Colony.* London: Zed, 1988, pp. 27–50.

———. "'Gender' and Global Capitalism." In Leslie Sklair (ed.), *Capitalism and Development.* London: Routledge and Kegan, 1994, pp. 107–122.

Mies, Maria, and Vandana Shiva. *Ecofeminism.* London: Zed, 1993.

O'Sullivan, Sister Helene M.M. (ed.). *Silk and Steel: Asian Women Workers Confront Challenges of Industrial Restructuring.* Published by the Committee for Asian Women, Hong Kong, 1995.

Mies, M., V. Bennholdt-Thomsen, and C. von Werlhof. *Women: The Last Colony.* London: Zed, 1988.

Omvedt, Gail. *Reinventing Revolution: New Social Movements and the Socialist Tradition in India.* New York: M.E. Sharpe, 1993.

Plumwood, Val. *Feminism and the Mastery of Nature.* New York: Routledge, 1993.

Shiva, Vandana. *Staying Alive: Women, Ecology, and Survival in India.* New Delhi: Kali for Women, 1988.

Shiva, Vandana, Vanaja Ramprasad, and Radha Hollar Bhar. *NAVDANYA: Renewing Diversity and Balance Through Conservation.* New Delhi: Research Foundation for Science Technology and Natural Resource Policy, 1994.

Viezzer, Moema, et al. *Com Garra E Qualidade: Mulheres em economias sustentaveis: Agricultura e extrativismo.* Sao Paulo: Rede Mulher, 1992.

Waring, Marilyn. *If Women Counted: A New Feminist Economics.* San Francisco: Harper and Row, 1988.

Werlhof, C. von. "The Proletarian Is Dead: Long Live the Housewife!" In M. Mies, V. Bennholdt-Thomsen, and C. von Werlhof, *Women: The Last Colony.* London: Zed, 1988, pp. 168–179.

Political and Legal Contexts

Reclaiming Women's Human Rights

Rosa Briceño

Introduction

The Universal Declaration of Human Rights protects everyone "without distinction of any kind such as race, colour, sex, language . . . or other status" (Art. 2). Furthermore, "everyone has the rights to life, liberty, and security of person" (Art. 3) and "no one shall be subject to torture or to cruel, inhuman or degrading treatment or punishment" (Art. 5). Therefore, we, the undersigned, call upon the 1993 United Nations World Conference on Human Rights to comprehensively address women's human rights at every level of its proceedings. We demand that gender violence, a universal phenomenon which takes many forms across culture, race and class, be recognized as a violation of human rights requiring immediate action.

This is the text of a petition campaign launched in November 1991 and sponsored by more than 1,000 groups who gathered nearly half a million signatures from 124 countries by the time of the Second World Conference on Human Rights (Vienna, 1993). The campaign reflects the emergence of a global movement for women's human rights that gained momentum in the early 1990s and that has been enormously successful in gaining international recognition that "women's rights are human rights."

As a result of this movement, remarkable progress has been made in making women's human rights visible and in expanding the international human-rights agenda to include some gender-specific violations. The greatest gains have been achieved in the area of violence against women. A few years ago, violence against women was not even considered a human-rights issue, much less one requiring international attention. By 1996, women could argue at the local and national level that the United Nations has recognized this human-rights abuse and has mandated state action on it. This will not automatically end these violations, but it provides a valuable tool toward their elimination.

This essay offers an introduction to the rapidly growing field of the global movement to transform and reclaim the vision of human rights for women; it is meant to provide but a "grand tour" or overview of this movement—identifying key issues, moments, accomplishments, and challenges ahead.

Vision and Reality

The concept of human rights gained prominence in the modern era with the establishment of the United Nations in 1948. Coming out of the horrors of World War II, the international community recognized the need to protect individuals from abuses on that scale and agreed on "the promotion and encouragement of a respect for human rights" as a central purpose of the organization (United Nations Charter). On 10 December, 1948, the General Assembly proclaimed the Universal Declaration of Human Rights "as a common standard of achievement for all peoples and all nations." Its vision, "the advent of a world in which human beings shall enjoy freedom of speech and belief and freedom from fear and want," resonates deeply in all of us.

The Universal Declaration remains the most widely recognized statement of the rights to which every person on the planet is entitled. Despite this broad agreement and the significance of the human-rights concept in international and national politics, its meaning remains controversial. Major international human-rights instruments shy away from a definition of human rights, contenting themselves with listing the rights and freedoms that they seek

Table 1. Human Rights Guaranteed in Main International Treaties

Right to self-determination
Right to nondiscrimination
Prohibition of apartheid
Right to effective remedy for violations
Prohibition of retroactivity for criminal offenses
Prohibition of imprisonment for contractual obligations
Right to procedural guarantees in criminal trials
Right to life
Right to physical and moral integrity
Prohibition of torture and of cruel, inhuman, and degrading
    treatment or punishment
Prohibition of slavery, forced labor, and trafficking in
    persons
Right to recognition of legal personality
Right to liberty and security
Prohibition of arbitrary arrest, detention, and exile
Right to freedom of movement and residence
Right to seek asylum
Right to privacy
Right to freedom of thought, conscience, and religion
Right to freedom of expression
Right to freedom of peaceful assembly
Right to freedom of association
Right to marry and to found a family
Right to protection of motherhood and childhood
Right to a nationality
Right to work
Right to food
Right to social security
Right to enjoy the highest standard of physical
    and mental health
Right to education
Right to participation in cultural life

Source: Tomasevski, 1993, p. 47.

to protect. (Table 1 offers a list of human rights guaranteed in major international treaties.) Despite attempts by scholars to define the concept, "the content of human rights and the nature and extent of obligations assumed by states is not definitive. It may vary from one historical moment to another. Human rights is a dynamic concept and this fact has significant implications for women" (Butewga, 1995, p. 27).

Since the drafting of the Universal Declaration in 1948, women have had to fight vigorously to be included in the human-rights vision. As Tomasevski (1993) recounts, an early draft of the declaration opened with "all *men* are brothers." This reflected the gender insensitivity within the Commission on Human Rights, which was drafting the declaration, even though it was chaired by Eleanor Roosevelt, and despite the efforts of its female members. The Commission on the Status of Women (CSW) effectively opposed this exclusionary language.

The final text of the Universal Declaration reaffirms the United Nations Charter's postulate of the equal rights of women, stating that "all *human beings* are born free and equal in dignity and rights" (Art. 1) and that "*everyone* is entitled to all the rights and freedoms set forth in this Declaration, without distinction of any kind, such as race, colour, sex, language, religion, political or other opinion, national or social origin, property, birth or other status." (Art. 2). Eleanor Roosevelt and the Australian and Latin American women who fought for the inclusion of women in the declaration and for its passage, clearly intended that it would address the problems of women's subordination (Bunch, 1991). There is, however, a wide gap between the vision and the reality.

Since the beginning of the 1990s, there has been increasing recognition that the international human-rights movement has benefited men more than women, and some encouraging steps have been taken toward reversing a long pattern of neglect of women's human rights. This progress has been the result of the activism and lobbying efforts of women's-rights activists from around the world who have taken on the challenge of reclaiming the vision of human rights for women. "The concept of human rights, like all vibrant visions, is not static or the property of any one group; rather its meaning expands as people reconceive of their needs and hopes in relation to it" (Bunch, 1991, p. 4). In this spirit, feminists in the 1990s have been defining human rights to reflect the specific experiences of women at all stages of their lives and to transform human-rights thinking and practice so that it takes better account of women's lives.

Transforming the concept of human rights means going beyond its late-twentieth-century form. Let us for a moment put on the traditional lenses that have been worn to shape and develop human-rights policy since the late 1940s. What do we see? First, we see a male. Second, we see a male being arbitrarily arrested and tortured. And third, we see a male being arbitrarily arrested and tortured somewhere in a developing country. Wearing a different pair of lenses—gender lenses—women's human-rights activists have launched a critique of this dominant perspective. As suggested below, it is a perspective that not only violates the "universality" and "indivisibility" concepts embedded in the human-rights vision but also has contributed to slowing down the promotion and protection of women's human rights.

Accepting the Male As the Norm

Universality is a central tenet of the human-rights vision. Universality implies that human rights and the mechanisms designed to actualize them should be equally available to men and women. In practice, the prevailing view has been that "universal" human rights are best protected through the norm of nondiscrimination, broadly formulated in the Universal Declaration of Human Rights, which entitles all to the rights and freedoms set forth therein, "without distinction of any kind," including distinctions based on sex. This general norm barring sex-based discrimination is enshrined in the declaration's two implementing covenants of 1966 (the Political Covenant and the Economic Covenant) and reinforced in all major human-rights instruments.

Women are bringing into question this "unisex" approach to human rights. There are some aspects of life that are common to women and men, and, clearly, women should be accorded equal opportunity in those areas. But there are also actions or omissions that impinge on the inherent dignity of women in the context of the Universal Declaration that are not experienced by men. Because traditionally, human-rights thought and practice have accepted the male as the norm and the point of departure, legitimate concerns of women that lacked a male norm or experience have been considered irrelevant to the human-rights framework. The result has been an absence of guarantees for fundamental rights and freedoms when women are the actors most affected (Butewga, 1995).

A particularly clear example is gender-based violence against women in all of its manifestations. As the Global Campaign for Women's Human Rights highlighted in the 1993 World Conference in Vienna,

> more women die each day from various forms of gender-based violence than from any other type of human rights abuse. This ranges from female infanticide and the disproportionate malnutrition of girl children, to the multiple forms of coercion, battery, mutilation, sexual assault and murder that many women face in every region of the world, throughout their lives, simply because they are female (Bunch and Reilly, 1994, p. 4).

Yet, only recently has violence against women been recognized as a human-rights issue and one urgently requiring attention from the international human-rights community.

A feminist human-rights policy would pursue the collective interests of humanity in a way that integrates the perspectives of men *and* women. Indeed, an enumeration of human rights that reflects the realities of most women's lives would look very different from the current one. It would focus, for example, on freedom from gender-based violence, on autonomy within the family, on conditions suitable for healthy reproduction, and on adequate food and shelter (UNIFEM, n.d.).

This blindness has been exacerbated by the insistence in traditional human-rights theory on a division between public and private responsibility. In general, human-rights abuses are considered violations under the various human-rights instruments if a state party to those instruments can be held responsible for abuse. This raises several obstacles for the realization of women's human rights because, often, private individuals, not the state or its agents, perpetrate human-rights violations against women. This is true primarily in cases of violence against women, as well as discrimination against women in employment and the acquisition of land rights.

Activists for women's human rights have challenged this public/private split as a politically constructed barrier that has been used to justify inaction by the state and continued subordination of women. Moreover, it is selectively applied. As noted in the report of the 1991 Women's Global Leadership Institute that met at Rutgers University: "The human rights community has proven willing to stretch the boundaries of state responsibility to accommodate the concerns of men," taking on, for example, the phenomenon of "disappearances" in Argentina, murder of indigenous rubber tappers in Brazil, and racially motivated hate crimes—all abuses perpetrated by private individuals (Center for Women's Global Leadership, 1992, p. 17).

Neglecting Economic, Social, and Cultural Rights

The indivisibility and interdependence of human rights is another central tenet of the human-rights vision. It was first expressed in the Universal Declaration of Human Rights, which treats all the rights and freedoms guaranteed therein as equal. These include civil and political rights—such as the right to life, liberty, and security of person (Art. 3), to equal protection of the law (Art. 7), and to freedom of thought, conscience, and religion (Art. 18)—as well as economic, social, and cultural rights such as the right to work, to just and favorable conditions of work, and to protection against unemployment (Art. 22), the right to an adequate standard of living, including food, clothing, housing, and medical care (Art. 25), and the right to education (Art. 26).

The indivisibility of human rights was reaffirmed in the declaration of the Second World Conference on Human Rights, which states that "the international community must treat human rights globally in a fair and equal manner, on the same footing, and with the same emphasis (Vienna Declaration, 1993, paragraph 5). Despite the rhetoric, in practice the international human-rights community has privileged civil and political rights, which are often regarded as *the* human rights. At the same time, it has neglected the promotion and protections of social, economic, and cultural rights, thereby ignoring some of women's most pressing concerns.

These developments were the result of political self-interests of the states and of Cold War politics. As Tomasevski (1993) recounts, soon after the adoption of the Universal Declaration in 1948, work began on translating its principles into a formally binding treaty. This turned out to be a long and conflict-ridden task. In the context of the Cold War, governments were not able to agree on a treaty. Positions of Communist and developing countries emphasized the protection of social and economic rights. Western and developed countries gave preference to the protection of civil and political rights, which, they argued, could be expressed in law, judged in court, and assessed by the United Nations Human Rights Commission (Ashworth, 1993).

The end result was the adoption of two separate treaties: the International Covenant on Civil and Political

Rights (the ICCPR, also known as the Political Covenant) and the International Covenant on Economic, Social and Cultural Rights (the ICESCR, or Economic Covenant). These constitute, together with the Universal Declaration of Human Rights, the International Bill of Rights. Civil and political rights have received the lion's share of attention by the international human-rights community. The Political Covenant receives more resources than other human-rights instruments and has more effective implementation mechanisms; it therefore plays a predominant role in the practice of human rights globally (Bunch and Reilly, 1994).

The hierarchical importance given to civil and political rights has slowed progress in the recognition and protection of women's human rights because "much of the abuse that women experience is part of a larger socio-economic web that entraps women, making them vulnerable to abuses that are not solely political or caused by states" (Bunch, 1991, p. 4). Some of the most urgent concerns of women's day-to-day existence involve the denial of economic, social, and cultural rights, including access to employment and credit, to adequate food and housing, and to education and health care.

These concerns include the calamitous impact of structural-adjustment policies imposed by multilateral agencies, such as the World Bank and the International Monetary Fund (IMF), that have affected women most harshly. Who is responsible for violations of women's human rights attributable to these policies? Advocates of women's human rights are wrestling with this issue (Schuler, 1995). They are also increasingly aware of the need to challenge the false belief that women's human rights are not an issue in developed countries. This is especially evident when violence against women, other gender-specific violations, and the denial of economic and social rights are brought into the picture.

CEDAW: A Landmark in Women's Human Rights

The adoption in 1979 of the United Nations Convention on the Elimination of All Forms of Discrimination Against Women (CEDAW) constituted a landmark in the history of women's human rights. Until then, there was no convention that comprehensively addressed women's rights within political, economic, cultural, social, and family life.

The drafting of CEDAW was orchestrated by the Commission on the Status of Women (CSW), which has been a major player pressuring for increased attention to women's issues within the United Nations. As of 1 April 1997, 157 countries had ratified the convention. As of late 1996, the United States and Switzerland were the only democracies that had not ratified it. Appendix A offers a summary of CEDAW circulated by activists for women's human rights pressuring for U.S. ratification of the measure.

CEDAW moves beyond the sex-neutral norm that requires equal treatment of men and women, usually measured by how men are treated, to recognize the fact that the nature of discrimination against women and their distinctive gender characteristics are worthy of a legal response. The convention draws a distinction between de jure and de facto rights. In this regard, it recognizes not only present discrimination but also past discrimination and introduces the concept of corrective measures to overcome the effect of past discrimination. It provides for measures through which affirmative action and women-centered development-policy measures can be legitimized to ensure de facto equality for women.

Unlike other human-rights treaties, which are usually limited to the conduct of the state or its agencies, CEDAW specifically obliges state parties to take all appropriate measures to eliminate discrimination against women by any person, organization, or enterprise. Recognizing that women are subject to pervasive and subtle forms of discrimination, it binds state parties to seek to modify cultural patterns of behavior and attitudes regarding the sexes and attempts to impose standards of equality and nondiscrimination in private as well as public life (UNIFEM, n.d.).

CEDAW also makes a strong case for the indivisibility of human rights. The preamble recognizes "that in situations of poverty women have the least access to food, health, education, training and opportunities for employment and other needs" (paragraph 8). It entitles women to equal enjoyment with men not only of civil and political rights but also of economic, social, and cultural rights, and it mandates both legal and development-policy measures to guarantee the rights of women in *all* areas of life.

While all international law suffers some for lack of enforcement, oversight of CEDAW is especially weak. Among the specific problems: (1) unlike its mainstream cousin, the Human Rights Committee, the Committee on CEDAW, the body of experts that monitors CEDAW, has neither the necessary resources nor the authority to investigate individual or group claims of violations; (2) there is no complaints procedure by which individual women or groups could seek international remedies for violations of the convention; (3) the reporting is the state's responsibility, and women are excluded from the process; (4) governments often do not submit reports, and, when they do, they are rarely self-critical; (5) the interpretation of the articles tends to be left to the governments, which often results in narrow definitions of rights and limited analyses of problems and remedies (Tomasevski, 1993).

Advocates and the CEDAW Committee cite the issue of reservations to this convention as a major obstacle. To facilitate the process of undertaking human-rights obligations, governments are allowed to reserve the right not to apply a specific part of a treaty by submitting a reservation at the moment of ratification. Not only is CEDAW the human-rights convention with the most substantive reservations, but some of these reservations are contrary to the very aim of the convention. As of late 1996, CEDAW contained no internal mechanism to reject reservations considered to be inconsistent with the object and purposes of its obligations.

Tomasevski (1993) offers a list of CEDAW's main provisions to which reservations have been submitted by specific countries, noting that most reservations concern nondiscrimination in family law—the very sphere where rights for women are most crucial—and women's citizenship and legal capacity. Many governments argue that they cannot implement parts of the convention because to do so would interfere with existing religious or cultural law. As Butewga notes, what is insightful is that some states entered reservations on articles they purport to accept in the context of other human-rights instruments. "It is, therefore, pertinent to ask whether cultural and religious or other relativism in human rights theory [is] in the interests of the protection of the human rights of women" (Butewga, 1995, p. 34).

By elaborating the meaning and scope of discrimination against women, CEDAW has provided a vital tool to advocates of women's human rights. At the same time, these advocates recognize that CEDAW has yet to realize its promises for women. One concrete way in which activists are working to strengthen the convention is by calling for the adoption of an optional protocol to CEDAW.

Toward this end, they have targeted the Commission on the Status of Women, which is responsible for drafting such a formal mechanism and has the optional protocol in its agenda. The optional protocol would provide a complaints procedure and an inquiry procedure, by which the CEDAW Committee could decide on its own initiative to investigate reports of widespread human-rights violations against women.

Global Movement for Women's Human Rights

Women have always played an active role in the struggles for human rights around the globe. Two have been awarded the Nobel Peace Prize in recognition of their human-rights work: Daw Aung San Suu Kyi in 1991, in absentia and while enduring detention, for her contribution to human rights in Burma; and Rigoberta Menchu, from Guatemala, in 1992 for her work on indigenous rights. Women's contributions to human-rights movements have been best documented in Latin America, where organizations founded by female relatives of those who disappeared under military dictatorships came to form the backbone of human-rights groups. The most famous of these groups, the Mothers of Plaza de Mayo in Argentina, became known internationally for its unprecedented impact on the need for democratic governance.

Women's rights traditionally have been treated as separate and not taken as seriously by human-rights organizations and governments. This attitude is reflected in the fact that when the United Nations resolved to hold its Second World Conference on Human Rights, its proposed agenda did not mention women or any gender-specific aspects of human rights. "Yet, by the time the World Conference ended in Vienna in June 1993, gender-based violence and women's human rights emerged as one of the most talked about subjects, and women were recognized as a well-or-

ganized human rights constituency" (Bunch and Reilly, 1994, p. 2).

The movement for women's human rights can be traced back to the United Nations Decade for Women (1976–1985), which facilitated the proliferation of women's nongovernmental organizations (NGOs) in Third World countries as well as the further establishment of U.N. programs with respect to the advancement of women. The decade ended with the Third United Nations World Conference on Women (Nairobi, 1985), which brought greater awareness about the obstacles to women's advancement and generated a new momentum for collective action at the international and regional levels. The *Forward-Looking Strategies,* the conference document, placed greater emphasis on the deeper structural and institutional changes required in societies everywhere for women to achieve full equality. The Third World Forum on Women, Law, and Development, held as part of the parallel NGO activities, identified two key strategies for action: using the law as a resource for women's empowerment and creating regional women's-rights networks (Schuler, 1986).

At the same time, violence against women began to emerge as one of women's most urgent concerns around the world and a powerful unifying issue. In 1981, women's groups in the Latin American region held the first *Encuentro Feminista de Latinoamérica y el Caribe,* which declared November 25 International Day Against Violence Against Women. This date was chosen in memory of the three Mirabal sisters who were brutally murdered in 1960 by the Trujillo dictatorship in the Dominican Republic. This and subsequent feminist meetings in the region played a forceful role in articulating the need for an effective response to violence against women. Whether women's organizations had been focusing on reproductive health, economic development, or legal reform, the problem of violence against women became increasingly visible as a major obstacle to the well-being and advancement of women across the board. Within the United Nations, United Nations Development Fund for Women (UNIFEM) responded to this trend by making violence against women a major program area. CEDAW and the CSW developed a number of recommendations, including a Draft Declaration on Violence Against Women in December 1991, which was finally adopted by the General Assembly in December 1993.

As noted by Chiarotti (1993), an activist for women's human rights in Argentina, the greatest obstacle to advancing this work has been persuading human-rights groups to take gender as an important variable, as important as class and race. Gender-specific concerns have been treated as marginal by the male leaders of human-rights organizations, although the situation has improved somewhat, thanks to the growth of the women's movement and the increasing presence of women in human-rights organizations.

By 1990, women's groups around the world were meeting locally, regionally, and internationally to articulate the connections between women's rights and human rights. During that year, for example, regional consultations were held in Thailand, India, and Indonesia under the auspices of the Asia Pacific Forum on Women, Law, and Development "to gather experiences on the link between women's rights and human rights in order to develop an alternative conceptual framework and strategies" (APWLD, 1992, p. 9). These consultations emphasized the need to restructure the system and created task forces to explore in depth the connections between human rights and "women and democracy," "women's economic rights," "women and religion," and "violence against women." At the same time, ISIS Internacional, an international women's network based in Santiago, Chile, produced an anthology on women's human rights that explored some of these connections (ISIS Internacional, 1991).

International human-rights organizations, as well, established new projects and initiatives to focus on abuses of the human rights of women. Amnesty International conducted studies and issued reports on rape in detention as a form of torture and other gender-based abuses perpetrated by state actors. The Women's Project of Human Rights Watch broke ground in reporting on such matters as domestic violence in Brazil.

The Second World Conference on Human Rights (Vienna, 1993) provided a unique opportunity to make concerns about women's human rights and transformative perspectives more visible. It also became a natural vehicle for women to organize globally around these issues. A highly effective global strategy was the launching of an annual campaign of "16 Days of Activism Against Gender Violence" linking November 25 (International Day Against Violence Against Women) with December 10 (International Human Rights Day). The idea was developed by international participants in the First Global Institute on Women, Violence and Human Rights, organized by the Center for Women's Global Leadership at Rutgers University, New Brunswick, New Jersey, in 1991.

Targeting the 1993 Second World Conference on Human Rights, the campaign initiated a petition drive calling upon the conference to "comprehensively address women's human rights" and to recognize gender violence "as a violation of human rights requiring immediate attention." The petition, originally sponsored by Center for Women's Global Leadership and the *International Women's Tribune,* was circulated through dozens of women's networks in 124 countries. By the time of the conference, more than 1,000 sponsoring groups had gathered almost a half million signatures (Bunch and Reilly, 1994).

At the same time, regional movements for women's human rights worked at influencing the conference agenda at the regional preparatory meetings. Bunch and Reilly's (1994) account of the Global Campaign for Women's Human Rights gives a good sense of the numerous organizing efforts by women's groups around the world to put women's concerns on the conference agenda. For example, Women in Law and Development in Africa (WiLDAF) or-

ganized a series of subregional meetings that highlighted human-rights issues of concern to women in Africa that required immediate attention, including violence against women, the use of culture and tradition to deny women their fundamental rights, the impact of economic structural-adjustment programs on the human rights of women, and the failure of the human-rights community to provide human-rights education to the people (Butewga, 1995).

In Latin America, women's-rights leaders from the region gathered at a women's human-rights conference called *La Nuestra* (Ours) and prepared a 19-point agenda to be submitted to the regional meeting. Asian women also drew from these efforts, as well as from their own regionally specific initiatives, to influence the preparatory process, despite frequent opposition from their governments. In their Declaration for the Bangkok Regional Conference, for example, they strongly defended the universality of human rights, which was being questioned as a Western construct by various Asian governments.

At the parallel NGO meeting in Vienna, more than 60 of the panels and workshops specifically addressed women's human rights. Among the strategies that captured the most attention was the Global Tribunal on Violations of Women's Human Rights. Thirty-three women from 25 countries testified on specific issues in each region within five broad areas: human-rights abuse in the family, war crimes against women in conflict situations, violations of women's bodily integrity, violations of social and economic human rights of women, and gender-based political persecution and discrimination. The tribunal provided a powerful venue to expose the violations of human rights that women suffer and to protest the failures of the international human-rights movement to protect women.

Many of the groups involved in these actions also joined forces in lobbying governments—both at the final preparatory meeting held in Geneva in April 1993 and at the actual conference. The women's caucus represented a loose coalition that crossed long-established divisions—not only along First–Third World lines but also joining women working in government, in nongovernmental human-rights and women's organizations, and in U.N. agencies. In spite of the differences, tensions, and disagreements, they worked effectively for the inclusion of text on women that was accepted by governments in Geneva almost without reservation, a process which virtually assured its passage at the full conference in Vienna. This work formed a basis for many women to continue working together at subsequent world conferences.

The movement maintained its momentum and worked to affirm and expand the gains made at the Vienna World Conference on Human Rights (Vienna, 1993) during subsequent world meetings: the International Conference on Population and Development (Cairo, 1994), the World Summit for Social Development (Copenhagen, 1995), and the Fourth World Conference on Women (Beijing, 1995). In Cairo, activists for women's human rights fought to establish recognition of their reproductive rights and health in the face of massive opposition from the Vatican and religious fundamentalists. Although the press focused on the issue of abortion, what was at stake was women's control over their reproductive functions and the means necessary to achieve that control. Women insisted that their health and bodies could not be the vehicles through which governments met demographic targets and other policy objectives with no consideration for their fundamental human rights. At the World Summit for Social Development (Copenhagen, 1995) activists emphasized the links between women's human rights and social and economic justice.

As women made inroads in the recognition of their human rights, the opposition forces also became stronger. In the United States, for example, the Far Right turned into a fervent champion of human rights, trying to prevent the United States from sending a delegation to the 1995 Beijing conference on those grounds. The reaction of the U.S. ambassador to the United Nations Madeleine Albright, that one cannot in the name of human rights sabotage a conference that is about the human rights of half of humanity, expressed perfectly the sentiment felt by many women at the time.

At the final preparatory meeting for the Beijing conference, about 35 percent of the document was in "brackets"—governments had not been able to reach agreement on the language—including most of the human-rights section. At the insistence of the Vatican, even the use of the word "gender" in the document was contested. Moreover, throughout the process, women's groups faced numerous barriers to their effective participation, including a sudden change of site for the NGO forum a few months prior to the meeting, which threatened their ability to be effective at the conference.

Gains for Women

Given the degree of invisibility of women's human rights at the beginning of this process, the progress has been remarkable. Both the *Vienna Declaration* and *Programme of Action* (the final document adopted by consensus by the 171 governments participating in the 1993 Second World Conference on Human Rights) and the *Platform for Action* (the document adopted by the 181 governments participating in the 1995 Fourth World Conference on Women in Beijing) significantly expanded the international human-rights agenda to include gender-specific violations; they are powerful tools that can be used to pressure local and national authorities to live up to the international standards to which they have agreed.

The *Vienna Declaration* devotes several pages to treating the human rights of women as a priority for governments and the United Nations and cites examples of gender-specific abuse as human-rights violations. The *Platform for Action* recognizes the lack of respect for and the lack of promotion and protection of the human rights of women as one of the critical areas of concern in future agendas and reaffirms and extends the commitments to promote and protect women's human rights. Key areas where gains in women's human rights have been made include:

Universality of Women's Human Rights

The universal nature of human rights was a contentious issue in Vienna and again in Beijing. It is an issue especially crucial for women, whose human rights are often eclipsed in the name of culture, religion, or tradition. The *Vienna Declaration* explicitly stated for the first time that "the human rights of women and the girl child are an inalienable, integral and indivisible part of universal human rights" (paragraph 18) and reaffirmed the universality of human rights regardless of differences in political, economic, and cultural systems (paragraph 5).

In Beijing, women found themselves lobbying for the removal of the term "universal" as a modifier of human rights in several parts of the draft document, where it could have been regarded as a shorthand for "universal recognition or acceptance" as opposed to their "universally binding character." This cynical move was spearheaded by the Vatican and supported by its allies, such as Malta, Argentina, Benin, and Honduras. Ultimately, the use of "universal" as a modifier to "human rights" was eliminated in the final document, retaining only the strong language from Vienna that the human rights of women and the girl child are part of universal human rights.

Debates about the role of religion and culture emerged in many parts of the *Platform for Action* in reference to sexuality, reproductive health and rights, inheritance, and adherence to human-rights norms. One of the most extensive debates took place over text that sought to condition the promotion and protection of women's human rights on the basis of culture and religion (paragraph 9). The women's human-rights caucus supported language that placed human rights, rather than "the full respect for various historical, cultural and religious backgrounds" as the primary concern. The final document is an odd mix of the two positions. It uses *Vienna* (paragraph 5) as its basis by placing human rights as the primary concern, while, at the same time, it recognizes the importance of national sovereignty and maintains that the "significance of national and regional particularities and various historical cultural and religious backgrounds must be borne in mind."

Overall, the *Platform for Action* asserts the universal and holistic nature of the human rights of women. At the same time, there were a large number of reservations and other interpretive statements that pose real obstacles for the implementation of the *Platform*. Some delegations placed reservations on efforts to utilize broad definitions of reproductive health and rights, sexuality, and the family; others entered reservations to text that did not conform to Islamic law. The Vatican put forward its interpretation of much of the *Platform,* arguing that it expressed "exaggerated individualism." As Fried (1995) notes in her report on the work of the human-rights caucus at Beijing, "the extensive reservations by governments to language on religious and cultural grounds reflect an ongoing debate about the human rights of women which could not have been resolved in Beijing, and lay out the contours of future collaborations and confrontations" (Fried, 1995, p.3).

Integration and Implementation of Women's Human Rights

Both the Vienna and the Beijing conferences urged governments, institutions, and intergovernmental and nongovernmental organizations to intensify their efforts for the protection and promotion of the human rights of women and the girl child. Governments were to develop their implementation strategies in consultation with relevant institutions and nongovernmental organizations.

Both conferences also stated that the human rights of women should form an integral part of the United Nations human rights activities, including the promotion of all human rights instruments relating to women. They made a strong call for strengthening CEDAW, through universal ratification by the year 2000, reducing the excessive number of reservations to the convention, and calling on the Commission on the Status of Women to expedite its consideration of the optional protocol.

The realization of women's human rights requires not only that CEDAW and gender-specific mechanisms be implemented but that, in addition, other human-rights treaties, such as the ICCPR and the ICESCR, be used to benefit women. In this regard, a challenging aspect in all of these documents is the recognition that the equal status of women and the human rights of women should be integrated into the mainstream of U.N. system-wide activity (Platform for Action, paragraph 231) and that these issues should be regularly and systematically addressed throughout relevant U.N. bodies and mechanisms. Despite the stated commitments about gender integration, there is little reference in these documents to which institutional tasks, specific agencies, and actors will be responsible for such integration.

Violence Against Women

The *Vienna Declaration* sounds a historic call to bring about the elimination of violence against women in public and private life: "Gender based violence and all forms of sexual harassment and exploitation, including those resulting from cultural prejudice and international trafficking, are incompatible with the dignity and worth of the human person, and must be eliminated" (paragraph 18). It further states that "violations of the human rights of women in situations of armed conflict are violations of the fundamental principles of international human rights and humanitarian law. All violations of this kind, including in particular murder, systematic rape, sexual slavery, and forced pregnancy, require a particularly effective response" (paragraph 38). Finally, it calls upon the United Nations General Assembly to adopt the draft declaration opposing violence against women.

Following the Vienna Conference, some important gains were made at the international and regional policy levels toward addressing gender-based violence, including the adoption of the United Nations Declaration on the Elimination of Violence Against Women (December 1993); the appointment by the Commission on Human Rights of a United Nations Special Rapporteur on Violence Against Women (March 1994); the adoption of the Inter-American Convention on the Prevention, Punishment, and Eradication of Violence Against Women (April 1994); and the appointment of a liaison on women's human rights at the United Nations Center for Human Rights to initiate the task of integrating gender into the United Nations' human-rights machinery.

The *Platform for Action* from the Beijing conference reaffirms and extends the commitments to "combat and eliminate all forms of violence against women in private and public life, whether perpetrated by or tolerated by the State or private persons" (paragraph 225). A debate took place at Beijing concerning rape in situations of armed conflict. The human-rights caucus opposed proposed language that threatened to make rape a human-rights offense only if it could be shown to be "systematic." While the term "systematic" rape was retained, the final *Platform for Action* document strongly articulates that "rape in the conduct of armed conflict constitutes a war crime and under certain circumstances it constitutes a crime against humanity" (paragraph 147); it also calls on governments and international agencies to respond to this problem more effectively.

Health and Reproductive Rights

Echoing the commitments made at the International Conference on Population and Development (Cairo, 1994), the Beijing *Platform for Action* reaffirms that reproductive rights "rest on the recognition of the basic right of all couples and individuals to decide freely and responsibly the number, spacing and timing of their children . . . and to have the information and the means to do so, and the right to attain the highest standard of sexual and reproductive health. It also includes the right to make decisions concerning reproduction free of discrimination, coercion, and violence, as expressed in human rights documents" (paragraph 96).

The women's human-rights caucus lobbied for the recognition of sexual rights in the *Platform* as an essential component of a woman's right to determine her reproductive life and fundamental to her empowerment. As expected, the concept of sexual rights as human rights was extremely contentious. While it was not explicitly articulated in the final document, it is contained in a broad statement in it that "the human rights of women include their right to have control over and decide freely and responsibly on matters related to their sexuality, including sexual and reproductive health, free of coercion, discrimination and violence" (paragraph 97).

The inclusion of language explicitly proscribing discrimination on the basis of sexual orientation (such as in paragraph 48) was also debated until the final moments. While it was ultimately decided not to include any references to sexual orientation in the document, the discussion reflected support from all regions of the world (including Slovania, South Africa, Brazil, Colombia, Jamaica, Norway, the European Union, the United States, Canada, New Zealand, and Australia) for the recognition that discrimination against women on the basis of sexual orientation is a barrier to their full enjoyment of human rights.

Challenges Ahead

The gender blindness of the Bosnia peace accords negotiated in Dayton, Ohio, in 1995 and the absence of women at the table—even though women suffered horrible violations, including rape as a weapon of war, constitute the majority of the refugees, and have been the loudest voices for peace in the region—illustrate how far the international human-rights community has yet to go to include the perspectives of women and guarantee their dignity. This situation, happening on the heels of the Fourth World Conference on Women, reminds us that the gains made will be realized only through women's vigilant oversight.

One of the challenges is to work to assure the implementation of the promises made to women in recent world conferences, from the local to the global level. Toward this end, women's groups around the world are engaging in a variety of follow-up activities to the Beijing conference. These include political, legal, and educational strategies such as conducting "report back" meetings, launching campaigns to elicit concrete commitments from governments, developing national strategies to monitor the implementation of the *Platform,* lobbying governments to support the optional protocol to CEDAW, producing popular versions of the *Platform,* and organizing advocacy workshops. Many of these strategies are being shared via the World Wide Web.

Another key challenge is to address economic, social, and cultural inequities that women suffer by promoting and protecting social and economic rights more effectively. There is a growing consensus about the need to do more in this area. The challenges ahead include lobbying for a faster pace in the development of an optional protocol to the ICESCR, contributing to the development of jurisprudence and human rights indicators in this area, and pressuring international human-rights organizations to include economic, social, and cultural rights as an integral part of their mandates.

First World–based activists for women's human rights face the additional challenge of bringing human-rights language and practice home to their own backyards. In the United States, for example, activists must work to expose the hypocrisy of a government that presents itself as a champion of human rights in the world while it has failed to ratify major human-rights covenants, including CEDAW.

Organizing globally is important because global actions are key to generating the necessary pressure for making changes in the international human-rights system. At the same time, gains at the international level are powerful tools for women to use at the local level, to create local pressure and as a recourse to seek redress of women's human-rights violations.

One challenge that the global women's human-rights movement faces is finding new ways of working together after the recent string of U.N. conferences, which provided unique opportunities and resources for groups to come together, to strategize regionally and globally, and to build coalitions and a common agenda. After the Beijing meeting, the global movement for women's human rights faces the challenge of finding new venues for working together. It also must continue to address issues related to diversity, leadership, and power within the movement. These issues have surfaced at various points during the struggles of the past years (Schuler, 1995). It is time for the movement to engage openly in self-criticism and to develop new approaches stemming from lessons learned. Finally, there is the question of resources. The role of donor agencies in facilitating the activities of the global movement for women's human rights in the years ahead cannot be underestimated.

References

Ashworth, Georgina. *Changing the Discourse: A Guide to Women and Human Rights.* Thinkbook no. 9. London: Change, 1993.

Asia Pacific Forum on Women, Law, and Development (APWLD). *Women's Rights/Human Rights: Asia-Pacific Reflections.* Kuala Lumpur: APWLD, 1992.

Bunch, Charlotte. "Women's Rights As Human Rights: Towards a Revision of Human Rights." In Charlotte Bunch and Roxanna Carrillo (eds.), *Gender Violence: A Development and Human Rights Issue.* New Brunswick: Center for Women's Global Leadership, Rutgers University, 1991, pp. 3–18.

Bunch, Charlotte, and Niamh Reilly. *Demanding Accountability: The Global Campaign and Vienna Tribunal for Women's Human Rights.* New Brunswick: Center for Women's Global Leadership, Rutgers University, and UNIFEM, 1994.

Butewga, Florence. "International Human Rights Law and Practice: Implications for Women." In Margaret Schuler (ed.), *From Basic Needs to Basic Rights: Women's Claim to Human Rights.* Washington, D.C.: Women, Law, and Development International, 1995, pp. 27–39.

Center for Women's Global Leadership. *Women, Violence, and Human Rights: 1991 Women's Leadership Institute Report.* New Brunswick: Center for Women's Global Leadership, Rutgers University, 1992.

Chiarotti, Susana. Interview by Rosa Briceño. In Susana Fried and Rosa Briceño (eds.), *International Campaign for Women's Human Rights, 1993–1993 Report.* New Brunswick: Center for Women's Global Leadership, Rutgers University, 1993, pp. 23–27.

Dairiam, Shanti. "The U. N. Convention on the Elimination of All Forms of Discrimination Against Women." Cairo: NGO Forum, International Conference on Population and Development, 1994. Mimeographed.

Fried, Susana T. "Report of the Women's Human Rights Caucus at the Fourth World Conference on Women,

Beijing, 1995." New Brunswick: Center for Women's Global Leadership, Rutgers University, 1995. Mimeographed.

ISIS Internacional. *La Mujer Ausente: Derechos Humanos en el Mundo.* Serie Ediciones de las Mujeres no. 15. Santiago: ISIS Internacional, 1991.

Schuler, Margaret A. (ed.). *Empowerment and the Law: Strategies of Third World Women.* Washington, D.C.: OEF International, 1986.

———. *From Basic Needs to Basic Rights: Women's Claim to Human Rights.* Washington, D.C.: Women, Law, and Development International, 1995.

Tomasevski, Katarina. *Women and Human Rights.* London: Zed, 1993.

United Nations Development Fund for Women (UNIFEM). "CEDAW and Women's Human Rights." New York: UNIFEM, n.d. Leaflet.

Women in Law and Development in Africa (WiLDAF). *The World Conference on Human Rights: The WiLDAF Experience.* Harare: WiLDAF, 1993.

Violence Against Women

Charlotte Bunch
Roxanna Carrillo
Rima Shore

Introduction

The 1990 *Human Development Report* (UNDP, 1990) broke new ground by defining human development as a process of widening the range of people's choices. The 1994 *Report* called for a new development paradigm that puts people at the center of development and "empowers people—enabling them to design and participate in the processes and events that shape their lives." More specifically, it recognizes that "not much can be achieved without a dramatic improvement in the status of women and the opening of all economic opportunities to women" (UNDP, 1994, p. 4).

Yet, throughout the world, violence against women has been an obstacle to efforts aimed at enabling women to exercise greater choice. This issue has come onto the world agenda from the direct experiences of women at the grass roots. Development workers report problems of gender-based violence emerging in the midst of everything from income-generating projects to leadership development. At a 12-country workshop on women's nonformal education held in China in 1989, participants who were asked to name the worst aspect of being female almost unanimously responded that it was fear of male violence (Bradley, 1990). The threat and the reality of violence loom over women's everyday lives, coloring their ability to participate in development projects, to exercise democracy, and to engage in a variety of other pursuits. Further, gender-based violence cannot be eradicated in isolation because it is, simultaneously, an economic, political, and cultural issue illustrating the interconnections of development, human rights, and peace in women's lives.

The 1994 *Report* elaborated on the meaning of human development by introducing the concept of "human security" as a necessary condition for world peace, noting:

The world can never be at peace unless people have security in their daily lives. Future conflicts may often be within nations rather than between them—with their origins buried deep in growing socioeconomic deprivation and disparities. The search for security in such a milieu lies in development, not in arms (UNDP, 1994, p. 1).

Human security is based on the conditions that allow people to exercise choice safely and freely, with some confidence that the opportunities they have today will not vanish tomorrow. The concept of "human security through development" holds great promise for women in the Third World, whose lives epitomize the insecurity and disparities that plague the late-twentieth-century world order. As the report further notes: "In no society are women secure or treated equally to men. Personal insecurity shadows them from cradle to grave. . . . And from childhood through adulthood they are abused because of their gender" (UNDP, 1994, p. 31).

In this context, gender-based violence emerges as a primary human-security concern for women and, thus, as a crucial development challenge. Women in the North and the South live with the risk of physical harm in ways that have no direct parallels for men. The experience and fear of violence are threads in women's lives that intertwine with their most basic human-security needs at all levels—personal, community, environmental, economic, and political. In virtually every nation, violence or the threat of it, particularly at home, shrinks the range of choices open to women and girls, limiting not only their mobility and their control over their own lives but, ultimately, their ability to *imagine* mobility and control over their lives.

This essay is based on a background piece commissioned by the United Nations Development Program (UNDP) for preparation of the *Human Development Report 1995*. It reflects the views of the authors and not necessarily those of UNDP or the Human Development Report Office.

Experts tell us this is exactly the aim of violence against women. It may not be the conscious intention of each batterer, of every rapist, but, as research on gender violence mounts, it points ever more forcefully to the conclusion that, as a social phenomenon, violence against women is not about sex; it is not even about conflict. It is about control. It is not an aberration; rather, it is an extension of the ideology that gives men the right to control women's behavior, their mobility, their access to material resources, and their labor, both productive and reproductive (Bunch, 1990, p. 491; Heise, 1994, p. 2).

Violence against women remains a steep barrier to securing human-centered development goals. It narrows women's options in almost every sphere of life, public and private—at home, in school, in the workplace, and in most community spaces. It limits their choices directly by destroying their health, disrupting their lives, and constricting the scope of their activity, and indirectly by eroding their self-confidence and self-esteem. In all of these ways, violence hinders women's full participation in society, including participation in the full spectrum of development efforts.

Violence against women also compromises the healthy development and well-being of children and families. Children who witness or share the violence unleashed against the women they rely on not only suffer traumatic stress, they are also socialized, at an early age, into accepting violence as a legitimate or inevitable means of resolving conflict. The 1994 *Report* recommends a global compact for human development, urging international commitment to attaining the most important targets for minimal human security: universal primary education, adult literacy, primary health care, family planning, safe water, an end to malnutrition, and financial credit for all. All of these goals hinge on women's ability to exercise and realize their options.

Defining Gender Violence

When the International Women's Year World Conference convened in Mexico City in 1975, the proceedings reflected a general awareness that wife abuse was problematic and that women would profit from more family counseling and more responsive family courts. As the international women's movement gained strength, public awareness of the dimensions and impact of the problem. Discussions at the World Conferences on Women in Copenhagen (1980) and Nairobi (1985) recognized domestic violence as an obstacle to equality and an intolerable offense to human dignity—see paragraph 231 of the *Forward-Looking Strategies,* the final document of the Nairobi conference. In 1985, the United Nations General Assembly passed its first resolution on violence against women, advocating concerted and multidisciplinary action, within and outside the U.N. system, to combat domestic violence.

In 1993, the General Assembly adopted the United Nations Declaration on the Elimination of Violence Against Women, defining this phenomenon and recommending measures to combat it. This document defines violence as:

> Any act of gender-based violence that results in, or is likely to result in, physical, sexual or psychological harm or suffering to women, including threats of such acts, coercion or arbitrary deprivations of liberty, whether occurring in public or private life. Violence against women shall be understood to encompass but not be limited to:
>
> Physical, sexual and psychological violence occurring in the family and in the community, including battering, sexual abuse of female children, dowry-related violence, marital rape, female genital mutilation and other traditional practices harmful to women, nonspousal violence, violence related to exploitation, sexual harassment and intimidation at work, in educational institutions and elsewhere, trafficking in women, forced prostitution, and violence perpetrated or condoned by the state. (United Nations, 1993).

This was a landmark document for three reasons:

1. It situated violence against women squarely within the discourse on human rights. The declaration affirmed that women are entitled to equal enjoyment and protection of all human rights and fundamental freedoms, including liberty and security of person, and freedom from torture or other cruel, inhuman, or degrading treatment or punishment.
2. It enlarged the concept of violence against women to reflect the real conditions of women's lives. The declaration recognized not only physical, sexual, and psychological violence but also threats of such harm; it addressed violence against women within the family setting, as well as within the general community; and it confronted the issue of violence perpetrated or condoned by the state.
3. It pointed to the gender-based roots of violence. The declaration reflected the fact that gender-based violence is not random violence in which the victims happen to be women and girls: The risk factor is being female.

Women's groups and researchers have built upon this definition, extending and refining the notion of gender-based violence. They have stressed that such violence causes physical and psychological harm in a way that perpetuates and reinforces the subjugation of women, whether this is the conscious intent of the perpetrator or not. They have noted that this harm is socially tolerated, in part because the victims are female (Heise, 1994, pp. 46–47). Moreover, they have enlarged the notion of violence against women to include both harm by commission and harm by omission, including life-threatening deprivation of nutrition, health services, and other critical resources and arbitrary deprivation of liberty, forced isolation, and extreme control—abusive treatment that denies women the degree of autonomy and self-determination that adult men in their communities take for granted.

Dimensions of Gender-Based Violence

Researchers have made strides in gauging the dimensions of gender-based violence, despite problems of data collection. In nearly every country, crime statistics grossly underreport violence against women, particularly battery and sexual assault within the family setting. Only a small percentage of these crimes is reported to the police, and studies that ask women to identify themselves as victims underestimate physical and psychological violence within intimate relationships (United Nations, 1989).

Large-scale epidemiologic surveys of sexual assault have emerged in industrialized countries. Survey data compiled in the 1980s by the United Nations Statistical Office indicate that approximately one in four women in developed regions has been hit by an intimate partner. These are representative statistics: Belgium, 25 percent; Canada, 25 percent; Netherlands, 21 percent; Norway, 25 percent; New Zealand, 17 percent; and United States, 28 percent.

Studies of gender-based violence in industrialized countries have documented disturbing patterns, such as these: In the United States, three to four million women are battered each year, but only 1 in 100 cases of domestic violence is ever reported (United Nations, 1991a, p. 67). Domestic battery is the single most significant cause of injury to women, more than car accidents, rapes, and muggings combined (National Center on Women and Family Law, 1988). Between one in five and one in seven U.S. women will be the victim of a completed rape in her lifetime (Kilpatrick et al., 1992). In Austria, in 59 percent of 1,500 divorce cases, domestic violence was cited as a cause of marital breakdown (United Nations, 1991a, p. 67). In the United Kingdom, according to the Home Office, assault takes place in one in three families; serious assault occurs in one in five families (United Nations, 1991a, p. 67).

Data collection has begun in many Third World countries as well. Some of these studies are small and rely on convenience samples; however, most draw upon probability samples with a large number of respondents. These studies indicate that gender-based violence is widespread, though not universal. One review of 90 peasant and small-scale societies found significant levels of violence against women in 74 societies; the remaining 16 were described as "essentially free of or untroubled by family violence" (Levinson, 1989). Based on research conducted in the North and the South, it is becoming clear that gender-based violence:

- affects women throughout the life cycle. It can extend from prebirth and infancy (sex-selected abortion and infanticide) through old age (violence against widows and elder abuse) (Bunch, 1990).

- affects women of every nation, ideology, class, race, and ethnic group. Researchers have documented communities where gender-based violence is minimal; on this basis, they conclude that it is not inevitable—a crucial finding. However, these examples are the rare exception, not the rule. On the whole, the data support the conclusion that violence or the threat of violence affects almost all women (United Nations, 1995).

- is exacerbated by poverty but cannot be solved exclusively through economic remedies. Violence against women appears to be associated with poverty and poverty-related stress; some studies suggest that wife abuse, for example, is more prevalent among the poor and unemployed (United Nations, 1995). But evidence from the First World indicates that many women living in relatively secure economic circumstances cannot extricate themselves from situations in which they are abused. In many cases, these women's income, social status, rights to child custody, and self-esteem hinge on their continuing cohabitation with the abuser.

Only in the 1990s have more than a very few researchers paid serious attention to violence against women, and there is an urgent need for more and better data collection on a national and international basis. In addition, statistics on violence that are published by international bodies should, whenever possible, include sex-disaggregated data.

Homicide

Statistics from places as diverse as Bangladesh, Brazil, Canada, Kenya, Papua New Guinea, and Thailand demonstrate that more than half of all murders of women were committed by present or former partners (United Nations, 1989). Political, social, and cultural conditions place some women at especially high risk of murder. Domestic violence can also provoke murder by women. Studies from the First and Third Worlds show that when women murder men, it is often in self-defense, often after years of steady, increasing abuse (Browne, 1987). Some of the cultural patterns that place women and girls at high risk of murder are:

Female infants. Infanticide is believed to account, in part, for skewed male–female sex ratios in many developing countries. One study in a remote region of southern India found that 58 percent of deaths among female infants were due to infanticide, usually within seven days of birth (George, Abel, and Miller, 1992, cited in Heise, 1994).

Rape victims. A study showed that nearly half of all women murdered in Alexandria, Egypt, were rape victims, murdered by a relative to cleanse family honor (Heise, 1994).

Dowry brides. Dowry—the payment of money and gifts to the bridegroom's family—has been made illegal in many countries, but the practice continues. A bride who fails to bring an adequate or promised dowry may be subject to beatings or even death. In India, official statistics suggest approximately 5,000 dowry deaths per year and undoubtedly underestimate the actual problem. Data for some urban areas of India indicate that one in four deaths of women age 15 to 24 is from "accidental burns," a medical euphemism for dowry-related murder and suicide. Dowry practices (in the form of betrothal gifts or "bride-price" payments) have been linked to violence against women in other countries as well, including Bangladesh, China, and Papua New Guinea.

"Wayward" wives. In many cultures, it is possible for a man to kill an allegedly unfaithful, disobedient, or willful wife and be absolved on the grounds of honor. The legal defense based on "honor" and "provocation by the victim" often requires little or no evidence and results in unduly short prison terms for wife-murder, even in cases involving premeditation.

Suicide

Suicide by women is often associated with other forms of violence against them. For example, 1990 World Bank estimates that battered women are 12 times more likely than other women to attempt suicide. A cross-cultural study, drawing on data from Africa, South America, and several Melanesian islands, established marital violence as a leading cause of female suicide (Counts, 1987). (In the United States, there is mounting evidence that domestic violence may be the single most important precipitant for suicide attempts by girls and women. One study found that domestic violence precedes one-fourth of suicide attempts by all women and half of all attempts by African-American women.) In extreme cases, women have apparently resorted to mass suicide. As recently as 1991, there were reports of mass suicides in rural China among women forced or sold into unwanted, often violent, marriages (Heise, 1994).

Domestic Battery

Studies suggest that the abuse of women by intimate male partners is a pervasive phenomenon: they show that from one-fifth to more than one-half of all women have been physically assaulted by a spouse or male intimate in their lifetime. Most of the victims have been beaten many times; many have been subject to escalating abuse over a substantial period of time. Most often, domestic violence takes the form of physical assaults by a husband or male partner; however, other members of the extended family, including co-wives and in-laws, may commit violence (United Nations, 1995).

These beatings often result in serious injury. A large-scale study conducted in Brazil in 1990–1991, for example, found that well over one-third of domestic-battery cases involved serious injury, usually committed by the aggressor's feet and fists (Human Rights Watch, 1991). The great majority of women who are beaten also suffer emotional and psychological abuse; many report that this can be even more damaging than physical assault.

It is difficult to compare the incidence of domestic violence among countries, because the problem is so widely underreported, and cultural barriers to revealing details of private relationships affect reporting rates differently in different societies. A recent report by the World Bank summarized the results of 35 studies of wife abuse in a variety of countries. The results cannot be directly compared, because different methodologies were used. They leave no doubt, however, that domestic violence occurs at extremely high rates in many cultures, exhibiting many forms of social organization. Studies in Mexico, Chile, Papua New Guinea, and Korea indicate that the great majority of wives—two-thirds or more—have experienced marital violence. Other researchers report significant levels of domestic violence in Brazil, Kuwait, Kenya, Thailand, Nigeria, and Uganda (Heise, 1994).

Rape and Sexual Abuse

Sexual assault and domestic battery are not mutually exclusive. Indeed, research shows that physical brutality often accompanies or culminates in sexual violence. Most rapes go unreported, however, particularly in cultures that place great symbolic importance on physical purity and subject rape victims to social ostracism or even murder to cleanse a family's honor. Rape statistics may also underestimate the incidence of sexual assault because they reflect different, sometimes narrow, definitions of rape. For example, in most countries, the law does not recognize marital rape as rape.

Statistics on rape from industrialized and developing countries suggest remarkable similarities: between one in five and one in seven women will be the victim of completed rape during their lifetimes (Heise, 1994, p. 5). Data compiled from eight countries representing both the industrialized and the developing worlds show that most of the time, perpetrators of sexual assault are known to the victim (United Nations, 1995).

Women are especially vulnerable to sexual abuse in the context of war and state militarism and in other political, social, and cultural upheavals. In the history of international law, sexual abuse has been largely invisible. Forced impregnation, for example, has never been explicitly recognized or redressed as a war crime by any international convention or tribunal. In the 1990s, efforts have been made to create a framework that addresses the full range of human-rights abuses that women suffer (Goldstein, 1993).

There have been few population-based studies on the sexual abuse of children and adolescents; however, those that exist suggest that it is widespread. *The World's Women 1995: Trends and Statistics* reports that, ". . . across diverse continents and hemispheres, a large subset of known sexual assaults (40 to 60 percent) are committed against girls 15 and younger) (United Nations, 1985, p. 2). According to the same source, national random surveys from Barbados, Canada, the Netherlands, New Zealand, Norway, and the United States show that 27 to 34 percent of women report sexual abuse during childhood or adolescence.

There is also substantial indirect evidence of the sexual abuse of children, including alarming numbers of very young girls being treated for sexually transmitted diseases (STDs). A 1988 study (Kisekka and Otesanya, cited in Heise, 1994) conducted in Zaria, Nigeria, found that 16 percent of female patients treated for STDs were under age 5.

Deprivation of Nourishment and Health Care

According to the World Health Organization, girls in many developing countries receive less nourishment than boys. They are breast-fed for shorter periods of time, receive fewer calories, and suffer from malnutrition (leading to death or to mental or physical disability) at higher rates than boys. In rural Bangladesh, malnutrition has been found to be almost three times more common among girls than boys (Bhatia, 1985). Girls also receive less health care. Studies show boys outnumbering girls at diarrheal treatment centers, although both sexes get diarrhea with equal frequency. For all of these reasons, the mortality rate among girls is one to four times higher than among boys in 43 of the 45 countries for which data are available (Heise, 1992).

Violence Related to Culture and Tradition

In many parts of the world, in both developing and developed countries, women suffer physical and psychological damage from practices rooted in culture and tradition:

Female Genital Mutilation (FGM)

Female genital mutilation—the ritual cutting and removal of all or part of the clitoris and other external genitalia—affects an estimated 85 million to 114 million girls and women in the world today (Toubia, 1993). At least two million girls are at risk of being subjected to this procedure each year, about 6,000 each day. In some cultures, girls undergo FGM in infancy; in others, it may not occur until adolescence. But most commonly, girls experience FGM between the ages of 4 and 8. FGM is practiced in 26 African countries; in a number of countries, such as Djibouti, Mali, Sierra Leone, Somalia, and large areas of Ethiopia and the Sudan, nearly all women are affected. FGM is also practiced in some minority communities in some Asian countries and by some immigrant groups in Europe, Canada, Australia, and the United States (Toubia, 1993). Female genital mutilation causes pain, trauma, and frequently severe physical complications, including bleeding, infection, infertility, and even death. It doubles the risk of maternal death during childbirth.

Servile or Mercenary Marriage

Women's ability to control their own economic, sexual, and reproductive lives is linked to their ability to make mature decisions about marriages. Cross-cultural research suggests that forced marriage is prevalent. Many countries have signed international conventions requiring free and full consent of both spouses to marriage. In some of these countries, however, legal conventions are routinely flouted; forced marriage and the marriage of minors, who cannot give legal consent, are the rule. In extreme instances, such as traditional Gambian weddings, the bride is not present and may not even know that the marriage has taken place until after the ceremony. A growing form of mercenary marriage is the expanding market for "mail-order" brides, promoted by agencies that broker "marriages," usually between Third World women and men from industrialized countries. A study of Thai women trafficked by German marriage brokers shows that the agency offers men a three-month trial period during which they can exchange brides as often as they like. The report documented cases of women who wanted to return home but could not because the agency would not release their passport and airline tickets ("All About the Mail Order Bride Business," 1989).

Child Marriage

In many parts of the developing world, significant numbers of brides are younger than 15. (Countries that are party to the United Nations Convention on the Elimination of All Forms of Discrimination Against Women have specified a legal minimum age for marriage above 15, but only in cases in which parents withhold consent.) The minimum marriageable age for women in many countries, including Chile, Ecuador, Panama, Paraguay, Sri Lanka, and Venezuela, is 12. In some regions, such as northern Ethiopia and some parts of Somalia, girls are married off even earlier. In other countries, legislation has set the minimum age at 16 or above, but traditional practices often persist. In Yemen, most girls marry between the ages of 12 and 15, despite a law forbidding marriage of girls under 16. Young brides typically have arranged marriages. Many are traumatized by adult sex and forced to bear children before their bodies are fully mature.

Preference for Male Children

In some parts of the world, including China, India, and South Korea, access to amniocentesis and ultrasound are sufficiently widespread to allow selective abortion and the consequent skewing of male-female sex ratios. The preference for sons stems from the social and cultural stigma attached to having girl children and the considerable costs of marrying off a daughter.

Eating Disorders

In Western industrialized societies and Japan, many young girls and women compromise their health to conform to cultural standards of beauty. Significant numbers of young women in these countries suffer from anorexia, a psychological disorder characterized by self-starvation, sometimes leading to death; bulimia, a related disorder, is characterized by binge eating accompanied by purging and also destroys the individual's health (Gordon, 1990, cited in Heise, 1994).

All of these practices—some rooted in culture, others in religious practice—are harmful and potentially life threatening for millions of girls and women around the world.

Violating Women's Human Rights

The wide range of criminal and discriminatory practices that constitute gender-based violence cause tremendous suffering for individual women and girls and exact incalculable social costs as well. There are fewer women living on Earth today than we would expect, based on well-established ratios of women to men. In countries where both sexes receive similar care, there are about 105 females for every 100 males, reflecting women's biological advantage. But in South Asia, West Asia, North Africa, and China, men significantly outnumber women: There are about 94 women for every 100 men. The bottom line: Roughly 90 million women who should be alive today are "missing" due to the effects of gender discrimination (Klasen, 1994). This is a staggering statistic—nearly twice the generally accepted figure for military and civilian casualties in World War II.

Women who survive gender-based violence, particularly domestic violence and rape, have been shown to suffer from numerous resulting conditions, including AIDS and other sexually transmitted diseases, depression, alcohol and drug dependence, posttraumatic stress disorder, and a high rate of accidents. Studies of women's health indicate that 90 percent of homicides and intentional injury and 60 percent of cases of depression and posttraumatic stress disorder are attributable to rape or domestic battery. The dependents of these victims suffer as well (Heise, 1994).

There are obvious parallels between violence against women and other recognized forms of human rights abuse. Battery of women is a form of *torture,* with many of the same characteristics as other types of torture, and it often includes *imprisonment* in the home, whether physically or psychologically enforced by sexual terrorism and fear. For example, the Congress in Colombia passed a law in 1992 to make *secuestro* (confinement) of a wife a crime in response to men locking up their wives to prevent infidelities, especially when men travel. Women are held involuntarily in *slavery* for prostitution and pornograpy, and even domestic servants are sometimes beaten, raped, and locked into their bosses' homes. Women face *terrorism* in the form of sexual assault on the streets and in jobs, where sexual harassment can be a condition for receiving a paycheck. Rape has been recognized by the rapporteur of the United Nations on torture as a form of torture when it is performed by police officers or by agents of the government and has become visible as a *war crime* through the tragedies of the former Yugoslavia.

Women's groups around the world have argued that freedom from violence, including domestic violence, is a fundamental human right. They advanced the case most powerfully at the Second World Conference on Human Rights in Vienna in 1993 (see Bunch and Reilly, 1994). To counter critics' claims that violence against women does not constitute human-rights abuse because it is not perpetrated by the state, the Global Tribunal on Violations of Women's Human Rights in Vienna documented numerous cases in which police failed to protect, prosecutors failed to prosecute, and the government tolerated and even promoted laws that discriminate against women. They showed that the state is often complicit in women's victimization and must be held accountable for its failure to take significant measures against such violence.

The link between gender-based violence and human rights is not a matter of rhetoric. Situating violence against women within the discourse on human rights creates a normative framework, in which governments can be held accountable for combatting violence against women. It also places it on the human-rights-policy agenda; thus, if governments link foreign aid and policy to human-rights concerns, they should take into account the state's treatment of women, including policies on violence against women. Moreover, recognizing women's rights as human rights can lead, and in many nations it has led, to more preventive measures, more and better services for victims, and stronger sanctions against perpetrators.

Destroying the Roots of Violence

Recent cross-cultural studies on family violence and rape, drawing on data from 90 societies throughout the world, suggest that four factors, taken together, are strong predictors of the prevalence of violence against women in a society: economic inequality between men and women; a pattern of using physical violence to resolve conflict; male authority and control of decision making; and restrictions on women's ability to leave the family setting (Levinson, 1989).

These factors may seem self-evident; however, they are often obscured by pervasive myths about violence against women: that lording power over women is an inherent part of "maleness" and an acceptable exercise of male prerogative; that it has the salutary effect of relieving male tension during periods of stress; that it reflects a natural male tendency toward sexual aggression; that it reflects women's inferiority and their desire for men to dominate them; that it is an inevitable and permanent feature of male–female relations. In contrast, the research shows that violence against women is a function of socially constructed norms of acceptable behavior. It can be reduced and eliminated only through fundamental social change and attitudinal change at all levels. This requires formal and informal education, the effective use of media, and a clear commitment from the government not only to condemn such practices legislatively but, even more important, to ensure that the legislation is enforced.

Key Challenges

Confronting gender-based violence raises certain problems that set it apart from other kinds of human-rights abuse. For example, when the perpetrator of violence is an outsider, it is usually possible to mobilize community support to fight it. But how can people be persuaded to stop accepting or condoning violence committed by their own friends and relatives?

Further, many human-rights abuses are amenable to legal remedies within the present structures of the law. But the law is not a neutral force in patriarchal societies. How can power holders be motivated to acknowledge, and ultimately relinquish, their own stake in practices that reinforce male control?

These questions point to three key challenges that should inform policy and practice: working toward community responsibility and state accountability, deconstructing traditional power structures and cultural assumptions, and engendering sustainable human development.

Working Toward Community Responsibility and State Accountability

In looking at strategies to eradicate violence against women, one underlying question stands out: What does it take to make local communities and national governments decide that it is their responsibility to prevent such violence? Related ones are: How could world resources be mobilized and a commitment of technical expertise be amassed against this scourge on a massive scale such as is represented at the World AIDS Conference held each year? When will consciousness be raised against gender-based violence as a threat to public health comparable to the national campaigns against smoking in some Western countries?

Meeting this challenge requires a profound change in public attitudes away from individual blame toward community accountability for gender-based violence. In most instances, the public attitude is that the individual woman is responsible for such violence—either because she did something to deserve it or because she did not remove herself from a violent situation. Even those who blame the individual male perpetrator, which is more likely to happen if that person is not a member of the victim's family or ethnic group, rarely see prevention of such violence as their responsibility. Significant change will occur only when the community is seen as responsible for the violence that it has promoted or tolerated by failing to intervene.

A critical element in any strategy to eliminate violence against women is a community's decision to not tolerate violence—by strangers or by its own members. As a U.N. expert group recommended, acceptance within the community of the responsibility to shame persons publicly who commit such violence can be an effective preventive measure. Strategies that empower individual women are important to reducing women's vulnerability to violence, but they are not sufficient by themselves. Even women who are able to gain enough control over their lives to avoid situations of domestic violence remain vulnerable to violence in the public sphere.

Communities need models and mechanisms for dealing with violence against women that occurs within their own boundaries. This effort must be built into community-development work. Moving toward community responsibility requires strengthening community-based organizations—especially, but not exclusively, women's

organizations. These groups may need assistance in incorporating into their policies and training, strategies to eradicate discrimination generally and gender-based violence specifically. Development initiatives should reinforce structures, control mechanisms, or associations—whether formal or informal—that are capable to delegitimizing violence as a means of conflict resolution in the family or in the community.

Creating a sense of public responsibility for gender-based violence also involves examining the ways in which societal institutions condone such violence either actively or by passively looking away. Schools, religious institutions, workplaces, advertising, social clubs, communications media, and families must be challenged about their collusion in the perpetuation of violence against women. In this process, the establishment of standards of state responsibility and of mechanisms for international monitoring can play an important catalyzing role.

Some argue that the state should not intervene in such private matters, but the state is explicitly or implicitly involved in gender violence at all levels, by laws and policies that encourage or discourage such violence and by the efforts it exerts to implement those measures:

> Gender relations are already regulated by states, through fiscal arrangements, social security, immigration law, and marriage and family law, established religion, military service, and executed through all the statutory instruments, administrative procedures, and legal and judicial processes, as well as the executive and elective bodies. It is the duty of "good governments" to enforce respect for women's human rights within them (Ashworth, 1992, p. 22).

State accountability for actively seeking to eradicate violence against women is based on the state's universally recognized responsibility to respect and ensure the fundamental human rights of all individuals in its territory.

Deconstructing Traditional Power Structures and Cultural Assumptions

The subjugation of women, including violence in many forms, is so common in societies and so deeply entrenched in cultural and religious traditions that it has eluded analysis as human-rights issue. Yet, as noted above, gender-based violence parallels other forms of abuse that are clearly and consistently included in the human-rights discourse.

Despite the evident parallels, there remains strong resistance to treating violence against women as human-rights abuse. Among the major barriers are cultural and religious assumptions that reinforce and justify existing power structures. Of course, no culture or religion intends to be abusive; the oppressive nature of many of their assumptions lies beneath the surface. Deconstructing and demystifying the oppressive aspects of such practices, as part of the ongoing process of reinterpreting basic principles, is a crucial step toward eradicating violence against women.

Historically, religion and culture have proven extraordinarily adaptive. Most belief systems have been revised over time to accommodate new understandings and new values that emerge in human society. In the words of an African woman:

Traditions are highly sacrosanct and untouchable where women are concerned. Still, I have seen traditions change during my lifetime. The change was so easy and smooth when the men took the initiative. Change, however, requires a lot of pain and hard work when it is initiated by women (Halim, 1994, p. 22).

Numerous cultures offer examples of traditions, including customs harmful to women, that have been changed or died out. For generations, women (and some men) in Sudan endured painful cutting to acquire face marks, a traditional sign of beauty and an indicator of tribal affiliation, but this tradition has disappeared. The binding of women's feet in China is another example of a widely practiced custom that has been successfully abolished.

While culture and religion remain barriers to the eradication of violence against women, efforts are being made to counter their influence. A key element in the elimination of gender-based violence is the principle of the universality of human rights, reaffirmed in the United Nations Declaration on the Elimination of Violence Against Women and in the *Vienna Declaration* and *Program of Action* at the Second World Conference on Human Rights (Vienna, 1993). These rights are cited there as irrevocable, and no custom, tradition, or religious consideration is to be invoked to limit their enjoyment.

Increasing recognition of the power of education and the media to challenge societal norms and transform cultural values has fostered efforts in many parts of the world to remove gender bias and gender stereotyping from school curricula and teaching materials; to integrate gender awareness training, parenting skills, and nonviolent conflict resolution into school curricula; and to provide gender-awareness training to teachers and educators, including teaching them to recognize signs of abuse. The human security of women, and the well-being of their dependents, will rely in large measure on the recognition that societies can alter traditions without sacrificing their identity or stability. It will hinge on the ability of religious and cultural communities to emphasize the core values of love and justice over the traditions of subjugation of women. And it will depend on the incorporation of those values into the legal frameworks of national and international bodies.

Engendering Sustainable Human Development

A growing body of evidence confirms what development workers have long known: Legal and economic discrimination against women (including limitations on their right to own and manage property) impedes the process of sustainable human development. For example, a recent United States Agency for International Development

(USAID) report (Freeman, 1993) found that issues of legal status blocked women's participation in approximately half of all AID development projects. Experts on violence against women stress that it is linked to inequality in all spheres and thus cannot be addressed through isolated and fragmented strategies; rather, coordinated and comprehensive approaches, supported by resources at all levels of the national and international community, are needed.

There has been less acknowledgment of the ways that violence against women inhibits development efforts. There is growing awareness, inside and outside the development community, that our planet's most urgent problems cannot be addressed effectively if we fail to tap the energy and creativity of half of its population. But if they are continually victimized by the violence or the threat of violence, women cannot take full advantage of educational or employment opportunities, and they cannot participate fully in public life, including development.

A study for the United Nations Development Fund for Women (UNIFEM) (Carrillo, 1992) outlines as an investment in the future of development, directions for programs and policies aimed at eradicating gender violence. It suggests that development agencies can make a key contribution by highlighting the obstacles that gender violence places in the path of development and identifying means of countering this in all phases of the project cycle. The paper goes on to outline possible interventions, such as strengthening women's ability to identify and combat violence, taking time to work with husbands and community members when violence erupts, training personnel in an awareness of this issue, and documenting gender violence when it occurs within the development process.

To the extent that development work empowers women—there is growing consensus that it must—it may sometimes create new forms of violence against them. When a development projects puts the key to the village water pump in a woman's hand, or income from an orchard in her pocket, or business shoes on her feet, it upsets long-held, deeply embedded attitudes about male–female relations in ways that may unleash new violence against her. The key may be confiscated, the trees chopped down, the shoes destroyed, and the woman herself risks assault or even murder. This backlash is hardly theoretical. In India, a revolving-fund project of the Working Women's Forum nearly fell apart when its most committed participants began dropping out due to increased levels of domestic violence; in Peru, an action research project with women workers in the electronics industry could not succeed until project leaders addressed the problems the women were experiencing (Carrillo, 1992).

Attacks against women's development efforts may target individuals. In a particularly dramatic case, a woman who led a successful government-sponsored women's development program in Rajasthan, India, was gang raped—at home, in front of her husband—by men in the community who were enraged by her efforts to organize against child marriage (Mathur, 1992). The incident was hushed up until a group of women in the state agency that employed the victim exposed the case. Backlash may target an entire project. In Bangladesh, opponents of an economic-empowerment program involving women cultivating mulberry trees to feed silk worms destroyed the program by burning down all of the trees. The project had been supported by loans from a large development bank in Bangladesh. When the women appealed to the bank for another loan to replant the trees, they were turned down because they lacked the capital to repay the original loan.

Clearly, the answer is not to bow to intimidation, not to deny women access to economic and political power, not to scuttle their human rights, but, rather, to pursue a vision of sustainable human development in which the prevention of gender violence and the provision of services to its victims are integral components of every strategy and every initiative. The planning and implementation of every project must incorporate measures to prevent and combat violence against women, and to involve everyone in the community, male and female, in that effort.

Lending policies may also indirectly lead to heightened violence against women. There is growing evidence the structural-adjustment policies may, by decreasing public services and increasing unemployment and despair, make women more vulnerable to violence and increase its incidence. Multilateral and bilateral funders must take full account of the impact of their lending practices on women's social and economic opportunities, and on their hopes of attaining human security.

Conclusion

While violence against women remains so pervasive that hopes for its eradication may seem remote, there has been some movement in that direction. Perhaps most encouraging is the growing recognition that violence against women constitutes abuse of their human rights—a principle that is reflected in the international conventions and declarations that address this topic.

In requiring states to regulate private as well as public gender-based violence, the United Nations Convention on the Elimination of All Forms of Discrimination Against Women (CEDAW) provides a working forum for the process of eradicating it and promoting gender equity in the international community and at every level of national life. Its General Recommendation No. 19 notes:

> Traditional attitudes by which women are regarded as subordinate to men or as having stereotyped roles perpetuate widespread practices involving violence or coercion, such as family violence and abuse, forced marriage, dowry death, acid attacks [when acid is poured on women's faces], and female circumcision. Such prejudices and practices may justify gender-based violence as a form of protection or control of women. The effect of such violence on the physical and mental integrity of women is to deprive them of equal enjoyment, exercise, and knowledge of human rights and fundamental freedoms.

Women from around the world testified at the Global Tribunal on Violence of Women's Human Rights in Vienna in 1993, documenting numerous cases in which states that failed to prosecute gender-based violence and noting that many even promoted law and practice to encourage it. In December 1993, the United Nations General Assembly adopted the United Nations Declaration on the Elimination of Violence Against Women, recognizing that physical, sexual, and psychological violence against women in public or private, and also threats of such harm, are human rights violations.

At the same time, many governments have taken initiatives against gender-based violence: supporting research; strengthening the legal framework for criminalizing, prosecuting, and punishing crimes against women; retraining law enforcement personnel; providing refuge and services to victims; providing treatment for offenders; and developing more effective policies and programs. Nongovernmental organizations have also been active in these areas and also have sponsored or supported initiatives for gender-violence prevention. The cause of human security has much to gain from the growing recognition that women's rights are human rights. Perceiving gender-based violence as a life-threatening, intolerable breach of human rights opens the door to dismantling the assumptions and attitudes that entrench it.

Finally, there is increasing awareness that violence against women is a critical development issue, and this offers hope that more women will live healthy, productive lives, largely free of violence or the threat of violence. Development agencies are acknowledging more frequently that, unless they confront the massive and deeply entrenched problem of gender-based violence, the "empowerment of women" will be a hollow slogan. Sustainable human development must rest on the shared conviction that unless women and men resolve together to not tolerate violence against women in their homes and their communities, development will not be human, and it will not be sustainable.

References

"All About the Mail Order Bride Business," *Nation,* April 18, 1989.

Ashworth, Georgina. "Women and Human Rights," background paper for Development Assistance Committee—Women in Development (DAC-WID) Expert Group on Women in Development of the Organization for Economic Development (OECD), Paris, 1992.

Bhatia, Shushum. "Status and Survival," *World Health,* 1985, pp. 12–14.

Bradley, Christine. "Why Male Violence Against Women Is a Development Issue: Reflections from Papua New Guinea." New York: United Nations Development Fund for Women (UNIFEM), 1990. Mimeographed.

Browne, A. *When Battered Women Kill.* New York: Free Press, 1987.

Bunch, Charlotte. "Women's Rights As Human Rights: Toward a Re-Vision of Human Rights," *Human Rights Quarterly,* vol. 12, no. 4, 1990, pp. 486–498.

Bunch, Charlotte, and Niamh Reilly. *Demanding Accountability: The Global Campaign and Vienna Tribunal for Women's Human Rights.* New Brunswick: Center for Women's Global Leadership, Rutgers University, and UNIFEM 1994.

Carrillo, Roxanna. *Battered Dreams: Violence Against Women As an Obstacle to Development.* New York: UNIFEM, 1992.

Counts, Dorothy. "Female Suicide and Wife Abuse: A Cross-Cultural Perspective," *Suicide Life Threatening Behavior,* vol. 17, 1987, pp. 197.

Freeman, Marsha A. *Human Rights in the Family.* International Women's Rights Action Watch (IWRAW), University of Minnesota, 1993.

Goldstein, Anne Tierney. *Recognizing Forced Impregnation As a War Crime Under International Law.* New York: Center for Reproductive Law and Policy, 1993.

Halim, Asma Mohamed Abdel. "Tools of Suppression." In Center for Women's Global Leadership (ed.), *Gender Violence and Women's Human Rights in Africa.* New Brunswick: Center for Women's Global Leadership, Rutgers University, 1994, pp. 21–29.

Heise, Lori. *Fact Sheet on Gender Violence.* New York: International Women's Tribune Center/United Nations Development Fund for Women, 1992.

———. "Violence Against Women: The Hidden Health Burden," discussion paper no. 255. World Bank, Washington, DC, 1994.

Human Rights Watch. *Criminal Injustice: Violence Against Women in Brazil.* New York: Human Rights Watch, 1991.

Kilpatrick, D.G., C.N. Edmunds, and A.K. Seymour. *Rape in America: A Report to the Nation.* Arlington: National Victim Center, 1992.

Klasen, Stephen. "'Missing Women' Reconsidered," *World Development,* vol. 22, no. 7, 1994, pp. 1061–1071.

Levinson, David. *Violence in Cross-Cultural Perspective.* Newbury Park: Sage, 1989.

Mathur, Kanchan. "Bhateri Rape Case: Backlash and Protest." *Economic and Political Weekly,* October 10, 1992.

National Center on Women and Family Law. *Information Package on Battered Women.* New York: National Center on Women and Family Law, 1988.

Toubia, Nahid. *Female Genital Mutilation: A Call for Global Action.* New York: Rainbo, 1993.

United Nations. *Violence Against Women in the Family.* New York: United Nations, 1989.

———. *Women: Challenges to the Year 2000.* New York: United Nations, 1991a.

———. *The World's Women 1970–1990: Trends and Statistics.* New York: United Nations, 1991b.

———. *United Nations Declaration on the Elimination of Violence Against Women.* New York: United Nations, 1993.

———. *The World's Women 1995: Trends and Statistics.* New York: United Nations, 1995.

United Nations Development Program (UNDP). *Human Development Report 1990.* New York: UNDP, 1990.

———. *Human Development Report 1994.* New York: UNDP, 1994.

Women in Transition to Democracy

Beatriz Schmukler
Maria Elena Valenzuela

Introduction

The democratic transition in the Southern Cone (Argentina, Chile, and Uruguay) and Brazil took place during the 1980s, a period when the military governments there lost their predominant political position and were thus unable to prevent the succession of civilian rule (Cavarozzi, 1992). The influence that the military had during the transition created nondemocratic conditions in the electoral process in Uruguay and Brazil. However, the political opening meant the end of an era of repetitive military coups in Argentina and Brazil and a process of increasingly successful negotiations between the military and civilians to guarantee a future democratic consolidation.

The democratic transition made apparent the paradox of authoritarian relationships and gender and ethnic inequalities in state, political, and private institutions. This political opening, along with specific historical conditions, helped Latin American women start their own struggle against authoritarianism and marginalization. Their struggle was possible because of their awareness of their subordination within the political parties, the state, the family, the community, and local organizations.

To understand the role of women during that period, it is necessary to explain the concept of transition we are using. We are thinking of simultaneously political and social processes of democratization. This situation occurred during the period between the end of the military regimes and the consolidation of democracies (between the end of the 1970s and the end of the 1980s). The opening of the political process meant the consolidation of political parties, the development of new political forces and alliances, and new modes of representation. It also meant a process of social democratization—the struggle for equality within private and public institutions, the family, the school, the political parties, the social organizations. One of the main roles of women during this period was to reveal the authoritarian mechanisms of institutions. Their social and political participation provided evidence of the "disappeared" people at a time when military repression did not allow political organizations to act openly. Women denounced discrimination against them within political parties and won a quota system in elections. Women's groups became aware of violence within the family, and participating women underwent a subtle process of awareness of their limits of authority within the families. Awareness helped many women push for a more flexible sexual division of labor within the family, work outside the household, and participate in their communities as a new social right.

Analysts are ambivalent about the consequences of the transition for women and have not reached a clear conclusion as to whether women have been able to introduce specific principles into the democratic regimes. Women's organizations, particularly those involved in human rights, had a strong profile after the mid-1970s, depending on the country. Women's-rights groups and those responding to the economic crisis emerged more strongly during the transition. Although women did not obtain significant representation in the political system, they developed their own organizations around issues of political, individual, and social rights. Throughout the human-rights movements, women were leaders of groups of families and relatives of prisoners and those who had disappeared, playing a significant role during the transition. In the cases of Argentina and Chile, they created powerful mothers' and grandmothers' associations, which persisted during the 1980s. The middle-class feminist movement was able to consolidate an agenda around individual- and social-rights issues and to support the activities of poor women. In addition, they gained state positions running women's offices in national and federal governments. The women's movements in the Southern Cone and Brazil—the *pobladoras* (low-income residents), human-rights, and feminist groups in Chile; the women's-rights, social-rights, and grass-roots women's movement in Brazil; the housewives, self-help, feminist, and human-rights organizations run by women in Argentina; the housewives, human-rights, working-feminist, and political women's groups in Uruguay—were all developed on the bases of defending their autonomy, working to achieve internal cooperation and solidarity, and struggling against hierarchical relationships.

Often, they defended their autonomy in the face of isolation and lack of representation in political parties. However, the defense of their autonomy helped the women in these organizations who did not have previous political experience develop a new sense of their personal values and rights and learn political, leadership, and managerial skills. In Uruguay, autonomy did not isolate women from the popular movement. Autonomous groups linked their activities with women in the political parties, and many of the women in autonomous groups were members, albeit not leaders, in political parties. They defined themselves as members of the popular, trade-union, and political movement (Prates and Rodriguez Villamil, 1989).

The transition helped women create and consolidate a gender identity in both the working and the middle classes. It also helped develop an alliance of women across social classes, which allowed the enforcement of a social-rights policy that benefited both women and poor families (Caldeira, 1992). In addition, the increasing social and political participation of women accelerated democratic changes in the authority systems of families, which favored more equal gender and generational relationships.

Women in the Transition: Success or Failure?

The weak representation of women in formal politics ran parallel to the incorporation of gender consciousness in the formation of new identities of women. This seeming contradiction creates many doubts and ambiguities about the success of women's movements in this period.

The way this success is rated depends on different emphases in the concept of democracy. Some authors have expressed disappointment that women have not been able to increase their political representation or, what is worse, in some cases stop the decline. From this perspective, democracy means the expansion of citizenship of women, minorities, and other marginal social actors in the political system. Barrig (1992), expressing an important concern among feminists in Latin America, believes that the weak political assimilation of women reflects a lack of confidence on their part to integrate, or even coordinate, with broader local organizations. Although women trained themselves in a democratic exercise within their organizations, Barrig is worried about their connections with the state, the local governments, and the political system as a whole. Integration might bring about future changes in the agenda and modes of representation within political institutions, which, in turn, incorporate more women (Barrig, 1992).

This view of the women's movement emphasizes the need for the integration of women beyond their small groups. The enclosure of women within their own organizations might also create the reproduction of other forms of isolation of women within the private sphere, arising from their fear to expose themselves and to work with other local organizations or political parties. Such isolation has, in fact, had consequences on women's participation as citizens. Usually, the women have more of a sense of belonging to local women's organizations that stress their identities as neighbors and not so much as members of the same nation. Membership in these organizations gives a sense of security, but the lack of broader involvement emphasizes women's vulnerability and limits their agenda and their possibilities of representation in the political system.

This concept of democracy as an expansion of citizenship for marginal actors emphasizes the need for integration in preexistent formal organizations, mainly male-oriented ones. This leaves many questions to be answered: What are the terms of the incorporation of marginal actors into formal organizations? In the case of women, what does incorporation mean regarding preserving and representing their specific identities within organizations? Would women be able to participate in the struggle for equal rights of members through strategic political action? Schield (1992) stresses the need to achieve equality and not merely formal recognition in order to become a citizen. She accurately concludes that real incorporation is a way to participate in the definition of who is able to engage in strategic political action.

An alternative definition of democracy sees it as a process in which women address and fulfill their specific needs. Valenzuela (1995) is concerned with the need for women to achieve social and political identities before they can be meaningfully incorporated into the political system. She thinks that the autonomy of women's organizations should help women prepare their own agenda and develop their own sense of politics. Women, says Valenzuela, "decided to create their own organizations that would prepare the women's agenda without the interference of party structures nor any other male organization" (Valenzuela, 1995, p. 63). The development of separate and autonomous organizations also implies the redefinition of hierarchical relationships and the creation of cooperative goals and structures of solidarity within these organizations.

Valenzuela's arguments foster reflection upon the character of preexistent institutions. The transition was not a process of restoring a set of democratic institutions that existed previous to the military regimes. On the contrary, most of the countries going through the transition had not previously built real democratic institutions that guaranteed equal rights and duties of members, independent of their sex and race. During the three decades before the transition, popular political participation was not oriented toward building up democratic representation or promoting more legitimate institutional mechanisms of decision making. Nor did electoral participation serve those purposes since it was oriented mainly toward disrupting the legitimacy of authority of both civil and military regimes (Cavarozzi, 1992).

During the transition, conservative tendencies also undermined institutional democratization. In Brazil, for instance, the expansion of political and social rights coexists with a disrespect for individual, civil, and human rights, including a weak justice system. Defenders of human rights failed to protect the legitimacy of the notions of individual and human rights in a national polemic where even the human rights of prisoners were discussed. In fact, conservative factions also challenged women's rights, trying to force women to return to their traditional role of consolidating the strength and stability of the family, which was, and is, seen as necessary to preserve the nation against increasing crime rates.

The transition was not then a smooth path of political and institutional democratization. In any such period, political conditions may exist for citizens to struggle for new collective arrangements to defend equal rights and duties of members of social institutions (O'Donnell and Schmitter, 1986). However, such actors have to be created by the development of new social identities among members of political and private institutions. The construction of new social identities may enable new collective arrangements of rights and duties in addition to building up new sets of values and interpersonal practices of gender and authority. Women's movements during the military regimes and during the transition tied together the struggle for women's rights with the struggle against all modes of authoritarianism and the positive construction of new models of democratization of popular organizations. There are examples that show that, in spite of not being formally integrated into political organizations, women have been able to influence political discourses. Ackelsberg and Shanley (1992) have shown this in the human-rights movement of the Mothers of Plaza de Mayo, which disrupted the traditional distinctions between private and public spheres in Argentina and insisted on the importance of human rights during the transition even when the democratic government, fearing the erosion of its relationship with the military, tried to undermine those issues.

The transition took place under a major crisis in the systems of political representation in Argentina and Brazil (Paramio, 1992). The frustration of the popular sectors in these countries about the capacity of parties to represent their interests contrasts with the reorganization of political parties in Chile during the post-Pinochet era (after 1990), which showed a greater responsiveness of the Chilean political system. This, in turn, influenced the desire of women to emerge as a social force and to develop their own agenda in order to negotiate with the political parties, creating fluid channels of communication and participation within the political system (Valenzuela, 1992).

Even if integration is absolutely necessary for women to become competitive in the political system, the need to (1) build a constructive politics based on urgent solutions to increasing poverty, (2) enforce a social-rights policy, and (3) struggle for individual and collective rights in institutions has given a distinctive direction to the political action of women. The different interpretations of the success or failure of women's autonomous movements is related to the importance that one gives to the democratization of institutions allowing women to become new political actors. In this process, women construct their identities on the basis of diverse social-class experiences and the common grounds of their gender history. This common experience has awakened in women substantial concern about those institutions that reproduce discrimination and equality, such as the family and the school. Valenzuela (1995) stresses the importance of women's democratic struggles within institutions in private and public spheres, such as the family and the educational and legal systems.

Social Democratization

We have explored the concept of democracy as an expansion of social equality and a liberalization of social institutions. This liberalization implies the struggle of new or marginal members to negotiate rules that will increase their equality, strategic action, power, and leadership. It also takes into account the impact that women's new identities might have on the authority systems of institutions, particularly on schools and families, as Valenzuela (1987) has said, where women have had historical influence, although they were secondary authorities. The increasing participation of women during the transition expanded their personal freedom within the household and increased the presence of mothers or mothers' groups demanding and negotiating equal treatment of poor children and the organization of day-care facilities oriented toward democratic values and teaching methods.

These forms of social participation imply a struggle for a new type of citizenship that incorporates not only the idea of belonging but also the achievement of equality (Schield, 1992). It includes participation in organizations that have difficulties being recognized in any political system of representation, since they do not have quotas in political parties or "natural" ways to participate in elections.

It is also worth pointing out some dangers and contradictions in these first steps of the women's movements. There is the danger of isolation and bureaucratization of the women's popular movement when it becomes more centralized (Barrig, 1992). The goals of creating cooperative relationships of solidarity in organizations might become utopian and frustrate the growth of newly born associations. The apparent controversy between integration and autonomy might lead to a new modus operandi combining autonomy and strategic influence.

Early in the emergence of democracy, there was a conflict within the women's movement between autonomy and integration. Women in the middle classes were responding weakly to the elected government's demands to participate in social policy. Women in the popular sectors were forming their local organizations to respond to the economic crisis as members of families in poor communities. Yet, at the same time, women of both the middle and the popular sectors were reconstructing their gender identities and developing an appropriate sense of politics for these new identities.

The image of mothers struggling in the public arena always reminds us of traditional roles. Yet, during the military regimes and during the transition, women-as-mothers were protagonists of conservative, human-rights, and antirepressive state movements and autonomous community action. Mothers, resisting the military regimes or opposing the adjustment policies of the transitional governments, were exposed to manipulations of external agents. However, maternal social and political action, when it has not been the target of paternalistic policy and when it has been autonomous from manipulative external agents, has been one of the paths for women to defend their rights and to develop a creative mode of political action that reflects their feminine experience but does not trap them into traditional femininity.

The Women's Movement During the Transition

During the 1980s—a decade of a strong political and economic crisis—all Latin American countries witnessed the appearance of many forms of collective protest, especially in urban areas. These have been called new social movements; they express new popular interests and practice new ways of political participation. These movements challenged the state's economic and political models and called into question the authoritarian and hierarchical ways of doing politics.

In this context, women organized themselves to confront issues such as human rights, economic survival, political participation, and gender issues. The women's movement was, therefore, a collection of diverse groups with different patterns of organization and different goals. Among the main elements that fostered the development of the women's movement were the Catholic Church, which urged women to participate in community struggles and fostered the creation of mothers clubs and housewives' associations among the poor; the Left, which took up the intensive organizing of the urban popular classes, leading some women militants to work with neighborhood women's groups; the military regimes, which allowed women to organize while still repressing other sectors of civil society; and the political liberalization process, which later increased the political space available to women's organizations.

Feminism originally had been labeled as too elitist and too hostile to men to be appropriate to Latin American social and political reality, especially by those who considered feminism as the product of contradictions existing in highly industrialized countries but not in the underdeveloped societies. Yet, as political exiles from Europe, the United States, and Mexico returned with new feminist ideas and as the United Nations Decade for Women (1976–1985) internationalized the content of the feminist agenda, feminism mobilized women from different sectors, pervading all the women's movements with a gender perspective. This was an alliance that produced a feedback with mutual benefits. While feminist ideas contributed to the shaping of the social agenda of the women's movement, the movement provided a mass base.

Human-Rights Organizations

Human-rights organizations were composed predominantly of women who had little previous political experience, who were mothers and grandmothers of the disappeared, who were politically repressed, and who called upon their maternal roles and family voices. They developed public activities to oppose the military and denounce the invasion of the private sphere of the family by governments, which, despite their public commitment to preserve traditional family values, used state terror to maintain political control. Even though these organizations were not seeking to break the barrier between public and private, they did so, establishing at the same time a new pattern that broke the traditional role of women in politics.

Economic Survival Organizations

The structural-adjustment policies designed to cope with the debt crisis brought heavier demands on women, who had to devote their energies to providing consumer goods and the basic services their families needed. This invisible face of the adjustment policies provoked a significant increase in women's work both within the home and outside because it forced women to play a compensatory role in relation to the disruptions taking place in the labor market and in society at large.

The deteriorating economic situation led large contingents of poor urban women to initiate collective strategies for survival, designed to satisfy the basic needs of their families. Women formed popular economic organizations, including subsistence and craft workshops, soup kitchens, and programs for collective shopping. These groups consisted mostly of housewives who were trying to meet their families' basic food and survival needs under a state that had forsaken its role as benefactor. Even though these new economic organizations created by women had survival as a principal objective, they rapidly became promising centers for political organization and the development of gender identity. The new roles assumed by women also had important effects in the development of a women's social movement. They brought attention to areas of conflict that had previously been ignored, and they questioned the idea that class contradictions were the only focus of social conflict.

Unlike the middle-class-based feminist movement, these women's organizations did not propose the end of gender discrimination. Changes in women's lives caused by the economic crisis, such as new tasks and responsibilities, led to changing attitudes, a greater sense of personal worth, and the creation of a sense of gender identity. Women's experience of leaving their homes, making contact with other women who were suffering the same problems, and discovering their own unsuspected capacities and abilities had an important impact on their lives.

The experience of these organizations shows that even though poor women were seemingly organizing themselves around their families' needs, they were also negotiating, and sometimes challenging, power relations in their daily lives. Yet another action they were taking through this collective process of creating a new identity was challenging hegemonic (dominant) discourses about gender and about development and politics and developing a critical perspective about the world in which they live. The experience of these groups forces us to reinterpret the politics of need. The very notion of what constitutes a survival strategy must be expanded and reconceptualized to include noneconomic dimensions.

Feminist Groups

Contemporary feminism emerged in the mid-1970s, in the midst of a repressive period, during the military regimes. It was born as an intrinsically oppositional movement. It challenged not only patriarchy but also its paradigm of male domination: the militaristic state. Feminists, as a consequence, joined forces with other opposition groups in denouncing social, economic, and political oppression. State repression and class warfare were instrumental in shaping a feminist practice distinct from that of feminist movements in other regions.

Social movements aimed at the vindication of women's rights are not new. During the 1920s and 1930s, upper- and middle-class women organized to demand political and civil rights, but after the right to vote was obtained in 1949 there was a long period of feminist silence. The second wave of feminism, in the 1970s, was born from the Left. These were women who had been activists of left-wing political parties, who became aware that women and their issues were invariably relegated to a secondary position within progressive political parties and revolutionary movements. Feminist consciousness was thus fueled by multiple contradictions, and, although some feminist groups broke with the Left organizationally, they did not do so totally, in terms of ideology. All of them retained a commitment to social change even as some continued, linked to their political institutions, to struggle against sexism within the parties.

In the 1970s and early 1980s, women mobilized themselves for the antidictatorial struggle more than for gender demands, but the majority of them evolved through these activities to incorporate a feminist perspective. By the mid-1980s, when the process of democratization was already in full swing, most of them had consolidated a feminist movement. From that moment on, feminism spread rapidly to the popular sectors, destroying the myth that feminist concerns reflect only the interests of middle-class women. Shantytown groups with clear feminist tendencies were created, even though, in most of these popular organizations, women defined themselves in terms of their domestic roles and were interested primarily in the struggle for survival.

The women's movement played a significant role in the transition to democracy. In Chile, for example, before the 1988 plebiscite that defeated Pinochet, women mobilized and organized themselves at the grass-roots level, and it was through these mass mobilizations that the quantitative superiority of the opposition to the military government was made obvious. Between 1988 and 1990, when the democratically elected President Patricio Aylwin took power, women organized to introduce women's issues in the agenda for democracy. They organized through the Concertación Nacional de Mujeres por la Democracia (National Union of Women for Democracy), an autonomous women's coalition in support of the Concertación Nacional de Partidos por la Democracia (National Union of Parties for Democracy), which campaigned through 1989 to win the presidential and legislative elections in December of that year.

The unexpected emergence of feminist movements and importance of women's issues and women's organizations in the period of transition to democracy are the consequences of a combination of factors. The political crisis of authoritarianism gave rise to social movements and to demands for more participatory politics. The transition was marked by an ideological openness and flexibility, born of the desire to break out of the polarized political situation of the past. This provided a fertile ground for feminist social criticism.

Contemporary feminism is one part of a larger, multifaceted, socially and politically heterogeneous women's movement. What makes this feminism so special is its focus on the intersection of gender oppression with other, more local, forms of oppression and domination.

A Feminist Proposal for Democracy

In their approach to politics, feminist groups argued that it was necessary to redefine the concept of democracy, since democracy had never existed for women. Accordingly, they proposed, the struggle for democracy should include the struggle for women's liberation; otherwise, patriarchal structures will not be eliminated. The argument implies that there is an authoritarian pattern behind political and personal relations, and that both structures, therefore, must be democratized. It is in this context that the Chilean feminist movement coined its slogan, "Democracy in the nation and in the home," seeking not only more equality for women but also a transformation of political and day-to-day relations. Without ignoring the problem of social inequality, feminist demands in all of the countries analyzed identify expressions of inequality in a broader context, focusing their attention on key social institutions that reproduce discrimination: the family, the educational system, the political parties of all ideologies, the state apparatus, and the legal system.

Feminists or Politicians?

While there is extensive feminist commitment to a project of wider social change, as well as general agreement on the need to link women's struggle to other social struggles, feminists have been faced with a debate about whether the strategy should come from within or from outside the political system. Some feminists seized the political space made available to women by parties during the period of democratization and promoted innovative changes from within; others continued to organize at the grass-roots level, fostering critical and organizationally and ideologically autonomous feminist politics.

The difficult relationship between social movements and political parties led the various women's organizations to consider maintaining their independence, even as their members remembered with apprehension the way in which they were co-opted once they won the right to vote. At first, this led some women to remain loyal to both their parties and feminism, thereby producing a "double militancy." However, because of the lack of institutional channels for representation and participation, this double militancy was later reoriented as women tried to incorporate feminism into the party structures.

Some parties, mainly those from the Left, produced positive action plans, establishing quotas for women. The proportion of women in leadership positions, however, remains low, with no large differences among parties of the Right, center, or Left. This has provoked a debate between party activists and independent or autonomous feminists, dividing the women's movement. One position holds that neither capitalism nor socialism alone could eliminate women's oppression and that, consequently, women's specific demands should be articulated in a movement outside, and independent of, all existing political parties. The other position asserts that women should follow a flexible strategy, one that would permit them to influence political structures from inside.

Incorporating the Women's Agenda

The active mobilization of women's organizations and their capacity for protest made these groups increasingly legitimate in the context of the democratization process. The parties of the Left, center, and Right became interested in women's issues, which they incorporated into their platforms in varying degrees. The return to democracy also made them aware of the electoral importance of women, and parties started to court them. Women saw themselves as an important political force; they became conscious of their contribution to the electoral struggle and less willing to continue performing secondary tasks at the margins of power. Thus, the feminist movement successfully influenced the political agenda of the democratization period by incorporating issues with a clear gender perspective.

Experience has shown that the state can be a mechanism for social change or for social control in women's lives. Feminist political practice is concerned with discerning the difference, seizing available political spaces, and avoiding excessive institutionalization or depoliticization of the feminist agenda.

The identification of, and advocacy on behalf of, women's issues is a controversial theme that divides many institutions, including church and government, and prompts a new system of alliances and tensions among political actors. The women's movement has become such an actor, as much for the role it has played in the antiauthoritarian struggle as for the potential for change that the process of redemocratization brings. However, even as the political parties and state institutions have incorporated some women's issues, other demands of the women's movement have met strong resistance.

Feminism has been linked to the political Left since its inception. Even though it later claimed independence, it still maintains a strong ideological tie to the Left. However, other social sectors also make gender demands, providing the women's movement a high degree of heterogeneity in its composition (cutting across social classes and ideological divides) and system of alliances. Common gender identity has allowed the creation of women's organizations that cover a wide political spectrum. However, the tendency of the political system to give greater importance to economic demands could undermine the effectiveness of these groups, since they advance their demands from a different perspective.

The democratic transition has benefited from women's proposals to democratize politics, but, in the end, it has helped restore the role of traditional political organizations more than it has helped the organization of women. Traditional political organizations have shown a relatively open attitude toward the incorporation of some gender demands, but they have tended to exclude women from the system of power and the areas of political and economic decision making.

Moreover, in some countries where the Catholic Church played a progressive role in the authoritarian period, it is now exercising a veto power over the progressive demands regarding women's and family issues.

The political importance of the women's movement during this period does not lie so much in its capacity to mobilize large numbers of people, but rather in the fact that women have reinforced the pro-democracy movement and brought about greater participation of sectors that would otherwise remain excluded from the political system.

The traditional question of whether autonomously organized groups of women are more successful in the struggle for power than those who have joined preexisting structures remains. While radical feminist groups have insisted on remaining independent of the parties, important numbers of women who define themselves as feminists have joined the parties with the purpose of struggling for greater spaces within existing power structures. The absence of other mechanisms of participation or influence, such as lobbies, over the political system excludes forces without representational capacity from the political process. Thus, autonomous feminist groups, as well as other women's organizations, are largely isolated from the political system. The traditional view of women in politics is that they are not interested in competing with men. This has led to the existence of women's departments in the parties and to an emphasis on women in public roles as mere attendants to the wives of political leaders.

Even though there is a growing consciousness and common identity of gender problems, feminism is still perceived by large segments of the political spectrum as an antimale movement that does not represent the majority of women. In addition, the persistence of traditional values and attitudes concerning the role of women in society has made the work of feminist pressure groups more difficult and, at the same time, has limited women's space in the political system.

References

Ackelsberg, Marta, and Mary Lyndon Shanley. "From Resistance to Reconstruction? Madres de Plaza de Mayo, Maternalism, and the Transition to Democracy in Argentina," paper presented at the 17th Congress of the Latin American Studies Association, Los Angeles, 1992.

Barrig, Maruja. "Violence and Economic Crisis: The Challenges of the Women's Movement in Peru," paper presented at the Conference on Women and the Transition from Authoritarian Rule in Latin America and Eastern Europe, Berkeley, December 3–4, 1992.

Caldeira, Teresa. "Justice and Individual Rights: Challenges for Women's Movements and Democratization in Brazil," CEBRAP and Department of Anthropology, State University of Campinas, Brazil, December 1992.

Cavarozzi, Marcelo. "Beyond Transitions to Democracy in Latin America," *Journal of Latin American Studies,* vol. 24, no. 3, 1992, pp. 665–684.

———. "La Política: Clave del Largo Plazo Latinoamérico," paper presented at the 17th Congress of the Latin American Studies Association, Los Angeles, 1992.

O'Donnell, Guillermo, and Philippe Schmitter. *Transitions from Authoritarian Rule: Tentative Conclusions About Uncertain Democracies.* Baltimore: Johns Hopkins University Press, 1986.

Paramio, Ludolfo. "Consolidación Democrática y Desafección Política." Instituto de Estudios Sociales Avanzados. Madrid: CSIC, 1992. Mimeographed.

Prates, Susana, and Silvia Rodriguez Villamil. *Los Movimientos Sociales de Mujeres en la Transición a la Democracia.* Serie Documentos Ocasionales. Montevideo: Grupo de Estudios sobre la Condición de la Mujer en el Uruguay, 1989.

Schield, Veronica. "Struggling for Citizenship in Chile: A 'Resurrection' of Civil Society?" paper presented at the 17th Congress of the Latin American Studies Association, Los Angeles, 1992.

Valenzuela, Maria Elena. *La Mujer en el Chile Militar.* Santiago: CESOC, 1987.

———. "Women and the Democratization Process in Chile," paper presented at the Conference on Women and the Transition from Authoritarian Rule in Latin America and Eastern Europe, Berkeley, December 3–4, 1992.

———. "Women in the Democratic Transition: The Chilean Case." In Paul Drake and Ivan Jaksic (eds.), *The Struggle for Democracy in Chile, 1982–1990.* Lincoln: University of Nebraska Press, 1995, pp. 161–187.

Refugee and Displaced Women

Sima Wali

Introduction

This essay discusses the refugee crisis in a post–Cold War world, especially as it relates to displaced women and children. Particular emphasis is given to refugee and displaced women from Muslim countries. In addition to highlighting gender inequality that contributes to their plight as refugees, this essay examines donor policies in the West; and in particular, U.S. humanitarian and development-assistance programs that exacerbate gender inequality by not factoring in women as active participants in the rebuilding of their societies. It concludes with recommendations for how donor policies can be reconstructed to move away from primarily supporting relief efforts to enabling women to play a greater part in the building of democratic societies upon return to their homelands.

This essay also seeks to draw attention to the pervasive and egregious human-rights violations against women and girls during war. The human face of tragedy is largely depicted in the personal accounts of women and girls who have been forced to flee their homelands under dire circumstances. These circumstances involve women making the most difficult decision of their lives under dangerous conditions and in hostile environments. Imagine a woman suddenly experiencing the loss of her husband and adult male children to war and who is unable to protect her shattered family or herself, who is suffering from hunger and hunger-related diseases, whose home has been destroyed, and who is either threatened by, or subjected to, sexual violence. In such a situation, when she decides to flee she does so under extreme conditions when she and her children are risking additional tragedies, and she is often subjected to multiple rape during all stages of her plight. Once she has crossed the border, fearful and isolated in a foreign land where she is dependent on the protection of the United Nations High Commissioner for Refugees (UNHCR) and the host country, she lacks adequate food and access to education for her children. During the prolonged period of her life as a refugee (often lasting more than a decade), she experiences lack of health care, food, relief supplies, and adequate travel and identity documentation. In effect, once she crosses borders she loses all identity by losing her country, her family, her emotional stability, her health, and her dignity. She may even have to agree to an early marriage for her girl child and the conscription of her sons into resistance forces. If she is fortunate to have her family intact, she is forced to bear numerous children to replenish the lost male population while she is undernourished and traumatized by displacement. Her emotional trauma is exacerbated by her isolation and lack of political voice in an environment where she is unable to understand the circumstances that led to her uprooting. She is devoid of hope of rebuilding her own life and her shattered family.

This scenario is described to help the reader come to some general understanding of the human consequences of war and civil upheaval. When the human face of war is graphically depicted, the impact is so profound on those who were unfamiliar with the circumstances of uprooted women that they often remain debilitated in the trauma rather than move beyond it to seek durable solutions. The suffering of war victims continues, with few remedies for comprehensive solutions. Although some advances have been made in alleviating this human tragedy—by courageous individuals in relief and humanitarian-assistance programs—they have been short term and sporadic rather than sustainable and comprehensive approaches.

This essay does not enter into the details of the human consequences of war but rather seeks to shift the mode of thinking and approach common to the field of uprooted people. This shift demands transcending the mode of thinking that centers on compassion fatigue or viewing the situation of victims as the "normal" consequences of war and, instead, inciting the collective conscience of humanity to bear responsibility for the circumstances that lead to war and conflict. In short, humanitarian assistance is not only a humanitarian but a political imperative as well.

World peace rests on collective humanity challenging war and international conflict at its frontiers. It is even more important not to segregate humanitarian assistance from political and economic realities or remedies, in order that peace and stability may become a practical possibility rather than mere hope. Without such a shift in approach, responses to the prevailing human tragedy will remain within the realm of humanitarian assistance, offering temporary relief rather than long-term and durable solutions.

The Shifting World Order and Forced Migration

The last decade of the twentieth century has been extremely violent, with the dismemberment of nation-states and the crippling of humanitarian and political response mechanisms. As low-intensity warfare (military struggle on a small-scale and using conventional weapons) spreads across the globe, people are becoming forcibly uprooted, thereby contributing to the existing massive refugee influxes across international borders. The world community is no longer willing or able to systematically respond to this accelerated humanitarian crisis, especially as the internal economies of individual nations face recession. Furthermore, the West, especially the United States, is still pondering its own position as the sole remaining superpower in a changing world order.

The history of humankind is a history of migrations. Yet, the discourse on migration is highly charged and divided and often disregards the phenomenon of forced migration and its impact. Although assistance to refugee, displaced, and returnee people falls under the general category of humanitarianism, it is most often addressed solely in its economic and political contexts. In the 1990s, we are witnessing unprecedented numbers of forcibly displaced persons. In the 1970s, there were approximately 2.7 million refugees worldwide. In the 1980s, that number grew to 8.2 million. According to conservative statistics by the UNHCR, there are 20 million refugees and an additional 25 million people displaced internally within the borders of their own countries (Van Rooyen, 1993). These statistics do not include the 1 million refugees forced from Rwanda by tribal and ethnic warfare in 1993 and 1996 (United States Committee for Refugees, 1994).

In total, approximately 45 million people—a population roughly the size of France—80 percent of whom are women and children, have been forced from their homelands. The challenge is what to do with the new refugees being created every day and whether to write off an entire nation's worth of people as the world community strives for a new economic and political order. Resolving the problem of forced migration and the problem of women in this context is not only a humanitarian issue but a political imperative with which the world community must contend in the coming century.

The impact and legacy of the Cold War and the end of the East/West conflict—a process that started in early 1990—can no longer be disputed as having major implications for the forcibly uprooted. On a positive level, the world community has empowered the United Nations to negotiate conflicts and assist in the return of the world's refugees. Ironically, the end of the East/West conflict has become a major tragedy for many developing nations. In the absence of a bipolar balance of power, developing nations formerly belonging to a particular "sphere of influence" can no longer leverage power for protection and resources (United States Committee for Refugees, 1993). Contrary to widespread expectations, the end of the Cold War has not brought an end to conflict but has actually fostered violence as the aftermath of a power vacuum created by collapsing economic systems. Refugees and immigrants are bearing the brunt of heightened xenophobic and isolationist attitudes in the West generally, and quite visibly in the United States.

Although prior to the demise of Communism the Soviet Union had a major role in the creation of refugee and displaced people, it did not take the responsibility to adopt immigration, refugee, or development-assistance policies either for reparation or humanitarian purposes. This essay offers suggested solutions for the United States, since it is the only remaining superpower in the post–Cold War era. On a global scale, the United States has not only created refugee and displaced people, it has concomitantly devised policies to bear responsibility for its role. Although the United States remains the only nation with progressive asylum and immigration policies, during the mid-to-late 1990s its policies regressed some in response to growing public and right-wing discontent. In the post–Cold War era, it is increasingly apparent that European and other emerging powers are unwilling or unable to take on leadership roles in global conflicts without their ally, the United States. In this regard, the United States is also the major provider of the bulk of foreign aid and military assistance across the globe.

Media coverage, especially television, following the 1993 bombing of the World Trade Center in New York City has preyed on public fears by portraying negative images of Muslims and Islam. Such practices have contributed to a change in U.S. asylum policies, further restricting asylum seekers' entry into the United States. The effects of such policies, which endanger international protection for genuine asylum seekers by limiting their right to asylum and international protection, have been overlooked by policy proponents. Moreover, this regressive asylum policy has seriously infringed upon refugee and displaced women's right to asylum and to protection from gender-based persecution.

This concern has led many refugee advocates to address directly the issue of gender-based types of persecution, including the practice of targeting women and girls, who are physically unable to protect themselves, for rape campaigns; the demand for sexual favors in exchange for food supplies and documentation; and forced prostitution. As a result of their efforts, the Refugee Women and Children Protection Act was passed into law in the Foreign Relations Authorization Act in 1994, permitting refugee and displaced women subjected to gender-based persecution to seek asylum in the United States. Passage of the act has focused more attention on gender-based types of persecution, which have been largely ignored by immigration opponents and the media. Moreover, the Immigration and Naturalization Services (INS) adopted the Consideration for Asylum Officers Adjudicating Asylum Claims for Women in May 1995.

There is growing recognition that women represent at least half of the global population and are responsible for approximately two-thirds of all hours worked (UNDP, 1995). Civil conflict, war, generalized violence, and other human-rights abuses skew population demographics, resulting in populations whose majority comprises women, children, and the elderly. Under such conditions, men who are conscripted or join active military campaigns die in larger numbers than under peaceful conditions, leaving adult women solely charged with the responsibility of taking care of children and the elderly in situations in which there are insufficient resources to satisfy basic needs.

The legacy of war is commonly accompanied by two main population dynamics: a large number of widowed and handicapped women (in Cambodia, for example, it is estimated that one in every 30 women is handicapped); and many handicapped and/or orphaned children (Subcommittee on Immigration and Refugee Affairs, 1991). These changes have important implications for women as caretakers, who are further burdened by their physical vulnerability, making them susceptible to the most egregious forms of human-rights violations.

When female refugees and displaced persons become the target of large-scale terror campaigns, human-rights violations in the forms of mass/multiple rapes, abduction, torture, forced pregnancy, unhealthy birth-spacing, domestic violence, and trafficking in women and young girls have lifelong negative impacts on people and nations, depleting the victims' dignity and energy for remaking their own lives as well as the lives of their communities in asylum and upon return.

When asked, refugee and displaced women identify gender-based persecution as a priority concern for themselves and for their communities. Refugee Women in Development (RefWID), an international agency focusing on refugee, displaced, and returnee women, has, in its domestic and international work, been systematically approached by uprooted women with this problem. In response, RefWID has focused its advocacy and programmatic efforts on raising awareness to this widespread phenomenon. Moreover, the issue of violence against uprooted women has, in the 1990s, drawn much attention in international symposia, such as the United Nations Fourth World Conference on Women (Beijing, 1995).

Through dialogue, symposia, and advocacy, uprooted women seek to empower themselves to address their problems by changing community attitudes toward women violated by gender-based types of persecution. This awareness has resulted in rethinking and redefining programs and responses to meet uprooted women's special needs. For example, uprooted women activists are calling for programs to provide crisis-intervention services during the early stages of plight to rehabilitate women and girls who have been subjected to gender-based persecution. Another priority for refugee and displaced women is their immediate involvement in the planning, development, implementation, and evaluation of programs and services affecting their lives. Typically, refugees receive relief assistance but no community-organizing or community-development support, even though experience has shown that participation in community-based projects enables victims of war to take control over their lives by restoring their communities and bringing some sense of normalcy to their abnormal conditions. This is primarily due to the fact that the field of refugee relief and assistance is structured to provide care and relief assistance effectively to large numbers of refugees crossing international borders. While the critical condition of refugees merits emergency relief and assistance, little attention is paid to the long-term development needs of communities to rehabilitate. There is some effort by the UNHCR and its implementing agencies to adopt long-term development approaches during the early stages of refugee crisis intervention. International relief and development agencies are becoming increasingly aware of the unique needs of uprooted women, who commonly defer their own needs to those of their family and larger community and are left with fewer relief and development resources as a result. However, uprooted women often lack access to institutional-development assistance and to all-important resources. Traditional programs are not designed to empower refugee and displaced women; inadvertently they foster dependence on international humanitarian and development-assistance agencies.

Much of the violence targeting refugee and displaced women is instigated in the name of culture, ethnicity, nationalism, and religion. It is estimated that women and children constitute 80 percent of the refugee and displaced population—a substantial subsection that itself merits special attention (Rogers and Copeland, 1993). Although they are most vulnerable to pervasive and outrageous forms of human-rights violations, their protection has regressed enormously. Two factors contribute to this phenomenon: the massive increase in the number of refugee and displaced populations, which is crippling humanitarian assistance machinery, and the global economic decline, coupled with restructuring policies (countries) unable or unwilling to finance humanitarian assistance machinery. The United Nations as a whole, and the UNHCR, a branch of the United Nations, in particular, decry its rapidly shrinking budget.

The post–Cold War environment has given rise to newly emerging forms of human rights specifically targeting women and girls. For example, since time immemorial the violation of women and girls has been dismissed as a by-product of warfare; however, mass rape campaigns against them as a calculated objective of warfare have been rampant in the 1980s and into the 1990s. The example of mass rapes of Cambodian and Vietnamese women and girls and, in the early 1990s, ethnic breeding (the forced impregnation and rape of women and girls by opposing factions) of Bosnian Muslims are stark reminders of such campaigns. Public violence against women who do not conform to cultural practices and norms is also becoming widespread, especially in the Muslim world. In effect, the world is waging war against its women. This has important implications, not only for how to protect refugee and displaced women and girls but also for how to help them rebuild their lives and families. Widespread abuse against uprooted women and girls constitutes a new set of human-rights abuses in the late twentieth century for which new solutions need to be devised. Advocates for women in general and uprooted women in particular have successfully decried the declining status of women and girls at a number of international conferences, which has generated cooperation from international institutions.

Muslim Women

Approximately 80 percent of the total world refugee population is Muslim, including the Afghans, Azerbaijanis, Bosnians, Kurds, Palestinians, Iranians, Iraqis, Burmese, Somalis, Sudanese, and Tajiks. Women and children make up almost 75 percent of this total Muslim population (International NGO Working Group on Refugee Women, 1995). Contemporary theopolitics in the Muslim world pits women as political pawns in the ideological warfare gaining momentum among extremist Islamists. A closer examination of the status of Muslim women and girls reveals that they are bearing the brunt of ethnic breeding and public violence. Such forms of violence against women are left unchallenged because there is little awareness in the West of their situation. On the sociopolitical front, growing dissension between Islam and the West further complicates the situation of women: In this context, the mistreatment of Muslim women and girls is dismissed as cultural rather than treated as a universal form of human-rights abuse. The debate has been taken up by Muslim women activists who challenge the notion of cultural relativism as justification for human-rights violations, but the lack of networking between Muslim and Western women contributes to their isolation. A major cause of the high incidence of violence against refugee and displaced women is the lack of appropriate and effective legal and legislative mechanisms to punish perpetrators. Furthermore, there is little financial support to upgrade the capacities of agencies for refugee and displaced women to organize their communities on human-rights issues. The status of refugee and displaced women can be understood only by examining in general context the historical, political, economic, and social realms of violence against women. Although violence against women is a global phenomenon, it takes its most severe form when directed against refugee and displaced women.

Traditional societies have little regard for gender equality, especially when resources are scarce and competition for access to social, political, and economic opportunities is the province of men. Power and control of resources are likewise the entitlement of men. The results of such male control of resources and repressed economies particularly affect refugee, returnee, and displaced women. Their condition and status invariably decline during war and civil conflict, as resources are shifted to meet the needs of male soldiers. Upon repatriation, it is once again the men who have priority access to resources and employment as paybacks for their political participation during war and upheaval.

Most of the world's refugees are generated by, and concentrated in, developing countries that fiercely subscribe to patriarchal societal models. Women in traditional societies are socialized as second-class citizens. This is particularly true for women in Muslim societies, where a wave of political resurgence to Islamic law and practice aims to restrict the minimal progress achieved for women and girls. Western relief agencies, including U.S. nongovernmental organizations (NGOs) such as International Rescue Committee, CARE International, and Catholic Relief Services, continue to respond in these situations as if gender biases do not exist, perpetuating the human-rights abuses directed against women.

U.S. NGO Policies and Practices

U.S. NGOs, as major implementors of the policies of the UNHCR, play a big role in providing refugee relief and development assistance. Two major transitions have occurred in the refugee field: a growing recognition of the need to move from relief to development; and a growing recognition of the trauma among refugee women and girls. The resulting presence of new players in the refugee field, such as the United States Agency for International Development (USAID) and development-oriented NGOs, has added challenges and opportunities for refugees and displaced women, some of which are briefly described here.

U.S. Funding Possibilities. USAID provides grants to development and relief agencies and is increasingly stipulating that its multiyear grants to U.S. NGOs be matched with nongovernmental sources. Typically, U.S. NGOs with strong institutional and managerial capacities are recipients of these grants because they have significant experience with governmental grant-writing procedures. USAID funding to large U.S. NGOs discourages smaller organizations with credible program experience in the field. U.S. NGOs headed by women addressing the acute problems of women and children are almost nonexistent and remain marginalized within the array of NGOs.

Repatriation and Development Assistance. Development agencies are becoming increasingly involved in refugee and displaced people's preparation for return and redevelopment. Although most development agencies are competent at implementing traditional development-assistance programs during times of peace, they have little or no experience working with refugees, returnees, or displaced people.

Furthermore, they are not familiar with the problems of transition from relief to development. Development for war victims who have high levels of stress due to uprooting and trauma has not been sufficiently understood by these agencies. When development agencies with good intentions, yet little regard for partnership and equity with local institutions, become involved in the field, there is danger that the "refugee dependency syndrome" will once again take effect.

Lack of Service Coordination and Cohesion. Provision of services for refugee and displaced people in the field may be divided between international NGOs and local agencies without planning, coordination, or cooperation. Often, new players in the field "reinvent the wheel" and/or provide services that inadvertently overwhelm the local agencies. As a case in point, Croatian agencies in the 1990s complained about being overwhelmed by international NGOs providing trauma relief services, while in reality what the local agencies needed most was technical assistance to upgrade their own capacities. These agencies were managed by expert psychiatrists and psychologists who needed specialized assistance to build their long-term organizational capacity to exist beyond the crisis stage.

Fast-Paced Development Assistance. Development agencies providing assistance to refugees to rebuild war-torn communities and nations must recognize that the normal pace used in traditional development-assistance projects must be altered in favor of a fast-paced development approach: They must make use of all of society's human resources. In the case of refugee and displaced people, this means actively involving the women and youth in development-assistance projects. This method will ensure that war-torn societies will quickly rehabilitate themselves and move toward reestablishing community-building efforts.

Sociopolitical Shifts in the Muslim World

In the Muslim world, lack of economic equity and social justice has given rise to growing resentment. The Islamists believe that the present (1990s) world system has kept the Muslim world unjustly marginalized politically and economically. This position contrasts with Muslim achievements and past glories. Therefore, the Islamists are seeking a just worldview that, in part, restores the old order of things. They believe that the Western worldview has been left unchallenged by the Muslim world, resulting in many Muslim governments' co-optation by Western hegemony.

Islamists believe that intrusive Western ideas such as liberalism, socialism, and secularism have no place in their societies; they have called for political resurgence to Islam and Islamic values, and they are implementing the Islamic solution, *Al-halul Islami,* to counter such heresies. In actuality, this male-led fundamentalist resurgence movement has little to do with Islamic values or principles, nationalism, or the imperatives of activism but is based, instead, on the power of religious groups to impose their will by attacking Western hegemony at the place where it is most alien to Muslim identity. In this context, women's liberation has caused much contention between moderate and fundamentalist Muslims.

The state of affairs in the Muslim world in the last decade of the twentieth century is characterized by rising tensions among secularists, moderates, and Islamists and is complicated by rising tensions between Islam and the West. Much negative attention has been paid to uprisings in the Muslim world, while little connection has been made between Islamic political resurgence and the role of the West in perpetuating the problem. The end of the East/West conflict, the West's policy toward Bosnian refugees, and the U.S. invasion of Iraq are points of contention for Islamists. Muslims believe that the fundamental problem lies in a general misunderstanding of Islam in the West and in the growing need to fill the ideological vacuum created following the demise of Soviet Communism.

The resurgence of religious extremism compounding global conflicts is not exclusive to the Muslim world, but it is perceived as synonymous with Islam. Knowledge of Islam in the West is minimal; the lack of it significantly exacerbated by the media. Misperceptions in the West that Islam is necessarily fundamentalist, antifeminist, and undemocratic contribute to the widening gap in Western–Islamic relations.

The conditions and status of refugee and displaced women cannot be segregated from the larger socioeconomic, political, and cultural contexts. In examining the root causes of women's second-class status in society, it is equally important to understand their historical basis. It is safe to say that uprooting and forced migration have resulted in more meager resources for women than for men. As a consequence, their survival rate, nutrition, education, skill development, and psychosocial well-being fall far below those of men.

In societies where a political resurgence of Muslim practice is observed, women and girls have been used as political pawns to portray the ideological virtues of Islam. For example, the practice of veiling, seclusion from public spaces, and lack of access to education are defended as necessary for upholding cultural norms and practices. The argument of cultural relativism, widely used by the international donor community as a reason not to act, in effect buys into the male political agenda of keeping women subservient and victimized. In fact, the refugee and displaced world is rapidly becoming institutionalized along patriarchal models, inadvertently supported by the West and by donor policies.

Western Donor Policies and Practices

In light of these developments, an appropriate role for international donors is to support U.S. NGOs that have strong backgrounds in working with refugee and displaced women and women in Islam. The role of international donors should be to support refugee and displaced Muslim women's efforts by maintaining dialogues and networks and providing organizational and community-development training and technical assistance. Consequently, long-term structural policies must institute the framing and shaping of culturally appropriate strategies in collaboration with the local Muslim women's agencies and the support of their governments. Such a funding policy would aim to establish democratic processes in which women are equitably involved.

Repatriation, Institution Building, and the Role of Women

The end of the East/West conflict has resulted in the world community mandating the United Nations and its entities to negotiate peace and end conflict in order to enable refugees who have languished in refugee camps to return home. This movement is known as repatriation. The UNHCR designated the 1990s as the Decade of Repatriation, in which it is estimated that between seven million and nine million refugees will be returned to their countries of origin. In the mid-1990s more than two million people have already been returned, with two million more on the move.

Repatriation has gained momentum partly for economic reasons; the refugees who were used as political bargaining chips by superpowers have little utility now. However, there are no guarantees that the return of refugees will be voluntary nor that measures will be taken to ensure safety once U.N. peacekeepers leave a country. The issue of safety has often been taken for granted, with disastrous results; the international community must watch these movements of people more closely. They and the receiving governments should also pay closer attention to other factors inherent in repatriation that are commonly overlooked, such as the fact that a large segment of the returnees comprises handicapped women and children and the elderly. Upon returning home, these disadvantaged repatriates find a weak social system, ill prepared to offer them the services and support they require. The task of caring for them falls to more-able-bodied women, to whom this role has traditionally been relegated, yet without the support of resources from the larger social structure. To dispel such disparities, it is essential that efforts to rebuild the infrastructure in nations emerging from war be tied to efforts to rebuild the social structure. Displaced and returnee women, who constitute the majority of national populations in postwar environments, are also a nation's primary resource to rebuild the ruptured value systems and social structures. Traditional structure-rebuilding policies do not take this factor into consideration. This situation is further compounded when constitutions and legislative systems of nations emerging from war conditions do not integrate language and policies that affirm democratic principles applied to both genders—women as well as men. Policies as well as aid programs must be revisited to ensure that newly emerging nations are based on pluralistic democratic models.

With massive return comes the issue of nation building and the nurturing of newly emerging democracies. Repatriation is enhanced by the promise of rebuilding and rehabilitation. However, little effort has ever been made during the decades of refugee life to prepare refugee women for the daunting task of development. Governments, international relief and aid agencies, indigenous NGOs, the United Nations, and multinational corporations must devise and implement new policies and practices that take into account the unique needs of women refugees in the early stages of uprootedness. Their physical and legal protection must be upheld in tandem with training, education, and skill-building programs. Also needing to be addressed directly are the emotional consequences of the flashbacks that many women experience upon resettling in areas where they were originally subjected to torture and abuse. An all too familiar lack of physical protection in refugee and displaced life commonly results in sexual violence and other human-rights abuses against women and girls.

Gender-based persecution is organized and rampant. It must be viewed in the larger context of less obvious forms of structural violence. As noted above, a common oversight in the building of newly emerging democracies is the linkage between infrastructure development and the rebuilding of social systems. Years of warfare and conflict have ruptured social infrastructures, resulting in widespread violence, the breakdown of law and order, and a shattered value system incapable of holding citizens accountable to the larger community and society. Violence, lack of education, and use of drugs are common among young male adults, who often lack family and community structures. Although women suffer from trauma syndrome, they are the ones responsible for rebuilding their families' and their communities' shattered lives and for creating conflict-free and compassionate societies for their children. The role of women in such societies is critical to reestablishing a value system and assuring the education of children. Their emotional and mental health must receive its due attention to integrate women fully into the process of rebuilding functional societies.

It is also important that attention be accorded to leadership development of refugee and displaced women, to train and strengthen their organizational rebuilding efforts. Donors can play an important role in supporting these efforts by holding their grantees accountable for designing and developing programs that support refugee and displaced women's capacities for community organizing and institutional development. This must be done in refugee camps and other places of refuge.

Supporting the organizational development efforts of refugee and displaced women early in the process of return is crucial. The international community has concentrated on addressing structural issues and on establishing legal and legislative frameworks for these new societies. While this is valuable, more attention must be given to rebuilding *social* structures, including a greater emphasis on gender equity. When the United Nations and other international entities support newly emerging democracies by helping them develop their constitutions and legislation, they must ensure the inclusion of language at all levels and strata of governance to protect basic human rights of both female and male citizens. This level and type of integrated approach to humanitarian and development assistance is especially critical in societies that are governed by Muslim laws and practice.

The signing of the Peace Accords has transformed the Palestinian society from a government in exile into an internationally recognized state. This transformation signals a new era and challenge for the Palestinian state—the challenge of transforming a resistance-oriented movement into a functional civil society. A newly emerging dimension is the challenge confronting Palestinian women where formerly politically active women have been relegated to lesser roles by their own patriarchy during the post-resistance era. Palestinian women are lamenting that they have been assigned to the private spheres of their homes while their male counterparts have been assigned to lucrative public posts as political paybacks for similar roles during the resistance against Israel. Little effort has been made, however, by male decision makers to continue to use formerly politically active Palestinian women by retraining them in organizing their own communities to provide critically needed human services that the new government is unable or unwilling to undertake. The situation of Ethiopian, Central American, and other returnee women is similar to the Palestinian case where women receive crumbs off the table of the male patriarchy. It is therefore critical that women's movements in general and women in the Free World in particular recognize this phenomenon and advocate for equal treatment of women by lobbying their own governments to earmark development assistance funds toward women's projects. The mistake of Bosnia should be avoided where the Peace Accords left women out of rehabilitation and reconstruction schemes even though it was women who constituted the majority of the highly traumatized. This oversight galvanized women activists in the United States to call for particular attention to Bosnian women. Their efforts led President Clinton to earmark $5 million toward women's projects.

Similarly, Afghan women have not remained immune from the traumas of warfare and other atrocities. In their uprootedness, they and their children have contended with the breakdown of their social fabric and value system, rising drug addiction, violence against women and girls, and hunger and death by starvation. However, despite the disproportionate trauma Afghan women in exile suffer, they are making contributions to family and community life. In spite of dire hardships, women and women-headed households hold much hope for eventual return. It is in this spirit that the international community must support the leadership efforts of Afghan women refugees. An effective strategy is political leveraging of economic- and humanitarian-assistance programs. Assistance to nations must be provided on the basis of upholding and supporting the efforts of women at all times.

It is important for the international community and international donors to support the efforts of women everywhere to develop strategies and responses for rebuild-

ing environments that will foster nontraumatized, nonviolent children. Women of different ethnicities and religious backgrounds are receptive to the development of coalitions to work together on issues that concern their children. In supporting such efforts, Western donors are not advocating socioreligious strategies but, rather, are empowering indigenous women to communicate, network, and receive training to define and implement strategies that they themselves originate and carry out. In this context, it is particularly important for donors to play a role in financing the peace process and supporting pluralism. International policies can be proactive and effect a supportive role for women by ensuring gender equity; by failing to do so, they may inadvertently foster inequity in the societal struggle for democracy, peace, human rights, and pluralism.

Support also must be given to help establish an environment and mechanisms for a popularly elected leadership that supports democratic and human rights for all, women as well as men.

Donor policies must support grass-roots and community-based groups in the best interests of people from all strata of society based on democratic and pluralistic principles. International aid and relief agencies must integrate programs (education, training, and services) for women. Mechanisms in support of civil institutions actively managed by a strong independent sector providing social, human, and development services must be upheld. The empowerment of people, especially women, depends highly on such models.

Role and Responsibilities of the Media

As crises erupt, international media play a major role in depicting the distant humanitarian horrors. Commonly, media images during these crises include grotesque pictures of maimed women and children, as in the case of the Kurds, Bosnians, Croats, and Rwandans, to name just a few. Advocacy efforts must be supported to educate the mass public to dispel myths and misperceptions of refugee and displaced people in general and of women in particular. For example, media projects depicting the stories of refugee and displaced women would help raise public awareness and bridge the informational gap that often relegates the innocent victims of war to mere statistics. Public media campaigns put a human face on the blatant inhumanity of man toward man—and woman—that exists today on a global scale. Journalism must be conducted in responsible ways with appropriate follow-up reporting of the conditions as needed. Too often, continuing crises are forgotten because the media become involved in covering new crises. The situation of the Afghan refugees, the Rwandans, the Somalis, to name a few, continues to exist and to haunt its people, but international print and television news media coverage has waned.

Some Recommendations

The following donor policies would help counter growing negative public and media misperceptions regarding refugee and asylum seekers in the United States and abroad. The recommendations do not imply that their implementation would resolve all aspects of uprooted women's problems. They would, however, ameliorate their most salient needs and provide them with the hope that the world has not forgotten and abandoned them.

1. U.S. and European foundations can be the necessary link in supporting small and reputable U.S.-based NGOs, especially those working on behalf of refugee and displaced women, by strengthening their institutional capacities and committing funds to match USAID grants. The benefits of this funding strategy are twofold: It will encourage small NGOs to apply for U.N. and governmental funds, and it will create a private-public partnership in the field of refugee women and development assistance between U.S. and European foundations and local entities.

2. Since the issue of human rights is highly relevant to refugee and displaced women, it is important that NGOs with strong backgrounds in training such women receive financial support. While the traditional approach of monitoring individual human-rights abuses is essential, efforts should also be made to support the efforts of local agencies to promote their causes on their own behalf with U.N. entities, international NGOs, and human-rights groups. This approach will establish a working model of partnership between these groups and local refugee and displaced women's agencies consistent with democratic and pluralistic principles. Furthermore, it will challenge the token involvement of refugee and displaced women in international and national conferences initiated by First World institutions.

3. U.S.-based and international NGOs with successful institutional and management structures must develop programs to share their own institutional-development approaches with local refugee and displaced women's agencies. Such a working relationship will not only create a valuable link between U.S.- and European-based NGOs and local agencies, but it will also dispel the traditional inequitable model of relief and development assistance that views refugee and displaced women in terms of their needs rather than their capacities as resources for development.

4. This strategy must also be instituted by U.S.-based NGOs to strengthen their knowledge base of felt priority needs of indigenous women's community-based NGOs enabling them to strengthen their long-term institution-building efforts. This approach will also serve to forge strong and viable partnership between Northern and Southern NGOs.

Moreover, the above strategy will help alleviate existing tensions between Northern and Southern NGOs where Southern women are calling for equal representation with Northern women's NGOs in the forging of policies and programs affecting them. Southern women are concerned that their institution-building efforts are overlooked due to their lack of access to financial resources concentrated in the North as well as the lack of adequate understanding of their issues by Northern donors. This is particularly true for women in the Muslim world where a trend toward Islamic extremism is directed against women.

Conclusion

It is time for international donors to assess their policies and their impact on refugee and displaced women. In so doing, donors must address the level and extent of their funding policies regarding refugee and displaced women and their institutions. The major challenges that international donors confront are: (1) institutional development of Western NGOs juxtaposed with that of local refugee and displaced women's agencies, and (2) avoidance of perpetuating the dependency of local refugee and displaced agencies on Western NGOs.

Too often, in addressing the problems of forcibly uprooted people, efforts are ostensibly directed toward economic and structural frameworks, thus overriding individual and collective human suffering. Lack of comprehensive international laws on human-rights asylum further compounds the problems of uprooted people in general and of women in particular. Wounds of human dignity are central to the suffering of war victims, with the offenses on decency being connected to the feeling of complete powerlessness that follows from organized violence (Arcel, 1994). Moreover, the post–Cold War environment has given rise to newly emerging forms of human-rights abuses specifically targeting women and girls (Wali, 1995). Thus, any effort to alleviate such problems must address primarily the psychosocial and emotional recovery of these wounds as well as the accompanying shame. Shame and loss of dignity are common to all female war victims regardless of age, ethnicity, religion, or culture. Addressing these as well as the more conventional problems previously noted by international organizations will help shape and define the future development of equitable partnerships between Western NGOs and local refugee and displaced women's agencies, as well as the establishment of democratic institutions. If donors aim to influence and support Western policies to foster democracies, then it is imperative to support NGOs that work to empower refugee and displaced women to upgrade their own organizational and civil-institutional development.

Donor policies aimed at supporting the humanitarian- and development-assistance efforts of refugee and displaced women will help fill the gaps created by Western foreign policy in areas in which it was involved in financing the wars that create such women. Such a campaign in itself will result in the emergence of local agencies that can provide critical human services for refugee and displaced women and pave the way for creating strong local institutions. Donor policies also can help address the myths and growing public misperceptions about refugee and displaced women in general and about Islam in particular, in Muslim and non-Muslim societies, and empower victims of war, especially refugee and displaced women, to initiate dialogue with American women and the larger public through education campaigns to help dispel those existing myths and misperceptions.

References

Arcel, Libby T. (ed.). *War Victims, Trauma and Psycho-Social Care.* Zagreb: European Community Task Force, 1994.

International NGO Working Group on Refugee Women. "Statement submitted to the United Nations Commission on the Status of Women," 39th session, New York, March 15–April 4, 1995.

Rogers, Rosemarie, and Emily Copeland. *Forced Migration Policy Issues in the Post–Cold War World.* Medford: Fletcher School of Law and Diplomacy, Tufts University, 1993.

Subcommittee on Immigration and Refugee Affairs. "Cambodia: Towards Peace and Relief," staff report prepared for the use of the Subcommittee on Immigration and Refugee Affairs, Committee on the Judiciary, United States Senate. 102nd Cong., 1st sess., November 25, 1991.

United Nations Development Program (UNDP). *Human Development Report 1995.* New York: UNDP, 1995.

United States Committee for Refugees. *World Refugee Survey 1993.* Washington, D.C.: United States Committee for Refugees, 1993.

United States Committee for Refugees. *World Refugee Survey 1994.* Washington, D.C.: United States Committee for Refugees, 1994.

Van Rooyen, Rene. "Challenges and Opportunities in the Decade of Repatriation: The Role of NGOs." Conference, Georgetown University, Washington, D.C., June 16–17, 1993.

Wali, Sima. "Muslim Refugee, Returnee, and Displaced Women: Challenges and Dilemmas." In Mahnaz Afkhami (ed.), *Faith and Freedom: Women's Human Rights in the Muslim World.* London: I.B. Tauris, 1995, pp. 175–183.

See also Barbara Swirski and Manar Hasan, "Jewish and Palestinian Women in Israeli Society"

Women and War

Mary E. Haas

Introduction

War, the failure of more peaceful means of interactions, has as its goal the gaining of dominance over the wealth of a region by controlling the factors of production, land, labor, and capital. Throughout most of history, wars were localized events of short duration fought by small numbers of professional military forces rather than the population as a whole. It didn't make too much difference who won the war since one ruler's policies were about the same as another's. The population continued living at a subsistence level and paid tribute in taxes or labor. Rulers were not the representatives of the people or even liked by the people, and military power, not wealth or ideology, created the empire. In medieval kingdoms wealth was measured as the production of the land, and the rulers provided protection for the peasants, who worked the land, through diplomacy and warfare. This obligation of the ruling class extended to all of its members, including the women, who often ruled and led in battles.

Since then, the advancement in weapons has created a world in which international war is so costly in the destruction of life and resources that the national leaders try to settle differences through diplomacy rather than war. Whereas heads of states may no longer fight in battles, they need the military to protect the nation's people and wealth. This fact places the members of the military in potentially powerful positions. History reveals that the military frequently assumes political power when its leaders believe it is necessary. Since the end of World War II in 1945, there have been more than 130 formal wars (Nordstrom, 1991).

The vast majority of these wars are civil in nature and do not involve major world powers. In essence, the definition of war has been changed, but the goal remains the power to control a region. One of the major changes in warfare has been the targeting of the civilian population by a military force. The use of "dirty war" tactics places women in the center of the conflicts; they become the target of those seeking to destroy the existing culture and to replace it with a culture of fear. Since the 1980s, 90 percent of all war casualties are civilian. Because of this, Nordstrom (1991) argues that the definition of war, including who can declare it and who can determine the terms ending it, needs to be redefined. War, she proposes, must be defined by the participants—including the victims, who are in large part female—because those with the power to define war are those with the power to control war and to end war.

Therefore, in examining the roles of women and war it is necessary to consider the roles women play in preventing and ending conflicts as well as in preparing for and taking part in armed fighting. Since most of the nations of the world are creations of colonial powers or treaties ending World War II rather than creations of the people within their borders, most developing nations have a short history, and their rationales for existing can be easily attacked. Additionally, some of the nations continue to exist only because they are maintained and supported by other, more powerful, nations rather than by their own inhabitants. Since the end of World War II, most warfare has been internal or civil wars among classes of people or nationalistic and ethnic groups who may have a long history of distrust, exploitation, and violence. If, as politicians claim, there is truly a movement toward democracy in the world today, all citizens, not merely a formal military or police force and certainly not a local individual or group with a private cache of weapons, need recognition as the power base of government.

Warfare always impacts the lives of women, forcing them to be either victims, reactive participants, or proactive participants. Throughout history, women have been victims of war and have reacted when conditions demanded immediate or emergency behaviors. As long as there are wars, women will continue to be among their victims, but, in the modern world, women are making personal decisions and assuming more proactive roles throughout society: Military careers are a proactive option for many women, as are other private and governmental roles that affect policies concerning war. Women who are becoming proactive concerning war believe that human dignity, justice, and equality are the important values on which to build a society.

In the developing world, 30 years of war (1945–1975) in Vietnam provide the best documented record of women and modern warfare. Proactive roles such as soldier, guerrilla fighter, peace activist, politician, and work in the medical and social-service professions are recorded and supported with research evidence from nations throughout Asia, Africa, and Latin America. Adequately verified documentation of women in other wars in the developing world is difficult to find. Details about individual events or interviews by journalists form the majority of the data. References about women participating in warfare, when available, reveal many common trends with the Vietnamese experiences. By combining the isolated data from developing nations with historical experiences from the developed world, it is possible to generalize and put forth the beginnings of a theory of women and war and to speculate on some of the causal factors behind the actions of women concerning war and warfare.

Table 1. Women's Roles in War

Victims	Reactive Roles	Proactive Roles
Widow	Refugee	Soldier/combatant
Orphan	Emergency fighter	Soldier/supporter
Disabled	Emergency defender	Guerrilla fighter
Rape victim	Lookout/informer	Journalist
Prisoner	Rescuer	Recruiter/propagandist
	Prostitute	Politician
		Peace activist
		Intelligence officer
		Medical worker
		Social worker
		Protester

The model put forth here describes three general classifications of the roles women have been and are playing in warfare. Table 1 also identifies the specific roles subsumed by the classification headings: victims, reactive roles, and proactive roles. The causal factors for the model are postulated to be directly related to the value societies, particularly the power groups of the society, place upon individual members of the society. The explanation begins with a discussion of the role of victims and ends with the proactive roles. By so doing the discussion also presents the numerically largest group of women first and the numerically smallest and most complex of the roles last. The three categories reflect an increasing ability of women to take actions to protect not only themselves but also people and things they personally value. Within the discussion of each category, relationships and hypothesized causes are presented first followed by specific, verifiable examples of women's behaviors that support the hypotheses. Even

though all categories of roles and their behaviors can be documented throughout history, most examples come from the developing world since the end of World War II. The actual numbers of women in each of the categories changes over time as more women and men come to believe in the correctness of the motivating values necessary to elicit the behaviors of the specific roles. What is emerging throughout the world is a slow but progressively greater acceptance of the proactive roles for women. Unfortunately, in some areas of the world vast numbers of women continue to be made the victims of warfare, reacting to situations rather than having personal control over their fate.

Women As Victims of Warfare
A victim role is one in which an innocent person receives suffering because of actions taken by others. Wives, mothers, sisters, and children are at risk of becoming victims in war. "Victim" is the most ancient and enduring category of roles for women associated with warfare; statistics reveal that, in many of the recent wars, committing terrorist acts against civilians is a dominant activity. The mid-1990s discussion of designating rape as a war crime was an outgrowth of the acceptance of individual human rights that in years past were not considered as important a value as human survival and the survival of the family. The discussion also reflects a growing recognition of the importance of extending human rights to all citizens, including women and children.

The most obvious consequences to victims of war are clearly illustrated in the media. The immediate physical pain of injury and permanent disabilities move the viewing public and prompt humanitarian relief. Three additional consequences for victims are less immediate but have a profound impact on individuals, the society, and the culture for generations. The deaths of large numbers of workers perpetuate poverty and present the very real possibility of even more people in poverty because the society has a larger than normal number of widows, orphans, and unmarried young women to provide for causing economic strain on the traditional families, extended families, and work force. Additionally, among some societies, particularly those that are Muslim, victims of rape are held in social and emotional disgrace and even blamed for not stopping the rape. Closely related to this is the existence of outcast children that are not fully accepted by a society because they are the offspring of mixed racial or ethnic unions. Lastly, there are both personal and societal conflicts stemming from the conflict between the value of the individual, her physical and emotional needs and wants, and the value given to the immediate and long-term survival of individual families. Performing physical labor is not as difficult as supporting or caring for the emotional needs of others who have undergone the stresses of war. For many the end of a war does not provide the conclusion hoped or worked for and instead requires women to abandon or drastically change their personal expectations for the future. Societies expect women to be care givers and not to ask for assistance from the society for their own wants.

The roles of victims are particularly devastating in the developing world, where societies have strict divisions of labor. In such societies, a child's or a woman's very survival often depends on family members because social services are not available from private or governmental organizations. Immediately after the death of a son or a father, the war victim may be able, through hard work, to assume the labor of caring for herself and her dependents. When a patriarchal society loses large numbers of males, the care of elderly parents and young children falls to families with fewer than a normal number of workers. Such a situation results in increased poverty within the society and is often accompanied by increased violence against women. Individual emotional satisfaction is sacrificed for family survival. Women may locally assume new roles and advance their own social positions as they pick up the work of men and the prestige associated with the new role. However, as the next generation matures with its normal number of males, decisions concerning whom to educate and whom to advance at the cost of family wealth and laborers are often made on the basis of old traditions: The eldest son is selected to receive the benefits of the family's sacrifice; the new prestige earned by women disappears.

The Vietnam War was the first war covered by television, which nightly brought the real experiences of the war into living rooms of civilians. The availability of information makes it much more difficult for a government to control or explain away events. People tend to respond to the visual news media, believing its accuracy. When people throughout the world see hundreds of thousands of refugees in camps in Cambodia, Zaire, and Tanzania fleeing from unrest and persecution in their own nations and hear journalists and social workers describe the medical and starvation potentials, they recognize the failure of governments to solve problems of basic needs and services. This results in greater pressures being placed on governments and on international organizations, and prompts individuals who have not been politically active to respond. The result is usually that the life-threatening physical needs of refugees are attended to quite rapidly; political solutions, if addressed, come much more slowly. However, when the press is excluded, much less help and political pressure are forthcoming and the number of victims increases, which allows unrest to continue and resentment to smolder beneath the surface, planting the seeds for revenge at a later date. Confronting problems of resentment between classes and ethnic groups by means other than police and military force does not appear to be a high priority for most governments. So the potential for truly democratic, representative governments suffers and the seeds of civil wars continue to grow.

The Soviet actions in Afghanistan between 1979 and 1988 went relatively uncovered in the media. Most people know of the military struggle between Afghan rebels and Soviet troops; few know that there were many demonstrations in cities in which unarmed civilians were shot and imprisoned. Nazifa Afghan recalled that many of the girls who demonstrated were taken to the Pul-i-Charkhy jail in Kabul, Afghanistan, where they were tortured and raped, and that many come out of the jails pregnant. She also told of seeing many dead bodies of women who were raped. Because of the great shame and severe consequences prescribed by their religion against those who are raped, she said, the people did not know if the women had been murdered by the soldiers or had committed suicide. Had she been raped, she said, she would never have returned home to bring shame on the family but would have killed herself instead (Hayton-Keeva, 1987).

Another war that was not covered immediately by the international media was the Khmer Rouge attack on the Cambodian people in 1977. Here again, rape was commonly used against women who had no weapons with which to defend themselves. Many of these women also had to make the difficult decisions to stop nursing infants or to abandon or deny food to some children so that they could physically continue to flee and help other children escape. Medical teams on the Thai–Cambodian border caring for refugees reported the women had to overcome both emotional and physical traumas from their experiences (Hayton-Keeva, 1987).

Civil wars have very high death rates. In Vietnam, there are large cemeteries filled with graves of unidentified soldiers. In many of the villages of the North, only 50 percent of the soldiers returned after the war. Sandra Collingwood, who served as a volunteer in the South with Catholic Relief and International Voluntary Services, recalled that every Vietnamese person with whom she had contact was directly affected by the war, by the loss of either a parent, a sibling, or another relative in the immediate or extended family (Walker, 1985). During the war, the Communists talked of the "Three Postponements": love, marriage, and childbirth. After the war, the combination of deaths, reeducation camps, disabilities, old age, and miscarriages and embryo deaths (attributed to the chemical dioxin in Agent Orange) kept many from having the families they wanted. American nurses who saw civilian casualties recalled that most were women and children; many had lost arms and legs, which would have made them unacceptable as marriage partners. There were 131,000 widows and 300,000 orphans in Vietnam after the war; estimates for abandoned offspring of U.S. GIs have run as high as 15,000 (Haas, 1991). In the Vietnamese tradition, children placed in orphanages are not necessarily without parents but often are members of families unable to continue to care for them. Catholic and Buddhist nuns cared for increasingly more children as the war lengthened, most of whom were in poor physical condition when they arrived at an orphanage. Among the infants, 60 percent died before reaching three months in age, and many were deformed because their caretakers lacked the time to properly rotate their position throughout the day.

As the war progressed, more and more villages and

farms were destroyed as a direct result of bombing by American forces. Hundreds of thousands of victims became refugees within their own nation fleeing to the larger cities of the South. To care for their families, women peddled soup or rice from sidewalk stands or opened small businesses that catered to the Americans. The resale of American items on the black market became a common practice as did prostitution (Hayslip, 1990).

Following the war, the Vietnamese people continued to suffer because of distrust among their own people and isolation within the international community. It took 20 years, even into the 1990s, before a few groups of individuals and the governments could begin the healing process.

Women As Reactors

Reactors reach their decision to act quickly, often without consideration of the consequences. Being a mother or older sister brings with it an emotional responsibility toward caring for the survival of the family, which prompts women to acts of bravery and sacrifice regardless of the personal risks. History is filled with accounts of wives and daughters assuming the positions of fallen males or undertaking high-risk actions to obtain assistance in desperate situations. Women also endure great physical hardships laboring at multiple jobs and for longer hours so their loved ones can survive. Becoming a refugee and making the decision to leave loved ones to increase the chances for the survival of some of the family is still another reactive behavior undertaken by millions since the end of World War II.

All societies condition members to react in certain ways. Societies that train women and children to be obedient, rather than independent thinkers, actively condition their young children to quickly and unquestioningly react to opportunities that they have been told will serve their family and friends. Thus women steal, bluff, and tell lies in emergency situations even though such behaviors are not generally thought acceptable. Parents, governments, and revolutionary groups tell children how to behave and reward children with praise for exhibiting exemplary behaviors while chastising, threatening, and severely punishing unwanted behaviors (Hayslip, 1990).

Experiencing a single act of injustice that impacts them personally during their early teens motivates many young women to react and become fighters. Such women experience their seemingly comfortable world (traditional society) so greatly disrupted that they turn in hatred and grief to seek revenge on the government they blame for failing to protect, or for disrupting, their world. Revolutionary forces often claim to support those values that the people believe are the "best" aspects of their traditional cultures and point out the government's role in the destruction of traditional values. Striking back through acts of violence and revenge provides the young reactors with a sense of power and satisfaction. The women quickly form strong bonds of loyalty to revolutionary forces with whom they struggle and develop a new, larger family of revolutionaries. Having entered the struggle as reactors such

women often spend the remainder of the war or their lives among the fighting forces. These women often train and become ferocious guerrilla soldiers or spies, propagandists, or recruiters of other women. If they marry, it is to soldiers, and children from such unions are usually left in the care of others so the mother can return to the revolutionary forces.

The reactor role, while it may be thrust upon an individual quickly, is not limited to a short demand on the energies of a woman, nor to any particular age group. If and when an individual has time to consider the consequences of reactive behavior, she may make the decision to assume a new, proactive role. Or, having performed the reactive behavior, she may be forced to permanently join a military group to survive or to protect family members from retaliation by the government.

When she was 15, Vo Thi Mo reacted to the destruction of her village and family farm by American bombs by quickly deciding to join the National Liberation Front (NLF) in Vietnam. She served until the end of the war as a sniper and officer in the elite, all-female C3 fighting company in the tunnels near Cu Chi, and she organized a spy ring. She won many medals for killing the enemy and taking part in dangerous missions. When interviewed many years after the war, Mo still exhibited the fearless desire to fight the enemy and face death that was typical of the young, female, Vietnamese guerrilla: "The first time I killed an American, I felt enthusiasm and more hatred. I thought I would like to kill all the Americans to see my country peaceful again. . . . The Americans considered the Vietnamese animals; they wanted to exterminate us all and destroy everything we had" (Mangold and Penycate, 1985, p. 237).

Le Ly Hayslip recalled that, as a young child, she was encouraged by the NLF forces to steal hand grenades, bullets, and first-aid kits from the South Vietnamese troops. In the evenings during meetings conducted by the guerrilla education agents, the names of children who had accomplished these tasks were listed, and they received verbal praise. The children were also told that if they were caught stealing grenades, they should pull the pin and die for the cause. Hayslip remembers that one of the young girls in her village did as instructed, taking her own life and that of a sibling (Hayslip, 1990).

In the Afghan-Soviet conflict, Afghan women sometimes used their bodies as decoys to halt the progress of Soviet tanks by lying down on the road or by dancing long enough for rebel forces to isolate and capture Soviet weapons and soldiers. When there were few men, the women acted as lookouts so the men could deploy in the best positions for fighting (Hayton-Keeva, 1987).

Reactive behaviors are dangerous and often personally costly. As a response to Soviet soldiers coming to take a group of young Afghan men away for the army, one woman reacted by calling out, "You can't take him!" and the soldiers responded by shooting her in the chest and then shooting and killing her two children (Hayton-Keeva, 1987, p. 18). In the last days of the Vietnam War, children from families who worked for Americans or whose fathers served in the military forces of South Vietnam were brought by their mothers to the few orphanages from which Americans, Australians, and Europeans adopted children and placed them among the true orphans. These mothers reacted by giving away their children in the hope that they were protecting them.

Women As Proactive Participants in Warfare

Women of the ruling class have long been accepted as, and expected to be, fighters, politicians, diplomats, peace activists, and medical-care givers: (1) Khawla Bint al-Azwar fought with the prophet Mohammed; (2) new discoveries from archeological digs and examinations of ancient manuscripts verify that Chinese women were generals and official members of the military as early as the Shang Dynasty (Segal et al., 1992; Li, 1994); (3) mythical and legendary female warriors are a part of the folklore of many nations; (4) Roman chronicles record the deeds of women fighters. Even if the female warriors of folklore are myths, their deeds are praised and celebrated by cultures and they are approved as role models to youth through stories, festivals, and monuments. It should not be surprising, then, to find women from all classes of society seeking to become proactive about warfare.

History books usually fail to acknowledge the acts of female fighters, just as they tend to omit women's other accomplishments in male-dominated societies. Societies that are patriarchal often allow large numbers of women to be members of the military and even combatants when there is a shortage of male fighters and when survival of the society or cause is a risk. However, the number of females who remain in the military after the fighting is over is quickly and greatly reduced. Having used the women during an emergency to remain in power, governments return to viewing women's role as nurturing children, not fighting and protecting resources. In the past, many women preferred or accepted this role and often failed to inform their own children about their military actions. Therefore, knowledge of women's contributions to nations through military service is largely lost in history once the women die. This appears to be changing some; once a society or some portion of it advances economically beyond the subsistence level, women can seek employment beyond the home. They begin to place a higher value on things other than the survival of the family—things that include both service to the community or nation and individual gratification in personal accomplishment.

Most nations that allow women in the military do so from a patriarchal perspective, limiting service to single women in noncombat positions and limiting the rank women can attain (Isaksson, 1988). Such procedures effectively keep women from leadership roles and the potential power such roles bestow. Often, women are told that such policies are necessary to protect them from injury and death. However, history clearly shows that warfare has injured and killed many women who have not had any training in how to protect themselves and that women have performed well in military positions. In the late twentieth century, women's reasons for formally joining a national military force vary greatly. Many believe that their service in the military contributes to a more just and, ultimately, less violent society, while others see the military as their best opportunity for education, prestige, and/or economic advancement. Others act from what they believe to be a citizen's patriotic obligation to protect, serve, or repay her nation.

Proactive Decision to Join a National Military Force

In the twentieth century, women throughout the world are playing increasingly proactive roles as members of the military. Modern military forces require many more people to work at supplying troops and keeping records. In effect, a modern military force increases the total demand for personnel, while, at the same time, it reduces the percent of the forces who are combatants. This fact provides the opportunity for more women to serve in the military at times other than emergencies. Regardless of the level of economic development, nations, especially those with small populations, are opening military service to women, including such diverse nations as Algeria, Argentina, Brazil, Burma, Chile, China, Cuba, Ecuador, Egypt, Guatemala, Honduras, India, Indonesia, Ireland, Jordan, Kenya, Kuwait, Libya, Malaysia, Mozambique, Nicaragua, Nigeria, Pakistan, Palestine, Panama, Peru, the Philippines, Senegal, Sierra Leone, Singapore, Somalia, Sri Lanka, Sudan, Taiwan, Tanzania, Thailand, Turkey, the United Arab Emirates, Venezuela, Vietnam, Zambia, and Zimbabwe. Often, the training for the women is as physically demanding as that given the men. Most of the nations restrict the jobs of women to areas of noncombat, while an additional group limits women's activities to the medical or the civil-police corps. The actual number of women in the military services of any nation in the developing world is small, and the percent of the total force is often only a fraction of 1 percent (Isaksson, 1988).

Each nation has its own unique needs for military forces and responds accordingly. When deciding to include women as official members of their military forces, nations seem to consider the most recent experiences of women's service, the immediate need for soldiers, and the projected military needs more than political ideology and traditional practices. Even nations that have had socialist revolutions, complete with proclamations and laws supporting the equality of women, have reduced both the presence and positions of women in the military during periods of peace, seemingly contradicting revolutionary doctrine. Practices in China, Vietnam, and Kuwait serve to illustrate the variety of women's participation in the military in developing nations.

During the Chinese revolution, many women were combatants with the communist forces. As the need for military forces in China decreased after 1949, the roles and numbers of women in the Chinese military changed. Chinese women did take part in the Korean conflict, but only as cultural workers, nurses, doctors, and telephone operators. Today, Chinese women make up about 4.5 percent of the total military personnel. The military is a popular career choice for women because it opens opportunities for education, training, and better jobs in the future. Most Chinese women in the military serve in traditional female roles and support positions in headquarters, hospitals, research institutions, and communications facilities; there are no female ground-combat forces or pilots. After 1980, when ranks were instituted in the Chinese military, some women were awarded officer ranks (Li, 1994).

In Vietnam during the war, the governments of both North and South Vietnam officially provided ways for women to be members of the military. In the South, women served only in policing roles. In the socialist North, the entire range of military roles was open to women. Some women traveled down the Ho Chi Minh Trail to serve as combatants and medical workers, and even those serving in the medical corps were trained as fighters of last resort in an evacuation of troops (Haas, 1991).

After Kuwaiti women fought alongside men to defend and take back their nation from Iraq during the Persian Gulf War in 1991, Kuwait formally opened its military to women, as did the United Arab Emirates. The small areas and populations of these nations and the perceived need to have a defense force that frees as many men as possible for fighting have prompted the admission of women into the military, despite the fact that women are predominately limited to domestic roles in both countries.

Protester, Politician, Peace Activist

One way to keep from being a victim of war is to approach the problem before violence occurs. Women, especially those who have been the victim of war, are likely to speak from their sorrow and pain to demand the end of suffering for people. They seek to stop the killing and torture by military, police, and paramilitary groups alike—by protesting to draw attention to a problem, pressuring for reform from grass-roots organizations, or seeking office and use of the power of the office to bring about change. Nations that restrict the franchise deny women the election option.

Proactive peaceful protests are not without consequences, some of which are personally costly and violent to those seeking peaceful reform. Activism for peace tends to be a choice made by older, more educated, and more reflective women. Examples of women who are proactive for peace are found among all classes in many nations of the developing world. These women seek to care for the next generation by giving them a better quality of life that is free from the oppression of human rights and provides for justice and equality.

Organizations such as Women in Black protest violence between Israelis and Palestinians in demonstrations every Friday at major intersections in Israel. Early in their efforts, they held a three-day conference in Switzerland titled "Give Peace a Chance." Well-known women such as Israeli journalist Yael Dayon, Palestinian professor/politician Hana Mikhail-Ashrawl, and others voiced their concerns for the future of the people and continue to meet and seek long-term solutions to the problem of violence. Such conferences often provide the first opportunity for women from various groups to share their views and concerns. Hana Mikhail-Ashrawl explains that women are drawn together by their common concerns for human life and sustaining life, a caring attitude that seems to be different from the attitude of men who seek power (Mitchell, 1991).

Similarly, Protestant and Catholic women in Northern Ireland, who culturally have been encouraged to stay out of politics and follow the decisions of their men, join together to bring about grass-roots changes. The most famous of these groups is the Peace People, whose founders, Mairead Corrigan and Betty Williams, were presented the Nobel Peace Prize in 1976. Through the pain of the senseless death of a sister and a son, these women, like others, join forces to stress the need of poor people to work together to solve the common problems of poverty and deal with bread-and-butter issues and the role of education in removing hatred and bigotry. They want the next generation to create its own solutions to these problems (Mitchell, 1991).

Peace groups are not limited to those nations that allow and respect the freedom of speech. Women peacefully protest in nations where they know their actions may well lead to beatings, violence, and imprisonment when they express the need for peace and basic human needs. In El Salvador during the late 1970s and early 1980s, women organized large demonstrations demanding to learn the fates of their children and husbands and advocating an end to death squads, and they returned to protest even after members of their group were arrested (Mitchell, 1991).

During the Vietnam War, Vietnamese women of all political persuasions demonstrated against the various governments established in the South. Many were not members of the National Liberation Front and sought religious freedom, the right of opposition parties to run in the 1971 presidential election, and an end to political corruption and indiscriminate or unfair imprisonment. Madame Ngo Ba Thanh, the founder of the Third Force Women's Committee to Defend the Right to Live and a graduate of Columbia University Law School, was imprisoned and nearly died on a hunger strike (Haas, 1991). More Buddhist nuns than monks took their lives by self-immolation in protest of the Diem, Ky, and Thieu governments. The Reverend Mother Thich Nhu Hue, who, according to Buddhist procedures, had the power to give permission for self-immolations, said she granted permission to only a few of the many who made the request because such an act had to be a consciously chosen act of sacrifice by an adult who truly believed in reverence for life and not an act of an impulsive, radical youth (Haas, 1991). Violently taking one's own life for peace is done to shock people and force the guilty to think, which is why it is a powerful weapon for peace.

Against the Soviets in Afghanistan in 1981, women's protests were exceedingly costly in human life. Female students were killed by machine gunning as they demonstrated, chanting, "Allah-o-Akbar" (God is great). Demonstrations against the Soviet-supported government were mounted by people who had only their numbers and their voices as weapons, and thousands were shot as helicopters, jets, and tanks were used against them (Hayton-Keeva, 1987).

Civil wars and insurrections are the predominant form of warfare in the modern world among the developing nations. Under these conditions, large numbers of women—as high as 30 to 40 percent of guerrilla forces—are combatants. [In the final offensive in Nicaragua against the dictator Anastasio Somoza Debalye in 1979, 30 percent of the fighting force were women (Harris, 1988)]. When the struggles continue for years, the women often rise to leadership and command positions. They are dedicated to their cause and willing to do what is necessary to attain victory. They leave children and families for long periods of time, lead double lives, are imprisoned and tortured, move to remote regions and live under difficult physical conditions, and commit acts of terrorism. They also become political activists, propagandizing, educating, and recruiting new members. What prompts women to begin such lives is often a belief that it is the only way to attain justice and equality in a nation where only a very few have wealth and power and suppress the vast majority of the population. Women who become revolutionaries, like those who become peace activists, are often the victims of poverty and personal injustice. For some, it is the culmination of a childhood that has trained them to assume the role of the next generation of fighter. For others, the awareness of injustice comes later in life, and the decision is finally made out of frustration with a government that seems not to care and is riddled with corruption and bribery.

For the guerrilla fighters, the end does appear to justify the means. The fears and feelings of doubt and guilt they encounter on their first mission tend to leave as they are reinforced by satisfaction and the success of each new mission. When difficult situations develop during missions, the women act to protect the immediate needs of their new friends and concentrate on the end goals of their struggle. Revolutionary women include the poor, members of the middle class, and, less frequently, members of the very wealthy class. Older women also become proactive revolutionaries, but they are more likely to serve the cause by caring for the children and the injured and by cooking and washing for the fighters than by actually fighting.

During the Vietnam War, women were encouraged to become proactive fighters, and a number of different forces were established that provided them with the opportunity to serve in a variety of locations. Women constituted one-third to one-half of the troops of the Communist Peoples Liberation Armed Forces (PLAF) and 40 percent of its regimental commanders. All women volunteered for membership in the PLAF. When not in combat, they assisted in harvesting, medical training, and building projects in the South to help win converts to the cause. In the regional fighting units, the percentage of women who were full-time fighters was higher than for the PLAF in general. Another group of women were members of the militia, or local self-defense units, fighting only when their area was attacked. It was the responsibility of the militia women to keep the villages fortified with trenches, traps, and spikes. Perhaps the most widely known female fighter was General Nguyen Thi Dinh, who joined the anti-French resistance forces at the age of 17 in 1937 and eventually became the deputy commander in chief of the PLAF. General Dinh developed and used tactics that became widely adopted by the PLAF. Her tactics combined both military and political actions. Political demonstrations were organized at locations and times that resulted in blocking the path or diverting troops so that a small guerrilla force could successfully attack military targets (Haas, 1991).

Interviews with women who have experienced conflicts and war reveal the typical training, dedication, and trials of proactive female guerrilla fighters. Nuha Nafal, a politically passionate member of the Palestine Liberation Organization (PLO), speaks as one born a member of the political-opposition group (Hayton-Keeva, 1987). She resents the term "terrorists" being applied to forces that she sees as seeking legitimate rights. She points to the continuing suffering of mothers and sisters after the death of male family members, the many Palestinian women jailed as political prisoners since 1967, and the fact that, since 1948, Palestinian women have fought side by side with men. She declares her intent to return to the Middle East with her sons and daughters when they are old enough to fight and to fight with them. She says that her hair, body, and language are Palestinian, and she wants her death to also be Palestinian. Karla Ramirez, who grew up in San Salvador not knowing the poverty of most Salvadorans until she was in high school, illustrates a woman's gradual conversion to fighter (Hayton-Keeva, 1987). While in college at the age of 18, she made the decision to work to bring a decent life to her people and started to learn about and assist the Popular Revolutionary Bloc (PRB). She describes seeing the bodies of friends and relatives who were tortured and killed by the military forces. Eventually, she feels compelled to move to the mountains and join a group of about 950 PRB *compañeros,* 250 of whom are women. She explains her training and feelings concerning fighting. It is difficult to kill. At times, she cries and prays, asking God to forgive her actions since they are performed out of desperation with a government that will not reform and work for all of the people. She explains that, in thinking about her murdered friends, the safety and needs of her *compañeros,* the people waiting to be liberated, and, most of all her son, she is able to morally justify her killing for a greater good. Ramirez believes that her feelings are shared by most women and explains that she is not afraid to die for her beliefs because, in dying, she would leave her heart to the people and to her child.

Conclusion

The concept of "developing nation" is basically an economic one, measured by such variables as per-capita income, gross national product, and the use of electricity. Colonialism, with its goal of a one-sided economic and territorial control, left most of the former colonies with economies that continue to be exploited for the benefit of just a few people, the former colonial nation, or the developed world as a whole. Their raw materials continue to be sent to nations whose economies are much more powerful and whose producers and consumers seek to pay the lowest prices. The labor force of developing nations is exploited, too. Most developing nations are products of the collapse of colonialism and the rise of ethnic nationalism and democracy that have accelerated since World War I. Two important characteristics of many of these nations, their physical borders and governmental structures, are outgrowths of the actions of the colonial powers. Often, these remnants of colonialism have not served the people well, and many groups within the borders of nations have sought, and continue to seek, to change them.

In the process of controlling a region, colonial powers destroyed the existing social systems and much of the traditional culture. People were told that their ways were inferior or wrong and were compelled to adopt the practices of the colonial people. A colonial education system selected the most promising students and trained them in the new language, religion, and history that supported the colonial economy. This educated group took positions in government and business, settling in urban areas, while the majority of people remained subsistence farmers. Thus, at independence, many of the new nations were governed by people who, while nationals by birth, tended to be colonials in their values and behavior. Such a situation is potentially destabilizing because it accentuates differences among the socioeconomic classes. Coupled with the multiethnic characteristic of many of the nations, this creates a situation in which democracies are replaced by dictatorships or the military takes control to prevent or stop unrest.

Returning nations to civilian control fails more times than it succeeds; the result is that many developing nations are controlled by those who use military and police force against their own citizens. This situation does not create loyalty to the central government; rather, it encourages people to give their loyalty to other groups or causes. Thus, it is not surprising that many nations are engaged in ethnic class strife or are constantly on the verge of disintegrating into chaos. Such situations impact women greatly, making them and those for whom they care victims of economic and political oppression. As we have seen, such conditions give rise to political unrest, oppressive reactions, and revolt. Women find that, in such situations, protesting groups and revolutionaries welcome them as active leaders and supporters more readily than do the institutional governments, with their colonial values of mercantilism and male superiority and dominance. If this scenario continues in the developing world, then more nations will experience civil revolts, and women will become increasingly more proactive in their roles in these revolts. In examining women's choices to work for peace or to become warriors, Boulding (1988) concludes that for highly motivated social activists there is a fine line between those who select violent and nonviolent actions. Since history shows that nonviolent peace activism has done little to improve the economic position of the poor, a greater number of women could be expected to select the more violent proactive roles.

If, on the other hand, the politically powerful are sincere in bringing about democratic reforms in government and successful in improving the lives of average people, thereby creating a better situation for children and the elderly, women may well work in the more peaceful proactive roles and be supporters of the government and reforms that improve the lives of all people. Then only a few women will assume proactive roles in a professional military as needed by their government, and the victim and reactive roles will be greatly reduced or eliminated. If a reform government fails to improve the condition of its citizens, then women will not continue to support what they consider a failing reform government. Women, in even larger numbers than in the past, will assume the proactive roles to gain their goals and the number of women in the victim and reactor roles will continue to be large.

References

Boulding, Elise. "Warriors and Saints: Dilemmas in the History of Men, Women, and War." In Eva Isaksson (ed.), *Women and the Military System.* New York: St. Martin's, 1988, pp. 225–246.

Haas, Mary E. "Women's Perspectives on the Vietnam War." In Jerrold Starr (ed.), *Lessons of the Vietnam War.* Pittsburgh: Center for Social Studies Education, Unit 9, 1991, pp. 1–32.

Harris, Hermione. "Women and War: The Case of Nicaragua." In Eva Isaksson (ed.), *Women and the Military System.* New York: St. Martin's, 1988, pp. 190–207.

Hayslip, Le Ly, with Jay Wurts. *When Heaven and Earth Changed Places.* New York: Plume, 1990.

Hayton-Keeva, Sally. *Valiant Women in War and Exile: Thirty-Eight True Stories.* San Francisco: City Lights, 1987.

Isaksson, Eva. *Women and the Military System.* New York: St. Martin's, 1988.

Li, Xiaolin. "Chinese Women Soldiers: A History of 5,000 Years," *Social Education,* vol. 58, no. 2, February 1994, pp. 67–70.

Mangold, Tom, and John Penycate. "Vo Thi Mo: The Girl Guerrilla." In Tom Mangold and John Penycate, (eds.), *The Tunnels of Cu Chi.* New York: Berkeley, 1985, pp. 228–240.

Mitchell, Pat. *Women in War: Voices from the Front Lines,* vols. 1–2. New York: Filmmakers Library, 1991.

Nordstrom, Carolyn. "Women and War: Observations from the Field," *Minerva: Quarterly Report on Women and the Military,* vol. 9, no. 1, Spring 1991, pp. 1–15.

Segal, Mandy Wechsler, Xiaolin Li, and David R. Sigler. "The Role of Women in the Chinese People's Liberation Army," *Minerva: Quarterly Report on Women and the Military,* vol. 10, no. 1, Spring 1992, pp. 48–55.

Walker, Keith. *A Piece of My Heart.* New York: Ballantine, 1985.

Gender Justice

Suma Chitnis

Introduction

Modern societies have depended on legislation as a powerful instrument of social justice and reform. Human-rights activists and social reformers all over the world have consistently struggled to obtain legislation to support and secure their demands. The ultimate step in this direction is to have the desired rights and reforms incorporated into the constitution by which the country is governed. Underlying this confidence in legislation and in the provision of constitutional rights is the faith that rights supported by law and provided for in the constitution are defended by governments, particularly by governments considered to be responsible to the people they govern.

As the struggle for gender justice gains strength, and measures for the institution of equity and justice for women are instituted around the world, some global patterns in the persistence of the subordination of women are beginning to surface. One of these is the fact that the assertion of ethnic identities and the emergence of religious fundamentalism almost invariably lead to a revival of traditions and practices that restrict and subordinate women. Typically, practices such as female circumcision, or the use of the *chaddor* (veil), are revived—probably because their powerful image has a value in underlining ethnic identity or the authority of religion. In such situations, demands for gender justice, as defined by rational, secular, and humanist norms, are rejected as irreligious, and the revival of traditions that restrict women is upheld in the name of God. Thus, the struggle for gender justice gets distorted into a contest between the sacred and the profane.

This discussion of gender justice takes as its key referent India, a nation that places great importance on religion within its culture. In India, legislation has been used extensively to change, even eradicate, oppressive traditions that deny liberal, democratic, and egalitarian values and expectations. This strategy of using legislation to bring about social reform and social justice started toward the end of the eighteenth century with the introduction of the

British legal system under colonial rule. It is significant that initially it was used primarily in the interests of gender justice. Moves for legislation, aimed at obtaining justice for women, came from educated Indians. Influenced by European values and liberalism, they were appalled by oppressive Hindu customs and traditions such as child marriage, the practice of female infanticide, *sati* (the self-immolation of widows on the funeral pyres of their husbands), and the denial of the right of remarriage to widows, so they reached out to legislation as an instrument for reform. The British were not particularly willing to concede the demands for legislation made by these reformers. Unlike some others who colonized in order to establish and extend the supremacy of their religion, the British had come to India primarily as traders. Conscious that interference with existing cultural and religious practices would offend their subjects and create situations that would adversely affect their commerce and trade, the British were careful not to intervene even when traditional practices were grossly at odds with the liberal values they advocated and upheld. In fact, they went to the extent of curbing missionaries who were aggressive in the propagation of Christianity. Toward the beginning of the nineteenth century, the reformers had persuaded the British government to legislate against *sati* (1829). From then on, a number of laws aimed at improving the status and rights of women were passed.

After India acquired independence from British colonial rule in 1947, the equality of women as citizens was underlined in the country's constitution, which identified women as "a weaker section of society," traditionally marginalized and denied opportunities, and, therefore, eligible for special assistance and support for advancement. This provision was made with a view to ensure that affirmative action on behalf of women would not be hindered by the criticism that special privileges were unequal. In the decades since independence, there has been a great deal of legislation aimed at improving the status and situation of women. It covers matters as diverse as marriage, divorce, family support, adoption, property rights, dowry, domestic violence, rape, prostitution, "indecent" representation in advertisements, sterilization, sex-discrimination tests, education, employment, and conditions at the workplace. The decade of the 1980s stands out as the golden era, with laws aimed at giving women equal rights.

The Distance Between Law and Reality

While India's track record with respect to legislation toward gender justice has been impressive, Indian experience reveals factors that render much of the legislation ineffective. To start with, it has been found that laws are ineffective unless they are clearly focused, unambiguously phrased, and purposively geared to the objective for which they are instituted. The point may be illustrated with reference to the practice of dowry—the custom by which the parents of a girl are expected to give the bridegroom and his relative large gifts at the marriage and to continue to do so afterward. This practice is the source of tremendous oppression of women. Brides are routinely harassed by their husbands and in-laws with the criticism that their dowries are inadequate and poor. They are continuously ill named, even tortured, with demands to bring more. This torture often leads to suicide and, not infrequently, murder, resulting in what are known as dowry deaths.

As may be imagined, there has been strong and persistent agitation for legislation against this practice. In 1961, the Dowry Prohibition Act was passed, but it is full of loopholes and contradictions. Its basic flaw is that it lays down a narrow definition of dowry as "property given in consideration" of marriage and as "a condition of marriage taking place." It excludes "presents" in the form of cash, ornaments, clothes, and other articles. It does not cover money asked for and given after marriage.

Similarly, the efficacy of legislation against rape is seriously affected by the fact that it leaves ample room for arguments over whether the victim "consented" to intercourse or was "submitted" to it, whether there actually was penetration of the vagina or whether what happened was "merely a violation of modesty." Legislation against rape was first enacted in 1860 as Section 375 of the Indian Penal Code. The latest amendment to this law, passed in 1983, continues to contain many loopholes.

Careful examination of reform legislation suggests that the weakness in the legislation is often due to the fact that those who legislate are not necessarily committed to the cause the legislation is expected to serve. Consequently, the instrument they produce lacks sharpness. One often senses a measure of tokenism in social-reform legislation; it is almost as if the bodies empowered to legislate use legislation for political mileage—a measure with which to gain popularity or silence opposition.

On the other hand, the best of reform laws may become ineffective because of poor implementation. For instance, India legislated against the practice of female infanticide in 1870, more than a century ago, but in some parts of the country the law was never implemented. Conse-

quently, the practice continues to survive. So, too, the practice of child marriage, which was banned by law in 1929.

Poor implementation, in turn, may occur either because of sheer inefficiency or inadequacies in the administration or because those who are responsible for the interpretation or administration of laws are not adequately committed to the other values the laws seek to implement or uphold. The latter has been heavily responsible for the limited efficacy of legislation aimed at gender justice in India. The weight of tradition is so strong that neither the women who are wronged nor their close relatives seek legal redress easily. Worse yet, conservative judges—firmly committed to traditional notions about a woman's virtue or a woman's place—interpret the behavior of women who are victimized or violated, and of those who victimize or violate them, in ways that perpetuate oppressive traditions and negative concepts of gender justice (Sathe, 1995a).

Taking their cues from this, social-reform activists have begun to realize that it is not enough to organize action for the enactment of legislation. Rather, organized pressure for gender justice has to be used at every point to ensure that (1) injustices against women are reported and legal action is initiated, (2) the judgments are fair, and (3) the judgments given are eventually honored by the full imposition of the sentence decreed. Above all, activists and others have recognized that legislative action by itself cannot dislodge deep-rooted traditions and customs; it has to be backed by a firm effort to change.

The Growth of Religious Fundamentalism

As activists fight the obstacles named above, a new phenomenon has emerged: the growth of religious fundamentalism. Since it and its impact on politics and government are a global phenomenon, the Indian experience should be relevant to other countries. To understand this experience, it is helpful to know the basics of the shape of religious plurality in India, how the British government handled it, and the implications for independent India.

Religion in Politics and Government

The majority (83 percent) of Indians are Hindus. Muslims (11 percent), Sikhs (2 percent), Christians (3 percent), Jains (0.5 percent), Buddhists (0.7 percent), and "others" (0.5 percent) constitute significant minorities. Although these percentages are small, they add up to substantial numbers because of the size of the Indian population (nearly one billion). These religious communities have coexisted in India for centuries. Each has made a distinctive contribution to the country's variegated culture. Their diversity has also been the source of conflicts at the local, regional, and national levels.

As far as they could help it, the British did not interfere with the religious practices of their Indian subjects. But they were shrewd administrators, and, as their rule in India stabilized, they consciously moved to use India's religious plurality to implement a policy of divide and rule. They used available religious denominations to classify the population for administrative purposes. This was done and emphasized in such a manner that religious divisions grew to be deep and violent.

The British had tried to introduce Anglo-Saxon jurisprudence in India. Beginning in 1858, under what was known as the Queen's (Queen Victoria's) Proclamation, the British Parliament initiated a process of law reform for India by administering such codes of law as the Indian Penal Code, the Indian Contract Act, the Criminal Procedure Code, the Civil Procedure Code, the Indian Evidence Act, and the Transfer of Property Act. These were uniformly applicable to all Indians. However, in matters relating to adoption, marriage, separation, divorce, family maintenance (obligations similar, but not identical, to alimony), and intestate succession, which they referred to as matters of "Personal Law," the British decided to administer justice on the basis of the religious laws and practices of the different religious communities of the country. Most of the religious laws covered by the Personal Law are patriarchal and disadvantageous to women. For instance, they permit polygamy and unilateral (male) right to divorce and allow women only very limited rights of inheritance.

Prior to British rule, the religious laws at least had the merit of being administered in a manner that allowed their adaptation to changing societal needs. Local religious bodies and community authorities were responsible for the interpretation of justice. To legitimize their roles as interpreters and sustain their authority, these administrators had to be flexible and dynamic in the interpretation of religious codes. Although the British had decided to honor the laws of India's different religions in their administration of the Personal Law, they entrusted the administration of justice entirely to the British courts. Far more centralized than the local bodies, these courts did not have the same sensitivity to local traditions, customs, and changing needs, nor did they allow themselves the same freedom and flexibility in interpreting the religious laws. Instead, they emphasized basic religious texts in a purist fashion. Inevitably, Personal Law lost its dynamism and became rigid and stagnant.

This movement had been taking shape from what is known as the Sepoy Mutiny of 1857, but it acquired a firm identity with the establishment of the Indian National Congress in 1985. It is commonly accepted that the British cleverly used the religious and other differences between their Indian subjects to implement a policy of divide and rule. This policy certainly helped in stalling the independence, but the nationalists finally succeeded in wresting political freedom from the British in 1947. However, the policy resulted in lasting damage to Indian unity. By the time the country acquired independence in 1947, the division between the Hindus and the Muslims had become so irreconcilable that the country had to be partitioned to create Pakistan as a separate Islamic state. Partition, accompanied by severe violence and dislocation of populations, was a traumatic experience. India came out of the trauma as a democratic state firmly committed to secularism and to restoring the confidence of the minorities that had been shattered by events accompanying partition. Thus, not only does the constitution of independent India guarantee full freedom in the practice of religion, it further accords religious minorities the right to practice their cultures, with an assurance of state protection.

The Politicization of Minority Rights

In independent India's democratic government, these minority rights have, from the beginning, become a politically contentious issue. As political parties as well as the government woo the minorities for political support, some sections have found it advantageous to assert their religious identities for political gain. In retaliation, a section of the majority Hindu community has started a countermove to assert its Hindu identity. Over the course of the last decade these moves have intensified. That this is less a matter of religious faith and more a matter of using religion for political power is evident in the fact that, where politics within the majority community becomes factional, the factions assert their religious identities in terms of sectarian Hinduism. The constitutional commitments to secularism as well as to gender justice and to the advancement of women are being set aside in the interests of political expediency, even by the state. Unfortunately, this is facilitated by a structural flaw in India's legal system whereby the traditional codes of its different religious communities continue to guide the Personal Law.

The Demand for a Uniform Civil Code

Undeterred by the attitude of the British, Indian social reformers and nationalists pressed for reform on behalf of women. In 1930, the All India Women's Conference demanded a Uniform Civil Code to replace the religious codes by which the Personal Law was administered. Soon after, this demand was placed as a priority issue on the agenda of the Indian National Congress (INC), but the division between Hindus and Muslims hampered support for this move. Within the INC, some Hindus apprehensive that a Uniform Civil Code might be too radical because it called for reforms in the personal laws of all the different religions practiced in the country, suggested that the plans for this code be temporarily kept in abeyance and a less ambitious objective taken in hand. They pointed to the many regional and other variations in the Hindu Personal Law, and suggested to start by developing a Common Code for all Hindus. Thus, by the time India acquired independence, the move for a rational, secular, Uniform Civil Code was already in trouble. Reformers had hoped that the constitution of independent India would nevertheless provide for it, but it didn't: Opposition from a section of the Muslim minority helped place it in Part IV of the constitution as a "directive principle" of state policy. Directive principles are expected to influence state policies of governance and legislation, but they are not enforceable. During the first few decades of independence, the move for a Uniform Civil Code assumed lower priority than other competing issues of development. The growth of religious fundamentalism and the misuse of minority rights by vested political interests combined to convert the issue of gender justice into a pawn in political dealings between the government, religious authorities, and politicians. The Shah Bano case and the *sati* incident described below illustrate this point.

The Shah Bano Case

In 1975, Shah Bano, the 65-year-old wife of a successful Muslim lawyer, was forced to move out of their home because of marital conflict. For the next two years, her husband paid her 200 rupees per month for support, then he stopped. Unable to support herself, in 1978 Shah Bano sued him for maintenance under Section 125C of the Criminal Code, which was intended to prevent vagrancy due to destitution. While Shah Bano's application for maintenance was being processed, her husband divorced her, using the triple *talaq* (the provision by which a husband can terminate his marriage contract by saying "talaq"—which means that the contract is terminated—three times in a row to his wife). In keeping with the requirements of the *Shariat* (the religious code of the Muslim), he also placed 3,000 rupees in court toward payment of the *mehr* (the amount that, as a part of the marriage contract in a Muslim ceremony, is accorded as maintenance for the wife), agreed upon at the time of their marriage.

Meanwhile, the magistrate ruled that Shah Bano was entitled to maintenance, but he fixed the amount at 25 rupees per month (approximately one U.S. dollar); Shah Bano appealed to the High Court, which raised the amount to 179.20 rupees per month, a figure closer to the 200 rupees her husband had paid each month for the first two years.

Upon this, Shah Bano's husband appealed to the highest legal authority, the Supreme Court. His argument was that, now that he had divorced Shah Bano, the matter under dispute was an issue of maintenance for divorce; it fell under Muslim Personal Law and it could not be treated as a criminal case of destitution under Section 125C of the Criminal Code, as the high court had done. He produced a statement from the Muslim Personal Law Board to support his argument that the *Shariat* does not require the husband to pay maintenance for more than three months (the period of *iddat*) after divorce. Thus, by the time it came before the highest judicial authority in the country, the case had been removed from the space available to Muslim women under criminal law and placed squarely in the domain of Muslim Personal Law. Further, it was being asserted that Muslim Personal Law did not require a husband to pay maintenance to his wife for more than three months after giving her a divorce.

Shah Bano's counsel produced two verses from the Koran to prove that the provision of maintenance was considered "the duty of the righteous." On the basis of this evidence, in 1985 the Supreme Court upheld Shah Bano's right to maintenance.

If the judgment had stopped there, the issue would have concluded within the boundaries of Muslim law. But because Shah Bano's husband had raised the issue regarding whether or not her case could be tried under Section 125C, and probably because the Supreme Court wanted to emphasize the country's secular obligations, the judgment went on to affirm that Section 125C of the Criminal Code "cuts across the barriers of religion and transcends Personal Law." It firmly criticized the way women had been treated in both Hindu and Muslim law and urged the government to enact a Uniform Civil Code. It went on to say that the establishment of such a code was essential for national integration (Kishwar, 1985).

The ruling angered Muslims, who thought that it transgressed minority rights. For instance, although the judgment criticized both Hindu and Muslim law, the wording made it easy for Muslims to interpret it to mean that the judges were saying not only that the Muslim law was bad, but also that the Muslim community preferred unjust laws and that the state would, therefore, have to "impose" justice on the Muslims. Muslim religious leaders proclaimed that the judgment was against their community. On the other hand, Hindu chauvinists applauded it, arguing that it supported their contention that Muslim law was unfair to women. At the same time, feminists and secularists were upset that the court had not had the courage to judge the issue as a matter of criminal law and that it had entertained Shah Bano's husband's appeal that the matter be judged under Muslim Personal Law.

In the seven years after Shah Bano had first petitioned for maintenance in 1978, her case had received considerable publicity and generated sympathy from feminists, human-rights activists, the press, and others. Secular Muslims were in the forefront of those who supported her. In at least two cases similar to Shah Bano's, the courts had decreed payment of maintenance. This was accepted by the Muslim community without protest. Shah Bano could have received simple justice as well, but the political climate of the country had been completely vitiated by communalism by the time her judgment was delivered, and the Supreme Court used it as an opportunity to drive home some strong points about secularism, national integration, and a Uniform Civil Code.

The judgment was delivered in 1985, five years after the assassination of Prime Minister Indira Gandhi by a Sikh triggered a steep rise in communal riots. After the assassination, anti-Sikh communal sentiment ran high. A major Hindu fundamentalist political party took advantage of this situation to attempt to displace the ruling Congress Party. The leaders of the fundamentalist party decided that the best strategy was to further whip up communal sentiments and make it difficult for the Congress Party to govern. Having decided upon this course, they discovered that the easiest way to draw the majority Hindu masses into their party was to promise them that Hinduism would be restored to its pristine glory if their party was brought to power. With this new emphasis on power for Hinduism, old communal disputes reappeared and conflicts flared up.

Babri Masjid and Ram Janmabhoomi in Communal Opposition

One such conflict related to a nineteenth-century dispute over the site of Babri Masjid, a Muslim mosque in the city of Ayodhya, the legendary home of the favorite Hindu deity, Ram. For many years, Hindus had complained that Muslims had vandalized the ancient Ram temple in Ram Janmabhoomi (Ram's birthplace) at Ayodhya and had built Babri Masjid over the ruins. They said that remains of the Ram temple were visible, claimed the right to worship at the site, and demanded permission to rebuild the temple. The Muslims resisted. In 1984, while the court had yet to deliver its judgment in the Shah Bano case, the Vishwa Hindu Parishad (VHP), a Hindu communal organization, led a 200,000–strong march to Ayodhya to liberate the shrine. By the time the Supreme Court issued the Shah Bano judgment in 1985, the VHP had been performing fire rituals all over the Hindi-speaking belt of the North to mobilize support for the demand that a shrine be built on the site of the mosque. Alarmed by this, the Muslims formed the Babri Masjid Action Committee to defend the status quo.

In the heat of these events, the Shah Bano judgment came to be seen as communal, and in favor of Hindus. Muslim *ulemas* (scholar priests) issued a widely publicized *fatwa* (proclamation) that the judgment was against the teachings of Islam. Syed Shahbuddin, a leader of the Babri Masjid Action Committee, launched a petition campaign against the judgment, quickly garnering 300,000 signatures. Within days after the *fatwa* was issued and Syed Shahbuddin's petition drive began, communal tensions intensified into a national upheaval.

In August 1985, a Muslim League member of parliament (MP) named Banatwala introduced in parliament a bill that sought to exclude Muslim women from Section 125C of the Criminal Code. The Congress Party government decided to oppose the bill and briefed a Congress MP, Arif Ahmad Khan, a Muslim, to argue against it in parliament. In an effort to find his own compromise between his loyalty to the Congress Party and to his religion, Ahmad Khan instead pleaded for a humane interpretation of the *Shariat.* In December 1985, Syed Shahbuddin, encouraged by the success of his signature campaign, stood for election to the State Assembly. He defeated his Congress opponent despite the fact that the latter was a Muslim, the secretary of a powerful organization of the Ulemas, and had had 200 *ulemas* canvassing for him.

In early February 1986, the magistrate before whom the VHP had first placed its petition claiming right to worship at the shrine within the precincts of Babri Masjid declared that the shrine be opened to Hindus for worship. While the VHP celebrated this with victory processions, Muslims held mourning processions. Riots between Hindus and Muslims spread to, and escalated in, different parts of the country and spilled over to Pakistan. Alarmed by these events, the government retreated from its earlier assurance that Muslim women would continue to be covered

by Section 125C of the Criminal Code; it announced that it would consider reviewing the judgment and would introduce a bill in parliament similar to the lines of the Banatwala bill.

There was widespread criticism of the announcement that it amounted to a declaration that the government was willing to sacrifice the rights of Muslim women in order to save itself. This attitude was no different from, and perhaps worse than, that displayed by the British government in the matter of legislative reforms on behalf of women during colonial rule. Since a government bill carries somewhat greater weight than a private bill in the Indian system, Banatwala withdrew his bill to make way for the government bill. On February 25, 1986, a bill, somewhat misnamed as the Muslim Women's Protection of Rights to Divorce, was introduced. It stated that a Muslim husband's obligation to maintain his wife ended within three months of *iddat* and that the woman's family (presumably parents) or the *wakf* (Muslim religious trust) would thereafter be responsible for supporting her.

There was massive opposition to this bill in the press, public meetings, demonstrations, protest marches, and lobbying, but it never came together as a united front. Protests were organized separately by various groups, representing a plurality of loyalties, viewpoints, and interests. Efforts were made to bring all of these groups together into a coalition for a combined rally on March 7, 1986, a day before International Women's Day, but the coalition was loose and short lived. Moreover, the women's groups of the two main Communist parties had a separate coalition and a separate demonstration later, on April 18.

Quite apart from the fact of competing loyalties, however, there were different views on how the issue was to be resolved. A view widely held was that the kind of maintenance to be made available to women would differ according to the social class, caste, and community to which a woman belonged. This implied that women were not a single category for whom basic human rights could be unequivocally defined; it was a major setback for gender justice.

Many Muslims who had been neutral on the issue of reform in Muslim Personal Law, and some who had strongly supported a Uniform Civil Code, now became ardent supporters of the bill. As Muslims of different shades of political commitment thus came together to support the bill, they started to organize themselves on communal lines, regardless of whether they supported the bill or opposed it. For instance, having decided to restrict its membership to Muslims, one of the most prominent of these organizations, the Committee for the Protection of the Rights of Muslim Women, specifically formed to oppose the bill admitted Muslim men but excluded non-Muslim feminists. The issue of gender justice came to be drowned under other issues—secularism, fundamentalism, and the obligation to uphold Asian values vis-à-vis the onslaught of Western feminism.

In the midst of the communal tensions generated on

its account, the bill was passed in parliament on May 6, 1986, after a night-long debate and became the Act for Muslim Women's Protection of Rights to Divorce. The Congress Party had instructed all of its Members in Parliament to vote for the bill; it was the first time in the 40-year history of democracy in India that the ruling party had taken such a step.

Two Concepts of Secularism

On the issue of secularism, the government took shelter in the fact that the connotation of the term had never really been specified for official purposes. Liberal Indians had generally assumed it to mean that religion should govern only the relationship between an individual and his or her god and not the relationship between man and man, man and woman, or woman and woman. The 1986 act does not fit in with this liberal concept of secularism. Rajiv Gandhi, who was prime minister of India when the bill was passed, announced that the state did not support this concept either. He firmly proclaimed the alternative view that secularism is the right of every religion to coexist with another religion, and stated that the government acknowledges this by allowing every religion to have its own secular laws.

Although the discrepancy between the passing of the bill and the constitutional commitment to secularism may seem to be resolved by this explanation, the prime minister's statement reaffirmed the British practice of religion-based family laws and ran counter to Articles 13, 14, and 15 of the Indian constitution, which guarantee all Indians equality before the law and prohibit discrimination on the basis of sex, religion, or race. It also made mockery of the fact that the constitution recognizes women as a "weaker section of society," takes full cognizance of their subordination through oppressive customs and traditions, and offers them protection with the promise that the state will make special efforts to enable them to advance toward equality.

By supporting the bill and indicating that religion could define relationships within the state, a government that claimed to be secular and committed to gender justice and the advance of women had yielded to religion. This opened the floodgates for religious fundamentalism and for the continued exploitation and marginalization of women in the name of God, as may be seen from the sensational case of Hindu fundamentalism described below.

Sati and Its Practice

One of the most outstanding features of the polytheistic Hindu religion is the importance it accords to goddesses. Archaeological evidence traces worship of the mother goddess in India far back to the Mohenjodaro civilization that flourished in the third millennium B.C. This respect for womanhood is combined with reverence for women as mothers and extremely demanding ethical norms and expectations regarding women as wives. Although marriage rites emphasize the partnership between husband and wife, a woman is expected to regard her husband as her lord and master and to serve him as her deity. The epitome of this concept of a virtuous wife is the *sati*—the wife who, upon widowhood, voluntarily immolates herself on the funeral pyre of her dead husband. (The term also applies to the act of self-immolation itself.) The courage for this extraordinary act is supposed to come with the *sat* (grace), which, upon the death of her husband, enters a wife who is truly virtuous. One of the most alarming features of the politicization of religion in India is the attempt to revive this custom, which was abolished in 1829.

Sati's aura came from the fact that it was so heroic as to be exceptional. Rules on how *sati* was to be performed were clearly specified. Among them was the strict requirement that the act be purely voluntary, performed by the widow in full possession of her judgment, and not under the influence of drugs or hypnotism. A *sati* was expected to walk up to, and mount, the funeral pyre on her own. As may be expected, a woman who performed *sati* was not only respected but deified. A temple was built, or at least a memorial stone placed, at the site where she had performed the act of self-immolation, marking the site as a sacred spot for worship.

The practice of *sati*, though infrequent, was found all over the country. But, in the state of Rajasthan, it took a distinctive form as the wives of Rajput princes and nobles, defeated in war, performed *johar*—or *sati* en masse—choosing death to being captured by the enemy. Worship at *sati* temples continues to be a common practice in the state of Rajasthan, and once a year Rajasthani women celebrate an annual *sati* festival.

The rich and vivid descriptions left behind by some British officers reveal how awesome and emotionally overpowering an act of *sati* could be. They also indicate the immense conviction, courage, and confidence with which widows performed the act, and the tremendous faith and sense of devotion with which their relatives and others supported them. At the same time, these descriptions also reveal occasions on which the act was far from voluntary, when helpless women stupefied by drugs, chantings, and exhaustion were literally pushed to their death. Not infrequently, this was done with a view to clearing the path of another for the right to succession. Some widows performed *sati* to escape the harsh treatment meted out to widows in those times.

Records show that, although they were horrified by the practice since it was first reported to them in 1787, the British authorities did not venture to interfere with *sati* until 1805, when the governor-general of India finally sought advice on whether the practice could be abolished, whether the administering of intoxicants to widows could be prevented, and whether women who were not capable of deciding for themselves could be forcibly rescued. He was advised that the practice could not be abolished because it was founded on the religious texts of the Hindus. However, it could be prevented if the woman was pregnant, under puberty, the mother of an infant for whom there was no one else to care, in menstruation, or stupefied with drugs. Above all, the texts clearly mentioned that the act had to be completely voluntary. The governor-general took seven long years to act upon this information: In 1812, he sent out a circular stating that, like all religious practices, *sati* should be allowed where countenanced by religion but would be considered illegal if the woman was (1) under compulsion, (2) intoxicated, (3) pregnant, (4) or had a child under 5 years old for whom there was no one else to care.

This 1812 circular was not particularly effective. During the years 1813–1816, only 10 illegal *satis* could be prevented out of a total of 400 that occurred in the presidency of Bengal. Among the illegal *satis* were six girls between the ages of 8 and 12. Disturbed by this, the British government decided to seek fresh advice and to reconsider its policy. In 1817, the body appointed for the purpose made new probes and came up with the bold conclusion that *sati* was *not* sanctioned by the *Shastras* (the religious authority of the Hindus). It quoted eminent religious authorities such as Pandit Mrytyunjaya Vidyalamkara, chief Pandit of the Supreme Court, to prove that the practice was merely a custom that had acquired the sanctity of a religious rite. It advised that it would not, nevertheless, be appropriate for the British government to stop the practice outright, but, on the basis of its findings, it suggested some important administrative measures: (1) that the circumstances under which *sati* was to be considered illegal be extended; (2) that relatives of the widow who intended to commit *sati* be required to give notice of her intentions to the police; (3) that they be required to prevent the act if the widow did not meet the requirements of legal *sati* and that they be made liable to criminal prosecution if they failed to do so; (4) that the police be required to read out all relevant rules to the widow who had decided to perform *sati* and that they do so in the presence of witnesses. It further recommended that protection be provided to the widow who withdrew at the last moment and that relatives who subsequently maltreated her be punished; (5) that any property that may have devolved on a widow after her husband's death should go to the government if she was burned or buried alive. The government implemented all but the last two recommendations, rejecting those on the ground that it was unwilling to interfere with the family life and property of the widow.

Declaration of *Sati* As a Crime

As though in protest of the government's revised policy, the incidence of *sati* increased visibly, from 442 cases in 1816 to 839 in 1818. Social reformers pointed to this fact and pleaded for more rigorous legislation, without much success, for 10 years. Finally, in 1829, after a great deal of consultation within and outside government, the Sati Abolition Act was enacted. It unequivocally declared the practice of *sati* illegal and punishable by criminal courts.

As anticipated, the Hindu orthodoxy objected to the act. They sent a petition to the governor-general in India and to the British Parliament. In 1830, orthodox Hindus in Calcutta formed the Dharma Sabha, a religious organization, to campaign against the act. In 1932, the Privy Council dismissed the petition. The orthodox kept up their protest, and, finally, in 1846 the British government bowed to these pressures and revised the act in such a manner that the distinction between voluntary and forced *sati* was reintroduced. It justified this step as an expression of its commitment to freedom of religion for its Indian subjects. In 1860, the age of consent for voluntary *sati* was placed at 18 years.

Thus, the legislation against *sati* remained somewhat unsatisfactory. But the British government was rigorous in the implementation of the legislation available. By the beginning of the twentieth century, the practice had almost disappeared. Or so it was believed.

Sati in Contemporary Times

In September 1987, exactly two centuries after British officers had first reported the practice of *sati,* and a century and a half after the institution of the Sati Abolition Act in 1829, the country was rocked by media reports that Roop Kanwar, a young Rajput widow, had performed *sati* in the presence of thousands of spectators and that no one had prevented her from doing so. That Roop Kanwar was barely 18 years old and had been married for only a few months was awful enough. What made the incident even more difficult to accept was that she was an educated girl who had completed high school, her husband was a science graduate, and her father-in-law, in whose home she dwelt as a member of a joint family, was a schoolteacher. Deorala, the village where the incident occurred, is by no means a backward community. It is a small town with a population of about 13,000 located about 30 minutes off the Delhi-Jaipur highway. In the mid-1990s, it had five primary schools and three secondary schools.

As details of the event trickled in, it was clear that this was no private self-immolation. It had been witnessed by the entire town. Later, over a 12-day period following the act, more than 500,000 people visited the site. There were highly conflicting reports on whether Roop Kanwar's act was purely voluntary or not. However, the facts available left no doubt that it was a well-organized event. There were other disturbing details. Roop Kanwar's "decision" to perform *sati* was announced in Deorala, but her parents, living in a neighboring village, were not informed. Local authorities knew of the announcement but did nothing to prevent the event. Hearing that the press was on its way, the organizers had moved up the event. When the reporters arrived, they were abused and manhandled. Government representatives failed to visit Deorala until three days after the self-immolation. Despite sensational press reports, the government took two weeks to make a statement. No attempts were made to arrest members of Roop Kanwar's family or the doctor attending her, despite the fact that the law clearly indicated that this could be done.

Immediately after the event, the site of the self-immolation became a place of pilgrimage. Stalls selling food, the paraphernalia for *puja* (worship), mementos, and audio cassettes of devotional songs sprang up. Parking lots appeared, and loudspeakers were installed to convey instructions from a central tower. Sophisticated technology was employed to replay the act of *sati* at the site, complete with a blazing funeral pyre. Roop Kanwar's father-in-law and other prominent men from the village, together with some influential persons from the city of Jaipur, formed an organization called the Sati Dharm Raksha Samiti (Organization for the Defense of the Religioethical Ideal of *Sati*). They also formed a trust for the offerings and donations that poured in.

Following tradition in the matter, a *chunni* (veil) ceremony (for the ritual cremation of her *chunni* at the site of the event) was announced for the 10th day after the self-immolation. Women's organizations in the nearby city of Jaipur successfully petitioned the courts to stop this ceremony, but the state government did little to implement the court order. Police stopped vehicles at a certain distance from the site of the *chunni* ceremony but allowed people to walk up to it. The route was lined with police officers dressed in civilian clothes to prevent the crowds from being offended. State-level politicians of different parties went to the site to pay their respects to Roop Kanwar.

Sati As a Symbol of Traditional Virtue

As the national debate on these happenings heated up, several disturbing realities about the manner in which the powerful practice of *sati* was being glorified, to serve vested political interests, grew visible. Supporters extolled *sati* as the symbol of the rich, courageous, traditional, rural values of womanly virtue, and portrayed members of the women's groups protesting *sati* as upper-class, urban, Westernized, materialistic women who claimed to represent a section of the population from which they were emotionally and in every other way distanced. Statements of this kind had also been made in connection with the Shah Bano issue, but, this time, they were much more sophisticated and subtle. Head priests of the major Hindu temples across the country announced that *sati* represented not only a noble Rajput tradition but also a noble Hindu tradition. They also declared that Hinduism was in danger and reminded people that matters such as *sati* came under their jurisdiction, not the state's. Extreme right-wing Hindu nationalist organizations, such as the Shiv Sena in Maharashtra, supported the glorification of *sati*.

Undeterred by this criticism, journalists, feminists, and human-rights activists kept up their protest and probed further into the issue. As they did so, they stumbled upon the startling fact that the revival of the *sati* tradition was part of a strategy, consciously and carefully being employed by local Rajputs, in a bid for power. During British rule, the region in which Deorala is located consisted of small, independent, princely states. At independence, these states were abolished and the region was brought under the authority of the Indian government. While former princes and then nobles thus lost authority

as rulers, the land reforms introduced by the Indian government further deprived them of the wealth and power they had earlier enjoyed. In a firm bid to regain their status, they organized themselves under the banner of the chivalric Rajput tradition. In this tradition, Rajput men had defended Hinduism on the battlefields, and Rajput women had defended the Hindu ideal of a virtuous wife by courageously performing *johar* if their men lost in war. In the organization that the former princes of the Deorala region established, this tradition was merged with a new fundamentalist Hinduism. Activities to express and assert this combination of the old and the new Hinduism were frequently conducted. The ancient *sati* temple at Jhunjhunwala in the region was rebuilt and expanded, and the annual *sati* festival then received new support. Rallies and demonstrations, at which thousands of volunteers dressed in saffron robes performed *lathi* drills. (A lathi is a stick, about five feet long and two inches in diameter. It is used as a simple but effective weapon. The use of a lathi is a skill and an art which has to be carefully learned. The police frequently use lathis to beat and disperse mobs. In this particular context the lathi drill was expected to make much the same kind of statement that is conveyed by a parade of armed policemen or army personnel.)

As this evidence of the organization of a new Hindu fundamentalism in aid of political power for the Rajputs surfaced, other events began to fall into place. In 1954, an incident of *sati* had taken place in the Deorala area. In 1983, a campaign to popularize the ideology of *sati* was launched in Delhi by an organization named Rani Sati Sarva Sangh, funded by Rajputs from Rajasthan. A procession organized by this body was thwarted by Delhi feminists, but the processionists were, nevertheless, able to raise slogans stating that Hindus should have the right to commit, worship, and promote *sati*. Among other slogans was one that said: "We, the daughters of India, are not flowers but fiery sparks." In 1985, the police had intercepted a *sati* at Jaipur. In 1987, only four months prior to Roop Kanwar's self-immolation, the police had prevented a woman from committing *sati* in the Deorala region again. This incident had been watched by about 30,000 people.

Thus it was clear that the event was part of a movement that was steadily gathering strength. Details from the descriptions of the Roop Kanwar self-immolation add up to support this hypothesis. Roop Kanwar's in-laws were well-to-do Rajputs with influential connections that would likely have linked them to the organization promoting Rajput chauvinism. Their failure to inform Roop Kanwar's parents of her decision to perform *sati* suggests that they did not want anyone to hold her back. Press reports about the event say that Roop Kanwar was escorted to the funeral pyre by saffron-robed, sword-wielding young men. The garb and demeanor of these young men were alarmingly similar to those of the volunteers who demonstrate on behalf of the Rajput organization in Delhi mentioned above. If all of this evidence points to the possibility that Roop Kanwar's self-immolation was engineered to aid the politics of Rajput chauvinism, it is also evident that such politics does not take cognizance of the constitutional commitment to provide women special protection for advancement. The politicians who visited Deorala to pay homage to Roop Kanwar did not seem to be bound by these constitutional commitments either. For that matter, they may not even have had any deep commitment to Hinduism. Their primary concern seems to have been to ensure the Rajput vote.

Having failed to take strong action in Deorala, the government of Rajasthan decided to make amends by passing a strong anti-*sati* act in 1987. So did the national government the same year. However, neither government allowed the bills proposing these acts to be adequately debated and discussed. Both were passed hurriedly, as though the governments were desperate to appease the feminists and human-rights activists, yet afraid to face the fundamentalist lobby, and not really interested in the substance of the issue.

Not unexpectedly, both the state and the national acts have serious flaws. Earlier legislation had consistently treated the woman who attempts to perform *sati* as a victim. In contrast, both of the 1987 acts consider her an offender, punishable with imprisonment ranging from six months to five years. In doing so, they equate *sati* with suicide. Both acts also treat as offenders all those who take part in the event, regardless of whether they do so as sightseers or as organizers. By equating sightseers with organizers, the acts miss the point that organizers are practically murderers. Finally, both acts shift the burden of proof on the accused. The Rajasthan act goes so far as to lay it upon the woman as well. Thus, though it was well intended, the new legislation seems to be completely insensitive to the manner in which *sati* functions.

Religion and the State

Both the 1986 Act for Muslim Women's Protection of Rights to Divorce and the glorification of Roop Kanwar's performance of *sati* have set the clock for gender justice in India back by more than a hundred years. However, that may prove to be only a partial victory for the fundamentalists. Both Shah Bano's plea for maintenance and Roop Kanwar's performance of *sati* raised issues that touched the mass of Indian women in a manner that abstract statements on gender equity and justice could never have done. The media exposure the two cases received was so extensive and of such good quality that discussions of the issue reached far and wide. Feminists and human-rights activists gained a reference point for further action. Moreover, the events between the time that Shah Bano filed her petition for maintenance in 1978 and the passing of *Sati* Abolition Acts by the government of Rajasthan and the government of India in 1987 have revealed some basic truths about the relationship between religion and the state in India, about secularism, and about the working of fundamentalism. These revelations should be valuable in defining the further course of action for the movement for gender equality and gender justice in India.

By far the most important outcome of the Shah Bano case and the Roop Kanwar event is the exposure of the fact that religion continues to be central to the politics of governance in India. Up to the 1980s, the official stand had been that the problem of religious communalism was largely resolved with the partition of the country at independence. The events between 1978 and 1987 proved that not only does communalism continue to exist, but it also has acquired a new dimension and a new strength because of the politicization of minority rights, so generously provided in the constitution.

Another important outcome is the explication of the concept of secularism that occurred as the government justified its acceptance of the move to deny Muslim women justice under Section 125C of the Criminal Code. After independence, human-rights activists and feminists had functioned with the confidence that the constitutional commitment to secularism was an assurance that religion would be separated from politics. While supporting the bill for the protection of Muslim women in divorce, the Indian government made it clear that it did not espouse this classic liberal democratic view of secularism. Rather, as Prime Minister Rajiv Gandhi told critics of the bill back then, the Indian state "defines secularism as the right of every religion to co-exist with another religion." This definition gives religious bodies the right to make their own laws or at least affirms that all religions have the right to representation within the law.

The revelation of the centrality of religion to the politics of government is disconcerting. Nevertheless, it is valuable to have facts in the open. It was hitherto believed that the constitutional promise of secularism provided a firm frame of liberal commitments within which battles for gender justice, as well as for other human rights, were to be located. With the realization that this frame does not exist, the battle lines can be redefined and redrawn more realistically.

Both the Shah Bano and the Roop Kanwar cases have revealed some of the critical features of the functioning of religious fundamentalism. Feminists and others who have to deal with religious fundamentalism in the course of their struggle for gender justice can take some valuable cues from what has been exposed.

Both events reveal that one of the first steps a community takes when it decides to assert its religious identity is to revive traditions and practices that subordinate and control women, probably because such a move helps illustrate discipline within the community. The revival is so engineered that women from the fundamentalist lobby themselves come forward to speak for, and celebrate, this revival. This happened in both the Shah Bano and the Roop Kanwar cases. The events also revealed that, rather than deal with feminist arguments at face value, those who oppose gender justice resort to discrediting feminists. In the Roop Kanwar case, Indian feminists were maligned, discredited, accused of being Westernized, and labeled as colonialists, cultural imperialists, materialists, and even capitalists. Meanwhile, the views and voices of fundamentalist women are counted as representing the "real" women. As evidenced in the glorification of *sati,* one of the key strategies used by fundamentalists seems to be to skillfully appropriate the right to speak for the entire spectrum of views, feelings, and opinions on the issue in contention, despite the fact that they generally represent only one end of the spectrum. Meanwhile, the issue of the subordination and oppression of women is systematically sidetracked by other issues that are created. In both the Shah Bano and the Roop Kanwar cases, issues of Hindu versus Muslim, Indian versus Western, traditional versus modern, spiritual versus materialist, and holy versus unholy clouded the basic issues of an old woman abandoned without support by her former husband and a young girl allowed to commit suicide or perhaps even pushed to her death—both in the name of religion (Sathe, 1995b).

Conclusion

This essay has offered a glimpse of the obstacles to the efficacy of the constitution, the law, and the instruments of gender justice. It has illustrated how precious traditions are, how deep they run, and how they are exploited by vested interests. As in any other society, oppression, exploitation, subordination, and marginalization of women in India are rooted in social structures, ranging from the family, kin, and community to the state. They permeate interpersonal relationships at every level and in all spheres of a woman's life. Their shape differs from subculture to subculture and within subcultures, and their expressions change through time. But they persist, sometimes hidden, sometimes visible. As religious and cultural fundamentalism are used as instruments to political power, bodies with vested political interests find it useful to reaffirm tradition, which often results in women being denied equality, freedom, and justice in numerous spheres of their lives. Despite the obstacles women face, the constitution and the law continue to be the most powerful instruments of gender justice. The challenge for those who use these instruments is to sort out the traditional and cultural elements really worth preserving from those that are not, to define the direction of change, to identify the interests that obstruct the desired change, and to forge ahead.

References

Agnes, Flavia. *State, Gender, and the Rhetoric of Law Reforms.* Bombay: SNDT Women's University, 1995.

Dhagamwar, Vasudha. *Law, Power, and Justice.* New Delhi: Sage, 1992.

Kishwar, Madhu. "Pro Women or Anti-Muslim? The Shah Bano Controversy," *Manushi,* vol. 6, no. 2, January/February 1985, pp. 4–13.

Kumar, Radha. *The History of Doing.* New Delhi: Kali for Women, 1993.

Sathe, Satyaranjan. *Towards Gender Justice.* Bombay: SNDT Women's University, 1995a.

———. "Uniform Civil Code: What, Why and How," *Toward Secular India,* vol. 1, no. 1, January/March 1995b, pp. 31–42.

Sex-Role Ideologies

Machismo in Latin America and the Caribbean

Violeta Sara-Lafosse

Introduction

The term "machismo" is often considered synonymous with sexism; that is, feelings of superiority by individuals of one sex (male) in behaviors toward the other (female). During the 1950s, with the publication of the study by Stycos (1958) of families in Puerto Rico, machismo entered research in the social sciences for the first time (Stycos, 1958, p. 45). Subsequently, Lewis (1964) employed it in the introduction to the life histories of a Mexican family, and de Hoyos and de Hoyos (1966) attempted to define it as they analyzed behaviors by Mexican men who alienated their wives.

However, none of these authors sought to provide an explanation for this form of masculine behavior, which comprises the man's desire to take sexual advantage of women, the failure to assume responsibility for the consequences of such actions, and the self-praise for sexual exploits within the subculture of the peer group.

If one considers machismo a form of sexism, it is necessary to be precise about other forms of sexual discrimination. One of the best-known forms is patriarchy, a term that evokes an ancient past and a perspective that has been maintained with varying degrees of rigor in both Western and Eastern cultures. In Western cultures, patriarchy achieved its highest point during the second half of the nineteenth century, in the Victorian era. At that time, a high infant-mortality rate obliged women to withdraw from the productive sphere to devote themselves exclusively to childrearing. Simultaneously, this demanded from men a redoubling of efforts to support their families. In this way, the role of the exclusive housewife came into being, making women totally dependent economically on their husbands, who thus became absolute patriarchs.

The patriarch is a father. The interest he has for his children, as a means to achieve permanence on Earth, leads him to be concerned about their survival and education. The patriarch has a double interest in his wife: She provides him with the necessary sons (and their concomitant care), and she is a source of sexual pleasure. In consequence, she is treated in condescending ways, often as a minor.

Machismo and patriarchy are both forms of sexist behavior, for both treat the woman as a sexual object and thus as a person subject to domination. The fundamental difference lies in the fact that the patriarch becomes responsible for the children he begets from that woman, while the macho man is not interested in them and does not recognize them as his offspring. He considers the children the concern of the woman.

This irresponsibility toward the children is a basic feature of machismo, a term that arose in Latin America and has no equivalents in other languages. Therefore, those who use it elsewhere tend to use it, erroneously, as a synonym for sexism.

The Emergence of Female-Headed Households in Latin America

Single-parent families led by women often present multiple problems economically and in the socialization of the children. However, in many of these families, the problematic situation of the mothers was already present before the father left, and it is necessary to know the specific family structure to understand the motivations of fathers who desert their homes. It also must be noted that the daily absence of the father is not necessarily a sign of parental abandonment in economic and educational terms. Finally, our understanding of the situation requires an examination of the laws that spell out the obligations of fathers and that consider as an unlawful act the father's failure to provide family support, especially for meals. The forms in which justice is administered to the violators is often extremely lenient; consequently, those who break these laws go unpunished with great frequency.

Family studies back to 1970 demonstrate that the phenomenon of female-headed households is widespread in Latin America and is associated with poverty (Buvinic, 1990). Research in the 1980s reported that, in the Caribbean, children whose parents could not provide them with the required meals and lodging could be found in many asylums (Massiah, 1984).

Prior to the father's desertion, these families present a great degree of instability (Guzman, 1991). According to studies conducted in Peru, "the status of the family is centered around the figure of the father or what he can demonstrate by himself or spend outside the home" (Castro de la Mata, 1972, p. 62). The father spends the largest part of his income drinking with friends and buying clothes for himself. These fathers forcibly demand attention to their desires but are not willing to give much, materially or emotionally, in exchange. They are the first to have their needs satisfied, and often only the leftovers go to other family members. They do not need rebellion by any members of their family as a condition to abandon them. "Generally, another woman attracts his attention and he goes with her, leaving his family behind" (Castro de la Mata, 1972, pp. 62). This reason for abandonment is corroborated by another study, which affirms that "the man forms another family parallel to his cohabitation or union with the [current] woman. Thus, it turns out that the desertion does not take place to avoid economic burden" (Chueca, 1986, p. 10). Similar findings have been reported in Mexico City through psychological studies, according to which many families are in an unstable situation, constantly threatened by possible disintegration due to the man's refusal to assume even a minimum of responsibility for his marital duties and family duties in general (Castro de la Mata, 1972). It must be remembered also that behind each "macho" man in Latin America, there are women willing to accept his behavior, which includes sexual advances and even violence.

An additional feature of abandoned families is the informal nature of the couple's conjugal arrangement. Most of these families are based on partnerships established only on a verbal or implied common agreement. In consequence, the children face the sociological condition of being illegitimate. Most of the women who head families appear in census data as unmarried mothers with children. Also included within this category are women who have had a child or children as a result of seduction or deception and who have not succeeded in living together with the child's father, as well as those who have offspring as a result of rape. Rates of illegitimate children in Latin America are high, ranging from 18 percent in Chile to 71 percent in Panama (Covarrubias et al., 1981).

The problems concerning the socialization of children from households headed by a single parent are multiple, and they vary according to the gender of the child who lives without a father. Generally, single mothers tend to overprotect their children and to discourage their independence. Children in female-headed households do not perform in school as well as children in two-parent families, and they are more likely to repeat their school year or abandon their schooling altogether.

Living in a single-parent home appears to affect the intellectual development of boys: Their academic performance suffers more than that of girls from one-parent homes, and this is most evident in boys who were less than 5 years old at the time their father abandoned them (Hoffman, 1988, p. 234). The absence of the father also affects boys' sex-role identification. Since the sexual model to be imitated is scarcely present in daily life, the child gains a greater identification with the mother, who is the person who maintains direct and affective contact with the child for the longest time. In this situation, the boy learns masculine roles through requirements that are rather negative—that is, how he must *not* behave in order to be a man. This compulsive masculinity leads boys to evince greater anxiety than girls regarding their identification with sexual roles (Muñoz, 1983, p. 109).

Older boys from homes with absent fathers exhibit behavioral patterns of exaggerated masculinity, which is manifested through harshness, self-sufficiency, rebelliousness, and sexual shamelessness. This "compensatory masculinity" predominates among boys of low social class with absent fathers, who live in settings where the peer group, whose members face similar problems concerning their identification with sexual roles, provides the substitute masculine models (Shaffer, 1979, p. 506). The most serious consequence of this situation is that fatherless boys tend to be more aggressive and run the risk of becoming juvenile delinquents. The absence of a "masculine authority" has been invoked as a possible explanation for the antisocial behavior of some of these youths (Delpino, 1990, p. 76).

The various findings described above, all of them narrowly linked with masculine behaviors of child desertion, have enabled researchers to assess not only the detrimental situation caused by this masculinity, but also the social relevance of a masculine feature held in substantially low esteem by this group (fatherhood) and its replacement with a less humane behavior (that of the macho) as a result of compensatory masculinity.

Taking stock of the damage, we find that homes in extreme poverty led by women suffer from the father's desertion after a sequence of chronic violence, infidelity, or insecurity within a fragile institutional framework: the informal union of the couple and the illegitimacy of the children. Likewise, the egocentric and irresponsible behavior of the father, with its negative effects upon the psychosocial development of the sons, who lack father roles and end up developing deficient sexual identification and aggressive and antisocial behaviors, has been demonstrated.

The father's desertion is considered a social deviation in most cultures: It is seen as an act of marginal character, it is viewed with disapproval, and it is given sanctions by different groups and institutions in every society. Moreover, it is considered a marginal situation, in the same manner as other social deviations. However, the widespread presence of this phenomenon in Caribbean and Latin American societies has contributed to its becoming a subculture for a substantial proportion of society. This internalized and legitimized pattern affects not only family institutions but also those concerning law and order. The subculture it creates must be labeled machismo to differentiate it from the dominant culture of patriarchy which, in contrast, develops an authority based on fatherhood. Both cultures are distinct and stand in contrast regarding the relationship of fathers with their offspring. Both are equally sexist because they share the assumed superiority of men over women.

Few are the social scientists who have sought to define machismo. Successive definitions have enriched its content but not its precision. For some, machismo emphasizes independence, impulsiveness, physical strength as the "natural" form to resolve disagreement, roughness as the best form of relations with women, and force as the most frequent form of relationship with the weak or the subordinate (de Hoyos and de Hoyos, 1966, p. 104). For other scholars, machismo is used to label the cult of virility whose characteristics are aggressiveness and high intransigence in male–male relations and arrogance and sexual aggression in male–female relationships (Stevens, 1973, p. 122). Both descriptions highlight the value assigned to physical attributes linked to the body in terms of force and aggression. Valuing the macho in the literal and biological sense alludes to the animal side of human nature and excludes rationality in most relationships. Machismo, understood and practiced in this manner, amounts to nothing less than a degradation of the human condition.

In other cases, social scientists engage in a definition of machismo that emphasizes its subjective face, which enables it to be presented as a form of cultural expression:

> The particular form in which each man or woman suffers or lives his/her machismo. Thus, for some men it is natural to have several lovers; for some women it is natural to accept that their husband or lover beats them; for others it seems natural for a man to deliver one-third of his salary to the home and to spend the rest with his friends. The term "mother" for women in those conditions means an unwanted pregnancy or a new child she won't be able to feed (Gissi, 1975, p. 365).

The dimension of irresponsibility vis-à-vis the expenditures linked to supporting a family (the children) and the display of violence is frequently present in other authors addressing the topic, as in the reality described by Lewis (1964): the use of violence to resolve any differences or to "educate the children," frequent beatings of the woman, a high incidence of abandonment of wife and children, incest, rape, adultery, and bigamy. In Mexico, references are also made to the initial appearance of machismo and the reasons behind it:

> Machismo appears from early years among the mestizo children. . . . The Mexican's machismo is at heart no other than the insecurity of his own masculinity, the baroque expression of his virility, the distance from a diffused paternity. . . . The Mexican macho spends the greatest part of his income to highlight his attributes as "macho" (Lugo, 1985, p. 42).

Those who have conducted macrosocial studies of the family in Latin America note that the machismo complex originates in the large numbers of illegitimate births and irresponsible paternity, in the form of extramarital relations, spouse abandonment, and single motherhood. This macrosocial fact is reflected in the consensual union and in the child's uncertain or deficient socialization (Covarrubias et al., 1981).

Some authors report that machismo is not found in certain places in Latin America. Regarding indigenous communities that occupy the high plateaus in the Andes, it is affirmed that the farmer *(comunero)* is not *machista,* since "Indians do not seek sexual conquests as a validation of their masculinity; sexual conquest does not add shine to the individual's reputation. The exploitation of one sex by the other finds little sympathy within the bounds of the community" (Wolf, 1959, p. 223). In a study conducted in Lima, Peru, of migrants from Andean communities, a concurrent and revealing judgment can be found: "Men from Lima are devoted to conspicuous consumption, they are very irresponsible and many times cause the breakdown of their families" (Golte and Adams, 1987, p. 87). It can be affirmed that, in Andean communities, the female-head-of-household phenomenon is not caused by the father's desertion, nor is it as frequent. According to 1990 data from the United Nations Development Fund for Women (UNIFEM), 37 percent of the poorest families in Lima were headed by women, while the figure was much less (23 percent) for families in the rest of the country.

It is not clear from the existing data, however, the proportion of female-headed households in which women seek to raise children by themselves or find themselves in such a condition because of particular social arrangements (for example, the child's father is married to someone else, the father does not wish to be married, the father is unknown, or the mother became pregnant by contraceptive error and could not have an abortion). Qualitative studies are needed to gain deeper knowledge of this situation.

Machismo and the Law

The social order is affected by the machismo subculture and, in particular, by the entire judicial apparatus, which often operates as an institutional force that maintains the masculine behavior of child desertion and the difficult situation of the female head of household. An observer of this condition affirms the long-standing nature of this problem in Peru:

> The failure to comply with his duties toward the children is not a problem that emerged recently; already in the decade of the 1950s there was a growth in the failure to comply with obligations regarding meal allowances (Aldave, 1987, p. 7).

The author presents data that enable him to detect the lack of sanctions for those who fall behind on their obligations.

Despite the fact that the percentage of defaults in meal allowances is very high in Peru, the population in prison for committing such a crime is extremely low. This can be due among other causes to judges being reluctant to apply punishment for these crimes (Aldave, 1987, p. 7).

This lack of punishment regarding the failure of fathers to provide family allowances has also been detected in other countries.

> Mexican men are absolutely irresponsible regarding their fatherhood. The abandonment of the mother and children is a misdemeanor not generally punished. . . . [T]he judicial system becomes a key player in the oppression of women. . . . [I]t ensures impunity for men and irresponsibility toward women (Lugo, 1985, p. 47).

Likewise, other aspects of complicity by judges have been observed: one concerns the amounts assigned to cover the basic needs of each child; another is the discrimination that judges evince in setting this amount in cases of consensual unions: In decisions made in one court, two women heads of household in consensual unions received settlements of between 2 and 10 U.S. dollars per month, respectively, for two children in the first case and three children in the other. The only woman who was married received about 15 U.S. dollars per month for three children (Delpino, 1990).

The judicial practice informs us that the basis for meal support is human solidarity and that, as such, this support fulfills a social function. The purpose of the legal obligation is to enable persons to survive. The right to meal allowances seeks to address the problem of maintaining life in the interest of society (Aldave, 1987). Therefore, the abandonment of a family is an act against life.

The information examined thus far enables us to affirm that the behavior of irresponsible fatherhood exists and is preserved in Latin America and the Caribbean because social norms that value fatherhood have lost meaning in several sectors of the population—at the individual level, by not being internalized by social subjects; and at the collective level, by institutions that do not enforce compliance to legal norms mandating family support. This explains why women who head households can rarely have their rights respected and why those who study the problem do not identify the abandonment of women as a cause of major sociocultural problems.

The Historical Origin of Machismo and Women's Headship of Family

What do specialists have to say about this paradoxical fact of irresponsible fatherhood? An immediate response, which does not contemplate the negative effects of rejecting fatherhood, tends to deny universal validity to this norm or to argue, in a pejorative way, that certain groups or racial

or cultural sectors are distant from the norm of responsible fatherhood by their very nature. There are also those who consider that this norm can be pushed aside in situations of economic crisis such as those facing some Latin American countries.

One of the indicators used by William Goode, a pioneer in the study of illegitimacy in Latin America and the Caribbean, is the rate of out-of-wedlock children. He analyzed these rates in different countries and cultures throughout the world and reached the following conclusions: Some evidence, still scant, suggests that a modest or low rate of illegitimacy (4 percent) has been present throughout the Western world for centuries. This pattern also existed in Japan until the tenth century and decreased afterward, albeit gradually (Goode, 1964). The United Nations (1965) showed rates of illegitimacy of 1 percent in China; among African countries, it was 3.4 percent in Morocco and 0.08 percent in Algeria. In contrast to these low rates of illegitimacy, Goode observed the rates in Latin America, where the lowest rate was 17 percent in Chile; the highest, 71 percent in Panama. In the Caribbean, the lowest figure was 29 percent for Puerto Rico; the highest, 72 percent for Jamaica (Goode, 1964). These rates would have been lower if the legitimacy of traditional marriages of native cultures had been recognized in those areas where ancestral values have been maintained. These high rates led Goode to examine studies about indigenous communities in the New World. These studies demonstrated that a tolerance of illegitimacy had never existed in those societies, leading to a hypothesis that this pattern of higher rates was a result of the European conquest of the New World, which affected all of the countries of Latin America and the Caribbean (Sara-Lafosse, 1984). Illegitimacy was widespread among colonial cities in the seventeenth and eighteenth centuries (Lavrin, 1992)

Information on illegitimate children—defined as those from parents who were not married at the time of birth, regardless of whether the child was recognized or legitimized after birth (United Nations, 1986, p. 104)—is difficult to obtain and thus is reported within relatively long intervals. The most recent data (1986) indicate that illegitimacy rates are increasing. In the Caribbean, rates go from a low of 54 percent in Belize to a high of 85 percent in Antigua and Barbuda. In Latin America, they go from a low of 26 percent in Uruguay to a high of 54 percent in Venezuela (United Nations, 1986, p. 857).

The Conquest and Machismo in Latin America

The small number of Spanish migrant women to the New World affected the patterns of sexual selection and promoted unions between Spanish men and Indian or Black women; the few White women and their descendants enjoyed great social status (Lavrin, 1995). The relationship between the Spanish men and the indigenous women must be considered as part of the conquest of the New World. Violence, an intrinsic element in the conquest, "must, from this perspective, be seen as the rape of indigenous women, a violent act that seeks to subjugate and oppress" (Burkett, 1985, p. 128). After the period of conquest (1492 to the early 1800s) was over, violence continued in other forms of oppression. These are embedded in the demand for labor, legal as well as illegal, imposed by the Conquistadores, now in charge of *encomiendas* (large tracts of land as well as the native population residing on them). The demands by the Spaniards for women to provide personal services created a new element in Latin American society: "If we had no other proof than the number of children that indigenous women had with those masters, it could seem that the service provided also included sexual duties" (Burkett, 1985, p. 132).

It is evident that the Spanish Crown was "concerned about the treatment indigenous women received in the houses of Spaniards, especially because these men did not allow the marriage of their servants, with the purpose of controlling them and keeping them available for sexual activities" (Burkett, 1976, p. 20). In 1541, it is documented in Lima that the Spanish king warned that he had been informed that Spaniards

> kept in their houses large quantities of Indian women to carry out with them their evil desires, and to remedy this it would be advisable to order that no Spanish man should have in his home a suspicious Indian woman, neither recently pregnant nor pregnant, except for those who are needed to cook or to provide domestic service (Burkett, 1985, p. 134).

In 1569, in a document addressed to the Audience of Quito, the king condemned the fact that "in order to have the service of Indian women, it is accepted that they be in sin and without becoming married and prevented from having their husbands take them, which is a form of slavery"; in Bogota, a royal letter dated 1606 ordered the governor of the New Kingdom of Granada to ensure that "Indian women lending service in the house of Spaniards be not detained and be free to marry" (Gutierrez, 1963, p. 294). These commands were ineffective despite their frequent expression during the colonial period. "The problem was widespread, which made it difficult to carry out the task of family acculturation of the Indian woman" (Gutierrez, 1963, p. 294). The threat of suffering violence in Spanish hands was still present at the beginning of the nineteenth century, as evident in accounts made during that period (Portocarrero, 1986).

The mixture of races in Latin America, as in other places under the effects and consequences of conquest, was inevitable. At the beginning, it was not only unavoidable but also fostered. During the fifteenth century, Spanish men married the daughters of the Indian nobility as part of a policy promoted by the Spanish Crown to facilitate pacification (Stein and Stein, 1979, p. 61). The situation of the mestizos (Spanish-Indian offspring) was varied and uncertain. The descendants of marriages with noble Indian women were incorporated into the Spanish group, while the others—the largest number by far and the fruit of sporadic relations—were viewed as beneath contempt. From early colonial times (i.e., the 1600s), Indians and Spaniards were prohibited by the Spanish Crown from marrying one another (Cotler, 1978). At the same time, the Spanish men did not have access to enough women of their own race—the ratio of immigrant Spanish men to immigrant Spanish women during the colonial period was nine to one—which by force pushed them toward racial mixing.

Amerindian women soon learned that the children of European fathers could not be considered Indians and thus were not subject to conscription, Indian taxes, and the many other prohibitions imposed on the conquered population (Stein and Stein, 1979). Although the father did not legitimate or recognize his offspring, the son had progressed in social position relative to his mother, dragging her along in this ascent.

The patterns of sexual behavior we have discussed became stronger within colonial society and were expressed in, among other forms, the reproduction of a set of illegitimate children, generally of mixed origin (Mannarelli, 1991). So it can be affirmed that "Indian women must have provided personal services in the house of the Spaniards, which included social obligations, thus producing thousands of bastard children" (Blondet, 1993, p. 95). The presence of illegitimate, or bastard, children is a fact during the entire colonial period. Peruvian historian Pablo Macera, in a study of the eighteenth century, observed the large number of names of illegitimate children that appear in parish registers and cited the report by Friar Bernardo Serrada, who said that, in Cuzco, "of all children baptized in a year, one-fourth at most were legitimate children." Macera concluded that "it can be supposed that this proportion was also representative of the other cities of the viceroyalty" (Macera, 1977, pp. 337).

It was the Indian woman who became responsible for the mestizo. For some scholars from Chile and Argentina, "there is a double configuration, the European family and another in which the mother is in charge of the offspring" (Covarrubias et al., 1981, p. 345). Finally, Macera himself detected by the nineteenth century an issue that concerns us here: "[T]hose who maintained non-married unions were not fathers. . . . [T]hey were not preoccupied with the education of their children" (Macera, 1977, p. 340). Macera is careful not to address the topic in connection with the large peasant groups in the rural areas.

The Situation in the Caribbean

The presence of women heading a family and responsible for their children was a characteristic of the everyday life of families in the West Indies from the very first days of its colonial history (Massiah, 1984). In the Caribbean, under the slave regime, the most basic conditions required by a legally married couple to maintain and share their home were inaccessible to slaves. Christian marriage was incompatible with the slave system, since either member of the couple could be transferred or sold at any time. The son of the slave woman inherited the same social condition of his mother, thus automatically becoming property of her owner. The efforts to promote procreation frequently took the form of special permits for the mother and for the foremen or administrators of the plantation. In this way, the central function of the mother and the invisibility of the father were legitimated (Massiah, 1984). Slavery undermined the status of the man as head of the family much more so than that of the woman. This happened not only in the Caribbean but also in the South of the United States, where Whites, even during the twentieth century, opposed "ambitions" by Blacks to obtain certificates of marriage and divorce (Goode, 1964, p. 46).

An issue scarcely treated by those who study the historical process of conquest and the colonial period concerns the form in which children became socialized. What is known is that a large proportion of mestizos were raised by their mothers. However, many of them knew their fathers before the abandonment, and such experience left an indelible mark. The sons of Spanish men and Indian women found themselves in an anomalous situation. The subjugation by which his father held his mother made the mestizo feel jealousy and even hatred toward his own father; at the same time, it provoked a desire to be like him (Corredor, 1962). A psychological study about contemporary mestizo male youths found that, "since identification is an unconscious process, the male youth identifies with partial traits of his father, with attitudes that he subsequently reproduces, even after having criticized him" (Ruiz and Cánepa, 1986, p. 15). Individuals respond, that is, through well-known psychological mechanisms that tend to perpetuate the attitudes of their fathers, and then they themselves teach their sons how to behave when they become adults (Castro de la Mata, 1972). This helps explain, according to Castro de la Mata, how the reproduction of despotic families is made possible by those who suffered within them as children.

The precarious nature of the social norm regarding responsible fatherhood and the leniency of competent authorities regarding infractions were also found during colonial times, as were voices of alarm, such as that of a Spanish visitor in 1786 who stated: "Having observed a relaxation of the customs in which one lives and the mistreatment that husbands give their women, it is necessary to establish a careful control of family customs and to impart the necessary punishment" (Gutierrez, 1963, p. 358–359). Friar Pedro de Aguada, a clergyman of this time, declared to the king:

> The state of dissolution is so great among the Spanish in terms of living lustily and carnally that it shocks and astonishes me. And judges and lawmen provide very little remedy to this disorder and dissolution. . . . I have never seen a single punishment by the law on this issue (Gutierrez, 1963, p. 185).

This writer also observed that the authorities can do little because "the officers of your Highness and the governor are the first to consent and take advantage of this" (Gutierrez, 1963, p. 186).

It was in the interest of the conquerors and masters to prevent the development of a native system of social control, be it family or community based, because this posed a potential threat to their domination. There were some indigenous "New World" communities that were able to continue to be internally coherent along social and cultural lines and to ensure compliance with norms of legitimacy. "These communities can be found, for example, in the most elevated lands of the Peruvian Andes" (Goode, 1964, p. 53).

Future Perspectives

Finding a solution to machismo will be easier, and more successful, the more the general public comes to learn about the historical origins of father desertion and the consequent massive presence of female heads of household. It will also help to learn about the psychological and social mechanisms that have enabled deserting fathers to persist across generations.

Examinations of the factors producing these social conditions can facilitate the struggle against these forces, especially machismo, which acts as a subculture at both the individual and the social levels. As with any other cultural ill, education must be the primary treatment. In universities, especially in the field of law, future judges and prosecutors must be trained more thoroughly and effectively to understand the problem and the importance of social sanctions in order to prevent and eradicate machismo.

In earlier schooling, it will be necessary to create conditions propitious to a deeper understanding among girls and boys through the elimination of sexual stereotypes. In Latin America, these conditions can be facilitated through coeducational settings. Machista subculture feeds upon, and becomes strengthened in, segregated educational environments (Sara-Lafosse et al., 1989). Such coeducational school settings must be supported with comprehensive teacher training and nonsexist textbooks.

Other aspects more specific to family desertion require an extensive and persistent global education effort concerning the rights of children, especially their right to a family environment that satisfies their affective needs and enables each child to develop a secure and balanced personality. In this sense, the importance of the educational and role-modeling presence of the father must be underscored. This supposes, at the same time, information for women that revalues women, dissemination of information about women's basic rights as human persons, and knowledge that questions a merely functional identification of motherhood or sexuality.

The situation of machismo in Latin American society of the late twentieth century is linked to population growth and the difficulties in reducing it. Efforts to limit population growth must be accompanied by educational measures to combat machismo and elevate the status of women. Within this effort, more needs to be learned about the origin of machismo and its impact on the lives of families so scholars and the public alike may stop considering it natural to the masculine condition.

Internalization of new models of gender behavior will require a change in the mind-set of authorities in the judicial system, in relation particularly to the basic rights of children and the duty fathers have to fulfill equitably their obligations to their children. The right to food support (by the father) addresses the need to protect life and thus benefit society (Aldave, 1995). On the latter point, legislative modifications are necessary to implement the norms concerning human rights, the constitutional mandates, regulations, and other legal principles of lesser importance that address the obligations of fathers toward their children.

The *Plan for Action* of the International Conference on Population and Development in Cairo (Naciones Unidas, 1994), in its reference to single-parent households, recommends that

> Governments should adopt measures to ensure that the children receive adequate financial support from their fathers, and among other matters, to monitor whether laws regarding payments of meal allowances are being fulfilled. Governments should consider the possibility of modiying their laws and policies so that men may fulfill their responsibilities and provide financial support for their children and their family (Naciones Unidas, 1994, p. 27, para. 4.28).

All of these institutional changes will need to be transmitted and diffused through the mass media and embedded in a wider conception of society and the individual. It must be understood that aspirations and accomplishments in political spheres are also aspirations and accomplishments in everyday life. In the life of groups such as the family—a basic unit that should protect the comprehensive development of each individual—the relations among its members must become more harmonious and just.

References

Aldave, Cecilia. *Situación Cualitativa de la Mujer en Relación al Abandono Infantil—Perspectiva Legal.* Lima: Departamento de Ciencias Sociales, Catholic University of Peru, 1987.

———. "L'abandon de famille et les auteurs sociaux impliqués: Le système juridigue face aux problèmes socio-cultureles au Pérou." M.A. dissertation, Criminology Program, Catholic University of Louvain, Louvain-la-Neuve, 1995.

Blondet, Cecilia. *Mujeres Latinoamericanas: Perú.* Santiago: Facultad Latinoamericana de Ciencias Sociales, 1993.

Burkett, Elinor. "La Mujer durante la Conquista y la Primera Epoca Colonial," *Estudios Andinos,* vol. 5, no. 1, 1976, pp. 4–25.

———. "Las Mujeres Indígenas y la Sociedad Blanca: El Caso del Perú del Siglo XIV." In Asunción Lavrin (ed.), *Las Mujeres Latinoamericanas: Perspectivas Históricas.* Mexico City: Fondo de Cultura Económica, 1985, pp. 121–152.

Buvinic, Mayra. *La Vulnerabilidad de los Hogares con Jefatura Femenina: Preguntas y Opciones de Política para América Latina y el Caribe.* Santiago: Comisión Económica para la América Latina, 1990.

Castro de la Mata, Renato. "Un Intento de Clasificación de la Familia Peruana." Ph.D. diss., Universidad Peruana Cayetano Heredia, Lima, 1972.

Chueca, Marcela. "Madres Jefas de Hogar, Mujeres en Abandono Permanente." Lima: Faculty of Social Work, Catholic University of Peru, 1986. Mimeographed.

Corredor, Berta. *La Familia en América Latina.* Bogota: FERES, 1962.

Cotler, Julio. *Clases, Estado y Nación en el Peru.* Lima: Instituto de Estudios Peruanos, 1978.

Covarrubias, Paz, Mónica Muñoz, and Carmen Reyes. "Población y Familia." In *Estudios de Referencia sobre Educación en Población para América Latina.* Santiago: OREALC-UNESCO, 1981, pp. 341–363.

De Hoyos, Arturo, and Genevieve de Hoyos. "The Amigo System and the Alienation of the Wife." In Bernard Farber (ed.), *Kinship and Family Organization.* New York: John Wiley, 1966, pp. 102–115.

Delpino, Nena. *Saliendo a Flote la Jefe de Familia Popular.* Lima: Fundación Friedrich Naumann, 1990.

Gissi, Jorge. "El Machismo en Chile," *Mensaje,* no. 241, 1975, pp. 364–370.

Golte, Jurgen, and Norma Adams. *Los Caballos de Troya de los Invasores.* Lima: Instituto de Estudios Peruanos, 1987.

Goode, William. "Illegitimacy, Anomie, and Cultural Penetration." In William Goode (ed.), *Readings on the Family and Society.* Englewood Cliffs: Prentice-Hall, 1964, pp. 38–55.

Gutierrez de Pineda, Virginia. *La Familia en Colombia: Transfondo Histórico.* Bogota: Universidad Nacional de Colombia, 1963.

Guzmán, Gloria. "La Familia Expulsora." Paper presented at the workshop on Work, Family, Development, and Population Dynamics in Latin America and the Caribbean. Santiago: Comisión Económica para la America Latina (CEPAL), 1991.

Hoffman, Lois, Maria Scott, Elizabeth Hall, and Robert Schell. *Developmental Psychology Today.* New York: McGraw-Hill, 1988.

Lavrin, Asunción. "Introduction: The Scenario, the Actors, and the Issues." In Asunción Lavrin (ed.), *Marriage and Sexuality in Colonial Latin America.* Lincoln: Nebraska University Press, 1992.

Lewis, Oscar. *Los Hijos de Sánchez.* Mexico City: Fondo de Cultura Economica, 1964.

Linton, Ralph. *Estudio del Hombre.* Mexico City: Fondo de Cultura Economica, 1965.

Lugo, Carmen. "Machismo y Violencia," *Nueva*

Sociedad, no. 7, 1985, pp. 40–47.

Macera, Pablo. *Trabajos e Historia,* vol. 3. Lima: Instituto Nacional de Cultura, 1977.

Malinowski, Bronislaw. "Parentood, the Basis of Social Structure." In Marvin Sussman (ed.), *Sourcebook in Marriage and the Family.* Boston: Houghton Mifflin, 1963, pp. 40–48.

Mannarelli, Maria Emma. "Sexualidad y Desigualdades Genéricas en el Perú del Siglo XIV, *Allpanchis,* vol. 22, no. 35–36, 1990, pp. 225–248.

———. "Las Relaciones de Género en la Sociedad Colonial Peruana: Ilegitimidad y Jerarquías Sociales." In Maria del Carmen Feijóo (ed.), *Mujer y Sociedad en América Latina.* Buenos Aires: Comisión Latinoamericana para las Ciencias Sociales (CLACSO), 1991, pp. 63–107.

Massiah, Joycelin. *La Mujer como Jefe de Familia en el Caribe: Estructura Familiar y Condición Social de la Mujer.* Paris: United Nations Educational, Scientific, and Cultural Organization (UNESCO), 1984.

Muñoz, Mónica. "Ser Hombre y ser Mujer." In Paz Covarrubias, Mónica Muñoz, and Carmen Reyes (eds.), *Crisis en la Familia?* Santiago: Instituto de Sociología, Catholic University of Chile, 1983, pp. 91–115.

Naciones Unidas. *La Mujer en el Mundo.* New York: ONU, 1992.

———. *Plan de Acción de la Conferencia Internacional sobre Población y Desarrollo.* New York: United Nations, 1994.

Portocarrero, Gonzalo. "La 'Idea Crítica': Una Visión del Perú Desde Abajo," *Los Caminos del Laberinto,* no. 3, 1986, pp. 3–14.

Ruiz Secada, Rosa, and María Angela Cánepa. "Los Jóvenes del Cono Norte." Lima: CIPEP, 1986. Mimeographed.

Sara-Lafosse, Violeta. "Crisis Familiar y Crisis Social en el Perú," *Revista de la Universidad Catolica,* no. 15–16, 1984, pp. 92–112.

Sara-Lafosse, Violeta, Carmen Chira, and Blanca Fernández. *Escuela Mixta: Alumnas y Maestros la Prefieren.* Lima: Fondo Editorial, Catholic University of Peru, 1989.

Shaffer, David. *Social and Personality Development.* Los Angeles: Brooks/Cole, 1979.

Stein, Stanley, and Barbara Stein. *La Herencia Colonial de América Latina.* Mexico City: Siglo XXI, 1979.

Stevens, Evelyn. "Marianismo: La Otra Cara del Machismo en Latinoamérica." In Ann Pescatello (ed.), *Hembra y Macho en Latinoamérica: Ensayos.* Mexico City: Editorial Diana, 1973, pp. 121–134.

Stycos, J. Mayone. *Familia y Fecundidad en Puerto Rico.* Mexico City: Fondo de Cultura Economica, 1958.

United Nations. *UN Demographic Yearbook 1965.* New York: United Nations, 1965.

———. *UN Demographic Yearbook 1986.* New York: United Nations, 1986.

Wolf, Eric. *Sons of the Shaking Earth.* Chicago: University of Chicago Press, 1959.

See also Noemí Ehrenfeld Lenkiewicz, "Women's Control Over Their Bodies"

Islam and Women's Roles

Golnar Mehran

Introduction

In the mid-1990s, more than one billion Muslims live throughout the world, half of whom are women. A close examination of the impact of Islam as a set of religious beliefs and as a political ideology is, therefore, essential in understanding the role and position of women in Muslim countries. Presenting itself as a social and political alternative in the final decades of the twentieth century, Islam is playing an increasingly important role in determining the possibilities provided for, and the limits imposed upon, women. This essay aims at explaining how religion, sociopolitical transformation, and women as active agents shape the lives of women in Muslim nations. It also discusses how feminism is understood and interpreted by the institution of religion, the state, and the women themselves in the Islamic world.

The Islamic Understanding of the Role and Rights of Women

The Muslim world is marked by extreme variation. The realm includes such diverse countries as Indonesia, Bangladesh, Pakistan, Iran, Afghanistan, Turkey, Sudan, Senegal, and Mali, as well as countries in the Arab world. Variation, however, is not limited to geography; it also includes political economic, racial, ethnic, and linguistic diversity and can include urban, rural, and nomadic populations (Beck and Keddie, 1978). Despite such differences, Muslims share in common their belief in the Qur'an (Koran) as the word of God as revealed by the prophet Muhammad, the rulings of which are central to religious beliefs and behavior. As the ultimate source of Islamic jurisprudence, the Qur'an has long determined every aspect of life, including gender status, throughout Muslim lands.

This is not to say that pre-Islamic tradition as well as contemporary indigenous customs have had no influence on the rules and regulations shaping female existence in the Islamic world. Nor does it negate the impact of the degree of economic development and modernization or the nature of the state on the position of Muslim women in each country. What is implied is that, regardless of existing geographic, economic, political, social, and cultural variations, the rulings of the Qur'an and Islamic law remain in force (Esposito, 1982) and continue to exert lasting influence on the public and private lives of women in the Muslim world.

The general rulings of Islam regarding female rights and responsibilities are clear and may be found in the verses of the Qur'an. What is interesting, creating even further variation among countries, is how the very same verses are interpreted in different contexts at different times. A review of Islamic laws drawn from the Qur'an can shed light on the legal position of Muslim women.

Islam came to the fore in seventh-century Arabia, a tribal society marked by patriarchy and disregard for female rights. Pre-Islamic Arabia witnessed the buying, selling, and inheriting of women as slaves, unlimited polygamy, female circumcision, and infanticide of baby girls. The status of women changed radically upon the advent of Islam as a new religion with a reformist nature.

Under the rule of Islam, women were granted a legal status along with certain rights and duties. According to Islamic law, women are entitled to own and inherit property and enter a trade or business. They are allowed to keep their father's name after marriage. Their sexual rights are recognized in the marriage contract, the unfulfillment of which may be cause for divorce initiated by the woman. According to Islamic law, it is the duty of the husband to provide his wife with shelter, food, and clothing, while the woman is not required to do housework or contribute financially (Walther, 1993, p. 59). Furthermore, Islam does not prevent women from becoming socially and politically active. Granting such rights to women when they were traditionally treated as merchandise to be bought and sold and even inherited before the advent of Islam in Arabia has long been regarded as a great advantage enjoyed by women under Islamic law. Wearing twentieth-century glasses and looking at the rights and responsibilities of Muslim women today, however, does not present such a favorable picture.

Contemporary Muslim women living in countries that abide by the Islamic law inherit half what a male heir does. The evidence of two women is regarded as equivalent to that of one man. Islam grants the husband the right to have up to four wives provided that he can treat them all equally as stated in the Qur'an. The justification behind polygamy in Islam is the concern of the prophet Muhammad for the widows and daughters of men who fought for Islam and were killed in battle during the territorial expansion of the faith. Allowing men to have more than one wife is, in fact, presented as protection and security for otherwise abandoned women. In Shi'i Islam, "temporary marriage" for a certain period of time is also allowed for men. The reason stated for such a marriage is the satisfaction of the man's sexual desires in a legal context while he is not married or is married but away from his wife, especially since adultery is punishable in Islam for men and women alike. Although the child resulting from such a temporary bond would enjoy equal rights with children from a permanent marriage, the woman cannot claim any rights upon the termination of the temporary contract.

Islamic law permits divorce, since marriage is regarded as a contract, not a sacrament. A husband has the right to divorce his wife, while the latter has to have good cause and a strong case, such as male impotence, abandonment, mental disorder, or cruelty, to ask for divorce. The same law applies to all women regardless of their social ranking. Since the father is the legal guardian of the children in Islam, the mother has the right of custody for boys only until the age of seven and for girls only until they are of age. (Being "of age" varies in different Muslim countries.) Usually, mothers cannot remarry without losing their children. A Muslim woman is not allowed to marry a non-Muslim unless he converts to Islam, while a Muslim man may marry a woman of either Jewish, Christian, or Zoroastrian faith. The restriction imposed on women is due to the fact that Islam regards the mother as the transmitter of faith to her children.

Yet another difference between men and women under Islamic law concerns remarriage after death or divorce. A woman must wait three months after divorce before she can remarry, to ensure that she is not expecting a child from her previous marriage. Men, however, can remarry immediately after divorce. On the other hand, a woman must wait four months and 10 days after the death of her husband before remarrying, during which she should observe a period of mourning. Within the public realm, the Muslim woman has the right to education and employment, yet in traditional societies she is not allowed to be a political ruler or a judge because of what is perceived to be her "emotional" nature. Furthermore, in most Islamic countries, a woman can work outside the home and travel only after obtaining the written permission of her male guardian.

The most visible and controversial mark of living in a Muslim society is veiling. The covering of women has many implications, ranging from control of female sexuality to the separation of the male (public) from the female (private) domain with a clear division of labor between the two spheres. The invisibility imposed upon women though veiling has its roots in the belief that female sexuality is a source of temptation and seduction to men. Women should, therefore, be controlled, covered, and secluded to the private realm to prevent their disruptive potential and thus protect them from any threat to their honor. Although veiling has been used to restrict female conduct and mobility and has often led to the exclusion of women from public life, the interpretation of separate spheres as "sexual apartheid" is a simplistic view of the intricate existence of women in Muslim countries and their active participation in the affairs of their societies. This has not always been the case in modern times, especially among women who have personally chosen to return to the veil. To gain insight into the reasons behind veiling and the justification for its practice, it is necessary to look at the roots and history of veiling and its role, real or symbolic, throughout time.

Modern scholars point to the prevalence of veiling in the pre-Islamic period, thus disclaiming it as a Muslim custom or tradition. According to Keddie (1991) and Walther (1993), veiling and seclusion developed in the pre-Islamic Near East among Persians, Assyrians, Babylonians, and noble ladies of Mecca as a symbol of class distinction among urban upper-class women to distinguish them from slaves, protect them from strangers, and show that they did not have to work. Veiling of "respectable" women also existed in the Greco-Roman world, among Athenians and Mediterranean women in Christian societies. The veil, as a symbol of honor and rank, is thus believed to be a "class and urban phenomenon." The wearing of the veil was not practiced by low-ranking, hard-working Bedouin women of early tribal Arabian society. Extensive veiling of Muslim women at a later time might, therefore, have been an imitation of customs practiced in Muslim-conquered lands such as Persia. Even today, strict veiling is not practiced by nomadic and rural women who have to perform hard manual labor and toil for survival.

Scholars also argue about the extent to which the Qur'an stresses veiling as practiced in the late twentieth century. Quranic verses demand covering for the believing women, especially the wives of the prophet Muhammad (Mutahhari, 1987), ordering them to veil their bosoms and hide their ornaments. Today, veiling is interpreted as total coverage of the female body except the hands, feet, and face (although some women also cover their faces). In general, Muslim women should be veiled in public and not use cosmetics or perfume in mixed gatherings.

What can we conclude about the role of woman in Islam? What is her status as ordained by the Qur'an and Islamic law? There is no single answer to this question. The position of the Muslim woman is the source of heated debate among traditional Islamists, reformist Islamists, and feminists, be they secular or Islamic. The traditionalists are clear about male–female differences and insist upon distinct sex roles based on what is perceived as the inherently different nature of men versus women. The majority of fundamentalists view women as basically emotional, sentimental, and weak humans whose primary duty is in the home. The most extreme example is the case of Afghani women who have been forced out of the public sphere by the victorious forces of the Taleban in Kabul in the late 1990s. This is not to say that all Islamists oppose female participation in the public domain. In fact, a close examination of the 1979 Iranian revolution clearly proves the opposite. What is implied is that traditional Islamists place a high premium upon domesticity and motherhood, after the fulfillment of which the woman may explore other territories.

Traditional Islamists take every verse of the Qur'an and every ruling of the Islamic law as a source of protection and security for women, who are regarded as minors to be controlled and protected by superior men. Blaming pre-Islamic tradition as the real source of female subordination, they view the position of Muslim women as a superior one compared to the status determined for them by other religions and ideologies. Women are regarded as precious and sensitive beings and compared to flowers and pearls; their protection is obligatory for the Muslim society. Thus veiling, protecting them from the harsh realities of the outside world by restricting them to the domestic sphere, and providing for their food, shelter, and clothing are all measures to ensure a smooth existence for "delicate" creatures that are unable to fend for themselves otherwise. The woman will, in turn, nourish and nurture the family and try to please her father, brother, husband, or son, whoever may be the head of the household. From this point of view, the woman has yet another responsibility: maintaining the honor, reputation, and status of the family. From a traditionally Islamic point of view, a man's honor depends to a great extent on the virginity of his sisters or daughters, the fidelity of his wife, and the sexual abstinence of widows and divorcees in his immediate family. If, for whatever reason, family honor is not respected and female sexual purity not abided by, the religious institution, as well as the legal system, in Muslim countries allows the man to impose any sanction deemed appropriate at the time. The woman is thus regarded as the "guardian of tradition" and the preserver of family honor.

The paradox of viewing woman as a minor to be protected as well as a cherished creature to be put on a pedestal and given the torch of family honor is clearly justified by Islamic reformists. According to them, female inferiority is only in relation to her physical strength and weaker anatomy. Otherwise, she is capable of participating in all spheres of public social life. Liberal, modernizing reformists try hard to accommodate the rulings of Islam with the realities and requirements of modern times. That is why they act as apologists, returning to, interpreting, and reinterpreting the Qur'an and Islamic jurisprudence to answer the needs of rapidly changing Muslim societies.

What the reformists are, in fact, doing regarding female status is walking on a tightrope. On the one hand, they cannot relinquish women from the demands of tradition and free them from the double binds of security and control. On the other hand, they are aware of, and responsive to, the requirements of changing times, demanding women to be exemplary wives and mothers as well as active participants in all walks of life. What reformists have offered Muslim women is justification for their unequal treatment and a double burden of traditional expectations and modern demands. The product of reformist attitude and behavior is a lonely woman who has lost the secure pillars of male patriarchy and protection without having replaced it by a viable alternative that can offer her support during periods of uncertainty.

Yet a third response to the question of female status in Islam comes from the feminist point of view. Feminism in this context is a secular movement to achieve equality for women. The feminist point of view, with its roots in the social movements of the 1960s and 1970s in the West, regards the position of the Muslim woman as a totally subordinate one, marked by inferiority and exploitation. Viewing the veil as the ultimate symbol of female subordination, secular feminists in the West and in Muslim countries have long attempted to abolish it and alter the rulings of Islamic law that they believe treat women as half-human by viewing them as physically, mentally, and intellectually inferior to men. While regarding the dictates of Islam as enlightening and liberating for seventh-century Arabian society characterized by patriarchal and tribal values, feminists have opposed their practice in modern times. The feminist stance is explored more fully later in this essay. Suffice it to say that debate about the status of women in Islamic countries has entered the realms of biology, psychology, sociology, economics, and even politics. The "woman question" has become an ideological issue in which every political group is pushing its own set of beliefs and trying to gain popularity through the publication of treatises about the "right" place of women in society.

Such diversity in viewpoints about women adds yet another dimension to the aforementioned variations in Islamic countries. These, in turn, are influenced by the various stages of development/modernization in each country. The next section is devoted to in-depth discussion about how variation in political ideology, the stage of economic development, and the composition of the social structure directly affects the role of women in different countries.

Social Transformation and the Changing Role of Women

Muslim women have been treated differently, according to the political ideology of the ruling group and the socioeconomic structure of their societies. Although women have continuously contributed to the shaping of their lives, this section deliberately highlights women's changing social, political, and economic contexts as opposed to their own active role. To look at social transformation as a key element in determining the role of woman may seem contradictory to what was originally stated in the essay—that the Qur'an and Islamic jurisprudence are the major determinants of female status in Muslim lands. Yet it is an undeniable fact that the Qur'an and the law have been interpreted and reinterpreted to suit the needs and exigencies of each period. Claiming to have practiced "real" Islam, the heads of states have repeatedly used the rulings of the faith to justify their acts. Seldom has a leader acted totally indifferent, if not hostile, to the framework determined by Islam, with the exception of Kemal Ataturk, the founder of the Turkish Republic in 1923. In the majority of contexts, political rulers have sought the help of religious leaders to add a touch of Islam to their rulings in order to legitimize their actions in the eyes of the common people who continue to remain ardent believers.

The two major determinants of Muslim women's role and status are the political ideology of the rulers and the economic structure of the society. Whenever the two fit together and reinforce each other, a clear pattern emerges in which the position of women is determined by law and strengthened by social policies and cultural norms. At other times, when, as a result of expansion, reform, or revolution, the economic realities of the society do not match the ideology of the rulers, women are forced to live in an ambivalent situation burdened with contradictory messages and conflicting realities. A historical look at female status immediately before the advent of Islam and its transformation throughout time (see Mernissi, 1991) will further clarify the above contention and illustrate the contradictions women have faced at all times.

The ambivalence in female status was evident in Arabia before the advent of Islam. In the tribal/pastoral society, marked by patriarchal values that bestowed an inferior position on women, female infanticide was widespread and women were treated as material goods to be bought, sold, and inherited. At the same time, certain women contributed to the social and economic affairs of their society; the most distinguished example is Khadija, a strong and wealthy merchant for whom the prophet Muhammad worked before he married her.

The rulings of the Qur'an bestowing a higher status on women and the realities of the Islamic world at the time of conquest and expansion reversed, to a certain extent, the subordinate position of women. Some women became active and visible in public life in the early period of Islam (Walther, 1993). A study of the lives of A'isha (among Sunnis) and Zeinab (among Shi'is) illustrates their active role in shaping the history of early Islam. A'isha's participation on the battleground and Zeinab's active role in recounting the Shi'i struggle are symbols of the militant role of women in the early period of Islam, shattering myths about the passive role of Muslim women throughout time.

The gradual expansion of Islamic lands through territorial conquests and contacts with other peoples and civilizations fundamentally altered the status of Muslim women. The transition to an agrarian lifestyle and the rise of urban life led to a new division of labor among men and women in which men remained active in the public sphere and women were relegated to the private domain, restricted to household chores and childbearing. Furthermore, the deep impact on Arabs of the Byzantine and Persian empires' attitude and practice of veiling and secluding elite women led to the imitation of the lifestyle of the new converts and the seclusion and veiling of Muslim women. Notwithstanding periods of time during which women acted as powerful leaders during the rule of the nomadic Seljuks, Mongols, Mamluks, and early Safavids (Keddie, 1991), more fundamental and widespread change had to wait for growing female consciousness and legal and economic reforms in the modern era.

Contradiction continues to mark the lives of Muslim women in contemporary times. It rises from social transformation, through gradual reform or more often through radical upheavals; changes in the stage of development and modes of production; alteration in the political ideology of the state (Kandiyoti, 1991); and, last but not least, fluctuating relations with the West.

The realization by Muslim rulers in the nineteenth century that, ever since the end of the golden age of Islamic science, philosophy, and medicine (from the eighth to the thirteenth century), they had faced military, technological, and scientific defeat at the hands of the mighty West tilted them toward reform within the military and educational arenas. Opening Western-style military academies and schools at home and sending local students abroad led to significant transformation in the lives of Muslims. The exposure to Western science and technology, behaviors, and lifestyles had a deep and lasting impact, first on the members of the ruling class and the elite and later on a significant portion of the population. The East–West contact gradually led to changing roles among women, whose lives could no longer remain as before. This led to even more contradictory messages for women, caught between the demands of tradition and the expectations of modernity.

The elite women of the ruling class were the first to respond and/or accommodate to change brought about by the West, but it was not long before middle-class women also tasted the flavor of modern times. Attempts at modernization and economic development by twentieth-century governments were supplemented by the adoption of the capitalist mode of production. Women entered into wage labor for the first time, followed by active participation in public life, mostly in institutions and professions connected with education and health.

Slowly but surely the engines of modernization—mass education and employment—began to affect the social and cultural institutions of traditional societies. Muslim societies became divided into the modern and the traditional sectors, each with its own customs and lifestyles, resulting in a cultural dichotomy within the population. Debates raged regarding the "appropriate" place of women in the home, or in the modern factories, schools, and hospitals that desperately needed their labor. Whether women were allowed to contribute to the social and economic affairs of their society, prohibited from participation in the public domain, or held in limbo due to fluctuating demands and contradictory expectations depended very much on state ideology.

The sociopolitical ideology of the ruling group continues to exert influence on the status of women in each country. Traditional states, such as Saudi Arabia, have clear guidelines concerning the limitations imposed on women. Complete segregation of the sexes and a clear-cut division of labor in the private as well as the public sphere are the hallmarks of countries that are headed by traditional leadership. On the other hand, modern state ideology, which is often more secular in nature, such as Turkey and Tunisia, calls for using the labor and ability of women in all spheres of life and poses far less hindrance to their active participation in the world outside the home.

In this context, value-laden terms such as "traditional" or "modern" are defined in relation to how much freedom is granted to women to participate in public life. Thus a traditional state ideology is one that adheres to the principle of segregation and regards the woman's role as first and foremost a maternal and domestic one, whereas modern ideology is characterized by laws that ensure equal opportunities for women and encourage their active participation in the public domain.

The ideological variety leads to practical variety: at one end of the continuum, the establishment of coeducational centers of learning, the banning of the veil, the outlawing of polygamy, and legalization of female voting in Turkey as early as the 1930s; at the other, the public segregation of the sexes, the prohibition of female voting, and the illegality of female automobile driving in certain countries of the Persian Gulf. In the middle are states that fluctuate between granting equal rights to women in peaceful times to surrendering them at times of political unrest in order to appease religious opposition. Egypt and Algeria in the 1980s and 1990s are examples of such states.

The more complicated case is a state ideology that is both modern and traditional. A clear example is the Islamic Republic of Iran. Known as a fundamentalist state that advocates a return to Islamic roots, postrevolutionary Iran was long identified as an antimodern, backward-looking country. The reaffirmation of Islamic laws such as polygamy and compulsory veiling was often cited as further evidence of the return to the medieval past in the late twentieth century. It took some time for both Iranian leaders and scholars in the field of Iranian studies to realize that one thing is certain: One cannot swim against the stream of industrialization and modernization for too long. Even more important, one cannot ignore the active role of women as determiners of their own destiny. Neither the rulers of Iran nor those who have tried to "formalize" the Iranian experience could have predicted the resistance—active or passive—of the Iranian female population against dictates that they leave the public domain and remain in the seclusion of the private sphere. No force could stop the Iranian women who had actively participated in the revolutionary process, had fought for the independence of the country, and had helped reconstruct a war-torn country to return to the kitchen and be content. The new Iranian woman has fought, and continues to fight, for her rights in the private realm of the family as well as the public spheres of education and occupation.

It is due to the perseverance and resistance of Iranian women that in the 1990s we witness the attempt of leading female political and ideological figures in Iran to bring about more opportunities for their sisters in the legal, educational, economic, and social spheres. Such amendments to the law as more rights for women in the case of divorce, the requirement that husbands seek permission from their first wife to marry a second wife, compensation for services rendered at the marital home in cases of unjustifiable divorce by the husband, and further opportunities for pursuing a variety of specializations in higher education are all the results of Iranian women's struggles.

The female experience in postrevolutionary Iran is a clear example of living with contradictions. Women living in times of rapid social change, such as in Palestine or Algeria, learn to live with duality. Life in transition has neither the security of tradition and repetition nor the protection of established institutions of modern times. It is a daily struggle marked by instant decisions and trial-and-error acts. On one side of the coin are instability and unforeseen change, while on the other side are innovation and creativity. Modern history has illustrated over and over again that women can build and maneuver and lay the foundations for the future at times when state ideology is zigzagging between tradition and modernity, conservatism and revolutionary fervor.

The discussion to here has focused on the role of outside factors—namely, the rulings of Islamic jurisprudence and the dictates of the social, political, and economic order—in determining the status of Muslim women. The next section highlights women themselves as the main actors in transforming their role and position. It demonstrates clearly that women are not passive recipients of rights and services who merely react or adapt to social change, but active contributors to their society who, through their daily struggles and choices, help shape their own lives.

The Women's Response

Whereas, in the past, feminist consciousness and expression appeared to be the exclusive right of elite women whose lifestyles permitted them to dare to be different, the onset of social upheavals, radically transforming order and the status quo in the modern period, led to more vocal expression by popular-class women. It is true that educated, professional women are still the spearheads of change, yet popular movements and grass-roots activities have their own way of introducing new role models hitherto unknown. Together, and in spite of their differences, Muslim women are trying to create a new identity for themselves.

Whatever the result may be, it is the process of formulating an alternative identity and the struggle to build something novel that are of utmost significance for gender studies in Islamic countries.

Two groups of women are active in determining female status in the Muslim world: the secularists and the Islamists. Although the term "feminist" is used for both (Islamic feminist, Muslim reform feminist, secular feminist), using the term "feminist" in this discussion for the secular advocates of women's rights; "Islamist," for those who propound a return to the Islamic roots; and Badran's "gender activist" (Moghadam, 1993, pp. 160, 167), when referring to both, seems more appropriate. Whereas secularists such as Huda Sharawi and Halide Edip have been active since the early years of the twentieth century, Islamists have gained prominence since the mid-1970s. Bound by a love–hate relationship, both groups are committed to the cause of female empowerment, attempting to restore the rights of Muslim women, yet each in their own way and through their own worldview. It is important to understand the debate between the two groups in order to assess the possibility of future dialogue and cooperation. Secular feminists and Islamists are usually regarded as "ideological adversaries," mainly because of their disagreements on issues of utmost concern to women. A close examination of their worldviews may determine whether their hostility toward one another is irreversible. The reaffirmation of structural differences and the discovery of new realms of cooperation are important for social redefinitions of gender in the Islamic world.

Just as the name indicates, secular feminism is a secular ideology treating religion as a personal matter that should be relegated to the private sphere and not mixed with public and political issues. As a movement concerned with the status of women, secular feminism struggles to ensure equal opportunities for women in the domain of the family as well as in the economic and political spheres. Secular feminism has basically been popular among modern middle- and upper-class professional women whose Western liberal education has led them to question tradition and call for gender equity. The long-standing opposition of the secularists to what they perceive as the patriarchal rulings of Islamic law and veiling as a form of "social control," as well as their disdain for traditional customs, has earned them such labels as "Western feminists" and "rootless imitators" of foreign ways. Secularists are blamed for negating the indigenous, authentic identity of the Muslim woman and attempting to model her after the Western ideal. Given such a negative view of secular feminists, is there any possibility of a coalition—even if temporary—with Islamists during periods of transition? To answer the above question, one must also study Islamism and its position on women.

Referred to as political Islam, militant Islam, radical Islam, Islamic revival, or fundamentalism, Islamism may be defined as a political-ideological-cultural movement advocating return to Islamic tradition in the face of modern challenges in the late twentieth century. Islamism is the "reconstruction of the moral order that has been disrupted or changed" (Moghadam, 1993, p. 135) as a result of modernization, industrialization, urbanization, and Westernization in the modern era. Various scholars point to the underlying causes for such disruption, including socioeconomic dislocation brought about by rapid modernization resulting in cultural duality and a growing gap between the Westernized elite and the popular classes; increasing disillusionment with political rulers and a crisis of legitimacy of the state; the growing power of the West undermining the independence and national sovereignty of Muslim countries; an increased gap between the traditional petty bourgeoisie and bazaar merchants and the Westernized upper-middle class engaged in the modern sector; weakening of the patriarchal family and the breakdown of the security provided by the extended-family structure; a growing sense of anomie resulting from the dehumanizing and degrading dimensions of modernity; and feelings of sociocultural malaise and crisis of national identity vis-à-vis the technological and military supremacy of the West (Keddie, 1991; Moghadam, 1993).

Presented as such, Islamism is a reaction, a remedy, a retreat, a return when faced with crisis, breakdown, dislocation, tension, frustration, anomie, grievance, and malaise. Yet, painting Islamic revival as a reactive position portrays only part of the reality. Islamism is also an active search for alternatives; it is a conscious and deliberate choice as opposed to an automatic reaction. This active quest has focused mainly on the question of identity; as such, it is first and foremost a cultural act, seeking a cultural alternative. Formulating a new, strong, and independent identity for Muslims has indeed been the mission of Islamist movements in recent decades.

There are many reasons for the focus on identity. One of the most important may be the fact that Muslims have not been able to offer an alternative economic model or technological know-how to compete with that presented by the West and Japan. Since the golden age of Islamic civilization, Muslims have, for the most part, remained imitators of Western ways and consumers of Western goods and ideologies. Given their inability to replace Western ways in science, technology, economy, and education, Muslims consciously decided to present an alternative in the one sphere in which they had retained their pride and dignity—namely, the realm of culture. Not only can Muslims find a sense of pride and security in their cultural heritage, they can also present it as something that truly belongs to them and is not a variation on the Western theme. Thus the response to external challenge and internal deterioration is the strengthening of identity and cultural integrity. It is true that the institution of culture alone cannot face the modern challenges of rapid urbanization, overpopulation, rural–urban migration, and industrialization, but it may provide a refuge in which Muslims can seek comfort and security. As such, it is the single remaining vestige of authenticity in a world where every aspect of Muslim life has been questioned and challenged. Since not all dimensions of the cultural resistance and reconstruction are tangible, Islamists have focused on the most visible one—the role and position of women.

One of the key dimensions of the process of identity formation, and probably the most controversial one, has been the role and position of women. Indeed, the "woman question" has become what Keddie (1991) refers to as a "politico-ideological" question. Not restricted to the realm of tradition and culture, female dress in the contemporary Muslim world has also become "the badge of ideology." While some argue that the Islamists' emphasis on the status of women and their obsession with the issue of proper clothing for Muslim women have their roots in the failure of the Islamic revivalist movement in offering a viable economic alternative and rescuing the Muslim world from the cycle of dependence, others point to the importance of the woman question as the symbol of what contemporary Islamist ideology stands for. Let us explore the Islamist position on women and discuss its inherent contradictions.

In their endeavor to retrieve traditional culture and create a new identity, Islamists have selectively and meticulously combined the old and the new. The coexistence of the traditional and the modern presents a real paradox for those who view it from outside. The contradiction presents itself most in the composition of the Islamist supporters, both male and female, and the means used to reveal the message of Islamic revival. A prevailing myth at the onset of the Islamist movement in the 1970s was that its supporters were mainly uneducated rural migrants and urban marginals who had experienced only the negative side of modernization. In Iran, the Islamic revival found its most vocal adherents among young, middle-class, urban professionals who had enjoyed the benefits of modernization through the educational and occupational structure. Together with the traditional petty bourgeoisie composed of shopkeepers, artisans, bazaaris, and the urban poor, they make up the strong support triangle of Islamists. As in other Muslim countries, those who have benefited most from Westernization and modernization—the educated upper class and the modernizing liberal bourgeoisie—continue to oppose the rise of Islamism.

The same contradiction exists regarding the female supporters of Islamism. The popular expectation was that only women who were at the margins of the modernization-industrialization-urbanization experience would seek refuge in Islamic revivalism. Furthermore, it was believed that the traditional stance of Islamism on women would not attract the beneficiaries of Westernism and liberalism. Yet, many middle-class women, including young and urban professionals, joined the ranks of revivalists. A combination of political, economic, cultural, and psychological reasons may explain women's conscious choice to return to the rulings of Islam. In the economically dependent and politically unstable Muslim countries, Islamism, especially if it is still the voice of opposition and not the ideology of the state, offers a theoretically self-reliant politicoeconomic alternative that rarely fails to stimulate nationalist sentiments. As conscious and conscientious members of the society who seek to become active contributors to social reconstruction as opposed to passive recipients of Western leftovers, educated women find cultural independence and integrity in Islamism. Furthermore, having struggled every step of the way to seek education and employment, in spite of family disapproval and traditional restrictions that prohibited women from participation in desegregated centers of instruction and work, "first-generation" urban, educated, professional women can relax in an Islamic environment that also facilitates the educational and professional advancement of their daughters.

It was also believed that women exposed to, or accustomed to, wearing Western-style clothing and gathering in mixed groups would not easily and/or willingly reject or give those opportunities up. However, in some countries where compulsory veiling and sex segregation are not legally enforced, such as Turkey and Algeria, some Muslim women choose to keep or put on the veil and seek single-sex alternatives. A close examination of veiling as a symbol reveals that covering for women is no longer only a personal and private matter but has become a political statement and an ideological stance.

Veiling has represented different things at different times. While the veil symbolized high status in the pre-Islamic Near East, it gradually became identified as the distinguishing mark of Islam, so that any form of covering was associated with Muslim women. With the onset of Westernization and secularization in the modern period, veiling became the sign of tradition, male domination, and low socioeconomic status while Western clothing symbolized modernity, female liberation, and elite standing. In the early twentieth century, Muslim women had to choose between tradition, symbolized by wearing the veil and remaining isolated from political life, and modernity, which included active participation in the social, political, and economic spheres. The choice was imposed by secular governments that used the image of women dressed in Western clothes as a symbol of the nation's modernization. Such imposition impeded the public activity of traditional women who chose to remain loyal to the dictates of their religion. In the late twentieth century, veiling is no longer seen as a sign of backwardness but rather as a facilitator that enables women coming from traditional religious backgrounds to pursue their interests and dreams without the anxiety and pressure imposed on their mothers who sought to break tradition, while adhering to their religious beliefs. The veil is no longer a hindrance today. Muslim women do not have to choose between tradition and modernity; they can have both at the same time. Furthermore, in the age of Islamism as an ideology of discontent, veiling has become a mark of political activism and militancy. Women in prerevolutionary Iran, Algeria, Turkey, and Palestine have returned to the veil to protest the status quo and resist Western cultural penetration. The veil has become an identifier, labeling one as Islamist or secular depending on her choice of Islamic covering or Western dress. Thus veiling has gradually transformed from a personal choice and a religious and cultural symbol to a value-laden political and ideological issue.

Going beyond the veil as the most visible sign of Islamism, let us examine the stance of this religiopolitical ideology regarding the role and position of women. The issue of gender is of paramount importance for Islamists, male or female, who aim at creating a new moral order in which women act as the main transmitters of the desirable sociocultural values and the socializers of the next generation. In this ideal society, women are to be symbols of modesty, purity, and religiosity—the preservers of community morality. They are to stand against Western cultural penetration and raise the flag of Islamic revival. They are even regarded as repositories of national, cultural, and religious identity (Moghadam, 1993). They are the ones who can lead the community to the Islamists' ideal, the formation of an Islamic state. While Islamism exalts women to such a high position, however, it also teaches the Muslim woman that she should know her limitations. She has to internalize the belief that men and women are biologically and psychologically different and that their natural distinctions bring about different sex roles and responsibilities for them. She has to understand and accept that domesticity and motherhood should be her priorities; she has to fulfill her family and domestic role as an exemplary mother and housewife. Here the difference between the Islamist and the feminist point of view regarding the status of women (and thus the emphasis on not using the term "Islamic feminist") is clarified. Whereas feminism stresses "woman's essential worth" and her "common humanity," which do not depend on her relationships with others, the Islamist viewpoint emphasizes her role as a wife, mother, sister, and daughter, guarding the family honor at all times.

The ideal Muslim woman has to fulfill other roles as well. She is not to be domesticated like her mother and grandmother before her. She has to assume sociopolitical responsibilities. She has to advocate the cause of Islam through active participation in public life. She is to be educated and is allowed to work outside the home in occupations that suit her unique biological and psychological makeup as a physically weaker and more emotional creature. An outsider may think that there is a role conflict between the private and the public demands made on women. According to the Islamists, there is no ambiguity. The double burden poses no conflict because the priorities are clear: first home, then outside. Such prioritization is fully internalized by the Islamist woman and she is, therefore, not torn between double roles and double expectations. By knowing her priorities, she can adopt simultaneously a traditional and a modern identity. The veiled doctor, professor, lawyer, engineer, physicist, computer scientist, or artist presents the most contradictory image. Yet the contradiction is only in the eyes of the outsider. Inside, there is a sense of pride to have been able to maintain tradition while taking on the modern challenge. The new Muslim woman enters the age of computers and satellites, using modern science and modern institutions while clinging to her religion and her veil. She does not have to forfeit one in order to gain the other.

The future discourse on women in the Muslim world has to include both the secular feminist and the Islamist dialogue to portray the reality accurately. The challenge today faced by all those who struggle for the betterment of women's lives in the Muslim world is to initiate communications between secular feminists, inspired by Western experience and "indigenous feminists" (Ahmed, 1992; Kandiyoti, 1996). It may very well be that a third, less extreme, more balanced redefinition of gender will enter the picture in the future. Such an alternative ideological framework may be better able to select the best in indigenous and Western cultures and create something new. More important, the third way may be more rooted in the experience of women in the popular classes as opposed to the secular or traditional urban elite. As such, it may address the real needs and aspirations of the common woman as opposed to the political ambitions among the elite.

Conclusion

An examination of the role of women in Muslim countries shows clearly that, although the rulings of Islam on women are clear and binding, social transformation and the interplay of political, economic, and cultural forces in each society play a crucial role in determining the status of women in different countries at different times. The structure of the economy and whether the society is a nomadic/pastoral, agrarian, or urban one; the nature of the state and its adherence to a conservative, reformist, or revolutionary ideology; and the dominance of a traditional or modern/liberal culture—all are external factors that have continuously shaped women's role in Muslim lands. Meanwhile, reformists, apologists, Islamists, fundamentalists, revivalists, modernists, and feminists have interpreted and reinterpreted the Qur'an and Islamic law to determine what "real" Islam has ruled for women. Yet, there is another determining factor, an internal one, whose impact should not be undermined—the women themselves. Both their characteristics, such as socioeconomic status, religious background, and level of education and occupation, and their degree of participation in, and contribution to, social affairs determine how women view themselves and how they are treated in Muslim societies.

It is ironic that while women have had to face many obstacles in their struggle against various forms of patriarchy and male domination throughout Muslim lands, in the late twentieth century they were the first group to whom the Islamists turned in their modern contestation. Basically ignored in the process of nation building and relegated to an inferior position behind the doors of the private domain, Muslim women are now expected to symbolize a nostalgic past and an ideal future. The new Muslim woman is, in fact, asked to become a cultural warrior in the Islamists' struggle. Although the fervor of Islamic revival has led to its classification as a political and revolutionary movement, it should be noted that contemporary Islamism is first and foremost a cultural act, at the core of which lies a quest for identity. Muslim women are placed at the center of the process of identity formation, this time not only in their domestic capacity but also as active agents in the cultural struggle.

Given the new role and responsibility of women, how should Islamists and secular feminists act? Can they work together or are they eternal enemies? Is there a common ground for coalition, even if short term and focused on a specific issue? An examination of modern national liberation movements in Islamic countries illustrates that women have usually aligned with men to fight against a foreign power or depose a domestic dictator, then have returned to the kitchen upon victory. Seldom have Muslim women independently and simultaneously raised the cause of gender equity and female empowerment, believing that it is of secondary importance in the national struggle and should be attended to at a later stage. Yet, experience has shown that the second phase will never come. The Iranian and the Palestinian experiences clearly illustrate that the woman question should be raised constantly throughout national movements so women can gain equity and liberation during the struggle as well as afterward. Consciousness raising regarding the necessity of perpetual activity may be an issue on which Islamists and secular feminists agree and cooperate. Other issues around which possible coalitions among Islamists and secular feminists can be built include reforming family law to meet the needs and demands of wives and mothers; providing equal opportunities for female students at all levels of education; seeking expanded occupational choices for women; providing various forms of assistance for working women such as day-care centers; supervising and ameliorating the working conditions of female laborers; and providing legal, economic, and social aid for poor, illiterate women. Although the context of cooperation is clear, are coalition and dialogue possible? The answer depends on the situation of each country.

Dialogue is possible in Muslim countries such as Iran that have overcome the bitterness of violent confrontation and/or witnessed consolidation of Islamists' state power and may be prepared for debate regarding points of mutual concern. Dialogue is more difficult in transitional societies in a state of upheaval. The often violent struggles for power in Algeria, Palestine, or Afghanistan, for example, do not leave much room for coalition building on common points.

The major challenge facing secular feminists and Islamists who want to facilitate dialogue and coalition building is the conscious and active elimination of stereotypes and simplistic presentations of the "Other." The Islamist portrayal of countermodels as alienated, Westernized, immoral, and decadent beings, and the secularists' presentation of the veiled Muslim woman as an ignorant, regressive, and passive creature, are neither helpful nor true. The black-and-white presentation of female reality as two hostile camps, one good and one evil, only further divides and weakens women in their struggles for empowerment. Many secular women may seek not to cooperate with the state feminism of Islamic fundamentalist countries since it is considered an ideological pillar of the ruling group. Likewise, Islamist women may choose not to align themselves with what they perceive as secular elitism among modernizing feminists. Yet, each group can struggle to create a democratic environment in which the two ideologies are presented to the public, leaving women free to choose what they regard as more appropriate. Imposing ideological monopoly will only close the doors of communication, and this is one thing women living in Muslim countries cannot afford.

Maybe the answer lies in women themselves forming grass-roots organizations in which they can discuss issues of pivotal importance to them and seek personally meaningful solutions to real problems instead of joining the rank and file of state feminist organizations or elitist institutions that seek to consolidate their own power. Women, especially those who are educated, have begun forming their own self-reliant units; it may take more time for women from the popular classes to find the time in their daily toil, or feel the need, to express their demands in a more organized format.

Muslim women have begun seeking their rights, the most important of which is the right of identity. They want to determine who they are and what they seek to become through their own efforts. They are engaged in the enlightening, if difficult, task of self-exploration, the result of which is the affirmation of an independent identity, like the one expressed in the words of this Iranian university student:

> My grandmother was forced to take off her veil during the reign of Reza Shah. So she burned her hair to avoid the shame. My mother was taught in Mohammad Reza Shah's schools that it was shameful to be covered—veiling was a sign of backwardness. I had to put on the veil to avoid the shame of becoming a sex object. It seems that the history of Iranian women is the history of shame. Yet I am all and none of the above. I am someone new. I am a new generation of Iranian women. I choose consciously. I choose despite the limitations and the restrictions. I am a New Iranian Woman (Mehran, 1992, p. 20).

References

Ahmed, Leila. *Women and Gender in Islam.* New Haven: Yale University Press, 1992.

Beck, Lois, and Nikki R. Keddie (eds.). *Women in the Muslim World.* Cambridge: Harvard University Press, 1978.

Esposito, John L. *Women in Muslim Family Law.* Syracuse: Syracuse University Press, 1982.

Kandiyoti, Deniz (ed.). *Women, Islam, and the State.* London: Macmillan, 1991.

———. *Gendering the Middle East.* Syracuse: Syracuse University Press, 1996.

Keddie, Nikki R. "Introduction: Deciphering Middle East Women's History." In Nikki R. Keddie and Beth Baron (eds.), *Women in Middle Eastern History: Shifting Boundaries in Sex and Gender.* New Haven: Yale University Press, 1991, pp. 1–22.

Mehran, Golnar. *Female Education and Identity Formation in the Islamic Republic of Iran.* Los Angeles: Von Grunebaum Center for Near Eastern Studies Publications, University of California, Los Angeles, 1992.

Mernissi, Fatima. *Women and Islam: An Historical and Theological Enquiry.* Oxford: Blackwell, 1991.

Moghadam, Valentine M. *Modernizing Women: Gender and Social Change in the Middle East.* Boulder: Lynne Rienner, 1993.

Mutahhari, Murteza. *On the Islamic Hijab.* Tehran: Islamic Propagation Organization, 1987.

Walther, Wiebke. *Women in Islam: From Medieval to Modern Times.* Princeton: Markus Wiener, 1993.

Modernity and the Mass Media

Ila Patel

Introduction

The role and status of women in relation to men have gradually changed since the beginning of the twentieth century. The mass media have played a decisive role in perpetuating gender stereotypes and in maintaining the status quo. With the introduction of new communication technologies, the media have acquired a global dimension. The very nature of technology enables the transmission of uniform media messages and images across regions. Although the form and content of the mass media have changed dramatically in the past few decades, the media in modern society have continued to play a significant role in the production and transmission of patriarchal culture. The relationship among modernity, the mass media, and patriarchy is, however, far more complex. The mass media function in the larger system of patriarchy and capitalism that controls media structures and organizations and represents women as subordinate.

In the 1960s and 1970s, the mass media in developing countries were assigned the role of modernizing traditional societies. Exposure to the media was perceived as an important indicator of modernity. The mass media repeatedly engage in redefinition of modernity and individuals. They influence the cultural domain in two ways. On the one hand, the mass media provide a large proportion of society with entertainment. With the proliferation of "women's genres"—soap operas, melodramas, women's magazines, and so on—women have also emerged as important consumers of mass entertainment. On the other hand, the mass media in contemporary society are increasingly responsible for the construction and consumption of social knowledge and meanings that people draw on to make sense of their world and to act upon their social reality. Thus, the mass media play an important role in setting the agenda for public opinion by selecting themes, items, and points of view that tend to reinforce patriarchal culture.

The way in which men and women are portrayed in the mass media is a reflection of reality. The media, however, do not represent reality as it is lived; instead, they choose some aspects of reality and re-present them to us such that we become recipients of the media's selective interpretation of the social reality. They selectively pick up certain existing realities and censor the others. By highlighting only selected aspects of existing attitudes, values, behaviors, and images in a particular context, by projecting them as larger than life, and by continual repetition of images and messages, the media influence reality and shape our perception of it.

The media, however, are only one element in the construction of women's marginality in culture. Feminist analyses of the media, culture, and society (Gallagher, 1992) produce much more complex understandings of the cultural dimensions of power and equality. The relationship between the media and reality is dialectical. Culture is not a static system, but an ever-evolving process that constantly changes to accommodate emergent alternative and oppositional meanings, values, and practices (Williams, 1977). The mass media, particularly the print and broadcasting media (television and radio), also provide some space for construction of new meanings and images. Although the progressive discourse is often co-opted in the mass media and reconstructed to confirm the hegemony of dominant social classes and to reproduce gender relations, women's groups and some media professionals continue to challenge this co-optation in the context of a larger system of patriarchy and capitalism that controls the mass media and subordinates women.

Although feminist activists and academicians concerned with the "woman question" have critically analyzed negative portrayals of women in the media, research literature related to developing countries is uneven across regions. This essay attempts to give a global picture of the representation of women in the mass media of developing countries. The subject has received far more attention in India than in any other developing country. Some research is reported from Asian countries; less from Latin America and Africa, partly due to the unavailability of much international research literature on women and mass media in India and partly due to my familiarity with Indian scholarship and strategies for change. The assumption throughout is that, despite apparent differences in societal context within which mass media function, there are striking similarities in the portrayal of women across divergent media systems in developing countries.

The discussion in this essay is divided into four sections. The first highlights feminist approaches to the study of media content; the second examines continuity and changes in the portrayal of women in the mass media; the third highlights strategies for changing the images of women in the mass media; and the final section summarizes emerging trends in the portrayal of women in the context of late-twentieth-century media developments.

Feminist Approaches to Media Content

Contemporary women's movements in North America and Western Europe have played an important role in shaping the academic agenda of feminist communication research. A feminist critique of media content and its implications in the construction of gender has been an important part of feminist cultural politics. The first international review of research and action, initiated in the 1970s by the United Nations Educational, Scientific, and Cultural Organization (UNESCO), drew our attention to striking similarities in the negative portrayal of women in the mass media (Ceulemans and Fauconnier, 1979; Gallagher, 1983) and to women's lack of decision-making power in media organizations throughout the world (Gallagher, 1981). However, a disproportionate volume of this research on women and the media was found to be from the developed countries.

In developing countries, feminist communication research on media content, images, and representation has grown since the early 1980s. It was the United Nations International Decade for Women (1976–1985) that provided political impetus worldwide to initiate research and action for changing women's portrayal and participation in the mass media. Compared to Western feminist scholarship in the field of communication, however, communication research on Third World women has remained fragmentary and descriptive. Despite the proliferation of women-in-development literature since the early 1980s, gender in the field of mass communication has remained a marginal area of inquiry among feminist scholars (Steeves, 1993). Hence, empirical and theoretical work in the area of women and the mass media in developing countries is limited and uneven across regions. Media research on women in developing countries has relied heavily on the feminist paradigm for communication research developed in the West. Despite limitations of the Western feminist approaches in the study of the content, images, and representations in the portrayal of women in mass media in the Third World, it is necessary to understand these approaches.

Gallagher (1992) highlights three strands in feminist research in this area. The focus of feminist scholarship in the 1970s, conducted mostly in North America and Asia (Japan, Korea, and the Philippines), was on quantitative content analyses of sex roles and media stereotypes. This research documented the invisibility of women in various media forms and highlighted how media images reinforce the negative portrayal of women in terms of behaviors, aspirations, psychological traits, and so on. Gallagher argues that such a juxtaposition of "positive" and "negative" media images of men and women is problematic because there is often a tendency to define "positive" images of women in "masculine" terms. Furthermore, such an approach ignores the ways in which the audience reads and reinterprets media content. Despite the limitations of this approach, it condemned and drew our attention to sexism in the media and provided impetus to feminist research on media content.

In the 1980s, feminist film criticism, based on qualitative European perspectives and methodologies that use psychoanalytical, semiotic, and poststructural frameworks in analysis, contributed to broadening our understanding of how the media construct definitions of femininity and masculinity and how images of women "reflect" or "distort" reality. This textual-analysis approach has been criticized for neglecting the dialectical relationship between the media and culture that constructs the notion of "women." The question is not merely one of examining whether the media reflect or distort images of women, but of exploring how images and meaning of femininity and masculinity in the media are socially constructed within the broader context of patriarchal social relations.

More recent feminist media criticism, which has drawn from cultural studies, shifts our attention from the text to the context of reception, in which the audience plays an active role in producing and negotiating textual meanings (Gallagher, 1992). Construction of textual meanings is an integral part of social and power relations in society, which are constantly contested and negotiated by the audience. Gallagher argues that recent feminist media criticism, heavily influenced by poststructuralist and postmodern theory, places far more emphasis on the autonomy of the audience reading of the text and on validating audience "pleasures" and ignores the fact that women as audience are positioned within a cultural system that reproduces particular representations of "femininity" and "masculinity."

Gallagher (1992) concludes that recent feminist media research and criticism give us useful insights in understanding cultural dimensions of power and equality, but she cautions against the apolitical trend in late-twentieth-century Western feminist media studies to focus on the microlevel while ignoring the issue of power and broader political and economic concerns. She argues that, in order to formulate relevant policies and strategic judgments, it is important—from the point of view of feminist action and politics—to situate microlevel work within the political-economic context of media development and to examine macrolevel forces impinging on the media as institutions and structures in society.

Because there are some commonalities in women's oppression around the world, and in women's position in relation to culture and communication, analytical and heuristic frameworks developed by Western feminists can be useful in understanding the representation of women in the mass media in the Third World. However, the recent discourse of feminist media scholars, based on a postmodern view of society, may not be appropriate for most Third World women, who have not yet experienced the "modern" life and who do not have access to mass media.

The challenge before the feminist communication researchers in the Third World is to evolve conceptual frameworks that situate the mass media in the changing political-economic context while taking into consideration the feminist agendas of Third World women and the differences and specificities of their experiences in a given cultural system. What is needed is a broader conceptual approach that examines how gender identities are constructed in the mass media through media production, media text, and audience reception within the changing global context of new communication technologies and rapid commercialization of the mass media.

Continuity and Changes in the Portrayal of Women in Mass Media

The limited reach of the mass media among a majority of the Third World women is the result of factors such as illiteracy, inaccessibility, lack of respite from household chores, inconvenient program schedules, and traditional mobility restrictions that inhibit women from going out to such places as the movie theater. Nevertheless, the role of the mass media in perpetuating patriarchal norms and ideology is pervasive worldwide.

It was the comprehensive study by Gallagher (1983) of the portrayal of women in the mass media that drew our attention to demeaning and derogatory media images of women around the world. This study was conducted in both developed (Australia, Austria, Canada, Denmark, Finland, Germany, Japan, New Zealand, Norway, Sweden, Switzerland, and the United Kingdom) and developing (Brazil, China, Colombia, Iran, Jamaica, the Philippines, Puerto Rico, Senegal, and Venezuela) countries. It concluded that, except in the case of government-controlled media in socialist countries (for example, China), the media underrepresent or misrepresent women and their concerns, use them in advertising as a commodity, and present traditional stereotyped images of women as passive, dependent, and subordinate to men. Subsequently, while the media portrayal of women has been written about and discussed by women's groups and researchers in numerous forums, there has been no systematic review of the literature on that subject in developing countries.

Enough is known, though, to present a general picture of the continuity and changes in the portrayal of women in each medium. Although the images of women and the types of messages transmitted through the various mass media—print (newspapers and magazines), broadcasting (radio and television), and films—are not very different, it is worthwhile to examine how each reinforces existing gender ideology. (Although radio is inexpensive and easily accessible to the vast majority of the illiterate population, it is a neglected medium of study among feminist scholars and activists and so is not included in this discussion.)

This discussion draws heavily from the Indian literature. It is assumed that, despite apparent differences in media development and content in divergent societal contexts, the Indian situation throws light on how the media reinforce gender and social relations and, at times, challenge them.

Women in Films

In the milieu of widespread illiteracy, film is the most important medium for entertainment. Popular cinema is an extremely potent medium since it combines sight, sound, motion, drama, and messages to capture audience attention and influences us at the subliminal level through powerful images. Narratives of various successful genres—such as family and social dramas, romance, vendetta sagas, or mythological stories—around familiar conflicts and their resolution in family and society, create myths that infiltrate the unconscious world of the collective psyche and reinforce patriarchal ideology. Popular cinema reinforces myths that exist about women in society by presenting a woman as what she represents for men and not in terms of what she actually signifies.

Popular cinema is an integral part of popular culture and becomes the distorted mirror of modern society. The representation of women in popular cinema has been a major issue of debate among Western feminist scholars in communication. In fact, feminist film theory has made a significant contribution to the understanding of how the mass media construct definitions of femininity and masculinity (Gallagher, 1992). In developing countries, however, there have been few similar efforts to build a feminist film theory or to examine the question of women in cinema through systematic research. In India, analysis of Indian films by feminist film critics and researchers is often related too closely to the specific film narratives; still, a closer look at images of women in Indian films will give us some insights into understanding how the powerful medium of films has helped redefine femininity and masculinity in the changing context of modernity. (There are 15 official languages in India; this discussion is based on Hindi films, which capture the largest audience nationwide.)

Tracing the changing versions of idealized femininity over 75 years in mainstream films, and since the 1960s in the New Wave (art) cinema, Rao (1989) shows that although images of women in Hindi films have changed from goddesses to dream girls to the "new" women, the heroines still project patriarchal norms and values. The form and content of Indian popular films have changed over the years with technological advances in cinematography and social milieu, but the traditional mythical female characters of the ideal women have continued as archetypes and are reinforced even today in characterizations of modern women in Indian cinema.

Until the late 1960s and early 1970s, the Indian popular cinema projected dichotomous images of women. Women were depicted primarily in their relationship to men. In the context of gradual transition from traditional to modern society, traditional women were characterized in the "good" role of mother, wife, sister, and daughter and portrayed as demure, submissive, passive, and self-sacrificing, upholding the traditions and accepting patriarchal norms and authority. "Modern" women were portrayed in the "bad" role of vamp, as Westernized and highly sexual, displaying independence and initiative in their relationships with men outside marriage. However, neither the tradition-bound women nor the "modern" women questioned male dominance within the patriarchal family or society. Thus, the two apparently contradictory images of women essentially highlight the hero's masculinity by reinforcing his control over women.

However, with the trends toward increasing commercialization of popular Indian films in the 1980s, and increasing sex and violence in the context of growing social unrest, there has been a significant change in the screen images of men and women. The romantic hero of the 1960s and early 1970s has been replaced by the angry young man who takes up the cause of fighting for the downtrodden against the corrupt social system. The gory violence by the hero is often romanticized as the reaction to injustices to the women (mother, sister or lover) in his life. The heroine's role in such narratives is made secondary to the hero's fight against society. The vamp who was used in earlier films to represent the dark side of modernity has more or less disappeared from popular Hindi cinema. The heroine is now portrayed as both a seductive dancer-singer enticing the hero and as a "good" modern woman upholding patriarchal traditions. Such characterization of a heroine has created alternative spaces within the song-and-dance sequences that enable her to express her sexuality. However, the heroine's role within a film's main narrative is made decorative and secondary to build up the macho image of the hero. Thus, the heroine's femininity is defined within the boundaries of patriarchy. She continues to need the protection of a macho hero and does not rebel against his dominance.

With growing commercialization of the film industry, there is a growing trend toward visual representation of women to enhance mass appeal of the films. The body of a woman is constructed on the screen from the eyes of the male "gaze" that objectifies her as a commodity. On the one hand, the depiction of violence against women has moved away from psychological and subtle forms manifest in denial and negation of women's individuality and identity in the earlier Hindi films to the use of direct physical force. Illustrating female victimization through rape or attempted rape scenes has become a new form of exploitation of the female body. On the other hand, seductive dances combine the rhythm of Indian and Western popular music and dance forms to create a synthetic and glamorous image of a heroine to attract the male audience. Thus, rape sequences and seductive dances are becoming increasingly common in the film narratives, projected to serve as a voyeur's delight rather than enrage and invite public debate.

Despite increasing participation of men and women in the work force, work is an underdeveloped theme in commercial Indian films. On the basis of an analysis of several Hindi films, Kishwar and Vanita (1987) highlight the diminished importance of women's work outside the home in film narratives and examine how the middle-class ideal of a domesticated woman is reinforced in divergent work roles for men and women.

In general, the work life of a hero usually exists to feed the film narratives of romance, family melodrama, and violent conflict, while women are shown mostly as working only when compelled by circumstances. Even educated middle-class women are shown without any occupation. When they do work, they are concentrated in stereotyped jobs such as typist, secretary, schoolteacher; only occasionally do they appear as a lawyer or doctor. The working middle-class woman, in general, is a young woman who is doing a job while waiting to get married. By and large, women are shown to work only in the absence of a male breadwinner, a father, a brother, or a husband. These women are presented as unfortunate victims who are sacrificing their own interests to support the family. It is only when she tries to rebel against her role as a wife and mother or chooses to work to assert her independence that the hostility against a working woman surfaces.

In the case of poor women, work is represented as an economic necessity, and their work lives are either romanticized or sensationalized: they are depicted as victims of poverty and sexual harassment. Sexual molestation is presented as a dominating reality of poor working women. While catering to the voyeuristic impulse of the audience, the molestation episodes obscure harsh working conditions and injustices and mask hostilities toward her for not conforming to the ideal of house-bound domesticated women.

In creating new archetypes of modern Indian women, commercial films also draw upon the progressive films and the women's movement. The new woman is shown as a "strong" character—educated, articulate, independent, and capable of taking the initiative in a relationship with a man. She signifies "good" modernity. However, the bright new image of "modern" women in the popular Indian films is superficial. To keep the audience emotionally involved, commercial cinema has created certain gender stereotypes with considerable ambiguity; underlying the spurious concerns for women's oppression in the mainstream cinema, deeply entrenched ideals of femininity are disguised in the glossy images of "liberated" women.

Characterization of women in women-centered commercial films is ambivalent. In imitating the role of a macho hero, she appears neither credible nor powerful as the "strong" woman character. She continues to be depicted as submissive in her romantic relationships with men. Furthermore, women who protest against the institution of marriage and patriarchal oppression have been ossified into new stereotypes. They are depicted either as home wreckers in the role of "other" women or as irrational and hysterical wives abandoning their villainous husbands. Even when a woman sets out to find herself as an individual, eventually she is shown as finding solace in motherhood or in another romantic relationship. Occasionally, she takes up a career as a consolation prize for the broken marriage or relationship, but not for defining her identity.

With the emergence of alternative cinema (often known as the New Wave, or art, cinema) since the late 1960s, there has been some effort to bring women to the center of film narratives to move away from the traditional stereotypes of the Indian woman and to characterize her as a person with distinct identity by projecting her as a strong and often dominant character. The redefinition of femininity so produced and portrayed in fact highlights the tension between "modernity" and "tradition" (Mazumdar, 1991).

Women are often used in New Wave films as symbols of resistance and victims of exploitation. The emphasis is on replacing the "myth" in popular films with "reality." However, realistic cinema can also create new myths about women through powerful and controlling narratives and cinematography (Laxmi, 1986). To establish visual authenticity of filmic reality, the penetrating gaze of realism often exploits the sensuous female body and uses women as ideal symbols to represent a social "issue" (Rao, 1989). Representation of women in New Wave cinema is also constrained by dominant ideological discourses on women that perceive the woman question as only a gender war within the framework of liberal feminism (Mazumdar, 1991).

It has been only in the late twentieth century that women have begun working in commercial and New Wave cinema in India as directors and writers. To what extent have they succeeded in shifting the perspective on women and related topics in films? At one level, female directors have created a much-awaited constructive space in films for promoting feminist ethos and views. But the terrain they have opened is still uneven and patchy. In the prevailing economic and cultural context of commercial films, female directors are mostly allowed to deal with "women-oriented" issues, such as the family, romantic relations, maternal relations, and so on, which are thought to be traditionally the domain of women (Gupta, 1994). Some of the female directors reflect more sympathetic understanding of women characters, but there has not been a significant shift in the roles assigned to women or the boundaries of gender stereotypes in the films they have directed.

In short, while there has been an increasing concern with women-centered issues in both art and commercial films in India, both kinds of films tend to use images and issues from each other that reinforce the mythical portrayal of the new Indian woman in the guise of modernity to ensure mass appeal.

Women in Television

Television is a medium of entertainment as well as of education and information. The penetration of new communication technologies in the 1980s and 1990s has widened access to television in developing countries. With the growing awareness among development planners and media professionals about the "woman question," there has been a significant increase in the number of television programs focusing on women and women's issues. How are women and their concerns reflected on television?

On the basis of comprehensive analyses of portrayals of women in the mass media across many countries of the world, Gallagher (1983) has shown that, not only are women and their concerns misrepresented and underrepresented on television, women's images on television consistently follow traditional stereotypical patterns and are very often derogatory. Have things changed since that study was reported? They have, though not for the better.

Seldom has the total output of television been studied in order to understand gender portrayal and the treatment of women's concerns in television programs. Feminist research on television production and audience reception is also very limited in developing countries. Nevertheless, available research on the representation of women in television gives us some insights into understanding the construction of gender discourse in television.

A pioneering study of the content of television programs on Indian television (focusing on one- to two-minute spots and continuity programs for 15 days) revealed that the state-controlled television reflected common sex-role stereotypes embedded in Indian cultural, religious, and political traditions (Krishnan and Dighe, 1990). Television affirmed a limited definition of womanhood that confined women in the home, the most private of all social spaces, and denied them spaces in the public sphere, traditionally the domain of men. The authors argue that the limited representation of women in public television in India is part and parcel of a hegemonic process that supports women's subordination in society. Furthermore, television's heavy reliance on commercial films brings all of the problems of commercial cinema into public television, which was expected to educate and inform.

In recent years, there has been a tremendous increase in television programs in which the text provides continuity by way of narrative or characters or themes or situations. An overwhelming number of these are fictional programs that hook large audiences through the narrative of the story. Among teledramas, soap operas have become one of the most popular forms of entertainment around the world. Compared to Western scholarship, feminist research on soap operas in developing countries is very limited. (There is substantial communication research on Latin American soap operas, known as *telenovelas.* However, as most of the research is in Spanish or Portuguese, it is not easily accessible to the wider English-speaking academic community.)

The Anglo-American soap opera appears to be less rooted in the domestic world. However, the *telenovela,* the

Latin American variation of the soap opera, differs in its narrative structure. Romance and melodrama associated with the women's world are key features of Latin American telenovela. A study of telenovelas in Chile in the 1970s by Maria de la Luz Hurtado reveals that the shows follow traditional role models of men and women, endorse a romanticized ideal of motherhood, and promote the traditional double moral standards for men and women in sexual matters (Frey-Vor, 1990a). Since then, there have been considerable variations in the form and content of telenovelas in Latin America, although romance and melodrama have continued to be key features.

To reach the wider audiences and commercial sponsors, telenovelas, like soap operas, also use many elements of the popular patriarchal culture. Hence, they are less likely to break away from the dominant patterns of representation of women. More recent audience research on telenovelas indicates an increasing viewership by men and children in the family (Frey-Vor, 1990a). However, further research is needed for understanding the changing nature of patriarchal ideology and the portrayal of gender in telenovelas and how men and women as audiences perceive these programs.

Soap operas are also used in many developing countries for imparting pro-developmental messages. For example, in Mexico the private television network Televista has produced pro-developmental telenovelas on family planning, adult literacy, child care, and female equality. In India, the public television network Doordarshan introduced the first development-oriented soap opera, Hum Log, in 1984; it emphasized the themes of family planning, the status of women, family harmony, and family welfare. Although there is considerable research on the effects of such soap operas in attaining their educational-developmental goals, critical analysis of gender issues in these programs is limited. A study of the effect of Hum Log on the Indian audience reveals that exposure to such pro-social television soap opera did not make viewers more aware of women's status issues (Brown and Cody, 1991). In addition to pro-developmental soap operas, several women-centered teleserials have appeared on public television. Some have addressed issues pertaining to women's subordination in society, but most generally treat the "woman question" in a superficial manner and fail to challenge unequal gender relations in society.

Television is not merely a medium for entertainment. It also transmits public knowledge and shapes public opinions on a wide range of development issues. Television programs on family planning, health and hygiene, child care, home care, foods and nutrition, and so on are often targeted at women as a general or specific audience. Portrayal of gender roles and activities in such programs on public television in India was not found to be radically different from the fictional television programs (Krishnan and Dighe, 1990).

Television has emerged as an important source of news, and visual presentation has become an important area of news coverage. However, feminist media criticism has neglected this important genre of news, current affairs, and other information-oriented programs (Gallagher, 1992). Research on television news in India shows a quantitative and qualitative underrepresentation of women in national and regional television news (Krishnan and Dighe, 1990; Media Advocacy Group, 1994). Given the celebrity orientation of political news, women figure mostly as wives, mothers, or daughters of well-known leaders. Otherwise they are depicted as a passive audience, victims of calamity or accident, or beneficiaries of various welfare programs. To understand the complexity of gender issues in television news, feminist research needs to go beyond the simplistic content analysis of the news text and visuals and examine how television news is produced, articulated in the text, and received by the audience.

With increasing commercialization of television broadcasting, there has been a considerable spurt increase in television commercials. Many of the gender differences that appear in television programs also are reflected in television advertising. The impact of advertisements is not due only to their verbal content, but also to their use of sexual symbols and images—it is the images that we see that provide significance to the product and reinforce patriarchal ideology. Although women's groups and feminists have often raised their voices against images of women in television advertising, systematic research in this area is limited.

In India, content analysis of roughly 186 television commercials, repeated several times during July 1986 on the public television network, reveals that women were utilized in advertising to sell products to both male and female consumers through their two-dimensional role: as caretakers of the household and the family and as decorative sex objects (Krishnan and Dighe, 1990). Although women were featured in various categories of commercials, they tended to be prominent figures in the commercials for grooming aids, household goods, and foods, while men were prominent in the advertisements for medical aids and agricultural, industrial, and electronic goods. Women were featured predominantly as housewives; men were either depicted as professionals (scientists, doctors, executives), or their occupations were unspecified. The activities of male and female characters in advertisements varied according to the type of product being advertized. Women were shown as engaged in cooking, feeding children, serving, caring, and so forth; men's activities were either of an outdoor nature or related to their occupations. On the whole, the latent messages of the commercials were more demeaning than the manifest messages. Commercials urged women to enhance their appeal to men or gain men's approval by using the product concerned. Further empirical research could detect similar trends in television advertising in the other developing countries.

Women's issues are no longer invisible on public television. The increase in women-centered programs on television gives us an impression that the media have at least responded to women's demands for better representation in television programs. However, there appears to be only a repackaging of women's images on television in the changing cultural and economic context.

The new woman in television programs is now an excellent wife and a mother who happens to work, but her success is not gauged by her achievements in the workplace. She is still defined in terms of her relations with men, marriage stability, motherhood, and family. On the other hand, with the emergence of women as an important group of the audience, television commercials have co-opted some of the images of "modern" women and created a new cultural "type" of assertive and ambitious women who are profitable consumers. The liberated woman is presented as a sensual person who is primarily involved in the work of purchasing appliances, accessories, and cosmetics. This new cultural stereotype of a woman is produced by television commercials essentially to serve the commodity markets of the capitalist economy. However, the changing images of women in television programs and advertising in the context of globalization and privatization of television are not adequately captured in media research.

This does not imply that no effort is made on television to present alternative perspectives on women's struggles and problems. The space within mainstream television is often used to create new representations of women, to project their experiences, struggles, and hopes, and to voice their concerns. However, such space is limited and constrained by the existing media structures. A closer look at the medium suggests that alternative discourse on gender is accommodated within the existing parameters of patriarchal discourse to reconstruct dominant values.

Women in the Print Media

The reach of the print media, newspapers, and magazines is far more limited in developing countries due to widespread illiteracy. Nevertheless, the print media play an important role in influencing public opinions and setting the agenda for what is construed as news. The 1983 Gallagher study pointed out that women and women's issues find comparatively little space in newspapers and that newspapers and magazines generally reinforce sex stereotypes. However, over the years the situation has somewhat changed.

In the 1980s, the general apathy among newspapers and periodicals toward women's issues has given way to some awareness and better coverage. Earlier, women and their issues or problems rarely figured on the front page of a newspaper, and women were predominantly depicted as victims of atrocities. In the 1990s, women are more "visible" in the mainstream print media, where they figure side by side with the old stereotyped sexist images and the back-page pinups. Although there has been a gradual increase in the space devoted to selected women's issues over the years and a noticeable decline in the overtly sexist and antiwomen items and articles, there is considerable ambiguity in representation of women and women's issues in the newspapers and magazines.

A quick review of literature by the Women's Feature Service (1993) on women in the print media in India during the decade 1980–1990 chronicles the emerging trends. Popular magazines, women's magazines, and a few newspapers that carry a women's page continue the tradition of defining the "women's world" in terms of beauty tips, recipes, fashions, home decoration, and such. Occasionally, some serious articles on the status of women or related issues are thrown in, but often they are sacrificed for encroaching advertisement space or an emphasis on "light" articles to attract readers. On the other hand, women are featured in articles as film celebrities or as successful urban professionals. This coverage of high-profile women depicts them as successful individuals within the existing structures even as it keeps their domestic and reproductive roles intact. Furthermore, women's magazines, through stories, jokes, and trivial headings for serious articles, contribute to the negative representation of feminists and the women's movement. The independent woman is often portrayed as an angry woman disruptive to the patriarchal order, antifamily, and antimale. Such negative portrayal marginalizes her demands and her viewpoints.

Journalistic articles and advertising in magazines show an interesting accommodation of feminism (Women's Features Service, 1993). In advertising, there is a reference to a "new" woman from the middle class, who is assertive and independent, who manages a nuclear family, and who works outside the home. These new archetypes are used to carve niches for specific products in the market. Often, film personalities are used to create attractive images of sensuous, independent, yet vulnerable women. The new stereotypes of the middle-class women in print advertising show how the imagery of the women's movement (women as articulate, independent, and capable) is used by the system to construct new identities.

A cursory look at research on women in the print media indicates that there have been some significant changes in presentation of women and their concerns in the Indian press in the late twentieth century (Balasubrahmanyan, 1988; Prasad, 1992; Joseph and Sharma, 1994). The space given to the coverage of women appears to be more than what it was earlier. A study of women's issues in the English-language newspapers indicates that such space has been carved out as a result of the urban women's movement and the liberal, reformist stance of the English-language newspapers (Joseph and Sharma, 1994). Nevertheless, in the hierarchy of news and news values, serious articles on women still form only a small part of the entire coverage. The print media continue to emphasize events rather than processes. Newspapers give selective coverage of women's issues and often present distorted feminist views (Balasubrahmanyan, 1988).

On the basis of research studies of the press in English and regional languages (Bengali, Gujarati, Hindi, and Tamil), Joseph and Sharma (1994) demonstrate that women's issues are usually given significant coverage in newspapers only when they fit dominant norms of what constitutes news. With the shift in the strategy of the women's movement from single, issue-oriented, highly visible public campaigns centered on atrocities against women, to low-key, grass-roots activism, there has been some dilution in the media's coverage of women's issues. In addition to changes in strategy, other factors affecting the press coverage of women's issues include changes in the presentation of news and views in the context of the expansion of electronic media and new consumer orientations of the Indian economy, and increasing participation of media women in the "hard" news areas of politics and economics.

So there is some visible space in the coverage of women and women's issues in the print media—and beyond the count of the words, stories, and column inches, the underlying perspective is often superficial, simplistic, and sensational. Nevertheless, progressive media professionals continue to work within these spaces to change the mainstream print media.

Although construction of gender differs somewhat in different media, the preceding discussion shows that women and their concerns have received attention in the mass media within the changing societal context. Media images and messages about women, however, appear to be contradictory in nature. Old gender stereotypes coexist with new characterizations of gender identities. At the same time, the mass media co-opt serious discourse on gender and trivialize gender issues. To examine the complex relationship between gender and power in the media, we also need to understand media production and media reception. Further systematic communication research can shed light on how the media contribute to constructing gender identities in the Third World.

Strategies for Change

Given the wide reach of the mainstream mass media and their structural linkages with the business and commercial interests through ownership and control, it is not easy to change the subtle and sophisticated ways in which women are portrayed. Nevertheless, the contemporary women's movement has played an important role in challenging the demeaning images of women in the mass media by creating alternatives and by intervening in, and negotiating with, the mainstream media.

Creating Alternatives

It was the Third World Conference on Women at the end of the United Nations Decade for Women (Nairobi, 1985) that provided the impetus to change the portrayal of women in the mass media and to increase women's participation in media industries. For raising feminist consciousness, women's movements in the developed and developing countries underscored the need for alternative media, based on the experiences and aspirations of women. Thus, alternative media were conceived to demystify the demeaning representation of women in the mainstream media and to empower women with the tools to project themselves.

Since then, considerable efforts have gone into creating alternative media systems as part of the global networking for consciousness raising. Women, individually and collectively, have captured a large space that has opened up in alternative media and have made significant contributions to providing an ongoing critique of the mainstream media. The creation of innovative and alternative means for self-expression has been an empowering experience, both for the women developing and using them and for the audiences and groups they reach and interact with while creating the communication.

In alternative media, efforts are made by individuals and organizations in India and elsewhere to develop successful communication that makes people subjects and participants rather than objects or targets of communication. Experiments have been undertaken in India to demystify communication technology and to empower women at the grass roots with the tools to project their issues and problems in their own voices. One is the Center for the Development of Instructional Technology, which has organized training workshops with women in participatory video production. Another is the video unit of the Self-Employed Women's Association (SEWA), which also trains women in the urban informal sector in video production and uses their programs to educate, organize, and train other women workers and mobilize public opinion (for a discussion of alternative practices in various media in South Asia, see Bhasin and Agarwal, 1984; Kapoor and Anuradha, 1986). The built-in interactivity of the alternative media allows a privileged communication with the audience. Alternative media for empowerment have used a variety of modern and traditional media forms, ranging from films, videos, and print material, to storytelling, street theater, folk dances, songs, and puppetry.

In addition, communication and documentation centers, women's news agencies, feature services, newsletters, cooperatives, and organizations have emerged to produce and distribute women's work in the alternative media. Feminist publishing houses have also emerged in the Third World to make the work and concern of feminists visible. Some examples are Cuarto Propio in Chile, Kali for Women in India, Kalayanamitra in Indonesia, and Tigress Press and Genderpress in Thailand.

The alternative media have made contributions, but their reach is limited. The challenge they face in the 1990s is marginality: Audiovisual and print materials developed in the alternative media are not for mass distribution since they are used primarily by progressive groups and individuals who are already sympathetic to women's issues. Systematic research on alternative media is required to understand the extent to which it has succeeded in bringing profound changes in media policies and in altering existing gender biases in the mainstream mass media.

Interventions in the Mainstream Media

Left to itself, the mainstream media will continue to pay lip service to women's issues and reinforce gender stereotypes. For a gender perspective to permeate the mainstream media, it is crucial to negotiate a space within the existing media system. An overview of some of the negotiation strategies used by women's groups and activists, particularly in the Indian context, offers useful pointers to similar initiatives in the other developing countries.

Few developing countries have well-articulated media policies; within the policies, seldom is attention given to gender issues. A UNESCO report on the end of the United Nations Decade for Women shows that only half of the 95 member states had formulated specific policies and guidelines requiring the media to promote advancement of women and that their effectiveness was doubtful (Krishnan and Dighe, 1990). Critical media-policy research is a neglected area of inquiry in feminist communication research in developing countries, so it is difficult to assess the extent to which media policies attend to gender issues. Nevertheless, women's groups, activists, and academicians have played important roles in challenging the existing media policies that reinforce gender stereotypes and in pressuring governments to formulate regulations concerning the portrayal of women in the mass media.

In the early 1980s, efforts to change the portrayal of women in the mainstream media (print, films, and television) in India took the form of direct protest actions against specific instances of sexism and obscenity and public consciousness-raising through the general media critique (Bhasin and Agarwal, 1984; Balasubrahmanyan, 1988). Some national commissions and committees on women also took up the issue of sexism in the media and attempted to pressure the government to change media policies and to regulate the media. The Committee on the Portrayal of Women in Media, constituted in New Delhi in 1983 by professionals from different walks of life, played an important role in monitoring and analyzing the depiction of women in the media and mobilizing public opinion by disseminating its recommendations through seminars and newspapers. In collaboration with the Center for Women's Development Studies (New Delhi), a leading women's studies center, it drew up the guidelines for the nonsexist portrayal of women in television images and presented it to a panel appointed by the government to devise a policy for television. The panel incorporated the guidelines in its report and suggested the government enact specific laws against stereotyped representation of women and their concerns on public television. It was in response to the growing feminist criticism of sexism in mass media that the Government of India enacted the Indecent Representation of Women (Prohibition) Act, 1986 (Act no. LX of 1986). Based on this Act, the government formulated the rules in 1987 that prohibit the indecent portrayal of women in any form through commercial advertisements, publications, writings, or any other manner. However, these two measures are full of loopholes. The concepts of "indecent" and "obscenity" are value laden and cannot be clearly defined. Hence, interpretation of these terms is open to misuse and controversy. This Act is primarily concerned about "indecent representation" of women in commercial advertise-

ments and does not express the same concern about sexism and violence against women in the contents of television serials, films, and film-based programs that are shown in public television.

Over the years, more professional women have entered the mainstream mass media, particularly print, films, and television. Unfortunately, that by itself does not ensure that they will succeed in changing the media's approach to, and treatment of, gender issues and concerns. Similar to other professions, women are, by and large, low profile, undervalued, and conspicuous by their absence in decision-making positions. Nevertheless, progressive men and women have been working within the mainstream media system toward promoting gender-sensitive approaches in productions. Although women producers and writers are often confined to dealing with "women's issues" (marriage, divorce, rape, dowry, and family violence) and are constrained by the organizational structures of media organizations, a few women have used this space given to them in the mainstream media by treating their subjects with feminist awareness (Balasubrahmanyan, 1988; Anand, 1993). Interventions in the popular mainstream media also enable women to establish a direct link with the large audience. Thus, despite the possibility of co-optation of their voices, media professionals are increasingly taking the risk of working within the constraints of the media organizations to promote an alternative perspective on gender issues.

Since the late 1980s, there have been some changes in the strategy of the women's movement in India to promote women's concerns and perspective in the mainstream media (Anand, 1993, 1994). In the 1990s, media professionals and activists have been receptive to the idea of regular and sustained monitoring in order to intervene in the mainstream media. There is more openness among the media advocacy groups to build new alliances and reframe strategies in a manner that enables them to influence policy formulation and media practitioners.

The mainstream media have the power to reach wider audiences. Interventions can go a long way in changing media content, in using the media to influence public opinion on the "woman question," and in giving wider exposure to alternatives and innovations in communication. Engagement in the mainstream media can co-opt feminist voices into the discourse of dominant culture, but that is a risk that women need to continue to be willing to take to penetrate the mainstream media through deliberate and planned efforts. Otherwise, new representations of women in the mainstream media will remain an elusive dream.

Conclusion

Women and their concerns are no longer invisible in the mass media as they were in the 1960s and 1970s. However, shifting definitions of femininity and masculinity appear to continue to be shaped by prevailing patriarchal norms and culture in the Third World.

At one level, the media are engaged in producing these

shifting definitions within the context of popular culture and articulating them in different ways in different media. In fact, there has been an increasing commodification of women in the mass media as the latter have become increasingly commercialized. At another level, the media appear to be depicting some positive and progressive representations of men and women, while accommodating and delegitimizing feminism and feminist issues. Construction of a stereotype of the "liberated" men and women in the mass media is, however, devoid of progressive meanings as the new gender identities are co-opted within the dominant discourse. Moreover, stereotyped images of women (old and new) are not confined to one medium, but repackaged and reinforced through all media—films, videos, television, magazines, and newspapers. The multimedia construction of women and the commonalities in the images graphically illustrate how media structures are linked to market forces.

The mass media can serve as a major source of education and information about ideas and issues central to any progressive social movement. However, given the pessimistic context of "tokenism" to women and their issues in the media, all of those concerned about improving the status of women need to be vigilant in the struggle against co-optation and marginalization. It remains to be seen whether positive changes taking place in the portrayal of women in the media in developing countries are simply cosmetic or part of a substantial transformation in consciousness.

References

Anand, Anita. "Moving from the Alternative to the Mainstream for a New Gender Perspective," *Intermedia*, vol. 21, no. 1, 1993, pp. 54–56.

———. "Mainstreaming Women's Concerns," *Voices*, vol. 2, no. 1, 1994, pp. 6–8. Special Issue: "Women and Communication: The Power to Change."

Balasubrahmanyan, Vimal. *Mirror Image: The Media and the Women's Question*. Bombay: Center for Education and Documentation, 1988.

Bhasin, Kamala, and Bina Agarwal (eds.). *Women and Media: Analysis, Alternatives, and Action*. New Delhi: ISIS International, 1984.

Brown, William, and Michael Cody. "Effects of a Prosocial Television Soap Opera in Promoting Women's Status," *Human Communication Research*, vol. 18, no. 1, 1991, pp. 114–142.

Ceulemans, Mieke, and Guido Fauconnier. *Mass Media: The Image, Role, and Social Conditions of Women: A Collection and Analysis of Research Materials*. Paris: United Nations Educational, Scientific, and Cultural Organization (UNESCO), 1979.

Frey-Vor, Gerlinde. "Soap Opera," *Communication Research Trends*, vol. 10, no. 1, 1990a, pp. 1–16.

———. "Soap Opera," *Communication Research Trends*, vol. 10, no. 2, 1990b, pp. 1–12.

Gallagher, Margaret. *Unequal Opportunities: The Case of Women and the Media*. Paris: UNESCO, 1981.

———. *The Portrayal and Participation of Women in the Media*. Paris: UNESCO, 1983.

———. "Women and Men in the Media," *Communication Research Trends*, vol. 12, no. 1, 1992, pp. 1–15.

Gupta, Shubha. "Women Feature Film Makers: Changing Course?" *Voices*, vol. 2, no. 1, 1994, pp. 27–29. Special Issue: "Women and Communication: The Power to Change."

Joseph, Ammu, and Kalpana Sharma (eds.). *Whose News? The Media and Women's Issues*. New Delhi: Sage, 1994.

Kapoor, Shushma, and Anuradha. "*Women and Media in Development*" a report of the South Asian Regional Workshop, Delhi, March 11–24, 1986, Center for the Development of Instructional Technology and FFHC/AD, Food and Agriculture Organization.

Kishwar, Madhu, and Ruth Vanita. "The Labouring Woman in Hindi Films," *Manushi*, nos. 42–43 September/December 1987, pp. 62–74.

Krishnan, Prabha, and Anita Dighe. *Affirmation and Denial: Construction of Femininity on Indian Television*. New Delhi: Sage, 1990.

Laxmi, C.S. "Feminism and the Cinema of Realism," *Economic and Political Weekly*, vol. 21, no. 20, 18 January 1986, pp. 113–115.

Mazumdar, Ranjani. "Dialectic of Public and Private Representation of Women in Bhoomika and Mirch Masala," *Economic and Political Weekly*, vol. 26, no. 43, 26 October 1991, pp. WS81–WS84.

Media Advocacy Group. "Women and Men in News and Current Affairs Programmes." New Delhi: Media Advocacy Group, 1994. Mimeographed.

Prasad, Nandini. *A Pressing Matter: Women in Press*. New Delhi: Friedrich Ebert Stiftung, 1992.

Rao, Leela. "Women in Indian Films: A Paradigm of Continuity and Change," *Media, Culture ,and Society*, vol. 11, no. 4, 1989, pp. 443–458. Special Issue: "Indian Media and Mass Communication Research."

Steeves, Leslie. "Creating Imagined Communities: Development Communication and the Challenge of Feminism," *Communication*, vol. 43, no. 3, 1993, pp. 218–219.

Williams, Raymond. *Marxism and Literature*. Oxford: Oxford University Press, 1977.

Women's Feature Service. "Towards a Positive Portrayal of Women in the Media: A Review of Literature, 1980–1990." New Delhi: Women's Feature Service, 1993. Mimeographed.

See also Ella Habiba Shohat, "Gender, Nation, and Race in Film and Video"

Gender, Nation, and Race in Film and Video

Ella Habiba Shohat

Introduction

Recent years have witnessed a terminological crisis swirling around the term "Third World," now seen as an inconvenient relic of a more militant period. Some have argued that Third World theory is an open-ended ideological interpolation that papers over class oppression in all three worlds while limiting socialism to the now nonexistent Second World. Three-worlds theory not only homogenizes heterogeneities, masks contradictions, and elides differences, it also obscures similarities—for example, the common presence of "Fourth World" (indigenous) peoples in both "Third World" and "First World" countries. Third World feminist critics such as Nawal el Saadawi (Egypt), Vina Mazumdar (India), Kumari Jayawardena (Sri Lanka), Fatima Mernissi (Morocco), and Lelia Gonzales (Brazil) have explored these differences and similarities in a feminist light, pointing to the gendered limitations of Third World nationalism. Exploring issues in national cinema such as post–World War II economic decline and geographic borderline changes, and redefining gender roles from a feminist perspective, leads to a complex formulation of the Third World and its relationship to the First World. Both are implicated in the "postcolonial" era of global circulation of goods and principles, and the First World–Third World struggle takes place not only between nations but also within them.

While hegemonic Europe may clearly have begun to deplete its strategic repertoire of stories, Third World peoples, as well as First World minoritarian communities, women, and gays/lesbians have only begun to deconstruct and tell their tales. For the "Third World," this cinematic "counter-telling" basically began with the post–World War II collapse of the European empires and the emergence of independent nation-states. Third World feminists, for their part, have participated in these counternarratives while insisting that colonialism and national resistance have impinged differently on men and women and that remapping and renaming must take into account these fissures and contradictions.

National film production can represent a gauge of the social changes in the Third World and the evolving roles women assume to control the creation of their images. Although relatively small in number, women directors and producers in the Third World already played a role in film production in the first half of the century: Aziza Amir, Assia Daghir, and Fatima Rushdi in Egypt; Carmen Santos and Gilda de Abreu in Brazil; Emilia Saleny in Argentina; and Adela Sequeyro, Matilda Landeta, Candida Beltran Rondon, and Eva Liminano in Mexico. However, their films, even when they focused on female protagonists, were not explicitly feminist in the sense of a declared political project to empower women in the context of both patriarchy and (neo)colonialism. In the postcolonial, or postrevolution, era, women, despite their growing contribution to the diverse aspects of film production, remained less visible than men in the role of film direction. Furthermore, Third Worldist revolutionary cinemas in places such as China, Cuba, Senegal, and Algeria were not generally shaped by an anticolonial feminist imaginary. Yet, they can be considered vanguard movements in the repositioning of women. While a pronounced feminist agenda in First World cinema can be discerned, Third World films, in contrast, present the participation of women as far from central. Their growing production since the 1980s corresponds to a worldwide burgeoning movement of independent work by women. New low-cost technologies of video communication—not structural changes in the patriarchal and neocolonial system that exerts influence over the film-production economy—facilitated this increase.

Beyond the "Global Sisterhood" Discourse

Any serious discussion of feminist cinema must engage in the complex question of the "national." Third Worldist films, produced within the legal codes of the nation-state, often in (hegemonic) national languages, recycle national intertexts (literatures, oral narratives, music) and project national imaginaries. But if First World filmmakers have seemed to float "above" petty nationalist concerns, it is because they take for granted the projection of a national power that facilitates the making and the dissemination of their films. The geopolitical positioning of Third World nation-states continues to imply that their filmmakers cannot assume a substratum of national power.

Already during the 1970s, Third World film and culture challenged White feminist film theory and practice of the First World metropolises. What I would call "post–Third Worldist feminist" writings and films continue to refuse a Eurocentric universalizing of "womanhood" and even of "feminism." Eschewing a discourse of universality, such feminisms claim a "location," arguing for specific forms of resistance in relation to diverse forms of oppression.

Aware of White women's advantageous positioning within (neo)colonialist and racist systems, feminist women of color (both in the First and Third Worlds) have not premised their struggles on a facile discourse of global sisterhood but have often conducted them within the context of anticolonial and antiracist struggles. From the outset, feminists of color have been engaged in analysis and activism around the intersection of nation/race/gender. While still resisting the ongoing (neo)colonized situation of their "nation" and/or "race," post–Third Worldist feminist cultural practices also break away from the narrative of the "nation" or "race" as a unified entity so as to articulate a contextualized history for women in specific national and racial contexts.

A discussion of Ana Maria Garcia's documentary *La Operación* (Puerto Rico, 1983), a film that focuses on U.S. sterilization policies in Puerto Rico, for example, reveals the historical and theoretical aporias of such concepts as "the female body" when not addressed in terms of race, class, and (neo)colonialism. Whereas a White "female body" might undergo surveillance by the reproductive machine, the dark "female body" is subjected to what might be called a *dis-reproductive* apparatus operating within a hidden racially coded demographic agenda. Women of color who reside outside the First World encounter specific oppressive mechanisms, and, in this film, policies concerning birth and population control consequently undermine the international standing of Puerto Rico as they suppress Puerto Rican women's control of their bodies.

Third World cinema has privileged issues of national identity. Third Worldist films by women filmmakers assume that revolution is crucial for the empowerment of women—that the revolution is integral to feminist aspirations. Sarah Maldoror's short film *Monangambe* (Mozambique, 1970) narrates the visit of an Angolan woman to her husband who has been imprisoned by the Portuguese, while her feature film *Sambizanga* (Mozambique, 1972), based on the struggle of the MPLA (Angola's liberation movement) in Angola, depicts a woman coming to revolutionary consciousness. Heiny Srour's documentary *Saat al Tahrir* (The Hours of Liberation) (Oman, 1973) privileges the role of female fighters as it looks at the revolutionary struggle in

Oman, and his *Leila wal dhjab* (Leila and the Wolves) (Lebanon, 1984) focuses on the role of women in the Palestine liberation movement. Helena Solberg Ladd's *Nicaragua up from the Ashes* (United States, 1982) addresses the role of women in the Sandinista revolution. Sarah Gomez's well-known film *De Cierta Manera* (One Way or Another) (Cuba, 1975), often cited as part of the late 1970s and early 1980s Third Worldist debates about women's position in revolutionary movements, interweaves documentary and fiction as part of a feminist critique of the Cuban revolution. From a decidedly prorevolutionary perspective, the film deploys images of building and construction as a metaphor for the need for further revolutionary changes; it dissects and analyzes macho culture within the overlaid cultural histories (African, European, and Cuban) in terms of the need to revolutionize gender relations in the postrevolution era.

Already in the late 1960s and early 1970s, in the wake of the Vietnamese victory over the French, the Cuban revolution, and Algerian independence, Third Worldist film ideology was crystallized in a wave of militant manifesto essays—Glauber Rocha's *Aesthetic of Hunger* (1965), Fernando Solanas and Octavio Getino's *Toward a Third Cinema* (1969), and Julio Garcia Espinosa's *For an Imperfect Cinema* (1969)—and in declarations and manifestos from Third World film festivals calling for a tricontinental revolution in politics and an aesthetic and narrative revolution in film form. Within the spirit of a politicized auteurism, Rocha called for a "hungry" cinema of "sad, ugly films," Solanas and Getino called for militant guerrilla documentaries, and Espinosa called for an "imperfect" cinema energized by the "low" forms of popular culture. The resistant practices of such films are neither homogeneous nor static; they vary over time, from region to region, and in genre, from epic costume drama to personal small-budget documentary. Their aesthetic strategies range from "progressive realist" to Brechtian deconstructivist to avant-gardist, tropicalist, and resistant postmodern. In their search for an alternative to the dominating style of Hollywood, such films shared a certain preoccupation with First World feminist independent films that have sought for alternative images of women. The project of digging into "herstories" involved a search for new cinematic and narrative forms that challenged both the canonical documentaries and the mainstream fiction films, subverting the notion of "narrative pleasure" based on the "male gaze."

Beyond the "Unitary Nation" Discourse

As with Third Worldist cinema and First World independent production, post–Third Worldist feminist films and videos conduct a struggle on two fronts, at once aesthetic and political, synthesizing revisionist historiography with formal innovation. Considering the political agenda, the early period of Third Worldist euphoria has since given way to the collapse of Communism, the indefinite postponement of the devoutly wished "tricontinental revolution," the realization that the "wretched of the earth" are not unanimously revolutionary (nor necessarily allies to one another), the appearance of an array of Third World despots, and the recognition that international geopolitics and the global economic system have forced even the "Second World" to be incorporated into transnational capitalism.

At the same time, the tropes of a unitary nation often camouflage the possible contradictions among different sectors of Third World society. The nation-states of the Americas, of Africa, and of Asia often "cover" the existence not only of women but also of indigenous nations (Fourth World) within them. Moreover, the exaltation of the "national" provides no criteria for distinguishing exactly what is worth retaining in the "national" tradition. A sentimental defense of patriarchal social institutions simply because they are "ours" can hardly be seen as emancipatory. Indeed, some Third World films criticize exactly such institutions: *Xala* (1990) criticizes polygamy and *Finzan* (1989) and *Fire Eyes* (1993) critique female genital mutilation, while films like *Allah Tanto* (1992) focus on the political repression exercised even by a pan-Africanist hero like Sekou Toure, and Sembene's *Guelwaar* (1992) satirizes religious divisions within the Third World nation. All countries, including Third World countries, are heterogeneous, at once urban and rural, male and female, religious and secular, native and immigrant, and so forth. The view of the nation as unitary muffles the "polyphony" of social and ethnic voices within heteroglot cultures. Third World feminists especially have highlighted the ways in which the subject of the Third World nationalist revolution has been covertly posited as masculine and heterosexual.

Subverting conventions in aesthetic presentation allows post–Third World feminist filmmakers to reinforce the advances gained by their political agenda. Previous Third Worldist films often favored the generic (and gendered) space of heroic confrontations, whether set in the streets, the casbah, the mountains, or the jungle. The minimal presence of women corresponded to the place assigned women both in the anticolonialist revolutions and within Third Worldist discourse, leaving women's home-bounded struggles unacknowledged. Women occasionally carried the bombs, as in *Battle of Algiers,* but only in the name of a "nation." More often, women were made to carry the "burden" of national allegory—the woman dancing with the flag in *Battle of Algiers,* the Argentinean prostitute whose image is underscored by the national anthem in *La Hora de las Hornos* (The Hour of the Furnaces), the mestiza journalist in *Cubagua* (1987), as embodiment of the Venezuelan nation—or scapegoated as personifications of imperialism (the allegorical "whore of Babylon" figure in Rocha's films). Gender contradictions were subordinated to anticolonial struggle: Women were expected to "wait their turn."

A Tunisian film, *Samt al Qusur* (The Silence of the Palace) (1994) by Moufida Tlatti, a film editor who worked previously on major Tunisian films of the postindependence *cinema jedid* (New Cinema) generation and who has now

directed her first film, exemplifies some of the feminist critiques of the representation of the "nation" in the anticolonial revolutionary films. Rather than privilege direct, violent encounters with the French, necessarily set in male-dominated spaces of battle, the film presents 1950s Tunisian women at the height of the national struggle as restricted to the domestic sphere. Yet, it also challenges middle-class assumptions about the domestic sphere as belonging to the isolated wife-mother of a (heterosexual) couple. *The Silence of the Palace* focuses on working-class women, the servants of the rich pro-French Bey elite, subjugated to hopeless servitude, including at times sexual servitude, but for whom life outside the palace, without the guarantee of shelter and food, would mean even worse misery—of, for example, prostitution. Although these women are under the regime of silence about what they see and know about the palace, the film highlights their survival as a community, an alternative family. Their emotional closeness in crisis and happiness and their supportive involvement in decision making show their ways of coping with a no-exit situation. They become a nonpatriarchal family within a patriarchal context. Whether through singing, as they cook once again for an exhibitionist banquet, or through praying, as one of them heals one of their children who has fallen sick, or through dancing and eating in a joyous moment, the film represents women who did not plant bombs but whose social positioning turns into a critique of failed revolutionary hopes as seen in the postcolonial era.

Other contemporary feminist works look at national identity in a critical perspective; some do so within a diasporic context. The voice-over and script of *Measures of Distance* narrates a paradoxical state of geographical distance and emotional closeness. The textual, visual, and linguistic play between Arabic and English underlines the family's serial dislocations, from Palestine to Lebanon to Britain, where Mona Hatoum has been living since 1975, gradually unfolding the dispersion of Palestinians over very diverse geographies. The foregrounded letters, photographs, and audiotapes call attention to the means by which people in exile maintain cultural identity. In the mother's voice-over, the repeated phrase "My dear Mona" evokes the diverse "measures of distance" implicit in the film's title. Meanwhile, background dialogue in Arabic, recalling their conversations about sexuality and Palestine during their reunion, recorded in the past but played in the present, parallels a shower photo of the mother, also taken in the past but looked at in the present. By manipulating the soundtrack and by capturing the various modes and devices of representation to affect temporality, the film narrative accentuates the multiplicity of women's images across generations.

Measures of Distance also probes issues of sexuality and the female body in a kind of self-ethnography, its nostalgic rhetoric concerning less the "public sphere" of national struggle than the "private sphere" of sexuality, pregnancy, and children. The women's conversations about sexuality leave the father feeling displaced by what he dismisses as "women's nonsense." The daughter's photographs of her nude mother make him profoundly uncomfortable, as if the daughter, as the mother writes, "had trespassed on his possession." To videotape such intimate conversations is not common practice in Middle Eastern cinema or, for that matter, in any cinema. (Western audiences often ask how Hatoum won her mother's consent to use the nude photographs and how she broached the subject of sexuality). Paradoxically, the exile's distance from the Middle East authorizes the exposure of intimacy. Displacement and separation make possible a transformative return to the inner sanctum of the home; mother and daughter are together again in the space of the text.

In Western popular culture, the Arab female body, in the form of the veiled bare-breasted women who posed for French colonial photographers or the Orientalist harems and belly dancers of Hollywood film, has functioned as a sign of the exotic. But rather than adopt a patriarchal strategy of simply censuring female nudity, Hatoum deploys the diffusely sensuous, almost pointillist, images of her mother naked to tell a more complex story with nationalist overtones. She uses diverse strategies to veil the images from voyeuristic scrutiny: Already hazy images are concealed by text (fragments of the mother's correspondence, in Arabic script) and are difficult to decipher. The superimposed words in Arabic script serve to "envelop" her nudity. "Barring" the body, the script metaphorizes her inaccessibility, visually undercutting the intimacy verbally expressed in other registers.

Rewriting the Exotic Body

Exile can also take the form of exile from one's own body. Dominant media have long disseminated the hegemonic White-is-beautiful aesthetic inherited from colonialist discourse, an aesthetic that exiled women of color from their own bodies. Until the late 1960s, the overwhelming majority of Anglo-American fashion journals, films, television shows, and commercials promoted a canonical notion of beauty that White women (and, secondarily, White men) were the only legitimate objects of desire. In so doing, the media extended a long-standing philosophical valorization of whiteness. European writing is replete with homages to the ideal of White beauty, implicitly devalorizing the appearance of people of color. For Gobineau, the "white race originally possessed the monopoly of beauty, intelligence and strength." For Buffon, "[Nature] in her most perfect exertions made men white." Fredrich Bluembach called White Europeans "Caucasian" because he believed that the Caucasus mountains were the original home of the most beautiful human species.

Gendered racism left its mark on Enlightenment aesthetics. The measurements and rankings characteristic of the new sciences were wedded to aesthetic value judgments derived from an Apollonian reading of a de-Dionysianized Greece. Thus, Aryanists like Carl Gustav Carus measured the divine in humanity through resemblance to Greek statues. The uratic religion of art, meanwhile, also worshiped

at the shrine of whiteness. Clyde Taylor, Cornel West, and bell hooks, among others, have denounced the normative gaze that has systematically devalorized non-European appearance and aesthetics.

The hegemony of the Eurocentric gaze, spread not only by First World media but even at times by Third World media, explains why *morena* women in Puerto Rico, like Arab-Jewish (Sephardi) women in Israel, dye their hair blond; and why Brazilian television commercials are more suggestive of Scandinavia than of a Black-majority Country; and why Miss Universe contests can elect blond "queens" even in North African countries; and why Asian women undergo cosmetic surgery to appear more Western. The mythical norms of Eurocentric aesthetics come to inhabit the intimacy of self-consciousness, leaving severe psychic wounds. A patriarchal system contrived to generate neurotic self-dissatisfaction in *all* women (whence anorexia, bulimia, and other pathologies of appearance) becomes especially oppressive for women of color by excluding them from the realms of legitimate images of desire.

Set in a Hollywood studio in the 1940s, Julie Dash's *Illusions* (1982) underlines these exclusionary practices by foregrounding a Black singer who lends her singing voice to a White Hollywood star. Like Hollywood's classic *Singin' in the Rain* (1952), *Illusions* reflexively focuses on the cinematic technique of postsynchronization, or dubbing. But while the former film exposes the intraethnic appropriation whereby silent-movie queen Lina Lamont (Jean Hagen) appropriates the silky dubbed voice of Kathy Selden (Debbie Reynolds), *Illusions* reveals the racial dimension of constructing eroticized images of female stars. The film features two "submerged" Black women: Mingon Dupree (Lonette McKee), invisible as an African-American studio executive "passing for White," and Esther Jeeter (Rosanne Katon), the invisible singer hired to dub the singing parts for the White film star (Lila Grant). Jeeter performs the vocals for a screen role denied her by Hollywood's institutional racism. Black talent and energy are sublimated into a haloed White image. But by reconnecting the Black voice with the Black image, the film makes the Black presence "visible" and, therefore, "audible," while depicting the operation of the erasure and revealing the film's indebtedness to Black performance. But if Gene Kelly can expose the injustice and bring harmony in the world of *Singin' in the Rain,* Lonette McKee does not have such power in *Illusions.* While she is far from being a "tragic mulatto" and is portrayed with agency, she is still struggling to rewrite her community's history.

Many Third World and minoritarian feminist film and video projects offer strategies for coping with the psychic violence inflicted by Eurocentric aesthetics, calling attention to the sexualized/racialist body as the site of both brutal oppression and creative resistance. Black creativity turned the body, as a singular form of "cultural capital," into a canvas of representation. A number of late-twentieth-century independent films and videos—notably, Ayoka Chenzira's *Hairpiece: A Film for Nappy-Headed People*

(1985), Ngozi A. Onwurah's *Coffee Colored Children* (1988), Deborah Gee's *Slaying the Dragon* (1988), Shu Lea Cheang's *Color Schemes* (1989), Pam Tom's *Two Lies* (1989), Maureen Blackwood's *Perfect Image?* (1990), Helen Lee's *Sally's Beauty Spot* (1990), Camille Billop's *Older Women and Love* (1987), and Kathe Sandler's *A Question of Color* (1993)—meditate on the racialist/sexualized body in order to narrate issues of identity.

In contrast to the Orientalist harem imaginary, all-female spaces have been represented very differently in feminist independent cinema. Documentaries such as Attiat El-Abnoudi's *Ahlam Mumkina* (Permissible Dreams) (Egypt, 1989) and Claire Hunt's and Kim Longinotto's *Hidden Faces* (Britain, 1990) examine female agency within a patriarchal context. Both films feature sequences in which Egyptian women, speaking together about their lives in the village, recount in ironic terms their dreams and struggles with patriarchy. Through its critical look at the Egyptian feminist Nawal el Saadawi, *Hidden Faces* explores the problems of women working together to create alternative institutions. Elizabeth Fernea's *The Veiled Revolution* (1982) shows Egyptian women redefining not only the meaning of the veil but also the nature of their own sexuality. And Moroccan filmmaker Farida Benlyazid's feature film *Bab Ila Sma Maftouh* (A Door to the Sky) (1988) offers a positive gloss on the notion of an all-female space, counterposing Islamic feminism to Orientalist fantasies.

Conclusion

While the media can destroy community and fashion solitude by turning spectators into atomized consumers or self-entertaining monads, they can also fashion community and alternative affiliations. Just as the media can "exoticize" and disfigure cultures, they have the power to offer countervailing representations and to open up parallel spaces for antiracist feminist transformation. The alternative spectatorship established by the kind of film and video works discussed in this essay can mobilize desire, memory, and fantasy, where identities are not only the given of where one comes from but also the political identification with where one tries to go.

References

Alexander, Jacqui, and Chandra Talpade Mohanty (eds.). *Feminist Genealogies, Colonial Legacies, Democratic Futures.* New York: Routledge, 1976.

Anzaldua, Gloria (ed.). *Making Face Making Soul/ Haciendo Caras: Creative and Critical Perspectives by Women of Color.* San Francisco: Aunt Lute, 1990.

Armes, Coy. *Third World Filmmaking and the West.* Berkeley: University of California Press, 1987.

Bowser, Pearl (ed.). *In Color: Sixty Years of Minority Women in the Media, 1921–1981.* New York: Third World Newsreel.

Burton, Julianne (ed.). *Cinema and Social Change in Latin America.* Pittsburgh: University of Pittsburgh Press, 1990.

Cham, Mbye, and Claire Andrade-Watkins (eds.) *Blackframes.* Cambridge: MIT Press, 1988.

Charkravarty, Sumita. *National Identity in Indian Popular Cinema.* Austin: University of Texas Press, 1993.

Gabriel, Teshome. *Third Cinema in the Third World.* Ann Arbor: University of Michigan Press, 1982.

Grewal, Inderpal, and Caren Kaplan (eds.). *Scattered Hegemonies: Postmodernity and Transnational Feminist Practices.* Minneapolis: University of Minnesota Press, 1994.

Jayawardena, Kumari. *Feminism and Nationalism in the Third World.* London: Zed, 1986.

Minh-ha, Trin. *Woman, Native, Other.* Bloomington: Indiana University Press, 1989.

Mohanty, Chandra, Ann Russo, and Lourdes Torres (eds.). *Third World Women and the Politics of Feminism.* Bloomington: Indiana University Press, 1991.

Moraga, Cherrie, and Gloria Anzaldua (eds.). *This Bridge Called My Back: Writings by Radical Women of Color.* Latham: Kitchen Table, Women of Color Press, 1991.

Naficy, Hamid, and Teshome Gabriel (eds.). *Otherness and the Media.* Langhorne: Harwood, 1993.

Shohat, Ella, and Robert Stam. *Unthinking Eurocentrism: Multiculturalism and the Media.* London: Routledge, 1994.

Demographics and Health

Women and Health

Cynthia Myntti

Introduction

The World Health Organization (WHO) defines health as a state of complete physical, mental, and social well-being and not merely the absence of disease or infirmity (Cook, 1994). More recently, WHO has proposed a definition of women's health as

> not only physical well-being, but also exercising more control over their lives and relationships, and having the information and the resources to take responsibility for their own health and that of their family (World Health Organization, 1995, p.1).

Women in the Third World live in such varied circumstances that it is difficult to generalize about their health status. What do upper-class, educated women in Cairo share with rural agricultural workers in the Kathmandu valley or with women living in shanty towns of Rio de Janeiro? This essay summarizes commonalities and differences in Third World women's health, by reviewing the major health problems affecting women and their determinants, then addressing the psychological aspects of women's health, health care, and, finally, local action to improve women's health.

A few words first about the data available: Women's health in the Third World has not received the attention it has deserved from the international scientific community, and we know much less than we should about the distribution and causes of ill health among women (Holloway, 1994). There are a number of reasons for this.

First, the international public health community has focused overwhelmingly on "tropical diseases," those problems mainly infectious in origin. Scientists have not been particularly interested in investigating the differences in disease patterns between males and females, at least not until the pandemic of a major new infectious disease, HIV/AIDS.

Second, where women *have* been in the lens of analysis and action, it is their reproductive functions that have received most attention: pregnancy care to promote *child* survival, or fertility control to slow the rate of population growth. Neither reproductive approach has addressed the needs of women, as women themselves would define them.

Third, judging from research-spending priorities, the largely male medical community has had most interest in problems that affect men. Problems such as breast and cervical cancer are only now (mid-1990s) receiving attention, and mainly still in industrialized countries.

Fourth, to understand women's health problems in the Third World, one must examine causal factors that go far beyond the biomedical domain, including social, cultural, and economic issues such as poverty, and systematic discrimination, neglect, and abuse. The scientific community has been reluctant to investigate these root causes of poor health in women.

Health Programs and Their Determinants

The Life-Cycle Approach

A World Bank (1994) report describes Third World women's health problems using a life-cycle approach. This approach highlights age-specific problems while considering the cumulative effects of poor health and nutrition throughout women's lives. It is a useful place to begin the discussion of determinants of women's health in the Third World.

Discrimination can begin before birth; the selective abortion of female fetuses has become a well-publicized problem in India and China. Girls are born with biological advantages that are negated in many parts of the world by neglect. In some societies, most notably South Asia and the Arab world, girls receive less food and less health care than their brothers (Tekce, 1990; Merchant and Kurz, 1993).

By the time they are adolescents, their growth may be stunted from the combined effect of untreated illnesses and undernutrition. Yet, in some societies, females are married and begin their childbearing while they are still adolescents.

These roles place further physical demands on their not yet mature bodies. In other societies, where sexual activity begins before marriage, young women may be exposed to involuntary intercourse, sexually transmitted diseases, unwanted pregnancy, and unsafe abortion. Adolescence can also be a time for experimentation with harmful substances such as tobacco and drugs. Clearly, the period of adolescence poses specific risks to the health of young women, whether married or unmarried, urban or rural, rich or poor.

In societies where fertility is still relatively high—more than four children per woman—the reproductive years place additional burdens on a woman's health. The physical and emotional demands of pregnancy, birth, breast-feeding, and childrearing—besides the other responsibilities a woman has—can lead to what some researchers have called the maternal-depletion syndrome, with anemia being a major underlying problem. In South Asia, for example, more than 60 percent of women are anemic and underweight. The World Bank (1993) describes the causes as: inadequate food supply, inequitable distribution within the family, improper food storage and preparation, taboos against eating certain foods, and a lack of knowledge about certain foods.

For older women, the same report suggests that the

cumulative effects of a lifetime of nutritional deprivation, hazardous and heavy work, continuous childbearing, and low self-esteem [leave] them both physically and mentally frail, while widowhood and abandonment often leave them destitute (World Bank, 1993, p. 23).

With this overview from a life-cycle perspective, let us now look more closely at Third World women's productive and reproductive roles. It is here where differences among Third World women can be best understood.

Work and Health

If information is scarce on Third World women's health in general, the problem is even more acute when we examine the relationship between work and health. The fields of occupational health and worker safety are relatively new in the Third World. And, as epidemiologists argue, measuring the relationship between certain kinds of work and health is a complex undertaking anywhere in the world. In industrialized countries, researchers have been able to establish cause-and-effect relationships by using elaborate longitudinal data sets, with precise data on "exposures" and "outcomes." This kind of information is less available for all workers in the Third World.

The analysis of the relationship between work and health has been limited by another problem that is more specific to women: a narrow definition of what constitutes work. Most women's work stretches far beyond what is normally measured as work—that is, formal-sector employment. Women's work encompasses informal-sector production, such as trading or petty manufacturing at home, unpaid work in agriculture and animal husbandry, and household maintenance tasks. Each of these can affect women's health.

Across the Third World, women juggle these different types of work, producing additional stresses in their lives. The share of domestic maintenance tasks that men are expected to do are scant in comparison with their female cohorts. For upper-class women, the problem may be remedied by hiring poorer women to do the housecleaning, cooking, and child care. But for the vast majority, the balance is a daily and ongoing struggle.

A small number of studies have alerted us to health problems associated with domestic chores. In part, it is simply the long and strenuous working day of women. In Nepal, for example, combined economic and household work is estimated at 13.2 hours per day; in Bangladesh, 9.3. Men work fewer hours (Asian and Pacific Women's Resource Collection Network, 1990).

The hazardous nature of women's work can also affect health. In many parts of rural Africa and South Asia, women haul the family's water. Jars or tanks carried on the head can weigh up to 50 kilograms (110 pounds) and can damage the spines of the water carriers. Cooking also has risks. In many parts of Asia, women cook on open fires in poorly ventilated buildings. One study in India found that women were exposed to 100 times the acceptable level of suspected smoke particles, levels much higher than found in other family members (Chatterjee, 1991). In Egypt, one study found that burns accounted for 9 percent of the deaths occurring to women age 15–49. Two-thirds of those deaths were attributed to burns from kerosene cooking stoves (Saleh and Gadalla, 1986).

Moving into the realm of primarily unpaid but "economic" production, increasing attention is being paid to agricultural work hazards for women. One study in India (Batliwala, 1988) found an unexpected incidence of stillbirths, premature births, and deaths during the peak rice cultivation season. The researchers postulated that the squatting, bending, and long hours put undue burden on pregnant women. Another study in Malaysia has underscored the dangers of pesticides: Women sprayers complained about sore eyes, rashes, burned fingernails, and disruption in menstrual periods (Asian and Pacific Women's Resource Collection Network, 1990).

In many Third World countries, women work for an income in the informal sector, producing goods from home on an out-work basis, selling articles in the market, or using their own bodies to earn an income in construction or in commercial sex. Some of these occupations are the most exploited and most dangerous known. Some years ago, one of India's best-known women's organizations, the Self-Employed Women's Association (SEWA) of Ahmedabad, organized a successful loan program for its members. Repayment rates exceeded those of commercial banks. But when SEWA organizers investigated why a small number of loans had not been repaid, they found that the women—

desperately working to support their families through their pregnancies—had died. SEWA then set up a maternity-insurance scheme to allow self-employed women to take time off work, with an income, in the end stages of pregnancy.

Increasingly more young women are joining the industrial labor force in the rapidly developing countries of Asia and in the trade zones of Latin America. Women work in textile mills and in factories processing food, assembling electronics, manufacturing clothing, and packaging pharmaceuticals. They often work in poorly ventilated and cramped surroundings, for long hours and exploitively low wages. Although little scientific research has been conducted to document the effect of these conditions on the health of women workers, the workers complain of respiratory problems, eyestrain and deteriorating eyesight, menstrual irregularities, and repetitive strain injuries (Asian and Pacific Women's Resource Collection Network, 1990).

Sex, Reproduction, and Health

From the 1950s, when industrialized countries began to worry about population growth in the Third World, it has been difficult to discuss any other aspect of reproduction besides "high" fertility. Wealthy countries devoted considerable resources to setting up family-planning programs to publicize and distribute contraceptives in poor countries. The programs emphasized contraceptives, such as the pill and the IUD, for women. It became possible to control fertility without reference to male partners or even to sexual intercourse.

In more recent years, Third World women have raised their voices in strong criticism of population-control programs. Increasing numbers of women, even in high-fertility regions such as the Arab world and East and West Africa, do want to have fewer children, on their terms. But many have suffered abuses at the hands of overzealous family-planning providers and managers more concerned about increasing the prevalence of contraceptives than about meeting women's needs. Others have pointed out the absence of any understanding of the context in which women have sex and get pregnant. Indeed, the extent to which a woman can protect herself against involuntary sex, unwanted pregnancy, and sexually transmitted diseases is based on her ability to negotiate when, how, and with whom she has sex (Dixon-Mueller, 1993).

Finally, still other critics of the narrow emphasis on family planning have presented new epidemiological data to underscore the need to broaden the programmatic focus to other reproductive-health problems. For example, while women around the world experience the same rate of complications in pregnancy, 99 percent of deaths are estimated to occur in the Third World, because women do not receive the emergency obstetric care they need in time (Thaddeus and Maine, 1994). While early Safe Motherhood programs began assuming that women and their birth attendants were the problem, more recent study suggests a more complex picture, with the access to, and qual-ity of, emergency care in hospitals being major issues (Sundari, 1992). It is also important to note that approximately 40 percent of all pregnancy-related deaths in the world are due to complications from unsafe abortions (Coeytaux et al., 1993).

A virtual explosion of sexually transmitted diseases has come about through increased travel opportunities and money in the hands of some, the persisting double standard rationalizing male sexual pleasure, and increasing poverty, which forces increasing numbers of women to sell sex for money. The sequelae of infections in women are more serious than they are in men; they include pelvic inflammatory disease, which can cause debilitating pain and infertility. Some areas of sub-Saharan Africa are experiencing dramatic rises in secondary infertility attributed to untreated sexually transmitted diseases (Germain et al., 1992). Evidence is also growing that cervical cancer may be related to the sexually transmitted human papilloma virus (Zunzunegui et al., 1986).

We are only beginning to comprehend what the double pandemic diseases that are essentially transmitted sexually—hepatitis B and HIV/AIDS—mean for women in the Third World. For a number of years, the emphasis on "high-risk groups" limited a proper analysis of risk. We now know, for example, that monogamous women with partners who have many sexual partners are themselves at great risk of being infected. Biologically speaking, women are more than 10 times likely to be infected through unprotected vaginal intercourse with an infected man than vice versa (Berer and Ray, 1993). It is already clear that women not only will become the majority of infected persons in Africa and Asia, they also will be expected to care for other ill family members when hospitals can no longer cope with the numbers of sick people.

Two other problems that affect African women, in particular, must be mentioned here. When undernourished adolescents begin childbearing, their pelvises are often not big enough to allow uncomplicated births. Many die from obstructed labor. Still others are seriously torn in the birth process. Unless they are repaired, the resulting tears can cause fecal or urinary incontinence (World Health Organization, 1991b). Such women are often ostracized and forced into destitution.

An estimated 100 million women in Africa have been "circumcised," with two million new operations performed each year (Toubia, 1993). This traditional surgery varies from a superficial cut to the clitoris to severe excision and infibulation of the external genitalia. Stated reasons for the practice include: cleanliness, beauty, religious dictates, and protecting a girl's marriageability. Genital mutilation poses immediate risks of infection and serious long-term emotional and physical problems.

Psychological Aspects

Because international public health has been so focused on fertility, child survival, and infectious diseases, scant attention has been given to Third World women's psychological health. Women of all countries and classes share the stresses of balancing production and reproduction, or work and family responsibilities. Women are witness to, and often victims of, violence, whether in the home, the streets, or as part of civil war. The World Bank's *World Development Report 1993* ranks depression as fifth among problems affecting women in the world age 15–44, yet women in the Third World have few therapeutic options. And women are increasingly concerned about a related problem: the lack of adequate controls on the prescription of tranquilizers. Anecdotal evidence from many places suggests that physicians are giving women prescriptions of potentially addictive medications based on superficial diagnoses (Smyke, 1991).

As Western medicine takes the preeminent commercial place in health care in Third World countries, many traditional medical practices are losing favor. Some traditional practices, such as massage, take a holistic approach to the person and have an important comforting effect. Positive traditional practices may not have a substitute in modern medicine, so when they are abandoned, healthcare options are less rich.

Health Care

A large literature exists on health services in developing countries, especially on their financing. Macroeconomic forces are having a strikingly negative impact on the coverage, quantity, and quality of health services in many countries, particularly those in Africa and Latin America undergoing "structural-adjustment" programs as a condition of World Bank loans. These programs encourage governments to cut back on social spending, including health and education. Emergency obstetric services are reported especially hard hit, making it difficult for governments to reduce already high maternal mortality.

Less is written about services, from clients' or patients' points of view. The literature on quality of care in family-planning services suggests, however, that where class, educational, and gender differences separate the patient and the provider, respectful treatment, full understanding, and informed consent are often lacking (Bruce, 1990).

Local Action to Improve Women's Health

So far, this essay has tended to portray Third World women as passive recipients of health policies and programs, many of which have been conceived by medical specialists trained in the industrialized countries or defined by international aid agencies as part of development assistance. As noted above, these forces and others, such as structural adjustment, have been significant factors in shaping health care.

But it would be wrong to cast women as only helpless victims of global forces. Since the late 1980s, we have seen impressive advances of women into the health professions, and it is only a matter of time until the first cohorts are in senior and decision-making positions. A number of women scientists and activists have questioned the appropriateness for the Third World of certain reproductive tech-

nologies, such as the contraceptive implant Norplant, contraceptive vaccines (World Health Organization, 1991a), and the abortifacient RU486 (Kamal, 1992).

In addition, grass-roots activism around women's health concerns is in evidence on all continents. In Sao Paulo, Brazil, and Dacca, Bangladesh, for example, women have set up health centers run on feminist principles. In other places, notably Cairo, Egypt, and Jakarta, Indonesia, women have produced reference books similar in spirit to the classic *Our Bodies Ourselves.* In Nigeria, women's organizations are struggling to prevent and treat obstetric fistulae, and, all over Africa, women's organizations are devoting themselves to eliminating female genital mutilation.

Still other groups, such as SEWA in India, have organized to protect women's occupational health. In Thailand, women's organizations such as Empower and Friends of Women have publicized the exploitive and dangerous conditions of commercial sex workers.

Conclusion

Third World women say that they are suffering the worst aspects of underdevelopment and development at the same time.

The cumulative and debilitating effects of undernutrition, excessive and often dangerous work, and risky reproduction come up against woefully inadequate health services. Yet, in many places, unregulated industrialization and private medicine also pose new threats to women's health.

By virtue of their gender, all Third World women share two problems: the legacy of neglect of women-specific illnesses by the biomedical establishment and their vulnerability to violence, in private and in public. But the factors that determine their health are also strikingly different. Nations of the Third World have dramatically different political and legal systems and different approaches to health care, and, even within countries, women are further differentiated by class and by the resources they control or do not control for their health.

The International Conference on Population and Development (Cairo, 1994) and the Fourth World Conference on Women (Beijing, 1995) raised public awareness about many of the health problems Third World women confront. Nongovernmental organizations received new encouragement to fight for improvements in women's health, and, whether by advocacy to change laws and design better policies or by direct action to provide new models of care, they are making a difference.

References

Asian and Pacific Women's Resource Collection Network. *Health.* Asian and Pacific Women's Resource and Action Series. Kuala Lumpur: Asian and Pacific Development Center, 1990.

Batliwala, Srilatha. "Fields of Rice: Health Hazards for Women and Unborn Children." *Manushi,* vol. 46, 1988, pp. 31–53.

Berer, Marge, and Sunanda Ray. *Women and HIV/AIDS.*

London: Pandora, 1993.

Boston Women's Health Book Collective. *The New Our Bodies, Ourselves: A Book by and for Women.* New York: Simon and Schuster, [1996], c 1992.

Bruce, Judith. "Fundamental Elements of the Quality of Care: A Simple Framework," *Studies in Family Planning,* vol. 21, no. 2, 1990, pp. 61–91.

Chatterjee, Meera. *Indian Women: Their Health and Productivity.* Washington, D.C.: World Bank, 1991.

Coeytaux, Francine, Ann Leonard, and Carolyn Bloomer. "Abortion." In Marge Koblinsky, Judith Timyan, and Jill Gay (eds.), *The Health of Women: A Global Perspective.* Boulder: Westview, 1993, pp. 133–146.

Cook, R.J. *Women's Health and Human Rights.* Geneva: World Health Organization, 1994.

Dixon-Mueller, Ruth. "The Sexuality Connection in Reproductive Health," *Studies in Family Planning,* vol. 24, no. 5, 1993, pp. 269–282.

Germain, Adrienne, Judith Wasserheit, and King Holmes (eds.). *Reproductive Tract Infections: Global Impact and Priorities for Women's Reproductive Health.* New York: Plenum, 1992.

Holloway, Marguerite. "Trends in Women's Health: A Global View," *Scientific American,* vol. 271, no. 2, 1994, pp. 77–83.

Kamal, Ghulam (ed.). *Proceedings of the International Symposium on Antiprogestins.* Dacca: Bangladesh Association for Prevention of Septic Abortion, 1992.

Merchant, Kathleen, and Kathleen Kurz. "Women's Nutrition Through the Life Cycle: Social and Biological Vulnerabilities." In Marge Koblinsky, Judith Timyan, and Jill Gay (eds.), *The Health of Women: A Global Perspective.* Boulder: Westview, 1993, pp. 63–90.

Saleh, Saneya, and Saad Gadalla. "Accidental Burn Deaths to Egyptian Women of Reproductive Age," *Burns,* vol. 12, 1986, pp. 241–245.

Smyke, Patricia. *Women and Health.* London: Zed, 1991.

Sundari, T.K. The Untold Story: How Health Care Systems in Developing Counries Contribute to Maternal Mortality. *International Journal of Health Services,* vol. 22, no. 3, 1992, pp. 513–538.

Tekce, Belgin. "Households, Resources, and Child Health in a Self-Help Settlement in Cairo, Egypt," *Social Science and Medicine,* vol. 30, no. 8, 1990, pp. 929–940.

Thaddeus, Sereen, and Deborah Maine. "Too Far to Walk: Maternal Mortality in Context," *Social Science and Medicine,* vol. 38, no. 8, 1994, pp. 1091–1110.

Toubia, Nahid. *Female Genital Mutilation: A Call for Global Action.* New York: Rainbo, 1995.

World Bank. *World Development Report 1993.* New York: Oxford University Press, 1993.

———. *A New Agenda for Women's Health and Nutrition.* Washington, D.C.: World Bank, 1994.

World Health Organization. *Creating Common Ground: Women's Perspectives on the Selection and Introduction of Fertility Regulation Technologies.* Geneva: WHO, 1991a.

———. *Obstetric Fistulae: A Review of Available Information.* Geneva: W.H.O., 1991b.

———. Women's Health. "Improve Our Health: Improve the World," position paper, executive summary, WHO, Geneva, 1995.

Zunzunegui, M., M.C. King, C.F. Coria, and J. Charlet. "Male Influences on Cervical Cancer Risk," *American Journal of Epidemiology,* vol. 123, no. 2, 1986, pp. 302–306.

Women and Mortality Trends

Monica Das Gupta
Jacob Adetunji

Introduction

Females have a biological advantage in survival from the time of conception until old age. This advantage can be reversed by active discrimination against females, as is the case in much of South and East Asia. It can also be reversed under conditions of poor nutrition, high fertility, and high levels of maternal mortality. Under these conditions, the stresses of reproduction can tilt the balance of survival advantage in favor of males in the age groups of peak childbearing. In some situations, there also may be a pattern of disease that affects females more than males. These deviations from the norm of female survival advantage are discussed below, following a description of the basic advantage.

The Female Biological Advantage in Survival

At the time of conception, there are far more males conceived than females. The rate of spontaneous abortion and stillbirth among males is higher, however, due to a higher incidence of genetic damage (Hassold et al., 1983). As a result, there is a more balanced sex ratio among live-born children. Data on trends in the rates of miscarriage, stillbirth, and live birth by sex are rare, but they have been collected over centuries in Sweden. These trends show an interesting pattern. As the Swedish population moved from the mid-eighteenth century to the twentieth century, the proportion of males among the miscarriages and stillbirths came down and the sex ratio of live births rose from around 104.4 males to 100 females in 1751–1760, to 105.8 males per 100 females in the 1980s. This is attributed to improvements in the nutritional standards of the Swedish population, reducing the rate of miscarriages and stillbirths (Johansson and Nygren, 1991). We can expect to find similar trends elsewhere as economic and health conditions improve, making for better maternal health and better conditions of delivery.

During the first month after birth, and especially during the first week of life, males have higher rates of mortality than females. Among the causes are a higher rate of congenital deformity and a greater susceptibility to infections such as tetanus. This disadvantage wears off after the first few weeks of life, and the mortality rates during the rest of the first year of life and thereafter show a smaller advantage for females. This is reflected in higher rates of neonatal (the first month of life) mortality among males. For example, in Matlab Thana in Bangladesh in the period 1974–1977, there were 1.16 male neonatal deaths to one female neonatal death (D'Souza and Chen, 1980). It can also be seen in the estimates of perinatal mortality (currently defined as mortality from the seventh month of gestation) for developing countries, where a higher proportion of male than female deaths are due to perinatal mortality and higher rates of male disability arise from perinatal problems.

It follows that female life expectancy at birth is normally higher than that of males, as is the remaining expectation of life at any given age. Preston et al. (1972) have analyzed the lifetables and the causes of death for males and females for all of the countries and time periods for which reliable data on mortality by cause were available. Most of these data are from developed countries. They show that females commonly have lower mortality rates than males at any age. The same pattern can be seen in the lifetables of most developing countries in the 1990s, except those characterized by substantial discrimination against females.

These lifetables show that females had lower rates of mortality from most infectious diseases than males. For example, lifetables from England and Wales for the period 1871–1921 show that males had higher death rates than females from every category of infectious diseases (Preston et al., 1972). The differential can be seen not only in 1871, when mortality rates from infectious diseases were very high, but also in 1921, by which time mortality from infectious diseases was greatly reduced. Not only during infancy, but also in most age groups, males had higher rates of mortality from each infectious disease for which data are given.

Not all countries show the same pattern of sex differences by cause of death. In some Scandinavian settings, for example Sweden in 1920, female death rates from tuberculosis were higher than male death rates from this cause, although males had higher mortality rates from other infectious diseases. Numerous epidemiological factors account for these differences, including questions of living conditions. Women can sometimes be more vulnerable to some diseases because they tend to stay at home and tend the sick, leading to prolonged exposure to the disease. This was the reason given in official reports of the plague in northern India in the first decade of the twentieth century to explain higher rates of female mortality from the disease.

Males are similarly affected by their own living and working conditions, which contribute to higher male mortality rates from infectious diseases. Men are more exposed to the world outside the home and, therefore, to a wider range of sources of infection. Some diseases are also aggravated by occupational stresses. Males are also exposed to more accidents and violence than women because of conditions in the workplace and for behavioral reasons. This is reflected in higher rates of male mortality from accidents and violence in England and Wales in 1871 and 1921, a pattern commonly found across the world. Behavioral factors predisposing males to greater risk of accident are apparently at play even at very young ages, as is reflected in higher death rates from accidents among young boys than girls.

Estimates based on data from developing countries show higher male than female mortality from injuries. This is reflected also in morbidity rates (shown in Table 1). The morbidity estimates for the developing world are drawn up by the World Health Organization (WHO) on the basis of morbidity surveys in different countries, so these estimates should be viewed simply as estimates and not as actual measurements. They are presented as Disability-Adjusted Life Years (DALYs) lost. This is an estimate of the

total burden of morbidity borne by people as a consequence of having a disease; it is weighted by the age of the person. Thus, the total DALYs lost by a young child contracting polio are considerably higher than those lost by an older person contracting diabetes, since the younger person has more years to live with the disability. As in the case of mortality, males suffer more than females from injuries, disabilities contracted during the perinatal period, and communicable diseases.

Technological change has the most powerful influence on survival probabilities. Compared to this factor, any differences found between the sexes pale into insignificance. Technological change has far outstripped the effect of economic differentials on health conditions. Economic differentials have, of course, considerable influence on health. There are several reasons for this, including the fact that people with higher incomes can afford to live in better housing with better sanitary conditions, can afford to be better nourished, and can afford to make use of health-care services. Nevertheless, the effect of economic levels is swamped by that of technological change. Figure 1 shows that a national (based on international dollars) per-capita income of $5,000 was associated with a life expectancy at birth of below 55 years in 1900, which rose steadily to above 72 years by 1990. Life expectancy is correlated with per-capita income in each of these periods, but the entire curve has been shifted up by dramatic improvements in medical technologies.

Maternal Mortality
The main point in the lifecycle at which females may be placed at a biological disadvantage is during the childbearing years. The stresses of pregnancy and delivery take a toll on women. When combined with poor prenatal care and delivery conditions, this can result in elevated mortality rates, especially when women bear many children. In some societies with high or moderate fertility rates, female mor-

Table 1. Disability-Adjusted Life Years (DALYs) Lost per Thousand Population, by Cause and Sex, 1990

Cause	EME[1]		FSE[2]		Latin America		Sub-SaharanAfrica	
	Male	Female	Male	Female	Male	Female	Male	Female
Communicable/maternal	10.8	12	14.8	14.2	102.2	93.2	415.3	403
Infectious/parasitic	4.7	5.6	4.9	4.7	62.2	54.2	303.1	278.2
TB	0.3	0.1	1.9	0.3	6.8	4.8	29.6	24.1
STD	0.1	3.8	0.1	2.8	1.1	9.7	11.4	17.7
HIV	3.2	0.8	0.8	0.1	15.4	4.6	37.2	34.8
Diarrhea	0.3	0.3	0.7	0.6	14.1	12.4	62.4	56.7
Respiratory	3.2	2.9	5	3.7	15.3	13.4	64.4	59.6
Maternal		1.4		2.7		8.1		31
Perinatal	3	2.2	5	3.1	24.7	17.5	47.8	34.2
Noncommunicable	102.6	82.6	140.8	109.7	103.1	95.1	113.7	108
Malignant neoplasm	25.7	19.6	30.4	20.1	11.5	12.6	9	8.7
Cardiovascular	31.2	24.3	54.6	44.4	22.2	20.8	22.1	25.8
Injuries	20.2	7.9	45.7	11.5	52.9	16.7	76.8	30.9
All causes	133.7	102.4	201.3	135.5	258.2	205.1	605.8	541.8

Table 1—*continued*

Cause	Middle East		India		China		Other Asia	
	Male	*Female*	*Male*	*Female*	*Male*	*Female*	*Male*	*Female*
Communicable/maternal	135.7	157.4	161.1	189	38.9	51.2	127.4	123.5
Infectious/parasitic	70.5	77.1	92.9	100.5	20	24.2	75	67.3
TB	8.4	7.6	14.3	11	5.9	4.5	15.1	11.1
STD	0.1	2.6	1.2	7.8	0.1	6.1	0.2	3.7
HIV	1	.03	6.2	3.3	0		2.3	1.4
Diarrhea	29.3	32	31	35.1	3.5	4	22.9	20.3
Respiratory	31.5	34.4	35.4	39.5	10.3	12.6	30.2	27.4
Maternal		17.1		19.1		4.5		12.8
Perinatal	33.6	28.9	32.7	30	8.6	9.9	22.2	16
Noncommunicable	103	103.1	136.3	140.8	104.1	101.8	104.6	102.7
Malignant neoplasm	10.4	9.1	15.1	13.2	19.4	13.2	12.2	10.9
Cardiovascular	25.2	25.7	33.5	33.8	25.5	24.5	24.9	25.7
Injuries	49.9	24.3	33.6	29.1	33.8	25.3	44.9	13.6
All causes	288.6	284.8	331	358.9	176.8	178.3	276.8	239.9

[1]EME refers to the established market economies (all Organization for Economic Cooperation and Development countries except Turkey, plus several small high-income European economies).

[2]FSE refers to the formerly socialist economies of Europe.

Source: Adapted from Murray et al., 1994, pp. 112–135.

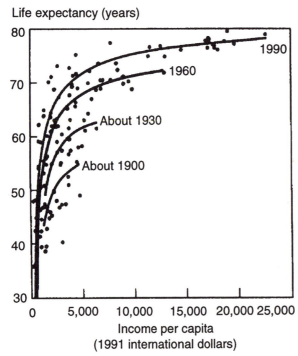

Life expectancy (years)

Income per capita
(1991 international dollars)

Note: International dollars are derived from national currencies not by use of exchange rates but by assessment of purchasing power. The effect is to raise the relative incomes of poorer countries, often substantially. For illustrative country comparisons and a more detailed explanation, see Table 30 in the World Development Indicators.

Figure 1. Life Expectancy and Income per Capita for Selected Countries and Periods. Source: Originally published by Preston, Keyfitz, and Schoen, 1972; adapted by World Bank 1993.

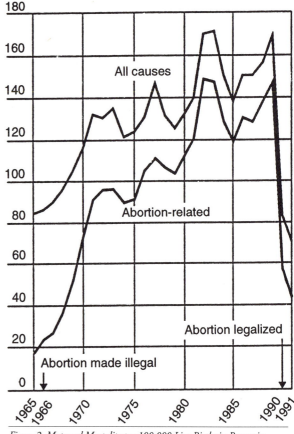

Maternal deaths per 100,000 live births

Figure 2. Maternal Mortality per 100,000 Live Births in Romania, 1965–1991. Source: Adapted from Stephenson et al., 1992, which used Romanian Ministry of Health data.

Table 2. Selected Indicators of Reproductive Health Among Women in Developing Countries

	Births Attended by Health Staff (%)[a] 1983–1989	Maternal Mortality per 100,000[b] 1988	Total Fertility Rate[c] 1991	Life Expectancy at Birth[d] 1991 Male	1991 Female
Sub-Saharan Africa					
Nigeria	40	750	5.9	50	53
Ghana	55	700	6.2	53	57
Kenya	28	600	6.3	46	49
Sudan	60	700	6.3	56	62
Senegal	41	750	6.1	46	49
Botswana	78	300	4.8	66	70
Latin America					
Brazil	95	230	2.8	63	69
Mexico	94	150	3.2	67	73
Argentina	N/A	140	2.8	68	75
Colombia	71	150	2.7	66	72
Peru	78	300	3.4	62	66
North Africa/Middle East					
Egypt	47	300	3	60	62
Tunisia	68	200	3.5	67	68
Jordan	83	200	5.3	66	70
Saudi Arabia	88	220	6.5	68	71
Asia					
China	94	130	2.4	67	71
India	33	550	3.9	60	60
Sri Lanka	94	180	2.5	69	74
Indonesia	49	300	3	58	61
Thailand	71	180	2.3	66	72
Philippines	57	250	3.6	63	67
Pakistan	40	600	5.7	59	59
Bangladesh	5	650	4.4	53	52

Sources: [a, b] UNDP, 1992, pp. 148, 150.
[c, d] World Bank, 1993, pp. 290–291, 300–301.

tality during the peak childbearing years exceeds that of males in the same age group. In India, in 1991, women had lower mortality throughout their adult years, except during the peak reproductive years.

Despite poor conditions of childbearing and high fertility, in many settings the female advantage in survival can outweigh these stresses such that female mortality rates remain lower than male mortality rates even during the peak childbearing years. For example, England and Wales in 1871 had high fertility and maternal mortality rates, but women had lower death rates than men at every age. Women's risk of maternal mortality was more than outweighed by men's greater susceptibility to infectious diseases and injuries, even during the reproductive ages.

The opposite effect can also be induced—namely, the female advantage can be negated by neglect of women's needs, either in the home or in the wider society. One example of a societal factor accentuating reproductive stresses on women is poorly designed legislation regarding abortion. Some countries have restrictive laws on abortion, which force women to resort to the poor-quality services of "backstreet"

abortionists. Romania provides a well-documented example of how this can affect women's mortality rates: When abortion was made illegal, female mortality rates soared; they fell when it was legalized once more (see Figure 2).

Inadequate provision and inadequate use of prenatal care also raises maternal mortality rates: There is a strong correlation between the proportion of deliveries attended by a trained person and the maternal mortality rate at the country level. Levels of maternal mortality are high in the developing world, especially in sub-Saharan Africa and South Asia (see Table 2), where the proportions of births attended by trained staff are lower than in the rest of the developing world.

The mother's physical condition and the age at which women give birth affect maternal mortality rates. In Bangladesh, many women begin childbearing in their early teens, well before reaching full maturity. Combined with undernutrition, this makes for high risks of maternal mortality during the teen years. Where female life expectancy at birth is lower than that of males, the situation is altered by a decline in fertility. In Sri Lanka, such a decline was accompanied by a crossover from higher male expectation of life to higher female expectation of life (Nadarajah, 1983). This was because the excess female mortality during the reproductive ages decreased as fertility declined, until female mortality was lower than male even at childbearing ages. Simply removing the stresses of reproduction enabled women's natural survival advantage to manifest itself.

Reproductive Health

There has been an increasing interest in the question of women's reproductive health, as evidenced in the strong commitment to improving women's reproductive health made at the International Conference on Population and Development (Cairo, 1994).

This concern arises from several roots. One is that levels of reproductive health are poor in many developing countries. Few studies exist that document reproductive health carefully; those that have been carried out show that high proportions of women suffer from reproductive health problems but few seek treatment. They may be isolated or embarrassed to discuss these intimate problems with a doctor. In some cultures, if a female doctor is not available, then male doctors have to treat women without necessarily being able to examine them properly. A variety of factors combine to make gynecological diseases a "woman's problem," which women fear they must bear in silence, with little support from others. Estimates of morbidity from sexually transmitted diseases (STDs) are far higher among women than among men (see Table 1).

Another source of concern with reproductive health is the fact that power relations within sexual unions are often such that women are at a special disadvantage in protecting themselves from sexually transmitted diseases, including HIV, the virus that causes AIDS. Men are often exposed more to such diseases because of extramarital relations—for example, in the course of labor migration. They may pass on these diseases to their wives. Unless both husband and wife seek treatment simultaneously, they can reinfect each other. Studies of female sex workers reveal that women find it difficult to enforce preventive measures, such as the use of condoms, with their clients.

Another source of concern is that the felt need to control population growth has resulted in family planning programs organized to maximize fertility control, often with little concern for the individual woman's health and welfare. The majority of modern contraceptive techniques are for the use of women rather than men. Even when a good technology exists, such as male sterilization (vasectomy), there seems to be an overwhelming preference in most countries for female sterilization. Although male sterilization is a far simpler operation than female sterilization, many men prefer not to undergo the operation. Other contraceptive techniques require more follow-up to monitor side effects than is generally carried out in most developing countries. Women are thus left with considerable burdens of contraception, which sometimes tell on their health.

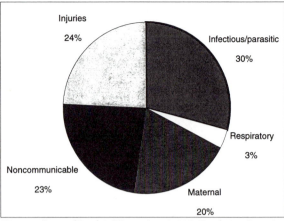

Figure 3. Causes of Death in Women Age 15–29 in Developing Countries
Source: Adapted from Murray and Lopez, 1994.

Women's reproductive health needs far more attention than it has received, but it is important to note that the totality of the burden of ill-health that women face as a result of discrimination far exceeds that arising from reproductive-health problems alone. In terms of mortality, maternal-related causes of death account for about 2 percent of all deaths among women in developing countries. Only 5 percent of total estimated morbidity among females is accounted for by maternal causes (Murray et al., 1994). Even if we include the reproductive-health problems of sexually transmitted diseases, this adds only another 1 percent to the total burden of morbidity attributable to reproductive-health problems.

Even in the peak childbearing age group of 15–29, with maximum exposure to reproductive-health problems, deaths related to maternity account for only 20 percent of total deaths in developing countries, substantially less than that from infectious and parasitic diseases (see Figure 3). These percentages shrink further in significance when we take into account the fact that death rates at these young-adult ages are far lower than at any other age, so the universe of which these figures are percentages constitutes a very small fraction of total female deaths. A similarly low percentage of total morbidity is related to maternity among women of overall childbearing age, 15–44 (see Figure 4).

It is possible that these estimates do not succeed in capturing the full extent of reproductive-health problems. It has been shown that women tend not to report such health problems (Bang et al., 1989). On the other hand, other health problems are also substantially underreported in developing countries (the estimates reproduced here do attempt to correct for that). It is apparent that, while focusing on women's reproductive health, attention should also be given to communicable diseases and other important disease groups.

Discrimination Against Women

We do not have reliable lifetables for most African and some Asian countries, but we do have estimates of life expectancy at birth by sex (see Table 2). The levels of life expectancy at birth tell us much about how women's lives are socially valued relative to that of males. Sub-Saharan Africa, for example, shows consistently higher female than male life expectancy at birth, despite the fact that levels of fertility and maternal mortality are high. Latin American countries also show the normal pattern of higher female life expectancy than male. This suggests that, in sub-Saharan Africa and Latin America, women are not neglected relative to men, at least not to a degree resulting in elevated mortality levels. South Asia stands in contrast to this situation, with considerable evidence of artificially raised levels of female mortality. The biological norm is that male life expectancy should be a few years lower than female life expectancy. So, when they are similar, as in South Asian countries, this indicates discrimination that overrides the biological advantage of women.

When comparing male and female life expectancy at birth in different regions of the world, we must bear in mind that the female advantage increases as life expectancy at birth rises. Given this, South Asia stands out as showing the least female advantage within the developing world, followed by the Middle East. Sub-Saharan Africa and Latin America show the largest female advantage over men. This suggests that women are discriminated against the most in South and East Asia, to a lesser extent in the Middle East, and perhaps not at all in Latin America and sub-Saharan Africa, at least as far as discrimination leading to excess mortality is concerned.

The same pattern of female advantage shows in the estimated levels of morbidity, measured in terms of DALYs (cf. Table 1). In China, India, and the Middle East, females suffer from substantially higher levels of communicable diseases than males. This contrasts with Latin America, sub-Saharan Africa, and the rest of Asia, where it is the males who suffer more from communicable diseases. China even has higher estimated morbidity rates from perinatal causes for females than for males, which is the opposite of the pattern in all the other regions and countries for which estimates are made. This could result from discrimination against newborn girls.

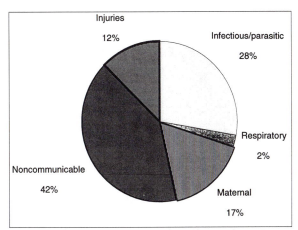

Figure 4. Causes of Death in Women Aged 15–44 in Developing Countries. Source: Adapted from Murray et al., 1994.

Another feature suggesting excess female vulnerability in South and East Asia is that while disability resulting from injuries is uniformly much higher for males than females in all countries and regions of the world, the gap between male and female rates is substantially smaller in the case of China and India than elsewhere. This suggests greater exposure of women to injuries in these South and East Asian countries, perhaps the result of poor domestic arrangements such as fire hazards in cooking or of greater physical violence against women.

There are many subtle ways in which women's health can be affected by how they are socially valued. For example, low status of women and neglect of their needs can exacerbate the effects of reproductive stress and slow down the pace of improvement in women's health when fertility declines and the physical toll of reproduction is reduced. Lack of care during the process of childbearing can take the form of inadequate nutrition and heavy workloads during pregnancy and breast-feeding, inadequate antenatal care, and poor conditions at delivery.

There is an apparent lack of clear correlation between fertility rates and maternal mortality rates. Jordan has low maternal mortality compared with India, although fertility is much higher. These anomalous findings result to a large extent from the conditions of pregnancy and delivery. One obvious and easily measured aspect of differences in conditions of delivery is the proportion of births attended by trained personnel: this is far lower in India than in Jordan or most sub-Saharan countries (see Table 2).

India provides a very good illustration of the adverse consequences of low female status on women's health. India's maternal mortality rates are similar to those of sub-Saharan Africa, which is surprising because India has lower fertility levels (and therefore lower exposure to maternal mortality) as well as a far higher number of doctors available per capita than in sub-Saharan African countries. Moreover, medical facilities are available throughout the country, including the rural areas. These facilities are widely used, and enormous sums of money are spent on private health care in the country, in addition to the widespread provision of public health care.

Yet the proportion of births attended by trained staff in India is low compared to most sub-Saharan countries, as already discussed. Although there is widespread utilization of public and private health-care facilities, such care is not lavished on pregnant women. In large parts of the country, young women lack the necessary autonomy to make their own decisions about their own health care. Such decisions are made by more senior members of the household. Pregnancy and childbirth do not tend to be treated as matters requiring much attention and expenditure of household resources, and the vast majority of deliveries are conducted by poorly skilled midwives. This is one major reason why Indian women have elevated levels of maternal mortality. Their sub-Saharan counterparts are better placed to use available household resources to look after their own health.

Inadequate nutrition also plays a part in reducing Indian women's capacity to cope with the stresses of pregnancy and lactation. Women's nutritional requirements rise during this period, and unless efforts are made to meet these additional requirements, they are subject to special nutritional stresses. To a large extent, the problem of undernutrition during pregnancy is inseparable from the problem of poverty and general undernutrition of the population in large parts of India. However, women are undernourished during pregnancy even in the parts of the country where levels of nutrition are high, because their lack of autonomy creates an obstacle to reallocating household resources to respond to their increased nutritional needs. For example, in a state such as Punjab, where levels of per-capita income and nutritional intake are high, both men and women are adequately nourished on average, except for women who are pregnant or lactating (Das Gupta, 1995). The nutritionally more dense foods women need to eat during this period are available in the local diet, but young women are not free to choose to consume them because of their low position in the household hierarchy. In large parts of the country, levels of poverty are such that nutritional levels are inadequate for substantial sections of the population, which further increases the probability of women being undernourished during reproduction.

East and South Asia are also the regions of the world that show the strongest evidence of son preference, as measured by the proportion of mothers who want their next child to be a boy rather than a girl (WHO/UNICEF, 1986). Middle Eastern countries register more moderate preference than South and East Asia, while Southeast Asian countries register moderate or no preference. Sub-Saharan African countries register moderate or no son preference, and Latin American countries mostly register no preference and occasionally even daughter preference.

The strong son preference of East and South Asia is reflected in excess female-child mortality, either before or after birth. Traditionally, unwanted daughters were killed or abandoned at the time of birth or neglected in such a way as to increase the risk of dying. Technological change has altered the methods of discrimination used, as people turn increasingly to sex-selective abortion as a way of eliminating unwanted daughters. Countries where people have better access to the technology, such as South Korea, rely almost entirely on sex-selective abortion. Elsewhere, as in China and India, people rely more on a mix of methods. In India, the most prevalent method is that of neglect, and studies indicate that the most important form of neglect is underutilization of health care for girls (compared to boys in the same households) (Wyon and Gordon, 1971; Chen, et al., 1981; Das Gupta, 1987).

Nor is there equal discrimination against all girls. The element of conscious choice in the discrimination is highlighted by the fact that it is daughters born into families that already have a daughter who suffer the excess mortality. First-born daughters have mortality rates comparable to that of their male counterparts. This is evident from the child mortality rates from northern India and Bangladesh (Das Gupta, 1987; Muhuri and Preston, 1991; Pebley and Amin, 1991). The same pattern is indicated by the sex ratios at birth in South Korea and China, which show a normal sex ratio for the first birth and increasingly masculine ratios thereafter. As fertility levels have come down in these countries, the total number of children desired by families has fallen more rapidly than the total number of sons desired, reducing further the space for tolerating daughters. This has led to intensified discrimination against girls at each birth order beyond the first. A net increase in excess female child mortality is evident from the rising overall sex ratios at birth in China and South Korea (Zeng et al., 1993; Park and Cho, 1995), and from increasing masculinity of the sex ratio of children in India (Das Gupta and Bhat, 1995). Thus, while fertility decline improves the survival chances of adult females, in settings with strong son preference it has reduced the survival chances of female children.

Mother's Impact on Child Mortality

Parents influence their children's life chances in many ways, including their prospects for survival, education, and occupation. Here we discuss two of the most direct ways in which the mother's status affects on the child's health: her own health conditions during pregnancy and delivery, and the effect of her educational status on her children's survival.

The implications of poor care during pregnancy and delivery for the mother's health have been discussed above. To a considerable extent, children are protected by their mothers from the impact of some aspects of poor conditions of childbearing, such as undernutrition during pregnancy and lactation. Mothers absorb much of their nutritional shortfalls, and their bodies give priority to the nourishment of the child. However, any infant who is not adequately protected by this biological mechanism and is born underweight has substantially reduced chances of survival, compared with normal-birthweight infants. Poor health care during and after delivery substantially raises the chances not only of maternal mortality, but also of child mortality.

Maternal education appears to have an impressive effect in improving child survival rates. Of all of the social and economic factors known to be associated with child survival, it has been found consistently to be among the most powerful—much stronger than the father's educational level or occupation. Hobcraft (1993) and the United Nations (1994) analyzing the data from a large number of developing countries, have found that, within each country, the probability of child mortality falls sharply as the mothers' education level rises, especially for women who are educated beyond primary school.

There is some controversy as to how much of the apparent relationship between mother's education and child survival is simply a reflection of the fact that better-educated mothers are likely to belong to higher-income households and be married to better-educated men. However, studies that have controlled for the effect of husband's education and household socioeconomic status reveal that the mother's education continues to have a powerful independent effect on child survival (Cleland and van Ginneken, 1988; LeVine et al., 1991; Jejeebhoy, 1995). The effect is particularly strong after the first month of life, when care-related factors become more important than congenital problems as causes of death.

The exact mechanisms through which maternal education may influence child survival are not fully understood. Quantitative and qualitative evidence suggests that schooling improves health care in various ways, including better hygiene and health management within the home and increased use of outside health services for preventive and curative care (Caldwell, 1979; Cleland and van Ginneken, 1988; LeVine et al., 1991). As with many other aspects of life, education is believed to improve people's effectiveness in obtaining higher-quality services from medical personnel and their ability to understand and follow the instructions given by health personnel.

Conclusions

To summarize, females have a basic biological advantage over males in terms of survival. This advantage is least manifest during the peak reproductive years in societies with high levels of fertility, because of the stresses associated with repeated pregnancy and lactation. This is exacerbated by poor conditions of reproductive health care, abortion, and delivery. Active discrimination against females reverses their survival advantage in some societies, notably those of South and East Asia. These societies show elevated levels of female mortality, especially among children. Discrimination against women can also slow down the pace of health benefits derived from having fewer children. The status of women also has clear implications for child survival. One example of this is that child survival is adversely affected if women lack autonomy and are poorly cared for during pregnancy and lactation. Similarly, child survival is improved by maternal education.

References

Bang, Rani, A.T. Bang, M. Baitule, Y. Choudhary, S. Sarmukaddam, and O. Tale. "High Prevalence of Gynaecological Diseases in Rural Indian Women," *Lancet,* 14 January 1989, pp. 85–88.

Caldwell, John. "Education As a Factor in Mortality Decline: An Examination of Nigerian Data," *Population Studies,* vol. 33, no. 3, 1979, pp. 395–413.

Chen, Lincoln C., Emdadul Huq, and Stan D'Souza. "Sex Bias in the Family Allocation of Food and Health Care in Rural Bangladesh," *Population and Development Review,* vol. 7, 1981, pp. 55–70.

Cleland, John, and Jeroen van Ginneken. "Maternal Education and Child Survival in Developing Countries: The Search for Pathways of Influence," *Social Science and Medicine,* vol. 27, 1988, pp. 1357–1368.

Das Gupta, Monica. "Selective Discrimination Against Female Children in Rural India," *Population and Development Review,* vol. 13, no. 1, 1987, pp. 77–100.

———. "Lifecourse Perspectives on Women's Autonomy and Health Outcomes," *American Anthropologist,* vol. 97, no. 3, 1995, pp. 481–491.

Das Gupta, Monica, and Mari Bhat. "Intensified Gender Bias in India: A Consequence of Fertility Decline," working paper no. 95–02, Center for Population and Development Studies, Harvard University, Cambridge, 1995.

D'Souza, Stan, and Lincoln Chen. "Sex Differentials in Mortality in Rural Bangladesh," *Population and Development Review,* vol. 6, no. 2, 1980, pp. 257–270.

Hassold, T., S. Quillen, and J. Yamane. "Sex Ratio in Spontaneous Abortions," *Annals of Human Genetics,* vol. 47, 1983, pp. 39–47.

Hobcraft, John. "Women's Education, Child Welfare and Child Survival." *Health Transition Review,* vol. 3, no. 2, 1993, pp. 159–175.

Jejeebhoy, Shireen. "Women's Education, Autonomy, and Reproductive Behavior: Assessing What We Have Learned," paper prepared for the Conference on the Status of Women and Demographic Change, East-West Center, Honolulu, Hawaii, 1995.

Johansson, Sten, and Ola Nygren. "The Missing Girls of China: A New Demographic Account, " *Population and Development Review,* vol. 17, no. 1, 1991, pp. 35–51.

LeVine, Robert, Sarah E. Levine, Amy Richman, F.M.T. Uribe, Clara S. Correa, and Patrice M. Miller. "Women's Schooling and Childcare in the Demographic Transition: A Mexican Case Study," *Population and Development Review,* vol. 17, no. 3, 1991, pp. 459–496.

Muhuri, Pradip, and Samuel H. Preston. "Effects of Family Composition on Mortality Differentials by Sex Among Children in Matlab, Bangladesh," *Population and Development Review,* vol. 17, no. 3, 1991, pp. 415–434.

Murray, Christopher, and Alan Lopez. "Global and Regional Cause-of-Death Patterns in 1990." In Christopher Murray and Alan Lopez (eds.), *Global Comparative Assessment in the Health Sector.* Geneva: World Health Organization, 1994, pp. 21–54.

Murray, Christopher, Alan Lopez, and Dean Jamison. "The Global Burden of Disease in 1990." In Christopher Murray and Alan Lopez (eds.), *Global Comparative Assessments in the Health Sector.* Geneva: World Health Organization, 1994, pp. 97–138.

Nadarajah, T. "The Transition from Higher Female to Higher Male Mortality in Sri Lanka," *Population and Development Review,* vol. 9, no. 2, 1983, pp. 317–325.

Park, Chai Bin, and Nam-Hoon Cho. "Consequences of Son Preference in a Low-Fertility Society: Imbalance of the Sex Ratio at Birth in Korea," *Population and Development Review,* vol. 21, no. 1, 1995, pp. 59–84.

Pebley, Ann, and Sajeda Amin. "The Impact of a Public-Health Intervention on Sex Differentials in Childhood Mortality in Rural Punjab, India," *Health Transition Review,* vol. 1, no. 2, 1991, pp. 143–169.

Preston, Samuel H., Nathan Keyfitz, and Robert Schoen. *Causes of Death: Life Tables for National Populations.* New York: Seminar, 1972.

Stephenson, Patricia, Marsden Wagner, Mihaela Badea, and Florina Servanescu. "Commentary: The Public Health Consequences of Restricted Induced Abortion—Lessons from Romania." *American Journal of Public Health,* vol. 82, no. 10, 1992, pp. 1328–1331

United Nations. *The Health Rationale for Family Planning: Timing of Births and Child Survival.* New York: United Nations, 1994.

United Nations Development Program (UNDP). *Human Development Report 1992.* New York: UNDP, 1992.

World Bank. *World Development Report 1993.* Oxford: Oxford University Press, 1993.

World Health Organization (WHO)/United Nations Childrens Fund (UNICEF). *Health Implications of Sex Discrimination in Childhood.* Geneva: WHO/UNICEF, 1986.

Wyon, John, and John E. Gordon. *The Khanna Study.* Cambridge: Harvard University Press, 1971.

Zeng, Yi, Tu Ping, Gu Baochang, Xu Yi, Li Bohua, and Li Yongping. "Causes and Implications of the Recent Increase in the Reported Sex Ratio at Birth in China," *Population and Development Review,* vol. 19, no. 2, 1993, pp. 283–302.

Reconceptualizing Risk:
A Feminist Analysis of HIV/AIDS

Lynellyn D. Long
Lisa J. Messersmith

Introduction

HIV/AIDS is a disease that affects women throughout the world through direct infection from sexual intercourse, needle exchange (intravenous [IV] drug use), or contaminated blood supplies. Women are also often the primary caregivers of those with HIV/AIDS. In many countries, HIV/AIDS is one of the leading causes of death in women of reproductive age. The incidence of reported HIV infections is increasing faster in women than in men; by the year 2000, it is projected that women will make up a majority of those infected. Throughout the world, HIV/AIDS affects primarily men and women in their most productive years and, therefore, has profound consequences for household and social welfare.

The high prevalence of HIV/AIDS is straining healthcare systems and financial resources in some regions, including much of sub-Saharan Africa, parts of Asia, Latin America, and the Caribbean, and the urban areas of many industrialized countries. Women in their traditional caregiving roles often become responsible for providing care and support to those infected. Many women must care for their ailing spouses, children, and grandchildren. HIV-infected women with children also have to cope with both the physical and the socioeconomic consequences of this disease for themselves and their families. Many elderly women in regions with a high HIV prevalence are raising orphaned grandchildren without the traditional economic support of their children.

HIV/AIDS is primarily transmitted heterosexually. Thus, an understanding of gender relationships, male and female sexuality, and sexual practices is critical for designing and implementing effective HIV/AIDS prevention and control programs. Traditional epidemiological approaches that solely target "high-risk" groups have proven inadequate in defining the scope of the epidemic and in designing effective prevention strategies (Schneider and Stoller, 1994). Some researchers argue for approaches that go beyond targeting particular groups of "high-risk" women

such as sex workers, IV drug users, and young adolescent girls (Standing, 1992; Ulin, 1992). HIV/AIDS must be understood within broader sociocultural and economic contexts; a comprehensive prevention strategy also requires a better understanding of gender relationships, roles, and responsibilities in all aspects of life.

Researchers have been forced to design new methodologies and theoretical approaches because the issues of prevention of this disease are so complex. They consequently observe that HIV/AIDS is reconfiguring social thought and reality (Herdt and Lindenbaum, 1992; Parker and Gagnon, 1995). For those infected, HIV/AIDS may change how they perceive themselves and their intimate as well as social relationships. The social and historical significance of HIV/AIDS extends beyond the physical consequences of the disease. Representations of HIV/AIDS associate intimacy, pleasure, and desire with death. To cope with their fears of AIDS, people often distance themselves from the disease through new social representations of it and stigmatization of the "Other." Women's inequality vis-à-vis men in many patriarchal societies makes them particularly vulnerable to this stigmatization.

The socioeconomic and biological complexity of HIV/AIDS has forced researchers to recognize the limitations of prevailing theoretical and methodological approaches. In response, some have adopted transdisciplinary approaches, action-oriented methodologies, and ways to include participation from those living with HIV/AIDS. Research on HIV/AIDS involving some of the most intimate social practices and relationships has also forced researchers to reexamine their own identities, roles, and relationships to their subjects.

Although HIV/AIDS is reconfiguring social thought, these changes have had relatively little effect on contemporary feminist theories and analyses. For fear of stigmatizing their own agendas and women as a group, many women's groups and feminist scholars have been reluctant to address HIV/AIDS and its consequences for feminist theory and practice. A few studies have considered the disease from development paradigms using gender and (women-in-development) WID perspectives and have made a significant contribution to our understanding (Schoepf, 1991; Gupta and Weiss, 1993). Women's groups have also played an important role in shaping public conceptions and responses to this disease. While some researchers have also considered feminist strategies of empowerment, primarily in U.S. settings, an analysis of HIV/AIDS and its effects globally on social life and thought from broader feminist perspectives has yet to be undertaken. Yet, an analysis of the social construction of HIV/AIDS for and by women could inform contemporary feminist analysis, while, in turn, contemporary social constructionist and materialist approaches to feminism could contribute to a more informed understanding of HIV/AIDS.

In this essay, we provide a framework for considering HIV/AIDS in women from gender and feminist perspectives. In the first section, we present a gender analysis of the demographic, epidemiological, and sociocultural factors of HIV/AIDS and women. In the second section, we consider governmental policies and actions, efforts of nongovernmental organizations (NGOs) to reduce the vulnerability of women to AIDS, and new prevention technologies. In the third section, we conclude with a discussion of how HIV/AIDS affects feminist theory and research, and how gender and feminist analyses, in turn, change our understandings of this disease and its representations.

Demographic, Epidemiological, and Sociocultural Factors

In 1993, the World Health Organization estimated there will be more than 13 million HIV positive women worldwide by the year 2000 (WHO, 1993). Currently, women account for up to 50 percent of all new HIV infections (WHO, 1995). As of December 1996, the cumulative number of adults and children living with HIV/AIDS reached an estimated 22.6 million, and the number of cumulative AIDS cases was reported to be approximately 8.4 million (UNAIDS and WHO, 1996a). UNAIDS estimates that there were over 3.1 million new HIV infections in 1996 or approximately 8,500 a day—7,500 of whom were adults and 1,000 of whom were children, and that the majority of newly affected adults were between the ages of 15 and 24 years. Women account for approximately 42 percent of the 21.8 million adults living with HIV and the ratio of women to men continues to rise (UNAIDS and WHO, 1996a).

The developing world has been hardest hit by the pandemic. More than 90 percent of the total number of adults infected worldwide live in developing countries. Sub-Saharan Africa has only 11 percent of the world's population but accounts for over 60 percent of all HIV infections (UNAIDS and WHO, 1996a). In addition, more women than men are living with HIV in Africa (UNAIDS and WHO, 1996a). Although the 1990s HIV prevalence increased more rapidly in Asia than in any other region, high-quality population-based seroprevalence (HIV positive)

surveys have not, as of 1996, been conducted. However, in Bombay, India, from the late 1980s to the mid-1990s, HIV prevalence reached 50 percent in commercial sex workers, 36 percent in STD patients, and 2.5 percent in pregnant women attending antenatal clinics (UNAIDS and WHO, 1996a). In Thailand, Malaysia, Myanmar, and India, HIV infection in IV drug users has increased faster than in any other group (United States Bureau of the Census, 1994). In Sao Paulo, Brazil, HIV prevalence among women STD clinic attendees increased more than fivefold between 1993 and 1994 (UNAIDS and WHO, 1996a). In the Caribbean, women represent more than 40 percent of all HIV positive living in the Caribbean (UNAIDS and WHO, 1996b).

HIV is one of several sexually transmitted diseases (STDs) that pose threats to women's health. According to the World Health Organization in 1966, more than 200 million sexually transmitted reproductive tract infections occur each year in women living in less developed countries. The significance of this statistic lies not only in the consequences of STDs but from the evidence that STDs are cofactors in HIV transmission. In fact, STDs are known to increase the risk of HIV transmission from two to nine times (Wasserheit, 1992). Effective diagnosis and treatment of STDs are, therefore, viewed as important and essential components of HIV prevention.

Women are more biologically vulnerable than men to STD and HIV infection (Schultz et al., 1987). Their genital epithelium is more vulnerable to trauma, and they have a larger area of exposure to the virus. For these reasons, transmission of HIV and other STDs from men to women is also more efficient than from women to men (Al-Nozha et al., 1990). One study found that the risk of contracting gonorrhea in a single act of intercourse in which one partner was infected was 25 percent for men and 50 percent for women (Hatcher et al., 1986). In addition, women are often asymptomatic for STDs such as chlamydia and gonorrhea and, therefore, less likely to seek treatment.

Sexually transmitted diseases that cause genital ulcers, such as chancroid, herpes, genital warts, and syphilis, facilitate the transmission of HIV because these STDs compromise the genital epithelium through which the virus can easily pass. The clinical course of genital-ulcer disease is complicated by concomitant HIV infection, which worsens ulcerations and retards the healing process.

There is also evidence that nonulcerative STDs, such as gonorrhea, trichomoniasis, and chlamydia, facilitate the transmission of HIV (Laga et al., 1991). These STDs cause inflammation, which, in turn, may produce microscopic lesions through which the virus may enter the body; and an increase in white blood cells to the genital tract. These white blood cells are vulnerable to HIV infection and can become sources of HIV in the body. Nonulcerative STDs may account for greater numbers of women becoming infected with HIV than ulcerative STDs due to their often asymptomatic presentation in women (thus, decreasing their likelihood of being treated) and their more common occurrence (Wasserheit, 1992).

Treatment of STDs in individuals infected with HIV has a higher failure rate than in those not infected because the former retain genital lesions and inflammation of the genital tract longer. Transmission of HIV from those who also have an STD is, therefore, easier.

Sociocultural factors can also increase women's vulnerability to HIV and other STDs. Women may not be able to prevent and/or control these diseases due to their lesser status vis-à-vis men in sexual decision making, including when and where to have sex and whether or not to use a condom. In addition, polygyny is associated with an increased risk of STD infection in Africa (Caldwell and Caldwell, 1990). STD sequelae, such as pelvic inflammatory disease, ectopic pregnancy, cervical cancer, infertility, and negative outcome on the fetus and neonate also have social as well as medical consequences for women, whose roles in developing countries are primarily those of mother and wife.

Many experts conceptualize HIV/AIDS epidemiologically in women as specific sets of risk groups, such as commercial sex workers, IV drug users, and, more recently, young adolescent women. In addition, women's vulnerability is subsumed within their roles as mothers, who transmit the disease to "innocent" children either in utero or through breast-feeding. A vast amount of research also focuses on women in their roles as commercial sex workers, who are treated as vectors, core transmitters, and reservoirs of infection to uninfected men and children. However, characterizations of high-risk groups are incomplete for several reasons: (1) all women who practice unsafe sex (penetrative sexual intercourse or oral sex without a condom) for whatever reason and regardless of the type of relationship are potentially vulnerable; (2) epidemiological risk groups at best describe the characteristics of those who have already been infected but cannot predict who is most likely to be infected; and (3) risk-group characterizations fail to recognize the underlying sociocultural and economic determinants and dynamics that place women at risk. For example, the focus on commercial sex workers in prevention and risk-assessment efforts failed to reveal that men were carrying the infection home to their wives and that heterosexual transmission is increasing. Since HIV/AIDS in women is often popularly associated with commercial sex and/or IV drug use, women who do not identify themselves in these ways may falsely assume that they are not at risk. Yet, increasingly, any woman who practices unsafe sex, including those believed to be in the "safest" relationships, such as monogamous marriages, are vulnerable. There is a need to go beyond the "mothers and whores" approach to reconceptualize the disease in ways that are more effective for HIV/AIDS prevention and that more clearly typify the multidimensional aspects of this disease in women.

Women throughout the world survive by exchanging sex for money, goods, and food. Women who work as commercial sex workers are particularly vulnerable to infection if their wages are low and they require large numbers of partners to support themselves and their families. Vulnerability to HIV is more marked in lower-class sex workers who work in brothels and bars.

Once thought to be an epidemic revolving around socially marginalized groups such as sex workers, homosexual men, and intravenous drug users, HIV/AIDS has reached "mainstream" society, infecting heterosexuals, children, and monogamously married men and women. Women in "monogamous" relationships make up an increasing percentage of those infected with HIV/AIDS. Monogamy itself could be considered a risk factor, because many women in "monogamous" relationships falsely assume they are safe. In many societies, being male means having several sexual partners, yet women are not expected to know and/or be informed about their partner's behavior. Using a condom in a married relationship is often considered a betrayal of trust; it can also be difficult because often women and men want to conceive. Thus, throughout the Third World and increasingly in industrialized countries, heterosexual transmission is becoming the predominant pattern of transmission.

Epidemiological, biological, demographic, and sociocultural factors combine to place women at higher risk than men for HIV/AIDS. The public-policy response to this pandemic has often been to regulate women's sexuality by encouraging sex workers to use condoms and counseling young girls to delay and/or abstain from sex. Such responses inherently discriminate against girls and women, who are expected to assume responsibility but may have little control over sexual decision making. Too often, policies and programs have not recognized the diverse reasons for girls' and women's vulnerability. Age, place of residence, class, ethnicity, cultural norms and ideals about gender, and particular life and historical experiences are all characteristics that affect women's and girls' vulnerability and their sexual practices. Research, policies, and programs are just beginning to address the multidimensional aspects of this disease in girls and women and to involve them in the process. In the next section, we outline a few promising, innovative programs and policy approaches that account for this diversity and that recognize the particular contexts of girls' and women's lives.

Programs and Policy Efforts to Prevent HIV/AIDS in Women

Although there are numerous innovative small-scale efforts to prevent HIV/AIDS in women, few have been evaluated or can demonstrate they would be effective if replicated on a large scale. Promising interventions fall into the following categories: (1) education and counseling; (2) condom promotion; (3) social and economic empowerment programs; (4) legal and human-rights reforms; (5) control and prevention technologies; (6) behavior change in men; and (7) integration of HIV/AIDS prevention into other health services and programs. These interventions focus more on the determinants of behavior, sexual beliefs, and practices than on targeting specific "high-risk" groups.

Education and Counseling

Peer-education and counseling programs have been organized to convey information and knowledge about safe sex, improve women's and adolescent girls' negotiating skills, and promote behavior change in women in several sites. In northern Thailand, for example, Cash (1996) designed a peer-education program for young female garment factory workers to share information about sexuality and sexual health, to improve their negotiation strategies, and to discuss condom usage. Cash argues that peer education is the most effective way to reach young women. Likewise, in South Africa, Mtshali (1994) observes that peer counseling may be the most effective way to convey information to adolescents because teenagers are more likely to listen to their peers.

Peer counselors, however, need to be trained to be effective and culturally appropriate. In Thailand, Cash (1996) found that teaching negotiating skills required practice, but that peer leaders tended to be very motivated and felt a sense of ownership in the program. Peer educators were also sensitive and committed to the particular needs of their peers. Women's counseling and focus groups may also provide a venue for conveying information about intimate aspects of sexual practices. In Brazil, Klein-Alonso (in press) organized women's focus groups that enabled the participants to role-play negotiation strategies and to confront their own stereotypes about HIV/AIDS.

Girls' basic education and literacy may also play a role in HIV/AIDS prevention, although hard data are not available to support this claim. In many societies, older girls in school may have to negotiate sexual favors in return for good grades and promotion and thus be exposed to greater risks. Delayed marriage may mean that women have a greater number of sexual partners. Yet, even minimal amounts of literacy may improve women's social status and grounds for negotiating safer sex and their ability to decontextualize and decode information from other sources, such as the radio. Improving women's educational, employment, and career opportunities also means that there is less need for young women to look to childbirth as the major route to maturity and personal fulfillment.

Educating parents and respected elders to talk openly about sex with their children may encourage intergenerational communication and a shared responsibility for prevention. In South Africa, Mtshali (1994) observes that the traditional extended-family system is still strong and that AIDS education and support systems should build on this system. However, this strategy has also proven difficult in many settings because of intergenerational difficulties in communication, parental denial, and cultural proscriptions against talking about sex.

Condom Negotiation

Educating and providing condoms for commercial sex workers may be successful strategies for preventing transmission between clients and sex workers provided sex workers themselves have a say in the process. Condom-only brothels wherein all sex workers use condoms with every client are becoming more numerous in Africa and Asia. The condom-only policy in many brothels has another benefit: a sense of solidarity and support among women working in common circumstances. Sex workers are a source of knowledge on sex practices and sexuality and can help to design effective interventions. They are sometimes in a better position than other women to negotiate condom use because they offer a service that has economic value in the tourist economies of many countries. In Thailand, for example, the government instituted an HIV/AIDS prevention (i.e., condom use) program for commercial sex workers when it realized that its tourist trade was being affected by rising rates of HIV/AIDS (Enloe, 1990). While women sex workers in upper-class establishments could afford to enforce this policy, those in lower-class brothels and bars and young girls in the poorest villages in northern and northeast Thailand were more vulnerable because they could be arrested for not using a condom. However, regardless of their economic situation, sex workers can be agents of behavior change among themselves and by helping to educate men.

Social and Economic Empowerment

Empowering women in various ways to negotiate safer sexual practices and even to refuse unsafe sex is recognized as critical to HIV/AIDS prevention and control. Strategies to empower women socially and economically include providing education and training, organizing support groups, promoting microenterprise development, facilitating access to credit, and improving employment opportunities. Empowerment strategies are aimed at improving women's status in many societies and, thus, imply cultural change. Such strategies are generally long term and must be tailored to specific social and cultural practices. Many empowerment efforts have been introduced by nongovernmental organizations (NGOs) and community groups that are well established and have local support and knowledge.

In many HIV/AIDS control and prevention programs, however, the concept of empowerment can conflict with the daily constraints women face within their households and communities. In addition, empowerment defined and promoted by outsiders can be a contradiction in terms. The definition of empowerment suggests that the individuals involved recognize and take responsibility for change rather than accede to change defined by "professional others" and "experts" from the outside.

Indigenous efforts to change inheritance, marriage, and property laws may have a positive effect on women's status and, therefore, on their ability to protect themselves. In Uganda, changes in property laws allow women to retain title to property and possessions on the death of their husbands. This change has meant that women are less likely to have to marry or exchange sex for financial stability. Discouraging the practices of levirate (compulsory marriage of a widow by her deceased husband's brother) and polygyny may also decrease both women's and men's risk of infection in some societies.

Control and Prevention Technologies

The male condom has been the primary means of HIV/AIDS prevention, and many interventions are predicated on increasing condom use. However, this method is problematic for women because its use is controlled by men. In addition, because condoms are often associated with illicit or commercial sexual encounters, both men and women often view condom use in noncommercial, intimate relationships as inappropriate and impractical. In a few regions, condoms are also not widely available.

The female condom has been introduced as a new prevention technology. Those who support its use cite its advantages over the male condom, including the fact that women, not men, wear the device, therefore increasing their control over protection. Women can also insert it before the sexual encounter takes place. Others believe that the device is awkward, expensive, and noisy, causing many women to discontinue its use. Furthermore, it is physically apparent to men, who must, therefore, consent to its use. Finally, as with the male condom, it does not always prevent conception as well as some sexually transmitted diseases.

Due to the disadvantages of both the male and the female condom, efforts are being made to develop a female-controlled microbicide, a chemical-barrier method to block HIV/AIDS, including types that would prevent transmission of HIV and other STDs while allowing for conception.

These new technologies are promising and potentially give women more control over HIV/AIDS prevention. Microbicides are more advantageous than the female condom because women can use these without their partner's knowledge and consent. However, it will be many years before these technologies are both affordable and readily available for Third World women. Like any barrier method, women will also have to learn how to use these new technologies effectively.

Behavior Change in Men

AIDS prevention programs in the Third World have traditionally focused on women, especially commercial sex workers. However, an increasing number of women who are not in the commercial sex industry are becoming infected through sex with their husbands and regular sexual partners. In most cultures, men control sexual decision making; often they have more than one lifetime sexual partner or more than one partner at any given time (including wives), despite the well-known potential health hazards of multiple partners. All of this leaves women vulnerable. Men and adolescent boys need to be educated about the consequences of their sexual practices not only for themselves, but also for their partners and children.

Integration of HIV/STD Prevention into Other Health Services

Women are less likely than men to seek health care for sexually transmitted diseases for three reasons: (1) they are more likely to be socially stigmatized in their communities and within health-care settings; (2) they may not recognize that they have an STD because they are often asymptomatic for many STDs; and (3) many women prefer to go to family planning and maternal and child health clinics, which may not offer these services. Treatment of STDs is a critical component of HIV/AIDS prevention for both women and men. Integrating STD/HIV prevention services into traditional health-care settings can increase women's access to STD treatment. A more comprehensive approach to women's reproductive health that includes sexual-health counseling may also improve the quality of care. There is some evidence that integrated family planning and STD/HIV prevention encourages women to protect themselves against both unwanted pregnancies and STDs (IPPF, 1994). Involving and training traditional healers in AIDS and STD prevention may also be an important part of creating an integrated service-delivery system (Mtshali, 1994).

Although the interventions outlined above can help reduce women's vulnerability to HIV/AIDS, they still do not reach many women because they and their societies are often slow in recognizing their vulnerability. Given scarce and diminishing resources, many programs can only afford to implement a single intervention at a given time or target only one "high-risk" group—even though it may be more cost effective over the long term to implement multiple interventions targeted at women and men of all socioeconomic classes and across the life span.

To address the pandemic, women with and without HIV/AIDS need to work together. Women with HIV/AIDS are not only an important source of information about the disease but also are catalysts for change in understanding the dynamics of HIV/AIDS transmission, prevention, and support. Women who are HIV negative are also deeply affected. Because this disease is straining local health-care systems, many women are bearing the burden of care and support in their households. The highest prevalence rates of HIV/AIDS among productive-age men and women also means that HIV/AIDS is affecting local economies, leaving fewer resources for HIV/AIDS prevention and care. Thus, women with and without HIV/AIDS share a common concern for mobilizing resources at local, national, and international levels for prevention and care.

Integrating HIV/AIDS in Feminist Analysis

In this final section, we address (1) how feminist theories and methodologies could contribute to understanding the dynamics of the epidemic; (2) possible reasons why there has been little feminist theory or analysis of HIV/AIDS; and (3) the potential for a feminist construction of HIV/AIDS analysis transforming social thought and theories of gender.

Toward a Feminist Theory of HIV/AIDS

HIV/AIDS is rarely addressed in feminist theory and research, excepting a recent study that looks at feminist strategies of empowerment (Schneider and Stoller, 1994). HIV/

AIDS, however, would seem to be an obvious topic for feminist inquiry and theorizing for several reasons. First, this issue addresses a critical feminist concern involving women's control over their own bodies. Issues of women's reproductive health are diverse, contested terrains that often reveal much about the social construction of gender in different locales. A number of scholars have conducted insightful feminist analyses on other aspects of women's reproductive health, including amniocentesis, pregnancy, menopause, abortions, and birth control (Ginsburg, 1990; Rapp, 1990; Martin, 1992). Like these other issues, HIV/AIDS speaks to the most intimate aspects of peoples' lives and relationships and the construction of their sexuality.

Second, HIV/AIDS itself is socially constructed (Sontag, 1988; Herdt and Lindenbaum, 1992), while political and economic considerations affect who gets HIV/AIDS, how it is transmitted and managed, and how it affects the person with the disease and his/her larger social network. HIV/AIDS is often constructed in the popular imagination as a disease of the "Other," which, as discussed earlier, increases many women's vulnerability; it also has been socially constructed by particular communities of experts, while other forms of knowledge and expertise that could contribute to understanding this disease have been largely ignored. As Sontag (1988) observes, "'Plague' is the principal metaphor by which the AIDS epidemic is understood" (Sontag, 1988, p. 132). She argues that diseases such as AIDS acquire meanings that stand for our deepest fears and inflict social stigma on the sufferer. At the same time, a strictly social-constructionist approach does not characterize the materialist aspects of HIV/AIDS. HIV/AIDS reveals socioeconomic and political disparities among classes, ethnic groups, locales, and men and women. This particular complex of issues—social constructionism coupled with materialist concerns—is at the heart of contemporary feminist theory and analysis.

Third, HIV/AIDS in women speaks to a feminist political agenda that celebrates diversity while demanding equal treatment. Specifically, HIV/AIDS is often the outcome of sexual and gender inequalities. Yet, the specific forms of discrimination reflect a diverse set of experiences, and the disease course itself can vary widely. Although much of the research through the mid-1990s seems to stereotype women, as victims or vectors, the experience of a heterosexual, middle-class North American woman with HIV/AIDS has little to say to a rural Burmese woman in a commercial sex establishment in northern Thailand. Likewise, many women who sleep with women have not been recognized as a potentially vulnerable group in certain situations. The diversity of experiences by class, ethnicity, locale, and gender identification, to name a few, of and for women are enormous and need to be further explicated.

Obstacles to a Feminist Analysis of HIV/AIDS

There are several complex and interrelated reasons for the relative absence of HIV/AIDS in feminist theory and research. First, early on, analyses of HIV/AIDS were primarily the domain of the medical, public health, and gay men's communities. The first two groups were largely concerned with managing the epidemic, preventing its spread, and finding a cure, while the third was initially concerned with the politics of HIV/AIDS—garnering resources, addressing issues of stigma, and promoting community survival. Even as all three groups recognized that heterosexual transmission predominated worldwide, women—primarily commercial sex workers, family planning advocates, and medical anthropologists—had a marginal voice at best in setting the HIV/AIDS agenda. To the extent that the research became increasingly medicalized, many women's groups and advocates and feminist scholars may have thought they had little to say about HIV/AIDS as a subject of inquiry.

Second, many Third World scholars as well as women's groups may choose to avoid HIV/AIDS in their work because the disease has been socially stigmatized by its association with commercial sex workers, "poor," "oppressed" women, and women of color. By failing to recognize their common interest in supporting the legitimate claims of other "stigmatized" groups of women, however, women are contributing to that stigmatization and their own vulnerability. Yet, many women of color also quite rightly point out that HIV/AIDS is not their specific concern; that women in most societies and situations are vulnerable. To the extent that HIV/AIDS in women has been seen as a problem for poor women or prostitutes, it has also been difficult to garner funding. Much of the research has been managed and controlled largely by men. Women who engage in HIV/AIDS research, programs, and policy dialogue for and about women may find themselves marginalized in their own institutions. They also may threaten the status quo by pointing out who is being most affected by this disease and by challenging the relevancy of certain kinds of predominantly Western male knowledge and expertise.

Third, many gender and WID specialists until the early 1990s were unwilling to take on this issue because they viewed it as outside their domain. Although household-level analyses and knowledge, aptitude, and practices (or KAPs) studies helped demonstrate the sociocultural and economic diversity of HIV/AIDS, the most interesting aspects of HIV/AIDS in women are often lost in these analyses, which are not reliable for addressing women's most intimate relationships, sexual practices, and bases for decision making about whether to engage in unsafe sex. By relying on KAPs studies and public focus groups, researchers re-create a domestic (private) versus public distinction in their avoidance of issues of intimacy and sexuality in their studies and programs. Likewise, WID and gender specialists may eschew HIV/AIDS for more public concerns, such as women's employment, even though the two may be interrelated. Traditional WID and gender analyses, however, may also lack the conceptual tools and frameworks to provide a broader understanding of this disease in women's lives. Merely adding women to the conventional analyses or disaggregating by gender does not seriously challenge the status quo. Gender itself as a category of analysis needs to be elaborated and explicated within particular locales, communities, and social networks and, if necessary, be contested as a framework of analysis.

A fourth and final reason that women's groups and scholars have avoided HIV/AIDS as a topic of concern is that it is enmeshed in many other issues around which women are losing ground politically and economically. The vulnerability of women to HIV/AIDS and its effects on their lives often reflect larger political and economic forces and struggles, including structural adjustment in developing economies, economic transformations in socialist societies, the feminization of poverty worldwide, the control over women's bodies, and war and conflict. HIV/AIDS in refugee women in Rwanda or the Balkans, for example, will be recognized as a priority only years hence when the women involved achieve some degree of safety and security. Likewise, women who have been fired or laid off due to an economic adjustment may have to address the immediate economic survival of their household first, and only secondarily their own health and welfare. As feminists, gender specialists, and women's groups struggle to find effective ways to address these issues, HIV/AIDS often becomes just another cofactor. The prevalence of the disease usually has to be quite high and has to affect mainstream society before it becomes a concern in its own right. Yet, if viewed more broadly, HIV/AIDS challenges the very construction of gender and sexuality in many different settings and may argue for different kinds of analysis and engagement in these concerns.

Transformation of Theories of Gender Through HIV/AIDS Analysis

Although gender and WID analysts and women's groups have made significant progress in building constituencies, garnering resources, and revealing gender disparities, there has been little analysis that recognizes how HIV/AIDS has transformed social thought and theories of gender. Contemporary feminist theory that combines social constructionism with materialist perspectives provides a powerful conceptual approach (see DiLeonardo, 1991). Feminist participatory research methodologies, including analyzing women's experiences, locating the researcher on the same plane as the subject, and doing research that directly benefits women, could also inform research and programmatic efforts on women and HIV/AIDS.

While a feminist analysis is needed to better understand the dynamics of HIV transmission, HIV/AIDS has the power to transform feminist thought to demonstrate how intimacy and sexuality affect social organization and the construction of gender. The disease itself cannot be controlled, prevented, or managed without a better understanding of the diversity of sexual practices and relationships across time and geographic location. It also cannot be controlled, prevented, or managed without a better understanding of the political, economic, and social conditions that constrain women's full equality.

Women in the Third World per se are not more vulnerable than women in industrialized countries, but unequal access to resources in either locale constrains women's ability to negotiate on an equal level with men about the most intimate aspects of their lives. Likewise, women who have access to resources but lack the information or self-esteem to determine their own lives may also be vulnerable. Who gets HIV/AIDS and how they are treated in a given society reveal the fault lines of that society. HIV/AIDS also reveals how gender inequalities, although structured in many different forms, increase all women's and men's vulnerability.

Conclusion

Women in the Third World have made progress in preventing and living successfully with HIV/AIDS. Their successes have often been small scale and locally specific, and sometimes they have been undertaken by women who are highly stigmatized within their own societies, such as commercial sex workers and widows. Yet, the varied experiences of women in the Third World can inform all HIV/AIDS prevention, control, and care efforts. Those living without HIV/AIDS have much to learn from those living with the disease since potentially all women and men are vulnerable.

From a feminist perspective, HIV/AIDS has the power to act as a vehicle through which feminists can recognize diversity yet unify women around a common political agenda. Even though women themselves may be unwilling or unable to change their own unsafe practices, the future of their children, familial networks, and communities often depends on their willingness to recognize and address gender inequalities in all their myriad forms. Such a perspective means that conceptions of feminism itself and feminists must go beyond symbolic construction and analysis to engage in the everyday, mundane political, economic, and social realities of women's lives. The particular scripts of women's lives must be considered, because these scripts help construct and maintain gender inequalities. However, by living with HIV/AIDS and confronting these traditional scripts, many women are beginning to assert their own autonomy and capacity to make a difference. Thus, HIV/AIDS provides a new, more complex understanding not only of gender inequalities but also of women's capacity for transformative political action.

References

Al-Nozha, M., S. Ramia, A. al-Frayh, and M. Arif. "Female to Male: An Inefficient Mode of Transmission of Human Immunodeficiency Virus (HIV)," *Journal of AIDS,* vol. 3. no. 2, 1990, p. 193.

Caldwell, J.C., and P. Caldwell. "High Fertility in Sub-Saharan Africa," *Scientific American,* May 1990, pp. 118–125.

Cash, K. "Women Educating Women for HIV/AIDS Prevention." In Lynellyn Long and E. Maxine Ankrah (eds.), *Women's Experiences with HIV/AIDS: An International Perspective.* New York: Columbia

University Press, 1996.

DiLeonardo, Michaela. "Gender, Culture, and Political Economy." In Michaela DiLeonardo, *Gender at the Crossroads of Knowledge: Feminist Anthropology in the Postmodern Era*. Berkeley: University of California Press, 1991, pp. 1–48.

Enloe, Cynthia. *Bananas, Beaches, and Bases: Making Feminist Sense of International Politics*. Berkeley: University of California Press, 1990.

Ginsburg, F. "The 'Word-Made' Flesh: The Disembodiment of Gender in the Abortion Debate." In F. Ginsburg and A.L. Tsing (eds.), *Uncertain Terms: Negotiating Gender in American Culture*. Boston: Beacon, 1990, pp. 59–75.

Gupta, G.R., and E. Weiss. *Women and AIDS: Developing a New Health Strategy*. Washington, D.C.: International Center for Research on Women, 1993.

Hatcher, Robert, F. Guest, G. Stewart, G. Stewart, J. Trussell, S. Cerel, and W. Cates. *Contraceptive Technology 1986–87*. New York: Irvington, 1986.

Herdt, G., and S. Lindenbaum. *The Time of AIDS: Social Analysis, Theory, and Method*. Newbury Park: Sage, 1992.

International Planned Parenthood Federation (IPPF). "Women in Brazil: Confronting AIDS," *Forum*, vol. 10, no. 2, December 1994, pp. 3–4.

Klein-Alonzo, L. "Women's Social Representation of Sex, Sexuality, and AIDS in Brazil." In L. Long and E.M. Ankrah (eds.), *Women's Experiences with HIV/AIDS: A Global Perspective*. New York: Columbia University Press, in press.

Laga, M., N. Nzilambi, and J. Goeman. "The Interrelationship of Sexually Transmitted Diseases and HIV Infection: Implications for Control of Both Epidemics in Africa," *AIDS*, vol. 5, suppl. 1, 1991, pp. S55–S63.

Martin, E. *The Woman in the Body: A Cultural Analysis of Reproduction*. Boston: Beacon, 1992.

Mtshali, N.A. "Transferability of American AIDS Prevention Models to South African Youth." In Beth Schneider and N.E. Stoller (eds.), *Women Resisting AIDS: Feminist Strategies of Empowerment*. Philadelphia: Temple University Press, 1994, pp. 162–169.

Parker, R.G., and J. Gagnon (eds.). *Conceiving Sexuality: Approaches to Sex Research in a Postmodern World*. New York: Routledge, 1995.

Rapp, R. "Constructing Amniocentesis: Maternal and Medical Discourses." In F. Ginsburg and A.L. Tsing (eds.), *Uncertain Terms: Negotiating Gender in American Culture*. Boston: Beacon, 1990, pp. 28–42.

Schneider, Beth, and Nancy Stoller (eds.), *Women Resisting AIDS: Feminist Strategies of Empowerment*. Philadelphia: Temple University Press, 1994.

Schoepf, B.G., E. Walu, W. Rukarangira, N. Payanzo, and C. Schoepf. "Gender, Power, and Risk of AIDS in Central Africa." In M. Turshen (ed.), *Women and Health in Africa*. Trenton: Africa World Press, 1991, pp. 187–203.

Schultz, K.F., W. Cates, and P.R. O'Mara. "Pregnancy Loss, Infant Death, and Suffering: Legacy of Syphilis and Gonorrhoea in Africa," *Genitourinary Medicine*, vol. 63, 1987, pp. 320–325.

Sontag, Susan. *AIDS and Its Metaphors*. New York: Doubleday Anchor, 1988.

Standing, H. "AIDS: Conceptual and Methodological Issues in Researching Sexual Behaviour in Sub-Saharan Africa," *Social Science and Medicine*, vol. 34, no. 5, 1992, pp. 475–483.

Ulin, P.R. "African Women and AIDS: Negotiating Behavioral Change," *Social Science and Medicine*, vol. 34, no. 1, 1992, pp. 63–73.

UNAIDS (Joint United Nations Programme on HIV/AIDS) and WHO (World Health Organization). *HIV/AIDS: The Global Epidemic*. Geneva: UNAIDS and WHO, December 1996a.

UNAIDS and WHO. *HIV/AIDS Situation in mid 1996: Global and Regional Highlights* Geneva: UNAIDS and WHO, July 1996b.

United States Bureau of the Census. "Recent HIV Seroprevalence Levels by Country," research note no. 15, December 1994.

Wasserheit, J.N. "Epidemiological Synergy: Interrelationships Between Human Immunodeficiency Virus Infection and Other Sexually Transmitted Diseases," *Sexually Transmitted Diseases*, vol. 19, no. 2, 1992, pp. 61–77.

World Health Organization (WHO). "WHO Estimate of HIV Infection Tops 14 Million." Press Release WHO/38. Geneva: Global Programme on AIDS, May 21, 1993.

———. "WHO Calls on Policy Makers to Reduce Women's Growing Vulnerability to HIV/AIDS." Press Release WHO/11. Geneva: Global Programme on AIDS, February 8, 1995.

Women's Nutrition Through the Life Cycle

Kathleen M. Merchant

Introduction

This essay describes nutritional problems faced by under-privileged women living in a context of poverty within less affluent nations. In general, women of more affluent nations and women with higher standards of living within less affluent nations face very different nutritional problems, generally characterized by imbalance through excessive intake and low levels of physical activity; some exceptions, such as eating disorders, often result from various forms of social and psychological stress. The biological role of women in reproduction and their consistently lower social status throughout the world makes women vulnerable to lower health and nutritional status in settings of poverty, particularly within less affluent nations.

Using a life-cycle approach, this essay examines women's nutrition in four broad stages: infancy and child-hood, adolescence, reproductive years, and later years of life. For each, social vulnerabilities particular to women are identified and the cumulative consequence of undernutrition is discussed. The section addressing the reproductive years includes consideration of the stress of frequent reproductive cycling and a discussion of the benefits and burdens of breast-feeding. The risks of malnutrition and their link to maternal mortality are examined as is the contribution of high fertility rates to increased risk of mortality, with consideration of the reduction in maternal mortality that might be achieved if women had authority over their own fertility. Next, the intergenerational implications of undernutrition and their role in perpetuating malnutrition are described. Finally, potential actions to reduce the nutritional problems of women living in poverty in less affluent nations are listed in the concluding section.

It is beyond the scope of this essay to address the influence of gender and socioeconomic injustices on the nutritional status of women in detail, but it is possible to describe ways in which the social context contributes to poorer health and greater undernutrition among females. The conditions mentioned below are by no means universal—they vary in their cultural patterns and combinations with each setting—but a consideration of their existence is essential in an analysis of the health and nutritional problems faced by women in any context where women face poverty and low social status.

Poverty and Social Vulnerabilities to Undernutrition Through the Lifecycle

The problem of undernutrition is created through social relationships rather than any biological effects of gender. Economic power plays a large role in facilitating or preventing access to adequate food and health care. A major underlying cause of malnutrition is poverty, and it is within settings of poverty that the social vulnerability of females—a product of their low social status—increases the risk for nutritional problems. The main biological vulnerability that women have for developing nutritional problems relative to men is a result of their role in reproduction. But reproduction alone will not compromise nutritional status. It is only in combination with the nutritional deprivation generated through socioeconomic relationships that conditions of malnutrition will develop from frequent episodes of pregnancy and lactation.

The social context can reduce or increase the impact of the biological burden of reproduction. In many cultures, there is a traditional period following the birth of a child during which a woman is expected to reduce her usual activities, sometimes referred to as a "lying-in" period. This social practice *reduces* the impact of the biologically rooted high energetic cost of lactation. Conversely, a social prac-

---

Portions of this essay appeared previously in Merchant and Kurz in M. Koblinsky et al. (eds.), 1993, and are reprinted here with permission from the publisher.

tice of food taboos prohibiting the intake of a nutrient-rich food during childhood, pregnancy, or lactation, *increases* the risk of a nutritional problem developing during these periods of high nutrient needs.

Infancy/Childhood: Preference for Males

The preference for males can express itself from birth onward through a reduced perceived need for, or delivery of, food, health care, and education to the girl child, particularly when resources are scarce. Additionally, children, particularly female children, may have important roles in household management, including physically demanding tasks such as collection of fuel and water, from a very young age.

Children are at risk for many nutritional deficiencies. Their high growth rates, small stomach capacities, and higher rates of illness due to less fully developed immune systems are partly responsible. During childhood, males and females have similar energy and nutrient needs, as illustrated by the grouping of males with females for the United States recommended dietary allowance (RDA) of vitamins and other nutrients up to the age of 10. Given that the nutrient needs are the same for the first 10 years of life, one might expect that children's vulnerability to nutritional problems and their health consequences would not differ by gender. Actually, girls have a biological advantage, in optimal environments, of 1.15 to one over boys in mortality rates (Royston and Armstrong, 1989). In spite of this, many countries exhibit higher mortality rates for girls. According to data from the World Fertility Survey reported by United Nations Population Fund (1989), among 1–2 year-olds in Turkey, the mortality rate of girls was 1.6 times that of boys; in Pakistan and Sri Lanka, it was 1.5 times higher; in Bangladesh and Trinidad and Tobago, 1.4 times higher; and in Colombia, 1.3 times higher; among 2–5 year-olds in Costa Rica, the mortality rate of girls was reported to be 1.7 times higher; in Syria, 1.6; and in Thailand and Pakistan, 1.5.

One implication of this evidence is that discriminatory behavior on the part of caretakers shifts the balance. A belief exists in some regions that young boys require more and better quality food than young girls. In families where the resources are highly constrained, the choice may be made to conserve by giving girls less, although, when food shortages become particularly extreme, the difference between the food intake of girls and boys may narrow. In many regions, girls will become members of other families upon marriage and do not bear the social responsibility for supporting their parents in old age. In these cases, they may be viewed as temporary and somewhat less valuable members of the family than the boys.

Along with equal access to food, equal access to health care and education are also key determinants of vulnerability to undernutrition. Specifically, frequent or severe episodes of illness can have an impact on nutritional status and, therefore, growth through anorexia (depressed appetite leading to a reduced intake) or increased need for certain nutrients. Additionally, nutritional deficiencies decrease resistance to infection. Education has an impact on the timing of childbearing and income-earning potential, both of which affect nutrition and health status. A dramatic difference in school-enrollment and literacy rates between girls and boys exists in many countries, particularly those countries with the highest infant-mortality rates.

A preference for male children is evident in many countries surveyed, which may or may not lead to preferential care for boys when resources are limited. In an effort to compile all available evidence of male preference in caretaking behavior, an annotated bibliography was produced by the World Health Organization (WHO) (1986). Differential breast-feeding practices (a longer duration for boys) were reported in Pakistan, India, and the Philippines. Evidence of unequal food allocation between male and female children was recorded for communities in Bangladesh, Mexico, and the Philippines. In the studies from Bangladesh and Pakistan, the gender differential was also observed in anthropometric (weight and height) data of the same children, as was the case in additional reports from communities in Jordan, Colombia, Saudi Arabia, Iran, Bolivia, Kiribati, Jamaica, and several regions of India. Seeking health care for male children appears to be more common than for female children in some communities of India, Nigeria, Egypt, Bangladesh, and Korea. The evidence from Korea is particularly illustrative: In the Kanghura Community Health Project, when measles immunizations were provided free of cost the proportions of boys and girls being immunized were almost equal, but when a small fee was charged the proportion of girls fell to little above a quarter of the children immunized. Other investigators also reported that female children are taken less frequently to health centers for care. And some studies have shown that the prevalence of stunting during the early years of life in developing countries is strikingly high among girls and boys. Stunting is usually caused by inadequate food, poor health, and other deprivations; the consequences of stunting among females are reduced work capacity and higher risk of compromised reproduction among women, such as difficult deliveries and lower birth weights. Given worldwide and regional variation, both culturally and in terms of available resources, evidence of discriminatory practices must be examined locally (WHO, 1986; UNFPA, 1989).

From available evidence, the disparity between girls and boys appears to be greatest for access to education, followed by access to medical care, then access to food. It is possible that the discrimination in favor of boys decreases as the item or service is perceived as more necessary for survival. For example, food is an immediate need for survival, medical care may be regarded as less so, and education may be regarded as nonessential, particularly for girls.

The intention here has been to identify the potential sources of social vulnerability for females during childhood that may increase the nutritional problems they face during this stage and later. The importance of food, medical care, and education for girls is obvious, but equal access may be an area in need of improvement in many settings. In parts of India, this has been recognized; television messages stress the equal food needs of girls and boys.

Adolescence: Early Reproductive Role

During adolescence, the circumstances of a young girl can be rapidly shifted from childhood to marriage and pregnancy, thereby eliminating opportunities for education and training as well as emotional maturation and physical growth prior to the physical and emotional demands of reproduction and other productive responsibilities of female adult life.

Developmentally, adolescence is a time of transition from external control (most often from parents) to internal control. It is recognized as a crucial and influential time for development of behavioral patterns, which include eating patterns and self care. Sources of information beyond family, such as the media (television, radio) become more important. Therefore, adolescence is potentially a good time for educationally based interventions.

There is limited information concerning the nutritional status of adolescents. Dietary intake and physical activity patterns are the major components of energy balance. The physical activities of work and play during this age period will vary tremendously; illness generally plays less a role at this time than earlier in life; and growth accelerates, resulting in significant nutritional needs, and then slows considerably.

The potential to realize many of the opportunities of adolescence, including continued education and optimal growth, is reduced if girls assume adult roles at an early age, especially if they begin having children during adolescence. Although growth begins slowing for females by approximately the age of 14, gains in linear growth, particularly of the long bones, are not complete until the age of 18, and peak bone mass is not achieved until the age of 25. The nutrient needs of pregnancy and lactation will be in addition to those of growth. Although the impact of the competing nutrient needs of pregnancy on the young mother's linear growth may be minimal (unless she is less than 13 years old), there is little information on how bone formation and calcium deposition will be affected in a young mother. Fetal growth is likely to be affected. The incidence of low-birth-weight babies is higher among young mothers. Adolescent mothers have a higher risk of developing anemia. Reduced growth through the stress of early pregnancy may have lifelong deleterious consequences.

Although legislation often exists prohibiting it, early marriage is a common phenomenon in many societies. Some of the most striking statistics have come from the World Fertility Survey, in which it was found that "25 percent of 14-year-old girls in Bangladesh, and 34 percent of 15-year-old girls in Nepal, were married, although the legal minimum age for marriage is 16 in both countries" (WHO/UNFPA/UNICEF, 1989). Formal education for the girl usually ends when she marries, and she experiences pressure to conceive to gain social status. Early marriage is not the only reason for early pregnancy: In certain societies, such as some Caribbean and African countries, adolescent pregnancy and childbirth are common outside marriage and regarded as a means of improving status, demonstrating fecundity, and attracting a new partner to provide support for each successive child (WHO/UNFPA/UNICEF, 1989). Some statistics may illustrate the magnitude of the problem: The percentage of women giving birth by age 18 in Africa was 28 percent; in Latin America, 21 percent; and in Asia, 18 percent; the percentage of first births to women 15–19 years old in Costa Rica was 44 percent; Mexico, 41 percent; the United States, 29 percent; the Philippines, 24 percent; Malaysia, 19 percent; and Jordan, 18 percent (United Nations, 1986).

Reproductive Years: Multiple Roles

Women have multiple roles to fulfill within their family and community, including the major biological role in the process of reproduction, and frequently spend a large proportion of their reproductive years pregnant or breast-feeding. Assuming that a woman is capable of reproducing for 35 years of her life, it has been estimated that, on average, a woman in Bangladesh spends 21.1 of these 35 years pregnant and/or lactating; a woman in Pakistan, 17.5 years; Senegal, 16.4 years; Kenya, 16.2 years; Thailand, 15.8 years; Peru, 13 years; and Haiti, 12.4 years (McGuire and Popkin, 1990).

In addition to reproduction, the social roles and responsibilities of adult life for women can be very physically demanding, including care for other family members, household management, food preparation, cleaning duties, use of health care, education, and supervision of children. Women frequently have other kin and community roles and, finally, paid or unpaid productive roles in agriculture, the marketplace, home production, factory, or other work activities. There is a growing recognition of the important role women play in food-chain activities—from food production and acquisition to preparation, storage, and consumption—and health maintenance.

Time-allocation studies have shown repeatedly that a woman's work day is longer than a man's and that women have less leisure or discretionary time available than men (McGuire and Popkin, 1990). This also has been shown for girls, who frequently spend more time in household maintenance activities than boys. When classified by reference activity levels established internationally, the measured energy expenditure of Bolivian Aymaras living in the Andean mountains within a small, rural agropastoral community was higher among adult women (heavy) than among adult men (moderate-heavy) (Kashiwazaki et al., 1995).

Heavy workloads increase women's food requirements. The amount of physical labor performed daily by women varies tremendously worldwide, depending on environmental and familial factors. Is firewood used for fuel? Is it nearby? Is the water source a well? Is the well nearby? Are there children, siblings, elders to help with the tasks? Too often the situation is similar to the following scenario:

Burkina Faso: "The whole trip took almost four hours"

Kalsaka's women used to get wood close to their compounds. The village committee had banned this, so now they walk for about five kilometers into the hills that form a backdrop to the village, where erosion, crusting and years of low rainfall have killed off many trees.

To see exactly what the job is like, we went out with the women on one expedition. We set out at 7:30. The women had already been to the wells twice for water. I used to imagine wood gathering was merely a matter of picking up sticks lying around. In fact it is a complex and energy-consuming operation.

On arrival at their destination, the women split up in all directions. Branches are attacked with machetes and hoes. As these are not very sharp, it can take a long time to hack through a single branch. Stumps are too thick to cut through. Usually the women leave them till the following year, when they are rotted enough to be pried loose. This is done by flinging as big a stone as they can lift at the top of the stump, then shaking until it comes out of the soil.

The women work with energy and considerable courage. Small babies are carried along and shaken at every blow of the machete. Young girls come along to help. The fittest women climb up trees, scamble up steep slopes of sharp scree, often in bare feet, and wrestle with shrubs perched on the edge of cliffs. Falls and injuries from cutting tools and stones are common.

The whole trip took almost four hours. The women do it two or three times a week (UNFPA, 1989, p. 8).

Time constraints may lead to infrequent meals, and exhaustion may lead to a reduced appetite. Given the low income, long hours worked, and multiple roles frequently fulfilled by women in settings of poverty, they are most likely to have trouble meeting their food needs and to be at risk for general undernutrition.

Nutritional Needs
Pregnancy, lactation, and menstruation increase women's requirements for various nutrients compared to their premenarcheal years. A thorough nutrient-by-nutrient discussion of needs during the reproductive years is given elsewhere (NAS, 1990, 1991). Although severe deficiencies of micronutrients result in the most dramatic and easily assessed consequences, it is important to remember that milder forms of these deficiencies may also have consequences: "The severely deficient persons represent index cases, or the tip of the iceberg, in the spectrum of nutri-

tional status within the population," (Buzina et al., 1989, p. 172). Mild-to-moderate deficiencies of iron, iodine, vitamin A, and energy may result in women from frequent reproductive cycling in a context of poverty and chronic deprivation.

Women have peak iron needs during the reproductive years. Iron deficiency anemia results from inadequate intake of iron-rich foods, as well as from excessive blood loss during events such as childbirth, hemorrhage, menstruation, and various parasitic infections. The consequences of iron deficiency anemia are severe (DeMaeyer et al., 1989). Physical-work capacity and resistance to fatigue and infection are decreased among anemic adults, as well as those with mild iron deficiency (Buzina et al., 1989). Conversely, an increased feeling of general well-being is reported following recovery from anemia. The danger of death due to hemorrhage, a condition that is not unusual following labor and delivery, is greatly increased among anemic women. Serious fetal consequences of iron deficiency anemia in pregnant women include increased risk of miscarriage and of low-birth-weight babies (DeMaeyer et al., 1989). The estimated prevalences of anemia among all women and pregnant women, ages 15–49 years, divided into categories of "developed" and "developing" countries are 14 percent and 59 percent, respectively (DeMaeyer and Adiels-Tegman, 1985).

The dietary intake of carotenoids and vitamin A is low among women in many regions of the world, although it varies tremendously (ACC/SCN, 1987). Vitamin A has important roles in growth, vision and the health of the eye, and immune response. The additional fetal growth occurring during pregnancy contributes to an increased need for vitamin A, leaving women more susceptible to vitamin A deficiency at this time. A clinical sign of deficiency is night blindness, which occurs more commonly among pregnant women. Depletion of vitamin A stores is a possibility among women who, although not pregnant at the time, have had many pregnancies in rapid succession. Many aspects of the immune response may be depressed if vitamin A deficiency is present. Even mild-to-moderate vitamin A deficiencies impair immunocompetence (Buzina et al., 1989). A reduced resistance to infection is particularly hazardous to women after labor and delivery, when the potential for infection is high. Infection following birth is one of the major causes of maternal morbidity and mortality (Royston and Armstrong, 1989).

Iodine deficiency is the most common cause of endemic goiter. A goiter is the enlargement of the thyroid gland generally due to a deficiency of iodine, which is required for the production of thyroid hormone. Pregnant women in regions with low iodine levels in the food supply are at high risk for developing goiter because of their increased need for this hormone (and therefore iodine) during pregnancy. Iodine is required for adequate brain development at crucial stages during fetal development. Cretinism, a form of permanent mental retardation that affects a fetus during gestation, is the most severe consequence of iodine deficiency in women. A less debilitating consequence for adults is lethargy. Low iodine intake has generally been observed in populations living in regions where the soil is poor in iodine content. Frequently, these are high mountainous regions.

Frequent Cycling/Depletion

"Too young, too old, too many, and too close." This statement from the UNFPA (1989) summarizes the problem of frequent pregnancies and periods of lactation that contributes to nutritional depletion of the mother and the high maternal mortality rates experienced in many settings of poverty. There is increasing evidence that, when pregnancies are too frequent, intake too low, and work demands too high, women do not have adequate time to recuperate from the nutritional demands and will show signs of nutritional stress such as loss of fat stores (Merchant, 1994). Specific nutrient deficiencies may occur or chronic undernutrition may be the outcome of such reproductive stress. A weakness of most research addressing maternal depletion has been a failure to accurately quantify reproductive stress and women's nutritional status, using instead broad and misleading indicators such as parity and maternal weight (Merchant, 1994).

Breast-feeding

There is a fallacious, but widespread, belief that the energetic costs of lactation place such a burden on undernourished women that it would be in their best interests not to breast-feed their infants. The problem with this reasoning is that it does not consider the benefits to women of such practices in their entire context. The well-being of mother and child are often inextricably linked. In a simple, narrow consideration of the nutritional energy balance of a woman, it would be possible to conclude that minimizing the amount of breast-feeding or choosing not to breast-feed an infant would result in a net benefit for the mother nutritionally. But when one considers what the consequences of such a decision or action would be on a woman, it is more difficult to justify such a conclusion, even when *only* the mother's well-being is considered.

If an undernourished woman were not to breast-feed her newborn, it is possible the infant would not survive. In that case, there is an immediate reduction in the energetic stress to the woman, but the energetic investment of the entire pregnancy is lost, not to mention any emotional investment. Additionally, if the woman wants or "needs" another child, she will now be faced with enduring another pregnancy (and the necessary energy costs) to "replace" this child. If the newborn was unwanted, and the death is not considered tragic, then withholding breast milk by the mother merely becomes a form of infanticide. Clearly, a much better solution to unwanted children would be to improve accessibility to acceptable forms of family planning and avert the pregnancy. Preventing the pregnancy or terminating it early would save the undernourished woman the nutritional investment of growing the fetus and supporting a pregnancy.

If the mother decides or is advised not to breast-feed, and if the newborn is to have a chance at survival, it must be fed with breast-milk substitutes, which are expensive if they are to be nutritionally adequate (for example, infant feeding formula). If the mother is undernourished, one can assume that this is a result of poverty and that, given the cost of infant feeding formula, purchasing it will most likely deplete family financial resources further, making it even more unlikely that the woman will improve her own diet and achieve adequate nutritional status. Additionally, the infant's susceptibility to diarrheal disease and respiratory infections is increased through a loss of the immunologic protection provided by mother's milk as well as the increased opportunity for contamination of the breast-milk substitute directly or through the method of feeding.

If the newborn becomes sick, there will be an additional burden of caretaking and any health care will be an additional expense—assuming such care may be available. To be sure, breast-feeding during the first four to six months of the newborn's life will place an additional energetic demand on the mother. But, unless extreme famine-like conditions prevail, the need can be met in the short term using maternal tissue stores that can be repleted through additional maternal consumption, either during the period of lactation (if possible) or following weaning of the infant. If the mother is extremely malnourished, it would be unlikely that she could carry the pregnancy to term or produce breast milk. A small increase in food for the mother to replete her tissue stores will be less expensive than the cost of formula for the newborn. A reduction in her physical workload during the period of pregnancy and lactation would also be beneficial if possible. If the conditions of food security are so marginal that neither of these adjustments is possible, then control of fertility through family planning is the most efficient short-term solution to increase the time period available for nutritional recovery from the period of pregnancy and lactation. Ultimately, improved socioeconomic status is the crucial change necessary to improve and protect the mother's nutritional status.

An important benefit to the mother resulting from exclusive breast-feeding during the first six months of her infant's life is the inhibition of ovulation and, therefore, safe, simple, effective, and automatic contraception. This is a direct and immediate benefit of lactation, initially. After six months and/or the introduction of other foods to the infant's diet, the effectiveness of amenorrhea resulting from breast-feeding is less reliable for an individual, but it is still present in many cases. One effect of the increase in the interpregnancy interval is an increased opportunity for nutritional recuperation of the mother. The contraceptive effect of breast-feeding is well-documented as are the advantages for adequate birthspacing. Therefore the practice of exclusive breastfeeding during the first six months of life generally has a net health benefit for the mother, particularly where access to other methods of family planning is limited. Additionally, it should not be forgotten that maximizing infant health and growth will have consequences in adulthood.

So, it is inappropriate merely to transform the energetic burden of lactation into another financial, emotional, or caretaking burden due to infant formula feeding, infant illness, or infant mortality. Ultimately, the cost-benefit analysis of any action relative to an individual must take into consideration the consequences of that action within the individual's socioeconomic context. It is fallacious to suggest that undernourished women should not breast-feed their infants. The more appropriate responses are to identify and remove the barriers to obtaining an adequate food supply and to make voluntary control of fertility a realistic option for women.

Maternal Mortality

A maternal death is defined by the WHO as "the death of a woman who is or has been pregnant during the previous 42 days," although deaths attributable to pregnancy and childbirth also occur outside of this time frame (Royston and Armstrong, 1989). Maternal mortality is a tragedy that shows one of the sharpest contrasts between affluent and less affluent countries. The lifetime chance of maternal death in North America is 1 to 6,366; in Africa, it is 1 to 21. The tremendous gap in this indicator of health is a result of the conditions of poverty. Not only are adequate health-care services and facilities essential for the appropriate response to medical emergencies of labor, delivery, and recovery from childbirth, but maternal mortality is a problem of such magnitude that it is crucial to consider the ways in which nutritional problems cause or aggravate dangerous situations of childbirth complications. It is becoming increasingly clear that many problems could be reduced in severity or prevented through adequate nutrition at earlier stages. Poor nutritional status contributes to at least three of four major causes of maternal mortality: hemorrhage (20–35 percent of maternal deaths), infection (5–15 percent), obstructed labor (5–10 percent), and eclampsia (15–25 percent). Women who die may have experienced several of these conditions. For example, obstructed labor can lead to the tearing of tissue, causing blood loss (hemorrhage), and, ultimately, an infection could set in and be recorded as the cause of death.

Adequate nutrition and health care in childhood can prevent or reduce the likelihood of such obstetric complications. A well-nourished mother with adequate iron status is much less likely to die from hemorrhage than a severely anemic mother, and a well-nourished mother will be better able to fight an infection than a malnourished mother. Additionally, well-grown women face obstructed labor much less frequently than women whose growth has been stunted—and pelvises malformed—as a result of chronic malnutrition. The nutritional link with eclampsia is poorly developed, but there has been some indication that high calcium intake reduces the likelihood of developing eclampsia. The nutritional links with maternal deaths resulting from unsafe abortions—generally as a result of infection or hemorrhage—are again found in the greater ability of a well-nourished woman to withstand hemorrhage relative to an anemic woman and the greater ability of a well-nourished woman to prevent and fight off infection relative to a malnourished woman. But the ultimate primary health-care solution to the complications of intentional termination of pregnancy is adequate access to effective contraceptive and family planning methods.

From these important relationships between conditions of extreme obstetrical emergencies and nutritional status, one can conclude that poor nutritional status is likely to be an underlying, hidden cause or contributing factor to the recognized clinical causes of maternal mortality.

Fertility

The actual risk of maternal undernutrition, morbidity, and/or mortality faced by any particular woman is also dependent on her fertility. The more times she is pregnant, the more opportunity for nutritional depletion and obstetric complications to occur. In Africa, high maternal mortality rates are compounded by high fertility. The average number of live births per woman is 6.4. But in rural Africa, it is quite common for a woman to have given birth to eight live babies and to have been pregnant several more times. If, at each pregnancy, such a woman has a one-in-140 chance of dying (calculated for a maternal mortality rate of 700 per 100,000), she has a lifetime risk of dying from pregnancy-related causes of at least one in 15. Comparable figures for North America are one in more than 6,000; for northern Europe, one in 10,000 (all above figures from Royston and Armstrong, 1989).

Maternal mortality rates are higher for adolescent women, for women over 40 years of age, for women with high parity (a large number of borne children), and for women with many pregnancies occurring in rapid succession. Women know of the burdens of frequent reproductive cycling through their life experience. They seek abortion, even though it may be illegal or unsafe, to avoid them.

The desire for limiting family size has been well documented. Recent survey data demonstrate that in developing countries, one in five births is unwanted and among countries outside of Africa at least half of all married women do not want any more children (World Bank, 1994a). Many studies show that abortion-related deaths account for a very large proportion of maternal mortality, more than 50 percent in some Latin American cities; more than 25 percent in Addis Ababa, Ethiopia; and more than 20 percent in rural Bangladesh. The desire of women to limit their own fertility becomes a marker of increased risk for maternal mortality. It has been shown that women with an unwanted pregnancy are less likely to seek prenatal care or to deliver with a trained attendant. Women who want no more children tend to be older, to have a higher parity, and to have a higher than average risk of maternal mortality. From the 1970 data, it has been estimated that one-fourth—25 percent—of maternal mortality worldwide could be prevented if women had the ability to prevent unwanted pregnancies; if the latter were the case, approximately 125,000 fewer women would die each year from causes related to childbirth.

Later Years: Marginalization

In their postreproductive years, in some settings, women can be particularly vulnerable to undernutrition due to a loss of health through aging coupled with a loss of social roles perceived to be valuable to the community. Women without secure financial or familial resources can easily be overlooked by the community and by programs targeted to benefit those who are worst off. The social vulnerability of being "postreproductive" and somewhat "invisible" in health programming and policy development increases the health risks of older women tremendously.

By the time a poor woman reaches the later years of her life, she is experiencing the cumulative effect of social vulnerabilities that started earlier in her life: the preference for males, early reproduction, and multiple roles, among others. Although the social position of women rises with age in some cultures, many older women may become more socially vulnerable as they become marginalized. The same may be true for an older man, but, given the low status of women worldwide, generally an aged woman has less power and economic independence than an aged man in the community. He may have been more able to acquire education, possessions, or status within his community, which are assets for survival. This is particularly true of highly patriarchal societies in which financial responsibility for a female traditionally lies in the hands of a male throughout her life. One illustrative case is Bangladesh: If a woman is abandoned or widowed and destitute she falls out of the realm of traditional protection and is left to her own devices for survival. Katona-Apte (1988) provides vivid descriptions of the factors that lead to destitution for women in Bangladesh. Although widowhood or abandonment can happen during the reproductive years, it occurs more frequently during later years of life. Soysa (1987) reports:

> As a widow, a woman suffers much indignity. If a woman survives her husband, it is believed to be the results of "karma"—her sins in a previous life. She is dependent upon the son's kindness for her support and she is often bereft of possessions, jewelery or fine clothes. More importantly, she eats sparingly, and fasts often because it is said to be unhealthy to eat much in this state of life. In fact, widowhood in the lower socioeconomic groups condemns women to beg for their food (Soysa, 1987).

Although very little information is available, one would suspect that chronic energy deficiency is an important problem for elderly women living in circumstances of poverty. Additionally, the elderly, male and female, have a tendency to have poorer intestinal aborption of some nutrients. Food intake decreases and, when combined with

problems of decreased absorption, may make the elderly vulnerable to specific nutrient deficiencies. Loneliness, isolation, poverty, depression, apathy, and debilitation may contribute to lower food intake. In addition, the prevalence of chronic disease is much higher as age increases. Evidence is accumulating that many chronic diseases have nutritional causes and consequences. As in childhood, the factors that predispose women to be discriminated against in the allocation of resources are the factors that increase their risk of undernutrition during old age. The specific impacts of frequent reproductive cycling on later health, as well as long-term depletion or undernutrition and how they all might be measured, are areas for future research.

Osteoporosis is one condition worth mentioning here because women are biologically vulnerable to developing it. A common definition of osteoporosis is "bone loss sufficient to bring about one or more symptomatic fractures with minimal trauma." Although epidemiological studies have shown evidence for a series of risk factors, the etiology of osteoporosis is not well understood. Following menopause, there is a dramatic loss in bone mass. This is believed to be due to the loss of estrogen production, although the role of estrogen in bone-mass maintenance is also not yet well understood. Nutrition is believed to play a role in the prevention of osteoporosis, and many believe that it plays a role in treatment, though that remains controversial. Calcium and vitamin D are particularly important for bone formation, and calcium loss is the main component of the loss in bone mass. Weight-bearing activity is also important for bone formation.

Although the prevalence of osteoporosis is very high in industrialized countries among postmenopausal women, there is little information regarding its prevalence in developing countries. The following risk factors have been identified: female gender, White or Oriental race, early menopause, advancing age, thinness, small frame, immobilization, sedentary lifestyle, poor dietary calcium intake, family history of osteoporosis, cigarette smoking, alcohol abuse, high sodium diet, caffeine consumption, childlessness, high protein intake, high phosphate intake, and the absence of breast-feeding. Many of these risk factors would be less common among developing-country populations, but one might expect the incidence of some of them to increase as these countries continue adopting Western lifestyles.

The most striking conclusion that can be drawn regarding women's nutrition at this final stage of the life cycle is that almost no information is available. This alone is a dramatic illustration of the invisibility of the elderly in developing countries to nutrition and health-care planners, policymakers, and providers. This is a reflection partly of the undervaluing of women beyond their reproductive years, and partly of the all-too-often short life span of individuals exposed to all of the health risks of poverty.

Intergenerational Effects of Undernutrition and Small Body Size

Nutritional problems are generally the consequence of

earlier problems and the cause of later problems. Therefore, they can rarely be assigned to a single stage of the life cycle, particularly as the consequences can be felt by later generations. For example, a cycle of suboptimal growth can be perpetuated across generations. Figure 1 illustrates this intergenerational cycle. Many social factors contribute to the less-than-optimal growth from conception to puberty. Indirectly, factors such as poverty, low social status, and lack of health care play a role. More directly, factors such as infrequent feeding (small stomach capacity), low energy density of food, high exposure to infection, reduced immunocompetence, and anorexia due to illness, both during pregnancy and early childhood, contribute to growth retardation. It is also important to recognize that behaviors are passed on intergenerationally, including behavioral patterns that contribute to growth retardation.

There is evidence that maternal size constrains fetal growth during the final stages of pregnancy. Therefore, small maternal size resulting from stunting during early childhood and/or from very young maternal age will constrain fetal growth beyond what it would have been had optimal childhood growth and/or pregnancy timing for the mother occurred. Compromised growth at early stages (gestation to three years) is particularly difficult to make up for at later stages, partly because growth is occurring at such an accelerated pace during this time period. In addition, due to the overwhelming environmental factors common in settings of poverty, it is unlikely that an initially poor start will be entirely overcome, resulting in a small adult stature. The females will continue in the cycle by producing offspring with a greater probability of having intrauterine growth retardation, and so the process cycles on.

This cycle of undernutrition must be broken by optimizing growth through better nutrition and health. The major window of opportunity for growth is during gestation through three years of age, although adolescent girls may also benefit from supplementation.

Nutritional Status of Women

Nutritional Indicators and Assessment of Women
Specific criteria for using indicators of body size (anthropometry) to assess the nutritional status of adult women are lacking. The choice of which indicator(s) to use will depend on the purposes of the assessment. Considerations of where the woman is in the reproductive cycle should be incorporated into definitions of adequate nutritional status. Otherwise, undernutrition in women may be underestimated. For example, the weight measurement of a pregnant woman includes the weight of the baby, the placenta, and additional support tissues such as extra blood and breast tissue, in addition to her usual body weight. After giving birth, if a woman begins lactation, her fat stores will not only support her own nutrient requirements, they will also be needed to support milk production during the early months of lactation. Therefore, the range chosen to define adequate nutritional status for her may need to be higher because she is supporting this energetically costly process. There also are difficulties in the anthropometric assessment of adolescents, mainly due to the variation in maturation rates that affects the timing of the pubertal growth spurt.

Worldwide data describing the prevalence of undernourished adult women are not available. But it has been suggested that the prevalence of low-birth-weight babies (less than 2,500 grams) is an indiret indicator. It has been estimated that maternal nutritional factors (specifically, low caloric intake, low weight gain during pregnancy, low prepregnant weight, and short stature, an indicator of childhood undernutrition) cause close to half of intrauterine growth retardation in rural developing country settings (Kramer, 1987).

Fetal growth is initially protected at the expense of maternal fat stores when there is energetic stress during pregnancy (Merchant, 1994). What degree of maternal malnutrition must occur before fetal growth is compromised is not known. Based on WHO estimates, 20 million babies (16 percent of all births worldwide) are born each year with low birth weight, and 18 million of these are born in developing countries (Kramer, 1987). The implication of this evidence is that the incidence of low-birth-weight babies attributable to maternal malnutrition would underestimate maternal malnutrition.

Latest Estimates of Malnutrition Among Women
Leslie (1991) reports that conservative estimates suggest that among the 1.1 billion women 15 years of age and older living in developing countries in 1985, more than 500 million were anemic due to iron deficiency, close to 500 million were stunted as a result of childhood protein energy malnutrition, about 250 million suffered effects of iodine deficiency, and almost two million were blind due to vitamin A deficiency. Information specifically addressing prevalences among women is difficult to find for vitamin A and iodine deficiencies, therefore the estimations are based on population prevalences, most likely leading to an underestimation of the deficiency among women alone. Additionally, conservative estimates suggest that more than 500 million of the world's people are chronically hungry, and that women are disproportionately represented among the hungry.

Conclusion
Women's nutritional status is crucial to their health, well-being, and productivity throughout their lives. Poor nutritional status arises from economic, political, social, cultural, and gender inequalities, as well as natural and human-made disasters. This essay is meant to alert the reader to the nutritional problems that are most likely to manifest themselves in women, particularly those women living in circumstances of poverty in less affluent countries.

The health and nutritional status of women can be examined across several dimensions, revealing long-standing inequities: a gender gap, resulting from a range of social and biological vulnerabilities; a maternal-child gap,

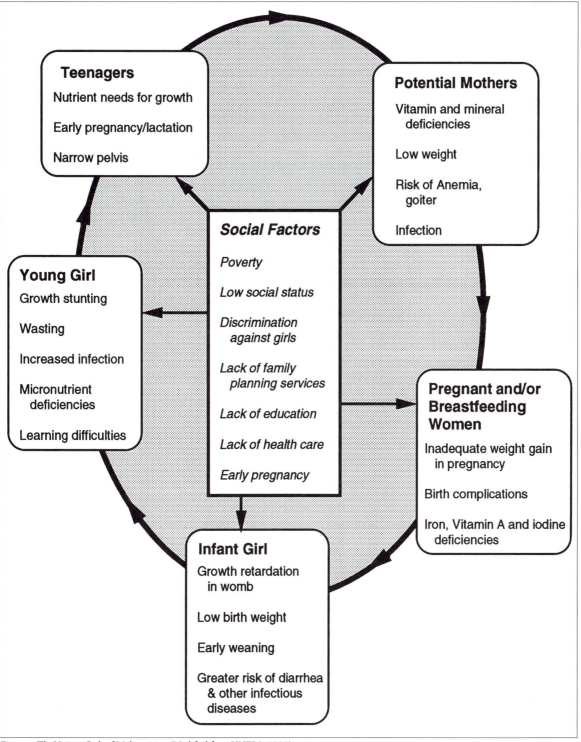

Figure 1. The Vicious Cycle of Malnutrition (Modified from UNFPA, 1989)

creating a view of women limited to reproduction, valuing their importance solely in terms of their ability to produce and maintain the health and nutrition of children; and an economic gap, existing between affluent and less affluent countries and among economic levels within countries, demonstrated by large differences in maternal mortality rates among these groups.

The United Nations Decade for Women (1976–1985) has played a large role in increasing the recognition of in-

equities and neglect in the areas of health and nutritional status. The series of conferences and papers commissioned through this forum launched a number of initiatives addressing problems women face in achieving health and obtaining adequate nutrition. An international dialogue has been stimulated and continues. The health and nutritional needs of women are being considered and even made priorities by some international health agencies, research institutions, and policymakers—not just to produce a larger baby or to deliver more and better breast milk (although these goals still dominate the discussions and objectives), but to have healthier women who do not face a high risk of death or debilitation from pregnancy or childbirth. And health issues of women beyond those related to their biological role in reproduction are receiving additional attention: Issues of genital mutilation, domestic violence, rape, depression, and sexually transmitted diseases such as AIDS, social/health issues that affect women to a greater extent than men in many settings, generally because of their lower status in social relationships, are receiving recognition and calls for action from society and particularly the health-care community (World Bank, 1994b).

The goal of improving women's nutrition will be accomplished only by working with women and their communities to define and act upon meeting their nutritional needs. Priority areas of need for action are listed in Tables 1 and 2.

Strengthening women through improvements in their nutrition and health status not only improves their own welfare but also is likely to be an effective avenue for promoting the health and welfare of those dependent on them. The intergenerational effects of malnutrition demonstrate that nutritional improvement throughout the life cycle is required to truly break "the vicious cycle of malnutrition."

References

ACC/SCN. "First Report on the World Nutrition Situation," report compiled from information available to the United Nations agencies of the ACC/SCN, World Health Organization, Geneva, 1987.

Buzina, R., C.J. Bates, J. van der Beek, G. Brubacher, R.K. Chandra, L. Hallberg, J. Heseker, W. Mertz, K. Pietrzik, E. Pollit, A. Pradilla, K. Suboticanec, H.H. Sandstead, W. Schalch, G.B. Spurr, and J. Westenhofer. "Workshop on Functional Significance of Mild-to-Modrate Malnutrition," *American Journal of Clinical Nutrition,* vol. 50, no. 1, 1989, pp. 172–176.

DeMaeyer, E., and M. Adiels-Tegman. "The Prevalence of Anaemia in the World," *World Health Statistics Quarterly,* vol. 38, no. 3, 1985, pp. 302–316.

DeMaeyer, E.M., P. Dallman, J.M. Gurney, L. Hallberg, S.K. Sood, and S.G. Srikantia. *Preventing and Controlling Iron Deficiency Anaemia Through Primary Health Care: A Guide for Health Administrators and Programme Managers.* Geneva: World Health Organization, 1989.

Gibson, Rosalind S. *Principles of Nutritional Assessment.* New York: Oxford University Press, 1990.

Kashiwazaki, H., J. Dejima, J. Orias-Rivera, and W. Coward. "Energy Expenditure Determined by the Doubly Labeled Water Methods in Bolivian Ayama Living in a High Altitude Agropastoral Community," *American Journal of Clinical Nutrition,* vol. 62, no. 5, 1995, pp. 901–910.

Katona-Apte, J. "Coping Strategies of Destitute Women in Bangladesh," *Food and Nutrition Bulletin,* vol.

Table 1. Summary of Nutritional Problems and Potential Actions

Period of the Lifecycle	Major Problem Leading to Undernutrition (in Addition to Poverty)	Priority Actions
Infancy/childhood	Preference for males	Mount public-awareness campaign (regarding equal nutritional needs of young girls relative to young boys)
Adolescence	Early reproductive role	Teach job skills; improve access to education; take action (through legislation and public awareness) to delay early marriage
Reproductive years	Multiple roles	Reduce women's workload; increase men's involvement in caretaking at the household level
	Frequent cycling/depletion	Increase access to methods of family planning; fund and conduct research for safe, healthy, acceptable contraceptive options
Later years	Marginalization	Increase availability of social services and support to elderly

Table 2. Specific Nutrient Deficiencies and a Selection of Possible Actions

Iron	Increase consumption of iron-rich foods and foods that enhance iron absorption; supplement women and children in regions of high prevalence of deficiency at their workplace, school, or health center; increase parasitic-control measures; seek means of appropriate food fortification
Vitamin A	Increase consumption of foods rich in vitamin A; supplement women and children in regions of high prevalence of deficiency at their workplace, school, or health center; seek means of appropriate food fortification
Iodine	Supplement women and children in regions of high prevalence of deficiency at their workplace, school, or health center; seek means of appropriate food fortification

10, no. 3, 1988, pp. 42–47.

Kramer, Michael. "Determinants of Low Birth Weight: Methodological Assessment and Meta-Analysis," *Bulletin of the World Health Organization,* vol. 65, no. 5, 1987, pp. 663–737.

Leslie, Joanne. "Women's Nutrition: The Key to Improving Family Health in Developing Countries?" *Health Policy and Planning,* vol. 6, 1991, pp. 1–19.

McGuire, Judith S., and Barry M. Popkin. "Helping Women Improve Nutrition in the Developing World: Beating the Zero-Sum Game." Technical Paper No. 114. Washington, D.C.: World Bank, 1990.

Merchant, Kathleen M. "Maternal Nutritional Depletion." *SCN News: A Periodic Review of Developments in International Nutrition Compiled from Information Available to the ACC/SCN,* no. 11, 1994, pp. 30–32.

Merchant, Kathleen, and K. Kurz. "Women's Nutrition Through the Life Cycle: Social and Biological Vulnerabilities. In Marge Koblinsky, Judith, Timyan, and Jill Gay (eds.) *The Health of Women: A Global Perspective.* Boulder: Westview, 1993.

NAS. *Nutrition During Pregnancy.* Washington, D.C.: National Academy of Sciences, 1990.

———. *Nutrition During Lactation.* Washington, D.C.: National Academy of Sciences, 1991.

Royston, E., and S. Armstrong. *Preventing Maternal Deaths.* Geneva: World Health Organization, 1989.

Soysa, P. "Women and Nutrition," *World Review of Nutrition and Diet,* no. 52, 1987, pp. 1–70.

United Nations. *Contraceptive Practice: Selected Findings from the World Fertility Survey Data.* New York: Population Division, United Nations, 1986.

United Nations Population Fund. *State of World Population 1989. Investing in Women: The Focus of the Nineties.* New York: United Nations Population Fund, 1989.

WHO. "Health Implications of Sex Discrimination in Childhood: A Review Paper and an Annotated Bibliography." *WHO/UNICEF/FHE* 86.2. Geneva: World Health Organization, 1986.

———. "Global Nutritional Status: Anthropometric Indicators Update 1989," *NUT/ANTREF/1/89.* Geneva: World Health Organization, 1989.

WHO/UNFPA/UNICEF. "The Reproductive Health of Adolescents: A Strategy for Action." A Joint WHO/UNFPA/UNICEF Statement. Geneva: World Health Organization, 1989.

World Bank. *Development in Practice: A New Agenda for Women's Health and Nutrition.* Washington, D.C.: World Bank, 1994.

———. "Women's Health and Nutrition: Making a Difference." Discussion Paper No. 256. Washington, D.C.: World Bank, 1994a.

Fertility Trends and Factors Affecting Fertility

Chloe O'Gara
Bryant Robey

Introduction

Since the mid-1960s, birthrates have shown a steady decline in most of Asia, Latin America and the Caribbean, the Near East, and North Africa. Fertility has declined not only in the cities and among more educated women, but also in rural areas and among the less educated. Behind these trends are important changes in Third World families and communities: rapid diffusion of new ideas and values, increasing education of girls, changing roles of women, and growing availability of family planning information and services. This essay examines recent fertility trends in Third World countries and discusses some of the origins and implications of those trends.

Fertility Levels and Trends

In the Third World as a whole, excluding China, the total fertility rate (TFR) has fallen from about six children per woman in the mid-1960s to about four in the mid-1990s, a decline of one-third. The TFR represents the number of births the average woman would have by the end of her childbearing years if she followed current age-specific fertility rates. Including China, where women average 2.5 births, the TFR of the Third World is estimated to be 3.5 children per woman. By comparison, in the developed world the TFR is about two children per woman.

These averages mask substantial differences in fertility among Third World countries. Findings from 44 recent (1985–1994) demographic and health surveys (DHS) showed that the TFR ranges from 7.4 children per woman in Niger and Uganda to 2.2 in Thailand (see Table 1).

Only one of 23 countries surveyed in regions outside of sub-Saharan Africa had a TFR of six or more; in 11 countries, women averaged between four and five children. In another 6 countries—Bangladesh, Brazil, Dominican Republic, Egypt, Peru, and Trinidad and Tobago—the TFR is between three and four children per woman; and it is below three children per woman in 5 countries—Colombia, Indonesia, Sri Lanka, Thailand, and Turkey (see Table 1).

Fertility is higher in much of sub-Saharan Africa. The TFR is above seven children per woman in 3 of the 21 sub-Saharan countries surveyed by the DHS—Mali, Uganda, and Niger. It is between six and seven children per woman in 11 of those countries, and between five and six children per woman in 4 of them. The TFR is below five children per woman in only three surveyed sub-Saharan countries—Botswana, Sudan, and Zimbabwe (see Table 1).

Comparison of the DHS data with data from the World Fertility Survey (WFS) conducted in the 1970s reveals how dramatically fertility has declined in many Third World countries. For example, in eight of 11 Latin American and Caribbean countries surveyed in both the 1970s and the 1980s, the TFR fell by at least one child per woman. Fertility declined by more than two children per woman between 1975 and 1991 in the Dominican Republic, and by nearly two children per woman between 1976 and 1990 in Colombia. In Asia, fertility has fallen substantially in Bangladesh, India, Indonesia, Sri Lanka, and Thailand, as it did earlier in South Korea, Taiwan, and elsewhere in East Asia. In India, the world's most populous country after China, the TFR declined from 4.3 in 1988 to 3.6 in 1992–1993, a 16 percent reduction in about five years.

The fertility decline in Bangladesh has been particularly striking. From a TFR of 6.1 reported in the 1975–1976 WFS, the TFR fell to 3.4 in the 1993–1994 DHS, a decline of 44 percent. This decrease occurred despite widespread poverty, high infant mortality, low women's status, little education, and other factors usually associated with high fertility. Popular interest in having smaller families appears to have grown as the population density has increased. Villagers have had less and less land to divide among their children, while cities are becoming overcrowded. The rising popular interest in family planning has been met with widespread contraceptive information and services available from government family planning programs in rural and urban areas.

In contrast, only a modest fertility decline has oc-

Table 1. Fertility, Contraceptive Use, and Breast-feeding Among Married Women of Reproductive Ages 15–49. Demographic and Health Surveys, 1985–1994.

Region	Total Fertility Rate[a] (TFR)	% Currently Using Contraceptives (Any Method)[c]	Median Duration (Months) of Breast-feeding[d]
Sub-Saharan Africa			
Botswana 1988	4.9	33	18
Burkina Faso 1993	6.9	8	25
Burundi 1987	6.9	7	24
Cameroon 1991	5.8	13	17
Ghana 1993	5.5[b]	20	21
Kenya 1993	5.4	33	21
Liberia 1986	6.7	6	17
Madagascar 1992	6.1	17	19
Malawi 1992	6.7	13	21
Mali 1987	7.1	3	18
Namibia 1992	5.4	29	17
Niger 1992	7.4[e]	4	21
Nigeria 1990	6.0	6	20
Rwanda 1992	6.2	21	28
Senegal 1992–1993	6.0	7	20
Sudan (Northern) 1989–90	4.7	9	19
Tanzania 1991–1992	6.3	10	22
Togo 1988	6.4	12	22
Uganda 1988–1989	7.4	5	19
Zambia 1992	6.5	15	19
Zimbabwe 1994	4.4[f]	48	N/A
Asia/North Africa			
Bangladesh 1993–1994	3.4	45	36+
Egypt 1992	3.9	47	19
Indonesia 1994	2.9	55	N/A
Jordan 1990–1991	5.6	35	12
Morocco 1992	4.0	42	16
Pakistan 1990–1991	5.4[e]	12	20
Philippines 1993	4.1	40	14
Sri Lanka 1987	2.7	62	20
Thailand 1987	2.2	66	15
Tunisia 1988	4.2	50	15
Turkey 1993	2.7[g]	63	12
Yemen 1991–1992	7.7	10	16
Latin America/Caribbean			
Bolivia 1994	4.8	45	18
Brazil 1986[h]	3.4	66	5
Colombia 1990	2.9	66	9
Dominican Republic 1991	3.3	56	6
Ecuador 1987	4.2	44	14
El Salvador 1985	4.2	47	15
Guatemala 1987	5.5	23	21
Mexico 1987	4.0	53	8
Paraguay 1990	4.7	48	11
Peru 1991–1992	3.5	59	17
Trinidad and Tobago 1987	3.1	53	6

[a] Based on three years preceding survey (women 15–49)
[b] Based on five years preceding survey
[c] Excludes prolonged abstinence
[d] Children less than three years old (any breast-feeding)
N/A Not available

[e] Based on six years preceding survey
[f] Based on four years preceding survey
[g] Based on one year preceding survey
[h] Women 15–44 years

Source: Demographic and Health Surveys Newsletter, 1995.

curred in Pakistan, although its culture, historical background, and women's status are similar to those in Bangladesh. In 1975, Pakistan's TFR was 6.3 children per woman, about the same as that in Bangladesh. Although Pakistan ranks higher than Bangladesh in most development indicators, fertility fell only to an estimated 5.4 in the 1990–1991 DHS. Despite evidence of popular interest in family planning, the Pakistani national family planning program has not provided information and services as broadly and effectively as the government in Bangladesh has. In Pakistan, only 12 percent of married women of reproductive age use contraception, compared with 45 percent in Bangladesh (see Table 1). It appears that in Pakistan many of those who want to have small families are frustrated in their efforts. Compared with Bangladesh, nearly twice as high a percentage of women as in Pakistan respond to surveys that they wish to avoid pregnancy but are not using contraception.

In North Africa and the Near East, fertility has fallen in all five countries where comparative survey data at two points in time are available—Egypt, Jordan, Morocco, Tunisia, and Turkey. For example, in Morocco the TFR declined from 5.8 children per woman in the 1979–1980 WFS to 4.0 in the 1992 DHS, a decrease of 31 percent. In Egypt, the TFR fell from 5.3 in 1980 to 3.9 in 1992, a 26 percent decline. However, in Yemen, where women's mobility and access to family planning are very restricted, the TFR stood at 7.7 in 1991–1992.

There are some indications that fertility is beginning to fall in sub-Saharan Africa as well. In Kenya, the TFR fell from 8.3 in the 1978 WFS to 5.4 in the 1993 DHS, a 35 percent decline. Although traditional Kenyan values and social structures in many communities continue to favor large families, rapid population growth has put pressure on agricultural land and caused migration to cities. Improvements in both the status and the educational attainment of women have brought new attitudes toward childbearing. More parents want to educate their children, while rising costs have made it more expensive to maintain and educate large families. These trends and changing fertility preferences have created more demand for family planning services, which has been met by family planning programs.

While few other sub-Saharan countries match the experience of Kenya, there are signs that fertility rates are beginning to fall elsewhere. Factors behind these changes include expansion of women's rights and education, changing economic pressures and options, improvements in child survival, evolution and erosion of traditions favoring high fertility, and increased acceptability and availability of contraception. In Botswana, the TFR declined 26 percent in just four years, from 6.5 in 1984 to 4.9 in 1988; in Zimbabwe, it declined 18 percent, from 6.5 in 1984 to 4.4 in 1994. Modest fertility declines were also evident in Ghana, from 6.1 in 1988 to 5.5 in 1993, and in Senegal, from 6.4 in 1987 to 6.0 in 1992–1993. There are not enough data to determine the consistency of these trends across all countries of sub-Saharan Africa.

Explaining Fertility Differences and Trends

Why has fertility declined so rapidly and dramatically in much of the Third World? This question may be investigated on two levels, first by addressing the direct factors affecting fertility, which demographers term the "proximate determinants," and then by examining the underlying social, cultural, and economic factors that affect fertility indirectly, through one or more of the direct determinants. These indirect factors include girls' education, urban or rural residence, women's status, family structures and responsibilities, women's legal and economic rights, and other influences on women's fertility desires and their ability to achieve those desires.

Proximate Determinants

Primary proximate determinants of fertility include: (1) the use of contraception; (2) women's age at first marriage; (3) postpartum infecundability from breast-feeding and lactational amenorrhea or sexual abstinence following childbirth; and (4) induced abortion. Other such determinants, whose impacts are of lesser magnitude, include levels of infertility, frequency of intercourse, and spontaneous abortion.

Use of Contraception. By far the most important proximate determinant of fertility is use of contraception. Dramatic increases in such use have largely accounted for the changes in fertility since the 1960s. In the mid-1960s, about 10 percent of married women of reproductive age in the Third World used contraception; in the mid-1990s, outside China, about 40 percent do. Among recently surveyed countries, contraceptive prevalence ranged from 4 percent in Niger to 66 percent in Brazil, Colombia, and Thailand. Including China, where the use of contraception is above 70 percent (on a par with the industrialized world), more than one-half of Third World women use contraception, for a total of more than 380 million women.

Differences in contraceptive prevalence—the percentage of married women of reproductive age using contraception, whether a modern or traditional method—explain more than 90 percent of the variation in fertility levels among Third World countries. In country after country, fertility is low where contraceptive prevalence is high, and fertility is high where there is little use of contraception (see Table 1).

The dramatic increase in contraceptive prevalence is due almost entirely to rising use of modern methods, chiefly voluntary female sterilization, oral contraceptives, and the intrauterine device. Outside China, about one-third of married women of reproductive age are using modern contraception, which is much more effective in preventing pregnancy than the traditional methods of withdrawal and periodic abstinence. It is also much safer than the folk methods of earlier times and than abortions in most Third World countries. While women have long sought to control their own fertility, until recently they had no choice but to abstain from sexual relations, have an abortion, or use ineffective, often dangerous methods to avoid pregnancy. Today, however, the great majority of women in the Third World have access to some form of effective contraception.

In addition to women's individual actions to control their fertility, throughout history governments, communities, families, and other social institutions have tried to manage population growth. Governments have adopted both pronatalist policies—most commonly offering praise or incentives for fertility, or discouraging urban migration or other changes that are associated with limiting fertility—and antinatalist policies. The latter are typically characterized by promotion of an "ideal" family size and widespread availability of modern contraceptives to achieve national demographic goals. Some countries have taken extreme measures, such as the "one-child family" policy in China.

Most governments support or allow widespread provision of contraceptive information, services, and commodities as part of national family planning or reproductive-health programs that serve women by helping them achieve their individual fertility desires. Studies show that many people quickly seek and use voluntary family planning information and services as soon as they become widely available and socially acceptable. This suggests a latent demand for family planning—that is, many women wished to have lower fertility than they were able to achieve without modern contraception.

The need for contraception is still unmet, on the basis of DHS data, *Population Reports* (Robey et al., 1996) has estimated that in 1996 about 100 million, or one-fifth, of women of childbearing age in the Third World did not want to become pregnant but were not using contraceptives. Demographers refer to such findings as indicators of an "unmet need for family planning."

Age at First Marriage. In all countries where fertility has declined, the rising age of marriage has played a role. This proximate determinant of fertility has been especially important in North Africa. However, the younger generation is marrying later than earlier generations in almost every country. For example, surveys show that a substantially smaller percentage of women age 20–24 in Third World countries were married before age 20 than was the case among women age 40–44. Also, the median age at first marriage—the age at which the number of married women equals the number of unmarried women—has risen almost everywhere. The DHS data indicate that the rising marriage age has played an even greater role than increasing use of contraception in Sudan and Morocco.

When marriage is delayed, women who become pregnant soon after marriage and continue to be pregnant or lactating for most of their reproductive years begin the childbearing process later and thus average fewer children than women who marry earlier. Later marriage is associated with broad social changes, modernization, and more autonomy and education for women, which also are associated with increased contraceptive use and smaller family size. As these changes progress, age at marriage becomes less important as a determinant of total fertility, and contraceptive prevalence becomes more important.

The effect on fertility of rising age at marriage depends in part on how delayed marriage affects the reproductive behavior of young, unmarried people. To the extent that it brings an increase in pregnancies and childbearing among young unmarried women, this would counter the decline in marital fertility, and would create new social problems for many of the unmarried mothers and their children.

Postpartum Infecundability or Sexual Abstinence. Breast-feeding is still an important proximate determinant of fertility in some parts of the world. Replacing the fertility impact of breast-feeding with modern contraceptives is well beyond the capacity of many of the world's health systems. Notwithstanding, breast-feeding was largely ignored by the population community for decades. Now, with the growing recognition that contraception is an essential, but not singular, element in demographic, sexual, reproductive, and economic changes, breast-feeding has become respectable again.

In some Third World countries, women breast-feed their infants for many months (see Table 1). This practice provides important benefits in child health and survival, as well as varying degrees of protection against pregnancy, depending on the age of the infant, frequency of suckling, and other factors. During lactational amenorrhea, breast-feeding provides protection from pregnancy through hormonal suppression of ovulation and implantation. Any breast-feeding provides more protection against pregnancy than no breast-feeding at all, but only "full" breast-feeding practices give as much protection from pregnancy as modern contraceptives for six months postpartum (unless menses resume). Full breast-feeding consists of frequent suckling without giving other foods or fluids to the baby on a regular basis. Such breast-feeding is not common, however; it is rarely practiced for longer than two or three months except in sub-Saharan Africa.

In many countries surveyed by the DHS, women practice postpartum sexual abstinence for two or three months, and for longer periods in some countries of sub-Saharan Africa.

In countries where use of contraception is low, as in sub-Saharan Africa and parts of Asia, postpartum fecundity remains a key determinant of fertility levels. DHS data show that, even in cities, traditional extended durations of breast-feeding are the norms and inhibit postpartum fecundity.

Breast-feeding is a suitable temporary child-spacing method for women who do not want to use modern contraceptives. To avoid pregnancy, breast-feeding must be used in conjunction with other methods at six-months postpartum or following resumption of menses. In populations of sub-Saharan Africa where breast-feeding is prolonged but is not supplemented with contraceptive use fertility remains high.

Preserving and improving the quality of breast-feeding to maximize postpartum amenorrhea reduces pressure to increase immediate contraceptive use among postpartum women; it gives women added time without fear of pregnancy to make informed choices about contraception. In addition, the economic benefits of breast-feeding are substantial, both in reduced food costs and reduced health care for mother and baby. Finally, in many cultures, the respect accorded a breast-feeding woman translates into assistance with domestic and market work.

Inclusion of breast-feeding as a family-planning option is an important feature of comprehensive reproductive health care for women. Providers of health and family-planning services need training in lactation management to be able to teach and support women who want the physical, emotional, and economic benefits of breast-feeding for themselves and their children.

Induced Abortion. Data on numbers of induced abortions are difficult to obtain and of questionable reliability because the practice is illegal in most Third World countries and disparaged almost everywhere. Thus, while abortion is a proximate determinant of fertility, estimating the impact of induced abortion on fertility rates is difficult. The prevalence of abortion tends to rise in the early stages of fertility transition and to fall as contraceptive availability and use increases. Every induced abortion does not represent a birth averted. Depending on the stage of pregnancy, rates of spontaneous abortion, stillbirths and low-birth-weight infants, and other factors, births averted vary across populations.

An estimated 36 million to 60 million women have induced abortions every year. Of these, 21 million to 34 million are in developing countries. Estimates of total abortion rates in Latin America are 1.4 to 2.0 per lifetime; in East Asia, 1.3; and in South Asia and Africa, fewer. Between 20 and 25 percent of pregnancies are aborted in East Asia (about seven million births annually in the 1980s) and Latin America. In China, one in three pregnancies ends in abortion; in the former Soviet Union and parts of eastern Europe, 40 to 50 percent of pregnancies do. Abortion is believed to play a smaller role in fertility reduction in South Asia and Africa. However, limited hospital data from Africa and Asia suggest that rates of abortion rose steeply in the 1970s and 1980s.

In the majority of developing countries, women do not have ready access to legal and safe abortion services. Abortion is practiced in every society in the world and has been for most of recorded history. However, by the mid-1990s, only 21 percent of countries allowed abortion. Restrictions on women's access to legal abortion in the Third World vary from prohibition under any circumstance, including to save the life of the mother (10 countries), to specific criteria that qualify a woman for a legal abortion, including saving the woman's life; preserving her physical or mental health; rape or incest; fetal impairment; and, least common, preserving her economic or social well-being.

In addition, while illegal abortions are dangerous almost everywhere, in many developing countries legal abortions are not as safe as they should be. Maternal mortality due to unsafe abortions is estimated to be 180,000 annu-ally, or more than one-quarter of maternal deaths, mostly in the developing world.

Reliance on abortion, particularly where it is illegal, dangerous, or expensive, testifies to the gap between women's reproductive desires and the realities of their lives. As effective contraception becomes more widespread in a country, abortion rates fall; however, safe abortion will always be needed as a backup for contraception, as some women become pregnant from contraceptive failure. Most important, safe, legal abortion is an essential component of adequate reproductive-health services for women because of the high mortality associated with illegal abortion.

Infertility, Frequency of Intercourse, and Spontaneous Abortion. These three determinants together probably account for as little as 4 percent of variance in fertility rates. But the most significant of the three, infertility, appears to be increasing, and the circumstances surrounding it reveal gaps in preventive and reproductive health care for women. Primary or secondary infertility affects as many as one-third of women of reproductive age in parts of sub-Saharan Africa. Infertility is often a significant personal tragedy. Furthermore, in many societies, bearing children is still essential for a woman's status as a worthy adult. Although such social norms are changing, failure to bear children changes a woman's stature and roles in her family and society.

High prevalence of sexually transmitted diseases (STDs), early intercourse (teenagers seem to be more susceptible to chlamydia), and unprotected intercourse are some of the factors behind infertility. Unsafe abortions contribute as well, but probably not to the extent commonly assumed. All of these are associated with women's lack of power and autonomy in their reproductive lives, with the scarcity of readily available contraceptives, with the absence of prophylactics designed for and controlled by women, and with limited treatment and diagnosis of STDs.

Other factors may also marginally decrease women's fertility. Falling sperm counts in many areas are attributed to environmental factors; these same factors may reduce women's fertility. Finally, a male partner's sterility is often denied, and a woman assumed to be infertile.

Except for periods of postpartum abstinence, divorce, and separations due to migration or other causes, the frequency of intercourse seldom accounts for much variance in models of fertility. Patterns of sexual intercourse may be more significant for young unmarried women, but information is scarce about their sexual behaviors and less understood than for married women.

Spontaneous abortion probably contributes relatively little to differences in fertility rates. Causes are poorly understood, and data concerning prevalence are limited.

Underlying Factors Affecting Fertility

Social, cultural, political, and economic factors affecting women's lives influence fertility indirectly, through one or more of the proximate determinants. Because they are indirect, their relationships to fertility are not easily measured; nevertheless, these and other underlying factors are critical determinants of fertility. They are important to understand because they are the motivating factors that influence women's fertility choices.

Women's Education. Women's education is highly correlated with reduced fertility. Education decreases fertility by increasing women's economic options, information about and access to family planning, status, fertility preferences, and life choices. More educated women tend to use contraceptives and delay marriage.

Norms for smaller families are taught in some schools and modeled by many teachers. Schooling may directly change fertility preferences and practices when the curriculum addresses sex education (including contraception), maternal and infant health, and nutrition. Perhaps more significant, schooling can provoke value shifts about fertility, sexuality, personal autonomy, competence, and self-efficacy. For girls who have been restricted to their homes for most of their lives, school can be an experience of radical change and exposure to a larger world. School socialization can teach girls that there are systems of knowledge, values, and authority outside of their households; that some knowledge, skills, and earning power are accessible to them; that they have rights and responsibilities to public authorities and to their community as well as to their families; and that they can make some choices without direct control or guidance from their families. When schooling conveys literacy, numeracy, and confidence in the face of new tasks, a woman's access to contraceptive information and services is facilitated. Educated women are able to use family planning methods more effectively.

Schooling is associated with later age at marriage, due in part to restrictions on attendance of married and pregnant girls. Those restrictions are being lifted in many countries as education of girls becomes a priority. Changing regulations about enrollment will not eliminate this correlation, however. Girls from traditional families are more likely to be withdrawn from school before they reach puberty to avert inappropriate sexual advances and to prepare for marriage. Bride-price and dowry systems also affect parents' willingness to invest in girls' education and give impetus to strategies for minimizing expense or maximizing wealth through marriage.

Educational attainment in part reflects a girl's social and economic status, which, in turn, is positively correlated with lower fertility in many developing countries. Girls from poor families often drop out of school because their labor is needed at home or their families cannot supply school fees.

In many countries, a woman's level of education marks not only her status in society, but also her social and economic stature in the home. Educated women have more power in their households, more discretionary control over their own and their husband's income, and more control over their reproductive choices and can negotiate more effectively with spouses. These outcomes of education are significant for women's effective contraceptive use, for

maintenance of their own health and well-being, and for their ability to successfully rear healthy children.

Infant mortality is significantly lower among the children of better-educated women. Among families at the lower end of the income spectrum, women who have more education also have more income and use the income more effectively to feed their children well. Malnutrition is lower among these children, and disease is managed more effectively by educated mothers: They immunize more regularly and seek medical assistance for their children sooner, more frequently, more appropriately, and more effectively than do less educated women. Most educated women can expect their children to survive; thus their desire to bear more children to compensate for excess mortality is reduced.

The relationships between education and fertility are not linear. Thresholds of effect vary across countries and cultures. Differences in educational systems, skills mastered, dropout rates, and local gender politics suggest that there are many factors in the educational experiences of girls that lead to reduced fertility. There are multiple pathways from female education to smaller families.

Residence. In most countries, fertility is lower in cities than in rural areas. Three of the four proximate determinants of fertility operate more strongly in urban than in rural settings to lower fertility: contraceptive use is higher, age at marriage is higher, and abortion is much more readily available. Breast-feeding practices contribute marginally less to spacing births in cities than in rural areas.

Underlying factors associated with family planning are also clustered among urban residents. Women living in cities are more likely to be educated, part of the modern sector of the economy, and thus more exposed to attitudes and lifestyles that favor smaller families. Living conditions in cities are more crowded, it is more expensive to raise children, and the economic value of children to an urban family typically is less than to families in the countryside. Family planning information and services generally are more accessible and varied in cities than in rural areas.

Neither having a formal education nor living in a city, however, is a prerequisite to wanting a small family. Women with little education in some Third World countries have lower fertility than educated women in other countries. The DHS data make clear that fertility has fallen among less educated women as well as among the more educated—and in rural areas as well as in cities. Indeed, fertility rates could not have fallen so profoundly since the mid-1960s without the widespread participation of rural women and of women with little or no formal education, because they are in the vast majority.

The pattern of fertility decline by education level and residence can be explained as follows (as discussed in Robey et al., 1992): The transition from high to lower fertility typically begins with high-status women who are educated and who live in cities. Lower fertility then spreads to other groups and eventually becomes the norm throughout society. In Colombia, for example, women with a secondary education already had low fertility (TFR of 2.5 children per woman) in the 1970s, whereas women with no education had high fertility (TFR of 6.4). By 1990, however, the TFR of women with no education had fallen to 4.9, while that for women with a secondary education had remained steady at a low 2.4 children per woman. Thus, almost all of Colombia's overall fertility decline during this period was the result of a substantial narrowing of fertility differences among women of different educational attainment.

This pattern explains why countries at different stages of overall fertility decline display different relationships of fertility to education level and residence. In countries where the fertility transition is just beginning or has yet to begin at all—as in much of sub-Saharan Africa—fertility is high among all groups, the educated as well as the less educated. For example, in Uganda, according to the 1988–1989 DHS, women with no education averaged almost eight children, and the fertility of women with a secondary education was nearly as high, at about seven children per woman. As the transition from high to low fertility in a country progresses, and more educated women begin to reduce their fertility before others with less education do, the fertility differences by education levels often become wide. This describes many Third World countries in the 1990s.

Differences in fertility by education shrink again as the whole society approaches a small-family norm. In such countries as South Korea, Sri Lanka, and Thailand, as in Europe and the United States, fertility is low or moderate among less educated as well as more educated women.

Access to Information. The influence of radio and, increasingly, television in Third World countries has accelerated the fertility decline. Unlike past generations, today millions of people—uneducated as well as educated, rural as well as urban—have direct access to the rest of the world through the mass media and to the new ideas, values, and behavior models carried therein. Together with the greater availability and promotion of modern contraception, the widespread reach of modern communication has contributed to rapid diffusion of the small-family norm.

Education, residence, access to information, and status all increase women's power to make informed choices about their reproductive health and family planning. Having information, making good choices, and carrying out plans is empowering to women. That empowerment leads, in the long run, to more opportunities in life and to reduced fertility.

While voluntary contraception gives women greater freedom of reproductive choice and life options, involuntary or uninformed contraception robs them of control over their lives. Abusive and unethical attempts by some governments to manage population growth by coercion, misinformation, or incentives too valuable to be rejected by many clients have been documented. Among such were early efforts by the Indian government to encourage sterilization that violated citizens' rights to informed choice and were concentrated on lower-caste groups. Gender bias in these violations of human rights was secondary initially;

men were targeted for vasectomies, to be replaced subsequently by women after tubal ligations became available.

Since sterilizations, IUDs, and abortions are more readily administered to women, the latter are the targets of most abuses. Reports of involuntary sterilizations, insertions of contraceptive devices, and surreptitious or compulsory abortion circulate periodically in many countries, particularly where service providers are given incentives for reaching target numbers of clients. Some of these reports are based on good evidence, others on misunderstandings or ignorance. As their education, power, and access to education increase, women are less vulnerable to these abuses and more able to exercise informed reproductive choice; nevertheless, governments that have put a priority on population control—aggregate numbers of births or contraceptive acceptors—at the expense of women's improved reproductive health and choice have raised the specter of violations of human rights, abuse of power, and misuse of contraceptive technology in family planning programs. Fortunately, for the vast majority, family planning increases their freedom of choice and autonomy.

Women's Status. Socially constructed gender expectations assign women many responsibilities and few rights in many communities. Comparisons with the responsibilities and rights of men show that many women are greatly overburdened and relatively powerless. In these circumstances, few women can effectively control their reproductive lives. They are socialized from birth to support their families and sacrifice for others; seldom do they have or use their personal power, social support, or resources to do otherwise.

Women of low status are less able to make and act on decisions about their fertility and health, negotiate with spouses, control their time and economic resources to get contraceptives when needed, and use contraceptives effectively. It is difficult to quantify and measure the effects on fertility at a population level of women's empowerment or changes in women's status. Nevertheless, analysts consistently find that, as women's status rises, their fertility falls. This is due primarily to increased family planning among women with higher status. Higher-status women also have more reliable access to safe abortion procedures and are more likely to delay marriage, especially if they pursue education.

On the other hand, high fertility is clearly a determinant as well as a consequence of women's status. Having many children constrains women's ability, both individually and in the aggregate, to improve their status. Although bearing many children is greatly respected in many cultures, can be personally satisfying and empowering, and can be critical to a woman's stature and social acceptance, in the modern world children are not the economic resource they once were, and values are changing in many communities. Economic realities are leading to pragmatic choices to limit family size. Men as well as women are making these choices, but empowered women lead the way in most families. Rearing healthy children, educating them, and preserving a degree of autonomy and income potential are more widespread and attainable goals for women today than a few decades ago.

Son Preference. The low status of women is no more clearly revealed than in son preference. Particularly in some Asian cultures, sons are still valued over daughters. Modern technology allows parents to know the sex of a fetus, and women of higher parity appear to be aborting girls in some countries. In Korea, for example, the ratio of boys to girls for third births in 1989 was 185 to 100 rather than the expected 106 to 100 ("Ultrasound Effects," 1995). The impact of selective abortion on total fertility is not yet significant and may never be; these practices are most prevalent among middle- and high-income families who limit family size in any event. Nevertheless, modern technology is being used by women, men, and families to manage women's reproductive lives and perpetuate inherently self-deprecating denials of the value of unborn daughters.

Fertility Decline and Women's Lives

Much of the research about how lower fertility changes women's lives is qualitative, and much of the analysis conceptual, but findings are remarkably consistent across cultures. Quantitative models and estimates have proved elusive (McCauley et al., 1994).

Women who are pregnant or lactating almost continuously throughout their reproductive years have very few options. Maintenance and survival for themselves and their children are all absorbing. As female-headed households around the world increase, these realities are becoming even more stark for women living alone with many children.

Every additional child adds significantly to the hours a woman works in her home; this is not true for fathers in most cultures. Research shows that most women eliminate their own leisure or sleep time to find the hours they need to sustain their families each day. Having fewer children and starting childbearing later gives women time and opportunity to mature, to conserve their own health, to educate themselves, to earn income, to invest in their futures, to optimize their children's health, and to more fully contribute their talents and energies to their communities and families.

Later initiation of childbearing, lower fertility, longer life expectancy, and better health have expanded women's possibilities. More education, increased migration by women and men, improved communications, and diversity of employment and income-generating opportunities are concomitant trends that also expand women's horizons and choices. Nevertheless, relative to men and in absolute terms, most women remain poor, poorly educated, and with very limited choices. One-third of the world's children are malnourished, a reflection of their mothers' poor nutrition in pregnancy, absolute lack of resources, and lack of control over household income. These realities persist although in every developing country women work longer hours than men—and still earn significantly less.

The demographic revolution has brought new opportunities and potential to women, but the revolution is only beginning. Future generations of women in every society will need to answer difficult questions about social reproduction, the meaning and purpose of education, redefinition of gender roles, relations between aging parents and (grand)children, production and income disparities, and continuing problems with development and distribution of wealth.

There are signs that fertility has still further to fall in the Third World. The strongest direct determinant of fertility is contraceptive use, and it is rising. The strongest underlying factor in fertility choices is the education of women, and that, too, is rising. As education and contraceptive use rise, so will women's desire for smaller families and their ability to achieve that desire.

Surveys show that many women continue to have more children than they would prefer, while many others are experiencing mistimed births (too soon after a previous pregnancy). A number of women continue to desire lower fertility than they actually achieve. Many of them, however, face obstacles in using contraception. Studies suggest that major obstacles are the opposition of husbands, concerns about health and side effects, lack of knowledge, and the unavailability of a range of effective methods. Presumably, more male involvement, better communication between spouses, and improvements in the quality and availability of family planning programs will help reduce unmet need levels. As family planning program coverage grows, and programs offer better information and more choice of contraceptive options, delivered with more concern for the interests and situations of women, more women will be able to achieve their desires to have fewer children, and fertility will fall further.

Other proximate determinants are also moving in the direction of fertility reduction. Women's age at marriage and at first birth is rising in many countries and although this trend typically levels off when women's median age at first birth is in the mid-20s, there are many countries where the current median age for first births is still in the teens. With continuing urbanization and modernization, it is reasonable to expect that women's age at first birth will continue to rise in many countries with high fertility rates.

Concern about abrupt declines in breast-feeding has been addressed in most countries with education and support programs for breast-feeding in increasing numbers of health services. The significant drop in infant and child mortality that has been achieved in many countries through immunization, oral rehydration therapy, and improved breast-feeding is another important underlying factor that is expected to continue to confirm women's desires for smaller families by ensuring the survival of more children to adulthood. Policies ratified at the International Conference on Population and Development (Cairo, 1994) and at the Fourth World Conference on Women (Beijing, 1995) endorse improved and more complete reproductive-health services for women. Over time, policies that acknowledge women's rights may lead to improved safety and expanded legality concerning induced abortion services, as well as better postabortion care when problems occur.

Among underlying factors, the acceleration of girls' education since the mid-1980s has been remarkable; despite economic hardships in many developing countries, both enrollments and grade attainment of girls are expected to increase. Strong commitments by developing-country governments, women's organizations, private voluntary organizations, and the international donor community to sustain female education initiatives reflect growing awareness of the many positive social, political, and economic outcomes associated with educating women and reducing fertility. With increased knowledge, skills, and earning power, women around the world are increasingly empowered actors in their nations, communities, and families.

Although patterns of migration and settlement change over time and vary among countries, the aggregate trend toward urbanization is expected to continue, bringing with it smaller families. The reach of the mass media and other information channels will continue to expand, bringing small-family norms, contraceptive information, and awareness of family planning services to more women.

Increased education, smaller families, and recognition of the rights of women in international policy forums and national legal codes all support the empowerment and improved status of women around the world. Local and regional variations in progress are likely to remain significant for decades, however.

Prospects for fertility trends will be driven by the sexual behaviors of today's young adults and children. As the twentieth century draws to a close, there is little evidence overall of significant increases or decreases in fertility among young adults; however, changes in marriage patterns, residency, household headship (increasing numbers of female-headed households), and patterns of employment (a greater increase among women than men), may have far-reaching implications for childbearing and childrearing in the next generation. One-tenth of the world's population are girls age 10–19. They will drive the world's fertility rates in the first decades of the twenty-first century. We know too little about whether, how, and why their sexual and family planning practices will differ from their mothers'. Too few countries are designing reproductive-health services and family planning programs to ensure that all of these young women will be able to make and carry out informed choices.

References

Bruce, Judith, Cynthia B. Lloyd, and Ann Leonard. *Families in Focus: New Perspectives on Mothers, Fathers, and Children.* New York: Population Council, 1995.

Demographic and Health Surveys Newsletter, vol. 7, no. 1, 1995.

Freeman, Ronald, and Ann K. Blanc. "Fertility Transition: An Update," *International Family Planning Perspectives,* vol. 18, no. 2, June 1992, pp. 44–50, 72.

Hong, Sawan, and Judith Seltzer. "The Impact of Family Planning on Women's Lives: Toward a Conceptual Framework and Research Agenda," working paper no. WP94–02, Family Health International, Research Triangle Park, North Carolina, 1994.

Mason, Karen O. *The Status of Women: A Review of Its Relationships to Fertility and Mortality.* New York: Rockefeller Foundation, 1984.

McCauley, Ann P., Bryant Robey, Ann K. Blanc, and Judith S. Geller. *Opportunities for Women Through Reproductive Choice.* Population Reports, Series M, no. 12. Baltimore: Johns Hopkins School of Public Health, Population Information Program, 1994.

Robey, Bryant, John Ross, and Indu Bhushan. *Meeting Unmet Need: New Strategies.* Population Reports, Series J, no. 43. Baltimore: Johns Hopkins University School of Public Health, Population Information Program, 1996.

Robey, Bryant, Shea O. Rutstein, Leo Morris, and Richards Blackburn. *The Reproductive Revolution: New Survey Findings.* Population Reports, Series M, no. 11. Baltimore: Johns Hopkins School of Public Health, Population Information Program, 1992.

Ross, John A., and Elizabeth Frankenberg. *Findings from Two Decades of Family Planning Research.* New York: Population Council, 1993.

"Ultrasound Effects," *Economist,* vol. 336, no. 7926, August 5, 1995, p. 34.

United Nations, Department for Economic and Social Information and Policy Analysis, Population Division. *World Abortion Policies 1994 Wallchart.* New York: United Nations, 1994.

Weinberger, Mary Beth. "Recent Trends in Contraceptive Behavior." *Proceedings of the Demographic and Health Surveys World Conference,* vol. 1. Columbia: IRD/Macro International, 1991, pp. 555–574.

Westoff, Charles, and Akinrinola Bankole. *Unmet Need: 1990–1994.* Demographic and Health Surveys Comparative Studies no. 16. Calverton: Macro International, 1995.

See also Carol Vlassoff, "Women and Contraception"

Women and Contraception

Carol Vlassoff

Introduction

Modern demography has tended to view contraception in purely contemporary terms, as something that began in the late nineteenth century, on the heels of the "demographic transition" and closely linked to advances in medical science. Ethnographic accounts and ancient historical texts, on the other hand, clearly indicate that this is a false and ethnocentric interpretation of history. Contraception did not begin with modern technology but has existed, in a variety of forms, throughout the ages. Moreover, it has always been intimately bound up with gender roles and relations.

This essay begins with a historical review of birth control practices and their relationship to gender roles and responsibilities. Modern contraceptive methods, which are becoming increasingly sophisticated and effective, are then discussed, as well as their implications for women's autonomy and control over their reproductive behavior. Against this background, population policies in developing countries are considered, and the challenges they have posed for women, the major clients of family planning programs.

A central premise of this review is that women's perspectives have been neglected in the design of contraceptive technology and family planning programs, despite the fact that so many programs have exclusively targeted women. The result is that many such programs have had limited success. Improved quality of health care to women and female-controlled contraceptive methods will not only improve family planning outcomes but will also empower women to protect themselves from sexually transmitted diseases.

A Historical Perspective

Contraception is not a new phenomenon. It has existed throughout the entire span of human history (Himes, 1963, p. 3). In this section, contraception and its links to gender relations are reviewed from a historical perspective, based on the pioneering work of Himes (1963) and McLaren (1990).

Anthropological studies confirm that abortion was prevalent in most early societies. Other practices, such as delayed marriage, prolonged breast-feeding, taboos upon sexual intercourse during specified periods, and infanticide seem to have been important in limiting population growth. The sophistication of early fertility-control practices is demonstrated by the Thonga, a South African tribe that prescribed sexual abstinence following delivery until the child began to crawl. Also, although intercourse could be resumed, conception was to be avoided until the child was weaned.

Semine non immisso, probably referring to coitus interruptus, or withdrawal, was used in preliterate societies throughout South, Central, East, and North Africa. Female-controlled methods of contraception were also documented. Among the Dahomey in West Africa, an intravaginal plug made of crushed root was recorded as a contraceptive method. In Central Africa, the vagina was plugged by cloth or chopped grass, often with fatal consequences for the woman. While little is known about gender relations in these societies, an ambivalent attitude may well have existed. Women seem to have been responsible for fertility control, yet blamed for contraceptive failures and abortions. For example, among the Riffian hill tribes in North Africa, women attended a weekly market where they received contraceptive advice. Men acknowledged this as a purely female event and did not interfere. Yet, a woman found practicing birth control could be killed.

The native Indians of North America attempted to limit family size by magic and potions made of roots. Abortion, as well as delayed intercourse and prolonged breast-feeding after childbirth, were also used. In South America, douching with lemon juice and other solutions was documented among women in Guyana and Martinique. In Dutch Guyana (now Suriname), a seed pod that acted as a kind of "vegetable condom" was inserted into the vagina before intercourse (Himes, 1963, pp. 18–19).

In the South Pacific, a variety of contraceptive methods were used, including withdrawal, abortion, and potions to induce sterility. While the efficacy of practices to induce sterility is unknown, there is some evidence that the substances employed may have damaged women's reproductive organs so that conception was prevented or hindered. An unusual method, seminal expulsion, possibly achieved by straining downward while squatting or by certain bodily movements, was also widely reported in Australasia, especially among unmarried girls.

In Indonesia, several methods were employed, including the inducement by midwives of a retroflexed uterus (a condition of the womb in which its body is bent backward at an angle with the cervix), especially after abortion or childbirth. This painful procedure had mixed success in preventing conception. In Sumatra, vaginal suppositories were used, at least one of which contained tannic acid, which was later used as a spermicide in the West.

Infanticide played an important role in limiting family size in early societies. Children were not considered fully human until they had received formal tribal recognition through special initiation ceremonies and, hence, could be disposed of as parents saw fit. This public acceptance of infanticide, and the view that children were the property of parents, was also common in Europe throughout much of its history.

Ancient Times

In ancient Egypt, references to recipes for pessaries that could be inserted into the vagina to prevent conception are found in papyri dating from around 1850 B.C. Some of these, such as crocodile dung mixed with paste, or honey that has a clogging capacity, may have had contraceptive effects. Even more significant was the mention, in ancient Egyptian writings of around 1550 B.C., of the use of "gum arabic" on vaginal tampons. This product was later to be employed in modern contraceptive jellies in the United States and England. Breast-feeding was also recognized and used as a birth-spacing method. All contraceptive methods employed in ancient Egypt seem to have been female, rather than male, dependent.

Among the ancient Hebrews, several methods of birth control were used, but, as with other ancient peoples, the prevalence of these practices is unknown. They included "embryotomy," the destruction of the embryo in cases where the mother's life was endangered by childbirth; withdrawal; vaginal sponges; violent postcoital movements; removal of semen from the vagina (although how this was done is unclear); and the drinking of root potions.

The ancient Greeks were preoccupied with scientific explanations of reproduction. They tended to denigrate women's essential procreative power as secondary to the male role in producing the seed, mounting the woman, and begetting the child. Begetting children was seen as the main purpose of marriage, and it seems that a woman who did not produce offspring was divorced and sent back to her father's home. Although the dominant power in procreation was attributed to men, the responsibility

for preventing conception seems to have been borne by women.

References to coitus interruptus are found in the ancient Greek writings, but the more frequent references to "pessaries, plugs and potions . . . make it clear that limitation of births was primarily a woman's business" (McLaren, 1990, p. 26). Certain herbal potions were also recommended for their contraceptive properties, including abortifacts and recipes for producing temporary sterility, and references were made to barrier methods, such as applying cedar gum to the genitals and inserting various kinds of suppositories into the vagina prior to intercourse.

In spite of widespread contraceptive practices, abortion was frequently resorted to in ancient Greece. Techniques such as drinking strong potions, inserting sharp objects and abortive substances into the uterus, and various physical exertions were used, but their harmful effects on women's health were also widely recognized. While abortion was tolerated, it was performed only with the husband's consent. Otherwise, there was a suspicion that women might turn to abortion to conceal adulterous affairs. Another means of limiting family size was the practice of abandoning or "exposing" children—that is, leaving them outdoors, exposed to the elements, to die—especially those who were deformed or had resulted from extramarital relations and rape. Exposure seems to have been more common for female children, and the right to expose a child rested with the father.

Although we tend to associate the Roman period with sexual excesses, research shows that sexuality and procreation were treated with caution and concern. During most of the Roman period, childbearing was considered the sole purpose of marriage, and childless wives could be divorced. Motherhood enhanced a woman's status, fertility was praised, and sons were essential for the continuation of the family line. Nonetheless, state efforts to encourage high fertility never succeeded, and contraception seems to have been widely practiced. As with the Greeks, abortion, infanticide and exposure, particularly of female children, were employed. The Romans, like the Greeks, saw women as primarily responsible for contraception, and similar methods were mentioned, with the addition of breast-feeding, which appears to have been more frequent among Roman wives. Douches with mixtures of vinegar and brines, not unlike those employed in the early twentieth century, were also used.

We know little about contraception in the early Eastern civilizations. In India, abortion seems to have been a common method of fertility control. Potions based on ingredients such as flowers, roots, and sour rice water were used to prevent conception, and a variety of other methods were recommended in the *Kama-Sutra* (Vatsyayana, 1984) of the early fourth century and later Sanskrit texts. These included coitus obstructus, the forcing of ejaculation into the bladder by pressing on the base of the urethra with a finger, smearing the vagina with honey and clarified butter, and inserting tampons of rock salt mixed with oil. All of these methods have a rational basis and may have been fairly effective. Salt, in particular, is an excellent spermicide, and mixing it with oil would also impede the progress of sperm. However, the salt must have been painful for the woman, especially if she had vaginal lesions.

References to contraception are infrequent in ancient Chinese writings, as great importance was placed upon fertility and large families. Several early works contain prescriptions for potions to prevent conceptions. In early Japanese texts, references are made to sophisticated contraceptive techniques, including a hard condom made of buffalo horn or tortoise shell, a leather condom, tampons of oiled bamboo paper, and retroflexion of the uterus, as practiced in Indonesia.

Early Christianity and Islam

Contraception and abortion seemed to fall increasingly out of public favor from the fourth century onward, along with the spread of religion and a decline in the status of women. During the early Christian period, this conservatism became increasingly pronounced: Marriage was considered necessary for procreation, and childbearing was viewed as the only rationale for sexual relations. Abortion and contraception, seen as two sides of the same coin, were considered sinful since they implied that pleasure, rather than procreation, had motivated the sexual act. Celibacy was considered more worthy than marriage, and women were deemed useful only as a vehicle for reproduction. St. Augustine (A.D. 354–430), among the most influential of the early Christian philosophers, reflected, "If it is not to generate children that the woman was given to the man as a helpmate in what could she be a help for him?" (St. Augustine, 1963, p. 14.22).

In spite of their lower status, women had a choice with respect to their life course: They could choose to remain celibate and avoid marriage and childbearing altogether. McLaren (1990) points out that this may have been appealing to many women: "Marriage and the bearing of children had traditionally been policed by men to serve their purposes. Now women at least had a legitimate alternative" (McLaren, 1990, p. 8). Women continued to be blamed for unwanted births and abortions, and contraception was forbidden. Those who used it were labeled adulteresses and murderers.

The medieval church praised procreation and high fertility, but in reality people continued to have small families due to late marriage, high infant mortality, and economic pressures in rural areas. Coitus interruptus seems to have been widely practiced in the early medieval period, but, from the eleventh century, Greek medical texts were reintroduced and a wide variety of birth-control methods became known. Abortion, infanticide, and exposure of children continued. Monasteries provided a refuge for abandoned children.

Unlike most Christian physicians, Islamic medical scientists studied birth control over the centuries, and contraception was permitted. Heavily influenced by the Greeks, the measures recommended were largely scientific, focusing on contraception and the protection of women's health rather than abortion. In the eleventh century, the great Islamic scientist Avicenna prescribed contraception for women suffering from diseases of the uterus or weakness of the bladder or for whom childbirth would be dangerous. The methods included withdrawal, strenuous exercise and sneezing after intercourse, avoidance of coitus during certain periods, smearing tar or other substances in the vagina before and after intercourse, and anointing the penis with spermicidal substances.

The economy of western Europe improved from the sixteenth century onward but poverty and landlessness prevailed. In the eighteenth century, it experienced rapid population growth, and families made deliberate efforts to control their fertility. With advances in medical science and improved understanding of the reproductive process, male obstetricians began to replace female midwives. Ironically, the advantages accruing to upper-class women who could afford male doctors were negligible, because doctors were no better equipped than midwives to perform deliveries. Dangerous Caesareans and "therapeutic abortions" were prevalent in Europe toward the end of the eighteenth century. With the growth in medically performed abortions, the state increasingly policed women's use of them, and abortion was made a statutory crime in France and England at the turn of the century.

Early Modern Times

Contraception seems to have been widely employed to limit family size in early modern Europe. Withdrawal, sponges, tampons, and condoms made of animal skins and bladders were frequently mentioned. Both Catholics and Protestants praised domesticity, and the concept of "love" in marriage and the family came into vogue. Contraceptive use in eighteenth-century Europe is often attributed to this closer relationship between couples, including the increased responsibility and concern of men for their wives' health. McLaren (1990) highlights two gender issues that were probably even more important in explaining fertility control during this period: women's independent decision making on the use of birth control, and female solidarity as women began to share experiences in contraception and to explore new methods.

Over the period 1870–1920, the United States and western Europe experienced a dramatic decline in fertility rates, known as the "demographic transition." Marked decreases in marital fertility during this period are documented: in the United States fertility declined by 50 percent between 1800 and 1900, and similar drops occurred in western Europe. The demographic-transition theory holds that in traditional society high fertility and mortality prevailed, but that advances in medicine and hygiene caused mortality to begin to fall, leading to rapid population growth as a result of continuing high fertility rates. Society countered this imbalance with the widespread adoption of birth control. This interpretation, as McLaren (1990) notes, is extremely narrow since it ignores the long history of fertility regulation that preceded this transition.

Thomas Malthus's *Essay on the Principle of Population* (1798) advocated family limitation as the sole means of avoiding environmental and social degradation. While Malthus opposed the use of contraception, his theory facilitated public discussion of population control and paved the way for nineteenth-century birth control advocates such as Annie Besant in England, Alice B. Stockham in the United States, Helène Stocker and Marie Stritt in Germany, and Madeleine Pelletier in France.

The impact of industrialization in Europe had a mixed effect on women's status and fertility. Initially, the demand for cheap labor encouraged the employment of women and children, stimulating high fertility among the working classes. Toward the end of the nineteenth century, as men began to compete for jobs and to oppose female and child labor, the sexual division of labor into "private" domestic roles for women and "public" wage work for men was adopted by middle- and upper-class families. Growing emphasis was placed on the "quality" rather than "quantity" of children: Whereas in the past children provided extra hands for the family economy, the abolition of child labor reduced the necessity of large families, and social pressure favoring children's education meant that greater costs were attached to high fertility. Although societal norms lauded women's domestic role, the private reality was usually different. Women continued to undertake jobs they could do at home, such as piecework, to supplement their husbands' wages.

Early-twentieth-century feminists such as Emma Goldman and Margaret Sanger in the United States and Marie Stopes in England argued strongly that contraceptive use should be promoted among the poor. Their argument was based on eugenic principles: The upper and middle classes were controlling their fertility through contraception while the poor continued to multiply. Sanger was prosecuted and eventually jailed for her outspoken defense of contraception. In 1921, Stopes opened a birth control clinic in London. Sanger's and Stopes's preoccupation with racial purity, which persists in some segments of society even today, was especially common among English and American upper classes at that time. Echoes of this philosophy sometimes emerge in population policies that encourage selected groups to multiply and others to restrain their numbers. Nonetheless, Sanger and Stopes promoted an important feminist message: that women had a right to enjoy sex without worrying about pregnancy and that women's and children's health could be protected by having fewer births. For the first time, women's right to sexual happiness and health were put at the forefront of the fertility-control debate.

The Medicalization of Contraception

It was only in 1827 that the modern scientific explanation of reproduction was finally established and, with it, recognition of the existence of life at conception. This revelation led doctors to condemn abortion at any stage, and, for the first time, a clear distinction was drawn between abortion and contraception. Nonetheless, a surge in abortions took place in Europe and North America during the nineteenth century in an effort to stem rapid population growth. Although doctors provided contraceptive services in clinics, notably the fitting of cervical caps and diaphragms, the medical profession largely ignored birth control during this period.

Experimentation with new methods of birth control took place in the early twentieth century. Surgical sterilization became available, although it was used mainly to prevent conceptions among the mentally ill. The rhythm method became more popular with improved understanding of ovulation, but it continued to be unreliable in practice. Research on the intrauterine device (IUD) also began in the 1920s, and chemists produced a range of spermicidal creams and jellies. Hormonal research on the prevention of ovulation eventually led to the development of oral contraceptive pills and their approval by the United States Food and Drug Administration in 1960. The medical profession began to take a serious interest in contraception for the first time with the appearance of the pill:

> Biochemical and hormonal contraception appealed to doctors' idea of "real" medical science and complemented their view of the necessity of scientific experts' managing births. Medical scientists' desire to sanitize reproduction was made clear in their embryology texts; the metaphors employed to describe conception and development had customarily been drawn from common, earthy domestic activities (sowing, baking and fermenting), but now doctors spurned such lowly associations and drew their metaphors from the science of engineering. . . . Similarly, doctors who were still embarrassed to fiddle with a messy cream or floppy rubber contraception were happy to distribute a pill, a product of scientific research, a "preventive medicine" that was simply prescribed (McLaren, 1990, p. 240).

The entry of women, including those with children, into the labor force reinforced the need for fail-proof contraceptive methods so that women could plan their families with confidence. The most reliable methods, the pill and the improved IUD, were female controlled, assuring that contraception, and the risks attached to it, remained squarely in women's domain. Women thus became increasingly dependent on medical science, an area dominated by males. Lobbying groups and doctors strongly protested the large number of illegal abortions resulting in women's deaths; consequently, abortion was legalized in the United States in 1973. Yet, control over abortion rested with the medical profession, not with women themselves.

During the 1970s, women began to voice concern over the risks attached to modern fertility-regulation methods, including cancer and blood clotting from pills and infections from IUDs. The Dalkon Shield, a defective IUD, was particularly dangerous, causing a range of pelvic infections, miscarriages, birth defects, and maternal deaths. Whereas feminists in the 1960s saw contraception as an answer to women's oppression, those in the late 1970s became increasingly skeptical about the "medicalization" of childbirth and contraception (McLaren, 1990, p. 257). Even in the 1990s, dependence on medical intervention and monitoring characterizes the more effective contraceptive methods: hormonal contraceptives, including the Norplant implant, injectables containing progestin, and the contraceptive pill; IUDs; and sterilization. Those requiring less frequent medical intervention—barrier methods (the diaphragm, cervical cap and sponge, spermicidal jellies, creams, and foams, and condoms), natural family planning (abstinence during the fertile period), and withdrawal—all tend to be less effective in actual use than the medically controlled ones.

Contraception, Population Policy, and Women's Status in Developing Countries

Early Population Policies

Population policies in developing countries have been motivated mainly by a preoccupation with population growth. In the late 1940s and early 1950s, the United States began to increase its contacts with developing countries in order to procure raw materials for its industries. Rapid population growth was seen as a potential threat to the stability of industrialized nations that were increasingly promoting birth control at home. Slogans such as "yellow peril" and "population bomb" became popular, and American private foundations began to promote an international strategy of population control. The language employed by such programs resembled that of late-twentieth-century disease-control programs today. The image conjured up is one of pathology: Births were seen as potentially harmful—much as unhealthy cells could multiply and become life threatening. Fears of the consequences of uncontrolled population growth were similar to those associated with epidemics or plagues.

The United States Agency for International Development (USAID) assumed leadership in the population field with the appointment of Dr. R.T. Ravenholt as head of its population branch in 1966. Ravenholt's philosophy became the guiding principle for most population assistance to developing countries: "Reduce fertility by means of the direct provision of family planning services and the development of new and better contraceptive technology, regardless of social context" (Hartmann, 1987, p. 105). Contraceptive devices, especially the IUD and condom, were distributed en masse to developing countries, which were asked to recognize their own "unmet needs" for family planning services.

Because women as childbearers were seen to have the main responsibility for contraception, and perhaps because female methods were considered more feasible in developing countries, women became the focus of fertility research and control efforts. Up to the early 1970s, these programs

had limited success because of their one-sided nature and their failure to take into account the cultural values and constraints of the women and couples whose behavior they sought to change. It became increasingly apparent that women in developing countries had sound reasons for wanting relatively large families. In most developing areas, a married woman's status within the family was closely related to her ability to bear children and to continue to do so until a satisfactory number was attained. Economic considerations often encouraged large families: Children could work on the family farm, provide support in old age, and inherit property. Children also provided emotional support and fulfilled many social and religious obligations, such as the continuation of the family line and performance of rituals at the time of parents' death. Also, as long as infant mortality remained high, couples were understandably unwilling to adopt birth control.

Faced with this reality, demographers began to focus more on factors influencing fertility, such as preferences for large families; high infant mortality; the economic value of children, including old-age security and the work children contributed; and inheritance and lineage issues. The status of women, and its relationship to birth control, received relatively little attention, or attention only within the context of how to change high-fertility behavior. For example, it was increasingly recognized that female education was an important determinant of contraceptive use: Educated women were more likely to adopt family planning than those with little schooling. In addition, research showed that children whose mothers were educated were less likely to die. Hence, education favored lower fertility on the one hand and lower infant mortality on the other.

Many researchers thus became preoccupied with the question of what it was about education that elicited these changes. Was it, for example, because education taught girls better hygiene, or to accept modern ideas and technologies? The debate was distinctly utilitarian. The implication was that birth control could be realized without having to wait for female education to be widely adopted if the key to the link could only be found. Similarly, by focusing on the education–fertility link, other aspects of gender inequalities were ignored. These investigations diverted attention from female education as an important goal in itself to the search for an intermediate solution largely unrelated to women's welfare.

As demographic research increasingly concentrated on local factors affecting fertility control, the metaphors used for population programs also became more humanistic. These included "family welfare" and "family planning" that conveyed the philosophy of later age at marriage, later entry into childbearing, and birth spacing. This approach was accompanied by the argument that family planning was good for women's health, as repeated pregnancies and deliveries put them at risk of anemia, infection associated with unsafe deliveries, and abortions due to unwanted pregnancies. Hence, greater interest in maternal and child health and their integration with family planning programs started to emerge. Accordingly, in the late 1980s international development agencies began to broaden their focus to "reproductive health," a more encompassing concept that included not only fertility regulation and maternal–child health, but also sexually transmitted diseases, reproductive tract infections and cancers, and larger family and couple-centered issues.

The Evolution of Women's Reproductive Rights

Within the international family planning discourse, the recognition of women's rights has been a long and difficult process. A beginning was made with the United Nations Declaration on the Elimination of Discrimination Against Women in 1967, which recognized that women have equal civil rights but that these were to be exercised "without prejudice to the safeguarding of the unity and harmony of the family, which remains the basic unit of any society" (United Nations, 1973, p. 39). As Dixon-Mueller (1993, p. 23) notes, however, the sanctity of the family was preserved in this statement and, with it, a subtle endorsement of patriarchal ideologies. In 1968, the United Nations International Conference on Human Rights in Teheran declared that couples should have the right to decide on the number of children they wanted and their spacing, and the 1968 Economic and Social Council meetings also recognized the crucial link between population and the status of women (Dixon-Mueller, 1993, p. 63).

The first World Population Conference (Bucharest, 1974) was a landmark in the debate on population and development, involving for the first time government representatives with different perspectives on population matters. A central issue was whether fertility control, in the absence of economic development, was a realistic or desirable goal. Whereas the U.N. *World Population Plan of Action* endorsed strenuous population policies with the goal of reducing population growth toward replacement-level fertility by the year 2000, developing countries argued that this "neo-Malthusian rhetoric" ignored the real underlying causes of underdevelopment—poverty and the exploitation of poor countries by the rich. India's argument, that "development is the best contraceptive," became a popular and central slogan of the conference. The *Plan of Action* that resulted echoed this sentiment, emphasizing the key role of economic development in fertility reduction. Women's concerns and gender relations were overshadowed by these broader ideological questions (Dixon-Mueller, 1993).

The United Nations Convention on the Elimination of All Forms of Discrimination Against Women, adopted by the United Nations General Assembly in 1979, and entered into force as a convention in 1981, affirmed the equality of men and women and women's right to individual freedoms such as voting and to social entitlements such as employment and education. It introduced specific rights for women, including the right to maternity leave and child-care benefits, recognized the need for affirmative action to redress the gender gap in opportunities and treatment, and urged countries to recognize that gender dis-

crimination in the name of culture or tradition is unacceptable.

The Second International Conference on Population and Development (Mexico City, 1984) was marked by the conservative position of the U.S. which, echoing the policy of U.S. President Ronald Reagan of free markets, individual initiative, and nonintervention, emphasized "sound economic policies" as the key to development and challenged the view that rapid population increase was an obstacle to economic growth. The U.S. policy statement made no mention of women's rights or gender inequalities. Despite this, other countries moved the *World Population Plan of Action* further ahead in highlighting the links between high fertility and the low status of women, and recommendations were made to improve economic, cultural, and social inequities. Interestingly, whereas the 1974 Population Conference was marked by a split in ideology between Northern industrialized and Southern developing countries, the position of countries at the Mexico conference was one of common opposition to the rightist U.S. position and confirmation of family planning as important to the realization of empowerment for women.

The Third World Conference on Women (Nairobi, 1985) marked the conclusion of the United Nations Decade for Women. A great deal of literature was generated on the social and economic condition of women and on gender differences, which led to the strengthening of networks of nongovernmental organizations (NGOs) and sparked the formation of women's groups, especially in developing countries. The Nairobi *Forward-Looking Strategies* recognized that women's right to control their fertility was essential to their enjoyment of other rights. The Second World Conference on Human Rights (Vienna, 1993) reaffirmed the centrality of individual rights in all population and development activities, and women's human rights received particular attention. Nonetheless, it was not until two more recent U.N. forums—the International Conference on Population and Development (Cairo, 1994) and the Fourth World Conference on Women (Beijing, 1995)—that women's health, reproductive, and sexual rights were fully tackled.

Despite considerable opposition from religious bodies, notably the Holy See and certain Muslim groups with fundamentalist positions, the action plans agreed upon in Cairo and Beijing were groundbreaking in terms of the compromises made and the recommendations reached. In Cairo, these included recognition that gender equity and women's empowerment are critical to their development and to their reproductive health; commitment to women's reproductive-health care throughout their lives within primary health care; recognition that unsafe abortion is a major public-health problem requiring action by governments and others; support for sex and reproductive-health education for adolescents; and recognition of male responsibility in the area of reproductive and sexual health. In Beijing, further progress was made with respect to the rights of children and adolescents to health information and services, privacy, confidentiality, respect, and informed consent; the responsibility of governments in the prevention of unsafe abortion and in the decriminalization of abortion; sexual rights, especially women's right over their own sexuality and fertility; and the overriding responsibility of states to promote human rights despite cultural, ethnic, and religious objections and differences.

Who Controls Contraception

The importance of understanding the perspectives of women who seek family planning and of providing high-quality services has been increasingly recognized by population researchers and program planners. The work of the Population Council has been important in developing a framework for assessing the quality of family planning care (Bruce, 1990) that includes choice of contraceptive methods, patient–client information exchange, provider competence, interpersonal relations, mechanisms to encourage continuity or follow-up, and appropriate constellation of services. This framework is being used in operational research studies to evaluate family planning activities; clearly, however, considerable commitment and resources on the part of developing-country governments will be required to bring about and sustain higher-quality services.

Bruce (1987) has highlighted several important feminist issues relevant to the acceptability of fertility-regulation approaches. These include the gender balance of responsibility and risk, referring to who uses the method and how this is affected by power relations, as well as the risks and benefits to male and female contraceptive users. These considerations, often complex, are extremely relevant to which methods are preferred and why. For example, women who do not wish to have more children but whose partners object to contraception may prefer methods that are female controlled and nondetectable. However, many of these, including hormonal contraceptives, IUDs, and sterilization, entail more health risks than male-controlled methods. These risks tend to be portrayed by family planning protagonists as considerably lower than the risks involved in childbearing. While this is true on the aggregate level, Bruce (1987, p. 364) points out that the argument is unacceptable from the perspective of the individual woman because those seeking to use contraception and those seeking to have children are different groups of women. A woman wanting to become pregnant is unlikely to weigh the risks attached to having a child against the risks of using contraception.

Another key issue is whether the use of a contraceptive method is controlled by the woman or by the health provider. Most modern methods require intervention by a health provider and at least one visit to a family planning facility. For women, this usually entails obtaining the consent of others, not least their husbands. In addition, women must submit to physical examinations, often by males, and this, in many parts of the world, is unacceptable. On the other hand, the long-term efficacy of some provider-dependent methods or the one-time insertion and removal of others may provide considerable peace of mind to women, relieving them of the need for constant attention to self-administered methods.

Bruce also discusses the issues of the reversibility of contraceptive methods, their side effects, and the need for these to be fully understood by female clients. Unusual or prolonged bleeding episodes and pain or discharge caused by certain methods may cause embarrassment and discomfort to women. These concerns can be considerably reduced through adequate counseling and follow-up. Those who choose to be sterilized should be fully informed about the implications of this method, including its possible irreversibility. Another issue around which much debate has centered is the provision of incentives to potential sterilization acceptors and health staff. Incentives may mean that women's free choice is jeopardized because of coercion, economic pressures, or inadequate information.

The HIV/AIDS epidemic has brought to center stage the role of barrier methods in avoiding infection during intercourse. The condom is the most effective in preventing contact with body fluids. In many countries, it is common for males to have more than one sexual partner, a practice that is culturally acceptable, even to married women. For females, too, sexual freedom is customary in many areas, but faithfulness among married women is expected and valued in most societies.

Females worldwide are an extremely vulnerable group for HIV/AIDS for reasons associated with gender inequity. Sex workers, for example, who insist on condom use may lose clients who, in the short term at least, ensure economic survival. Condoms have been widely associated with prostitutes and "dirty" practices and, therefore, are not widely favored within couple or other stable relationships. Hence, in many areas, the rate of HIV infection among women is high and rapidly rising. The impact on rural African women has been devastating as they watch their partners die, realizing that their own death and, in some cases, that of their offspring is not far behind. The tragedies associated with HIV/AIDS are many, as women contract the infection because cultural taboos and powerlessness prevent them from questioning their husbands about other sexual contacts.

Several feminist issues linked closely with contraception are relevant here. In many parts of the world, it is considered immodest for women to see the male sexual organs, even during intercourse. In such situations, women cannot suggest the use of condoms or ask their partners about previous sexual contacts. Even where women can discuss these matters with their partners, their desire for children may preclude the use of condoms. Many such women are young and childless and, in order to attain their rightful status in life, require children. They therefore risk HIV infection to become pregnant. How can women whose partners are infected with HIV become pregnant with a minimum of risk of acquiring the virus? As the twentieth century closes, health experts have no clear or consistent answers to this question.

Because of a lack of attention to the improvement of female barrier methods in the past, many women find themselves in a precarious situation that is costing their lives. The recently developed polyurethane female condom may have considerable potential as a female-controlled method of protection from HIV and pregnancy. As of 1996, it was still being tested for efficacy and acceptability in developing countries, but early results have identified some user dissatisfaction, especially among men. It is costly compared to the male condom (about three times the price in the United States); cases of breakage have been reported; sometimes the penis enters around rather than inside the sheath; and it can be noisy during intercourse. The impact of the female condom on women's negotiating power may, however, be more important than the efficacy of the method as such because it may motivate men to use protection themselves.

In many parts of the world, men will probably have the final say on use of female condoms for some time to come. But because women can wear the product themselves, they may be more successful at negotiating its use than they are with male condoms. . . . But providing technology is only the first step to action. Women need to know how to take advantage of this new technology. Efforts must begin now to help women gain the skills and the confidence to ensure that condoms—male or female—are used to safeguard their own health (Liskin and Sakondhavat, 1992, p. 707).

More research of this kind, focusing on female-controlled methods and their relationship to sexual negotiation and the empowerment of women, is urgently required.

Follow-up is another important aspect of quality of care. If women are not assured counseling and help if methods are unsatisfactory, they will hesitate to use them. The lack of provision for follow-up to Norplant implants in many countries has been strongly criticized. The most glaring case is in Indonesia, where about two-thirds of the world's Norplant insertions had occurred (more than 2 million women by the mid-1990s). The Population Council documented numerous cases in which women could not obtain removal on request; also, while many women were past the five-to-six-year utilization mark when the capsules should be removed, up to one-quarter of them may be unable to obtain removal because of the lack of adequately trained personnel.

While abortion is not, strictly speaking, a method of contraception, it has always been used by women to prevent unwanted births, and it remains an important fallback method for terminating unwanted pregnancies. "Menstrual regulation" (manual vacuum aspiration) has been legalized in many countries, including many developing nations. Mention should also be made of an important new early abortion technology (mifepristone), in combination with prostaglandin analogues, which can be used very early in a pregnancy until the ninth week from the last menstrual period. It is highly medicalized in that it is administered under close medical supervision, and several follow-up visits are required. As of the mid-1990s, it was being tested in many countries, including the United States, where final approval by the FDA is under litigation by the end of 1997. Some feminists argue that rules concerning the administration of the drug are unnecessarily restrictive and leave women just as dependent upon the health system as surgical abortions did. With continuing advances in medical technologies such as amniocentesis and RU486, ethical issues will continue to fuel debates between feminists and the medical world.

Toward a Gender Approach to Contraception

This review has shown that women, no matter what their culture or status, have always been concerned with controlling fertility. Early reasons for limiting family size, including high infant mortality, pain during childbirth, danger to the mother's health, and economic constraints, were strikingly similar to those later expounded by Malthus (1798) and subsequent family planning protagonists. Efforts to stem the growth of population have not, in general, been thwarted by women's resistance to birth control; rather, they have frequently taken narrow, top-down approaches in their design and implementation. Population programs have been motivated by concern more over high population growth than over the health and well-being of women in developing countries. Women were the vehicles through which broader objectives of these programs were to be realized. Only when programs failed, or were less successful than hoped, were they modified to more humanistic approaches. These programs could have benefited greatly from more two-way communication developed in collaboration with women and their families, understanding how women view their reproductive roles, their bodies and their needs, and arriving at compromises and common strategies for action.

Feminists have played an important role in bringing to light the shortcomings of the past, and these are gradually being recognized by international donors and national family planning programs. Concern is now being expressed for the quality of care received by women in family planning and health services, and dialogue is improving between women's advocacy groups and reproductive-health programs on integrating women's perspectives into research on contraceptive technology.

Women's groups have been strong advocates for better barrier methods of contraception, including microbicides, as well as for more extensive distribution of these by international agencies. This pressure is finally having an effect, especially in light of the growing concern over the HIV/AIDS epidemic. In December 1995 approximately 250,000 female condoms were provided to various developing countries as part of a pilot project, and there has been increased funding for barrier methods. The Rockefeller Foundation has also supported, with other donors, an international effort toward a "woman-centered agenda,"

"Contraception 21," intended "to revitalize contraceptive research and development" ("Program," 1995, p. 1). The agenda emphasizes improvement in the quality of reproductive-health care and research in three main areas: improved vaginal methods, especially those that protect against sexually transmitted diseases; menses inducers; and expanding male contraceptive choices and responsibilities.

There is growing recognition of the need to expand and intensify research on female-controlled methods of birth control, as well as on the role of men in family planning. Female-controlled methods can help free women from dependence on others and can give them a measure of confidence in their ability to control their own reproduction. They can help women negotiate with their partners about family size and the timing and spacing of children. Nonetheless, cooperation from male partners and their involvement in reproductive decisions and in the use of contraception are essential to this process.

Finally, family planning efforts should take into account existing sexual practices, patterns, and partnership arrangements of the populations they serve. While this may sound like simple common sense, many programs have failed to consider these and have focused on specific target groups, especially married women, with little consideration of practices prevailing in the larger society of often unstable families and single but sexually active adolescents and men. On the other hand, programs that begin from a gender perspective—with an understanding of the sexual relations within a given cultural context—are more likely to suggest appropriate and feasible interventions, including the sharing of responsibility by males and females for contraceptive decision making and action.

References

Bruce, Judith. "Users' Perspectives on Contraceptive Technology and Delivery Systems: Highlighting Some Feminist Issues," *Technology in Society,* vol. 9, 1987, pp. 359–383.

———. "Fundamental Elements of Quality of Care: A Simple Framework," *Studies in Family Planning,* vol. 21, no. 2, 1990, pp. 61–91.

Claro, Amparo, Lezak Shallat, and Kathleen Vickery. "Round Table on RU486," *Women's Health Journal,* vol. 2, 1993, pp. 29–59.

Dixon-Mueller, Ruth. *Population Policy and Women's Rights: Transforming Reproductive Choice.* Westport: Praeger, 1993.

Hartmann, Betsy. *Reproductive Rights and Wrongs.* New York: Harper and Row, 1987.

Himes, N.E. *Medical History of Contraception.* New York: Gamut, 1963.

Liskin, Laurie S., and Chuanchom Sakondhavat. "The Female Condom: A New Option for Women." In Jonathan Mann, Daniel J.M. Tarantola, and Thomas W. Nelter (eds.), *AIDS in the World.* Cambridge: Harvard University Press, 1992, pp. 700–707.

Malthus, Thomas. *An Essay on the Principle of Population As It Affects the Future Improvement of Society, with Remarks on the Speculations of Mr. Godwin, M. Condorcet, and Other Writers.* London: Johnson, 1798.

McLaren, Angus. *A History of Contraception: From Antiquity to the Present Day.* Oxford: Blackwell, 1990.

"Program for Appropriate Technology in Health (PATH)," *Outlook,* vol. 13, no. 2, 1995, pp. 1–8.

St. Augustine. *The City of God Against the Pagans.* Cambridge: Harvard University Press, 1963.

United Nations. *Human Rights: A Compilation of International Instruments of the United Nations.* New York: United Nations, 1973.

Vatsyayana. *Kama-Sutra: The Classic Hindu Treatise of Love and Social Conduct.* Translated by Richard F. Burton. New York: Viking Penguin, 1984.

See also Lynellyn D. Long and Lisa J. Messersmith, "Reconceptualizing Risk: A Feminist Analysis of HIV/AIDS"

Women's Control over Their Bodies

Noemí Ehrenfeld Lenkiewicz

Introduction

The right of women to "control their bodies" has different meanings representing various standpoints. The subject of the body has been a field rich in theoretical discussion, philosophical argumentation, and religious writings in almost all cultures; the debate has been particularly explicit in the Western world. As Connell remarks, two opposing conceptions have dominated discussions of the body in recent times:

> In one, which basically translates the dominant ideology into a language of biological science, the body is a natural machine that produces gender difference. In the other, which has recently swept the humanities and social sciences, the body is a more or less neutral surface or landscape on which a social symbolism is imprinted (Connell, 1994, p. 9).

The body, in particular woman's body, has also been a fertile arena of discussion for several scientific disciplines. Hence, throughout history science has appropriated woman's body, giving to it a variety of meanings, values, and cultural environments. Michel Foucault, in his revolutionary *History of Sexuality* (1977–1987), initiated a modern rethinking of the development of sexuality as he demonstrated how the concept of sexuality had changed through different historic periods and cultural settings. Before Foucault, sex and sexual concerns were the domain of psychiatry, which tended to operate with binary approaches, emphasizing the concepts of "normal" and "abnormal" sexual behavior. Despite the many changes, we still find today in the medical field health professionals who locate women's sexuality/gender/sexual identity in an outmoded frame of analysis.

"Sex"—defined here as the biological bases that characterize females and males and make them compatible reproductively—is a term that seems simple at first glance. The most striking conditions for assuming gender differences are women's fertility, birth, and child care. Among the biological features, different brain organization has to be included. Through the years, scientists have conducted a substantial amount of research seeking to explain the ontogeny and phylogeny of sex differences. This effort has had two undesirable consequences: First, many similarities between men and women were ignored; and second, this approach was used to argue that one of the sexes—the male sex—presented some clear advantages over the other. From that arose the distinction between the "strong sex" and the "weak sex," which presented men as the dominant sex and women as the passive sex. This representation led in turn to the attribution of different gender roles, with men dominant over women in many respects, including in the area of sexuality, so deeply rooted in many cultural values.

The idea, practice, and conviction of male dominance are present in different degrees in almost all societies, and are reflected in their laws and customs. In some societies, this domination becomes tantamount to the ownership of women. It is evident that male dominance leads to one of the most frequent, least studied, and most hidden problems of women's health: violence against women. Such violence occurs across cultures, through all social classes, from infancy to adulthood, and is expressed in a broad spectrum of behaviors: battery, rape, sexual abuse, and selective infanticide, to name but a few. These produce varying degrees of damage but often create an unfortunate impact on women's health.

Control of the Woman's Body by Men and the State

All societies have had to make arrangements, including the provision of laws and moral precepts, to regulate and organize erotic life. In both Western and Eastern cultures, with the West probably the more concerned about this need of regulation, attitudes concerning sex, sexuality, reproductive behavior, and sexual behavior of societies at large, and minorities in particular, became essential components of the social and cultural construction of masculinity and femininity in the evolution of humankind. As Weeks observes, "Gender, the social condition of being male or female, and sexuality, the cultural way of living out our bodily pleasures and desires, have become inextricably linked" (Weeks, 1986, pp. 45–46).

The relationships between sexuality and reproduction are deeply rooted in the subconscious of any woman living in a male-dominated system. Such a system is embedded in one's daily experience, through generations, learned in early childhood, in which boys and men make the rules of the game and girls and women become subordinate to them. Hence, a woman has to struggle in a male-dominated society to individualize herself regarding almost all of her decisions, among them those concerning sexual life and reproductive health, including when to have sexual intercourse and with whom, when and whether to use contraceptives, whether to have an abortion, when to adopt a baby.

The social actors most strongly supported by the state to assume responsibility for women's reproductive health are the health ministries, related health institutions, and health workers or physicians. Religious and cultural beliefs also strongly regulate women's freedom toward any reproductive choice. The state, too, has an interest in directing these decisions in order to control social reproduction, which also affects the semiprivate domain of family life, the domestic domain, intimacy, and, of course, the decisions made about sexual and reproductive practices. Beginning in the 1970s, feminist movements and women's associations in many countries have created an awareness of women's subordinated condition concerning these and other decisions. Many decisions regarding these practices, in consequence, are increasingly contested by women.

Since the beginning of the twentieth century, the medical profession has become one of the most powerful agents in women's health and life decisions. In many societies, to be a physician means that, beyond the specific health knowledge in which he or she has been trained, this person has special status—a higher one—over the rest of the community. In fact, in institutionalized medicine, the treatment of patients seems to be based on a hierarchy, in which doctors prescribe what they consider advisable and patients receive the prescription without questioning it.

Since most physicians traditionally have been male, the doctor–patient relationship is clearly an unequal one. In terms of reproductive health, many abuses have been committed in the name of "state interest in women's reproductive health." The prescription and distribution of family planning methods, in many cases, respond more clearly to a population-control strategy than to women's health requests and needs. There is a wide range of quality (including complete absence) of services offered and, even more important, there is an absence of ethical considerations in many countries.

In a changing and uneven world, where women have increasing access to education, healthcare, and laws ensuring equal rights, it seems that men will have to change: They are losing power, and must therefore learn to negotiate.

Reproductive choice includes the practice of sexual rights and the condition of freedom to make decisions about sexual and reproductive matters throughout life. This entails autonomy for women, defined as the ability to make independent decisions. It is imperative for women to have access to education, employment, and healthcare, among other basic conditions.

For women, making a single decision about their sexuality is part of a process in which a whole set of personal and familial circumstances, institutions, social rights, and legal status are linked in a complex social network.

For example, Mexico's constitution clearly states that every woman has to be "informed" and has to make a "free" decision about her reproductive life. Yet, information is exactly what women do not have, for many and complex reasons. And freedom is possible only if a woman is able to have autonomy and, in actual practice, exert satisfactory control over her body. Paradoxically, one of the main mechanisms by which the dominant society has control over women's decisions and bodies—information and access to it—is exactly the element that could help most to promote women's autonomy, through promoting their reproductive health.

Feminist groups all over the world are making women aware of this situation, and each day more and more women demand their right to decide freely about their bodies.

Control over Women's Bodies in Islamic Societies

It is important to consider other societies in addition to Western ones. In countries of the Middle East, there is wide variety in women's social conditions, reproductive rights, and reproductive health. The deficient Western references to many of these cultures as the "Islamic or Muslim countries" is a simplistic reduction that shows ignorance and sometimes prejudice; it is dangerous in terms of understanding other people's—and women's—conditions and rights. In the Islamic context, there is a tension between the egalitarian view, in which believers are judged solely according to merit, and the inegalitarian elements that clearly define different roles for women and men (Makhlouf, 1994). While traditional interpreters emphasize some religious passages that give women a lower valuation than men and so define a clear subordinated role for women, others interpret the religious precepts in a more egalitarian manner. Unfortunately, many of the Middle Eastern countries reflect the traditional conception in their approaches to gender issues, allowing women little, if any, control over their own bodies and over decisions that affect their lives. Data available from the Middle East as a whole indicates a dramatic and clear disadvantage for women in almost all indicators of reproductive choice, with related health and social consequences: high maternal mortality and morbidity, high fertility rates, low education levels, and limited access to the labor market. Some harmful customs are widespread, such as arranged marriages at very young ages, the male-determined legal and religious permission to repudiate a wife and to have multiple wives, the social pressure to bear children, the preference for sons, and a domestic unequal provision of nutritional and health care that favors males from childhood onward. All of these cultural practices leave women little space to exercise any sexual or reproductive rights. The adoption of constraining reproductive policies by states is often used to justify their position on gender and reproduction (Hedley and Dorkenoo, 1992; Toubia, 1993).

There are a variety of practices that have negative impacts on women's health but are nevertheless accepted by different cultures, the most offensive involving genital mutilation. Circumcision and infibulation are the most outstanding examples of how a male-dominated culture can affect women's health in its attempt to control female lives. These traditional practices, which some authors consider a clear exercise of gender violence, are also linked to child marriage (Hedley and Dorkenoo, 1992; Toubia, 1993).

It is estimated that 85 million to 114 million girls and women in the world have undergone genital mutilation, most of them in Africa (Heise, 1994). In Sudan, a representative case, genital circumcision and infibulation are commonly accepted practices, and frequently a women has to undergo the infibulation process more than once. Men have total power and control over women's bodies and rights. Purity—along with motherhood—is the highest value of women; social regulations often make it impossible for women who have not been circumcised to marry. Many women in Islamic countries, being products of their own socialization, report agreement with these customs, and most of them also profess a "satisfactory" sexual and marital life (Lightfoot-Klein, 1989).

Women's rights are far from being uniform throughout the world. There is, however, a common element: almost all of the variations and combinations of beliefs, religious codes, laws, and political strategies are thought, made, and implemented by men whose goal is to control women's sexuality and reproductive rights through the control of women's bodies.

Women's responses to this objective are quite different depending upon various factors, including what society they live in, whether or not they have their own resources, and whether they are literate or illiterate. In almost all countries, government policies related to gender are somehow inconsistent or ambivalent. There are advanced and "modern" countries that still strongly condemn divorce or abortion. Women as individuals must try to define and trace their own strategies, using ambiguous policies to skillfully avoid laws, and manipulating rules and codes for their own benefit. But there are also some women who are more involved in politics and who engage in overt opposition of such laws. What seems clear is that in order for the legal status of women to change, women themselves will have to assume the initiative and influence political agents or else assume power directly to change unfair policies.

Sexuality, Education, and Socialization

Although it is generally accepted that knowledge is power, schooling often presents a particular facet in regard to gender representations, sexuality, and sex education. Developing countries, despite the enormous variety and complexity of their political, social, and cultural patterns, share a conservative gender status in terms of the values that define family life and family formation, values that are transmitted through schooling both in what is covered and in what is ignored in the curriculum.

The role of the state in different countries of the Middle Eastern region as well as the incursion of "modern" states into the domain of the family presents substantial limitations to women's rights (Kandiyoti, 1991). Throughout the Muslim world there is a conservative ideology regarding the social roles of women, idealizing their importance as mothers and stressing allegedly innate gender differences that make work outside the home unsuitable for women (Esposito, 1995). Almost all conservative values and restrictive moral, cultural, and educational constraints that stress innate gender differences and emphasize women's traditional roles, such as motherhood, have an undesirable impact on women's status by removing from them the ability to control their own bodies and lives. Sociocultural behaviors are expressed through child marriages, men's ownership over women, polygamous practices, and male decisions on divorce. The recent introduction in some countries of population policies that include reproductive control oppresses many women. The "value" of women is stressed in severe control of their sexuality and capacity for motherhood and in the extreme social value attached to chastity and purity.

In Latin American countries, control of women's sexuality reflects a variety of cultural influences; however, these countries share some strikingly similar attitudes toward sexual mores, gender roles, sexual behavior patterns, and the lack of rights among sexual minorities (i.e., gays, lesbians, bi-sexuals). In this region, a double moral standard has been established through many decades, due mainly to the influence of the Catholic Church. It is true that Catholicism is not a homogeneous institution but the deeply rooted lessons about moral standards that place women in permanent conflict: women learn that premarital sex is a sin, but it is expected that men will have sexual experience by the time they marry, and that women will have to "learn" from men. Modern contraception is forbidden because sex is for reproductive goals not for pleasure, though, given the formidable threat of AIDS, some sectors of the Catholic Church are allowing the use of condoms. This double standard always favors male dominance over women's decisions and status.

Mexico is one of the countries considered a "success" in terms of family planning. The government has developed national programs for population control, family planning, and healthcare, which try to improve women's conditions by reducing fertility rates and providing educational services for reproductive health. Almost all of these programs view family planning services as the main strategy for reaching the goal of lowering the fertility rate to one child by the year 2000. Unfortunately, little has been advanced in other areas, such as nationwide sex-education programs for young people and special educational programs and services for marginalized people such as the various ethnic groups and sexual minorities. Almost all of the existing programs target women, except for some recent programs for men that have focused on vasectomies. Mexican official institutions, such as the National Population Council and, more recently, the Ministry of Health, faced serious problems in developing sex education programs, in particular programs that introduced sex education content linked to family planning strategies. Conservative and religious groups criticized those programs severely, raising the obsolete arguments that sex education belongs strictly to the family domain, and that such programs promote an early onset of sexual practices, among teenagers and thus lead to promiscuity.

Some nongovernmental organizations (NGOs) and other private institutions have made remarkable efforts to include information about sexuality in the official textbooks of the formal educational system; some NGOs have created their own programs and tried to reach students attending formal schooling and urban youths. The Mexican state, however, still upholds a very conservative position. Although the question of introducing sex education into the school curriculum was first raised in the 1970s, in the mid-1990s sex education was limited to a discussion of reproductive systems, omitting information on sexual life, sexual practices, desires, and sexual minorities. Absent are a gender viewpoint and a wider conception of sexuality that could promote at least a safer as well as a healthier and happier sexual life for students and future citizens. Tensions between a conservative state position and social forces that disagree with and confront state policies are present in many other countries of the region. The Mexican Catholic leadership allowed for the first time in 1997 the introduction of public information spots in the mass media such as television and radio about the need for sexual information and education, but the messages made clear that adolescents must speak with their parents on the subject. This strategy once again places on the family the responsibility for sex education. Studies by some NGOs and the Family Planning Unit of the Health Ministry show that it is imperative to include sexual education in family planning programs if they are to reach fertility-rate goals and create effective services for reproductive health.

Tensions between conservative state positions and social forces confronting them are present in several other countries in Latin America. Argentina—considered with Uruguay and Chile to rank among the most "modern" and liberal countries—is an interesting case: While gender relations are much more equal than in many other countries (probably due to an early participation of women in the labor market as well as to high educational levels), many contradictions are linked to women's status and reproductive health. Abortion is illegal except in very few circumstances but it is a frequent practice and women have access only through private services; the procedure is always performed clandestinely. Divorce became legal only in June 1987, with the first democratic government following the military dictatorship. During the first month after the legalization of divorce, two million divorce petitions entered the civil courts. Sex-education programs are absent at schools, except for some information about sexually transmitted diseases, including AIDS.

Sexual feelings and ethics, the language and cultural representations of sexuality, and the concepts of intimacy, control of the body, and autonomy for both genders are dangerously absent. This leads to a misunderstanding of the wide range of human sexuality and reinforces gender differences based mainly on genital characteristics and arbitrary sexual roles. This clearly is not enough if we want to create a culture that favors a "gender dialogue" rather than a "war of the sexes."

Control of the Body: Adolescents and Their Problems

In developing countries, especially in Africa and Latin America, adolescents represent an important sector of the population. In Latin America, there is a great concern about the reproductive health of the adolescent population, not only because of its impact on women's lives but also because it is one of the most important variables affecting fertility rates. Family planning programs have reached other age groups, but they still have had no impact on youth practices. Almost all countries lack comprehensive policies to target young women. Women of all ages face conditions of unwanted pregnancies, abortion, and sexually transmitted disease.

It is common to think of adolescence as the golden years, in which both genders have the capacity to live life to its fullest. It is also a common belief that adolescents are able and willing to change deeply rooted values of society, transforming it with new and revolutionary ones. But adolescent girls and boys are actually quite conservative, and they reproduce family, social, and cultural values as a well-learned script.

Since the mid-1980s there has been an increasing interest among researchers and health-policy decisionmakers in the reasons for and the contexts under which adolescents engage in unprotected sexual intercourse since this action has undesirable outcomes for boys and girls, especially regarding sexually transmitted diseases and unwanted pregnancies. Adolescent fertility represents 12 percent of the total fertility rate in Mexico (Ehrenfeld, 1994), 18 percent in Nicaragua, and even more in some African countries (United Nations, 1989). Girls are affected by these unwanted pregnancies because they not only tend to complete their pregnancies but are also abandoned by their boyfriends, face opprobium of family and peers, and often have to drop out of school before completion to take care of the baby. Women without instruction have on average five to six children in Latin America, a much greater figure than for those having completed secondary education. Thus, one of the most serious problems facing young mothers is that they have limited education, and this condition will continue through life.

Adolescence is that life period in which gender identity is almost fully shaped and boys and girls "act out" what they have acquired through their socialization. This learning is a complex process that is rooted in the family, where sexual roles are more or less spontaneously acted out and where boys and girls learn how they are expected to behave; this process is also strongly influenced by the mass media and peers.

But there are different groups of adolescents: those who belong to rural or urban areas, those who enter into productive life at very young ages, those who have no family at all and are much more influenced by their peers while living on the streets. Most adolescents in low-income groups face the problems, moral codes, and obligations of adults. For those who drop out of school, there is limited hope for readmission. Among this population, reproductive life starts very early, and in rural areas, early pregnancies are frequent.

Adolescents represent almost one-fourth of the total population in Mexico, and young women of ages 10–19 were 34.4 percent of the total female population in 1990. A study focusing on young women patients in public hospitals in Mexico (Ehrenfeld, 1994) indicates that low-income women have their first sexual intercourse around 15–16 years of age; these girls are about 17 years old when they seek health care from a public hospital because they are pregnant or have had an abortion. More than half of these young women live in informal, consensual unions, which is a common and socially accepted way to start a family in Mexico; fewer than 30 percent are married, and 16 percent are unattached. A remarkable fact within this population is that the average time between first intercourse and the initiation of pregnancy is five months. These girls were not attending school at the time they became pregnant, and, with few exceptions, they were not working; they were at "home." Interviews with these adolescents revealed that they got pregnant because, "He asked me to give him a child," or, "He wanted me to get pregnant." A high proportion knew about at least three modern contraceptive methods, but few of them used any form of protection at first intercourse (more than 85 percent never used a contraceptive) (Ehrenfeld, 1994). These findings are similar to those found by Langer et al. (1996) for other Latin American countries.

Girls are educated to be passive; they wait for and respond to the boy's requests but seldom initiate a sexual encounter. In focus-group discussions with peers, young women expressed ideas that highlighted their self-perception as women, their beliefs and fears about family planning methods, their social role, their knowledge and representation of their body, and their self-esteem. These young women clearly stated that boys are responsible for "taking care of them" in terms of using some contraceptive method. They are not able to suggest that their partner use a condom; among the reasons frequently cited are the fear that making such a request would drive the boy and his love away or that the boy could interpret this as a signal that the girl is an "experienced woman" in sexual affairs.

Sexual roles are so deeply rooted that it is easy to perceive the fear and anguish young women experience if considered promiscuous—"bad women"—just because they seek protected sex. Moreover, in many countries, double moral standards in sexual mores can lead adolescents to perceive that becoming a mother has a higher social value than does seeking a healthy, protected sexuality at their ages. For many adolescent girls, to become a mother means to reach the status of "woman," to get social recognition, to acquire a higher value vis-à-vis boys or men.

Closely linked to these feelings is the sad fact that girls often do not have knowledge about their bodies. They are not informed about the pre-ovulation period, do not understand their own anatomy, and, in some cases, do not even know what or where a vagina is. They know even less about sexual desire, intimacy, and self-confidence in a boy–girl relationship. Young women are severely pressed to behave according to young men's expectations and requests.

Many adolescent pregnant girls seem to have no concept of autonomy and control of their own body in terms of their reproductive choices. They refer to the pregnancy and more specifically to the baby as "this that has happened to me" or "once *it* is born. . . ." For many of these adolescents, pregnancy is a "state" in which they have had little participation. In interviews, most stated that the present pregnancy was undesired and unplanned, particularly because they had just initiated their sexual life and had engaged in sexual intercourse at their partner's request (Ehrenfeld, 1994).

In contrast, men and boys have free access to a woman's body, and this attitude is strongly supported in Latin American societies through explicit male gender roles: Boys are expected to be aggressive, and sexually very active, to achieve sexual experience before marriage, to be possessive, and usually to define the rules of the game regarding sexuality.

In Latin America, undesired and unplanned pregnancy is often the result of undesired and unplanned sexual intercourse without contraceptive protection. But the most common finding is that the newborn is fully accepted by the new mother. Motherhood is a gender role widely accepted by adolescent mothers, by the family, and also by society. Every actor in this conservative society plays the role that is expected: young men engage in sexual experiences in order to become "knowledgeable"; young women should not engage in sexual activity—this is not moral—but they become mothers and reproduce the overwhelming "values" of maternity. Those women who choose either to postpone or to cancel maternity are viewed with suspicion.

Abortion carries the most undesirable consequences for women's health. It is the clearest expression of an unwanted pregnancy. Women do not make this decision easily; usually they do so for economic reasons (they cannot feed and take care of the new baby) or contraceptive failure. In developing countries, abortion is associated with higher maternal mortality because in almost all countries it is illegal. It has to be performed under secrecy, with high risk for women's health. It is estimated that between 36 to 53 million abortions were performed in 1987 in the entire world. Among them, 26 to 31 million were illegal and between 10 to 22 million were clandestine; most of the abortions were performed in the Third World. In Latin America, estimates of the number of induced abortions range from 2.7 million to 7.4 million a year (Paxman et al., 1993).

Almost all social scientists as well as feminist groups consider meeting women's need for safe abortion health services worldwide an urgent public-health concern. In Latin American countries, along with the national governments themselves, one of the agents most resistant to legalizing abortion is the Catholic Church. Gender inequality and male-dominant social roles are clearly shown in the abortion question: a man may impregnate a woman against her wish; it is the woman who has to face the outcome of her lack of control over her body.

Unwanted pregnancy is a condition that millions of women face during their life span. A study by Ehrenfeld (in press) comparing pregnant adolescents with a peer group that had induced abortions indicates that adolescents in the latter group were one year older, had more secondary school education, or were attending nonformal education courses compared to those who had not had abortions. Even the decision about whether or not to have an abortion was made under clear pressure, exerted by the girl's mother or boyfriend against the girls' will.

It is important to stress that women who seek abortion services are not only young women, but from all ages. In Latin American countries, with the exception of Cuba, abortion is illegal.

Adolescents are in the dawn of their lives, yet many lack the conditions necessary for a healthy and comprehensive foundation for their gender identity. One of the important and final goals of the process of identity formation should be that women become "first class citizens," with a full set of personal, institutional, and social conditions to enable them to enjoy autonomy and freedom for life choices and decisions.

Regarding African and Middle Eastern countries, sexuality control combines with other factors to have a strong impact on women's status. Most of these countries are more rural than urban, their family-planning policies have been implemented more recently than in many Latin American countries, and they present many cultural constraints to women. Child marriage, which is expected to occur around 12 years of age, is a common practice that marks "adolescent" status. Adolescence is a particularly unhappy, painful, and stressful period in African and Middle Eastern countries because all women have some degree of genital mutilation, and sexual intercourse to get women pregnant during the

first year of marriage becomes a complex process. Almost all decisions about family size, sexual life, and reproduction are in the hands of men. In Ghana, the decision about family planning is strongly influenced by the male spouse although it depends on his educational level (Ezeh, 1993). In Nigeria, large families are culturally desirable and because men play a dominant role in family and society, men's reproductive attitudes affect their wives' behavior (Isiugo-Abanihe, 1994).

Control of the Body and Sexual Violence

Adolescents are particularly vulnerable to physical violence, especially sexual abuse. The vast majority of the perpetrators are male, and, while boys are victims too, a complex network of institutional mechanisms, legal systems, family traditions, and even educational systems shapes gender roles according to behavioral standards that are socially recognized and thereby reinforced. As Heise (1994) observes: "In many societies children learn that males are dominant and that violence is an acceptable means of asserting power and resolving conflict" (p. 2).

Violence as a gender problem is a one-sided entity: men exert control and women suffer its consequences. In some societies, violence and the expression of sexual aggression seem to be as deeply rooted as gender roles are. Despite the lack of statistical data from national surveys, and other sources, women's efforts to achieve legal responses to the problem of domestic violence are gaining visibility. In Mexico, for instance, the office of the attorney general declared in 1996 that only 1 of 10 sexual crimes are reported by the victims, 90 percent of whom are women aged 13–17, and 1 in 10 women becomes pregnant due to rape.

Sexual abuse and rape are extreme demonstrations of gender domination exerted by males; they constitute a forced invasion of the body, an expropriation of privacy with severe health consequences for the victim. In a study by Ehrenfeld (1995), 90 percent of those women who made legal complaints and were referred to a hospital for medical services were aged 13–17, and 20 percent of the sample became pregnant due to the rape. These figures are higher than those reported by official sources but similar to those of other clinical researchers. In this study, though many raped women reported that the rape did not have any impact on their health, they had faced many obstacles to present a legal complaint. They had to admit to their family that they had been abused. They had to make this fact known within their community. They had to struggle with the legal bureaucracy. Throughout this long process, they stated, they had lost weight, they could not eat, they suffered insomnia, they did not dare to go out in the street. And still they denied any health consequences "except this vaginal infection. . . ."

Many other young women did not dare to admit that they had been raped until at least some time had passed. Many of those who had a partner and a regular sexual life reported experiencing serious sexual disorders. It was usually somebody emotionally close to the woman who encouraged her to report the rape and press charges. When asked for the main reason they reported the rape, the women always cited someone other than themselves: "So he will not do this to other women," or "He has to be punished; he cannot just go around abusing other women."

These young women seem to perceive themselves as not important enough to justify the police report. This shows that they have difficulties with, and very little knowledge of, their bodies. They do not consider their health as a whole state of well-being: Sexual health is an absent perception. Sexuality belongs to the private domain, so when it becomes public because a rape induces a woman to make it public, she continues perceiving that the damage to her sexuality is a private matter, or that it is less important than other health areas.

Most of the women who become pregnant due to rape in almost all Latin American countries, decide to keep their babies, despite their being informed that under this circumstance the law allows an abortion to be performed at a hospital. In the 1994 Ehrenfeld study, some of the women said, "The baby is not guilty for what happened," or, "Anyway, I always wanted to have a baby," or, "I like babies. . . ." Once more, motherhood as a gender role is so powerful that it overrides the conditions in which this pregnancy occurs.

This issue should be explored in other cultural contexts. Depending on legal regulations concerning abortion, women may interrupt their pregnancy more easily and at earlier stages in some countries than others. In many developing countries it seems that young women (most of them adolescents), victims of sexual abuse, rape, and other forms of violence, do not see themselves as persons with rights, as citizens with access to a legal framework that guarantees that they will not be discriminated against for any reason. If for normal adolescents it is difficult to have a sense of autonomy, for a sexually abused girl, such autonomy is even less likely. She has really been expropriated from her own body, and is left with deep wounds that may never vanish. Despite the lack of data about this problem, it can be surmised that sexual abuse of children and adolescents is a widespread fact.

Some Recommendations

The exploitation of gender roles that have adverse effects on women's sexual and reproductive health must end. It seems obvious that education will be an important means in this effort. It is important to teach children of both sexes that their gender differences can be used as complements, not as weapons of control. Educational strategies must be formulated to create changes in intersexual attitudes and behaviors. The investment in time—possibly several generations—money, and creative efforts may be daunting, but these educational goals are not unrealistic. There are societies where aggression is not the main axis around which sexual games are learned and later acted out in adolescence and adulthood.

Women in developing countries must begin to acquire and use the skills needed to start changing and improving their condition. In many cases they will have to create alternative strategies to those provided by the state and utilize, if necessary, loopholes in existing laws that discriminate against women. In cases where these measures prove ineffective, it is possible that women will have to "take the law into their own hands." The challenge is great, but feminist groups throughout the world are fighting and winning more battles against the male-dominated status quo.

References

Altorki, Soraya. "Women and Islam: Role and Status of Women." In John Esposito (ed.), *The Oxford Encyclopedia of the Modern Islamic World.* New York: Oxford Academic Press, 1995, pp. 322–326.

Connell, Robert. "Bodies and Genders," *Agenda: A Journal About Women and Gender,* no. 23, 1994, pp. 7–18.

Consejo Nacional de Población. *Programa Nacional de Población 1995–2000.* Mexico City: Consejo Nacional de Población, 1995.

Dixon-Mueller, R. "The Sexuality Connection in Reproductive Health," *Studies in Family Planning,* vol. 24, no. 5, 1993, pp. 269–282.

Ehrenfeld, Noemí. "Educación para la Salud Reproductiva y Sexual de la Adolescente Embarazada," *Salud Pública de México,* no. 36, 1994, pp. 154–160.

———. "Female Adolescents at the Crossroads: Sexuality, Contraception and Abortion." In *The Determinants of Induced Abortion in the Developing World: Results of a WHO/HRP Supported Research Initiative.* Geneva: World Health Organization, in press.

Ehrenfeld, Noemí. *Violencia y Violación: Su impacto en la salud de la mujer.* Mexico City: Colegio de México, 1995.

Esposito, John (ed.). *The Oxford Encyclopedia of the Modern Islamic World.* New York: Oxford Academic Press, 1995.

Ezeh, Alex. "The Influence of Spouses over Each Other's Contraceptive Attitudes in Ghana," *Studies in Family Planning,* vol. 24, no. 3, 1993, pp. 163–174.

Ford Foundation. "Reproductive Health: A Strategy for the 1990's," program paper of the Ford Foundation. New York: Ford Foundation, 1991.

Foucault, Michel. *Historia de la Sexualidad,* vol. 1: *La Voluntad de Saber.* Mexico City: Siglo XXI Editores, 1977.

———. *Historia de la Sexualidad,* vol. 2: *Historia del Uso de los Placeres.* Mexico City: Siglo XXI Editores, 1986.

———. *Historia de la Sexualidad,* vol. 3: *La Inquietud de Sí.* Mexico City: Siglo XXI Editores, 1987.

Hedley, Rodney, and Efua Dorkenoo. *Child Protection and Female Mutilation: Advice for Health, Education, and Social Work Professionals.* London: Forward, 1992.

Heise, Lori (with Jacqueline Pitanguy and Adrienne

Germain). "Violence Against Women," discussion paper no. 255. Washington, D.C.: World Bank, 1994.

Isiugo-Abanihe Uche C. "Reproductive Motivation and Family-Size Preferences Among Nigerian Men," *Studies in Family Planning*, vol. 25, no. 3, 1994, pp. 149–161.

Kandiyoti, Deniz. "End of Empire: Islam, Nationalism, and Women in Turkey." In Deniz Kandiyoti (ed.), *Women, Islam, and the State.* London: Macmillan, 1991, pp. 22–47.

Langer, A., and Kathryn Tolbert. "El aborto en México: Un fenéómeno escondido en proceso de descubrimiento." In Kathryn Tolbert, Noemí Ehrenfeld, and Marta Lamas (eds.), *Mujer: Sexualidad y Salud Reproductiva en México.* Mexico City: The Population Council, 1996, pp. 289–315.

Lightfoot-Klein, Hanny. "The Sexual Experience and Marital Adjustment of Genitally Circumcised and Infibulated Females in the Sudan," *Journal of Sex Research,* vol. 26, no. 3, 1989, pp. 375–392.

Makhlouf Obermeyer, Carla. "Reproductive Choice in Islam: Gender and State in Iran and Tunisia," *Studies in Family Planning,* vol. 25, no. 1, 1994, pp. 41–51.

Paxman, J., A. Rizo, L. Brown, and J. Benson. "The Clandestine Epidemic: The Practice of Unsafe Abortion in Latin America," *Studies in Family Planning,* vol. 24, 1993, pp. 205–226.

Toubia, Nahid. *Female Genital Mutilation: A Call for Global Action.* New York: Rainbo, 1993.

United Nations. *Adolescent Reproductive Behavior: Evidence from Developing Countries,* vol 2. New York: United Nations, 1989.

Weeks, Jeffrey. *Sexuality.* London: Routledge, 1986.

See also Nelly P. Stromquist, Molly Lee, and Birgit Brock-Utne, "The Explicit and the Hidden Curriculum"

Women and Migration

Pierrette Hondagneu-Sotelo

Introduction

Colonial economic regimes in the Third World were once based primarily on the extraction of wealth to core nations through large-scale, plantation agriculture, mining, or agrarian farming. Today, Third World nations are primarily integrated into the global economy through capital and labor mobility. Migration, both domestic and international, is an integral part of structural transformations occurring in developing societies, and women are key players in this movement.

Throughout the Third World, women migrate to cities to seek work in manufacturing, in export assembly, in informal-sector (i.e., unregulated) services and commodity production, and in commercial activities. They also go to cities to pursue education, to reunite with family members, and to flee aspects of rural life they may find oppressive. Throughout the twentieth century, women have predominated in domestic rural–urban migration in Latin America, the Caribbean, and parts of Asia, and now, on the eve of the twenty-first century, they are increasingly migrating to cities in Africa. Women have also been well represented among movements of political refugees and exiles from Cuba, Vietnam, and Chile. And women from various Third World nations have been key players in large-scale labor migrations from Southern Hemisphere to Northern Hemisphere countries, going to the United States, Canada, France, Germany, Britain, and Switzerland. Even Japan is becoming a receiving nation of immigrants: Women from various Asian countries and Brazilian women of Japanese heritage are migrating to Japan.

A retrospective of the migration literature in various disciplines obscures women's participation. In spite of all of the "women on the move" throughout the twentieth century, with few exceptions, research strategies focused relentlessly on men. Initial scholarly attempts to focus on women migrants were met, at best, with indifference; at worst, with vitriolic hostility. In a 1976 article commenting on one of the first conferences on women and migration, the promi-

nent anthropologist Anthony Leeds opined that "the category 'women' seems to me a rhetorical one, not one which has (or can be proved to have) generic scientific utility," and he decried this focus as "individualistic, reductionist, and motivational" (Leeds, 1976, pp. 69, 73). Leeds argued that focusing on women would deflect attention away from structural processes of capitalist exploitation, and that in itself is telling, since it encodes the assumption that women do not act in economic or structural contexts. Androcentric biases, assumptions that women are "too traditional" and culture-bound, or that they only migrate as family followers or "associational migrants" for family reunification figured heavily in the literature.

While women were traditionally ignored in the migration literature, women migrants have received more scholarly attention since the 1980s. The international women's movement, the subsequent growth of women's studies programs and feminist scholarship, and policymakers' and academics' renewed interest in migration account for this turnabout. As one commentator observed of the burgeoning scholarship, the topic of immigration and women "has mushroomed" (Pedraza, 1991, p. 304). After decades of neglect and absence, women are "in" in the migration literature. By the early 1990s, there was enough research on the topic of women and migration to yield two substantive, prominently placed review essays in sociology (Pedraza, 1991; Tienda and Booth, 1991), as well as several edited volumes (Brettel and Simon, 1986; Gabaccia, 1992) and numerous monographs and case studies.

Feminist scholarship has shown that gender—that is, the social and cultural ideals, displays, and practices of masculinity and femininity—organizes and shapes our opportunities and life chances. The concept of gender as an organizing principle of social life has encountered resistance and indifference in various social science disciplines. Feminist scholars tell us that the same is true of immigration scholarship conducted in such disciplines as economics, sociology, and even anthropology.

While there has been more attention placed on women and migration, much of the scholarship remains mired in an "add and stir" approach. Women are "added" as a variable to be inserted and measured, so that women's migration is examined with respect to fertility and compared with men's employment patterns. Gender as a set of social relations that organize immigration patterns is generally ignored, taken into consideration only when women are the focus (as though men were without gender). Research strategies that do more than simply "add" women to the picture would enhance the field. While it is beyond the purview of this essay to focus on men and migration, a truly gendered understanding of migration requires that we use gender as an analytical tool relevant equally to the study of men's and women's migration. This is as yet a relatively nascent stage in migration studies (Grasmuck and Pessar, 1991; Hondagneu-Sotelo, 1994). The first part of this essay focuses on women and how competing theoretical approaches view women and decision-making processes concerning migration; the second part examines some of the important empirical contributions to the study of migration and African, Asian, and Latin American women, briefly summarizing the demographic composition of particular migrations and paying particular attention to how migration affects women's participation in the labor market and gender inequality within the family and society.

Why Do Women Decide to Migrate?

The question "why do women migrate?" assumes that women possess the will and the means to make and enact important decisions about migration. But the two competing theoretical frameworks in migration scholarship operate on decidedly different assumptions and, hence, come up with different answers to this question.

Derived from neoclassical economics and congruent with modernization theory, the orthodox perspective, alternately called the equilibrium or "push-pull" model, posits migration as an individual response to negative "push" factors at the point of origin and positive "pull" factors at the point of destination. At its most extreme, this perspective casts the individual migrant—usually a man—as a purely self-interested economic agent, an actor who computes present income with potential earnings in alternative locations. In this view, migration is seen as a matter of individual choice, with men cast as risk takers and women as the guardians of tradition and stability. Accordingly, women are seen as economically irrelevant and unproductive, and men are seen through a prism of voluntarist assumptions, calculating and acting outside contingencies of history and political economy. The conditions that give rise to "push" and "pull" factors are not investigated but are assumed to derive from distinct, unconnected societies or, more distortionally, from universal conditions.

When studies of women and migration operate under some variant of this framework, they assume that migrant women are either responding to the same sorts of "push" and "pull" factors as men are, or are purely dependent, "associational" migrants who follow their husbands or fathers. But as Thadani and Todaro (1984) point out, "even the associational migration of women may be induced, in part, by the expectation of urban employment and the dislocation of their traditional economic activities" (Thadani and Todaro, 1984, p. 44). Working from this observation, they suggest that what is needed is a special theoretical framework for analyzing the migration of women in the Third World. In their view, theoretical explanations must account for income and employment opportunities in the cities of destination, as well as the urban marriage market and cultural constraints on women's mobility. With respect to marriage, they consider the migration of married women and the migration of women in pursuit of improved marital prospects. Once again, it is telling that, when scholars have studied male migration, they have not posited the importance of men's marital choices and constraints on migration. Although both women and men marry, the paradigm defines only women by marriage.

Macrostructural approaches to the study of migration developed in opposition to the neoclassical model and redirected the research focus to the structural and historical factors that make labor migration possible. Discussion has focused, for example, on how foreign investment in Third World countries disrupts established economic structures and generates emigration, on how capital mobility from core to periphery induces labor migration in the opposite direction, and the significance of labor migration for capital accumulation in the societies of immigrant destination. Unlike the orthodox model, the structural model conceptualizes migration as a phenomenon internal to one global "world system," not as the movement between two autonomous spheres. Research informed by macrostructural and comparative-historical approaches illuminates how broad structural factors induce and support migration, offering a necessary corrective to the orthodox views by providing the missing "big picture" focus.

The mass migrations and entrance of women into export assembly firms in the Caribbean, along the U.S.–Mexico border, and in parts of Asia have been examined through the prism of the macrostructural model. In a study of the *maquiladoras* (low-wage assembly plants) along the U.S.–Mexico border, Fernandez-Kelly (1983) underscored the significance of northward labor migration of young, single women to staff these plants. She noted the gender-specific nature of economic development at that time, reporting that "85 percent of those working in the export-manufacturing plants along the Mexican border are female" (Fernandez-Kelly, 1983, p. 209). Other scholars have reached similar conclusions about migrant women and development in Latin America, Asia, and the Caribbean. Sassen-Koob (1984) extended the structuralist perspective to the study of women and migration by noting the link between the recruitment of Third World women into service and manufacturing jobs in two different but related sites: export-processing zones in the Caribbean and Asia, and manufacturing and service jobs in cities such as New York and Los Angeles.

These contributions have enriched our understanding of women and migration, but they have replicated a basic problem with macrostructural approaches: In explaining the origins of migration and the functions that labor migrations play in the development and maintenance of modern capitalism, the social dimensions of immigration are neglected. What is generally missing from studies informed by the macrostructural perspective is a sense of human agency or subjectivity. Rather than human beings, migrant women and men are portrayed as homogeneous, nondifferentiated objects responding mechanically and uniformly to the same set of structural forces.

In the late 1970s and early 1980s, immigration researchers addressed some of these issues by focusing on more intermediate units of analysis, the household and migrant social networks. These efforts, however, remained flawed by several unexamined assumptions. The analyses presented the household as a unified collectivity, ignoring divergent and conflicting interests within households. Thus, they continued to obscure gender and generation as social relations shaping immigration. And while such studies highlighted the importance of social networks in encouraging migration and showed the assistance these provide to women as they move and resettle, they did not explore how gender may regulate these social networks. Migrant networks channel and support migration, but the opportunities available to women and to men within the networks may differ (Boyd, 1989; Hondagneu-Sotelo, 1994). Women and men may develop, and use, distinct networks of support, and single women may or may not enjoy access to the network resources that married women do (Hondagneu-Sotelo, 1994; Repak, 1994). Similarly, particular household strategies may favor women's migration over men's or vice versa (Grasmuck and Pessar, 1991; Toro-Morn, 1995).

Women's migration patterns take diverse forms. In the following section, I review some of the key studies of Third World women and migration with the proviso that we must guard against reifying generalizations. Each continent comprises a wide diversity of cultures, nations, economies, and geographies, and, even within particular communities, women may experience heterogeneous migration trajectories.

Africa

It is only fairly recently, in the late twentieth century, that women are participating in labor migration in Africa. Men had traditionally predominated in rural–urban migration because they were vigorously recruited for work in plantations, railroads, and mines and even as domestic servants. In some nations, employers and government authorities devised elaborate systems of labor control that would deliberately keep women and children in rural hinterland areas so as to avoid the creation of permanent, Black communities in White areas (Stichter, 1985; Tienda and Booth, 1991). The women "left behind" farmed, traded, engaged in petty commodity production, and became effective heads of households.

Since the 1960s, women have increased and men have decreased their participation in the rural–urban migration flows in Africa (Stichter, 1985; Tienda and Booth, 1991). Most of these new female migrants are married women who are joining husbands in cities, but more unmarried women are also migrating (Stichter, 1985; Wilkinson, 1987). Yet, while women are more likely to migrate to cities than before, men still predominate in international migration in Africa.

Sub-Saharan Africa is one area where women traditionally remained "back home" while their men migrated. Under racial apartheid in South Africa, men's migration was circular—that is, the men lived in the city of destination only while employed, and, when not employed, they returned to their rural or village place of origin. The system was not uncommon elsewhere in the continent. Although women themselves were not on the move, they played a key role in maintaining this system. Circular labor migration is characterized by the physical separation of employment and family home residence, as well as by the physical separation of the costs of maintaining and renewing labor. That is to say, the male migrant workers receive the resources necessary for daily subsistence or maintenance in the city of destination, but these same workers are "renewed" in their place of origin, as it is the women who take on the responsibility for intensive, small-scale, subsistence farming. Women's labor has thus been twofold: They are actively supporting their own family's daily reproduction and maintenance, but they are also indirectly subsidizing a system of labor migration that benefits primarily urban elites, many of whom are White, with access to workers who subsist on extremely low wages.

Although staying "back home" and assuming myriad economic and family responsibilities has been a struggle for women, it has traditionally afforded them some semblance of autonomy and independence. What happens when they migrate to cities? New migration patterns have emerged in sub-Saharan Africa. In a study of migrant women in Lesotho, Wilkinson (1987) found that migration did not enhance, but rather diminished, women's position vis-à-vis men. As young women have acquired education, many have aspired to clerical, industrial, or technical jobs, but more often than not they do not find the jobs they seek. In the city, women work commonly as vendors, commodity producers, and prostitutes. In this context, migrating to the city actually increased women's dependence on men and intensified their subordination (Wilkinson, 1987). In the city of Maseru, Wilkinson found that women's only job opportunities were in the informal sector, while migrant men took jobs in the formal sector or the public sector. Even when migrant women have better employment opportunities and increased incomes, their relative economic position does not necessarily improve dramatically since they often remain dependent on men to purchase their goods.

In the second half of the twentieth century, a growing number of women have migrated to West African towns and cities. In a review of case studies of women and migration in southern Africa, East and Central Africa, Zambia, and West Africa, Tienda and Booth (1991) note that the impact of migration often depends on whether a woman is married or single at the time of migration:

> For women who moved with or to join husbands, migration transferred male control over women's labour and earnings from rural to urban areas, adapting hierarchical relations to an urban setting . . . [so that] wives in poor families were unable to challenge the patriarchal relations of redistribution. . . . By contrast, single women were able to improve their social position by increasing their autonomy via economic exchanges. For them the city provides the only alternative to assured subordination in the village. Because single and married women shared the economic constraints imposed by their relegation to informal activity, neither realised substantial improvements in material well-being (Tienda and Booth, 1991, p. 60).

Married women often lose much of the independence and control that they maintained over their own households "back home."

There is not much literature on African women and international migration. African women participate less in international migration than do African men, but when they do their status may be enhanced. Unlike in the African continent, in Asia, Latin America, and the Caribbean most migrants to the city are women. This is due to particular patterns of agricultural development and to labor demand for women to fill positions as domestic workers and assembly workers. Asia, Latin America, and the Caribbean are where most new immigrants to the United States originate. Third World women from these continents are key actors in both internal and international migration.

Asia

Asian nations' link to the global economy is solidified by their provision of sites for multinational corporations' manufacturing and assembly activities and for U.S. military bases. These developments are largely responsible for attracting primarily Asian women, not men, to the cities. There are, however, some significant exceptions, such as South Asia, where men outnumber women in the cities due to patterns of land tenure and agriculture (Pedraza, 1991, p. 310).

In the Philippines, women from the rural provinces migrate to cities, and many of them are employed as domestic workers and prostitutes in areas around U.S. military bases. Although prostitution is illegal in the Philippines, women sell sexual labor as "hospitality women" in bars, massage parlors, and night clubs. In this line of work, the migrant women are routinely abused economically,

sexually, and physically. Philippina women also migrate to cities, to other Asian countries, and to Canada for employment as paid domestic workers.

Research conducted by Wolf (1990) in Java, and by Ong (1987) in Malaysia examines the experiences of young, single women who migrate from rural areas for work in the new export-processing zones. In Malaysia, the combination of a decline in peasant farming and the introduction of Japanese manufacturing assembly plants has induced the rural–urban migration and proletarianization of young, single women. Ong's (1987) study found that these "neophyte factory women" marry later than their nonmigrant peers and that, although they earn low wages, they typically contribute about half of their earnings to their family of origin and thus gain greater power in the family. These transformations, however, are not without their contradictions, as the women suffer hallucinations and "spirit attacks" on the shopfloor, which Ong interprets as challenges to capitalist discipline and the male authority imposed on them in the factories.

Typically, one thinks of migrant workers supporting their families back home with their new incomes earned in the city. But in a study of young migrant women from rural Java working in Indonesian assembly factories, Wolf (1990) found that the women's earnings are so low that they are, in fact, subsidized by their rural, peasant families. She concludes that the peasant economy is subsidizing the factories. The young women contribute very little money to their families, as they must spend a good deal of their earnings for transportation and lunch expenses. While their earnings are modest, the young women derive satisfaction and status from their employment, and they used part of their income to purchase cosmetics and other small consumer goods and save for dowry. Wolf reports that many of these women

> exhibited an air of assertiveness compared with their peers who had never worked in a factory. Their makeup, nail polish, and in a few cases long pants were statements of modernity. . . . Many stated defiantly that they, not their parents, would choose their future mate (Wolf, 1990, p. 42).

Asian women have a long history of migrating to the United States, and while Korean, Chinese, and Vietnamese women have predominated in recent decades, in the past Chinese and Japanese women did. The latter two groups began coming to the United States in the late nineteenth and early twentieth centuries, respectively, but their migration was hindered by deliberate U.S. policies of racial exclusion.

Back in the beginning, Chinatowns were largely "bachelor" communities, and immigrant daughters from poor, peasant families in China supplied sexual labor in cities such as San Francisco, where there were few women (Cheng, 1979). These young women were usually sold by poor families, and daughters accepted their sale and subsequent migration out of filial loyalty. Chinese patriarchal ideology forbade the migration of "decent" women, and United States policies also prevented the migration of all Chinese after 1882. According to Cheng, the experiences of these young migrant women were not heterogeneous, as prostitution was organized in different ways. Some of the women gained autonomy and control over their lives, while others were virtual slaves.

In *Issei, Nisei, Warbride* (1986), Glenn examines the migration and domestic work experiences of three generations of Japanese women in the San Francisco Bay area. The first generation of Japanese women began arriving in the early twentieth century, rapidly transforming the ratio of Japanese men to women from twenty-five to one at the turn of the century to two to one by 1924. Many of these women were "picture brides," and, once in the United States, they worked triple shifts doing their own housework, and working in family farms and businesses and as paid domestic workers. The 1924 Immigration Act excluded all immigration from Asian countries, effectively curtailing the growth of Asian-American communities in the United States. Although the 1952 McCarran-Walter Act established token immigration quotas, it was only the 1965 amendment to the immigration act that allowed more Asians to immigrate. According to Glenn (1986), the primary struggle for Japanese immigrant women and their daughters—and perhaps for all subordinate racial-ethnic immigrants—has not been with fighting gender subordination within the family, but rather with struggling against a racist society and an exploitive, stratified labor system. In this regard, Japanese-American women used the family as a source of protection. Although gender and generational conflicts within the family existed, they were overshadowed by societal oppression.

Latin America and the Caribbean

In the second half of the twentieth century, urbanization accelerated in Latin America, and it is young, single women who have predominated in rural–urban migration. Many of these women come from poor, peasant families, and they were drawn to work as domestic servants in Latin American cities. Their migration served as a "demographic safety valve" for rural communities that could not absorb the labor of these women (Tienda and Booth, 1991, p. 63). Since the mid-1960s, women in Mexico and the Caribbean have also been drawn to cities to work in export assembly production (Fernandez-Kelly, 1983). Married women also migrate to cities in Latin America, where many work in relatively unskilled, informal-sector jobs as street peddlers and doing home assembly. Finally, Latin American women of various nationalities, but especially Mexican, Cuban, Dominican, Salvadorean, and Guatemalan women, also migrate to the United States, where many of them find work as paid domestic workers or nannies (Grasmuck and Pessar, 1991; Hondagneu-Sotelo, 1994; Repak, 1994). Puerto Rican women's migration to the U.S. mainland, particu-

larly the East Coast, is intimately tied to U.S. patterns of investment in Puerto Rico (Toro-Morn, 1995).

Even in the colonial period and the nineteenth century, young women going from Latin American rural communities to cities often found jobs as domestic workers in private households. In the late twentieth century, as live-out working arrangements became more common, domestic-work relations became more contractual and less personalistic. For poor, relatively unschooled women from the countryside, domestic work still provides one of the few means of economic survival and urban exposure, but it does not necessarily provide a stepladder to better jobs. Migrant women who are able to leave domestic service do so because they marry and become homemakers or because they make a lateral socioeconomic move to informal-sector vending.

Particular patterns of industrialization along the U.S.–Mexico border and in the Caribbean have also stimulated women's rural–urban migration. In 1965, in an effort to stave off rising unemployment, the Mexican government instituted the Border Industrialization Program (BIP), a program designed to generate the infrastructure and legal conditions to successfully attract foreign manufacturing investment along the northern border with the United States. Although the BIP was originally intended to occupy primarily male migrant workers, the plants employed predominantly young, single women because of employer preferences and the growing population of single women without the financial support of husbands or fathers. During the 1970s, women accounted for 75 to 90 percent of the workers at the border assembly plants and many of them were young, single, childless women drawn from the interior of Mexico (Fernandez-Kelly, 1983).

Research conducted by Fernandez-Kelly (1983) in the *maquiladora* assembly plants reveals that the employment of migrant daughters, wives, and mothers has neither eroded domestic patriarchy nor enhanced women's positions in families. Husbands and fathers, when they are present, still maintain family authority, and employed women still take primary responsibility for domestic household chores. One arena of control that does appear to be affected by women's employment in the *maquiladoras* is women's enhanced spatial mobility and concentration in new public spaces; Fernandez-Kelly reports that in their leisure hours, the young women congregate in discotheques and the all-female *cervecerías* (beer bars).

In the 1960s and 1970s, it was primarily single women who migrated and sought wage work in Mexico, but, with the economic crisis of the 1980s, financial needs propelled married women with small children into the labor force, and many of these women were former migrants to the city. The availability of many types of unregulated, informal-sector work did, and does, allow married women to continue to assume childrearing and domestic responsibilities in the home.

Male migration to the United States is another factor that has accelerated women's migration and employment in Mexico. Meager remittances and the prolonged absences of migrant husbands, starting during the bracero program (1942–1964) and continuing in the 1970s–1980s, often propelled women to migrate to cities and villages for jobs in agricultural occupations previously defined as male, in service-sectors jobs, and in newly created occupations in which labor is exclusively or primarily female, such as jobs in packinghouses, *maquiladoras,* shoe, clothing, and textile manufacturing, and home assembly work.

Have Mexican women's increased levels of employment, made possible largely through rural–urban migration, challenged traditional patterns of domestic patriarchy? While women's incorporation into the paid labor force has eroded men's positions as sole family providers, women's labor does not necessarily signal women's emancipation, in either the public or the domestic sphere. As is the case in the United States, women in Mexico earn significantly less than men do, and they are subject to occupational segregation and sex discrimination and harassment. And, just as in the United States, the inferior earnings of Mexican women are often justified by patriarchal assumptions that men deserve wages large enough to support their families and that women are necessarily secondary or supplementary earners and thus deserve less pay.

When women's employment occurs in regions of Mexico marked by U.S.-bound male migration, patriarchal gender relations show more receptiveness to change. Gail Mummert's (1988) research in the rural villages surrounding Zamora, Michoacán, where young single women entered wage labor in the mid-1960s underlines how women's employment in commercial agriculture, in conjunction with male wage labor in the United States, has brought about rapid and significant changes in gender relations. Female employment has shifted from being a stigmatized rarity to a virtual rite of passage, an expectation for all young, single women that will allow them to escape household drudgery and isolation. Initially, recruitment efforts focused on obtaining parental permission by assuring parents that their daughters would be properly chaperoned. Today, young women are no longer cloistered in the home but walk about with their friends, with whom they take buses outside the village to the fields and packinghouses. While the first generation of female workers turned over their entire earnings to their parents, young women now either keep all of it or contribute only part of their pay checks to their families and spend the rest as they please, usually on clothes and cosmetics. Even when working daughters compliantly turn over their salary to parents, young women have acquired more decision-making power in the domestic arena. Courting patterns are more open, and in less than two decades there has been a shift away from patrilocality to matrilocality or neolocality, as mothers and daughters pool their earnings to purchase the new couple land and thus avoid the subordination a newly married daughter suffers when living with parents-in-law.

It would be misleading to overstate these transformations in domestic patriarchy. Although women may "take the reins" during men's long-term absences by making decisions, securing employment, and venturing beyond the domestic sphere, their behavior often remains subject to strong double standards. Sexual infidelity, for example, may be tolerated in a man, who is allowed to maintain a *casa chica* (i.e., a second woman and children), but a woman left alone by her sojourning husband is often under the watchful eyes of other villagers, and any contact with an unrelated man may be cause for suspicion. Yet, young women's acceptance of these double standards may be eroding. Rosado's (1990) research suggests that, in the context of men's migration northward, women's beliefs about traditional double standards of sexuality are weakening. One of the young, single working women Rosado interviewed in a rural, migratory village of Michoacán discussed the issue of spousal sexual infidelity with her friends and her mother and then concluded: "I don't believe as my mother does that you must put up with your husband forever if he cheats and turns out bad" (Rosado, 1990, p. 67, translated by Hondagneu-Sotelo).

What happens once the women "left behind" migrate? Part of my research has looked at Mexican women who eventually migrate to the United States to join their husbands (Hondagneu-Sotelo, 1994). Although there are structural inducements to this pattern of migration, it is men's authority within families and men's access to migrant network resources that favors the husbands' initial departure. Yet, their departure rearranges gender relations in the family: As women assume new tasks and responsibilities, they learn to act more assertively and autonomously. This new sense of social power and, later for another cohort of migrant wives, additional access to women's network resources enable the wives to migrate.

Once the families are reunited in the United States, migration and resettlement processes elicit transformations in patriarchal gender relations. During the spousal separations, women often learned to act independently, and men, in some cases, learned to cook and wash dishes. In other instances, they began to concede to their wives' challenges to their authority. These behaviors are not readily discarded when the spouses reunite.

This egalitarianism is indicated by the emergence of a more egalitarian household division of labor and by shared decision-making power. With the diminution of patriarchal gender relations, women gained power and autonomy and men lost some of their authority and privilege. These gains and losses were reflected in the women's near-unanimous preference for permanent settlement in the United States and in the men's desire for return migration—a finding that echoes Pessar's (1986) data on Dominicans in New York City.

Migrations of political refugees are complex and varied; here I wish to discuss briefly the experiences of female refugees who have come to the United States fleeing Latin American political regimes of the "Left" (Cuban women) and of the "Right" (Chilean women). In Chile, the 1973 military coup and overthrow of the democratically elected, socialist president, and the repressive aftermath, forced many trade unionists, activists, and leftists to leave for

Europe, Mexico, Canada, and other nations, including the United States. Among working-class Chilean exile families in San Jose, California, Chilean women may have fared better than the men. The Chilean women's acquisition of new skills in employment, language, and driving, together with their income earned in the Silicon Valley electronics industry, improved the women's self-confidence and domestic authority. By contrast, the Chilean men lost much of the public status and life meaning that they had previously derived from political and trade-union activities.

Cuban women who came to Miami fleeing Fidel Castro's regime were primarily from the middle and upper classes. Familiar explanations for the Cubans' success in Miami emphasize the class privileges and human-capital resources that they brought with them and the politically motivated actions of the U.S. government in offering substantial resettlement assistance to those fleeing Communism. But Cuban women also played an important part in building Cuban economic success in the United States through their (Fernandez-Kelly and Garcia 1990) work in building family businesses. While the women momentarily gained in autonomy and resources, they "surrendered" these benefits to preserve traditional, patriarchal notions of manhood and womanhood. Family economic power ultimately strengthened their allegiance to family patriarchy.

Conclusion

More than a quarter of a century ago, Boserup (1970) called attention to women's role in economic development. Her findings indicated that geographical moves failed to improve women's employment and that gender inequality was further exacerbated in urban contexts. In the time since Boserup's study was published, Third World women have increased their participation in intranational and international migration. The reviews of African, Asian, Latin American, and Caribbean women's migration reveals mixed outcomes. On the one hand, migration may deepen women's subordination. Many Third World migrant women who go to cities or other nations find limited employment opportunities and are relegated to informal-sector work. They develop income-earning strategies, preparing and selling food and beverages on the street or assembling products in their homes while they look after their children. Yet, research conducted in a variety of locales also finds women who take greater control over the purse strings and who have greater say in family decision making than they did prior to migration.

Future research on migration and Third World women could be enhanced if it follows at least two new directions. First, investigations should adopt approaches that consider both macrostructural arrangements and living arrangements as they are actually experienced by migrant women on a daily basis. By following this route, research will continue to unveil gender and generational hierarchies and dynamics within households. As I suggested at the outset, this will involve conceptualizing men as gendered actors.

Another promising direction would be the conceptualization of migrant women acting within the "transnational community." As the mobility of both labor and capital redefines the meaning and effectiveness of nation-state entities, researchers are realizing that it is no longer viable to focus on only the sending or the receiving area of migration. International migrants create their own social field of interaction, which may be bidirectional. Although this "transnational community" approach is gaining in popularity, we still need gendered analyses built into migration studies. Achieving a truly gendered understanding of Third World women and migration will require weaving conceptual tapestries in which the multiple strands of migrant women's voices remain clear.

References

Boserup, Ester. *Woman's Role in Economic Development.* London: Allen and Unwin, 1970.

Boyd, Monica. "Family and Personal Networks in International Migration: Recent Developments and New Agendas," *International Migration Review,* vol. 23, 1989, pp. 638–670.

Brettel, Caroline B., and Rita James Simon (eds.). *International Migration: The Female Experience.* New Jersey: Rowman, 1986.

Cheng, Lucie Hirata. "Free, Indentured, Enslaved: Chinese Prostitutes in Nineteenth Century America," *Signs,* vol. 5, 1979, pp. 3–29.

Fernandez-Kelly, Maria Patricia. *For We Are Sold: I and My People.* Albany: State University of New York Press, 1983.

Fernandez-Kelly, Maria Patricia, and Anna Garcia. "Power Surrendered, Power Restored: The Politics of Work and Family Among Hispanic Garment Workers in California and Florida." In Louise Tilly and Patricia Gurin (eds.), *Women, Politics, and Change.* New York: Russell Sage, 1990, pp. 130–149.

Gabaccia, Donna (ed.). *Seeking Common Ground: Multidisciplinary Studies of Immigrant Women in the United States.* Westport: Praeger, 1992.

Glenn, Evelyn Nakano. *Issei, Nisei, Warbride.* Philadelphia: Temple University Press, 1986.

Grasmuck, Sherri, and Patricia R. Pessar. *Between Two Islands: Dominican International Migration.* Berkeley: University of California Press, 1991.

Hondagneu-Sotelo, Pierrette. *Gendered Transitions: Mexican Experiences of Immigration.* Berkeley: University of California Press, 1994.

Leeds, Anthony. "'Women in the Migratory Process': A Reductionist Outlook." *Anthropological Quarterly,* vol. 49, no. 1, January 1976, pp. 69–76. Special Issue: "Women and Migration."

Mummert, Gail. "Mujeres de Migrantes y Mujeres Migrantes de Michoacán: Nuevo Papeles Para Las Que Se Quedan y Las Que Se Van." In Thomas Calvo and Gustavo Lopez (eds.), *Movimientos de Población en el Occidente de Mexico.* Mexico City:

Centre d'Etudes Mexicaines et Centramericaines; Zamora: Colegio de Mexico, 1988, pp. 281–295.

Ong, Aihwa. *Spirits of Resistance and Capitalist Discipline: Factory Women in Malaysia.* New York: State University of New York Press, 1987.

Pedraza, Silvia. "Women and Migration: The Social Consequences of Gender," *Annual Review of Sociology,* vol. 17, 1991, pp. 303–325.

Pessar, Patricia. "The Role of Gender in Dominican Settlement in the United States." In June Nash and Helen Safa (eds.), *Women and Change in Latin America.* South Hadley, MA: Bergin and Garvey, 1986, pp. 273–294.

Repak, Terry. *Waiting on Washington.* Philadelphia: Temple University Press, 1994.

Rosado, Georgina. "De campesinas inmigrantes a obreras de la fresa en el valle de Zamora, Michoacán." In Gail Mummert (ed.), *Población y Trabajo en Contextos Regionales.* Zamora: Colegio de Michoacán, 1990, pp. 45–71.

Sassen-Koob, Saskia. "The New Labor Demand in Global Cities." In Michael P. Smith (ed.), *Cities in Transformation.* Beverly Hills: Sage, 1984, pp. 139–171.

Stichter, S. *Migrant Labourers.* Cambridge: Cambridge University Press, 1985.

Thadani, Veena N., and Michael P. Todaro. "Female Migration: A Conceptual Framework." In James T. Fawcett, Siew-Ean Khoo, and Peter C. Smith (eds.), *Women in the Cities of Asia: Migration and Urban Adaptation.* Boulder: Westview, 1984, pp. 36–59.

Tienda, Marta, and Karen Booth. "Gender, Migration, and Social Change," *International Sociology,* vol. 6, 1991, pp. 51–72.

Toro-Morn, M.I. "The Family and Work Experiences of Puerto Rican Women Migrants in Chicago." In H.I. McCubbin, E.A. Thompson, and J.E. Fromer (eds.), *Resiliency in Ethnic Minority Families,* vol. 1: *Native and Immigrant American Families.* Madison: University of Wisconsin Press, 1995, pp. 277–294.

Wilkinson, C. "Women, Migration, and Work in Lesotho." In J.H. Momsen and J. Townsend (eds.), *Geography of Gender in the Third World.* Albany: State University of New York Press, 1987.

Wolf, Diane L. "Linking Women's Labor with the Global Economy: Factory Workers and Their Families in Rural Java." In Kathryn Ward (ed.), *Women Workers and Global Restructuring.* Ithaca: ILR Press, Cornell University, 1990, pp. 25–47.

Marriage and the Family

Women in the City

Sônia Alves Caliô

Introduction

This essay explores women's perceptions of the city. It uses a feminist perspective to consider to what extent the city fulfills women's wishes and dreams. The city, the product of society and a strong reflection of its contrasts, reproduces on a larger scale the social discrimination experienced by women who suffer from the same time-and-space fragmentation in their own personal lives. The city structure affects woman's availability: the place where she lives and the place where she works are extremely important since these affect all of the areas she must travel. As the main administrator of the family's daily life, she is always running against the clock trying to conciliate the demands of her job, her housework, school, bills to be paid, and the occasional needs of family members in times of illness or other crisis. Her right to enjoy the city as a citizen is dependent on her social role.

Women and City Public Places

It is not the same for a woman to go into a public space and use it as it is for a man: Strolling in a park, going to the movies, visiting a bar, or just walking in the street requires self-determination and self-confidence, which for women are often in short supply. This means a woman cannot enjoy public areas as freely; she must generally restrict her use of these areas to perform conventionally assigned roles (Coutras and Fagnani, 1982). A solitary woman visiting a park is more likely to be approached by a man than a women (i.e., wife/mother) visiting a park with her children.

What does this mean? What should we think of a city that does not accept a woman in its various spaces, unless she plays the role(s) determined for her by society? How should we view public areas, places of collective use, the so-called outdoors as opposed to the "indoors" area? (Engeu and Save, 1974; Dumont and Franken, 1977).

What areas of the city are specified as appropriate for women? When a woman is not at work, she is in places such as the market, health centers, schools, playgrounds.

Women use public areas that are seen as an extension of the home (Pouchol and Severs, 1983; Caliô, 1991).

Today, many women are no longer at home cleaning and cooking all day—they go to the hospital, to school, to the bank, to shopping centers, to the dentist, to the doctor's office, and so on. These are public places in which women engage in the sexual division of labor necessary to maintain the social system. In this way the city functions as an extension of the home: women's use of public places is an extension of their housework.

Interesting in this light is an examination of the impact of technology upon city life. Electric and electronic appliances, cars, and other new products are present today in women's lives. For most women, this technology represents the possibility of working against the clock and finishing housework faster. By optimizing their working time, women have increased their productive capacity. Tasks that formerly required much of their time can now be performed faster, enabling women to do other activities (Fagnani and Coutras, 1977; Coutras, 1978; Coutras and Fagnani, 1982; Fagnani, 1982).

In the city, a woman who works outside the home is usually within the limits of women's work ghettos, an extension of her so-called natural activities. A married woman with or without children is also a wife, or a mother and wife, who works a double workday, working outside against the clock in order to have time enough to perform her daily household tasks. Does the single woman without children have more opportunities to enjoy her life in the city? Not always. Prohibitions, exclusions, and physical and moral violence are present in her daily life, making her feel afraid to be in the street. Is a woman who works at home more protected? Would she do better living within the boundaries of home and private life in this world of patriarchal laws that govern urban life? (Groupe Espace, 1977; "Revue Sorcières," 1978; Rogers, 1979; Guillot and Neyrand, 1985; Pelletier, 1987).

The City and the Consumer Woman

The city as a place where goods are manufactured and consumed welcomes the consumer woman. As a housewife, her work is not recognized by dominant society as part of social production. But when a woman is a consumer, the city welcomes her by continuously creating new needs and making her the target of its marketing strategies. Although the woman is frequently not allowed by society to make decisions about her body—regarding abortion, for instance—she is exhibited in publicity ads on walls and outdoor signs. She is consumed as a symbol in commercial transactions. Is this—as a sex symbol—the only way she can freely enjoy the city?

The City: Inhuman and Forbidding for Children

In the late twentieth century most areas of the city are highly valued and have become expensive commodities. In developing countries very few people can afford to possess a part of the city in which they live. City land is increasingly expensive and this further shapes the profitability of all of its areas. Green and leisure areas are seen by developers as generally unproductive. Their preservation consequently is made difficult by real-estate speculation. Such areas are also affected by traffic, and they are often slighted by the city public administration. Green and leisure areas are continuously thus disappearing, making the city less hospitable for children; this, in turn, affects women's lives since child care is their responsibility. What to do with children? Where can we take them? Where can we leave them when we go to work? In the streets? Locked up at home? How to protect them from the violence of the city? Fear and insecurity keep many women away from the street. Increasingly in developing countries, particularly in Latin America, many stay at home absorbed by television soap operas (Lima, 1989).

Yet it is in the city, where technical, economic, political, and intellectual innovations occur, that women show changes. Today, rather than working at home producing consumable goods, women are engaged in the provision of services in the public sphere because technology has changed and relieved the load of their work at home. Women are increasingly participating in the servicing processes required in the big cities. The city offers her a job (although her work is devalued), giving her the opportunity to leave the private area of her house, discover the working world, meet other people, and expand her space and social life. This is the hidden side of the city—a side traditionally forbidden to women but at the same time desired.

Citizenship and Women's Demands for Reclaiming Urban Spaces

Women and men experience problems in the city differently. Women's roles in the reproduction of the labor force and in their family make their presence fundamental and necessary in social fights aimed at improving city services and the standard of living. Women, especially poorer women, are strongly affected by poorly provided city services that significantly increase the difficulty of their responsibility to perform housework and participate in community life (Massolo and Ronner, 1984; Suremain, 1987; Ucles, 1987).

In the context of Latin America, most public policy proposals to date have not offered any serious alternative to improve the standard of living for women. As a move to counter discrimination against women such policies should use strategies that take into account women's unique characteristics and conditions, thereby integrating women as a fundamental group.

Citizenship for women requires government to examine in depth the specific issues and take measures to: develop special protection systems against violence, provide health policies, fight against job discrimination, create new opportunities in different areas of the economy, ensure educational and professional training, encourage income-generating activities by facilitating credit, create specific housing policies, and so forth. This means that government must make decisions to help fathers, mothers, and society to share responsibilities. This requires a fundamentally greater participation of the government in the socialization of housework (i.e., making these activities collective or available through the market place), which is basically in the hands of women today. Policies addressing housework would greatly facilitate women's management of daily life (Wekerle, 1985).

Working toward these goals are popular movements, many of which are headed specifically by women, to secure better housing, a better standard of living, day nurseries, nonsexist education, equal pay, health services, and occupational training. Women's police headquarters, refuge houses, legal centers, social support institutions, community groups, and cultural groups, as well as knowledge about sexuality, contraception, and abortion, are important resources in the struggle for women's right to citizenship.

The Consolidation of the City

During the past decades, the growing urbanization process of Latin American cities has reduced the already inadequate services in general and social services in particular. This has exacerbated efforts to receive the permanent influx of underprivileged people. Since part of this urbanization process has occurred through the occupation of areas where social services and social security are not available, to create the structures necessary to make these areas suitable for living has been the task of women (Feijóo, 1987).

Since women are involved in the family's daily survival, they emerge, in a way, playing the role of administrators of various aspects of urban life. They participate in important movements in the city consolidation process: fighting for housing, basic services, social services, a better environment. By bringing the family and neighbors together, and by encouraging relations of mutual solidarity, women participate in the process of community development and these activities often become an expansion of their housework. Thus women's needs become collective needs, and public policies that deal with these needs require women's participation in the construction or reconstruction of new communities or established neighborhoods. In this way, the lives of both the women and the city can be democratically led.

The present economic crisis in Latin America makes city poverty worse. The numerous "economic adjustment" policies made by the governments (which have reduced the role of the state in the provision of social services) have caused deterioration in the living conditions of the inhabitants and reduced the number of jobs available, the purchasing power, and public expenditures in basic services. Women must work very hard to face this seemingly permanent lack of services. (Feijóo, 1987).

There is another important issue to be examined here, one that concerns women's struggles for better environmental conditions (i.e., against pollution, poor sanitation, contaminated water, floods, etc.). The growing environmental deterioration of urban areas has contributed greatly to increasing difficulties in the daily lives of poor people and to the present moment, it has been women who have been in charge of this fundamental task.

The Democratization of Public Areas

Changes have occurred slowly in the democratization of urban public space. Some places that had been for the exclusive use of men, where men would gather and talk and where women were not accepted, began to be occupied by women, and they began to organize them differently, through sports, parties, dancing, and film show activities. Their design and function also began to change: the sports court became also a dancing salon, classrooms served also as sewing and painting workshops and meeting rooms. There were many discussions about collective and voluntary work in the community, such as digging ditches for laying water pipes, covering courtyards with roofs so that they could have multiple uses. Many places were transformed by those women who had not been accepted there before (Calió, 1992).

The presence of women in leisure areas where once there had been only men brought up fundamental women's issues: a woman is someone who needs her identity as a citizen, to have her survival guaranteed as a human being who has wishes, desires, and aspirations. As designers of their physical environment, women have changed these areas by discussing and elaborating on rules for the community, by learning to be responsive not only to other people but to themselves as well.

Developing a Plan from the Women's Point of View

The proposals presented below are the results both of discussions held within the women's movement throughout Brazil and of many other discussions that have taken place between women and administrators in local communities. Four basic issues emerged from the discussions: housing, security, social services, and income-generating activities (Calió, 1992).

Housing

Housing is part of a greater issue involving social inequalities and the standard of living. Thus, a housing shortage cannot be seen solely as the result of low supply and high demand but also concerns the distribution of services in general. The availability of housing is directly associated with the standard of living.

Women occupy all social classes, but those in the lowest social classes participate in the labor market earning very low salaries; since their conditions are worse than those of men in the same social stratum, women face serious problems with housing. This is the reason that most of the poorest women live miserably in slums (Feijóo, 1987).

Although houses in poverty make life difficult for every member of the family, it is especially difficult for women who are in charge of the management of the household. Living poorly means walking long distances to fetch water, putting the garbage out, taking children to school and to the doctor by long walks or time-consuming public transportation, etc. It also means infestations of rats and cockroaches, no drinking water, and diseases that will not receive professional treatment.

The discussions resulted in the following proposals:

To elaborate policies involving "women and housing" and to establish housing programs directed toward women supporting large families.

To reexamine housing financing policies, which traditionally require an income report and an income higher than the average salary of women. Current policies eliminate the possibility of housing for women who have lower incomes. Moreover, it is known that within housing programs there are many financing institutions that discriminate against women by saying that they are "high-risk clients."

To encourage and set up housing associations for developing projects with good prospects for obtaining financial support from international institutions. These associations should be involved in buying land for the community, organizing themselves with their own rules and with women in their administration, assuring property rights, and creating an architectural design appropriate to women's needs, with leisure and common areas for the community (particularly laundries, restaurants, and collective kitchens).

To ensure that housing programs consider women not only as "priority clients" but also as participants when decisions must be taken. Government technocrats and building designers will finally understand the differences between "house" and "home" by taking into account the needs of women, children, and old and sick people who are rarely present during planning.

To assure women's representation in housing programs through community associations.

To offer credit facilities both to build new

housing and to improve old, deteriorating houses.

To have owner-builder programs and collective voluntary work, with women trained in building techniques so they can participate not only with their time but also with their expertise.

To adopt measures that foster private initiative and provide houses for underprivileged people.

Security

The issue of security has been present in all the meetings with women. Children, people without cars, and old people need security. Women face additional safety threats and develop common fears. Always mentioned are fears of sexual violence and rape. Since lack of security is associated with "empty and dark places," in which no public activities take place, special emphasis has been given to revitalizing life in low-income communities (Castelain, 1982).

In cases where the community government has responded to the main demands of the women's movement by providing such facilities as a Women's Defense Police Headquarters, a 24-hour Social Aid Service at the Police Headquarters, and an Association for Women Victims of Violence, security was discussed as a far-reaching matter concerning the whole city and the following items were suggested for action:

a high taxation for abandoned lands where violence normally occurs;

a building code requiring that buildings position their facade towards the street so as to avoid the long blank walls where assaults normally occur;

a revitalization of community life by offering its inhabitants various possible functions—housing, culture, entertainment, shops, etc.—thus avoiding the total desertion of isolated residential areas at night;

an establishment of a public lighting plan, particularly in dangerous areas such as train stations, bus stops, parks, and public squares;

a systematic pruning of trees, especially along dark streets; and

a placement of police units in dangerous areas.

Social Services and Income-Generating Activities

The purpose of the proposed plan was to create a better distribution of services (day-care centers, health centers, leisure areas, etc.) in order to avoid the situation in which some communities offer many such services while others offer none. With this specific goal in mind, the following points were made during the discussions:

Since good social services (shops, education, health, leisure areas) offer a better daily life, they must be as close to women's houses as possible. This means "to interfere deliberately" in women's daily life, bringing relief to their burdensome housework, enabling them to leave home, and increasing their

time available to be involved in public issues.

Income-generating activities as well as social services specifically designed and managed by women should be encouraged by stimulating commercial and service activities (the so-called backyard work). For instance, associations and bazaars and small shops could help to sell different products (handicrafts, clothes, food, gardening supplies, stationery, etc.).

Some Suggested Social Services

Collective laundries. Based on public incentives (tax incentives, financing, technical training), these laundries would provide women with an opportunity to generate income. This suggestion would not entail the creation of an additional public service.

Restaurants and/or collective kitchens (similar to collective laundries).

Child-care facilities in public leisure areas (squares, parks, clubs, social centers).

Recreation equipment specially designed for children and appropriate to the country's climate, with adequate standards of security. Recreational toys made of iron or other metals can become extremely hot under the sun, and those made of certain types of wood can splinter and hurt the children; these materials should be avoided. Also to be avoided are toys that do not stimulate creativity.

Toilets and drinking fountains especially designed for children should be provided.

The recreational spaces in parks and squares should be built in areas easily supervised by adults. Normally, these areas are not designed with this in mind and parents have to run after their children all the time. If we take into account the fact that most Brazilian families have at least two children and that the recreation areas typically are not safe enough for them, we can imagine how "pleasant" a walk in such a park might be.

Good lighting systems should be installed in leisure areas so that they can be used in the evenings for various activities (parties, sports, etc.) and these places, formerly dangerous and threatening at night, could become a source of pleasure after sunset as well.

Concluding Comments

An urban environment responsive to needs that women have as a result of the various roles they presently play and, at the same time, responsive to the desire for flexibility, equality, and autonomy in their lives remains a dream under construction. This essay has identified some of the requisite components. It has dealt with broad issues; the precise fitting to particular cases and milieus must be a local invention.

References

Calió, Sônia A. "Relações de Género na Cidade: Uma contribução do pensamento feminista a Geografia

Urbana." Ph.D. diss., University of Sao Paulo, 1991.

———. *A Mulher e o Espaço Urbano.* Santo André: Assesoria dos Direitos da Mulher, Prefeitura Municipal de Santo Andre, 1992.

Castelain, C. "Insecurité Urbaine et Langage de Femmes." Proceedings of a national seminar on "Femmes, Feminisme et Recherches," Toulose, 1982, pp. 577–582.

Coutras, J. "Femmes et Transport en Milieu Urbain," *Revue Internationale de Recherche Urbaine et Regionale,* no. 3, 1978, pp. 432–439.

Coutras, J., and J. Fagnani. "Les Femmes dans Leurs Espaces Habituels en Milieu Urban," *Analyse de l'espace,* no. 4, 1982, pp. 1–11.

Dumont, M., and E. Franken. "Et si la Ville Etait à Nous Aussi?" *Cahiers du GRIF,* no. 19, 1977, pp. 44–62.

Engeu, C., and J. Save. *Structures Urbaines et Reclusions des Femmes.* Paris: Temps Modernes, nos. 333–334, 1974, pp. 1736–1750.

Fagnani, J. "Impact de la Motorisation sur le Mode de Vie des Femmes." Proceedings of a national seminar on "Femmes, Feminisme et Recherches," Toulouse, 1982, pp. 727–739.

Fagnani, J., and J. Coutras. "Transports et Femmes," *Cahiers du GRIF,* no. 19, 1977, pp. 99–107.

Feijóo, María del Carmen. "La Mujer y el Habitat Popular," *Boletín de Medio Ambiente y Urbanización,* no. 20, 1987, pp. 25–31.

Groupe Espace. "Espace des Femmes. Espace pour Femmes," *La Lune Rousse,* no. 2, 1977, pp. 4–8.

Guillot, C., and Neyrand, G. "Le Sexe de l'Espace," *Revue Espaces et Societés,* no. 46, 1985, pp. 55–69.

Lima, M.S. *A Cidade e a Criança.* Sao Paulo: Editorial Nobel, 1989.

Massolo, A., and L. Ronner. "La Participación de las Mujeres en los Movimientos Sociales Urbanos." In CEPAL (ed.), *La Mujer en el Sector Popular Urbano.* Santiago de Chile: Comision Economica para la America Latina, 1984.

Pelletier, L. "Au Sujet des Espaces Feminises," *Cahiers de Geographie du Québec,* vol. 31, no. 83, 1987, pp. 177–187.

Pouchol, M., and M. Severs. *Travail Domestique et Pouvoir Masculin.* Paris: Editorial du Cerf, 1983.

"Revue Sorcieres," *Espaces et Lieux,* no. 14, 1978.

Rogers, S. "Espace Masculin, Espace Feminin: Essaie sur la Différence," *Études Rurales,* no. 74, 1979, pp. 87–108.

Suremain, M.D. "Las Mujeres y las Energías Domésticas en los Barrios Populares de Bogotá," *Boletín de Medio Ambiente y Urbanización,* no. 18, 1987, pp. 109–117.

Ucles, M.L. "Las Nuevas Tecnologías Urbanas, Crisis y 'Nuevas Ciudades' en América Latina," *Boletín de Medio Ambiente y Urbanización,* no. 18, 1987, p. 2–3.

Wekerle, G. *La Planification Urbaine: Comment la Mettre au Services des Femmes.* York, Ontario: Department of Environment, University of York, 1985.

Allocation of Labor and Income in the Family

Gale Summerfield

Introduction

The image of the wife at home cooking, cleaning, and caring for the children while her husband ventures out to earn the income that brings food into the home is a two-dimensional stereotype that is remarkably resilient among people inclined to generalize about women's roles. Although women typically do prepare the food in families in developing countries, they also often grow it or earn it through activities such as small-scale production and trading (Boserup, 1970). While they do bear the children and provide most of the care for them, they also provide for the family by increasing their workload in the face of losses accompanying economic restructuring or environmental degradation. Items formerly bought, such as bread or clothing, are more often made at home, and vegetables are grown in tiny urban gardens instead of being purchased; as water and fuel become harder to obtain, the women of the household get up earlier and walk farther in search of what is needed. If there is a universal truth about the gender division of labor within the family, it is most likely that women work more hours than men, especially in developing countries, and that development policies have often placed more demands on women's time. For women's workload to be reduced, either there must be advances in appropriate technology and more appropriate development policies, or men must work harder.

Many types of families exist in the world today, from cooperative, where decisions are made democratically and work is shared, to authoritarian, where decisions are made by a single, head-of-household decision maker who may be an "altruistic" dictator or simply dictatorial (Becker, 1981; Sen, 1990). Female-headed households with no adult males present are increasing rapidly (Bruce, 1989). In recognition of this variety, the treatment of the family as a "black box" has been criticized in recent literature for ignoring interaction and allocation among the individual members (Sen, 1990; Ferber and Nelson, 1993; Blumberg, 1994; Folbre, 1994). Sen characterizes the family as a "cooperative-conflict" in which individual members may care very much for

one another and work toward common goals but simultaneously may have conflicting interests, especially over the division of work and income.

This essay examines the variation in the division of labor and income within the family in developing countries. In the first section, the presentation focuses on whether a family is more appropriately treated (by researchers and policymakers) as a fully cooperative unit, as supported by Becker's work (1981), or as a cooperative-conflict, as supported by the work of Sen (1990). Becker's theories, presented briefly in the first section, have had a substantial impact on development policies and provide a defense for those who claim that giving credit, land title, or other support to women will undermine family values (since the family is viewed as fully cooperative). The argument is persuasive because families do cooperate and many people strive to ensure the stability of their families. Development policies that have taken this view, however, have often failed or had unanticipated and inequitable impacts; a few examples of these problematic policies are presented. The first section next examines the areas of conflicting interests, followed by a discussion of the appropriateness of considering both cooperation and conflict in families. The second section provides an illustrative case of how intrafamily division of labor and income are changing in response to economic reform in a village in China. Then, drawing on the model of family as a cooperative-conflict and the case study, factors that influence allocation of labor and income within the family are identified. The final section presents conclusions and addresses policy implications of the analysis.

The Model of the Family

The Fully Cooperative Family in Theory

Despite the growing incidence of women-headed households without a husband's contribution, the majority of people in developing countries live in families with at least two adults present for at least part of their life cycle, and many view the institution of the family with husband and wife as the core as something that needs protecting in the face of change. The cooperative characteristics of families have been well documented (Blau and Ferber, 1986). People marry for many reasons—love, security, power; one reason is economic. Families, even in the city, consume and produce; in addition to subsistence-oriented production such as vegetable gardening, they produce "so-called z-goods, that is, commodities produced by combining market-purchased goods with labor—child care is an example" (Haddad et al., 1994, p. 12).

Becker (1965, 1981) was one of the first well-known economists to examine the formation of households and interaction within the household, and his theories have had a strong impact on development policies. (Other economists such as Margaret G. Reid had written earlier on topics about households but had not received much attention; see Blau and Ferber, 1986, p. 37.) Becker states that households form to produce goods, such as "children," that are not obtained through market transactions and to gain from the comparative advantages of the individual members in producing other goods more cheaply than they could be acquired elsewhere. "Though some of these goods could be produced by the market, the ability of spouses to monitor each other's behavior, and their ability to use loyalty to obtain certain ends, minimizes transaction costs" (Haddad et al., 1994, p. 10).

The gains from specialization provide potential improvements in well-being for all members (Blau and Ferber, 1986). Men may be relatively better at earning an income to support the family (partly because better jobs are open to them in the framework of that particular society), and women may be relatively better at tending to the needs of reproduction (again, much of the difference is due to the institutional and cultural setting). By taking advantage of the relative opportunity costs through total or partial specialization, the couple can increase the total prosperity or advantages of the household.

In addition, other economic reasons, such as economies of scale, public goods, and externalities, may give people incentives to form households (Blau and Ferber, 1986, pp. 45–46). Housing and food tend to be cheaper for families than for people living on their own, reflecting the scale economies in these areas; public goods, such as watching a television show or enjoying a pleasant room in the house, are items that are not diminished by one spouse's consumption (wealthier families are more likely to have the goods mentioned in this example, but some very poor people now have televisions as can be observed in the slums of Bangkok); adults may get more enjoyment from watching their children grow up together than they would separately, and they may gain pleasure from the success of the other spouse, reflecting externalities.

Although Becker examines some sources of conflict within the household, such as the behavior of spoiled, "rotten kids," he determines that such conflicts are avoided or resolved by the ability of the altruistic benefactor of the household to impose his or her preferences through the implicit threat of withholding resources. Under the assumptions of this model, the altruist's preferences become the preferences of the household, but the altruist must control more resources than any other member or group of members of the household (usually a characteristic of the male head of household). Family decisions are treated as if made by a single unit regardless of whether that unit reflects unanimous agreement among the members of the family, or the decree of the altruistic dictator (a term that has never been welcomed by feminist economists).

Cooperation in Practice: Problems with Development Policies

Development policies have often failed to achieve their goals of improving well-being of both women and men because of the limitations imposed by viewing the household as a completely cooperative unit. In addition to increasing women's workload, as occurred with structural-adjustment programs referred to above, policies that overlook the conflicting interests of family members often create incentives for conflict within the family. In Cameroon, for example, women were reluctant to work in the rice fields controlled by their husbands. Although their work could increase the return to the household, their husbands would receive and control the income. The wives preferred to work on their own plots growing sorghum. When projects try to increase production of cash crops for exports, the earnings are typically controlled by the husbands, who sometimes resort to beating their wives to get more help with the crops (Haddad et al., 1994, pp. 37–38). The prevalence of domestic violence is sufficient to warrant a model that can accommodate noncooperative behavior in families.

Many other unsuccessful policies in Africa incorrectly assumed that labor and income were pooled within the family (Whitehead, 1990). In Sri Lanka, the Mahaweli irrigation project resulted in lower food consumption by poor households related to differential gender control of cash crops; a project for female fish smokers in Guinea suffered from overlooking the roles of males as well as females (both examples reported in Jaquette, 1993, pp. 50–51).

Policies designed to improve the nutrition and education of children are especially likely to be affected by which adult in the family controls a transfer; numerous studies have shown that women more consistently spend money they control on these items (Blumberg, 1994). Even so, both mothers and fathers may agree on decisions that give preferential treatment to sons: Educational fees that are instituted as part of structural adjustment often mean that daughters are kept home from school; new opportunities for families to earn income often lead to demands that the daughters stay home to earn money rather than "waste" their time on education. The closer financial ties that grown sons keep with their families in many societies make the education of boys a preferable investment for their families (Summerfield, 1994).

Perceptions of Individual Interests

Conflicting interests, such as the maximization of current earnings for the family by keeping a daughter out of school despite the long-run cost to her, can exist within a family, and yet those bearing the costs may not perceive their personal loss. Some women, especially in developing countries, cannot even respond to questions about their personal interests and needs versus those of the larger family unit; they perceive their own interests as fully congruent with that of the rest of the family (Sen, 1990; interviews). Even when traditions are physically harmful to them or their daughters, mothers frequently take an active role in perpetuating the traditions. Extreme examples are the former practice of foot-binding in China and the genital mutilation in some African countries or among immigrants from these countries living in other areas; women are often instrumental in carrying out the painful rituals, which can inflict permanent damage to young girls, to assure that their daughters will marry and have a family (Papanek, 1990). The perception of what constitutes well-being is strongly influenced by opportunities available in the context of the society and may differ from an outside view of the elements of well-being.

> Insofar as intrafamily divisions involve significant inequalities in the allotment of food, medical attention, health care, and the like (often unfavorable to the well-being—even survival—of women), the lack of perception of personal interest combined with a great concern for family welfare is, of course, just the kind of attitude that helps to sustain the traditional inequalities. . . . It can be a serious error to take the absence of protests and questioning of inequality as evidence of the absence of that inequality (or of the nonviability of that question) (Sen, 1990, p. 126).

This discussion does not imply that the solution to family inequities must be the dissolution of the family. On the contrary, the above presentation has focused on some of the gains women receive from living in families, and women in developing countries often express their concern about the effect of economic reform in destabilizing the family. This section, however, provides a caveat that gains cannot be assessed merely by interviewing the members of families because perceptions based on social context can mask serious forms of inequality. Addressing this inequality through development policies and community groups may actually strengthen the family, especially in the process of economic transformation, rather than weaken the institution (Jaquette, 1993, p. 47).

Conflicts over the Allocation of Labor and Income

Because women are disadvantaged in society, the assumption of the total equality of decision making within the family is not reasonable. If there is at least one other adult present (for example, a legal or common-law spouse or an adult sibling), the probability of disagreement over major decisions exists. Wife and husband may have conflict over decisions about how to spend their income but still unite in giving preferential treatment to sons, such as sending them to school or feeding them better than the daughters. The mother may comply with a decision that hurts her daughter even though she does not agree because she does not have the means of altering the decision. Domestic violence is far too common for a fully cooperative model to be more than a special case.

It is easy to document examples of violence against women. In India, hundreds of men every year throw kerosene on their wives and burn them in a vicious effort to extract more dowry payments from the women's parents; at times, the mother-in-law participates in this extortion. In both India and China, millions of women are "missing" because they have been denied medical care and adequate nutrition (especially common forms of neglect of young girls), because they have been abandoned or killed at birth, and increasingly because parents choose to use ultrasound to identify the sex of the fetus and abort those that are female. In Cambodia, a generation brutalized by arbitrary and immediate executions finds wife-beating rampant. These extreme examples from Asia are indications of a global problem.

The institutional framework of society reinforces traditional forms of discrimination against women in most countries. Frequently, laws guarantee the husband the right to make decisions for the family. Even in matrilineal societies, women are often subordinate to their brothers; although property is handed down through the mother's side of the family, the arrangements do not give women enough power for the system to be called "matriarchal." Relatively little of the work done by women goes through the market; because most of their hours are spent on "unpaid" work, their efforts are often subjectively evaluated by other family members as worth less than the explicit wage contribution of the husband (their own perceptions may reinforce this low valuation).

The unequal opportunities in society are reflected in unequal bargaining power within the family about who is going to do the work and how income will be divided. Throughout the world, women generally work longer hours than men; "the widest gaps are in Africa and in Asia and the Pacific, where studies show that women average 12 to 13 hours more work a week than men" (United Nations, 1991, p. 82). A woman in the Philippines obtains a small loan to open a stall selling local foods; she gets up at 4 A.M. to prepare breakfast for her family, buy the necessary inputs, and prepare the food for the day's sales. After cleaning up and feeding her family again in the evening, she finally goes to bed at midnight. A rural woman in Kenya awakes before the rest of the family to make the first long walk of the day to fetch water. She then prepares breakfast and wakes the other family members. Throughout the day, her work continues. She usually spends several hours on trips to get water. Even then, the water is often contaminated, but there is no clean source available to her. Her eight children whine that they are thirsty, but she tells them to wait. At last, she closes her eyes long after other family members have fallen asleep (World Bank, 1989). For these women, sleep is among the scarcest of resources; their long days attest to their lack of bargaining power and limited alternatives in allocating the heavy workload they face.

Although much of the mainstream approach to the advantages of forming a cooperative household rests on gains from specialization according to lower opportunity costs, this specialization can also result in disadvantages and conflict. Women are assumed to be relatively more efficient in housework, but, regardless of whether this is true, the relative costs change if a woman finds a job outside the home. As the costs change, husbands frequently refuse to help more with the housework because it is "women's work" (Blau and Ferber, 1986, pp. 46–47). Conflict can easily arise over the division of labor in the household. Even if working outside the home does not necessarily translate into lighter workloads within the household for women, it does often give them more control over income, which can increase their bargaining power within the household and allow them more choice in buying goods for the household. A woman in Mexico City, for example, reports that her husband only buys shoes or clothes for the children when he feels like it, not when they are needed. She has to buy them from her work (Benería and Roldán, 1987, p. 122). Consistently, studies show that women tend to spend more on the nutritional and educational needs of their children and other family expenses (Benería and Roldán, 1987; Senauer et al., 1988; Haddad et al., 1994).

Cooperative Conflict

Despite the conflicts in families, most people in developing countries support the concept of family and resist threats to its integrity. Therefore, the relationships within the family are appropriately characterized by the combination of cooperation and conflict (Sen, 1990). The family works cooperatively to increase total advantages but it experiences conflict in how to allocate labor, leisure, opportunity, and income among family members. At times, total advantages are also affected by the eventual distribution within the family as illustrated by the example from Cameroon, referred to above, in which women resisted working to increase total production because they did not expect to share in its benefits.

The variation in family arrangements is large. Pooling of income is least common in sub-Saharan Africa, but many spouses in families that superficially appear to pool income resort to different degrees to hiding some of their earnings from the other spouse, a behavior that reinforces the view that families that do not completely pool their income constitute the majority in the world. The average contribution by other family members to the pool differs even in the countries that are geographically close. In Taiwan, it is common for an unmarried daughter who works for an export-processing plant to remit 70–80 percent of her earnings to her family, but, in Indonesia, the daughter usually retains her earnings unless there is a crisis in the family (Bruce, 1989; Wolf, 1990).

A matrilineal tradition, as observed in Laos and northeast Thailand, usually carries more of a partnership role for the woman in marriage than the strongly patrilineal traditions of nearby countries such as China. Women's right to inherit property clearly reduces the incidence of female infanticide or neglect of young girls observed in China, India, Bangladesh, and Pakistan. Seclusion in countries with Islamic traditions has reinforced the strong division of labor in which women are responsible for essentially all of the housework and child care, with limited means of earning an income. Because men can easily divorce their wives in this system, some countries such as Iran have traditions of both dowry and bride price, where the bride price is agreed upon at marriage but paid only if the woman is divorced.

The intrafamily bargaining process is dynamic. Regardless of the extent of potential gains from intrahousehold specialization and trade, the prospects change as family members come and go and as more opportunities open up to women outside the home. When a woman can get an education and work at a job that pays well, the cost of staying home increases.

Variation in Sex Role Division of Labor and Income: A Case Study

A case study of a village in rural China is presented here because the rapid pace of economic reform in China can be expected to have widespread effects on the traditional division of labor in the countryside. Since the economic reforms began in the late 1970s, Dongyao Village in north China has changed in many ways, and the division of labor and income within the family has reflected these changes (for a more comprehensive analysis of the impact of the Chinese reforms on women, see Summerfield, 1994). In 1992, the population of the village was 1,014, with a labor force of 510 composed equally of women and men. Annual per-capita income had increased from 45 yuan before the reforms to 1,100 yuan (Li, 1993, p. 9). Although nonfarm opportunities had opened up for men and unmarried women, most middle-age married women engaged in agricultural work on the family plot (agriculture was de-collectivized in the early 1980s). The average age of women classified as farmers or housewife/farmers was 38; their average education level was 4.46 years, and they worked 14.3 hours per day on average—longer than other groups (Li, 1993, p. 14).

Traditionally in rural China, men did the skilled work and women the unskilled, but the definitions of what is considered unskilled are changing as men move into industrial work and women take over the tasks formerly performed by men. The tasks associated with the more skilled work in agriculture are increasingly considered "unskilled" as they become women's responsibility (Li, 1993, p. 39). The government's exhortations for men to share housework have had less impact in the countryside than in the city, and women still do most of the domestic work. While husbands may have time for a game of *mahjongg* with friends, wives rarely do (Li, 1993, p. 14).

Women did most of the housework before the reforms and still do; their unchanged duties include cooking, washing, cleaning, child care, care for in-laws, and feeding the livestock (Li, 1993, p. 38). Some tasks that only men performed before are now done by either men or women; these include storing grain and vegetables, processing food, trading produce in state or farmer markets, and building walls of houses using bricks. Water carrying was formerly men's chore, but the success of the reforms has permitted families to have a pump well in their yards and so eliminated this task (Li, 1993, p. 39). Most families have bought appliances such as televisions, washing machines, and even a few VCRs; these new treasures are operated mainly by the men. These changes mean that the share of household labor provided by women has increased as the opportunity to earn more income from the family farm has increased and, simultaneously, more opportunities in rural industry have opened up for men and young women.

The increased labor does not necessarily mean that women are worse off; indeed, they are responding to incentives that enable them to earn more income and retain their earnings for the family's use. At this stage, women face a heavier burden than may be expected in the future since some of the chores now performed within the family will be offered by the market, as they are in the city—for example, ready-made clothes and canned food. Changing to purchasing these commodities instead of relying on women's unpaid work, however, will require a significant shift in attitudes and decision-making power within the family.

When asked about who makes decisions in the family, most villagers indicated that the husband did, but women are taking an increasingly active role in decision making (Li, 1993, pp. 32–34). Most young women still follow the tradition of relocating to their new husband's village and moving in with his extended family; in such circumstances, the father-in-law is typically the head of the family and orders the others about. Nuclear families, however, are becoming more common, and, in these, the husband and the wife both make decisions even though the husband is viewed as the head of the household and has more authority. Even in the extended family, there are varying degrees of participation by the members of the household, but the tendency for the patriarch to make crucial decisions has not been effectively challenged. Traditional attitudes about women's inferiority linger, and women are prohibited from even watching the digging of wells or the setting of the main beams in a house for fear they will bring bad luck (Li, 1993, p. 41). With such forms of bias against women, they cannot be expected to have equal bargaining power in making household decisions. Attitudes are changing, however, as evidenced by one family that proudly announced spending one-third of its income on sending the daughter to college (Li, 1993, p. 45).

The families in this village in China demonstrate aspects of cooperation (they voluntarily divide labor to maximize family income by having the husbands work in the new industries while the wives take over agricultural duties) and of conflict (the wife does the least-skilled jobs, doesn't get to operate the new appliances, and has less leisure than the husband). The women in the village tend to work longer hours on a daily basis because, in addition to income-generating activities, they do most of the housework and take care of children and in-laws. And the division of labor and income within the family is in a process of change as women take over the skilled agricultural jobs formerly done by men, as young women get more opportunities in industry, and as daughters receive more education.

Factors That Influence the Division of Labor and Income Within the Family

A single case study is useful in providing an empirical example of the division of labor and income, but a broader perspective is required to illustrate which aspects of the case are representative. This section utilizes the model of the family as cooperative-conflict and draws on the Chinese case and other studies from developing countries around the world to identify factors that influence the division of labor and income within households or families. Some of the factors are related more to how families are organized and how they can gain from specialization or scale, such as the size of the household and whether it is located in rural or urban areas; some more strongly affect the relative bargaining power of the members, such as the impact of government policies for structural adjustment, for controlling fertility, or for equality of employment opportunity.

Size of Household

The number of people living in the household can be expected to affect the gains from specialization and economies of scale. There is no single pattern of family or household that accurately describes the living situations of women in developing countries. All families change over time: In addition to the models of the extended family, the nuclear family, and the increasingly common female-headed household, a grandparent may make an extended visit of more than a year, a cousin may live with the family to attend a nearby school, a child may go off to college, another may migrate to a bigger city but still send home remittances, or a friend may rent a room from the family (Bruce, 1989; Tinker, 1993). Smaller families are becoming more common, especially in the cities. When there are more people, the potential for economies of scale and specialization in certain tasks increases, but there are also costs to coordinating the different interest groups that emerge. Apparently, as income increases, people are choosing increased privacy over the gains associated with specialization and scale, and the gains from these factors are smaller in cities.

These family and household patterns do not dictate a particular type of intervention on the part of the government or development groups. They are a response to economic conditions, and are based on culture and tradition. Indeed, the doubling up of families in apartments in Latin American cities is attributed to the austere conditions associated with structural adjustment (Montecinos, 1994). While it does not appear to be part of the government's or nongovernmental organizations' sphere to legislate or promote size of families (except for population policies regarding fertility), an awareness of the types of living arrangements can be expected to have an impact on the effectiveness of policies. Female-headed households, for example, are more likely to be poor, but women living in male-headed households can also be subject to the "international feminization of poverty" if their opportunities in society are reduced or if they become de facto heads of household because their husbands migrate to other areas for long periods of time (Summerfield and Aslanbeigui, 1992).

Region of Residence

A second obvious factor influencing the division of labor in the household is whether the members earn their income mainly through agriculture, industry, or services. In the countryside, the family is more likely to produce commodities than purchase the items through a market, and the women are more likely to be the ones producing the items. As people in developing countries continue migrating to the city in large numbers, more commodities are purchased rather than made; even if the husband does not help with the housework, this change potentially could reduce the workload of the woman. Bargaining power again enters into how any gain in leisure time is allocated within the family. In the transition to a market economy in Poland, women needed to spend less time standing in lines to buy food and other commodities. This potential gain in leisure was quickly usurped by their husbands, however, who reduced the amount of time they spent helping with housework because the wives no longer "needed" it (Leven, 1994).

In addition to providing gains, rapid urbanization has created numerous problems for both genders associated with urban slums and pollution. These problems resulted in the emphasis on rural development among specialists since the 1970s (some say over-emphasis; see Tinker, 1993, p. 65). People are choosing to migrate to the cities for a variety of reasons, and, for the most part, countries have treated this move as an individual decision rather than make laws to restrain it. Some policies, however, distort individual decisions, such as the urban bias in subsidies on food and rent. There are many contradictory impacts of urbanization on women, and the impact on the division of labor or income within the household is just one aspect of a more complex whole.

When a policy addressing an issue involves an increase in family labor, the allocation of that labor clearly plays a role. For example, one focus of environmental policy in the latter twentieth century has been reducing the use of fuelwood in rural areas of developing countries. The promotion of some types of energy-efficient stoves designed to reduce the need for fuelwood ignored the increased time demands on women for collecting the fuel for the new stoves. These women already worked long days with no leisure. The result, of course, was that the stove projects were often unsuccessful because time is a scarce resource (Tinker, 1987).

Opportunities Outside the Household

A critical area for policy and program intervention to facilitate improving women's bargaining power over the allocation of scarce resources within the family involves the impact of opportunities outside the home on what happens within it. The ability to get an education or find employment, for instance, exerts a strong influence on a woman's bargaining power within the family.

Amount and Source of Income. In making decisions about spending the family's earnings, the source as well as the amount of income matters. Higher income typically indicates an improvement in well-being for family members; they can all benefit from having a new house or appliances, and wealthier families will usually provide better for their children's nutrition and education. Although a woman's well-being may improve as her husband's earnings increase and she is able to share consumption of more goods, she generally has more control over the spending of her own income. Earning wages in her own name also is expected to increase her bargaining power within the household, partly because it provides clear evidence of her contribution to the family's well-being and partly because it improves her "breakdown position"—what will happen if all attempts at cooperation fails (Sen, 1990; Summerfield, 1994). Since women tend to spend a larger share of their income on the children and family than men do (Senauer et al., 1988; Bruce, 1989), poor families benefit when the mother has earnings to allocate to improving the health of her children.

Peers reinforce the inclination of women to spend on their children by rewarding them with the label "good mother." Fathers, on the other hand, are often expected by their peers to partake more in the life of the community, ranging from frequenting coffeehouses and movies to gambling and going to bars; these activities are often combined with business opportunities and are part of the male networking apparatus; in other cases, they are straightforward means of increasing the man's well-being without regard for, or even at the expense of, other family members.

Although studies show a positive impact from the mother's earnings on children's well-being, men may reduce their contribution to the family as women's income goes up. Some women, however, respond to this potential loss by hiding part of their income from their husbands and thus keeping the husbands' contributions higher. Observation of such behavior strengthens the view of the family as a unit containing both cooperation and conflict.

Existence of Alternatives. Even when women are full-time homemakers, the existence of opportunities outside the home has an impact on their bargaining power within the family. The availability of alternatives to complete dependency on her husband strengthens the woman's breakdown position if she is dissatisfied with conditions at home. These alternatives can also affect the perception of the wife's contribution to a marriage; when women have the possibility of earning money, they are often perceived to be more valuable even though their unpaid work has also contributed to the family's well-being (Blumberg, 1994). They may also perceive themselves to be worth more, and this change in attitude affects their bargaining power as well (Sen, 1990).

Government Policies

Although government officials often claim that what goes on within the family is outside the scope of their control, government policies frequently do have direct and indirect impacts on the family. Most governments have policies about population intended to alter family decisions; in addition, they have laws about sex, domestic violence, and property ownership (not all of the laws are enforced with equal commitment, of course). Some governments still give the husband full right to control all of the wife's earnings, to be the sole owner of property, and to make all decisions for the family.

Structural Adjustment. Government policies that are not directly aimed at women can have a tremendous impact on them. Structural-adjustment programs (SAPs) that began in the late 1970s in response to the debt crisis have had strong effects on women, especially in Latin American and African countries. As governments privatized national industry, cut payrolls, liberalized trade, and reduced government spending, these countries lost more than a decade of growth and investment. In the 1990s, some of them were again seeing small positive growth rates, but the human costs of the restructuring process have been high. In the face of male unemployment and falling wages for those employed in Latin America, women have entered the labor market in increasing numbers. Simultaneously, the demand for women workers increased because trade-liberalization policies favored export-processing industries (such as garment, textiles, and electronics) that traditionally employed a high proportion of women. These jobs require few skills and pay the minimum wage or below. As women entered the labor force in Latin America, men did not rush to help at home. Women's outside work may have prevented more serious loss from the adjustment process, but it did not lead to a sense of increased well-being in countries in the throes of recession. The women's workday lengthened without any obvious improvement in their circumstances of living.

Not only are more women working outside the home, women's work at home has intensified. Researchers have documented how women in Mexico and Chile filled the slack of economies undergoing similar reforms to prevent the family's level of consumption from falling too low (Benería, 1992; Montecinos, 1994). They organized soup kitchens, grew vegetables in pots or tiny gardens, and produced goods at home that would otherwise have been purchased.

Family Planning. Most governments have policies aimed at slowing the growth rate of their population or at least providing access to birth control methods. The one-child policy in China since 1979 has achieved notoriety for its severity, but it has reduced the rate of population growth; the fertility rate, however, had already been cut in half between 1950 and 1978 (falling from six to three children per woman) with much less coercive methods that included an emphasis on increased employment and basic education of women (Banister, 1991). Bangladesh has allocated significant resources to a successful program focused on information about, and availability of, birth control methods. Kerala, a large state in India, has achieved similar success with policies aimed at education and employment of women instead of birth control methods. The choice between policies stressing birth control or employment and education indicates a decision about the allocation of scarce resources; when a government puts funds into birth control campaigns, it usually reduces funds available to other areas. Providing opportunities for women may be the most effective and least coercive method of slowing the rate of population growth in a country (Sen, 1994).

Government policies, whether directed at women or not, typically have differential impacts on them as illustrated by the structural-adjustment programs and family planning policies discussed above. Because their policies often have negative impacts on women, government policymakers have a responsibility to address these problems. Some responses are more appropriate than others, however. The Chinese government has made the use of ultrasound to detect the sex of the fetus illegal because so many families were choosing to abort female fetuses; this approach could achieve some small success, but it does not address the heart of the matter that creates strong incentives for son preference. People just find ways to get around the laws and continue to use the ultrasound tests for this purpose.

In the late 1980s, the World Bank responded to a degree to criticism that its structural-adjustment programs were detrimental to the health of women and children because the provisions to these groups were the first to be cut by government officials: It began to publish analyses addressing the need for a social safety net during adjustment. Critics have charged that actions have not been forthcoming (Turshen, 1994); still, the commitment to include social conditions among the requirements for loans reduces the tendency of officials implementing adjustment programs to impose most costs on those least able to resist.

Women's Solidarity Groups

Feminist movements and other forms of community support groups can also have an impact on decision making within the household. Women's groups can help women get access to resources from the community and the government, learn how to read and write, and understand basic nutrition that enables them to improve the health of their children. Even government-sponsored groups such as the

Women's Union in Vietnam and the Women's Federation in China play a role beyond popularizing government policies: They can organize projects to relieve poverty and help women find employment, among other things. The Tianjin branch of the Chinese association, for instance, has helped middle-age women who are laid off from their jobs develop small businesses and market handicrafts.

In Kenya, an extensive government-assisted network of more than 16,000 community groups aids women with goals of increasing women's income, promoting saving, and improving health (World Bank, 1989). In Mexico, *Los Niños* aids women in setting up urban gardens and in studying about nutrition. In Bangladesh, credit programs, such as the Grameen Bank and the Women's Entrepreneurship Development Program, have helped women contribute to the family's well-being, and women's groups, with the aid of NGOs, have helped women set up ventures in joint farming, pond fisheries, rice mills, brick making, export-oriented garments, and frozen-fish processing (World Bank, 1990, p. xix).

Thus . . . there is an interplay between the familial and extrafamilial: "The scene of women's advancement seems to be the household and . . . the household's perception and evaluation of women's role, its hierarchy, its monetary and nonmonetary sources of power." But key in changing the dynamics within the household are extrafamilial experiences which permit women an opportunity to see themselves differently, to become discomforted with their subordinated status, and empowered to confront and transform the aspects of family and income relations that oppress them (Bruce, 1989, p. 988).

Conclusions and Policy Implications

Three key conclusions about the division of labor and income within the household or family can be drawn from the above analysis: (1) the family is most accurately represented as a cooperative-conflict in which some actions are performed cooperatively to the benefit of all members, and others involve conflict and trade-offs among members (fully cooperative families are a special subcategory of the collective model); (2) women have fewer opportunities in society, and this is reflected in unequal bargaining power in the home—the result is that women tend to work more hours than men in countries throughout the world; and (3) the division of labor and income within the family is a process that changes over time, not something set at the beginning of a marriage that then remains stationary. This division is influenced by societal changes, such as types of employment open or closed to women, as well as by factors within the family.

Government policies, projects of international development agencies, and NGO activities are all more likely to achieve their goals with fewer unexpected negative impacts on women and girls if those involved in the design and implementation of the programs (optimally including some of those affected by the programs) keep in mind the dual aspects of families as cooperative-conflicts. Program designers can search for ways to support families and still be attuned to the needs of individual members; the data suggest the efficacy of aiming development policies at women in addition to men in the household. The policy aspect is especially relevant in the mid-1990s, with the economic success of the East Asian countries. Playing on this success, in 1994 the leadership of Singapore announced that the "Asian family values" of its development model included targeting all development programs to male heads of households (Hoong, 1994, p. 14). The approach suggested in this essay contrasts with the Singapore model by considering aspects of conflict as well as cooperation in families and by recognizing that the recipient of a transfer often influences how the transfer is used. Especially in light of the empirical research that demonstrates the effectiveness of channeling finances through the mother in order to improve the health, education, and status of the children, policymakers need to consider carefully the focus of development strategies and programs and take advantage of more variety in implementation options in efforts to achieve their goals.

References

Banister, Judith. "China: Population Changes and the Economy." In Joint Economic Committee, *China's Economic Dilemmas in the 1990s*. Washington, D.C.: United States Government Printing Office 1991, pp. 234–251.

Becker, Gary. "A Theory of the Allocation of Time," *Economic Journal,* vol. 75, no. 299, September 1965, pp. 493–517.

———. *A Treatise on the Family.* Cambridge: Harvard University Press, 1981.

Benería, Lourdes. "The Mexican Debt Crisis: Restructuring the Economy and the Household." In Lourdes Benería and Shelley Feldman (eds.), *Unequal Burden: Economic Crises, Persistent Poverty, and Women's Work.* Boulder: Westview, 1992, pp. 83–104.

Benería, Lourdes, and Martha Roldán. *The Crossroads of Class and Gender: Industrial Homework, Subcontracting, and Household Dynamics in Mexico City.* Chicago: University of Chicago Press, 1987.

Blau, Francine, and Marianne Ferber. *The Economics of Women, Men, and Work.* Englewood Cliffs: Prentice-Hall, 1986.

Blumberg, Rae Lesser. "The Intersection of Family, Gender, and Economy in the Third World," *INSTRAW News,* no. 20, first semester 1994, pp. 36–44.

Boserup, Ester. *Woman's Role in Economic Development.* New York: St. Martin's, 1970.

Bruce, Judith. "Homes Divided," *World Development,* vol. 17, no. 7, July 1989, pp. 979–991.

Ferber, Marianne, and Julie Nelson (eds.). *Beyond Economic Man: Feminist Theory and Economics.* Chicago: University of Chicago Press, 1993.

Folbre, Nancy. "Children As Public Goods," *American Economic Review,* vol. 84, no. 2, May 1994, pp. 86–90.

Haddad, Lawrence, John Hoddinott, and Harold Alderman. "Intrahousehold Resource Allocation: An Overview," policy research working paper no. 1255, World Bank, Washington, D.C., 1994.

Hoong, Chua Mui. "The Asian Family," *Strait Times* (Singapore), 1994, p. 14.

Jaquette, Jane S. "The Family As a Development Issue." In Gay Young, Vidyamali Samarasinghe, and Ken Kusterer (eds.), *Women at the Center: Development Issues and Practices for the 1990s.* West Hartford: Kumarian, 1993, pp. 45–62.

Leven, Bozena. "The Status of Women and Poland's Transition to a Market Economy." In Nahid Aslanbeigui, Steven Pressman, and Gale Summerfield (eds.), *Women in the Age of Economic Transformation: Gender Impact of Reforms in Post-Socialist and Developing Countries.* London: Routledge, 1994, pp. 27–42.

Li, Bongmin. "Changes in the Role of Rural Women Under the Household Responsibility System: A Case Study of the Impact of Agrarian Reform and Rural Industrialization in Dongyao Village, Hebei Province, North China," research paper no. 113, Land Tenure Center, University of Wisconsin, Madison, June 1993.

Montecinos, Verónica. "Neoliberal Economic Reforms and Women in Chile." In Nahid Aslanbeigui, Steven Pressman, and Gale Summerfield (eds.), *Women in the Age of Economic Transformation: Gender Impact of Reforms in Post-Socialist and Developing Countries.* London: Routledge, 1994, pp. 160–177.

Papanek, Hanna. "To Each Less Than She Needs, From Each More Than She Can Do: Allocations, Entitlements, and Value." In Irene Tinker (ed.), *Persistent Inequalities: Women and World Development.* Oxford: Oxford University Press, 1990, pp. 162–181.

Sen, Amartya K. "Gender and Cooperative Conflicts." In Irene Tinker (ed.), *Persistent Inequalities: Women and World Development.* Oxford: Oxford University Press, 1990, pp. 123–149.

———. "Population: Delusion and Reality." *New York Times Review of Books,* September 22, 1994.

Senauer, Benjamin, Marito Garcia, and Elizabeth Jacinto. "Determinants of the Intrahousehold Allocation of Food in the Rural Philippines," *American Journal of Agricultural Economics,* vol. 70, no. 1, 1988, pp. 179–180.

Summerfield, Gale. "Effects of the Changing Employment Situation on Urban Chinese Women," *Review of Social Economy,* vol. 52, no. 1, Spring 1994, pp. 40–59.

Summerfield, Gale, and Nahid Aslanbeigui. "The Feminization of Poverty in China?" *Development,* vol. 4, 1992, pp. 57–61.

Tinker, Irene. "The Real Rural Energy Crisis," *Energy Journal,* vol. 8, 1987, pp. 125–146.

———. "Women and Shelter: Combining Women's Roles." In Gay Young, Vidyamali Samarasinghe, and Ken Kusterer (eds.), *Women at the Center: Development Issues and Practices for the 1990s.* West Hartford: Kumarian, 1993, pp. 63–78.

Turshen, Meredith. "The Impact of Economic Reforms on Women's Health and Health Care in Sub-Saharan Africa." In Nahid Aslanbeigui, Steven Pressman, and Gale Summerfield (eds.), *Women in the Age of Economic Transformation: Gender Impact of Reforms in Post-Socialist and Developing Countries.* London: Routledge, 1994, pp. 77–94.

United Nations. *The World's Women 1970–1990: Trends and Statistics.* New York: United Nations, 1991.

Whitehead, Anne. "Rural women and Food Production in Sub-Saharan Africa." In Jean Dreze and Amartya Sen (eds.), *The Political Economy of Hunger,* vol. 1. Oxford: Clarendon, 1990.

Wolf, Diana. "Linking Women's Labor with the Global Economy: Factory Workers and Their Families in Rural Java." In Kathryn Ward (ed.), *Women Workers and Global Restructuring.* Ithaca: ILR Press, Cornell University, 1990, pp. 25–47.

World Bank. *Kenya: The Role of Women in Economic Development.* Washington, D.C.: World Bank, 1989.

———. *Bangladesh: Strategies for Enhancing the Role of Women in Economic Development.* Washington, D.C.: World Bank, 1990.

Female-Headed Households

Sonalde Desai
Samia Ahmad

Introduction

While family remains a basic unit of social and often economic organization in most societies, social science research as well as public policies often fail to recognize the diversity of family forms throughout the world. It is often assumed that families, whether nuclear or extended, are headed by men and rely primarily on a male breadwinner for economic support. However, it has been recognized increasingly that a significant minority of families do not fit into this model and, hence, that public policies must be devised to meet their special needs. This issue has become particularly salient with changes in family structure that often seem to accompany economic development and industrialization.

Definition and Prevalence

Recent research—often grouped together as research on female-headed households—has documented the pervasiveness of female-supported families and analyzed the determinants and consequences of this complex phenomenon. The term "female-headed households" is used to refer to two types of households: (1) those in which no adult males are present or in which women have been established as household head through some sort of family-level consensus (de jure female-headed households); and (2) those in which a majority of economic support for the household maintenance is provided by women (de facto female-headed households). It is relatively easy to ascertain de jure female-headed households through use of census or household-survey data, and much of the past research has focused on ascertaining the characteristics of this type of household.

De jure female headship varies across countries, but a sizable minority of the households in all countries appear to be headed by women. Demographic and health surveys (DHS) conducted in 26 countries between 1986 and 1990 show the percent of female household headship ranging from 9 percent in Mali to 45 percent in Botswana. A large number of countries show rates of female household headship between 15 and 25 percent. Relatively little data are available to calculate trends over time, and data quality of surveys conducted at different points in time vary. In 18 of the 26 countries, for instance, the percent of households classified as female-headed increased between the 1970s and the 1980s, whereas in eight of the 26 countries there was a marginal decline in female headship.

In contrast to de jure household headship, de facto female headship is more difficult to determine. A variety of factors contribute to this difficulty: (1) economic contributions of housework, particularly such expenditure-saving activities as collection of firewood and processing of food grains, are rarely seen as being productive; (2) women often work on family farms or in family enterprises, and the income from these activities is usually attributed to a male family member, reinforcing his position as the household head; (3) female household headship is often dynamic in nature, with women contributing a greater proportion of income during economic hardship but relinquishing the job of primary breadwinner to a male when feasible. Thus, cross-sectional measures of female headship often underestimate the number of families who experience female headship at some point in time.

Choice of de facto or de jure measure of household headship can make a remarkable difference in the percent of households classified as female headed. A study in Peru reports that when using a self-reported (de jure) household-headship definition, 17 percent of the households were classified as female headed, but, when a definition based on whether men or women work the greatest number of hours in economic activities was used, de facto female-headship rates increased to 29 percent of all households (Rosenhouse, 1989).

Although few studies give clear measures of de facto household headship, a review of available data indicates considerable contributions by women to household maintenance. A review of this literature by Bruce (1995) shows some startling findings:

Community studies in Nepal find that women contribute an average of 27 percent of household monetary income; however, when economic value is added to all the goods and services they produce—gathering wood for fuel, collecting and toting water, preparing food—women's average economic contribution to the household rises to 50 percent. Women in farming households in Thailand contribute an average of half of their households' economic resources when these are calculated to include domestic consumption derived from home production. An extensive study of how women and men use time in the Philippines indicates that women's share of market income is a third of men's on average; when the economic value of women's home production is added to the equation, women's economic contributions to the household exceed men's by about 10 percent. Analyses of data from Ghana indicate that, in terms of market hours of work, 33 percent of households with children were maintained primarily by women in 1988 (Bruce, 1995, pp. 27–28).

Research on de facto female household headship remains in the formative stages; the subsequent discussions will focus primarily on determinants and consequences of de jure female headship.

Routes to De Jure Female Household Headship

Female-headed households around the world differ considerably with respect to the nature of female headship and the routes through which they become female headed. The patterns and nature of female headship are strongly influenced by demographic antecedents that vary considerably across countries. Three major routes to female household headship are: (1) widowhood; (2) single parenthood through divorce, out-of-marriage childbearing, or such other tenuous marital relationships as visiting unions or polygynous marriages; and (3) absence of a male breadwinner for such socioeconomic reasons as migration, unemployment, or disability.

Widowhood

Demographic data show that, in almost all countries in the world, women have longer life expectancies at birth than men. Hence, it is not surprising that the probability of being widowed is higher for women than for men. For example, in 1991, 91 percent of Egyptian men above age 65 were still married, compared to only 34 percent of the women (Naguib and Lloyd, 1994).

Widowhood occurs at both younger and older ages and is often an important cause of female headship. In countries like Bangladesh and Indonesia, widowhood at younger ages is a common phenomenon, caused by a large spousal age difference. In Bangladesh, women marry at a very early age (average at first marriage was less than 17 years) and average spousal age difference exceeds seven years. In Indonesia, although age at first marriage is slightly higher (20 years), the spousal age difference is similar to

Bangladesh (United Nations, 1990). Consequently, women are often widowed at relatively early ages and are often left with young children to raise. In Bangladesh, nearly 35 percent of the women age 45–49 do not have a spouse, compared to only 3 percent of men. Comparable figures for Indonesia are 33 percent and 6 percent (United Nations, 1992).

Differences in male and female mortality have other implications as well. For instance, widowhood tends to be particularly high in the aftermath of a war. In Vietnam, the 1989 census (General Statistical Office, 1991) showed only 81 males for 100 females age 40–44. This high mortality among males, not surprisingly, results in high rates of widowed and single women. About 10 percent of Vietnamese women age 40–44 are widowed. At the same time, declining overall mortality results in diminished probability of widowhood. For example, the percent of single women age 65 and above in Egypt declined from nearly 80 percent in 1980 to about 66 percent in 1991 (Naguib and Lloyd, 1994).

Single Parenthood

Single parenthood through divorce or out-of-wedlock birth plays an important part in creating female-headed households in some parts of the world, but the role of these factors is less important in other areas. For example, as the *United Nations Demographic Yearbook* (1992) covering years 1980–1985 indicates, nearly 14 percent of women age 25–44 in Mauritania and nearly 9 percent in Ethiopia are divorced, whereas less than 1 percent in Pakistan, China, India, and Haiti are. Divorce rates vary considerably across different regions of the world, with higher rates seen in Africa and Latin America than in Asia.

Polygamous marriages form another route through which many women become household heads. Polygamous marriages are common in sub-Saharan Africa, particularly in West Africa. Often, each co-wife maintains her own household and may not have a coresident husband. Consequently, a considerable part of women's lives may be spent as a household head, living without a coresident partner. For example, DHS data show that about 34 percent of currently married women do not live with their husbands in Ghana; the corresponding percentage is 26 percent for Botswana, 22 percent for Kenya, 20 percent for Senegal, and 29 percent for Zimbabwe. The financial relationship between husbands and wives in polygynous unions varies by region but in many regions, men carry only a limited responsibility for providing for their wives and children. In many parts of middle and West Africa, polygyny is associated with separate spousal budgets, with the basic childbearing unit being a mother and her dependent children. Hence, polygyny intensifies the economic burdens on women for children's upkeep, with the wife assuming the responsibility of providing food for herself and her children, although the husband often contributes toward such other expenses as school fees. Studies show that food costs can be substantial, amounting to about two-thirds of the total expenditures of Ghanaian households (Boateng et al., 1990). Moreover, women's claims to their husband's resources for household maintenance tend to decrease with the number of wives and children in the union. It is important to note that although separate spousal budgets are pervasive in West Africa and show little evidence of declining, they are rarely found in polygynous marriages in eastern and southern Africa (Caldwell et al., 1992).

Entry into nonformal unions may be another route into female household headship. A large number of women in different parts of Latin America live in unions that are common-law unions, often called consensual marriages. For example, in the early 1980s, the percent of women in consensual unions among all married women was 68 percent in Haiti, 40 percent in Nicaragua, 26 percent in Trinidad and Tobago, and 13 percent in Brazil (United Nations, 1990).

The emergence of different types of family forms across different regions is intimately tied to changes in political and economic situations of these regions. Common-law marriage in Latin America and the Caribbean provides an interesting example. During the early phases of Latin American colonization, a large number of men relative to women migrated to the New World. As a result, many soldiers brought indigenous women to their beds. Their power over these women was reflected in a lower level of commitment to women and children from these consensual marriages than to scarce and precious Spanish women and their legitimate offspring. Marriage was also a determining factor in the inheritance of land and dispensation of tribute rights by the Crown (*encomiendas*), but outside the Spanish elite marriage lost much of its significance for intergenerational transmission of property. Today, although some consensual unions are just like formal marriages except that spouses simply wanted to avoid paying the relatively high costs of marriage, others are not so stable.

Consensual unions are significantly more likely to dissolve than formal marriages (Goldman, 1981; Rao and Green, 1991). Additionally, the nature of the union seems to be related to the partners' degree of commitment to the marriage. Stycos (1955) quotes a male respondent in Puerto Rico as saying, "If the wife does not turn out well, one separates easily from her. If she turns out to be a fine worker and faithful, one may marry her later" (Stycos, 1955, p. 117). This supports the finding by Goldman and Pebley (1981) that a substantial proportion of long-lasting informal unions are later legalized into formal marriages. Similarly, Potter and Ojeda de la Pena (1982) point out that the decision to legalize consensual unions in Mexico hinges on the couple's perception of the stability of the union. This suggests that while the choice of formal versus consensual union is affected by the socioeconomic status of the partners, an element of trial arrangement is also involved in selecting a consensual union. Additionally, the choice of the type of union seems to be related to the bargaining power of the partners involved. Based on fieldwork

in Brazil, Rao and Green (1991) suggest that women rarely choose consensual union but end up living in such unions because their partners will not agree to a formal marriage or cannot afford it.

Nonmarital childbearing also plays an important role in creating female-headed households. The percent of all births that take place outside marriage varies across different regions of the world, with a high proportion of nonmarital births in industrial countries and in Latin America but relatively few nonmarital births in parts of Sub-Saharan Africa and Asia. However, considerable diversity exists within each region. DHS-based research shows that a large number of women are sexually active before marriage in many developing countries: Nearly 80 percent of women in Botswana have a nonmarital sexual experience before age 20, and 42 percent of never-married women age 15–24 have had a child; corresponding figures are 61 percent and 16 percent for Kenya and 21 percent and 6 percent for Colombia. For the United States, the figures are 68 percent and 6 percent (Alan Guttmacher Institute, 1995). At the same time, nonmarital childbearing is virtually nonexistent in countries like Mali and Burundi. Although some unmarried mothers maintain close ties to their partners, a large proportion of unmarried mothers often end up becoming household heads with little financial support from their partners.

However, it is important to note that not all divorced, separated, or single women become heads of households. A woman's decision regarding where and with whom she lives is likely to depend on three factors: her individual preference, economic circumstances, and the alternative living-arrangement options open to her. Research in Santiago, Chile, shows that many single mothers lived with other family members since they could not afford a dwelling of their own and, consequently, did not form female-headed households (Buvinic et al., 1992). Tabulations from demographic and health surveys for Ghana, Kenya, Mali, and Senegal also indicate that, although women in each country spend substantial amount of their reproductive life without a resident partner, they spend relatively little time heading their own households. For example, although 33 percent of the reproductive life of women in Senegal is spent without a resident partner, only 2 percent of their life is spent as household heads (Lloyd and Duffy, 1995). This suggests that many of these women are absorbed in other families. However, the implications for women and their children of this subfamily arrangement remain unclear.

Moreover, the role of women's economic independence in increasing single parenthood and, consequently, the proportion of female-headed families is far from clear. The experience in the United States shows that divorce rates soar during the years when employment opportunities and real wages for women are also growing. There is relatively little research on the relationship between opportunities for earning income independent of one's family and divorce rates in non-Western countries, but it seems likely that, if higher incomes are associated with higher rates of divorce or single parenthood, then women in female-headed households—at least those households that become female headed by choice—may well be financially more independent than women in male-headed households, although, as discussed below, their overall household income may be lower.

Absence of Male Breadwinner

Male- and female-headed households are rarely identical. Whereas most male-headed households contain adult women who contribute substantially to household maintenance through their participation in wage economy, production for domestic consumption, and other unpaid domestic activities, relatively few female-headed households contain employed males. For example, research in Ghana by Lloyd and Gage-Brandon (1993) indicates that nearly 56 percent of the female heads of the household were the only workers in their household, while only 8 percent of male heads of the households were. This brings us back to the definitional issue. In most cultures, as long as a husband is present and employed, few wives will be characterized as household heads. It is only in families without an employed adult male that women are characterized as household heads.

Male employment, male migration, and the nature of economic development are closely linked. These links are particularly relevant to the emergence of female-headed households in southern Africa. European colonization of sub-Saharan Africa led to a great demand for native labor. In many areas, colonial administration imposed "head taxes" that forced African men into the labor force to earn necessary cash. In countries of eastern and southern Africa, young men were the first to accept wage employment on White-owned farms. This led to sex-selective migration, resulting in a high incidence of female-headed households. Interestingly, although colonial authorities encouraged male migration, they often discouraged female migration. Women in colonial towns in Zambia, Zimbabwe, Kenya, and Tanzania were occasionally rounded up by the police and forcibly returned to native reserves. This colonial legacy continues to influence contemporary Africa, particularly in countries close to South Africa, where wage labor in mines continues to attract many immigrants. In 1986 in Botswana, for example, there were only 59 men per 100 women age 25–34 (United Nations, 1992), the other men having migrated in search of work, particularly to South Africa. Consequently, it is not surprising that as many as 17 percent of women in Botswana remain unmarried at age 45–49. Moreover, nearly half of the children in Botswana grow up in female-headed households (Lloyd and Desai, 1992).

Forced migration also results in a high percentage of female-headed households. With a dramatic increase in the refugee population around the world, a large number of families are being separated, leaving many mothers of young children as involuntary heads of their households. Young widows with children are heavily represented among refugee populations from Cambodia and Vietnam, for example.

Economic Conditions of Female-Headed Households

Research on female-headed households, particularly research based on the American experience, has tended to suggest that female-headed households are more likely to be poor than male-headed households, hence, the term "feminization of poverty" (Pearce, 1978). Research in this area has attributed the economic vulnerability of female-headed households to a variety of factors (Buvinic, 1990).

First, female-headed households are more likely than male-headed households to have a smaller number of adults contributing to the household's economic welfare. Research on urban Brazilian households shows that, on average, male-headed households contain 1.2 earners, whereas female-headed households contain 0.9 earners (Barros et al., 1993). Moreover, since male-headed households typically contain the spouse of the head as well as the head himself, these households can also rely on a larger number of adults for such nonmarket economic contributions as food processing and agricultural production for domestic consumption. In spite of their smaller size, female-headed households often contain a large number of children and have a greater dependency burden than other households. An extensive review by Buvinic and Gupta (1994) of studies from such diverse countries as Botswana, Ghana, Kenya, Malawi, Brazil, Pakistan, and Peru show relatively more dependents both young and old in female-headed households (for an exception, see Visaria and Visaria, 1985).

However, a greater dependency burden does not necessarily lead to poverty if households receive economic support from absent fathers. It is when income transfers from fathers to children are disrupted that female-headed households are more likely to fall into poverty. Studies show a variety of situations under which a disruption of paternal financial support for children occurs, including out-of-marriage childbearing and often accompanying low levels of father–child bonds, women's abandonment by their partners, and disruption of traditional systems of family governance that enforced income transfers from fathers to children.

Second, women are more economically vulnerable than men since they are usually more likely to be concentrated in poorly paid jobs, particularly in the urban informal sector. They are less likely to hold positions in the urban formal sector, particularly in administrative or managerial positions. Even when doing the same job, they are often less likely to be paid the same wage, and they are more likely to experience unemployment (UNDP, 1995; World Bank, 1995). Moreover, in most developing countries, women receive less education than men, which in turn increases their economic vulnerability and reduces their wages compared to men.

Third, women who become household heads may be more vulnerable to poverty than other women since they may be more constrained by child-care demands and, consequently, may have to accept lower-paying jobs. In societies where land inheritance is patrilineal, married women may have access to their husbands' land, but widows may lose that access if they do not have a son. Widowed women's loss of access to land has been recorded in research in India (Desai, 1994) and in other parts of South Asia.

Fourth, women often become household heads through processes that themselves lead to poverty. War may result in both female household headship and destitution for those households. Poverty and low employment opportunities are often associated with high incidence of consensual unions, which are more likely to dissolve than formal marriages.

The preceding review suggests that the economic conditions of female-headed households are likely to depend on the route through which households became female headed and on social and economic conditions of the society. On balance, however, an extensive review of 61 studies linking the gender of the household head to poverty suggests that in a majority of the situations, households headed by women are more likely to be poor than households headed by men (Buvinic and Gupta, 1994):

> Thirty-eight of the sixty-one [studies] reached this conclusion by using a variety of poverty indicators: total or per-capita household income, mean income per adult equivalence, total or per-capita consumption expenditures, earnings of the head, access to services, and ownership of land or assets, among others. Fifteen other studies found that poverty was associated with only certain types of female heads, or that the association emerged only for certain poverty indicators. Eight of the sixty-one showed no empirical evidence for the hypothesis of the greater poverty of female-headed households (Buvinic and Gupta, 1994, p. 8).

In spite of many instances in which female heads of household are more disadvantaged than other women, there are also instances in which they are more advantaged than other women in comparable situations. Setting up and maintaining one's own household requires financial resources. Research in Chile shows that a substantial portion of single mothers live as subfamilies within their extended families because they lack the resources and access to housing to set up their own households (Buvinic et al., 1992).

Moreover, women-headed households that emerge out of traditional customs that have been sanctioned by society are less likely to be poor than women-headed households that emerge out of situations that are not sanctioned by their societies (Buvinic, 1990). Examples of these types of households are independent households of wives of polygynous men in West African societies or households of women in matrilineal societies of sub-Saharan Africa. Wives of migrant husbands who are left behind as household heads are less likely to be poor than other wives if they continue to receive regular remittances.

Methodological Considerations
Pioneering work by Visaria and Visaria (1985) in India on

the economic conditions of female-headed households points out a major methodological consideration: Although female-headed households are often poorer than male-headed households, they are also smaller, at least in two Indian states that Visaria and Visaria studied. Thus, although these households were disadvantaged in terms of absolute income, they were not disadvantaged in terms of per-capita income. Buvinic and Gupta (1994) attempt to sort out the confusion created by two different poverty measures by reviewing results from 61 studies. Of the 61 studies they review, only eight found no link between gender of the household head and poverty. A detailed review of these eight studies revealed that even these studies showed that in terms of total household income, female-headed households are uniformly poorer than male-headed households, it is only when per-capita adjustment is undertaken that these households are relatively better off.

Additionally, when per-capita-expenditure measures of poverty are used, the link with gender of the household head is considerably weaker than when per-capita-income measures are used. Development economists vastly prefer consumption measures to income measures as indicators of well-being since consumption measures are less likely to be distorted by income underreporting and seasonal variability in earnings. Unfortunately, a focus on consumption is particularly biased when studying implications of gender composition of the household since female-headed households often tend to produce many items for home consumption, often at the expense of leisure. Indeed, a study in Ghana by Lloyd and Gage-Brandon (1993) reports that although female-headed households are not poorer than male-headed households in terms of per-capita consumption, women heads of the household work nearly 47 hours per week (counting both market and nonmarket work) compared to 37 hours for male household heads or 40 hours per week for women who are not household heads.

In addition to a higher work burden and lower incomes, female-headed households also seem to suffer in terms of access to housing, credit, land, and other resources (Buvinic and Gupta, 1994). Much of the research has focused on economic consequences of living in female-headed households and ignored some of these other consequences. Hence, systematic data on other disadvantages suffered by female-headed households remain limited.

Consequences of Female Household Headship for Children
The possibility that living in female-headed households has a negative impact on children is one of the most persuasive rationales for targeting government programs toward female-headed households. Curiously, however, research on the relationship between the gender of the household head and the welfare of the children provides mixed results. While on the one hand the poverty associated with living in a female-headed household seems to have a negative effect on children, on the other hand female-headed households often have expenditure patterns that favor children.

Moreover, the effect of living in a female-headed household is different for different types of child outcomes, particularly children's nutrition and education.

Children's Nutrition

Buvinic and Gupta (1994) found 18 studies that examined the relationship between living in female-headed households and children's nutritional status: Roughly half reported positive effects and half reported negative effects. Studies reviewed by Buvinic and Gupta (1994) in such countries as Zimbabwe, Brazil, and the Philippines found the negative effect of living in female-headed households was due to differences in income, race, education, and standard of living associated with female household headship. In contrast, studies in Kenya, Malawi, Guatemala, and Chile found a beneficial effect of female household headship on children. It is hypothesized that this positive effect is associated with gender differences in expenditure patterns: Women may be more likely than men to spend a larger portion of their income on items related to children—particularly food. Women in male-headed households are often unable to carry out their preferences; in female-headed households, they may be more likely to follow them since they do not have to contend for power with men. Moreover, this shift in expenditure may be more apparent in poorer families than in better-off ones, either because investments in children yield greater returns at lower levels of income or because there are fewer competing alternative investments than in higher-income households (Kennedy, 1992).

Thus, although female-headed households are poorer, their incomes may be spent in a way that favors children more than those in male-headed households, reducing the potentially negative impact of female household headship on children's nutrition.

Children's Education

A relatively small number of studies has directly linked the gender of the household head to children's educational outcomes. Of the 10 studies examined by Buvinic and Gupta (1994), four found positive effects of living in female-headed households on children, and six reported negative effects. Female-headed households affect children's education in two contradictory ways. On the one hand, female-headed households have greater need for domestic labor as well as income, resulting in greater demands on children to stay at home and care for younger siblings or for participation in wage work. On the other hand, female-headed households, more than male-headed households at comparable income levels, make an effort to invest in children's future welfare and encourage school enrollment. Thus the ultimate impact of the gender of household head on children's education depends on the relative magnitude of these two effects.

It has also been argued that in female-headed households where women have the decision-making power, the distribution of resources to children is more even-handed and girls and boys are favored equally. If this hypothesis is true, gender disparities in children's education are likely to be lower in female-headed households. However, not enough systematic research has been conducted in this area to draw any generalizations.

Social Consequences

Much of the research on the consequences for children of living in female-headed households has focused on economic disadvantages rather than on other social consequences of living in male-absent families. To some extent, this can be attributed to a general belief that, besides conferring income and social status on the family, fathers play only a limited role in raising children. Moreover, the strength of the family tie does not depend on residential location. Children of divorced parents may see their fathers regularly; children in matrilineal societies may find substitute fathers in the form of their maternal uncles. Nonetheless, separate residences, particularly residences separated by long distance, are likely to reduce contact between fathers and children.

Does regular contact with father and a close relationship between father and child confer any special advantages to children? Conventional wisdom suggests that fathers play an important role in socializing and training their children, particularly adolescent sons, and, consequently, that their absence is likely to have negative implications for their children. A study of 8-year-old children in Barbados found that children who performed better in school had more involved fathers than those who did not perform as well, regardless of whether their fathers lived with them (Russell-Brown et al., 1992). Nonetheless, due to a paucity of empirical data, it is not possible to draw any definitive conclusions regarding the social importance of paternal presence for children in developing countries.

Policy Considerations

It is now widely recognized that a sizable minority of the households in many countries are headed by women and are often economically vulnerable, and public policy attention has focused on strategies for protecting these households. This focus on the needs of female-headed households in the developing and other countries has been a direct result of the feminist movement in identifying the magnitude of this population and in mobilizing political support around its needs. Three types of policy interventions have been discussed:

Focusing on Male Responsibility for Children. A large number of households become female headed through divorce or single parenthood. As discussed above, the poverty of female-headed households is often exacerbated by responsibilities for young children, resulting in high dependency ratios. Thus, one strategy is to focus on developing and strengthening institutional mechanisms to ensure children's access to paternal as well as maternal resources. The magnitude of this problem differs around the world, as does the attention it receives from both feminist activists and governments. For example, a focus on paternal responsibility has gained center stage in the United States in discussions of deadbeat dads who do not fulfill their child-support obligations. In contrast, a highly complicated legal system in Mexico remains insensitive to the economic needs of children following divorce or separation of their parents. A review of child-support legislation and compliance in different parts of the developing world paints a dismal picture. Even though Islamic law requires divorced fathers to contribute to their children's upkeep, in Malaysia only 50 percent do so. In Botswana, Lesotho, Mozambique, Swaziland, Zambia, and Zimbabwe, fathers are ordered by a court to pay child support only if it has been shown that they have failed to provide any support at all; fathers are not liable in cases of intermittent, unreliable support. In Chile, a study found that 42 percent of fathers of children born to adolescent mothers were not paying child support six years after their child's birth (Buvinic et al., 1992).

Payment of child support by fathers who were not married to the mothers of their children is even rarer than that by divorced fathers, particularly since the customs surrounding the establishment of paternity vary. In many countries, such as Botswana, Guatemala, and Zambia, the onus is on mothers to furnish proof of their child's paternity. In Zimbabwe, by contrast, a man must furnish proof that he could not have been the father.

Ensuring Women's Access to Livelihood, Particularly Land. In many countries, women are unable to obtain livelihoods independent of male family members. This is particularly problematic in South Asia, where women's access to land is limited. Widows often lose access to their husbands' land if they remarry, if they are viewed as being unchaste, or if they do not have a son. Moreover, in Islamic tradition, a substantial portion of a dead man's land is inherited by his blood relatives rather than by his widow. Thus, widows often lack the primary resource—land—required for survival in an agricultural society. Daughters rarely inherit land in their own right, and both customary practice and the legal system discriminate against women's inheritance of agricultural land. In addition, women are often unable to obtain credit or employment without support from male family members. Women's movements in many countries have pressed for legal as well as social reforms to address some of these issues. Some very creative institutions like the Grameen Bank in Bangladesh and Samakya in India have emerged to help meet women's credit needs.

Targeting Female-Headed Households for Assistance. Government programs can be specifically devised to provide assistance to female-headed households. In the United States, households headed by low-income women with young children are eligible for financial aid and medical insurance through the Aid to Families with Dependent Children (AFDC) program; in Chile, job training, health services, child care, and housing are provided to female-headed households by *Servicio Nacional de la Mujer;* in India, households headed by women receive a priority for participation in Integrated Rural Development Program (IRDP) schemes. These programs have been successful to varying degrees: The Chilean program is perceived to be quite successful; the American program has been successful with respect to providing basic necessities to families but has come under political attack and its fate remains uncertain; the Indian program has not been particularly successful due to its reliance on meeting targets (Buvinic and Gupta, 1994).

A major debate revolves around the desirability of targeting special programs to female-headed households versus providing some services or subsidies to all households. This debate is centered on the following questions:

1. What is the size of the population to be targeted? When female-headed households are a minority, targeting may be successful. In societies where female-headed households form a large portion of the population, it may be more efficient to deliver such services as nutritional supplementation or credit schemes to all households, not just female-headed ones.

2. Can female-headed households be uniquely identified? Any system based on preferential treatment of a subgroup must contend with possible leakages to the nonpoor, so targeting can be successful only if female-headed households in a society can be easily identified. For example, setting up pension schemes for widows in rural India has not been difficult since it is easy to identify widows. In contrast, in societies where common-law marriages or polygamy are common, it may be difficult to distinguish truly female-headed households from male-headed households and, therefore, to target female-headed households for subsidies.

3. To what extent will government subsidies provide incentives for household disintegration, leading to a higher rate of formation for female-headed households? If there is a causal effect, is it still a politically acceptable policy? In the United States, it has been argued that the availability of AFDC to single mothers hastens marital disruption and prevents unmarried parents from marrying. As of the mid-1990s, the magnitude of this relationship was unknown, but even a possibility that it exists has provided a powerful platform for lobbying against AFDC. At the same time, supporters of AFDC argue that the subsidies received under this program are not large enough for anyone to change their lifestyles radically. Based on generally accepted definitions of poverty, families on AFDC are likely to remain poor, but, for children in these families, these subsidies are enough to spare them destitution and to provide basic necessities.

This discussion highlights the complexities of targeting government subsidies to female-headed households, and it seems likely that public policies in this area will differ across countries and cultural settings. At the same time, it appears that gender-specific interventions directed at both male- and female-headed households may particularly benefit female-headed households. For example, rural women in different parts of the world often spend considerable time fetching water or collecting fuelwood. Time demands on female household heads are particularly heavy (Lloyd and Gage-Brandon, 1993). Hence, any interventions that increase or at least preserve the supplies of water and fuelwood in rural areas, thereby reducing time pressure on women, will particularly benefit women who are the sole heads of their household.

Conclusion

The preceding review suggests that a large number of households in developing countries are headed by women and will continue to be headed by women in the foreseeable future. Whereas most male-headed households usually also contain an adult woman, few female-headed households contain an adult man. This asymmetrical pattern, combined with women's disproportionate responsibility for child rearing and gender inequalities in the labor market results in considerable financial as well as time pressure on female-headed households.

Only a few developing countries have developed concrete public policies for assisting female-headed households, and even there, public policies are limited to direct public assistance. While this targeted assistance can be invaluable is some instances, as recent debates regarding welfare reforms in the United States suggest, direct public assistance in not enough to lift female-headed households permanently out of poverty. This points to a need for changes in institutional structures that strengthen children's claim to support from both father and mother and increase women's access to livelihood.

References

Alan Guttmacher Institute. *Women, Families, and the Future: Sexual Relationships and Marriage Worldwide.* New York: Alan Guttmacher Institute, 1995.

Barros, Ricardo, Louise Fox, and Rosane Mendonca. "Female-Headed Households, Poverty, and the Welfare of Children in Urban Brazil," Joint Program on Female Headship and Poverty in Developing Countries working paper, Population Council/International Center for Research on Women, New York, 1993.

Boateng, E. Oti, Kodwo Ewusi, Ravi Kanbur, and Andrew McKay. "A Poverty Profile of Ghana, 1987–88," Social Dimensions of Adjustment in Sub-Saharan Africa, working paper no. 5., World Bank, Washington, D.C., 1990.

Bruce, Judith. "The Economics of Motherhood." In Judith Bruce, Cynthia Lloyd, and Ann Leonard (eds.), *Families in Focus: New Perspectives on Mothers, Fathers, and Children.* New York: Population Council, 1995.

Buvinic, Mayra. "The Vulnerability of Women-Headed Households: Policy Questions and Options for Latin America and the Caribbean," Joint Program on Female Headship and Poverty in Developing Countries working paper, Population Council/International Center for Research on Women, 1990.

Buvinic, Mayra, and Geeta Rao Gupta. "Targeting Poor Woman-Headed Households and Woman-Maintained Families in Developing Countries: Views on a Policy Dilemma," Joint Program on Female Headship and Poverty in Developing Countries working paper, Population Council/International Center for Research on Women, New York, 1994.

Buvinic, Mayra, Juan Pablo Venlenzuela, Temistocles Molina, and Electra Gonzalez. "The Fortunes of Adolescent Mothers and Their Children: The Transmission of Poverty in Santiago, Chile," *Population and Development Review,* vol. 18, no. 2, 1992, pp. 269–297.

Caldwell, John C., I.O. Orubuloye, and Pat Caldwell. "Fertility Decline in Africa: A New Type of Transition," *Population and Development Review,* vol. 18, no. 2, 1992, pp. 211–242.

Desai, Sonalde. *Gender Inequalities and Demographic Behavior: India.* New York: Population Council, 1994.

General Statistical Office. *Vietnam Population Census—1989: Detailed Analysis of Sample Results.* Hanoi: General Statistical Office. 1991.

Goldman, Noreen. "Dissolution of First Unions in Colombia, Panama, and Peru," *Demography,* vol. 18, no. 4, 1981, pp. 659–680.

Goldman, Noreen, and Anne R. Pebley. "Legalization of Consensual Unions in Latin America," *Social Biology,* vol. 28, no. 1–2, 1981, pp. 49–61.

Kennedy, Eileen. "Effects of Gender of Head of Household on Women's and Children's Nutritional Status," paper presented at a workshop: Effects of Policies and Programs on Women, Washington, D.C., January 16, 1992.

Lloyd, Cynthia B., and Sonalde Desai. "Children's Living Arrangements in Developing Countries," *Population Research and Policy Review,* vol. 11, no. 3, 1992, pp. 193–216.

Lloyd, Cynthia B., and Niev Duffy. "Families in Transition." In Judith Bruce, Cynthia Lloyd, and Ann Leonard (eds.), *Families in Focus: New Perspectives on Mothers, Fathers, and Children.* New York: Population Council, 1995.

Lloyd, Cynthia B., and Anastasia J. Gage-Brandon. "Women's Role in Maintaining Households: Family Welfare and Sexual Inequality in Ghana," *Population Studies,* vol. 47, no. 1, 1993, pp. 115–132.

Naguib, Nora Guhl, and Cynthia B. Lloyd. *Gender Inequality and Demographic Behavior: Egypt.* New York: Population Council, 1994.

Pearce, Diana. "The Feminization of Poverty: Women,

Work, and Welfare," *Urban Change and Social Change Review,* vol. 11, no. 1, 1978, pp. 28–36.

Potter, J.E., and N. Ojeda de la Pena. "Dissolution of First Unions in Mexico," paper presented at the annual meeting of the Population Association of America, San Diego, 1982.

Rao, Vijayendra, and Margaret E. Green. "Marital Instability, Inter-Spouse Bargaining and Their Implication for Fertility in Brazil," paper presented at the annual meeting of the Population Association of America, Washington, D.C., 1991.

Rosenhouse, Sandra. "Identifying the Poor: Is 'Headship' a Useful Concept?" Living Standards Measurement working paper no. 58, World Bank, Washington, D.C., 1989.

Russell-Brown, Pauline, Patricia L. Engle, and John Townsend. "The Effects of Early Childbearing on Women's Status in Barbados," Family Structure, Female Headship, and Maintenance of Families and Poverty working paper series, Population Council/

International Center for Research on Women, New York, 1992.

Sycos, J. Mayone. *Family and Fertility in Puerto Rico.* New York: Colombia University Press, 1955.

United Nations. *Patterns of First Marriage: Timing and Prevalence.* ST/ESA/SER.R/111. New York: United Nations, 1990.

———. *Demographic Yearbook, 1990.* New York: Publishing Division United Nations, 1992.

United Nations Development Program (UNDP). *Human Development Report 1995.* New York: UNDP, 1995.

Visaria, Pravin, and Leela Visaria. "Indian Households with Female Heads: Their Incidence, Characteristics and Level of Living." In Devaki Jain and Nirmala Banerjee (eds.), *Tyranny of the Household: Investigative Essays on Women's Work.* New Delhi: Shakti, 1985, pp. 50–86.

World Bank. *World Development Report 1995.* New York: Oxford University Press, 1995.

Childrearing Practices

Robert G. Myers
Judith L. Evans

Introduction

A review of literature dealing with childrearing practices yields solid support for the following four conclusions: First, children, in whatever setting, have general physical, social, and emotional needs that require responses from others. Second, the specific ways in which these general needs manifest themselves and the childrearing practices adopted to meet these needs differ widely from place to place and from caregiver to caregiver, influenced by physical and social contexts, by beliefs, values, and norms, by available technologies, and by the characteristics and knowledge of particular caregivers. Third, in a rapidly changing world, it is difficult for cultures to adjust their norms and practices to fluctuating conditions. This results, more frequently than in the past, in beliefs, values, norms, and practices that do not fit well with actual conditions. These can work against the sound rearing and development of children. Fourth, rapid change has produced a move away from so-called traditional and family-centered practices toward placing greater responsibility for childrearing in institutional settings outside the family. As these trends and specific changes in practices associated with them are judged, it is important not to equate "modern" with "good" and "traditional" with "outmoded" or "bad," or vice versa. Rather, if we are to retain the good practices from traditional systems and to develop quality child care in response to the major changes thrust upon us by industrialization and so-called modernization, we will need to be much more systematic in our assessments and much more open to potential advantages of both the new and the old than we have been in the past.

Before entering directly into the review that substantiates this line of argument, and in a desire to avoid perpetuating stereotypes or promoting outmoded concepts, it is important to clarify three points. First, although the childrearing topic is set here within the general theme of "marriage and the family," it is important to recognize that a great deal of childrearing has always occurred outside the home and family in both formal and informal arrangements. It is common, for instance, for the larger community of which families are a part to play a significant role in raising children. This is not only the case for traditional cultures but is increasingly common today as more and more children are being cared for during long periods of the day in specialized centers that are not family run and have nothing to do with marriage ties (Olmsted and Weikart, 1994). While marital and family status certainly affect childrearing (and sometimes vice versa), so do many other factors. Thus, to consider childrearing only within the context of marriage and the family would be to limit both description and understanding of the process.

Second, childrearing is a productive activity. Women's activities are often classified as productive or reproductive. This classification is unfortunate, in our view, because it does not adequately represent the complexity of women's lives. Rather than distinguish these two roles, assigning economic value to one and not to the other, it seems more useful to place women's activities on a continuum. At one end of the continuum is work outside the home that is paid. This work, normally classified as productive, also supports reproductive activity. The next point on the continuum would be reserved for work that women do at home for which they earn money. Next comes support that women provide to other family members who are earning income, such as providing food to those working outside the home, and production for family consumption, such as caring for small agricultural plots. Toward the other end of the continuum are such tasks as maintenance of the household, including gathering firewood, collecting water, cooking, and the like, and childcare. If these household chores, normally classified as reproductive, were not performed by women as their duty to the family, they would have to be paid for, and, if women did not care for children, this service would also have to be bought, in which case the services would be classified as productive activities.

Because marriage and family activities tend to be associated with the reproductive role of women in society, it is important to stress that childrearing is an economically productive activity as well. Ironically, when "childrearing" is redefined as "child care" because it occurs outside the home and/or is paid, it is recognized as an economically productive activity and is included in national accounts. However, when childrearing occurs in the home and is unpaid, it is not. But whether or not children are raised primarily in the home and whether or not a caregiver is paid, childrearing constitutes a productive as well as a reproductive activity because it affects the ability of a child to make a productive contribution to the family and increases the child's chance of becoming productive in later life. Those who are involved in rearing children are performing a service to society that has economic as well as social value. Today's children are tomorrow's citizens and workers who will determine the general course of society.

Third, childrearing is not, nor should it be seen as, an activity restricted to women. A growing literature shows that fathering as well as mothering is important to the childrearing process (Lamb, 1987) and that the absence of fathers can leave important areas of a child's development unattended (Zoller Booth, 1995). We will return to this point when discussing who participates in childrearing.

Having cautioned the reader against a narrow and stereotyped view of childrearing as a reproductive role to be carried out exclusively by women in the context of the family, we will now proceed to discuss childrearing, what it means, who it involves, how it is changing, and what is required in terms of programming and policies to provide a healthy environment for all children.

The Childrearing Process

Childrearing, or the process of bringing up children, can be described in terms of practices and activities that reflect a society's response to the survival and developmental needs of both children and the society. Individual practices are influenced by cultural norms or patterns that have evolved over time and that, in turn, are grounded in cultural beliefs and values. They are influenced by the physical, social, and economic context as well as by the types of technology available to, and associated with, childrearing. The choice among possible childrearing practices, the manner in which they are carried out, and the results seen in children will also depend on the character and knowledge of the individual(s) who are responsible for childrearing. We draw here upon a useful framework for studying the cultural regulation of the micro-environment of the child labeled the "developmental niche" and developed by Super and Harkness (1986). The framework relates childrearing practices to changing settings, customs, and beliefs. To understand the process of childrearing, then, it is necessary to examine the basic needs of children and the general practices required to meet them; who is responsible for childrearing; differences among, and changes in, physical, social, economic, and technological characteristics of the environments in which childrearing occurs; and the cul-

tural beliefs, values, and norms that have evolved (and are constantly changing) to guide and ground choices among the available options in the form and content of childrearing. It is necessary also to identify how these various dimensions of childrearing are interwoven and how changes in any one dimension can produce changes in others, sometimes to the benefit and sometimes to the detriment of the child, the family, or the culture.

Children's Needs and Common General Practices
Children everywhere have certain basic needs that must be met if they are to survive, grow, and develop. For this reason, it is possible to specify general areas of childrearing practices that will be common to all societies. For example:

To meet needs that will guarantee the child's physical well-being, childrearing must include provision of proper attention to mothers during pregnancy and at birth; provision for the child of an adequate environment for sleeping; protection from harm; provision of shelter and clothing, feeding, and bathing; provision of safe places to play and explore; and prevention of, and attention to, illness.

To promote the child's psychosocial well-being, childrearing must include affection and nurturing activities that allow a child to form attachments, to develop emotional security and a sense of mastery and confidence, and to seek some independence.

To promote the child's mental development, childrearing must include interaction, stimulation (things to look at, touch, hear, smell, and taste), opportunities to explore the world, appropriate language stimulation, and play.

To facilitate the child's social development, childrearing must provide opportunities to acquire values, exposure to the wisdom of the culture, and the chance to practice socially and culturally related skills. In most cultures, this will include opportunities to learn to cooperate and to share by participating in group activities with members of the family and community.

Although these four general areas have been separated conceptually, in practice they come together. Attention to a mother during pregnancy affects mental and social development as well as physical development. Providing affection and nurturing influences physical as well as psychosocial development.

Specific Practices
At a very general level, children's needs are similar across all cultures and will require common areas of activity. But the specific ways in which these needs are manifested and the specific practices developed to meet these needs will vary widely from place to place. Examples of specific practices could be provided for each one of the general categories and areas of activity mentioned above, but, with space limitations, we focus on practices in two areas, feeding and protecting from harm, in order to illustrate variation.

Feeding is a common practice to all cultures, but the age-old practice of breast-feeding contrasts with the more recent practice of bottle-feeding. Feeding on demand, which is still characteristic of most rural cultures, contrasts with the scheduled feeding brought on by adjustments to city life. Active feeding contrasts with letting young children feed as they wish or with "bottle-propping" (allowing an infant to hold a bottle by herself). Weaning at three months, or nine, contrasts with weaning at two years.

The practice of protecting a child from harm occurs in many cultures through constant carrying, by parents, or siblings, or others who help mothers (Hewlett, 1987; Paolisso et al., 1989). In other cultures, children are placed in a crib, a cradle, a hammock, or a playpen for long periods of time (Kotchabhakdi, 1987).

As we discuss the various dimensions that influence childrearing and look at changes along these dimensions, we will return to these examples.

Who Cares for Children?

The activities chosen to meet children's needs are dependent in part on who cares for the child, which, in turn, is influenced by the context, and cultural beliefs, and the realities of daily life. The selection (or assignment by circumstance) of the person (or people) who take responsibility for raising a child is important because people differ with respect to their knowledge about childrearing, their experience, their beliefs, and their level of self-confidence. These differences will affect the choice of practices and the development of children.

As indicated at the outset, childrearing does not always fall to parents and may occur outside the circle of immediate family members. In Zambia, a study that included a diverse sample from throughout the country found that 16 percent of the children were being cared for by someone other than their own parents (Chibuye et al., 1986). Other examples include the collectivized childrearing of Israeli kibbutzim or Chinese communes. And, although legal and social responsibility may remain with the immediate family, in some cultures it has always been common to assign responsibility for daily childrearing activities to nonfamily members—wet nurses, servants, or neighbors, for instance. In a sense, the growth of childcare centers has formalized what were previously informal patterns of childrearing outside the home. Because nonfamily individuals sometimes bring to the childrearing task a very different knowledge base, set of beliefs, and cultural background from those of family members, their practices can be very different from those that are, or would be, applied by family members. These discontinuities can foster cultural disruption, help integrate children into changing cultures, or both, depending on the particular circumstances.

In most societies and in most cases, the family is the primary unit given responsibility for raising children. Even when childrearing is held closely within the family, however, there is considerable variation in who takes responsibility for childrearing, depending in part on who is present in the family but also on cultural norms for assigning childrearing roles, on the degree to which other family responsibilities compete with childrearing given the particular economic and social conditions in which the family lives, and on the psychological makeup of the available caregivers. Only rarely is childrearing assigned exclusively to the mother, although that may be the case during the first months of life. In parts of Andean Peru, for instance, a child is considered to be part of the mother until it is baptized—usually during the first weeks after birth. During that period, the mother provides exclusive care. Until weaning, others may be allowed to see or talk to the baby only for brief periods, but even then they are not to be effusive (Ortiz and Souffez, 1989).

In many cultures, grandmothers or mothers-in-law play a major role in upbringing. This is particularly true for the first child. The mother is not yet considered capable of providing proper upbringing on her own. This arrangement provides informal parental education, but care by grandmothers or mothers-in-law is also linked to maintaining kinship ties and power relationships within the broader family and society, and to the necessity for young women to work while older women are available to provide child care. It may even be linked to a sense of obligation to grandparents to whom a child is "fostered" (Dare and Adejomo, 1983).

Siblings have played an important role in upbringing, particularly when mothers must farm or work outside the home. In some cases, placing a child in the primary care of siblings has been shown to have a negative effect on the child's welfare. Shah et at. (1979) report such an effect for rural Maharastra in India. In other instances, the presence of older siblings to help out is seen as an important aid to child development or as a deterrent to illness. Paolisso et al. (1989) describe increased diarrhea among young children during periods when siblings must be in school and are not available for caring.

The role that fathers play in childrearing differs across cultures. (For a cross-cultural perspective on the role of fathers in upbringing, see Lamb, 1987; Evans, 1995.) At one extreme, one can cite the increase in male single-parent families in the United States or the case of fathers in the Aka group of Pygmies in West Africa who have been observed to spend as much as 20 percent of their time holding and interacting with their infant children (Hewlett, 1987). At the other extreme are cases in which a high percentage of fathers are absent from the home, either because they have migrated to work—as in rural Swaziland (Zoller Booth, 1995) or in rural Ecuador (Roloff et al., 1995)—or because of an increasing divorce rate.

Even when fathers are present in nuclear or extended family arrangements, however, it is common to find that fathers and young children have little contact. This was one of the findings of an 11-country study of 4-year-olds carried out by the High/Scope Foundation in conjunction with the International Association for the Evaluation of Educational Achievement. (The countries were Belgium, China, Finland, Germany, Hong Kong, Italy, Nigeria, Poland, Spain, Thailand, and the United States.) This study showed that fathers in all 11 countries averaged less than one hour per day alone with their child as contrasted to more than five hours alone with their child by mothers in the same countries (Olmsted and Weikart, 1994). An Indian study of the girl child (carried out in 10 states) showed that girls shared less than 1 percent of their play time alone with their fathers. The study concluded that, "It is our observation in the field (and supported by earlier findings from several studies on socialization available as theses and dissertations) that the Indian father has abrogated his responsibility of parenting. The tasks of providing for food, education, and marriage are in a sense the economic duties of the father, but beyond what is the basic minimum, the father steps out of the scene, surrendering his socialization role and losing the opportunity to develop emotional closeness with his children" (Anadalakshmy, 1994, p. 66). A study of Lao childrearing practices indicated that fathers in Laos do not get involved in any child care until the child is age 3–4 (Phanjaruniti, 1994). When fathers are present, it is common for them to be assigned the role of disciplinarian and to serve as a role model for their sons as they approach and enter school age (Evans, 1994). Fathers also influence the childrearing process indirectly by assuming (or not) the role of economic provider and by providing (or not) emotional support to mothers.

Research, mostly carried out in the North, suggests that when fathers are a significant part of a child's life from birth, the children score higher on intelligence tests and do better in school than children whose fathers are less involved. According to Lamb (1987), the level and the type of a father's involvement seem to be more important in determining positive effects on child development than the amount of time that a father spends interacting directly with the child. Further, that involvement seems to be determined not only by the time available to fathers outside their work routines, but also by their motivation (influenced in part by norms related to masculinity), by levels of confidence related to knowledge and skills (often low because fathers do not have prior experience or time alone with young children), and by the degree of support provided within the family. (In many families, women do not want husbands to be more involved in childrearing because it further undercuts their own status.) The results of studies of father absence vary from "no effects" to detrimental effects, related variously to length of absence, the age of children at absence, whether or not economic support continues during absence, the degree of extended-family support provided, and cultural limitations on women as they try to fill the gap left by absence (Lamb, 1987).

The Influence of Environment on Childrearing Practices

The wide variation in the physical, social, economic, and cultural settings in which childrearing occurs will have a major influence on who brings up children and on the practices they follow.

The Physical Environment

The climate and the geography of an area will help define both children's specific needs and the practices adopted to meet those needs. In a review of literature on this subject, John Whiting (1981) notes that the manner in which infants are cared for during their waking and sleeping hours has been substantially influenced by the physical environment, the temperature of the coldest month of the year being the most important factor. He also concludes that there is not sufficient evidence to indicate whether these variations in methods of infant care have any enduring effects.

Continuing the focus on feeding and protection, consider the following examples of how specific childrearing practices and techniques used must be adapted to varied physical settings, even though the needs they are designed to meet and the ends they serve are similar across them. Both an Eskimo (Canada) and Yoruba (Nigeria) mother may carry their babies and breast-feed on demand. But for the Eskimo mother to do so requires wearing a heavy, warm, very large, and loose garment that allows the baby to be carried on her back and swung around, inside the garment, at the time of feeding. A hood on the garment channels air to the baby while it is on the mother's back at the same time that it helps to keep the mother and baby warm. The Yoruba mother, faced with a tropical climate, needs only to sling the baby on her back, open to the elements. In both cases, the baby is in direct touch with the mother. In the Eskimo case, however, the baby cannot benefit from the sight of other people and of the world in general while being carried. Other means are required for providing that kind of stimulation and social interaction.

The Social and Economic Environment

The physical environment helps determine settlement patterns, which will also affect childrearing choices and patterns. The conditions for childrearing in a nomadic society differ from those in a settled society, and, among those who are settled, conditions in dispersed settlements will differ dramatically from those in concentrated areas. Further, small concentrations in rural towns provide different conditions from large concentrations in major cities.

Childrearing practices will also be influenced by variations in forms of economic livelihood and the division of labor. In some societies, when both parents directly share work obligations, as among the Aka (Hewlett, 1987), it improves the possibility that both will share childrearing tasks. However, this sharing of childrearing tasks is not usually the case. Studies from Jamaica, India, and the United States indicate that, when a woman works outside the home, this usually does not relieve her of childrearing responsibility nor lead to greater involvement of men in childrearing. Particularly in urban areas, when women work outside the home, the demand for and use of institutional care increases.

Practices are also influenced by social organization at the level of family, community, and the larger society. Factors that have an effect on childrearing include divisions by caste and class; whether a culture is patrilineal or matrilineal; the social definition of men's and women's roles; the degree of solidarity and hierarchy in communities; family size and composition; and organization into extended, polygamous, nuclear, or single-parent families. For instance, extended and polygamous families, in which the kinship groupings are large, patterned, and complicated, will provide a wider range of potential caregivers than either a nuclear or a single-parent family in which there is isolation from other kin. At the same time, polygamy can bring jealousies among women and conditions that negatively affect childrearing practices, and in-laws are not always supportive.

Beliefs and Values, Norms and Practices

Beliefs help rationalize, justify, and explain particular practices that have grown up over the ages or that are carried out by individuals. Beliefs merge with values in giving meaning to practices by defining the kind of child (and adult) a particular society seeks to produce. Beliefs and values also provide the backdrop against which cultural norms are established, representing how particular cultures define the way in which their children should be reared.

Cultures are guided, and distinguished by, their beliefs about what happens both in this world and in an afterlife. These beliefs may arise from long and practical experience in particular conditions that daily life presents or they may represent attempts to deal with the unknown (such as, a belief in the power of the evil eye).

In Ethiopia, a belief, derived from concrete experience, that a baby should be small at birth, has persisted in rural areas. Accordingly, food taboos that keep weight increases down during pregnancy commonly lead to the birth of what the World Health Organization (WHO) would define as "low birth-weight" babies. The origin of this apparently irrational belief lies in the fact that, at one stage, rickets was prevalent among teenage girls in Ethiopia, affecting pelvic growth and making it dangerous to give birth to a large child. With a change in conditions, however, the belief, originally well grounded and related to a value placed on the life of the mother so that another child could be born, had not changed as recently as the late 1980s (Negussie, 1988). The belief and the associated practice persists despite the fact that low-birth-weight babies tend to be less active and attentive and relatively more irritable, and despite a demonstrated relationship to infant mortality and to delayed or debilitated development.

In a less tangible vein, newborns are considered to have different relationships to the spirit world. In many cultures, a newborn is considered a gift from god (Evans, 1995), to be treated with the utmost respect and indulgence in the early years. In Bali, Indonesia, "the newborn are treated as celestial creatures entering a more humdrum existence and, at the moment of birth, are addressed with high sounding honorific phrases for gods, the souls of ancestors, princes and people of a higher caste" (Mead, 1955, p. 40). By way of contrast, in northeast Thailand,

"newborn infants are usually wrapped and placed in a basket lined with a blanket, close to the mother for a few days. Parents, relatives and neighbors usually do not openly express their enjoyment or admiration of the baby for fear that the spirits might take the baby away. Relatives usually say aloud, 'What an ugly baby he is,' in order to deceive the spirit" (Kotchabhakdi, 1987, p. 20). In societies in which a high level of infant mortality was, or is, common, beliefs have arisen that help explain and help people accept a young child's death. For instance, the newborn is believed to remain in a pure state for some period. If the child dies during the early period, it returns to the spirit world, having avoided the trials and tribulations of a temporary stay on Earth (Scheper-Hughes, 1985). When children are looked upon as a human creation and responsibility is assigned to people, "letting go" is more difficult.

Most cultures have a set of beliefs about how children develop physically, mentally, and socially. In some cultures, a child is believed to be fragile; in others, the child is perceived to be hardy. Therefore, handling practices differ. In some cultures, children are believed to be capable of learning in their early years, but not of being taught. This affects the choice of teaching methods: direct instruction, as contrasted with learning by observation and imitation, for example. Likewise, people in some cultures believe it necessary to take an active role in helping a child learn to sit and walk, while those in others do not.

Beliefs merge with values to define the kind of child (and adult) a particular society seeks to produce. Some cultures want children to be obedient; others foster a questioning child. Some tolerate aggressiveness; others value a quiet submissive child. Some stress individualism; others have a collective orientation and promote strong social responsibility. A study of Thai ways of childrearing (Amornvivat and Khemmani, 1989) identified the following key values: obedience and respect for elders, diligence and responsibility, generosity, honesty, gratitude, being economical and self-reliant.

Cultural norms or patterns are derived from experience and tied to beliefs and values, and they influence the choice of how to respond to children's needs. They represent the generally accepted styles and types of care expected of caregivers in responding to the needs of children in their early months and years. Normative patterns represent a culture's means of ensuring behaviors that will maximize the survival, maintenance, and development of the group or culture as well as of the child. Norms apply to behaviors expected at various points in a child's life. For example, societies differ in expectations for parental and community behavior with respect to such activities as birthing practices, how a child should be named, how an infant's death is handled, when to stop breast-feeding, what games to play, how a child should be punished, and so on.

An example of how beliefs, values, and norms may operate is provided in a study of an Ecuadoran community in which a belief that women should not be aggressive combined with a value placed on women's submissiveness led to differential norms guiding the length of time that boys and girls should be breast-fed. Girls were weaned earlier than boys. which had an impact on differential growth and development. The earlier weaning for girls was related to a higher mortality rate for girls prior to age 5 (McKee, 1980).

To balance this example, one could point to many norms, embodied in traditional wisdom, that seem to have a salutary effect on the survival and development of young children. For instance, a norm that young children should participate at an early age in work and ritual activity (as well as play and formal education) has many positive consequences. It provides learning that occurs in context and in naturally occurring sequences. Children are required to take on tasks that are increasingly complex, and they are provided with direct and repeated monitoring. This process builds confidence and competence and inculcates social responsibility.

While general agreement about what should be done in raising children may exist for the culture as a whole, these norms may or may not be followed by individuals, who are affected also by their particular circumstances and who bring different beliefs and knowledge to the task. Sometimes within a culture deviation from the norm leads to ostracism. In other instances, there is considerable latitude in terms of adherence to cultural patterns. In times of rapid change, norms often blur and they are hard to define.

Changes in Childrearing

Changes in the Environment

Perhaps more important than the fact that contexts within which childrearing occurs are different is the fact that contexts are changing constantly, sometimes slowly and almost imperceptively, allowing adjustments to occur along the way, sometimes rapidly and with potentially disastrous consequences because adjustment is difficult. For instance, in the Western world, important changes in childrearing practices and patterns accompanied the Industrial Revolution in the eighteenth century. In the predominantly agricultural and rural society that preceded industrialization, children were usually found within an intact, extended family and socialized to a relatively limited and unchanging world with agreed upon community values. The rural setting provided space to explore and a stimulating environment. The responsibility for childrearing lay clearly with women, whose work usually permitted them to breast-feed and to attend directly to the child in the early years. Families were often large, and older children were expected to help with child-care tasks. Children quickly entered an adult world and, in a sense, did not have a separate childhood as they do today.

The rural conditions of the eighteenth and nineteenth centuries in the West, and in most parts of the developing world of today, should not be romanticized. Rural life is demanding, and survival is threatened constantly by disease and often by lack of food. Still, for those rural children who manage to survive their early months, their development is less problematic than it is for many of their peers in new urban environments. The childrearing practices that evolve for socialization in a rural environment are not suited to cities.

With industrialization and migration to the cities come changes in values, in living conditions, in family arrangements, and in working patterns. The new conditions bring with them both a need to care for children of working mothers and a need to provide adequate stimulation for children within a restricted physical environment. They call for new parenting skills and a different kind of socialization process.

Childrearing continues to be affected by population growth, changes in fertility patterns and in economic policies, geopolitical shifts, and changes in technology. While population growth and the related pressure on land have helped fuel migration to cities, declining fertility has meant smaller families, resulting in fewer children to care for, but also fewer siblings to help with childrearing. A declining infant and child mortality rate means that more attention needs to be paid to strategies of childrearing focused on child development rather than on child survival. This basic shift in outlook does not come easily.

Economic shifts, including longer-term trends toward inclusion of all people in a monetary economy, cash cropping, and industrialization, and the changes since the 1980s associated with neoliberal adjustment policies, have had profound effects on the form and content of childrearing. We see increases in unemployment, reductions in social expenditures, and the need for poorer families to turn toward self-generated employment in the informal sector, either as a supplement to formal-sector earnings or as their sole source of support. As a result, there is a greater presence of women in the paid labor force, changing occupational distributions within households, and a growth of female-headed households. While women have entered the labor force in greater numbers, they have often done so in informal-sector jobs that are too frequently insecure, require long hours of work, and carry with them no social benefits. These trends have posed painful dilemmas for women with childrearing responsibilities, often forcing them to leave their children alone during the day.

Since the late 1980s, the global Cold War has given way to local hot wars and conflicts among and within nations. As a result, the displacement of families has become a problem of major proportions. An important part of that problem is on how to rear children under such conditions.

Technology has brought new dimensions to everyone's lives. A communications revolution has helped create the global village. Transistor radios now reach into most corners of the world, and even television is reaching rural areas to a degree not thought possible. The educational revolution, with its emphasis on literacy and schooling, has dramatically altered socialization patterns and defined new childrearing needs. For instance, participation in schooling requires much greater emphasis on cognitive skills related to abstract reasoning. The arrival, perhaps intrusion, of schools into rural areas has brought competition with indigenous educational forms. And schooling competes with domestic work that has been such a central part of children's lives over the centuries. Technology has also affected childrearing by making available such products as bottles and cribs and plastic toys.

Revolutions in transportation and organization have occurred as well. Buses not only help rural people visit cities and migrate to them, they also aid periodic or permanent return to villages, where new ideas and modes of behavior can be displayed. Developments in communications have facilitated the outreach of business and governmental organizations to villages. There is, then, not only a shift to the city with accompanying changes, but also a reach of the city into rural areas. In addition to the introduction of new technologies, there is a changing sense of community and a confusion about responsibilities and loyalties. The changes bring uncertainty about old values and ways of doing things, including rearing children. The implication of all of this is that even children who remain in rural areas increasingly live in national and global cultures, even as they are rooted in a local culture sometimes unsure about its own roots and directions.

Changes in Beliefs and Values, and Norms and Practices

With changes in the environment come changes in beliefs, values, and norms, all of which evolve as the contexts and specific needs of people change. When societies were more or less isolated from one another and there were few outside influences, what one generation passed on was similar to the way the next generation raised its children, and there was relative stability of values, beliefs, and norms. Changes could be incorporated without losing the traditional. However, few cultures have remained relatively isolated and intact. Most are vulnerable to outside influences—schools, mass media, missionaries, returned migrants—that challenge established beliefs, values, and norms. The juxtaposition of the old and the new can leave cultures disorganized and groups of people at a loss in terms of their values and beliefs. In the jargon of modern-day psychology, these cultures might be classified as "dysfunctional": They cannot define norms and no longer provide children with the grounding, stability, and vision that was found within traditional belief systems.

In the struggle for identity and in the desire to be modern, some cultures have completely cast off their traditions, or think they have. Yet both rural and urban, or indigenous and industrialized, cultures are recognizing that adopting "modern" ways does not always offer the best means of providing what is desirable for children to grow and develop. As a result, people are seeking to identify and recapture traditional values. There is an increasing awareness that much of what existed within traditional cultures was positive and supportive of growth and development, for the individual and for the society. Conversely, some traditional beliefs and practices that may have had a logi-

cal origin are recognized today as harmful to a person's health and well-being and are being changed.

As noted, the general needs of children do not change. For children to survive, grow, and develop, they still need to be nourished, to avoid disease and accidents, to be nurtured, and to learn the ways of the world. But against the background presented of changing environments, beliefs, values, and norms, it is clear that many childrearing practices must change and are changing. This is true for rural families who are asked to assimilate new ways but resist changes that may, indeed, be necessary for their children to function in the transitional world or multiple worlds that surround them. It is true for young, single-parent mothers who draw upon their memory of how they were brought up even though conditions have changed drastically. It is true for those who migrate to cities. All need to make adjustments to a new environment. To protect a child in an urban setting, for instance, it may be necessary to keep the child indoors a large part of the time. If a child must be indoors in a confined space rather than in the open countryside, new ways of stimulating the child must be created. If a family cannot grow its own crops, it must rely on what is available in a market. Thus feeding practices change. If a mother must work and cannot take the child along, breast-feeding may not be possible. If a child is now expected to do well in school rather than do well tending animals, there is a need to be socialized to new ways.

In general, childrearing practices change slowly, despite pressures for change and despite increasing access to information that supports alternative practices. As new demands are made on families, sound and solid practices that worked well previously may be set aside. These include such practices as breast-feeding, massaging, carrying, and storytelling, which are being replaced by new practices such as bottle-feeding, bathing in a bathtub, putting children in playpens, and watching television.

Childrearing practices do not stand alone. In general, they tend to cluster as a result of the interrelatedness of children's developmental needs, environmental conditions, and the fact that practical choices made to satisfy one need may limit what it is possible to do to satisfy another. For instance, the choice of breast-feeding on demand requires a kind of physical proximity to the mother at all times that scheduled bottle-feeding does not. That requirement affects carrying and sleeping practices. The physical closeness of mother to child also permits, or reinforces, a kind of nurturing that is different from that employed when there is greater physical separation.

Changes in Who Cares for the Child

We have noted that with the decrease in the number of extended families, the childrearing role of grandparents and in-laws and near relatives is diminishing. We have noted that changes in context and family structures lead more often than in the past to the absence of fathers. In addition, both the need for children to attend school and the tendency toward smaller families make it less feasible for siblings to participate in child care in a major way. Moreover, in urban areas, the sense of community is often very weak and neighbors are often unfamiliar people. Leaving children with the neighbor becomes more difficult, and the traditional tendency for childrearing to be a responsibility of all community members is lost.

A result of the above shifts is that women, who are under more pressure to work outside the home, but who, in most cultures, continue to be assigned the responsibility for childrearing, have fewer options than before for help with the childrearing task. It is not surprising, then, that the demand for institutionalized forms of child care should be increasing in most parts of the world and that a major share of childrearing is being assigned to nonfamily members working in nonfamily institutions.

Indeed, institutionalized forms of child care have grown by leaps and bounds in the last three decades of the twentieth century. (An exception to this seems to be found in many parts of the former socialist world, where the number of formal institutions of child care has dropped as economies have changed.) These include preschools, creches, centers for integrated attention to the child, nurseries, and home day care, each with its own characteristics. Some of these institutions are formal governmental systems of care, often created as a worker's benefit. Others are considered to be nonformal in the sense that the people who run the centers are not part of the government payroll, even though their center may receive some sort of subsidy from the government.

A review of literature dealing with women's work and child care (Myers and Indriso, 1987) suggests that, to be judged adequate by women, child-care centers, whether formal or informal, should be affordable, accessible, flexible (particularly with respect to hours), run by trusted and accountable persons, and meet minimum quality standards. Many centers do not meet such standards, which, in part, explains why it is sometimes possible to encounter a supply of available places that are unused, in the presence of an unsatisfied demand for such places.

Working women's expectations with respect to the need for adequate child-care arrangements are backed by a literature regarding the actual and potential effects of institutionalized care on children. If an institutional setting provides a warm, loving, safe place, child care outside the home can provide children with the types of experiences that foster healthy growth and development. This is particularly so if previously they were reared exclusively in a home environment that did not provide these conditions. If, however, centers are custodial places of poor quality, children may suffer developmentally and be worse off than peers who are cared for at home (Kagitcibasi, 1996). Thus the quality of a child's experiences in the child-care setting and at home is more important than whether the child is involved in programs of care outside the home.

There is some evidence that programs of child care and/or preschooling can be of particular benefit to children from impoverished or socially disadvantaged homes and for girls living in conditions that put them at a disadvantage. Such children will be more likely to attend school and perform well than their peers who have not had a similar opportunity (Lal and Wati, 1986). They will also be less likely to repeat grades or leave school early. In the best of circumstances, one might expect that quality programs of early education will lead to social and economic benefits for the participating individual and society as well.

Implications for Policy and Programming

It is clear that the balance of responsibility for childrearing is shifting. Institutionalized forms of child care are becoming more prominent, and family responsibility is being reduced. This has been true for many decades as government-run institutions (i.e., schools) have taken over the educational function beginning at about age 6. The shift to government responsibility for young children is now occurring for children during their preschool years, and all signs point to this trend continuing.

As the care for young children continues to shift to preschools, creches, home day-care centers, and other institutions, often under the auspices of the state, it is important to be sure that a process of assessment is in place that will inform in the best way possible policy formulation and program development by various levels of government, by private organizations, and by international funders. Such an assessment process will involve several steps, running from micro (child) level to the macro (political) level:

Assess the Situation for Children. It is important to assess children's needs in a given context. Are the children living in a relatively stable environment or is their world shifting on a daily basis? Do they have adequate nutrition? Are there gender differences in the way children are cared for?

Know the Current Child Care. It is important to understand the current child-care situation. Who is caring for the children? Has there been a shift in this as a result of recent economic and social changes? Where are children being cared for and under what conditions? Are mothers the only ones caring for children? To what extent are older siblings involved? If more and more girls are attending and remaining in school, who is caring for the youngest children?

If children are being left at home alone when mothers are involved in income-generating activities outside the home, and older siblings are in school, then alternative child-care systems need to be developed. This might involve the development of neighborhood day care, in which care is provided in the home of someone nearby. If a greater number of children require care, then perhaps it is appropriate to create a child-care center.

Provide Support for Caregivers. It is important to understand the caregiving environment. What kind of supports do the current caregivers require? Are these adequate? One place to begin is to look specifically at the situation of the caregiver, primarily mothers, in the community. What supports do they have and what do they need? What resources do they have available to them, human and financial? What kinds of stress are they under? What are the demands on their time?

Learn About the Current Childrearing Practices. Another place to look is at the childrearing practices themselves: It is important to assess whether or not current childrearing practices are adequate to support children's growth and development. What other forms of childrearing are required to meet children's needs?

As socialization for parenthood through experience with younger siblings and under the strict supervision of parents has diminished, there is an ever greater need for parental education programs that provide parents with better parenting information and skills. Childrearing practices and beliefs should provide the starting point for such programs. Data on the culture gathered by anthropologists, psychologists, and the community itself can be used to identify specific areas that should be stressed in a program and to develop positive parenting modules.

The education of future parents can begin in the late years of primary school or in secondary school. Parental education can occur in health centers in conjunction with monitoring during pregnancy and lactation. It can also be included in literacy courses. However, it is not only the content that is important. It is also critical to be aware of the process used to convey the information to parents. If the information builds on strengths parents already have, the new information will be incorporated much more easily than if parents feel or are led to believe that they are inadequate. It is best to begin by asking: What are the local childrearing practices and beliefs that can be built upon to strengthen childrearing practices?

Facilitate the Provision of Child Care Outside the Home. If alternative forms of childrearing are required, a question to ask is: What resources are available within the community to create supports for caregivers and/or alternative child care? Are there untapped resources within the community that could be used in providing better support for children's development?

It is important for the community to understand the need for a focus on child care. Those to be involved in the program (sponsors, community members, and parents) must come to believe in the importance of childrearing and to understand the importance of their role in the process. Further, since parents are the policymakers in their children's lives, their involvement in the design and operation of alternative child-care options is critical. The greater the level of parent and community involvement, the more likely it is that the program will meet family needs and become an integral part of community life.

In the development of alternative forms of child care, it is increasingly important to provide quality programs and avoid substandard programs for the disadvantaged; to work against the tendency to expand programs so rapidly that the system cannot keep up, leading to poor-quality programs for the poor; to see ways to decentralize programs so that a culturally and geographically appropriate response can be made to demands for institutionalized care and real community participation can be increased; and to assure that programs are accessible, affordable, and flexible, are run by trusted and accountable persons, and meet minimal quality standards.

To have a positive impact on children, programs do not necessarily need to have a child-development or parent-education focus. Policies that uplift the lives of family members, particularly mothers, and support the community, produce indirect benefits for children. Programs that give women additional disposable income have indirectly affected children because the women tend to use these new resources to enhance their children's health and education. Thus a variety of social-support-system efforts can be put into place to improve the context within which children live.

Beyond the community, there is a need to assess the climate of support for childrearing concerns within the larger sociopolitical arena. There are certain preexisting conditions that enhance or hinder the likelihood of success of any effort to support childrearing. Political commitment at all levels should include an appropriate supportive policy, adequate budget allocations, and a willingness to make structural reforms where necessary.

One question to ask is: What is required, in terms of national policy and support, to provide for children's healthy growth and development? Legal changes are called for in many countries, for example, to enable working women to dedicate full time to the care of their children during the first year of life without fear of losing their jobs or benefits. Legal provisions that promote greater participation of men in childrearing are also needed.

There is an additional need to recognize childrearing and child care as productive. On the one hand, this means that people, mostly women, who are entrusted with the care of children in institutions should be properly compensated. The tendency to exploit women in child-care positions by paying substandard wages and by withholding benefits is directly related to the lingering belief that a woman's place is in the home, that it is natural for women to care for children, that anybody can do it—and, therefore, that a woman's services in helping to rear children should be voluntary. A step toward recognition of child care as a productive activity would be to include this work in national statistics related to economic productivity.

The eve of the twenty-first century is a time of major change in childrearing beliefs, norms, and practices. If we are to retain the good practices from traditional systems and make needed shifts to respond to the major changes thrust upon us by industrialization and so-called modernization, we will need to be much more open than we have been in the past to recognizing both the wisdom of experience and the necessity of change.

References

Amornvivat, Sumon, and Tisana Khemmani. *Thai Ways of Child Rearing Practices: An Ethnographic Study.*

Bangkok: Faculty of Education, Chullalongkorn University, 1989.

Anadalakshmy, S. (ed.). *The Girl Child and the Family: An Action Research Study.* Madras Department of Women and Child Development, Indian Ministry of Human Resources Development, 1994.

Chibuye, P.S., Mary-Ann Mwenda, and Connie Osborne. *CRZ/UNICEF Study on Child Rearing Practices in Zambia.* Lusaka: Zambia Association for Research and Development, 1986.

Dare, Glenda, and D. Adejomo. "Traditional Fostering and Its Influence on School-Related Behaviour in Nigerian Primary School Children," *International Journal of Child Development,* vol. 15, no. 1, 1983, pp. 16–19.

Evans, Judith L. "Childrearing Practices in Sub-Saharan Africa: An Introduction to the Studies," *Coordinators' Notebook,* no. 15, 1994, pp. 25–41.

———. "Men in the Lives of Children," *Coordinators' Notebook,* no. 16, 1995, pp. 1–15.

Hewlett, Barry. "Intimate Fathers: Patterns of Paternal Holding Among Aka Pygmies." In Michael E. Lamb (ed.), *The Father's Role: Cross-Cultural Perspectives.* Hillsdale: Lawrence Erlbaum Associates, 1987, pp. 295–330.

Kagitcibasi, Cigdem. *Family and Human Development Across Cultures: A View from the Other Side.* Hillsdale: Lawrence Erlbaum Associates, 1996.

Kotchabhakdi, Nittaya J. "The Integration of Psycho-Social Components of Early Childhood Development in a Nutrition Education Programme of Northeast Thailand," paper presented at the Third Meeting of the Consultative Group on Early Childhood Care and Development, Washington, D.C., January 12–14, 1987.

Lal, S., and R. Wati, "Non-formal Preschool Education—an Effort to Enhance School Enrollment," A paper prepared for the National Conference on Research on ICDS, February 25–27, 1986. New Delhi, National Institute for Public Cooperation in Child Development (NIPCCD). Mimeographed.

Lamb, Michael E. (ed.). *The Father's Role: Cross-Cultural Perspectives.* Hillsdale: Lawrence Erlbaum Associates, 1987.

McKee, Lauris. "Ideals and Actualities: The Socialization of Gender-Appropriate Behavior in an Ecuadorian Village." Ph.D. diss., Cornell University, Ithaca, 1980.

Mead, Margaret. "Children and Ritual in Bali." In Margaret Mead and Martha Wolfenstein (eds.), *Childhood in Contemporary Cultures.* Chicago: University of Chicago Press, 1955, pp. 40–51.

Myers, Robert G., and Cynthia Indriso. "Women's Work and Childcare," paper presented at a workshop: "Issues Related to Gender, Technology, and Development," New York, February 26–27, 1987. New York: Rockefeller Foundation, 1987. Mimeographed.

Negussie, Birgit. *Traditional Wisdom and Modern Development: A Case Study of Traditional Peri-natal Knowledge Among Elderly Women in Southern Shewa, Ethiopia.* Stockholm: Institute of International Education, University of Stockholm, 1988.

Olmsted, Patricia P., and David P. Weikart (eds.). *Families Speak: Early Childhood Care and Education in 11 Countries.* Ypsilanti: High/Scope Press, 1994.

Ortiz, Alejandro, and M. France Souffez. *Patrones de Crianza Infantil en el Area Rural Andina.* Lima: Centro Latinoamericano de Estudios Educativos, 1989.

Paolisso, Michael, Michael Baksh, and J. Conley Thomas. "Women's Agricultural Work, Child Care, and Infant Diarrhea in Rural Kenya." In Joanne Leslie and Michael Paolisso (eds.), *Women, Work, and Child Welfare in the Third World.* Boulder: Westview, 1989, pp. 487–496.

Phanjaruniti, Somporn. *Traditional Child Rearing Practices Among Different Ethnic Groups in Houphan Province, Lao People's Democratic Republic.* Vientiane: UNICEF, 1994.

Roloff, Gerardo, Pilar Nuñez, and Juan Vásconez. *Los Niños del Ecuador: Prácticas de Crianza en Zonas Rurales y Urbano Marginales.* Colección Prácticas de Crianza no. 3. Santafe de Bogota: Consejo Episcopal Latinoamericano, 1995.

Scheper-Hughes, Nancy. "Culture, Scarcity, and Maternal Thinking: Maternal Detachment and Infant Survival in a Brazilian Shanty Town," *Ethos,* vol. 13, no. 4, 1985, pp. 291–317.

Shah, P.M., S.R. Walimbe, and V.C. Dhole. "Wage-Earning Mothers, Mother Substitutes, and Care of Young Children in Rural Maharashtra," *Indian Pediatrics,* vol. 16, no. 4, 1979, pp. 167–173.

Super, Charles, and Sarah Harkness. "The Developmental Niche: A Conceptualization at the Interface of Child and Culture," *International Journal of Behavioural Development,* vol. 9, 1986, pp. 545–569.

Whiting, John. "Environmental Constraints on Infant Care Practices." In Ruth H. Monroe, Robert L. Monroe, and Beatrice B. Whiting (eds.), *Handbook of Cross-Cultural Human Development.* New York: Garland, 1981, pp. 155–179.

Zoller Booth, Margaret. "Children of Migrant Fathers: The Effects of Father Absence on Swazi Children's Preparedness for School," *Comparative Education Review,* vol. 39, no. 2, May 1995, pp. 195–210.

Lives of Middle-Aged Women

Judith K. Brown

Introduction

In a publication such as *Network News: A Newsletter of the Global Link for Midlife and Older Women,* we can catch varied glimpses of the middle-age woman in the Third World. In Latin America and the Caribbean, her vulnerability is noted as follows:

> Women currently at midlife may be faced with an even greater disparity in the male/female ratio by the time they reach age 65. Taking into account the traditional disadvantages women have suffered in education, employment, and health, it becomes apparent that an important segment of the population of Latin America and the Caribbean is potentially at great risk in the coming years (*Network News,* 1987, p. 1).

But an entirely different picture emerges from the other side of the world, where older Chinese women are awarded considerable authority:

> The Chinese government has established the street committee [composed of older men and women] as the main link between the leaders of the central government and ordinary people. . . . From keeping track of family planning practices of neighbors to mediating family disputes, . . . the committees also distribute grain and food coupons, notify parents of vaccination schedules for their children, take care of hygiene, running water and public security. They even set up bicycle parking lots . . . (*Network News,* 1991a, p. 24).

From vulnerability to exercising authority, varied roles are described for the middle-age woman in the Third World.

But what about her role in traditional societies?

Anthropological researchers who conducted the early cross-cultural investigations of older women found that there was little descriptive data available in the ethnographic literature and their studies often combined information on aged women, on middle-age women, and on menopause. Today these are viewed as separate though related topics and there is a considerable anthropological literature on all three subjects. (For recent anthropological studies dealing specifically with menopause, see for example Beyene, 1989; Kaufert and Lock, 1992; Lancaster and King, 1992; Lock, 1993; for studies of middle-age women, see Datan, Atonovsky, and Maoz, 1981; Davis, 1983; Boddy, 1992; Counts, 1992; Kerns and Brown, 1992; Lee, 1992; Raybeck, 1992; Sacks, 1992; Solway, 1992.)

No doubt the study of middle-age women in the Third World as well as in the accounts of anthropologists has been hampered by the difficulty in defining "middle-age." Chronological age is not a useful index. In many societies, the ages of adults are approximations, because there are no written birth records. Furthermore, since there are societies in which the end of child bearing and rearing is negotiable by means of adoption and fosterage, the end of child care also does not provide a useful definition. To avoid these inaccuracies, the following definition is suggested for the cross-cultural data presented below: middle-age women (matrons) are women who have adult offspring and who are not yet frail or dependent. (Also see Brown, 1982, 1985, and 1992.)

As women reach middle-age in non-industrialized societies, three kinds of changes take place in their lives. First, they are often freed from the cumbersome restrictions they once had to observe when younger. Second, the changes may

Some sections of this chapter appeared before in Judith Brown, "Lives of Middle-Aged Women," in Virginia Kerns and Judith Brown (eds.), *In Her Prime: New Views of Middle-Aged Women.* Copyright 1992 by the Board of Trustees of the University of Illinois. Used with permission of Westview Press.

include eligibility for special statuses, and thus for recognition beyond the confines of the household. Third, matrons are expected to exert authority over specified younger kin. This may involve the right to extract labor or the right to make important decisions for the younger person.

In many societies, middle-age women are freed from exhibiting the deferential and even demeaning behavior they had previously been expected to display to the senior generation or to their husband. Although some aged persons may still require expressions of respect, at middle age women have become the active senior generation and now may receive the deference they once had to display. Furthermore, menstrual customs no longer apply, nor do the restrictions that may be imposed on pregnant or lactating women. In societies that demand great propriety in younger women, and in which their conduct is narrowly prescribed, many of the rules are lifted for matrons. Older women may interact informally with men who are nonrelatives; they may be allowed to drink too much on ceremonial occasions; they may use foul language, dress immodestly, and even urinate in public.

One aspect of this major change is the greater geographic mobility that older women are often allowed. Domestic chores are reduced or are performed by younger women. Child care has ceased or can be delegated. And so commercial opportunities, the hospitality of relatives living at a distance, and religious pilgrimages may provide an opportunity to venture forth from the village. Even in societies where travel is dangerous and where young women need protection, older women may be granted a certain immunity and therefore enjoy freedom of movement. In some societies, all younger women are restricted to the household and its courtyard because they are believed to be sexually voracious and in need of constant supervision. Here, older women are the supervisors, because they are believed to be beyond involvement in sexual escapades that would bring dishonor to the entire kinship group. In many societies in which the young bride must move into the household of her husband and his family, she is at first mistrusted. It is as the mother of grown sons that she is at last viewed as assimilated into the kin group she had entered at marriage. Only then is she finally considered above disloyalty. Thus, once a woman is the mother of adult offspring, she is no longer encumbered with elaborate rules of conduct concerning menstrual customs, modesty, and display of respect; nor is she confined.

A second major change brought on by middle age is the eligibility that older women have for special statuses and the possibility these provide for recognition beyond the household. Societies vary in the number of special positions that their female members can occupy. In some, such opportunities are restricted and there is only the vocation of midwife. Other societies make possible a variety of special offices: curer, leader of girls' initiation ceremonies, holy woman, guardian of the sacred hearth, matchmaker. Such positions are typically not filled by younger women, sometimes because the exercise of spiritual power is considered

harmful to a nursing child or to the baby a pregnant woman is carrying. Sometimes the incompatibility is the result of the great demands of child-care and subsistence activities, which when undelegated, allow neither time nor energy for ritual roles. Further, in many societies the belief that menstruation is defiling, disgusting to the spirits, or dangerous makes women ineligible for dealing with sacred matters until later in life. Even a midwife evokes more confidence if she has given birth to many children and has attended at the birth of many more.

A third major change brought on by middle age is a woman's right to exert authority over specific younger kin. (See, for example, Burbank, 1994; Dickerson-Putman and Brown, 1994.) In many societies, an older woman is expected to be leisured, and it reflects unfavorably on the entire family if she is not. Also, the daughters and/or daughters-in-law will be reprimanded by the woman's sons and will be the object of disapprobation by the community.

In many societies, older women administer the production of food, and its processing, preparation and preserving. They oversee food distribution. Within the household, the matron may have absolute control over who eats and when. A recalcitrant daughter-in-law can be starved into submission or even poisoned. In traditional societies, hospitality and the authority to distribute food to nonhousehold members have significant political and ritual implications. By providing or withholding food, older women can influence the celebration of a ceremony or the meeting of a political council.

The authority of older women also finds expression in shaping important decisions for certain members of the junior generation: what a grandchild is to be named, who is ready to be initiated, and who is eligible to marry whom. In some societies, older women have specific responsibilities in the material exchanges that solemnify a marriage. In societies that maintain very separate worlds for men and women, older women are the go-betweens. The mother of the groom may be the only member of his kin group who actually sees the potential bride and converses with her, before formal negotiations take place (Mernissi, 1975). The report of the older woman is crucial to the arrangements, although the groom's father is officially in charge.

Perhaps most significant are reports by many ethnographers concerning the tremendous influence that mothers exercise over and through their grown sons. Often a man's relationship with his mother supersedes that with his wife. In some societies, the husband is typically very much older than his wife. Early in the marriage, the age discrepancy favors the husband, but as he grows aged and feeble, his middle-age wife gains in power. This power is further enhanced when combined with the allegiance of grown sons, the old man's successors.

The Age Hierarchies of Women

Sixty traditional societies were the subject of a recent study (Brown et al., 1994), that focused on the relationship between older women and their younger female kin, and found this relationship shaped by three variables: (1) the division of labor by sex in subsistence activities (what members of a society do to make a living, such as hunting, herding, horticulture); (2) the cultural rule for postmarital residence (where a married couple is expected to set up housekeeping, whether with the groom's family, the bride's family or on their own); and (3) the cultural rule for reckoning descent (whether the individual is viewed as descended from the father's paternal male ancestors, the mother's maternal female ancestors, a combination of these lines, or from the ancestors of both parents and both sexes).

In some nonindustrialized societies, women are the major breadwinners. When the very survival of the group depends upon the output of female labor, older women dominate their younger female kin in the interest of productivity. The middle-age woman's managerial role often is carried out so deftly that it goes unremarked by the ethnographer. However, Murphy and Murphy (1974) provide a full description of women's work among the Mundurucú of South America. Traditionally in this society, each household of up to 50 people was under the undisputed direction of the oldest woman. She also organized the cultivation and processing of bitter manioc, the staple of the Mundurucú diet. The production of flour from bitter manioc was laborious and involved several sequential tasks that had to be coordinated and rotated among the women of the household. Furthermore, the household's activities had to be coordinated with those of other households in the village. According to Murphy and Murphy, the women cooperated easily with no friction, and the work was administered quite unobtrusively by the elder women.

The status inequality imposed by this division of labor by age (Mitchell, 1994) contains the promise for its younger members that they will become administrators. In their years as laborers, the junior women have ample time to acquire the skill needed for the senior managerial positions. A woman gradually builds up her authority by creating obligations among her juniors (Lambek, 1992; Sinclair, 1992). In a sense, she earns the right to delegate the tasks she formerly performed.

The relationship between a middle-age woman and her younger female kin may be that of a mother and daughter, mother-in-law and daughter-in-law, or senior wife and junior wife, among others. A single older woman can fulfill numerous kinship roles simultaneously. However, in any one society only some of these relationships will be of actual importance, because societal rules, concerning which relatives should share a household, will bring certain relatives into close contact, while separating others.

When these rules dictate that a bride must move into the household of her husband's family at marriage (patrilocal residence), particularly if she must marry a man from a community other than her own (community exogamy), the mother-in-law/daughter-in-law relationship becomes emphasized, while the relationship between a mother and her married daughter becomes tenuous. Patrilocal residence

makes it possible for the mother-in-law to provide supervision of her daughter-in-law. This monitoring is particularly significant for societies in which men and women lead separated daily lives, because custom dictates that men spend their time with men and women with women. (The demands of child-care and subsistence activities separate the sexes to a certain extent in all societies, yet in some it is as if men and women inhabit two different worlds; Whiting and Whiting, 1975.)

In some patrilocal societies, once women move into the household of the family of the husband at marriage, they are "lost" to the family and household of their birth. As Freedman (1967) has noted for traditional Chinese society, a woman's allegiance to the family and household of her husband is questioned until she proves herself worthy of trust by producing sons, and thus has blood ties to her husband's family.

The lives of women in these societies are characterized by discontinuity. As a bride in the household of her husband and his family, she is in a position of servitude, separated from her own kin, living under the authority of her mother-in-law, and treated as an outsider even by her own husband. It is the birth of sons that raises her status. As these sons mature, she becomes the imposing and respected mother-in-law, with authority over her daughters-in-law and with power to exert influence upon and through her grown sons. The latter often are more deeply attached to her than to their young wives (see Gallin, 1994).

A full description of this metamorphosis is provided by Roy (1975), whose report is based upon the detailed life histories of some 50 upper-class Bengali women. The young bride, homesick for her village and family, restricted to life in purdah (seclusion) among strangers, compelled to demonstrate her industriousness and respect for her husband's kin, eventually becomes the matron-mother. Kenyon (1994) provides a parallel description for women in central Sudan. And Mernissi (1975) gives a strikingly similar account for traditional Moroccan Muslim society: living in the household of her husband's family, the young bride is tutored and protected by her mother-in-law, for whom she performs daily, prescribed deference rituals. When she becomes a mother-in-law, the submission of her daughters-in-law is required by law.

In societies that confine and restrict junior women, the bride often is wedded at an extremely young age, and the marriage actually may not be consummated until years later. The mother-in-law becomes a surrogate parent who completes her daughter-in-law's upbringing and she may even orchestrate the sex life of the young married couple (Vatuk, 1992). Although the older woman who imposes her authority on the daughter-in-law was once similarly subjugated by her own mother-in-law, this is not simply an example of "identification with the aggressor"; rather as noted below, such behavior is motivated by the need to assure the legitimacy of her son's offspring.

When postmarital residence is other than patrilocal, the mother-in-law/daughter-in-law relationship is not emphasized. For example, when a married man is required to live in the home of his wife's family (matrilocal residence), a woman and her daughters share a household all of their lives. Here the mother/daughter relationship predominates and a women's daughters-in-law reside elsewhere in the homes of their respective mothers. The mother's and mother-in-law's authority is very much reduced in many societies (as in the United States), in which the married couple typically sets up a new home separate from both sets of parents (neolocal residence). This arrangement, relatively rare in the nonindustrialized world, provides younger women with a certain amount of autonomy.

Most non-industrialized societies practice patrilocality and also patrilineal descent. This means that the individual is viewed as descended from the father and the father's paternal male kin and is allied with relatives who share these male links to the ancestors. Whereas patrilocal residence makes it possible for a mother-in-law to supervise and restrict her daughters-in-law, patrilineal descent provides the motivation for that monitoring.

When descent is patrilineal, a problem arises concerning the certainly of paternity, creating a special need, on the part of the patriline, to control and monopolize the sexual behavior of women. Thus Gaulin writes: "There is an inherent sex-bias in parental certainty: females will always know their offspring but males can only elevate their parental certainty by monitoring their mates" (Gaulin, 1980, p. 229). In some societies, such "monitoring" requires the confinement of all women and girls of childbearing age, making them economically dependent and thus subservient to the patrilineage. For older women in patrilineal societies, only the children of their sons are culturally defined as their descendants. The offspring of their daughters are relinquished to the descent groups of the sons-in-laws. Since her sons' children are her descendants, the older woman has a stake in ensuring their legitimacy; therefore she participates in the "monitoring" noted by Gaulin above, supervising and restricting the behavior of her daughters-in-law.

On the other hand, in societies practicing matrilineal descent (in which the individual is viewed as descended from the mother and her female maternal ancestors), the parental certainty of a woman's descendants is inevitably assured, since they are the children of her daughters and not those of sons. And in societies practicing bilateral descent (such as the United States, in which the individual is viewed as descended from both the parents and their male and female ancestors), the concern with a daughter-in-law's virtue does pertain (Kerns,1992), yet it is less extreme, since the descendants of an older woman can include children of both sons and daughters.

A Point of Comparison: The American Matron

Unlike matrons in nonindustrialized societies, older women in industrial societies, such as the United States, do not experience the many positive changes brought about by middle age. First, middle age does not usher in a life of fewer restrictions because these are negligible anyway. Second, industrial societies do provide the possibility for recognition beyond the household, either through volunteer activities in the community or through a career, but such opportunities are open to all women, regardless of age. Spiritual power, delivering babies, and curing are not associated with older women but with various specialists in industrial societies, and the specialists are likely to be men. Third, the coresident, multigeneration family is the exception in industrial societies like the United States. Often great distances separate an older woman from her offspring. The support given by adult children, so central to the definition of middle-age women cross-culturally and so crucial to their relatively privileged position in nonindustrialized societies, in American society is often reduced to occasional telephone conversations, letters and brief holiday visits. Although younger kin may seek advice or may be manipulated covertly, the latter is disapproved as meddlesome behavior and not viewed as a maternal right. Both the negative valuation our society places on intrusiveness by the senior generation and the fact of geographic separation vastly reduce the dominance American middle-age women exert over their younger kin. The control of food is also not the older woman's prerogative. Caterers, vending machines, restaurants, and fast-food emporiums have undermined the family meal, trivialized food, and reduced its political and religious significance.

In American society, being in charge, exerting authority, and assuming managerial responsibility are roles for middle-age women that remain largely underutilized in the workplace and that are not esteemed within the family. Yet in certain nonindustrialized societies, matrons are expected to manage the cultivation of food and to exert political and domestic power. The cross-cultural evidence suggests that industrial societies may be wasting the potential of their middle-age women.

What about middle-age women in the Third World? Data for this group of women is beginning to be compiled. Perhaps the report from a workshop concerning mid-life and older women held in Jamaica provides a tentative and the two-edged answer:

> Many Jamaican women have developed money-earning schemes and crafts and services that contribute to the economic growth of their communities. Others maintain farms and provide food for local consumption and export.
>
> Jamaican women are also the mainstay of health and voluntary systems. And yet, Jamaican women, in particular older women, are under-valued and under-supported. (*Network News,* 1991b, p. 25.)

Third World middle-age women seem to be secure in the knowledge that their productivity furthers the well-being of their families and of their communities. Unfortunately, appreciation for their considerable contribution is not always forthcoming.

References

Beyene, Yewoubda. *From Menarche to Menopause: Repro-*

ductive Lives of Peasant Women in Two Cultures. Albany: State University of New York Press, 1989.

Boddy, Janice. "Bucking the Agnatic System: Status and Strategy in Rural Northern Sudan." In Virginia Kerns and Judith K. Brown (eds.), *In Her Prime: New Views of Middle-Aged Women.* 2nd ed. Urbana: University of Illinois Press, 1992, pp. 141–153.

Brown, Judith K. "Cross-Cultural Perspectives on Middle-Aged Women," *Current Anthropology,* vol. 23, no. 2, 1982, pp. 143–156.

_____. "Introduction." In Judith K. Brown and Virginia Kerns (eds.), *In Her Prime: A New View of Middle-Aged Women.* South Hadley: Bergin and Garvey, 1985, pp. 1–11.

_____. "Lives of Middle-Aged Women." In Virginia Kerns and Judith K. Brown (eds.), *In Her Prime: New Views of Middle Aged-Women.* 2nd ed. Urbana: University of Illinois Press, 1992, pp. 17–30.

Brown, Judith K., Perla Subbaiah, and Therese Sarah. "Being in Charge: Older Women and Their Younger Female Kin," *Journal of Cross-Cultural Gerontology,* vol. 9, no. 2, 1994, pp. 231–254.

Burbank, Victoria K. "Women's Intra-Gender Relationships and 'Disciplinary Aggression' in an Australian Aboriginal Community," *Journal of Cross-Cultural Gerontology,* vol. 9, no. 2, 1994, pp. 207–217.

Counts, Dorothy A. "*Tamparonga:* 'The Big Women' of Kaliai (Papua New Guinea)," In Virginia Kerns and Judith K. Brown (eds.), *In Her Prime: New Views of Middle-Aged Women.* 2nd ed. Urbana: University of Illinois Press, 1992, pp. 61–74.

Datan, Nancy, Aaron Antonovsky, and Benjamin Maoz. *A Time to Reap: The Middle Age of Women in Five Israeli Subcultures.* Baltimore: Johns Hopkins University Press, 1981.

Davis, Donna. *Blood and Nerves: An Ethnographic Focus on Menopause.* St. John's, Newfoundland: Memorial University of Newfoundland Press, 1983.

Dickerson-Putman, Jeanette, and Judith K. Brown. "Women's Age Hierarchies: An Introductory Overview," *Journal of Cross-Cultural Gerontology,* vol. 9, no. 2, 1994, pp. 119–125.

Freedman, Maurice. *Rites and Duties, or Chinese Marriage.* London: London School of Economics and Political Science and G. Bell and Sons, 1967.

Gallin, Rita. "The Intersection of Class and Age: Mother-in-Law/Daughter-in-Law Relations in Rural Taiwan," *Journal of Cross-Cultural Gerontology,* vol. 9, no. 2, 1994, pp. 127–140.

Gaulin, S.J.C. "Sexual Dimorphism in the Human Post-Reproductive Life Span: Possible Causes," *Journal of Human Evolution,* vol. 9, no. 3, 1980, pp. 227–232.

Kaufert, Patricia, and Margaret Lock. "'What Are Women For?': Cultural Constructions of Menopausal Women in Japan and Canada." In Virginia Kerns and Judith K. Brown (eds.), *In Her Prime: New Views of Middle-Aged Women.* 2nd ed. Urbana:

University of Illinois Press, 1992, pp. 201–219.

Kenyon, Susan. "Gender and Alliance in Central Sudan," *Journal of Cross-Cultural Gerontology,* vol. 9, no. 2, 1994, pp. 141–155.

Kerns, Virginia. "Female Control of Sexuality: Garifuna Women at Middle Age." In Virginia Kerns and Judith K. Brown (eds.), *In Her Prime: New Views of Middle-Aged Women.* 2nd ed. Urbana: University of Illinois Press, 1992, pp. 95–111.

Kerns, Virginia, and Judith K. Brown (eds.), *In Her Prime: New Views of Middle-Aged Women.* 2nd ed. Urbana: University of Illinois Press, 1992.

Lambek, Michael. "Motherhood and Other Careers in Mayotte (Comoro Islands)." In Virginia Kerns and Judith K. Brown (eds.), *In Her Prime: New Views of Middle-Aged Women.* 2nd ed. Urbana: University of Illinois Press, 1992, pp. 77–92.

Lancaster, Jane, and Barbara King. "An Evolutionary Perspective on Menopause." In Virginia Kerns and Judith K. Brown (eds.), *In Her Prime: New Views of Middle-Aged Women.* 2nd ed. Urbana: University of Illinois Press, 1992, pp. 7–15.

Lee, Richard. "Work, Sexuality, and Aging among !Kung Women." In Virginia Kerns and Judith K. Brown (eds.), *In Her Prime: New Views of Middle-Aged Women.* 2nd ed. Urbana: University of Illinois Press, 1992, pp. 35–46.

Lock, Margaret. *Encounters with Aging: Mythologies of Menopause in Japan and North America.* Berkeley: University of California Press, 1993.

Mernissi, Fatima. *Beyond the Veil: Male-Female Dynamics in a Modern Muslim Society.* Cambridge, MA: Schenkman, 1975.

Mitchell, Winifred. "Women's Hierarchies of Age and Suffering in an Andean Community," *Journal of Cross-Cultural Gerontology,* vol. 9, no. 2. 1994, pp. 179–191.

Murphy, Yolanda, and Robert Murphy. *Women of the Forest.* New York: Columbia University Press, 1974.

Network News. "Focus on Midlife and Older Women in Latin America and the Caribbean," *Network News: A Newsletter of the Global Link for Midlife and Older Women,* vol. 2, no. 2, 1987, pp. 1–2.

_____. "Older Women in China: Community Service and State Patrols," *Network News: A Newsletter of the Global Link for Midlife and Older Women,* vol. 6, no. 1, 1991a, pp. 24.

———. "Workshop on Mid-life and Older Women Held in Jamaica." *Network News: A Newsletter of the Global Link for Midlife and Older Women,* vol. 6, no. 1, 1991b, pp. 24–25.

Raybeck, Douglas. "A Diminished Dichotomy: Kelantan Malay and Traditional Chinese Perspectives." In Virginia Kerns and Judith K. Brown (eds.), *In Her Prime: New Views of Middle-Aged Women.* 2nd ed. Urbana: University of Illinois Press, 1992, pp. 173–189.

Roy, Manisha. *Bengali Women.* Chicago: University of Chicago Press, 1975.

Sacks, Karen Brodkin. "Introduction: New Views of Middle-Aged Women." In Virginia Kerns and Judith K. Brown (eds.), *In Her Prime: New Views of Middle Aged Women:* 2nd ed. Urbana: University of Illinois Press, 1992, pp. 1–6.

Sinclair, Karen. "A Study in Pride and Prejudice: Maori Women in Midlife." In Virginia Kerns and Judith K. Brown (eds.), *In Her Prime: New Views of Middle-Aged Women.* 2nd ed. Urbana: University of Illinois Press, 1992, pp. 113–137.

Solway, Jacqueline. "Middle-Aged Women in Bakgalagadi Society (Botswana)." In Virginia Kerns and Judith K. Brown (eds.), *In Her Prime: New Views of Middle-Aged Women.* 2nd ed. Urbana: University of Illinois Press, 1992, pp. 49–58.

Vatuk, Sylvia. "Sexuality and the Middle-Aged Woman in South Asia." In Virginia Kerns and Judith K. Brown (eds.), *In Her Prime: New Views of Middle-Aged Women.* 2nd ed. Urbana: University of Illinois Press, 1992, pp. 155–170.

Whiting, John W. M. and Beatrice B. Whiting. "Aloofness and Intimacy of Husbands and Wives: A Cross-cultural Study," *Ethos,* vol. 3., no. 2, 1975, pp. 183–207.

Theoretical and Practical Aspects in Gender and Aging

Elsa M. Chaney

Introduction

In the late twentieth century, we are beginning to note a change in attitude toward older women, both in scholarly research and in the popular press. At least a decade before Betty Friedan's significant contribution to the discourse, *The Fountain of Age* (1993), began its climb to a modest best-sellerdom, there were signs that a paradigm shift was already under way, one that rejected the stereotypes of aging and the aging process. Moreover, this shift in thinking is taking place not only in the industrialized countries, but also in some countries in the developing world.

No doubt part of the explanation for the refocusing is the sheer number of older persons—defined in official documents as those who are age 60 and older—and their growing proportions in the populations of many nations. In Latin America and the Caribbean region, the number of persons 60 and older grew from about 8.8 million in 1950 to about 31 million in 1989, according to a study by the Economic Commission for Latin America and the Caribbean (ECLAC, 1989). By 2025, elders in the region will number an estimated 96 million.

Any attempt to define who is "old" is difficult because age is a cultural and social, rather than a chronological, concept. The statistical sources used here define midlife quite arbitrarily as beginning at age 55 and older age as commencing at 75. Among many ethnographers, the convention has been to take the person's, or the major informant's, own view on whether he or she is old; people in preliterate cultures often do not know their age. Moreover, even though there are more elderly in the developing world as populations age, elder status may be attained as early as 40 or 45 years in some cultures.

Since the fertility rates in the Latin American region until recently remained high, *proportions* of older persons did not register alarming gains. Now, however, percentages in older age groups also have begun to show rapid increases. In 1980, for example, older persons represented 6.5 percent of the total population in the Latin American region;

by the year 2025, their proportions are projected to be 12.7 percent, according to the Pan American Health Organization (PAHO, 1990, p. 137).

While the projections for Asia and Africa are smaller, the *numbers* of older persons there also are increasing rapidly. Table 1 shows the number of women age 60 and older per 100 males by world region in 1970 and 1990. Although the table shows a slight decrease of older women in relation to older men in Africa and Asia, in every country the numbers of both older women and older men increased during the two decades.

Certainly, we should not exaggerate the shift in attitude noted above. Despite talk of the ripening and the wisdom that maturity brings, gerontologists still talk about the disengagement and withdrawal of the elderly from active life. We now have a category of the "oldest old"—those age 85 and older—to express this idea, distinguishing those who have passed a borderline into infirmity and dependency from the "young elderly."

The Independent Elderly: A Paradigm Shift

Until the 1990s, both in the industrialized West and in the Third World, a youth culture dominated the mass media and advertising. Now awareness is growing that the increasing numbers of older people must be taken into account, illustrated not only by debates on health care for the elderly but also by the growing market for products and services targeted to older persons, particularly to the majority of those who will remain independent at least until their later years.

This change in focus on the aging process means adjusting our lens to take in others besides the minority of sick, frail, and institutionalized elderly. The latter continue to be of major concern in both the industrialized and the developing worlds, and we cannot ignore the needs of many older persons, particularly those "oldest old." However, we are beginning to include in the picture the far greater numbers of healthy, vigorous, and independent older persons

Table 1. Women Age 60+ per 100 Men by World
Region, 1970 and 1990

Region	Number of Countries	1970	1990
Developed	34	129	136
Africa	52	122	119
Latin America/ Caribbean	30	117	119
Asia/Pacific	41	108	107
Totals	157	119	120

Source: Calculated from United Nations, 1991, pp. 22–25.

in nearly every country. Certainly, we are not yet at the stage of a "revolution" in relation to how we view old age, either among scholars or the general public. Nevertheless, a shift in focus is taking place at many levels:

Among medical practitioners and persons in the caring professions, there has been a shift from the "medicalization" of old age, with its overemphasis on senescence and chronic illness, to notions promoting the already strong trends toward healthy aging through prevention, diet, and exercise.

Among psychologists, it's a shift from an almost universal "psychologizing" of older persons—attributing their problems to their own depression, forgetfulness, and despair—to notions of mental well-being and lifelong learning.

Among sociologists, gerontologists, and health researchers, the change has been from an exclusive emphasis on "humanitarian support" or a "welfare" approach (in Britain, "compassionate ageism") and its preoccupation with caregiving, the burdens on the carers, and the complex management of nursing and retirement homes, to the notion of "productive aging": the view that older people should not be shunted off to retirement or old-age homes but should remain in their communities, where they still have valuable contributions to make. There is a growing consensus, if not yet a majority view, that older persons are a valuable resource for development.

In developing countries, the vociferous demands of younger population cohorts for education, health services, jobs, food, and shelter deflected attention from problems faced by the burgeoning numbers of older persons, particularly where elders are isolated and unorganized. There has been a tendency in traditional societies to assert: "We don't have any problems with our elderly; they stay with their families who take care of them."

Nevertheless, the notion that most older persons—healthy, vigorous, experienced, independent—might represent an opportunity rather than an insupportable burden is attractive to those few policymakers in developing countries who realize that their population pyramids are becoming heavier and heavier at the top and that there are fewer and fewer adults of working age at the base to support the older generations. Involving elders in solving the problems of their own age peers, and in collaborating in the development enterprise, may be an idea whose time not only has

come but is overdue. And in the industrial countries, as the mothers of the second wave of feminism themselves become "older." Perhaps we can anticipate greater interest in what contributions older women might make to family and community, work, politics, and the feminist movement.

Looking for the Invisible Older Woman

Except for studies on the diseases of old age, particularly psychological ills, not a great deal of information on midlife and older women has been published. Most health researchers don't deal with women's health from either a feminist or a development perspective, and feminist and other scholars writing on women so far have demonstrated little concern for their elder sisters. Betty Friedan has been a notable exception, but her concerns do not extend beyond domestic boundaries. Barbara Macdonald's assertion at an early National Women's Studies Association conference, widely quoted ever since, that women's studies has kept older women "outside the sisterhood" (1989, p. 6) also holds good for international feminism. Stoner's 1989 bibliography on Latin American women, for example, lists 136 entries under the headings of Fertility and Birth Control, but only 17 references to older women, all under Medicine, subhead Gerontology, and all dealing with illness and disease.

In Latin America, an interesting exception to the focus on women's ill health has been the militantly feminist Latin American and Caribbean Women's Health Network, whose *Women's Health Journal,* published in English and Spanish, displays the highest professionalism. Its second issue of 1991 included a 20-page insert on older women, "Disregard or Dignity? Approaching Old Age."

Initial interest in the potential of older women in the developing world is found, not surprisingly, among anthropologists and ethnologists, though even in most of this literature one must sift carefully to extract the nuggets of information on women elders since only occasional studies are focused directly on women, much less on those who are beyond their childbearing years. An article by Brown (1982) was one of the first to focus directly on older women in other cultures and is widely cited as the pioneering effort. Since then, only a few studies of women at midlife and older ages in any society have been published, and still fewer on older women in cross-cultural perspective. Among them are Walker's history of the crone (1985) and three collections: Giele (1982), Kerns and Brown (1985), and Katz and Monk (1993). The doyen of research on older women, Helena Znaniecka Lopata, took her classic studies of widows (1987b) beyond the United States with works on widows in the Middle East, Asia, and the Pacific (1987a) and a synthesis work (1995).

What the research in anthropological and ethnological sources reveals is that, in most countries of the world, women are living significantly longer than their own grandmothers, mothers, and male partners. Often they head their own two-, three-, or even four-generation households, and they are proving to be far more resilient and independent in old age than expected. In their later years, women gain prestige in many societies and their wisdom and experience are highly valued.

The ethnological and anthropolgical sources highlight the increased freedom of action that elderly status confers on women. In many cultures, women become much less subject to male authority and may engage in a variety of activities not permitted to younger women (Brown, 1982, p. 144). Nor are they required to exhibit the same degree of deference to males, and they are not held to the same rules of modesty in dress and action. They also may ignore some of the language taboos, and they begin to speak freely and with authority. In addition, they may travel from home much more freely than younger women.

This means that older women potentially have more possibility to challenge men in the councils, to engage in commerce and trade beyond the boundaries of their villages, and to collaborate in organizational activities. It is, therefore, appropriate for scholars to begin to explore what this new aging paradigm holds for women, and for practitioners to remember that women do not "somehow wither as they slide into their later years," as Giele (1982: p. 1) so aptly puts it. She documents the new theories on human development that go well beyond childhood and adolescence to midlife and older ages. Other scholars concur, suggesting a new developmental phase for older women, a time in later life that presents tremendous opportunities for learning and doing. In Latin America, this period of life has a name: the "Third Age."

Older women have more fun, too, according to several sources. They can go to the coffeeshops, where younger women never venture, smoke with the men, and go to the men's sweat lodges, drink as much as they like on ceremonial occasions, and engage in aggressive, even indecorous, behavior: there is a notable lessening of social constraints on the behavior of older women, and they are allowed greater latitude in social interaction especially with male age peers. By the time a woman is 60, she may be seen casually chatting with a group of men guzzling a beer at a public festival, things forbidden to younger women (Sokolovsky, 1982, p. 122).

In 1990, a first "sifting" of scholarly sources for the Women's Initiative of the American Association of Retired Persons was conducted (Chaney, 1991). The resulting abundant evidence from ethnography, anthropology, history, sociology, and psychology supports the following propositions:

1. Almost all preindustrial societies, past and present, were (are) structured so that older persons, including women, could (can) acquire prestige, authority, and material contributions to their well-being. There is an "amazing unanimity" in the findings that women's lives improve with the onset of middle age (Brown, 1982, p. 143).

2. Not all of the elderly in agrarian, hunting, and herding cultures uniformly acquire the same degree of prestige and power simply by growing older;

empowerment appears to depend, at least to some extent, on personal characteristics and initiative.

3. Opportunities to gain empowerment at older ages tend to diminish with modernization and development. However, deterioration of older peoples' position in industrial societies may be a stage; when societies reach a certain level, the situation of older persons often improves.

4. Older women's position at older ages improves less than the position of elderly men; nevertheless, in many cultures women's status increases substantially over the life span and remains high throughout old age.

5. Organization may be an important key to empowerment of older persons, hastening the day when both preindustrial and newly industrial societies extend certain benefits to older persons.

Empowerment of Older Women

Since empowerment is so important to older people's ability to attain status and contribute to the development of their societies, this key concept needs to be defined. In an early attempt at examining the fate of aging cross-culturally, Simmons (1945) suggests that empowerment includes access to food and other assets and resources, and the increase in, or acquiring of, authority over others. Others suggest that information confers power: When old people have knowledge that the young consider important, the position of the older generation in the community is enhanced.

For older women, the notion of empowerment means increased freedom to choose new roles, occupations, pastimes, and friends, not only because they are freed from the responsibility of raising children but also because they are no longer held back by formal taboos and/or the informal strictures of traditional cultures that often surround women in their childbearing years.

The anthropological and sociological literature is full of reports of women and men, at various points in their life course, as moving into new and different statuses. In preindustrial societies, elderly status typically opens up a series of roles generally reserved to those of advanced years, and these roles confer prestige.

The available roles for women appear to increase dramatically with age. From the constricted role of daughter, in which her conduct is closely watched because her purity must be guarded against the day of her marriage, women assume the perhaps even narrower roles of wife and mother; after that, however, they progress at midlife and older ages to a wide variety of attractive options.

This is not to say that women in their childbearing years, if they also participate in economic production and have some control over resources, do not earn esteem. But their activities center on their domestic roles, which take priority even when they are combined with agricultural production and perhaps some artisan work, horticulture, and/or market trading and commerce.

In contrast, at midlife women in these cultures can look forward to fulfilling roles endowed with much greater prestige and authority and to handing on to others—often their daughters-in-law—the onerous parts of their domestic and farming activities.

[The] "forty-year-old jitters," the "empty-nest," and the "rolelessness" which have been ascribed to the middle-aged women in Western society do not apply [to women in many societies]. Overwhelmingly, the cross-cultural evidence indicates positive changes. Middle age brings fewer restrictions, the right to exert authority over certain kinsmen, and the opportunity for achievement and recognition beyond the household (Brown, 1982, p. 143).

Older women in rural areas would appear to have an edge over women in the cities in the potential to attain greater status. Their new roles include those of councillor in societies where women elders also rule; shaman, *curandera* (curer), midwife, and other roles associated with childbirth and health care; mythkeeper or repository of the culture and history of the ethnic group; witch, who sometimes combines medical knowledge with broader powers of divination and control; master artisan, who knows the traditional stitchery, weaving, and basketry patterns and techniques handed down from ancient times; administrator of the family food supply and other resources, including, in some cases, land. Each of those roles encompasses a variety of activities, and when women reach elderly status they can fulfill more than one role—making the lives of women of more advanced years in preindustrial societies potentially rich indeed.

In contrast, older women in urban areas may find it more difficult to gain prestige. Activities in the rural areas associated with certain valued roles are no longer open to them. Just as their sisters in developed societies, they must often "grandmother" from a distance. Cheap plastic products take the place of handcrafted traditional articles of daily or decorative use, and the elder master artisans have no one to whom they can pass on their skills. Doctors and medical services, even if just at the local clinic, supersede women's midwifery and curing arts. Nor are the wise counsel and advice of the elders any longer valued because it is the young who learn how to manipulate the systems of the new technological setting—indeed, the young may need to translate and interpret the modern experience for their elders. In the rural peasant household, the young know that they must learn from elders in order to survive. In the urban or modernized rural milieu, the old depend on the young. Television, movies, and the fascination of the town itself blunt the younger generation's interest in indigenous lore. Modern living has little need for peasant or herding women to share their knowledge and experience.

Whyte (1978) found in his review of the status of women in 186 cultures of preindustrial societies that women in more complex cultures tend "to have less domestic authority and less independent solidarity with other women" (Whyte, 1978, p. 173) but that their informal influence and joint partnership with males may increase. But for another view, see Coles's (1990) account of urban Hausa women who, despite rules of strict seclusion, exercise a great deal of power and control over both males and females among their kin networks, through midwifery, trading, practice of herbal medicine, and property owning. An interesting parallel is provided by the sixteenth-century cloistered Poor Clare nuns in Salvador, Brazil, who carried on business affairs and managed landed estates through intermediaries (Soeiro, 1978).

There is an additional reason that an exploration of the roles that older women fulfill in traditional (and other) societies can be valuable: Women's performance of valued activities through which they gain prestige and authority provides evidence that other women also can do so and demonstrates that neither sex nor age implies any inherent genetic inferiority. The achievements of older women show us that biology—while it may determine sex at birth and age status in the inexorable march of the life span—is not destiny.

Older Women in International Development

There appears to be a growing interest on the part of researchers to study older persons' potential contributions to international development and on the part of development agencies to assess how the contributions elders already are making might be enhanced.

Some of the most imaginative work on the potential of elders to contribute to development has been carried out by Helen Kerschner, a pioneer thinker and doer in the field of "productive aging." Kerschner is a founder of the American Association for International Aging and creator of its Senior Economic and Enterprise Development Program, which fosters innovative income-generating projects for older persons in the United States and other countries. The Britain-based organization Help-Age International sponsors many income-generating projects throughout the world in which older women are particularly active. These initiatives are important. Many older women are poor; even though the majority are taken in by their families, they are often expected to make a contribution in cash and to carry their share of the workload. Another pioneer organization emphasizing older women's potential is the International Federation on Ageing (IFA), founded under the aegis of the American Association of Retired Persons (AARP) to promote cross-cultural interchange and understanding on aging. Under IFA auspices, several monographs on older women's activities and perspectives have been published from time to time.

Older persons have long been valued as experts in Peace Corps development projects. My own experience bears out the significant contributions older women can make. In the late 1970s, I was paired with a feisty, wise, practical, and endlessly energetic "elder" partner on a women-in-development project—a North American who related much better than I to the vigorous matriarchs who were the backbone of the small farmer economy in the hills of Jamaica. In the same project, an older Peace Corps volunteer, her hair threaded with plenty of gray, sped along the mountain roads on her moped to the astonishment and delight of the local people. She also was a hit with the older women, able to interact with greater ease and confidence than most of the 19 young Peace Corps volunteers on the project.

Sonia Nofziger, an executive of Volunteers in Overseas Cooperative Assistance, an organization that sponsors many older persons to work with cooperatives and farmers around the world, puts it this way:

> A 30-year-old may be able to deal with technical matters but does not have the life experience that older people with 20 to 30 years of experience bring to a project. Older people have qualities that give them the credibility they need. A lot of people are retiring early, and they have attained a certain level of affluence— and they want to give something back.

The American Association of Retired Persons has been one of the most important agencies in the "healthy aging" approach both in the United States and in addressing some of the particular problems facing older women in the Third World. In the 1980s AARP carried out numerous initiatives in information, publications, and conferences in relation to older women's health, socioeconomic status, and prospects for contributing to international development.

In 1987, the AARP Women's Initiative and its International Activities Unit signed an agreement with the Pan American Health Organization (PAHO) for a series of activities designed by Irene Hoskins, editor of *Network News*. The first effort, a three-day seminar in 1988, brought together at PAHO headquarters in Washington, D.C., 30 leading scholars and practitioners from Latin America and the Caribbean for a first meeting on midlife and older women in the region. Working in close collaboration with these AARP/PAHO activities has been the Dominican Republic's alternate representative to the United Nations, Ambassador Julia Tavares de Alvarez, who has become known in that body as one of the chief advocates for the elderly.

Lee Sennott Miller planned this first conference and carried out a study pulling together what was then known about older women's health and socioeconomic situation. Presentations appear in the conference proceedings published in both Spanish and English (Pan American Health Organization and American Association of Retired Persons, 1989; Sennott Miller, 1989).

Subsequently, the AARP Women's Initiative and the IFA cosponsored a follow-up workshop, "Coping with Social Change: Programs That Work," as an official preconference to the 1989 World Congress on Gerontology held in Mexico. Practical programs on health, shelter, income security, and new roles for older persons were explored, and a proceedings was published in Spanish and English. In 1993, the two groups again sponsored an official preconference workshop, "Midlife and Older Women in the Emerging Democracies of Central and Eastern Europe," as part of the Fifteenth World Congress on Gerontology held in Budapest, with presentations on Hungary, Armenia, Romania, and Poland.

Another effort the AARP Women's Initiative has been its collaboration with the Division for the Advancement of Women (DAW), the administrative arm of the United Nations Commission on the Status of Women, which resulted in the DAW's first-ever expert group meeting on older women, a long-postponed follow-up to the panels on the topic at the Third World Conference on Women (Nairobi, 1985). The five-day experts' seminar, chaired by Alvarez in Vienna in 1991, brought together scholars and activists from Canada, Ghana, India, Jamaica, Malaysia, Mexico, Nigeria, the Philippines, and Poland to work with several hundred representatives of member states on economic, social, cultural, legal, and health issues related to older women. The DAW sponsored some of these same experts for a panel on "Contributions of Older Women to Development" at the 1993 Budapest World Congress on Gerontology.

Other researchers, advocates, and innovators who highlight older women's issues include Atunes Filho in São Paulo, who founded and directs the theatrical group *Novo Horizonte* (New Horizon) for older (over age 55) actors; Mabel Bianco in Argentina, president of the Fundación para el Estudio y la Investigación de la Mujer (Foundation for Study and Research on Woman); and Adelina Brenes Blanco, pioneer gerontologist who holds the first doctorate in gerontology in Costa Rica and speaks and writes extensively on older persons in her capacity as director of social service in her country's social-security agency.

Other activists are Luz Esther Rangel, a physician and coordinator of geriatrics and gerontology in Mexico's Sistema Nacional para el Desarrollo Integral de la Familia (National System for the Integral Development of the Family); and Betsie Hollants, founder of *Vejez en México: Educación y Acción* (Old Age in Mexico: Education and Action), the pioneer in Latin America in her work on the problems of older age.

From the fact that in the citations the names of these and several other persons keep recurring, it is evident that since the late 1980s, a core group of researchers on women and aging has been forming—researchers who know each other and have had the chance to exchange ideas in a series of venues. The flip side is that, outside these core pioneers, little work is being done on women and aging except for the burgeoning literature on older women and health. While these efforts are welcome, researchers and activists need to broaden their horizons to include the many other facets of women's lives after age 60. The "Third Age" need not signal an entry into senescence, but the beginning of a different— active, productive, and satisfying—older age.

References

American Association of Retired Persons and International Federation on Ageing. *Coping with Social*

Change: Programs That Work. Proceedings of a Conference, Acapulco, Mexico, June 1989. Washington, D.C. Women's Initiative, American Association of Retired Persons, 1990.

Brown, Judith K. "Cross-Cultural Perspectives on Middle-Aged Women," *Current Anthropology,* vol. 23, no. 2, 1982, pp. 143–156.

Chaney, Elsa M. *Older Women and Empowerment: Cross-Cultural Views.* Washington, D.C.: International Activities Unit, American Association of Retired Persons, 1991.

Coles, Catherine. "The Older Woman in Hausa Society: Power and Authority in Urban Nigeria." In Jay Sokolovsky (ed.), *The Cultural Context of Aging: Worldwide Perspectives.* New York: Bergin and Garvey, 1990, pp. 57–81.

Economic Commission for Latin America and the Caribbean (ECLAC). *The Elderly in Latin America: A Strategic Sector for Social Policy in the 1990s.* Santiago: Division for the Advancement of Women, ECLAC, 1989.

Friedan, Betty. *The Fountain of Age.* New York: Simon and Schuster, 1993.

Giele, Janet (ed.). *Women in the Middle Years: Current Knowledge and Directions for Research and Policy.* New York: Wiley Interscience, 1982.

Katz, Cindi, and Janice Monk. *Full Circles: Geographies of Women over the Life Course.* London: Routledge, 1993.

Kerns, Virginia, and Judith K. Brown (eds). *In Her Prime: A New View of Middle-Aged Women.* South Hadley: Bergin and Garvey, 1985.

Lopata, Helena Znaniecka. *Widows.* Vol. I: *The Middle East, Asia, and the Pacific.* Durham: Duke University Press, 1987a.

———. *Widows.* Vol. II: *North America.* Durham: Duke University Press, 1987b.

———. *Current Widowhood: Myths and Realities.* Newbury Park: Sage, 1995.

Macdonald, Barbara. "Outside the Sisterland, Ageism in Women's Studies." *Women's Studies Quarterly,* vol. 17, nos. 1 and 2, 1989, pp. 6–11.

Pan American Health Organization (PAHO). *Health Conditions in the Americas.* 2 vols. Washington, D.C: PAHO, 1990.

Pan American Health Organization and American Association of Retired Persons (PAHO/AARP). *Mid-life and Older Women in Latin America and the Caribbean.* Washington, D.C.: PAHO, 1989. Also published in Spanish by PAHO in 1990 as *Mujeres de Edad Mediana y Avanzada en America.*

Sennott Miller, Lee. "The Health and Socio-economic Situation of Midlife and Older Women in Latin America and the Caribbean." In *Mid-life and Older Women in Latin America and the Caribbean: A Status Report.* Washington, D.C.: PAHO, 1989, pp. 1–125.

Simmons, Leo W. *The Role of the Aged in Primitive Societies.* New Haven: Yale University Press, 1945.

Soeiro, Susan A. "The Feminine Orders in Colonial Bahia, Brazil: Economic, Social, and Demographic Implications, 1677–1800." In Asunción Lavrin (ed.), *Latin American Women: Historical Perspectives.* Westport: Greenwood, 1978, pp. 173–197.

Sokolovsky, Jay (ed.). *Aging and the Aged in the Third World,* part 1. Studies in Third World Societies publication no. 22. Williamsburg: 1982.

Stoner, K. Lynn. *Latinas of the Americas: A Source Book.* New York: Garland, 1989.

United Nations. *The World's Women 1970–1990: Trends and Statistics.* New York: United Nations, 1991.

Walker, Barbara G. *The Crone: Woman of Age, Wisdom, and Power.* San Francisco: Harper and Row. 1985.

Whyte, Martin King. *The Status of Women in Preindustrial Societies.* Princeton: Princeton University Press, 1978.

Women and Production

Women in Agricultural Systems

Agnes R. Quisumbing

Introduction

This essay provides an overview of women in agricultural systems. It discusses women's roles in different farming systems—their participation in agricultural production throughout the world, their degree of control over productive resources, and gender-specific constraints to increasing their incomes and welfare. It examines the implications of generally lower levels of human and physical capital for productivity and wages. And it reviews how gender-specific constraints and control of productive resources determine whether women benefit from the introduction of new agricultural technologies and commercialization.

There is substantial variation in the gender composition of the agricultural labor force. Women make up a high proportion of the agricultural labor force in sub-Saharan Africa, followed by East Asia and Southeast Asia. Women also account for a significant portion of agricultural employment in some European countries, notably Hungary, Portugal, and the former Yugoslavia. In contrast, the agricultural labor force in Latin America, North Africa, and the Middle East is predominantly male, although this may reflect undercounting of women's participation. In Latin America, for example, women have had a traditionally important role in peasant farming systems and are becoming increasingly significant in export agriculture. Estimates adjusted for undercounting suggest that, on average, women are 46 percent of the agricultural labor force in sub-Saharan Africa, 45 percent in Asia, 40 percent in the Caribbean, and 31 percent in North Africa and the Middle East (Dixon, 1982).

While a breakdown of agricultural employment into categories of farm owners and managers, unpaid family workers, and hired workers shows that farm owners and managers account for the majority of the agricultural labor force across countries, females are a minority among farm owners and managers and are usually classified as unpaid family workers. Moreover, although wage labor is the dominant category in many countries, women's participation as hired laborers is less than men's, partly due to underreporting but also to women's domestic responsibilities and cultural restrictions on participation in work outside the family farm.

In addition to agricultural production, food processing, and market work, rural women also perform household production activities such as child care, food preparation, and provision of fuel and water. When costs are imputed to time spent on home production, and these are included in the computation of full household income, women contribute 40–60 percent of full household income in rural areas (Goldschmidt-Clermont, 1987).

Farming systems can be classified into three general types: (1) extensive, land-surplus systems; (2) intensively cultivated, labor-surplus systems in which most farms are of average size; and (3) dualistic systems with different factor intensities between large and small farms, the former usually mechanized, the latter labor intensive (Boserup, 1970; Lele, 1986). These types have usually been associated with Africa, Asia, and Latin America, although changes in relative resource endowments, such as increasing population pressure on limited land in Africa, have caused these associations to break down. These systems coexist with different family-structure "types": the polygamous societies of Africa, in which there is less congruence between women's interests and their husbands'; and the monogamous extended/nuclear family type in Asia and Latin America, with agricultural decision making primarily by men.

A different typology of farming systems focuses not only on labor input by men and women but also on access to means of production, participation in decision making, and control over the outcome of the productive process (Deere and León, 1982). Taking these factors into account enables distinctions to be made among male and female farming systems, in which only men or women provide the labor and control the productive process; patriarchal family systems, in which both men and women provide labor but men control decision making and the outcome of production; and egalitarian family systems, in which both men and women participate in production, decision making, and control over the outcome of production.

Unlike the unified "family farm" in Asia, in most of sub-Saharan Africa, households hold several granaries, or purses, controlled by men or women depending on different but complementary responsibilities to the household (Dey, 1985). Women tend to be involved more heavily in the production of traditional food crops, such as swamp rice in the Gambia or maize in Kenya, while men contribute more labor to cash crops: irrigated rice in the Gambia and sugarcane in Kenya. While evidence suggests that women are increasingly involved in cash-crop cultivation (Saito et al., 1994), the traditional patterns of specialization can be traced to customary rights and obligations of men and women. In most African societies, women have a traditional obligation to produce subsistence food crops for home consumption and to perform household maintenance activities such as fetching fuel and water. Traditionally, men have cultivated cash crops, the sale of which provides cash to meet nonsubsistence obligations to wives and children. Men are obligated to provide the land and to be responsible for housing, taxes, ceremonial and religious obligations, and at least part of the school fees. In most patrilineal societies, men are responsible for surplus accumulation, usually in the form of cattle. This is linked to longer-term security, which may often be achieved through the exchange of cattle for additional wives (with concomitant increments in household labor supply) and through sale of cattle in times of crop failure (Kumar, 1987). In many African countries and ethnic groups, both men and women also have the right to cultivate a personal field on their own from which they meet certain obligations to the household and their personal expenses.

One big difference between Asian and African agriculture is the importance of hired labor in Asia. Unlike in Asia, where both male and female casual laborers account for a significant proportion of labor input, particularly in paddy rice, the casual labor market in much of Africa is still relatively small and mostly male. Another major difference is the large contribution of male family labor in Asian farming systems. Plow agriculture is the norm, and men typically provide the labor in land preparation, with women providing auxiliary labor in hand operations, such as planting (particularly transplanting rice seedlings), cultivation, and crop care (notably weeding).

In Latin American agriculture, there is a gender division of labor in both industrialized crop production and peasant farming. In the large-farm sector, women are hired as wage laborers for such nonmechanized tasks as coffee harvesting and cotton picking, and they work with men as members of migrant family laborers doing piecework. Regional differences in the tasks men and women perform depend on the local supply of male and female wage labor and the substitutability of male and female labor in the large-farm sector.

In the peasant subsector, the significance of women's participation in family farming systems varies widely across Latin America, with ethnicity and across regions. Women's agricultural participation is much more important in the Andean countries and Central America than in the Southern Cone. Participation can also vary widely within a given country: The gender division of labor is responsive to a number of technical variables—specific crops and tasks and labor intensity of the activity—and social characteristics of the peasant household and the woman herself. There is also substantial variation between cultural values (such as "machismo") and actual fulfillment of role obligations across cultures and within social classes in a particular culture. Although cultural ideals categorize house work as "women's work" and field work as "men's work," in practice, one can observe men and women working together in the fields, not necessarily demarcated into exclusive sex-specific tasks. However, there are certain tasks reserved for men, usually those involving animal traction and farm machinery. Flexibility in the gender division of labor is influenced by cultural factors (the strength of Hispanic versus Indian traditions), social class (whether income comes from land or wages), labor market conditions, and the degree of market integration of the peasant economy. When men work off the farm, women have taken on more male-dominated tasks.

The above does not adequately reveal the heterogeneity and dynamism of farming systems all over the world. For example, the growing prevalence of female-headed households is due to very different reasons in Africa, Asia, and Latin America. In Africa and parts of Asia, the migration of men to cities leaves women in charge of farming. In Latin America, the predominant rural–urban migration flow is composed of young single females in response to declining incomes and increased poverty in the rural areas. Their ability to find unskilled jobs motivates them to migrate to urban areas at a very young age. In the rural sector, however, poor peasant women assume greater responsibility for agricultural production as men leave the household in search of temporary wage work. It has been argued that dualistic, bimodal patterns of development in which growth is concentrated in cities and export enclaves, such as those experienced in Africa and Latin America, are conducive to individual migration, while broad-based rural growth, as in the successful East Asian economies, induces family rather than individual migration (Lele 1986). Moreover, wars and civil strife, in which men participate as combatants, have also led to increasing numbers of female-headed households. To the extent that development policies contribute to changes in farming systems and family structures, policymakers need to evaluate the unintended social costs and benefits of agricultural policy.

Gender-Specific Constraints in Agriculture

Despite women's importance in agricultural production and their significant contribution to full household income, they usually have lower levels of physical and human capital compared to men. These disparities may persist due to legal, social, and institutional factors that create barriers for women.

Land Rights

There is wide diversity in laws governing women's rights to land and other property. Property and inheritance rights may be governed by the Civil Code, personal laws of religious communities, or customary laws. Different systems often exist in the same country, so individuals have some choice in the disposition of property. Most religious laws, however, discriminate against women in land rights. In Hindu law, women do not have the right to own, acquire, or dispose of property, although they may have the right to some personal property. Despite legal reform in India, loopholes in the legislation prevent equalization of inheritance rights. In particular, laws reforming women's right to succession in India do not mention agricultural lands and tenancy laws, such that in a majority of Indian states, legislation gives female heirs a very low priority in the list of heirs to agricultural lands. In Islamic law, women's land ownership rights are explicit, but they do not always control de facto land use. Furthermore, Islamic law limits the share to female heirs to one-half of that given to males. Daughters may also receive their share of the inheritance in the form of household goods and jewelry, but not in land or farm implements.

In some Southeast Asian countries, customary law gives women independent land rights. In Indonesia, the Philippines, and Thailand, women can own, inherit, acquire, and dispose of property in their own right. Among ethnic groups that have bilateral kinship, which account for the majority of the Indonesian population, customary law (*adat*) gives sons and daughters shares to property from both parents. Despite provisions for equality of inheritance between sons and daughters in civil law, actual practice may differ. Philippine law gives women the right to hold property in their own name, but income from the wife's premarriage property is considered conjugal income whose use is subject to the husband's consent. Although tenancy law gives first priority to the wife, and then the children according to age, in succession of tenancy rights, sons typically receive more land and nonland assets in farming households than do daughters.

Under customary law in many African countries, women usually had use rights to some land; it was allocated to women from their husbands and natal families based on their position within a kinship group and, in particular, on their relationship to a male relative (father, brother, husband). These rights entitled women to farm the land, often in exchange for labor on their husbands' and other family plots. However, these indigenous customs have been modified by Western colonization, which introduced private ownership by individual registration of land, and which often discriminated against women in titling. Furthermore, since women obtained land rights usually through a male relative, there was no guarantee that they would retain these rights in the case of death or divorce. Various studies from Kenya, Malawi, Nigeria, and Zambia summarized in Table 1 have shown that female-headed households (FHH) have smaller landholdings than do male-headed households (MHH). FHH's area cultivated ranges from 31 to 69 percent of land cultivated by MHH. Family size is also smaller in female-headed households, due to the absence of the male head. As a result, land cultivated per person does not differ much between MHHs and FHHs, although this ratio is lower for the latter.

Absence of formal land rights and smaller land sizes cultivated by women may be critical since land is usually needed as collateral in credit markets. A farm household survey in Kenya and Nigeria (Saito et al., 1994) found that more male than female heads of households, and more male than female farmers, were able to exercise their land rights fully. The ability of women to exercise the full range of land

Table 1. Size of Holdings by Gender of Farm Manager or Household Head, Selected Countries

Country and year	Area cultivated		Family size		Area per person	
	Male	*Female*	*Male*	*Female*	*Male*	*Female*
Kenya 1973[a,b]	1.8	1.2	N/A	N/A	N/A	N/A
Kenya 1989[c,d]	2.6	1.7	8.6	8.0	0.3	0.21
Malawi 1983–1984[c,d]	1.3	0.9	4.9	4.0	0.26	0.22
Nigeria 1989[c,d]	2.6	0.8	7.6	4.9	0.34	0.16
Zambia 1986[a,d,e]	6.8	3.0**	3.5	1.7***	1.94	1.76
El Salvador 1988[f]	0.78**	0.49**	5.3**	4.8**	0.15	0.10

Sources: Kenya 1973 from Moock, 1976; Kenya 1989 and Nigeria 1989 from Saito et al., 1994; Malawi 1983–1984 from Segal, 1986; Zambia 1986 from Sikapande, 1988; El Salvador 1988 from Lastarria-Cornhiel, 1988

[a] Area in acres

[b] By gender of farm manager

[c] Area in hectares

[d] By gender of household head

[e] Family size in adult-equivalents

[f] Area in manzanas; data refer to cooperative members

**Significant at $\alpha = 5\%$

***Significant at $\alpha = 1\%$

rights—to be able to sell or mortgage the land—may be essential to the functioning of land markets. While these rights may evolve from systems of communal control to individualized rights in response to commercialization and population pressure, in more monetized and market-oriented economies land-tenure security and the exercise of these rights may have a positive productivity impact.

Governments have attempted to change the distribution of property rights to land through agrarian-reform programs, resettlement schemes, or land-development schemes. However, where these policies were implemented without consideration of women's role in farming systems, or without explicit inclusion of women as beneficiaries, they may have led to a deterioration of women's land rights. In some African countries, government interventions have reinforced, rather than remedied, traditions against women in land rights. In Kenya, Saito et al. (1994) found that more men than women have received land from government allocations but also that the land allocated by the village tended to be more equitably distributed between men and women in both Kenya and Nigeria, possibly because these involved the distribution of use rights.

Formalization of customary rights may also have unintended consequences for women's traditional rights to land. In matrilineal areas of Malawi, where women have inheritance rights to customary land, the restriction, until the early 1990s, of burley tobacco–growing to leasehold land has led to the registration of tribal lands in the name of the husband, with no protection for the woman in the event of divorce (Mkandawire, Jaffee, and Bertoli, 1991). Resettlement projects and irrigation schemes have also altered the distribution of land rights. Two examples discussed below are the Jahally-Pacharr irrigated-rice project in Gambia and an irrigated-rice scheme in northern Cameroon (the SEMRY Project). In Burkina Faso, planned settlement areas managed by the Volta Valley Authority vested all title to land in the male household head, such that women had to clear land outside the settlement to establish individual plots (Saito et al., 1994).

Neither were women's land rights considered in the majority of land reforms in Latin America. Deere's (1987) review of 13 Latin American agrarian reforms found that the majority have not produced significant numbers of female beneficiaries or even given attention to gender as a beneficiary category. The available national-level data suggest that only in Cuba do women represent a significant number of agrarian-reform beneficiaries, accounting for 25 percent of production cooperative members and 14 percent of workers on state farms.

Legal, structural, and ideological mechanisms have excluded women from Latin American agrarian-reform programs. Since social custom dictates that if both an adult man and an adult woman reside in a household, the man is considered its head, the only women who could possibly benefit from an agrarian-reform program would be widows or single mothers without an adult male in the household. Moreover, many agrarian reforms have ben-

efited only the permanent agricultural wage workers employed on estates at the moment of expropriation, excluding from membership the often large seasonal labor force, of which women make up a large proportion. The use of point systems that give priority to peasants with more education, larger family size, and farming experience, as in Colombia, also discriminates against rural women, who have less education than men and whose farming experience is likely to be downplayed because of cultural stereotypes. Ideological norms governing the proper sexual division of labor—the man's place is in the fields, the woman's place is in the home—are also a barrier to women's benefiting from agrarian reform in Latin America. Even where female heads of households are explicitly included as potential beneficiaries, they may have lower land allocations than male household heads. In El Salvador, among cooperative members on former large estates, male household heads had significantly larger areas allocated than female heads (Lastarria-Cornhiel, 1988). Lastly, agrarian reform may introduce changes in traditional patterns of land rights that were formerly more equitable to women, such as the bilateral pattern of land inheritance in the Andean highlands. Whereas women would have had independent rights to land under traditional inheritance customs, these are replaced by agrarian reform laws in which only men have succession rights.

Even if men and women were deemed equal beneficiaries of land reform, differences usually exist between formal and real land rights, as the experience of China suggests (Wazir, 1987). Although the marriage law guaranteed equal rights of men and women, it was not forcefully implemented. Neither did formal equality of land ownership under the land reform loosen the hold of traditional values, although having land rights increased the bargaining power of peasant women within their households. When agriculture was collectivized, women were discriminated against since they received lower work points than men and their pay was usually added to the husband's pay packet. The unequal allocations under collectivization affected the reallocation of collective land after the 1978 agricultural-sector reforms. Under the production-responsibility system, land for individual cultivation was allocated according to the size of the household labor force and the number of work points—restoring the bias against women in land rights.

Tools

Female farmers generally own fewer tools than male farmers (see Table 2). In a Gambian irrigated-rice project area, 8.2 percent of men owned a plow, compared to none of the women. Less than 1 percent of women owned a seeder, weeder, or multipurpose cultivation implement, compared to 26.9 percent, 12.4 percent, and 18.1 percent of the men, respectively (Von Braun et al., 1989). A household survey in three districts of Kenya found that female farmers owned fewer farm tools and equipment than male farmers (Saito et al., 1994). When costs were imputed to these tools, based

Table 2. Farm Tools and Equipment by Gender of Farmer

Implement or Type of Technology	Men	Women
Gambia 1987[a]		
(Percent owning the tool)		
Plow	8.2	0.0
Seeder	26.9	.6
Weeder	12.4	0.2
Multipurpose implement[b]	18.1	0.4
Kenya 1989[c]		
(Value of tools and equipment as percent of		
male farmers' tools and equipment)		
Kakamega district	100.0	6.8
Muranga district	100.0	132.3
Kilifi district	100.0	64.1
Overall mean	100.0	17.8
El Salvador 1988[d]		
(Average number of farm tools)		
Cooperative members on large estates	2.0	0.8
Tenant beneficiaries	3.2	2.7

[a]Von Braun et al., 1989
[b]An animal-drawn implement convertible for plowing, weeding, and groundnut lifting
[c]Saito et al., 1994
[d]Lastarria-Cornhiel, 1988

on their market price, the value of tools owned by women, on average, amounted to only 17.8 percent of the value of male farmers' tools and equipment. While the value of tools owned by female farmers was higher than men's in the Muranga district, this was an exception; the value of tools owned by women compared to those owned by men was as low as 6.8 percent in the Kakamega district. Since farm capital contributes positively to yields, female farmers are more likely to have lower yields than male farmers. Among cooperative members organized from large estates in El Salvador, and among tenant-beneficiaries, female-headed households had fewer farm tools than male-headed households, although this difference was significant only among cooperative members (Lastarria-Cornhiel, 1988) (see Table 2).

Extension Advice

In spite of women's important role as farm managers and agricultural workers, whether as family or hired laborers they have seldom been beneficiaries of extension systems. Traditional agricultural extension based on only one crop often fails to consider the crops and activities in which women are involved. Community or rural extension, on the other hand, covers the broad spectrum of women's activities, but the range of tasks covered may limit the time devoted to any single task. Furthermore, extension systems in many developing countries are overloaded; ratios in Africa, Asia, the Near East, and Latin America range from one agent to 2,000 or 3,000 farmers. In contrast, in Europe and North America one extension agent serves 300–400 farmers (Saito and Spurling, 1992). In addition, women are underrepresented among extension agents. Even in regions with a long tradition of female farming, such as Africa, in the early 1990s, only 11 percent of extension staff, and 7 percent of field extension staff, were women. Moreover, while female extension workers may be trained in agriculture, they are mandated to give advice on home-economics subjects. This may constrain the delivery of agricultural extension messages to female farmers, who may be restricted from interacting with male extension agents, particularly in Muslim countries, and who prefer to interact with female agents.

Evidence from a number of African countries suggests that male farmers have greater contact with the extension service than female farmers. Moreover, male-headed households are more likely to ever have contact with an extension worker than female-headed households are. Similar patterns are reported for land-reform beneficiaries in El Salvador, with male-headed households having significantly higher access to technical assistance than female-headed households (Lastarria-Cornhiel, 1988).

Credit

Access to credit, both formal and informal, has important implications for the ability to attain a stable standard of living and undertake productive activities. Where agricultural production requires heavy equipment or cash inputs, credit makes possible the purchase of these inputs before income gains are realized. Credit is particularly important during the planting season, when seeds and cash inputs have to be purchased, and in the lean season before harvest as stocks for consumption are depleted. Credit may also be essential to smoothing consumption in case of crop failure or drought.

Female farmers have less access to credit than male farmers. Collateral requirements, high transactions costs, limited education and mobility, social and cultural barriers, and the nature of women's business limit women's ability to obtain credit. Property that is acceptable as collateral, especially land, is usually in men's name, and the types of valuables women have, such as jewelry, are often deemed unacceptable by formal financial institutions. The transactions costs involved in obtaining credit—transportation costs, paperwork, time spent waiting—may be higher for women due to higher opportunity costs from foregone activities. Indeed, in rural Kenya, the distance to a bank is a significant determinant of women's probability of obtaining credit but does not affect men's borrowing behavior (Saito et al., 1994). Women's lower educational levels relative to men, their lack of familiarity with loan procedures, and social and cultural barriers may constrain their mobility and interaction with predominantly male credit officers or moneylenders. Exclusion from local groups, such as farmers' groups, may prevent women from receiving both credit and extension advice, particularly if the extension worker plays an important role in credit delivery. Women also tend to be involved in the production of relatively low-return crops that are not included in formal-sector lending programs. Moreover, women may be credit constrained because their role as primary caregivers and the health risks associated with childbearing lead to intermittency in employment, which makes them risky clients for banks.

Domestic Responsibilities

In most societies, women's role as primary caregivers may limit the time they spend outside the home or in market work. In Africa, women may spend as much as two hours a day on child care, three hours on food preparation, and two hours fetching fuel and water. In rural Asia, food-processing activities take up two to three hours daily; in rural Bangladesh, women may spend as long as six hours a day fetching water. In rural Peru, such household tasks as cooking, washing clothes, and fetching wood and water are primarily the mother's responsibility (Deere and León, 1982). Pregnancy and child-care activities reduce women's participation in market activities, although daughters or other female adults in the household often substitute for mother's time in housework. The substitution of daughters' time in housework, if at the expense of time spent in school, may have unfavorable consequences for the next generation's human capital. Cultural traditions of seclusion may also limit women's participation in activities outside the home.

The extent of rural women's multiple responsibilities may impose time and energy constraints on their participation in programs designed to increase their incomes and on their willingness to adopt new technologies. It may be critical to increase their productivity in existing tasks—bearing in mind constraints created by their biological needs and social roles—rather than to expect them to sacrifice many of their existing functions. This highlights the importance of increasing women's productivity in both agricultural and home production.

Productivity and Wages

Women's limited control over productive resources and gender-specific constraints have important consequences for agricultural productivity. Econometric evidence on gender differences in agricultural productivity points to the importance of investing in women by increasing their human capital, through education and extension, and by increasing their access to physical and financial inputs. The econometric evidence encompasses managerial efficiency of male and female farmers, their adoption of new crops and technologies, their contribution to agricultural output as laborers, and returns to schooling in rural areas.

Technical Efficiency of Male and Female Farmers

Numerous studies have found that, where male and female farmers manage separate plots, as in many African farming systems, plots controlled by women have lower yields than plots farmed by men. Studies analyzing the technical efficiency of male and female farmers have suggested that this is due to lower levels of input use on women's plots, and not to inherent managerial differences between male and female farmers (e.g Moock, 1976; Saito et al., 1994). For example, studies from Kenya, Nigeria, Thailand, and Korea indicate that female farmers are equally as efficient as male farmers, once individual characteristics and input levels are controlled for (Moock, 1976; Saito et al., 1994; Jamison and Lau, 1982). However, these studies did not take into account the relationship between input application and farmer characteristics. Since more-educated farmers are more likely to adopt new technologies and apply modern inputs, the contribution of such inputs may be overstated and that of education underemphasized. Alternatively, the consequences of underinvesting in women in rural areas may be underestimated. Among land-reform beneficiaries in El Salvador, men and women do not have significant differences in production value per unit of land; access to land was more important than the gender of the household head in determining production, income, and family labor allocation (Lastarria-Cornhiel, 1988).

More recent literature suggests that the asymmetry of roles and obligations within the household, particularly in African farming systems, may underlie women's lower yields. A study in Burkina Faso (Udry, 1996) found that the yield differential between male and female plots planted to the same crop within the same household was attributable to significantly higher labor and fertilizer inputs per acre on plots controlled by men. If the household had common objectives and preferences, factors of production would be allocated efficiently to the various productive activities of the households. The consistently higher input intensities on men's plots suggest misallocation of household resources, which is traced to stronger incentives for individuals to achieve high output on their own plots, as well as imperfect labor-allocation processes within the household.

Technological Adoption by Male and Female Farmers

In some countries, especially where new technologies are associated with "male" crops or activities, women may be less likely to adopt new crops or technologies. This may matter more in some cultures than others: Female decision makers are less likely to adopt coffee in Kenya and livestock in Tanzania but are equally likely to have cattle in Kenya and to grow coffee or cocoa in Cote d'Ivoire (Appleton et al., 1991). Where plowing is a traditional male activity, women may be less likely to use oxen in cultivation. However, women may more readily adopt technologies related to tasks they perform, particularly if the extension agent is female. Evidence from Kenya suggests that female farmers are equally likely to apply technical advice from extension agents and are even more likely to adopt relatively complex practices such as top dressing, chemical use, and stalk-borer control (Bindlish and Evenson, 1993). In general, more-educated farmers, and more-educated women, in particular, are more likely to adopt new technologies. Increasing the educational level of female farmers by giving them universal primary education may have higher marginal effects on the probabilities of adoption than increasing the educational level of all farmers, due to the generally lower levels of female education in Africa. This is because universal primary education stimulates early adoption by female farmers, whom other women are more likely to imitate than male farmers. The significance of gender-specific-copying effects highlights the need not only for female extension agents to work with female farmers but also for female contact farmers to be chosen. The importance of cooperatives and extension services in many of these studies emphasizes the need for provision of support infrastructure to rural areas. Farmers with larger areas cultivated and higher values of farm tools are also more likely to adopt new technology. To the extent that female farmers may have less education, have less access to land, and own fewer tools, they may be less likely to adopt new technologies.

Productivity of Male and Female Workers

Labor productivity is affected by differences in the division of labor across crops and tasks, and in the availability of male and female hired labor. In countries where women are culturally constrained from participating in farm work (particularly on others' farms, as in India), women tend to have lower marginal products of labor—that is, an additional unit of women's labor contributes less to output than an additional unit of men's labor.

Limited access to complementary inputs also reduces women's labor productivity. In Gambia (Von Braun et al., 1989), women in a centralized pump-irrigation area exhibited lower average labor productivity than men in the same crops and broad technology groupings because of reduced access to labor-saving implements and more constraints on their time, which allowed them to cultivate only smaller pieces of land, leading to a smaller scale of operation with less efficiency.

Returns to Schooling of Rural Men and Women

The rate of return to schooling, or the "return to schooling," is the percentage increase in wages or earnings due to an additional year in school, an indicator of the gains to increasing educational attainment. Although returns to schooling for both men and women are higher in nonagricultural occupations, income increases associated with more years in school are significant in dynamic agricultural settings where modern technologies have been introduced. In traditional agricultural societies, wages may depend more on local market characteristics than on an individual's human capital. In India in the 1970s, the rate of return to an additional year of schooling for both men and women was not statistically different from zero. However, more recent evidence in West Bengal finds positive and significant returns to male and female schooling (Mukhopadhyay, 1991). An additional year in school increases female wages by 3.5 percent, while an additional year increases male wages by 1.6 percent. The increased rate of return may be associated with the diffusion of modern rice varieties, which require farmers to make informed decisions regarding the timing and quantity of inputs used.

In the Philippines, where modern rice varieties are widely adopted, rates of return to schooling are positive and are not significantly different between men and women in the agricultural labor force. Male rates of return to an additional year of schooling are between 8 to 10 percent, and female rates of return are about 7 percent (Behrman and Lanzona, 1989).

Limited employment opportunities for women in rural areas may lead to low or insignificant rates of return to schooling. However, increases in agricultural productivity may provide additional justification for increasing schooling attainment in rural areas. In rural Peru, for example, education increases the contribution of both men and women to agricultural production (Jacoby, 1992). For the average male farmer, with 3.5 years of schooling, an additional year of schooling would increase his contribution to agricultural output by 11 percent. The corresponding increase for the average female farmer with 1.7 years of education would be 16 percent.

To summarize, returns to schooling are higher, and individual characteristics more important in determining wages, in areas where agriculture has benefited from technological change. In contrast, in areas where labor-market participation of women is limited or constrained by cultural factors, an additional year of schooling does not raise women's wages significantly, and women receive lower wages than men. Women's lower income-earning capacity in these settings may be one reason why parents invest less in their daughters' education.

Technological Change, Commercialization, and Women's Welfare

Agricultural development is associated with technological change and commercialization. Technological change, which increases the productivity of *all* factors of production, means that the same output can be produced with fewer inputs, or more output can be produced with the same inputs, although the nature and the composition of inputs may change (Von Braun et al., 1989). Commercialization, on the other hand, is a shift from subsistence to greater market orientation, through either an increase in the marketed surplus of a subsistence crop or a change in crop mix. While increased commercialization may occur with or without technological change, the latter rarely occurs without commercialization, at least in the factor-input side (modern seeds, fertilizers, hired labor), if not on the product-output side (Von Braun et al., 1989). Men and women of the same socioeconomic class may not benefit equally from technological change and commercialization, due to initial differences in men's and women's involvement in agricultural fieldwork and nonfieldwork, especially domestic work and child care; the extent of their control over, and the patterns of distribution of, household earnings and expenditures; and the extent of their direct access to productive resources, especially land.

Consequences of Technological Change

Technological change involves not only the adoption of a specific type of innovation but also the adoption of other associated components. For example, the adoption of fertilizer-responsive, high-yielding rice and wheat varieties with short growing periods has often been accompanied by increased use of tractors and direct seeding, a shift from traditional methods using draft animals and transplanting rice seedlings. Furthermore, the impact of any technological change can not be discerned in the period immediately after adoption, since the diffusion of agricultural innovations is a long-term process. Some of these long-run adjustment effects may involve the movement of labor from agriculture to nonagriculture. Since the majority of the poor—and women—in Africa and Asia derive incomes from labor on their own and others' farms, the employment effects of new technologies are important factors determining changes in incomes and welfare. This is illustrated by the adoption of irrigated rice and high-yielding varieties (HYVs) or modern varieties (MVs) in Asia and Africa, as well as the introduction of other crops in Africa.

When there is a growing supply of landless women's labor, women will benefit only from productivity increases that are accompanied by an increased demand for women's labor. In addition, whether technical change benefits women depends on their control over resources. Women in farm households who have some control over income from land will benefit from any type of technical change in agriculture. This is because they will reap the returns from increased productivity of both household labor and land (ignoring intrahousehold distribution). For women in landless households whose only resource is labor, neutral or labor-using technical change in agriculture will raise demand for their labor as well as their productivity, but labor-saving technical change will reduce their employment opportunities (Unnevehr and Stanford, 1985).

Adoption of Modern Rice Varieties in Asia. The impact on women of the adoption of modern rice varieties depends on nonagricultural earnings opportunities, existing land-tenure patterns, and the gender division of labor in rice farming prior to the introduction of the new technology. In the Philippines, Indonesia, and parts of India, the increased labor requirements for transplanting and weeding, which are traditionally women's tasks, were met by hired female laborers from landless households, as women in landed households have tended to move out of agricultural labor into more lucrative nonagricultural occupations (in the Philippines and Indonesia) and to provide supervision rather than labor in farm production (Unnevehr and Stanford, 1985). In West Bengal, where the new technology uses a higher proportion of male-to-female labor, women from land-owning households spent less time in agriculture and more on domestic activities or leisure (Mukhopadhyay, 1991).

After the widespread adoption of MV technology, however, nonfarm income sources may contribute more to increases in rural incomes than rice income, as the experience of Taiwan and Japan suggests. Movement of men into nonagricultural enterprises on either a part-time or a full-time basis may increase women's contribution to agricultural production. In Japan, movement of men into nonagricultural enterprises has made rice farming predominantly a woman's activity, while in Indonesia, female agricultural wage rates increased in the 1980s due to increased demand for labor in the nonagricultural sector.

To a great extent, the benefits from adoption of high-yielding varieties of rice in Asia have been captured by those with secure land rights. Since labor is geographically mobile across regions, growth in agricultural productivity tends to benefit owners of fixed factors, such as land. However, possible increases in income inequality have been mitigated by land-reform programs in rice, the growth of nonfarm income sources, and the provision of support services that assured relatively egalitarian access to inputs, such as networks of effective farmers' associations, subsidized irrigation, and availability of credit.

Since wages in rice tend to be equalized across regions, the possibility that the adoption of modern varieties will continue to raise incomes and wages for landless workers is limited. It is unlikely that the spread of such technology will generate the upward pressure on real wages—and increased wages of women in the rice sector—similar to the early experience in East Asia. Population pressure on limited land, and slow absorption of surplus labor into nonagricultural enterprises, have tended to depress real wages. With the exception of Japan, South Korea, Malaysia, and Indonesia, where there has been strong growth in the industrial and nonagricultural sectors, agricultural wages in most of Asia have been stagnant since the 1970s.

Population pressure may also be exacerbated by the effects of the adoption of modern varieties of rice. Evidence from India suggests that the latter has positive effects on fertility (Mukhopadhyay, 1991), although increases in female wages and female education depress fertility. In West Bengal, higher yields of the modern, compared to traditional, varieties are associated with higher fertility, perhaps because of the relatively lower use of women's labor in the modern varieties and women's withdrawal into home production activities that do not conflict with childbearing. Moreover, increases in income in West Bengal have been invested in larger families rather than in children's schooling. In contrast, in the Philippines, landed women from households adopting the modern varieties have moved into nonagricultural occupations in which they have greater control over their own incomes. Thus, while the new agricultural technology may have increased the income of farm households in both India and the Philippines, the difference in its impact on women is partly due to cultural differences in women's roles. Raising female educational attainment and female wages (from an increase in labor demand) is, therefore important, to reduce fertility and increase child schooling. However, given the limited labor absorptive capacity of agriculture, sustained increases in female wages will have to come from the expansion of nonagricultural employment opportunities.

Introduction of Irrigated Rice in Africa. Many agricultural development projects have also tried to introduce irrigated-rice technologies in rice-farming areas in Africa. However, Dey (1985) found that the impact of these rice-development projects on women is ambiguous, because of the unsuitability of agronomic and agroclimatic conditions, inadequate extension services for women, loss of women's access to rice land, increase in women's labor input, and limited input supplies, marketing, and other support services for women. In many cases, the development programs repeatedly failed to take into account differences in social structure and the roles played by women in African rice cultivation. Many technologies introduced were inappropriate for existing conditions, and inadequate extension services meant that suitable technologies were not introduced to women. In many irrigated-rice projects, sickles and threshing machines were introduced for men's irrigated rice but were not available to women.

Many rice-development projects deliberately handed over improved land to male household heads, even though women had previously been exclusively responsible for rice cultivation and often had independent rights to rice land. The possibility of double cropping also changed the traditional allocation of time across seasons, the dry season being customarily reserved for other activities like field maintenance and ceremonial activities. In addition, many of the new technologies increased demand for labor in situations in which traditional tasks were being performed with very low-level, time-intensive technologies. Where the introduction of the programs had detrimental effects on the economic security of women, the success of the programs themselves was jeopardized.

Two classic cases are the introduction of irrigated rice in Gambia (Von Braun et al., 1989) and Cameroon (Jones, 1983). Technical difficulties and a lack of understanding of farming systems led to unsuccessful early attempts to introduce irrigated rice in Gambia. In the Jahally-Pacharr Smallholder Rice Project, attempts were made to preserve women's customary role in rice farming by giving women land titles (long-term leases) to irrigated rice areas. Nevertheless, the introduction of centralized pump irrigation transformed rice into a male-controlled crop. As yields per unit of land increased from 1.3 to 5.9 tons, the share of women's rice fields in the total dropped from 91 percent (of swamp rice fields) to 10 percent (of fully water-controlled rice fields). Less of the new rice crop was sold for cash than expected: The high-technology rice became a communal food crop under the control of the compound head, while traditional rice remained a mixed private (for cash sale) and communal food crop. This represents a shift from rice being a woman's individual crop to a communal crop under the authority of the male compound head. Moreover, the higher yielding technologies with higher variable costs became the responsibility of men. As women's control of rice production declined, they reduced their labor input in rice and increased it in the cultivation of upland crops.

Traditionally, women in northern Cameroon spent most of their time on sorghum fields allocated to them by the head of the compound. However, in the SEMRY irrigated-rice project, even when women were assigned rice plots the proceeds were controlled by men. The higher labor requirement for rice reduced the time available for, and spent in, sorghum production. Following the harvest, rice was retained for home consumption, and, from sales of the surplus, men gave their wives a lump sum in cash to compensate them for their labor. Jones (1983) found that women were willing to work more days in their husbands' rice fields only if they were compensated more. While wage rates for rice were higher than the returns from any alternative uses of the women's time, women apparently did not have the freedom to choose the option of wage labor, and they had little incentive for labor input in rice.

The failure to consider women's roles in the design of rice-development projects has had three detrimental consequences: (1) a loss of adaptive efficiency from not taking women's operational knowledge into account; (2) a reduction of women's bargaining power within the household, as well as an increase in work; and (3) lower adoption rates due to women's lack of access to technology and training and to a failure by the proponents of the technology to address women's time constraints (Dey, 1985).

Introduction of Other Crops in Africa. In other attempts to introduce new technologies in other African countries, where only men had direct access to the improved technologies, there was either a loss of women's crop-production role to men or an improvement in crops grown by men and not previously grown by women (Kumar, 1987). While there may have been a short-run decrease in women's incomes and an increase in their labor input, long-run effects were not necessarily unfavorable. When cocoa was introduced in western Nigeria, it became a men's crop, and women's traditional obligation to provide labor was extended to cocoa, subject to bargaining and remuneration. However, much greater reliance was placed on male hired labor. Eventually, women decreased their time in food production but increased it in food processing, cloth making, and trade. While self-subsistence in food declined, food security did not.

In another case, women's incomes dropped initially when families migrated to a resettlement area devoted to improved cotton and sorghum in Burkina Faso, but later increased (Kumar, 1987). In the resettlement area, women did not have traditional rights to land, and their food crops were not included in the program rotation. Husbands did give their wives informal authority to cultivate part of the bush fields. After five years, however, families had paid off initial debts, family size had increased with the in-migration of additional family members, cash income was seven times higher, and grain yields were two to three times higher than in the home villages. Much of the increase in yields was used to guarantee consumption and to hire labor. The increase in hired labor enabled women to decrease their labor inputs in common fields, expand their private grain fields, and become more involved in trading and livestock activities.

These two cases highlight the importance of relieving labor constraints through the adoption of labor-saving technology in tasks women traditionally perform, such as food processing or fetching fuel and water, or through the ability to hire labor. In both of the above cases, the ability to hire labor enabled women to diversify into more profitable nonagricultural enterprises. Although much of the literature on Africa argues that the introduction of new technology only increases women's labor inputs, taking time away from their other responsibilities such as child care, this view does not recognize that women's tasks are time-intensive because of lack of tools and complementary inputs. Women can increase the efficiency of their time use if appropriate labor-saving technology is introduced.

Effects of Commercialization

While concerns have been raised that increased commercialization reduces production of food for home consumption, with detrimental effects on nutrition, a number of studies by the International Food Policy Research Institute have found that the increased income due to commercialization increases household calorie consumption (Kennedy, 1989; Von Braun et al., 1989; Bouis and Haddad, 1990). However, actual impacts on women's income, women's and men's time allocation and women's access to land and off-farm employment opportunities vary widely.

Commercialization of Rice in the Gambia. The Gambian irrigated-rice project discussed earlier was estimated to have increased income by 13 percent in the lowland villages. In these villages, women's income is spent mostly on food. If income were held constant, the shifting of women's private crop and female-controlled communal crop to male control would have decreased women's incomes, leading to lower food expenditures, and lower calorie consumption. However, the net effect of the new technology on calorie consumption was positive since the increase in income more than compensated for the decline in calorie consumption due to the shift in crop control.

However, this project's experience suggests that the selection of a "woman's crop" for promotion does not necessarily mean it will benefit women, especially if women do not have access to cash inputs required to take advantage of the new technology. Neither can projects guarantee protection of traditional production arrangements while fundamentally changing the nature of production. Project experience also suggests that women are more involved in independent cash-crop production than generally acknowledged and would benefit from increased access to inputs. Finally, it may be important to promote not only a single crop but also rice and upland crops to reduce competition between men and women in rice production.

Shifts from Maize to Sugarcane in Kenya and the Philippines. There were major differences in the gender impact of shifts from maize to sugarcane production in Kenya (Kennedy, 1989) and the Philippines (Bouis and Haddad, 1990). Although incomes of smallholder sugarcane farmers substantially increased in both areas, the effects on women's involvement in agricultural production and nutritional status were markedly different, due to fundamental differences in the gender division of labor in agriculture in both farming systems and relative resource scarcities. Maize employs considerably more family than hired labor per hectare in both Kenya and the Philippines, but sugar cultivation depends mostly on hired labor in the latter. Moreover, agricultural production in Kenya is more intensive in female labor. In Kenya, female adult labor accounts for one-half of family labor in maize, while in the Philippines, it is less than 15 percent. On a per hectare basis, women contribute less labor in sugarcane in both Kenya and the Philippines.

In absolute terms, the effects of the shift from maize to sugarcane cultivation were profoundly affected by relative resource scarcities. In the Philippines, labor is abundant relative to land. When sugar, which uses less female labor, replaced maize, there was an absolute decrease in the demand for female labor in agriculture. Women in sugarcane households were able to increase calorie intakes, due to increases in income, and reduce activity levels, due to the lower demand for female labor, and in doing so, improve their nutritional status. In Kenya, where labor is a binding constraint and sugarcane production took place on previously fallow land, there was a marginal increase in the demand for female labor in agriculture. In this labor-constrained environment, the increased demand for female labor was met by women working longer hours and increasing their energy expenditure. While sugarcane production increased incomes and calorie intakes, the increase in calorie intake was used by women to increase the energy intensity of their activities rather than to improve their own nutritional status.

Impact on Women's Welfare

Normally, technological change and commercialization stimulate agricultural growth, improve employment opportunities, and expand the food supply, all of which are central to the alleviation of poverty. However, those instances in which the poor have failed to reap the benefits of commercialization were due mostly to inelastic demand for food, adverse institutional features, or unfavorable policy environments.

Some of these adverse structural and institutional features need to be considered more explicitly to avoid unfavorable impacts on women. The introduction of technology may change the nature of cropping arrangements and the division of labor, and detrimental impacts may be exacerbated by constraints in access to credit and hired labor (in labor-scarce societies). The full range of constraints under which women operate needs to be understood when development programs and policies are targeted by gender, and when untargeted new technology is introduced in rural areas (Binswanger and Von Braun, 1991).

Some of these constraints are dictated by a country's relative resource scarcities: In a labor-scarce economy, the introduction of technologies that increase the demand for labor in agriculture may have adverse impacts on time spent on other tasks, bargaining within the household, and women's nutritional status. In a labor-surplus economy, technological change that increases labor productivity should be matched by an increase in demand for labor, to avoid unemployment. It is important that relative prices of land, labor, and capital reflect the true cost of technologies adopted, since distorted prices may promote technologies inconsistent with a country's resource endowments. In surplus-labor situations, potential increases in income and wages due to technical change will be limited by population pressure and the availability of nonagricultural employment opportunities. Such sustained increases in female wages (accompanied by increases in female education) would reduce fertility and ease population pressure on limited land.

Women's access to fixed factors, especially land, is another important determinant of the distribution of the benefits of technical change. Women's independent land rights have been crucial to their capturing the gains from technological adoption and irrigation expansion. Another institutional feature affecting the gender impact of technical change is household structure, which varies widely across countries and cultures. In households with asymmetric rights and obligations, as in much of Africa, there is no guarantee that technical change will benefit women. In fact, most case studies of commercialization show that,

despite income increases and general improvements in nutrition, women have had a much reduced role in the new technologies in commercialized crops. Whether household welfare and women's welfare are at cross purposes, however, depends on the degree of income pooling in the household and the degree to which development programs can protect, or strengthen, women's rights to productive resources and incomes.

References

Appleton et al., "Public Services and Household Allocation in Africa: Does Gender Matter?" (unpublished manuscript). Oxford: Center for African Studies, Oxford University, 1991.

Behrman, Jere, and Leonardo Lanzona. "The Impact of Land Tenure on Time Use and on Modern Agricultural Technology Use in the Rural Philippines," paper presented at a conference, "Family, Gender Differences, and Development," Yale University, September 4–6, 1989.

Bindlish, Vishva, and Robert Evenson. *Evaluation of the Performance of T&V Extension in Kenya.* World Bank technical paper 208, Africa Technical Department Series. Washington D.C.: World Bank, 1993.

Binswanger, Hans, and Joachim von Braun. "Technological Change and Commercialization in Agriculture: The Effect on the Poor," *World Bank Research Observer,* vol. 6, no. 1, 1991, pp. 57–80.

Boserup, Ester. *Woman's Role in Economic Development.* New York: St. Martin's, 1970.

Bouis, Howarth, and Lawrence J. Haddad. *Effects of Agricultural Commercialization on Land Tenure, Household Resource Allocation, and Nutrition in the Philippines.* Research report no. 79. Washington, D.C.: International Food Policy Research Institute, 1990.

Deere, Carmen Diana. "The Latin American Agrarian Reform Experience." In Carmen Diana Deere and Magdalena León (eds.), *Rural Women and State Policy: Feminist Perspectives on Latin American Agricultural Development.* Boulder: Westview, 1987, pp. 165–190.

Deere, Carmen Diana, and Magdalena León. *Women in Andean Agriculture: Peasant Production and Rural Wage Employment in Colombia and Peru.* Women, Work, and Development Series no. 4. Geneva: International Labor Office, 1982.

Dey, Jennie. "Women in African Rice Farming Systems." In International Rice Research Institute, *Women in Rice Farming: Proceedings of a Conference on Women in Rice Farming Systems.* Brookfield: Gower, 1985, pp. 419–444.

Dixon, Ruth. "Women in Agriculture: Counting the Labor Force in Developing Countries," *Population and Development Review,* vol. 8, no. 3, 1982, pp. 539–566.

Goldschmidt-Clermont, L. *Economic Evaluation of Un-*

paid Work in the Household: Africa, Asia, Latin America and Oceania. Women, Work, and Development Series no. 14. Geneva: International Labor Office, 1987.

Jacoby, Hanan. "Productivity of Men and Women and the Sexual Division of Labor in Peasant Agriculture of the Peruvian Sierra," *Journal of Development Economics,* vol. 37, nos. 1 and 2, 1992, pp. 265–287.

Jamison, Dean, and Lawrence Lau. *Farmer Education and Farm Efficiency.* Baltimore: Johns Hopkins University Press, 1982.

Jones, Christine. "The Mobilization of Women's Labor in Cash Crop Production: A Game Theoretic Approach," *American Journal of Agricultural Economics,* vol. 65, no. 5, 1983, pp. 1049–1054.

Kennedy, Eileen. *The Effects of Sugarcane Production on Food Security, Health, and Nutrition in Kenya: A Longitudinal Analysis.* Research report no. 78. Washington, D.C.: International Food Policy Research Institute, 1989.

Kumar, Shubh. "Women's Role and Agricultural Technology." In John Mellor, Christopher L. Delgado, and M.J. Blackie (eds.), *Accelerating Food Production in Sub-Saharan Africa.* Baltimore: Johns Hopkins University Press, 1987, pp. 135–147.

Lastarria-Cornhiel, Susana. "Female Farmers and Agricultural Production in El Salvador," *Development and Change,* vol. 19, no. 3, 1988, pp. 585–615.

Lele, Uma. "Women and Structural Transformation," *Economic Development and Cultural Change,* vol. 34, no. 3, 1986, pp. 195–221.

Mkandawire, Richard, Steven Jaffee, and Sandra Bertoli. "Beyond 'Dualism': The Changing Face of the Leasehold Estate Subsector in Malawi," (unpublished manuscript). Bunda College, University of Malawi, and Institute for Development Anthropology, Binghamton, New York, 1991.

Mollel, N.M. "An Evaluation of the Training and Visit (T&V) System of Agricultural Extension in Muheza District, Tanga Region, Tanzania." M.S. thesis, University of Illinois, Urbana-Champaign, 1986.

Moock, Peter. "The Efficiency of Women As Farm Managers: Kenya," *American Journal of Agricultural Economics,* vol. 58, no. 5, 1976, pp. 831–835.

Mukhopadhyay, Sudhin. "Adapting Household Behavior to Agricultural Technology in West Bengal, India: Wage Labor, Fertility, and Child Schooling Determinants" discussion paper no. 631, Economic Growth Center, Yale University, New Haven, 1991.

Saito, Katrine, and Daphne Spurling. "Developing Agricultural Extension for Women Farmers," discussion paper no. 156, World Bank, Washington, D.C.,1992.

Saito, Katrine, Daphne Spurling, and Hailu Mekonnen. "Raising the Productivity of Women Farmers in Sub-Saharan Africa" discussion paper no. 230, World Bank, Washington, D.C., 1994.

Sikapande, E. "An Evaluation of the Training and Visit (T&V) System of Agricultural Extenson in Eastern Province, Zambia." M.S. thesis, University of Illinois, Urbana-Champaign, 1988.

Udry, Christopher. "Gender, Agricultural Production, and the Theory of the Household," *Journal of Political Economy,* vol. 104, no. 5, 1996, pp. 1010–1046.

Unnevehr, Laurian, and M. Lois Stanford. "Technology and the Demand for Women's Labor in Asian Rice Farming." In International Rice Research Institute, *Women in Rice Farming: Proceedings of a Conference on Women in Rice Farming Systems.* Brookfield: Gower, 1985, pp. 1–20.

Von Braun, Joachim, Detlev Puetz, and Patrick Webb. *Irrigation Technology and Commercialization of Rice in the Gambia: Effects on Income and Nutrition.* Research report no. 75. Washington, D.C.: International Food Policy Research Institute, 1989.

Wazir, Rekha. "Women's Access to Land As Owners and Workers: A Stocktaking for Five Asian Countries," report prepared for the Food and Agriculture Organization of the United Nations, Rome, 1987. Mimeographed.

Women's Experiences As Small-Scale Entrepreneurs

Susana T. Fried

Introduction

Women's experiences as small-scale entrepreneurs have received increasing attention in the late twentieth century. This stems, in part, from the recognition of the importance of women's economic contributions, especially in the informal sector—as microentrepreneurs, domestic workers, home-based subcontractors, and sweatshop workers—to family survival throughout the world. Most women's small-scale enterprises, especially microenterprises, are located in the informal sector, either because government regulations exempt businesses below a certain size or because these enterprises intentionally evade government regulations. One may even regard the informal sector as composed primarily of microenterprises, and this has led to a tendency to equate one with the other.

As support for microentrepreneurial efforts has gained approval among governments, multilateral agencies, and foundation funders, so has the observation of women's extensive participation in programs that support self-employment and the initiatives of small-scale enterprise. Women's increasing employment in both the formal and the informal sector and the growing dependence upon women's income-earning capacity set the context for understanding women's experiences as small-scale and microentrepreneurs and the evaluation of support for these efforts.

This essay explores women's experiences as small-scale entrepreneurs in a variety of locations throughout the world. It begins by setting out a broad context from which to view women's small-scale entrepreneurship. Economic restructuring, the informal sector, and women's work in these contexts are discussed, followed more specifically by sections on the location and the strategies of women working in microenterprises. The essay also presents an overview of women's experiences as small-scale entrepreneurs, based upon research conducted by scholars, development practitioners, and feminist organizers throughout Africa, Asia, the Caribbean, and Latin America.

A brief history of microenterprise assistance follows, highlighting the implications of various types of support and the different perspectives on development that drive such programs. Next comes a discussion of dominant development paradigms and issues related to evaluating these efforts. Finally, I suggest that a critical assessment of microenterprise from the perspective of women's experiences requires that we clearly examine both the philosophy and the strategies of microenterprise assistance organizations and the sociopolitical context in which they are set.

Economic Restructuring

To assess women's experiences as small-scale and micro-entrepreneurs, one must first comprehend the circumstances that women of varying cultures, races, and economic classes face as economic actors. In particular, it is necessary to examine the factors that circumscribe women's economic and political opportunities. The structure of opportunities and constraints in the economy, culture, and society are founded upon frequently unexamined assumptions and prescriptions about what it means to be female or male. Women and men, as individuals or groups, enter societal arenas (labor markets, political institutions, and the like) with differential power, resources, and access to entrepreneurial opportunities. We can call this "gendering," and it entails intersecting processes that construct and reinforce the marginalization of some groups through categories of "difference" (meaning at its basic level identity categories different from the "norm" based upon the White, Western, middle-class male), including gender, race, culture, and class. Economic patterns are also "gendered." For instance, the rising reliance upon "contingent" workers, who may appear to be self-employed, among business enterprises of all sizes and locations assumes a substantial cadre of women workers willing to accept often irregular and unsafe employment conditions. Some refer to this as a "feminizing" of the labor force, a phrase that might misconstrue the historical importance of women's paid and unpaid work in all parts of the world.

At the broadest level, the balance between formal- and

informal-sector employment opportunities shifts as a result of economic restructuring, macroeconomic policies, and microeconomic strategies. Economic restructuring here refers to the shifting location and structure of industry worldwide and the resulting changes in the balance of power between labor and capital. Such changes are based upon industrial transformations that result from technological developments, such as the shift from steam-based production of textiles and steel to the creation of chemical and electrical industries in the late nineteenth century, or the late-twentieth-century growth of "postindustrial" sectors based upon computers and telecommunications. New forms of production entail geographical shifts in investment as well as transformations in the manner in which labor is deployed. The *maquiladora* system is a good case in point. First, the number of *maquiladoras* at the U.S.–Mexican border rose from virtually none in 1964 (the year before the border-area industrialization program legalized the establishment of the "offshore plant" in Mexico) to more than 1,900 in 1991, in part because of the vastly lower cost of land and labor in Mexico. Second, this locational shift to seek greater profits and cut costs relies upon a work force that is primarily young women who are paid low wages and work for only a short period of time in a particular factory. At the level of individual nations, economic restructuring brings with it varying patterns of business and trade cycles, as well as debt burdens and changes in economic and social policy.

Two elements of the patterns created by economic restructuring are particularly important for understanding the form that gender structuring of the economy takes and the opportunities and constraints women face in the economy. First, throughout the world, women's labor-force participation has increased in both the formal and the informal economy. This general increase results from several factors. One is rapid urbanization, which in some regions, particularly Latin America, has been led increasingly by women seeking wage labor (Berger, 1989). Another is the decreasing value of currency in many countries, or the decreasing value of wages, which has necessitated more income earners in a household for survival.

Changes in production have transformed the work that women once did at home, like food preparation or clothing manufacturing, to work for pay in the marketplace. Such technological movements have contradictory impacts on women, since they create some opportunities while eliminating others. For instance, even jobs with long hours and minimal pay provide women with independent income, however inadequate it may be for supporting themselves and their families. The increasing employment of women, alongside the decreasing employment of men, generates unstable patterns of gender relations, while creating some prospects for women's increased autonomy even though much of the work women find is contingent and low paid.

Second, one can conceive of informalization itself—and, therefore, the increase in women microentrepreneurs—as a gendered process, upon which global economic and political restructuring rely. Historically, women's work has often been informal and small scale, when visible at all. Indeed, this is tied to the general position of labor in any political economy, and the location and placement of women within this structure. For instance, governmental policies in most countries generally have not favored labor. However, even in those cases of what might be regarded as a labor-capital-government compromise, policies have functioned to subsidize male incomes, creating a labor elite. Most often, this shifts revenue out of job creation and social welfare strategies, but it also assumes the unpaid and invisible family and community maintenance work, sometimes called social reproduction, done primarily by women. In other words, much of women's informal economic activity takes place in conjunction with women's responsibility for assuring family survival. Such obligations lead women to engage in multiple ventures, of which small-scale production, commerce, or personal-service provision are frequently central. Studies indicate that women all over the world must engage in "income packaging" or "livelihood strategies" of mixing income from work in the labor market, support from family members, and government aid in order to survive (Escobar, 1989; Downing, 1990; United Nations, 1991; Downing and Daniels, 1992; Spalter-Roth et al., 1994).

The Informal Sector

Many studies on the impact of lending practices and economic crises indicate that cutbacks in government employment, the reduction of vital social services, and the elimination of price supports and subsidies on primary goods and services have brought about decreases in women's economic opportunities and a deterioration in the health and well-being of women and their families. Such changes have also severely decreased employment in the formal sector, and they help to explain the burst of informal activity observed throughout the so-called First and Third Worlds. Not only does it appear unlikely that the formal sector will expand quickly and extensively enough to create jobs for all of those who seek them, but in many cases the formal sector is contracting. Hence, the increasing discussion about structural unemployment, and the corresponding need for new macro- and microlevel policies, has led to a situation in which microenterprise assistance has assumed growing importance.

In many communities, regions, and nations, poverty rates are growing rapidly while unemployment rates in the formal sector similarly multiply. At the same time, the informal sector continues to expand. For instance, a 1988 estimate for 19 selected countries noted that the informal sector accounts for from 31 percent of the labor force in high-income countries to 51 percent in low-income countries (Turnham, 1993). Such a large informal economy confounds traditional theories of development, which suppose a progressive formalization of the labor force, along with a disappearance of the informal economy. Yet, even high-income countries retain a significant informal sector, and it appears that informalization is growing rather than decreasing in importance. Such observations demand that government planners and makers of development policy grapple with responding to labor markets that are at odds with their presuppositions and with empirical data that contradict their basic theoretical assumptions.

The "informal sector" or "informal economy" is an ill-defined and controversial term. It comprises a wide range of unregulated and extralegal activities, generally involving work for pay that does not come in the form of wages, and employment conditions that are not regulated by local, state, or national government. "Informality" describes not only the relation of the enterprise to the state but also the relation between employers and workers, many of whom are likely to be family, and between buyers and sellers. In this way, informality characterizes both production processes and distribution networks. Small-scale enterprises, on the other hand, refer to a larger category of business ventures, which includes microenterprises, but does not carry the assumption of informality.

The informal sector is least problematic if we restrict it to focus on microenterprises as the primary economic units that constitute it: microenterprises. The size of the informal sector and the extent of microentrepreneurship vary extensively from region to region. The sectoral location of microenterprises also varies, but the greatest concentration of these enterprises is in commerce and trade. There is disagreement about the contribution of microenterprises to the gross domestic product (GDP), their growth and job-creation potential, and the possibility of "graduating" informal enterprises into the formal economy (see Rakowski, 1994). Some argue that microenterprises constitute an important and vital source of job creation and economic development with the capacity for expansion and, therefore, provide an important contribution to national economic development. Others contend that the overwhelming bulk of microenterprises are fragile and marginal and are not likely to survive long enough to add greatly to the economic health of communities or nations.

Legal and regulatory practices play a significant role in determining the size and extent of the informal sector. The extent of informality may be very high in countries, sectors, or areas where regulation is lax or conformity to regulation is low, as in Peru, and very low in places where regulation is tight, as in North Korea. The cost of conforming to regulations will also have an impact on the extent of informality. High costs of conforming to regulations may create incentives for formal-sector enterprises to subcontract to small, informal ones. Such regulatory practices make it clear that informality does not imply backwardness or lack of development.

Indeed, it appears that the rise in informal economic activities corresponds to economic restructuring in the late twentieth century. In these often marginal economic initiatives, one sees the intersection of race, class, culture, and gender. For instance, a study of microvendors in La Paz, Bolivia, found that women are more likely than men to be involved in the sale of basic goods along sidewalks or in market stalls. These women microvendors are more likely to be migrants from rural areas and to have a lower educational profile than economically active women in general (Escobar, 1989). A number of studies have found that the quality of microenterprise employment is hierarchically organized by gender as well as by age, educational level, skill, and length of time in residence. Thus, young women who are recent migrants may be more likely to become domestic servants, while middle-age women who are heads of household are more likely to be self-employed or working in small shops in commerce and industry (Turnham, 1993). In cultures that explicitly restrict women's mobility in public, women's economic activities will most likely be home based. Given their restricted mobility, they rely upon others to sell their products and purchase the goods they need for production. As of the late 1990s, little of the research on women in the informal sector intensively addresses the intersection of gender, race, class, and culture in women's experiences in microenterprise. Differences among women and the impact of cultural differences, therefore, can only be inferred. The research that does exist, along with more extensive information about the intersection between race, gender, culture, and class, indicates that the category of "women microentrepreneurs" is differentiated in ways that replicate the greater or lesser access to resources and opportunities in society in general.

Such a nuanced framework is particularly crucial when we move from the description of women's experiences in small-scale enterprises to an evaulation of microenterprise assistance. To understand women's experiences as small-scale entrepreneurs and the potential impact of credit and training programs, these activities must be placed in their economic and cultural context, while specifying the differences and links among microenterprises, small-scale enterprises, and the informal economy.

Women in Microenterprises

While women may be involved in small-scale enterprises in greater and greater numbers, the explanations for this increase vary from place to place. Both macrolevel and microlevel conditions are important in accounting for the different social, economic, and political circumstances women face as small-scale entrepreneurs: the general level of economic well-being and political conditions in any particular location, as well as the construction of the formal and informal sectors, and the corresponding educational and economic opportunities for women.

Women's participation in the informal sector varies by region, but generally it is extensive. A survey of women around the world indicates that more than half of economically active women in sub-Saharan Africa and southern Asia and a third in northern Africa and the rest of Asia are self-employed (United Nations, 1991). Such economic activity also varies by industrial sector. For example, two-thirds of informal-sector service production in African countries like the Congo and Zambia is done by women, except in transport, where women are almost completely excluded (United Nations, 1991).

In looking at women's work in the informal sector, several factors are quite clear. First, in most regions of the world, particularly Latin America, domestic service is the predominant location of women's informal work. (The question of where domestic service is located is contested. Some argue that it is informal-sector work because of the characteristics of the work—its casualness and lack of regulations being primary considerations. Others argue that it is waged work and not an economic unit, in any case. Berger [1989] deals with this debate at length.) Second, the regulatory environment may either enhance or constrain women's opportunities for self-employment. In Belize, as in a number of other countries, a woman must have a man's signature in order to start a business and obtain a loan. Such requirements not only constrain women's capacity to earn income through self-employment but also serve to limit women's autonomy and freedom of movement.

For poor women in urban areas, two options predominate: domestic service and microenterprise. For the poorest women, the primary option among various forms of microenterprise is microvending. Microvending and domestic service have a number of common characteristics that make them forms of employment to which women with the fewest resources turn: They are both easy to enter, require little or no capital or savings, provide immediate returns, and draw upon skills that most women acquire early in their lives. Requiring long hours for minimal pay, they are both also labor intensive, unorganized, and generally unregulated, making them fragile and insecure for income generation. The ease of entry is matched by the ease of exit, which may, in fact, work to the detriment of women workers. This helps account for the "casualness" of women's labor, and the low pay for women's work, each of which then exacerbates the other.

The Location of Women's Microenterprises

The prominent features of informal-sector initiatives include: small size of operations, reliance on family labor and local resources, low capital endowments, labor-intensive technology, limited barriers to entry, a high degree of competition, an unskilled work force, and acquisition of skills outside the formal educational system. Informal enterprises may be usefully classified along two lines: the industrial sector of which it is a part (manufacturing, services, or commerce); and the characteristics of the enterprise itself (casual work, temporary and seasonal activities, or precarious activities like street vending). Unstable or temporary work must be distinguished from more or less stable enterprises like small stores, manufacturing operations, and traditional artisanry. Finally, enterprises may be organized as subcontracting firms, which some consider to be a form of disguised wage employment (Berger, 1989). Such distinctions are particularly important in discerning the sustainability of women's enterprises, their job-generation capacity, and their potential as part of broader local development initiatives. All

three, in other words, will most likely be greater if the activity involves creating a small manufacturing workshop rather than selling vegetables on the street.

When examining gender differences in microenterprises, one sees the replication of the larger pattern of sex segregation in the labor market. For instance, women's microentrepreneurial activities are more likely to be in the service or commerce sector than men's. In addition, women's microenterprise initiatives are more likely to be concentrated at the survival end of the survival-security-growth continuum. Such "sex typing" in entrepreneurial activity is prevalent throughout the world, but the precise nature of the gender-based division or segmentation varies by region and culture. Regional and cultural variations in women's informal enterprises stem from multiple factors ranging from laws, policies, and regulatory mechanisms, such as the ability of women to own property independently, to kinship patterns and family practices. For instance, West African women dominate long-distance trade, while women in East Africa are more often relegated to local trade. Some explain this by the greater freedom and mobility that West African women have because they live in matrilocal societies, where the extended family shares domestic responsibilities (Downing, 1990).

Women engaged in "traditional" activities are in the least-remunerative enterprises. Traditional economic activities are those, like vending and personal services, in which women's initiatives have tended to be clustered, or manufacturing activities like dressmaking and pottery making that are highly labor intensive. A study of microvending in Bolivia found that women vendors are more likely to sell household staples and agricultural products, while men are more likely to sell nonessential consumer products (Escobar, 1989). It also found that the high proportion of women in the informal sector is not simply the result of a lack of employment opportunities in the formal sector, but reflects the difficulty many women face in managing their multiple responsibilities and the discrimination they face in the informal and formal economies.

The gender segmentation of informal enterprises is matched in the formal sector. The very notion of the informal sector might be called into question by looking at the way in which the category of gender cross-cuts the formal–informal distinction. Indeed, the notion that the economy is divided into a formal and an informal sector is complicated by the fact that women's work appears to be quintessentially "informal" (insecure, low–paying, poorly regulated), whether in the formal or the informal sector (Scott, 1991). In nearly all regions of the world, women's informal businesses have lower sales revenues, lower asset bases, and smaller profit margins than men's. In some countries in Africa, one potentially perverse effect of the economic crisis is a change in relative income and profit levels between women and men. In these cases, women may be earning more than men, but this is primarily caused by a decrease in men's income rather than a substantial increase in the returns to women's labor.

Women's participation in the labor force, whether formal or informal, is countercyclical: It increases during periods of economic crisis, as unemployment among men increases and the need for additional income rises. As household incomes decline, families increasingly depend on women as income producers. While more women seek to engage in market-oriented economic activities, and women's participation in microenterprises increases, these enterprises become more concentrated in a few subsectors, leading to greater competition. All of this results in lower remuneration for such economic activity.

Some channels of communication, purchasing, and distribution between producers and consumers are dominated by women, while others are controlled by men. This segmentation limits women's access and purchasing patterns to more "traditional" industries and products (Downing, 1990). In their study of women entrepreneurs in southern Africa, Downing and Daniels (1992) found that women's firms are concentrated in more-traditional and less-dynamic product markets. As an example, women's manufacturing enterprises are primarily located in beer brewing, dressmaking, knitting, crocheting, and grass and cane work. Men's manufacturing activities, on the other hand, are concentrated in more technologically sophisticated areas that tend to have more growth potential, like construction, welding, auto repair, radio and television repair, and brick or block making. In addition, women are more likely to buy from other women, when possible, and women sellers are more likely to reach other women as purchasers. In this way, different goods are dominated by women or men throughout the process of selling to purchasing final products. Women are also more likely to sell directly to the consumer, while men's firms are more likely to produce intermediate goods sold as inputs to other firms (Downing and Daniels, 1992). This gender segmentation is reinforced and articulated through backward and forward linkages.

These gender differences are well documented and frequently result in lower returns to women's enterprises than men's (Downing, 1990). In part, this can be explained by women's limited access to resources, resulting in start-up capital that is a fraction of men's, a reliance on more rudimentary technology, and fewer workers in each enterprise. The link between women's household and marketing activities also accounts for some of the disparity, since women's trading activities often grow out of their daily tasks and household responsibilities and utilize basic equipment and premises that are readily available to them.

Women's Entrepreneurial Strategies

The objectives of women microentrepreneurs range from survival to security to growth (Grown and Sebstad, 1989; Downing, 1990), but the overwhelming concern of most is survival: These women tend to be very poor, and their ventures fragile. Thus, the purpose of these tiny economic initiatives does not correspond to a neoclassical notion of the dynamic entrepreneur who seeks growth and profit

(Berger, 1989). Instead, the concern of the majority of women in the informal sector is to generate enough resources to maintain themselves and their dependents while accommodating their other activities. As a cause and a result, women tend to cluster in the most fragile and least lucrative enterprises and sectors and to remain in these sectors because they are easy to enter and they make it feasible for them to manage their multiple responsibilities. Women, therefore, seek initiatives that allow them to bring their children to the market with them, if they are vendors, or have their children assist them in production of the goods they are making. For the same reason, women seek enterprises that can be home based, so they can continue to look after their children and perform their household responsibilities.

More generally, women's participation in the informal sector relies upon, and allows for, a blurring of household and market elements in women's daily activities—with mixed results. At best, this means that women can intersperse household responsibilities and income requirements in a manner that makes security possible for themselves and their families. At worst, such a blending of activities leads to extreme exploitation of women's labor and to overwork, illness, and death. The self-employed, or those employed in informal enterprises, frequently work long hours for little pay, and they work under the most difficult of conditions without protection of labor legislation, health care, training, or representational organizations. In addition, while women must bear the costs of initiating their enterprises, the income from these initiatives is frequently controlled by other family members (Greenhalgh, 1991).

To the extent that enterprises are home based and women mix household and market work, the opportunities for greater autonomy are circumscribed. In a larger sense, microentrepreneurial activity is marked by a lack of separation among consumption, production, and household sustenance. Such patterns appear to be quite different for men. Their relative lack of responsibility for child care and household maintenance means that men are more willing to enter riskier ventures and are able to reinvest profits back into their economic activities.

The result of these gender differences, or the gender structure of microenterprise opportunities, is a significant disparity in the growth, income, and profit potential of women's and men's small-scale enterprises. Such inequities constitute the impetus for a number of microenterprise assistance programs, which seek to provide credit at reasonable rates to small entrepreneurs and to reduce the disparity in women's and men's access to resources.

Assistance to Women's Small-Scale Enterprises

Governmental and nongovernmental organizations (NGOs) and agencies that address poverty and unemployment are increasingly looking to projects that assist women's microentrepreneurial activities. In part, this is a result of sustained criticisms by feminist organizers and development planners about the perverse impacts of traditional development projects, which have often displaced women from economic activity and promoted men's control over production and distribution. In fact, some organizations set out explicitly to address the inequities in the gendered division of labor and the need to develop programs that increase women's economic opportunities and income-generating potential. This focus on microenterprise assistance is also a response to the increasing importance of informal-sector enterprises and the inability and unwillingness of formal lending institutions to serve the needs of microentrepreneurs.

History of Assistance

Among agencies assisting informal entrepreneurial efforts, there is no one definition of what constitutes a microenterprise. They generally agree that their target is sole proprietorships, partnerships, or family businesses with fewer than 10 employees. The characteristics of small-scale enterprise initiatives of the poor generally mean that they have extremely limited, if any, access to the commercial banking sector.

At the same time, small and microenterprises can clearly utilize small loans, particularly since a lack of capital constitutes one of their major barriers to growth and success. In fact, despite the reluctance of commercial banks to lend to small-scale ventures, repayment rates, even at market interests rates, are often high (Bennett and Goldberg, 1993). Thus, nonprofit efforts to create credit sources for microenterprises have the potential for becoming self-sustaining over time. The Working Women's Forum in India was initiated in the 1970s as a cooperative banking system to provide credit for poor women. By the mid-1990s, it had reached nearly two million women and had a credit recovery rate of more than 95 percent, a figure not uncommon among nonprofit lenders (Working Women's Forum Report, 1994).

Women's World Banking, the Grameen Bank, ACCION International affiliates, and many others similarly report near-perfect repayment rates. Clearly, these high repayment rates contradict the impression of most lending institutions that the poor, and poor women, are not good credit risks. In fact, many poor women have a long history of borrowing, from family or local moneylenders, who generally charge extremely high interest rates. Many women in villages and communities are long-term participants in local savings associations and mutual support networks (sometimes called *roscas* or rotating savings and credit associations). Indeed, informal-sector enterprises often rely upon informal credit providers who are generally either moneylenders or wholesalers or "middlemen" who provide goods on credit and who tend to charge high interest rates, thereby draining the capacity of marginal initiatives to expand. Given this history, it is clear that there is a great need for microenterprise assistance programs.

Since the late 1960s, self-employment as an antipoverty and economic-development strategy through the creation of microenterprises has been used extensively in Asia, Africa, and Latin America and the Caribbean. Agencies providing assistance to microenterprises come from a range of institutional sources: government programs that receive support from international assistance programs; international organizations with local affiliates, like ACCION Internacional or the affiliates of Women's World Banking; or local, grass-roots organizations that function independently of affiliation to international NGOs, although these are likely to rely upon international and government funding sources. There is no typical microenterprise that receives funding. Working Women's Forum listed 109 occupations of the nearly two million members of its cooperative bank, including street vendors, hawkers, and service specialists, *beedi* (tobacco) rollers, handicraft workers, fisherwomen, embroiderers, and migrant workers.

The increasing popularity of small-scale and microenterprise projects is largely a consequence of their market orientation, which matches the goals of donors and governments for "privatization" of social-development policies regardless of the empirical evidence about which programs generate the most assets for participants over the long term. In fact, microenterprise assistance can take a range of forms, which vary from the so-called minimalist model that only provides credit, to the more extensive "credit-plus" model, which includes social as well as financial intermediation. In fact, the most far-reaching programs seek also to bring about changes in women's perceptions of themselves and their roles in their families and communities, and may even promote individual and group empowerment. For example, in a 1993 interview, Women's World Banking president Nancy Barry described the core principles of the organization as reliance upon local, self-sustaining initiatives that regard women as dynamic economic agents, and market-oriented strategies that "create transformational institutions and relationships to open up whole systems to people who have not had access" (Howells, 1993, p. 32). In fact, the evaluation of microenterprise assistance for women requires careful attention to the form or model of the program and the existence and extensiveness of social as well as financial intermediation.

Types of Assistance

In addressing the inability of the formal sector to keep pace with the bulk of those seeking employment, microenterprise assistance programs include training, credit and lending, and support services. Programs draw upon existing individual and community self-help strategies, particularly the dynamism of the informal sector, savings associations, and mutual support networks. Many of the programs focus on or target women, and, even if they do not, they generally include a significant proportion of women. In part, this is a result of the clustering of women in the most precarious ventures in the informal sector; it is also a consequence of the recognition among donors that funding women's development efforts is money well spent.

Among the forms of assistance to microenterprise, women are most likely to take advantage of informal mechanisms. In part, this results from women's concentration in the low end of the socioeconomic spectrum, as well as their need for flexibility in loan-repayment terms. The more assistance programs use methods with which women are familiar, such as informal credit processes, the more likely that women will turn to and trust these programs.

The shift in the focus of assistance from large-scale development projects to smaller-scale self-reliance support has evolved over a period of time, pushed by critiques of the development assistance process, particularly in its consideration of gender issues. An evolution of perspectives has undergirded gender and development planning that has accompanied changes in economic and social policy frameworks for development in general. Microenterprise assistance falls clearly within this pattern. The shifts in emphasis and ideology have corresponded to more general shifts in Third World development policies, ranging from modernization policies of accelerated growth, to the basic-needs approach that addresses the need for redistribution, to more recent compensatory measures associated with structural adjustment. Project designs and perspectives have often relied upon a notion of women as passive recipients of policy and the importance of women's roles as mothers. Other attempts to recognize the importance of women's productive activity continue to reinforce the gender division of labor by viewing women's income-generating activity as less important than men's.

The assumptions underlying the "income-generating" approach downplay the importance of women's economic activities by creating projects that draw almost exclusively upon "traditional" skills. Such projects also inadequately account for the importance of women's reproductive work in supporting poor families and poor communities. The "income-generating" model of development for women has, moreover, failed in many instances to raise income levels because traditional activities command little remuneration in the market. In some cases, such projects add to women's daily burdens and multiple roles without reasonable payback.

These observations, along with a growing recognition of the distinct importance of women's participation in the informal economy, especially as self-employed workers or as microentrepreneurs, has led governments and nongovernmental assistance agencies to consider projects and research that focus on supporting women's economic activities as integral to family and community survival—which has, in turn, reinforced the popularity of microenterprise assistance.

Methodologies and Mechanisms of Assistance

Methodologies and mechanisms used by assistance agencies differ, as does the size of the individual programs. To some extent, the philosophy of assistance will be connected to the type of assisting organization (governmental or nongovernmental, international or local). The organizational philosophy may be more or less participatory and incorporate a range of strategies of financial and social intermediation. (For a more detailed discussion about the concepts

of financial and social intermediation in the context of the experience of World Bank enterprise development and financial services to women in Asia, see Bennett and Goldberg, 1993). As noted above, a number of program models exist, ranging from "minimalist" to credit-plus ones. Some, like ACCION affiliates, focus on providing credit to already existing businesses. Other programs, like FINCA (the Foundation for International Community Assistance) and other village savings schemes, build upon existing informal credit mechanisms, like savings groups, *roscas* and indigenous support systems and networks. While groups disagree about many aspects of assistance, there is broad consensus about the need for program flexibility and informality in the provision of credit and services (Otero and Rhyne, 1992). Programs function best when they build upon existing networks of savings and mutual support. Furthermore, some efforts have successfully provided women with the means to expand their enterprise activity and develop control over the assets and income generated while also reducing their multiple responsibilities through the provision of child care and other services.

Credit Strategies

Credit is the most popular form of microenterprise assistance, as a way to counteract both the lack of access microentrepreneurs have to formal banks and the very high rates commanded by local moneylenders. Forms of provision range from direct lending to small enterprises to initiating cooperative producer groups. In all cases, however, they contrast with the development strategies based upon investing in labor-intensive industries that are likely to create jobs for women. To make loans available to those with little or no collateral, many organizations use what are called "solidarity groups," which provide support for entrepreneurs and stand in as collateral while the group collectively guarantees the loans. This approach was pioneered by the Grameen Bank in Bangladesh and ACCION International affiliates throughout the Americas and has been extremely popular across all program types. Most researchers and practitioners agree that the creation of solidarity mechanisms, or "social intermediation," provides crucial support to microentrepreneurs and vital links to other economic institutions while it also creates the conditions for broader social and political participation (Bennett and Goldberg, 1993).

Local, grass-roots initiatives are frequently more capable of constructing these links because they can build upon already existing networks and trust. In fact, the process of social intermediation itself creates links between disadvantaged groups and the institutions charged with the delivery of services (Bennett and Goldberg, 1993, p. 35). Such links increase access to resources and strengthen opportunities for sustained security or growth of women's economic ventures.

Capacity of Assisting Organizations

Lending agencies vary widely with regard to their ability

to serve women's specific needs. Bank lending schemes rarely include more than 20 percent female participation (Berger, 1989). Banks generally operate these as public-relations efforts and as social programs rather than financial projects. In contrast, parallel projects, like many of the ACCION affiliates or the Self-Employed Women's Association (SEWA) in India, lend directly to the poor. These tend to be more accessible to women, particularly since they are more likely to provide alternatives to collateral in property or goods and to preexisting savings associations and information networks more extensively. In fact, projects that use solidarity groups tend to include the greatest number of women.

Yet, while some large projects exist, most parallel projects are small in scale and, therefore, lack sustainability. Poverty-focused development banks, such as the Grameen Bank in Bangladesh or BancoSol in Bolivia, began as small, private, grass-roots efforts and have been successful in reaching women. These banks build upon traditional savings schemes of low-income communities and "scale-up" to operate more extensively. Since formal banks do not lend to those they consider high risk, especially those without collateral, these forms of financial intermediation are especially important. Nongovernmental organizations are developing a greater capacity to "intermediate" resources. In some cases, this has involved the conversion of NGOs into commercial banks, like BancoSol, Grameen, and SEWA, which provide credit to the poor and give them a place to save. This process can be transformative because it empowers people by giving them access to money and knowledge.

The gender impact will, of course, vary, depending upon whether projects reach women as a primary goal or offer services to women as a secondary activity. In addition to discrimination, there are a number of reasons why women have so little access to credit. One is that low-income women often develop a "portfolio" of activities and engage in a diverse set of economic activities to reduce their risk. They therefore are more likely to increase the number, rather than the size of the enterprises in which they are engaged. (Downing, 1990). Another reason is that it is often difficult to disentangle the business from household activities in general, particularly for self-employed women and home-based enterprises (McKee, 1989). Agencies that are concerned with assisting women's small-scale enterprises must recognize this and build mechanisms to support women's diverse activities.

Evaluating Microenterprise Programs

Programs for small-scale enterprises in the informal sector can be viewed as falling into three categories, based upon their central goals: (1) enterprise development; (2) poverty alleviation; and (3) social transformation, empowerment, and community development. While extensive overlap exists in the services offered and program strategies, philosophical differences are evident. The enterprise-development approach is primarily concerned with making

women's enterprises more secure and getting women to enter higher-growth sectors. It is most often targeted toward existing businesses, through marketing and business development assistance, along with the provision of small loans at market interest rates. The success of this approach depends on reaching female entrepreneurs whose existence is somewhat secure and who are, therefore, able to focus on accumulating profits.

Poverty-alleviation programs focus on providing low-income women with the means to move from a concern with survival to security. This requires access to baseline resources for women who enter the informal sector with small-scale enterprises. Poverty alleviation may involve integrated programs, which include credit along with policy and management, and may link program participants to other social institutions. In contrast to the enterprise-development approach, agencies focused on poverty alleviation are more likely to encourage women to pursue "safe" sectors as opposed to higher growth–higher risk businesses. While safe sectors—those in which women are traditionally concentrated—may be the least lucrative, they contain fewer barriers to entry and are often easier to combine with household responsibilities.

The third approach takes social transformation, empowerment, and community development as its goals. Agencies that profess these goals often provide access to basic services, along with income generation and enterprise development. For these organizations, a concern with enterprise development is linked to a broader social, political, and economic critique. As a result, they often support cooperative as well as individual efforts. Running throughout these programs is a concern with helping people meet basic needs while increasing self-esteem. Such processes promote efforts to find collective solutions to problems in the community. Women's participation in microenterprises is, therefore, seen as a vehicle for broader social change. While concerns with survival and security are not ignored, social-transformation strategies attempt to link the problems that individual women face to the problems of women as workers more generally. Social-transformation strategies highlight structural factors of exclusion that have led to the preponderance of women and marginalized groups in the population of poverty.

While one might be tempted to regard these three approaches to microenterprise assistance as progressing from less to more intervention, it is more useful to see each of them as having benefits and drawbacks and serving different groups of female microentrepreneurs. For instance, one might criticize the enterprise-development organizations as too narrowly concerned with bringing poor women into the economic mainstream without challenging the structural roots of women's poverty. Yet, these organizations are the most likely to become self-sustaining and will, therefore, continue to provide credit and services to women who may not be reached by other organizations or sources of low-interest credit. Indeed, the ability to function as economic agents is itself empowering for many women. Social-transformation groups, on the other hand, often have a difficult time constructing a clear program focus and must constantly face the difficulty of raising funds to continue operations. Yet, some such organizations, such as SEWA in India, function very successfully on a large scale. The commitment of SEWA to social change is connected to the fact that it was initiated with trade-union support as a private, grass-roots project to serve poor women.

Some general observations can be made about the experience of assisting women's small-scale, informal enterprises through training, support, and lending. The smaller the loans available, the more "informal" the intermediary, the higher the participation of women (McKee, 1989). The most successful organizations have institutionalized in a manner that increases women's participation and maximizes the impact of financial assistance on the security of women's small-scale enterprises. Many agencies involved in assisting these enterprises retain a commitment to connecting survival and financial security with women's participation in broader social and political networks. In many of these NGO-supported efforts, however, the success achieved depends upon an intensity of effort that cannot easily be "scaled up."

At their best, microenterprises and self-employment provide women with the means for survival, security, and growth. They may lead to changes in women's perceptions of themselves and their roles in their families and communities and to increased individual and group empowerment. However, strategies that focus solely on the individual entrepreneur have limitations for broader change: While self-employment or owning and operating a microenterprise may help individual women, it does not provide solutions to the deep problems of structural inequality—among nations and between different categories of people. Moreover, women's participation in the informal sector as microentrepreneurs may not give them the opportunity to be active agents in a public process, since much of their work is home based and the businesses they operate tend to be the most fragile and insecure. Furthermore, women's responsibilities are generally increased by self-employment, as are their working hours, making participation in community or political activity difficult, if not impossible.

Finally, the perceived "autonomy" resulting from self-employment may not, in fact, indicate that women really control their time or income. As evidence of women's lack of autonomy, a recent study (Greenhalgh, 1991) observed a pattern of male or family control of women's businesses and prior financial claims on the profits from these businesses. Greenhalgh argues that economic relations are embedded in social relations and patterns of domination or control. The fact that women may be responsible for assuring their family's survival does not necessarily confer upon them control over the income or assets they earn. As a result, Greenhalgh observed, the income women earn from self-employment is not theirs to distribute. Rather, their husbands or family networks determine the use to which the money will be put.

Therefore, the analysis of women's informal enterprise activity must consider broader social relations and cultural assumptions about women's appropriate roles. Even in the case of market women in Ghana, who are reputed to be the most economically independent in the developing world, family obligations take precedence over business growth. In this way, the autonomy of Ghanaian market women is circumscribed by their social embeddedness and financial obligations to kin. Such observations make it clear that assistance programs must consider women's obligations and responsibilities so that self-employment means more than greater burdens and longer hours of work for little additional income.

On the other hand, assisting agencies that place political and social goals at the center of their programs run the risk of getting so immersed in those goals as to be ineffective at helping participants gather the resources they need to survive. If this occurs, organizations will not be able to reach large numbers of beneficiaries nor contribute meaningfully to aggregate employment and economic growth. Moreover, supporting women's microenterprises may exacerbate the tensions women face when they work long hours for little return—without challenging the structural conditions that generate women's widespread poverty and extensive (often sole) family responsibilities.

The evaluation of support to women's small-scale activity ultimately relies upon a decision about the relative importance of individual change within the context of wider opportunities for mobilization and social transformation. As a form of local development, supporting women's small-scale enterprises is vitally important. In fact, there may be larger income-multiplication effects of women's expenditures, since women tend to spend more of their money on locally produced food and enterprise inputs (Downing, 1990). But because microenterprise assistance programs tend to remain very small, they have limited impact on low-income communities in terms of job creation or community investment. Moreover, they do not necessarily challenge gender hierarchies or the gendered division of labor. In fact, women's microenterprise initiatives are concentrated in segments of the economy that reinforce rather than challenge the gendered division of labor. Still, many microenterprises exist as crucial elements of survival and security strategies. Assistance projects must proceed with a clear understanding of the benefits and restrictions of microenterprise, particularly since the gendered impacts of development are still open to debate.

Conclusion

Women's overall economic opportunities will largely be the result of the structure of the division of labor in each region and, in particular, the way in which the division of labor relies upon and inscribes differences of gender, race, class, and culture. In every location, some work and entrepreneurial activities will be promoted as "appropriate" for women of particular races and cultures while others will be proscribed. With regard to gender, development planning and practice have generally relied upon and reinforced a gendered division of labor that may serve women's immediate needs but does not address broader questions about labor-market segmentation and gender-based discrimination. Policies and practices must focus on efficiency and effectiveness with regard to economic gain, including increased access of women and their families; if we hold an ideal of development strategies as democratic, we must also consider the broad implications for enhancing women's opportunities to participate in the construction and implementation of policies and programs that affect their lives and, one might hope, help them gain greater control over productive resources.

One possible evaluation framework is to ask whether and how policies and strategies might move from viewing program participants as recipients with needs to citizens with rights. Such policies and practices would not only improve people's livelihoods and their opportunities for economic gain but would also increase their opportunities to participate in public life and decision making. Unless microenterprise assistance programs are designed to ensure women's control of their own labor, while at the same time confronting the causes of women's poverty and the need for political mobilization, such efforts will address the symptoms of such poverty without adequately grappling with the causes and persistence of women's marginalization.

References

Bennett, Lynn, and Mike Goldberg. "Providing Enterprise Development and Financial Services to Women: A Decade of Experience in Asia," technical paper no. 236, World Bank, Washington, D.C., 1993.

Berger, Marguerite. "An Introduction." In Marguerite Berger and Mayra Buvinic (eds.), *Women's Ventures: Assistance to the Informal Sector in Latin America.* Hartford: Kumarian, 1989, pp. 1–18.

Downing, Jeanne. "Gender and the Growth and Dynamics of Microenterprises," GEMINI Project working paper no. 5, Development Alternatives and GEMINI Project, Bethesda, 1990.

Downing, Jeanne, and Lisa Daniels. *The Growth and Dynamics of Women Entrepreneurs in Southern Africa.* GEMINI Technical Report no. 47. Bethesda: Development Alternatives and GEMINI Project, 1992.

Escobar, Silvia. "Small-Scale Commerce in the City of La Paz, Bolivia." In Marguerite Berger and Mayra Buvinic (eds.), *Women's Ventures: Assistance to the Informal Sector in Latin America.* Hartford: Kumarian, 1989, pp. 64–76.

Greenhalgh, Susan. "Women's Informal Enterprise: Empowerment or Exploitation?" working paper no. 33, Population Council, New York, 1991.

Grown, Caren, and Jennifer Sebstad. "Introduction: Toward a Wider Perspective on Women's Employment," *World Development,* vol. 17, no. 7, 1989, pp. 937–952.

Howells, Cynthia. "Women's World Banking: An Interview with Nancy Barry," *Columbia Journal of World Business,* vol. 28, no. 3, 1993, pp. 21–32.

McKee, Katherine. "Microlevel Strategies for Supporting Livelihoods, Employment, and Income Generation of Poor Women in the Third World: The Challenge of Significance," *World Development,* vol. 17, no. 7, 1989, pp. 993–1006.

Otero, Maria, and Elizabeth Rhyne (eds.). *The New World of Microenterprise Finance: Building Healthy Financial Institutions for the Poor.* Hartford: Kumarian, 1994.

Scott, Alison MacEwen. "Informal Sector or Female Sector? Gender Bias in Urban Labour Market Models." In Diane Elson (ed.), *Male Bias in the Development Process.* Manchester: Manchester University Press, 1991, pp. 103–132.

Spalter-Roth, Roberta, Enrique Soto, and Lily Zandniapour. *Micro-enterprise and Women: The Viability of Self-Employment As a Strategy for Alleviating Poverty.* Washington, D.C.: Institute for Women's Policy Research, 1994.

Turnham, David. *Employment and Development: A New Review of Evidence.* Paris: Development Center, Organization for Economic Cooperation and Development (OECD), 1993.

United Nations. *The World's Women, 1970–1990: Trends and Statistics.* New York: United Nations, 1991.

"Working Women's Forum Report," *GROOTS Network News,* vol. 1, no. 4, 1994.

Women in the Informal Sector of the Economy

Benson Honig

Introduction

The "informal sector" is a broadly defined term that encompasses a ubiquitous and heterogeneous phenomenon in the Third World: the labor of small firms and individuals, often with minimal resources, operating in the "gray" areas with respect to legality and employing a wide range of activities and services. The 1971 Nairobi conference organized by the International Labor Organization (ILO) provided the following characteristics of informal activity and still serves as a guide for scholars and practitioners alike: ease of entry; reliance on indigenous resources; family ownership of enterprise; small scale of operation; operation in a semipermanent or temporary structure or in a variable location; skills acquired outside the formal education system; and operation in unregulated and competitive markets (ILO, 1972).

The concept of the informal sector has been contested on theoretical grounds as failing to transcend models based on economic dualism (Moser, 1978). Some advocates of the informal sector insist that it should be considered a permanent component of national economic activity; examples are the neoliberal model advocated by the World Bank and the research of the Peruvian Hernando de Soto (1989). They seek to strengthen linkages between the two sectors, thus encouraging growth and promotion from micro to small business activity. Another approach focuses on structural failures resulting from unemployment, migration, and class, ethnic, and gender inequalities, such as the work by the International Labor Office (Tokman, 1978). Structuralists maintain that the informal sector is a result of insufficient demand for labor, a type of hidden unemployment. A third line of research examines issues of class and gender along neo-Marxist perspectives (Bromley and Gerry, 1979). This argument focuses on the exploitation of labor in the informal sector by the owners of capital, who are said to design linkages that enhance their control to minimize wages in both sectors. Further neo-Marxist analysis examines relationships that result from the internationalization of capital.

The role of state regulation provides one of the most important perspectives in defining the informal sector. Global economic competition and the resulting restructuring of labor markets highlight the function of unregulated labor in both developed and developing countries (Castells and Portes, 1989). These new markets are characterized by their heterogeneity, as well as their linkages with formal-sector firms that help to reduce costs in the face of global competition. The expansion of informal markets is thus characterized by increasing horizontal networks amid the growing trend toward a reduction in centralized hierarchical industrial models. In this view, governments are seen to ignore illegal labor practices and occasionally promote informal labor markets to solve social problems and otherwise increase political influence and create economic growth.

Despite wide-ranging controversies among the aforementioned groups, considerable attention and investment continue to sustain research in the field. Almost by definition, accurate statistics regarding the size and growth of the informal sector are unavailable, as the activities are unregulated and bureaucratically hidden. Studies suggest that it is not only the largest location for urban employment in Africa and Latin America but also the fastest growing, with rates exceeding 10 percent annually. Estimates of urban informal-sector employment in Latin America range from 33 percent to 57 percent of the labor force; in Africa, from 40 to 60 percent; and in Asia, upward of 20 percent. Across all Third World countries, approximately 30 percent of economic transactions are said to be informal (USAID, 1990). Much of this growth has resulted from the lack of formal-sector employment to keep pace with rapid population growth. Structural-adjustment policies prevalent in developing countries have also acted as a catalyst for informal-sector growth by reducing government and parastatal employment. To many, the informal sector remains the only silver lining in an otherwise grim economic prospectus.

Informal Labor and the Role of Women

Because much of the economic literature examines formal-sector employment in the Third World, the role of women is often undervalued because they have limited formal opportunity and so participate less. As economic actors, however, women clearly deserve considerably more attention. In much of Africa, for example, their role as agriculturalists provides the primary source of economic production and vitality (Boserup, 1970). So, too, in the informal sector, where women are making contributions far in excess of their numbers. While their efforts all to frequently go unnoticed in examinations of the marginal productivity of wages or per-capita indicators of gross domestic product, studies of the informal sector consistently highlight the importance and contributions of women.

The high participation level of women in the informal sector remains one of the most salient and universal justifications for assisting the sector. Women are found in disproportionately higher numbers in certain occupations that often lend themselves to informal activity. In most studies, women represent more than half of all informal-sector activity, far exceeding their role in formal-sector employment (Berger, 1989; Everett and Savara, 1994). They are often less prepared than their male counterparts due to lower levels of education, reduced access to capital, and limited information. Further, opportunities for women tend to provide less remuneration, a fact, exacerbated by additional unpaid household labor, that presents a pattern of subordination (Elson, 1992): Many normative and structural elements of society direct and constrain the activities of women's informal labor.

Low barriers to entry, minimal capital and training requirements, and flexibility in both time and location that allow for child care are a few of the explanations often cited for the preponderance of women in the informal sector. Petty trading, food services, textile production, and domestic service represent a few of the occupations typically dominated by women. Whether determined by caste, convention, or limited opportunity, such activities demonstrate the high degree of segmentation by gender in informal labor markets.

Family roles are often responsible for female marginalization in the informal sector. Traditional norms and social systems worldwide place primary responsibility for childrearing and sustenance on women. These roles are structured such that productive activity, whether it is economic or house work, is considered "women's work" and fails to carry the requisite credibility necessary for advancement and growth. In addition, practical requirements often limit activities to those that can take place inside or close to home. Such activities, where competition is greatest and markets are weakest, tend to offer the lowest financial rewards. Additionally, social constraints restrict ownership, leadership, and even attendance at organizational meetings, further marginalizing women and limiting their occupational choices.

Women's economic activities are in marked contrast with the more viable high-growth enterprises dominated by males, such as automobile repair, construction, and small-scale manufacturing, in which economic opportunities, including linkages with the formal sector, create a more hospitable environment. In part because of the increased requirements of both capital and expertise, male-dominated occupations enjoy higher profits and greater long-term stability (Buvinić et al., 1989; Honig, 1994). However, even when men and women are engaged in similar activities, gender plays a strong role in determining the nature and economic potential of the enterprise. In agriculture, for example, women in many countries are relegated to selling and producing subsistence crops, while the more lucrative cash-crop industry is run primarily by men. Where women engage in cash-crop production, there are often restrictions on property rights and on the final determination and distribution of earnings. Thus, the informal sector itself is observed to be highly gender-stratified, with many of the resultant consequences mirroring formal-sector occupational constraints and biases in favor of males. Addressing the issue of gender segmentation requires a multidimensional approach, since altering only a single component of the economic environment is unlikely to yield permanent opportunity in the face of widely held biases and prejudices. Careful analysis is necessary to understand the nexus between gender, institutional and macro policy, and normative household customs and relationships.

Regional Characteristics of Women in the Informal Sector

While there are many parallels regarding the role of women in the informal sector worldwide, regional variations are still discernible (Lubell, 1991). In Africa, where many families are living the most marginal existence, often under threat of starvation, the business climate is particularly harsh. Informal activities in the rural areas, where most of the population live, consist primarily of petty trading in village markets and the simple production of handicrafts and household items such as baskets. As a result of the limited infrastructure, the lack of electricity, poor transportation, and the unavailability of raw materials, there is less room for growth, or "graduation," from microbusinesses employing household members into small businesses employing outside workers. A particular constraint lies with many traditional female roles in Africa that promote child marriage (which reduces school participation), limit property ownership, and delegate subsistence farming and household and child care to the exclusive domain of the female. One of the more profitable ventures in the African informal sector is meal production and beer and spirit production and sales. In a number of African cultures, this activity is traditionally relegated to women, providing an essential source of livelihood for hundreds of thousands of them (Birks et al., 1992; Lopez and Muchnik, 1994).

In Latin America, the infrastructure is comparatively sophisticated, and poverty somewhat less debilitating. Transportation capacity allows for a greater exchange between urban and rural areas, encouraging the production of a wider range of products, including craft items designed for export and the tourist trade. While the role of women in many Latin American societies continues to be restricted, there tends to be a higher level of political and social awareness, resulting from a long tradition of cooperative and social movements. Many of the most successful microenterprise-development schemes originated in Latin America, where they have been particularly effective at organizing and assisting female microentrepreneurs.

The role of women in Asia is particularly heterogeneous, varying greatly depending upon the country. In general, however, the region enjoys the most sophisticated and extensive infrastructure in the Third World, allowing for the goods produced in rural areas to be distributed easily on a worldwide basis. Informal activity is extensive, particularly in India and China. The major constraint faced by women consists of traditional and legal restrictions, including caste and religious beliefs, that artificially limit growth and ownership possibilities. Extreme poverty is also a severe problem in a number of countries. Micro-lending programs targeting women in India, Bangladesh, and Pakistan have been particularly successful in alleviating poverty and promoting basic economic self-reliance. Unfortunately, they have been less successful at expanding microbusinesses into small businesses, with the attendant growth in employment and incomes.

Worldwide, while formal-sector firms are often in a position to manage their operating environments and markets, informal-sector businesses rarely enjoy that advantage. During standard business cycles, for example, larger businesses may be positioned to redeploy their capital or production. Not so for informal-sector microentrepreneurs, who seem to disappear and reappear with unusual frequency corresponding to the demands and economic trends of already weak developing markets. In the informal sector, family life and economic life are blurred, resulting in an amalgamation of activity that makes close evaluation and assistance all the more difficult. Women, acting a dual role as focal points for both domestic and market activities, are particularly vulnerable to economic swings. When formal-sector wages are reduced or otherwise unavailable, women frequently hire themselves out as labor to the informal sector, where wages are unregulated and relatively low. Many assistance programs have recognized the vulnerability of such women. Citing the importance of equity, efficiency, and growth potential, they have designed projects that address the specific needs and requirements of low income women and their families.

Assistance to the Informal Sector

Support for the informal sector has focused on three main types of activities: technical, socioeconomic, and training. The technical component of assistance involves both macroeconomic and technology policy. For example, governments and nongovernmental organizations (NGOs) can promote the importation and distribution of essential raw materials and inputs or may work toward the enhancement of a popular product or technology, such as the design of

a more efficient stove or other consumer product. Socioeconomic programs focus on the formation of groups and cooperatives, including handicraft zones and microbusiness incubators, and access to credit and financial assistance. Training is occasionally designed to improve the production processes, but it more typically concentrates on the management practices of microbusinesses and often involves development of bookkeeping and marketing skills. In rare instances, the formal educational system adopts curricula designed to promote and enhance the informal sector; however, it has been shown that basic literacy and numeracy have a major effect on the productivity of microenterprise (Fluitman, 1989).

More recently, the ILO has used a multidimensional approach: Rather than concentrate on only one or two aspects of informal activity, such as credit and training, it supports a number of issues relevant to developing and enhancing microenterprise, including nonformal and informal training, credit, marketing, production techniques, cooperative and community building, child care, and transportation.

Because of the characteristics of the informal sector, attempts to evaluate the effectiveness of various assistance programs have been limited. Among the difficulties are the reluctance and inability of informal workers and owners to utilize written records (due, in part, to fears of taxation and governmental intervention), the seasonality and fluctuation of activity, and family and household unpaid labor. In many cases, projects are instituted on a shoestring budget and are incapable of effectively monitoring outcomes on a longitudinal basis. Women are particularly difficult to monitor because of the nature of the informal work in which they are typically engaged. Unpaid household work often results in limiting market activities, particularly for retail traders. As a result, many women are unable to appear at the same market location from day to day, creating a formidable obstacle for those attempting to research their situation. When familial responsibilities require them to work out of their homes, women's long working days reflect an enigmatic mixture of domestic and income-generating activities that seem to defy demarcation and study. Their activities tend to be further constrained by comparatively higher degrees of illiteracy. Childrearing and household responsibilities often require the temporary cessation of activity, confounding even straightforward research regarding the outcomes of specific interventions. Thus, assisting women in the informal sector requires extensive study and careful monitoring, as well as interventions at both the macro and micro levels.

Training Programs

One of the most comprehensive attempts to evaluate training programs for women was conducted across three African countries by the ILO, which maintains a strong leadership role in training for the informal sector (Nubler, 1992). The study utilized a quasi-experimental approach, including the examination of both participants and a small control group. The training program itself was conducted in Nairobi, Kenya; Dar es Salaam, Tanzania; and Abidjan, Ivory Coast; it consisted of a methodology based on pictures, stories, and role playing that attempted to consider the specific social and environmental conditions of women in the informal sector. It was developed for women with low levels of education and literacy and designed to cover three topics: the enterprise and its environment, marketing, and financial management. The ILO findings indicated that participants in the three countries valued the different components in unique ways. In Kenya, for example, women valued training more highly than advisory services. This probably reflects the heterogeneity of the informal-sector environment and the variation from country to country that makes universal approaches problematic at best. Examinations of the rate of retention indicated that stories relating to the environment were most memorable, followed by marketing examples. The rate of retention for financial management was extremely low—and, unlike the other components, was highly correlated with educational level. The difficulty in providing useful financial management training has been identified in a number of studies. Microentrepreneurs in Jamaica were found to ignore advice relating to bookkeeping and cost evaluation, citing lack of time and interest and concerns regarding government intervention (Honig, 1994).

Although financial management repeatedly appears problematic, informal-sector participants appear more receptive to a number of other aspects of business training. Most respondents indicated that they had made major changes to their businesses as a direct result of the training program. These actions ranged from new skills to bookkeeping to new credit policies. Between 33 percent and 66 percent (depending on the country) expanded their business activities after completing the seminar. Eighty percent of the Kenyan and Tanzanian women and 40 percent of those from Ivory Coast reported increases in sales after the seminar, with many adding new products and employees. A cost-benefit analysis conducted on the entire training exercise, evaluated all costs, including the opportunity costs. Opportunity costs reflect the earnings that the participants would have earned while they were attending the seminar, and are an important factor in determining the actual costs of such programs. The resulting evaluation demonstrated increases in business profitability, and was found to provide positive returns to the costs invested in the project.

Despite reports that suggest informal activity is constrained primarily by resource and market shortages, there are some indications that individual decisions also play an important role in the comparative success of informal-sector activities. Technological choice can make an important difference, above and beyond capital and market constraints. In one study, for example, dressmakers, who employed a range of different technologies and market strategies, were compared and evaluated. While some women were found to be using multiple basic sewing machines, others incorporated more technically complex machinery of equal cost, such as serging machines. The women who employed the more complex technology were more competitive in the retail and mass market trade, due to the higher quality of their finished products (Honig, 1996).

Credit Services

Credit services continue to be the overwhelming focus of informal-sector assistance programs by both governments and NGOs. A wide body of research suggests that there is a persistent shortage of microbusiness credit, which acts as one of the most serious constraints to microenterprise expansion. In most developing countries, large banking institutions represent the primary, and often the only, source of short-term credit to businesses. The comparative size or scale of the loans processed by banks and the collateral requirement for such loans virtually excludes the informal-sector microentrepreneur. Women are particularly marginalized because, in many countries and cultural systems, they are forbidden, or otherwise unlikely, to own land and are thus denied the primary source of collateral. Illiteracy due to historic low educational participation further limits their ability to complete a legal transaction. Finally, they are often engaged in the smallest businesses with the lowest profit margins and highest risks, making them unlikely traditional borrowers.

This dearth of institutional credit has been recognized by bilateral and multilateral institutions such as the ILO, USAID, CIDA (the Canadian Agency for International Development), and the World Bank. Since the 1980s, they have undertaken to develop, promote, and support credit institutions whose primary directive is to provide credit to microenterprise. Examples include USAID's Pisces II program operating in Costa Rica, Kenya, Egypt, and the Dominican Republic; PRODEM (*Fundación para la Promoción y Desarrollo de la Microempresa*) in Ecuador; and in India, SEWA (Self-Employed Women's Association) and the Working Women's Forum (Singh and Chakraborti, 1991). As a potential source of intervention, credit programs continue to be viewed in a favorable light. Their attraction is enhanced by the relative ease with which they are implemented and monitored. Compared with the expense and length of loosely coupled formal and nonformal education, lending small sums of money leads to a rapid and efficient project cycle. Monitoring is also relatively simple—consisting mainly of reported default rates in the portfolio accented with numerous examples of successful case studies to send back to program directors abroad.

A number of organizations, of which the Grameen Bank in Bangladesh is the most widely noted, have organized credit institutions that effectively assist poverty-stricken women in obtaining credit and starting businesses in the informal sector (Hossain, 1988). They operate by organizing women into small groups and encouraging them to help one another, effectively leveraging the human capital available to them. Loans start out very small, sometimes less than $10.00, and are repeated at relatively frequent intervals in growing amounts, as new businesses prosper. The development of similar social resources has been cited as the primary vehicle in explaining the success of ethnic enterprise in the United States, particularly among Asian and Hispanic business owners. In developing countries, advocates suggest that the high proportion of female clientele (85 percent in the case of the Grameen Bank), accounts for the unusually low default rate—typically less than 3 percent of all loans issued. USAID missions in a number of African and Latin American countries have reported that women have a significantly higher repayment rate than men.

The apparent compatibility between women and microcredit schemes probably reflects traditions in many existing cultural systems, as women in many countries have historically maintained various types of rotating credit associations. The Nigerian practice of *esusu* has been documented in Africa as early as 1843. Slaves are thought to have brought this practice to the British West Indies, where lending groups are referred to as *aus* or *susu* in Trinidad and Barbados and "partners" in Jamaica. It is not surprising that such institutions have endured. All loans are subject to varying levels of uncertainty—borrowers might die, inflation may dilute the value of money. Part of the interest demanded by lending institutions reflects the uncertain nature of these contractual arrangements, which are essentially future promissory notes. Friends and family have the best information regarding potential borrowers and are positioned to lend money with a minimum of bureaucratic transaction cost (usually at no interest, with no formal written agreement). Women engaged in selling merchandise and produce on a regular basis have an intimate knowledge of their colleagues and have been known to maintain lending pools comprising more than 1,000 accounts. Thus attempts to formalize women's informal finances and to integrate them into the broader economic structure may compete with or supplement preexisting and highly efficient practices.

While microcredit schemes have been shown to be very effective at reducing absolute poverty levels of the women involved, there has been less evidence demonstrating enterprise growth into the more established small-business sector. Theoretically, access to credit markets should make a business more competitive, as a result of the ability to purchase raw materials in higher quantities at a lower price, as well as to invest in newer, more efficient, and profitable technologies. Targeting microcredit often precludes the few individuals who might be able to make this transition, as they must utilize their limited profits for survival.

A study of five different credit institutions in Jamaica highlights the importance of gender in maintaining an effective program (Honig, 1994). Those institutions that attempted to promote mid-level firms, lending in the U.S. $1,000.00 range, found very few women who met their defined criteria. Although more women walked in the door than men, the majority of female applicants were engaged in the most marginal informal-sector businesses and were viewed by the loan agents to be such poor risks as to be ineligible for formal application. To fulfill mandates of gender equality, many women who might otherwise have been disqualified appeared to be pushed through the loan process. Even so, three of the five organizations were unable to maintain a clientele of at least 50 percent females. Of particular note was the preponderance of male lending agents employed by the five different agencies. Only one agency relied primarily on female agents, and that institution was also the most successful at attracting qualified female clients. This is particularly significant because the organization that employed the highest proportion of female loan agents was the only agency utilizing loan agents who had not themselves completed a college education. This finding suggests that the background and gender of the loan officer may be of considerable importance in determining the clientele of the agency. Field interviews supported this premise. Among all of the agencies, the loan officer occupied a pivotal position regarding recommendations. Examination of the loan-approval process of all five showed that a recommendation for approval by a loan officer was acted upon 80–95 percent of the time.

Evidently, the strategy already employed by a few microlending organizations, of utilizing women as operations managers, demonstrates an important model that should be applied more widely, not only for microlending but also to banks normally targeting a more affluent clientele. Failure to do so may retard the participation rate of women. Empirical support of this proposition is evidenced by a worldwide evaluation of USAID efforts that indicated that women were much less likely to engage in transformation and expansion programs than were men; it suggested that distance to the home, unfamiliarity with the lender, and the lack of collateral requirements were the main obstacles to equality of participation (USAID, 1990).

Finally, there is a growing belief that credit programs alone are unlikely to assist the lower stratum of the informal sector, which women tend to predominate. Evidence from Latin America and Africa suggests that weak demand is a greater constraint than the unavailability of credit (Holt and Ribe, 1991). One study in Ecuador demonstrated that, while the effects of credit seemed to be small, female participants gained considerable efficiency (Buvinić et al., 1989). The conclusion was that women profited from the credit program by carving out additional time during the day to manage household and childrearing duties, a product of the increasing returns to their economic activities. Such a finding calls into question normative criteria for project assessment and suggests the importance of monitoring the role of female labor in the informal sector from a multidimensional perspective. This is a particularly acute problem because of the large number of studies demonstrating segmentation according to gender (Everett and Savara, 1994). Women are often relegated to working in informal-sector occupations, in which mobility, despite education, is extremely limited.

Formal Education

While formal education is normatively involved in the traditional pursuit of academic certification, a few attempts have been made to influence vocational and informal-sector activity in developing countries. Because of the high cost of maintaining traditional vocational education and the difficulty in providing environmental relevance and economic efficiency, vocational education has more recently fallen out of favor. Where it was instituted, it reflected a highly gender-biased structure, in which a minority of girls participated in highly feminized vocational occupations.

One of the most underutilized roles of formal education in developing countries is to teach practical skills and promote microenterprise role models in the primary classroom. A properly implemented self-employment curriculum provides exposure to the largest educational segment, as comparatively few students, and even fewer girls, are able to participate in secondary education. Secondly, cost differentials are minimized, since the level of instruction is typically the most basic, utilizing available materials and technologies. Most important, such a curriculum offers an opportunity to alter prejudices and misconceptions regarding the social status of informal-sector activity, as well as gender roles within that context (King, 1989). Unfortunately, there have been few attempts to introduce such a curriculum; it is an area unfamiliar to most educational experts, advisers, researchers, and consultants. In East Africa, Kenya has pioneered this subject, introducing successful microenterprise case studies in primary readers and teaching basic bookkeeping and management skill training to rural primary students. Due to the long delay of any anticipated outcomes regarding this curriculum, effective evaluation is unobtainable as of the late 1990s.

Macro Policy and the Role of Government

The subject of the regulatory environment is necessarily complicated, and a full review is beyond the scope of this essay. Briefly, the legal framework covers areas such as access to property, minimum wages, building codes, taxation, tariffs, international trade, and licensing (Lubell, 1991). It is important to recognize that much informal-sector work is a direct result of the avoidance of formal-sector labor laws (de Soto, 1989). In a sense, informal firms are already operating against established state policy, so creating additional laws is unlikely to alter conditions without enforcement. Compelling workers in the informal sector to comply with legal and administrative codes would be viewed as politically untenable in most Third World countries. In many cases, compliance would dramatically reduce profitability or otherwise force businesses to close (ILO, 1990). Thus lack of regulation is likely to continue to define informal activity. In other matters, however, the state has greater control over macro policies that influence and structure the informal labor market. Reducing tariffs on essential raw materials and developing a distribution system can be of immense benefit to many informal activities. Employing social marketing that promotes gender equity over a broader occupational range may serve to reduce some of the observed gender segmentation. Organizing and perhaps subsidizing child-care and preschool institutions would have a radical effect on women's productivity.

At the state level, regulations involving the informal sector will require specialized enforcement to have any effect. For example, unless labor commissioners are willing to enter into homes, they are unlikely to ameliorate the conditions of many informal working women. Compliance in the informal sector might be encouraged through an incentive program. If microentrepreneurs are offered credit, training, access to cooperatives, or other marketing assistance from the government, they might be in a better position to respect labor and tax laws. Such a trade-off, in effect, is what motivates compliance in the formal sector.

Another important role the state can play in supporting the informal sector regards the technological choices available to microentrepreneurs. Research and the promotion of more effective technologies offers the possibility of expanding the productive efficiency of the entire sector. If linkages between the formal and the informal sectors are promoted, enhanced, and well monitored, there exists a potential for considerable economic growth and social development.

One area of macro policy that offers significant potential assistance concerns the role of institutions. Organizations need to be developed that coordinate public policy with the demands and outcomes of informal-sector workers. Because of the fragmented nature of informal activity, it is difficult to identify the full implications of particular regulations on these economic subsectors, so much public policy follows a "wait-and-see" approach. Further, the informal sector can benefit from a consolidation of interests, by, for example, importing raw materials and parts in larger and more economically efficient quantities. Distribution systems are often primitive and noncompetitive, providing unfair advantages to larger, more established firms and brokers. The resulting profit-taking contributes to marginalization, scarcity, and poor rates of return to capital and labor in the informal sector. Macro policy can contribute toward enhancing the competitive nature of the market for intermediary goods and services.

The Future of Women in the Informal Sector

Population growth, increasing economic inequity on a global scale, and fiscally conservative policies such as structural adjustment are likely to expand the informal sector well into the twenty-first century. The meager progress toward gender equity in the last two decades of the twentieth century suggests that Third World women will continue to play a dominant role in the maintenance and expansion of low-cost labor, particularly in less regulated informal activity. In addition, the expanding roles of women as primary providers of child care, health care, and welfare highlight how important strengthening women's economic opportunity will be to Third World nations. As more women continue to expand their role vis-à-vis the informal sector, competition and an oversupply of labor threaten the income-generating potential of their activities. Only by increasing productivity can this be ameliorated, which emphasizes the need for improved technology, higher skills, and access to credit and distribution markets. Because little is known regarding the long-term effects of various policies and interventions, further longitudinal research and study are imperative.

As we enter the new millennium, there is an unusual worldwide consensus among governments that their capacity to manage, employ, and regulate economic activity is limited. Although the pendulum is likely eventually to swing in the opposite direction, the harsh reality faced by women and other marginalized populations of the Third World is that external resources will be in short supply.

There are a number of reasons to expect continued growth and participation of women in the informal sector. First, the historical record strongly demonstrates sustained long-term informal growth as populations in developing countries grow and industrialization increases (Castells and Portes, 1989). Second, as economies in developing countries attempt to negotiate the dual sword of structural adjustment and market liberalization, informal-sector activities provide many women with supplemental sources of income generation in the face of reduced government subsidies and assistance. Finally, economic restructuring increases the demand for women's labor and so makes opportunities available that were previously nonexistent. It is, therefore, critical to develop global networks that share information and assist in the promotion and development of low-cost strategies designed to assist women in the informal sector. Evidence suggests that effective linkages between the formal and the informal sectors can serve to direct, support, and strengthen productive capabilities, thus enhancing the welfare for those involved. Governments, institutions, and assistance agencies have an obligation to examine the implications of such integration in terms of their effects on families and specifically on women.

Growing political participation, particularly by women, is likely to result in increasing attention at the state level to the development of informal activity as a national resource. Women in many parts of the world have demonstrated successful cooperation and interrelatedness in institutions of trust, often eclipsing the capability of males. This trust, if properly organized and harnessed, provides women in the informal sector with a margin of advantage exceeding any assistance or activity conducted to date. Much needs to be learned regarding these characteristics. Increasingly, gender is regarded, along with class and ethnicity, as an essential explanatory variable in the amelioration of poverty. There is increasing recognition that additional research is needed that transcends the examination of gender from a dichotomous perspective and explores the role of women embedded in both society and the household. The presence or absence of men, relationships of power and position, and educational, legal, and resource opportunities—all contribute to the marginalized role of women in the informal sector. Only after understanding these dynamics can we expect income promotion and support projects to become demonstrably effective.

References

Berger, Marguerite. "Introduction." In Marguerite Berger and Mayra Buvinić (eds.), *Women's Ventures: Assistance to the Informal Sector in Latin America*. West Hartford: Kumarian, 1989, pp. 1–18.

Birks, Stace, Fred Fluitman, Xavier Oudin, Bernard Salomé, and Clive Sinclair. *Skill Acquisition and Work in Micro-Enterprises: Recent Evidence from West Africa*. AFTED Technical Note no. 4. Washington D.C.: World Bank, 1992.

Boserup, Ester. *Woman's Role in Economic Development*. New York: St. Martin's, 1970.

Bromley, Ray, and Chris Gerry. *Casual Work and Poverty in Third World Cities*. Chichester: Wiley, 1979.

Buvinić, Mayra, Marguerite Berger, and Cecilia Jaramillo. "Impact of Credit Project for Women and Men Microentrepreneurs in Quito, Ecuador." In Marguerite Berger and Mayra Buvinić (eds.), *Women's Ventures: Assistance to the Informal Sector in Latin America*. West Hartford: Kumarian, 1989, pp. 222–246.

Castells, Manuel, and Alejandro Portes. "World Underneath: The Origins, Dynamics, and Effects of the Informal Economy." In Alejandro Portes, Manuel Castells, and Lauren Benson (eds.), *The Informal Economy: Studies in Advanced and Less Developed Countries*. Baltimore: Johns Hopkins University Press, 1989, pp. 1–37.

De Soto, Hernando. *The Other Path*. New York: Harper and Row, 1989.

Elson, Diane. "From Survival Strategies to Transformation Strategies: Women's Needs and Structural Adjustment." In Lourdes Benería and Shelley Feldman (eds.), *Unequal Burden: Economic Crisis, Persistent Poverty, and Women's Work*. Boulder: Westview, 1992, pp. 26–48.

Everett, Jana, and Mira Savara. *Women and Organisations in the Informal Sector*. Bombay: Himalaya, 1994.

Fluitman, Fred. *Training and Work in the Informal Sector*. Geneva: International Labor Organization, 1989.

Holt, Sharon, and Helena Ribe. "Developing Financial Institutions for the Poor and Reducing Barriers to Access for Women," discussion paper no. 117, World Bank, Washington D.C., 1991.

Honig, Benson. "Education and the Microentrepreneur: A Study of the Jamaican Informal Sector." Ph.D. diss., Stanford University, Stanford, 1994.

———. "Education and Self-Employment in Jamaica," *Comparative Education Review*, vol. 40, no. 2, 1996.

Hossain, Mahabub. *Credit for Alleviation of Rural Poverty: The Grameen Bank in Bangladesh*. Washington, D.C.: International Food Policy Research Institute, in collaboration with the Bangladesh Institute of Development Studies, 1988.

International Labor Organization (ILO). *Employment, Incomes and Equality: A Strategy for Increasing Productive Employment in Kenya.* Geneva: International Labor Office, 1972.

————. *Informal Sector and Urban Employment.* Geneva: International Labor Office, 1990.

King, Kenneth. "Training for the Urban Informal Sector in Developing Countries: Policy Issues for Practitioners." In Fred Fluitman (ed.), *Training for Work in the Informal Sector.* Geneva: International Labor Organization, 1989, pp. 17–38.

Lopez, Elisabeth, and Jose Muchnik. "Food Related Micro-Enterprises and the Current Crisis." In M. Kaddar and P. Gerbouin-Rerolle (eds.), *Micro-Enterprises: What They Are and What They Can Be.* Paris: International Children's Center, 1994.

Lubell, Harold. *The Informal Sector in the 1980s and 1990s.* Paris: Organization for Economic Cooperation and Development (OECD), 1991.

Moser, Caroline O. "Informal Sector or Petty Commodity Production: Dualism or Dependence in Urban Development?" *World Development,* vol. 6, no. 9–10, 1978, pp. 1041–1064.

Nubler, Irmgard. "Cost-Effectiveness of Management Training in the Informal Sector," training discussion paper no. 101, International Labor Office, Geneva, 1992.

Singh, Sunila, and Sreemati Chakraborti. "Credit Extension and Women's Organisations: Two Case Studies in the Informal Sector." In Chetana Kalbagh (ed.), *Women and Development: Women, Employment, and the Workplace,* vol. 1. New Delhi: Discovery, 1991, pp. 223–235.

Tokman, Victor. "An Exploration into the Nature of Informal-Formal Sector Relationships," *World Development,* vol. 6, no. 9–10, 1978, pp. 1065–1075.

United States Agency for International Development (USAID). *Women in Development: A Report to Congress by the U.S. Agency for International Development.* Washington, D.C.: Office of Women in Development, Bureau for Program and Policy Coordination, USAID, 1990.

Women's Labor Incomes

Zafiris Tzannatos

Introduction

Nearly 50 percent of women around the world are in the labor force, and they constitute approximately one-third of all workers. Yet women's work pays relatively little: Most of it is not paid, and, even for paid work, women's earnings average two-thirds of men's. Women earn less than men in practically all countries, including the industrialized ones. Overall, no more than one-fifth of the world's wages accrue to women.

There are many reasons why women's work is so poorly rewarded compared to men's. Most of women's work follows the traditional division of labor and is performed within the household or, in rural areas, is unpaid work on the family farm. Such work includes fuel collection, gardening, food preparation, childrearing, and care for other household members. When women become part of the labor force, they often engage in unpaid work as a member of a self-employed family. Finally, when women do earn money, they typically do so in low-paying occupations. Yet even in those occupations, women are commonly paid less than men doing comparable work. A woman's relatively lower remuneration often intensifies over her career as she loses out on promotions, seniority, and the corresponding wage increases because her family responsibilities encourage or require intermittent employment. Later in life, women often withdraw from the labor force earlier than men, and before they realize their potential earnings, often because of gender-biased pension provisions.

A host of social, economic, and legal factors are responsible for the patterns that lead to low remuneration of female work. From a social perspective, from birth boys and girls are usually treated differently within the family. While boys are generally expected to become breadwinners and household heads when they grow up, girls are usually raised to be caretakers and workers within the family. Girls are often given less education and are raised with different expectations than men. As adults, as spouses or partners or parents, women and men perceive their household roles and tasks very differently. When in the labor market, men and women, and also their employers, follow various social norms. Although women may be constrained in what they want to do, employers may also be precluded from offering women employment and pay comparable to those given to men. Men's work and women's work are often perceived to serve different purposes and thus merit a difference in pay: Men are assumed to need a wage intended to support a family, while women earn "secondary" wages on the assumption that their earnings constitute only a supplementary income to the household.

Economically speaking, women's low earnings can be attributed to lifetime choices between work and family formation (from the viewpoint of labor supply) and to employment discrimination (from the viewpoint of labor demand). Since women usually have a greater role than men in caring for the family, they may invest less in their own education and may work for shorter periods and in occupations that require fewer hours or less effort than men. This, combined with interruptions in labor-force participation, limits women's access to better jobs and promotions. Employers, in turn, may invest less in nurturing women's skills through training or education because women are expected to drop out of the labor force while they are raising young children or, in many circumstances, to stop all work outside the home once they are married. Alternatively, women's lack of development of their own human capital may well be caused by discrimination, to the extent that women are excluded from certain jobs and thus do not realize as much return from their education as men or because women are paid less than similarly educated men.

From a legal perspective, labor legislation and family and tax laws often intensify the gender division of labor within the household and perpetuate differences in earnings between women and men. In some cases, labor laws preclude women from working in certain occupations or industries, specify different working conditions for women and men, or impose overtly different rates of pay for women and men. In other cases, family law gives differing power and responsibility to men and women within the household, especially with respect to who owns and controls family resources and how bequests can be made. Finally, tax laws often penalize the earnings of additional workers within the family, given the structure of allowances and taxation of supplementary earnings. In nearly all cases, the primary wage earner within the family is a man and the earnings of the secondary earner, often the wife, face high marginal tax rates. This induces women not to work and reinforces traditional family norms and gender stereotyping in the labor market.

At the early stages of development, when much of women's work is unpaid and only a small part of the economy is monetized, lower pay for women, though lamentable, has a negligible effect on economic efficiency and social welfare. But as more economic activity involves paid work, the unjustified underpayment of women leads to investments in human capital that favor men and are biased against women. This bias deprives women and the economy from reaping the full gains of work and adversely affects women's welfare, leads to lower growth, and perpetuates poverty. In addition, low incomes for women change consumption patterns within the family in ways that are disadvantageous to children because women tend to spend more of their incomes on children than do men.

Though there is no general agreement about the reasons for women's low incomes, whether they are individually or socially determined and whether they are "justified" or not, economic development tends to improve women's access to jobs and to increase women's earnings. However, the process is neither linear nor always obvious, especially in the short run. Policies that affect the legal framework and create incentives for equal opportunities and treatment of the sexes in the labor market and within the household can accelerate this process and contribute to increased incomes and social welfare for women. This does not necessarily imply that women's pay will become equal to men's, at least in the foreseeable future. Equality in the labor market will not be achieved as long as there are inequities in other spheres of private and public life.

Why Are Women Paid Less Than Men?
Women's earnings in developing countries average two-thirds of men's earnings (see Table 1). This statistical regularity was also common in most industrialized countries until the 1980s (Blau and Ferber, 1986). Coincidentally, the Bible, in Leviticus 27:1–7, states that the value of a woman shall be assessed at three-fifths of the value of a man.

There are many reasons why women's work, when paid, is nearly always rewarded less than men's work. In the absence of effective legislation, women can be subjected to wage discrimination—that is, they can be paid less than men in the same job. Also, women can face job discrimination in the form of employment segregation—that is, by being precluded from holding certain jobs. This discrimi-

nation reduces average women's pay for two reasons. First, it reduces pay directly because women are typically excluded from the better-paying jobs. Second, it reduces pay indirectly because women tend to become concentrated in certain fields, where their relative abundance depresses wages. In addition to the wage discrimination and employment discrimination that take place in the labor market, the causes of gender wage differences include discrimination before entry into the labor market if boys and girls are not treated equally within the family, such as when girls receive less education than boys. Finally, the gender division of labor among adults within the family, as well as various legal and institutional aspects of society, can reinforce the tendency of women's wages to be lower than men's.

Wage Discrimination
Economists have attempted to distinguish what, if any, part of the gender pay gap is justified (Blinder, 1973; Oaxaca, 1973). "Justified" differences are those caused by differences in the measurable qualifications of women and men. For example, if women tend to have less schooling, fewer skills, or less work experience than men, their pay could be expected to be lower by the amount that these differences affect productivity. An "unjustified" pay gap is one that is not explained by such differences in work qualifications. The unjustified gap is often called "the upper bound of discrimination" because it can be caused wholly by discrimination or partly by discrimination and partly by other differences that have not been accounted for. The interpretation of the unjustified gap is that the labor market rewards a given characteristic, such as education, in a discriminatory way—that is, it rewards the characteristic differently depending on whether the worker is male or female.

It is possible to break down the gender pay gap into justified and unjustified components based on regression analysis. Separate earnings functions for women and men are estimated in which the logarithm of wages or salaries (the dependent variable) is regressed upon a constant term and representative human capital variables such as education and work experience and on other variables such as urban/rural location, marital status, hours of work, and so on. Then the justified part of the gender pay gap is derived by comparing differences in the means of the independent variables, and the unjustified part by comparing differences in the regression coefficients of these variables.

This breakdown of the gross gender wage gap into justified and unjustified components typically reveals that one-third of the gap is justified, or caused by differences in women's qualifications, and the rest is unjustified—that is, cannot be accounted for by such differences (see Table 1). In other words, most of the gender wage gap is attributed to differences in the way the labor market rewards women's productive capabilities compared to men's, though some productive characteristics of women are often rewarded as much as, if not more than, those of men (an extra year of education usually increases women's earnings more

Table 1. Female-to-Male Relative Earnings (%) and Percentage of the Gender Pay Gap Unaccounted
for by Differences in Amount of Human Capital Held by Women and Men

Country	Year	Relative Pay	% of Relative Pay Unexplained*	Wage Measure
Latin America				
Argentina	1985	65	74	Monthly, Buenos Aires
Bolivia	1989	63	85	Weekly
Brazil	1989	70	81	Hourly
Chile	1987	71	115*	Weekly
Colombia	1988	84	92	Weekly
Costa Rica	1989	81	95	Monthly
Ecuador	1987	66	62	Hourly
Guatemala	1989	77	54	Monthly
Honduras	1989	81	151*	Weekly
Jamaica	1989	58	114*	Weekly
Mexico	1984	85	72	Weekly
Nicaragua	1978	43	67	Fortnight
Panama	1989	80	86	Monthly
Peru	1990	84	81	Hourly
Uruguay	1989	74	77	Monthly
Venezuela	1989	77	86	Weekly
Asia				
India	1988	56	74	Urban, employees
Indonesia	1990	54	62	Urban, employees
Japan	1988	59	44	Hourly wages
Korea	1988	51	53	Manufacturing
Malaysia	1984	69	93	Employee annual earnings
Philippines	1988	76	127	Urban, hourly
Thailand	1989	90	85	Employees
Middle East/Africa				
Jordan	1991	78	74	Annual
Tanzania	1980	86	4	Urban, manufacturing, monthly

Sources: Based on individual country studies summarized in Tzannatos, 1995a.
*A figure more than 100 percent means that working women are more qualified (e.g., in terms of education) than men and, had this not been the case, the wage gap would have been greater than the observed in practice (that is, greater than 100 percent).

than it does those of men; Psacharopoulos and Tzannatos, 1992, p. 206).

Employment Discrimination

Women's earnings may be lower than men's not because similar women and men are paid differently in the same job, but because women and men are engaged in different jobs that are rewarded differently. Though the difference in men's and women's types of employment usually declines over time (one increasingly sees women working as managers or doctors and men working as secretaries or nurses), occupational and industrial segregation remains significant in the labor market. Types of jobs that are predominantly female usually pay lower wages than those categories that are dominated by men. Across the labor market, there is a high negative correlation between the percentage of women in an occupation and the average wage in an occupation.

The effect of gender-based occupational differences on female earnings can be estimated by extending the previous analysis of Blinder (1973) and Oaxaca (1973) as proposed by Brown et al. (1980). This is done by estimating equations that predict occupational attainment and then using the coefficients from these equations together with the coefficients of conventional earnings functions to break down the total gender wage difference into explained and unexplained intra- and interoccupational components. The unexplained interoccupational component is attributed to occupational segregation. Though this approach is more comprehensive than the previous one based entirely on earnings functions, the effects of occupational segregation are often underestimated because the occupational groups are too aggregated.

Employment differentials typically explain up to one-third of the gender wage gap, particularly in developed countries (Chiswick et al., 1975; Treiman and Hartmann, 1981; Zabalza and Tzannatos, 1985). In Israel, employment differences between women and men have a positive effect on the relative earnings of females (Hallerstein and Neumark, 1993). Smaller estimates are reported for developing countries. In Latin America, employment differences account for little, if any, of the gender pay gap (Psacharopoulos and Tzannatos, 1992; Winter, 1994).

These summary results of discrimination in the labor market suggest that, though a statistically significant part of the gender wage gap remains unjustified from a conventional market perspective, factors other than pay and employment discrimination are also relevant. What appears as discrimination in the labor market may to a considerable extent be attributable to discrimination that occurs before entry into the market. This discrimination includes how children are treated within the family, how adult rights and responsibilities vary within the family, gender-biased institutional factors such as societal norms and legal structures, and how women choose between market and home work, regardless of whether such choices are determined individually and freely or by laws and society.

Discrimination Prior to Labor-Market Entry

In developing countries, unpaid work absorbs a much larger percentage of total time than paid work. If this unpaid work were to be incorporated into national accounts, the world's gross domestic product would increase by one-quarter. Most unpaid production is carried out by women, who devote a much larger fraction of their time to unpaid activities than do men. Women are responsible for up to three-quarters of the food produced in the developing world although they constitute only one-third of the world's wage labor and a mere one-fourth of the industrial labor force. Women are perceived to be primarily responsible for domestic work, including reproduction. In the more extremes cases, women's roles outside the family or household are practically nonexistent. The ways women and households respond to external economic forces such as prices, transfers, and government services, and the resulting labor-market decisions and outcomes, depend to a considerable extent on the gender division of labor at home.

These perceptions affect allocation of resources within the household. In some social settings, when resources such as food, health, education, and so on are scarce, they are allocated preferentially to boys. In fact, different allocations among boys and girls can be pervasive even in households with higher incomes. In India, discrimination against female infants, in the form of sex-selective abortions and unequal allocation of health care, is prevalent even among the better-off propertied people of the northwestern plains.

The effect of health and nutrition on labor productivity and wages is well documented. The adverse effects of females' unequal share of these benefits show up in the labor market. Iron-deficiency anemia in female tea workers in Sri Lanka and cotton-mill workers in China decreases their productivity. In some rural areas, the female labor force can be reduced by one-fifth because of various disabilities. The effects of underinvestment in girls' education on the lifetime

labor-market outcomes of women cannot be overstated; for every 100 boys in primary school in the 1990s there are only 85 girls. Family or preentry discrimination creates different gender-based labor-market outcomes even in the apparently competitive rural labor marketby affecting the relative position of buyers and sellers of labor.

Inequality Within the Family and Gender-Biased Norms

Women also face inequality in the control of family income. Who controls and spends the income is often an important factor in allocating household resources and determining individual welfare. There are substantial differences in how resources are used within households, and these differences are often determined by differing rights and norms for women and men (Ott, 1992). Inequality within the family can exacerbate inequalities in the labor market: When women have a weak bargaining position at home, there will be unequal outcomes from marriage and labor (Cigno, 1991). In addition, women behave differently than men within the family in that they may lack a sense of having interests separate from those of the family; women may see their self-worth as deriving from sacrifice (Sen, 1990) or from group, as opposed to individual, interests (Folbre, 1993).

Women usually have less education and more constrained mobility than men, and this reduces their labor price, occupational choices, and earnings. In addition, women often face legal or social exclusion from particular activities, and employers often use norms of female inferiority to justify prescribing less favorable terms in women's contracts (Bardhan, 1983). Thus, women may have lower earnings than men partly because their opportunities are constrained by family arrangements and partly because of social conventions that restrict their access to certain types of employment.

The absence of effective rights to ownership of land and the lack of educational opportunities can reduce women's earnings and lead to inferior contracts for women within the family. In turn, this can lead to a lower effective labor-supply price for women and, as a result, lower wages for women than for men. Even women who could be judged to be rich by virtue of a wealthy marriage are vulnerable in the case of desertion, separation, divorce, or death of a spouse. In parts of Asia, divorced or separated rural women work as agricultural laborers on the farms of their well-off brothers and in-laws; widows beg after the deaths of their husbands (Agarwal, 1995). Nevertheless, maintaining dependence on the family reduces risk, and gender-biased familial norms can continue as long as the family remains the only available option and where other alternatives, such as insurance, are absent (Binswanger and Rosenzweig, 1984).

Women's Intermittent Labor Force Participation

The link between women's family responsibilities and employment is evident in all countries. Unmarried women

Table 2. Female-to-Male Relative Earnings (%) Corrected for Hours Worked, by Marital Status, and Age

Country (by Marital Status)	Married	Single
Germany	57	103
United Kingdom	60	95
United States	59	96
Austria	66	97
Switzerland	58	94
Sweden	72	94
Norway	72	92
Australia	69	91
Country (by Age)	25 Years	45 Years
Brazil	75	63
Colombia	94	70
Malaysia	82	61
Indonesia	81	60
Venezuela	92	70

Sources: Constructed from Blau and Khan, 1992; Sedlacek et al. 1993.

participate in the labor force at rates that approach or even equal those of men, and young women tend on average to work more hours than older women. Among married women, those with young children (especially preschool-age children) work less than childless women or mothers whose children are grown. On average, fewer women than men participate in the labor market and, when they do work for pay, they work fewer hours than men. Also, women work for fewer years overall, and the discrepancy is greatest during the time when they are forming families. Thus, women's connection to the open labor market is weaker than that of men (see comparative studies on Latin America and Asia by Psacharopoulos and Tzannatos, 1992; and Horton, 1996).

The relationship between this weaker connection and women's wages is shown in Table 2, though data limitations force the adoption of different classifications for industrial and developing countries. In industrial countries, female earnings are on average 60 to 70 percent of men's earnings. In contrast, the average of single women's earnings nearly equals that of men. Relative earnings vary also in developing countries, this time classified by age group. Young women's earnings are generally 80 percent to 90 percent of men's but decline to 60 to 70 percent by the age of 45 when most phases of family formation and interruptions in employment have taken place. The experience in industrial and developing countries taken together suggests that, as women age, marriage and family formation in general, take a toll on women's earnings.

It is possible that part of the decline in women's relative earnings is caused by women being shifted to dead-end occupations by discrimination. There is also the question of employer's expectations; employers often assume that all women will leave work and start families at some point in their lives, and those women who do not do so cannot signal this in advance. But part also can be caused by women's expectations of intermittent participation, along with their low investment in human capital and self-selected occupational choices. Of course, women's expectations can also be affected by existing employment or wage discrimination.

From a market perspective, some loss in wages is likely when a worker interrupts her career. Intermittent participation reduces wages because of productivity decline caused by depreciation of existing skills, smaller lifetime accumulation of work experience, and less specific training by employers. This loss in productivity can be anticipated and taken into account when women's initial career expectations are formed. Women's expectations result in investing less in human capital or investing in areas where costs and depreciation of investment are lower.

Intermittent participation has been found to be a significant cause of the gender wage gap; in some cases, it can statistically account for the whole gap. A study in Britain showed that as much as 85 percent of the pay gap between married women and men could be labeled discriminatory if the women had the same employment experience as men. However, when actual work experience was included, the discriminatory, or unjustified, component of the gender wage gap was reduced by more than half (Zabalza and Tzannatos, 1985). Similar results are reported for the United States: tenure and labor-market experience seem to be rewarded equally among female and male workers, and atrophy rates from intermittent labor-force participation are not greater for women than men (Polachek, 1995). The evidence from developing countries on the effects of intermittent female participation is scanty. One study for Taiwan, China, estimated that three-quarters of the gender wage gap is explained by such lifetime decisions; a conventional adjustment based on the women's education and experience explained only one-quarter of the gap (Kao et al., 1994).

Though women's lifetime decisions appear to explain much of the gender wage gap, women's choices may be significantly constrained by their greater family responsibilities and limited mobility. These constraints reduce the supply price of women and are themselves the result of social norms. They result in a relatively inelastic supply of female labor in certain occupations or industries in the sense that women may keep supplying their labor even at lower wages. Such restricted alternatives for women can give employers bargaining power and lead to different pay scales for female and male workers. The interplay of these factors can be seen in rural areas, where norms affecting occupation and migration result in different patterns of male and female employment and earnings. These differences affect gender equality, women's welfare, and, more generally, overall poverty and economic growth.

Trade Unions

Unions have not always been sympathetic to women workers' problems. Women have often been seen as less permanent workers who are prepared to accept low pay and poor employment conditions. In some cases, unions have even opposed equality legislation, as in the case of abolishing the "marriage bar," the requirement that women leave their jobs when they marry. Women's role as wage earners was considered to be detrimental to their role as mothers.

The net effect of unions on the relationship of female wages to male wages is uncertain. The differential effect of unions on female and male wages can arise in two ways: first, from different unionization rates; second, from the ability of unions to influence wages in some sectors but not in others. On the one hand, unionization increases women's earnings more than those of men. On the other hand, women workers are less likely to be unionized, and this increases the wage gap.

There are no studies on the effects of unionization in developing countries. In any case, informal sector employment dominates formal employment in developing countries and unions are not that important. For some industrial countries for which estimates exist, the evidence is mixed. The net effect of unionization on gender-based wages is found to be small in some countries, such as Canada (Doiron and Riddell, 1994), but significantly positive in others, such as the United Kingdom (Nickell, 1977). In Canada, female-to-male annual earnings of full-year, full-time workers increased smoothly from about 58 percent in 1967 to 70 percent in 1991. Shapiro and Stelcner (1987) attribute the female wage gains between 1971 and 1991 roughly equally to an improvement of female workers' qualifications and a reduction in discrimination. Doiron and Riddell (1994) incorporate in their analysis the effect from the increase in the female unionization rate and the decrease in the male unionization rate that took place in the 1980s. Their estimates suggest that the gender pay differential in the nonunion sector makes a larger contribution to the gender pay gap than does that in the union sector; had it not been for union effects, the gender gap would have increased by 7 percent in the 1980s.

The Effects of Legislation

Legislation that attempts to correct for women's underpayment can be divided into three broad types. First, there is equal-pay legislation, which requires that women and men in the same or similar jobs be paid the same. Second, there is comparable-worth or pay-equity legislation, which requires that women and men be paid the same for work of equal value, irrespective of whether they work in similar employments. Third, there is affirmative action, which aims to increase the representation of women in jobs in which they are usually underrepresented and for which pay is usually higher than for typical female jobs. In addition to these more or less direct labor-market policies, women's wages (and employment) can be affected by other types of legislation, such as protective legislation or family and tax laws.

Equal pay in the same or similar jobs is simple and relatively easy to enforce; it is easy to determine that male and female teachers should be paid the same. Such legislation enhances economic efficiency by requiring that similar workers are treated similarly. There are instances of this type of legislative changes in developing countries. But in developed countries where wage discrimination was overt, the effects of "equal pay for the same or like work" legislation have been impressive. In Britain, Australia, New Zealand, Canada, and Greece, female pay used to be prescribed in collective agreements with reference to, and often as a percentage of, male pay.

The institutionalized underpayment of women was based on two philosophies. First is the idea that men typically support families and should, therefore, be paid a "family" wage—that is, a wage that can support a man and his dependents. Second is the idea that women are typically working for themselves as single women or, if married, to supplement family income. These ideas reflect the then predominant family arrangements and their associated norms. At the same time, setting wages in this way allowed employers to have lower labor costs in the short run. When differential pay rates for women and men workers were outlawed by equal-pay legislation in the aforementioned industrialized countries during the 1960s and 1970s and a single wage rate was adopted, the gender pay gap was reduced in some cases by as much as one-third.

The success of this type of legislation should not be overrated. The impressive effects of equal-pay legislation occurred only in countries where collective bargaining was widespread and wage discrimination overt, that is, on statute. In this case, enforcement took the form of "a stroke of the pen." Employers could not avoid its implementation, though some adjustment, unjustified or justified, took place. For example, in Canada following the equal-pay legislation, some employers deliberately reclassified jobs to avoid applying the principle of equal pay "in the same or like work." In other cases, the previously "male" and "female" wage rates were replaced by rates for "heavy" and "light" work (MacDonald, 1994).

Equal-pay policies are harder to administer when employment is segregated and the principle of "the same or like employment" does not apply. For example, if there is an industrywide collective agreement that specifies the wage rate of, say, semiskilled workers in the car industry, it makes little difference what these semiskilled workers do, and such broad collective agreements can lead to wage equality across different groups of workers. However, it can make a lot of difference if collective agreements are specific to skilled workers in mechanical tasks or upholstery production, because the former are usually undertaken by men and the latter by women. In this case, gender pay differences will be driven by segregation in employment rather than by unequal pay in the same employment. In fact, since overt wage discrimination is often illegal, firms can react to equal-pay policies by diversifying labor demand so that jobs become segregated between women and men, and the principle of comparison cannot apply.

Comparable-worth legislation, unlike the "equal pay for the same or like work," requires that equally productive workers are paid the same. Job-evaluation schemes can be undertaken to examine whether a cook is, for example, "worth" as much as a carpenter. Though it is true from a demand side that the two types of work may require the same level of training, effort, responsibility, and so on, it is also possible that cooks (mostly women) are in more abundant supply than carpenters (mostly men). Under these conditions, the labor market will assign lower wages to cooks than carpenters. This will not be "fair," since two workers who have comparable human-capital characteristics will be rewarded differently. But enforcing equal pay between these two types of work could ignore differences in preferences (the supply side) and can introduce an inefficiency in the form of higher labor costs for cooks and, ultimately, higher prices overall. The evidence in the United States shows that there have been adverse employment effects following the adoption of the comparable-worth principle, and there are doubts whether women as a group benefited (Faundez, 1994).

"Comparable worth" defined in terms of knowledge and skill requirements, mental demand, level of accountability, working conditions, and so on may end up as an arbitrary exercise: Job ranking is often based on a total score calculated from assigning these factors "points" that are subsequently somewhat aggregated. For example, attempts in Britain at the time the Equal Pay Act was introduced in the early 1970s were soon abandoned, since in some cases they showed that women's market wages were actually higher than those suggested by a job-evaluation scheme. Similar attempts, though, uncovered some interesting comparisons in the United States (Bergmann, 1986). Secretarial jobs, for example, were ranked ahead of the higher-paid delivery-truck drivers mainly because of the mental demands on the former. Nurses' work received a greater evaluation than that of civil engineers, based on the required ability to motivate and persuade, the higher mental demands, and, above all, the diverse and adverse work conditions under which nurses operate. Overall, job-evaluation results correlated well with men's wages but suggested that wages in female-dominated jobs should be higher than the actual ones.

Affirmative action can move the most qualified workers from the discriminated group into higher-level jobs, where they can be more productive than members of the nondiscriminated group. This can be enforced through prescribing employment quotas. However, this type of legislation can be hard to justify, as its overall effects are difficult to predict. Quotas may induce an inefficiency if workers who secure employment through them are less qualified than those who would have been selected in the absence of the policy. In addition, workers who benefit from the policy may have lower incentives to invest in their own human capital as long as they can find employment as a consequence of employers' obligation to satisfy numerical goals.

In addition to labor legislation that explicitly aims to equalize opportunities or outcomes between female and male workers, other types of regulation can directly or indirectly affect women's earnings in the labor market. Family law and strong social conventions in some countries prevent married women from work. This was the case in the United Kingdom and the United States until the end of World War II (Goldin, 1990) and in Japan and Korea until the late 1970s and early 1980s (Horton, 1996). In Korea, although it is no longer permissible to discriminate against married women in promotion and dismissal, discriminatory hiring is still not punishable by law. In Indonesia, discrimination against married women in hiring is still permitted. Such provisions result in women workers being initially channeled into dead-end jobs, being paid low wages, and being the first to be laid off. This reduces women's incentive to stay in the labor force, as well as their incentive to invest in their own human capital. The treatment of the family and married workers can, therefore, reinforce the traditional specialization of labor within the household and perpetuate women's underpayment relative to men.

The treatment of the family as a tax unit often creates disincentives for women to work. Joint taxation tends to keep women at home more than separate taxation because of the higher marginal taxes that the earnings of the secondary worker, usually the wife, attract. Separate taxation increases women's incentives to work, can result in greater participation by the husband in domestic activities and a smaller share of household responsibilities for the wife, and can eventually increase women's relative earnings.

Divorce laws often do not oblige fathers to effectively support their children, and women often lack the financial ability to get access to courts. When legal provisions enable a more equal sharing of child care and responsibilities among parents, women's constraints on work are eased. In Sweden, for example, authorities make serious attempts to identify fathers and make them share the financial cost of childrearing (Gustafsson, 1994). This also increases the price of children to men and reduces fertility, with resulting beneficial effects on women's welfare and potential labor-market outcomes, including higher earnings.

Maternity legislation can increase female costs of

employment in two ways. The first is by the mere interruption of women's employment during maternity leave and the need for employers to find a temporary substitute. The second occurs when legislation prescribes that the costs of maternity pay should be borne by the employer rather than paid for from a social insurance fund or general taxes. This tends to increase the costs of employing women, which may be addressed by employers attempting to lower women's wages, or limiting recruitment to childless women. Therefore, it is usually better to pay maternity benefits out of social insurance funds to which all workers contribute, or general revenue. In addition, such schemes should not increase the overall cost of labor to the point that they result in a significant reduction of employment. Such adverse effects usually affect disproportionately the employment prospects of the poor, while the benefits of policies tend to accrue to the better-off workers.

Protective legislation, such as prohibition of night and hazardous work, tends to predate equity-promoting legislation. In its early applications, generally before the 1960s, it was based on the notion that female workers are younger and more vulnerable than males. Protective legislation was seen as more valuable to women than legislation to ensure equality, given the exploitive conditions that women (and children) faced during the early stages of industrialization. In addition, occupational and environmental hazards can have different effects on women than on men. The mere fact that women tend to be employed in unregulated, low-wage positions, such as food vendors, petty traders, domestic workers, or agricultural laborers, can expose them to hazards (pesticides and other toxic chemicals, extreme temperatures, excessive noise) or require undue physical effort. The effects of unsafe working conditions intensify in the case of pregnant women. Protective legislation, though desirable in principle, can in practice be overdesigned and may restrict women's choices or increase labor costs to employers. In fact, some regulations are considered incompatible with equality (Mason, 1988). Restrictions on overtime or night work or exclusions from certain occupations can unduly reduce women's ability to compete effectively with men, for example.

This discussion suggests that labor-market interventions are only one policy instrument, and their design becomes complex in the presence of other social policies. Equal pay and equal treatment can be easily pursued both on theoretical and practical grounds, but the picture becomes more complicated in the case of comparable worth and affirmative action. When other policies are in force, the ultimate effects of labor regulation depend on family-law provisions, tax structures, social insurance, and social assistance. It is important that legislation be carefully designed and timely. This is particularly relevant in the emerging national and international economic conditions, since one side effect of the "feminization" of the labor force is "flexibilization" of labor contracts and increasing difficulty in enforcing labor legislation. In fact, legislation even

in advanced countries reaches successively fewer women workers because of certain preconditions and qualifications, such as minimum hours of work or earnings, or the increasingly casual nature of contracts. The problems of enforcement in developing countries are even more acute.

Wage Differences over Time

Women's educational attainment has increased over time, both in absolute terms and also relative to men. Their employment experience has increased as well, through greater participation in the labor force. Increases in women's education and work experience over time imply that relative female-to-male wages should rise in the long term.

The evolution of relative wages in developed countries confirms that women's wages have increased over time, though the narrowing of the gender wage gap has been intermittent and slow (except when equal-pay legislation proved effective). Cohort effects can mask the direct effect of women's increased human capital on their pay. When female participation increases, this can lower the average female wage because the increase most likely comes from the entry into the work force of relatively young and inexperienced workers, who typically earn less than more-senior workers. When the growth rate of the female labor force declines, as women approach high participation rates, their relative wages start increasing. In fact, the convergence of gender wages in the United States in the 1970s was masked by such compositional effects and did not show up in the statistics until the late 1980s.

There is also a tendency for women's wages to increase in developing countries (see Table 3). In six of the seven Asian countries studied in Horton (1996), women's relative earnings improved over time. (No long-term data were available for India.) The gender wage gap also declined in some Latin American countries in the 1980s (Winter, 1994). Overall, the improvement in female wages relative to male wages has occurred primarily because of women moving into more lucrative employment rather than their being paid more relative to men in the same sector (Tzannatos, 1995b).

However, the entry of women into what might have previously been a high-paying job does not necessarily imply full wage gains. The increase in the labor supply to those jobs may result in a fall in the wages for professional workers. This has happened in industrialized countries as the teaching, banking, and other clerical fields became feminized.

Conclusions and Strategies

In all countries, the average wages of women are lower than those of men. However, unequal wages do not necessarily mean that there is discrimination. Conversely, equal wages do not rule out discrimination. Social scientists disagree about the extent to which women's jobs are paid less because of discrimination through employer practices, union preferences, and legal asymmetries; the cause could also be a relatively abundant supply of female labor. Furthermore, abun-

Table 3. Relative Female Earnings (%) in Selected Countries over Time

Country	Year	F/M wage	Year	F/M wage	Annual Change
Brazil	1981	50.2	1990	53.6	0.7
Chile	1980	68.0	1987	71.0	0.6
Colombia	1984	67.2	1990	70.2	0.7
Costa Rica	1980	90.0	1989	97.0	0.8
Cote d' Ivoire	1985	75.7	1988	81.4	2.4
Honduras	1986	65.0	1990	68.0	1.1
Indonesia	1980	38.5	1992	54.4	2.9
Korea	1984	41.9	1988	51.0	5.0
Malaysia	1973	56.8	1984	69.3	1.8
Philippines	1978	70.9	1988	80.0	1.2
Thailand	1980	73.5	1990	79.8	0.8
Venezuela	1981	87.0	1990	93.0	0.7

Sources: Constructed from Horton, 1996, for Indonesia, Malaysia, and Korea; Winter, 1994, for Chile, Costa Rica, and Venezuela; and Tzannatos, 1995a, for remaining countries.

dant labor in certain occupations can be the result of women's preferences or of choices made to accommodate conflicting responsibilities. Either way, the case for legislation requiring equal pay in the same or like work and mandating equal employment opportunities can be adopted without reservation for reasons of economic efficiency and social justice. The case for comparable-worth provisions or affirmative-action programs is less clear on identification, measurement, enforcement, and efficiency grounds.

Women's labor incomes are greatly affected by women's intermittent participation in the labor force. Women are disadvantaged in the labor market because they spend less time at paid work than men and because of gender-specific constraints within the household. Women's looser reliance on paid work gives them less incentive to invest in their own human capital and eventually lowers their earnings. Ultimately, if a choice is to be made as to which partner will be employed and which one will drop out of work, the nonworking partner will almost automatically be the woman. Thus, even well-functioning labor markets will not necessarily produce equal outcomes between women and men as long as women must split their responsibilities between the home and the market while men continue to specialize in market work. Under these circumstances, equality-promoting policies in the labor market alone can have limited, or even adverse, effects on growth unless they address simultaneously the different production and reproduction roles of women and men.

Within-family and societal discrimination are responsible for fundamental gender differences in the labor supply and in labor demand. Preentry discrimination and impediments to the smooth transition from unpaid to paid work can exercise a significant adverse effect on women's work outcomes that can be even greater than the effect of employer discrimination. Though it is not always easy—or, indeed, desirable—to affect the intrahousehold allocation of resources or decision making, policies in education, health, population, and public finance (revenue/tax and social assistance/insurance) can provide gender-neutral incentives or reduce gender stereotypes and inequalities in a socially desirable way.

Increasing girls' education is perhaps the most important factor for improving women's lifetime opportunities for work. Literacy rates are lower among women in most regions, with the lowest indicators found in Africa. Families underinvest in women's education either because of societal norms or because they are unable to predict whether the women will need to work and what their occupational requirements will be. Both factors restrict the demand for girls' education. Later on, women find themselves in a disadvantaged position in the labor market, even though they may have the same aspirations as men. More part-time programs and continuing-education options can ease the constraints women face to upgrade their qualifications when they have formed their families. Education also affects reproductive outcomes. Female education raises demand for family planning and promotes effective contraceptive use (Subbarao and Raney, 1993). This reduces fertility and alleviates some of the burdens women bear because of family responsibilities.

Investments in infrastructure, if properly appraised to include the costs and benefits that occur in the invisible part of the economy, can alleviate women's house work. For example, Tanzanian women use up to 20 percent of their caloric intake for basic chores, including collecting water and preparing meals (Tinker et al., 1994). Investment in infrastructure reduces the amount of time women must spend on such basic tasks and enhances their choices and outcomes in the labor market.

As a general rule, policies in the labor market should treat reproduction and unpaid work as recognized economic activities. But the benefits should not be financed by individual employers. Putting responsibility for such benefits onto employers increases women's labor costs—and the benefits do not reach those women in the informal sector, who most likely need them most. One way of addressing this issue is for governments to encourage women and men to share responsibility for childrearing by adopting legislation allowing either parent to qualify for the leave and benefits associated with having a child. Such legislation can be supported by changes in the tax system to ensure treatment of workers within the household and a marginal tax rate on the earnings of additional workers in the household low enough to avoid creating a disincentive to women's participation in the labor force. These policies can be complemented by legislation that requires absent parents to pay child support.

While the reasons women earn less than men range from personal choices to institutionalized discrimination, the experience of many developed countries indicates that the income differential is not necessarily inevitable and can be diminished. Progress has been made in many countries in bringing women's wages closer to parity with men's wages. However, the overall earnings gap between women and men will not change without social changes, policy innovations, and improvements in the economic environment.

References

Agarwal, Bina. "Gender, Property and Land Rights: Bridging a Critical Gap in Economic Analysis and Policy." In Edith Kuiper, Jolande Sap, Susan Feiner, Notburga Ott, and Zafiris Tzannatos (eds.), *Out of the Margin: Feminist Perspectives on Economic Theory.* London: Routledge, 1995, pp. 264–294.

Bardhan, Pranab K. "Poverty and Rural Labor Markets in India: A Survey of Research," working paper no. 54, Rural Employment Policy Research Program, International Labor Organization, 1983.

Bergmann, Barbara. *The Economic Emergence of Women.* New York: Basic Books, 1986.

Binswanger, Hans, and Mark R. Rosenzweig (eds.). *Contractual Arrangements, Employment, and Wages in Rural Labor Markets in Asia.* New Haven: Yale University Press, 1984.

Blau, Francine, and Marianne Ferber. *The Economics of Women, Men, and Work.* Englewood Cliffs: Prentice-Hall, 1986.

Blau, Francine, and Lawrence Kahn. "The Gender Earnings Gap: Learning from International Comparisons," *American Economic Review,* vol. 82, no. 2, May 1992, pp. 533–538.

Blinder, Alan S. "Wage Discrimination: Reduced Form and Structural Estimates," *Journal of Human Resources,* vol. 8, no. 4, 1973, pp. 436–455.

Brown, Randall S., Marilyn Moon, and Barbara S. Zoloth. "Incorporating Occupational Attainment in Studies of Male-Female Earnings Differentials," *Journal of Human Resources,* vol. 15, no. 1, 1980, pp. 3–28.

Chiswick, Barry, J. Facklar, J. O'Neil, and Solomon Polachek. "The Effect of Occupation on Race and Sex Differences in Hourly Earnings," *Review of Public Data Use,* 1975.

Cigno, Alexandro. *Economics of the Family.* Oxford: Clarendon, 1991.

Doiron, Denise J., and W. Craig Riddell. "The Impact of Unionization on Male-Female Earnings Differences in Canada," *Journal of Human Resources,* vol. 29, no. 2, 1994, pp. 504–534.

Faundez, Julio. *Affirmative Action: International Perspectives.* Geneva: International Labor Office, 1994.

Folbre, Nancy. *Who Pays for the Kids? Gender and Structures of Constraint.* London: Routledge, 1993.

Goldin, Claudia. *Understanding the Gender Gap: An Economic History of American Women.* Cambridge: Cambridge University Press, 1990.

Gustafsson, Siv. "Single Mothers in Sweden: Why Is Poverty Less Severe?" In Katherine McFate, Roger Lawson, and William Julius Wilson (eds.), *Poverty, Inequality and the Future of Social Policy: Western States in the New World Order.* New York: Russell Sage Foundation, 1995, pp. 291–326.

Hallerstein, Judith K., and David Neumark. "Sex, Wages, and Productivity: An Empirical Analysis of Israeli Firm Level Data," discussion paper no. 9301, Maurice Falk Institute for Economic Research, 1993.

Horton, Susan. *Women and Industrialization in Asia.* London: Routledge, 1996.

Kao, Charng, Solomon Polachek, and Phanindra V. Wunnava. "Male-Female Wage Differentials in Taiwan: A Human Capital Approach," *Economic Development and Cultural Change,* vol. 42, no. 2, 1994, pp. 351–374.

MacDonald, Martha. "Social Security Policy and Gender," paper presented at the Gender Symposium, World Bank, Poverty and Social Policy Department, Washington, D.C., 1994.

Mason, Mary Ann. *The Equality Trap: Why Working Women Should Not Be Treated Like Men.* New York: Simon and Schuster, 1988.

Nickell, Stephen. "Trade Unions and the Position of Women in the Industrial Wage Structure," *British Journal of Industrial Relations,* vol. 15, no. 2, 1977, pp. 192–210.

Oaxaca, Ronald L. "Male-Female Wage Differentials in Urban Labor Markets," *International Economic Review,* vol. 14, no. 1, 1973, pp. 693–709.

Ott, Notburga. *Intrafamily Bargaining and Household Decisions.* Berlin: Springer Verlag, 1992.

Polachek, Solomon. "Human Capital and the Gender Earnings Gap: A Response to Feminist Critiques." In Edith Kuiper, Jolande Sap, Susan Feiner, Notburga Ott, and Zafiris Tzannatos (eds.), *Out of The Margin: Feminist Perspectives on Economic Theory.* London: Routledge, 1995, pp. 61–79.

Psacharopoulos, George, and Zafiris Tzannatos. *Women's Employment and Pay in Latin America: Overview and Methodology.* Washington, D.C.: World Bank, 1992.

Sedlacek, Guilherme, Leah Gutierrez, and Amit Mohindra. "Women in the Labor Market." Washington, D.C.: World Bank, 1993.

Sen, Amartya. "Gender and Cooperative Conflicts." In Irene Tinker (ed.), *Persistent Inequalities: Women and World Development.* New York: Oxford University Press, 1990, pp. 123–149.

Shapiro, Daniel M., and Morton Stelcner. "The Persistence of the Male-Female Earnings Gap in Canada 1970–1980: The Impact of Equal Pay Laws and Language Policies," *Canadian Public Policy,* vol. 13, no. 4, 1987, pp. 462–476.

Subbarao, Kalanidhi, and Laura Raney. "Social Gains from Female Education: A Cross-National Study," discussion paper no. 194, World Bank, Washington, D.C., 1993.

Tinker, Anne, Patricia Daly, Cynthia Green, Helen Saxenian, Rama Lakshminarayanan, and Kirrin Gill. "Women's Health and Nutrition: Making a Difference," discussion paper no. 256, World Bank, Washington, D.C., 1994.

Treiman, Donald, and Heidi Hartmann. *Women, Work, and Wages: Equal Pay for Jobs of Equal Value.* Washington, D.C.: National Academy Press, 1981.

Tzannatos, Zafiris. "Economic Growth and Equity in the Labor Market: Evidence and Policies for Developing Countries." Washington, D.C.: World Bank, 1995a.

———. "Growth, Adjustment, and the Labor Market: Effects on Women Workers," paper presented at the Fourth Annual Conference of the International Association for Feminist Economics, Tours, 1995b.

Winter, Carolyn. "Working Women in Latin America: Participation, Pay, and Public Policy." Washington, D.C.: World Bank, 1994.

Zabalza, Antoni, and Zafiris Tzannatos. *Women and Equal Pay: The Effects of Legislation on Female Employment and Wages in Britain.* Cambridge: Cambridge University Press, 1985.

See also Haneen Sayed and Zafiris Tzannatos, "Sex Segregation in the Labor Force"

Sex Segregation in the Labor Force

Haneen Sayed
Zafiris Tzannatos

Introduction

When women are in the labor force, they usually perform different tasks (jobs) and work in different sectors (industries) than men. The conditions of their employment are, on average, inferior to those faced by men and are often atypical—part-time, temporary, or casual work, work in the home, and subcontracting. In terms of occupations, for example, nearly two-thirds of women in manufacturing are categorized as laborers, operators, and production workers while relatively only a few can be found in the administrative and managerial positions predominantly held by men. Globally, female workers are concentrated in a limited number of industrial sectors: More than two-thirds of those in garment production are female, accounting for almost one-fifth of the total female labor force in manufacturing. With respect to employment status, the majority of family workers are female and often unpaid (International Labor Organization, 1985; United Nations Development Program, 1995; World Bank, 1997).

These patterns suggest that women and men in the labor market are concentrated in different sectors and that, even when they are in the same sector, they carry out different tasks at different levels of responsibility. This labor-force phenomenon is called sex segregation; it occurs when female and male workers are employed in "compartmentalized" activities that usually lead to different rewards and different career opportunities even though workers may have comparable labor-market attributes. Segregation can manifest itself in industries (more males are found in some forms of manufacturing as compared to more females in social services), occupations (managerial versus clerical jobs or skilled versus unskilled manual labor), and even in location (with men usually more mobile and women more homebound). In fact, the term originally applied to racial segregation in jobs and residential location.

As a concept, segregation breaks down into two types: horizontal and vertical. Horizontal segregation exists when women and men are employed in different occupational groups such as professional and technical vocations. Vertical segregation exists when men and women work in the same occupational group, such as teachers, but men do the more skilled, more responsible, and better-paid activities while women perform complementary activities. For example, within the teaching profession, the majority of headmasters are men; in healthcare, the majority of nurses are women. Though this distinction is followed in much of the literature on segregation, treating vertical and horizontal segregation as separable—that is, as mathematical concepts suggesting two unrelated dimensions—is not always appropriate, for often the same social processes lead to both vertical and horizontal segregation.

Vertical segregation is more complex than horizontal segregation in that it involves many different aspects of inequality, such as differences in skills, responsibility, pay, status, and power. Such elements differentiate all occupations (for example, managers and professionals) and not just activities within specific occupational groups. Unfortunately, cross-occupational comparisons have been lacking; until this gap is closed, our understanding of segregation in the total labor force will be limited. In practice, even simple approaches to vertical and horizontal segregation have been curtailed by the availability of data. Most studies of horizontal segregation tend to use a few large occupational groupings; most studies of vertical segregation examine only a limited range of occupations.

When sex segregation reflects the interaction of constrained choices of women resulting from gender-biased demands for child care, exclusion strategies of male workers, discrimination by employers, and legislation, it becomes a clear policy issue. This is because it entails both inequality in the treatment of women and men and inefficient processes in the labor market, which subsequently result in the underutilization of women's labor (inefficiency), lower wages for women (inequality), lower levels of output (welfare loss), and social injustice (inequity). This is not to say, however, that the integration of the sexes in the workforce should be a social objective per se. Policies can reduce impediments to women's employment and enable women to exercise their own choices on the same basis as men.

Measuring Segregation

Segregation is usually measured in terms of outcomes—for example, in terms of unequal shares of men and women within occupations. A commonly used measure is the Duncan index (Duncan and Duncan, 1955), an index of employment dissimilarity between any two groups of workers. In the case of women and men, the Duncan index is calculated by finding out the absolute difference between the sectoral employment ratios of women and men to their respective labor force and dividing their sum by two. Absolute differences are used because, by definition, the excess female and male labor in some occupations or industries is exactly equal to their respective shortages in the other sectors. In other words, the sum of the differences, if conventionally calculated, will be zero. This also explains why the resulting sum is divided by two: The formula involves a double counting of excess female/male labor in some sectors with a corresponding shortage of female/male labor in the rest of the sectors.

The minimum value of the index is zero; it occurs when women and men have identical employment distributions across sectors—that is, when the percentage of women in each sector is the same as the percentage of women in total employment. The maximum value, unity, occurs when there is complete dissimilarity (no women and men work in the same sector). Since the index varies from 0 to 1, it is often expressed as a percentage and is often wrongly assumed to show the percentage of workers who should be reallocated across sectors to achieve the same employment distributions of women and men across sectors, a problem we will return to shortly: The percentage of workers who should be reallocated depends upon the value of the index and the total size of, and the share of, women in the labor force.

A feature of the index is that its change over time can be decomposed into changes of the sex ratio within sectors as well as changes in the size of the sectors. For example, a decline in the index can come either from some equalization of the sex ratios within sectors or from different growth rates of employment sectors over time. The former, called the sex ratio effect, captures changes in segregation within sectors. The latter, the structural effect, relates to changes in the employment concentration of women and men in specific sectors.

The value of the index depends on the number of sectors for which information is available and the way in which these sectors are defined. For example, the agricultural sector may be severely underestimated because women's farm work often goes unrecorded in official statistics. Had "invisible" women workers been included in the labor force, the

Table 1. Employment Dissimilarity (Duncan Index)[1] by Industry
(regional averages; N = number of countries)

	N	Early Duncan (1950s–1960s)	Late Duncan (1980s–1990s)
All Workers[2]			
Africa (excluding North Africa)	4	0.2772	0.2509
East Asia/Pacific	9	0.2723	0.2182
South Asia	3	0.1487	0.2041
Eastern/Central Europe	5	0.2105	0.2604
Rest of Europe	15	0.3448	0.3028
Middle East/North Africa	6	0.3761	0.3995
Americas	19	0.4519	0.3656
Total	61	0.3458	0.3058
Employees Only[3]			
Africa (excluding North Africa)	4	0.4559	0.3003
East Asia/Pacific	9	0.3368	0.2332
South Asia	3	0.2827	0.2677
Eastern/Central Europe	5	0.3234	0.3141
Rest of Europe	15	0.3711	0.3216
Middle East/North Africa	6	0.4250	0.3781
Americas	19	0.4602	0.3258
Total	61	0.3948	0.3097

[1]The calculation of the index is based on seven industries—agriculture, mining, manufacturing, construction, utilities, transport, and services—as tabulated in ILO, 1990.
[2]Wage employment, self-employment, and family work.
[3]Wage employment only.
Sources: Tzannatos, 1995.

value of the index might be different. Within agriculture, occupations are often grouped broadly; for example, no distinction is made between farmers and farm workers. In this case, the value of the index would be lower, especially in countries where agriculture is important. It is, therefore, more meaningful to examine changes in segregation over time than to concentrate our attention on the value of the index per se, given that the definitions of the labor force and occupational classifications do not change significantly between two periods of time.

An examination of data worldwide shows that, for all types of workers across the globe, employment dissimilarity by industry as measured by the Duncan index declined on average from 0.35 to 0.31 in the period between the 1950s–1960s and the 1980s–1990s (see Table 1). But employment dissimilarity among employees only, albeit higher initially (0.39 in the earlier period), declined more quickly and in the 1990s stands at the same level as the dissimilarity for the whole labor force (0.31).

Similarly, occupational dissimilarity among all workers registered practically no change between the two periods under consideration (0.39 percent and 0.38 percent), but the dissimilarity among employees declined from 0.44 percent to 0.40 percent (see Table 2). One explanation for the more rapid decline in segregation among employees compared with all workers may be that market forces are less discriminating against women than are noneconomic

factors: It may be more difficult for women to break sex stereotypes in self-employment and family work that is undertaken more often at the village or community level.

The relatively slow decline in dissimilarity over time masks significant cross-country changes. Since the 1960s, many sectors have changed from being overrepresented by females to being underrepresented by them (overrepresentation is defined as a state in which the share of women in an employment sector exceeds the share of women in total employment; the converse holds for underrepresentation). Among 86 countries for which information exists in both the 1950s–1960s and the 1980s–1990s, the industrial sectors that changed in that time from being "female overrepresented" to being "male overrepresented" were agriculture (from 25 to 18 countries), manufacturing (from 30 to 26 countries), and utilities (from 6 countries in the early period to no country in the late period; see Table 3). The number of countries in which women were overrepresented in the mining and the services sectors in both periods remained the same (3 and 71 countries, respectively). Neither was a change in services. The number of countries in which women were overrepresented in commerce increased from 60 to 70. These 27 net changes (7+0+4+6+10+0) were the result of a total of 63 reversals (see Table 3, last two columns).

With respect to employment status, women were initially overrepresented in self-employment in 7 of the 75 countries for which information exists in both periods but

Table 2. Employment Dissimilarity (Duncan Index)[1] by Occupation
(regional averages; N = number of countries)

	N	Early Duncan (1950s–1960s)	Late Duncan (1980s–1990s)
All Workers[2]			
West Africa	2	0.2258	0.2301
East Asia/Pacific	9	0.3169	0.3190
South Asia	2	0.2011	0.1958
Eastern/Central Europe	1	0.2655	0.2623
Rest of Europe	13	0.4068	0.4022
Middle East/North Africa	5	0.3566	0.4623
Americas	11	0.5191	0.4469
Total	43	0.3860	0.3804
Employees Only[3]			
West Africa	2	0.3807	0.2909
East Asia/Pacific	9	0.4052	0.3257
South Asia	2	0.4783	0.3528
Eastern/Central Europe	1	0.2548	0.2714
Rest of Europe	13	0.4254	0.4165
Middle East/North Africa	5	0.4072	0.4926
Americas	11	0.5442	0.4553
Total	43	0.4421	0.4030

[1]The calculation of the index is based on seven occupations—professional, administrative, clerical, sales, services, farming and production.
[2]Wage employment, self-employment, and family work.
[3]Wage employment only.
Sources: Tzannatos, 1995.

Table 3. Women's Overrepresentation by Industrial Sector and Number of Countries in Which Changeover Occurred

N = 86	Early Period (1950s–1960s)	Late Period (1980s–1990s)	Over to Under	Under to Over
Agriculture	25	18	9	2
Mining	3	3	2	2
Manufacturing	30	26	13	9
Utilities	6	0	6	0
Commerce	60	70	1	11
Services	71	71	2	6

Source: Tzannatos, 1995.

only 4 in the later period. Women's overrepresentation also declined among family workers (from 57 to 54 countries). Among wage and salaried workers, women's overrepresentation increased by one country, from 45 to 46. These seven net changes (1+3+3) mask the fact that a reversal took place in no less than 38 cases (see Table 4).

Although the degree of sex segregation has been declining over time, employment differences may arise in new forms. The introduction of new technologies usually changes not only gender employment patterns but also social relations among the people. A division of labor by sex gets reconstituted with shifts in production methods and technology and redistributes power, resources, and authority in ways that keep women in subordinate positions. In this shift, a new division of labor by sex often becomes more disadvantageous to women, as may be the case in rural Java (Wahyana, 1994). These changes may not show up in numerical measures of segregation that defeminize female jobs (nurses, telephone operators, flight attendants) and feminize some formerly male jobs (bank tellers, teachers). Defeminization usually affects more jobs than feminization (Cohn, 1985). New forms of segregation can occur with women employed predominantly in part-time work with high turnover rates, short job tenures, and low trade-union membership. These forms of segregation are typical of workers with a weak commitment to employment careers or a strong commitment to family responsibilities that lie outside the market, and such workers are usually women.

There seem to be two trends in employment segregation over time. On the one hand, a larger proportion of women move into lower-skill jobs. There is some evidence that self-employment among women outside agriculture has increased, as has contract work at home in Asia (Lim, 1994).

On the other hand, increasingly more women engage in work traditionally performed by men. In general, growth and the opening up of new employment opportunities, combined with the increasing educational achievements of women, should decrease the value of the Duncan index.

Some evidence points to the fact that an increase in segregation is not necessarily synonymous with poorer opportunities or inferior labor-market outcomes for women. In Puerto Rico, segregation patterns have been found to be consistent with median annual earnings of women that are quite close to those of men (Presser and Kishor, 1991). An explanation for this is that women are offered more opportunities for upward mobility when production is organized around strictly segregated occupations than when women and men work together. Under the first arrangement, women will be required to supervise other women and get better jobs than they would otherwise have, while in male occupations some men simply have to accept jobs that are low status if they work solely with other men. On similar grounds, horizontal desegregation does not unambiguously represent an improvement in the labor-market position of women: The feminization of previously male dominated jobs can be associated with deteriorating employment conditions due to the increase in *total* labor supply that, in turn, depresses wages for both women and men in those jobs. This may also lead to an increase in vertical segregation if men move up to top positions.

A recent empirical study of women in developing countries also found that, despite increases in segregation over time, women's earnings increased relative to men. A possible explanation for this is the decline in women's employment in agriculture, where wages are typically very low, and the increase in job opportunities in manufactur-

Table 4. Women's Overrepresentation by Employment Status and Number of Countries in Which a Changeover Occurred

N = 75	Early Period (1950s–1960s)	Late Period (1980s–1990s)	Over to Under	Under to Over
Wage employment	45	46	9	10
Self-employment	7	4	5	2
Family workers	57	54	5	7

Source: Tzannatos, 1995.

ing and services, especially in the public sector. Thus an increase in women's relative wages can occur although women may still tend to be employed in female-dominated occupations. This does not rule out an increase in employment discrimination, but rather relates to the issue of whether women's wages need necessarily decline during the early stages of industrialization or whether women workers suffer disproportionately more than men workers during implementation of policies for economic adjustment and stabilization.

From Segregation to Misallocation

A common interpretation of the Duncan index is that it refers to the proportion of either women or men who would have to be transferred from one sector to another to obtain equal proportions across sectors. This is clearly not the case, since the proportion of women who would have to change sectors is always less than the proportion of men because there are fewer women working than men. The index does not even refer to the total number of workers who would have to change sectors. The index is simply the ratio of actual to potential reallocations if no woman and man worked in the same sector at the current size and sex composition of the labor force. Consequently, the index is independent of the size and sex mix of the total labor force. The number of workers who would have to be reallocated is equal to twice the value of the index times the total number of male (female) workers times the share of female (male) workers in the total labor force. It is the value of this product that, if divided by the total labor force, will provide the percentage of all workers that should be reallocated to achieve equality in the employment distributions of women and men (Tzannatos, 1990).

The percentage of the labor force that should be reallocated, rather than the Duncan index per se, can be used as an indicator of labor misallocation and potential efficiency losses. In countries in which historical data exist, the

percentage of workers that should be reallocated increased while the Duncan index declined. For example, in Britain between 1900 and 1980 the index decreased by more than 10 percent, whereas the percentage of workers that should be reallocated increased by 3 percent. For the United States, the corresponding changes were a 20 percent decline and a 25 percent increase (Tzannatos, 1990). In addition, changes in the Duncan index in Latin America since the 1960s correlate weakly with changes in the percentage of workers who should change sectors. The reason the decline in the value of the Duncan index has not reduced the percentage of the labor force that would have to be reallocated to achieve the same employment distribution between women and men is that the share of women in the total labor force has also been increasing over time.

Estimates for 11 countries in Latin America suggest that as much as 12 percent to 37 percent of the total labor force should be reallocated to achieve occupational equality between women and men (see Table 5). As women constitute typically approximately one-third of the labor force, this amounts to around half of all female workers, not an insignificant number. Further analysis of data revealed that the percentage of required reallocations to achieve occupational equality among the self-employed and family workers has a tendency to increase in Latin America over time. In contrast, reallocations among employees declined in most countries, and, in those in which they increased, the increase was small. This suggests that the paid/formal labor market responds to changes during economic development more quickly than the informal sector. An explanation for this may be that social factors and norms are a more binding constraint in the informal sector than in the open labor market.

Economic Effects of Segregation

If employment segregation were solely the result of employer discrimination and women were excluded from some occu-

Table 5. Effects of Within-Industries Elimination of Occupational Segregation

| | | % Change in GDP | | | Reallocations (% of Labor Force Who Would Have |
Country	Year	Female Wages	Male Wages	(gross domestic product)	to Change Occupation to Achieve Gender Wage Equality)
Argentina	1987	38	−9	4	25
Bolivia	1989	50	−9	6	28
Brazil	1980	96	−8	9	23
Chile	1987	41	−6	3	18
Colombia	1988	46	−8	5	20
Costa Rica	1989	35	−6	3	18
Ecuador	1966	59	−13	9	37
Guatemala	1989	25	−6	2	14
Jamaica	1989	61	−8	8	28
Uruguay	1989	30	−8	3	16
Venezuela	1987	24	−6	2	12

Source: Adapted from Tzannatos, 1991a.

pations and crowded in others, then, by virtue of the distorted relative labor supply across occupations, pay would be higher in male-dominated occupations and lower in female-dominated occupations than it would be under perfectly competitive conditions. The effect of occupational segregation upon female wages relative to male wages is discussed in the essay "Women's Labor Incomes" elsewhere in this volume. It is shown there that, after controlling for women's labor-force characteristics, interoccupational differences in pay explain little of the gender wage gap: Most differences arise from intraoccupational differences. This analysis has two shortcomings. First, part of the intraoccupational differences can be the result of discrimination that has fed back into women's occupational choices and appears to be justified incorrectly on statistical grounds. Second, this type of analysis tells us little about either the divergence of male and female wages from the competitive level of wages or the welfare loss (reduction in total output) arising from the misallocation of the labor force.

A method to evaluate the wage and output effects of employment segregation has been proposed in the context of racial segregation (Bergmann, 1974). In a gender context, the method is based on the assumption that, had women and men had the same supply characteristics (human capital and preferences) and had they been treated equally in the labor market, they should have the same occupational distribution within industries, assuming we take the existing industrial distribution of employment as given. The result is an equalization of female and male wages within industries as women abandon the crowded low-paying occupations (where wages should increase) and enter the high-paying male occupations (where wages should decline). At the same time, assuming conventional marginal productivity conditions, there will be efficiency gains—that is, increases in total output. The reason for this: Productivity in the female occupations is low due to crowding, and, when women leave them to enter male occupations, their productivity will increase, resulting in a net change in output that will be positive.

The results of this exercise are suggestive about what can happen in the long run when (1) women and men are equally endowed with human capital, (2) there is no employer discrimination, (3) family constraints are equally binding upon women and men, and (4) the gender-specific effects of social norms and other institutional factors have withered away. Though subject to a series of qualifications (Tzannatos, 1988), simulations suggest that women's wages can increase significantly with generally little loss in male wages, partly because there can be significant output (GDP) gains (see Table 5). In other words, a reduction in segregation is not a purely redistributive issue; the "size of the pie" increases, with women claiming a bigger share. In fact, given that the economy grows over time, men's wages need not decline in absolute terms—a point worth noting because, with zero-sum gains, the losers (in this case, men) may devise strategies for forestalling equality. This phenomenon has been emphasized by feminist economists and is

known in welfare economics as "the reversal rest" (those losing from economic change can bribe some potential winners and forestall the move toward a situation in which they would have lost more). In any case, the mere size of labor reallocations required to achieve complete equality (see Table 5) confirms that these estimates refer to the long run, given that the annual *flows* to the labor force that are usually the main avenue for changes of existing labor-market patterns are only a fraction of the labor force *stock*.

Theories of Labor Segregation by Gender

Many theories have been put forward in an attempt to account for sex segregation; none of them, however, offers a complete account for its levels and persistence. Theories can be grouped into two broad categories: those that adopt an individual's perspective and fall broadly into neoclassical theory of marginal productivity and human capital, and those that adopt a structural-analysis approach to labor markets and can be collectively referred to as institutional models.

Neoclassical Views

In the neoclassical framework, both labor-supply and labor-demand factors are relevant even though due emphasis is placed on women's preferences and women's choice to invest less in education and training as well as to spend a smaller proportion of their adult years in the labor force than men. Thus gender differences in labor supply and individual choices are usually considered to be the main cause of women's differential placement in the labor market, although neoclassical economists neglect to examine how choices are made and whether preferences are individually or collectively determined. By the neoclassical line of reasoning, men and women enter the labor market with different tastes and qualifications, such as education and formal training. Once women are in the labor force, their desire to find jobs that do not conflict with their domestic obligations leads them to engage in flexible employment, often in self-employment, outcontracting, and supportive family work. They also make frequent changes in employment status to adapt to the changing family cycle due to childbearing and -rearing when time constraints are particularly binding. Women in Lima, Peru, who have no children or only one child tend to work in the wage sector, whereas women with two or more children are predominantly engaged in self-employment (Herz and Khandker, 1991). Data from 79 developing countries display an unmistakable pattern of high fertility (completed number of births per woman), a low share in paid employment, and a high percentage working at home for no pay (see Table 6). Women's occupational outcomes can be the result of three factors: (1) gender-biased decisions within the household that result in women being less qualified than men at the point of entry into the labor market, (2) women's compromise between work at home and in the market, or (3) labor-market imperfection such as discrimination.

Segregation can, therefore, be traced in part to women's dual responsibilities at home and in the market.

Table 6. Fertility and Women's Labor-Force Status

Number of Children	Women in Paid Employment (%)	Women in Unpaid Positions (%)
More than 7.0	11	47
6.1–7.0	17	32
5.1–6.0	25	27
Fewer than 5.0	30	18

Source: Adapted from Dasgupta, 1995.

Their engagement in unpaid/family work results in both lower participation in the labor market than men and intermittent employment patterns. Because women tend to leave the labor force at the time of marriage, employers may restrict their access to various job ladders; consequently, women end up in dead-end positions or jobs with limited mobility. Empirical evidence shows that it is profitable for firms to discriminate in hiring and pay practices against individual women workers, even those who they expect will never drop out if they expect intermittent labor-force participation of women on average (Lazear and Rosen, 1990). Since overt wage discrimination is often illegal, firms can react by diversifying labor demand. Jobs can be segregated between women and men so that the principle of comparison cannot apply, or women may be required to be more productive than men, if they are to be employed in the same jobs, as a compensation to employers for higher turnover costs. Employers may also offer less training to workers who are less permanently a part of the labor force.

The consequence of all this is that a woman who would be as productive and stable an employee as a man is deprived of equal opportunities because of statistical discrimination: Firms do not know in advance the productivity and commitment of individual workers and use stereotypes as inexpensive screening devices (Blau and Ferber, 1986). Statistical discrimination in hiring practices can cause occupational differentials (men in good jobs, women in bad jobs) and be responsible for the underpayment of women relative to men. These gender employment differentials become self-fulfilling for both employers and women workers; and, once embedded, the disparities are extremely slow to change. Competition can drive discriminatory employers out of the market in the long run, provided that some employers do not discriminate, or discriminate less than others (Becker, 1957). Still, this tendency seems to be small or at times not obvious because not all sectors are competitive.

Women can also face fewer job opportunities in the labor market than men because of social or legal exclusions from particular activities or from the labor force overall. In many countries, marriage is seen as incompatible with work; women withdraw from work after they get married either through peer pressure or employment regulations or gender-biased tax treatment of the family. The effect of

norms on patterns of female employment cannot be underestimated. In fact, they are often more important than economic factors. In a study of more than 100 countries in the 1980s, religion was the single most important explanatory variable for female participation in the labor market (Psacharopoulos and Tzannatos, 1989).

Social or employer discrimination due to marital status or in the form of restricted access to certain types of employment can also result in a rational defense strategy of underinvesting in women's education. It would be irrational for women to invest in areas in which the probability of employment is small, the costs of education are high, and skills rapidly depreciate—in science versus liberal-arts subjects, for example. The resulting loss in women's productivity, benefits, and career opportunities can be anticipated by women and taken into account when investment in human capital is made. The end result of this lack of investment is that women work in different areas and also bring in lower earnings than men.

Institutional Perspectives

The institutional models, which are alternative theories to the neoclassical models, locate the source of occupational sex segregation in employers' discriminatory practices or their attempts to prevent worker solidarity (Doeringer and Piore, 1971; Gordon et al., 1982). In the internal-labor-market model, firms hire workers from the open labor market for entry-level jobs while filling the remainder of jobs internally as workers progress up promotion ladders by acquiring job-related skills, many of which are firm specific. This creates an internal labor market within the firm that puts an emphasis on firm-specific skills; a large proportion of jobs are filled from internal sources. Firms take the occupational category as the decision unit, establishing pay rates for each category with some allowances for seniority and merit considerations, and link jobs together with promotion ladders. In this setup, the employer wants to make sure that workers within each job category are as similar as possible. If it is believed that men and women differ in their productivity-related characteristics, then discrimination is likely to result in men and women being channeled into different jobs.

In a similar way, the dual-labor-market model emphasizes the distinction between primary and secondary jobs.

Primary jobs have high levels of firm-specific skills and thus pay higher wages, have good promotion opportunities, and emphasize long-term relationships between workers and firms. In the secondary jobs, firm-specific skills are not as important, with the result that these jobs pay less, offer relatively fewer promotion opportunities, and have higher rates of turnover. Primary jobs are more likely to be located in monopolistic, unionized industries; secondary jobs, in competitive industries. The segmentation of the labor market produces both pay and productivity differences between men and women due to unequal access to on-the-job training. According to this model, feedback effects magnify any initial productivity differences as women rationally respond to lower incentives by self-selecting themselves into the secondary sector through constrained occupational choices and lower investment in human capital; thus, labor-market discrimination is not wholly the outcome of conscious deliberate acts by employers. Once men and women are channeled into different types of entry jobs, internal regulations and management practices can perpetuate differences in productivity, promotion opportunities, and pay (Roos and Reskin, 1984). The firm may have an interest in perpetuating these "divide and rule" practices to benefit from segmentation of its work force by sex (or race) because it prevents workers from seeing their common interests.

Feminist Theories

Feminists view the human-capital and internal/dual-labor-market theories as inadequate since neither incorporates the concept of gender into its analysis or gives it a central place. While there is no single feminist theory of sex segregation, a common point of departure for feminist analysis is to develop a theory that does not consider sex segregation a mere byproduct of employers' strategies to maximize profits.

A Marxist-feminist framework analyzes the conditions under which women and men sell their labor power because they tend to be different (Beechey, 1987), as well as the social construction of the family and the education and training system. This analysis also questions the conventional ideology of the family that asserts that women's primary responsibilities are to be housewives and mothers. Finally, this framework calls for the analysis of gender within the labor-market process itself, in particular along the lines of the "deskilling" hypothesis—the simplification of tasks and the substitution of unskilled for skilled labor. Women are drawn into unskilled, low-paying jobs in the course of capital accumulation that enables employers to accrue profits.

Radical feminists add patriarchy as an important determinant of segregation. Patriarchy is defined as a set of social relations with a material base, characterized by a hierarchical relation between men and women in which men are dominant and women subordinate. Male workers, unions, and employers all play a role in maintaining occupational segregation. The division of labor by sex in the workplace and at home is both universal and historical. This hierarchical structure is at the root of women's low social status. Based on anthropological explanations, stratification by sex came about with the increasing productivity, specialization, and complexity of society, such as the development of widespread agriculture and private property. In precapitalist patriarchy, men controlled women's and children's labor in the family and gradually learned the techniques of hierarchical organization and control. With the emergence of economic systems of wider exchange, larger production units, a division between the public and private spheres, and a state apparatus, the direct personal system of control over women's labor expanded into societywide indirect and impersonal institutions. Capitalism threatened patriarchy, however, in that it destroyed old institutions and created new ones such as the labor market. How was male dominance reconstructed?

One argument is that men, as capitalists, created hierarchies in the production process in order to maintain their power (Hartmann, 1976). Sex segregation in the labor market is the main mechanism that puts men in a superior position because it enforces and reinforces lower wages for women. Low wages keep women dependent on men and encourage women to marry. Married women must perform domestic chores for men and the family. Men are better off because of higher wages in the labor market and work done for them by women in the house, but this domestic division of labor weakens women's position in the labor market. The circle is complete—patriarchy and capitalism form an interlocking system.

In this view, labor market behavior of men, as employers and workers, is governed by their desire to maintain patriarchal privileges at home. Two mechanisms give rise to both segregation and lower wages for women (Strober, 1984). First, male employers set wages and working conditions and allow male workers to decide which occupations they will take first and which will be left for women. Second, in deciding which jobs to take and which to leave for women, male workers maximize their economic gain by comparing alternative occupations available to them. Therefore, jobs become male or female jobs not simply because they have inherently female characteristics, such as teaching or nursing. Whether a particular occupation becomes male or female depends on the range of occupations available at the time.

Despite differences, many feminists would agree that there is an ideology of gender, a normative/cultural system that lies behind and guides men's and women's behavior. In doing so, it separates the roles and activities of the sexes, including their occupational roles. Examples of such a system are assumptions that the sexes are different from each other in character, temperament, and capability, that women are naturally suited to be mothers, supporters, and caregivers, or that what is feminine is of lower prestige than what is masculine.

Finally, some feminist authors have attributed the emergence of segregation in the labor market under capitalism as an attempt to control sexuality. Under feudalism, sexuality was controlled within the family. The separation of home and workplace under capitalism, however, led to anxiety about mixing the sexes at work with the potential of producing illegitimate children. This, in turn, led to calls for prohibiting the employment of adolescent girls, for example, in the nineteenth century, when the employment of children was otherwise routine. A measure short of the outright prohibition of young women's employment was therefore segregation at the workplace. This "prophylactic" explanation of segregation is supported by statistical analysis that displays a significant negative relationship between the degree of workplace segregation and illegitimacy rates in early capitalist Britain; it, in turn, led to the view that segregation was prompted by social concerns and had nothing to do with the labor force as such (Humphries, 1987). Such concerns die hard: Even in mid-twentieth-century Britain, the post office was content to employ both male and female clerks only so long as individual offices were single-sex establishments.

Given that it is difficult to assess which part of sex segregation arises from what particular factor, how can these competing theories be evaluated? For example, in addition to historical factors, women's labor market outcomes can be reinforced by certain expectations by prospective women workers. Women may expect their future role to be more likely within the family than in the labor market. In this case, lower investment in education by parents or women themselves appears to be a rational rather than a constrained choice since the period when the monetary benefits to education accrue, during employment, is likely to be shorter than for men.

Given the traditional division of labor by sex in the family, the human-capital model provides an explanation for occupational sex segregation in terms of women's optimizing behavior: Women will choose occupations characterized by flat-earnings profiles, that is, in which loss of experience due to an interruption of employment and subsequent loss in earnings do not result in great loss of earnings. The neoclassical theory's prediction is that women in predominantly male jobs would not enjoy greater benefits ("returns") to each year of experience than women in predominantly female jobs. However, the earnings of women in predominantly female jobs do not always depreciate less during periods of time spent out of the labor force than do the earnings of women in predominantly male jobs; in addition, women who have discontinuous work histories are not more likely to be in a predominantly female occupation than are women who have been employed continuously. In the United States, the evidence does not always support the claim that it is rational for women planning intermittent labor-force participation over the life cycle to choose traditional female occupations or that women with continuous employment have an occupational distribution similar to men (England, 1982). Though this finding suggests that labor-demand factors (discrimination) are more important than labor-supply factors (women's own choices), in Britain lower rates of pay for women do not derive from job characteristics but from the conditions under which women supply their labor (Rubery and Tarling, 1988).

With respect to alternative theories, there is little agreement in the empirical literature on the reasons segregation occurs or how the segregated occupations or "compartments" should be defined. This has become more evident in industrialized countries: Women are no longer overrepresented in unskilled occupations, making it difficult to link female-dominated occupations directly to the secondary sector of the labor market, as segmented-labor-market theories would suggest. In more recent work, the case is made that working women are not a homogeneous group and that explanations of sex segregation must differentiate between part-timers and full-timers and between public- and private-sector jobs (Humphries and Rubery, 1992). Divisions within the female labor force are also important. The distinction between primary ("good") and secondary ("bad") segments in the labor market becomes blurred when the "bad" includes service and clerical occupations that are predominantly female as well as skilled blue-collar occupations that are predominantly male. Finally, more women are obtaining long and stable job tenures that are cyclically insensitive. It is, therefore, inappropriate to classify women as a single group when they can differ markedly in occupational status.

The Marxian approach is also incomplete, because deskilling is not always obvious in the case of women. For example, in some industries in the public sector where many women are found, conditions are usually better than in the private sector. Another argument here may be that women's increasing representation in professional categories in the health and education sectors is an extension of stereotypical female roles within the family such as nurturing and caring. At some point, the analysis should include the role the state takes in affecting the overall pattern of occupational segregation.

Government Policies

When segregation is simply the result of a legal exclusion of women from certain activities (for example, as a result of protective legislation), the abolition of such provisions when they are no longer justified can be effective in reducing segregation. In cases of employer discrimination, legislation that requires employers to treat women and men equally with respect to hiring, training, promotion decisions, and other employment practices can be introduced. In fact, most countries have passed national legislation concerning equal employment opportunity or have ratified the relevant ILO conventions.

The effectiveness of legislation depends on whether it is enforceable and to what extent it is enforced. Women workers and especially women applicants rarely know that they have been discriminated against. This reduces the effectiveness of equal-employment legislation. A proposition that has been put forward to counter the informal nature of the selection process by companies requires that unsuccessful job applicants be informed of the identity and background of the successful applicants. The realization of such a proposition is, of course, limited for reasons of confidentiality and cost. Requiring employers to prove that discrimination has not taken place may result in an avalanche of claims since any unsuccessful applicant can potentially launch a complaint at little cost once her candidacy is rejected. On the other hand, if the onus of proof that discrimination has occurred falls on the workers, it may involve substantial costs to women, which may include potential retaliation and eventual dismissal.

Affirmative action is another policy instrument that can reduce discriminatory hiring practices or the failure to promote women workers. Affirmative action can take the form of setting goals for increasing women's representation in certain areas, with a deadline agreed upon by employers, workers' representatives, and governments. Affirmative action is an integral part of U.S. legislation, but its effects and its future are debatable. U.S. firms subject to such legislation tend to hire more women than those not so subject, but the overall effect of such policies on women and economic efficiency has yet to be established. Affirmative action does not have strong proponents in the largely social-democratic regimes in Europe; in fact, it was outlawed by the European Union in 1995. In countries in which it is applied, such as in Sweden, it is voluntary. Still, it is in Sweden where measurable labor-market characteristics between women and men show the least divergence, a fact that can be attributed more readily to consistent policies addressing reproduction and class issues than an isolated focus on labor-market outcomes with uncertain distributional results (Tzannatos, 1991b).

Conclusion

Sex segregation is difficult to measure, and existing statistics do not represent adequately the complex economic and social processes that generate it. Nonetheless, sex segregation is clearly a visible and important dimension of gender inequality in the labor market. Female workers are concentrated in sectors that are characterized by lower status, lower pay, and employment insecurity. But the causes for this association are less clearly established. Since sex segregation in the labor force is a relatively modern phenomenon that appeared with the emergence of the market, it is tempting to attribute it to the operations of the market. However, much of the observed differences reflect or are still affected by the historical division of labor within the family. The market, in turn, seems to be reinforcing this sex segregation. The fact is, segregation remains an enduring feature of the labor market whether it arises from economic forces or from social institutions outside that market.

The cross-country evidence reviewed in this essay suggests that economic development contributes to a reduction of the gender employment differentials, a fact that is increasingly acknowledged in the literature (Glyn, 1992). One factor that has enabled more women to spread throughout the labor force is education. Gains in women's educational attainment augment their productivity and enable them to compete more effectively with men in the labor market. Higher levels of education are also associated with higher levels of labor-force participation, which in turn provides women with greater work experience and skills and reduces statistical discrimination as employers' expectations about women's longevity in the labor force become more optimistic. All of this justifies more in-service training for women. For developing countries, the elimination of gender differences in education is a priority policy area.

Employment discrimination that leads to segregation is still practiced in many countries and is not illegal. Such discrimination can take the form of differential hiring, training, promotions, and firings for women and men. In some countries, employers make sterilization certificates, pregnancy tests, marital status, and other screening practices a condition for employing women in production. In others, discrimination can arise directly or indirectly from legal provisions such as gender-specific clauses in family, inheritance, taxation, social insurance, and other laws. These practices give rise to economywide inefficiency and a feedback effect that influences worker and family expectations at the micro level. The economy suffers from static inefficiency (lower levels of current output) as well as from dynamic inefficiency in the form of underinvestment in women's human capital and lower future rates of growth. Removing impediments to women's work is an increasingly pressing issue: As female labor-force participation rates rise, a greater proportion of the labor force is affected, and efficiency losses from the underutilization of women's labor increase.

Public policies targeting the labor market can have an effect but should not focus exclusively on labor legislation. Discrimination within the family and prior to entry in the labor market is an important determinant of gender differences in the labor market. There is often an implicit acceptance and a de jure enforcement of the requirement that it is women who bear, or should bear, the responsibility of children. Women's employment patterns are shaped by norms and legislation that revolve around the notion that women can work while bringing up their children. Public policy can address gender biases within the household by encouraging men to share equally with women the responsibility for childrearing. Absent parents should be obliged to pay child support. Maternal leave beyond what is required for a woman to recuperate after she gives birth has already been replaced by parental leave open to both parents in some countries. Such legislation needs to be supported by changes in the tax system to ensure equal treatment of individuals within the household. The marginal tax rate on earnings of additional workers in the household should be low enough to avoid creating a disincentive to "secondary" workers, who are usually women.

The persistence of segregation creates little optimism that legislative attempts can effect a timely change in long-established, gender-biased employment structures. The response of women and men to economic development is partly conditioned by prevailing norms and institutions. Social norms are not abandoned easily or quickly if they are maintained by strong sanctions. Individuals or groups may conform to them out of fear of loss of reputation rather than because such norms improve their economic well-being. There is, however, an emerging body of empirical evidence that suggests that economic development is associated with increasing competition and less-intense segregation. Institutions that may have served the economy and society well for a number of years become dysfunctional and are changed to accommodate emerging market realities. Economic development changes the relative importance of industries, technical progress affects the relative importance of occupations, and globalization of production leads to international integration that alters the structure and level of domestic labor demand and employment opportunities (Joekes, 1987). If it is true that no country can successfully industrialize without relying on a significant expansion of female labor for its labor-intensive manufacturing or services sectors (Moghadam, 1995), it is difficult to predict how gender employment differentials will be affected by these global changes. Segregation as measured by economists, however, seems to be on the decline.

References

Becker, Gary. *The Economics of Discrimination.* Chicago: University of Chicago Press, 1957.

Beechey, Veronica. *Unequal Pay.* London: Verso, 1987.

Bergmann, Barbara. "Occupational Segregation, Wages, and Profits When Employers Discriminate by Race and Sex," *Journal of Political Economy,* vol. 79, no. 2, 1974, pp. 294–313.

Blau, Francine, and Marianne Ferber. *The Economics of Women, Men, and Work.* Englewood Cliffs: Prentice-Hall, 1986.

Cohn, Samuel. *The Process of Occupational Sex-Typing.* Philadelphia: Temple University Press, 1985.

Dasgupta, Partha. "The Population Problem: Theory and Evidence," *Journal of Economic Literature,* vol. 33, no. 4, 1995, pp. 1879–1902.

Doeringer, Peter B., and Michael J. Piore. *Internal Labor Markets and Manpower Analysis.* Lexington: Heath, 1971.

Duncan, Otis, and Beverly Duncan. "A Methodological Analysis of Segregation Indices," *American Sociological Review,* vol. 20, April 1955, pp. 210–217.

England, Paula. "Failure of the Human Capital Theory to Explain Occupational Sex Segregation," *Journal of Human Resources,* vol. 17, no. 3, 1982, pp. 358–370.

Glyn, Andrew. "Inequality and Stagnation," *New Left Review,* no. 195, 1992, pp. 71–95.

Gordon, David, Robert Edwards, and Michael Reich. *Segmented Work, Divided Workers: The Historical Transformation of Labor in the United States.* Cambridge: Cambridge University Press, 1982.

Hartmann, Heidi. "The Historical Roots of Occupational Segregation: Capitalism, Patriarchy, and Job Segregation by Sex," *Signs,* vol. 1, no. 3, 1976, pp. 137–169.

Herz, Barbara, and Shahidur Khandker. "Women's Work, Education, and Family Welfare in Peru," discussion paper no. 116, World Bank, Washington, D.C. 1991.

Humphries, Jane. "The Most Free from Objection . . . The Sexual Division of Labor and Women's Work in the Nineteenth Century England," *Journal of Economic History,* vol. 47, 1987, pp. 929–949.

Humphries, Jane, and Jill Rubery. "The Legacy for Women's Employment: Integration, Differentiation, and Polarization." In Jonathan Michie (ed.), *The Economic Legacy.* London: Academic Press, 1992, pp. 236–255.

International Labor Organization (ILO). *Women in the World of Work: Statistical Analyses and Projections to the Year 2000.* Geneva: ILO, 1985.

———. *Yearbook of Labour Statistics: Retrospective Edition on Population Censuses 1945–1989.* Geneva: ILO, 1990.

Joekes, Susan. *Women in the World Economy.* Oxford: Oxford University Press for INSTRAW (United Nations International Research and Training Institute for the Advancement of Women), 1987.

Lazear, Edward, and Sherwin Rosen. "Male-Female Wage Differentials in Job Ladders," *Journal of Labor Economics,* vol. 8, part 2, 1990, pp. S106–S123.

Lim, Lin Lean. "Women at Work in Asia and the Pacific: Recent Trends and Future Challenges," paper presented at the International Forum on Equality for Women in the World of Work: "Challenges for the Future," International Institute of Labour Studies, Geneva, June 1–3, 1994.

Moghadam, Valentine. *Economic Reforms and Women's Employment in the Middle East and North Africa.* Helsinki: World Institute for Development Economics Research (WIDER), 1995.

Presser, Harriet, and Sunita Kishor. "Economic Development and Occupational Sex Segregation in Puerto Rico 1950–80," *Population and Development Review,* vol. 17, 1991, pp. 53–85

Psacharopoulos, George, and Zafiris Tzannatos. "Female Labor Force Participation and Education." In George Psacharopoulos (ed.), *Essays in Equity, Poverty, and Growth.* New York: Pergamon, 1989, pp. 266–285.

Roos, Patricia, and Barbara Reskin. "Institutional Factors Contributing to Sex Segregation in the Workplace." In Barbara Reskin (ed.), *Sex Segregation in the Workplace.* Washington, D.C.: National Academy Press, 1984, pp. 235–260.

Rubery, Jill, and Robert Tarling. "Women's Employment in Declining Britain." In J. Rubery (ed.), *Women and Recession.* London: Routledge, 1988, pp. 47–75.

Strober, Myra. "Toward a General Theory of Occupational Sex Segregation: The Case of Public School Teaching." In Barbara Reskin (ed.), *Sex Segregation in the Workplace.* Washington, D.C.: National Academy Press, 1984, pp. 144–156.

Tzannatos, Zafiris. "The Long-Run Effects of the Sex Integration of the British Labour Market," *Journal of Economic Studies,* vol. 15, no. 1, 1988, pp. 1–18.

———. "Employment Segregation: Can We Measure It and What Does the Measure Mean?" *British Journal of Industrial Relations,* vol. 28, no. 1, 1990, pp. 107–111.

———. *Potential Gains from the Elimination of Labor Market Differentials.* Regional Studies Program Report no. 10. Washington, D.C.: Latin American Technical Department, World Bank, 1991a, pp. 4.1–4.14.

———. "Reverse Racial Discrimination in Higher Education in Malaysia: Has It Reduced Inequality and at What Cost to the Poor?" *International Journal of Educational Development,* vol. 11, no. 3, 1991b, pp. 177–192.

———. Economic Growth and Equity in the Labor Market: Evidence and Policies for Developing Countries. Washington, D.C.: World Bank, 1995.

United Nations Development Program (UNDP). *Human Development Report.* New York: UNDP, 1995.

Wahyana, Juliani. "Women and Technological Change in Rural Industry: Tile Making in Java," *Economic and Political Weekly,* vol. 29, April 1994, pp. WS19–WS33.

World Bank. *World Development Indicators.* Washington, D.C.: World Bank, 1997.

See also Zafiris Tzannatos, "Women's Labor Incomes"

New Industrial Labor Processes and Their Gender Implications

Martha Roldán

Introduction

This essay revisits the interest in feminist analysis of industrial dynamics, which examines the relationship between forms of work organization and the creation of gender hierarchies. This interest is related to two late-twentieth-century intertwined developments: the implementation of flexible new industrial labor processes (NILPs) as key features of the manufacturing reconversion in progress in advanced and less developed economies, and the emergence of a distinct phase of internationalization, or globalization, of the world economy. (For the debate on the new phase of internationalization versus the globalization of the world economy, see Ruigrok and van Tulder, 1995.) This essay focuses primarily on one region, Latin America, and explicates developments in one country, Argentina.

The restructuring of the advanced industrial economies along neoliberal lines, after the Golden Age of the post–World War II period (1950–late 1960s) shows unequal diffusion of the NILPs, productive and social heterogeneity, and conflict (Coriat, 1988, 1992a; Marglin and Schor, 1990; Boyer, 1996). Simultaneously, international flows of financial capital, trade, and direct foreign investment, mainly among the same developed economies, have accelerated significantly (Hirst and Thompson, 1996). The semi-industrialized countries of Latin America have also been affected by this transformation. During the 1980s and 1990s, the application of structural adjustment programs to secure payment of the external debt was followed by neoliberal economic policies purportedly to promote a new phase of growth based on tradable goods and services. These policies, applied by the governments of the area following the recommendations of international agencies such as the World Bank, and the International Monetary Fund, among others, have not met with success. (For different views on these policies see Williamson, 1990; Azpiazu and Nochteff, 1994.) The recovery in the growth of the domestic product, financial flows, and direct foreign investment derived from those recommendations have not meant the emergence of new, stable regimes of accumulation, better levels of social homogeneity, cohesion, or social equality.

On the contrary, industrial growth and exports, if attained, are often accompanied by the disarticulation of the national productive apparatus, balance of trade and payments deficits, economic polarization, and new social exclusions, although these negative effects differ according to the mode of application of those neoliberal principles in each individual country. (See Azpiazu and Nochteff, 1994, for Argentina; Hirata et al., 1994, for a comparison of specific aspects in Mexico, Brazil, and Argentina; Pérez Sáinz, 1994, for Central America.) In the industrial sector, in particular, the search for competitiveness, based on the application of NILPs, of which the Just-in-Time/Total Quality Control (JIT/TQC) Japanese model is perhaps the best-known, has been sui generis and unequal, besides contributing to the increasing productive and social dualism of the region.

The above scenario associated with the new phase of internalization of the world economy, brings old and new pertinent questions to the theoretical-political feminist debate. The old concern with labor processes includes its reformulation around new organizational forms. What is the origin of productivity earnings and what are the mechanisms of implicit discipline characterizing the Japanese JIT/TQC model? In what measure does its implementation incorporate women into the work force as was formerly done on the Fordist assembly lines? In which firms, industrial sectors, and levels of subcontracting, under what conditions, and in what technical capacities are women being employed? To what degree do those processes bring about the eradication or re-creation of gender hierarchies at the heart of industry and through what mechanisms? What are the implications for gender dynamics outside the confines of manufacturing? What are the practices and ideologies employed by the same actors, the enterprises, the unions, and women and men workers in the above-mentioned context of economic crisis and social exclusion?

The *new* questioning connects the formulation of public policies to the goal of social equity sensitive to gender. This goal requires intervention from a feminist perspective in a broader interdisciplinary debate focusing on the new theorizations about economic growth and the new phase of internalization of the world economy that incorporate NILPs and "congenial" or associated institutional networks. The debate offers a variety of positions, such as the transition from a Fordist regime of accumulation and mode of regulation to neo- or post-Fordism (Aglietta, 1979; Lipietz, 1987, 1994; Coriat, 1988) or post-Taylorism (Stankiewicz, 1991a, 1991b); or from productive paradigms of mass production to flexible specialization (Piore and Sabel, 1984) to lean production; or between forms of social organization in manufacturing; or from Just-in-Case to Just-in-Time. The existing research about new forms of work organization and institutional networks diffused in Latin America originates from economists associated with the French School of Regulation (for a critical examination of the Regulation School, adhered to by this author, see Brenner and Glick, 1991).

That more encompassing feminist vision of engendered solidarity, comprising all working sectors, could utilize the examination of NILPs in the manufacturing micro reality as: (1) a crucial axis for thinking about the relation between these phenomena, patterns of industrial growth, and the new features of the world economy, and (2) the basis for an elaboration of alternative national, regional, and international public policies based on social ethic principles. A reference to the institutional context and regulatory changes (of the industrial, technological, educational, and labor-related public policies, among others) could then be instrumental in securing the degree of economic expansion needed for overcoming exclusionary practices and attaining engendered social equity.

This intra- and interdisciplinary conceptual reformulation is still pending in the Latin American region. In the interim, this essay presents a summary of examples of the construction of NILPs of the JIT type in Argentinean manufacturing. It explores the adaptation of JIT rationality in distinct industrial contexts and its links with gender dynamics, situating both in the context of changes in the model of industrialization, of labor systems, and of principles of societal structuration. The essay consists of five main sections. Section One sets forth in detail aspects of the evolution of feminist research in the fields of economics and sociology of labor processes. Section Two examines the evolution of the debate about the nature of the JIT/ TQC Japanese model at micro and macro social levels. Section Three places the debate in the Argentinean context: the rupture of import-substitution industrialization (ISI), its social agreements, and current industrial restructuring. Section Four presents some field-research findings. The final section elaborates some implications from micro and macro social levels in the context of increasing international heterogeneity.

Labor Processes, Gender Hierarchies, and Industrialization Patterns

How does one apply a feminist vision to the theme of labor processes? Roldán (1992, 1993a) has observed that the polemic, whether from the perspective of economics or sociology, is usually carried out in terms of the relation between capital and labor, as if the gender of the social agents would not influence the practices, representations, and implications of the organization of labor processes in manufacturing. The empirical feminist literature shows, on the contrary, that as soon as the construction of concrete labor processes is explored, the latter lose their apparent gender neutrality. Roldán (1993b), among others, found that men and women tend to participate in distinct labor processes and to utilize different physical technologies inside and outside the manufacturing sphere. Moreover, the gender division of labor becomes interwoven in the social and technical divisions of labor. Hence, each form of labor organization and technical innovation becomes engendered—that is, defined as either masculine or feminine.

Feminist reflection on the nexus between forms of organization of the industrial labor process and the creation of gender hierarchies has commonly had as an empirical referent the experience of women's incorporation into industrial manufacturing in "rigid" Taylorist/Fordist lines. These processes are characterized by fragmentation of tasks, with separation of the conception from the execution of work; specialization of functions; and the transition from "assigned" to "imposed" time of work performance according to the speed and rhythm of the assembly line. This phenomenon has been studied at all levels—from the head offices of transnational corporations (TNCs) to domestic subcontracting (about this last point, and for the Mexican case, see Benería and Roldán, 1987). In broad terms, this perspective revealed that there did not exist an unequivocal relation between industrialization, gender, and employment. Gender divisions—by which labor positions become masculine or feminine—can vary in the course of a firm's history, but the sexual hierarchies, in terms of salary, labor conditions, training, and professional promotion generally persist throughout it. Furthermore, it has been demonstrated that not only is the firm's strategy important, but union activity, the conditions of specific labor markets, and male and female workers' own perspectives and reactions are also crucial for explaining gender divisions.

It is useful to distinguish here between the perspectives of economics and the sociology of labor processes. Contributions from the economics of labor processes include Phillips and Taylor (1980), Elson and Pearson (1981), and Benería and Roldán (1987); from the sociology of labor processes, Cockburn (1983, 1985), Game and Pringle (1984), Lobo (1991), and Guzmán and Portocarrero (1988); and from the socioeconomics of labor process, Roldán (1992, 1993a, 1993b, 1994a, and 1994b). Each model of the capitalist labor process implies a distinct logic of extraction of surplus value and of economic and productive efficiency. Therefore, different types and levels of technical skills and competencies are required from production workers, who are also subject to "typical" mechanisms of discipline and control. In this fashion, the incorporation of women into rigid, Taylorist/Fordist lines obeys management's engendered economic and political logic, but trade unions and male and female workers, in different degrees, generally contribute to that construction of gender.

The logic of the enterprise has been emphasized above all by the economists, who argue that firms hire women on the basis of preexisting gender differences (so-called gender traits), making the employment of women in place of men and for identical jobs more profitable. This is because the cost of feminine labor is lower than masculine labor, in terms of salaries, work incentives, and social-security costs, because women are more productive (the "nimble fingers" argument), or both. The gendered economic rationality is complemented by the political logic, or logic of control. The women, by their presumed or real docility, patience, and lack of union experience, constitute, at least in the enterprise's ideology, a more easily manipulable labor force than their masculine counterparts.

The advantages for men in firms and for the enterprise itself that derive from the subordinate incorporation of women have been emphasized mainly by the feminist sociologists, who focus usually on the logic of discipline and control of the Taylorist/Fordist lines. With differences of emphasis and nuance depending on author, this disciplinary approach elaborates the fundamental thesis of Braverman (1974) about a necessary relation between the "scientific management" characteristic of Taylorism, the separation of conception from execution of work, the fragmentation of tasks, and the deskilling and control of labor. The thesis becomes feminist through the introduction of patriarchy or gender relations, and, therefore, masculine interests, into the analysis of the relations between capital and labor.

The most important finding from these studies is the discovery of gendered subjects operating within the confines of the workplace, thus overcoming the theoretical barrier derived from the conception of unequivocal agency based on social class. The manufacturing sphere is redefined as a crucial setting in which gender relations are constructed and renegotiated, and in which the expressed interests are not uniquely those of capital, but also of the dominant masculine gender.

This conceptual progress finds specific limits in the implementation of NILPs—that is, when it focuses on the unit production/valorization of capital. The pioneering sociological studies do not analyze the transformations of the organization of labor processes, either at the theoretical or the empirical level. This presents a problem when the logic of the economy of time (i.e., there is no idle time) and control (Coriat, 1988), typical of each production model, changes and, with it, the forms of the regulation in which they are embedded at the mezzo and macro levels. This is the case with the new forms of JIT/TQC labor organization, or the Japanese model, that is considered next.

The Debate over the Japanese JIT/TQC Model

There is no unitary vision of the JIT/TQC labor processes: The representations vary according to academic discipline and the theoretical perspective of the particular researcher. This summary is based primarily on principles of microeconomic sociology.

In a broader sense, the label of the JIT (Just-in-Time) with TQC (Total Quality Control) system commonly refers to a combination of interdependent practices related to a form of organization of the industrial labor process for mass production in small batches; supplier and customer relations (subcontracting linkages); and commercial networks. The classical example of the application of JIT as a production concept is assembly-line work aimed at instantaneous production with a minimum "waste" and, ideally, perfect quality.

JIT production refers to work that is done only when necessary, in the necessary quantity, and at the moment in which it is necessary. The object is to eliminate all superfluous resources in the process of production, to implement the idea of "lean" production by means of the reduction of stock, such as raw materials, components, labor, and so forth, to the least amount necessary to meet unforeseen situations. That goal combines with *kaizen*, the constant search for new forms of eliminating superfluous expenses and increasing quantity and quality of the production. JIT processes imply a continuous improvement. The crucial point of TQC is a despecialization of the function of quality control: In this model quality control is accomplished with production itself, eliminating the need for quality inspectors. JIT production is more efficient when flows of materials are simple and unidirectional. This is generally accomplished through production layouts that require multiskilled (or polyvalent) and/or multitasked workers who rotate operations and tasks to carry out preventive maintenance and quality control.

Economic Rationality of the Japanese System

What are the advantages associated with the implementation of JIT, its institutional foundations, and its implications? The most central economic ones are: (1) a more efficient use of capital by drastically reducing the levels of inventories and the costs associated with storage and materials handling, and (2) increases in productivity and quality. Productivity gains under Taylorist/Fordist systems derive from an extreme form of division of labor, plus mechanization. In contrast, productivity gains and higher quality under JIT systems originate from the reduction of horizontal and vertical divisions of labor (hence, the need for multiskilled workers); from forms of workers' involvement; and from the application of the principle of *kaizen*. A multiskilled workforce allows the firm to become more efficient through collective learning and adaptation to changing circumstances. Task rotation reduces the porosity of the working day and increases workers' output, which is much higher than in typical Western factories. Production workers in JIT systems are required to assume responsibility and to mobilize their capacities in pursuit of higher productivity and quality goals in naturally cooperative ways (Coriat, 1992).

Discipline and Labor Control

In broad terms, JIT practices require a much higher degree of worker involvement with the firm than is typical of Taylorism/Fordism. Such involvement is sustainable only if workers and managers identify strongly with the company. This requires low rates of labor turnover and a process of employee "incubation" that can last several years. The incentive structures firms develop to ensure workers' involvement include benefits covering health and housing plans; systems of pay according to seniority and, in some cases, participation in profit sharing; and the guarantee of job security for life for core workers. JIT techniques, in turn, are imposed and reproduced through a battery of control mechanisms, some of which are embedded in the same labor process through *kaizen* and teamwork, while others range from company unions' compliance to green labor recruitment, an extremely rigorous selection procedure seeking workers without a history of trade-union militancy or labor struggles, recently graduated from high school or university, and preferably from rural areas. All of these factors make workers more vulnerable to informal and paternalistic forms of control. Group pressure also imposes discipline on workers. Work teams control their own members even to the point of regulating absenteeism (Fucini and Fucini, 1990).

Social Agreements, Productive Dualism, and Labor-Market Segmentation

New theorizations about growth and the new phase of internalization of the world economy show the relation between productive systems and the mode of regulation that makes this growth possible. The JIT system is associated with a peculiar institutional frame different from that of Taylorism/Fordism:

> The institutions for the representation of the interests of labor in Toyotism seek to secure conformity and obedience in cases in which one cannot achieve authentic consensus. The idea of adversarial interests between employees and employers and the notion of negotiating conflicts openly at the level of industrial relations are foreign to the system; in contrast to this, the institutions and procedures of Taylorism/Fordism suppose the possibility or inevitability of conflicting interests between capital and labor (Jurgens et al., 1993, p. 50).

It is not accidental that JIT was established in Japan following the defeat of trade unions and working-class struggles in the 1950s. As a system of work organization, it is based on the assumption of cooperation and commitment between the enterprise and its employees that is reflected in a level and content of "social agreements" different from Fordism (Coriat, 1992). According to Lipietz (1994):

Bargaining under Toyotism is not bargaining about the distribution of the general growth in productivity to the whole society. Distribution affects only those employees who have a right to the quasi-rents that belong to a given firm—the labor aristocracy club. It is a very particular form of negotiation *where there is not a generalization of the distribution of income*. It's evident that this creates macro-economic problems because there are no systematic forms of redistribution of increases in productivity to the whole society (Lipietz, 1994, p. 52; italics mine).

It is not surprising, then, that the literature indicates that the JIT/TQC model is based on a dual economy and contributes, at the same time, to the consolidation of a segmented labor market, in which it is estimated that less than 30 percent of the Japanese labor force is in the category "privileged," with high wages and stable employment. In the same way, the acceptance of the principle of *kaizen* and elimination of waste explains why JIT labor systems do not favor high levels of domestic employment, at least in central firms. At peak periods, production targets are achieved by using various forms of numerical flexibility, compulsory overtime, temporary employment, and certain types of subcontracting. The security of the workers in central firms, primarily men, is based on the insecurity of the many workers with temporary contracts in large companies or in smaller subcontracting firms.

In general, the theories of the JIT complex ignore the gender dimensions of the Japanese system. Characteristics such as employment for life and social benefits based on the enterprise accrue only to central workers (men) in large corporations. The objective of obtaining low absenteeism and responsible workers is achieved by recruiting personnel who are not distracted by domestic obligations. This, in the context of a patriarchal society, translates into a preference for male workers. Women are excluded from central Japanese firms and typically are concentrated in the peripheral labor force. However, women are not necessarily excluded from Japanese firms abroad and are hired by the automotive firms in Japan to a limited degree, together with Korean immigrants, when young Japanese men refuse to work in assembly line jobs.

Evaluation of the Japanese Model

Microsocial Level. The nature of the NILPs at a micro-manufacturing level constitutes a contested terrain. Two interpretations compete in the field of theory and prescriptions for related policies.

The first is what I call the "benevolent vision" of the new labor processes (Wickens, 1987; and some of the economists of the Regulation School and the majority of the articles included in *IDS Bulletin,* 1993). According to this perspective, the restructured capitalist relations between capital and labor should be cooperative, assuming that only workers with high levels of skills and in control of their work will guarantee the efficient and high-quality production necessary to achieve competitiveness in demanding world markets. Some of the characteristics already enunciated seem to support this view: the reduction of vertical and horizontal divisions of labor; the demand for high levels of formal education for assembly-line workers; the rotation of tasks; and workers' polyvalence, teamwork, and involvement with the firm. In this scheme, investment in flexible automation is taken for granted.

Second, we have the conflictive perspective. This position, supported by the workers' own perception and by some researchers who echo the view of this sector, rejects the supposed emancipating nature of JIT practices, and questions their origins and corollaries in terms of employment, skills, and workers' control (Dohse et al., 1985; Tomaney, 1990; Brenner and Glick, 1991; Jurgens et al., 1994). The criticism of the JIT system is corroborated by the perceptions of the workers themselves (Kamata, 1972; Parker and Slaughter, 1988; Garrahan and Stewart, 1992). Hyper-Taylorism (i.e., exaggerated Taylorism), management by stress, and similar concepts constitute the crucial points of workers' representations of structural restructuring.

Meso- and Macrosocial Levels. Questions about the nature of the Japanese model must also be related to meso and macro institutional levels to fully realize its potential implications when applied abroad. According to the Regulation School, during the Golden Age of capitalist expansion following World War II (from the 1950s to the mid-1970s), and with national variations, Taylorist/Fordist systems were associated with a mode of development called Fordism, which was characterized by mass production and mass consumption. The latter was supported by a regulatory framework that ensured that mass production could be absorbed by mass consumption and that inaugurated a continuous cycle of productivity increases, which were reflected in wages of workers in Fordist industries, and general income redistribution. In this scheme, the welfare state and the coding of work/labor relations by means of "social agreements" played a crucial role (Coriat, 1992). (For characteristics of the Fordist regime, see Coriat, 1988, 1992; Lipietz, 1994; for discussion of peripheral Fordism or sub-Fordism, see Lipietz, 1987, 1994.) This Fordism took on distinct characteristics with national variations in the advanced industrialized countries in North America and Europe, and sui generis varieties arose in the periphery, Brazil, Mexico, Argentina, Uruguay, Colombia, during the period of import-substitution industrialization (ISI).

One such national case is the collective bargaining system in the United States. During the period 1948–1979, collective agreements between labor and management in the Big Three automotive factories followed the Fordist setup, becoming one of its basic pillars. This was accomplished to the extent that its norms, based on contractual agreements at firm and branch levels, sought a true Keynesian policy of sustained demand that was constituted at micro and meso levels (Coriat, 1992a). To this effect, rules concerning the determination of direct salaries, with yearly automatic increases adjusted to inflation and unemployment insurance, were extended by means of "target agreements" to the whole industry. This process led to a strong homogeneity and uniformity of work conditions and salaries among all workers of a given industry and, in so doing, diminished competition among them, even though not all branches of industry and all categories of workers were similarly covered (Coriat, 1992a).

Toward a New Feminist Reflection on the Institutional Field

Features of JIT-related institutional arrangements in Japan that run counter to some major features of Western social agreements include: company unions operating in an assumed "naturally cooperative and harmonious" environment; worker involvement sought through incentive schemes applying to each individual company only; lack of a significant macroeconomic role for unions leading to income redistribution; and the furtherance of productive dualism and labor market segmentation (Coriat, 1992a). This raises two questions: What regime of accumulation and what type of society can adopt and, at the same time, be influenced by the Japanese system? What are the consequences and implications of its broad acceptance upon gender? Up to the mid-1990s, feminist research had not addressed these dimensions.

Nevertheless, it is crucial to reflect on the forms of productive dualism and labor-market segmentation associated with the Japanese model and on its adaptations in the context of contemporary restructuring at the world level. Not by chance, local versions of the Japanese work rules could be imposed in the United States and in other countries only in the 1980s, a period characterized by growing economic crisis and unemployment. The reforms of this decade changed the above-described Fordist social agreements. They also helped advance present-day dualisms according to industrial sector and company, while "contributing to segment and to differentiate the living conditions and reproduction of different categories of workers who occupied the national space" (Coriat, 1992a, p. 219). The new labor contracts constitute an example of defense bargaining, whereby wage laborers are made to pay the price of adjustment, and benefits obtained are precarious: some professional training, along with minimum unemployment insurance and employment defense for privileged workers. This low-level transition also took place, with national features, in other developed countries in the West, the United Kingdom and France being leading examples (Coriat, 1992a; Lipietz, 1994). Germany and Sweden, on the other hand, took a different path, with union arrangements at branch and macro levels, respectively.

Therefore, it is important that the analysis of new engendered labor processes be situated in the context of the political economy of each country concerned and especially in the context of institutional networks that facilitate or obstruct the implementation of those same systems. From this perspective, the dismantling of ISI and its pseudo-Fordist social agreements in Latin America can be seen as one possible, though not necessarily the only, route for the "hybridization" of the Japanese model in the region.

The Case of Argentina

Since the end of the 1970s, Argentine society has experienced fundamental transformations linked to the dismantling of the import-substitution model and its regulatory framework, including the sui generis Fordist social agreements that promoted growth with socioeconomic homogeneity. During the 1976–1990 period, the implementation of structural-adjustment policies and selective deregulation of the economy favored industrial concentration and the growth of transnational corporations and large domestic economic groups (Azpiazu and Nochteff, 1994). Simultaneously, these measures gave rise to the productive heterogeneity characteristic of the late twentieth century, even though they did not modify totally the regulatory scheme of the import-substitution phase.

Kosacoff (1993) synthesizes the industrial transformations in place in Argentina in the early 1990s as a process of regressive restructuring and growing structural heterogeneity. The regressive nature is attributed to two factors: (1) the inability of the industrial sector to rescue the positive contribution of the previous phase—engineering capabilities, knowledge and skills, and human resources—and the transfers of income derived from the reconversion process without generating any dynamic comparative advantages by the enterprises that benefited from this process; and (2) the concomitant deterioration of health, housing, education, and basic services that threatens social equity and systemic economic competitiveness. The growing heterogeneity is produced by distinct practices (such as those related to product design, process innovations, and overall functional reorganization of firms, from top management to shopfloor workers in sub-contracting chains). There coexist the backwardness and stagnation of many firms, with the growth and modernization of a few others. While it is true that empirical evidence of successful microeconomic performance is abundant, Kosacoff (1993) concludes that it has not been powerful enough to define a new path of growth for the economy as a whole. The economy, in turn, is becoming increasingly based on the service sector, particularly in low-paying activities.

What has happened in the meantime within the system of labor relations inherited from the import-substitution industrialization stage? During its apex, centralized collective bargaining fulfilled an essential role in the distribution of income. After much attack by the military government, only in 1988—under a democratic regime—did the labor unions recover; but, at that time, they struggled more for political gain than for better salaries, a tendency that reflects the weakening of workers relative to new actors such as businessmen's associations and the important presence of service-sector unions.

In the period 1990–1995, the implementation of structural-adjustment programs to honor external commitments, and the removal of barriers to the full and free operation of markets, would create, according to government sources, the conditions required for a new stage of accumulation. To overcome the recession of the late 1980s, economic agents would need to invest and restructure in accordance with standards applying in global markets. To this purpose, substantial changes in economic and labor regulation were introduced (Azpiazu and Nochteff, 1994). However, the government's expectations have not been fulfilled. Growth in the gross national product (GNP) in the 1991–1994 period and the reduction of inflation are well-known facts, but these indicators cannot be taken as signs of the emergence of a new, stable regime of accumulation based on tradable goods and services.

Changes in the regulatory framework have accompanied these trends, including labor legislation concerning employment, right-to-strike norms, workplace accidents, health plans, and collective bargaining regulations. Trade unions in the mid-1990s are allowed to bargain collectively at more decentralized—even plant or company—levels, and negotiations over wage increases can take place only in case of productivity increases. Trade unions in the mid-1990s are allowed to bargain collectively at more decentralized, even company or plant levels and negotiations over wage increases can take place in case of productivity increases only. But unions have lost their role at the macroeconomic level; they are no longer capable of influencing the dynamics of the regime of accumulation or of reproducing the high degree of socioeconomic homogeneity that supported their action during the ISI stage. On the contrary, there is a clear trend toward social polarization, expressed in increasing unemployment and precarious living conditions for broad segments of the population.

There are no precise data about the degree of diffusion of the Japanese model within Argentine industry, nor is there information about its possible ties with the dominant gender division of labor. The available evidence from various studies (Novick, 1991; Kosacoff, 1993; Roldán, 1993a, 1993b, 1994a, 1994b) suggests that practically the entire manufacturing sector underwent modification in the organization of work, even though there are few, if any, "pure" examples of the JIT/TQC technologies discussed previously. Research findings by Roldán (1994a, 1994b) indicate the existence of a continuum of engendered organizational forms that spans the spectrum from "high" or dynamic JIT to crisis JIT with different gradations in between. Only TNCs and large domestic firms competing in regional or world markets take the high route.

The key question to be considered is: What are the connections between the new labor-process dynamics and gender hierarchies? This can be explored by considering skill requirements and mechanisms of discipline and control. Some examples are available from the metal-mechanic sector, comprising, among others, the automotive, steel, metallurgical, and electronics industries. The firms observed in a study carried out in Argentina were classified into five categories: masculine (those with more than 95 percent male production personnel), predominantly masculine (75–95 percent male), mixed (25–75 percent male or female), predominantly female (75–95 percent female), and female (more than 95 percent female) (Roldán, 1993a and 1994a).

The Nature of Work and Gendered Skills

The optimistic scenario regarding the necessarily harmonious relations between capital and work is based on the assumption that there is a necessary tie between the reduction of vertical and horizontal divisions of labor, high levels of skills and control by labor, and efficient, high-quality production. No empirical evidence could be found to support this thesis in the work practices categorized as male or female. On the contrary, dominant restructuring trends suggest the slow formation of a multiskilled, largely masculine working class with its own internal segmentations made up of a male center (with higher-level technical polyvalence) and male and female multiskilled and/or multitasked peripheries.

The Male Center and Periphery. In broad terms, the firms continue to apply the norms of gender composition prior to the reconversion, be they male or predominantly male (in the steel, metallurgical, and autopart examples), but there also are examples of masculinization of some female-mixed enterprises. Regarding skills, the evidence does not support the benevolent view of the nature and level of technical polyvalence, despite a reduction in the vertical and horizontal divisions of labor. This general assertion must be qualified according to whether the firm observed approaches high-level JIT (i.e., firms with clear growth strategies and competitiveness in external markets) or crisis JIT (those that have adopted only "common sense" equivalents of the JIT model as a means of survival). The assertion also varies by industrial subsector, section observed (with or without use of new information technologies), and degree and type of divisions of labor being established, they themselves in constant flux (Roldán, 1994a, 1994b). Observed patterns in car, auto-part, and metallurgical firms, however, show the opposite tendency: the slow erosion of old craft skills that gave senior workers a higher degree of control under preexisting work arrangements, with no simultaneous re-creation of complex skills.

At first glance, the firms under study show a high degree of cooperation and social peace, with the workers obeying the new work rules, including additional responsibility for quality control and involvement, while at the same time guaranteeing greater efficiency and productivity. But this success has been attained without the re-creation or recomposition of high-level technical skills and without assigning workers greater control. Hence, the conflict perspective seems more appropriate to define the emergent configurations.

The Female Periphery. Women are not necessarily excluded from enterprises that adopt JIT/TQC. The experience of several firms in the metal-mechanic, automotive, and auto-part industries illustrates how the preexisting gender composition of the work force, the conditions of the labor market, and management's own evaluation of the trade-offs between the advantages and disadvantages derived from hiring women (retraining costs and real or presumed male staff or union opposition) are crucial for women's placement in the labor force. Two automotive firms located in an interior city of Argentina, one established during the 1950s and the other established in 1992, which asserts its strict endorsement of JIT principle, are illustrative. In the older firm, there is a process of masculinization, continuing the tendency present before the reconversion. The former jobs of women have been transferred to smaller subsidiary enterprises (module assembly) or have been masculinized as part of the new multiskill emphasis. The new firm has preferred to hire women in its JIT cells, creating a mixed composition of men and women in production; but the higher-level technical positions continue to be male. In the electronic assembly line, the female composition prior to the reconversion has not been modified.

Several factors favor the masculinization—or the very selective integration of women for specific sections or cells—of the firms reportedly working toward high-level JIT. For one thing, the requirement of a trained labor force with a low level of absenteeism and the absolute readiness to work overtime without advance notice and during weekends favors hiring men. This allows firms to avoid the vagaries of women's reproductive and domestic responsibilities, as well as the conflict among workers and labor unions regarding mixing men and women in productive spheres. Protective legislation for women in effect operates against women. Finally, in a context of economic crisis, unemployment, underemployment, and the absence of union protection, management can subject skilled and flexible male workers to minimum wages and strict control without fear of immediate problems.

Much the same can be said regarding the nature and technical level of newly defined female jobs; a conflict perspective seems to apply here too. However, the multiple skills of men and women are not strictly comparable because hierarchies are re-created through a two-tier flexibility (i.e., work tasks with two degrees of flexibility) within cells and pseudo-JIT lines. Among auto-part firms, signs of feminization of multitask cycles can be observed. Women are incorporated into some JIT cells and lines. However, men and women seldom perform similar routines. Women's routines (or work cycles) often require fewer skills than those performed by men. Also women keep jobs that may be skilled but that are not recognized as such by collective agreements still in force. High productivity, for instance, may be disguised as a gender trait, a domestic skill that does not deserve to be paid. In short, the fact that women may constitute a task-rotating personnel in high-level JIT firms is not synonymous with high-level multiskills and concomitant labor process control.

The restructured processes evince one constant: the re-creation of gender asymmetries within industry. In this manner, even though the patterns are complex and the gendering of JIT cannot be understood as a single dynamic, the experiences of the observed firms suggests that, inasmuch as companies implement high-level JIT forms—which demand high-level technical polyvalence, including preventive maintenance (keeping machines and tools in good shape), quality control, and other functions, besides the basic productive tasks—women are losing the comparative advantage they held previously as a result of their low wages, high productivity, and docility. Their lack of technical education, along with their need for additional training and its costs, presents a significant obstacle for the placement of women in higher-level positions.

Given the obvious lack of fit between the gendered labor reality and theory, it is appropriate to ask: How is capital-labor accommodation attained? A key element, highlighted by managers as well as workers at the firms studied, is the weakness of the labor movement in the face of unemployment. In fact, from the perspective of male and female workers alike, JIT restructuring depends upon the reinforcement of regressive labor practices compared to those prevailing during the ISI stage. The incentives offered by firms in terms of wages, training, and labor stability are minimal given the changes in labor legislation discussed previously. Variation exists, of course, depending on the type of enterprise, its policy regarding human resources, and the agreements that it may have signed with its union regarding salary increases and other bonuses for increased productivity.

The firms studied are employing a number of strategies to ensure the obedience, if not the consensus, of their production personnel. The mechanisms for technical control through mechanization and automation and the "personal" types of control through patriarchal authority and actual and potential sexual harassment are important. The experience of some of the firms also shows how ideological persuasion and the expensive development of a company's ideology can become a sine qua non for the achievement of worker involvement. In the larger firms and those more likely to invest in the "gestation" of their work force (awareness-raising talks by consultants and supervisors), one can see a wide divergence in the world visions of the enterprises and those of its workers derived from their own experience in the JIT line. However, with increasing unemployment, fear of being laid off is a sufficient force to secure workers' engagement—a far cry from a harmonious system that might justify the label of "negotiated capitalism," used by the French School of Regulation.

Conclusions

While the shift to new labor processes has benefited a segment within the manufacturing sector, it seems to have acted to the detriment of most workers. Only a minority of men and women are associated with a successful restructuring. New principles for societal reorganization on the basis of a heterogeneity within emerging polarization seem

to be the distinguishing features of the contemporary trends.

Current industrial restructuring in Argentina shows numerous paths that fit neither the traditional Fordist model nor the classical Japanese model. Kosacoff's (1993) assertion that there is an increasing heterogeneity in the industrial sector is supported in our case-study situation. From the perspective of labor processes, this heterogeneity is associated with the industrial branch (particularly with car, auto-part, steel, metallurgical, and electronic sectors): the size of, and resources available to, each enterprise; the type of product to be manufactured; its market (national or international); and the type of competitive strategy regarding costs or quality.

Regarding female participation in these new labor processes per se, the choice the firms make in terms of competitiveness based on quality or simple price of their production and training costs is important. But gendered patterns and hierarchies of the past are resilient. Thus, the re-created gender asymmetries are largely congruent with older norms. But the context has changed substantially, with destroyed social agreements and overall working-class fragmentation. As in the past, the firm's decision regarding the gender composition of its work forces is paramount. But under the new processes, the firms operate with total freedom concerning the masculinization and feminization of their personnel, without male agreement at the individual or the union level. Local labor-market conditions can be an important force to facilitate workers' demands, such as for greater technical education for women. But in the JIT scenario, female workers have lost the capacity to negotiate, not so much because they are women but because they are part of a fragmented and weakened working class.

The question that can be asked is: What type of industrialization do we want? JIT systems demand a change in the norms of work to permit the maximum and free use of labor power and to secure workers' involvement. Moreover, it is difficult to identify a single instance of JIT practices equally beneficial for male and female workers. How, then, to reconcile the interests of the firms with those of the workers? What types of nationally regulated, equity-oriented industrialization are possible in the new phase of internationalization of the world economy? Finally, we must ask ourselves whether a new type of social sciences—interdisciplinary in nature—can help narrow the gaps between countries and regions and engender new utopias that might lead the way to truly emancipatory work practices.

References

Aglietta, Michel. *A Theory of Economic Regulation: The U.S. Experience.* London: New Left Books, 1979.

Azpiazu, Daniel, and Hugo Nochteff. *El Desarrollo Ausente: Restricciones al Desarrollo, Neoconservadorismo y Elite Económica en la Argentina: Ensayos de Economía Política.* Buenos Aires: Flacso/Tesis, 1994.

Benería, Lourdes, and Martha Roldán. *The Crossroads of*

Class and Gender: Industrial Homework, Subcontracting, and Household Dynamics. Chicago: University of Chicago Press, 1987.

Boyer, Robert, and Daniel Drache (eds.). *States Against Markets: The Limits of Globalization.* London: Routledge, 1996.

Braverman, Harry. *Labor and Monopoly Capital: The Degradation of Work in the Twentieth Century.* New York: Monthly Review Press, 1974.

Brenner, Robert, and Mark Glick. "The Regulation School and the West's Economic Impasse," *New Left Review,* no. 188, July/August 1991, pp. 45–119.

Cockburn, Cynthia. *Brothers, Male Dominance, and Technological Change.* London: Pluto, 1983.

———. *Machinery of Dominance: Women, Men, and Technical Know-How.* London: Pluto, 1985.

Coriat, Benjamín. *El Taller y el Cronómetro: Ensayo Sobre el Taylorismo, el Fordismo y la Producción en Masa.* Mexico City: Siglo XXI Editores, 1988.

———. *El Taller y el Robot: Ensayos Sobre el Fordismo y la Producción en Masa en la Era de la Electrónica.* Mexico City: Siglo XXI Editores, 1992a.

———. *Pensar al Revés: Trabajo y organización en la empresa Japonesa.* Mexico City: Siglo XX Editores, 1992b.

Dohse, Knuth, Ulrich Jurgens, and Thomas Maisch. "From 'Fordism' to 'Toyotism'? The Social Organization of the Labor Process in the Japanese Automobile Industry," *Politics and Society,* vol. 14, no. 2, 1985, pp. 115–146.

Elson, Diane, and Ruth Pearson. "The Subordination of Women and the Internationalisation of Factory Production." In K. Young, C. Wolkowitz, and R. McCullagh (eds.), *Of Marriage and the Market: Women's Subordination in International Perspective.* London: CSE Books, 1981, pp. 144–166.

Fucini, Joseph, and Suzy Fucini. *Working for the Japanese.* New York: Free Press, 1990.

Game, Ann, and Rosemary Pringle. *Gender at Work.* London: Pluto, 1984.

Garrahan, Philip, and Paul Stewart. *The Nissan Enigma: Flexibility at Work in a Local Economy.* London: Mansell, 1992.

Guzman, V., and P. Portocarrero. *Una Nueva Mirada, Crisis, Mercado de Trabajo e Identidad de Género.* Lima: Centro de la Mujer Peruana Flora Tristan, 1988.

Hirata, Helena, Michel Husson, and Martha Roldán. "Restructurations Productives et Changements dans la Division Sexuelle du Travail et de l'emploi: Argentina, Brésil et Mexique." In *Amérique Latine Démocratie et Exclusion Futur Antérieur.* Paris: Harmattan, 1994.

Hirst, Paul, and Grahame Thompson. *Globalization in Question.* Cambridge: Polity Press–Blackwell, 1996.

Humphrey, John (ed.). Institute of Development Studies Bulletin. *IDS Bulletin,* vol. 24, no. 2, April, 1993.

Special Issue: "Quality and Productivity in Industry: New Strategies in Developing Countries."

Jurgens, Ulrich, Thomas Maisch, and Knuth Dohse. *Breaking from Taylorism: Changing Forms of Work in the Automobile Industry.* Cambridge: Cambridge University Press, 1993.

Kamata, Sutoshi. *Japan in the Passing Lane.* New York: Pantheon, 1972.

Kern, Horst, and Michael Schumann. "Limits of the Division of Labour: New Production and Employment Concepts in West German Industry," *Economic and Industrial Democracy* (Sage, London), vol. 8, no. 2, 1987, pp. 151–170.

Kosacoff, Bernardo. "La Industria Argentina: Un Proceso de Reestructuración Desarticulada." In Bernardo Kosacoff (ed.), *El Desafío de la Competitividad: La Industria Argentina en Transformación.* Buenos Aires: Comisión Económica para la América Latina/Alianza Editorial, 1993, pp. 11–67.

Lipietz, Alain. *The Crises of Global Fordism.* London: Verso, 1987.

———. *El Posfordismo y sus Espacios: Las Relaciones Capital/Trabajo en el Mundo.* Serie Seminarios Intensivos de Investigacion. Documento de Trabajo no. 4. Buenos Aires: Facultad Ciencias Economicas, Universidad de Buenos Aires, 1994.

Lobo, Elizabeth. *A Classe Operária Tem Dois Sexos: Trabalho, Dominação e Resistencia.* Sao Paulo: Editora Brasiliense, 1991.

Marglin, Stephen, and Juliet Schor (eds.). *The Golden Age of Capitalism: Reinterpreting the Postwar Experience.* New York: Oxford University Press, 1990.

Novick, Marta. "Nuevas Tecnologías de Gestión y Acción Sindical: Métodos Japoneses de Producción en la Industria Argentina," *Estudios del Trabajo* (Buenos Aires), no. 1, 1991, pp. 77–111.

Parker, M., and J. Slaughter. *Choosing Sides: Unions and the Team Concept. A Labor Notes Book.* Boston: South End, 1988.

Pérez Sáinz, Juan Pablo. *El Dilema del Nahual: Globalización, Exclusión y Trabajo en Centroamérica.* San Jose: FLACSO, Programa Costa Rica, 1994.

Phillips, Anne, and Barbara Taylor. "Sex and Skill: Notes Towards a Feminist Economics," *Feminist Review,* vol. 6, 1980, pp. 79–88.

Piore, Michael, and Charles Sabel. *The Second Industrial Divide.* New York: Basic Books, 1984.

Roldán, Martha. "El Debate Sobre Procesos de Trabajo, Crisis y Reestructuración Industrial en los 90: Hacia una Nueva Representación Androcéntrica de las Modalidades de Acumulación Contemporáneas?" *Estudios del Trabajo* (Buenos Aires), no. 3, 1992, pp. 85–124.

———. "Industrial Restructuring, Deregulation, and New JIT Labour Processes in Argentina: Towards a Gender-Aware Perspective?" *IDS Bulletin,* vol. 24, no. 2, April, 1993a, pp. 42–52.

————. "Nuevos Desafíos a la Teoría y Práctica de la Investigación Sociológica Feminista en la Década de los Noventa." In GRECMU (ed.), *Mujeres y Trabajo en America Latina.* Madrid: Iepala Editorial, 1993b, pp. 27–79.

————. "Critical JIT Restructuring in a Cluster Context: Autopart Manufacturing and Gender Relations in Vittoria City, Argentina," working paper, International Labor Organization, Geneva, 1994a.

————. "Un Debate Pendiente: Reconversión Industrial, Desregulación y Nuevos Procesos de Trabajo Flexibles en la Contexto Latino Americano de los 90: Hacia una Perspectiva Sensible al Género?" In B. Bustos and G. Palacio (eds.), *El Trabajo Femenino en America Latina: Los Debates en la Década de los Noventa.* Guadalajara: Instituto Latinoamericano de Servicios Legales Alternativos, Universidad de Guadalajara, 1994b, pp. 101–137.

Ruigrok, Winfried, and Rob van Tulder. *The Logic of International Restructuring.* London: Routledge, 1995.

Stankiewicz, Francois. "Las Estrategias de las Empresas Frente a los Recursos Humanos." In Francois Stankiewicz (ed.), *Las Estrategias de las Empresas Frente a los Recursos Humanos: El Post-Taylorismo.* Buenos Aires: Humanitas, 1991a, pp. 21–62.

Stankiewicz, François (ed.). *Las Estrategias de las Empresas Frente a los Recursos Humanos: El Post-Taylorismo.* Buenos Aires: Humanitas, 1991b.

Tolliday, S., and Jonathan Zeitlin. "Introduction: Between Fordism and Flexibility." In S. Tolliday and J. Zeitlin (eds.), *The Automobile Industry and Its Workers: Between Fordism and Flexibility.* Cambridge: Policy, 1986.

Tomaney, John. "The Reality of Workplace Flexibility," *Capital and Class,* no. 40, 1990.

Wickens, Peter. *The Road to Nissan.* London: Macmillan, 1987.

Williamson, J. "What Washington Means by Policy Reform." In J. Williamson (ed.), *Latin American Adjustment: How Much Has Happened.* Washington, D.C.: Institute for International Economics, 1990.

Women and Home Work

Elisabeth Prügl

Introduction

At the beginning of the twentieth century, the term "home work" conjured up images of tired and overworked women cowering over their sewing machines in crowded and filthy apartments, surrounded by sick children who helped out with simple tasks. This image of home workers informed campaigns and social policies in Europe and North America to combat "the evils of industrial home work," which were thought to defile motherhood, undermine the achievements of the working class, and spread germs to consumers. At the time, it was widely understood that home work was a remnant from a previous mode of production and that it was on its way toward extinction. Yet, almost a century later, home work thrives. According to the International Labor Organization (ILO), home work is not only widespread but probably on the increase in countries around the world (International Labor Conference, 1994). New images of home workers have appeared in a new historical environment. In addition to being portrayed as exploited, disguised wage workers, home workers appear in the public imagination as artisans who maintain national traditions, microentrepreneurs devising creative survival strategies, and professionals taking advantage of new technologies to escape the routines of office work.

Since the identity of home workers is very fluid in rhetoric of the 1990s, this essay provides a lengthy discussion of definitions and identities. I argue that the category "home work" is meaningful if it provides a basis for empowering practice, and I use the term in this sense. Statistical evidence from several countries indicates an increase of home work since the 1980s. Home work fulfills a distinct function in a political and economic environment where production processes are restructured to achieve cost savings and flexibility and labor market regulations are being dismantled. Home workers may be the prototypical workers of the future. They are organizing, and their struggle for protection may provide a gauge of things to come.

Questions of Definition

Home work is commonly defined as work carried out at home for pay. In addition to industrial home workers and artisans in a putting-out system (i.e., artisans who sell their product to a merchant who provides raw materials), this definition has allowed some surveys to include skilled professionals, such as consultants, doctors, and lawyers; those who work from home as a base, such as real-estate agents; and those who take work home from the office, such as executives and teachers. For example, government surveys in the United States and Great Britain have included these workers in their statistics, seeking to gauge futuristic predictions of work moving back into the home. But for the majority of home-based workers, this definition is not particularly meaningful because it lumps together a group of very privileged professionals with a group of extremely vulnerable workers. A category called "home worker" defined in this way casts its net too widely and, therefore, cannot provide a basis for political action. A definition of "home work" is useful only if it makes sense from the perspective of those included in the definition and if it provides a guide for empowerment.

In most countries around the world, home workers occupy a vulnerable position in the economy. They are largely unskilled and concentrated in manufacturing or in the informal sector. The ILO Convention on Home Work of 1996 more accurately captures this phenomenon:

"Home work" means work carried out by a person, to be referred to as a home worker,

(i) in his or her home or in other premises of his or her choice, other than the workplace of the employer;

(ii) for remuneration;

(iii) which results in a product or service as specified by the employer, irrespective of who provides the equipment, materials or other inputs used, unless this person has the degree of autonomy and [of] economic independence necessary to be considered an independent worker under national laws, regulations, or court decisions.

The definition adds to an emphasis on the place of work the criterion of economic dependence. Home workers work for an employer (who could be an intermediary) and are dependent on this person for their livelihood. They have little opportunity for profit or loss and rely on the work giver for access to raw materials and to the market. Thus the definition captures the unequal power relationship between home workers and their work givers and becomes politically relevant.

Commonly, a distinction is made between industrial home workers and crafts producers or artisans. Industrial home work is integrated into an industrial organization of production with a detailed division of labor in which a merchant or manufacturer provides designs and workers mechanically execute the designs. In contrast, crafts producers or artisans draw on their own skills and know-how to make goods from start to finish. While their products may be marketed by merchants or middlemen who are motivated by profit, the process of production itself has not been rationalized along capitalist principles.

Industrial home work is particularly common in the labor-intensive phases of electronics, footwear, and garment production. The typical industrial home worker is a seamstress who gets precut materials from a manufacturer or an intermediary, sews the garments together at home, and gets paid by the piece upon returning the finished products. Similar arrangements are common in a large variety of industries. In addition to sewing garments and shoe uppers, embroidering, and knitting, industrial home workers have been found to assemble artificial flowers, aquariums, jewelry, pens and pencils, toys, cartons and wood boxes, auto parts, dog collars and chains, rugs, wigs, suspenders, luggage tags, burglar alarms, ball-bearings, brass fittings, bird cages, electrical switches, light shades, and umbrellas. They package cloth, sweets, sunflower seeds, metal sponges, bath plugs, bolts, medical supplies, party masks, pencil sharpeners, screws, and straws. They peel shrimp, brussels sprouts and onions, roll cigarettes and incense sticks, polish plastic, and do quality-control work. Furthermore, in the United States, "industrial" home work is entering the service sector, and home workers process insurance claims, input data, and type.

There is broad agreement that industrial home workers are disguised wage workers. They do have to follow instructions of their work givers closely and are dependent on them for raw materials and designs. They have little opportunity for profit or growth. But their dependent status is disguised because they often own the tools needed for production (such as sewing machines and needles), they are not directly supervised, and they are free to determine the time and the place at which they want to carry out their work.

In contrast to industrial home workers, crafts producers are often portrayed as proud artisans, heroically independent in the face of the capitalist steamroller and upholding a venerable tradition. The image of the independent artisan harks back to preindustrial days when skilled craftswomen and craftsmen supplied a variety of everyday goods. However, comparing female crafts producers today to their preindustrial sisters is misleading. While many do rely on traditional skills, they have often adapted them to meet the demands of a global market. Furthermore, degrees of independence among crafts producers vary considerably. Women, in particular, are often integrated into crafts production as unpaid family helpers or as workers in a putting-out system. Their work is considered an extension of household activities, and often they are said only to "help out" or pursue a hobby. Merchants, middlemen, or husbands serve as intermediaries to the market, often an export or tourist market. Thus, more often than not, female crafts producers are unpaid family laborers or disguised wage workers performing tasks according to specifications of work givers, and dependent on merchants or husbands for raw materials and for the marketing of their products (Nash, 1993).

In rural Turkey, for example, women in farming households weave carpets that their husbands and fathers sell to intermediaries or dealers. Production of carpets is organized in three different ways: (1) Independent producer households provide their own inputs and designs and, in theory, are free to sell their carpets to any trader. However, in practice, households producing independently often buy yarn on credit from a trader and are then under obligation to sell the finished carpet to the same trader. (2) Under the putting-out system, intermediaries provide yarns, designs, and sometimes also looms. Weaving households are paid by the piece for the finished carpet. (3) Under the workshop production system, mostly young women work in a space provided by the intermediary. Even in this case, intermediaries pay male household heads directly rather than the weavers. Far from being independent crafts producers, the weavers face two types of subordination: They are subordinate to merchants, traders, and middlemen, and they are subordinate to males in the household (Berik, 1987).

The case of lace makers in Narsapur, India, provides a particularly striking example of the similarity between industrial home work and crafts production. Lace making was introduced by missionaries at the turn of the twentieth century and developed into a multimillion-rupee export industry, taking advantage of skills taught to women. To use the term "crafts producer" for lace makers is misleading on two counts. First, it conjures up images of independence, but lace makers are embedded in a putting-out system in which they receive all raw materials from traders and sell back their finished products to the traders. Second, lace makers do not produce whole pieces but operate along an invisible assembly line resembling an industrial division of labor. Some women produce lace flowers only and even specialize in certain kinds of flowers; others join together various patterns; and a third group fixes lace borders to pieces of cloth and joins pieces of lace with cloth to produce tablecloths or pillow cases. The finishing of lace products is always done in the houses of exporters and traders (Mies, 1982).

Not all home workers fit the image of disguised wage workers. Crafts producers as well as industrial home workers sometimes assume characteristics of the self-employed. This is particularly true in cases in which home workers have been able to acquire great skill or in which they have been able to hold on to a form of independence rooted in traditional practices. For example, in Narsapur, those women who know how to assemble the lace to make whole pieces sometimes produce on their own and try to sell their products in the market. These women then become microentrepreneurs, but their position in the market remains vulnerable.

In some cases, home workers move back and forth between self-employment and dependent home work. For example, seamstresses in the suburbs of Mexico City assemble clothes at home for merchants. In times of peak demand, they draw on relatives and neighbors for help, effectively becoming intermediaries and contractors themselves. When there is little demand, they make clothing for people in the neighborhood, turning themselves into self-employed artisans (Alonso, 1983).

The indeterminate and fluctuating work status of home workers has led some advocates to argue that all home workers should be covered by labor laws, regardless of whether they are self-employed or wage workers. Industrial home workers and artisans, disguised wage workers, and the self-employed all are vulnerable to exploitation; therefore, their legal employment status should not involve considerations over whether they are worthy of protection.

This demand is a response to existing confusion in labor law and public perception over whether home workers are "real workers." Because most countries do not have home working laws, home workers have to show that they are workers in order to receive protection. Employers and middlemen like to argue that home workers are not workers but housewives who take in home work as a hobby or to kill time. This allows them to pay less than minimum wages and deny home workers legally mandated benefits such as sick leave, maternity leave, and social-security payments. Workplace health and safety standards are ignored. Arguments that portray home workers as self-employed obscure the vulnerable status of these workers and deprive them of protection (International Labor Conference, 1994).

In sum, the category "home worker" is best understood to include industrial home workers and female crafts producers who may combine characteristics of disguised wage workers and the self-employed. Both industrial home workers and crafts producers operate from a disadvantaged position and struggle to assert their livelihood under difficult economic conditions. Both are integrated into the capitalist economy from a position of weakness and share a number of economic and political interests. It makes political sense to unite industrial home workers and crafts producers under the label "home worker."

Home Work Is Women's Work

The vast majority of home workers around the world are women. In half of the (mostly European) countries for which I found national statistics, at least 90 percent of home workers were women. In Latin America and in Asia, case studies document that home workers are disproportionately female as well. Women predominate in home-based garment production in Mexico, Argentina, Venezuela, Sri Lanka, and the Philippines. In Brazil, Peru, and Thailand, 89 percent or more of those who work at home in the garment industry are female. Other industries in which home work is heavily feminized are bidi rolling (*bidis* are a type of cigarette) in India and artificial-flower making in Thailand. Female home workers predominate in the stitching of shoe uppers in Italy, Spain, Colombia, Uruguay, and Mexico, and in carpet weaving in Turkey, Nepal, and Iran (Prügl, 1992).

Jobs are sex typed, and home work is no exception. The fact that home work is so heavily feminized reflects social constructions of home work as women's work. To some extent, the category "home work" is self-generating because home work is defined as women's work. When men work at home for pay, they tend not to call themselves home workers, but craftsmen, entrepreneurs, businessmen, consultants, freelancers, or self-employed. Men do not seem to engage in small-scale home-based operations, but it has been observed that men take over when a woman's home-based work expands and becomes a profitable business. Thus home work appears as women's work by definition in patriarchal cultures where women are relegated to the least remunerative occupations.

Feminized constructions of home workers vary in different cultural contexts. In many industrialized and urban cultures, home workers are constructed as housewives, and home work becomes a practice that allows women to care for children and perform their household duties while earning much needed income. Indeed, studies from Mexico show that women take on home work during certain phases of their life cycle, usually when they are first married and have young children. They work outside the home before marriage and after their children leave the household. The pattern of the "housewifized" home worker—a term coined by Maria Mies—is also prevalent in Europe, North America, and Australia, where national statistics show home workers to be much more likely to be married and to have dependent children than their office or factory counterparts or than women in the labor force in general. Similar patterns have been found among garment home workers and artificial-flower makers in Thailand, garment and textile home workers, *bidi* rollers, and food processors in India, and among all home workers in Hong Kong (Mies, 1982; Benería and Roldán, 1987; Karnasuta, 1987; Singh and Kelles-Viitanen, 1987).

In rural areas, home work provides additional income to farming. It is flexible enough that it can be adapted to changing labor needs during the farming cycle. Rural home workers are often more likely to perceive themselves as "peasant" women who engage in home work during their "leisure time." Home work appears as one female task in a traditional gender division of labor, and young girls often start home work at an early age as part of their gender socialization. This is the case among carpet weavers in rural Turkey and Iran, for example (Berik, 1987).

But the notion that women are housewives invades rural areas as well. Indeed, it is not uncommon that well-meaning middle-class women introduce home work to women in rural areas who eagerly adopt it as an income-earning strategy. But the image that it is a leisure-time activity for housewives follows them. Consider the following quote of a female village leader in Thailand:

> Artificial flower making activities done by the housewives here as home working, though with low pay, are good for them in the sense that they don't have to waste time doing nothing or playing cards. In the case of playing cards, if they lose, they also lose their temper. When their husbands return home from work and see their wives being in bad mood [*sic*], they usually leave home for alcoholic drinks or for the gambling den. With the introduction of vocational training especially in artificial flower making, these housewives have changed. Working among colourful flowers, they look cheerful. When their husbands return and see their wives in joyful mood [*sic*], they feel happy and help them at work instead of leaving the house. Beauty [*sic*] flowers have become "addicts" [*sic*] for women — they make women "beauty-loving" and attract men to refrain from liquors and gambles (quoted in Karnasuta, 1987, p. 9).

The introduction of artificial-flower making to farming women in this village has thus acquired ambiguous meaning. Rural women adopted the practice as a new income-earning strategy. Yet, in the process, they are being redefined as idle housewives whose work is nothing but a new hobby.

Another culturally distinct construction of home workers exists in areas where women live in seclusion. In Pakistan, for example, home work provides a source of income for secluded women. Young women, in particular, are less likely to be allowed to work outside the home because it is understood that their reputation has to be guarded closely. Thus, 15–20-year-olds form the largest part of the home-working population in Lahore, where they sew, embroider, and string flowers. Their prime motivation for home work is to save for a dowry (Shaheed and Mumtaz, 1983).

Home work is thus not only a politically useful category, but also one that, in culturally distinct social constructions, emerges as an occupation of women. Home workers may be seen as housewives, farmers, or women in seclusion, but they are always women. The identification of home with the female sphere can become an organizing principle for female workers whose work has remained invisible.

Incidence of Home Work and Trends

Official counts of home workers are rare and fraught with problems of definition. Some include those working at home as well as those working from home as a base. Few distinguish between dependent employees and the self-employed. Some refer to home workers who have registered with state authorities only. Thus the numbers of home workers in different countries cannot be compared, and they represent approximations only. Yet, the figures can indicate trends within countries. In the following, I compare only those figures that draw on similar definitions.

In Europe and North America, home work increased during the 1970s and 1980s. In Great Britain, the 1968 National Census recorded 1.1 million home workers; the number increased to 1.5 million in the 1971 Census and to 1.68 million in the 1981 National Survey on Homework. In the Federal Republic of Germany, home work increased by almost 8 percent between 1983 and 1987. France, Italy, Greece, the Netherlands, Spain, Switzerland, and the United States reported similar trends. In the province of Quebec in Canada, the number of registered home workers in the clothing industry increased from 823 in 1974 to 1,626 in 1980 and 1,714 in 1981. Even in the former Soviet Union, home work became an increasingly attractive option for a government that faced labor shortages and for women and pensioners who preferred to work at home. Between 1971 and 1988, the number of home workers in the Russian federation increased from 75,000 to 184,550 and in the Soviet Union from 56,000 to 316,000 in local crafts production between 1975 and 1988.

It is not clear whether these trends are sustained in the 1990s. Figures from Germany indicate that home work contracted there is possibly a result of increased subcontracting to Eastern European countries. Industrialized Asian countries with easy access to cheap labor abroad reported decreases in the number of home workers in the 1980s. Japanese government statistics show that the number of home workers dropped from 1.8 million in 1973 to 1 million in 1985. Government figures from Hong Kong show a reduction in the number of home workers from 50,000 in 1986 to 20,000 in 1989.

In contrast, studies from newly industrializing and developing countries in Asia, Latin America, and the Middle East document that home work is pervasive. While government statistics in these countries do not count home workers, the proliferation of studies on the issue may indicate a growth in the phenomenon. Home work is widely documented in the garment industry, leather and footwear industries, artificial-flower making, and *bidi* rolling in South Asia and carpet weaving in the Middle East, and in a wide variety of handicrafts. In Brazil, estimates are of 450,000 to 700,000 home workers in the garment industry in 1980. Seamstresses made up 7.1 percent of the female labor force in the informal sector, second only to domestic workers, who constituted the large majority— almost 70 percent (Abreu and Sorj, 1993). In 1985, the

Cámara Venezolana de la Industria del Vestido (a society to promote the garment industry) estimated that there were 65,000 home workers in the garment industry in Venezuela, accounting for about 45 percent of all workers (Lacabana, 1987). In Chile, home workers made about 60 percent of women's and children's clothing and 30 percent of men's clothing in the early 1980s. In the 1970s, home workers accounted for about 30 percent of the labor force in the garment industry in Mexico, and, in Peru, home workers and clandestine workshops were responsible for almost half of the garments produced in the country (Crummett, 1988). In Thailand, 423,600 home workers were employed in the ready-made-apparel sector in 1986, according to an estimate by the Thai Association of Textile and Clothes. This compares to a total of 683,000 workers in wearing-apparel establishments with 10 or more employees (Karnasuta, 1987). In the Philippines, the number of home workers was between 450,000 and 714,000 in 1981, accounting for 25 to 39 percent of employment in wearing apparel (Pineda-Ofreneo, 1982).

Home Work in the Global Economy

The growth of home work in the 1980s must be understood in the context of a globalization of production and markets. In the 1970s, management in center economies relocated parts of the production process to countries in Asia, Latin America, and North Africa to save labor costs. While unemployment soared in center economies during the 1970s, the import of manufactured goods increased significantly. Assembly workers and industrial home workers employed in footloose manufacturing industries in the center lost their jobs to a largely female work force in free-trade zones in developing countries.

The restructuring of production took on a different character in the 1980s, and the way home workers were inserted into the new organization of production took a new form. In addition to saving direct labor costs, companies sought to become more flexible. Managers argued that higher levels of flexibility would allow them to more quickly respond to changes in demand, increase their earnings, and save costs. One way in which large companies sought to achieve this was by subcontracting labor-intensive phases of the production process to small firms and by employing contingent workers, including home workers, who could be easily hired and fired. Firms that subcontracted did not have to pay wages when they had no orders, and the wages they paid to contingent workers were usually lower than those paid to full-time factory workers. Furthermore, resorting to subcontracting and contingent workers allowed firms to circumvent unions and the high costs of negotiated wages, including costs derived from employment guarantees. To remain globally competitive, companies around the world began to adopt these strategies.

Throughout the twentieth century, the "Fordist" factory was the paradigmatic unit of production in the center, and import-substituting industrialization sought to emulate it in the periphery. In this factory, semiskilled operatives subject to detailed supervision mass-produced standardized consumer goods. Bargaining between employers and unions, social-welfare systems providing a safety net, and Keynesian demand management typically complemented this way of organizing production, securing factory workers the means to buy the consumer goods they produced. The global restructuring of manufacturing entailed a full-scale attack on these modes of regulating the economy.

State policies often facilitated the efforts of businesses to save costs and gain flexibility. In the 1970s, policymakers in industrial countries as well as in the World Bank and the International Monetary Fund (IMF) came to reject Keynesian orthodoxy and put in its place neoliberal prescriptions to regain economic growth. In the neoliberal view the industrial and social policies instituted after the Second World War were thought to have created structural rigidities that impaired the free play of market forces and were responsible for the global economic crisis of the 1970s. The solution to economic problems was to eliminate these rigidities through the deregulation of both product and labor markets. In the 1980s, in the face of a growing debt crisis, international agencies imposed neoliberal orthodoxies on countries in the periphery through structural-adjustment and stabilization programs. These programs discouraged bargaining between workers and employers, prescribed cuts in welfare programs, and were antithetical to Keynesian economic management. To create an investment climate conducive to foreign capital, governments in developing countries deregulated their economies and dismantled existing labor protection. Some pursued active policies to encourage subcontracting with small firms and home-based workers in rural areas. For example, the government of Indonesia promoted linkages between large-scale modern industries and small-scale home-based firms as "father and son" relationships. Similarly, the government of Cyprus encouraged subcontracting linkages as a new model of development (Mitter, 1990).

In center economies, the process of restructuring led to a polarization of the working class into a relatively secure and protected minority, on the one hand, and a fragmented and relatively unprotected majority, including home workers, on the other. The rise in the number of home workers in Germany, the Netherlands, Great Britain, and the United States in the 1980s reflects efforts of firms to enhance their flexibility and service domestic "spot markets"—markets in which demand changed rapidly and required companies to adapt with virtually no lead time.

A distinctive feature of peripheral economies is the large number of petty commodity producers in urban and rural areas who struggle to make a living producing clothing, shoes, furniture, and a variety of crafts. The new movement toward flexibility and subcontracting increasingly recruits these workers to produce for national and global markets. At the top of the subcontracting chain may be small producers, large national companies, or multinationals producing for export. For example, garment home workers in Indonesia and Mexico largely produce for a national market (Boris and Prügl, 1995). But in the Philippines, Sri Lanka, and India, home workers produce garments for export, and exports have coincided with a rise in the number of home workers (Pineda-Ofreneo, 1982; Singh and Kelles-Viitanen, 1987). Subcontracting chains linking home workers to export have also been found in Taiwan and outside the garment industry in Mexico (Benería and Roldán, 1987).

The requirements of a global market have also led to a strengthening of decentralized crafts production. To satisfy Western consumer and tourist demand for indigenous crafts, merchants and traders often organize putting-out systems. The hammock industry in the Yucatan province of Mexico provides an example. Traditionally, people in the Yucatan wove hammocks from locally produced agave fiber for their own use and for exchange in the local market. Today, factory-spun cotton cord has become popular, and wholesalers in Merida, Yucatan's capital, have come to dominate the hammock industry. They employ small-town middlemen to distribute cotton cord to women and children. These weavers produce hammocks that the traders export or sell to tourists (Littlefield and Reynolds, 1990).

Similar examples have been reported from other states in Mexico involving the weaving of scarves, embroidery of traditional blouses, and production of pottery and wood carvings (Nash, 1993). In a village in Ecuador, intermediaries came to organize the production of sweaters after Peace Corps volunteers introduced handknitting in 1965. They maintain contact with suppliers and traders, who export most of the sweaters produced (Gladhart and Gladhart, 1981). In the United States, rural crafts producers similarly knit sweaters and paint wood buttons for an international clothing company. They make designer pillows, reproduce museum antiques, weave baskets, knot fringes on woven products, and market products through large companies.

The need to earn foreign-currency income, the dogma of export-oriented industrialization, and the desire to preserve indigenous crafts have led some governments in the periphery to support artisans, offering training, credit, and marketing support. The Sri Lankan government, for example, began to support crafts producers after independence in 1947 to counteract the effects of colonial policies that fostered imports from Great Britain over local production. In 1990, 40,000 out of a total of 55,000 artisans in Sri Lanka were registered with the National Crafts Council and sold their products through its affiliates. Sri Lankan government programs communicate with home-based piece-rate workers in the putting-out system, encourage their organization, train them, and guarantee a market for their products. In the changing economic environment of the 1980s, the government of Sri Lanka set up what it calls "export-promotion villages." Villages are constituted into a company, and villagers, together with the Export Development Board, buy shares in the company. The latter pro-

vides a marketing outlet for home-based and other producers and functions as an intermediary for subcontracting arrangements (Samarasinghe, 1993).

Various governments recognize the potential of carpets to earn foreign exchange. The Turkish government has actively fostered carpet exports since 1963. In 1966, it created an institution to promote carpet weaving for export. It organizes cooperatives, trains carpet weavers, and buys carpets from cooperatives and merchants. Carpets made up 11 percent of all manufacturing exports in 1963, and, while most carpets are still sold in the national market, their export share increased to 14 percent by 1982 (Berik, 1987). In Iran, handwoven carpets accounted for 54 percent of total nonoil exports in 1987 (Boris and Prügl, 1996).

In sum, home workers have a distinct role to play in a global economy in which producers scout the globe for cheap and flexible labor and in which consumer tastes have become increasingly open to foreign goods, creating an international market for handcrafted items. In a restructured manufacturing sector that prizes flexibility and subcontracting, home workers become the prototypical workers, available on demand, cheap, and easily fired.

Wages and Working Conditions

There is broad evidence that home workers earn very low wages. They generally make less than minimum wages, less than workers with similar jobs in offices or factories, and often less than other workers in the informal sector. (For sources of data presented below, see Prügl, 1992, pp. 323–332.) Big layers of profits from home work are apparently siphoned off at various stages of subcontracting. Rural home workers in the Philippines, for example, received 10 cents for sewing a baby dress that sold for $15.00 in the United States (Pineda-Ofreneo, 1982).

In the mid-1980s, home workers' wages were reportedly below the legal minimum among unregistered home workers in the Canadian women's garment industry as well as among home workers in toy and button manufacturing in Great Britain, and substantially below the legal minimum in all areas of industry in India, Thailand, and Mexico City. In Sri Lanka, no home-based worker interviewed in a 1989 study earned more than 500 rupees per month, which was well below the official poverty line of 700 rupees. At her rate of pay, a woman in the rural Philippines making rush baskets for a subcontractor would have had to work 44 hours a day to earn enough to cover daily living expenses for her family.

In the 1980s as well, home workers made less than comparable factory or office workers in Austria, Great Britain, the United States, and Pakistan. Studies from the late 1980s and early 1990s show home workers to make considerably less than their factory or office counterparts in Hong Kong, Japan, Spain, Sri Lanka, and Thailand, in Lima, Peru, and in Rio de Janeiro, Brazil. In southern India in 1983, most home workers in the textile industry lived right at the poverty line while workers in textile factories had incomes well above it.

Because their work is highly irregular, the monthly and annual income differentials between home workers and their counterparts in the formal and informal sectors are much starker than hourly differentials. Home workers are paid by results and thus do not have any income during periods in which they are not given work. They are the first to lose work if an enterprise experiences a reduction in demand. Therefore, home workers take advantage of periods of peak demand, put in long hours, including nights and weekends, and sometimes draw on help from family members and children. Figures on hourly pay do not account for this hidden labor. Furthermore, hourly wages do not take into account home workers' expenses, such as rent, lighting, thread, oil, and machine maintenance and repair, which cut their earnings even further.

The working conditions of home workers have been a cause for concern, particularly because of their health implications. Home workers often live in crowded quarters, and their work space tends to be badly ventilated and badly lit. There are reports of home workers losing their eyesight because of insufficient lighting. Home workers and their families breathe in fumes from glues, tobacco dust, and lint, and they frequently develop asthma, tuberculosis, and allergies. Furthermore, inadequate equipment leads to back pains, tension, and nervous disorders.

Home Workers Organize

Throughout much of the twentieth century, unions have argued that home workers cannot be organized because they are isolated and dispersed. As a result, many unions have favored a ban on home work as the only means of eliminating the low wages and dismal working conditions often associated with it. However, feminists and women concerned about the welfare of the poor have sometimes organized home workers quite effectively. For example, at the beginning of the twentieth century, women in Christian unions were able to organize about 10,000 home workers in Germany into the Association of Homeworkers. The British Homeworkers' League and the Homework Protective League in the United States provided a medium for home workers to lodge complaints about low wages and undesirable working conditions. Similarly, feminists today are resorting to innovative strategies to ameliorate the conditions of home workers (Boris and Prügl, 1995).

In England, nonprofit organizations such as the Leicester Outwork Campaign and the West Yorkshire Homeworking Group use community-organizing strategies to approach home workers. They run a hotline and provide information about employment rights, eligibility for benefits, and access to services. They help home workers with grievances against their employers; organize social outings and afternoon teas, which allow home workers contact with each other; and lobby governments to improve legislation and create public awareness about home work.

In India, nonprofit groups organizing women in the informal sector count home workers among their constituencies. Most vocal in speaking on behalf of home workers is SEWA, the Self-Employed Women's Association of Ahmedabad in the state of Gujarat. In the words of SEWA organizers, the association's approach to home workers combines "struggle and development." On the one hand, SEWA seeks to ensure that existing labor laws are enforced and that home workers get the same rights as factory workers. It has lobbied the Indian government for legislation to ensure that debates over employment status no longer deprive home workers of their rights, and when such legislation failed to materialize, put considerable effort into ensuring passage of the 1996 ILO Convention on Home Work. On the other hand, SEWA acknowledges the importance of income-generating strategies in an economic context in which jobs are rare. It runs a bank that provides easy access to credit for its members' productive ventures. Furthermore, it has created a number of cooperatives that help artisans reduce their dependence on traders and contractors. Cooperatives are also a fallback for women whom traders and agents victimize because of their organizing activities. In addition, SEWA provides training to artisans to improve their skills and make their products more marketable.

In other parts of Asia, the ILO is spearheading projects that seek to replicate SEWA's approach. In the Philippines, the effort has led to the creation of a national network of home workers, known by the acronym PATAMABA, which has chapters in 22 provinces and a membership of several thousand. PATAMABA has worked to increase the visibility of home workers and has gained the cooperation of the Philippine government to improve the enforcement of labor laws applying to home work and clarify home workers' eligibility for welfare benefits. PATAMABA also provides funds to home workers to form cooperatives and for other economic projects. Similar efforts are being pursued in other Asian countries, including Thailand and Indonesia.

Home worker organizers have realized that international contacts are useful for their national struggles in a world in which markets are increasingly integrated and global competition sets the agenda for national politics. Under the leadership of SEWA and the West Yorkshire Homeworking Group, they have formed a global network, called HomeNet International, that aims to facilitate the exchange and dissemination of information about home-based work and about organizations of home-based workers. The struggle for an ILO Convention on Home Work became a catalyst for HomeNet's global efforts. The network established a productive alliance with international trade union federations and coordinated a highly successful campaign, which helped convince governments to vote for the ILO Convention at the 1996 International Labor Conference held in Geneva, Switzerland.

Home worker organizers also maintain contacts at regional levels. In Asia, SEWA took a lead, and the ILO facilitated meetings between organizers in the Philippines, Indonesia, Thailand, and India. The West Yorkshire Group, with support from the European Union, spearheaded the formation of a Europe-wide network. Originally drawing strong interest especially in Southern Europe, this European Homeworking Group eventually included activists from virtually every member state.

Despite predictions to the contrary, home work has not died out but is flourishing around the world. Home workers are not simply passive, exploited workers but are organizing at a global level. As flexible work becomes an increasingly preferred pattern of employment, home workers are setting an important precedent for the way in which labor relations can be organized in the context of a flexible work environment.

References

Abreu, Alice Rangel de Paiva, and Bila Sorj (eds.). *O Trabalho Invisível: Estudos sobre trabalhadores a domicílio no Brasil.* Rio de Janeiro: Rio Fundo, 1993.

Alonso, José. "The Domestic Clothing Workers in the Mexican Metropolis and Their Relation to Dependent Capitalism." In June Nash and M. Patricia Fernandez-Kelly (eds.), *Women, Men, and the International Division of Labor.* Albany: State University of New York Press, 1983, pp. 160–172.

Bènería, Lourdes, and Martha Roldán. *The Crossroads of Class and Gender: Industrial Homework, Subcontracting, and Household Dynamics in Mexico City.* Chicago: University of Chicago Press, 1987.

Berik, Günseli. *Women Carpet Weavers in Rural Turkey: Patterns of Employment, Earnings, and Status.* Geneva: International Labor Office, 1987.

Boris, Eileen, and Elisabeth Prügl (eds.). *Homeworkers in Global Perspective: Invisible No More.* New York: Routledge, 1995.

Crummett, Maria de los Angeles. "Rural Women and Industrial Home Work in Latin America," working paper no. 46, World Employment Program Research, International Labor Organization, Geneva, 1988.

Gladhart, Peter Michael, and Emily Winter Gladhart. "Northern Ecuador's Sweater Industry: Rural Women's Contribution to Economic Development," working paper no. 81–01, Michigan State University, East Lansing, June 1981.

International Labor Conference, 82nd sess. *Report V (1) Home Work.* Geneva: International Labor Office, 1994.

Karnasuta, Kattiya. *Homework in Developing Countries: A Case of Thailand.* Bangkok: National Institute of Development Administration, 1987.

Lacabana, Miguel Angel. "Trabajo A Domicilio En Paises En Desarrollo: El Caso De Venezuela." paper prepared for the Conditions of Work and Welfare Facilities Branch, International Labor Office, Geneva, 1987.

Littlefield, Alice, and Larry T. Reynolds. "The Putting-Out System: Transitional Form or Recurrent Feature of Capitalist Production?" *Social Science Journal,* vol. 27, no. 4, October 1990, pp. 359–372.

Mies, Maria. *The Lace Makers of Narsapur: Indian Housewives Produce for the World Market.* London: Zed, 1982.

Mitter, Swasti. "Homeworking: An Evaluation in a Global Context." Prepared for the Conditions of Work and Welfare Facilities Branch, International Labor Office, Geneva, 1990.

Nash, June (ed.). *Crafts in the World Market: The Impact of Global Exchange on Middle American Artisans.* Albany: State University of New York Press, 1993.

Pineda-Ofreneo, Rosalinda. "Philippine Domestic Outwork: Subcontracting for Export-Oriented Industries," *Journal of Contemporary Asia,* vol. 12, no. 3, 1982, pp. 281–293.

Prügl, Elisabeth. "Globalizing the Cottage: Homeworkers' Challenge to the International Labor Regime." Ph.D. diss., American University, Washington, D.C., 1992.

Samarasinghe, Vidyamali. "The Last Frontier of a New Beginning? Women's Microenterprises in Sri Lanka." In Gay Young, Vidyamali Samarasinghe, and Ken Kusterer (eds.), *Women at the Center: Development Issues and Practices for the 1990s.* West Hartford: Kumarian, 1993, pp. 30–44.

Shaheed, Farida, and Khawar Mumtaz. *Invisible Workers: Piecework Labour Amongst Women in Lahore.* Islamabad: Women's Division, Government of Pakistan, 1983.

Singh, Andrea M., and Anita Kelles-Viitanen (eds.). *Invisible Hands: Women in Home-Based Production.* New Delhi: Sage, 1987.

The Impact of Structural Adjustment and Economic Reform on Women

Gale Summerfield
Nahid Aslanbeigui

Introduction

Since the early 1980s, economies throughout the world have experienced rapid transformation. Some countries changed because of the inability to service their international debt. Others transformed from centrally planned systems to market-based economies. Regardless of the cause, change was fundamental. One would be hard pressed to find any area that has not been touched by these events. This essay focuses on the gender impacts of the economic reforms in developing countries.

The main components of the reforms, commonly referred to as economic restructuring or transformation, were remarkably similar in developing countries regardless of the motivation for the changes: reduction in government spending, privatization of state firms, more emphasis on economic efficiency, and trade liberalization, including greater exchange-rate flexibility and reduction of tariffs and other trade barriers.

Restructuring in Latin America and Africa was propelled by the debt crisis and dissatisfaction with the slow growth rates associated with import-substitution industrialization (which relied on tariffs and quotas to reduce imports to promote domestic production). The International Monetary Fund (IMF) and the World Bank frequently set a package of reforms, called structural-adjustment program (SAPs), as a condition for obtaining loans. Most East Asian economies restructured voluntarily based on the success of the export-oriented model in Japan and Hong Kong (South Korea and Taiwan, for example) or on the desire to move from socialist planning to a more market-based economy (China). The changes in centrally planned economies were more extensive than in other countries, but the key reforms were the ones described above. In the Third World, the experiences of China, Chile, and Nicaragua indicate the similarity in the changes induced by SAPs and by the transition from socialism, and they demonstrate the extensive scale of the effects on women. This essay focuses on both the obligatory structural-adjustment programs and voluntary restructuring.

Asia, Latin America, and Africa have fared quite differently between 1980 and 1995. While the economies of Latin America and Africa suffered declines in per-capita incomes over the 1980s, many of the Asian economies flourished. By the early 1990s, China had the fastest-growing economy in the world; its gross national product (GNP) grew 13 percent in 1993 alone. While both the percentage of the population living in poverty and the absolute numbers of the poor fell in East Asia, they increased in sub-Saharan Africa, Latin America, the Caribbean, North Africa, and the Middle East. In South Asia, the percentage of the population living in poverty fell while the number of the poor increased (World Bank, 1993, pp. ix–x).

Regardless of region, women have gained less and lost more than men in all countries examined (Benería and Feldman, 1992; Aslanbeigui et al., 1994; Sparr, 1994). Women entered the labor market in increasing numbers in Latin America as export processing grew, but they were restricted to low-wage jobs with few long-term promotion possibilities. Simultaneously, many women who had held jobs in the formal sector were pushed into informal-sector employment, such as selling food on the street or cleaning houses. In the high-growth Asian economies, some women gained from new opportunities, but they received less than a proportionate share of the benefits. In all cases analyzed, women did, however, bear a disproportionate share of the costs of restructuring. In addition to working more outside the home, especially in the informal sector, women cushioned the costs by doing more unpaid work within the home, such as growing vegetables and making clothes rather than buying them. Problems created by cuts in government spending on health care and education have serious, long-run implications for girls and women.

In the following sections, key aspects of the restructuring process in the different regions are summarized and the effects on women analyzed. Regional similarities notwithstanding, there are significant differences among countries. These are introduced by examining critical issues in a few different countries. One large country from each of three regions is used as the main case for that region: Mexico is the case for Latin America and Zambia for Africa because these countries have tried to implement a range of structural-adjustment policies as prescribed by the World Bank; China is the case for East Asia because it undertook similar, though more extensive, reforms to move voluntarily to a market-based economy. China also offers the contrast of an economy that has responded to the reforms with rapid growth. Finally, conclusions about global trends are presented.

Restructuring Policies

The global phenomenon of restructuring began primarily in response to the credit crunch faced by heavily indebted countries, such as Mexico, Brazil, and Argentina, in the early 1980s. Many developing countries had borrowed heavily since the 1970s when funds from higher oil prices instituted by OPEC were deposited in U.S. and European banks. The banks funneled these "petrodollars" to borrowers in developing countries. The eager flow of petrodollars during the 1970s suddenly dried up at the beginning of the 1980s as commercial banks realized the risky nature of their loans. Anti-inflationary policies that led to higher interest rates in countries such as the United States increased the debt repayment (interest and principal) of developing countries. Recession reduced demand for exports from developing countries that were needed to earn foreign exchange necessary for debt repayment. Falling prices for commodities such as cocoa and misuse of borrowed funds by developing countries made matters worse (Darity and Horn, 1988; Sachs, 1989).

Mexico teetered on the brink of default, other Latin American debtors appeared almost as shaky, and U.S. banks realized that they could collapse if the major debtors jointly defaulted. No longer did the old saying hold that, If the U.S. sneezes, the less developed countries get pneumonia. The links between the more developed and the developing worlds guaranteed that all economies would suffer tremendously if the debt crisis were not resolved.

At this point, in the early 1980s, the IMF and the World Bank stepped in. Rather than completely bail out the private banking system, they demanded that commercial banks also provide increased credit to resolve what appeared to be a short-term problem with cash flow. When the problems in Mexico persisted for several years, the IMF and the World Bank decided that lack of liquidity was not as central as they had thought earlier. They thus began to demand that countries adopt fundamental, structural reforms as a condition for further loans.

Restructuring was designed to reduce the twin deficits—government budget deficits and deficits in the balance of payments accounts (trade and capital flows)—and to promote the growth of per-capita income though not necessarily the equality of its distribution. Governments cut their payrolls by removing public employees and selling off national enterprises. They reduced subsidies for education and health care, instituted or raised user fees, and encouraged private provision of these services. Those countries plagued by high rates of inflation adopted tighter monetary policies. Tariff barriers were reduced, currencies devalued, and foreign investment encouraged. In sum, countries focused on two policy areas: privatization and trade liberalization.

During the mid-1980s, restructuring was based on the general dictates of neoclassical economic theory, which considered neither country context nor meso decisions, such as the structure of government expenditure and programs for the social sectors (see UNDP, 1990, p. 43). The IMF and the World Bank specified only economic targets—for example, deficit reduction; they did not consider impacts on the poor in general, on women and children in particular, or on the environment. The universal application of this model was responsible for much of the suffering associated with the transformation. The poor ultimately bore much of the burden for insensitive policies made by wealthy bankers, businessmen, and government officials. Women and children found their entitlements shrinking so they were especially vulnerable to nutritional deficiency. This increased their need for health care at the same time that subsidized provision of these services was being reduced. Future opportunities for girls were compromised as relatively more girls than boys were kept home from school because their parents could not afford the fees or needed the help at home (Cornia et al., 1987; Benería and Feldman, 1992; Aslanbeigui et al., 1994).

The need for additional loans to meet interest payments forced most governments to acquiesce to the conditions. Food riots frequently greeted the price hikes for staples when subsidies were removed. Observers wondered whether the policies destabilized democratic governments. Some adjustments to the model were made eventually, but until the end of the 1980s the lending agencies denied responsibility for how the cuts were made.

In the late 1980s, the IMF and the World Bank stated that they could impose restrictions not only on the amount of government spending but also on how it would be cut. Powerful finance ministries would no longer be immune, and the vulnerable areas of health and education would no longer be required to bear the brunt of the cuts; structural-adjustment conditions would have to include a social safety net. That was the announced change in policy. In practice, change took place at a slower pace.

Latin America

Mexico is at center stage in the Latin American transformation process. Led by economists, it has been trying to satisfy the rules set up by the international lenders, the IMF and the World Bank, who have helped assure its success. But success has not come easily. Real GNP declined 3.1 percent between 1982 and 1988 (Benería, 1992, p. 84). Between 1987 and 1993, the annual unemployment rate was a stable 18 percent while estimates put the underemployment rate in the 25–40 percent range. It is not surprising, therefore, to note that consumer purchasing power has declined by more than 35 percent between 1982 and 1990 or that the deindustrialization induced by economic reforms has led to a decline in gross domestic investment (Barkin, 1990, p. 101; United Nations, 1990).

The Chilean experience following the overthrow of the government of Salvador Allende in 1973 was similar. Chile began reforms similar to SAPs much earlier than other Latin American countries. Sluggish growth attributed to lack of incentives under socialism, the rise in oil prices, and the drastic drop in price for copper exports motivated reform. By the end of the 1970s, liberalization policies had "substantially dismantled the mixed economy and one of the most progressive social welfare systems in the region" (Montecinos, 1994, p. 161). The continuation of reforms in the 1980s imposed much of the adjustment costs on workers as wage rates fell and unemployment grew. Targeting of welfare policies toward the poorest in society facilitated government budget cuts but left many slightly less poor people without resources. By the mid-1980s, 45 percent of the population was estimated to be living in poverty (Torche, 1987).

In general, then, the 1980s was a lost decade for development in Latin America. In the 1990s, after years of negative growth rates in per-capita income, high unemployment, and cuts in social services, many Latin American countries have finally begun to experience modest positive growth rates.

Africa

Although African countries were not among the major debtors, for some, the debt burden was heavy. Domestic policies had not succeeded in stimulating rapid growth, and most countries suffered the marks of a recent history of colonialism. During the 1980s, commodity prices dipped significantly, making export of primary commodities less profitable, at the same time slowing growth and making repayment of foreign loans more difficult. Although most African debt was financed by governments or international agencies, the countries had to vie with those in Latin America for the shrinking pool of available credit. They thus found themselves compelled to adopt restructuring to qualify for more loans.

The human costs of restructuring are most apparent in African nations. Despite assistance, poverty is increasing more rapidly than in other regions as countries experience the effects of wars, disease, falling commodity prices, and structural adjustment; between 1990 and 2000, the number of people living below the poverty line in Africa is expected to grow from 270 million to approximately 400 million (UNDP, 1990, p. 61). The social impact of restructuring is particularly visible in sub-Saharan Africa. Most of the countries that exhibited a decline in the Human Development Index (a composite of literacy and education, life expectancy, and per-capita income) between 1970 and 1990—Sudan, Angola, Benin, Zambia, and Mozambique—are found in this region (UNDP, 1993, p. 103).

Here, government budget cuts have typically meant reductions in education, health care, and other social services, despite the fact that, as poverty and malnutrition increase, health care needs escalate. Although some nongovernmental organizations (NGOs) began programs to help, their accomplishments in health care were slowed by problems of coordination and standardization of efforts, accountability, and training (Turshen, 1994).

Although international agencies realized that rural and urban conditions differ drastically in countries throughout the world, they underestimated the interaction between the city and countryside in sub-Saharan Africa. In particular, while the rate of urbanization is one of the highest in the world, the frequent movement between rural and urban areas and the intertwining of families blur the rural–urban distinction when estimating the effects of restructuring on people's welfare. This blurring of the rural–urban dichotomy stands out in Zambia, where the World Bank, to help the rural poor, supported agricultural diversification and promoted export crops. Subsidies on maize were to be removed in the late 1980s to promote market incentives for suppliers as well as to reduce government outlays. An initial attempt to remove the subsidies in 1987 was halted, however, after urban riots. In 1989, price controls on all crops except maize were eliminated, and subsidies on maize and fertilizer were reduced. As prices were liberalized, the cost of agricultural inputs and transportation climbed, but the price of maize did not keep up. The agricultural sector felt the squeeze of price changes more than it benefited from them (Geisler, 1992).

The World Bank has also frequently advocated targeting in sub-Saharan Africa. To reduce government subsidies, only the poorest are targeted for benefits. Poverty, however, is so deep and widespread in many of the countries that this artifice hurts more than it helps. In places such as Ethiopia, where targeting was used to allocate food aid during famine, some families starved one child, often a girl, to qualify for the target group (Sen, 1992).

Overall, the 1980s were a painful decade of adjustment in Africa, with few success stories and much belt tightening by already deprived people.

Asia

East Asian countries have, in contrast, become models of growth for the rest of the world. Singapore, South Korea, Hong Kong, and Taiwan have influenced economic strategies throughout the world far beyond what might be expected by their size. China has emerged as an awakened power; more than one billion people have been directly affected by its transformation process.

Restructuring in most East and Southeast Asian countries was undertaken voluntarily rather than at the behest of the IMF or the World Bank (for South Asia, see Benería and Feldman, 1992; for the Philippines, see Floro, 1994). These countries enjoyed better human capital and less foreign debt than those in other regions. With Japan as a model, they moved toward export promotion.

Although Korea was heavily indebted in the 1970s,

most of its loans were negotiated under a regime of fixed interest rates. Soaring interest rates in the United States, therefore, did not affect this country as much as it did in nations in Latin America. When emphasis on heavy industrial output led to disappointing growth rates in the late 1970s, South Korea resumed its earlier emphasis on light industry and exports. The country devalued its exchange rate and took steps to balance the domestic budget, which had never been as imbalanced as those in Latin American countries had. Wages were kept low as growth rates climbed.

The Chinese case is particularly important for two reasons: Reforms there have resulted in striking changes that directly affect almost one-fifth of the world's population; and China represents a Third World case of the transition from socialist planning to a more market-oriented economy. China's transition to a market economy, adopted voluntarily after the death of the nation's Communist leader Mao Zedong in 1976, began from a much different position than that in most developing countries. Rather than foreign debt, Mao's legacy included widespread primary education (but very little university-level training), a good basic health-care system, and a modest industrial base. Inefficient state enterprises, however, required increasing amounts of government funds to achieve output targets, and agricultural production barely kept pace with population growth. The examples of high incomes in Taiwan and Hong Kong further motivated reform.

The 1978 reforms initially focused on rural areas. They transformed agriculture from communal to household production, granting long-term rights to land (15–30 years). Families were allowed much greater decision-making power in production, including diversification of products. With the disbanding of the communes, most subsidized health care and educational systems were also eliminated. To speed per-capita growth, the government introduced the one-child policy, which instituted rewards and punishments to persuade couples to have a single child.

With rural reforms successfully under way, officials pushed for urban reforms in the mid-1980s. To improve efficiency, firms began to hire more people on contract and used incentives, such as piece-rate wages and bonuses, that tie reward to effort. Although massive layoffs were not allowed, managers identified workers who were considered "surplus" and who might be dismissed in future cost-cutting measures. In the late 1980s, many of these workers, mostly women, were sent home, sometimes with base pay but no bonuses, which had become an increasingly important component of wages. At the same time, opportunities for self-employment opened up, and foreign investors were courted. Foreign investment and trade were encouraged by establishing the Special Economic Zones for export processing.

After the Tiananmen Incident in 1989, during which scores of prodemocracy demonstrators were killed by the government, officials pulled back from economic reforms to solidify their political power, but reforms were in full swing by 1992. The changes unleashed a wave of entrepreneurial talent that has astonished observers throughout the world. The average annual growth rate of per-capita income between 1980 and 1993 was 8.2 percent (World Bank, 1995, p. 162).

Impact of Economic Restructuring on Women

Reforms have presented opportunities for, as well as inflicted costs upon, women. In all countries we examined, women have traditionally experienced some form of discrimination. The formation and growth of women in development as a field of study in the 1970s are due, at least partly, to the often adverse effects of economic development on women.

Have the reforms of 1980 to 1995 been gender neutral, or have women borne most of the burdens associated with them? Have the changes reduced or exacerbated traditional forms of discrimination? How are women responding to the changes?

Latin America

SAPs have affected women in Latin America in various ways: by increasing the unemployment of the male members of their households; by cutting social services and food programs; and, somewhat positively, by increasing women's employment opportunities in export-processing work. Since Mexico has played a key role as an international model in the implementation of its SAPs, it is the focus of this section, with supplemental material from other Latin American nations.

Employment and Labor-Force Participation. Since the Mexican government began to privatize production and increase exports in the 1980s, women in rural and urban areas have had different experiences. Rural areas were hard hit in terms of job opportunities, as small farmers were displaced when land was bought by the wealthy. Most new opportunities were in the informal sector and export-processing work (which provided more jobs for women than for men) in the cities. Both demand and supply factors thus increased rural-urban migration by young women to the point where women outnumber men in the cities; they are outnumbered by men in the countryside.

No discussion of the changes in Mexico can ignore the controversial *maquiladoras* (export processing plants) along the U.S. border that have provided jobs for many of the migrants; by the mid-1990s, half a million workers were employed in this region. This may be a relatively small fraction of the total Mexican labor force, but more jobs have been created in the Mexican *maquiladoras* than in most other countries' export-processing zones. Export-processing work also creates demonstration effects of the benefits of trade liberalization for domestic industries and thus has impact beyond the numbers of workers explicitly employed in the zones (for more analysis of export-processing work, see Lim, 1990; Ward, 1990).

Although the *maquiladoras* were set up in the mid-1960s, the restructuring of the 1980s built on and extended the program. The success of the *maquiladoras* played a central role in Mexico's inclusion in the North American Free Trade Agreement (NAFTA) ratified in 1994. Over 70 percent of the jobs in the first decade were held by women. This imbalance was not the result of a deliberate effort on the part of government to provide jobs for women. Rather, it was an effect of the particular industries that operated export-processing plants—textiles and electronics, which traditionally hire female workers.

The dominance of women's work has proved controversial because official unemployment statistics—mostly reflecting unemployed men—were not reduced by the job openings. Interestingly, unemployment statistics for women also did not fall because of the growing unemployment in other areas of the economy and women's increasing labor-force participation in Mexico. Many women who enter the labor force remain there even if they lose the jobs; also, more women may be attracted to the labor force than the number of jobs available. In part because of official pressure to hire more men and in part because the *maquiladoras* diversified into areas traditionally dominated by men as the zone matured, women's employment share in the *maquiladoras* has fallen since the mid-1980s.

Critics attribute the growth in the rate of labor-force participation by Mexican women to the *maquiladoras*. While the opportunities in the zones surely have attracted some women who would not otherwise have been in the labor force, they cannot take all of the blame or credit. Changing demographics reflect more women entering the labor force in the 1980s throughout most of the world. Reasons range from the need to have additional income earners in the family during the austerity associated with restructuring to the choice made by better-educated women to seek formal employment. Moreover, many women were previously already working in the informal sector, but their work was not always counted in the official statistics.

In Mexico, problems associated with the increased labor-force participation of women, in addition to resistance to any change in family structure, include officials' concern about unemployment statistics and researchers' concern about the increased workload of women. Mexican women face the double burden common to most working women, but structural adjustment has placed additional demands on their time (Benería, 1992). Economic necessity and falling wages mean that both husband and wife must work to attempt to maintain their standard of living; the wife's work may be prompted by her husband's job loss and his lack of alternative employment. Many, especially older, women have to find informal-sector work at low pay. In this environment, increased employment opportunities for women do not necessarily translate into increased well-being. The *maquiladora* jobs have the advantage of bringing women together in the workplace, but they provide low-level jobs at meager pay with little opportunity for advancement.

Nonwage Work and Intrahousehold Bargaining Power. Women in Latin America have played a key role in protecting their families from the negative impacts of reforms by shouldering the extra burdens. The importance of the family has grown as the need to work together to survive has increased (Elson, 1991; Benería, 1992). Some of the most creative responses to the crisis have been documented for Chile (Montecinos, 1994).

Women in Chile became active in new organizations sponsored by NGOs to aid the family. Programs such as growing vegetable gardens, restoring agricultural land, and preparing food with "magic" energy-efficient cooking pots were designed to ensure adequate nutrition for poor families. Women emerged as leaders in these activities, and their participation often translated into forms of political participation within an environment that suppressed open political opposition by men.

The economic value of women increased during the 1980s as they sought to meet their families' needs through domestic labor, barter, begging, and sometimes stealing. Soup kitchens opened where women cooperated in the preparation of food. Activities that increase women's economic value can be expected to increase their bargaining power within the household as well, but the climate of economic and social stress under President Pinochet's authoritarian regime created offsetting tensions that resulted in ambiguous impacts on well-being.

Targeting of social-welfare programs to the poorest groups inflicted social costs in Chile as it did in African countries. Schoolchildren were excluded from nutritional programs that were limited to those under the age of 6, and health-care programs available since the 1950s were discontinued in the late 1980s, leaving the population vulnerable to disease.

Sub-Saharan Africa

Women in sub-Saharan Africa have long had relatively high rates of labor-force participation compared to women in other developing countries. They produce much of the food for subsistence and have a long history as traders. SAPs in various countries of this region had the effect of pushing women more into low-paying, informal-sector work or frequently into work in which the form of pay was a small amount of salt or grain. The reforms also altered the distribution of benefits within the household by affecting sources of income.

Rural–Urban Interactions. In a sub-Saharan Africa, in contrast to countries like China where the rural–urban dichotomy is strong, the difference between rural and urban areas is fluid, and people flow easily between city and village. SAPs in this region often had stated goals of helping rural residents, who make up most of the population and live with high rates of poverty. Urbanization is occurring rapidly, however: By 1995, almost half the population in Zambia was living in cities (four to five million people). Not surprisingly, SAPs have not produced the intended results.

As part of the SAPs, prices of agricultural products were increased to provide an incentive for farmers to increase production. The price hikes have not elicited as much supply response as expected because of higher input prices (e.g., fertilizer). The area is so poor that farmers often have to buy food to supplement their own production. As consumers, they are hurt by the higher prices for grain; as workers, they have to work longer for less return. The pay for female farm labor has always been lower than that for men, and, since the SAPs were introduced, in the 1980s wages have fallen dramatically and are usually paid in-kind. A woman who previously may have earned a basket of maize for a day's work would have to work four days for the same basket of maize after one year of restructuring. In addition, the increased demands on her time often meant that she no longer would be able to grind the grain herself and would have to request processed maize meal, which is more expensive, further reducing the amount of maize that she would receive for a day's work. Urban residents often cannot afford the more expensive grain and are forced to switch to less-desirable staples, such as millet gruel. Wealthy urban residents, however, can afford to hoard maize.

Informal-sector activities have grown in importance for both rural and urban women. Rural women and men migrate to urban areas and perform low-paying tasks such as running errands or tending stalls where food or used clothing is sold. Since the reforms, the market for used clothing has grown in both urban and rural areas. People from the city frequently market used clothes in the countryside. When they reach retirement age, some people choose to return to their villages; this practice is more popular with men than with urban women.

In such an environment, the distinction between rural and urban poor becomes less clear because reverse migration is common and rural suppliers are also often consumers of the same products they are producing. The full implications of policies on both rural and urban areas need to be taken into account before the policies are implemented, but this was not the case with SAPs. Women have had to spend an increasingly large amount of their scarce time to help their families survive the restructuring process.

Intrahousehold Allocation of Income. Price reforms have had unanticipated impacts on women's control of income within the household in the context of traditional bias against women. Although men traditionally controlled the crops and the fields in which maize and other cash crops were grown, women usually received all or most of the return for less-valuable crops, such as soybeans in the north or groundnuts in the east, that they cared for as part of the crop rotation on these fields. They then used the money for household expenses. The alternative cash crops had risen in price enough to make them attractive to men. Soybeans are especially easy to substitute for maize because they require no fertilizer and minimal time to cultivate. At the new, higher prices, many farmers switched from maize to beans. Husbands frequently claimed the right to the income from this more-profitable crop. By ignoring the traditional patterns of discrimination in which husbands can control their wives' income, the reforms resulted in a loss of income that was controlled by women in the family (Geisler, 1992).

Much of the literature on women's poverty has focused on single heads of households (see Bruce, 1989), but many married women live in families marred by violence and conflict that may result in more severe impoverishment for them and their children than for their husbands. Women often prefer to receive the in-kind payments referred to above because they are harder for husbands to appropriate than cash, but some cash is still needed. At the same time that men are taking over earnings that had previously gone to women, the household's need for income to pay for food, schooling, and health care has increased.

Health Care. Cutbacks in health care have hit women especially hard because they are the providers for the children in the household and they also have extra requirements for food during pregnancy and lactation. Women have struggled to prevent serious malnutrition in their families during the reform process. Often, their struggles have been fruitless; many families can barely afford one small meal a day. Child malnutrition has increased especially in the North-Western Province of Zambia (Chipulu, 1990). In addition to the problems they face that are directly related to SAPs, these women also must eke out an existence in a region plagued by wars, drought, years of superpower politics, and exogenous shocks such as falling commodity prices. AIDS and resistant forms of malaria combine with conditions of endemic hunger to produce a greater need for health care by women and children at a time when less health care is available.

Asia

East Asia has had a refreshingly positive experience in terms of economic growth between 1980 and 1995. Thus, women in this region confront economic restructuring from a much brighter position than those in Africa and Latin America. The changes, however, have generated costs as well as benefits for women.

Labor-Force Participation and Employment. As in Latin America, most countries in East Asia have seen the labor-force-participation rates of women increase with trade liberalization because of the dominant role played by export processing. Singapore has chosen not to promote immigration as its labor force approaches full employment but rather to encourage women, especially educated women, to have more babies at the same time that they hold full-time jobs. The government's policy has created increased demands on women's time and put the burden of development squarely on their shoulders (Pyle, 1994).

In China, the labor-force-participation rate for women, which had been among the highest in the world, dipped slightly after the 1978 reforms. The changes have been quite different in rural and urban areas. Under the household responsibility system, the source of employment for rural women has changed from the collective to household production. Return to working within the confines of the patriarchal household has put women in a vulnerable position in terms of bargaining power within the family because their contribution is not as visible as when they work for an outside employer. Although the work points for all family members were usually given to the male head of household under the collective farming system, the woman's contribution was clearly delineated, but in family production the amount and value of individual contributions are less obvious. On the other hand, rising incomes and the ability to choose profitable types of production, such as raising pigs and chickens, provide positive counters to the negative aspects of the changes.

In addition, new opportunities outside the home have appeared in the township, village, and private nonstate enterprises (TVPs). These have grown rapidly since 1985; by 1994, they employed approximately 100 million people, about 40 percent of them women (Tan, 1994, p. 62). Although working conditions are frequently dismal, the TVPs are the fastest source of income growth in rural China and also provide about 15 percent of China's exports.

Change in the cities has been most dramatic in terms of the availability of consumer goods, from higher-quality vegetables to fancy clothing and jewelry. State firms still maintain excessive numbers of employees although they are under pressure to become more efficient. (Some experimental areas such as Tianjin have had more bankruptcies and layoffs than other cities.) Still, women are openly discriminated against in many jobs. This occurs partly because the cost of hiring a woman is marginally higher than that for a man based on benefits for childbearing and health care; partly it is a result of traditional discrimination that cannot be accounted for by higher costs or lack of education. For young, pretty, well-educated women, there are a number of new opportunities in joint ventures, hotels, and export processing. For older women who get laid off, the opportunities are fewer.

Self-employment is a new option under the reforms, but it requires some capital or credit, which remains less accessible to women. Despite these problems, women have found ways to open small businesses selling clothing, magazines, or food on the streets.

Education, Health Care, and Welfare. Education of girls in rural China since the reforms has been problematic. Schools have introduced fees, and the opportunity cost of sending a girl, who can now be earning income, to school has increased. The result is that many girls have been compelled to drop out of school, despite the legal requirement of nine years of education. "More than 80 percent of the 4.8 million school-aged children who dropped out of school in 1990 were girls, mostly from rural and remote mountainous areas and from minority groups, and there are still more than twice as many illiterate women as men" (*China Daily,* 1991).

In the cities, on the other hand, parents continue to send their children to school despite fee increases. The advantage of an education is much clearer in the urban context, and higher education has expanded since the reforms. Nevertheless, female graduates are discriminated against in their job search.

Health care has also become problematic. In some wealthy, typically urban, areas, the quality of health care improved during the 1980s. In much of the countryside, however, residents lost access to health care when the old system was dismantled. Although some private insurance providers have stepped in to fill the gap, the majority in the countryside still do not have access to medical care. From a combination of policies that give conflicting incentives (to enrich the family by having more sons but also to have only one child) in a context that values men more than women, families have elected to abort female fetuses, abandon female infants or hide them with relatives, and occasionally to commit female infanticide. Combined with neglect of young girls, these practices have resulted in an increased imbalance in the male/female sex ratio in China. While it is typical to have about 105 male babies for every 100 girls, much of China reports ranges of 116 male babies to every 100 girls (Kristof and WuDunn, 1994, p. 472). In other regions of the world, women usually outnumber men by the age of marriage, but in China there have traditionally been fewer women than men overall. This imbalance has contributed to the increase of another problem increasing during the reforms: the incidence of a slave market for wives. Growing numbers of women are abducted and sold to farmers who cannot or will not pay the "requisite" costs of a rural wedding. These costs have soared to the equivalent of about $2,000.00 compared to the "reasonable" $350.00 for an abducted woman (Kahn, 1993).

Conclusions About Global Issues

This essay has examined economic transformation in developing countries around the world. The transformation was frequently forced on the countries as SAPs and stabilization plans from the World Bank and the IMF. For other countries, similar reforms were undertaken voluntarily to strengthen the economy and integrate it into the world market, as occurred in the transition from socialist planning to a more market-oriented economy. The model of transformation has been rigidly similar across countries; culture, tradition, environment, and gender have rarely been considered. The emphasis since the late 1980s by the World Bank on targeting relief packages has protected the poorest women and children in some societies, but it has left most of the poor vulnerable to costs of adjustment. The results of the reforms and their impact on women vary from country to country. Still, some global themes emerge from analyzing the cases presented here.

Work. In most cases, women have had to work longer and harder both in the home and outside it, whether in the formal or the informal sector. Women's work has been pivotal in determining their families' ability to weather the difficult times of restructuring. The opportunity cost of women's time has, thus, been increasing. In most countries, so has women's labor-force participation, at least partly in response to the demands of economic change.

In socialist countries such as China, where women's labor-force participation already was higher than in the rest of the world, pressures to increase efficiency have translated into pressures to remove redundant workers from the payrolls. The opportunity to stay home is also an attractive one for some women. Thus, the rate of women's participation in the labor force in these countries is dropping to some extent.

Regardless of the country, however, the reforms have consistently led to increased discrimination against most women in hiring (young, attractive women may find some new opportunities) and to disproportionate numbers of women among those laid off, fired, or otherwise unemployed. In addition, most of the jobs available in the formal sector are low-level positions with few opportunities for promotion. A growing number of women have been pushed toward jobs at the lower end of the informal sector.

Capital and Credit. Many of the new opportunities made possible by restructuring are for self-employment in either the formal or the informal sector. These endeavors usually require some form of capital or credit, which women traditionally do not control or have accessible to them. NGOs and a few banks have provided some response to women's need for small loans, but such efforts have been inadequate. In some areas with a large overseas population, such as China, remittances from relatives also provide a source of investment capital.

Intrahousehold Allocation of Income and Bargaining Power. The increasing impoverishment of women internationally has often been associated with single heads of household. Nevertheless, women as members of male-headed households can also suffer increased impoverishment in terms of time and control of income under the restructuring. The family should be viewed as a cooperative conflict, combining elements of working for a common good and of trying to get more for oneself or one's children (for more on this view of the family, see Sen, 1990). The woman may work longer and harder both at home and in the labor force but not receive any increase in material or other goods to improve her well-being.

Education, Health Care, and Social Services. Cuts in investment in people through education, health care, and other social services will have negative repercussions for years. Both men and women are subject to the cuts, but reductions in services or increases in fees, combined with traditional attitudes about the inferiority of women, usually mean that women and girls will be pulled out of school sooner and sent to the doctor later than men and boys (Rose, 1994). The price hikes for food and the cuts in social services have meant that many people cannot get adequate nutrition, a problem especially serious for women who are pregnant or breast-feeding.

Tradition and Culture. All of the cultures examined have traditional forms of discrimination against women. Many manifestations of discrimination against women that have surfaced during restructuring are related more to this traditional bias than directly to the reforms. Reform policies, however, have not anticipated or adequately addressed the problems of gender bias and are, therefore, responsible for some of the damage.

Women's Actions

Women have not been passive recipients or victims in the reform process. They have found creative ways to use dwindling supplies to ensure the survival of their families. On an individual level, they have increased their work time both within the household and in the labor market. They have also participated, and taken leadership roles, in NGOs and organized for political action (Sparr, 1994).

International conferences sponsored by the United Nations, such as the International Conference on Population and Development (Cairo, 1994), the World Summit on Social Development (Copenhagen, 1995), and the Fourth World Conference on Women (Beijing, 1995), provide excellent opportunities for people concerned with the impact of policies on women to participate, meet others engaged in similar work, organize, and focus on continuing efforts in the future. To become equal players in the changing global economy, women will be compelled to organize at all levels, from small NGOs to national political office to international forums.

References

Aslanbeigui, Nahid, Steven Pressman, and Gale Summerfield (eds.). *Women in the Age of Economic Transformation: Gender Impact of Reforms in Post-Socialist and Developing Countries.* London: Routledge, 1994.

Barkin, David. *Distorted Development: Mexico in the World Economy.* Boulder: Westview, 1990.

Benería, Lourdes. "The Mexican Debt Crisis: Restructuring the Economy and the Household." In Lourdes Benería and Shelley Feldman (eds.), *Unequal Burden: Economic Crises, Persistent Poverty, and Women's Work.* Boulder: Westview, 1992, pp. 83–104.

Benería, Lourdes, and Shelley Feldman (eds.). *Unequal Burden: Economic Crises, Persistent Poverty, and Women's Work.* Boulder: Westview, 1992.

Bruce, Judith. "Homes Divided," *World Development,* vol. 17, no. 7, July 1989, pp. 979–991.

China Daily. "On Education," December 6, 1991.

Chipulu, P. "North-Western Area Development Project," nutrition working paper, Lusaka, 1990.

Cornia, Giovanni Andrea, Richard Jolly, and Frances Stewart (eds.). *Adjustment with a Human Face.* Oxford: UNICEF/Clarendon, 1987.

Darity, William, Jr., and Bobbie L. Horn. *The Loan Pushers.* Cambridge: Ballinger, 1988.

Elson, Diane. *Male Bias in the Development Process.* Manchester: Manchester University Press, 1991.

Floro, Maria. "The Dynamics of Economic Change and Gender Roles: Export Cropping in the Philippines." In Pamela Sparr (ed.), *Mortgaging Women's Lives: Feminist Critiques of Structural Adjustment.* London:

Zed, 1994, pp. 116–133.

Geisler, Gisela. "Who Is Losing Out? Structural Adjustment, Gender, and the Agricultural Sector in Zambia," *Journal of Modern African Studies,* vol. 30, no. 1, 1992, pp. 113–139.

Kahn, J. "Cultural Violence Against Women: An Uphill Battle for China Against the New Slave Trade," *Dallas Morning News,* May 19, 1993.

Kristof, Nicholas, and Sheryl WuDunn. *China Wakes: The Struggle for the Soul of a Rising Power.* New York: Times Books, Random House, 1994.

Lim, Linda. "Women's Work in Export Factories: The Politics of a Cause." In Irene Tinker (ed.), *Persistent Inequalities: Women and World Development.* New York: Oxford University Press, 1990, pp. 101–119.

Montecinos, Verónica. "Neoliberal Economic Reforms and Women in Chile." In Nahid Aslanbeigui, Steven Pressman, and Gale Summerfield (eds.), *Women in the Age of Economic Transformation: Gender Impact of Reforms in Post-Socialist and Developing Countries.* London: Routledge, 1994, pp. 160–177.

Pyle, Jean. "Economic Restructuring in Singapore and the Changing Roles of Women, 1957 to Present." In Nahid Aslanbeigui, Steven Pressman, and Gale Summerfield (eds.), *Women in the Age of Economic Transformation: Gender Impact of Reforms in Post-Socialist and Developing Countries.* London: Routledge, 1994, pp. 129–144.

Rose, Pauline. "Female Education and Adjustment Programmes: A Cross-Country Statistical Analysis," Gender Analysis and Development Economics working paper no. 5, University of Manchester, Manchester, 1994.

Sachs, Jeffry D. (ed.). *Developing Country Debt and the World Economy.* Chicago: University of Chicago Press, 1989.

Sen, Amartya. "Gender and Cooperative Conflicts." In Irene Tinker (ed.), *Persistent Inequalities: Women and World Development.* New York: Oxford Univer-

sity Press, 1990, pp. 123–149.

———. "The Political Economy of Targeting," discussion paper, Washington, D.C.: World Bank, 1992.

Sparr, Pamela (ed.). *Mortgaging Women's Lives: Feminist Critiques of Structural Adjustment.* London: Zed, 1994.

Tan, Shen. "Societal Changes and Women Employment in China," *Zhejiang Zuekan,* 1994, pp. 62–65.

Torche, Aristides. "Distribuir el ingreso para satisfacer las necesidades básicas." In Felipe Larraín (ed.), *Desarrollo Económico en Democracia.* Santiago: Centro de Estudios del Desarrollo, 1987.

Turshen, Meredith. "The Impact of Economic Reforms on Women's Health and Health Care in Sub-Saharan Africa." In Nahid Aslanbeigui, Steven Pressman, and Gale Summerfield (eds.), *Women in the Age of Economic Transformation: Gender Impact of Reforms in Post-Socialist and Developing Countries.* London: Routledge, 1994, pp. 77–94.

United Nations. *Trade Policies, Investment, and Economic Performance of Developing Countries in the 1980s.* New York: United Nations, 1990.

United Nations Development Program (UNDP). *Human Development Report 1990.* New York: Oxford University Press, 1990.

———. *Human Development Report 1991.* New York: Oxford University Press, 1991.

———. *Human Development Report 1992.* New York: Oxford University Press, 1992.

———. *Human Development Report 1993.* New York: Oxford University Press, 1993.

———. *Human Development Report 1994.* New York: Oxford University Press, 1994.

Ward, Kathryn (ed.). *Women Workers and Global Restructuring.* Ithaca: ILR Press, Cornell University, 1990.

World Bank. *Implementing the World Bank's Strategy to Reduce Poverty: Progress and Challenges.* Washington, D.C.: World Bank, 1993.

———. *World Development Report.* Washington, D.C.: World Bank, 1995.

Women's Employment and Multinational Corporation Networks

Jean Larson Pyle

Introduction

This essay discusses insights obtained from surveying the substantial body of research carried out between 1985 and 1995 on women's employment in multinational corporations (MNCs) worldwide. The first two sections show that it is timely and important to reassess the topic of women's employment in multinational corporation networks because significant changes have taken place in the global economy and in MNC strategies in that 10-year period. These changes, in turn, have had dramatic effects on the world's women.

The third section presents an overview of the research since the mid-1980s on women and MNCs by world region. It updates the first decade of research (1975–1985), outlining the ways in which earlier trends have continued and profiling new findings and lines of analysis. It shows how patterns of employment of women can differ by class, marital status, and ethnicity and illustrates how they are shaped by actions of international organizations, national governments, households, and the women themselves. It also highlights the similarities and differences in strategies of MNCs in developed and developing countries. It is suggested in the concluding section that, in light of the findings of this essay, the conceptual framework for analyzing women's employment associated with multinationals must be fundamentally altered—broadened and extended to include these trends and categories of analysis.

This essay draws on a wealth of detailed research by scholars from a variety of disciplines and countries. The studies cited are representative of the variety of careful, insightful research done in this field; the selection is not meant to be comprehensive. In addition, this essay does not discuss other issues of growing importance in the relationship between women and MNCs, including women as consumers and women as citizens who are adversely affected by environmental damage done by MNCs.

The Changing Global Economy: Increased Market Orientation and Broader-Based MNC Activity

Multinational corporations have been a presence in Third World countries for many decades, originally moving abroad in sectors involving agriculture, natural resources, and utilities. After World War II, many countries adopted an import-substitution-industrialization (ISI) strategy, which established protective barriers to encourage domestic production of needed products. It was widely used in Latin America and in some Asian countries, such as India. MNCs in many industrial sectors began production abroad as a way of circumventing these protective barriers and accessing markets.

During the same period, a few export-processing zones (EPZs) were established in Third World countries as special enclaves in which production could be established for export only. This allowed MNCs to produce abroad in lower-cost countries for export back to their home market. This type of process expanded in the 1960s as numerous countries began to adapt this approach into a national strategy—export-oriented, outward-looking development. Development policies were revised to encourage the production of manufactured goods or the provision of various service activities for export to the international market, often by multinational corporations. Governments provided MNCs with tax breaks, increased ability to repatriate profits, a favorable regulatory environment, and low-cost labor, often largely female.

Since the early 1980s, globalization of the world economy has accelerated. Two key changes have occurred that have had a substantial impact on the employment of women, particularly women in MNC production networks. First, increasing numbers of countries have moved more toward use of markets rather than government planning to coordinate their economies. Many of them have adopted the export-oriented development strategy. Often, countries have sought foreign direct investment by MNCs. This has provided an enlarged environment in which MNCs, particularly those in labor-intensive industries that hire largely female workers, can operate. Second, there has been a rapid expansion of productive and marketing activities of multinational corporations globally. Increasingly, multinational corporations have originated from more and more countries and have moved into new areas of the world. MNCs have accelerated and extended their cost-cutting strategies globally, with distinct and adverse impacts on the workers involved.

Each of these key changes is important for understanding the relationship between women and multinational employment networks in the world economy in the 1990s. First, substantial areas of the world have become significantly more market oriented since the early 1980s. Many previously socialist countries, such as the former Soviet Union and the countries of Eastern Europe, have allowed markets rather than planners to make most of their economic decisions. Capitalist industrialized countries have also moved further in the direction of market coordination. Third World countries heavily burdened by international debt or interested in obtaining loans have been pressured by international institutions such as the World Bank and the International Monetary Fund (IMF) into structural-adjustment programs (SAPs), which often involve production by MNCs for export as a way to raise revenue to repay the debt. In these and many other Third World countries, the World Bank has advocated export-oriented development since the 1980s as the appropriate strategy for economic development. Export-oriented development has been an integral part of globalization processes, much more compatible with them than ISI because of its fundamental orientation to production for export rather than production for internal markets that was the focus of ISI.

As illustrated in East and Southeast Asia since the 1960s, export-led development has been based heavily upon female labor. This is because women can be paid less than men and are generally considered more productive workers because of their abilities, patience, and reluctance to unionize. MNCs widely draw women into their workforces, even in areas where there are high rates of male unemployment and where women have little experience in the paid labor force. Thus, the movement toward more market-directed economies and the choice of an export-oriented development strategy have had a significant impact on women's labor-force participation in many areas. The spread of export-oriented development draws more women into production in the MNCs and their affiliated networks than other development strategies.

Second, corporations have become much more global in terms of origin, location, and function. The earlier view of MNCs in the international economy—that of MNCs from industrialized countries, largely the United States, siting manufacturing production in a few countries, such as Mexico or in East and Southeast Asia, to take advantage of lower-cost production, accessibility to markets, or political stability—is outdated. MNCs in the 1990s are involved in both manufacturing and service sectors and have originated from diverse areas: other industrialized coun-

tries, particularly Western Europe and Japan; the newly industrialized countries (NICs) of South Korea, Taiwan, Hong Kong, and Singapore; and other developing countries. They have greatly expanded the areas into which they locate to include additional tiers of developing countries and formerly socialist economies, and they have augmented operations in industrialized countries.

MNCs have developed and extended their strategies to combat increases in costs, particularly labor costs. Cost-cutting strategies have occurred at an accelerated pace and have involved both industrialized and developing countries. They affect all industries, from heavy industries (autos, steel, petrochemical) that hire largely male workers to more labor-intensive assembly industries (garments, electronics, shoes, toys, pharmaceuticals) that hire larger percentages of women. They have resulted in much controversial industrial restructuring. The impact has been particularly striking in labor intensive assembly industries that hire largely female production workers.

Cost-cutting has increasingly encompassed not only formal-sector employment (where individuals work for wages) but also the informal sector (where people are small-scale entrepreneurs or vendors or perform piecework in the household). Specifically, as labor costs have risen in an area in which MNCs have been producing, the corporations have decided to do one or more of the following: (1) automate, (2) increase the use of home working or subcontracting to other businesses within the country, or (3) move to another tier of low-cost countries, either as direct producers (via subcontracting) or in joint ventures in which the multinational and a local company jointly own the business. Repression of workers is often involved.

Each of these tactics has an adverse impact on the women and children involved in the original area. As MNCs in these labor-intensive industries automate, it is largely female production jobs that are replaced by machines. Males occupy larger percentages of more-technical job categories. As the labor force is informalized by the MNCs' increased use of subcontracting and home work, it is chiefly women and children who are employed in these informal sectors. They work for wages substantially below those for similar work in factories, in jobs that are even more precarious. As MNCs relocate, women in the new location gain employment while women at the original site lose their jobs. The newly unemployed are often unable to find other employment, since they gained no new skills at their old jobs that would be useful in economies that are evolving away from labor-intensive sectors.

In short, although MNCs have expanded their spheres of activity and, therefore, seek to hire more women workers in many sectors, changes in their cost-cutting strategies in such labor-intensive sectors can have an adverse impact on women workers.

The Importance of Reexamining Women's Employment in Multinationals

Although women's employment in MNC networks is a

small percentage of women's employment worldwide, it is increasingly important to reexamine for several reasons. First, the changes in the global economy and in MNCs discussed above are affecting growing numbers of women. As MNCs expand their presence, they employ an increasing percentage of women in developing countries in formal-sector wage work. As they enlarge their subcontracting networks, they affect increasing numbers of women working in the informal sector, including the household where home-based work is done. The converse is true too: If demand in the industry contracts, the subcontracting networks are the first to experience cutbacks in employment, often followed by reductions in formal-sector employment in MNCs.

Second, women's employment in MNC networks is a global issue, not simply a concern of those studying development in the Third World. Once we realize this and recognize the parallels that exist between the strategies of MNCs in developing and industrialized countries and their resultant impact on women, we can bridge the gap that often exists between women in Third World and women in industrialized countries. This provides a basis for the formulation of useful strategies for empowering women and improving their economic condition as well as for the construction of appropriate research agendas.

Third, this reassessment provides a clear focus on the critical issues of the 1990s. In the 1980s, the key issues were whether women were integrated into development or marginalized by it and whether their socioeconomic position was improved or further eroded. Women employed in MNCs were considered integrated into the development process; however, there was considerable controversy regarding whether women benefited from, or were disadvantaged by, such employment. As shown below, the issue is more complicated. The debate over this controversy has been replaced in recent years by a more complex analysis that addresses the dynamics and contradictions involved with female MNC employment. It suggests that, although women may initially regard employment in MNCs favorably because it provides an immediate source of income, over time they experience widespread adverse effects (lower wages relative to men's, health hazards at the workplace, harassment, lack of possibilities for advancement, and job insecurity and loss) (Ward and Pyle, 1995).

Fourth, this overview suggests the need for a broader approach for understanding the effect of MNCs on the largely female labor force. A model based exclusively on foreign direct investment and employment of women in MNCs captures only part of their impact. The analysis must be widened from its relatively narrow base of workers within MNC factories to include those women working in the extended networks of subcontractors and home workers that MNCs have increasingly developed. This broadening requires understanding the interconnectedness of spheres of work that are often examined separately: the formal and informal sectors, including the household. It highlights the integral relationship between MNCs and other firms and households. Clearly, the way MNCs structure or restructure work arrangements has a significant impact on women's work in the formal sector as well as the informal sector, including the home. An extended model that incorporates horizontal linkages (foreign direct investment, including joint ventures) as well as vertical aspects (the pyramid of levels of subcontracting and home work that underlie many corporations) may be more appropriate.

Women's Employment in MNC Networks by Region

This section traces the impact of MNCs on women's employment in the major regions of the world—Latin America, Asia, parts of Africa, and throughout the industrialized countries. It summarizes insights from research since 1985 and updates earlier findings which suggested that, although women did find employment in MNCs, their situation was generally unfavorable (low wages, poor working conditions, short job tenure). It highlights the parallels in MNC strategies within the developing world, and between the Third World and industrialized countries, and illuminates the differences that exist. Throughout are interwoven other themes that have received growing attention. One is the manner in which race and ethnicity, class, and gender are integral factors in how work is allocated and, therefore, in the resultant composition of the labor force. Another is the role of state policy in shaping economic development and the relationship between women and MNCs. A third is how the relation of women to MNC employment is shaped by the movement of labor as well as capital on an international scale. Earlier, the focus was largely on the international movement of capital to establish production in sites deemed favorable. Recent research has shown how the international movements of workers, from Third World to industrialized countries where they often work in sweatshops or perform home work, are also an important aspect of the MNC–female worker nexus.

MNCs in labor-intensive industries initially set up operations in the Third World in Latin America and Pacific Asia. Macroeconomic conditions in each area differed, affecting both the inflow of MNCs in labor-intensive industries and women's work. Export-oriented development was adopted in several Pacific Asian countries in the 1960s, much earlier than in most Latin American countries, which shifted from ISI to export-oriented development in the 1980s. This facilitated a larger and earlier influx of MNCs in labor-intensive industries in Asia as well as a wider spread throughout the region over time. Because of this, the Asian region provides a longer-term opportunity to examine many of the themes mentioned above.

The few incursions of MNCs into Africa, a continent that has experienced serious deterioration in the past three decades, are briefly surveyed followed by a discussion of the impact of MNCs on women in industrialized countries. The latter have had substantial inflows of immigrants and refugees from some less developed countries and have developed pockets of MNC production much like those in the Third World.

Latin America

Multinationals employing female workers in Latin America date from 1965 with the establishment of the *maquilas* or *maquiladoras* (export-processing plants in zones free of import and export taxes) in Mexico along the border with the United States. Since the mid-1980s, the MNCs have widened their activities in the region: The research examines their newer production networks and modes of operation in Mexico City, Peru, Colombia, Nicaragua, and Costa Rica—along with the older Mexican border industries and operations in Brazil.

This research indicates that patterns of treatment of female workers found earlier have continued and spread throughout the region. Among the findings it documents: the below-subsistence wage levels, even for women with extensive labor histories; the relatively low position of women on the occupational ladder, combined with virtually no possibility of promotion; the mixed feelings of female workers toward MNC employment; and the fact that employment provided by MNCs has not had a substantial impact on the high male and female unemployment rates.

Tiano's (1994) research shows that *maquilas* provide employment for women who would otherwise have to work in the informal sector. *Maquilas* have hired relatively well-educated young women, whose earnings are not simply supplemental but critical to the household economy. She finds that women have contradictory attitudes toward this work, needing the income but still believing in traditional sex roles regarding domestic responsibilities. Tiano finds no foundation for arguments that men's high unemployment rates may propel women into the *maquila* labor force or that *maquilas'* use of women may augment men's unemployment. At her household level of analysis, women took jobs in *maquilas* either because men's incomes were too low or because there were no men in their households.

Holt (1994) finds that although Mexican leaders since the early 1980s have viewed the *maquilas* as a model for Mexico's future, these industries have not had spillover benefits to other manufacturing industries. He traces the declining share of women as the *maquila* industry shifts away from lighter industry.

Wiegersma's (1994) research in Nicaragua shows how the government restructuring of the economy accompanying the shift from the socialist Sandinista government to the export-oriented, neoliberal UNO regime in 1990 had contradictory and largely adverse effects on women workers. Both the domestic sector and the EPZs were restructured. The state closed domestic sector textile and garment factories, decreasing female employment, while it retained domestic industries employing men. This occurred simultaneously with the privatization of state-owned garment factories in EPZs and the replacement of older female workers with younger women. Additional new employment accompanying the growth of garment and textile industries in the EPZs has not offset the losses for women in the domestic sector.

Benería and Roldán (1987) were among the first to study the rising use of subcontracting by MNCs. Their research in Mexico City extended the analysis of the effect of MNCs on women beyond formal, paid employment to the informal sector, where they examined the layers of subcontracting and home work arrangements that MNCs developed. They show how the types of work arrangements involved in subcontracting or home work blur the earlier distinction between formal and informal sectors. Subcontracting has become a favored strategy because it reduces costs. These female workers are paid less than those working in the MNC factories, and they have no benefits. The corporation can terminate them immediately without recourse in the event of an economic downturn. Subcontracting avoids protective legislation and taps sources of workers, such as married women and children, that may have not been utilized before and are unlikely to unionize. The authors discuss the contradictions that subcontracting and home work present for women: On the one hand, they offer married women and female heads of households the flexibility to combine their work in the labor force with household duties; on the other hand, they make these women more vulnerable economically.

Studies also reveal the use of additional types of strategies of MNCs to lower the cost of labor. For example, Truelove (1990) has studied agribusiness MNCs in the coffee industry in Colombia that employ male agricultural workers for seasonal work—and also have established mini-*maquiladoras* there to employ women at below-subsistence wages to produce shoes and garments for export. In this manner, the year-round work of the women subsidizes the male agricultural workers, who are unemployed only part of the year.

Caribbean

Puerto Rico was one of the first countries to offer an array of attractive incentives to foreign investors via Operation Bootstrap in 1948. Although the government was explicit over the next 25 years about its intention to reduce male unemployment, its development of export-processing zones consisting of textile, garment, and electronics manufacturing drew women into the labor force (Rios, 1995).

The pursuit of export-oriented development in the Caribbean intensified for part of the 1980s with programs such as the Caribbean Basin Initiative (CBI), initiated by the United States in 1984. This strategy was based on the establishment of free-trade zones (FTZs) to attract foreign industry. Factories in the FTZs produced chiefly garments and electronics and hired predominately female labor because they were low-wage, productive workers.

Recent research in this region documents the contradictions that women working in MNC networks face. French (1994) relates that women's employment in FTZs in Jamaica involves low wages, no job security because of vulnerability to shifts in the international market, workplace health hazards, and few unions. The government does not enforce labor legislation because it fears that corporations will leave if it does so. Safa (1995) shows that similar adverse characteristics prevailed in Puerto Rico and the Dominican Republic; nevertheless, the employment of women has led to their becoming major contributors to household income and to a weakening of men's roles as the primary earner in many households.

The Caribbean was the initial location for a new type of MNC global activity—service-sector jobs such as banking, airline reservations, and telemarketing—which were studied by Rios (1995). Research by Velvington (1993) illuminates how MNC jobs are structured by ethnicity as well as gender: In the factory he examined in Trinidad, 82 percent of line workers were female whereas all floor supervisors were male. Furthermore, 71 percent of the female line workers were Black and 25 percent East Indian, while 62 percent of the male supervisors were White and 38 percent East Indian.

New research in this region also reveals the failure of the strategy to provide a broad employment base, the precariousness of female employment in MNCs, and the resulting impact on women's and children's lives. MNC employment provided only a portion of jobs needed. Although women's employment in such Caribbean factories grew substantially, it did not significantly affect their overall labor-force participation rates or reverse the rising unemployment among them. Further, in recessionary times, enterprises that hired women were the ones most likely to be closed. This is a particular problem in this region, which, due to its external dependence, has been severely affected by world economic crises. In addition, although wages in Caribbean countries were low they were still higher than in Mexico or some Asian countries. Therefore, employment could be shifted internationally accordingly.

These effects of MNCs on women were especially devastating because a high percentage of households are headed by women. The ability of Caribbean women to provide for themselves and their dependents has eroded since the mid-1980s, forcing them increasingly into informal-sector activities.

Asia/Pacific Asia

As a whole, recent research on the differential effects of MNC employment by gender in Asia provides a complex picture of: (1) how women have been, and continue to be, important for the economic development of nations in this region as more adopt export-led development strategies and set up factories for the manufacture of electronics, garments and textiles, shoes and footwear, toys, and plastics, and the assembly of other consumer products; (2) how gender, class, and ethnicity interact in shaping the composition of the work force; (3) the sometimes contradictory— yet largely negative—effects of MNC employment on women and how these effects persist or change over time depending on whether the MNCs are automating, relocating to new countries, or increasing their layers of subcontracting and home work; and, (4) how the state affects the relation between MNCs and women and the close relationship between the household and the MNCs.

The Importance of Women in Regional Export-Led Development. Women have played a key and largely unrecognized role in the rapid economic and social development of the region, particularly Southeast Asia. They remain the preferred type of worker for direct employment in labor-intensive MNC industries and for employment in the informalized layers of subcontractors and home workers MNCs establish, even when male unemployment is high, because they work for relatively low wages, have higher productivity, and are reluctant to organize.

Women were important from the 1960s for the NICs of South Korea, Taiwan, Hong Kong, and Singapore. Several authors have noted that the flow of women into the Taiwanese labor force facilitated what has been called that country's economic miracle. For example, Li (1985) notes that "the mass entry of women into the labor market provided the 'manpower' needed for rapid industrial development and sharply boosted women's share of employment in every industry" (Li, 1985, p. 4). They provided a low-cost supply of labor, needed at a time of remarkable national economic growth.

Female workers have been important since the 1970s for the ASEAN-4 (Philippines, Malaysia, Indonesia, and Thailand, where ASEAN is the Association of South East Asian Nations), and they continue to be promoted as an attraction for foreign investors: State development agencies in Thailand and Malaysia actively advertise the availability of female workers to attract foreign direct investment. A guide for investors distributed in early 1990 by the government of Thailand says:

> Those foreign companies locating in Thailand benefit through the use of Thai female labour. . . . Throughout Thailand females are found by many companies to be manually skilled, keen to work and have the patience to work for long hours at repetitive activities (Thailand Board of Investment, 1987, p. 38).

Another group of Asian countries (Sri Lanka, Bangladesh, and areas of China and India) began establishing export-led development and export-processing zones in the 1980s, relying heavily on the labor of women and often children. MNCs producing in these countries originate from industrialized nations and other Pacific Asian nations such as Japan and/or the NICs, relocating to them when labor shortages result in rising wages in the home country. Production may be by joint venture or by locally owned businesses that are subcontractors of MNCs.

The Effect of Marital Status, Class, and Ethnicity. Although the composition of the workforce of the MNCs remains largely young, single women, some variation has occurred over time. Married women have entered this workforce in countries such as Thailand, Singapore, and the Philippines. The way this pattern has evolved depends not only on the preferences of MNCs and rising labor shortages but also an the preexisting relations of domination in the particular nation. In Thailand and the Philippines, where cultural traditions have allowed women a strong economic role, female employment is important because of low male incomes and high unemployment. Child care is provided by extended families. In contrast, patriarchal norms in Taiwan have restricted women's employment in the formal sector to the period prior to marriage (Gallin, 1990). Married women's economic participation in Taiwan has been chiefly in informal networks such as subcontracting and home work established by MNCs (Arrigo, 1985). This parallels events in the industrialized country in the region, Japan, where patriarchal norms have constrained women's labor-force participation.

Ethnicity as well as class and gender shape the structure of the MNC workforce in this region. Arrigo's research (1985) in Taiwan reveals that managers and union leaders in MNC factories were predominately mainland Chinese military men while the workers were native Taiwanese, most of them women. Standing (1989) studied the entrance of lower-class Bengali women into wage work in Calcutta. Traditionally, cultural norms prevented these women from any employment except teaching. However, with rising male unemployment, the increased preference of employers for women in light of strong union organization among men, and the reorganization of production via subcontracting and home work, these women were pushed and pulled into wage work.

The Contradictory Effects of Employment in MNCs on Women. A general overview of recent literature reveals that the effects of MNC employment on women in Pacific Asia are somewhat mixed, yet largely negative. Such employment may have contradictory effects. In the short run, employment in MNCs may provide greater material benefit to the women and their families than existing alternatives. Some have argued that although wages are low in MNCs relative to industrialized countries and working conditions more adverse, in many areas MNCs offer conditions and pay that are relatively better than other local employers. However, the adverse impact is clear in the longer run, when these women experience work-related deterioration of health, the lack of benefits, the absence of opportunities to gain skills and advance in the job hierarchy, and loss of jobs due to their health, work-related stress, or strategies of automation, retrenchment, or relocation of the MNCs. Negative aspects are often obvious even in the short run in the form of long hours, forced overtime, increased production quotas or speedups, poor working conditions or housing, stress, and harassment from management.

These general effects vary according to how long the country has been pursuing export-oriented growth, conditions within the country (such as culture or worker resistance), and the particular corporate cost-cutting strategies adopted (automation, relocation, subcontracting). As mentioned above, some countries have been slower to adopt the export-oriented strategy than others and, therefore, have only recently become sites for MNC production networks. Simultaneous with their adoption of such strat-

egies, other countries that utilized this strategy in earlier decades have been consciously moving from labor-intense sectors to more high-tech industries.

For those in the NICs who have not lost jobs to automation or relocation of MNCs, working conditions and absolute wage levels have improved somewhat over time due to worker resistance. Nonetheless, conditions of female employment in MNCs vary even within the NICs. Phongpaichit (1988) contrasts Singapore with South Korea. According to her analysis, in Singapore there has been a dramatic increase in female employment and a narrowing of the male/female wage gap since the 1960s whereas in South Korea women have remained a more peripheral workforce, with relatively short working lives. She attributes the more favorable situation in Singapore to the tighter labor market there and to state support for some child care and for the upgrading of skills.

However, although conditions may have improved in some NICs, adverse conditions persist in the ASEAN countries, resulting in continued worker efforts to resist and unionize. Furthermore, conditions are the worst in the latest group of countries to adopt export-oriented growth (Sri Lanka, Bangladesh, areas of China and India). As the MNCs move into the latest tier of countries that are pursuing export-led development, they adopt the same exploitive pattern as when they established operations in the NICs.

Wage differentials between men and women have persisted over time and appear to be substantially due to discrimination. According to Gannicott (1986), Taiwanese women earned two-thirds of what males earned in 1982. His econometric work reveals that these differentials persist, even when women have apparently the same productivity as men and are performing the same level of work.

With respect to the impact of corporate cost-cutting strategies on women, an overview of research on Pacific Asia puts into sharp relief the adverse effect of automating, relocating, and establishing networks of subcontractors and home workers. As MNCs move from one country to another, women in the original country lose their jobs while those in the new area gain them. This can have serious effects on the women originally involved, particularly if they were in low-skill, dead-end jobs in a country, such as South Korea, that has been structurally reorienting its export-led economy toward products that involve higher technology. A company closure in Masan Free Export Zone (Mafez), where the work force was 75 percent female, focuses attention on what has been called an industrial ghetto for women. Attracted by employment in the MNCs, these women left school early and had few skills transferable to more high-tech jobs when companies closed. Such displacement is widely expected to be the trend throughout the labor-intensive industries in garments, footwear, and toys in South Korea. The paradox is that South Korea has had somewhat tight labor markets. This misuse of female labor is not only inequitable, it is inefficient, in addition to inequitable.

Automation has affected male and female employment differentially by its reduction of jobs that are primarily female. The percent of the electronics workforce that is female has been reduced. This trend is expected to involve the garment industry as it automates in the future. Salih and Young's (1989) study of the semiconductor industry in Malaysia reveals that, although there was net growth in employment in this industry in the 1977–1984 period, the proportion of females decreased. The direct labor component (largely female) fell while there was increased use of more skilled labor, technicians, and engineers (male). In the 1970s, women made up more than 90 percent of the total workforce. However, by 1986 they were only 71.9 percent of the workforce in free-trade zones in Penang. With the trend toward automation and more-skilled labor, the proportion of women is expected to fall further. This process also has ethnic implications. The percent of *bumiputeras* (native Malaysians) is correspondingly falling since the *bumiputeras* were also placed in the lowest-level production jobs.

Subcontracting and home work have spread throughout all tiers of countries in the region. Jayaweera (1994) discusses the rapid spread of subcontracting through Sri Lanka, pointing out that it reinforces women's subordinate and marginalized position in the labor force.

Other tactics of labor shedding and wage discrimination that MNCs are employing in the region are mentioned by Salih and Young (1989). Corporations cut back employment as demand falls by compulsory shortening of both working hours and days, which immediately reduces the incomes of working women. In addition, corporations are using an emerging strategy called recontracting, a simultaneous process of hiring and firing: Women who are retrenched are rehired at a dramatically lower wage on a short-term contract, the latter obviating the need for compensation in the case of future retrenchment. Last, there are a variety of tactics to force resignation "voluntarily," such as unreasonably high production quotas, harassment by managers, and limited-term "special offers" for severance pay.

The Impact of State Policy on the Relationship Between Women and MNCs. Many governments in this region have long been active in controlling the MNC workforce via a variety of tactics, including specifically designed family policies aimed at shaping the workforce over time. Arrigo (1985) has examined "the MNCs means of political control of women's labor, specifically the role of the Taiwan government, labor unions, security agencies, control of public media, the factory personnel offices and MNC-Taiwan government relations" (Arrigo, 1985, p. 79). Noting parallels at that time between Taiwan and South Korea and the Philippines and Malaysia in terms of the military dictatorships in each country that maintained the climate for foreign investment, she shows that even exemplary government labor legislation can be simply a lever the government uses to extract profits and control in the government–MNC power struggle, "rights" that the government and its supported unions will then consent to forego.

Jayaweera (1994) reports that the state in Sri Lanka assisted MNCs by restricting the activities of trade unions, by lax enforcement of legislation on working conditions, and by withdrawing from the International Labor Organization (ILO) Convention on Night Work, thereby ensuring the availability of low-cost female workers 24 hours a day. Rosa (1995) surveys tactics used by governments in Malaysia, the Philippines, and Sri Lanka to prevent organizing by workers. These were often utilized in conjunction with corporate repression that involved violence, fear, and murder.

Stivens (1987) examines the complexities and contradictions of state policies toward economic development, women, and families in Malaysia. Growth of the industrial sector depended upon the large increases in female labor-force participation in all ethnic groups. However, the Malaysian government put increasingly heavy pressure on women to maintain levels of family welfare via remittances (sending some earned income to other family members) and household duties (maintenance and care of family members in the household), thereby reducing its need to underwrite base levels of family welfare. In addition, the government pressed young women to have more children to ensure an adequate supply of labor for the future. This was done by adopting a pronatalist policy in an astounding reversal of family planning practice in 1984. These state policies are contradictory, however, because increased responsibility for household tasks and increased numbers of children tend to reduce women's labor-force participation.

Africa

The presence of MNCs in Africa has been localized to South Africa, countries in North Africa, and Mauritius. Women's work is a crucial resource in African food production, but women's participation in formal-sector employment is negligible compared to men's. Industrial development has been limited in Africa. Where urbanization has occurred, women are involved largely in the informal sector in petty commodity production or in trading activities (rather than in informal-sector activities attached to the formal sector).

MNCs in South Africa received widespread notoriety in the past because of their presence in a country that long practiced a form of institutionalized racism called apartheid. The MNCs drew on Black males as their chief source of workers. This left the women to commute from areas such as the Black township of Soweto into the cities to work as domestics or to maintain households of children and older people on the very poor land granted Blacks. Although Black women were finally allowed (in 1985) to become factory workers, the last group to be tapped, only a very small percentage of them work in MNCs.

Mauritius was the first country in Africa to provide free-trade zones in promoting its export-led development strategy. Textiles and garments are the primary industries, replacing the formerly dominant but now nonexistent electronics industry. These industries hire largely female workers, utilizing networks of home workers and locating factories throughout the country to avoid unionization efforts. Factories have poor working conditions, forced overtime, and often increased quotas; employees burn out within a few years.

Proximity to Europe and their well-behaved workforces have made the North African countries of Algeria, Morocco, and Tunisia attractive sites for production. Although Algeria's development was capital intensive and, therefore, did not draw women into the labor force, modern clothing factories were established in Morocco. Both men and women were employed, but female labor was preferred, and the employment of women increased rapidly. Tunisia has been a supplier for the clothing industry in France.

Industrialized Countries: Western Europe and the United States

Analysis of the effects of the increased globalization of MNCs in the world economy and their impact on women must include a look at what is occurring within industrialized countries as well as developing nations. First, MNCs follow the same sets of strategies in Europe and the United States as in the developing world, with similar effects on their workers. Second, the contradictions such employment presents for women also exist in these countries. Last, an analysis integrating industrialized and developing countries completes the picture of the international mobility of both labor and capital. It highlights the fact that women working in MNC networks throughout the world face the same MNC cost-cutting strategies and have similar problems. Analyses focusing only on women in developing countries miss this critical point and often implicitly depict women in Third World countries as low-cost labor undermining jobs in industrialized countries. This section examines MNCs in Europe first, including their spread into peripheral areas and eastern Europe, and discusses the role of the state in shaping the composition of their workforce. It then examines cost-cutting trends and labor-intensive industries in the United States.

MNCs that often produce in the Third World countries have also established operations in countries in the western European periphery, such as the Republic of Ireland, Spain, and Greece, and have subcontracted production to firms in eastern Europe. Elson's (1989) study of the European textile and clothing industry reveals a growing variety of forms of internationalization of production that are more predominant and complex than the foreign direct investment used by the firms in the past. These include coproduction (firms from western Europe producing with firms in eastern Europe), offshore processing (sending raw materials or components out of the country for processing by firms that may not be subsidiaries), and subcontracting (western European firms contracting for supplies of specific items from independent firms in other countries). These forms of internationalization rely primarily on female workers in western European periphery countries, heavily on women in eastern Europe, and somewhat on women in North Africa.

In addition to this regionalization of production, MNCs have extended their production networks in core European countries, using as low-cost sources of labor the flows of immigrants from eastern Europe and outside the region.

The national government in this region can be a major influence in shaping the composition of the MNC labor force, sometimes in surprising ways. Hadjicostandi (1990), in discussing the foreign capital that has invested in garment manufacturing for export in northern Greece for 25 years, outlines how the government sponsored the strategy that structured production to take place both in the factory and at home, under a piece-rate system, and has employed largely women. The Republic of Ireland has had one of the world's most sophisticated approaches to attracting foreign investment. Unlike many other governments, its state-promoted export-oriented development has been targeted to the provision of male employment, explicitly in the 1970s and implicitly thereafter. Although this has reduced the types of MNCs Ireland could attract and, therefore, constrained its growth, the goal of providing male employment was deemed more important.

MNCs in the United States that have not automated and are in industries that still seek a low-cost (low-wage, manageable, productive) workforce pursue the same range of strategies outlined with respect to the Third World—relocation and subcontracting. They can relocate to another country, resulting in loss of female jobs in the United States, and they can move production within the United States to areas such as the South, the Plains, or urban centers that have lower-wage workers, causing regional disparities in unemployment. This trend has been augmented by local and regional governments in the United States that employ strategies similar to governments of Third World countries to attract capital. They offer an array of financial incentives, a plentiful workforce, and an attractive location.

In addition, MNCs can increasingly utilize subcontracting, sweatshops, and home work within the United States. They often tap pools of immigrants or refugees as low-wage workers. Such people have relocated to the United States for a variety of reasons involving economic and political hardship. They are part of the international flow of labor in the MNC capital-labor nexus.

Fernandez-Kelly and Garcia (1992) show that MNCs are present in the United States in the same major industries examined in the Third World—garments and electronics. Contrary to the widespread impression that these are declining industries, the authors found that this was not the case in the 1970s and 1980s. In the 1980s, the apparel industry employed more than the automotive, steel, and electronics industries combined in the United States. Firms in this industry in the United States hire the same type of worker (female, largely young, and single) as in the Third World, and for the same reasons. The clothing industry has restructured rather than declined, with the demise of larger firms and a proliferation of smaller ones. Corporate strategies have been altered in the same ways as in the Third World, with the enlargement of subcontracting and home work, thereby increasingly informalizing the labor force.

In addition, the hiring strategies of these MNCs have evolved in distinct patterns involving ethnicity, class, and gender. Garment manufacture is largely female, and overwhelmingly minority women—91 percent of women garment workers in the Los Angeles area are minority, with 71 percent of them Hispanic. Employers distinctly prefer Hispanic women, especially foreign born, to other women.

Conclusions

This essay has reexamined the effect of MNCs on women in light of major shifts in the global economy with respect to national macroeconomic policy and regarding the operation of MNCs: (1) the trend toward increased marketization of economies and widespread pressure to adopt more export-oriented strategies for development, often based on production by MNCs, which expand the opportunities for MNCs and, therefore, for women's employment; and (2) the proliferation of MNCs from expanded numbers of countries, their movement into more widespread areas of the world, and their increasing use of cost-cutting strategies of automation, relocation, and subcontracting in industrial sectors that are labor intensive, each of which has an adverse impact on female workforces involved in the original country.

The essay has highlighted the fact that, although there are differences within and between areas, due to the impact of culture, ethnicity, development strategies, and other state policies, there are many similarities in the effect of MNCs on women globally. Research since the mid-1980s has increased our understanding of the parallels that exist internationally in the impact on female employment over time of changing MNC strategies (automating, relocating, and informalizing by subcontracting and home work); the sometimes contradictory effects of MNC employment on women; the mediating effect of state policies; the way in which the workforce is shaped by the interaction of class, race, and gender; and the direct relationship between the household and MNCs.

It is clear MNCs are affecting increasing numbers of women globally as they move into new areas, expand in existing areas, and extend their growing subcontracting and home work networks. The effects are being felt by women in the formal as well as the informal sector, including the household. Understanding how this happens becomes additionally important with the passage of world trade agreements, such as the Uruguay Round of the General Agreement on Trade and Tariffs (GATT). This agreement, passed by 124 nations, reduces worldwide tariff and nontariff barriers in many sectors and establishes the World Trade Organization (WTO). This U.N. agreement itself may have contradictory effects on women globally. On the one hand, it will free trade and increase the possibilities for MNCs in labor-intensive industries to move internation-

ally. On the other hand, as certain trade restrictions are gradually eliminated, it will remove the motivation for MNCs to move production facilities into additional tiers of Third World countries to circumvent present quotas on the amount of exports permitted from each country.

This overview of MNC trends in developing and industrialized countries has shown that women's employment in multinational networks is a global issue rather than simply a concern of particular developing countries or women in isolated pockets in industrialized nations. Recognizing the many parallels that exist in processes regarding women and MNCs both in developing and in industrialized countries can bridge the gap that often exists between women in different parts of the world. It also makes it clear that strategies for change must be developed not only locally but also internationally, with women in industrialized countries and those in developing nations recognizing the similar ways in which they are adversely affected by MNCs. Among the avenues for this are ILO conventions regarding workers, such as one adopted in 1995 to standardize and improve treatment of home workers.

This overview can assist in updating the debates regarding the effects of MNCs on women in the 1990s. It is now recognized, for example, that the effects of employment cannot be discussed solely in dichotomous terms (whether women benefit by it or are disadvantaged by it) because such employment has contradictory effects, providing an income opportunity that might otherwise not exist but with low wages and often adverse working conditions. Also recognized is the need for a broader framework of analysis for understanding women's employment in MNCs. Such an updated framework must include acknowledgment that the impact of MNCs includes direct employment in MNC factories and indirect employment in subcontracting and home work; comprehension of the parallels that exist internationally; and an understanding of the role of the government, as well as race, ethnicity, and culture, in mediating the relationship between women and MNCs.

Women are clearly of great importance as workers in MNCs' extended production networks. In many areas, they are also targeted specifically and encouraged, as reproducers of the next generation of workers, to increase their family size. And, as happened earlier in industrialized countries, women in developing countries are becoming increasingly critical to MNCs as the corporations seek to market their products in wider areas of the world. An agenda for future research regarding women and MNCs must be based on this broader framework of analysis and include all three aspects—women as producers, childbearers, and consumers.

References

Arrigo, Lourdes. "Economic and Political Control of Women Workers in Multinational Electronics Factories in Taiwan," *Contemporary Marxism,* vol. 11, 1985, pp. 77–95.

Benería, Lourdes, and Martha Roldán. *The Crossroads of Class and Gender: Industrial Homework, Subcontract-*

ing, and Household Dynamics in Mexico City. Chicago: University of Chicago Press, 1987.

Elson, Diane. "Women's Employment and Multinationals in the EEC Textiles and Clothing Industry." In Diane Elson and Ruth Pearson (eds.), *Women's Employment and Multinationals in Europe.* Basingstoke: Macmillan, 1989, pp. 80–110.

Fernandez-Kelly, M. Patricia, and Anna Garcia. "Power Surrendered, Power Restored: The Politics of Home and Work Among Hispanic Women in Southern California and Florida." In Louise Tilly and Patricia Gurin (eds.), *Women, Politics, and Change in America.* New York: Russell Sage, 1990, pp. 130–149.

French, Joan. "Hitting Where It Hurts Most: Jamaican Women's Livelihoods in Crisis." In Pamela Sparr (ed.), *Mortgaging Women's Lives: Feminist Critiques of Structural Adjustment.* London: Zed, 1994, pp. 165–182.

Gallin, Rita. "Women and the Export Industry in Taiwan: The Muting of Class Consciousness." In Kathryn Ward (ed.), *Women Workers and Global Restructuring.* Ithaca: ILR Press, Cornell University, 1990, pp. 179–192.

Gannicott, Kenneth. "Women, Wages, and Discrimination: Some Evidence from Taiwan," *Economic Development and Cultural Change,* vol. 34, 1986, pp. 721–370.

Hadjicostandi, Joanna. "'Facon': Women's Formal and Informal Work in the Garment Industry in Kavala, Greece." In Kathryn Ward (ed.), *Women Workers and Global Restructuring.* Ithaca: ILR Press, Cornell University, 1990, pp. 64–81.

Holt, Richard P.F. "The 1982 Reforms and the Employment Conditions of Mexican Women." In Nahid Aslanbeigui, Steven Pressman, and Gale Summerfield (eds.), *Women in the Age of Economic Transformation.* London: Routledge, 1994, pp. 178–191.

Jayaweera, Swarna. "Structural Adjustment Policies, Industrial Development, and Women in Sri Lanka." In Pamela Sparr (ed.), *Mortgaging Women's Lives: Feminist Critiques of Structural Adjustment.* London: Zed, 1994, pp. 96–115.

Li, K.T. "Contributions of Women in the Labor Force to Economic Development in Taiwan, the Republic of China," *Industry of Free China,* August, 1985, pp. 1–8.

Phongpaichit, Pasuk. "Two Roads to the Factory: Industrialisation Strategies and Women's Employment in South-East Asia." In Bina Agarwal (ed.), *Structures of Patriarchy.* London: Zed, 1988, pp. 151–163.

Pyle, Jean Larson. *The State and Women in the Economy: Lessons from Sex Discrimination in the Republic of Ireland.* Albany: State University of New York Press, 1990.

Rios, Palmira N. "Gender, Industrialization, and Development in Puerto Rico." In Christine E. Bose and Edna Acosta-Belen (eds.), *Women in the Latin American Development Process.* Philadelphia: Temple University Press, 1995, pp. 125–148.

Rosa, Kumudhini. "The Conditions and Organisational Activities of Women in Free Trade Zones." In Sheila Rowbotham and Swasti Mitter (eds.), *Dignity and Daily Bread.* London: Routledge, 1995, pp. 73–99.

Safa, Helen. "Gender Implications of Export-Led Industrialization in the Caribbean Basin." In Rae Lesser Blumberg, Cathy A. Rakowski, Irene Tinker, and Michael Monteon (eds.), *EnGENDERing Wealth and Well-Being: Empowerment for Global Change.* Boulder: Westview, 1995, pp. 89–112.

Salih, Kamal, and Mei Ling Young. "Changing Conditions of Labour in the Semiconductor Industry in Malaysia," *Labour and Society,* vol. 14, 1989, pp. 59–80.

Standing, Guy. "Global Feminization Through Flexible Labor," *World Development,* vol. 17, no. 7, 1989, pp. 1077–1095.

Stivens, Maila. "Family and State in Malaysian Industrialisation." In Haleh Afshar (ed.), *Women, State, and Ideology.* Albany: State University of New York Press, 1987, pp. 89–110.

Thailand Board of Investment. *Thailand: Investor's Guide.* Bangkok: Government Printing Office, 1987.

Tiano, Susan. *Patriarchy on the Line: Labor, Gender, and Ideology in the Mexican Maquila Industry.* Philadelphia: Temple University Press, 1994.

Truelove, Cynthia. "Disguised Industrial Proletarians in Rural Latin America." In Kathryn Ward (ed.), *Women Workers and Global Restructuring.* Ithaca: ILR Press, Cornell University, 1990, pp. 48–63.

Ward, Kathryn B., and Jean Larson Pyle. "Gender, Industrialization, Transnational Corporations, and Development." In Christine E. Bose and Edna Acosta-Belen (eds.), *Women in the Latin American Development Process.* Philadelphia: Temple University Press, 1995, pp. 37–64.

Wiegersma, Nan. "State Policy and the Restructuring of Women's Industries in Nicaragua." In Nahid Aslanbeigui, Steven Pressman, and Gale Summerfield (eds.), *Women in the Age of Economic Transformation.* London: Routledge, 1994, pp. 192–205.

Yelvington, Kevin A. "Gender and Ethnicity at Work in a Trinidadian Factory." In Janet Momsen (ed.), *Women and Change in the Caribbean: A Pan-Caribbean Perspective.* Bloomington: Indiana University Press, 1993, pp. 263–277.

Women and the Environment

Global Struggles for a Healthy Planet

Moema Viezzer
Thais Corral

Introduction

The story of how, in the closing years of the twentieth century, women of every class, color, and culture have emerged as a powerful, organized force for positive change around the world is a remarkable one. This essay focuses on the processes that formed women's visions and their influences on global issues, taking as a point of departure the analysis of women's participation in United Nations conferences, particularly the Earth Summit in Rio de Janeiro, Brazil, in 1992. This analysis will show that the field of women and the environment is more than a new entry in women's studies. The new circumstances that led to the various mobilizations, including the women NGOs' participation in world conferences, provide insights to bring about a change in the present model of socioeconomic development.

The question of women and the environment started to be incorporated in the women's struggles during the 1990s, after more than a decade of advocacy for equality in different spheres of human life. During the 1970s—at the beginning of the contemporary feminist movement—it was important for women to call attention to the fact that their bodies, as well as their lives in different classes, races, and cultures, were controlled; therefore, the first slogan adopted worldwide by the movement, which was growing rapidly everywhere, was: "Our Bodies, Ourselves." The outcome of this first period was the demand for recognition of women's basic right to control their own fertility and for the elimination of all kinds of violence against them.

The movement spread in all directions, including international institutions. One example of this phenomenon—very strong at the time in North America and Europe—was the convocation of the first U.N. conference of women, the International Women's Year World Conference (Mexico City, 1975). At this international gathering, it was recognized that discrimination against women all over the world deserved special attention by governments and multilateral institutions. From the deliberations of the 2,000 government delegates present, two documents emerged: the

Mexico Declaration on the Equality of Women and Their Contribution to Development and Peace and the *World Plan of Action for the Implementation of Objectives of International Women's Year.* As a result of women's action, the same year the United Nations voted a Decade for Women (1976–1985). These documents contained the guidelines that oriented numerous actions at both governmental and nongovernmental levels. These initiatives were evaluated at a mid-decade conference in Copenhagen in 1980. A fundamental outcome of that conference was the high visibility given to the Convention on the Elimination of All Forms of Discrimination Against Women (CEDAW) enacted the year before by the United Nations General Assembly.

The dissemination of "women's issues" was rapid with groups organized around the world as well as through networks and less formal special units at universities, institutions, and social organizations, many taking up specific issues. Demands arose for measures to be implemented by governments, international institutions, and development agencies. The international women's movement was blooming. The U.N.-Sponsored Third World Conference on Women (Nairobi, 1985), convened to evaluate the decade, demonstrated first of all that women had assumed their personal questions to be political. Fifteen thousand delegates from all over the world converged at the NGO (nongovernmental organization) Forum at Nairobi, bringing all sorts of questions and proposals. The evaluation of the decade revealed gaps and failures in the efforts to integrate women in development. Ten years later, at the Fourth World Conference on Women (Beijing, 1995), the women's movement demonstrated stronger networks and more shared interests than ever before. But, again, substantial gaps remained between intended policies and actions by governments.

The Rise of Women, Environment, and Development (WED)

It is possible to identify three phases in the understanding of "development" and its link to the conditions and status of women. (1) In the early 1970s, the analysis of women in development was best represented by Boserup (1970; see also Boserup and Lijencrantz, 1975). She asserted that women should be given access to technology as a precondition of benefiting from development. (2) Ten years later, the GAD (gender-and-development) perspective emerged, recognizing a much more complex framework within which development affected women. It emphasized the role that class and gender relations, as well as patterns such as culture, race, age, ethnicity, nationality, and history, play in this process (Benería and Roldán, 1987; Sen and Grown, 1987). (3) More recently, analysis of the effects of the economic and environmental crisis on entire populations of the Third World (aggravating the situation of women) has given rise to studies that have challenged the assumptions made by both Boserup and GAD. These studies state that women cannot benefit from a model of development that, in its essence, is grounded in the exploitation of women, of other races, of other countries, and of nature (Merchant, 1980; Mies, 1989; Mies et al., 1988; Shiva, 1988; Warren, 1987).

Mies points out that this perspective, which has not been addressed yet within the Marxist or liberal approaches, has come basically from the ecological movement, the alternative social movement, and the women's movement. Mies et al. (1988) suggest that the ecological approach has shown that a social model based on unfettered development of productive forces and unlimited growth will destroy the foundations upon which a socialist life might develop. There are limits to human action upon nature; nature is not limitless, which, in turn, affects humans because they are part of nature and are also limited.

The connection between the exploitation of women and of nature has been extensively addressed by Merchant. In her words: "We live under a world view and science that by reconceptualizing reality as a machine rather than a living organism, sanctioned the domination of both nature and women" (Merchant, 1980, p. 52).

Along the same lines, Mies et al., using the tools of materialistic analysis, state:

> Women and the subjugated peoples are treated as if they were means of production or natural resources such as water, air and land. The economic logic behind this colonization is that women (as means of production for producing people) and land are goods that can in no way be produced by capital (Mies et al., 1988, p. 5).

Stamp provides another important observation, analyzing the implications of development aid for women:

> Aid projects have been built in male dominated aid institutions and governments seek male knowledge in "local centers" of power knowledge, thereby unwittingly reinforcing male domination, disrupting the local power knowledge relationship, and alienating women from the development process (Stamp, 1990, p. 154).

Women come into the global debate on environment from a broad range of entry points. Their contribution takes a multiplicity of forms, from collecting garbage for recycling in the Brazilian city of Porto Alegre and fighting exploitive logging in the Amazon region by developing extractivist (that is, centered on the extraction of natural resources) activities as an alternative, to establishing the innovative Citizen's Clearinghouse to organize against hazardous waste in the western United States. In a number of countries, women have joined Women-in-Black Action, which was initiated by the Palestinian and Israeli women in Haifa and has become a worldwide movement of women for peace and against nuclear tests and violence.

About half of the world's food is grown by women. In Africa, women produce most of the food their families consume; in Asia and Latin America, they carry out key stages of producing and processing cash crops and are the main producers of vegetables, poultry, and livestock for the household. Women's knowledge of local social conditions, growing cycles, and other environmental aspects make them key in conservation. In India, a huge movement of peasant women has boycotted the market of genetically engineered seeds by maintaining seed banks on which local crops depend. They argue that the appropriation of seeds and food by transnational corporations will deprive local food growers and expose consumers to health risks from the new products.

Women of developing nations also have the primary responsibility for gathering fuel, food, and fodder from forest areas and for collecting and managing water. Women's traditional use of these resources has generally ensured their availability for future use. This perspective, born of everyday experience, differs from the priorities laid out by some environmental groups in the industrialized nations. As a woman from the Fiji Islands stated, ozone depletion and global warming are features that seem very far from the reality of mothers who see their children dying from drinking contaminated water, or large tracts of forests destroyed, or huge dams built; if technology displaces human labor, it is women who have to cope with the increased difficulties of day-to-day survival of their families.

This recognition has given Third World women the power to demand an equal voice in the fate of the earth: Since the mid-1970s, women have been bringing their unique life experiences, concerns, perspectives, and holistic analyses into the processes through which the United Nations, governments, international finance, transnational corporations, and public and private institutions shape policies.

The global women's environment movement had a starting point in two major U.N. actions: the Our Common Future Commission linking the environmental crisis to unsustainable development and financial practices that were worsening North/South inequities; and the Earth Summit, the informal name for the United Nations Conference on Environment and Development (UNCED), which took place in Rio de Janeiro, Brazil, in 1992.

Two events addressing the link between the global ecological and social crises were organized by women in preparation for the UNCED conference. Sponsored by the United Nations Environment Program (UNEP) and organized by the World WIDE Network of Women, the November 1991 "Global Assembly of Women and the Environment" brought together in Miami, Florida, more than 200 women who presented "success stories"—grass-roots initiatives that demonstrated that women are effective environmental leaders in solving a variety of problems in every region of the world. Their stories were considered tangible, sustainable, affordable, and replicable (IPAC, 1991; World WIDE Network, 1991).

Immediately following the assembly, the Women's Environment and Development Organization (WEDO) held in 1991 the First World Women's Congress for a Healthy Planet in Miami, which featured dozens of workshops and a tribunal of five female judges taking testimony from 15 experts who presented documented analyses of how the environment and development crisis affected and involved women. Attended by 1,000 women from 83 countries, the congress aimed at bringing women's perspectives to the discussions culminating in the drafting of UNCED's *Women's Action Agenda 21* and other official documents.

Women in Rio '92: Planet FEMEA

The Women's Caucus (a coalition of women's NGOs) and the advocacy methodology developed during the UNCED preparations in 1991–1992 were key to incorporating 120 recommendations and a whole chapter, "The Role of Women in Sustainable Development," in the official *Agenda 21* (UNIFEM, 1992). It proved to be so effective and popular that WEDO also organized women's caucuses during the preparatory meetings for the 1994 Sustainable Development of Small Islands Developing States Conference (SIDS), the 1994 International Conference on Population and Development (ICPD), the 1995 World Summit on Social Development, and the 1995 Fourth World Conference on Women.

During UNCED in 1992 in Rio de Janeiro, women's organizations, ranging from community groups to international networks, organized the first major NGO Forum ever held at a U.N. conference. They shared their unique life experiences, concerns, and perspectives in a colorful tent, Planet FEMEA. The work that took place there resulted in an action-oriented program that provided the grounds for the platform of the *Women's Action Agenda 21*.

Women's Action Agenda 21

This agenda made recommendations for practical steps that could be taken by the United Nations, national governments, industry, and NGOs on a host of related issues such as land rights and credit for women, women's health and reproductive rights, biodiversity and biotechnology, nuclear and alternative energy, environmental ethics, use of women's consumer power to protect the environment, and democratic rights (WEDO, 1992b).

The first item on the action agenda, "Democratic Rights, Diversity and Solidarity," highlights a code of practice that has shaped the women's movement. For women, only a society founded on the values of solidarity and diversity can right the wrongs and injustices that taint the world in which we live. Women have proposed to the United Nations a Code of Environmental Ethics and Accountability, based on principles of cooperation rather than competition, "which acknowledges the responsibility that accompanies power and is owed to future generations." They are critical of the current system, "the barren instruments (systems of national accounts) on which all major economic and environment decisions are made" and suggest that "governments agree to a timetable for implementation of full costs of accounting that includes environmental social costs—and assigns full value to women's labor in national accounting systems and in calculation of subsidies and incentives in international trade" (UNIFEM, 1992).

Women have always been a majority in the organized pacifist movement. They lead most of the movements in the world for improved quality of life that combat disposal of toxic materials, transport, and resource use. In the *Women's Action Agenda 21,* they urged "an immediate 50 percent reduction in military spending, with the money saved reallocated to socially useful and environmentally friendly purposes." Pragmatically, they proposed that "armies be used as environmental protection corps to monitor and repair damage to natural systems, including clean-up of war zones, military bases and surrounding areas, and . . . be available to assist citizens in times of natural and man-made disasters." Women also took a clear stand on the foreign debt and the rules of international trade, "rejecting the structural adjustment policies that shift the responsibilities of basic social services from governments to women without compensation or assistance" (UNIFEM, 1992).

In areas where women represent a majority of the labor force, their lives are particularly damaged by environmental destruction. The item "Women, Poverty, Land Rights, Food Security and Credit" on the *Women's Action Agenda 21* calls on "United Nations, governments and non-governmental organizations to cease discriminatory practices that limit women's access to land and other resources, to increase allocation of resources that enhance food security, and to provide appropriate technologies to reduce women's work" (UNIFEM, 1992).

The topic of "population and the environment" was one of those most debated by women during UNCED, where Planet FEMEA was responsible for coordinating the Population and Environment Treaty. Women objected to insinuations that population pressure is the chief cause of environmental degradation and countered that the true causes of the problem are "industrial and military pollutants, toxic wastes and economic systems that exploit and misuse nature and people" (UNIFEM, 1992).

The women also addressed issues of biodiversity and biotechnology, nuclear power and alternative energy, and science and technology transfer. *Women's Action Agenda 21* stressed one point in particular that guides women's action: their power as consumers.

> Aware that the power of the consumer is decisive in industrial planning and production . . . we will engage in campaigns supporting investment in environmentally sound productive activities and encourage initiatives to reduce fossil fuel energy use, over-consumption and wastes (UNIFEM, 1992).

The special nature of women's participation in UNCED has given the women's movement a new profile. The common thread running through these mobilizations resides in the women's will to give political force to a particular way of being and behaving in the world, founded on values that can usher in a new world order and that are both sensitive to women's autonomy and respectful of the environment.

Besides the official *Women's Action Agenda 21,* three major NGO treaties benefited from the full input and participation of women at the UNCED: the Treaty on Environmental Education for Sustainable Societies and Global Responsibility, the Treaty on Consumption and Lifestyle, and the Treaty on Population and Environment (CCIC, 1992). The treaties, agreed upon by the NGOs at the Global Forum, replicate the conventions and agreements that governments had made among themselves and serve as an alternative way of posing questions and issues that have been elaborated by the governments.

From Rio to Beijing

The implementation of these formal agreements has been carried out at different levels. Regarding the NGO treaties, various networks at national and regional levels have been created. The Treaty on Environmental Education for Sustainable Societies and Global Responsibility, for example, has been implemented among networks that already exist, such as ICAE (International Council for Adult Education) at the international level and CEAAL (Council of Adult Education for Latin America) at the regional level. In Brazil, as an outcome of the treaty, a Network of Organizations of Environmental Education has been created and has already promoted different actions agreed upon in the treaty.

The coalitions that organized Planet FEMEA coordinated the Treaty on Population and Environment, which was used to influence the agenda of the U.N.-sponsored International Conference on Population and Development (Cairo, 1994). The treaty was drafted to counteract the view that the population explosion was the main cause of environmental degradation. It established ethical principles that include respect for the human rights of women to decide the number of children they want and to have access to reproductive health services. It states:

> We condemn policies and programs, whether by governments, institutions and organizations, or employers, that attempt to deprive women of their freedom

of choice or the full knowledge or means to exercise their reproductive rights, including the right to interrupt unwanted pregnancies.

To assure full participation in the implementation of both the *Women's Action Agenda 21* and the treaties, WEDO has developed a methodology to highlight some of the main questions related to women. This methodology is embodied in the *Community Report Cards* (CRC) that have been translated into different languages (WEDO, 1992a). The methodology of the CRC has been used in various ways, at public hearings, on radio programs, and in workshops. It has proven to be a good start for different types of assessments and decisions.

In a 1994 meeting promoted by WEDO and REDEH (Network in Defense of the Human Species), an NGO in Rio de Janeiro, "Women, Environment and Development—Planning for Joint Actions," 35 representatives from Latin American countries gathered to evaluate developments since the UNCED conference and to promote new forms of joint action. The participants developed guidelines for action concerning women and the environment, including the following:

Methodological Principles. In principle and spirit, feminist methodology may be summarized as follows: It builds from women's knowledge and experience, thinking and acting locally and globally; it is participatory and endeavors not to reproduce authoritarian relations; and it is integrative and creative because it respects diversity.

Political Applications. Women are to access spaces for political decision making in order to argue environmental issues from the gender perspective, including the design of strategies for decision making for natural-resources management. Women are also to organize groups to pressure and negotiate with governments, states, and international organizations via lobbies. Further, women should design decision-making strategies for natural-resources management from a woman's perspective (considering the needs of women in their domestic, public, and community-organizing roles).

Integration of Local, Regional, and International Groups. Women are to organize regional networks and committees with women's groups working on environmental issues to exchange information and form pressure groups for gender and environment problems; women are also to reinforce relations via South–South networks (i.e., contact among the developing countries themselves).

Communication. Women are to make use of all available alternative or mass communications media to broadcast women's environmental message and actions and to undertake sensitization campaigns to curb the environmental impact of unsustainable development projects. For this purpose, they should use community, local, and national radios, television channels, and the printed press; they should use electronic mail for interchange among regional groups and to participate in international conferences.

Education and Training. Women are to train other

women and men to integrate and articulate the subjects of gender and environment, make decisions on proposals for intervention in the environment, design educational materials for training programs on gender and environment, and give prominence and due value to the various forms of popular culture, such as theater, songs, dance, rituals, folklore, and the like, in education and training programs.

Research. Women should carry out diagnoses from a gender viewpoint, in the form of both documents and fieldwork; from this perspective, they should identify those responsible for environmental problems and monitor development projects and their possible environmental and social impacts.

Strategies and Criteria for Negotiating Projects. Women should facilitate women's access to knowledge about institutions that fund environment projects; they should train women in methodologies they need to present, negotiate, execute, and assess environment projects.

Daughters of the Earth: The Environment and Development Collaborative Web

Under WEDO's coordination and with the participation of 80 women's organizations, the women's movement in ecology (commonly known as the Web) held the Second World Women's Congress for a Healthy Planet during the NGO Forum at Beijing in 1995. This congress reviewed what women had accomplished in the years 1991–1995 and what remained to be done in the face of conflict, violence, environmental degradation, poverty, sexual exploitation, rising fundamentalism, lack of political power, and a global economic system that subordinates human wellbeing to growth. Four female judges listened to testimonies from witnesses from different regions of the world on a range of issues and activism that revealed the struggles of women for environmental and social justice. WEDO (1995) cites some of the participants' statements as follows:

As one-tenth of the world's population, Chinese women are a vital force for environment protection. In the nearly three years since the first World Congress for a Healthy Planet, 30 million women planted 2.1 billion trees (Wan Shuxian, vice president, All China Women's Federation, p. 6).

As habitat and resource base disappear, many rural and indigenous communities are becoming environmental refugees. . . . If they enter the waged labor market, they tend to do so at the bottom. . . . Women are in a double bind: exposed to new vulnerabilities and dependencies while still obliged to fulfill existing responsibilities as caregivers of the communities (Marta Benavides, Instituto Internacional para la Cooperacion de los Pueblos, El Salvador, p. 7).

As consumers and investors, we women have the power and the leverage to change corporate behavior. Today, corporate social responsibility is discussed in the most conservative of forums. Companies will increasingly devote efforts to addressing social and environmental concerns—if consumers and investors keep the pressure on (Alice Marlin, president, Council on Economic Priorities, p. 7).

The principles of human rights should be extended to protect all people from risks to life or health arising from environmental damage, hazardous waste disposal, and air, water, or land pollution, whether from private acts or the acts of governments (Rosalie Bertell, International Institute of Concern for Public Health, p. 7).

The Web focused on women's action in both the First and the Third Worlds on some of the most critical issues for the twenty-first century: trade and the global economy; technology and communications; new militarism and new peace movements; health and healing; women's resistance strategies and sustainable alternatives; and sustainable consumption and livelihood.

New Issues

Two new topics have been added to the ecological movement, further increasing the range of issues that are addressed under this rubric.

Women's Perspectives on Peace and Militarism

Global arms and nuclear industries have drastically damaged Earth's ecosystems, bringing tremendous suffering. Indigenous peoples and communities have been targeted disproportionately, and women and children have suffered.

The logic of militarization and nuclearization in the name of social security has transformed all conflict situations into political wars of genocidal proportions. From international commerce and economic interdependence has evolved the military-industrial complex that sustains and fuels conflict situations between and within nation-states, be they in the name of nationalism, communalism, fundamentalism, ethnic assertion, or even "peacekeeping."

Cancer and Women

An area that attracts attention and, thus, support from women in developed countries is breast cancer. It is the most common cancer in industrialized countries and one of the most rapidly increasing cancers in developing countries. There are strong and compelling reasons to conclude that exposure to hormone-mimicking chemicals explains some of the increase of breast cancer. A campaign already taking place in some countries involves the organization of hearings in which the scientific community, the women's movement, and authorities try to come together to put forward some concrete measures.

Conclusions

Actions identified and implemented by women in the ecological movement are numerous and involve many issues dealing with a good and just life in general. From protection of the environment, ecology is now defined as protection of the environment and its inhabitants—particularly human beings. Rich and poor people make different demands on their habitat, yet across social classes women are emerging as particularly sensitive actors in response to problems that affect the world in which we all live.

The voluminous references concerning women in the documents issued at U.N. conferences are the result of women and men working together in the past few years. In the next stage, the Global Women's Environmental Movement plans to use a linkage strategy, in which local solutions impact global problems and governments are held accountable for their performance in various sectors. This strategy involves reiterating the need for greater recognition and support for women's concerns for a healthy planet. Darcy de Oliveira and Corral have written "Agenda of the Unacceptable," which synthesizes the visions of women:

The Disharmony of Earth
in the disruption of seasons
in the killing rains
in the poisonous dust of the wind
in the desolation of amputated forests
in the drifting of the poles
in the dispersion of the ozone aura
in the perversion of matter
in the wandering graveless wastes in the slaying

The Sorrow of the People
in the invisibility of Others
in the solitude of forgotten lands
in the unnecessity of the poor
in the price pegged to each gesture
in the ruthlessness of the markets
in the banning of meaning
in the timidity of Hope
in the denial of the Sacred
in the silence of Femininity

Life Gone Astray
in the soils become deserts
in the barrennes imposed upon wombs
in the industrialization of the cell
in the deprivation of desire
in the agony of species
in the withering away of diversity
in the hallucinations of Science
in the exile of Ethics (Oliveira and Corral, 1992).

The connections between the questions and issues posed by the women's movement and the environmental movement will deeply influence the future of civilization. The women-and-environment agenda, which has its roots in the history of marginalization of women and reflects other cultures besides the European, represents a real change in the principles that have oriented and marked the societies in which we live. Women have already begun to work for the transformations those principles demand.

References

Benería, Lourdes, and Martha Roldán. *The Crossroads of Class and Gender: Industrial Homework, Subcontracting, and Household Dynamics in Mexico City.* Chicago: University of Chicago Press, 1987.

Boserup, Ester. *Woman's Role in Economic Development.* London: Allen and Unwin, 1970.

Boserup, Ester, and Christina Lijencrantz. *Integration and Women in Development: Why, When, How.* New York: UNDP (United Nations Development Program), 1975.

Canadian Council for International Cooperation (CCIC). *International NGOs Forum at the Global Forum—Rio 92: NGOs Treaties.* Ottawa: CCIC, 1992.

Corral, Thais, and Rosiska Darcy de Oliveira (eds.). *Planeta FEMEA.* Rio de Janeiro: Brazilian Women's Coalition, 1993.

Dankelman, Irene, and Joan Davidson. *Women and Environment in the Third World: Alliance for the Future.* London: Earthscan, 1988.

International Planning Action Commission (IPAC) Steering Committee. *World Women's Congress for a Healthy Planet.* Official Report, 1991.

Merchant, Carolyn. *The Death of Nature: Women, Ecology, and the Scientific Revolution.* San Francisco: Harper and Row, 1980.

Mies, Maria. *Patriarchy and Accumulation on a World Scale: Women in the International Division of Labour.* London: Zed, 1989.

———. *Radical Ecology: The Search for a Livable World.* New York: Routledge, 1992.

Mies, Maria, and Vandana Shiva. *Ecofeminism.* London: Zed, 1993.

Mies, Maria, Veronica B. Thomsen, and Claudia von Werlhof. *Women: The Last Colony.* London: Zed, 1988.

Sen, Gita, and Caren Grown. *Development, Crises, and Alternative Visions: Third World Women's Perspectives.* New York: Monthly Review Press, 1987.

Oliveira, Rosiska Darcy de, and Thais Corral. "Agenda of the Unacceptable." Mimeographed, 1992.

Shiva, Vandana. *Staying Alive: Women, Ecology, and Development.* London: Zed, 1988.

———. *Staying Alive: Women, Ecology, and the Scientific Revolution.* New York: Harper and Row: 1988.

Stamp, Patricia. *Technology, Gender, and Power in Africa.* Ottawa: International Development Research Center, 1990.

United Nations Development Fund for Women (UNIFEM). *Agenda 21: An Easy Reference to the Specific Recommendations on Women.* New York: UNIFEM, 1992.

United Nations Environment Program (UNEP). *Global Assembly of Women and the Environment "Partners in Life."* Official Report. New York: UNEP, 1992.

Waring, Marilyn. *If Women Counted: A New Feminist Economics.* San Francisco: Harper and Row, 1988.

Warren, Karen. "Feminism and Ecology: Making Connections," *Environmental Ethics,* vol. 9, no. 1, 1987, pp. 3–20.

Women's Environment and Development Organization (WEDO). *Community Report Cards.* New York: WEDO, 1992a.

———. *Women's Action Agenda 21.* New York: WEDO, 1992b.

———. *News and Views,* vol. 8, nos. 3–4, 1995.

World WIDE Network. *Global Assembly of Women and the Environment: Partners in Life: Success Stories of Women and the Environment: A Preliminary Presentation in Anticipation of the Global Assembly.* Washington, D.C.: World WIDE Network, 1991.

Women's Role in Natural-Resources Management

Sara Ahmed

Introduction

Since the early 1970s, there has been considerable interest in the relationship between women, particularly poor rural women in developing countries, and the environment—that is, the natural resource base on which development depends. Initially, women were viewed as the primary victims of environmental degradation since they were the main users and providers of household biomass and subsistence needs (firewood, fodder, minor forest products, and water). Gradually, the debate recognized that women have a particular role to play in natural-resources management, because of their knowledge and privileged experience gained from working closely with their environment (Mies and Shiva, 1993). Consequently, women came to be seen as the solution to the development-environment crisis, as major assets to be "harnessed" in initiatives to conserve resources and as "fixers" of ecological problems (Braidotti et al., 1993). But ascribing to women the additional responsibility of being caretakers of the Earth, without, in turn, addressing their access to and control of natural resources, knowledge, information and decision-making systems, and the products of their own labor, means that the positioning of women as "naturally privileged environmental managers" is problematic (Agarwal, 1992).

This essay looks at the scope and nature of rural women's managerial roles and responsibilities in collective efforts to use, conserve, and manage community-based natural resources. It argues that while participatory community-resources management projects have, to some extent, been able to target women's practical gender needs—clean and accessible water, sanitation facilities, wood-saving stoves—they have yet to address strategic gender interests. Women's generally subordinate position vis-à-vis men, which stems from gender inequalities in the division of labor as well as sociocultural, political, and historical factors, determines their participation in decision-making regarding project design, implementation, operation, maintenance, and management. In other words, rural women are not a homogeneous unit: Their responses to environmental degradation are mediated by their different livelihood systems and strategies, which reflect multiple objectives.

Property Regimes and Women

"Resources may be defined as those components of an ecosystem which provide goods and services useful to [humankind]" (Gibbs and Bromley, 1989, p. 22). Natural resources exist either as nonrenewable "stocks," in which the physical quantity available for use is fixed (coal, mineral deposits) or "flows," which can be managed to yield goods and services sustainably (water, forests).

The concept of management when applied to natural resources, suggests control over decision making concerning the rational and productive use of resources. At the level of the community or village, management has to meet the needs of diverse resource users while ensuring that the rate of resource exploitation employed is sustainable in the long-term and that ecological degradation is minimal.

Prudent management depends upon a careful orchestration of resource-use policies and implementation practices guiding the conservation of natural resources. This includes a number of actors and agencies with conflicting demands and priorities. At base, conflicts stem from the nature of tenurial (property) regimes, which govern rights, or access to, and control over particular resources. In addition, economic and political interests, social structures, and institutional relationships of power all play a part in determining the operational context of natural-resources management.

There are essentially three broad categories of property regimes, each raising different implications for natural-resources management and the role of women:

State-Property Regimes. These regimes, in which ownership and control over resource use rest with the state, are particularly extensive in most developing countries. The increasing appropriation of resources in the name of the public or national interest—for example, the reservation of land for "developmental" purposes (missile sites), the creation of parks and sanctuaries, or the construction of large dams—has a serious impact on users, especially women, who are denied access to the resource or are affected by displacement and rehabilitation.

Private-Property Regimes. These are the most prevalent form of regimes governing the use and distribution of land. Despite numerous attempts at land reforms in developing countries, the concentration of land among a few accounts for increasing poverty and overuse of marginal lands by the landless, leading to further ecological degradation.

Women's lack of ownership of land or their inability to self-manage land they have inherited has several implications. Decision making on crops, access to credit from institutional and private sources, as well as access to information or technology on productivity-enhancing agricultural practices and inputs, are usually denied to women. Land titles are invariably in men's names since, officially, they are identified as heads of households, though they may have migrated a long time ago, leaving women as de facto managers. Even if a woman does inherit land, the ideological division of village space constitutes a strong barrier to her exercising effective control over her land. Not only is a woman's mobility defined by age, seniority, status, or whether she is a daughter or a daughter-in-law, but her ability to command the labor of other household members is also limited. Inevitably, many women relinquish, or are forced to relinquish, claims to parental land (Agarwal, 1988).

Livestock inheritance also reveals gender differentials, with men having predominant ownership rights over large stock such as cattle, while women usually own smaller stock, though they can inherit cattle from their parents. Decision making over the disposal of animals, or herd migration in pastoral societies, is the prerogative of the male, although women may influence such decisions.

Common-Property Regimes. These are similar to private-property regimes in that they exclude nonmembers from use and decision making. Property-owning groups are essentially ". . . social units with definite membership and boundaries, with certain common interests, with at least some interaction amongst members, with some common cultural norms, and often their own endogenous authority systems" (Bromley and Cernea, 1989, p. 15). Women have traditionally had access to the "commons" to meet household biomass and subsistence needs, but the growing privatization and statization of grazing lands and forests have impinged on their usufruct rights (Jacobson, 1992).

Natural resources are rarely held under any one property regime. It may be useful at this point to make distinctions among resource use, management, and control (ownership). For example, private-property regimes need the power of the state (laws) to ensure individual rights and prevent intrusion by nonowners. At the other end of the scale, rural communities who have had de facto use, or access to, resources around their villages (forests, grazing lands, water) are not necessarily the owners, except in the case of common-property regimes. The state, for its part, is unable to manage resources it technically owns—bureaucracies are underfunded, large scale, and managerially distanced from the resources in question (Murphree, 1993, p. 4). Thus, the breakdown of property regimes means that many resources are held under "open-access regimes": In the absence of any visible authority system, "everybody's access is (or becomes) nobody's property" (Bromley and Cernea, 1989, p. 19). Most rivers, lakes, oceans (beyond the territorial zone), and the atmosphere also fall under this regime; attempts to control the use of such resources face many institutional problems arising from the difficulty of imposing effective behavioral norms.

Approaches to Natural-Resources Management

Although resource exploitation has been going on for centuries, the concern to "manage" extensive resource degradation is a more recent phenomenon. Early approaches to natural-resources management in developing countries were, on the one hand, rooted in the discourse of nature conservation and protection through the creation of parks and sanctuaries to preserve wildlife. On the other hand, colonial rule imposed on its subjects a set of "scientific management" principles, which, although they did herald the advent of conservationism, also contributed to large-scale resource exploitation in the interests of imperial capital.

By the mid-1900s, ecological managerialism was dominated by neo-Malthusian concerns about population pressures on "spaceship Earth." Apocalyptic visions of the future—the doomsday syndrome—were pointed at resource issues in the Third World. Bioeconomic models (limits to growth) had significant political implications: Developing countries saw them as attempts to impose "technocratic globalism" on their path to development (Adams, 1990, p. 29).

However, it soon became clear that governments could not manage the environmental crisis on their own. The World Conservation Strategy, launched in 1980 by the International Union for the Conservation of Nature, blamed the weak sectoral approach of resource-management agencies, with their often overlapping responsibilities. Its calls for more public participation in development planning and management were reiterated through concepts such as "ecodevelopment" and, later, "sustainable development." Despite the focus on appropriate technology, basic needs, and self-reliance, the top-down rhetoric of participatory sustainable development ignores the hard reality of unequal sociopolitical structures, which deny marginalized people, such as tribals and poor women, the right to participate in such development. Moreover, professionalism implies specialization, and the so-called "tyranny of expertise" tends to ignore other knowledge systems, particularly the gendered domain of female knowledge.

The failure of the state to sustain environmental management through the development of decentralized community-governance systems has led to several alternative initiatives, usually facilitated by external agents like nongovernmental organizations (NGOs) or strong local leaders. Participation, and more recently women's participation, has also become a buzzword in this area, crucial to the success and sustainability of many such interventions. Although NGO approaches and methodologies have tended to be bottom-up, democratic, and flexible, they, too, raise questions about the scope and nature of women's participation: At what level is participation sought, by whom, and why?

The rest of this essay examines the extent to which rural women have the right and the ability to participate in the management of community-level water, land, and forest resources. Most of the examples drawn upon refer to resources that are state owned and/or open access, with attempts being made to foster partnerships among government, NGO, and community-level (village groups, cooperatives) actors through different projects and programs.

Management of Water Resources

Fresh water, the main source of potable water for human use, is a renewable, but vulnerable, resource. The availability of water is contingent on a number of interconnected factors, such as deforestation in catchment areas, which leads to surface runoff and depletion of groundwater. Growing populations and competing demands for water—for example, between agriculture, which accounts for 73 percent of global water use, industry, and leisure activities, such as watering golf courses—affect decisions about the distribution of water resources. In addition, pesticides and fertilizers from agriculture and untreated effluents from industry pollute surface water, and eventually groundwater, in the absence of effective regulatory mechanisms. The lack of sewage disposal and treatment facilities also contributes to water pollution. The conflicting demands on water sources and the interdependencies among water users suggest the need for a holistic approach to the sustainable management of water systems.

Although in many developing countries women play a primary part in water collection and use, their role as managers of water resources was officially recognized only with the launching of the United Nations Water Supply and Sanitation Decade (1981–1990). Women's knowledge about the location and availability of water sources and social customs governing the use of water and disposal or reuse of wastewater have provided valuable inputs for the design and implementation of water supply and sanitation projects. In addition, women are increasingly assuming community leadership roles in organizing groups for the harvesting of rainwater in drought-prone areas and in improving the quality of water through the chlorination of village wells.

Women's relationship with irrigation projects, however, is poorly documented or conveniently overlooked on the assumption that it is not significant. Most of the institutional literature on farmer participation in irrigation management is essentially gender blind, even when considering issues such as the transfer of power from large, inefficient irrigation bureaucracies to water-user associations.

Gender Issues in Irrigation Management

Despite the growing awareness that effective irrigation planning and management approaches need to take gender considerations into account, women water users' needs and interests are often not clearly understood. Not only do women lack institutional support—technical training and knowledge concerning irrigated agriculture are assumed to be a male domain—their access to and control of critical resources, such as land, is also limited. In many cases, women's role in agriculture is not officially recognized; farmers are viewed as male, even when female participation in agriculture is greater.

The largely negative impact of large-scale irrigation and hydroelectric-power projects on women's livelihood systems has been brought out by several evaluation studies undertaken by women. Jackson (1985) looks at the differential impact of the Kano River Project in Northern Nigeria on Muslim and non-Muslim Hausa women in the command (irrigated) area. The KRP is a gravity flow irrigation scheme that lies about 50 kilometers south of Northern Nigeria in the region called Hausaland. Designed to increase food supplies (particularly wheat and tomatoes for urban areas), provide employment opportunities, and improve the standard of living, the KRP also included plans to provide basic infrastructure to transform the area from isolated farmstead type of settlements to nucleated villages.

Over 95 percent of Hausa are Muslim; Muslim Hausa women are mainly petty commodity producers (producing snack-food for sale, for example). Despite restrictions on their mobility and contact with nonfamily men, Muslim women appear to have greater financial autonomy and the right to inherit land compared to non-Muslim women. Thus, they were able to maximize their income-earning opportunities in response to the increased demand for snack food from construction workers on the KRP.

Non-Muslim Hausa women, on the other hand, depend on land as the primary means of production, but lack both the right to inherit and the finance to purchase land of their own. Thus, marriage provides access to small plots of land allocated to them by their husbands on use rather than ownership basis. However, as a result of the KRP, non-Muslim Hausa women were being allocated smaller pieces of land and these were usually appropriated during the dry season by men who wanted to benefit from irrigation for commercial crops such as tomatoes.

On the whole, despite expanded economic opportunities for Muslim women, both groups of women were affected by the destruction of trees (loss of fuelwood), the increasing contamination of water sources, and the general disruption of social networks.

The absence of efforts to integrate women has been documented by studies of the Aitang Ai Hydroelectric-Power Project on Sarawak, Malaysia, and the Mahaweli Scheme in Sri Lanka (see Heyzer, 1987). In both cases, women have lost their traditional rights to land as a result of resettlement. New systems of plantation agriculture have eroded women's traditional equality with men in the sphere of production, relegating them to the lowest strata in the work hierarchy. The disintegration of community-based social networks has also had a negative impact on women.

However, attempts to integrate women in irrigation projects are problematic. Schenk-Sandbergen (1991) describes how gender concerns were included in a Dutch-aided small-scale irrigation project in the rain-fed *terai* (plains) of North Bengal, India. The main focus of the project, undertaken by the Bengal Directorate of Agriculture, was the implementation of five types of small-scale irrigation facilities: river-lift, deep tube wells, shallow tube wells, pump dug wells, and hand pumps.

Social norms in the area, such as purdah (seclusion of women), and the absence of local women's groups meant that gender sensitivity had to be imposed from above. Tackling women's practical gender concerns through involving them in the selection of hand pump sites has contributed to limited empowerment of some women by increasing their mobility. Those from marginally better-off households are able to work in their vegetable gardens because of the introduction of new crops that are not associated with purdah restrictions like the traditional crops (jute and rice).

Women who have been able to inherit tiny plots of land from their fathers have also received ownership of hand pumps. However, recruiting women as extension workers, appointing them to water-user committees, or increasing their access to training is a slow process. This is partly because of official rules and regulations (slowly changing) that discourage women from applying for government jobs, bureaucratic mind-sets about the "role" of women, and the fact that nonlandowners are not perceived as "beneficiaries."

Jordans (1991) looks at women's participation in another Dutch-aided project, the Delta Development Project (DDP) in southwest Bangladesh. Initiated in the early 1980s, the DDP sought to protect low-lying areas from floods by building embankments. Other project objectives included improving agricultural yields and standards of living. Since 1987, the project has focused on the landless, half of whom are women, looking for ways to enhance their participation in the construction and maintenance of embankments. With the help of a local NGO, Nijera Kori, women's committees have been formed. Recognized by the Bangladesh Water and Development Board, committees are appointed for a one-year term, during which time they are responsible for monitoring organizational and financial aspects concerning women's work, such as the distribution of wages.

However, maintenance work is heavy and difficult, particularly during the monsoons. The financial impact of the work is limited because of its short duration, and gains have to be weighed against increases in women's daily workload. Although women's public work in this tradition-ally purdah-bound area was gradually being accepted by the local community at the time of the study, women were not involved in water management per se: They were not considered in decision making regarding the distribution of irrigation water through water-user groups. There were no female extension workers, and agricultural research was directed at male farmers.

In contrast, Proshika, one of the largest NGOs in Bangladesh, has actively promoted its all-women village-level groups of 15–25 females to take up management of pump-irrigation schemes. Women have been trained to operate and maintain a variety of irrigation equipment. Though the program was expanding in the mid-1990s, social constraints on women's mobility had forced a few groups to wind up activities or pass them on to their husbands.

Cultural factors force women to depend on male relatives to harvest and collect crop-share revenues for the sale of water or to help them sell rice in the market, thus undermining their ability to control financial aspects of the operation:

> Forbidden to go out in the fields at night, women have to make special arrangements for sleeping at the pump-house to guard it or to supervise irrigation at night. They are likely to encounter male farmers' resistance in negotiating water contracts with them and in negotiating the site of the pump (FMIS, 1992, p. 4).

The Sreeramsagar Project in Andhra Pradesh, south India, illustrates a rather ingenious collaboration of men and women on a minor irrigation scheme. In the village of Hyderpet, women do the irrigating, leaving the men free for procurement and guarding. Because of limitations in the availability of water, households in the command area have been divided into five groups by the women, and each group is entitled to water once a week by rotation. Based on their turn, men from the group (one from each family) set out to patrol the system, removing obstacles along the way, while the women in the group take turns to irrigate their fields. In the village of Ananthram, the roles are reversed: Women guard the irrigation system and remove obstructions while men irrigate the fields.

Although men have a more visible role in community irrigation management, women are involved in informal decision making, at the household level, about *when* to irrigate. In addition, women do play a mediating role in resolving disputes between male irrigators. But for the most part, despite their role as wage labor in the construction of community/public irrigation facilities, women have only marginal access to participation in formal decision making. Membership in water-user associations or cooperatives is contingent upon ownership of land, which inevitably rests with the male head of household. Although there are a few instances of token representation of women on water-user committees—for example, female heads of households—their active participation is constrained by their limited numbers, their workload, and social norms. There is clearly a need for more research on the various agricultural, irrigation, and management activities performed by women and men and the interdependencies between them so they can both be made effective stakeholders in projects.

Water Supply and Sanitation

Women's role as primary water carriers and family health educators has led to a greater focus on them in water supply and sanitation projects than in irrigation projects. However, the emphasis has been largely on meeting their practical needs: improved or accessible water supply and sanitation facilities to reduce the time and energy women spend in collecting water and finding a private place for their daily ablutions. In addition, public-health education, with its almost exclusivist focus on women and children, remains top down and prescriptive, overlooking men's need to support and adopt improved hygiene practices.

Women's participation at all levels of water supply and sanitation projects, from decisions concerning site selection, to training in the operation and maintenance of new technology, to being members of local user committees, is considered important for ensuring the effectiveness and sustainability of projects. In many areas, women do have informal decision-making roles in the public management of traditional water sources, such as the collective cleaning of village wells and the culturally determined allocation of different water sources for different use purposes (bathing, drinking, cooking, washing). Often, these roles are not easily obvious to male technicians involved in the introduction of new facilities, partly because of communication taboos that restrict them from talking to local women and partly because such roles are not always acknowledged, by either the community or the researchers.

Although women are increasingly being involved in project planning, design, implementation, and maintenance, because of the benefits perceived they face a number of hidden, negative impacts. While the drudgery of water collection may have been reduced, women may be spending more time in the maintenance and management of the new facilities. Yacoob and Walker (1991) remark that in Rwanda, women were trained to operate and maintain a community water supply and sanitation project. After the implementation of the project, women found they were spending more time collecting fees than collecting water. Undoubtedly, the former task is less physically taxing, but there are inherent conflicts, particularly when users do not want to pay for services.

In another example, as part of a pilot project to improve hand-pump maintenance, the Integrated Sanitation, Water, Guineaworm Control and Community Health (SWACH) Project in Rajasthan, India, trained 24 rural women as pump caretakers and community change agents in the mid-1980s. The SWACH Project involves converting traditional step wells into draw wells or covering them on the top and installing a hand pump.

Women were trained in hand-pump maintenance on the assumption that they would be the most motivated to maintain them, since they have to bear the burden of frequent hand-pump breakdowns. Maintenance is also seen as an income-generating activity, albeit limited. Women work in groups of three as hand-pump mechanics, necessitating travel of about 12 miles per day.

An economic analysis of the project evaluating the costs of training men and women found that although training costs per pump are almost three times higher in the case of women mechanics, the breakdown rate and the duration of the breakdown before it is attended to are much lower in the female-tended system. However, if an economic value is ascribed to women's work (domestic, agricultural, community management), then it was found that the social-opportunity costs for women mechanics are high. Training and repairing pumps, while giving women confidence in a traditionally male-dominated area, puts time pressure on their other responsibilities, sometimes forcing them to hire labor for farm work.

Maintenance and management tasks require considerable time inputs, from both men and women, inputs that are "seldom measured and incorporated as the contribution of villagers to the total financing of a project, as if time is a free good which they can spend at will" (Wijk-Sijbesma, 1992, p. 5). In addition, the largely physical nature of women's work in this respect (keeping standposts and drains clean, collecting fees and reporting problems) means that they are not always recognized as having the status and authority of managers.

Evaluating women's managerial roles in a rural water and sanitation program in Tanzania, Chachage et al. (1990) note that, although women are represented on user committees, they do not always have the time to attend meetings. If they do attend, cultural norms prevent them from speaking, and they have no recognized status with the village government. Furthermore, male community-development workers lack communication skills essential to mobilizing women. Similar insights were gained from the evaluation of gender participation in an integrated water supply and sanitation project in Zimbabwe. Extension workers, council members, and beneficiaries, including women, had strong beliefs about what women should or should not do, often reinforcing gender stereotypes.

However, women have played a prominent leadership role on water committees in decentralized water supply schemes in Colombia. Part of the reason for this could be that women on the committees already had positions of status in the community, as teachers and political leaders, and so were more readily accepted when it came to tasks like tariff collection.

Strong agency support is a key factor in mobilizing women's participation, as the Kwale Water Supply and Sanitation Project in Kenya illustrates. Essentially a government effort backed by the World Bank and financially funded by the Swedish Development Cooperation Agency and the Kenyan government, the project sought to introduce sturdy village-level hand pumps in the south-coast district of Kwale. An NGO, KWAHO (Kenyan Water and Health Organization), was involved, primarily to train women as extension workers and facilitate participatory methods of work and communication. KWAHO, in turn, approached PROWWESS (Promotion of the Role of Women in Water and Environmental Sanitation Services, a United Nations Development Program [UNDP] agency) and later UNICEF, United Nations Children's Fund, PACT (Private Agencies Collaborating Together, an American NGO), and Wateraid (a British NGO). "Every agency involved was characterized by strong leadership willing to take risks, make long-term commitment and practice partnership" (PROWWESS/UNDP, 1991, p. 131).

The impact of such widespread institutional support is clearly visible at the grass-roots level. Joint male and female village-level committees (more than 125) have proved successful in raising money for maintenance of pumps, forming rules, resolving conflicts, and seeking additional information from project authorities through extension workers. Women have been gradually accepted in public decision-making roles and have gained confidence in themselves through the support of the community, particularly men.

Experience has shown that the introduction of low-cost water and sanitation technologies, accompanied by public-health education, is a necessary, but not a sufficient, prerequisite for improving the quality of life of intended beneficiaries, particularly women and children. Although women are becoming involved in the management of community infrastructure, there is a need to understand and strengthen their decision-making roles in the context of project design, planning, and operation. By placing new demands on women's labor and time, projects that fail to address strategic gender needs only reinforce existing gender inequalities.

Gender Issues in the Management of Land Resources

Land-use patterns support a number of different activities, including grazing, agriculture, and human settlements. In addition, forests occupy more than a quarter of the world's land area (World Bank, 1992, p. 57). Land-degradation problems range from soil erosion and salinization to deforestation and desertification. The causes vary, from individual or communal pressures on fragile agroecosystems (intensive farming practices) to the appropriation and exploitation of land and forests by states and commercial interests (ranching, logging). Rarely, except perhaps in the case of private landholdings, is only one agent responsible. Decision making about the management of land resources must take into account the diversity of uses and users.

Responses to land degradation are undertaken either at the individual farm level, through better farming techniques and the like, or collectively through a combination of community-based, government- or donor agency–funded, and NGO-fostered initiatives. The most pervasive intervention has involved the planting of trees on degraded land (wastelands) officially classified either as forests, but

with little tree cover, or as community grazing lands. Trees are also planted as a buffer against desertification. Such efforts are supposed to meet both environmental and social objectives since trees are an important source of savings, security, and subsistence for the poor (Chambers et al., 1993).

Women have gradually been recognized as significant assets in land-rehabilitation and forest-management efforts because of their perceived relationship with trees. The division of labor determines that, in most societies, it is women and sometimes children who collect firewood, fodder, and other forest products, often walking considerable distances with heavy headloads. Shortages of firewood or woody biomass, the main source of domestic energy in most rural and semiurban areas, affect dietary patterns. Food is cooked either for less time or less frequently in the day, which means that it is always women (and sometimes the girl child) who, despite greater energy expenditure in collecting firewood, eat less. In addition to subsistence needs, women also use forest products for a variety of income-earning opportunities, such as the brewing of beer in parts of southern Africa, which requires large amounts of firewood.

Management of Trees in Forests and Wastelands

While deforestation has been acknowledged as a serious problem, with widespread socioeconomic and ecological consequences, there seem to be a number of conflicting views as to its causes, the agents responsible, and what can be done to halt it. The dominant Malthusian perspective of national governments, forest professionals, and some donor agencies blames overpopulation and poor, ignorant peasants and tribals for overgrazing or overcultivation on fragile slopes, as well as cutting down trees to meet their basic needs. An alternative argument sees the overexploitation of tropical forests for commercial, industrial, and developmental purposes, such as construction of roads and dams, coupled with the systematic enclosure of the commons, as the main reasons for the marginalization of forest users and degradation of forest lands.

Not surprisingly, these two viewpoints have led to different responses, both to the forestry crisis and to the role of women in the management of forest resources. Community or social forestry projects, launched in the late 1970s, were essentially attempts by governments and donor agencies to involve local communities in the plantation and management of forests near villages. The need for community participation and decentralized management arose from the increasing failure of governments to meet afforestation targets (success rate of trees planted) or objectives (firewood and fodder needs). Many plantations were systematically destroyed, either by communities that had been excluded from decisions regarding the choice of species or indirectly by straying grazing animals and encroachments.

However, for forest officers accustomed to running things from above, the sharing of decision-making power and authority with local communities was both personally painful and institutionally difficult. In practice,

> Local people, assumed to be ignorant, were subjected to disciplinary technology: extension programs were designed to educate farmers, schoolchildren, women and cooperatives about the benefits of tree planting. Farmers were taught how to plant and lop trees correctly and how to manage forest plantations (Hausler, 1993, p. 86).

By the mid-1980s, as pictures of women carrying heavy loads of firewood across barren landscapes began to flood the media, forestry planners realized that it was important to involve women in plantations. Not only were they the primary victims of deforestation, but men and women made different uses of forest resources, and, therefore, they had different needs. Women's needs had inevitably been overlooked by a profession dominated by men, in terms of both research on suitable types of tree species and practice—the inability or unwillingness of male extension workers to talk to women in the community.

However, women could not simply be added on to existing social forestry programs, nor was it going to be easy to change embedded attitudes at the level of the community and in forestry bureaucracies regarding the role of women. Thus, most top-down initiatives have ended up paying lip service to women's participation, reducing it to another wage-labor component.

The Community Forestry Development Project (CFDP), launched in 1980 in Nepal, is a case in point. Funded by the World Bank, the UNDP and the Food and Agriculture Organization (FAO), it broadly sought to undertake reforestation and resource utilization through village-level *panchayats* (administrative councils) with varying degrees of legal control over forest land. Although women were not specifically referred to in the original project plan, it gradually became evident to implementors that women needed to be involved because of their relationship with the forests. For example, it was thought that since women generally graze animals, they must be made aware of newly planted areas, the need to protect them, and to switch to stall feeding.

While it was possible to target extension information to women, particularly in the complementary woodstove program, it was difficult to sustain women's participation on forest-protection committees or in training schemes. Employment of women under the project was also far from satisfactory, partly because of sociocultural norms and the nature of political power at the local level and partly because they had not been consulted at the project-design stage (Molnar, 1989). Collaborative forest-management systems in Southeast Asia have also been planned by centralized bureaucracies distanced from social realities and unable to respond to diverse community-defined needs. Attempts to reorient forest management on Java through experimental community-development strategies in the 1980s supported by the Ford Foundation have largely failed to involve women despite the gender sensitivity of planners. Forest Farmer Groups initiated in pilot agroforestry projects have excluded women and the landless because membership was based on land ownership (head of household). A few women, mostly widows, were able to obtain plots in selected pilot sites, but foresters tended not to select women for the program and most women were too embarrassed to ask to be included. Female heads of household were not given the same access opportunities to benefits; only men were allowed to sign contracts for access to the fruits and leaves of trees planted on state forest lands.

Social forestry programs in India have fared little better. A review of the literature on program evaluation suggests that women's main role has been that of wage laborers, and that, even in this limited sphere, women have not acquired new skills but have simply been doing weeding, watering, and preparing polythene bags for saplings. Thus, the benefits (income) they have gained "have not been so much the result of efforts to target them, but because poor women constitute a substantial part of the daily wage earners in the countryside" (Venkateswaran, 1992, p. 68). Employment is short term, inadequate, or underpaid and depends on the kind of land on which planting is undertaken and the choice of tree species.

The deployment of women extension workers has not always proved constructive. Unclear job descriptions have led many female extension workers to focus on men, using women only as an entry point to the village. Moreover, the use of women is not a guarantee of "natural" solidarity with poor rural women, for research shows that "whatever the class background of women fieldworkers, their primary reference group is likely to be their superiors or colleagues, not their clients" (Goetz, 1992, p. 14).

The failure of state-initiated community forestry has led to a shift in focus to farm forestry or agroforestry, either at the individual household level or through the privatization of the commons, in which a select group is granted exclusive rights over land once accessible to the community. Not only do such schemes exclude women because of their lack of ownership of land, they invariably involve extensive cash cropping, which, in turn, denies women access to biomass needs and often leads to ecological degradation.

Officials of the Kenya Woodfuel Development Program (KWDP), a multiagency effort to increase the supply of wood fuel through tree planting at the household level, soon realized that, although farmers were actively planting trees, there was still a firewood shortage. Social and cultural norms denied women, major collectors of firewood, access to trees on farmland, since trees demarcated the ownership of individual landholdings and were seen as important sources of income and investment. In most parts of southern Africa as well, trees are generally regarded as men's property. Men exercise the sole right to plant trees, as land-ownership disputes are resolved by adjudicating on the basis of male tree planting. Women generally collect firewood from indigenous tree species grown on agricultural land, such as *Sesbania sesban,* a bushlike plant that enhances soil fertility. Through dialogue with the community and cultural awareness programs, the KWDP was able to introduce *Sesbania* and similar species to new areas as soil-fertility-improving bushes rather than trees.

Over the years, many NGOs have become involved in tree-growing programs, acting as important links between farmers and governments and, in the process, creating decentralized forest-protection groups or cooperatives. Increasingly, women have been considered as a specific target group, either through the creation of separate women's projects or by addressing gender concerns in joint programs. However, despite the gender-and-environment rhetoric, many NGOs have often found it as difficult to include women in forest management as the state, for both organizational and social reasons.

In India, NGOs have played an important part in trying to integrate women in collective efforts at regenerating degraded wastelands through three characteristic forms of organization: mixed groups, exclusive women's groups, and separate groups for women that can eventually merge with village-level institutions. Experiences from the field suggest that targeting women exclusively can lead to male suspicion and resentment and that initial joint meetings with both men and women can help address apprehensions about separate groups. But conservative social structures dictate that women do need their own space for some time to express their views freely (Singh and Burra, 1993).

However, the most critical factor affecting women's participation is that of ownership and usufruct rights to the produce of the land they are developing or protecting. In the case of exclusive women's projects, such as the Bankura experiment in West Bengal, these are clearly spelled out (International Labor Organization, 1991). Landless and marginal peasant women were donated degraded land by private landowners who lacked the resources to develop it. Supported by the Center for Women's Development Studies and the ILO Program for Rural Women (Delhi), the women formed *samitis* (committees or registered groups) to organize the collection and sale of forest produce and undertake other income-earning opportunities, such as the rearing of silkworms on plantations. Since their inception in the early 1980s, *samitis* have grown in strength and number, receiving more offers of land for rehabilitation and development of enterprise activities organized and managed by rural women.

In June 1990 the Indian government initiated the Joint Forest Management (JFM) program as a response to the National Forest Policy of 1988 emphasizing people's, particularly women's, participation in managing their local forest resources. By 1993–1994, 14 states in India accounting for 72 percent of the country's 75 million hectares of public forest land had issued resolutions specifying the basis of partnerships between forest departments and local communities (Sarin, 1993). However, conflict continues to prevail between villagers and traditionally hostile forest bureaucracies as well as between men and women in several villages over the distribution of benefits (forest products), access to protected forest lands, and the membership of village forest management institutions.

Although groups vary in their size, structure, and method of functioning, there is little active or even symbolic representation of women on forest management committees in India. On the contrary, women have only been disadvantaged by so-called community protection: They have to walk farther to collect firewood, in many cases resorting to stealing from other forests and facing harassment from forest officials, villagers, and even other women. Thus many concerned individuals are asking whether joint forest management is turning women against women in the name of "participatory" or "community" management.

Desertification Control

Apart from planting trees in degraded forests and wastelands, rural women, particularly in southern Africa, have been planting trees to act as buffers against the encroaching desert. An evaluation of women's participation in desertification-control projects across six Sahelian countries (Monimart, 1991) maintains that there is no significant difference between the type of agency involved (bilateral, multilateral, or NGO) or type of project (large or small scale) in terms of integrating women. What is of critical importance is the role of the project management, particularly in supervisory and extension jobs. However, women are far less concentrated in these areas than men, and there is strong male resistance, from both villagers and project managers, to training for women. Not surprisingly, much of women's participation remains as unskilled and largely unpaid manual labor on project sites, undertaking arduous, time-consuming work in the absence of men, with few possibilities that their own land will be protected.

The Green Belt Movement in Kenya was established by Professor Wangari Mathai in the late 1970s through the National Council of Women as a campaign to halt desertification (Kamau, 1991). Green belts were planted around school compounds and landholdings by individuals and groups using seeds or saplings provided by the council. To ensure success, the movement maintains rigorous standards: Green-Belt promoters inspect the land to be planted and discuss maintenance of seedlings with potential participants. The council is responsible for overall monitoring, coordination, and evaluation, while community-based women's groups organize the daily management of the seed collection, quality control, and payment. Decentralized tree nurseries serve as important income-earning opportunities for women, particularly the young and disabled.

In addition to planting trees, women are involved in other soil-conservation works, such as bench terracing. The best-known example in this respect are the *mwethya* (cooperative self-help) groups in the hilly, arid areas of Katheka district, Kenya. Begun in the mid-1970s, the groups have 20–30 members, largely with marginal landholdings and limited access to capital. They work two mornings a week, 10 months a year, on the farms of members, determined by rotation. In addition to bench terracing, the groups undertake weeding, cultivating, mulching, and carrying manure.

> The *mwethya* groups enable women to share labor not only at critical times in terms of the agricultural cycle, but more particularly, on a regular basis to prevent the deterioration of the resource base on which agriculture depends (Thomas-Slayter, 1992, p. 816).

Women As Resource Managers: A Natural Responsibility?

This essay has shown that women do not have any intrinsic connection with the environment based on their biology (sex) or their domestic responsibilities, namely the collection of firewood, fodder, and water. Rather, women's relationship with the environment is mediated by the social construction of gender relations, "the social relations which systematically differentiate men and women in processes of production and reproduction" (Jackson, 1993, p. 1949). Women's access to resources is contingent upon a number of factors, including their economic background and local institutional structures that govern resource use. Control or ownership of resources, which is to be distinguished from the management of resources, is largely dependent on the nature of property rights and the cultural context in which they are exercised.

The high visibility of women in the use of communally held or based resources is not an indicator of their greater commitment to community management than men—their environmental responsibilities do not necessarily make them more or less environmentally friendly. The operational context of community-resource management varies considerably depending on the resource to be managed and on the relationship between the tenurial system under which it is legally owned and the social system through which it is to be managed. Any understanding of women's managerial roles and abilities needs to be located in the context of this relationship.

On a day-to-day basis, women are involved in decision making concerning resource management, sometimes concerning their own land in the absence of male family members, and sometimes, through informal channels, influencing decisions concerning community resources. However, the institutionalization of such decision-making processes, through the formation of user groups for most community-centered water and forest resources, often denies women formal, recognized participation in their management. Since membership of the group is open to the head of the household, this invariably means men, as land is registered in their name. The token representation of two women on management committees, usually from a socially significant family in the village, or, in a few cases, female heads of household, has limited significance. These women rarely attend meetings or feel confident about speaking out because of their limited numbers.

At another level, in the case of drinking-water and sanitation projects in which there are no specific property rights limiting access or defining control, women have usually been given often time-consuming management responsibilities with few incentives. Such work is in addition to their already numerous chores, and, since it is underpaid or poorly paid, it in not always valued by the community.

In the final analysis, projects and programs of natural-resources management need to take into account the different relations of men and women to resources, and to design management incentives that are not based on gender stereotypes and do not unconsciously reinforce gender inequalities. Management responsibilities must be sensitive to the local context; otherwise they will remain only symbolic attempts at enhancing women's participation. Finding realistic means through which women can exercise decision-making control over the resources on which they depend is one step in the process of empowerment.

References

Adams, William M. *Green Development: Environment and Sustainability in the Third World.* London: Routledge, 1990.

Agarwal, Bina. "Who Sows? Who Reaps? Women and Land Rights in India," *Journal of Peasant Studies,* vol. 15, no. 4, 1988, pp. 530–581.

———. "The Gender and Environment Debate: Lessons from India," *Feminist Studies,* vol. 18, no. 1, 1992, pp. 119–158.

Braidotti, Rosi, Ewa Charkiewicz-Pluta, Sabine Hausler, and Saskia Wieringa. *Negotiating for Change: Debates on Women, the Environment, and Sustainable Development.* London: Zed, 1993.

Bromley, Daniel W., and Michael M. Cernea. "The Management of Common Property Resources: Some Conceptual and Operational Fallacies," discussion paper, no. 57, Washington, D.C: World Bank, 1989.

Chachage, C.S.L., J. Nawe, and L.L. Wilfred. *Rural Water and Sanitation Programme in Morogoro and Shinyanga Regions: A Study on Women's Involvement in the Implementation of the Programme.* Dar es Salaam: Dar es Salaam University Press, 1990.

Chambers, Robert, Melissa Leach, and Czech Conroy. *Trees As Savings and Security for the Rural Poor.* Gatekeeper Series no. 3. London: International Institute for Education and Development, 1993.

FMIS. *Newsletter of the Farmer-Managed Irrigation Systems Network* (International Irrigation Management Institute, Colombo), no. 11, December 1992.

Gibbs, Christopher J.N., and Daniel W. Bromley. "Institutional Arrangement for Management of Rural Resources: Common Property Regimes." In Fikret Berkes (ed.), *Common Property Resources: Ecology and Community-Based Sustainable Development.* London: Belhaven, 1989, pp. 22–32.

Goetz, Anne-Marie. "Gender and Administration," *IDS Bulletin* (Institute of Development Studies, Sussex), vol. 23, no. 4, 1992, pp. 6–17.

Hausler, Sabine. "Community Forestry: A Critical Assessment: The Case of Nepal," *Ecologist*, no. 3, May/June 1993, pp. 84–90.

Heyzer, Noeleen (ed.). *Women Farmers and Rural Change in Asia: Towards Equal Access and Participation.* Kuala Lampur: Asia and Pacific Development Center, 1987.

International Labor Organization. "The Samitis of Bankura." In Annabel Rodda (ed.), *Women and the Environment.* London: Zed, 1991, pp. 137–141.

International Union for the Conservation of Nature and Natural Resources (IUCN). *World Conservation Strategy: Living Resource Conservation for Sustainable Development.* Gland, Switzerland: International Union for Conservation of Nature and Natural Resources, 1980.

Jackson, Cecile. *The Kano River Irrigation Project.* West Hartford: Kumarian, 1985.

———. "Doing What Comes Naturally? Women and Environment in Development," *World Development*, vol. 21, no. 12, 1993, pp. 1947–1963.

Jacobson, Jodi L. "Gender Bias: Roadblock to Sustainable Development," paper no. 110, Worldwatch Institute, Washington, D.C., 1992.

Jordans, E.H. *Survival at a Low Ebb: Women Farmers and Water Development in Bangladesh.* Wageningen, Netherlands: Wageningen Agricultural University, 1991.

Kamau, Wanjiru. "The Work of the Green Belt Movement." In Annabel Rodda (ed.), *Women and the Environment.* London: Zed, 1991, pp. 111–112.

Mies, Maria, and Vandana Shiva. *Ecofeminism.* London: Zed, 1993.

Molnar, Augusta. "Forest Conservation in Nepal: Encouraging Women's Participation." In Ann Leonard (ed.), *SEED: Supporting Women's Work in the Third World.* New York: Feminist Press, 1989, pp. 98–119.

Monimart, Marie. "Women in the Fight Against Desertification." In Sally Sontheimer (ed.), *Women and the Environment: Crisis and Development in the Third World.* London: Earthscan, 1991, pp. 32–64.

Murphree, Marshall W. *Communities As Resource Management Institutions.* Gatekeeper Series no. 36. London: International Institute for Environment and Development, 1993.

Poffenberger, Mark (ed.). *Keepers of the Forest.* West Hartford: Kumarian Press, 1990.

PROWWESS/UNDP. "People, Pumps, and Agencies: The South Coast Handpump Project." In Annabel Rodda (ed.), *Women and the Environment.* London: Zed, 1991, pp. 129–134.

Sarin, Madhu. "From Conflict to Collaboration: Local Institutions in Joint Forest Management." Joint Forest Management working paper no. 14, Society for Promotion of Wastelands Development and the Ford Foundation, New Delhi, 1993.

Schenk-Sandbergen, Loes. "Empowerment of Women: Its Scope in a Bilateral Development Project—a Small-Scale Irrigation Project in North Bengal," *Economic and Political Weekly* (Bombay), vol. 26, no. 17, April 27, 1991, pp. 27–35.

Singh, Andrea M., and Neera Burra (eds.). *Women and Wasteland Development in India.* New Delhi: Sage, 1993.

Thomas-Slayter, Barbara P. "Politics, Class, and Gender in African Resource Management: The Case of Rural Kenya," *Economic Development and Cultural Change*, vol. 40, no. 4, 1992, pp. 809–827.

Venkateswaran, Sandhya. *Living on the Edge: Women, Environment, and Development.* New Delhi: Friedrich Ebert Stiftung, 1992.

Wijk-Sijbesma, Christine van. "Drinking Water Supply and Sanitation Projects: Impacts on Women," *Women, Water, and Sanitation, Annual Abstract Journal*, no. 2. The Hague: International Water and Sanitation Center, 1992.

World Bank. *World Development Report 1992.* Washington, D.C.: World Bank, 1992.

Yacoob, M., and J. Walker. "Community Management in Water Supply and Sanitation Projects: Costs and Implications," *Aqua*, vol. 40, no. 1, 1991, pp. 30–34.

Women and Environmental Activism

Marta Benavides

Introduction

This essay examines the actions of women, particularly women of the nonindustrialized Third World nations, to live healthily and in a healthy environment. It reflects on the meaning of such concepts as environment, development, progress, and consumer and looks at conditions in the late twentieth century and how we arrived at them. In the process, it presents examples of effective work to achieve a healthy environment at local, national, regional, and global levels.

We live in the 1990s in the knowledge that we face an environmental crisis of such dimensions that we can predict the destruction of the human race and life as we know it on the planet. Across the world, soil is constantly eroding, the result in some countries of centuries of monoculture and export economies; in others, of agribusiness, cash crops, and the ever-increasing use of chemical fertilizers, pesticides, and herbicides; in others, of strip mining, the clearing of thousands of acres, the felling of trees for export purposes, and war. Water supplies are also drying, a usual consequence of erosion of the land. Potable water was a promise that did not materialize for the majority of peoples of nonindustrialized nations, and more of their water sources have not only dried but become polluted, a result of the dumping of toxic wastes into lakes and rivers, oceans, and the countryside. This is a reality that an increasing number of communities of color, impoverished communities in the industrialized nations, face as a daily fact of life.

Thousands of children and adults die of hunger every day in the nonindustrialized world, while in industrialized nations the number of elderly and of infants younger than one year of age dying as a result of malnutrition and lack of medicine has reached alarming levels. The air is polluted, the result of untreated toxic emissions from industries and cars. We talk of the impending warming of the Earth and its potential destructiveness, while U.N. studies affirm that the process has started, and the United Nations demands rigor-ous, urgent, and enforced controls of the causes behind this situation.

Death of humans and death of nature result from the same causes: abuse, disrespect, lack of recognition of the natural laws, and a persistent determination to continue to exploit human beings and the environment for economic profit. Profit and suffering are not equally distributed. Very few people profit; most suffer. How did it come to be this way? What we see today is the result of the process of globalization for profit that started more than 500 years ago. At that time, various processes of regionalization, present in varying degrees in the continents of the world, reached an economic level in Europe that demanded expansion. This expansion was for new trade, for economic gains that would enable the countries involved to become powers, to develop an effective military force that would enable them to move and gain control of raw materials and peoples necessary to consolidate their industrial development.

At the time, women and children were not considered fully human and thus did not have rights to property or decisions. Those with power mounted an aggressive witch hunt to do away with those who had special knowledge, such as medicine in its various forms, or who knew too much—those who knew how to think creatively and critically, that is, people with wisdom. Most of those hunted were women. In this way, the pursuers effectively destroyed leaders, knowledge, and wisdom, rendering those people remaining subservient to their ways of thought. They also pursued ethnic cleansing, as is evident in the history of the persecution and killing of the Moors and the Jews. In Christian Europe, the Roman Catholic Church supported and justified these actions and even contributed by way of the Crusades. The Church also purged those who thought and believed differently, as happened with Galileo, who stated that the Earth turned around the sun, yet, it blessed the adventures of Columbus in search of new routes for silk and spice.

It is this kind of understanding that became globalized

by force and gunboat diplomacy. If these people could think that women and children were not fully human, if they could destroy thinkers and wise persons under the guise of paganism (and in defense of the one true God), if they could persecute those different among them (to keep their businesses and gold), if they could militarily invade strategically located nations to gain control as they strove to Christianize, what wouldn't they do to exploit the people and control the natural resources of other continents? Imagine the fate of many cultures, the fate of spirituality and religion, the fate of women and men, the fate of Mother Earth. England needed timber for its navy to have the best battle and trade ships. Besides various raw materials, it specialized in trading the most important natural resource—that is, in the kidnaping, transportation, and sale of people as slaves.

The people and "discovered" continents became the raw materials for the industrialization of the European nations, and that is when the former's "development" began. This power, or "development," has always been dependent on assured "free" or cheap raw materials. It is this logic that has evolved into today's globalized condition of ecological, social, economic, and moral deterioration. It is these needs and priorities that have determined what profit is (and for what and for whom), influenced the domestic policies of industrialized nations, and decided the imposition of laws and military, even dictatorial, powers on the nonindustrialized nations. This is the reason we should hesitate to call ourselves "Third World." We are not underdeveloped, or developing nations—we are the direct result of the creation of the "First World."

In the last decades, the globalized situation of the world resulted in the "development" of high interest rates that brought most countries with dependent economies to bankruptcy through foreign or external debt. As a result of this and technological advances, as well as market competition, a new globalization has occurred. The new forms of internal and external relations have been regulated with the advent of the GATT (General Agreement for Tariffs and Trade) regulations, the WTO (World Trade Organization), and the three new economic world blocs that have replaced the "Big Seven" industrialized nations: NAFTA (North American Free Trade Agreement), the European Economic Community, and the Pacific Rim. At this point, the struggle is for control of food and for biodiversity, which includes the patenting of all forms of life, to be exploited when and how it is deemed necessary by those corporations holding the patents. Already some indigenous peoples are fighting the patenting of their genes.

Today it is understood that a better terminology for the two sides of this division of the world's nations is the "North" for those that are industrialized and have a relatively high standard—though not necessarily quality—of life, and the "South" for those nonindustrialized nations that generally have been formally or informally a colony of one or various nations of the North. We have been forced to live at the periphery, while the North has been the center. The people and the natural materials of the South have been seen as resources, and we have been used as such, so there is depletion as a result of abuse and careless use (Luxemburg, 1951). A few years before the fall of the Berlin Wall in 1989, it became rather clear that a new world order had evolved. The industrialized nations of the North and the impoverished nations of the South are economic and sociological concepts that deal with the quality of social conditions in the countries of the world. We recognize that, due to economic readjustments, there is an increasing South in the North and a small North in the South. We have to be mindful of the real meaning of development. There are destruction and war, inequality and injustice where there must be peace; there are disintegration and erosion where there should be health and wholeness; there are oppression, repression, violation of the right to be human where there should be freedom. This is, then, the background needed to understand the disabling and eroding environment we face today and to understand the unhappy fate of women and of the Earth (Boserup, 1970).

Development, defined as the making of a more advanced or effective state, and progress, defined as advancement in general, cannot be fostered if they depend on the exploitation of half of the human race—women—or on the destruction of that which sustains life itself. In this context, those people who live in affirmation of all and cooperate with nature should be the ones considered developed and progressive. As we reflect on the way women work, we are able to see that, over the course of thousands of years, they have learned from nature the ways of sustenance and have worked with nature to maintain health and to heal, knowing that the best medicine was food and a balanced life. They have developed a practice of wholeness, which is a more integral expression of permaculture, for the maintenance of their homesteads.

"Environment" is not only about natural surroundings, but about all that constitutes our life, and, for that matter, the mind-set, logic, or understanding of the world that guides our thinking, analysis, and actions. "Environment" is not only about ecology, but about our ways of being, about what is important in life, and the nature and meaning of healthy relations within ourselves and with others, starting with those closest and with that which provides our sustenance. It is about awareness of the partnership in which we live with each other and with nature, and the respect and humbleness in which we carry ourselves while we are on Earth. Would we live in a way that would enable us to leave a legacy of wisdom, of understanding, of affirmation and respect for diversity, of cooperation and collaboration among peoples, nations, and nature to guarantee quality of life for the seventh generation? Or will we be the mindless or mere consumer? The word "consumer" comes from the word "consume"—to destroy or expend by use, use up, spend wastefully. These words tell us the reality of what is going on; they give the definition of how we are. The way in which we exist in the world determines that, for us to have a full, secure, and meaningful life, we can and must do it only in partnership with others and with nature. The planet was not created for consumption, regardless of the fact that there are theoreticians, scientists, and theories that support such conduct (Singh, 1987). We must develop the appropriate language and understandings to describe the world we want. We have the power to name, thus to create.

Women of the South Shaping Their Environment

The Women of Color of North America at the First World Women's Congress for a Healthy Planet (Miami, 1991) made a statement to define a more integral concept of environment when they said that they believe that a women's environmental agenda will be nothing without an embodiment of economic justice, political democracy, and a respect for the contributions of all civilizations and cultures. They believe that we must be guided by principles of equity in the quality of life for all (WEDO, 1991). This position resonates with the concerns and aspirations of women and peoples from all over the world, as is evident by their efforts to achieve a plenitude of life. Many women do the work of healing the environment as they create new understandings and appropriate practices. We are naming ourselves, our ways, and our relations. Women are partners, facilitators, coordinators, and creators. It is not at all, as some people of the North believe, that women are just not interested in the environment and, thus, do nothing about it and continue to degrade it.

The work of women around the world reflects their efforts for the well-being of their families, their homes, and their communities, and their consistent efforts to shape their immediate environments in ways that affirm the best possible quality of life. All we must do is be mindful of all that women do, and it will be evident. For what else is it to keep a home, and keep everyone fed, clean, healthy, and growing, even while one has to suffer a double journey, unvalued work, or the stigma of racism or classism, not to mention sexism? What else is it to be a *comadre,* the godmother committed to take care of the neighbors' children; the *curandera,* who keeps and shares the knowledge of the medicinal plants to heal or nourish whoever needs it; or the *partera,* the midwife who cares for women as they bring a new baby to life? What else is it to carry the water in a clay basin on one's head or the firewood on one's back, or wash the clothes in the river, or carry the baby in one's womb, or give birth and raise the future farmers, scientists, factory and construction workers, and the young women who will give birth to future ones?

The women in the "Third World"—a high percentage of whom have lived for centuries in impoverishment generally due to colonialism—are the miracle makers, the ones who multiply the bread and the fish, the ones who heal by touching. These are the women who are now facing, because of the globalized economy, a new world order of economic and ecological crisis and devastation. They are working and searching to find partners among their community peers, and cooperating with other women nationally, or across borders, using their experience to develop recommendations and actions for a healthy planet, as the bases for their own fulfillment and health.

Women of all countries and from various backgrounds—small farms, indigenous communities and networks, grass-roots groups, consumer organizations, neighborhood leagues, and flower and horticulture clubs, as well as governments and universities, and women who are social and physical scientists, technicians, doctors, geographers, parliamentarians, lawyers, church members, bureaucrats, and bankers—have joined at local, national, regional, and international levels to work with determination to improve the quality of life for themselves and their families, for their community, for their country, and for the world. This can be seen concretely in the local, national, and regional participation of women's nongovernmental organizations (NGOs), and the movements culminating in the U.N.-sponsored Earth Summit (Rio de Janeiro, 1992), the Second World Conference on Human Rights (Vienna, 1993), the International Conference on Population and Development (Cairo, 1994), and the World Summit on Social Development (Copenhagen, 1995). Through these, women have created an environment of participatory governance for quality of life for themselves and for the healing of their societies and the planet. These women, often joined by men, want healthy environmental standards to be upheld in their factories in the provision of water, light, and ventilation, and similar standards to be applied to sewage and recycled waters. They are pressuring to prevent regular and toxic waste from being dumped in their communities or rivers; to stop desertification, the sale of public lands, and legislation that permits the sale or "urbanization" of wetlands; and to pass legislature that guarantees the right to clean water and air and asbestos-free schools. They are also applying pressure to legalize their right to health, to control their reproduction, and to have their rights and those of the girl child be recognized as human rights.

Women are uniting in efforts around the world to transform the conditions of the home, the community, and the workplace; to acquire access to water; to develop appropriate and compatible technologies; and to manage and recycle waste to create jobs, energy, and fuel production. They are also carrying out research that will impact biodiversity and bioengineering work and they are working to enact laws to bring the international financial systems under the control of people-centered systems and to value unpaid work.

Indigenous and peasant women are working to preserve medicinal plants and indigenous knowledge, as is occurring in communities aiming toward sustainability in El Salvador. In an effort to stop desertification, women university students are studying and documenting the increasing number of animal and plant species at the verge of extinction (Benavides, 1991, p. 95).

Winona LaDuke, an indigenous woman whom the Spirits call Thunderbird Woman and who lives on the White Earth Reservation in North America, is a good example of this coming of age. She tells us:

I am coming home. Coming home. That is, I believe the challenge we all face is making a home, restoring, building, investing in, and reclaiming a community, a destiny, a way of life. . . . That is the essence of becoming part of the Land. There are no more frontiers, and no greener pastures. This is what we are fortunate enough to have. . . . From here I try to look back, and look forward. I am looking at the community 20 years from now, and trying to do my part to make it what I dream it will be. In the future I plan to hear in my community primarily Ojibwa/Anishinabeg. . . . I plan to see a recovered traditional economy—from wild rice to buffalo, deer, maple sugar, and the ecosystem that supports them. . . . I see the restoration of our traditional religious institutions, and I see non-Indians, as well as Anishinabeg on this reservation, sharing common land, values, and language. . . . In short, I wrestle personally, politically, intellectually with how to restore my home, and how to come home (*Utne Reader,* 1995, p. 80).

In Bhopal, India (where a toxic gas leak caused hundreds of deaths in 1984), women continue to protest Union Carbide operations, educate nationally and internationally on the dangers they present, go to court about various related issues, and look for programs to heal those affected by the emissions. In Ecuador, women have organized to stop local factories' chemical emissions; indigenous women participate with their husbands, friends, and brothers to defend their rights to land, territoriality, and all resources within. In Puerto Rico, a free associated state of the United States, women and men, in coordination with churches and universities, have organized industrial-mission campaigns to press factories to respect U.S. environmental protection regulations because the factory emissions are causing skin, breathing, and hormonal diseases and disorders in women, men, and children. They also have sought to stop the construction of port facilities that were to bring the largest oil tankers to the island, which would continue to pollute and destroy the already deteriorated environment.

Thousands of women work to have a nuclear free South Pacific. They denounce the destructions of their islands and the pollution of the seas, and, with each new nuclear test, they know that more babies will be born without spinal columns. The nuclear-bomb tests in the area by foreign powers continue in violation of human rights and the international principles for the rights of nations. The coalition working in educational and political campaigns to have a nuclear-free region and future spans oceans and involves people of the South and the North.

In the United States, women and men of the United Farmworkers of America (UFW) have led successful strikes against grape and lettuce growers, who contaminate their crops with pesticides that poison the milk of breast-feeding women, causing sickness, deformation, and death among babies. They defend their economic and health

rights, demand environmental regulations, and campaign for the protection of consumers and their rights. Leaders of the UFW have crossed the border to inform people, trade unions, and the government to promote Mexico's support of environmental laws and the defense of the rights of Mexican men and women who work in the United States, especially as they relate to environmental hazards.

In India, the first major contemporary environmental movement was Chipco, which means "to embrace":

Though this resistance movement had predecessors in the 1800s, now in the late 60s women created the movement to resist logging in the Himalayan region. For over a decade, villages and communities protected a particular patch of land from being logged. The people brought food, and kept the children of those "embracing" the trees, until equipment and contractors had moved from the area, and the patch removed from the list to be logged. The people were conscious that forests on mountain tops are the source of one's water, that water is the source of life, and that it must be protected. This is a good example of protracted nonviolent action. (*Indigenous Women,* n.d., pp. 18–19.)

The Alternative Nobel Prize (Right Livelihood Award) was granted to this movement in 1987.

Environmental Movements and Processes Facilitated by Women

All of the activities presented so far as examples of women's actions for a healthy and better environment have resulted in movements and processes that have protected the quality of life of women in general. There is no way that the status and condition of women would be considered important if it were not for the work that women have been carrying out at various levels.

The 80,000 members of Kenya's Green Belt Movement (GBM), started in 1977, have planted more than seven million trees to prevent desertification and have set up more than 600 income-producing nurseries, in spite of the Kenyan government's direct interference. GBM has also documented and campaigned actively against official corruption, misuse of funds, and structural-adjustment programs and policies (SAPs), which leave the people without social safety nets.

Crucial actions have been initiated by individual women. Chee Yoke Ling of Friends of the Earth led effective mass protests in Malaysia against the use of poisonous pesticides and the dumping of radioactive waste. Laila Kamel of Egypt worked with a Cairo garbage-collector community in the Mokattam hills recycling garbage and generating jobs and educational opportunities for the people of this community. Tuenjai Deetes cofounded the Hill Area Development Foundation to increase the self-sufficiency of marginal communities, which, having fled ethnic wars and hardships in Burma and Laos, settled in the mountainous area of northern Thailand. The foundation promotes self-sufficiency for the marginal communities while protecting the natural resources against deforestation and soil erosion and respecting tribal culture.

Ester Yazzie is an indigenous Navajo woman who lives in the state of Arizona and uses her law-enforcement training to sustain traditional values. After uranium was discovered in their land, hundreds of Navajos worked in the open-pit and deep mines without being told about the dangers of radiation, and now they die of radiation-related diseases. They leave their widows and children with no support. Most homes are also contaminated since Navajos use the earth to build the traditional hogans (Navajo dwellings built of earth walls supported by timbers). Yazzie works with her people to get the uranium-mining industry to stop the exploitation of lands and peoples for nuclear-power generation, nuclear testing, and radioactive-waste dumping; to clean and restore all homelands; to end the secrecy about the nuclear industry and its dangers; and to provide full and fair compensation for damage to peoples, families, and communities. Her people share a vision for the future: Given the unity of humanity and the world, they appeal on behalf of future generations to those in the present to use sustainable, renewable, and life-enhancing energy alternatives. The need to live in harmony and respect for life, as a means of peace, leads people to discuss the concept of Gaia, the planet as a living thing (*Indigenous Women*, n.d.).

Leonor Briones, professor at the University of the Philippines, president of the Freedom from Debt Coalition, declared:

In 1988, the Philippines paid $2 billion a year interest on its [foreign] debt to the wealthy nations of the North—and received a pitiful $236 million in return. It's foreign aid in reverse . . . and we are only one of 50 countries that annually pay the North $50 billion more than we get . . . Many of our people live in absolute, grinding, unimaginable poverty. . . . We are paying $350,000 interest a day on a corrupt nuclear plant that is now mothballed because it was defective from the start. The leaders knew that the project was riddled with bribery; the whole world knew. Yet they gave the money. And now they insist that to pay back such fraudulent loans, our government must cut subsidies on food, education, health, and social services. . . . Our women hold four and five jobs at once, struggling to pay for medicines for their babies, begging the schools to let their children write exams. Our country is hemorrhaging its people, its financial and natural resources. . . . To get the foreign currency we need to pay our debt, we are pushed to encroach on the environment. Only one-fifth of our beautiful coral reefs are healthy. Fish stocks have plummeted by 50 per cent . . . (WEDO, 1991).

Moema Viezzer of Sao Paulo, Brazil, directs Women in Citizenship Action. This group struggles against hunger and poverty and in defense of life. It educates women and their impoverished communities about nutrition, food security, and how to create urban community orchards. It also studies the national and international policies that bring the women and their country to these conditions.

Marta Benavides of El Salvador carries out an initiative, the Women's Environment and Development Program, for people's land and human rights, environmental education, and ecological action for sustainability. In this process, each community participates in the creation of a healthy and natural environment. Sustainability is about how we treat ourselves, each other, and our environment, and about how we live, using and affirming our strengths and differences, sharing common goals and resources.

In all of these cases, we can see the need for dialogue and analysis, and the importance of committed, collaborative, and consistent action, which comes from the grass roots to impact on their communities and groups and on social institutions and governments agencies as well as on other NGOs.

Let us review the example of a multinational project that has a direct impact on future programs and commitments of women, churches, ecumenical groups, and NGOs. At the Mexican–U.S. border, 50 women of all races, representing impoverished communities, Protestant and Catholic churches, and women's groups in the United States and Canada and from Mexico, Guatemala, and El Salvador, gathered to learn about each other's situations and about the impacts of the *maquiladora* industry—factories of free-trade zones—on the quality of life of women, children, families, communities, nations, and the environment. The meeting was called and sponsored by Agricultural Missions of the Commission of Women in Development. The representatives learned that NAFTA recently signed by the United States, Mexico, and Canada, had already had resulted in severe detrimental consequences for some working people in the United States, as it depended primarily on Mexican "cheap" (low-wage) labor, especially that of young women. They learned about increased family violence, disruption of homes, impoverished communities, pollution of the environment, babies born without spines or eyes or with other deformations, and chronic illnesses of women—on both sides of the border. They learned that these are the result of a way of doing business that is a common practice across the United States, Mexico, Central America, and the Caribbean. Having visited, studied, and reflected on these conditions, the representatives at the meeting recommended that the churches study and monitor NAFTA for the purpose of taking effective action on behalf of both male and female workers and the environment. The representatives further recommended that actions in this regard be included in the work of churches in Canada and that the women's division of each denomination be educated about it, in the context and spirit of the Program of Justice, Peace, and the Integrity of Creation sponsored by the World Council of Churches. They challenged the churches to continue this type of work, to maintain close ties with one another, and to keep a spiritual dimension in the work on the environment. They also recommended that this challenge be shared with all of the partner communities and programs supported by sponsors in the NAFTA nations and in the rest of the world (National Council of Churches-USA, 1993).

Global Connections and Their Impact

The work and the impact of women in environmental processes and efforts show that they must become full partners in local, national, and global environmental plans and strategies for the new paradigm of sustainable social and human development to be a success. In this context, DAWN (Development Alternatives with Women for a New Era), a women's NGO from the South led by Barbadian Peggy Antrobus, states:

> Women must be seen neither as victims nor saviours but as people whose interlocking roles in reproduction and production in the household and in the economy offer them a special vantage point for addressing the issue of the current debates in development theory, policy and practice—the debates revolving around the role of States vs. Markets, as well as those revolving around issues of Population, Environment and Development (quoted in *DAWN Informs,* 1994).

DAWN works also with other women's networks of the South, and actively coordinates with women's networks from the North, which share these perspectives.

Gus Speth, United Nations Development Program (UNDP) administrator, shares his views on the relationship among women, development, and the environment:

> Sustainable human development is . . . people-centered . . ., participatory . . ., pro-poor and pro-nature. It gives highest priority to poverty alleviation, to environmental regeneration, and to job-led growth. And it recognizes that none of this is possible unless the status of women is elevated. Sustainable human development is an essential precondition to bringing human numbers into balance with the carrying capacities of nature and the coping capacities of societies (*UN-NGLS Go-Between,* 1993).

Women's lives and roles profoundly shape and affect every aspect of the environment. Women total more than half the population of the world. People are the most valuable natural resource of a nation; women can say that they are the planet's most valuable resource. Yet, women are overexploited and neglected; it is they who face higher illiteracy rates and who live in poverty in the highest numbers. The United Nations, in its Women, Environment, and Development Program, implemented with the support of its United Nations Development Fund for Women (UNIFEM), states:

> Although women grow, process, and market between 50 and 80 per cent of the food consumed in develop-

ing countries, governments rarely record these inputs or support them with financial credit, technologies, education or training—inputs which could raise productivity while safeguarding the environment and reducing the physical burdens involved. Overlooking the central role which women play in the economic and social life of their countries constrains national development and jeopardizes the natural resource base on which growth is founded. For millions of women in the developing world, the struggle for survival and environmental protection are inseparable. Women are among the first to suffer when land is degraded, when trees disappear, and when water supplies are polluted. As the main providers of food, fuel and water for their families, women are acutely aware of the need to protect their surroundings and to manage natural resources. In rural areas, where they spend many hours of every day fetching and carrying fuelwood and water, they know from harsh experience that the depletion of woodlands or watersources will eventually force them to walk farther afield in search of new supplies. Experience has taught them that soil erosion, caused by intense agriculture on fragile soils, will ultimately reduce the amount of food they can put on the table. . . . They do whatever is necessary to survive from day to day. Poverty and environmental degradation embrace in a daily downward spiral. . . . In the poorest communities of Africa, Asia and the Pacific, Latin America and the Caribbean, an often destructive quest for survival is fueled by high rates of population growth and the failure of policymakers to recognize obvious truth: That as long as women remain poor, fragile eco-systems will remain at risk (UNIFEM, n.d.).

It is clear that the U.N. findings and documents of the conferences and summits of the 1990s contradict the blessings heralded in the new world order of peace, prosperity, and progress for all peoples and nations following the fall of the Berlin Wall and of the socialist and communist camps. All of them—the Children's Summit (New York, 1990); the Earth Summit (Rio, 1992), where the intimate relation of poverty, development, environment, and women was clearly presented; the Second World Conference on Human Rights (Vienna, 1993), where women pressed for women's and children's rights to be accepted and affirmed as human rights; the International Conference on Population and Development (Cairo, 1994); and the World Summit on Social Development (Copenhagen, 1995), at which delegates worked on poverty alleviation and eradication, productive-quality employment, and social integration—have shown that the existing model of development has resulted in serious illness for humans and the planet.

The United Nations has begun to use Human Development Indexes: Each country must show its development as measured not only by the gross national product or per-capita income, but also by the literacy rate, housing, employment, education, health, transportation services, and environmental protection available to its citizens and by the possibilities for improved quality of life being created for future generations. This is a process to assure sustainable human and social development, which is an important way to follow up on, and interconnect, all of the international U.N. conferences, especially the Earth Summit. The United Nations produced the Women's Human Development Index in August 1995, in time for the Fourth World Conference on Women in Beijing.

The U.N. conferences and summits provide an excellent example of the partnership and collaboration among women of the South and the North, women and men in general, young and old people, peasants and city dwellers, officials of civil society and of government, and the importance and relevance of this work. Each of the U.N. meetings is preceded by preparatory commissions, and on these occasions 200–300 women and men representing organizations from 50–70 countries from all over the world come together daily to inform one another, develop analysis and strategy, and monitor and lobby the texts being worked line by line by government representatives. This process enables participants in the Women's Caucus at each U.N. conference to learn of the processes, plans, actions, and interest of governments and the United Nations. It also enables them to have an impact not only on global governance but also on the forging of a new society.

Bella Abzug, the cochair of WEDO (Women's Environment and Development Organization, a highly active NGO) speaking for the international women's caucus and addressing the official U.N. delegates at a preparatory commission of the 1995 Social Development Summit, summarized the way women are working in this process and the effect they are achieving:

> As women who seek an equal voice for women in international and national decision-making and democratic and human rights for all people, we have a simple test for evaluating progress. We ask, what is happening to women around the world? Women have been the shock absorbers for the dramatic changes the world has witnessed in the last five years and for the so-called structural adjustments that put corporate profiteering ahead of human needs. . . . In some parts of the world, women are losing whatever rights and benefits they had, and those who never had either are bearing the burdens of unpaid labor in factories owned by absentee, faceless owners. For many women and girls the alternative is exploitation and prostitution in the international sex market. We propose a major capitalization strategy for a people-centered economy. This would redirect World Bank and other financial institutions' resources into small-scale credit without collateral for self-employment. We propose to work for an international agreement on corporate responsibility that assures factory and field workers in the South, many of them women, of their right to decent wages,

social benefits, health and child care facilities, and working conditions free of industrial pollutants and other environmental hazards. We urge education and awareness strategies that teach the poor, both children and adults; that make technology an ally, not a job-destroying, health-destroying foe; that penalize male violence against women; that recast gender stereotypes which subordinate women and deny them an equal share in fate-of-the-earth decisions (Statement presented at PrepCom I, World Social Development Summit, 1995).

The Asia-Pacific Women Action Network, in the context of the Social Summit, submitted a formal declaration to be placed in the docket of documents. The way these women carried out their process of self-awareness and education, and lifted it to the international arena both for NGOs as well as officially, shows how helpful this process can be for the region, and participants, and for all the other areas of the world. It states, among other positions, that:

> The realities of women's lives are shaped by contextual and perceptual paradigms. A combination of market-driven, profit-oriented development and patriarchal values have placed the vast majority of our women in extremely marginalized and vulnerable positions. Women become the object of development, first as impoverished citizens of Asia-Pacific, and second by facing specific gender discrimination and violence. . . . Sustainable development, poverty alleviation, and social integration cannot be attained without the full participation and co-leadership of women at all levels of society and decision making. A gender perspective is one of the most forward looking and socially just approaches for social development. The different and irreversible impacts that structural adjustment policies, trade liberalization, and the trickle-down programs have on women have not been responsibly dealt with by our nations. . . . Women have always been and still are in Asia-Pacific primary food producers. Therefore, agricultural sustainability cannot be achieved without women's active participation in setting and implementing policy. Food security is a basic right, inextricably tied to sustainable agriculture. Food security is the access by all peoples at all times to food needed for a healthy life. It is in this context that food for domestic consumption must override agriculture products for trade. Food security can only be a reality when women's rights to land and resources are protected. . . . Women, in their role as primary food producers, have necessarily been custodians of biodiversity in all forms. They have maintained a relationship with nature without ownership and degradation. Modern agricultural biotechnologies have begun to displace women from their central role as food producers and gatherers of medicinal plants. . . . The corruption of traditional lifestyles, the depriva-

tion of productive land, lack of income opportunities and increasing violence have resulted in large-scale migration of women to already overcrowded urban centers where they live in abject poverty . . . (Asian-Pacific NGOs, 1994, pp. 68–69).

Vandana Shiva of India, a physicist, philosopher, feminist, ecologist, and director of the Research Foundation for Science, Technology, and Natural Resource Policy in Dehradun, India, has made a great contribution to the establishment of the connections between the personal and the global and to the recognition of the importance of local, national, and international work. She details in her writings the link among women, ecology, development, health, and the agricultural conditions of the "Third World" (Shiva, 1989, 1991a, 1991b, 1991c, 1993). Shiva is concerned with the protection of cultural and biological diversity. She argues that, unless we can put limits and boundaries on commercial activity and on new technologies, the violence against nature and against people will become uncontrollable. "The question I constantly ask myself is, 'What are the creative catalytic linkages that strengthen community and enable communities of people to exercise social and ecological control on economic and technological processes?'" (*Utne Reader*, 1995, p. 81).

Women met at the regional level to work on the development of the *Platform for Action* decided during the Fourth World Conference on Women (Beijing, 1995) and at similar meetings for the other U.N. conferences and summits. Women are learning from each experience to establish, lobby, and defend their positions and to work cooperatively with other women in their community, nation, region, and the world. During these meetings, women have analyzed and affirmed past agreements. They have developed quality benchmarks, bottom-line goals, commitments, and deadlines; they have also produced an economic analysis and demands for democratization, transparency, and accountability for U.N. programs and the Bretton Woods institutions: the International Monetary Fund (IMF) and the World Bank. They have disseminated the information and have been negotiating with their governments and other delegates. They are using the experience gained for local and national quests and legislation, and they are firmly defending what has been achieved.

It is important to see that the work that we, as women, are carrying out in favor of the environment is much more than mere activism. We are consciously and tenaciously bringing about a new world order of equality, equity, peace, justice, and people-centered development. What we are doing is not just saving the Earth but also working to establish an enabling environment in which we can reach our whole potential as human beings and live in harmony with a healthy natural environment: This is the exercise of our power. Women are doing this at national, regional, and international levels, attempting then to bring about the era of being in fullness, creating from all of these interrelations and work platforms an impact on national and international laws, agreements, accords, and practices.

We are intent not simply to live, but to achieve quality of life, and a life lived simply. Our task is to think and understand globally and work effectively locally so that we may have an impact both locally and globally. To be effective, we understand that we must learn self-knowledge and the universal principles of life. We recognize the need to work and participate cooperatively for human and social fulfillment on a healthy planet. We know that, in the interrelatedness of life, to be working for a healthy environment does not necessarily mean that we must work only on those issues directly linked to the health of nature and the planet; rather, we acknowledge that all of those people who, conscientiously and with love, work for lasting peace, for quality of life, for equity and equality, for justice, for the elimination of all forms of discrimination and exploitation, and for understanding among peoples are making a contribution to these ends. Together we are creating an environment and a climate that allow humans and the planet to live in harmony and be healthy. This, then, would be the enabling and healthier environment—and practices—that can result in an enriched and meaningful life for us, our societies, future generations, and our planet.

References

Asian-Pacific NGOs. *Breaking a Common Ground in the Pursuit of Alternatives.* Declarations of the Asian Pacific NGOs in the Preparatory Activities to the World Summit on Social Development, Manila: Asian-Pacific NGOs, June-October 1994.

Benavides, Marta. *La Mujer y Medio Ambiente en América Latina: El Papel de la Mujer.* Quito: CEPLAES, 1991.

Boserup, Ester. *Woman's Role in Economic Development.* New York: St. Martin's, 1970.

DAWN Informs. (St. Michael, Barbados). February 1994.

Indigenous Women, vol. 2, no. 1, n.d.

Luxemburg, Rosa. *The Accumulation of Capital.* London: Routledge and Kegan Paul, 1951.

National Council of Churches-USA. *Women Crossing Boundaries: Fighting Back.* New York: Agricultural Missions, National Council of Churches-USA, 1993.

Shiva, Vandana. *Staying Alive: Women, Ecology, and Development.* London: Zed, 1989.

———— (ed.). *Biodiversity: Social and Ecological Perspectives.* London: Zed, 1991a.

————. *Ecology and the Politics of Survival.* Tokyo: United Nations University Press, 1991b.

————. *The Violence of the Green Revolution: Third World Agriculture.* London: Zed, 1991c.

————. *Monocultures of the Mind.* London: Zed, 1993.

Singh, Narendra. "Robert Solow's Growth

Hickonomics," *Economic and Political Weekly*, vol. 23, no. 45, 1987.

UN-NGLS Go-Between, no. 42, November 1993.

United Nations Development Fund for Women (UNIFEM). *Women, Environment, Development.*

Educational pamphlet. New York: UNIFEM, n.d.

Utne Reader, no. 67, January/February 1995.

Women's Environment and Development Organization (WEDO). *Findings of the Tribunal: World Women's Congress for a Healthy Planet.* New York: WEDO, 1991.

Enabling Conditions for Change

Girls' Educational Access and Attainment

Margaret Sutton

Introduction

Since the early 1970s, the education of girls and women in the Third World has become an increasingly prominent item on the agenda of scholars of education and of international assistance agencies. Educating girls has lately been heralded as one of the most significant steps that can be taken to ensure social and economic development in poorer countries. This essay summarizes late-twentieth-century understandings of why the education of girls and women is considered important and what are seen to be the obstacles to furthering the goal of educating all girls and women. The discussion focuses on the education of girls and young women within the formal schooling system. Despite widespread international support for the idea that girls should be educated, female participation and achievement in schools around the world lags behind that of males. Thus, much of the growing body of literature on girls' education examines the questions of why more girls are not in school and what can be done to increase their numbers.

On the basis of these studies, governments, nongovernmental organizations (NGOs), and international assistance agencies have instituted programs and policies to heighten girls' participation in education. As is the case in many fields of Third World studies, analysis of the problems related to educating girls and documentation of the efforts made to redress them are heavily biased toward the views of Northern scholars and the donor community. The literature thus leans heavily toward questions that interest external policymakers, rather than addressing issues that might be raised by educators and activists in the Third World.

Education of Girls and Women As a Development Issue

The educational expansion that has taken place in the Third World in the second half of the twentieth century is nothing short of remarkable. In 1960, approximately 73 percent of children were enrolled in primary schools in the developing countries of Asia, Africa, and Latin America, 15 percent in the secondary schools, and 2 percent in tertiary, or higher, education. In 1990, these numbers had risen to 98 percent, 42 percent, and 7 percent, respectively. For postcolonial nations, the expansion of education was one of the earliest and most prominent steps taken to extend the rights of citizenship and the promise of development to formerly subjugated people. The concept of developing "human capital," or an educated workforce, largely through education, took hold in international assistance agencies and among many national officials. Both to meet social demand and to create a supply of educated workers, Third World governments built schools and trained teachers at a rapid pace.

As the Second Development Decade of the United Nations dawned in 1970, critical scholars began to make the case that, in the realm of education, as in other areas of social development, growth was not necessarily resulting in equity. Differential access to education by ethnicity, regional location, and gender was emerging as an all too common pattern. In 1970, more than 88 percent of the male school-age population was enrolled in primary schools while only 69 percent of females were so enrolled. The evident inequities in access to schooling as well as other promised benefits of development fostered the "basic needs" paradigm of development, which reached its height in the late 1970s (Streeten et al., 1981). From the "basic needs" perspective, development is a process not only of growth but also of equalizing benefits. Moreover, it recognizes that the existence of dire poverty and gross inequalities limits the absolute potential for economic growth. At the same time, the women's movement was gaining force and visibility in Europe and North America. With the United Nations Decade for Women beginning in 1976, the nascent field of women in development began to take hold in international assistance agencies and universities. Within the education sector, the new concern for women's social and economic status led to increased attention to the education of girls and women. Work on women's issues in other sectors, notably agriculture and health, recognized that increased knowledge and skills were either prerequisite or concomitant conditions for enhancing women's overall social and economic status.

Thus, research on girls' and women's education began to be accorded legitimacy and financial support from the donor community and associated scholars. In a 1981 bibliography of comparative education, only 39 of 3,080 entries dealt with girls or women and education (Altbach, Kelly, and Kelly, 1981). In a recent bibliography, which focuses on research published in the period 1979–1988, Kelly was able to identify 1,200 entries in English, French, Spanish, Portuguese, and German (Kelly, 1989, pp. l–9).

Increased scholarly concern with female education has been paralleled by a corresponding growth in programming by donors that concentrates on the education of girls. A report on World Bank activity (Herz et. al., 1991) identifies seven education projects initiated between 1972 and 1979 that included explicit efforts to enhance girls' education. By contrast, between 1980 and 1991, 31 World Bank education projects had components directed at improving girls' access to, and attainment in formal education. A similar concentration of efforts can be found in USAID (United States Agency for International Development) support to education, as well as the educational programming of UNICEF (United Nations Children's Fund), UNESCO (United Nations Educational, Scientific, and Cultural Organization), and the Swedish International Development Agency. The importance of educating girls has been underscored in international forums, including the Third World Conference on Women (Nairobi, 1985) that capped the United Nations Decade for Women and the 1990 conference in Jomtien, Thailand, on Education for All.

Despite widespread and growing concern for educating girls and women, in the majority of developing countries there is still a noticeable gender gap in educational participation rates and associated outcomes such as literacy. UNESCO (1991) estimated that 640 million women were illiterate in 1990, representing two-thirds of the world's illiterate population (see Figure 1). Given that 60 percent of the 100 million children worldwide who do not attend primary schools are girls, the literacy differential is likely to continue well into the twenty-first century.

The ABCs of Girls' Education

Three issues have received the largest share of attention in the literature on girls' education in the Third World: (1) the extent of girls' access to schooling and how it has changed over time; (2) the potential benefits of educating girls; and (3) the constraints on girls' full participation in education. These three areas constitute the ABCs of girls' education.

Access

The evidence that people have access to education is that they take part in educational programs. Thus, access to schooling is measured by participation rates. The gross enrollment ratio (GER) is the most widely available international measure of educational participation. It is the ratio of the total number of students enrolled at a given level to the estimated population of school-age children. Gross enrollment ratios can be computed for any level of education and separately for males and females. Unlike the net enrollment ratio (NER), which computes age-specific enrollment in different levels of schooling, the GER includes overage students. In countries such as China, which has special primary-education programs for adults, or in other countries with many overage primary-school students, the GER can and does exceed 100 percent. The NER gives a more accurate accounting of children's participation in schooling, but it is not available for many countries. Thus, the discussion that follows is based on GER calculations.

UNESCO data show that, for the developing countries of Africa, Latin America, and Asia as a whole, girls' GERs at the primary level rose from 58 percent in 1960 to 91 percent in 1990. At the secondary level, girls' GERs went up from 10 percent to 36 percent over the same period. The percentage of young women enrolled in tertiary institutions increased fivefold in the 30-year period, from 1 percent in 1960 to 5 percent in 1990. Figure 2 displays the trends in female and male enrollment at all three levels of education.

As is evident from Figure 2, male enrollment has in the past exceeded and continues to exceed female enrollment at all levels. Figure 3 illustrates the gender gap, the difference between male and female enrollment ratios at each level, over time. The gender gap at the primary level declined substantially from 1960, when it was 29 percent, to 1975, when it registered 19 percent, and stood at 14 percent in 1990. By contrast, at the secondary level the difference in participation rates between males and females rose from 9.5 percent in 1960 to 14 percent in 1975. Since the mid-1980s, this gap seems to be declining; it reached 12 percent in 1990. In higher education, the difference in participation between young men and women seems to be gradually rising, from 1.9 percent in 1960 to 2.5 percent in 1975 to 3.1 percent in 1990. The fact that the gender gap appears to grow at times or to disappear slowly at the secondary- and higher-education levels calls for closer analysis. In all likelihood, it reflects the increasing restrictions of gender and class roles upon individuals as they mature. Social sorting by gender and class at the higher end of formal education inevitably feeds back into the earlier levels of education, as young girls begin to comprehend the options that are available to them and the relevance of education to fulfilling socially sanctioned roles.

Aggregate enrollment figures obscure substantial variations in girls' and women's access to education both within and between countries. Participation rates vary across regions and by the income level of countries. Figures 4–7 show female and male gross enrollment ratios at the primary, secondary, and tertiary level for the countries of Asia (see Figure 4), Africa (see Figure 5), the Arab states (see

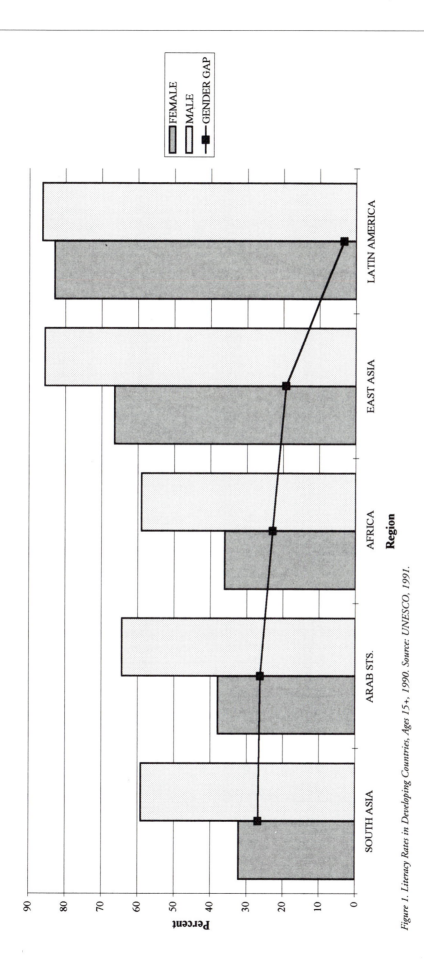

Figure 1. Literacy Rates in Developing Countries, Ages 15+, 1990. Source: UNESCO, 1991.

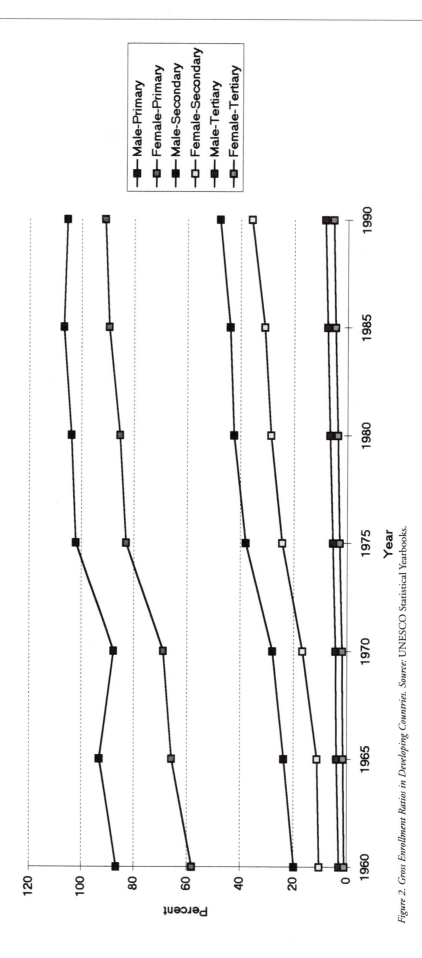

Figure 2. Gross Enrollment Ratios in Developing Countries. Source: UNESCO Statistical Yearbooks.

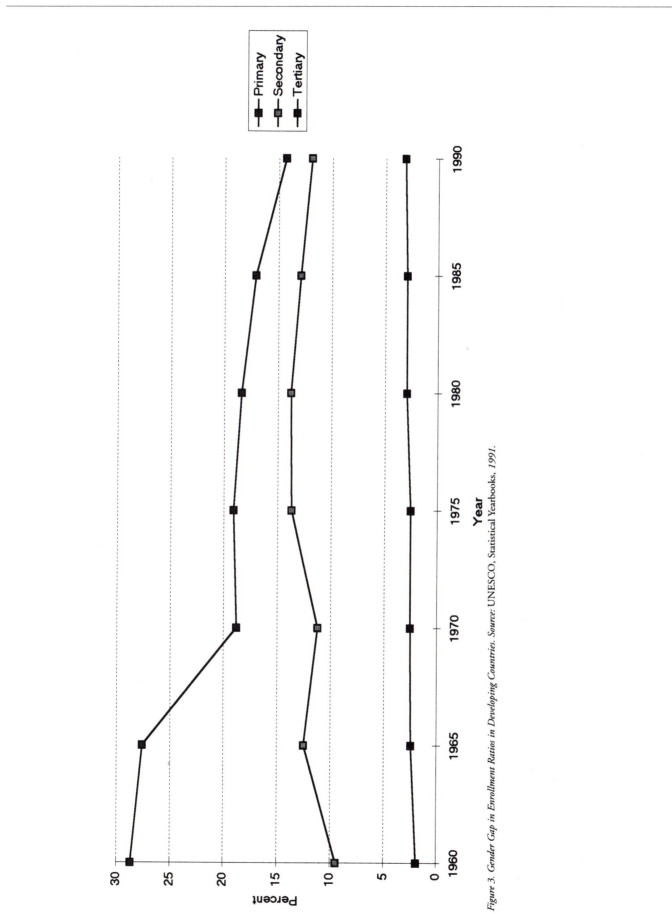

Figure 3. Gender Gap in Enrollment Ratios in Developing Countries. Source: UNESCO, Statistical Yearbooks, 1991.

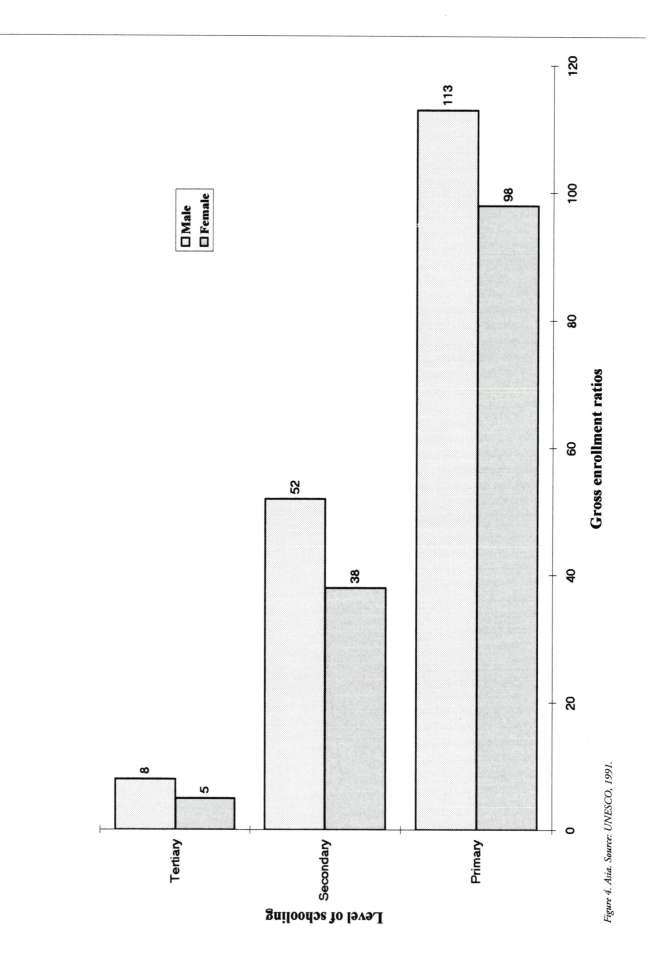

Figure 4. Asia. Source: UNESCO, 1991.

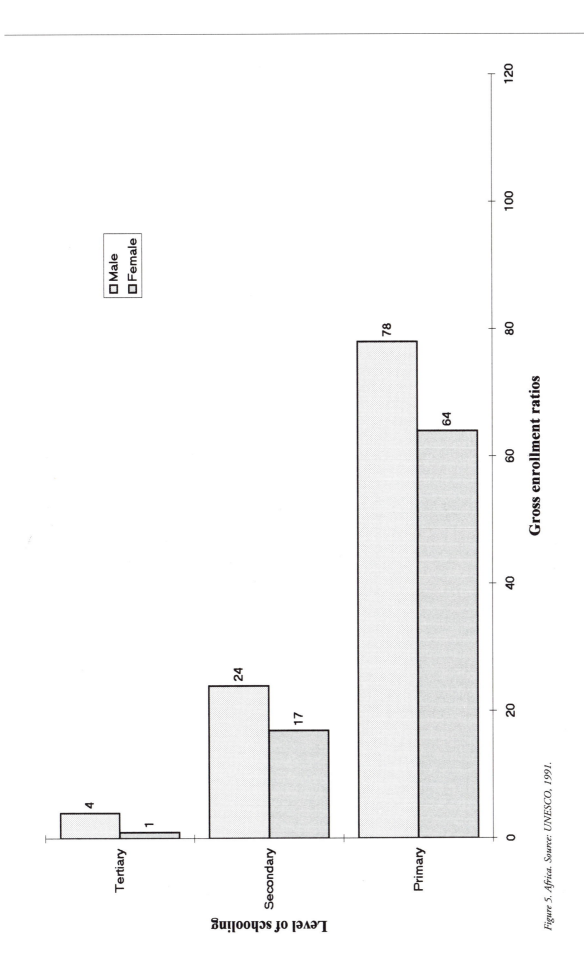

Figure 5. Africa. Source: UNESCO, 1991.

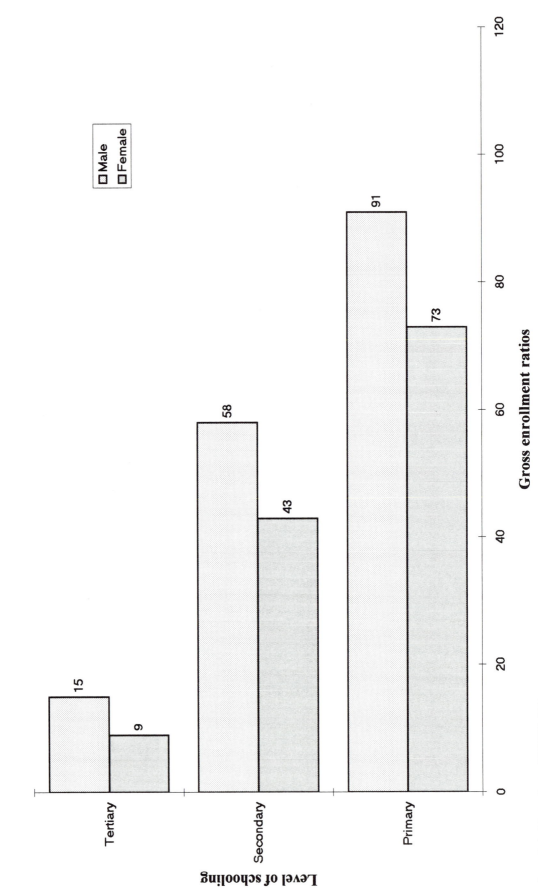

Figure 6. Arab States. Source: UNESCO, 1991.

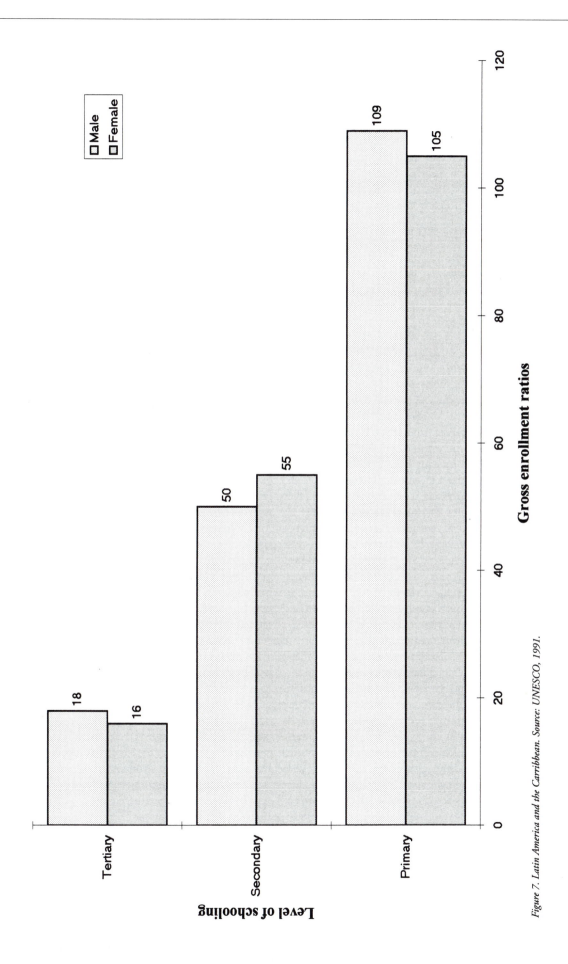

Figure 7. Latin America and the Carribbean. Source: UNESCO, 1991.

Figure 6), and Latin America and the Caribbean (see Figure 7). In Latin America and Asia, the primary GER for girls is near to or greater than 100 percent, indicating that there are, on average, sufficient school places for all girls as well as boys. In Africa and the Arab region, the primary GER for girls stood at 64 percent and 73 percent, respectively, in 1990. In these two regions as well as Asia, girls' primary enrollment was 14 to 18 percent less than boys'.

The near-parity of girls' and boys' primary enrollments in several countries of Latin America and the Caribbean reflects specific national circumstances in that region. At the secondary level, the unweighted average (computed on the number of countries rather than population per country) of girls' GER across countries (55 percent) exceeds boys' GER (50 percent) by 5 percent. Argentina, Chile, Colombia, Costa Rica, Cuba, Ecuador, El Salvador, Guyana, Jamaica, Nicaragua, Panama, Suriname, Trinidad, and Venezuela all report slightly higher enrollment of girls than boys at the secondary level; the relationship remains when participation is measured by age-specific, or net, enrollment ratios. Of the countries for which data are available for the 1990s, only Bolivia, Haiti, and Peru measure female secondary participation that is lower than that of males. Several factors undoubtedly contribute to this statistically anomalous phenomenon and deserve further study, including differences in migration patterns and wage-labor opportunities for men and women. An equally important set of questions could be raised about the intersection of gender with class and ethnicity in defining access to education and the rewards that accrue to it. Whatever causes lead to higher secondary enrollment of females in many Latin American countries, they do not seem to carry into the post-secondary level. Higher education in Latin America, as elsewhere in the Third World, is characterized by greater participation of males than females.

Some associations have been established between the income level of countries, measured by gross national product (GNP) per capita, and the amount of participation by girls in formal education. Studies by World Bank researchers (see King and Hill, 1993) have shown that girls' enrollment in primary schools is lowest in low-income countries, and highest in upper-middle-income countries. Girls' primary GERs were near 100 percent in lower-middle and upper-middle-income countries in 1985. For the lower-income countries, which are located primarily in South Asia and sub-Saharan Africa, girls' primary GERs were just over 60 percent in the same year. Not only levels of participation but also the size of the gender gap differ between income levels. There was an average 20 percent difference between girls' and boys' primary GERs in low-income countries in 1985. For lower-middle-income countries, the gap was less than 10 percent; in upper-middle-income countries, less than 5 percent (see King and Hill, 1993, p. 8). As is the case when looking at regional averages, these composite, cross-country averages provide a picture that is suggestive in outline but lacking in relevant detail. Several

countries stand as exceptions to the general income-level and regional trends. Madagascar ranked 12th from the lowest GNP per capita in 1987, yet showed 99 percent gross enrollment of boys and 95 percent of girls at the primary level. Secondary enrollment ratios, 20 percent for males and 19 percent for females, though not high, were nearly equal between genders. In Jordan, female and male gross enrollment ratios at the primary level exceeded 100 percent in the early 1980s, with 73 percent female and 75 percent male secondary participation at the same time. Sri Lanka, with a long history of free education through university level, is one of the handful of countries in which female participation at the secondary level exceeds that of males, with rates of 74 percent and 68 percent, respectively, in 1988.

To the extent that they obscure the differences between and within countries, gross enrollment ratios tell only part of the story of girls' and women's access to education. Another shortcoming of these indicators for assessing educational access is that they do not illuminate the differential attainments of males and females within a given level of education. When other statistics are brought into the picture, such as the net enrollment ratio, the distribution of enrollments at different grade levels and the percentage of children reaching the upper grades of primary school, more profound gender differences emerge. In Rwanda, for example, gross enrollment at the primary level reached 69 percent for boys and 66 percent for girls in 1988. However, UNESCO estimates show that only 43 percent of girls, compared to 50 percent of boys, will reach the eighth and final year of the primary cycle. Figures for primary-school retention in 24 African countries show, on average, 46 percent retention of girls, compared to 53 percent retention of boys (Hyde, 1993, p. 108).

These and related figures underscore the fact that in many Third World countries, not only do fewer girls than boys attend primary school, but more drop out before completing the primary level. The continuing differences in educational access between boys and girls reflects realistic assessments by girls and their families of the possibilities that education offers for young women. If women are unlikely to obtain the jobs or other social positions that require formal education, the motivation to persevere with schooling is undercut. Gender differences in educational participation have become of increasing concern to analysts and policymakers, who have found compelling evidence that the education of girls and women makes substantial contributions to societies and economies.

Benefits of Education

In 1979 and 1982, researchers at the World Bank published two important studies that highlighted the social and economic impacts of educating girls and women (Cochrane, 1979; Cochrane and Jamison, 1982). Both studies examined how the education of women affects fertility and child health; the second also established that education of women raises agricultural productivity. These relationships have been further analyzed and documented, as has the impact of women's education on earnings and economic development (Floro and Wolf, 1990; Herz et al., 1991; King and Hill, 1993). It appears indisputable that education of girls leads to lower fertility and improved child health. It is more difficult to determine the impact of education on economic productivity among women, due to discrimination in many labor markets.

Among the many potential effects that the education of girls and women may have on society, perhaps none is so extensively documented as the relationship between a woman's education and the number of children to whom she gives birth. The analyses generally concur that women with four to six years of schooling average one fewer birth than women with no schooling. For those with seven or more years of schooling, the number of children conceived drops even further, by two fewer pregnancies in Africa and three fewer in Latin America and Asia (Schultz, 1993, p. 77). Figure 8 summarizes fertility rates associated with levels of women's education.

The studies of education and fertility identify a number of plausible pathways by which more education of women leads to fewer pregnancies. Women with higher levels of education tend to marry later than those with less and to avail themselves of contraceptive technologies at much higher rates. In Latin America and Asia, women with seven or more years of education are twice as likely to use contraceptives as those with no schooling; in Africa, the rate of contraceptive use is tripled with seven or more years of schooling (Schultz, 1993).

Educated women, it has been shown, have not only fewer but also healthier children. Summarizing a number of studies on the relationship between mother's education and children's health, Schultz (1993) observes that each year of a mother's formal schooling is associated with a 5–10 percent reduction in child mortality. When other factors such as access to and affordability of health services and the availability of clean water are included in the equation, the education of mothers is consistently one of the two most powerful determinants of child health, clean water being the other. The effect of mother's education is so strong that, in many cases, it overrides access to health care as a determinant of child survival.

Less attention has been devoted to the impact of maternal education on children's life experiences than to their conception and survival. A partial exception to this rule are studies of intergenerational effects of education. Herz et al. (1991) report that seven studies covering eight countries all find strong associations between levels of parental education and the educational attainment of children. In three of these countries, Malaysia, Peru, and Ghana, the education of mothers exerts a greater influence over daughters' education; that of fathers, over sons' (Herz et al., 1991, pp. 20–21). In all countries studied, the mother's education is as important as or more important than the father's for determining educational outcomes of both sons and daughters.

The research concerning the economic benefits of educating women is more difficult to summarize than that on social benefits. This is partly due to technical issues; variations in econometric modeling make it difficult to compare studies done by different people in different countries. In addition, most analyses of economic returns to education rely implicitly or explicitly on wages received. Women's contributions to the economy through the informal sector, domestic production, and, in some cases, agriculture are rarely incorporated into these studies. Differential participation of women and men in the labor market and discrimination in pay also cloud the picture.

Two general approaches prevail in assessing the economic benefits of education, one macro and the other micro. On the macro level, growth in the economy, indicated by the GNP, can be regressed on levels of educational participation at prior time points. Benavot (1989) conducted such a study for 76 countries over the period 1965–1980. He found that, while the education of both genders had a substantial impact on economic growth, the education of girls was a stronger predictor of growth than that of boys; he also found that this effect was particularly powerful in the poorer countries of sub-Saharan Africa (reported in Floro and Wolf, 1990, pp. 10–12).

Microlevel analysis, more common in the economics of education, uses the "rate of return" approach. Put simply, rates of return compare lifetime earnings of an individual to investment in that person's education. Returns to education vary by level of education. In developing countries, returns to primary education tend to be the highest, followed by returns to secondary education, and then to higher education. This form of analysis distinguishes between private returns (the economic benefit realized by individuals in relation to their investment) and social returns (the benefits accorded to society in association with the total social costs, including government expenditure on an individual's education). Private rates of return provide a measure for inferring the individual's or family's willingness to invest in education. Social rates of return may be used by policymakers to determine which investments in education (at what level, for which populations) are most beneficial to the society as a whole.

The relative private returns to education between men and women vary across countries and by levels of education. For example, in 10 studies summarized by Schultz (1993), returns to each level of education were estimated to be lower for women than men in about half of the cases. Several economists have attempted to recalculate rates of return by gender using some correction for the difference in labor-force participation of women and men. The results of these analyses tend to show much higher returns to women than to men, especially at the secondary level (Herz et al., 1991, p. 14).

Finally, more recent analyses of the social and economic benefits of educating women have looked not only at the outcomes associated with different levels of female participation but also at how these benefits are affected by

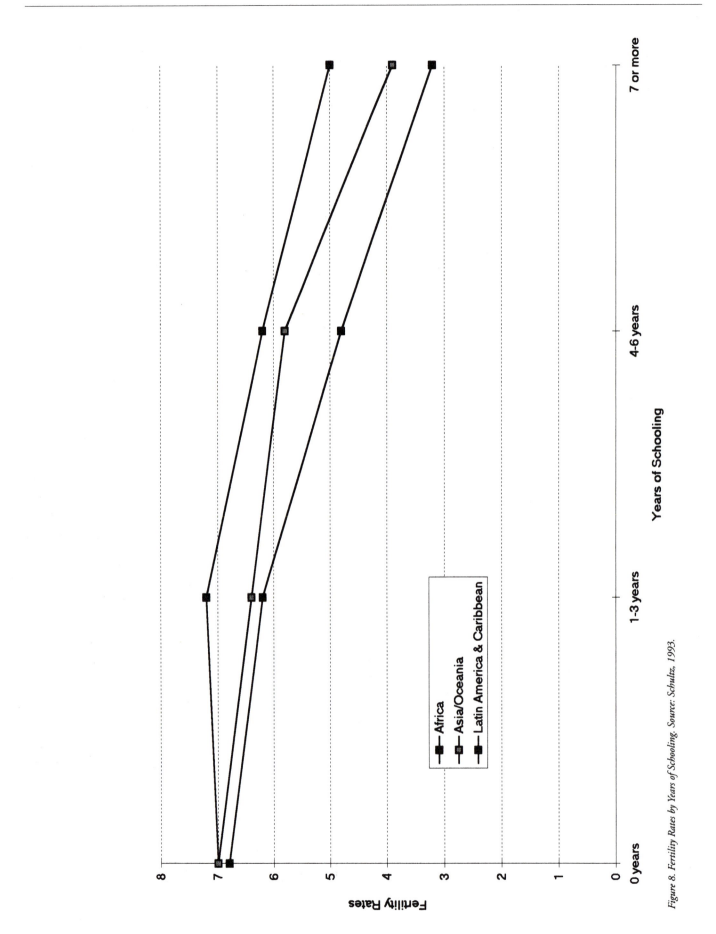

Figure 8. Fertility Rates by Years of Schooling. Source: Schultz, 1993.

differential educational attainment by girls and boys—in other words, by the gender gap in education. The results are informative. Looking at the established relationships between women's education and fertility, child health, and economics as they relate to differential participation, King and Hill (1993) observe:

> The levels of fertility and infant mortality associated with a particular level of female education are much lower, and GNP per capita and life expectancy much higher, in countries with greater equity between the sexes. Since these observations hold even for countries that have achieved gross female primary enrollment rates of 100 percent, the gender gap at the secondary level is apparently quite important too (King and Hill, 1993, p. 15).

In other words, it is not only the overall level of education in the female population that leads to expected societal outcomes like economic growth and lowered fertility. Equity in educational participation between females and males also determines whether beneficial results are obtained. In terms of the social benefits deriving from girls' education, equity at all levels of education appears to have a positive impact. This is a significant finding that suggests the need to look beyond girls' access to primary education, an issue that has gained increasing support from international assistance agencies. In order for girls to participate fully in primary education, they and their families must ultimately perceive that there are rewards for girls who persist into higher levels of education. Thus, the issue of attaining full participation of girls in basic education brings us face to face with questions of how gender roles are constructed in different societies.

Constraints to Education

The evidence of the benefits of educating girls and women has grown dramatically since the early 1980s, yet the enrollment of girls and women has not reached levels comparable to those of boys and men. What accounts for this disparity between the ideal and the actual? The most benign explanation is that educators do not have sufficient knowledge of the reasons that girls are out of school to formulate effective policies and programs to increase access for them. Consequently, educational researchers and policy analysts have devoted considerable attention to understanding the constraints to girls' and women's participation in education. Various factors have been identified as constraints to girls' participation. For present purposes, they will be classified as economic, household, sociocultural, and school-system factors. As the discussion will make clear, these factors are interactive and overlapping.

There are two important economic constraints to girls' educational attainment. The first is that women cannot always command the same wage, or even receive employment in the same occupations, as men. Such labor-market restrictions can make the returns to education lower for

females than males, thus depressing the demand from girls and their families for more education.

At the same time, girls spend more time in domestic labor than boys in most societies. Girls' work includes tending children, fetching fuel and water, cooking, garden and farm work, and marketing. Attending school takes time away from household work, creating an often rather high "opportunity cost" for families. Opportunity cost refers to the labor and income that are lost to the household when a child attends school. Numerous studies show that girls age 6–14 in many countries devote from 1.5 to 2.5 times as many hours to home and market production activities as do boys of the same age (King and Hill, 1993). Because of the important contribution that girls make to the well-being of their families and because of the greater expected earnings of males, household decision-making processes may place the education of sons over the education of daughters. In many societies in the Third World, girls enter primary school at a later age than boys and so may reach puberty before completing primary school. In all parts of the world, puberty is likely to be a period of what anthropologists call "cultural compression" in the lives of girls. With sexual maturity comes a host of specific restrictions on public activity and a focus on entering marriage. These expectations can push young women out of school. Moreover, in many societies girls "marry out" of their natal family. Parents can reasonably expect to attain greater direct financial benefit from their sons' education than from their daughters'. Sociocultural norms thus reinforce economic ones in lowering household demand for the education of girls. Household decision-making processes, in fact, can be viewed as the daily enactment of economic and sociocultural norms and practices.

Economic, household, and sociocultural factors shape the demand for education by girls and their families. Demand is also influenced by the supply of school places and the quality of education offered. Schools that are far from home and those that have few teachers overseeing many students can be perceived as threats to the safety or chastity of young women. Low-quality education offers little leverage over high opportunity costs. For girls as well as for boys, the potential for achievement offered by the available schools is certainly a major factor influencing individual decisions to enter, stay in, or leave school. The higher the direct costs, such as tuition, and the indirect costs such as uniforms and transportation, the more families are likely to choose education for sons over daughters.

Opening Schools to Girls

Based on a cumulative understanding of why it is important to educate girls and women, and what factors impinge on meeting this goal, governments have implemented policies and programs designed to enhance the participation of girls in schools. Some of these steps have resulted in documented improvements. In many cases, however, more careful evaluation is required to determine the efficacy of particular measures in specific contexts. At the same time, research and policy interventions have generally assumed that the challenge of educating girls can be met by attending to a rather narrowly defined set of questions. What is needed is a redirection of analytic and policy attention to the larger questions that are raised in the process of attempting to realize the promise of education for girls and women.

Steps Taken

The most immediate means for policymakers to address educational issues lie in the educational system. School-based interventions formulated by international assistance agencies and by governments to enhance girls' education have sought to ameliorate the social and economic factors that constrain girls from participating in schooling. Creating more schools closer to where children live is an obvious measure for reducing the indirect and opportunity costs of educating girls and ensuring a higher degree of safety than is afforded by long commutes. In some countries, including Egypt, Bangladesh, and Indonesia, studies have shown that girls' participation dropped in relation to long distances from schools. In other countries, like India, Ethiopia, and Nepal, having schools within a kilometer of home did not seem in itself to lead to higher enrollment of girls (Floro and Wolf, 1990, pp. 18–19). Despite this ambiguity about the magnitude of effects gained by creating local schools, it is an approach that has strong intuitive appeal and popular support.

Programs in some countries have focused on reducing direct costs of schooling by offering scholarships for girls. Such efforts have been undertaken in Bangladesh for general secondary education and in Nepal for training primary teachers in normal schools (Herz et al., 1991). These programs appear to have made an impact not only on individual beneficiaries but also more broadly on perceptions of the importance of educating girls. Directly or indirectly, they contribute to increasing the proportion of women in the teaching force. The presence of female teachers has been cited by parents as a determinant of their willingness to send daughters to school, although correlational studies have not found a compelling relationship between girls' participation and the percentage of women teachers.

In some countries, such as Pakistan and Saudi Arabia, the majority of schools are segregated by gender, with male and female schools under separate administrative authority. Even in countries in which schools are not formally or in large number so segregated, some argue that single-sex schooling may make an important contribution to improving access, retention, and achievement of girls. The idea has intrinsic appeal for those who view traditional sanctions against girls mixing with boys as a major constraint to girls' education. A few studies have shown that, in the Third World as in the United States, girls in single-sex schools appear to stay in school longer and achieve more highly than in co-ed schools (Tietjen and Prather, 1991, pp. 74–77). Although there are evident academic benefits to single-

sex schooling, creating such systems on a wide scale raises some problems, not the least of which is the cost of running a dual system. Donor attempts to direct large portions of their educational assistance to establishing girls' schools might also be met with resistance by some national policymakers, who may view such efforts as detrimental to boys. The content and outcomes of gender-segregated schooling also must be carefully considered. It is important for educators, parents, and societies at large to consider whether it is preferable to socialize girls and boys separately or together.

Governments and donors have begun to initiate public campaigns to raise awareness of the importance of educating girls. These campaigns are predicated on the belief that parental attitudes are a primary constraint to girls' schooling. In South Asia, UNICEF declared 1992 the Year of the Girl Child. One of the many activities supported in this year was a media campaign, using a cartoon character named Meena, that demonstrated ways in which educating girls was beneficial to themselves, their family, and society. Since the late 1980s, USAID projects in Mali, Malawi, Bangladesh, and Guatemala have adapted social-marketing campaigns, long a component of public-health programs, to influence parental attitudes concerning girls' education. These campaigns raise two sets of issues. The first is assessment: It is inherently difficult to determine the specific impact of media campaigns aimed at changing behavior. The second issue is more ideological. The campaigns in Bangladesh and Guatemala, as examples, both argued for educating girls in terms of the contributions that education would make to traditional female roles, such as wife and mother. For some, the end of getting girls into schools justifies the means of reinforcing traditional roles. For others, this approach seems destined to perpetuate sexist stereotypes. Because the efforts are so new, it is too early to determine whether this is so. Certainly, it merits future study.

One other approach to enhancing girls' education has gained increasing support from donor agencies, including UNICEF, USAID, the World Bank, and the Asian Development Bank, since the late 1980s: community schools. These vary substantially in their specific form and their position in the education system as a whole. In the province of Balochistan in Pakistan, community schools are a joint responsibility of a local NGO, the Society for the Community Support of Schools in Balochistan (SCSSB), and the provincial government. The SCSSB follows a well-defined process for bringing parents together, selecting a local woman to be a teacher, and supporting the community in establishing a school, on property the community donates. When the school has been functioning for three months, the new teacher receives training and the government assumes her salary. The society also conducts regular community visits to check on progress. From 1992 to 1995, more than 250 such schools were opened, enrolling more than 2,000 girls. Schools that are organized and run by local communities have great potential for altering the pattern of girls' educational participation, particularly in places where there has been parental resistance to sending girls to school. For these schools to last and to provide meaningful education, they require the careful, ongoing, and labor-intensive support that is provided by the SCSSB.

Unanswered Questions

The research conducted on girls' education since the early 1970s has contributed to our understanding of the issues involved. It has also led to thoughtful policy measures designed to improve the education of girls. As a result of this work, scholars, activists, and policymakers concerned with girls' education now face a new set of questions that must be addressed for our efforts to be effective and meaningful. Two areas in particular demand greater attention: the context in which girls live and learn, and the experiences of girls within the schools.

As noted earlier, girls may be entering schools in increasing numbers, but many do not attain more than a few years of schooling. In addition to the economic and social pressures for girls to leave school, some researchers are beginning to look seriously at the experiences of girls within the classroom, in order to understand ways in which girls may be discouraged from continuing their education. Odaga and Heneveld (1995) offer an initial analysis of classroom-level case studies of girls' education in Africa. The case studies document teacher preference for males, sexual harassment by fellow students and teachers, and overt discouragement of aspirations among the factors that need to be addressed if girls are to have a stimulating and satisfying educational experience.

Studies based on models of household decision making have been extremely important for formulating policies to overcome immediate constraints to girls' education (Floro and Wolf, 1990). By identifying the economic value of girls' labor and parental attitudes toward daughters' education, these analyses have encouraged policymakers to reduce the invisible costs of schooling for girls, to provide schooling for girls that is more acceptable to parents, and to encourage parents to view the education of girls as valuable. At the same time, this model of analysis tends to assume that the context in which schooling takes place is static. This assumption is erroneous. The growth of educational systems changes societies. It reformulates the skills and credentials available in the population. Schooling is explicitly designated as a transmitter of common national values and identities.

The expansion of schooling also changes gender roles. As more males than females attain the level of education required to enter government and industry, the structure of families and subsistence activities has changed. As some girls also attain requisite credentials, women take on social and economic roles that may conflict with prevailing norms concerning gender identity. Thus, the question of how to enhance girls' education is as much a sociocultural issue as it is an issue of household economics.

It is also a political issue. The process of education changes people, for better, or for worse, and most often in ways difficult to evaluate as either. For those in groups that have historically been marginalized by exclusion from the schooling system, access to education can raise aspirations and heighten awareness of social injustice. This process is perhaps most fully captured not by social science, but in literature such as Tsitsi Dangeremba's novel, *Nervous Conditions* (1988). Dangeremba portrays the social and emotional costs paid by two women who attain rare levels of education in preindependence Zimbabwe. Both must find their own balance between the female roles open to them and the desires and aspirations created through their learning.

As Stromquist (1989) has argued, it may be disingenuous to believe that states as presently organized will act to neutralize power differentials between genders, as well as those based on ethnicity or class. This point is underscored by the study conducted by Mbilinyi and Mbughuni (1991) of gender in education in Tanzania, a country that has explicitly endeavored to equalize education but failed to adequately address gender differences in the process. For external actors, such as foreign scholars and the donor community, it is difficult to exert pressure on governments to effect what are, in the end, political changes within their own countries. For these and other reasons, it is perhaps more comfortable for external actors to treat girls' education as an economic issue susceptible to technical solutions.

To fully realize gender equity in education, however, means changing the process so girls realize as much learning as boys, and changing the context in which education operates so women can benefit from their education to the same measure as men do. Such processes are slow, complex, and situation specific. They require, above all, an active dialogue concerning gender relations and roles. This is a process that can be encouraged, but not imposed, by actors outside the societies in which it takes place.

References

Altbach, Philip, Gail Kelly, and David Kelly. *International Bibliography of Comparative Education.* New York: Praeger, 1981.

Benavot, Aaron. "Education, Gender, and Economic Development: A Cross-National Study," *Sociology of Education,* no. 62, 1989, pp. 14–32.

Cochrane, Susan. *Fertility and Education: What Do We Really Know?* Baltimore: Johns Hopkins University Press, 1979.

Cochrane, Susan, and Dean Jamison. "Educational Attainment and Achievement in Rural Thailand." In Anita Summers (ed.), *Productivity Assessment in Education.* San Francisco: Jossey-Bass, 1982.

Dangarembga, Tsitsi. *Nervous Conditions.* London: Women's Press, 1988.

Floro, Maria, and Joyce Wolf. *The Economic and Social Impacts of Girls' Primary Education in Developing Countries.* Washington, D.C.: Office of Women and Development, USAID, 1990.

Herz, Barbara, K. Subbarao, Masooma Habib, and Laura Raney. "Letting Girls Learn: Promising Approaches in Primary and Secondary Education," discussion paper no. 133, World Bank, Wasington, D.C., 1991.

Hyde, Karin. "Sub-Saharan Africa." In Elizabeth King and Anne Hill (eds.), *Women's Education in Developing Countries*. Baltimore: Johns Hopkins University Press, 1993, pp. 100–135.

Kelly, Gail P. *International Handbook of Women's Education*. Westport: Greenwood, 1989.

King, Elizabeth, and Anne Hill (eds.). *Women's Education in Developing Countries: Barriers, Beliefs, and Policies*. Published for the World Bank. Baltimore: Johns Hopkins University Press, 1993.

Mazumdar, Vina. "Education and Rural Women: Toward an Alternative Perspective." In Aruna Rao (ed.), *Women's Studies International: Nairobi and Beyond*. New York: Feminist Press at City University of New York, 1991, pp. 38–54.

Mbilinyi, Majorie, and Patricia Mbughuni (eds.). *Education in Tanzania with a Gender Perspective*. Stockholm: Swedish International Development Authority, 1991.

Obura, Anna. *Changing Images: Portrayal of Girls and Women in Kenyan Textbooks*. Nairobi: Acts Press, 1991.

Odaga, Adhiambo, and Ward Heneveld. *Girls and Schools in Sub-Saharan Africa: From Analysis to Action*. Washington, D.C.: Africa Technical Department Series, World Bank, 1995.

Schultz, T. Paul. "Returns to Women's Education." In Elizabeth King and Anne Hill (eds.), *Women's Education in Developing Countries: Barriers, Beliefs, and Policies*. Published for the World Bank. Baltimore: Johns Hopkins University Press, 1993, pp. 51–98.

Streeten, Paul, with Shahid Javed Burki, Mahbub Ul Haq, Norman Hicks, and Frances Stewart. *First Things First: Meeting Basic Human Needs in Developing Countries*. New York: Oxford University Press, 1981.

Stromquist, Nelly. "Determinants of Participation and Achievement of Women in the Third World: A Review of Evidence and a Theoretical Critique," *Review of Educational Research*, vol. 59, no. 2, 1989, pp. 143–183.

Tietjen, Karen, and Cynthia Prather. *Educating Girls: Strategies to Increase Access, Persistence, and Achievement*. Washington, D.C: Office of Women in Development, USAID, 1991.

United Nations Educational, Scientific, and Cultural Organization (UNESCO). *Statistical Yearbook 1972*. Paris: UNESCO, 1972.

———. *Statistical Yearbook 1978*. Paris: UNESCO, 1978.

———. *Statistical Yearbook 1982*. Paris: UNESCO, 1982.

———. *Statistical Yearbook 1988*. Paris: UNESCO, 1988.

———. *Statistical Yearbook 1993*. Paris: UNESCO, 1993.

———. *World Education Report*. Paris: UNESCO, 1991.

The Explicit and the Hidden School Curriculum

Nelly P. Stromquist
Molly Lee
Birgit Brock-Utne

Introduction

Schools are often considered the main vehicle for achieving individual mobility and collective social progress. They are also considered institutions that promote the development of democratic ideals, given their practices based on universalistic social norms and meritocracy. Often forgotten is the crucial role schools play in maintaining dominant class and gender ideologies through both curriculum content and teaching practices. Also forgotten is the role that schooling has played as a Westernizing force in colonized societies, including the introduction of a sexual division of labor and a separation between the social and economic spheres occupied by men and women that were generally much stronger than any in the precolonial experience.

The knowledge and skills transmitted by schools are considered to be, if not value neutral, at least beneficial to both individual and society. Sober and deeper views of educational institutions, however, show that they are powerful ideological institutions that transmit dominant values, and function as mechanisms of social control. Schools transmit values that not only reproduce social class but also maintain gender structures. The formal school system contributes to the reproduction of gender inequalities through such mechanisms as selective access to schooling, the content of what is being taught and what is not and how it is taught, and the kinds of knowledge men and women (and boys and girls) get. School curriculum functions to legitimate the political order, and any curriculum change often involves changing the definition of knowledge held by dominant groups; thus these changes are often fiercely contested. Institutions such as UNESCO (United Nations Educational, Scientific, and Cultural Organization), the World Bank, and the International Bureau of Education affect the school curriculum in developing countries by promoting changes ranging from child-centered curricula to all-round development of the individual, but they have been mild and rather ineffectual in the promotion of gender transformation in the curriculum. A recent trend—

promoted in the *Education for All Declaration* (1990) and the *Platform for Action* (agreed upon at the Fourth World Conference on Women in Beijing in 1995) and endorsed by a large number of world governments and development agencies as well—is to reform the school curriculum so that it is gender sensitive and more conducive to sex-role equality; it remains to be seen to what extent such a goal will be attained.

This essay examines the school curriculum in the broadest sense of the term, including both the formal and the hidden curriculum. It discusses the conditions of the curriculum at elementary, secondary, and university levels in Asia, Latin America, and Africa. Since the curriculum cannot exist independent of its transmitters, the essay examines the issue of teacher training and identifies what might be considered the optimal content of educational programs designed to contribute to the improvement of the social relations of gender. Finally, the essay reviews successes, counterforces, and obstacles to the development of gender-sensitive curricula.

The Formal Curriculum

Also known as the explicit curriculum, the formal curriculum covers the knowledge and skills schools officially seek to transmit via their program of studies, courses, and textbooks. This transmission cannot be considered aproblematic, because some students reject school messages and values. Yet, most discursive and material representations are accepted by students. Few studies have explored the effects of textbooks on students—a question that would require careful longitudinal research tracing influences over time. Nonetheless, observers and educators consider that textbooks and curricular content leave lasting influences in our memories, as phrases and stories heard, read, and written about men and women condition our minds.

Research on education and gender in industrialized countries has made significant advances in documenting conditions and experiences in schools and in accounting for their diverse manifestations. This research has also begun to look at the interaction between gender, race, social class, and ethnicity, noting the times at which gender becomes salient and the times it does not play a significant role. In contrast, in developing countries, due to the limited resources and low interest of governments, this type of research has been much more limited. Yet, important aspects have been mapped.

Programs of study, particularly at the primary-school level, are commonly assumed to be value neutral, concentrating on the transmission of basic skills such as reading, writing, and arithmetic. A more detailed look shows that these skills cannot be separated from the content they transmit. In this respect, the content of most schooling portrays a world with clearly demarcated feminine and masculine roles, with women depicted in domestic and familial roles and men in professional, public engagements. One of the most researched aspects of schooling has been the textbooks. These are surprisingly similar across countries in their definition of women as followers, passive and self-sacrificing individuals, having immense loyalty to their families—but seldom involved in political or economic activities. Knowledge increasingly considered by feminists as necessary to gain a more critical view of the gendered world we live in is seldom touched by school textbooks; rather, there is an effort to define those subjects as "controversial" and thus to remove them from or to ignore them in the curriculum. Questions crucial to young women involving knowledge about their bodies and how to control them directly rather than submit to being controlled by others, about legal rights, about the general economic and political situation of women, and about the role of ideology in shaping social definitions of reality are first postponed because the students are "young minds," and later not covered because the curriculum "is already crowded."

Some of the new knowledge about the socialization that occurs through schooling advises planning school interventions by student age. During early childhood education, it is important to make teachers aware of their language use toward girls and boys (such as the frequent use of the diminutive toward girls), the distribution of seating in the classroom, and the types of games assigned to boys and girls. In primary education, much of the reproduction of stereotyped models occurs through reading. This period also marks the beginning of differential expectations regarding the physical and intellectual performance of girls and boys. Teachers also promote differential expectations regarding education and work. During secondary schooling, messages favoring domesticity for women prevail, and strong messages are conveyed about images, valorization, and expectations regarding the world of work. By the time of higher education, women are in teaching and feminine fields, and few women can be located in important research and teaching jobs (Ministerio de Cultura y Educación, 1992).

In many developing countries, given the poor access to schooling, and especially the weak representation of women as teachers (the case in many African and Asian countries) in rural primary schools and in secondary education, efforts for improvement have emphasized the issues of access and completion of schooling. Less frequent have been expressions of demand and pressure to modify the school curriculum (content and modes of instructions), despite its great importance to the schools' fulfillment of social reproductive functions. Although ostensibly gender neutral, classrooms and schools activate gender forces in ordinary and recurrent situations such as playtime, choice of friends, themes in student writings, and selection of courses when there is student discretion.

The Informal Curriculum

The informal curriculum refers to the no less important but "hidden" curriculum—the set of knowledge that is transmitted through the roles men and women play in the staffing of schools, the way teachers treat male and female students, and the manner in which adults interact with each other. Additional important aspects of the hidden curriculum are the expectations teachers have of boys and girls regarding occupational and family roles, the differential vocational advice given to boys and girls, the behavioral norms and disciplinary sanctions enforced at school, and the re-creation within the school of norms and values concerning masculinity and femininity by the peer group (Levinson, 1997).

Textbooks are also significant mechanisms for the transmission of the informal curriculum. In addition to certain subjects, knowledge, and skills, books disseminate sexual bias, prejudice, and discrimination through the ways in which men and women are depicted in stories and illustrations. By focusing on some and ignoring others, the identity of persons may be strengthened or weakened accordingly.

Social and political forces in the forms of particular interest groups and social institutions shape and influence choices of content and pedagogy related to the school curriculum. Very often, the curriculum functions to legitimate the existing political order; likewise, curriculum changes reflect the changing definitions of knowledge by dominant groups (Apple, 1977) and the changing educational culture of the global system (Meyer et al., 1992).

Latin America

Most of the work connecting gender and curriculum in Latin America has occurred through the content analysis of textbooks. Studies have been conducted in several countries, including Peru, Brazil, Colombia, and Argentina (see Stromquist, 1992). A study of fourth-grade textbooks in Brazil, covering the period between 1941 and 1975, found that men were given more frequent and prominent coverage in textual and illustrated materials; the books presented women as passive, emotional, and affectionate, while men were shown as enterprising, curious, endowed with authority, and occupying professional roles (Pinto, 1982). The study by Wainerman and Raijman (1984) of more than

100 reading textbooks used in grades 1–7 in Argentina during three different periods (1900–1910, the 1950s, and 1975–1980) found that, despite the passage of time, textbook illustrations and accounts of women and men emphasized a traditional sexual division of labor and practice. The textbooks had changed little, even though there had been significant changes in the labor force, such as the increased participation of women, their increase in professional roles, and the social and political struggles in which they had participated (Wainerman and Raijman, 1984).

A study of primary-school textbooks in Colombia found that they represented men and women in three dichotomies: men as public and women as private, men as strong and women as weak, and men as active and women as passive. The same study found that a recurrent theme was the family, a nuclear family in which the mother worked at home while the father worked in the office. Girls were portrayed as performing secondary activities and having a weaker intelligence than their brothers (Bonilla, 1978). A study in Peru found that the frequency of male illustrations and characters increased as the grade advanced. Thus, while in the first grade two out of three, about 65 percent, references were to men, by the sixth grade three out of four references, or 75 percent, were. The study also found that women and men were depicted with highly differentiated personality traits, occupational roles, and household tasks within the family (Anderson and Herencia, 1983). Working in the area of the hidden curriculum, an ethnographic study of a high school in Mexico found that, although boys and girls upheld conventional views of masculinity and femininity, these views were not fixed and that well-designed interventions by the school administration could create positive changes (Levinson, 1997).

Beyond the identification of stereotypes in school textbooks, there has been little work in Latin America on the questions of rewriting the textbooks and providing gender-sensitive training to teachers. Indeed, a comprehensive response by the state has been extremely rare. An exception has been Argentina, where a National Program for the Promotion of Equal Opportunity for Women in the Educational Area (PRIOM) was established in 1991 through an agreement between the Ministry of Culture and Education and the National Women's Council. This program, headed by a knowledgeable and charismatic leader, identified and implemented a comprehensive agenda: raising gender awareness and providing training for teachers, students, and administrative staff; providing special programs at all levels to improve the educational possibilities of women; providing vocational, professional, and educational advice to improve the participation and performance of women in the political, professional, and social spheres; introducing women's issues in textbooks to describe women's role not only in the family but also in history, economy, politics, and society; conducting research projects to attain a wide, systematic, and permanent knowledge of women's situation in education; and creating a data bank on problems facing women in the educational sphere (Ministerio de Cultura y Educación,

1992). PRIOM implemented its program in 20 provinces through agreements with provincial ministries of education and became linked with women's units in provincial governments, universities, and women's NGOs. The motto of the women's efforts in education in Argentina was, "revise received knowledge, consider the omitted knowledge." In 1995, just as PRIOM was moving toward integrating the new gender curriculum into ongoing efforts to improve the national curriculum in other aspects, the Catholic Church of Argentina attacked its work, accusing PRIOM's coordinator of trying to destroy the family and introduce homosexuality in schools. One of the main reasons for the Church's opposition was PRIOM's commitment to work on the question of sex education and women's control over their bodies. PRIOM continues to exist, but under a more cautious and less transformative leadership.

Another important case in Latin America has been that of Costa Rica, where the Ministry of Education and the Office for Women and the Family have produced a national policy for promoting the participation of girls in nontraditional areas of technical education. In Brazil, Ecuador, Chile, and Nicaragua, promising programs promoting girls' education are in place, but they are not comprehensive.

Feminist work on education in Latin America has favored the creation of more coeducational schools. Research in developed countries, particularly in Great Britain has indicated that coeducational schools and classrooms do not necessarily produce greater achievement in girls, as the mixed-gender environments do not eliminate the differential treatment that girls and boys tend to receive from teachers. The Latin American argument in favor of coeducation is not predicated on student achievement. Rather, it upholds coeducation as a more proper setting in which to socialize girls and boys, so that they learn to appreciate men and women as individuals and move away from social representations that link the genders to romance and sex. The pro-coeducational movement seeks a more natural environment for girls and boys so that they are socialized into norms of equality and mutual respect. The movement also seeks to introduce substantive changes in classroom practices so that a new environment is created; coeducation, therefore, is seen as part of the qualitative transformation of education in general.

Asia

Two common forms of perpetuating gender distinctions in the school in much of Asia have been textbooks and course assignment. Images depicting sexist bias in textbooks transmit values that reinforce the sex-gender division of labor that exists in numerous Asian societies. Despite this realization, many textbooks in Asia continue to portray debilitating roles and values for women. Confucian ideals include high respect for education, but at the same time assign domestic roles to women. A study of primary-school textbooks in China, conducted by the government, argued that significant improvements in advocating and portraying equality between the genders had occurred in Chinese textbooks (Chen et al., 1983), but a subsequent study conducted by an independent researcher, who compared U.S. and Chinese primary-school textbooks in language and reading, found considerably greater gender inequalities in China in the portrayal of occupational, political, and scientific roles (Shu, 1989). Sexual stereotypes have also been found in Hong Kong. Yau and Luk (1988) studied the contents of history and social-studies textbooks in junior secondary schools, and Au (1993) analyzed Chinese-language, social-studies, and health-education textbooks for primary schools in Hong Kong. Both studies found that the textbooks depicted women as passive and men as active and full of initiative; they also presented the domesticity of women compared to the political, professional, and intellectual leadership of men. A similar study of selected picture books and reading schemes used in kindergarten in Malaysia also found a sexist portrayal of female and male characters (Manan, 1991). The ideal woman portrayed in Korean textbooks tends to be a respectful daughter, a good mother, a sacrificing sister, and a loyal factory worker (Chung, 1994). Indian school textbooks also promoted traditional sex roles (Kalia, 1980): Instead of fostering a basic equality between men and women, the messages given to schoolchildren sanctioned the dominance of males. Females were often described in terms of beauty, obedience, and self-sacrifice; men, in terms of bravery, intelligence, and achievement.

The practice of channeling boys and girls into fields of study that are traditionally filled, respectively, by men and women—and thus of limiting both in their choice of subjects in the school curriculum—is a powerful mechanism to genderize fields of study and training. For a long time, it was a common practice for girls to choose home economics and needlework while boys opted for "male subjects" such as industrial drawing, metalworking, and woodworking. This practice has been abolished in curriculum reforms in Malaysia and Singapore. Today in Malaysian secondary schools, both girls and boys take a new subject called "Living Skills," which consists of cooking, sewing, woodworking, electricity, and electronics.

Women's socialization in a school setting that constantly portrays them in subordinate and dependency positions affects the level and form of participation at higher levels of education. In South Korea, the enrollment of men (52 percent of college-age men) is almost double that of women (28 percent) in colleges and universities, and the predominant field of study among women is home economics, followed by education, the arts, and medical science and health-related fields (Chung, 1994).

Parental influences operate together with the school. In many Asian societies, parents hold differential expectations and aspirations for their sons and daughters. Japanese parents, despite the substantial modernization of their country, view educating daughters as less important than educating sons, and the purpose of their education as preparing them to be better wives and mothers. Similar findings have been documented in the Philippines, especially among families of rural origins and those in which the father's education is low (Smith and Cheung, 1981).

Africa

During the colonial period, some African boys were provided with technical education in secondary schools or post-primary education institutions. Girls had no access to technical education during this time. After independence, the colonial educational system was abolished, and single education systems were established for all racial groups. But few things changed for girls. As Kalugula (1991) notes for Tanzania: "The colonial legacy of sex specificity in some courses was maintained. Girls were still denied access to technical education. And they still retained the exclusive access to domestic science education" (Kalugula, 1991, p. 153). The five-year plan for 1957–1961 (just before the end of colonial rule in Tanzania), stated:

> Because of the close association which has long existed between African women and the land, it is considered that more should be done to encourage school gardening at girls' middle schools and such gardening should be closely related to the domestic science course, since it is important to know how to grow food as well as to cook it.

This line of thinking was perpetuated and used to plan girls' curricula at the secondary-school levels as well. Girls in coeducational schools and in girls' schools had to take domestic science as a compulsory subject in the first two years of "0" level, but later in forms 3–4 they took it as an optional subject. Boys did not have to take this subject. The real turning point in the curriculum-reform process in Tanzania was initiated by former president Julius Nyerere when he formulated his "Education for Self-Reliance" policy and philosophy in 1967. This document (Nyerere, 1967) does not go into specific curricula or course programs but rather addresses the formulation of goals of education for independent Tanzania. A diversification of the secondary-school curriculum took place in 1972 as special emphases were identified: agriculture, commercial, technical/craft, and home economics. Practically all girls' secondary schools had to offer home economics. All technical schools were essentially boys' schools, and no boys' school had an option in home economics. Coeducational schools took a commercial emphasis. Only two of the 20 agricultural-emphasis schools were for girls; the other 18 were for boys.

There seems to be agreement in the literature of the newly independent states in Africa that increasing the quantity of education has overshadowed the work with quality, including the reconstruction of the curriculum (Beshir, 1974; Uchendu, 1979; Jansen, 1989). This has occurred in part because the legitimacy of the new states depended heavily on their ability to provide access to schooling for large parts of their populations, and this expansion often took place with a blind eye turned toward the gender question.

In Tanzania, textbooks continue to depict men and boys twice as often as girls. Content analysis of 18 textbooks written by the Institute of Curriculum Development in Tanzania found that girls are depicted serving others, in submissive behaviors, and preparing food (Kalugula, 1991). The most submissive behavior in the pictures showed a girl kneeling down before elders. The only people pictured in leading and white-collar jobs were men (Kalugula, 1991). An earlier study of books in Zambia (Tembo, 1984) also documented the presence of sexual stereotypes, with women frequently depicted in passive and secondary roles. In an article on the effects of the biased curriculum of the secondary schools in Tanzania on the education and future employment of girls, Katabaro (1991) concludes that, following independence, when educational planners talked about achieving self-sufficiency in local manpower by the year 1980, they "realistically meant male personnel. Until recently the question of gender differences never attracted the attention of the planners" (Katabaro, 1991, p. 85).

There is an urgent need for textbooks and learning materials to restore the best parts of African culture, history, and dignity after a long period of colonization and to provide these through an increased publishing capacity within the African countries themselves. Boulding (1976) gives a glimpse of the great variety of occupational roles African women played in earlier times. She tells of three successive reigning queens in one Transvaal tribe during the nineteenth century. In the Niger and Chad regions and in the Hausa territory, women founded cities, led migrations, and conquered kingdoms. Songhai groups still remember the names of celebrated female ancestors who governed them. In Katsina, Queen Amina became famous during the first half of the fifteenth century through her widespread conquests. She extended her influence as far as the Nupe, built many cities, and is deemed responsible for introducing the kola nut to the region. In northern Cameroon, it was often a woman who chose the site of a city, held the insignia of power, or governed the district. In West Africa, women administered the market system and the protolegal systems associated with it. European colonial administrators eventually came to forbid administrative market roles for women because they were considered inappropriate for the female sex by European standards. The most Western-oriented women were the ones who faced the greatest status deprivation as they were pushed to accept dependent roles by virtue of their adoption of Western lifestyles as the wives of professional men. The Nigerian Omu Okwei, the "merchant queen" of Ossonari (1872–1943), is an example of the success that could be obtained by women who did not permit themselves to be placed in Western-style sex roles. Omu Okwei amassed a large fortune and was elected the last market queen of Ossonari and chairwoman of the Council of Mothers (Boulding, 1976). During her lifetime, however, the supervision of markets was transferred from the Council of Mothers to the City Council under British pressure.

One of the few studies concerning the classroom experience of girls in Africa is Biraimah's (1982) study of secondary-school girls in Togo. She observed that teachers had little regard for the ability, character, and potential of female students and that their messages emphasized a gender division of labor alien to African culture, where women have always engaged in productive tasks inside and outside the home. Biraimah also found that, despite the teachers' low expectations of the female students, the girls themselves had high occupational expectations, which suggested that the school message was being rejected in some of the cases. There is a need for further research into the richness of African cultures and for this research to be translated into textbooks for African schools. But without researchers with a feminist perspective, the accomplishments and lives and women may be forgotten.

The assignment of a separate curriculum to the girls and the lower expectations attached to women's education have led, in many African countries, to the provision of fewer boarding schools for girls than for boys. This type of school, by enabling the students to engage full-time in their studies, has been instrumental in moving a larger proportion of students into higher education. Hence, the existence of fewer boarding schools for girls also operates to limit their educational mobility.

In the 1990s, the economic crisis in Africa does not create positive prospects for a redefinition of the curriculum and a rewriting of textbooks. The situation is exacerbated by concomitant pressures by the World Bank and the IMF to cut public expenditures in education and by bilateral donors to liberalize and privatize the textbook sector and to do away with national-curriculum-development centers.

Building an Antisexist Curriculum

From a feminist perspective, cognitive outcomes are only one result sought from the schooling experience; equally important are cultural outcomes. These include (1) human dignity, a sense of self-esteem and confidence; (2) social legitimacy, belonging to the group on an equal basis; and (3) personal and social power, the capacity to influence decisions and negotiate the transactions of everyday life (Wyn and Wilson, 1993).

Some feminists make a distinction between a nonsexist curriculum—one that is free of gender stereotypes and other forms of distortions—and an antisexist curriculum—one that seeks to destroy stereotypes and to build a new way of perceiving and establishing social relations between men and women. Obviously, eliminating stereotypes is only part of a much larger project. The progression in the development of a gender-sensitive curriculum is illustrated in Figure 1.

The issues to be covered in such a curriculum are numerous and imperative: characteristics of the patriarchal family and its role in domestic violence and transgenerational socialization; the social value of domestic work and its impact on the identity, health, and social participation of

Figure 1. Phases of Curriculum Change

Stage	Focus	Stimuli for Advances in Knowledge	Means	Outcome
1. Women's invisibility	Contribution by "true" thinkers and actors in history	To maintain standards of excellence	Legitimating existing knowledge	A stable curriculum and students as recipients of knowledge
2. Search for the absent women	Recovery of great women in all fields	To correct gaps in knowledge	Adding new information to existing paradigms	The incorporation of exceptional women
3. Women as a subordinate group	Understanding women's marginalization throughout history	To obtain justice	Questioning the ideological aspects of paradigms	Creating specific courses in women's issues
4. Establishment of women's studies	Describing women's experience	To produce new educational knowledge for cultural change	Legitimating and developing women's perspectives	Interdisciplinary courses on women's studies, valuing the students' experience in knowledge production
5. Multidisciplinary approaches	Questioning traditional disciplinary concepts and methodologies	To revise epistemological assumptions	Utilizing gender, class, and race as analytic categories	Greater theoretical production
6. Transformed curricula	Understanding the intersection of gender, class, and race; seeking knowlegde that captures the diversity of women's experience	To produce an inclusive conception of the diverse human experience	Transforming paradigms	Total integration of feminist perspectives in all disciplines

Source: Adapted from Bonder, 1994, pp. 70–71.

women; traditional and modern forms of sexual-reproduction control; the impact of mass media on the awareness and practices of women; power relations between the genders in politics, the economy, and the law; the participation of women throughout history; the need for decision-making, participation, organization, and negotiation skills; and the analysis of philosophical ideas across time regarding gender equality in society.

The curriculum work must be done in conjunction with the mass media to transmit images and messages that are not stereotyped but are more realistic and diversified regarding roles for women and men. The development of a gender-sensitive curriculum must also utilize advances of women's studies to remedy the persistent invisibility and distortions of women's experiences.

Specifically in the case of Africa, a rewriting of textbooks from a feminist perspective would mean not only making known the accomplishments of African women, but also examining the forces oppressing women, requiring them to work longer hours and for less pay than men, preventing them from obtaining equal rights with men. An analysis of the neocapitalist market structure whereby countries in Africa have been forced to grow cash crops to repay debts to the rich industrialized Western countries must find its way into textbooks in social science in African schools. Such an analysis will also have to include a feminist perspective looking at who among the Africans makes profits on the sale of cash crops—men or women? And who are the ones who have difficulties growing enough food for the family when the best soil is taken for cash crops? When the White experts come to teach industrial forestry, commercial fishing, and mechanized agriculture to Africans, whom do they teach? How much do they bother to find out first about the type of forestry and agriculture already going on in Africa and about who has been responsible for the care of forests and lands? A transformative gender curriculum must be multidisciplinary, link the private and the collective, and seek global transformation, and it must begin with the questioning of gender stereotypes.

Additional important issues that remain to be addressed are the formal polygamy in many African countries, the informal polygamy in all developing regions, the widespread presence and acceptance of machismo, the prevalence of informal marriages in Latin America, and the major role played by women in the informal sector of the economy in Africa.

Two specific curriculum subjects that need special attention are sex education and women's studies.

Sex Education. Sex education is increasingly important, not only because of its contribution toward the empowerment of girls and women but also because of the tremendous expansion of HIV/AIDS throughout the Third World. Sex education courses are severely underplanned. Even in countries where sex education is well established, such courses are brief and often superficial. In the United States, for example, most high school programs offer short courses of less than 10 hours, and fewer than 10 percent

of the students take comprehensive sex education courses of more than 40 hours or courses designed as a component within a health or sex program that spans both primary- and secondary-school levels (Stubbs, 1989). In many African and Asian countries, sex education also needs to address the cultural practice of sexual mutilation that exposes more than two million girls each year to needless pain and infection. Sex education courses have to be offered so that they question current norms regarding social relations of sexuality; otherwise they simply reinforce double standards of sexual behavior (Stubbs, 1989). They need to offer not only more information on anatomical and physiological features but also a new way to conceptualize affected and interpersonal relations between women and men.

Women's Studies. Women's studies programs operate not only to create new gender-sensitive knowledge in universities but also to develop crucial links between the educational system and universities. An important feature of these links is the provision of knowledge that can be incorporated into teacher training programs. Women's studies programs have expanded substantially since the mid-1980s, but their impact on education, especially on teacher training programs, has been modest. In building new content as well as more emancipatory ways of learning and teaching, much greater use should be made of women's studies.

One example of the fruitful role of women's studies derives from Tanzania. Feminist educators there, after extensive analysis of their educational system, presented demands leading to the creation of a "compulsory *basic life skills* course in all schools for all students that would combine domestic science, agriculture, household repair and maintenance, car and machine repair and maintenance, typing, and bookkeeping and would drop domestic science as presently constituted at all levels" (Mbilinyi et al., 1991, p. 26).

Training As a Critical Accompaniment to Curriculum Change

Development of gender-sensitive content in textbooks does not ensure its transmission. From a feminist perspective, both the content and the process of education are equally important, thus the need for a democratic experience in the classroom. Teachers are key actors in the transmission of knowledge, skills, and values in the school; they are an essential component of the learning process and central in any attempt to improve the treatment of gender issues in the schools. In many ways subconsciously, the products of their own socialization, both male and female teachers tend to hold lower occupational expectations about girls than boys and to believe that girls are intellectually inferior to boys, particularly in the area of mathematics. Teachers often act as reproductive agents of gender inequalities, and they are often unconscious of their discriminatory practices. As Bonder (1994) argues, it is not a question of blaming teachers, but neither is it one of exonerating them. There have been few studies of teacher practices and student outcomes in developing countries. A study of students in grades 4–5 in Pakistan found that girls had much lower math performance than boys. As the analysis differentiated between students of male and female teachers, students of male teachers had significantly higher achievement scores in mathematics than students of female teachers. However, when students with either male or female teachers in urban areas were compared to their counterparts in rural areas, the gap in mathematical achievement disappeared and was even reversed. The explanation that emerged was not that boys are more capable than girls of learning and teaching mathematics, but that girls in rural schools tend to be served by very poorly trained women teachers (Warwick and Jatoi, 1994).

In many countries, women are the majority of primary-school teachers, which makes their gender-sensitization training both easy and difficult. On the one hand, they have experienced many of the problems faced by the girl students in their classes. On the other, because gendered practices are the product of prolonged and ongoing socialization, many female teachers are not aware of their own subordination. Furthermore, teachers are immersed in a social and institutional system that does not foster questioning of the values and models transmitted through the average educational practices. Silveyra (1992), reporting on a study about the professional and technical training of women in nine Latin American countries, found that teachers are not aware of the discriminatory messages they transmit through textbooks and instructional methods. They see most problems as originating elsewhere, such as from poor educational facilities, demanding schedules, and difficulties in gaining access to training classes. There is, therefore, an urgent need to train teachers, not only to acquaint them with a new curriculum but to make female teachers recognize their own oppression. Especially needed, in the context of a more effective teacher role in sex education, is training to inform teachers about human sexuality and to help them reflect on their own experience.

Part of the teacher training (both pre-service and in-service) should be the development of feminist pedagogy. Although it takes many forms, a common principle in this pedagogy is the centrality of personal experiences. Thus the need for students to learn from their feelings and experiences and to integrate different perspectives and viewpoints is fostered. The students, therefore, are encouraged to become involved with the self in a continuing reflective process, engaged actively with the materials being studied and engaged with others in a collective struggle against all forms of oppression (Shrewsbury, 1987). Instead of individualism and competition, the gender-balanced curriculum stresses the need for collaboration and cooperation. Both the reward structure of the classroom and the content of the curriculum need to be altered so that students see each other as sources of help and support, share their learning experiences, and learn to be responsible for one another's learning (Sapon-Shevin and Schniedewind, 1991).

Noddings (1994) argues that schools must pay attention to the moral and social growth of the students and that students should have practice in caring. In a classroom dedicated to caring, students are encouraged to support one another, and opportunities for peer interaction are provided; the quality of that interaction is as important as the academic outcomes. One of the learning goals is to develop a caring community, and one way to achieve that goal is to include community or social services in high schools. Last, but not least, a gender-balanced curriculum should take into account women's ways of knowing (Belenky et al., 1986). A gender-balanced curriculum is one that accords respect and allows time for knowledge to emerge from first-hand experiences.

Teacher training is needed because an increasing number of teachers, especially in primary education, are women. Female teachers who are married face tensions between being mothers and being professionals. These tensions are reflected in everyday occurrences in the classroom and in teachers' limited participation in administrative and political activities at the school.

The training for teachers must enable them not only to be more sensitive to the treatment of female and male students, but also to address problems that increasingly affect girls and women, such as domestic violence, teenage pregnancy, and the girls' apparent lack of interest in disciplines dealing with math and science. Teacher training should also enable teachers to reflect upon their gendered identities in society and ways to alter them.

Counterforces and Obstacles

Liberal feminists argue that, by eliminating all sex biases and gender stereotyping in the curriculum, girls will opt for "male subjects" and improve their academic achievement in science and technical subjects. Radical feminists, in contrast, argue that we live in a patriarchal world—a system in which men maintain power over women. Socialization of girls and boys into sex roles, according to this perspective, is not merely a means to perpetuate misplaced attitudes and prejudices but a necessary tool in maintaining male dominance. Radical feminists further maintain that education transmits "male" knowledge and that schools tend to ignore or trivialize women's experiences in, and contributions to, society. Thus, a male-dominated subject such as history often is reduced to the exploration of wars and politics. Feminists criticize the dominant male intellectual traditions not only for their omission of knowledge about women's experience but also for excluding experiential learning, undervaluing qualitative research, and relegating subjective knowledge, insight, and the potential for experience-based change to secondary status. While the radical-feminist perspective would argue for more drastic and encompassing changes in curriculum, it would not oppose many of the initiatives proposed under a more liberal framework.

The curriculum in developing countries is greatly affected by policies endorsed by international agencies, policies that may create positive as well as negative consequences for gender issues in education. A number of international development agencies are promoting gender-sensitive curricula, mostly involing revisions along gender lines of new courses in primary-school programs. But less affirmative influences are occurring through efforts to make the schools more effective. This call for efficiency, especially from the World Bank, usually translates into content that emphasizes the basic skills (language and math) and is less concerned with issues affecting social relations. The World Bank's efforts to improve the efficiency of schools recognizes the importance of textbooks in increasing student learning; at the same time, this institution makes no reference at all to the possible content of textbooks or to who should design them.

Financial difficulties, as a result of pressures to reduce public expenditures in education, have created a climate of raw survival in many school systems, which is not conducive to exploration of changes in the treatment of gender. Curriculum changes require budgetary allocations to set up new curriculum teams, to pilot test materials, to produce textbooks and teaching aids, and to train teachers. Also unfavorable for gender-sensitive curriculum changes are the intentions of multilateral and bilateral development agencies to liberalize and privatize textbook production. While this could result in a set of publishers that are more progressive than many a Curriculum and Textbook Board, the problem lies in the cultural and political distance that is created between publishers in one country and the national plans or feminist pressures to modify the curriculum and textbooks along gender lines in other countries.

At the higher-education level, women's studies programs are growing considerably in Third World countries, as reflected in the many sessions on this topic offered at the NGO Forum at the Fourth World Conference on Women (Beijing, 1955). The creation of supportive women's networks has permitted the circulation of new knowledge and the application of new research methods.

The linkages between feminist educators and NGOs have been modest, despite the vitality and innovativeness demonstrated by many female NGOs in the design and implementation of their programs. This is an area that deserves more strategic attention.

Conclusions

Given the fact that an estimated 130 million school-age children worldwide—60 percent of them girls—are not *in* school, it is tempting to give exclusive priority to issues of school access. However, the content of schooling is equally crucial. This importance is evident in the fact that although school enrollment increased by two-thirds during 1975–95, women's labor-force participation rose only 4 points, from 36 percent to 40 percent, during the same period (UNDP, 1995), suggesting that gender-role perceptions have changed little.

School narratives are constructed in basic terms that contrast and oppose masculinity and femininity. The oppression of women by men is a nonissue in most textbooks. The formal curriculum, through textbook content and instructional dynamics, continues to promote the creation of gendered identities of asymmetrical nature. Therefore, textbooks should continue to be a prime target in strategies to modify the curriculum. Linked to a modified curriculum is the need to train and retrain teachers to recognize gender ideologies and alter them. Many international and national declarations have recognized the importance of working on the curriculum, but few agencies follow through.

The recommendations for curriculum change along gender lines are often accompanied by invocations to "develop materials sensitive to local culture." While at one level it may seem adequate to consider cultural differences and preferences, at another level this claim may serve as a loophole not to modify existing gender roles and conceptions. The creation of a new social reality by means of school interventions necessitates the recognition of the value of endogenous culture, but it also requires the courage to discard traditional elements that are conducive to the oppression of women, whether in the name of culture or religion.

In many developing countries, girls' lack of access to school and their inability to complete their studies at all levels constitute major problems to the attainment of women's equality. Nonetheless, we must also be conscious that access and retention need to be supported by an education that awakens in girls and women the desire to alter those secular and sacred aspects of their society that limit their lives.

References

Anderson, Jeanine, and Cristina Herencia. *L'image de la femme et de l'homme dans les livres scolaires peruviens.* Paris: UNESCO, 1983.

Apple, Michael. "Power and School Knowledge," *Review of Education,* vol. 3, 1977, pp. 26–49.

Au, Kit-Chun. "A Study of Gender Roles As Defined in Primary School Textbooks in Hong Kong," occasional paper, Hong Kong Institute for Asian-Pacific Studies, Chinese University of Hong Kong, Hong Kong, 1993.

Belenky, Mary, Blythe Clinchy, Nancy Goldberger, and Jill Tarule. *Women's Ways of Knowing: The Development of Self, Voice, and Mind.* New York: Basic Books, 1986.

Beshir, M.O. *Education in Africa: Two Essays.* Khartoum: Khartoum University Press, 1974.

Biraimah, Karen. "The Impact of Western Schools in Girls' Expectations: A Togolese Case." In Gail Kelly and Carolyn Elliott (eds.), *Women's Education in the Third World: Comparative Perspectives.* Albany: State University of New York Press, 1982, pp. 188–200.

Bonder, Gloria. "Los Estudios de la Mujer en la Universidad: Antecedentes, Perspectivas a Futuro y Aportes a la Actualización Curricular de los Otros Niveles Educativos." In Gloria Bonder (ed.), *Los Estudios de la Mujer en la Argentina: Avances y*

Propuestas para el Cambio Educativo. Buenos Aires: Programa Nacional de Promoción de la Igualdad de Oportunidades para la Mujer en el Area Educativa, Ministerio de Cultura y Educación, 1994, pp. 64–84.

Bonilla, Elsy. La mujer y el sistema educativo en Colombia, *La Educación*, vol. 2, pp. 37–47.

Boulding, Elise. *The Underside of History: A View of Women Through Time*. Boulder: Westview, 1976.

Chen, Zijun, Naizhen Xi, Guangshuo Huang, and Li Tao. *Study on Portrayal of Men and Women in Chinese Textbooks and Children's Literature*. Paris: UNESCO, 1983.

Chung, Ji-Sun. "Women's Unequal Access to Education in South Korea," *Comparative Education Review*, vol. 38, no. 4, 1994, pp. 487–505.

Cowen, Robert. "Cultural Hegemonies, Subsidiarity, and Schools: A Comparative Approach in the Social Construction of Educational Identities in Selected Asian Countries." In H.W. Lam and Y.W. Young (eds.), *Curriculum Changes for Chinese Communities in Southeast Asia: Challenges for the 21st Century*. Hong Kong: Chinese University of Hong Kong, 1993, pp. 173–180.

Jansen, Jonathan. "Curriculum Reconstruction in Post-Colonial Africa: A Review of the Literature," *International Journal of Educational Development*, vol. 9, no. 3, 1989, pp. 219–231.

Kalia, Narendra. *Sexism in Indian Education*. New Delhi: Vikas Publishing House, 1980.

Kalugula, Charles. "The Policy Implications of Sex Role Stereotypes in Texbooks for Girls in Tanzania." In Birgit Brock-Utne and Naomi Katunzi (eds.), *Women and Education in Tanzania*. WED-Report no. 3. 1991, pp. 152–175.

Katabaro, Joviter. "The Effects of Curricular Options on Girls' Education and Their Future." In Birgit Brock-Utne and Naomi Katunzi (eds.), *Women and Education in Tanzania*, WED-Report, no. 3, 1991, pp. 78–92.

Kinyanjui, Kabiru. "Enchancing Women's Participation in the Science-Based Curriculum: The Case of Kenya." In Jill Conway and Susan Bourque (eds.), *The Politics of Women's Education: Perspectives from Asia, Africa, and Latin America*. Ann Arbor: University of Michigan Press, 1993, pp. 133–148.

Levinson, Bradley. "Masculinities and Femininities in the Mexican *Secundaria*: Notes Toward an Institutional Practice of Gender Equity," paper presented at the 20th Latin American Studies Association meeting, Guadalajara, 17–19 April, 1997.

Manan, Shakila. "And Girls Will Be Girls: A Preliminary Study of Gender Bias in Kindergarten Curriculum and Classroom Practices," paper presented at the regional seminar on "Teacher Education: The Challenges in the 21st Century," Penang, Malaysia, November 1991.

Mbilinyi, Marjorie, Patricia Mbughuni, Ruth Meena, and Priscilla Ole-Kamaine. "A Historical Perspective on Women and Education in Tanzania." In Birgit Brock-Utne and Naomi Katunzi (eds.), *Women and Education in Tanzania*. WED-Report no. 3. Dar es Salaam: University of Dar es Salaam, 1991, pp. 1–34.

Meyer, John, David Kamens, and Aaron Benavot. *School Knowledge for the Masses: World Models and National Primary Curricular Categories in the Twentieth Century*. London: Falmer, 1992.

Ministerio de Cultura y Educación. *Programa Nacional de Promocion de la Igualdad de Oportunidades para la Mujer en el Area Educativa*. Buenos Aires: Ministerio de Cultura y Educacion, 1992.

Noddings, Nel. "An Ethic of Caring and Its Implications for Instructional Arrangements." In Lynda Stone (ed.), *The Education Feminist Reader*. New York: Routledge, 1994, pp. 171–183.

Nyerere, Juius. *Education for Self-Reliance*. Dar es Salaam: Government of Tanzania, 1967.

Pinto, Regina. "A Imagem da Mulher Atraves dos Livros Didáticos," *Boletim Bibliográfico de Biblioteca Mário de Andrade*, vol. 43, no. 3–4, 1982, pp. 126–131.

Sapon-Shevin, Mara, and Nancy Schniedewind. "Cooperative Learning As Empowering Philosophy." In Christine Sleeter (ed.), *Empowerment Through Multicultural Education*. New York: State University of New York Press, 1991.

Shrewsbury, Carolyn. "What Is Feminist Pedagogy?" *Women's Studies Quarterly*, vol. 34, no. 3–4, 1987, pp. 6–13.

Shu, Hangli. "Sex Role Socialization of Children in Chinese and American Cultures: A Comparative Study of Elementary School Language Textbooks," paper presented at the VII World Congress of Comparative Education, Montreal, 1989.

Silva, Renán. "Imagen de la Mujer en los Textos Escolares," *Revista Colombiana de Educación*, vol. 4, no. 2, 1979, pp. 9–52.

Silveyra, Sara. "Reunión de Consulta de los Grupos de Trabajo con las Especialistas Nacionales y Extranjeras." In *Estrategias para la Igualdad de Oportunidades de la Mujer*. Buenos Aires: Programa Nacional de Promoción de la Igualdad de Oportunidades para la Mujer en el Area Educativa, 1992, pp. 97–105.

Smith, Peter, and Paul Cheung. "Social Origins and Sex-Differential Schooling in the Philippines," *Comparative Education Review*, 1981, pp. 28–44.

Stromquist, Nelly P. (ed.). *Women and Education in Latin America: Knowledge, Power, and Change*. Boulder: Lynne Rienner, 1992.

Stubbs, Margaret. "Sex Education and Sex Stereotypes: Theory and Practice," working paper no. 198, Center for Research on Women, Wellesley College, Wellesley, 1989.

Tembo, L.P. *Men and Women in School Textbooks*. Paris:

UNESCO, 1984.

Uchendu, V. (ed.). *Education and Politics in Tropical Africa.* New York: Conch, 1979.

United Nations Development Program (UNDP). *Human Development Report 1995.* New York: UNDP, 1995.

Wainerman, Catalina, and R. Raijman. *La División Sexual del Trabajo en los Libros de Lectura de la Escuela Primaria Argentina: Un Caso de Inmutabilidad Secular.* Buenos Aires: CENEP, 1984.

Warwick, Donald, and Haroona Jatoi. "Teaching Gender and Student Achievement in Pakistan," *Comparative Education Review,* vol. 38, no. 2, 1994, pp. 377–399.

Wyn, Joanna, and Bruce Wilson. "Improving Girls' Educational Outcomes." In Jill Blackmore and Jane Kenway (eds.), *Gender Matters in Educational Administration and Policy: A Feminist Introduction.* London: Falmer, 1993.

Yau, Betty Lai-Ling, and Luk Hung Kay. *A Study of Gender Roles in Junior Secondary Chinese History and Social Studies Textbooks in Hong Kong.* Hong Kong: Institute of Social Studies, Chinese University of Hong Kong, 1988.

See also Charlotte Bunch, Roxanna Carrillo, and Rima Shore, "Violence Against Women"

Higher Education and Professional Preparation

Lisa Petrides

Introduction

This essay begins with the assumption that education is a basic human right and recognizes that higher education is a powerful agent for change. It examines the status of women in higher education in the Third World with regard to access and equity in both education and workplace opportunities and summarizes how present-day economic, cultural, and religious barriers that women face while in pursuit of higher education can subtly or overtly deny this right. It also looks at the influence of women's studies programs in institutions of higher education and how they have attempted to promote societal and structural change within the university and in the culture at large, especially with regard to the role and status of women in society.

The term "higher education" is used in this essay to mean any postsecondary, or tertiary, education, which is the most common use in development literature. This includes two-year nursing schools, four-year liberal arts colleges or professional schools, and research universities. Some countries differentiate the number of students in various types of higher-education institutions while others do not. For this reason, it is sometimes difficult to disaggregate enrollments in higher education, and, therefore, figures can be difficult to compare.

Finally, while this essay describes in general the condition of women in Third World countries in the 1990s, it also addresses specific regional differences and commonalities when possible, so as not to present a monolithic picture of higher education in the Third World.

Third World Women in Higher Education

In the 1970s, development agencies believed that the modernization of less developed countries would increase the participation of women in higher education, since there would be an increased demand for higher-level technical education. Western nations speculated that the benefits of modernization would trickle down to women and the lower social classes and might be an "antidote to social revo-

lution" (Mazumdar, 1994, p. 45). Boserup (1970) was one of the first theorists to argue that modernization would not increase the participation of women in the development process. She argued that the central role that women had held previously, in nonpaid work in the household production of goods, was displaced when the technological production of goods and services in the workforce became specialized. Consequently, the lives of women were not economically improved because modernization took over from them their only power—production in the household domain. In the mid-1980s, a study by the Committee on the Status of Women in India (Mazumdar, 1994) showed that development policies which rely on technological advances had, indeed, displaced female labor. Women did not receive the education necessary to fill these technical positions since the majority of women in higher education are found in humanities and education.

Just as modernization had the unintended effect of taking away power from women's lives, the entry of women into higher education did not produce societal transformation with regard to the changing roles of women. In some regions of the Third World, there was an underestimated resistance to lifting traditional restrictions imposed on women. So while the numbers of women receiving higher education have increased dramatically, those who argued that such education would be a panacea for the equality of women have been greatly disappointed. While some research argues that the illusion of equality is essential for people to believe that the current system is working, there have been few policy implications that address the removal of the structural barriers that still exist (Fagerlind and Saha, 1989). Women throughout the Third World still face enormous cultural and religious barriers, having primarily to do with their role and status in society. Even in Latin America, where women make up more than 50 percent of those enrolled in higher education, they still face discrimination and economic marginalization, which carries through to their treatment and experience in the workforce as well.

The benefits of education for girls and women in the Third World have been well documented in development literature and include lower fertility rates, healthier children and families, and increased participation in economic production. The education of Third World women has been profoundly affected by changing ideologies of development. From the mid-1960s through the early 1970s, technical, vocational, and higher education were promoted without any consideration for the involvement of women. In the 1970s, with the return to basic needs as the main objective of development, agencies focused on literacy and primary education and influenced Third World governments to work to eliminate illiteracy in their countries, with particular emphasis on the inclusion of girls.

Higher education is a much more costly endeavor to the Third World than primary or secondary education. Thus it has generally been found that the social rate of return on a country's investment in higher education is less than the rate of return for primary education. One exception is in Latin America, where there is a higher rate of return for higher education. This may account for the fact that Latin America, which has the highest enrollments in tertiary education in the Third World, has nearly equal representation of women and men in higher education.

It is important to acknowledge the ideological conflict between individual and social benefits of education, especially as they pertain to women. Policymakers have made the decision to focus on literacy and primary education for women and girls because this is where there are high social returns. Therefore, many countries of the Third World have come to believe that, in times of economic crisis and hardship, it is more worthwhile to invest in primary and secondary education than in higher education.

Rates of return to education (i.e., the benefits derived from investing in it) have been found to be higher for women than for men (Psacharopolous and Woodhall, 1985). Further, individual rates of return (pertaining to salaries and promotions) for higher education are greater than the social rates of return (concerning benefits for society as a whole, such as lower fertility and child mortality, higher national productivity) (Psacharopolous and Woodhall, 1985). It could be inferred that higher education has a greater individual payback for women than does primary education. However, the promise of individual payback for women becomes clouded when the social (the status of women) and economic (the stratification in low-paying fields of study) disincentives for women are considered. Without a demonstrated individual return, the incentive for women to continue in higher education is lessened.

The impact of higher education on Third World countries is extremely important because universities are often the training ground for a country's elite, and institutions of higher education have a nontrivial impact on political and ideological practices (Altbach, 1987). Additionally, it is the university graduates who will be at their country's technical intellectual forefront, especially with regard to global technological development.

The effects of higher education in the Third World on women have not been well documented. A study by Fernández (1993) found that women in Peru with higher education make two to three times the salary of women with less than primary education. But, for the most part, the economic effects of higher education on women's workforce participation and salaries are difficult to estimate. This is partly because figures for higher-education enrollments have only recently been recorded separately by gender—it has been difficult to convince educational researchers of the importance of studying women in higher education in the Third World. Even when enrollment data are disaggregated by gender, there are other confounding factors, such as type of institution and field of study. For example, UNESCO reported that, as of 1986, women were disproportionately enrolled in nonuniversity-based tertiary institutions (UNESCO, 1991). This means that the majority of women in the Third World received their higher education outside of the university, such as in teacher training institutions or nursing schools (Kelly, 1992). Women benefit little from postsecondary training in these types of technical fields.

These are critical obstacles in understanding the relationship between higher education and the workforce, since the social and cultural context in which women pursue higher education is uniquely different from that of men. For example, it was assumed that higher education for women would naturally lead to better employment opportunities for women. However, because women are still expected to do the majority of unpaid work within the household (such as providing emergency child care when children are home sick from school), women are more inclined to pursue higher education in fields that will allow them a greater degree of job flexibility. These types of jobs are often lower paying, part-time positions in the service sector. Therefore, it is important to conduct analyses that will lead to insight about how issues of education affect gender equality, especially since the relatively recent push for women in higher education provides a global setting in which gender intersects with sociocultural changes, such as class background, ethnicity, and religion.

The Status of Women in Higher Education

The proportion of girls enrolled in primary, secondary, and higher education has risen considerably throughout the Third World. However, there are still substantial gaps between the number of girls and the number of boys enrolled in formal education. In particular, it appears that the gap is narrowing in much of Latin America but growing in Africa and in parts of Asia. In general, girls are still more likely than boys to drop out earlier from the education pipeline. While in some countries girls make up 50 percent of primary-school enrollments, a smaller percentage continue on for secondary schooling, and even fewer pursue tertiary schooling. By the time young women enroll in some type of higher education, they are a significantly smaller percentage of the total numbers of students attending such institutions.

Figure 1. Gross Enrollment Ratios in Tertiary Education, by Sex, 1990

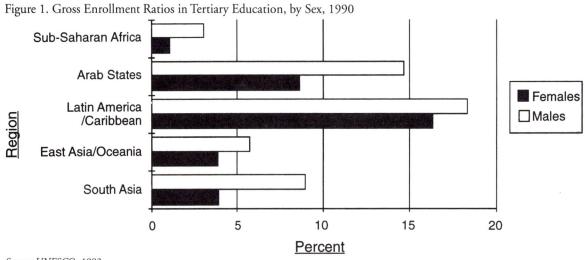

Source: UNESCO, 1993

The total gross enrollment ratio (GER) for tertiary, or higher, education in developing countries increased from 1 percent to 5 percent between 1960 and 1990 (UNESCO, 1991). Figure 1 illustrates the GERs—defined as the proportion of those 18–24 years of age who are enrolled in a tertiary institution—for females and males in the five regions of the Third World in 1990. Sub-Saharan Africa had the lowest GERs in the developing world, with 1.1 percent for females and 3.1 percent for males. The GER for females in South Asia and East Asia was 4 percent; for males in South Asia it was 9 percent; and for males in Eastern Asia, 5.8 percent. The figures for the Arab states were 8.7 and 14.7 percent, respectively, for females and males; for Latin America, 16.4 percent for females and 18.4 percent for males. Latin America is the only region of the Third World in which enrollments have nearly reached parity between men and women in higher education.

Figure 2 gives an aggregate picture of the change in total gross enrollment ratios for higher education for women and men in all developing and developed countries between 1980 and 1990. GERs for females in developing countries (without taking regional differences into account) showed a small but consistent increase, from 3.7 percent to 5.4 percent, between 1980 and 1990; and for males, enrollment rose from 6.5 percent to 8.4 percent. In developed countries, the GER for women increased from 29.3 percent to 39 percent between 1980 and 1990; for males, from 31.1 percent to 36.9 percent. These figures indicate that the rate of growth for females in tertiary education in developed countries is greater than the rate of growth for males, and, in fact, GERs for females have surpassed those for males. In developing countries, the rate of change for females is also greater than for males, but females still have a smaller gross enrollment ratio.

Figure 2. Gross Enrollment Ratios in Tertiary for Developing and Developed Countries

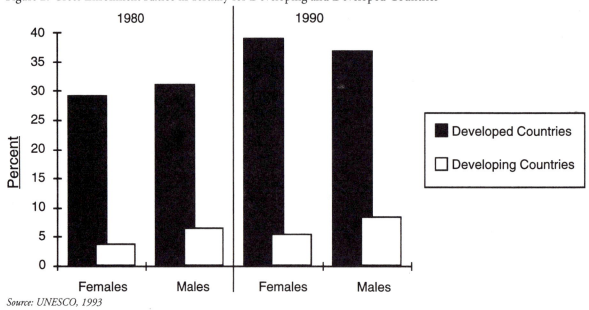

Source: UNESCO, 1993

Table 1. Completed Tertiary Education, by Gender, in Developing Countries

Region	Males in the Adult Population (%)	Females in the Adult Population (%)	Equal Attainment Index
Sub-Saharan Africa	0.70	0.20	0.29
Middle East/North Africa	2.70	0.70	0.26
Latin America/Caribbean	3.80	2.80	0.74
East Asia/Pacific	3.50	2.20	0.63
South Asia	1.70	0.70	0.41

Source: Kaneko, 1987, p. 26.

Enrollment ratios tell only one part of the higher-education story in the Third World since attrition for women is often higher than for men. Table 1 shows (by gender) the percentage of the total adult population of that region who have completed three or more years of postsecondary education. The Equal Attainment (EA) Index is the percent of women enrolled, divided by the percent of men. As this number approaches 1.0, gender parity in completion rates occurs (Kaneko, 1987). The EA Index is extremely low in sub-Saharan Africa and the Middle East/North Africa regions (0.29 and 0.26, respectively). This low index rate indicates that even though the gross enrollment ratio is 8.7 percent for women in the Arab states (see Figure 1), women have a much lower probability of completing their degree. The Latin American region is closest to gender parity (at 0.74) with East Asia next in line (at 0.63).

The two regions that have the greatest gender parity in completed tertiary education (see Table 1) have the smallest gap between female and male enrollments (see Figure 1). This seems to indicate that gender parity in completion rates for tertiary education is somehow correlated with gender parity in enrollments. One conclusion that could be drawn from this is that women are more likely to complete their educational training when they are more equitably represented in higher education. Future research may determine whether this is true, and if so, whether it might be because women are more comfortable in a situation when they are not the token minority.

The proportion of women to men is smaller in each subsequent level of higher education in all Third World countries. This means that fewer women will go on to complete graduate school, which is the training ground for faculty members. Graduate education for women accounts, on average, for less than 10 percent of the total, but these figures are difficult to come by since many countries do not disaggregate by gender. In India, where women are 38 percent of the total tertiary enrollment, they make up only 4 percent of those Indians pursuing doctoral education. In China, women make up 20 percent of enrollments in higher education but only 9 percent of enrollments at the Ph.D. level (UNESCO, 1990).

In 1990, women in all Third World countries accounted for less than 20 percent of all faculty in higher education, including lecturers, instructors, and assistant, associate, and full professors, and only 5 percent of the highest-ranked faculty, full professors. There are almost no female faculty in the math and science fields (1 percent or less in most countries) (UNESCO, 1993). There is also a paucity of women in senior full-time faculty positions, as opposed to lecturers and part-time instructors. Although no comprehensive study of publication rates of Third World women has been carried out, it can be argued that women face the problem of trying to advance in their careers by meeting publication rates while at the same time maintaining children and household duties.

In Pakistan, women are 12 percent of university faculty and only 4 percent of full professors, and they are not represented at all in engineering and agriculture. This percentage disparity is typical of all regions of the Third World, although in many countries there are no female full professors in any field. China has the highest percent of female faculty, at 24 percent, yet even there only 11 percent of full professors are women. Nigeria is at the opposite end of the spectrum: The female academic staff is approximately 1 percent of the total. As in the case of enrollments in higher education, the type of institution can mask issues of gender equity. While research has been scarce in this area, one study showed that women in India were 22 percent of the faculty at the less prestigious general education colleges and only 10 percent of the faculty at universities and research institutions (Gill, 1990).

Barriers to Higher Education

National primary- and secondary-education policies and enrollments set the tone for what happens in higher education. Because educational institutions are most often a reflection and product of their own society and culture, it is not unexpected that these institutions inherently serve to reinforce a gender bias in teaching, curriculum, textbooks, and administration (Conway and Bourque, 1993).

Nearly equal participation of girls and boys in primary and secondary education does not necessarily ensure the subsequent representation of women in higher education. A study in Kenya (Kinyanjui, 1993) showed that fewer than one-third of the high-school girls who take math and science courses pass tests that qualify them for higher education. The importance of effective primary and secondary math and science education for girls cannot be over-emphasized. Strong math and science skills are the best predictors of success for women in higher education and in the professional workforce. Studies of secondary enrollments for girls can be misleading unless the type of secondary schools that girls attend, which is an indicator of the level of math and science education they will receive, is also known. Low-quality schools, which girls tend to attend, offer far fewer math and science courses, so the girls' academic access is different from boys'. It also has been shown (Kinyanjui, 1993) that there are discrepancies in parental and teacher expectations for girls and boys. For example, girls are encouraged to drop out if they fail a grade while boys are encouraged to repeat such a grade.

It is essential that the local climate for learning is one in which girls have early access to good teachers, equipment, resources, and an environment that is free from negative stereotypes, gender-biased curriculum, harassment, and discrimination. Unfortunately, this is not the case in much of the Third World. Gender-biased primary and secondary textbooks are a perfect example: Young girls are confronted with negative stereotypes and a lack of beneficial female role models and female historical figures in most texts. A team of researchers in Kenya found that there were no role models of female leadership or of women engaged in nontraditional activities, such as driving a car or banking (Kinyanjui, 1993). The study also found no examples of women using mathematics, while men were shown as owners of land and businesses.

In Africa, young girls are already working in the home and there is little time for school. Only 14 percent of young women attend secondary school, and early marriage makes it difficult for women to even consider higher education. Additionally, single-sex schools for girls and women, which are required in many Muslim countries and in Africa, tend to be ill-equipped, and far away from rural homes, and they often have poor-quality teaching. Mobility is not easy for most families in the Third World; it is very difficult for girls to travel far from home.

Economic Barriers

The whole conception of education as an enabling condition of equality is premised on the belief that gender-based inequality in society was somehow rational, stemming from women's lack of qualifications equivalent to those possessed by men. Women were also presumed to have little power or authority because they were not engaged, as were men, in earning a wage and therefore being productive. With earnings came power, authority, and autonomy. Ignored were the effects of marriage, childbearing and childrearing on women (Kelly, 1989, p. 566).

Access to education has been the primary focus of educational development for girls and women in the Third World. Enrollment for women in higher education has, in some cases, tripled since the 1980s; however, the absolute numbers are still very small. In any thorough analysis of the condition of women in higher education, the first question that still needs to be asked is: Are all women allowed to enroll in institutions of higher education, or are there regulations that prohibit them from attending? The next question is: Are there cultural and economic barriers that prevent women from obtaining higher education? A third question is: If women do, indeed, have the opportunity to receive higher education, does the curriculum of secondary schools prepare girls for higher-education training and, if so, for what type of training?

A large majority of women who do have access to formal education in the Third World attend institutions based on patriarchal Western culture and development. Efforts to move away from the colonial past of higher education and Western development have been met with opposition and reluctance to change, similar to the way women have been denied social support in nontraditional subjects. In the countries in which higher education has begun to produce a shift in the attitudes of men and women, such as Peru, universities are blamed for provoking an uprising of revolutionary students during bad economic times and for creating expectations that society cannot fulfill. The education of women tends to be more adversely affected in times of economic hardship, when families, forced to prioritize resources, give more weight to a son's education than a daughter's.

Higher education often takes a back seat to primary education and issues of basic literacy, especially in countries, such as Nepal, where the majority of the population is illiterate. There is disagreement as to whether a national interest is better served by providing basic education for the majority of people or higher education to a relative few. In Francophone Africa, 39 percent of the educational resources go to the 2.4 percent of students who receive college education (Psacharopoulos and Woodhall, 1985). In many Third World countries, those attending institutions of higher education come predominantly from upper- and middle-class backgrounds. In Latin America, the emergence of national social and women's movements has enabled middle-class women to have access to universities, although women are channeled into higher education in fields of study such as arts and the humanities that are more highly represented in the lower economic sectors of the work place, thereby perpetuating economic discrimination. While it can be argued that women themselves make the choice to study in fields with less prestigious, lower-paying occupations, it has been suggested that socialization, access to relevant secondary math and science education, and stereotypical attitudes toward women's work are often the more salient factors (Conway and Bourque, 1993).

When the economy of a country takes a downturn and jobs become scarce, it is much more difficult for women to afford the expense of education since their education is often considered to be secondary to men's. In the debt crisis of the 1980s, Mexican professionals and academics lost 57 percent of their buying power (Arizpe, 1993). This produced a setback for universities because of the economic hardship felt by the educated middle class. A government may, therefore, think that there is less risk involved in development policies that put their main emphasis on primary and secondary education as opposed to higher education, thereby avoiding the expensive flight of educated professionals or the threat of political unrest.

Low enrollments at the tertiary level are a result of high attrition at the secondary level. However, there is a concerted effort and increased awareness of the importance of education in countries such as Cameroon, where one-fifth of the budget is spent on education, that are willing to invest heavily in basic and higher education. Government authorities believe that the seriousness of an investment like this will encourage a greater general acceptance of education, which will lead to sustained educational participation, especially for girls.

It has been shown throughout the Third World that poverty, debt crises, and other stagnations of the economy disproportionately affect women. For example, in times of economic hardship, women are often forced to terminate their education (Arizpe, 1993). Additionally, unemployment statistics in Latin America verify that women tend to be disproportionately laid off when jobs are cut. It has also been shown that there is a considerable disparity in wages for men and women who have the same amount of education, with women receiving roughly 50 percent of men's wages. In Brazil and Argentina, this gap is even more pronounced, particularly in jobs that require higher education (Rosemberg, 1993; Bonder, 1994).

The social, cultural, and religious histories of the regions of the Third World are deeply entwined with the economic barriers that prevent women from obtaining higher education. In fact, studies have found a number of variables, such as father's and mother's education, economic status, and class, significantly determining whether a woman will pursue higher education. While the general question of who has access to higher education is somewhat easy to determine, there is a dearth of research about the quality of education that women, relative to men, receive. This is where the question of equity in education becomes confounded. A closer look at the relatively large reported number of women in higher-education programs in many countries shows that the majority are in the lower-ranking colleges and professional schools, as opposed to universities. It is also important to look at the ranking of higher-education institutions with regard to the representation of women on the faculty. The aggregated numbers of female faculty may be deceptive due to the fact that the majority of female faculty are employed at teacher training institutions, which are predominantly female and lower ranked.

Cultural and Religious Barriers

Cultural and religious barriers play a major role in the degree of participation of women in higher education. Gender bias has traditionally served to limit access and equity for women and girls, to marginalize the role of women in society, and to devalue women's contribution to education, research, and reform.

The norms and societal expectations of many traditional cultures do not encourage women to become too educated lest their education become a social or financial liability in finding them appropriately educated husbands. Many countries will not support educational policies that call the role of women in the family or the status of women in society into question, because in most countries the primary locus of women's responsibility is still in the family and in the home. Additionally, families fear that the loss of traditional roles for women will negatively affect their society. For example, an Indian family with an educated daughter must have a larger dowry because women are expected to marry men with equal or preferably greater education (Kahle, 1985). Because education is often equated with personal wealth, not only would the husband's family have to be more wealthy, but the woman's family would have to increase the amount of the dowry to reflect her own education level as well.

If parents are supportive of college education for their daughters, it is often because college is seen as a way for women to increase their prospects of marriage, as opposed to a path to economic independence. On the other hand, in studies in which the women themselves were asked why they wanted to go to college, they reported not that they wanted to increase their prospects for marriage, but that they wanted to have career opportunities, advancement, and social mobility (UNESCO, 1990). In certain cultures, women who have college degrees are forced to retire from their jobs when they get married or pregnant. It is thought that women with higher education will not be good mothers because they will not respect the traditions of their culture. In Muslim countries, it is sometimes believed that there is a moral danger in educating women because educated women are likely to be promiscuous and will not make good wives (Okeke, 1989).

Education does tend to affect decisions women make with regard to marriage. Marriage patterns in Nigeria, for example, have begun to change due to the higher education of women: it is not uncommon for women there to wait to get married until they are 24 years old and have finished their university education. However, losing these most fertile years is still considered wasteful by many, divorce has increased among educated women, and women in Nigeria still do not have equal access to land and inheritance (Okeke, 1989).

The Muslim culture insists that women adhere to strict roles regarding their positioning in society, and in most Muslim countries women are less represented than men in the formal education systems. For example, in Pakistan, primary and secondary schools have always been segregated by gender, while general and professional universities have not. Additionally, women may receive a college degree but are then forced out of the public sector and into the home. This curtailment of rights is more common for middle- and upper-middle-class women who have begun to cast off their Muslim practices.

Studies throughout the Third World have shown that highly educated parents who are from higher socioeconomic classes have a greater desire to see their daughters also become well educated. In Nigeria, which is both a Christian and a Muslim country, a study was conducted at one northern university where Christian women constituted a two-thirds majority of the women, and Muslim women one-third. The study indicated that the Muslim fathers had more money than the Christian fathers. These findings suggest that economic and educational resources may neutralize the effect of religion by allowing women to have higher education (Stromquist, 1989).

Across the Third World, women who have acquired knowledge and skills in higher-education institutions have been naively optimistic in thinking that an increase in education would necessarily lead to greater equality in society or to a significant transformation of the division of labor in the home (Conway and Bourque, 1993). This becomes more obvious as women continue to make gains in higher-education enrollments. In Latin America, where women make up 50 percent of all students in higher education, they are still encouraged to pursue education that reinforces traditional roles of women.

Women with higher education tend to raise smaller families: In Peru, for example, they have an average of 2.7 children, compared to 6.2 children for those without such schooling (Fernández, 1993). It is still difficult for women who have any children to attend college, though, since it is rare to find day-care centers in institutions of higher learning.

Higher-education facilities, especially technical schools, are generally located in urban areas where women have less self-selected access due to safety considerations (UNESCO, 1990). Women in higher education are also subjected to sexual harassment in situations where there are no policies to deal with their complaints and where it is almost guaranteed that no disciplinary action will be taken against the offenders. So while higher education may be the first step, even the most important step, in ensuring social and economic equity for women in the Third World, much work still needs to be done to remove structural barriers.

The Effects of Gender Segregation on Employment in Higher Education

The political and economic conditions in the Third World are diverse. Some countries are in economic crises while others are relatively stable; some are only newly independent while others have been so for decades. They are socialist and nonsocialist, industrialized and agrarian, high income and low income. Yet, all have one thing in common: Women may enroll and attend formal higher education, but most women are tracked to nonscience fields and are, therefore, concentrated in the lower-paying, less-prestigious professions (Kelly, 1989).

Human-capital theory is a major argument used in Western-based economic theory to explain the difference in occupations and earnings between men and women and to justify how an increase in education results in an increase in economic production. It says that women, on their own accord, acquire less human capital because their tastes and desires are different from those of men (Becker, 1975). It assumes that men and women have equal access to educational and occupational opportunities but that women do not obtain comparable education since they do not intend to pursue the same high-paying positions that men pursue. For example, women may choose lower-paying careers that emphasize service over profitability, or flexibility of hours over prestige, because of their primary commitment to the family. This does not seem to be a plausible explanation for employment differences in the Third World since women are not afforded the same opportunities for equitable education, as shown in the examples above.

As the number of women in higher education has grown, so has gender segregation. In most countries of the Third World, as many as 90 percent of women in higher education are concentrated in less-professional courses that do not lead to specific careers. Instead they tend to pursue more generalized education in the humanities (or liberal arts), education, nursing, and service sectors in fields known as female ghettos. There is a dearth of women in science and technical fields; they account for only 5 to 10 percent of all students enrolled in math, science, and computer technology. Even in Third World countries where women make up 50 percent of the total higher-education enrollment, they are disproportionately overrepresented in humanities, health, and education (as is also true in developed countries). As a result, women are less represented in the higher-paying, higher-status occupations. Even in countries such as India, where the percentage of female doctors is increasing, they remain confined to the lower-paying, less-prestigious positions in medicine (such as family practice), with little opportunity for leadership.

In Botswana, rapid economic growth was accompanied by the demand for democracy and social justice after it gained independence from Britain in 1966. Since then, educational institutions have experienced a large increase in the number of female enrollments. Institutions of higher education, which had previously excluded women from their ranks, now enroll 49 percent women (Marope, 1994). However, gender segregation within fields of study continues: Women are 100 percent of the enrollment in nursing; they make up less than 10 percent of students in technical fields such as agriculture and engineering; and there are even fewer female faculty members in law and scientific fields (Duncan, 1989).

In Argentina, middle- and upper-class women have higher enrollments and completion rates than men in both secondary and higher education. Although there, too, women tend to go into certain disciplines, there are several professions, such as law and medicine, in which women are beginning to reach parity with men. Ironically, the most highly educated women in Argentina are those who experience the highest levels of wage discrimination. Female university graduates make 50 percent of the salaries of their male counterparts while the average female worker makes 57 percent of the male worker's wage (Bonder, 1994). This wage discrimination holds true in other parts of Latin America as well. In Mexico, where half of those enrolled in higher education are women, they are segregated in such fields as nursing and teaching. In Peru in 1982, 99.5 percent of the nursing graduates and 98.7 percent of those in social work were women (Fernández, 1993).

Why do so few women enter the math and science fields in higher education? Some argue that women do not see the need to do well in these fields and, therefore, do not choose to go into them. Others cite socialization and early educational preparation as reasons. Another critical element is geographic location: Often the more technologically advanced universities are located in urban centers, which are more difficult for women to access. Partly this is due to safety considerations: The high incidence of violence against women makes living and/or traveling alone in the city problematic. Partly it is a matter of economics. Because it is more expensive to live in cities than at home, many single women cannot study in urban areas. Additionally, the need for specialized materials and labs make science education more expensive than liberal arts education.

Because of discrimination in hiring, women encounter fewer employment opportunities in male domains, such as engineering, which is one reason they tend to enter less-specialized disciplines that give them a greater degree of flexibility in their profession and with their families. Women also face pressure to pursue education in fields in which their roles are already well defined. For this reason, education and health are popular fields for women because these fields will enhance their future childrearing and homemaking responsibilities by making them better mothers and managers of their own homes. It stands to reason that if women have a better chance of finding employment in humanities and nontechnical professions, that is where they will end up. Likewise, if there were a strong demand for women in engineering, they would pursue it. It can also be hypothesized that the inequity of earlier educational opportunity has predetermined their futures.

Women's Studies

The U.N.-sponsored Third Conference on Women (Nairobi, 1985) was instrumental in painting a global picture of women's lack of social, political, and economic power throughout the Third World. The conference encouraged governments in the Third World to make education policy that prohibits discrimination against women, and it gave impetus to the creation of dozens of women's studies programs in developing countries. These programs have served to introduce interdisciplinary academic and action-oriented research on women and development in higher education and to create women's studies courses for university students that analyze gender politics, translate materials into the national language, and incorporate knowledge about women and their research into higher-education curricula.

Women's studies programs in Asia, Africa, and Latin America have been the place where women in higher education have produced the greatest achievements. These include directing research that encourages development agencies to look at issues of gender, introducing feminist theory into the curriculum, promoting gender disaggregation of nationally kept higher-education figures, and conducting empirical studies of discrimination and enrollment disparities. Many countries have taken their women's studies agendas into mainstream culture—an example is India, where women's studies programs have addressed the role and status of women in society—and are assisting government agencies to make the necessary policy changes. Women's studies programs have introduced their respective countries to feminist theories and methodologies, and they were the early trailblazers in the critique of the adoption of Western education in Third World contexts. The analyses of the Western development model of education have produced a dichotomy in women's studies, especially in Asia and Latin America. The tension is experienced by those women's studies leaders who mistrust the Western industrial development yet have been educated by the very system that was built under and has been traditionally associated with colonialism. This has resulted in the call for women's studies programs in the Third World to insist on their own set of political and cultural priorities, which are not based on Western models of women's studies programs. For example, Narayan (1989) speaks to the tension for feminists who have to "be critical of how our culture and traditions oppress women and conflict with our desire as members of once colonized cultures to affirm the value of the same culture and traditions" (Narayan, 1989, p. 259).

Just as the regions of the Third World are not homogeneous, neither has the creation of women's and gender studies programs been the same across countries: Differences in the culture and the politics of these regions have influenced the conception and formation of these programs. In Latin America, where there has been a long-standing coupling between politics and education, the creation of women's studies has been initiated from outside of formal education, by women's nongovernmental organizations (NGOs) and activist groups. Women's studies programs in Africa have been formed in universities and focus more on theoretical issues, such as the inclusion of gender awareness in areas such as literature and history, than on public policy.

In India, the formation of women's studies has been primarily from the top down, by a special university commission that incorporated women's studies into higher education. Although the commission did not design the program, it did accept the ideology of the women's studies movement, which actively strives to change core values and perceptions of women's status in Indian culture. The stated objective of India's women's studies is to work for policy changes and to reexamine methodology and epistemology (the creation and process of knowledge legitimization) in social research by looking at social issues related to the status and role of women. It also questions the dependence on Western theories and methods in the social sciences, such as the concept of academic neutrality (Mazumdar, 1994) and has been helpful in identifying oppressive aspects of the culture, such as castes, that inhibit the use of education as a tool for social mobility.

In many Third World countries (and most countries of the world), women's studies programs have had to struggle for legitimization, power, and access to resources within their own academic institutions. Their challenges to traditional epistemologies may be discredited by the institution for being less than scholarly. This has resulted in a lack of concrete policymaking with regard to women at the institutional level.

Academic women have often been reluctant to organize and initiate programs on their campuses because they are in the tenuous position of trying to effect change while protecting their place in the academy. Because the universities of a country often operate independently from, and competitively with, one another, it has been very difficult for women's studies to work cooperatively with other programs (Bonder, 1994).

Conclusion

This essay has explored the regional gender politics in the Third World that have had an effect on the present condition of women in higher education. The increase in the numbers of women pursuing higher education has not been accompanied by any significant change in the role of women in the home; this is evidenced in workforce gender segregation, in which women are found in less prestigious fields and faced with discrimination in hiring and less pay for equal work.

This essay has acknowledged the influence of Western development, but this Western context, although it has been successful in many ways in its own right, is different from most Third World educational settings. In the context of the Third World, there are five main areas around which significant action can be taken. First, developing countries need to insist on creating institutions that accurately represent their own style of technological development and environmental concerns in the economic organization of their higher education programs. Second, there is a need to increase the leadership of women within these institutions. Governments of the Third World need to heed the United Nations' call for action to eliminate gender disparities in access to all areas of tertiary education by ensuring that women have equal access to career development, training, and scholarship; they also need to insist that education policymakers look at the structural and social barriers that prevent women from obtaining higher education that would enable them to improve their economic and social prospects. One way that this can be done is to ensure that women are actively participating in the decision-making process.

Third, higher-education institutions must actively work to change the social attitudes toward women's work and responsibility by not tolerating discriminatory practices, which lead to gender segregation in higher education and in the workforce. Higher education should also actively endorse and support equal education for girls at the primary- and secondary-school levels. This might include such things as partnerships between universities and secondary schools that could be used to plan activities for girls that would foster interest in math and science. Also, careful policy-driven research needs to be conducted that analyzes the structural barriers for women in math and science fields and then designs national objectives, which would, for example, offer technical courses in both rural and urban settings so that they are more accessible to women.

Lastly, there is a critical need for tools that accurately document the social and economic benefits of women's education. The most widely used standard of measure is a process that calculates individual and social rates of return on investment, based on market (workforce) production. This neoclassical Western economic view, which has influenced the rationale for Third World educational development, may not be the most accurate measure of the worth of investment in women's education inasmuch as it measures only the effect of education on market production. And even though the improved health and welfare of women and their families are attributed to education at all levels, this neoclassical view does not use women's nonmarket production to figure in family incomes or as part of the gross national product of a given country. This is why the total benefit of women's education in terms of economic returns for the individual and for society is unknown and why education is not as highly correlated with workforce participation for women as it is for men. Now is the time to make the best use of all of each country's natural resources and to begin to treat the investment in women as an investment in the human rights of all people.

References

Altbach, Philip. *Higher Education in the Third World: Themes and Variations.* New York: Advent Books, 1987.

Arizpe, Lourdes. "An Overview of Women's Education in Latin America and the Caribbean." In Jill Conway and Susan Bourque (eds.), *The Politics of Women's Education: Perspectives from Asia, Africa, and Latin America.* Ann Arbor: University of Michigan Press, 1993, pp. 171–182.

Becker, Gary S. *Human Capital.* New York: Columbia University Press, 1975.

Bonder, Gloria. "Women's Studies in Argentina: Keeping Feminist Spirit Alive," *Women's Studies Quarterly,* vol. 22, no. 3–4, 1994, pp. 89–102.

Boserup, Ester. *Women's Role in Economic Development.* London: Allen and Unwin, 1970.

Conway, Jill, and Susan Bourque (eds.). *The Politics of Women's Education: Perspectives from Asia, Africa, and Latin America.* Ann Arbor: University of Michigan Press, 1993.

Duncan, Wendy. "Botswana." In Gail P. Kelly (ed.), *International Handbook of Women's Education.* New York: Greenwood, 1989, pp. 3–24.

Fagerlind, Ingemar, and Lawrence Saha. *Education and National Development: A Comparative Perspective.* Oxford: Pergamon, 1989.

Fernández, Hernán. "Persistent Inequalities in Women's Education in Peru." In Jill Conway and Susan Bourque (eds.), *The Politics of Women's Education: Perspectives from Asia, Africa, and Latin America.* Ann Arbor: University of Michigan Press, 1993, pp. 207–216.

Gill, Veena. "In Two Worlds: Women Academics in India." In Suzanne Lie and Virginia O'Leary (eds.), *Storming the Tower: Women in the Academic World.* London: Kogan Page, 1990, pp. 178–193.

Kahle, Jane (ed.). *Women in Science: A Report from the Field.* Philadelphia: Falmer, 1985.

Kaneko, Motoshita. *The Educational Composition of the World's Population: A Data Base.* Washington, D.C.: World Bank, 1987.

Kelly, Gail P. "Achieving Equality in Education—Prospects and Realities." In Gail P. Kelly (ed.), *International Handbook of Women's Education.* Westport: Greenwood, 1989, pp. 547–570.

———. "Education, Women, and Change." In Robert Arnove, Philip Altbach, and Gail P. Kelly (eds.), *Emergent Issues in Education: Comparative Perspectives.* Albany: State University of New York, 1992, pp. 267–281.

Kinyanjui, Kabiru. "Enhancing Women's Participation in the Science-Based Curriculum: The Case of Kenya." In Jill Conway and Susan Bourque (eds.), *The Politics of Women's Education: Perspectives from Asia, Africa, and Latin America.* Ann Arbor: Univer-

sity of Michigan Press, 1993, pp. 133–148.

Marope, P.T.M. "Botswana Women in Higher Education: From Systematic Exclusion to Selective Engagement." In Suzanne Lie, Lynda Malik, and Duncan Harris (eds.), *The Gender Gap in Higher Education.* London: Kogan Page, 1994, pp. 24–32.

Mazumdar, Vina. "Women's Studies and the Women's Movement in India: An Overview," *Women's Studies Quarterly,* vol. 22, no. 3–4, 1994, pp. 42–54.

Narayan, Uma. "The Project of Feminist Epistemology: Perspectives from a Nonwestern Feminist." In Alison M. Jaggar and Susan R. Bordo (eds.), *Gender/Body/Knowledge: Feminist Reconstructions of Being and Knowing.* New Brunswick: Rutgers University Press, 1989, pp. 256–269.

Okeke, Eunice. "Nigeria." In Gail P. Kelly (ed.), *International Handbook of Women's Education.* New York: Greenwood, 1989, pp. 43–64.

Psacharopolous, George, and Maureen Woodhall. *Education for Development: An Analysis of Investment Choices.* New York: Oxford University Press. 1985.

Rosemberg, Fulvia. "Education, Race, and Inequality in Brazil." In Jill Conway and Susan Bourque (eds.), *The Politics of Women's Education: Perspectives from Asia, Africa, and Latin America.* Ann Arbor: University of Michigan Press, 1993, pp. 223–236.

Stromquist, Nelly. "Determinants of Educational Participation and Achievement of Women in the Third World: A Review of the Evidence and a Theoretical Critique," *Review of Educational Research,* vol. 59, no. 2, 1989, pp. 143–183.

United Nations Educational, Scientific, and Cultural Organization (UNESCO). *Women's Participation in Higher Education: China, Nepal, and the Philippines.* Paris: UNESCO, 1990.

———. *World Education Report.* Paris: UNESCO, 1991.

———. *World Education Report.* Paris: UNESCO, 1993.

Women and Literacy

Anita Dighe

Introduction

One of the serious problems of the world at the end of the twentieth century is women's illiteracy. According to UNESCO (United Nations Educational, Scientific, and Cultural Organization) estimates, there were 873.9 million illiterate adults in developing countries in 1990, of whom 567 million, or 64.9 percent, were women (UNESCO, 1993). Why are women illiterate in such disproportionately large numbers? There are a number of factors that impede access of poor women to literacy programs. These include poverty, family responsibilities, opposition by husbands, lack of time, physical distance to literacy classes, and lack of motivation, which results in part from the preceding problems. The obstacles to women's literacy are similar to those affecting access of the girl child to primary education and are fairly similar across cultures. While it is necessary to identify reasons that hinder women's access to literacy and to formal education, it is even more important to have a holistic understanding of the forces behind such constraining factors. Stromquist (1992) has underscored the contribution made by feminist theory and analysis to the understanding of the conditions of women in society and in education in order to grasp the exact nature of the phenomenon of women's illiteracy. According to Stromquist, patriarchal ideology has two mutually supportive components: the sexual division of labor and control over women's sexuality. This ideology has assigned women the roles of mothers and wives and assumed that these tasks are natural, and thus immutable, attributes of women. Women's "lack of time" is really a result of the sexual division of labor that has imposed on them numerous domestic duties. "Distance to the literacy center or school" is also a manifestation of control over girls' and women's sexuality as husbands and parents seek closer places that would enable them to have better supervision of wives' and daughters' sexual behavior and safety. The advantage of relying upon theory is that factors for women's and girls' participation or nonparticipation in educational programs are understood not just as a string of events but as systematically linked elements that provide a more holistic understanding of such participation.

The exclusionary mechanisms mentioned above have, over the years, led to a widening gender disparity in literacy in some regions of the world against a background of worsening economic crisis and increasingly severe national debt burdens. The deepening economic crisis and the remedial structural-adjustment programs (SAPs) in operation in many Third World countries have led to significant budget cuts in social sectors such as health and education and a deterioration in the living standards of large sections of the population. In the case of women, as greater demands are made on their time for subsistence living, they are less likely to have the time or the inclination to become literate. Given this scenario, and the growing international awareness of male–female literacy disparities that has been highlighted through world conferences, it is crucial to examine the kind of literacy relevant to the needs and interests of women. This essay argues that literacy programs must become a means for empowering women so that they can move away from the sense of powerlessness they experience as poor women and exercise more control over their lives and over the educational programs that are meant for them.

It is important to understand the extent of the problem of women's illiteracy in the different regions of the world.

UNESCO estimated that the total number of illiterate people in the world in 1990 was 905.4 million, of whom only 31.5 million lived in the developed countries and the rest (873.9 million) lived in the developing countries. The largest proportion of the world's illiterate population was concentrated in South Asia, East Asia/Oceania, and sub-Saharan Africa. Of the 873.9 million illiterates, 567 million, or 64.9 percent, were women. Table 1 provides a breakdown of the illiterate population age 15 and older, by sex and region; these data reveal that women constitute the majority of illiterates in all regions.

Not only are more women illiterate, but gender gaps

in literacy seriously disadvantage them. As shown in Table 2, the differences in literacy rates between men and women are quite large in South Asia (particularly India), the Arab states, and sub-Saharan Africa.

Rapid reductions in female–male literacy disparities have not been easy to achieve at the national levels because they are bolstered by social and cultural norms and practices that are difficult to change merely through legislative or administrative fiat. In many countries, progress has been made due to voluntary action at the community level, where nongovernmental organizations (NGOs) have often taken the lead. But unless gender disparities are reduced in the school system, particularly at the primary level, the problem of female illiteracy will not be solved. As is apparent from Table 3, the regions with low ratios of female to male literacy also have significant enrollment disparities at the first level of education.

International awareness of female–male literacy disparities has grown, as has the realization that improvement in the literacy rate of females relative to that of males is a precondition for improvement in the overall adult literacy rate. World attention has focused increasingly on the problem of adult illiteracy, particularly female illiteracy. Before considering the recent measures, it is important to understand the shifts in perspectives on the education of women and girls in the world in the last decades of the twentieth century.

Shifts in Perspectives on Education of Women and Girls

While the overall perspective on education in most Third World countries has been gender-blind, there has been a shift in perspective relating to women's literacy and the

Table 1. Estimated Illiterate Population (Millions) Age 15+ by Sex, 1990

	Male/Female	Female	Females As Proportion of Total
World Total	905.4	586.4	64.8
Developing countries	873.9	567.0	64.9
Sub-Saharan Africa	138.8	86.6	62.4
Arab States	61.1	38.4	62.8
Latin America/Caribbean	43.3	24.0	55.4
East Asia/Oceania	232.7	162.4	69.8
China	181.6	127.3	70.1
South Asia	398.1	242.2	60.8
India	280.7	173.6	61.8
Least-developed countries	148.2	89.0	60.1
Developed countries	31.5	19.4	61.6
North America	6.2	3.3	53.2
Asia/Oceania	1.0	0.7	70.0
Europe/Former USSR	19.7	12.6	64.0

Source: UNESCO, 1993

Table 2. Estimated Adult Literacy Rates (%) by Sex, 1990

	Male	Female	Difference in Percentage Points
World Total	82.2	67.3	14.9
Developing countries	77.0	56.2	20.8
Sub-Saharan Africa	59.5	35.6	23.9
Arab States	64.3	38.0	26.3
Latin America/Caribbean	86.4	83.4	3.0
East Asia/Oceania	88.2	71.9	16.3
China	87.0	68.1	18.9
South Asia	59.1	32.2	26.9
India	61.8	33.7	28.1

Source: UNESCO, 1993

Table 3. Enrollment (Millions) and Gross Enrollment Ratios in Primary Education by Sex, 1990

	Enrollment			Gross Enrollment Ratio		
	M/F	F	% F	M/F	M	F
World Total	610.9	280.5	46	99.0	104.8	93.0
Developing countries	498.5	226.0	45	98.4	105.2	91.2
Sub-Saharan Africa	58.1	26.0	45	68.3	75.4	61.2
Arab States	29.6	12.9	44	83.4	92.2	74.1
Latin America/Caribbean	75.8	36.8	49	106.9	108.6	105.1
East Asia/Oceania	196.1	92.1	47	119.4	123.2	115.3
China	122.4	56.6	46	125.3	130.4	119.7
South Asia	132.9	55.3	42	88.8	100.6	76.3
India	99.1	41.0	41	98.5	111.5	84.4
Least-developed countries	45.0	19.5	43	64.7	72.4	56.8
Developed countries	112.4	54.5	48	101.8	102.7	100.9
North America	32.4	15.3	47	106.3	109.8	102.8
Asia/Oceania	12.0	5.9	49	101.7	101.6	101.7
Europe/former USSR	62.7	30.7	49	101.9	101.9	102.0

Source: UNESCO, 1993

education of girls on the basis of certain research trends. In 1979, a book titled *Fertility and Education: What Do We Really Know?* (Cochrane, 1979) was published. Its purpose was to review the evidence on the relationship between education and fertility and to develop a model of the processes through which education operates. Since that book was published, a large number of studies have become available. In particular, results of the 1980 world fertility survey indicated that, averaged across 20 countries, a negative correlation was observed between level of education and fertility (United Nations, 1981). Studies on the impact of education on fertility attracted the greatest attention of scholars in the late 1980s and early 1990s, making possible several firm conclusions. For example, on the basis of research evidence from developing countries, Lockhead and Verspoor (1990) highlighted how a mother's level of schooling was highly correlated with infant and child mortality so that with the increase in her level of education, there was a decline in the infant and child mortality rate. By 1992, a case was being made for investing in the basic education of girls because it reduced fertility and child and maternal mortality, it prevented the spread of AIDS, and it had important environmental benefits (Summers, 1992).

In other words, more and more research evidence was being garnered to provide a rationale for promoting the basic education of girls and for making a case for investing in their basic education so as to arrest the rate of population growth in the long run. In the immediate future, besides reductions in fertility and mortality, basic education of girls was seen to ensure that an informed mother would send her children to school and make use of health facilities and services for the children, should the need arise. Basic education of girls was not regarded as an end in it-

self but as a means to an end, and that was clearly understood as the well-being of her family—her husband and her children.

Researchers have, however, differed in their interpretations. It has been suggested that the association of women's literacy with infant and maternal mortality is indirect and that there are other factors responsible. It has also been suggested that lower fertility in countries with high literacy rates is not so much a consequence of literacy as of the wider availability of schooling and hence the tendency for girls to marry later (UNESCO, 1993). Bown (1990) maintains that many of the so-called proofs of a relationship between women's education and various socioeconomic changes are based on schooling statistics rather than literacy data and that, while such a statistic is useful in the argument for promoting education of girls, the fact remains that very little is known about the outcomes of literacy for adult women.

A question that needs to be asked is: Why have policymakers and planners tended to ignore or underplay fundamental issues relating to illiteracy and the special educational needs of adult women? From a feminist perspective, this is because men are in positions of power and decision making, where patriarchal values prevail (Ramdas, 1990a; Stromquist, 1992). Male dominance in literacy policymaking has invariably resulted in "time-bound" plans and projects that do not come to grips with the structural problems that relate to attitudes toward women and women's education. But there is a more basic reason, and this relates to the very definition of what education stands for and what it can aim to accomplish. The policymakers and planners are also not paying attention to the more humanistic aspects of education that enhance critical faculties and abilities and are even empowering for those who participate in such educational programs.

Since the early 1990s, the concept of education for women's empowerment has been gaining ground among women's activist groups. According to this understanding, education has to become a means whereby women and girls begin to question their own beliefs about themselves and gradually develop self-confidence and a positive self-image so that they begin to appreciate their own capacities and potentialities. The term "empowerment" in the context of women's education is considered to be particularly relevant, referring to a range of activities, from individual self-assertion to collective resistance, protest, and mobilization, that challenge basic power relations. It is a process that is aimed at changing the nature and direction of systemic forces that marginalize women in a given social, economic, and political milieu. Literacy is regarded as an important skill whereby women acquire the ability to read, write, and compute and thereby gain access to knowledge and information hitherto denied to them.

Measures to Combat Women's Illiteracy

This shift in perspectives has been reflected in measures taken at the global level to deal with basic education of women and girls.

World Conference on Education for All (EFA)

"Education for All" has become the battle-cry of educationists in the 1990s. Events in different regions of the world have helped focus the attention of the world community on the importance of primary education and adult literacy in efforts to make the world population literate by the year 2000. The United Nations proclaimed 1990 as International Literacy Year, and in March of that year a meeting of historic significance took place at Jomtien, Thailand: Four U.N. agencies—UNESCO, UNICEF (United Nations Children's Fund), UNDP (United Nations Development Program), and the World Bank—teamed up to sponsor the World Conference on Education for All, which adopted two historic documents, the *World Declaration on Education for All* (UNESCO, 1992b) and the *Framework for Action to Meet Basic Learning Needs* (UNESCO, 1992a).

How was the issue of educational marginalization of women and girls presented in these documents? Ramdas (1990b) analyzed all of the draft documents, which were discussed at nine regional and three international consultations held prior to the world conference, and found that they lacked sensitivity to gender issues—despite the participation of feminists at some of these international meetings and their interventions to incorporate a gender perspective. Perhaps due to such interventions, women's issues got some visibility in terms of an addition here or a few lines there, but by and large these were cosmetic changes; for the essence of the original draft did not alter substantially. According to Ramdas, the main thrust of the background document that was presented at the Jomtien meeting was that there was no special focus on gender; gender was treated as just another disadvantage; and there was no attempt at deeper, more serious analysis of the underlying causes of the educational marginalization of women and girls.

The text of the two documents adopted by the EFA conference, which was claimed by one of the organizers to be a "product of a wide and systematic process of consultation" (WCEFA, 1990, p. i), did not reflect any major shift in focus and emphasis with regard to gender issues. Thus, while there was recognition of the problem that "more than 100 million children, including at least 60 million girls, have no access to primary schooling" and that of the more than 960 million illiterate adults, two-thirds are women, the overall tenor of the documents was to treat women as one of the disadvantaged groups, with priority placed on ensuring better access to basic education for girls while giving short shrift to the literacy needs of adult women. Since the documents reportedly reflected a worldwide consensus on the concept of basic education, it is likely that they played a significant role in setting the educational agenda of most Third World countries. It is, therefore, a matter of grave concern that gender issues were not adequately addressed in these documents. As rightly noted by Bown (1992), the Jomtien Conference on Education for All was largely hijacked by those whose exclusive interest was schooling for the young.

The Education for All (EFA) Forums

After Jomtien, the four principal sponsors of the world conference convened an International Consultative Forum on Education for All to serve as the international mechanism for consultation and information sharing among all of the broad constituencies represented at the world conference and concerned with its follow-up. The 60 forum participants met for the first time at UNESCO headquarters in Paris in December 1991. After reviewing the progress made by countries and agencies in implementing the conference's *Framework for Action,* the forum focused its deliberations on the prospects for achieving universal primary education. The second meeting of the forum, focusing on "quality education for all," was held in New Delhi in September 1993. The central theme was elaborated around four broad areas: (1) early childhood development, (2) improving schooling, (3) improving nonformal education programs, and (4) financing quality basic education. Gender equity was regarded as one of the transversal concerns that was to be addressed in each of the above areas.

Unlike at the Jomtien meeting, the education of girls and women did find a place on the agenda, but for a different reason: The problem of basic education got redefined, and nine of the most populous countries of the world were identified as those that also had problems relating to basic education. Statistics relating to these nine countries (see Table 4) showed that while each of these countries had high population figures, they also had comparatively lower levels of literacy attainment, especially in countries of South Asia. A case was made for investing in basic education of girls so as to arrest the population growth in the long run. A background paper titled "Financing Quality Basic Edu-

Table 4. Illiterate Population Age 15+ and Percentage of Illiteracy, 1990

Country	Illiterate Population (in Thousands)		Percentage of Illiteracy	
	Total	M/F	M	F
China	181,610	22.2	13.0	31.9
India	280,732	51.8	38.2	66.3
Egypt	16,492	51.6	37.1	66.2
Mexico	6,162	12.4	9.6	15.0
Brazil	18,407	18.9	17.5	20.2
Indonesia	20,899	18.4	11.7	24.7
Nigeria	28,723	49.3	37.7	60.5
Bangladesh	41,961	64.7	52.9	78.0
Pakistan	43,459	65.2	52.7	78.9

Source: UNESCO, 1993

cation" prepared by the World Bank (Tsang et al., 1993) made the rationale for women's education explicit: Women should be provided basic education for reasons of equity ("a society should take care of its most disadvantaged members") and efficiency ("investment in quality basic education for these populations can have a high rate of return"). Public investment in women's education is highly desirable because "there is substantial evidence from a wide range of countries that increased female education is linked to improved health, lower fertility, and other benefits, and that investment in female education has a high social rate of return" (Tsang et al., 1993, p. 3).

The approach advocated in the EFA documents indicates a narrow economistic view of education. This has probably been due to the predominance of the World Bank and its approach toward human-capital formation, according to which education is necessary for improving the skills and thereby the productivity of the human labor force. The economistic orientation to education has fostered strategies that are technomanagerial in nature. While barriers to education of girls and women are identified in these documents as family and cultural constraints, lack of time, shortage of female teachers, lack of child-care services, and the like, the strategies focus on materials production, curriculum development, training of teachers, strengthening managerial structures, and improving access to educational services. The underlying assumption is that it is these technomanagerial interventions that will ensure equitable access and participation of women and girls in the educational process. The EFA documents also create the impression that once the goal of education for all is achieved, the socioeconomic, cultural, and political inequalities in the Third World countries will disappear. Thus the larger structural issues of underdevelopment of a large number of Third World countries are not addressed at all (Tandon et al., 1989).

World Summit on Education for All
Although the World Bank was not one of the sponsoring organizations for the World Summit on Education for All

held in New Delhi in December 1993, its influence loomed large, for the economic argument for investing in the education of girls and women persisted. An issue paper for the panel on "Girls and Women's Education, Women's Empowerment, and Population Issues" prepared by UNFPA (United Nations Population Fund), cited findings of a World Bank study of 200 developing countries that "countries which allocated substantial resources to female primary education experienced higher economic productivity, lower fertility rates, lower infant mortality rates and improved levels of life expectancy for both men and women, compared to countries with lower levels of women's educational attainment" (UNFPA, 1993, p. 2).

It said that women's education delivered the highest return of any development investment and termed it "the sine qua non for the achievement of sustainable development" (UNFPA, 1993, p. 3). Once again, education of women and girls was promoted because it had an instrumental value rather than for any intrinsic value. Interestingly, the same paper recognized women's right to education as a basic human right and called education "perhaps the most important means for empowering women to exercise their rights in society and in the home" (UNFPA, 1993, p. 4). However, in the strategies enumerated for promoting female education, it was once again the education of the girls that was emphasized while that of the illiterate adult women was marginalized. Nor were any efforts made to elaborate on the strategies whereby education could empower women.

However, in the actual panel presentation on "Girls' and Women's Education, Women's Empowerment, and Population Issues," the issue of education for women's empowerment was central. The issues the all-women panel addressed were comprehensive in nature. It was recognized that the ideology of gender inequality is inculcated in both boys and girls at a very early age. Religion, mythology, cultural taboos, rewards, and punishments are all used to socialize girls to accept, and participate in, their own oppression and subordination. If education is to be of use to women, it must create an awareness of how gender as well as other socioeconomic and political forces are acting on them and preventing them from breaking free from a sense of inferiority that has been imprinted on them since childhood. For it is this internalization of their own subordination that has prevented them from recognizing their own strengths, innate intelligence, and skills and, above all else, from exercising their right to lead a life of dignity and social justice. If education is to empower women, the educational process must enable women to set their own agendas, articulate their problems, and find solutions through collective action, while at the same time their abilities to access new bodies of knowledge will redefine their own self-image as women and place a value on their productive role in society. If this happened, women would be able to make informed decisions regarding their lives and exercise their rights in society. Education for women would then become a much larger and more significant process of enhancing their self-esteem, ensuring an equal partnership for them in the development process, and equipping them to determine their own life.

Significantly, while education for women's empowerment received a major thrust during the presummit meeting, in the summit deliberations and in the *Framework for Action* that was adopted by the leaders of the nine high-population developing nations of the world, the issue was once again sidelined.

Education for Women's Empowerment: Challenges

The concept of education for women's empowerment has been gaining ground among women's activist groups. One of the earliest attempts to build on this experience and to provide meaningful education to poor women was made through the Women's Development Program of Rajasthan in India and, subsequently, since 1989, through the *Mahila Samakhya* (Education for Women's Equality), which has been in operation in some districts of Uttar Pradesh, Gujarat, Karnataka, and Andhra Pradesh.

These programs begin with an investigation of the socioeconomic reality by the women themselves, an examination of the problems faced by them, and a process of critical analysis leading to collective action against injustices suffered by them in the home, the workplace, and society. Literacy is not imposed on women; rather, they are allowed to seek literacy at a point when its meaning and value become evident to them. Literacy is thus not viewed as an end in itself, limited to the teaching of basic reading, writing, and numeracy skills, but as part of an overall strategy of empowerment. The educational process enables women to ask questions, seek answers, act, reflect on actions, and raise new questions. As women have collectively addressed themselves to problems of fuel, child care, and income generation, they have realized that the problems are linked at a very fundamental level to the question of access to authentic information and, hence, to literacy.

As women have gradually become empowered at the individual and collective levels, they have been able to address problems such as access to drinking water, payment of minimum wages, access to health services, ensuring functioning of the village school, and children's participation in education, and they have taken collective action against domestic and social violence. Every issue taken up by the women has resulted in an educational activity. When the *mahila sangha* (women's collective) decides to take up an issue for debate or action, it involves a systematic analysis of the problem, collection of necessary information, visits to the "block" or district headquarters, and collective planning on the course of action. *Mahila Samakhya* has given women a voice in the villages, provided legal and administrative support, and focused systematic efforts to improve women's access to available educational and developmental facilities. In this manner, the *Mahila Samakhya* approach has become an integral part of the strategy for mobilizing women for participation in development.

Elsewhere in South Asia, many NGOs and some government departments and ministries run educational programs for women and girls whose stated goal is to empower women. Batliwala (1993) visited organizations in South Asia and conducted in-depth discussions with field workers, decision makers, and researchers, building a conceptual framework of women's empowerment and analyzing the different strategies that are being used in the region to empower women. Sponsored by the Asia Pacific Bureau of Adult Education (ASPBAE) and the Food and Agricultural Organization's (FAO) Freedom from Hunger Campaign-Action for Development, Batliwala's study provides a South Asian perspective, with the expectation that similar experiences will be initiated in other parts of Asia and the Pacific.

UNESCO's initiatives in this direction also deserve mention. After the Jomtien conference, a project on skills-training literacy for women initiated by UNESCO's Regional Office of Education in Bangkok in seven countries of the Asia-Pacific region has become a means for empowering women. One of the outcomes of this project, in the training manual entitled *Educate to Empower,* shows the direction in which the UNESCO project has moved. An international seminar on "Women's Education and Empowerment" organized by the UNESCO Institute of Education (Hamburg) in January 1993 helped in building on some of these experiences and in carrying the debate forward. At this seminar, participants once again grappled with the theoretical base and the strategies and indicators of empowerment (Stromquist, 1995).

As this broad overview shows, the concept of education for women's empowerment has gained legitimacy and recognition in recent years. As a matter of fact, some governments and institutions have now included "empowerment" in their discourse and are actively seeking partnerships with women's groups that are increasingly focusing their attention on this approach.

While this is an interesting development, the question remains whether the state, which is intrinsically patriarchal in nature, would really promote education that would empower women. Two experiences from India highlight the response of the state as women got empowered in government-funded initiatives.

In 1992, a woman who worked as a *sathin* (village-level worker) in the Women's Development Program in Rajasthan was gang raped by men of a socially dominant community. One of the stated objectives of the program is to empower women; yet, when this active, articulate village woman campaigned against child marriages on directions from the state government, not only was she sexually assaulted and raped, the state turned a deaf ear to her pleas and to the women's groups that demanded justice.

The second example is from the Nellore district of Andhra Pradesh. As part of the literacy campaign there, village women came together in classes, where they discussed general problems of the village and also talked about the evils of excessive drinking. Their discussions focused on how, in many families, the men drank away all they earned and how women had to work and run the household on their earnings—and get beaten daily by their drunken husbands.

A lesson in the postliteracy (after the basic literacy phase) primer described an incident that had taken place in one of the villages in the Nellore district: The women had stopped the vending of *arrack* (country liquor) after two men died following a bout of drinking. This lesson had an electrifying impact: Women's committees were formed in several villages, and gradually the agitation against the sale of *arrack* engulfed the entire district. It was clearly a case of women becoming empowered and spearheading an agitation that elicited the support of political parties, voluntary organizations, women's groups, and civil-liberties organizations.

The movement suffered a setback, however, when the chief minister of Andhra Pradesh first branded the agitation "antigovernment," then ordered that the provocative lesson be expunged, and even threatened strict action against those government functionaries who supported the movement and worked actively for it. Elsewhere in the state, women were beaten, kidnapped, and terrorized.

These two incidents highlight the variable response of the state as women become empowered—from callous indifference in one case to blatant repression in the other. Given this experience, some have questioned the desirability of the women's groups working collaboratively with the government on any program whose stated objective is to empower women (Malika et al., 1993). While the advantages of the partnership are greater outreach, access to resources, and more stability, the fear on the part of the women's groups is that such a relationship invariably leads to their co-option—for an illusion is created that there is space for women's rights and that their demands will be met. In reality, however, the state represents the interests of the dominant class and caste groups and does not tolerate any collectivization of women's strength. If this is so, it becomes incumbent for women to organize as women and as workers first and then to forge links with the wider

class struggles (Malika et al., 1993). Interestingly, in the case of the anti-*arrack* agitation, it was the forging of links with mass organizations that sustained and even intensified the agitation so that very soon it spread to other districts of Andhra Pradesh. The rural women of Andhra Pradesh scored a major victory when, in early 1993, the chief minister did a complete about-face and announced that *arrack* would be banned in Nellore immediately.

Empowerment and Structural Adjustment

Due to the deepening of the world economic crisis, many developing countries of Africa, Asia, and Latin America have introduced sweeping economic reforms to meet the needs and demands of the global economy. Economic restructuring has been undertaken through a package of stabilization and structural adjustment policies of the International Montary Fund (IMF) and the World Bank. The primary thrust of these policies is on economic liberalization, privatization, and deregulation to pave the way for a global market economy. The increased interest in empowerment comes at a time when structural-adjustment policies are being implemented in many of the developing countries.

A review of the experiences of countries that have been affected by the adjustment policies and programs shows that while they have mixed effects on the national economies, the adverse effects are particularly damaging in human-development sectors such as education (Tilak, 1992). Tilak's review further shows that as income and living conditions are seriously affected during the process of adjustment, the demand for education declines. This is more true in the case of the disadvantaged poorer sections. Demand for education falls due to changes in the labor market—increased levels of unemployment, reduced levels of wage earnings and greater earnings differentials, and correspondingly, changing patterns of wage employment. Women, particularly poor women, bear the major brunt of the structural-adjustment policies. Some of the educational consequences of these policies are low enrollment and increased dropout rates at the primary level; increased costs of education due to privatization; and lower participation rates by women in literacy classes due to greater priority for meeting daily survival needs. Other consequences include increases in the drudgery of women's work as fuel, fodder, and water collection become more difficult; a decline in women's health and nutrition standards; retrenchment of women from the organized sector; and a higher demand for women as "cheap" labor through expansion of home-based and ancillary units. It is economic concerns, survival issues, and livelihood pressures that are going to affect the lives of poor women. Given the long-term nature of these macroeconomic changes, what kind of women's literacy would effectively meet the challenges posed by structural adjustment?

As of the mid-1990s, there is little empirical evidence to show what kinds of literacy strategies for women have been found to be effective in terms of enabling or fostering steady participation. But there are projects in different countries that are beginning to indicate a possible direction for future work. Chlebowska (1990) describes some of these projects, whose purpose is to lighten the workload of rural women so they have the time to engage in literacy activities. This is done with the introduction of appropriate small-scale, simple, intermediate, or village technologies. Chlebowska argues that the introduction of these technologies should be preceded by a detailed analysis by women of the gendered distribution of tasks and of the social and cultural norms that define such tasks. Training methods should encourage women to play an active role in the discussion of new technologies and enable them to learn how to use these technologies and undertake simple maintenance and repair tasks themselves.

Case studies from India (Mishra et al., 1994) show how a program to sustain literacy skills has to be integrated with technical skills so that they mutually reinforce and sustain each other. In Banda district of Uttar Pradesh, training of women as hand pump mechanics resulted in better maintenance of the pumps and propelled the troup toward producing a newsletter on issues linked to the lives of women in the area. Work with the rural women para-veterinarians in the Dungarpur district of Rajasthan, where veterinary services in remote rural areas were almost nonexistent, catalyzed a demand from them for a residential camp for sustaining their literacy. These and other skills-training programs, besides equipping women with relevant technical skills, have also raised gender issues and developed leadership among local women. With the increasing possibility of women entering the informal sector of the economy in large numbers, skills training is a vital component of literacy and postliteracy programs.

A growing number of NGOs have begun to think of initiating efforts to ameliorate the economic hardships experienced by the poor. PRIA (Society for Participatory Research in Asia) (1993) spells out strategic interventions the NGOs might consider to meet the new challenges. It is clear that economic activities have to be seen as a strategic intervention in strengthening the capacities of the organizations of the poor to begin to confront the forces that shape their lives. Among the challenges are ensuring that the poor, particularly women, have access to, and control over, natural resources; organizing pressure groups to influence public policy and legislation in favor of the economic enterprises of poor women; and improving the sustainability of subsistence economic efforts. The latter might include a variety of interventions, such as organic and sustainable agriculture, forestry and water harvesting, regeneration of the natural-resource base, and utilization of the local economy for meeting local needs. In the short run, development of a local subsistence base for the rural poor would be crucial to allow them to stand up to market pressures and state regulatory measures. In the long run, a regenerative economy would move away from both market and state, with local self-reliance in production and fulfillment of basic needs of communities (Omvedt and Gala, n.d.).

Conclusion

The above examples of projects and strategies suggest some general answers to our questions about what kind of literacy would be relevant to women's needs and empower them to meet the challenges they face. Clearly, successful literacy strategies for women would have to evolve in response to specific local contexts. While the critical role of education would be to empower women generally, special emphasis would have to be laid on empowering women about market and political institutions. Omvedt and Gala (n.d.) suggest that a beginning could be made with those sectors in which women have traditionally played a major economic role—such as vegetable and fruit production, milk processing, weaving, and textiles—by providing training and resources and gradually strengthening and modernizing these sectors within the total economy. Skills training, as part of literacy and postliteracy programs, would become one meaningful intervention. More important, poor women would have to be organized, and links simultaneously established with other affected groups, such as the workers in the formal and informal sectors, and with the peasant movement, the environmental movement, the women's movement, and the like to ensure that the interests of all weaker sections are safeguarded. For it is coalitions and alliances, possibly cutting across different women's organizations and political affiliations, that would help build a broad-based local and national movement to work as pressure groups.

Finally, any attempt at transforming of the structures of subordination would require long-term systematic interventions. The level of awareness of the various arms of the civil society about women's subordination would have to be raised through popular culture, the media, and formal and nonformal education. Male attitudes toward women would have to be changed not only in the classroom but also in the home, in the workplace, and in political spheres. Attitudes of planners, policymakers, and administrators at all levels would have to become more gender sensitive, and gender issues would have to be integrated into all development programs. Women's participation in the development process would have to be enhanced by involving them in the planning, management, and implementation of development projects and programs, making their voices heard in the definition of development and the making of policy choices.

The primary point is this: Any understanding of women's literacy would have to be all-encompassing to include the economic, political, legal, social, cultural, and all other aspects that impinge upon and shape the lives of poor women. Rather than an end in itself, literacy would have to be perceived and pursued as a means of empowering individual women and their organizations—a process inextricably linked to women's education that is ever evolving and whose long-term goal is to bring about not just gender but also class, ethnic, and racial equality.

References

Batliwala, Srilata. *Empowerment of Women in South Asia:*

Concepts and Practices. Delhi: Asia South Pacific, Bureau of Adult Education (ASPBAE), 1993.

Bown, Lalage. *Women, Literacy and Development.* London: ActionAid, 1990.

———. "Women's Literacy for Development," *Women's Literacy for Development.* Warwick: INCED (International Center for Education in Development), Paper 11, March 1992.

Chlebowska, Krystyna. *Literacy for Rural Women in the Third World.* Paris: UNESCO, 1990.

Cochrane, Susan. *Fertility and Education: What Do We Really Know?* Baltimore: Johns Hopkins University Press, 1979.

Lockhead Marlaine, and Adriaan Verspoor. *Improving Primary Education in Developing Countries: A Review of Policy Options.* Washington, D.C.: World Bank, 1990.

Malika and others. "Women's Development: What Is the State's Intention?" *Economic and Political Weekly,* vol. 28, no. 10, 1993, pp. 373–376.

Mishra, Renuka, Malini Ghose, and Dipta Bhog. "Concretizing Concepts: Continuing Education Strategies for Women." *Convergence,* vol. 27, no. 2–3, 1994, pp. 126–135.

Omvedt Gail, and Chetna Gala. "The New Economic Policy and Women: A Rural Perspective," n.d. Mimeographed, pp. 1–15.

Society for Participatory Research in India (PRIA). *Kriti.* New Delhi: PRIA, no. 2, January–June 1993.

Ramdas, Lalita. "Gender Issues and Literacy: An Analysis," *Convergence,* vol. 23, no. 4, 1990a, pp. 37–47.

———. "Women and Literacy: A Quest for Justice," *Convergence,* vol. 23, no. 1, 1990b, pp. 27–40.

Stromquist, Nelly. "Challenges to the Attainment of Women's Literacy." In Catherine Odora with Jan-Ingvar Loftstedt and V. Chinapah (eds.), *Women and Literacy: Yesterday, Today and Tomorrow.* Stockholm: International Institute of Education, 1992.

———. "The Theoretical and Practical Bases for Empowerment." In Carolyn Medel-Añoneuvo *Women's Education and Empowerment: Pathways Toward Autonomy.* Report of the International Seminar, UIE (UNESCO Institute for Education): Hamburg, 1995, pp. 13–22.

Summers, Lawrence. *Investing in All the People: Educating Women in Developing Countries.* Washington, D.C.: World Bank, 1992.

Tandon, Rajesh, Lalita Ramdas, R. Parikh, H. Swarup, O. Shrivastava, and B.K. Sinha. "Education for All as Determined by the Few?" New Delhi: PRIA (Society for Participatory Research in Asia), December 1989. Mimeographed.

Tilak, Jandhyala B.G. "Education and Structural Adjustment," *Prospects,* vol. 22, no. 4, 1992, pp. 407–422.

Tsang, M., assisted by A. Verspoor, P. Moock, M. Lockheed, and T. Moran. "Financing Quality Basic Education," Washington, D.C.: World Bank, 1993.

United Nations (UN). *Multivariate Analysis of World Fertility Survey Data for Selected ESCAP Countries.* Asian Population Series No. 49. Report and selected papers on the regional workshop and seminar (ST\ESCAP\151), ESCAP, Bangkok, 1981.

United Nations Education, Scientific, and Cultural Organization (UNESCO). *Framework for Action to Meet Basic Learning Needs.* Paris: UNESCO, 1992a.

———. *World Declaration on Education for All.* Paris: UNESCO, 1992b.

———. *World Education Report.* Paris: UNESCO, 1993.

———. *Education for All Summit of Nine High-Population Countries: Final Report.* Paris: UNESCO, 1994.

United Nations Population Fund (UNFPA). *Girls' and Women's Education, Women's Empowerment and Population Issues.* Rome: UNFPA, 1993.

Women's Participation in Science and Technology

Eva M. Rathgeber

Introduction

Women represented 36 percent of the global population engaged in formal-sector employment in 1990. However, in most countries they continue to be concentrated in lower-level and less well paid types of employment (ILO, 1993). Almost universally, female participation in modern science- and technology-based occupations has been remarkably limited. This is notwithstanding the important roles traditionally played by women in the development and management of technology-based tools and implements in households and in family autarkies. Although they have always been users and often developers of technology, women rarely have been recognized as central actors in science and technology.

As a result of social and cultural processes, most women were excluded from participation in science and technology as it emerged during the past centuries. Historians of science have provided evidence of female scientific activity, including physicians in the Middle Ages, mathematicians in early-modern Italy, and natural scientists in the nineteenth century (Mozans, 1991), yet female participation in science and technology was discouraged during most of modern European history. Britain's Royal Society, established in 1662, did not admit women until 1945, and by the mid-1990s only 2.9 percent of its fellows are female. Only seven women (3.5 percent of new Fellows) were elected in the period 1989–1993. By the early 1990s Britain's Royal Academy of Engineering has elected three women out of 901 fellows. The American National Academy of Sciences has 1,750 living members, only 70 of them female (Holloway, 1993).

For the most part, where women have been involved in science, their contributions have been minimized or overlooked, as occurred with British X-ray crystallographer Rosalind Franklin's work on the structure of DNA. She provided critical information that enabled James Watson and Franklin Crick to undertake research on the double helix for which they later won a Nobel Prize. By the early 1990s, only nine women had been awarded a Nobel Prize in science subjects, compared to more than 300 men.

Feminist theorists have emphasized the biased nature of science, pointing out that it is a human activity heavily influenced by prevailing social, political, and economic factors (Rosser, 1988). Some have argued that a feminist science would differ from "masculine" science because of fundamental differences in female perspectives and female approaches to problem solving. It has been pointed out that science, as commonly practiced, espouses an essentially male worldview and that female scientists who wish to succeed must of necessity work within this view or perspective. One object of continuing debate is the preeminence of scientific method with its emphasis on "rational" thought processes. Brush (1991) has argued that definitions of rationality should be expanded to include at least some aspects of other cognitive styles and ways of knowing. In this way, for example, greater emphasis might be placed on evidence from other sources, including different types of intuition. There has been little empirical work to explore whether, in fact, female scientists do work differently than male scientists or whether they bring different qualities or emphases to the scientific task (Sorensen, 1992). This would appear to be a critical area for further research.

Since the mid-1980s, a growing feminist concern with the exclusion of women from science and technology has led to determined efforts to correct gender imbalances. Nonetheless, attempts in North America to increase the numbers of female students in science and engineering programs have had mixed results. A study in the late 1980s showed that the interest of all American first-year college students in science and engineering subjects had declined by one-third during the past two decades (Task Force, 1988). An increasingly large number of places in American science training programs were being filled by foreign students.

Some critics have maintained that science and technology will take women's concerns into account only when

Table 1. Female Proportion (%) of Total Enrollment in Tertiary-Level Engineering and Medical and Health-Related Courses, 1985

Country	Engineering Courses	Medical and Health-Related Courses
Africa		
Cote d'Ivoire	3.6	32.2
Kenya	1.6	24.7
Senegal	14.0	34.6
Tunisia	9.6	52.2
Zambia	n/a	31.1
Latin America/Caribbean		
Chile	20.6	56.3
Colombia	26.5	56.6
Jamaica	n/a	41.3
Nicaragua	25.8	68.3
Asia		
Indonesia	16.4	32.0
Malaysia	14.0	47.0
Philippines	14.5	77.2
Sri Lanka	19.8	45.8

Source: Compiled from UNESCO, 1987.

more women become scientists and technologists. However, it is also argued that the current model of science and technology has been internalized by those (both women and men) who have gone through and succeeded in the system. For this reason, increasing the *numbers* of women scientists will not necessarily effect fundamental change in the *conception and practice* of science. It is apparent that if the practice of science and technology is to be reformed, it will be necessary not only for more women to be represented but also for a critical assessment to be undertaken of the underlying assumptions that guide the creation of scientific and technological knowledge.

Statistics

Participation rates of women in science in industrialized countries, as in developing countries, are relatively low. In Sweden, where equality between the sexes has been promoted longer than in most other countries, women accounted for only 11 percent of employed nonacademic scientists and engineers in 1985. In Japan in 1992, less than 8 percent of scientists and engineers were female, most of them employed in less prestigious scientific institutions. In the United States in 1992, the figures were considerably higher, although still far from representative of women's numbers in the general population: Approximately 22 percent of nonacademic scientists and engineers were female. However, while women constituted 36 percent of employed nonacademic scientists, they constituted only 8 percent of employed nonacademic engineers. In Great Britain, the numbers of women studying science and engineering were sufficiently low to warrant concentrated efforts to increase them in the 1980s. In 1988, science and mathematics became compulsory core subjects for all chil-

dren age 5–16, but there has been only a slow increase in the proportion of girls relative to boys choosing to study sciences. As in the United States, those girls who do choose to study sciences tend to concentrate in the biological sciences rather than in the natural/physical sciences and engineering. In 1991–1992 women made up only 27 percent of full-time science postgraduate students in Great Britain and 10–25 percent of postgraduates in engineering, mathematics, and the physical sciences (Committee on Women in Science, Engineering, and Technology, 1993).

Less is known about the experiences of female scientists in developing countries, but statistics suggest that trends parallel those in the industrialized countries, with a tendency toward even greater exclusion of women from science. Statistics are of varying reliability, but they confirm that the percentage of women enrolled in sciences and engineering is very small in most countries. Table 1 reveals significant differences in the proportion of women enrolled in tertiary-level (postsecondary) courses in engineering and in medical and health-related programs.

In all countries, the numbers of women enrolled in medical and health-related fields are substantially higher than those in engineering. However since the "medical and health-related" category used by UNESCO (United Nations Educational, Scientific, and Cultural Organization) does not differentiate between physicians and other types of health workers, such as nurses or physiotherapists, it is difficult to assess the extent to which women are enrolled in the most prestigious and potentially well-remunerated area of health studies—medicine. Regardless of the types of programs in which women are enrolled, however, it is clear that the category "medical and health-related" carries connotations of nurturing and social service, which are not immediately associated with "engineering." As discussed further below, even at an early age girls tend to express a preference for careers with a strong element of social service.

Table 1 shows that Central and Latin America have the largest proportion of female engineers and that the number of women of that region enrolled in the medical and health-related fields is also relatively high. It should be emphasized, however, that although relatively many women *study* in engineering and medical and health-related programs, this does not necessarily imply that they actually will *work* in those areas after graduation. With respect to Nicaragua, it should be noted that the participation of women in higher education, including science, was at an all-time high during the Contra war in the mid-1980s. Based on Table 1, it would seem that African women have the lowest participation levels in science programs. As discussed below, this may be due to continuing deficiencies and inadequacies in the teaching of science to girls in many African countries.

Table 2 presents data from select countries on the proportion of female students in higher education in natural science, engineering, and medical sciences and health. Again, the overall picture that emerges is one of low female enrollment in the so-called hard sciences, especially engineering.

In most of the developing countries included in Table 2, women make up less than one-half of the students enrolled in higher-education programs. Lesotho is an important exception. Unusually high female enrollment in higher education has been attributed to the tendency for young men to leave Lesotho to work in South Africa (Rathgeber, 1991). This began to change in the economic downturn in South Africa in the late 1980s and early 1990s and will no doubt be affected even more by the subsequent democratization and political restructuring of South Africa.

In Table 2, there are some surprising statistics with respect to the numbers of women enrolled in natural sciences and medical and health-related fields. For example, in eight of the 41 countries (Afghanistan, Argentina, Cuba, El Salvador, Nicaragua, Panama, Philippines, Singapore), 50 percent or more of the students enrolled in natural-science subjects are women, while 13 countries show more than 50 percent female enrollment in medical and health-related fields (Argentina, Barbados, Brazil, Cuba, Jordan, Laos, Lesotho, Madagascar, Malawi, Mozambique, Nicaragua, Panama, the Philippines). However, not a single country has more than 50 percent female enrollment in engineering. Cuba has the highest, at 32 percent. This low participation rate for women in engineering holds equally true in industrialized countries. In the United States, there are fewer women in engineering than in any other science-based profession, and the feminist movement has had limited impact in encouraging male members of the profession to reexamine their assumptions and/or biases (Hynes, 1992).

Education and Cultural Factors

Educational disparities at all levels between girls and boys are a primary cause of continuing female underrepresentation in science. In many parts of the world, girls continue to have substantially lower enrollments, especially at the secondary level, and, even in cases in which they are proportionately represented, the type of secondary education they receive commonly differs from that of boys. Some studies have found a stronger correlation between parental income and social status and school enrollment for girls than for boys (G. Kelly, 1984). As a result, during the 1980s girls were disproportionately affected by the imposition of structural-adjustment policies and education user fees in various countries, especially in Africa. Where poor parents have had to choose between educating sons and daughters, the preference usually has been for male education. Indeed, data from Kenya's Central Bureau of Statistics reveal that, by 1989, there was already a slight proportional drop in the numbers of girls attending secondary school. In Tanzania during the 1980s as structural adjustment policies came into effect, school enrollment rates for girls declined and dropout rates increased.

There is considerable evidence that girls receive less intensive training in science and mathematics at the primary and secondary levels. American, Canadian, and British studies have found that girls routinely get less attention from teachers and that teachers often give answers directly to girls while they give boys further information to enable them to solve problems for themselves. Male attitudes, particularly among peers, can have a negative effect on girls' aspirations in science. In a study of secondary-school students in Swaziland, Smith (1988) found that gender biases with respect to "appropriate" male and female occupations were held much more strongly by boys than by girls.

Although girls' science achievement levels frequently are equal to, or even higher than, those of boys in early primary school, they commonly drop in secondary school. Sociologists of education have compiled detailed evidence of classroom interactions between girls and boys and students and teachers in various settings. In a study in Great Britain, Alison Kelly (1985) noted that adolescent boys consistently undermined girls' efforts to participate in science classes by making disparaging remarks and frequently indicating that they considered science to be a male domain. There is evidence that girls' performance in science is more likely to remain stable in single-sex schools where, in the absence of male students, girls face less pressure to conform to predetermined "feminine" roles. Science teaching in single-sex schools can have a positive impact. Research from Nigeria showed that girls enrolled in science programs at the university level were more likely to have attended single-sex secondary schools (Erinosho, 1993).

In Kenya, the number of girls attending secondary schools increased in absolute terms during the 1980s, but they were more likely to attend unaided *harambee* schools (which receive little or no government financial support and consequently have inadequate teachers and facilities, especially for science (Kinyanjui, 1993). Even relatively prosperous girls' secondary schools often do not offer sci-

Table 2. Proportion (%) of Female Students in Higher Education in Three Fields of Study, 1985

	Women As Proportion of Total Enrollment in Higher Education	Women As Percentage of Field		
Country	Female	Natural Sciences	Engineering	Medical Sciences and Health
Africa				
Burundi	24	12	1	24
Congo	13	6	n/a	21
Egypt	28	33	13	39
Ethiopia	18	8	3	14
Gabon	29	16	5	42
Guinea	14	16	4	15
Kenya	26	14	2	25
Lesotho	63	30	n/a	69
Madagascar	38	37	7	74
Malawi	29	13	n/a	57
Mali	13	n/a	3	26
Mauritius	33	n/a	2	34
Morocco	19	27	11	31
Mozambique	23	30	8	51
Niger	18	8	n/a	18
Rwanda	14	6	n/a	15
Senegal	21	11	14	37
Sudan	37	33	9	36
Swaziland	40	6	n/a	23
Togo	16	6	n/a	23
Uganda	23	16	2	25
Caribbean				
Barbados	49	40	3	63
Trinidad and Tobago	43	47	14	n/a
Central and South America				
Argentina	53	62	12	56
Brazil	50	n/a	12	61
Cuba	55	52	32	58
El Salvador	43	77	10	23
Mexico	36	38	12	48
Nicaragua	57	71	26	68
Panama	58	50	26	79
Suriname	56	4	n/a	43
Middle East and Asia				
Afghanistan	14	80	11	31
Bangladesh	19	18	3	27
Indonesia	32	39	16	32
Jordan	45	n/a	15	53
Laos	36	n/a	15	53
Malaysia	44	41	14	47
Nepal	20	23	3	28
Philippines	54	67	15	77
Singapore	42	64	17	33
Sri Lanka	40	43	20	46

Source: Compiled from Gail P. Kelly, 1991.

ence courses, sometimes because they cannot find qualified female staff to teach them (Eshiwani, 1989). Analysis of the Kenya Certificate of Education Examination results for 1985 and 1986 showed that girls chose to sit for subjects that were less cost intensive to teach. Moreover, curricular changes in the mid-1980s reduced the amount of compulsory time spent on sciences from 17 to 12 periods per week, creating greater pressure on students to be self-motivated and to work independently outside the classroom. More than three-quarters of the girls who sat for math exams in 1985 and 1986 failed (Kinyanjui, 1993).

Girls tend to be more attracted to science if they see it as socially relevant. A British study found that girls were less interested in science that involved defense-funded work or animal experimentation (Wellcome Trust, 1994), and a Nigerian study revealed that "usefulness" was one of the three most important factors (together with personal interest and ability) influencing female university students' choice of a career in science (Erinosho, 1993). Research in Swaziland found that male secondary-school students were much more likely than girls to think that success in science will lead to a good job (Smith, 1988). Although girls were not uninfluenced by job prospects, they wanted to do something that was socially useful and involved contact with people. Finally, British research has shown that girls tend to perceive science as a male-dominated area with long working hours (Wellcome Trust, 1994). All of these studies point to significant attitudinal differences between male and female students at secondary and tertiary levels. In general, girls have stronger interest in people and social issues while many boys have an interest in tinkering and understanding the mechanical foundations of technology (A. Kelly, 1985). Nonetheless, science curricula tend to be structured in such a way as to appeal primarily to the interests of boys. It is evident that if greater numbers of girls are to be attracted into science, school science courses must be structured to appeal to the interests and tastes of both boys and girls. In most countries, these differences have not been taken into account in science-curriculum design.

There has been considerable research on the significance of cultural and socioeconomic factors in steering women toward science careers. An American study found that foreign graduate students from Africa and Asia tended to cite strong family pressure to enroll in science programs. Similar pressure existed for African-American and Asian-American students but not for students of European background (Bellisari, 1991). Foreign students from Africa and Asia also tended to see national development goals as an incentive for pursuing science-based careers. The governments of most developing countries have long emphasized the close links between science and technology and development; it is a message that apparently has been internalized by many young people. However, African research has shown that girls specializing in natural sciences tend to come from more-affluent socioeconomic backgrounds and to have better-educated parents with less strongly held notions of sex-role stereotypes (Erinosho, 1993). In this

sense, it is mostly girls coming from social elites who even have the option to contribute to national development through pursuit of careers in science and technology.

Erinosho (1993) found in a large Nigerian sample of female university students that two-thirds of their fathers and almost one-half of their mothers had completed tertiary education. Moreover, she found that the fathers of female science majors often had science-based professions or careers and had served as role models for their daughters. Conversely, research in Great Britain has indicated that female students enter university science courses because they like science or are good at it rather than as a result of specific encouragement from parents or schools (Wellcome Trust, 1994). As was the case in Nigeria, however, most girls in university science courses tended to come from middle- and upper-middle-class backgrounds. This suggests that there may be important differences as well as similarities in the experiences of women in developing and industrialized countries.

Most girls have few role models of successful female scientists. Sex-role stereotyping in school textbooks continues to be significant. Alison Kelly (1985) notes that girls are rarely depicted as active participants in science-textbook illustrations. More commonly, they are shown as participant observers or sometimes as amazed onlookers. With the exception of Madame Marie Curie, few, if any, female scientists are identified for students. In Africa, where women are key subsistence food farmers, texts rarely portray them as farmers. Moreover, in some cases, curricular expectations differ for men and women. A Kenyan study found that curriculum requirements varied for men and women training as junior-level agricultural officers in a technical institute (Bahemuka et al., 1992). Women were required to take hands-on assignments in home economics while men were required to take hands-on activities in agricultural engineering. It is clear that prevailing stereotypes of "appropriate" roles for girls and boys and men and women continue to inform the design of education programs at all levels, especially in countries where the influence of feminist thinking on curricular reform is at an early stage.

At the tertiary level, particularly in the physical sciences and engineering, female role models are rare. In most countries, the number of female professors in science and engineering programs is low—in 1985, women made up 13 percent of science faculty in the United States and 2 percent of engineering faculty (Task Force, 1988). Various studies have found that women tend to drop out of science and engineering graduate programs much more frequently than their male counterparts. One reason for this high attrition rate may be a lack of female role models.

Finally, in developing countries, girls are faced with a variety of factors that conspire to reduce female participation in higher education and in science and technology. These include higher opportunity costs for girls through the loss of their labor in the productive sector and in the household. Other factors include teenage pregnancy, as well as a range of socioeconomic, religious, and cultural influences that affect the parental decision to educate daughters. In some countries, legal constraints also make it difficult for women to work in certain industries or under certain conditions. In many African countries until the early 1990s, women were prohibited from working between the hours of 6 P.M. and 6 A.M. In other countries, Islamic law prohibits the employment of women in situations in which they will have to interact with men. Such obstacles create a further set of structural barriers to full participation of women in economic development.

Women in Science and Technology Careers

Women's experiences as scientists in the workplace usually differ from those of men. Statistics for the United States indicate that female scientists are more likely to be unemployed or underemployed and their salaries are lower than those of equally qualified males (White, 1992). By 1990, women accounted for only 4 percent of employed engineers and 30 percent of employed scientists, despite the fact that by 1986 women had earned 30 percent of all bachelors' degrees in science and engineering. Moreover, employed female Ph.D. scientists and engineers tended to be young. In 1990, 39 percent were under the age of 40, compared with 25 percent of their male counterparts (White, 1992).

British statistics reveal a similar trend. The proportion of female scientists decreases at the higher levels of appointment in industry, the civil service, and academia (Committee on Women in Science, Engineering, and Technology, 1993). The only scientific occupation in which women outnumber men is that of laboratory technician. Moreover, between 1980 and 1990 in Great Britain, the proportion of women employed in engineering actually fell. In the civil service, female scientists are also underrepresented. They constituted only 9 percent of senior scientific officers in 1992. In British universities in 1991, of 24,000 full-time academic staff, women accounted for only 15.5 percent in the biological and physical sciences, chemistry, mathematics, and computing, engineering and technology, and subjects allied to medicine.

Relatively little information is available about employment opportunities for female scientists in developing countries. Moreover, it is difficult to ascertain whether they are underrepresented because they are few in number or because employers discriminate against them, or perhaps both. For example, Ghana's Council for Scientific and Industrial Research employed 171 scientists, only 17 of them women and 9 of those in food research (Beoku-Betts and Logan, 1993), but, given the fact that the pool of women science graduates is low to begin with, that may be an accurate reflection of the proportion of women science graduates.

There is evidence that women in academic science careers in the United States tend not to be promoted to high academic rank and that, in general, their professional advancement is much slower than that of men (Cole cited in Brush, 1991). In 1990–1991, only 17 percent of women full-time faculty at American universities were full professors, compared with 44 percent of the men. In developing countries, the situation seems to be similar. In Cote d'Ivoire in 1987–1988, only 10 percent of full professors at the country's universities were women (Beoku-Betts and Logan, 1993). In the Chinese Academia Sinica in 1992, there were 286 women directors or deputy directors of research laboratories, accounting for 11.9 percent of the total (Guan Tao, 1992). Since the mid-1980s, there has been a steady, if modest, increase in the numbers of women scientists engaged in all sectors of research and industrial production. However, in China as elsewhere, female scientists are most likely to be concentrated in the biological sciences: In 1992, 47.3 percent of the Chinese Academy of Medical Sciences' research projects were led by women.

Both in developing and in industrialized countries, employment prospects for female scientists are negatively affected by the women's need to combine professional and home responsibilities. Since they believe that women will carry the major burden of child care and domestic work, employers often assume that women's commitment to science will be less intense than that of male colleagues. This can have an impact not only on the employer's decision to hire female scientists but also on the type of work to which women are assigned. It is clear that women's employment and promotion opportunities frequently are curtailed or hampered when they take time out for childbearing and rearing. Sweden has made an attempt to equalize the situation for men and women with historic legislation in 1994 that fathers must take compulsory leave after the birth of a child.

As noted by Arianrhod (1992), the mundane interruptions of daily domestic life, which are a reality for most working wives and mothers, are highly disruptive to the pursuit of science, especially scientific research, which demands long periods of uninterrupted laboratory time. This leads to the question of whether women are expected to make greater sacrifices than their male colleagues in order to succeed in science. Marriage and motherhood are difficult to reconcile with total dedication to scientific research. Although there are many examples of women who are managing to do so, it requires considerable organizational skill. This gives further credibility to the argument that neither the practice nor the content of science is neutral and reinforces the feminist appeal for a rethinking of the fundamental assumptions of science.

Learning Science and Technology on the Job

Since the mid-1980s, efforts have been made in many countries to integrate women into nontraditional, science- and technology-based careers. One successful example is the Jamaican Women's Collective, which has provided poor Jamaican women with skills in carpentry, masonry, and other construction areas. Another is the Sarvodaya movement in Sri Lanka, which has conducted water-pump-maintenance workshops for young women to teach them plumbing and welding techniques. Such initiatives seek not only to provide new sources of employment for poor women and to break down existing sex-role stereotypes but also to ensure that women share with men necessary everyday skills that are essential to the maintenance of communities. For example, the Sri Lankan project was developed after it became apparent that men did not place a high priority on the repair of water pumps since water was a female responsibility in most households. Since women's interests were more immediately involved, it was logical that women should acquire the necessary skills to repair the pumps themselves. Consequently, groups of young women learned theory and machine-shop work, as well as pump assembly, installation, monitoring, and repair. Women hand-pump technicians then became educators and change agents in their own communities.

Despite such success stories, it is relatively rare for women to be given technical training outside formal education systems. While women are often employed in industries undergoing technical change, they rarely are among the first beneficiaries of corporate efforts in the area of worker retraining and skills upgrading. A more common example is that provided by Roldán in her study of the introduction of some aspects of Japanese JIT (just-in-time) technology into two Argentinean factories, one small and the other medium size (Roldán, 1993). One of the characteristics of the JIT system is that it requires multiskilled workers who can rotate between jobs and carry out different functions. Roldán found that, in both factories, women were being kept in the same positions or even phased out while men were being given training to enable them to rotate between jobs and become multiskilled. Women were excluded because they did not have the basic technical competence considered necessary to become multiskilled. The firms considered it cheaper to invest in training men, even those with low bases of technical knowledge. In addition, Roldán studied an electronics plant with an all-female labor force. In that case, she found that although the male production manager was aware of Japanese techniques, he was not planning to introduce them into the plant because it would be too costly to provide technical training for women. Thus he was willing to forego the potential benefits of the technologies to avoid having to invest in technical training for women. Roldán suggests that, in Argentina, the introduction of new production technologies is serving to weaken the position of women in the labor force and leading toward a masculization of factory labor.

The computer explosion of the 1980s offers another example. Although facility with desktop computers is based on the mastering of typing, a skill that has been strongly associated with women throughout most of the twentieth century, computers have become another area of male dominance. Studies in the United States have shown that although boys and girls show equal interest in computers in early primary grades, girls' interest tends to decline after age 10 or 11 while that of boys continues to grow. Not surprisingly, the computer-games industry is aimed primarily at boys, with great emphasis on games of violence and destruction.

To a significant extent, the exclusion of girls and women from full participation in the computer industry reflects general sex-role stereotyping and the assumption that women are "not technical." Further, it is often assumed that women will not even desire such training both because it is "unfeminine" and because it is "too difficult." In an effort to break down such stereotypes and to encourage an interest among girls in the computer industry, which is projected to be a significant source of employment into the twenty-first century, a number of strategies have been employed. In Australia, weeklong computer holidays are organized annually for groups of young girls. Moreover, educational videos showing women successfully at work in the computer industry and questioning traditional stereotypes about female computer professionals have been shown to secondary-school girls with positive results. In the United States, some schools have experimented with the establishment of individual mentor programs while others have set up girls-only hours in computer labs.

On-the-job technical training frequently is acquired through formal apprenticeship or informal mentoring whereby an older worker teaches skills to a younger one. In most countries, both formal apprenticeships and informal mentoring tend to be heavily dominated by men. Apprenticeship programs, which grew out of the medieval trade guilds, have always been male oriented, and, while there are examples in many countries of efforts to provide opportunities for young female apprentices, these are difficult to legislate or enforce in the private sector, which often is a major source of apprenticeship opportunities. With respect to mentoring, given the separation of the sexes that occurs as part of childhood socialization processes in most societies, it is not surprising that experienced men prefer to train young men rather than young women, but this does mean that an important potential source of technical learning is less accessible to women. Again, it is difficult to legislate this matter since mentoring is commonly an unstructured and voluntary undertaking.

Finally, the profusion of new information technologies provides a rich source of instructional media for the imparting of science and technology knowledge and skills. These include computer-aided learning systems and transmission of training courses into factories by satellite, interactive video, and various other methods. If women are given equal opportunity to benefit from such courses, they should be able to acquire new technical skills and knowledge to enable them to compete effectively for better-paid positions in the industrial workplace. At the same time, however, older technologies, including radio, should be used systematically in developing countries to demystify aspects of science and technology for rural women. There is also much scope for the use of popular education materials and popular theater, songs, and dance to impart science and technology information.

Conclusion

Ultimately, it must be asked whether science and technology would be different if women had a greater voice. It seems evident that the problem goes beyond a mere increase in the numbers of female scientists and technologists. There is need for a complete reconceptualization and reorganization of the culture of science and technology. This must encompass the provision of space for alternative viewpoints and perspectives and the humanization of an area that traditionally has considered itself to be "neutral" and beyond the reach of immediate social, cultural, political, and economic manipulation. There is abundant historical evidence that science never really has been neutral, but the challenge lies in persuading practitioners of science to accept this as a given and to work from that basis rather than from the assumption of a value-neutral science.

There is also a need for science and scientists to move beyond the conceptual boundaries imposed by formal training and to recognize the significant contributions that have always been made, and continue to be made, by grassroots practitioners, many of them women. Science tends to be organized in an inherently hierarchical and elitist fashion. In developing countries, the extension of technology (in agriculture, health, or other areas) is almost exclusively one way. Scientists or technologists rarely examine existing agricultural or health practices with the thought that they can provide them with valuable insights or understanding about human survival strategies. Instead, there is always an assumption that knowledge held by the scientist or technologist is of a higher order than that held by the practitioner. This assumption is the foundation stone upon which most extension services have been built.

Finally, it can be argued that scientific and technological facts often are presented in deliberately arcane language that is aimed at creating "insiders" who have the training and expertise to understand and "outsiders" who do not. Women are frequently relegated to the status of "outsiders," and although they may have much to gain from understanding the mechanical details of a particular technology, it is rare that efforts are made to explain these in such a way as to be nonthreatening and comprehensible. Instead, women's lack of understanding is once again seen as an example of their intrinsically "nontechnical" nature. It is clear that we need to demystify and democratize scientific knowledge and involve women as agents of change.

References

Arianrhod, Robyn. "Physics and Mathematics, Reality and Language: Dilemmas for Feminists." In Cheris Kramarae and Dale Spender (eds.), *The Knowledge Explosion: Generations of Feminist Scholarship.* New York: Teachers College Press, 1992, pp. 41–53.

Bahemuka, Judith Mbula, Charles B.K. Nzioka, and Paul N. Mbatia. *Women Professionals in the Agricultural Sector: Kenya Case Study.* Development Studies Paper Series. Morrilton: Winrock International Institute for Agricultural Development, 1992, pp. 41–53.

Bellisari, Anna. "Cultural Influences on the Science Career Choices of Women," *Ohio Journal of Science,* vol. 91, no. 3, 1991, pp. 129–133.

Beoku-Betts, Josephine, and B. Ikubolajeh Logan. "Developing Science and Technology in Sub-Saharan Africa: Gender Disparities in the Education and Employment Process." In *Science in Africa: Women Leading from Strength.* Washington: American Association for the Advancement of Science, 1993.

Brush, Stephen G. "Women in Science and Engineering," *American Scientist,* vol. 79, no. 5, 1991, pp. 404–419.

Committee on Women in Science, Engineering, and Technology. *The Rising Tide: A Report on Women in Science, Engineering, and Technology.* London: HSMO, 1993.

Erinosho, Stella Y. *Nigerian Women in Science and Technology.* Dakar: International Development Research Center, 1993.

Eshiwani, George. "Kenya." In Gail P. Kelly (ed.), *International Handbook of Women's Education.* Westport: Greenwood, 1989.

Guan Tao. "Chinese Women As an Important Driving Force in Environment and Continued Progress," address at the International Symposium on Women's Role in Environment and Continued Progress, 1992.

Holloway, Marguerite. "A Lab of Her Own," *Scientific American,* November 1993, vol. 269, no. 5, pp. 94–103.

Hynes, H. Patricia. "Feminism and Engineering: The Inroads." In Cheris Kramarae and Dale Spender (eds.), *The Knowledge Explosion: Generations of Feminist Scholarship.* New York: Teachers College Press, 1992, pp. 133–140.

International Labor Organization (ILO). *World of Work,* no. 2, 1993.

Kelly, Alison. "The Construction of Masculine Science," *British Journal of Sociology of Education,* vol. 6, no. 2, 1985, pp. 133–154.

Kelly, Gail P. "Women's Access to Education in the Third World: Myths and Realities." In Sandra Acker, Jarquette Megarry, Stanley Nisbet, and Eric Hoyle (eds.), *World Yearbook of Education.* New York: Nicholas, 1984, pp. 81–89.

———. "Women and Higher Education." In Philip Altbach (ed.), *International Higher Education: An Encyclopedia.* New York: Garland, 1991.

Kinyanjui, Kabiru. "Enhancing Women's Participation in the Science-Based Curriculum: The Case of Kenya." In Jill Ker Conway and Susan C. Bourque (eds.), *The Politics of Women's Education: Perspectives from Asia, Africa, and Latin America.* Ann Arbor: University of Michigan Press, 1993, pp. 133–148.

Meena, Ruth. "The Impact of Structural Adjustment Programs on Rural Women in Tanzania." In Christina H. Gladwin (ed.), *Structural Adjustment and African Women Farmers.* Gainesville: University of Florida Press, 1991.

Mozans, H.J. *Women in Science.* Notre Dame: University of Notre Dame Press, 1991. Reprint, 1913 ed.

Rathgeber, Eva M. "Women in Higher Education in Africa: Access and Choices." In Gail P. Kelley and Sheila Slaughter (eds.), *Women's Education in Comparative Perspective.* Dordrecht: Kluwer Academic Publishers, 1991, pp. 47–62.

Roldán, Martha. "Industrial Restructuring, Deregulation, and New JIT Labor Processes in Argentina: Towards a Gender-Aware Perspective?" *IDS Bulletin,* vol. 24, no. 2, 1993, pp. 42–52.

Rosser, Sue V. "Good Science: Can It Ever Be Gender-Free?" *Women's Studies International Forum,* vol. 11, no. 1, 1988, pp. 13–19.

Smith, A.C. "Females in Science Courses in Swaziland: Performance, Progress, and Perceptions, *Swazi Journal of Science and Technology,* vol. 9, no. 1, 1988, pp. 65–82.

Sorensen, Knut H. "Towards a Feminized Technology? Gendered Values in the Construction of Technology," *Social Studies of Science,* vol. 22, no. 1, 1992, pp. 5–31.

Task Force on Women, Minorities, and the Handicapped in Science and Technology. *Changing America: The New Face of Science and Engineering.* Washington, D.C.: United States Congress, 1988.

United Nations Educational, Scientific, and Cultural Organization (UNESCO). *Statistical Yearbook 1987.* Paris: UNESCO, 1987.

Wellcome Trust. *Why May Women Science Undergraduates and Graduates Not Be Seeking to Take Up Careers As Scientists? A Scoping Study for PRISM.* London: Wellcome Trust, 1994.

White, Patricia E. *Women and Minorities in Science and Engineering: An Update.* Washington, D.C.: National Science Foundation, 1992.

Informal and Nonformal Education

Shirley Walters

Introduction

Informal and nonformal education focus attention on the question of knowledge and its social construction. Neither "knowledge" nor "ignorance" is a natural state; both are created socially. What constitutes knowledge and for whom different kinds of knowledge are appropriate vary from age to age. Powerful medieval kings saw no need for literary skills; these rested with a small group of specialized scribes. Over the ages, women have been cut off from many kinds of knowledge. At the same time, the kinds of knowledge and skills that women possess from their roles in child-rearing and domestic labor have been deemed trivial in many societies.

In any society, there is a social process for deciding what kinds of and whose life experiences are of sufficient value and importance to be systematized and theorized into a body of knowledge. The forms of valorizing knowledge vary enormously from one society and one age to another, from the received wisdom of a council of elders to sets of institutions like universities, libraries, and professional societies. And these decisions have everything to do with the power relations in a given society.

Informal and nonformal education are critically important areas for women's learning within male-dominated societies. Most women in the Third World are outside of formal economic, political, or educational structures. They often do not have easy access to education and training. It is only through informal and nonformal education that they can have an opportunity to learn new skills, develop different attitudes, or acquire new knowledge. It is also through informal education that indigenous knowledge, including cultural traditions, is passed on. But informal and nonformal education evade adequate definition because in many ways they are analytical constructs that give intellectual coherence to a range of activities that are integrally part of other forms of economic, social, cultural, or political life.

From the view of the educator, but in fact integral to the social fabric that produces and reproduces families, communities, and workplaces, informal education refers to educational activity that is unplanned, incidental learning. It occurs, for example, while organizing a community soup kitchen or rearing children. Another illustration relates to the many women who are involved in grass-roots groups in communities through savings clubs, burial societies, religious organizations, or community centers. The learning processes in these groups are not necessarily thought of as education. They are simply a part of what groups do and how they operate as they work to build democratic organizations and initiate change. A woman who is an active member of a grass-roots community center put it like this: "I finished school and I've taken night classes and I've been here and there, but actually the best education I've had was when I came to this center. Communication with these people is the best education there is." A fellow activist described her own experience as being one of having knowledge in her head "but no place to express it" until she "started getting involved in the community and with women" (Bingman, 1996, p. 219).

Nonformal education refers to planned educational activity that is usually, but not always, short term and noncertificated, such as nutrition education at a health clinic or driving lessons. It is often defined in terms of its relevance to the needs of the poor, its flexibility in organization and method, and its specificity of objective.

Informal and nonformal education for women are part of adult-education provision and practice and are distinct from formal education. In some ways this is a necessary distinction, since formal education usually has some form of state support, whereas informal and nonformal education occur within families, communities, or workplaces and receive limited, if any, support from the state. But from an educational viewpoint, the distinction is not necessarily helpful because the boundaries, particularly between nonformal and formal education, are blurred. As with all forms of education, adult education is concerned with the development of particular attitudes, values, skills, and knowledge among the adult population. "Adult education" is used here as the overarching term for informal, nonformal, and formal education for adult learners, particularly for women learners.

This essay focuses on the social purposes and dominant forms of adult education for poor, working-class, and peasant women in both the First and the Third worlds, whether informal, nonformal, or formal. I understand the terms "First" and "Third" Worlds as designating constructs referring to degrees of relative poverty or affluence, rather than geographical regions. This essay first situates adult education for women within their family, work, and community lives, then focuses on social purposes and forms of adult education. It describes feminist popular education, which is one of the most significant areas of nonformal and informal education for women. It will begin by situating adult education for women, and highlights important themes that confront educators working to empower women.

Situating Adult Education for Women

Adult education for women relates directly to their conditions and positions in society. It can be categorized along a continuum of whether the educational practices are consciously trying to change women's subordination or not. The form, method, and content will be influenced by the providers' political understanding of gender relations and whether they are aiming to conserve or transform gender power relations. In this essay, the focus is on feminist education, which is concerned with transforming these relations.

Since the mid- to late 1970s, new economic and political conditions have exacerbated the economic problems of less industrialized countries and the place of women within them. The World Bank introduced structural adjustment lending in 1979 to provide temporary balance-of-payment loans to assist governments in solving their financial crises while governments pursued changes in internal economic policies. Although the conditions imposed on each country differ, the general strategy of structural adjustment that developed in the 1980s and early 1990s resulted in the reduction in social-sector spending. Decreased public expenditure on education, health, and food subsidies means that increased costs must be borne by women, who work longer hours, look for less expensive food, spend more resources on basic health care, make difficult choices about which children will get an education and which will work to sustain the family economy, and face lower wages or fewer job opportunities as the wages in female-dominated industries decline. Women's labor and women's bodies often absorb the impact of reduced public-sector spending. Informal and nonformal education for women are situated within these economic and social circumstances, which are permeated by a patriarchal ideology.

Put simply, women in most societies lack institutional and decision-making power. They are seen as inferior to men because of their sex. Gender ideologies promote and reinforce patriarchy, which is defined as the social organization of the family, the community, and the state in ways that reinforce and perpetuate male power. The two main pillars of patriarchy are: first, the sexual division of labor that allocates men and women to different occupations and thus to different levels of prestige and reward; and second, the control of women's sexuality by men, which seriously constrains women's space and physical mobility and shapes conceptions of what women should be. These two pillars function in a mutually supportive manner, one justifying the other, making the combination seem totally natural, and rendering the questioning of either of them a formidable task, since economic benefits and deeply internalized norms of reality are then at stake. While the particular ways in which the division of labor and the control of sexuality manifest themselves are influenced in important ways by social class, technological levels of development, and religio-cultural norms, these connections represent the fundamental linkages in women's subordination in most societies.

Adult education that intends to change gendered power relations needs to challenge these two pillars of patriarchal ideology. It needs to be integral to women attaining real power as part of processes of economic, political, and cultural transformation. It therefore needs to relate to women's lives in their homes, communities, and workplaces. Women need to gain access to decision-making structures in all of these arenas. The political mobilization of women to struggle for economic and social justice is a vital aspect of changing their conditions and positions, and informal and nonformal adult education are integral to this.

Social Purposes of Adult Education

Defining the social purposes of adult education for women is critical to the development of a framework for analysis that captures how adult education is intertwined with all other social processes. This framework can be used to describe adult education that either conserves or transforms the status quo for women.

Wolpe (1994) has devised a classification of nonformal-education (NFE) programs that is sensitive to education within women's lives and that describes the social purposes of NFE. This classification is also relevant to informal education because it covers all aspects of women's lives. When using classifications, it is necessary to be cautious and to recognize that the categorizations can be disputed, particularly by those whose activities are being described. For example, women's participation in literacy classes is not necessarily viewed as part of a survival strategy by the women themselves.

The classification is: (1) educational strategies that help people survive the harsh conditions in which they live; (2) skilling for the informal sector of the economy; (3) skilling for the formal labor market, including training for the unemployed worker and reskilling people already engaged in the labor market; and (4) political and cultural NFE, which addresses civic issues, social justice, and people's participation in civil society.

Survival Strategies

Survival strategies are concerned with the reproduction of the family, which involves often interlinked economic, social, and cultural activities. Examples include informal and nonformal education relating to basic needs such as health care, nutrition, family planning, and literacy. Literacy programs for women illustrate the complexities of the interlinkages. Women in many Third World contexts are now actively involved in areas that men monopolized before. With the immigration of men to the towns to take up employment, women have been left in charge of agriculture and general home-improvement projects, and many of them feel the need to be literate as a way of coping with their responsibilities (Lind and Johnston, 1990, p. 66).

Cultural and other factors have led to a large gap between male and female access to education: UNESCO (1993) has estimated that two-thirds of the more than 900 million illiterate people in the world are women. In the literature, a central question is: What do women want from literacy and what sort of literacy programs should be offered to them?

Before addressing that, it is necessary to ask what is meant by literacy. The simple definition used by UNESCO since 1945 is that a person is literate who can both read and write a short simple statement on his or her everyday life. The definition encompasses the idea that literacy is a skill and links it with daily life, so that it hints at the notion of functionality, of literacy having to be used in a way that is relevant and useful. But this definition is inadequate. Sociolinguistic and anthropological studies have widened our understanding of literacy so that it is increasingly seen as social practices developed through participation in literacy events. This means that literacy is embedded in a set of social circumstances, that it is reinforced or blocked by parallel social activities, and that literacy acquisition cannot be divorced from literacy habits (Stromquist, 1992, p. 5).

Literacy is more than "just another skill"; it is bound up with oral and written communications in different languages and in different situations and contexts. It also includes numeracy. The acquisition of literacy, language, and numeracy skills is complex. For example, many women and men who have no schooling undertake business transactions on a daily basis and learn to know if they have received the correct change or not. Some nonliterates conduct complex business deals. How people acquire these skills is not understood well. New research into the social uses of literacy is exploring these issues in detail and should enhance our understanding of informal literacy learning (Prinsloo and Breier, 1996).

Additionally, the twentieth century has brought new means of communication, such as radio, television, and video, and it can be argued that they, too, can be "read." To be able to negotiate their way through life, people need to be able to "read the world," which means more than the learning and using of an alphabet.

The importance of literacy for women has been the subject of intense debate. People who are nonliterate are, in most instances, poor, but this does not imply that they are poor because they are nonliterate. They are poor because of the economic, social, and political relationships within which they are located. While this is true, based on a number of international case studies, Bown (1990, p. 4) reports on the significant social, economic, and personal effects that literacy can have for women. Among the social effects: Literate women are more likely to make use of child health-care techniques in the home and are more willing to present children for immunization. The economic effects include women's greater capacity to run small businesses and keep their own records. The personal effects of literacy on women include an increase in self-confidence.

The social and political context determines when and how literacy programs are meaningful for women. Although nonliterate women might like to become literate, relatively few manage to fulfill this wish because of overwhelming constraints. These include time-consuming duties, shortcomings in the design and delivery of the literacy program, and resistance from men.

Skilling for the Informal Sector

The informal economy is a critically important source of economic growth and employment for burgeoning Third World populations. This sector is growing in response to the globalization of the economy and its impact, including the structural-adjustment programs of the World Bank and the International Monetary Fund. Economic policies in the late twentieth century are leading to the downsizing of governments and their social and welfare programs and to growing numbers of jobless people. In order to survive, more and more people are engaged in the informal sector.

Women's engagement in this sector has a long history and includes such diverse activities as beer making, hawking fresh and prepared foods, and working in the sex industry. The informal sector can be differentiated into a subsistence segment and an enterprise segment (McGrath and King, 1995, p. 1), which also represents two corresponding types of self-employment. The enterprise segment is the upper tier of the informal sector where the self-employed may be thought of as microentrepreneurs; the subsistence sector is the much larger, lower tier where the self-employed are just surviving. It is this latter tier in which women are primarily involved.

Teaching skills, or skilling, for the informal sector most often occurs through informal learning, through active involvement on the job. In some parts of the world, for example in West Africa, there are highly formalized traditional apprenticeship systems. In Egypt, 83 percent of craftspeople acquire their skills through traditional apprenticeships. However, these seldom include women because the crafts are traditionally considered men's work. Training for the informal sector all too often has a male bias because women are seen as marginal or as invisible in the employment market. Women find themselves in forms of self-employment that require almost no training.

Where nonformal training is given for women in the informal sector, it often tends to be in the sphere of their traditional activities, such as sewing, knitting, and crocheting. The areas of training that researchers identify as being relevant relate to financing (either accessing credit or managing books), business skills, marketing, technological development, and organizational skills. Women in the informal economy rarely have ready access to these forms of training.

Skilling for the Formal Labor Market

Skilling of people for the formal labor market takes many forms, ranging from basic skills to technical work involving complex technological machines and instruments. Skilling may be directed toward people who have never worked in the labor market and have limited, if any, education. It may involve upgrading an existing labor force or skilling the unemployed labor force either to enter the existing labor market or to work more effectively in the informal sector.

Wolpe (1994, p. 17) notes that it is important to understand what constitutes "skill." Too often, skill is taken to refer to an accepted level of expertise and qualification in the labor process, such as a motor mechanic or an electrician. But what "skill" comprises is far more complicated than that and relates back to the issue of the social construction of knowledge discussed briefly at the beginning of this essay. Skill can refer to everyday knowledge of survival tactics. For example, a woman living in informal housing requires a range of competencies to feed, shelter, and clothe her children, though these are not normally recognized socially as constituting skills. Skill may be defined socially, and this may or may not be related to the labor process. For example, a woman who cooks meals every day for her family is not necessarily defined as having a skill. But a person (usually a man) who cooks for a salary, is defined as having a skill. The latter is called a chef, and the former a housewife who works in a way that is simply expected of her.

In countries like Britain, Australia, and the United States, skills learned in the home or in the community are sometimes being recognized toward further educational qualifications or for employment. This has been particularly important for poorly educated women entering or returning to work after raising children. Since the mid-1980s, there has been an explosion of access courses to facilitate entry into higher education. It is mainly women who have used these (Coats, 1994). While these opportunities are not widespread in Third World countries, they are being developed in places like South Africa and will in time assist women in developing countries to further their education.

The upgrading of skills, or reskilling, is mainly provided by employers. The extent to which this training is directed toward women at the lower levels of employment is likely to be limited. There has been in the 1990s a great deal of emphasis on training in the workplace to prepare workers to cope with changing, technologically driven labor processes. There is also an important critique of this emphasis—that the problem is not workers' levels of skills but the lack of jobs and of policies pushing toward full employment.

Upgrading for formal employment also takes place through "second chance" general-education programs. These programs are often state run and follow the curriculum of the schools. In many countries, for example in South Africa, the majority of the students are women who are upgrading their teachers' qualifications. The classes are usually run in the evenings and, not surprisingly, have high dropout rates. Women experience innumerable difficulties trying to combine familial responsibilities with studying. The threat of physical abuse in getting to and from the classes on unreliable transport is also a major disincentive for female students.

Skilling for the formal and informal sectors of the economy is affected directly by global economic and political developments. In parts of the world, trade unions and other popular groups have developed organizational and educational strategies to confront the appalling conditions of female workers, such as exist in the free trade zones of Malaysia or Mexico. These relate to the category of political and cultural education.

Political and Cultural Education

To change the social, economic, and political positions and conditions of women, women's movements and women's activism have grown stronger, more varied, and more accepted as indigenous expressions of women's interests. In the 1960s and 1970s, nationalists in the less industrialized world and Marxists worldwide often discredited women's movements as imperialist or bourgeois; they voiced concerns about the applicability of feminist ideas originating in Western Europe or North America. But some women in these societies have developed their own awareness of male domination and its relevant cultural forms, and in the 1990s in most countries there are vibrant groups of women vigorously inventing ways to respond to women's subordination. Simultaneously, there are strong indigenous movements to reinforce and maintain the gendered power relations as they are.

Growing recognition of the many national and community-based forms of women's activism and of the fundamentally plural nature of women's understandings of their problems has occurred simultaneously with increased international communication, and cooperation among women developed in ever-widening networks, committees, and caucuses. These groups, born of common interests and limited resources, have sustained international efforts to improve women's place in economic development, to mobilize opposition to violence against women, to encourage collaboration among scholars, and to facilitate hundreds of other causes defined by women as important to women and to the world. A critical part of women's organizing has been the informal and nonformal education that has been integral to it. The organizational and educational work that contributed to the U.N.-sponsored Fourth World Conference on Women (Beijing, 1995), is illustrative of this.

Political and cultural adult education takes place mainly within voluntary associations, including community-based organizations, nongovernmental organizations (NGOs), social movements, trade unions, religious organizations, and political parties.

Feminist Popular Education

Nonformal and informal education that is concerned with women's empowerment can be described as "feminist" whether it is concerned with workplace skilling, survival, or political action. "Popular education" is not a term used across the board, but it has gained an international currency and ascendancy among organizers and educators who have been influenced and inspired by the late 1960s writings of the Brazilian educationalist Paulo Freire (1972). It describes NFE practices that aim to challenge injustice and oppression and are variously called "community education," "radical adult education," "education for change," "liberatory education," "transformative education," and "education for empowerment."

It has a core orientation that can be described as a participatory, democratic, nonhierarchical pedagogy that encourages creative thinking that breaks through embedded formats of learning. It valorizes local knowledge and works collectively toward producing knowledge by starting from where people are situated and working to develop a broader understanding of structures and how they can be transformed. It strives to foster both personal and social empowerment. Feminist popular education obviously focuses particularly on the conditions and positions of women and the renegotiating of gender relations. But, given that gender is a social category, referring to the historically and culturally defined constructs of masculinity and femininity, feminist popular education must simultaneously engage with the ways in which the social categories of race, ethnicity, culture, age, social class, sexuality, and physical ability are implicated in constructions of gender.

Feminist popular education developed in the early 1980s as a conscious attempt to challenge the male bias in much of the popular-educational activities. It describes a host of nonformal and informal organizational and educational practices within NGOs, community-based organizations, and social movements in different parts of the world that consciously focus on changing the subordinate position of women.

There are a number of themes debated within the literature that have relevance for feminist pedagogy broadly and feminist popular education specifically. They resonate with complex, contested issues in feminist and educational theory. Some of these are: (1) consciousness-raising and the use of experience as a basis both for social analysis and for validating how women see their lives; (2) the acknowledgment of differences as a vital consideration in the design and facilitation of educational workshops; (3) social activism; and (4) the position of the educator. Before exploring these themes, some brief background on the literature on feminist pedagogy may be helpful.

Feminist pedagogy is part of, and an elaboration of, critical pedagogy. Critical pedagogy in general is concerned with transforming the position of the oppressed. Feminist pedagogy deepens and extends this with a particular focus on improving the position of women. It developed as a critique of critical pedagogy, which in the past largely ignored gender as a key social category.

The theoretical discussions of feminist pedagogy seem to have emanated mainly in conjunction with the growth of women's studies courses at colleges and universities in North America and Western Europe from the late 1970s. This has limited much of the discussion to pedagogy within formal educational institutions with students enrolled in a formal degree program. It has also constrained feminist pedagogy within the need to deal with the contradictions and difficulties of work in formal institutions. However, in South Africa, the Philippines, India, and elsewhere, innovative feminist popular education is taking place. Vivid descriptions of these practices are captured in Walters and Manicom (1996).

Defining what feminist pedagogy is, in practice, is not easy. On the one hand it speaks to the gendered character of the classroom and the curriculum. It is about teaching from a feminist perspective. But it is more than "good teaching" and is concerned with contributing to changing the subordinate position of women. The key intention of feminist pedagogy is providing learners with the skills to continue political work as feminists. Feminist pedagogy across the range of political perspectives echoes the struggles of its origins and retains a vision of social activism. It usually reflects this critical, oppositional, and activist stance.

Consciousness-Raising and the Use of Experience

Because of the power relations within society, women's knowledge is often trivialized and women lack confidence in what they think, feel, and do. It is a central task within popular education to raise women's consciousness of power relations within the society.

A fundamental aspect of consciousness-raising is reliance on experiences and feelings of women in order to affirm what they think, know, do, and feel. During the 1960s and 1970s, as women's movements grew around the world, this focus on women's own experiences came from a profound distrust of accepted authority and truth. The need to challenge patriarchal structures, which had defined common sense, meant that women had nowhere to turn but to themselves. Another aspect of consciousness-raising was the sharing of experience in a "leaderless" group, a process similar to testifying in the Black church in the United States and dependent on openness and trust among those participating. The assumption underlying this sharing of stories was the existence of commonality among women.

In Latin America, the consciousness-raising process has often begun with "daily lived experience." This emphasizes the importance of starting with the here and now, with the most immediate experience, including all aspects of life, at home, at work, in the community, in organizations. Through this, the many contradictions in society are made evident.

Consciousness-raising has also been closely linked to political action. It has been seen as both a method for arriving at truth and an impetus to organization and action. What was original in consciousness-raising was its emphasis on experience and feeling as guides to theoretical understanding, an approach that reflected the reality that women's subjectivities and the conditions of their lives were socially defined. However, at some stages and in certain contexts, consciousness-raising groups tended toward a loss of political perspective and too narrow a focus on the individual, to the detriment of political activism.

Some have pointed out the contradiction in an approach that treats experience and emotion as sources of knowledge, on the one hand, while acknowledging, on the other, that both experience and emotion are socially constructed. "Experience" is a highly contested concept within contemporary feminist theory. It is argued (Walters and Manicom, 1996) that privileging women's experience as a basis of knowledge tends toward an essentialist construction of "women," for that experience tends to be lodged in some foundational definition of women, such as reproductive capacity, or mothering, or the gendered division of labor. What about women who are differently located? How are the different claims to truth mediated? Appealing to "women's experience"—that is, tying experience to a particular identity—carries the danger of reproducing the ideological systems that define "women" and their world in particular and limiting ways.

There are further issues. When women's experience is given epistemological status as the standpoint from which an overall analysis can be built, there is the problem of accounting for the limitations in perspective that are assumed to be the consequence of an oppressed and excluded status. One of the rationales for promoting women's experience as a source of knowledge is to counter the masculine and class dominance of rational, abstract modes of analysis. But all experience and knowledge, it is argued, are mediated by language and discourses that shape what is possible to know and speak about. Experience is not a pristine category.

These complex questions translate into strategic issues for feminist popular educators. The resolution that is offered is that human beings are not completely shaped by dominant discourses, that we retain the capacity for self-critique and may challenge our own ways of feeling and knowing. If we gain distance, either physical or historical, and alternative perspectives on our realities, and if we recognize who we are and how we are positioned in the world, validation of feeling can be used to develop powerful sources of politically focused feminist education and action. An example of how this understanding has been translated into popular-educational practice is a set of carefully constructed exchange visits between people living in informal settlements in India and South Africa. Through physically gaining distance from their own situations in a structured program, they have been able to gain very different insights into their own situations and have since adopted different organizational strategies to secure housing for themselves.

Many feminist popular educators work actively to integrate the emotions, the intellect, the body, and the spirit to deepen consciousness, reconstruct the self-esteem of individual women, and build solidarity among them. Body consciousness, spiritual connectedness, and acknowledgement of the importance of emotions are all seen as integral to the consciousness-raising process. They all contribute data that are important reference points for the introduction of theoretical understandings, for the identification of commonalities and differences among women, and for building solidarity and a sense of a learning community.

The Question of Differences

Black women, poststructuralist feminist theorists, and critical Third World feminists converge in their critique of the concept of universal women's experience. This notion of a unitary and universal category of women has been critiqued for, among others, its racist assumptions. In addition, feminist theorists influenced by postmodernism have pointed to the need to consider the social construction of subjectivity and the "unstable" nature of the self. They argue that it is when individual selves are viewed as being constructed and negotiated that we can begin to consider what those forces are by which individuals shape themselves and by which they are shaped. The category "women" itself is challenged, seen more and more as part of a system of ideology. Critical Third World feminists point also to the imperialist assumptions behind the notion of universal women's experiences. Black women, lesbians, and women from the Third World ground such critiques on analyses of their experiences, which reveal that it is not only sexism that must be considered and dealt with by feminist theory and practice but also racism, homophobia, class oppression, and imperialism.

Investigation of the experiences of women leads to a view of the world that both acknowledges difference and points to the need for an integrated analysis and practice based on the fact that the major systems of oppression are interlocking. Acknowledging the reality of tensions and differences among women does not mean abandoning the goal of social justice and empowerment for all women, but it does mean recognizing that claims are contingent and situated and, at the same time, turning a critical eye on how our own histories and selves are constructed. "We need to work collectively to confront difference, to expand our awareness of sex, race and class as interlocking systems of domination, of the way we reinforce and perpetuate these structures" (hooks, 1988, p. 25). She and others still look to consciousness-raising as a key element in the educational process, but it is done with a more developed theory of pedagogy and an acute consciousness of the importance of difference.

A complementary methodological approach to the question of difference calls for coalition building. Difference is recognized and validated, but so is the need for mobilization around common goals. Some women argue that coalitions are necessary to successfully face the growing power of the Far Right, nationally and globally. Nevertheless, coalition politics does not remove the need for particular groups of women from different locations to organize themselves separately.

Social Activism

In addressing women's oppression, feminist pedagogy operates from the assumption that the status quo must change. While political activism is assumed in both Freire's work and the literature on feminist pedagogy, there is little discussion on what this might entail or what theories of social transformation underlie particular forms of social action. A prime source of contestation over the meaning of popular education is, of course, the political dimension, the implicit social vision. Who is the "subject" of popular education and what is the goal?

In the 1970s and the early 1980s, popular education tended to be linked to, and closely associated with, a clearly defined social movement, national liberation struggle, or women's movement, and a sense of the end point of "liberation." In the 1990s, particularly since the demise of formal socialism and the emergence of feminist and postcolonial critiques of the unitary narratives of such movements, social opposition has become more fragmentary. The building of civil society and of relations with states is much more the context of feminist popular education, although this obviously varies from country to country.

Broadly speaking, the politics of feminist popular education can be defined as the struggle against gender oppression. But, since gender has been understood increasingly as constructed in relation to race, class, and so on, feminist popular education has been working to integrate all aspects of power inequalities structured along social identities. It is, therefore, positioned to support the struggles of women in oppressed communities, rather than "women" in general. It is this that distinguishes feminist popular education from, for example, gender training or feminist pedagogy, both of which tend to be conducted for an undifferentiated category of "women."

The form that social activism takes clearly will be shaped by the specific context within which feminist popular education is operating. In many situations, particularly within formal educational institutions, learners participate as individuals. Social activism in these circumstances is limited in most cases to the development of critical consciousness among the individuals and to their individual interpretations of what they have experienced. In informal educational settings, particularly when participants come from political, cultural, worker, or other social organizations, the curriculum can include space for detailed planning and strategizing for collective social action. In certain situations, the line between education and social activism will be blurred. An example from South Africa is the organizational and educational efforts undertaken in 1993 and 1994 to draw up a women's charter that was to be integrated into the new democratic constitution. Activities included participatory research, mass rallies, discussions on the charter and its relationship to the constitution, production of popular media, and TV talk shows.

The Position of the Educator

The position of the educator is of critical importance in feminist popular education (Walters, 1993–1994). The recognition that people are shaped by their own experiences of class, color, gender, imperialism, and so on has powerful implications for pedagogy, in that it emphasizes the need to make conscious the subject positions not only of learners but of educators as well. Feminist theorists in particular argue that it is essential to recognize that we cannot live as human subjects without in some sense taking on a history. The recognition of our own histories necessarily implies articulating our own subjectivities and our own interests as we try to interpret and critique the social world. It requires educators to acknowledge their power and privilege, although this will vary according to circumstance. For example, those who have the power and authority to grade learners in formal programs will clearly be in a different position from educators in nonformal settings where grading is not an issue.

Some feminist writing points to the need to articulate and claim a particular historical and social identity and to build coalitions on the basis of a recognition of the partial knowledge of our own constructed identities. Educators and learners need to recognize and actively acknowledge differences while building solidarity in the quest for the empowerment of all women. Thus, feminist educators are involved in a particular form of Freirean dialogue with learners, recognizing differences and commonalities of experience and knowledge at different points. The role for feminist popular educators working across difference is to listen to women of other positions in the world, reflect upon their own privilege and oppression, and act, with these women, so as to transform themselves and society.

In educational programs, an important way of taking account of differences among the learners and the educators is to ensure that the group of educators is representative of social differences. While this strategy carries the danger of essentializing social categories, it can be argued that it still has validity in certain contexts. For example, in South Africa where there has been institutionalized racism and educators are dealing with the challenges of racism in a mixed group of Black and White people, then it is important to have both Black and White educators.

Conclusion

Feminist nonformal and informal education are concerned with challenging gendered power relations in the quest to attain equality for women. The form and shape of the educational practices are influenced by their social purposes and the particular political, social, economic, and cultural contexts within which they are embedded. But while there are specific differences, there are also common themes within feminist popular education that transcend national boundaries and challenge feminist educators in their work.

Within patriarchal societies, the educational needs of women are not met adequately through either the formal, the nonformal, or the informal system. In most instances, nonformal and informal education obtain little support from governments. One reason for this is that it is women who make most use of such education, and women's role in the economy is not acknowledged. Additionally, education that is likely to challenge male hegemony is unlikely to be encouraged. Also, it is most often only formal, certificated education in recognized educational institutions that is valued. Nonformal and informal education are mainly located within civil society, in communities and families, and at the workplace. Given many women's restricted mobility, it is this community-locatedness that makes nonformal and informal education particularly important for women's learning.

In a world with shrinking formal employment, growing numbers of female-headed households, and increasing impoverishment, the need for women to be affirmed has never been more urgent. Nonformal and informal education are integral parts of women's quest for the economic, social, political, and cultural survival and development of themselves, their families, and communities.

Nonformal and informal education are part of social processes that can either reinforce women's subordinate positions or challenge and transform them. Feminist approaches to such education are, therefore, critical to women's learning within male-biased societies if change is to occur.

References

Bingman, Beth. "Women Learning in Appalachian Grassroots Organizations." In Shirley Walters and Lizi Manicom (eds.), *Gender and Popular Education: Methods for Empowerment.* London: Zed/CACE, 1996.

Bown, Lalage. *Preparing the Future: Women, Literacy, and Development.* ActionAid Report no. 4. London: ActionAid, 1990.

Coats. Maggie, *Women's Education.* London: Society for Higher Education and Open University Press, 1994.

Freire, Paulo. *The Pedagogy of the Oppressed.* London: Penguin, 1972.

hooks, bell. *Talking Back.* Toronto: Between the Lines, 1988.

Lind, Agneta, and Anton Johnston. *Adult Literacy in the Third World.* Stockholm: Swedith International Development Authority, 1990.

Luke, Carmen, and Jennifer Gore. *Feminisms and Critical Pedagogy.* New York: Routledge, 1992.

Manicom, Lizi, and Shirley Walters. "Feminist Popular Education in the Light of Globalization." In Shirley Walters (ed.), *Globalization, Adult Education, and Training.* London: Zed/CACE, 1997.

McGrath, Simon, and Kenneth King. *Education and Training for the Informal Sector,* vol. 1. London: Overseas Development Administration, 1995.

Prinsloo, Mastin, and Mignonne Breier (eds.). *The Social Uses of Literacy.* Cape Town: SACHED Books, 1996.

Stromquist, Nelly P. "Women's Literacy and the Quest for Empowerment." In Jeannette Claessen and Lillian van Wesemael-Smit (eds.), *Reading the Word and the World.* Oegstgeest: Vrouwenberaad Ontwikkelingssamenwerking, 1992, pp. 51–70.

———. (ed.). *Women and Education in Latin America: Knowledge, Power, and Change.* Boulder: Lynne Rienner, 1992.

United Nations Educational, Scientific, and Cultural Organization (UNESCO). *World Education Report.* Paris: UNESCO, 1993.

Walters, Shirley. "Training Gender-Sensitive and Feminist Adult Educators in South Africa: An Emerging Curriculum," *Perspectives in Education,* vol. 15, no. 1, Summer 1993–1994, pp. 115–132.

Walters, Shirley, and Lizi Manicom. *Strategic Learning: Reflections on Gender in Popular Education.* London: Zed, 1996.

Wolpe, Annmarie. *Adult Education and Women's Needs.* CACE Publications. Bellville, South Africa: University of Western Cape, 1994.

Movements for Change

Women in Anticolonial Movements

Virginia W. Leonard

Introduction

This essay discusses the protest activities of women in Africa, Asia, and Latin America over the centuries. Ancient empires are not included; only European colonies created after the sixteenth century are considered. These areas are very different, and knowledge of them is usually limited to one region or country. This essay is offered as a starting point for comparative research.

At the outset, the terms "colonialism" and "imperialism," which are often used interchangeably, need to be defined. Colonialism is the occupation or settlement of a colony by a parent country, wherein the people are subject to or connected with that parent state. Imperialism is the rule or influence of a people over a foreign people or country.

This essay reflects the availability and quality of published research on women in resistance movements in Asia, Africa, and Latin America. Countries with complicated histories of colonization were left out: China had many colonizers (multiple spheres of influence), and Vietnam faced occupation and rule by the Portuguese, then the French, and, finally, the Japanese. In addition, the story of overall female resistance has barely been touched on. Working-class and peasant women are often missing from documentary histories. Their activities often went unrecorded or their movements were violently suppressed. Resistance to colonizers by women in some of the African colonies also has been overlooked until recently, making it difficult to include them in this essay. Data are lacking even when cases are extensively reported: How many women were involved, how many women were in the opposition, and who remained uninvolved?

Colonialism did not suppress nationalism but caused it. These regions learned nationalism—devotion to one's nation or a policy of national aspiration and independence—from the West. Since the Portuguese trading posts (entrepôts) in Africa were not true colonies, Latin America is the first of all three areas to have been colonized in the

modern period. The Amerindian resistance to the conquistadors included women and was followed by early-nineteenth-century movements for independence. Latin America is also different because miscegenation created a mixed-blood population with new cultural identities, and there were two main parent countries, Spain and Portugal, rather than many parent countries as in Africa and Asia.

There were also indigenous traditions of resistance to conquest. The Aztecs, Mayas, and Incas resisted Spanish conquistadors. Judith Van Allen (in Hafkin and Bay, 1976) argues that in the case of Igbo women in Nigeria, political action was part of their culture and Kikuyu women who joined the rebellion in Kenya had a ritual oath to back up militancy. Indigenous traditions, languages, and especially religions contributed to resistance movements against the colonizers. In French North Africa, nationalism often revolved around resurrecting a Muslim and Arabic culture, one that resulted in de-Westernizing women and men.

In the cases of Africa and Asia, there is a growing literature on the role of Western women. European and North American women (Black and White) who lived in these regions interacted with native women. The reaction of Western women who remained at home makes the subject more complex. English women at home and abroad were proud of the empire, and they demanded civic equality for Indian and African women. They mistakenly thought that female exclusion (purdah) was the dominant social relationship between men and women in India (Engels, 1989). The nationalism and racism of the imperialists affected English women, and they consequently looked upon African and Indian women as inferiors and unfortunates in need of saving (Burton, 1990).

This patronizing attitude explains the lack of credit given Westerners for educating colonial women in Africa, Asia, and Latin America, then bringing them into the public sphere, or otherwise improving their lot. Like many men, many Western women also condemned the veil, the harem, polygamy, pawnship, slavery, self-immolation of widows, female infanticide, clitoridectomy, ostracism of mothers of twins, and child marriage. They strongly advocated the education of Indian women.

The formal education of women by colonialists was especially helpful in most of these societies and allowed many women to advance to professional occupations and become economically independent. However, many writers dismiss these measures as having "little positive impact on women's lives" (Engels, 1989, p. 426). This statement slights the role of Western women as educators, missionaries, nurses, role models, and political activists (some of whom were freethinkers), anti-imperialists, socialists and communists (Chaudhuri and Strobel, 1992).

The Great Depression of the 1930s seems to have been a watershed for the colonial territories, as revenues from exports and agriculture slumped. The Europeans reduced their social services but not their taxes, leading to female rebellions against taxes in eastern Nigeria in 1929 and in India and elsewhere in the 1930s. The "civilizing" purposes of colonial administrations were brought into question. The economic crises and population growth in the countryside forced increasing numbers of peoples to migrate to the cities, creating a constituency for Western- and Muslim-educated nationalists. A social struggle joined a national struggle for independence. After independence was won, the nationalists, especially the men of Islam called traditionalists, halted the social struggle, to the detriment of women, especially in Africa.

During these struggles for independence and social reform, the language of kinship, family, and gender is used by both sides. Political discourse involves family relationships and sexuality, and since women are defined in terms of their relation to men more than men are, one finds females in opposition groups labeled as "loose" or "whores." The sexuality of opposition females is vilified as masculine, lesbian, impotent, or homosexual. One's comrades are considered "sisters" or "brothers." British women are seen as mother figures, and Indian women as their daughters. Countries can be identified with a mother, such as Mother India.

The status of women after an anticolonial movement provides an excellent measure of how radical the movement was. Prerevolutionary patterns, which often seem to reappear after the revolution, can actually be different, and gender roles must be examined carefully to detect these differences. While most sources attribute the return to former values and gendered hierarchies in the public sphere to male discrimination and prejudice, recent writers see it more as an effort to relieve the anxieties and insecurities produced by a revolution. This return to patriarchy props up the indigenous traditions, religion, and culture that resisted Western colonialism while new political and social structures are created. This brings comfort to males and to some women who then cooperate in the reimposition of restrictions (Wasserstrom, 1994).

Every anticolonial movement in Asia, Africa, and Latin America enlisted the support of women in various ways and would not have been successful without their help. Men in patriotic parties mobilized female activists to fight colonial powers and promised them equality. Once in power, however, men took the most powerful positions and reasserted traditional gender roles, toning down the implication of their egalitarian nationalist ideology, which, in many cases, was adopted from the Western colonizer.

Asia

India

More than nationalist movements elsewhere, Indian social and political movements exhibited a class and caste nature. Affiliation based on status, kinship, and personal alliances led to factions that merged into a greater nationalist movement. The Hindi, Christian, and Muslim leaders of the anticolonial movement in India came from the higher classes and castes. Although Indian women were secluded in the upper castes, their authority and responsibilities in the private domestic sphere were not regarded as inferior but were thought of as complementary to male public roles since they preserved caste purity. Indian women were seen by their compatriots as important and powerful persons whose sexuality had to be restrained, and the degree of restraint separated the upper castes from the lower.

Furthermore, Indian history contained examples of strong women. The upper-caste *rani* of Jhansi (Lakshmi Bai), for instance, took up arms against the British when they refused to affirm her regency and her adopted son's right to rule and moved to cut her funds. In the wake of the Sepoy Mutiny in 1857, she recruited an army of 14,000 men to fight, and women to work the batteries and carry food, ammunition, and water when the British successfully laid siege to Jhansi in 1858. The *rani* escaped on horseback, wore military dress, and died in battle while defending the town of Gwalior from British attack.

Hindu religion includes powerful goddesses and priestesses and recognizes the importance of women. Little wonder, then, that Indian women attacked British colonialism with its male-dominated religion as one of the major causes of inequality. Unless colonialism were ended, it would be hard to tackle inequality, which women also ascribed to custom and the patriarchal organization of the family in both India and the West.

The British believed that they had helped Indian women by not only reforming their rights to own property but also by passing laws ending child marriage, widow burning, polygamy, and female infanticide. Other British laws, however, supported male prerogatives: Husbands had conjugal rights even if wives had fled them; prostitution was legalized (this allowed for inspections for diseases that might infect the army); and women's suffrage was denied. Women perceived British rule as oppressing them as females and as Indians. The struggle for nationalism brought women, especially Muslim women, out of seclusion and created a national movement for their uplift and emancipation.

In 1905, Hindu middle-class women were mobilized for the Swadeshi movement (the boycott of foreign goods) in opposition to the partition of Bengal by the English. While a high-caste woman, Sarala Debi Chaudhurani, received praise and veneration for her role, British officials imprisoned and Hindu society rejected Nanibala Devi, a lower-class widow, for posing as a wife to help revolutionaries. Overall, the anticolonial activities of Indian women helped garner male support for the incipient women's movement, and as early as 1919 some men supported suffrage to show they favored equality more than British males did.

The women's movement consisted of urban, English-educated women of the new middle classes and castes; it did not use the term "feminist" because the nationalist question took priority over women's issues. Although foreign women helped initiate the women's organizations, these groups grew in strength because they coincided with the cultural heritage of Indian women as strong and powerful and the newer possibility of their being independent economically because of entrance into new professional careers (Liddle and Joshi, 1986).

In 1904, the National Social Conference began a separate Indian Women's Conference that met annually and organized women's educational programs. International events and persons affected the founding of three more organizations that became the largest and most influential in the women's movement.

In 1917, Margaret Cousins, Dorothy Jinarjadasa, and Annie Besant started the Women's Indian Association and linked it to the British suffragettes. This took place in the context of World War I, economic dislocation, expanding self-government, Mohandes Gandhi's movement for independence, and the anticolonialism of the United States. In 1925, the National Council of Women in India was founded by Lady Mehrbai Dorab Tata and Lady Ishbel Aberdeen as a branch of the International Council of Women. In 1927, the All-India Conference for Educational Reform was established with Margaret Cousins as chairperson; the *rani* of Sangli, Lady J.C. Bose, and Sarojini Chattopadhyaya as vice presidents; and the maharani of Baroda as president. In 1928, it was attended by Muslim women, changed its name to the All-India Women's Conference (AIWC), and appointed a Muslim president, the *begum* mother of Bhopal, Maimoona Sultana. The *begum,* ruler of a small central Indian state, promoted and financed the education of women in the Muslim community and throughout India. At first, conferences dealt with women's education, but women of the Communist movements and Indian National Congress pushed to adopt the political issues of sexual equality, votes for women, and reform of personal laws. The AIWC later brought in the Women's Indian Association and became the largest women's organization in India (Jayawardena, 1986).

In the early 1920s, women joined male nationalists in noncooperation, and in the 1930s they joined in rural protests, *satyagraha* (nonviolent resistance) campaigns,

revolutionary activities, and civil disobedience. During the civil-disobedience movement (1930–1931), women joined Indian National Congress committees, edited banned newspapers, demonstrated, picketed shops selling foreign goods, drugs, and liquor, and laid down in shop doorways, railway tracks, and roads. They also broke the Salt Laws by marching to the sea to make salt, thus flouting the government's monopoly on salt, and they burglarized salt factories. Rejecting nonviolence, some women became active in the Indian Communist Party and terrorist organizations in India and abroad. Ushanti Dange, married to the communist leader, organized militant strikes at textile mills. Others helped the *bidi* (cheroot) workers strike and organized female slum dwellers in the city of Bhajan Mandals. Thousands of women were imprisoned.

The entry of women into the struggle for independence created a mass movement. During World War II (1939–1945), women continued the anticolonial struggle, and they carried out relief efforts during the Bengal famine of 1942–1944. The All-India Student Federation set up a Girl Students Committee to organize political activities of young women all over the country; they were most active in Bombay, Bengal, and the Punjab. By 1941, there were 50,000 members, and some of their leaders were arrested. In 1942, left-wing activist women formed the Women's Self-Defence League (MARS), which worked for the release of Gandhi and other national leaders from prison, in addition to engaging in relief work during the famine. By 1944, MARS had 43,000 members (Jayawardena, 1986).

Male leaders differed in their views of these new activities. The more successful restricted them so that they were merely an extension of the women's private sphere in order to make their activities socially acceptable. Subhash Chandra (Netaji) Bose classified women as sisters and mothers since he was then living in celibacy and did not accept sexually active women (married) as equals. (Bose became infamous during World War II as India's only nationalist leader to affiliate with the Axis.) Gandhi restricted women to supportive jobs, and in the 1930s he had channeled the nationalism of women to picketing the drink and drug shops. He later found women better suited than men to passive resistance because of their "moral power," noting: "The role women played in the freedom struggle should be written in letters of gold" (Liddle and Joshi, 1985b, pp. 161–162). Jawaharlal Nehru, the most progressive of all, advocated equality in the family, education, and careers for women. Indian males began to understand that female participation countered British stereotypes of Indian males as oppressors of females and in need of Western civilizing; Britain had lost its moral cause to the national liberationists (Liddle and Joshi 1985b; Engels, 1989, pp. 432–433).

Muslim women in India were subject to the same forces as Hindu women were. In many ways, they were better off: They did not suffer from the same binding Hindu laws that forbade divorce and widow remarriage.

Muslim law allowed women to inherit property and denounced self-immolation of widows. The *begum* of Bhopal and other leaders overcame purdah in the course of their lifetimes to enter the public arena and, later, to address mixed audiences after first working with local women's associations. The All-India Muslim Women's Conference, established by the *begum* in 1916, went on record the next year in opposition to polygamy and received male criticism in return. Muslim women entered the *satyagraha* and non-cooperation movements, and they had male support in the independence movement for women's education and for suffrage (Jayawardena, 1986).

The British and Indian authorities were hesitant to club and arrest middle-class women during demonstrations in front of an urban population. Peasant women, however, were less fortunate. They were beaten and fired upon by the police in rural villages during a tax protest. The authorities attributed this "deviant" behavior of the protesters to either "loss of reason" or male instigation. When upper-class Indian women were arrested, authorities questioned their femininity and respectability and characterized their menfolk as henpecked or "unchivalrous." However, the nationalist movement was strengthened when a congressional committee began an inquiry in 1930 into police violence against women in Contai. The press publicized the inquiry and the Bengal superintendent of police instructed all officers to treat "lady *satyagrahis*" (nonviolent resisters) politely. After women began to mingle with men in marches in 1930, the police began to arrest women and appealed to men to keep their women at home. This approach worked for a while: The number of female demonstrators declined (Engels, 1989, pp. 433–435).

However, when Gandhi led the Salt March in 1931, an event that began the era of civil disobedience, he was arrested, and the British outlawed the Indian National Congress. Women emerged to replace the arrested male leaders. (The Indian National Congress, founded in 1885, allowed female membership.) Women repeatedly faced beatings, arrests, and bullets. Notable female leaders included Sarojini Naidu, who directed the salt protest after Gandhi's arrest, and Kamaladevi Chattopadhyay, who rallied women to surround Gandhi's male escort as British cavalry were about to ride them down; British troops refused to charge. Of the more than 80,000 persons arrested during the salt campaign, more than 17,000 were women (Liddle and Joshi 1985a, pp. 525–527, 1985b; Engels, 1989).

After independence in 1947, women gained equality under the law but were not treated as equals in practice. It took the power of Prime Minister Nehru to get laws assuring equality through Congress in 1955–1956. However, two measures failed: Housework was not included in the national accounting process, and the constitution did not guarantee equality in marriage. Even in the 1990s, the laws are not enforced by the state or by women, many of whom are poor and illiterate and do not know their legal rights (Liddle and Joshi, 1986).

Ironically, though Gandhi had successfully appealed to the lower castes, it was urban, middle-class women who gained equality after independence. Revisionist writers today emphasize the conservative nature of the gains for women. The subordination of women did not end after independence in spite of their achieving legal equality (Liddle and Joshi, 1985b; Jayawardena, 1986).

Africa

As in India, colonialism is the key to understanding the changing role of women in the many African cultures and states. As in India, too, colonialism combined with local African culture to keep women in a dependent state. The British, French, Portuguese, and Belgians used variations of indirect rule in sub-Saharan Africa, keeping African chiefs and headmen in place and using them to carry out colonial policies, surveillance, and taxation. The political power of women, where it existed, was dismissed and diminished. The Europeans were conscious of enforcing patriarchal rule because the chiefs and headmen often requested that they stop their wives and women from running away to the towns and becoming independent. European courts favored fathers over mothers, husbands over wives, and they granted custody of children to the father. In addition, the Europeans knew that the cheap labor of the women and children amounted to an agricultural subsidy for the labor of the males in mining and plantation agriculture. This cheap labor kept down costs of local commodities. Also, Europeans held to a sexual double standard and wanted women controlled to cut down on marital problems, prostitution, and venereal disease.

Black African women had to struggle against European administrators who were upset by what they branded "loose sexual practices," as well as by female political and economic power. Unlike most Asian women, African women did not live in seclusion unless they were wealthy Muslims. Colonial policies affected African women in contradictory ways, but the long-range effect was to bring them more under male domination. Under colonialism, African women lost political and economic status. For example, many African women once had pawns (persons who offered labor to pay off an economic or social obligation) and slaves (Hafkin and Bay, 1976).

African women often ran away to towns and entered liaisons with men of their choice, or they entered lucrative but illegal occupations such as prostitution and beer making. European and African males denounced women who remained independent of men and earned handsome incomes in their chosen occupations. Another place of refuge for a woman was the mission or convent. However, state law demanded that male guardians give their consent to these brides of Christ. The priests let them work in town as nannies so they could reimburse their families for the loss of any bride price or bride wealth (Schmidt, 1991).

In the 1960s, British West Africa was begrudgingly granted independence with little struggle. There, the British had prepared a political class of males to rule. Ghana set the precedent in 1957, and Nigeria became independent in 1960, Sierra Leone in 1961, and Gambia in 1965.

In contrast, the French were determined to retain their territories after 1945. Disaster and withdrawal in Indochina persuaded France to hand over power to political moderates in Tunisia and Morocco, which became independent in 1956. Algeria was a different story because it had been considered part of France, and the French began severe repression after 1945. French resistance in West Africa and Equatoria was mild in comparison. By 1960, all 12 territories had become nation states (only Guinea had voted for outright independence; the others were associated with France). Togo received independence in 1960, but a nationalist and socialist insurrection based on mass participation in Cameroon was militarily repressed by the French military aided by Muslim allies in the north.

Struggles for independence took place where British settlers occupied much of the land and considered the Africans to be savages—Kenya, Rhodesia, and South Africa. Settler minorities had governed Southern Rhodesia since 1923 and South Africa since 1910. The British East African countries of Tanganyika, Uganda, and Zanzibar produced mass movements resulting in independence in the 1960s.

In the lesser empires, the story is mixed. Italy lost its colonies during World War II. Ethiopia was liberated by the British in 1941; Libya gained independence in 1951; and a combined Somalia in 1960. The Belgian Congo continued to be exploited by the Belgians, but after police fired on crowds in the capital, the decision was made to grant independence by 1961. Belgian Ruandi-Urundi was divided and became independent Rwanda and Burundi. With little preparation for independence, all three countries suffered terrible civil wars. The Portuguese colonies of Angola, Mozambique, and Guinea-Bissau were scenes of mass participation in movements headed by male, and a few female, leaders considered "assimilated" (*assimilados*) to Portuguese civilization. Although influenced by radical and Marxist ideas, they had a respect for their own African civilization. It is in these peasant societies of Portuguese Africa that the contributions of women to anticolonial movements are extolled. The governments set up after independence are cited as models for more equal treatment of women (Hafkin and Bay, 1976).

Belgian Congo

The Belgians used taxation to support their Roman Catholic faith. Polygyny grew after World War II in rural areas because men needed plural wives to work the fields and increase the harvest revenues to fulfill the agricultural, health, and other *corvées* (forced labor). The elders and chiefs of tribes monopolized the women by paying higher bride prices to their parents, and they often lent the women to bachelors for a fee or an even higher bride price. The proliferation of polygyny pushed many men to the cities, where they frequented prostitutes or took concubines in pseudo-marriages, or "camouflaged polygamy" (Hunt,

1991, p. 140). Monogamous fathers of four children received family allocations and head-tax exemptions. Polygamous men had to pay an extra tax on each wife. In 1951, a law was passed to penalize polygamy and allow for the expulsion of polygamous men and their wives from urban residence. The Belgian authorities wanted to control the movements of the population and end polygamy and prostitution. In the 1930s, they had passed laws and taxes to control internal migration. African women's rights to visit or reside in a city were based on their husbands' permission or rights. Single women (divorced, widowed, or never married) were reluctantly given these rights and taxed as single males. It was thought that single women could afford the tax because they must be supporting themselves through the illegal means of prostitution or beer making. Such women were considered prostitutes even if they were not. This tax was criticized especially in the Muslim quarter of Usumbura, where honest single women, and the "camouflaged" plural wives who were claimed as single women to avoid the tax on polygamy, felt dishonored by the word "prostitute" (*malaya*). In 1955, the women began a protest and were supported by the men, who in 1957 also revolted against paying the polygamy tax. To soften this tax rebellion, the Belgian authorities exempted some women. The practice of surveilling women and men to carry out these taxes and the sexual labeling, abuse, and taxation of single women had resulted in a rebellion begun by women (Hunt, 1991).

Southern Rhodesia (Zimbabwe), Nyasaland (Malawi), and Northern Rhodesia (Zambia)

European plantations and mines needed laborers. A state system of migrant labor stripped the countryside of adult males, leaving behind women and children to fend for themselves. The older males who remained asserted control over women and monopolized their labor and marriages, receiving bride wealth and controlling kinship alliances. As wage earning for men became more important than food production, the social status of women was further undercut. African chiefs, headmen, and older men often profited from their connections with the colonizers at the expense of women. European males did not hold any women in high esteem, and their racial prejudice placed Black women on the bottom. Male administrators and missionaries blamed sickness, adultery, and venereal disease on African women.

Kenya

When nationalist leaders headed by Jomo Kenyatta were imprisoned in 1953, armed revolt against the British broke out to gain land and independence. Kenyatta headed the Kenya African Union, a political party connected to the secret society of Mau Mau. Peasant and urban men and women took separate oaths of secrecy in ritual ceremonies. Men were expected to be warriors, and women were to act as scouts, nurses, and spies (Itote Cartey and Kilson, 1970, pp. 111–117). The British army defeated them by 1956 and forced more than one million peasants into camps and designated villages.

Guinea-Bissau

The African Party for the Independence of Guinea and Cape Verde (PAIGC) succeeded in ending 90 years of Portuguese rule in 1974 after fighting a guerrilla war for nearly 12 years. It was the first guerrilla movement in sub-Saharan Africa to force a European power to withdraw. Related struggles were fought simultaneously in Angola and Mozambique against Portuguese rule. Women were consciously involved in these independence movements.

The emancipation of women was a goal of the PAIGC. Its socialist leaders believed that women needed to fight two colonialisms, of the Portuguese and of their male countrymen (Urdang, 1979). Women had to counter three local customs: polygyny, forced marriage, and divorce with no custody of children. Muslims made up one-third of the population of one million in 1960 and had the most restrictive rules for women.

At first, passive resistance, protests, and strikes were severely repressed by the Portuguese, resulting in counterviolence. The revolutionary PAIGC recruited mostly men as soldiers, which postponed any radical changes in the economy and in male–female relationships. In a country where 95 percent of the people were peasants, and women were denied freedom of movement, male mobilizers were sent to the peasant villages to recruit. When mobilizers spoke to the villagers about war against the colonizers and urged women to act to change their lives, women often joined the PAIGC. In turn, they mobilized others in their villages. They also fed the mobilizers and later the guerrillas, often their own menfolk, hiding in the forest. They cooked and prepared food, gathered supplies and water, and marched long hours to the hidden base camps. At great risk, they gathered information of Portuguese troop movements, and some women became trained as guerrillas (Urdang, 1979). Women also worked as nurses for both sides of the war; the Portuguese relied on African soldiers and nurses. Attracted by the promises of the PAIGC to work for their emancipation, women were some of the earliest supporters of the anticolonial movement. Their emancipation was part of the party ideology.

However, national independence did change the role of women. The PAIGC ordered all villages to elect a five-member council that included two women. Women in Muslim areas, with 30 percent of the population in the 1970s, made the greatest relative progress, leaving their secluded homes to attend school and serve on village committees and council meetings as socially acceptable officers of women's affairs. The state invested in the economy, began large-scale projects, interfered to control exports, and set prices for sale of crops. Production suffered, smuggling of rice, onions, and tomatoes by individuals across borders for higher prices increased, and thousands migrated to cities. Women had less money with which to purchase family food supplies. International aid propped up the social-ist state that increasingly distributed money to the capital of Bissau while the rural population became alienated from the PAIGC.

In 1980 "the rice coup" took place, when peasants and others backed a military leader of the PAIGC who took office by force and began to slowly abandon statist policies in favor of privatization of the economy. Elections were held in 1989 for regional councils and the National Assembly. Voters elected a single list of candidates, of whom 30 percent were women. Foreign aid continued to provide food and money. In 1991, a revised Constitution allowed a multiparty system, however PAIGC still dominated other parties in the 1994 elections (ACR, 1992; Bigman, 1993).

Women made up a majority of the 1.1 million population in 1996, four-fifths living in rural areas. The village economy still depended on controlled female labor, making polygamy an institution the PAIGC could not root out. The other two customs of forced marriage and taboo against divorce were whittled away by decrees, and laws after 1974, that the people's courts administered. Educational and training programs favored males, and as of 1996, twice as many boys as girls were enrolled in secondary school (Urdang, 1979; PRB, 1996).

Nigeria

Nigeria consisted of three different geographical areas and groups. In the Southeast were the Igbo (Ibo) whose women waged what they called the Women's War, 1929–1930. The British called the rebellion the "Aba Riots," to cover up what essentially was a protest against taxes at a time when profits from the palm-products trade were declining. The Igbo consider it a women's war because the women used traditional methods of "sitting on a man," that is, meeting at his hut, singing songs detailing their grievances against him, insulting his manhood, and banging on his hut with pestles for pounding yams until he repented.

In November 1929, thousands of Igbo women ridiculed and demanded the caps of office of the warrant chiefs of 16 native courts in Calabar and Owerri provinces. They were aroused by a rumor that women and their possessions were about to be taxed. They vented their fury on the Igbo men who served the British as warrant chiefs, who were perceived as corrupt. The women had acted in concert due to the strength of their grass-roots market network. They had met in market squares to solidify their demands. They signified war in the traditional manner, dressing in loincloths and carrying sticks wrapped with palm fronds, smearing their faces with ashes, and binding their heads with ferns to invoke the power of female ancestors. Their ancestors' power did not work when British police and soldiers fired on them on two occasions in mid-December, killing 50 and wounding another 50. British punitive expeditions razed villages well into the 1930s.

The war had been directed at African males who served as warrant chiefs in native courts, and became extended to the native administration. Native administrators had taken power and economic resources from the women as three British institutions—the colonial church and its mission schools, the government, and foreign investment—subordinated females by preparing males for wage-paying jobs and governance. During the "war," the women did not injure anyone seriously to the surprise of the British Commission of Enquiry report issued after the "war." Women had expanded their demands that native courts be abolished or allow women to sit on them, that women serve as district officers, and that the British return to their own country. Igbo men had sympathized with the women, but knew they would be shot if they had joined them. They were shocked when the British fired on the women, because the women had protested according to traditional rules that limited their actions. It had not appeared radical to Igbo men for women to lead a resistance movement using traditional methods of waging "women's war," and they were winning until the British opened fire.

Afterward, the British continued to bolster their own and Igbo male authority. In 1933 British reforms enlarged the native courts to include more judges who were no longer warrant chiefs, and increased the number of native court areas, benefiting the women somewhat, but women still had no institution to represent them as women (Van Allen in Hafkin and Bay, 1976).

Algeria

After 1956, the revolt against French rule begun by a few men became a national movement. Women assumed a significant role smuggling ammunition and terrorist bombs under their robes, serving as nurses and spies, hiding rebels and feeding them, and engaging in combat as members of the National Liberation Front (FLN). They marched in demonstrations, suffered torture and rape, and were refugees in camps in Morocco and Tunisia.

The most famous heroine was Djamila Bouhired, recruited by her brother to be a terrorist. Her trial in France became a propaganda boon for the Algerian revolutionaries. Djamila Boupacha was also from the middle class and had been educated in France. After her capture, she was brutally tortured. French liberals such as the artist Pablo Picasso and the writer Simone de Beauvoir lobbied for her freedom, which she received with independence in 1962. At the same time, however, an amnesty was granted her torturers (Gordon, 1972).

During the revolution for independence, the French tried to appease the women, and possibly their men, by promoting the shedding of the veil. They also tried to increase female enrollment past the 118,000 Muslim girls in elementary and secondary schools in 1958. Welfare services and film propaganda were aimed at women. All of these efforts failed as nationalism and Islam proved more attractive to women.

Women kept the veil and remained in seclusion until the militant revolutionaries needed couriers and spies. At first, wives were recruited, then widows and divorcees, and finally unmarried women. If the women needed to operate in European sections, they discarded their veils.

Independence leader Ahmed ben Bella and his revolutionary cadre recognized their valor and promised them equality and political participation in the new government. Women voted and served as delegates to the National Assembly.

Two years later, the revolution began to turn to traditional Muslim ways. Women were not to be seen in the streets, and former maids of the French lost their jobs and slipped into prostitution. Women in Western dress were taunted in the street. The economic depression after the war of independence, and the overthrow of Ahmed ben Bella by the army in 1965, created flux and uncertainty. Scarce jobs were given to men. Two organizations attracted women, the National Union of Algerian Women (UNFA) and the General Union of Algerian Workers (UGTA).

The Algerian revolution proved to be a significant force for change after more than 100 years of French rule (the French first occupied Algiers in 1830). Although it reasserted traditional Muslim values, the military was still committed to modernization and a mild socialism. Education was available to more women than before, and they had the vote and more opportunity. The nationalists resurrected a Muslim and Arabic culture to replace what they believed the French had contaminated, and women were encouraged to stay at home. They served as the litmus test to prove whether the revolution was traditional or Marxist. These tensions have made women's lives contradictory and confused, as depicted in the novels of Assia Djebar.

Latin America

The anticolonial movement in Spanish-speaking Latin America took place earlier than movements in Asia and Africa, possibly due to the earlier colonization of Latin America. More than elsewhere, this colonization led to the mixing of races and the systematic use of both forced and slave labor. Ironically, it is in Latin America that the symbolism of the native woman dominated by a masculine conqueror was used to repudiate European conquest and imperialism, but not colonialism.

This is because differences of race and ethnicity impeded anticolonial movements, ideology, and symbolism. Unlike the movements in Asia and Africa, the shedding of Spanish rule did not mean the emancipation of Amerindian women from "foreign" rule. Instead, whiter men and women continued to dominate them culturally, economically, and politically in independent nations. (The terms "anticolonialism" and "anti-imperialism" are also used by rebels in Nicaragua, Puerto Rico, and Cuba to refer to their struggle against their governments and the United States. The distinction between the two terms was made at the beginning of this essay.)

Amerindian Women

The Amerindian women who resisted Spanish colonialism served in numerous capacities: They cooked, carried food, and fought as soldiers in some battles. In Tlatelolco, Mexico, Indian women fought against the Spaniards by hitting them and throwing darts at them. They wore military insignias and fastened their skirts above their knees so they could pursue their enemy (León-Portilla in Hernández and Murguialday, 1993, p. 84).

At the end of every battle, they mourned the dead, and if the battle had been lost, they suffered rape and distribution to the conquerors as spoils of war. The Indian women's miscegenation with Spaniards led to mixed-blood mestizo progeny. By the end of the colonial era in the viceroyalty of New Spain (Mexico and Central America), 40 percent of the population was mestizo, 1 percent was White, and a small percentage was African (Hernández and Murguialday, 1993).

Amerindian women resisted by running away; aborting fetuses fathered by the conquerors; keeping up Amerindian traditions, religions, and languages; and creating an underground culture as well as an above-ground mestizo culture. For example, the cult of an Indian-featured Virgin of Guadalupe was diffused by Indian women throughout Mexico. In the early nineteenth century, Pope Pius X declared her the patroness of all Latin America.

Race and Ethnicity

African women in the colonial period were burdened by slavery and racism, yet were allied with the Spanish, French, and Portuguese conquerors. They, too, were scarce in numbers since Black males were preferred as slaves in tropical regions. Slave women had little choice but to enter into sexual liaisons with their masters, and a huge cohort of mulattoes appeared. A great part of a slave's life could be spent working for emancipation for herself and her children. In pursuit of freedom, slave women worked extra hours, sold their bodies, had children by masters so that those children would be lighter in color (resulting in more possibilities or even freedom), and ran away. Quilombo (Brazil) and Maroon (West Indies) communities of hundreds of runaway slaves, both Indian and African, existed in Brazil and the Caribbean islands during the colonial period.

The Creole fear of mixed bloods, Indians, and Blacks overthrowing the caste system and gaining civil rights thwarted unified movements of independence in Mexico, Brazil, and the Andean region. Upper-class Creoles remained allied with the mother countries as long as these mixed-blood groups threatened their privileges. In late-eighteenth-century Peru, for example, peasant rebellions were "essentially family enterprises" that failed to attract Creole adherents. Micaela Bastidas Puyucahua, wife of José Gabriel Tupac Amaru, leader of a 1780 rebellion in Tinta, Peru, worked hard to secure clerical, White, and mestizo support. Bastidas, a Catholic, acted to protect churches and churchmen. Although women were considered subordinate to men in the colonial period, the trial and execution of Bastidas along with her husband and other Indian women leaders suggests that the Spanish viewed women as autonomous and legally responsible in instances of deviance and rebellion (Campbell, 1985).

Independence

The eighteenth-century Enlightenment and American and French Revolutions produced men and women in Europe and North America who championed equal rights for men and women. The Benedictine monk Benito Jerónimo Feijóo advanced these ideas in Spain, and they became known in Latin America. Articles appeared in the press speaking of the necessity of educating women so they would be better mothers and housewives. However, these Enlightenment ideas were meant to apply to upper-class White women, and they got a cool reception in most of Latin America. Argentina was an exception: Argentine women, who were more educated than women elsewhere in Latin America, did foment independence by exposing their sons and daughters to ideas of the Enlightenment. One author credited them with molding "the generation of 1810" that achieved independence (Furlong, 1951, p. 262).

An American-born (Creole) elite replaced the French, Portuguese, and Spanish elites by 1822 in Latin America save in many of the Caribbean islands, Belize, and the Guianas. Elite-control made these revolutions more conservative than revolutionary. There was a major exception: Slave rebellions and efforts to break down the caste system after 1776 reflected the ideals of the eighteenth-century revolutions; Slaves and freedmen not only sought emancipation, as earlier rebellions of African-born slaves had, but also the transformation of society.

Black slave women joined in the successful Haitian revolution against their French masters that began in 1791 and ended in 1804. Victoria ("Toya"), who commanded troops, Marie Jeanne a la-Crete-a-Pierrot, and Henriette St. Marc, executed by the French in 1802, are just a few of the Haitian women rebels who fought (Morgan, 1984). In Brazil, Black and mulatto slave women became involved in the Tailors' Revolt in Bahia in 1798. It was a complicated social protest: The Blacks were fighting for emancipation, whereas the mulattoes hoped to end racial discrimination and make common cause with the White conspirators, who were nationalistic and anti-Portuguese. This divergence of goals of slaves, free Blacks, and Whites reflected status and racial divisions that impeded concerted political action against colonialism in Brazil throughout the colonial era.

Brazil

Brazilians achieved independence in 1822, at the same time as most of their Spanish-speaking neighbors. Brazil's revolution was relatively bloodless when Dom Pedro I refused to return to Portugal and became emperor of Brazil. He declared independence, convened a Constituent Assembly that ratified his act, and made sure the Portuguese military did not counterattack. Dom Pedro was bolstered in this decision by his intellectual Hapsburg-born wife, Leopoldina. She served as regent on several occasions while her husband was away from Rio de Janeiro. Her affection for Brazilians was well known, and their love for her as their empress helped check the republican movement. After her untimely death, attributed by the public to Pedro's public affair with Domitila Canto e Melo, he was forced to abdicate in 1833 in favor of his and Leopoldina's son, Dom Pedro II (Henderson and Henderson, 1978).

Spanish America

Women fought for independence against Spain in many ways. They served as combatants, guides, and spies; acted as propagandists and recruited their relatives and friends to the cause; hosted political salons (*tertulias*) where ideas of independence and revolution were discussed; and donated money, slaves, goods, and supplies to the rebel troops. They also served as quartermasters, seamstresses, nurses, carriers of water, and mourners and buriers of the dead. There are many stories of women taking their sons to a patriot officer and giving them to the revolutionary army in a semiceremonial manner.

In Guayaquil and Quito, Ecuador, Baltasara Calderón de Rocafuerte and Manuela Cañizares opened their homes to men and women who spoke of revolution. María Josefa Ortiz de Domínguez of Mexico City, a salon hostess, warned Father Miguel Hidalgo, chief revolutionary leader, of his impending arrest so he could escape in 1811. Nuns hid rebels, offered information, and provided refuge for revolutionary women. In Venezuela and Argentina, women worked in military hospitals. Women served as combatants in precursory revolts in Gran Colombia in 1781 and 1795. In the later revolt, and in 1797, women inspired slave rebellions (Reusmann, 1910; Cherpak, 1978).

The full-fledged independence movement in Latin America that began in 1810 involved women of all social classes on both the royalist and the revolutionary sides. Some participated for personal reasons and few expected their sex to make any legal, economic, or political gains. Women who supported the revolutionaries were able to violate tradition and still win popular approval, especially if they were connected to a great man. A nun, Juana de la Purísima Concepción, smuggled messages to her brother José Mariano Michelena in Mexico. Manuela Sáenz, a mestiza married to a British physician, became the mistress of Simón Bolívar, the liberator of northern South America. In addition to aiding the revolutionary cause, she is credited with saving his life.

Most of the women who fought as soldiers in Gran Colombia did so disguised as men. Those who fought openly did so while defending their towns or homes from invasion by the Spaniards. Simón Bolívar paid tribute to "our amazons" who had died fighting the Spanish to inspire his men to further struggle (Cherpak, 1978, pp. 221–222). In Argentina and Bolivia, women fought in battles, negotiated truces among patriot forces ready to fight each other, and delivered messages. In Mexico, Antonia Nava de Catalán led a group of women armed with machetes and clubs into battle; thereafter she was known as "the General." Juana Montenegro accompanied her husband's regiment into battle in Entre Ríos in 1814 and fought so ably that she received a citation and induction into the regi-

ment. Martina Silva de Gurruchaga of Salta received the nickname of "Captain of the Army" after she outfitted and provisioned a whole company of men for General Manuel Belgrano. Two women, in particular, captained military legions—Juana Azurduy de Padilla, an Indian schooled in a convent near her birthplace of Chuquisaca in Bolivia, and Juana de Arco of Argentina. Azurduy received praise and adulation from the patriot military generals for her intelligence, education, and valor (Reusmann, 1910; Furlong, 1951).

Camp followers of the rebel armies were usually lower class, mestizo, and wives and lovers of the foot soldiers. They rendered essential services such as preparing meals, caring for the sick, and taking up arms when necessary. They raised troop morale and cut down on desertion. The presence of camp followers irritated some officers, but orders from generals prohibiting their marching with royalist and rebel troops were ignored (Cherpak, 1978).

Cuba

The movement for independence in Cuba came later in the nineteenth century and culminated when U.S. troops defeated the Spanish in 1898. José Martí noted that popular campaigns succeed only when "they attract the heart of women" (Sarabia, 1991, p. 81). Cuban women wrote poems, songs, and books praising revolutionaries. In 1832, Emilia Casanova was exiled for her anticolonial politics. During the Ten Years' War, 1868–1878, women played the same roles as their counterparts elsewhere in anticolonial movements. Luz Vázquez hosted meetings of revolutionaries, and her daughter Adriana was a courier for them. Mariana Grajales Coello, a free Black who wanted to end slavery in Cuba, pushed her husband and 13 children to fight for independence. Several of her sons—the Maceos—rose to positions of command, suffered exile in Jamaica, and returned in 1895 when revolution broke out again. Evangelina Cisneros met with President William McKinley to get U.S. support for the revolutionaries. In addition, women filled political posts as prefects and subprefects; as such, they were responsible for the distribution, storage, and gathering of supplies and arms. Some received the rank of captain. Ana Betancourt de Mora helped her husband publish an underground rebel newspaper. In 1869, she asked the Chamber of Representatives of Guáimaro, the first town to fall to the revolutionaries, to extend civil rights to women and to abolish slavery (Henderson and Henderson, 1978; Bingham and Gross, 1985; Sarabia, 1991). Her request was ignored.

Consequences

Women suffered the consequences of their revolutionary activities. The famous spy Policarpa Salavarrienta ("La Pola") was executed in the public square of Bogota in 1817; in all, 44 women were executed in Colombia. Another 119 women were arrested and exiled, and others were forced to do public services such as feeding and mending the clothes of Spanish troops and sweeping the streets (Cherpak, 1978;

Henderson and Henderson, 1978). During the War to the Death (1813–1814), in Gran Colombia, royalist troops raped them, sacked and burned their homes, killed them, or forced them into prostitution. Many died along with their children when forced to migrate elsewhere. Some were forced to accompany the Spanish troops and carry baggage and do other chores. Rebel women were confined to their homes, forced to leave their towns, or sent to convents as punishment. Women fled war zones, and many evacuated the cities of Caracas and Cartagena in 1814 and 1815 for insecure lives in exile with their children to support. Wealthier women sought refuge in convents.

Misery and poverty were the fate of many women. They lost their loved ones—fathers, husbands, lovers, children—due to injury, disease, malnutrition, and death. The property of both royalist and rebel women was confiscated. Governmental insolvency and heavy debts prevented most widows and orphans from receiving pensions.

After the revolutionaries won, women were advised to return to their homes. Simón Bolívar wrote to his favorite sister, María Antonia, in 1826 that women should retire to their homes to take up domestic duties, keep out of politics, and maintain neutrality in public affairs (Cherpak, 1978). The upheavals and militarism of the postindependence period saw men reaffirming order and control in their family life.

The rights women had in colonial times to own, buy, and sell property, inherit and pass on wealth, petition the government, and enter into lawsuits were not amplified after independence and may have been restricted. Women made slight gains in two areas: education outside of the home, a process begun during the Enlightenment, and the right of association. Often these two were combined—for example, upper-class women in Buenos Aires received permission from the government to found the Beneficent Society in 1823, which established schools for girls as one of its many charitable causes. Women seemed to have adapted to the reassertion of patriarchy with barely a complaint. They returned to traditional concerns such as matrimony, motherhood, household management, religious life, and charity (Cherpak, 1978).

Conclusion

In Latin America, as in Asia and Africa, male leaders of anticolonial movements enlisted the support of women to win independence. These movements would not have been successful without that support. Male leaders were fearful of social unrest and sharing power with others. After winning independence, men denied and frustrated the egalitarian implications of their anticolonial and republican ideology by reasserting traditional gender roles.

The advances made by women after independence differ in these three regions of the world and in the individual countries within them. Independence was achieved in Latin America early in the nineteenth century at a time when women's movements were just beginning in the United States and Europe. Independence movements swept Asia and Africa after women had achieved many legal, social, and economic rights in the West. The influence of North American and European women on women and men in Asia and Africa in the twentieth century needs to be taken into account.

Women were patriots first and foremost, and independence was their priority. They recognized their men's need for control, at least in their homes, or felt that they had little recourse, and postponed the move to equality.

References

Bingham, Marjorie Wall, and Susan Hill Gross. *Women in Latin America: From Pre-Colombian Times to the 20th Century,* vol. 1. St. Louis Park: Glenhurst, 1985.

Burton, Antoinette. "The White Woman's Burden: British Feminists and the Indian Woman, 1865–1915," *Women's Studies International Forum,* vol. 13, no. 4, 1990, pp. 295–308.

Campbell, Leon G. "Women and the Great Rebellion in Peru, 1780–1783," *Americas,* vol. 42, no. 2, 1985, pp. 163–196.

Cartey, Wilfred, and Martin Kilson. *The African Reader: Independent Africa.* New York: Random House, 1970.

Chaudhuri, Nupur, and Margaret Strobel. *Western Women and Imperialism.* Bloomington: Indiana University Press, 1992.

Cherpak, Evelyn. "The Participation of Women in the Independence Movement in Gran Colombia." In Asuncion Lavrin (ed.), *Latin American Women: Historical Perspectives.* Westport: Greenwood, 1978, pp. 219–234.

Engels, Dagmar. "The Limits of Gender Ideology: Bengali Women, the Colonial State, and the Private Sphere, 1890–1930," *Women's Studies International Forum,* vol. 12, no. 4, 1989, pp. 425–437.

Furlong, Guillermo, S.J. *La Cultura Feminina en la Época Colonial.* Buenos Aires: Editorial Kapelusz, 1951.

Gordon, David C. *Women of Algeria: An Essay on Change.* Cambridge: Harvard University Press, 1972.

Hafkin, Nancy J., and Edna G. Bay. *Women in Africa: Studies in Social and Economic Change.* Stanford: Stanford University Press, 1976.

Henderson, James D., and Linda Roddy Henderson. *Ten Notable Women of Latin America.* Chicago: Nelson-Hall, 1978.

Hernández, Teresita, and Clara Murguialday. *Mujeres Indígenas Ayer y Hoy.* Managua: Puntos de Encuentro, 1993.

Hunt, Nancy Rose. "Noise over Camouflaged Polygamy, Colonial Morality Taxation, and a Women-Naming Crisis in Belgian Africa," *Journal of African History,* vol. 32, 1991, pp. 471–494.

Jayawardena, Kumari. *Feminism and Nationalism in the*

Third World. London: Zed, 1986.

Liddle, Joanna, and Rama Joshi. "Gender and Colonialism: Women's Organisation Under the Raj," *Women's Studies International Forum,* vol. 8, no. 5, 1985a, pp. 521–528.

———. "Gender and Imperialism in British India," *South Asia Research,* vol. 5, no. 2, 1985b, pp. 147–164.

———. *Daughters of Independence: Gender, Caste, and Class in India.* London: Zed, 1986.

Morgan, Robin (ed.). *Sisterhood Is Global: The International Women's Movement Anthology.* Garden City: Doubleday, 1984.

Pescatello, Ann (ed.). *Female and Male in Latin America: Essays.* Pittsburgh: University of Pittsburgh Press, 1973.

Reusmann de Battolla, Elvira. *Heroínas Americanas:*

Episodios, Anécdotas, Acciones Heróicas. Buenos Aires: Libreria Argentina de Enrique García, 1910.

Sarabia, Nydia. "La Mujer en la Historia Política Colonial de Cuba." In Jorge Núñez Sanchez (ed.), *Historia de la Mujer y la Familia.* Quito, Ecuador: Editorial Nacional, 1991, pp. 81–103.

Schmidt, Elizabeth. "Patriarchy, Capitalism, and the Colonial State in Zimbabwe," *SIGNS: Journal of Women in Culture and Society,* vol. 16, no. 41, 1991, pp. 732–756.

Urdang, Stephanie. *Fighting Two Colonialisms: Women in Guinea-Bissau.* New York: Monthly Review Press, 1979.

Wasserstorm, Jeffrey N. "Gender and Revolution in Europe and Asia, Part 2: Recent Works and Framework for Comparative Analysis," *Journal of Women's History,* no. 6, 1994, pp. 109–120.

International and Bilateral Aid Agencies

Kathleen Staudt

Introduction

In the global economy, technical and monetary resource transfers occur through concessionary grants, loans, and debt repayments. For the long haul of history, these transfers have privileged men: Men made decisions about transfers in official organizations; resources seemingly benefited (or burdened) men. In a world wherein men had material advantages relative to women, such patterns aggravated gender gaps in opportunities and resources. Women-in-development (WID) research and action emerged in aid agencies to address these concerns.

WID policies and offices were established as far back as the 1970s to increase women's access to development programs and projects. Pioneering aid agencies included those in Norway, Sweden, and the United States. Even the World Bank had a WID adviser in the 1970s, perhaps more window dressing than substantive at that time (Kardam, 1991). Many early efforts had a welfare orientation, focusing on maternal and child health, as opposed to equity, empowerment, antipoverty, or efficiency approaches. This essay relies on Moser's (1993) definitions of these orientations:

Welfare: providing relief aid provided directly to low-income women, who, in their engendered roles as wives and mothers, are seen as those primarily concerned with their family's welfare.

Equity: bringing women into the development process through equal access to employment and the marketplace, beyond their participation in concrete development projects.

Antipoverty: shifts from reducing inequality between men and women to reducing income inequality through income-generating projects on the belief that gender inequality is primarily due to proverty.

Efficiency: increasing women's economic participation on the assumption that increased income is automatically linked with increased equity.

Empowerment: increasing (women's) power in terms of the capacity of women to increase their own self-reliance and internal strength, the right to determine choices in life and to influence the direction of change, through the ability to gain control over crucial material and nonmaterial resources.

Early WID efforts were often separate, small projects, minimally funded in overall agency terms. In the 1970s, the United States Agency for International Development (USAID) spent 2 percent of its funding on WID; it subsequently doubled that after several years of WID advocacy to 4 percent (Staudt, 1985). More recently, the two most progressive national aid agencies can document almost 20 percent funding that, at least, integrates women (Jahan, 1995, p. 89). For aid agencies, generally, what patterns have existed from the 1970s to the mid-1990s?

This essay examines women's space in the aid relationship, with many illustrations from agencies such as the World Bank, the USAID, and European donor agencies (Staudt, 1994a and 1994b). Women have just begun to move from occupying a separate to occupying a mainstreamed space in many aid agencies, and this essay analyzes how we might understand that move in both *technical* and *political* ways. It also faults most agencies for a lack of documentation on budgets and on women's participation and voice in aid programs. We know very little about the effects of aid programs on women's lives.

The essay first sets forth basic terminology and the central question: Who decides aid agendas? It then briefly profiles 17 agencies and discusses their impacts on women's lives. It closes with strategies that emphasize the importance of political approaches that complement bureaucratic techniques in order to address the problem posed in the first paragraph: male preference in resource transfers.

Aid Rationale: But Who Speaks for Women?

In the best-case scenario, aid is transferred to develop countries, people, or economic sectors within countries. "Development" is a term more loaded than most. It generally refers to economic growth, equitable growth (with redistributive intentions built into transfers), and/or improvements in the quality of people's lives in areas such as education, health, and housing. The 1994 *Human Development Report* indicates that only 7 percent of aid is spent on human priorities such as education, health, and housing (UNDP, 1994, p. 7).

One might assume that decades of aid would have created more equality between the so-called rich and poor countries. Quite the opposite has occurred. According to the United Nations Development Program (UNDP), "more than three-fourths of the world's people live in developing countries, but they enjoy only 16 percent of the world's income—while the richest 20 percent have 85 percent of global income." Developing-country debt service to banks and aid agencies amounts to more than $1.8 trillion (UNDP, 1994, p. 14). Adding high-interest debt service, poor-to-rich-country resource flows *exceed* those from rich to poor countries. Nevertheless, the possibility exists that aid will, at best, temper the most virulent of market-generated inequalities and, at the least, avoid aggravating inequalities with ill-advised program agendas and conditions that advantage rich countries.

The sources of aid fall into two categories, each with disadvantages for recipient countries' sovereignty in this inherently dependent relationship. A central question is this: Who decides the aid agenda, its form, and related concessions? Even women-in-development programming is subject to that question.

An equally compelling question is this: Do recipient-country governments speak for women? Despite the transition to democracy since the mid-1980s, democracies exist in just a minority of countries, and in only a handful of those do female representatives and cabinet officials speak in more than token numbers (UNDP, 1995).

Bilateral aid, which is also known as foreign aid, transmits resources from one government to another government, both linked with special trading, historical, or ideological relationships. Resource transfers may be tied to trade or to foreign and domestic policy concessions, thereby undermining the sovereignty of recipient-country officials—that is, their ability to control other policies.

Multilateral aid comes from various sources, which award funds according to supposedly neutral development criteria rather than the foreign or commercial agendas of a single country. To the extent that a dominant economic ideology colors those awards, however, even multilateral aid may be tied to policy concessions, endangering national sovereignty.

Problems: Faulty Concepts, Questionable Democracies

Multilateral and bilateral aid staff have usually used historic conceptions of development oblivious to women's work and to gender inequity. Projections and measurements of economic growth made women's work or its underpaid (or unpaid) basis superfluous. Assumptions about nuclear families, complete with domesticized housewives and mothers, permeated aid conceptions and programs (Mies, 1986; Stromquist, 1994).

If donor conceptions and assumptions have been faulty, what about conceptions in recipient nations? Development discourse is global, albeit differentiated with market-driven and centrally planned alternative scenarios. Men monopolize politics in most countries, and their monopolization is more complete in finance and foreign ministries, where aid is negotiated and concessions made. Who speaks for women in such circumstances? Are national governments accountable to female citizens? The DAWN (Development Alternatives with Women for a New Era) collective exposes how some governments defend female subordination with cultural and religious rationales (Sen and Grown, 1987).

Questions about democratic accountability should also be asked about multilateral and bilateral aid. As Tendler (1975) pointed out in her classic study of aid, the users of aid funds in recipient countries lack political voice in the funder country. Jahan (1995, p. 45) makes a fundamental distinction between bilateral and multilateral aid in the politics of reporting relationships: Multilateral agencies report to their governing boards, which are composed of national appointees, shielding them from public accountability. Bilateral agencies report to elected officials in national representative bodies, enhancing public accountability to citizens and openness to the media. To the extent that bilaterals operate in democratic contexts, including those with critical masses of women's voices, we would expect greater responsiveness in mainstreaming women-in-development agendas.

The following section profiles aid agencies, most of which claim to have moved from a separate to a mainstreamed (or "integrated") approach. Agencies use a variety of technical approaches, but the most far reaching are enmeshed in political contexts wherein not only gender mainstreaming but also development discourses have been transformed to establish partnerships with developing countries, including women therein.

Profiling Agencies: From Separate to Mainstreaming?

Multilateral and bilateral aid agencies are far from a monolithic group. Each has its own history and special mission. The women-in-development offices also have complex histories and changing strategies, adapting discourses that are compatible with, yet a challenge to, their agency hosts.

Increasingly, WID offices within agencies have renamed themselves "gender offices" for a variety of reasons: to depoliticize a controversial effort, to demonstrate women's relation to men, and to pose a technical fix. A new name does not signify that mainstreaming has occurred—that the whole agency has stakes in, and shares responsibility for, responding to women as well as men. Jahan

(1995) also points out that the gender terminology does not translate well into a variety of languages used in recipient countries, where, after all, the impact should occur.

This section confines itself to brief treatment of 11 multilateral agencies in the U. N. family and six bilateral agencies in Europe and North America. It draws on interviews from 1992 and 1994 (Staudt, 1994a and 1994b) and a mix of academic analyses and consultant reports. The literature would be strengthened if better "mainstreamed" in academic writing, with its longer-term commitment to published knowledge.

Multilateral Agencies

In the multilateral group, the two largest agencies are the World Bank and the United Nations Development Program (UNDP). Both recruit few women professionals, especially at senior levels. Both agencies claim progress in gender mainstreaming. The bank is a postwar creation (1944); the UNDP was born in 1966.

The bank lends money for development projects and policy changes that have likely economic return and strong repayment prospects. It also has a long-term, low-interest grant-like concessionary arm, the International Development Association. It is a centralized organization, based in Washington, D.C.

The World Bank is most infamously known for its structural-adjustment lending of the early 1980s and thereafter, lending to strengthen market economies and limit government size. What this has meant in human terms is cuts to health and food-subsidy budgets, thus burdening many women on whose shoulders such responsibilities rest. UNICEF (United Nations Children's Fund) published a massive, two-volume, 10-case study on the human costs of adjustment (Cornia et al., 1987).

The bank's historic interest in women has been confined to population control, safe motherhood (a 1980s campaign), and, increasingly, girls' education as an investment payoff to produce more efficient economies with lower fertility rates. The bank is the exemplar of the "efficiency" approach to mainstreaming, eschewing other arguments as irrelevant in economic models. The bank also claims a hands-off policy in cultural and political interference in sovereign nations, yet critics say it sets policy conditions that represent clear political interference.

Gender watchers inside the bank count the number of basic project documents that mention women, pointing to rising figures to nearly a half with such mentions. Even so, reference to women has never guaranteed that implementation will result in projects that include women in documentable ways.

The World Bank publishes many books, papers, and reports—so many that people worry about its dominance of development discourse. Increasingly, it also publishes reports on women, which are written in a style that speaks to economists. Its annual *World Development Reports* focus on thematic issues. One in 1984 focused on population; another in 1990, on poverty alleviation, which included

attention to women. Can the bank's increasingly vigorous, if economist, attention to gender counterbalance the alleged damage done to women through structural adjustment? For some, the answer is a resounding no. For others, the bank is just too important in monetary and ideological terms to dismiss as hopeless.

The UNDP is a huge, multipurpose technical-assistance agency that has its own resident representatives in recipient countries and contracts with other U. N. affiliates. It officially established a WID, later changed to gender, unit rather late, in the 1980s. However, it sponsored the only comparative study of multilateral agencies that documented the lack of attention to women in the majority of projects studied: 56 percent (63 percent of total funding) affected women but had no provision for women's involvement; 12 percent mentioned women's participation; 5 percent were women's projects; and 27 percent had no interest for women (UNDP, 1985).

In early years, the UNDP seemed to concede a lead role to the United Nations Voluntary Fund for Women (UNIFEM). By the mid-1980s, the UNDP steered its own version of mainstreaming, apart from UNIFEM, which also turned from innovative (if sparsely funded) women's projects to strategic investments that would mainstream women in society and government (Anderson, 1990). Both UNIFEM and INSTRAW (International Research and Training Institute for the Advancement of Women) are the only women-specific development agencies in the United Nations.

The UNDP produces an annual *Human Development Report* (*HDR*). Unlike the economist orientation of the World Bank, the UNDP ranks countries based on a composite measure that documents achievements and government investments in human priorities such as health and education. Besides being a useful tool to undermine narrowly economist orientations to development, the *HDR* has always attended to women and gender equity. It produces "gender-adjusted" rankings for countries (i.e., gender gaps in literacy reduce ranks); the thematic focus of its 1995 report was gender. *HDR* legitimizes gender equity and transforms discourse.

The UNDP executive director has been a frequent and outspoken advocate of attention to gender. In speeches, James Speth has rarely missed the opportunity to talk about development as "pro-woman, pro-poor, pro-jobs, and pro-environment." Yet, of both the UNDP director and the *HDR,* one might wonder what impact such rhetoric and discourse have on everyday programming and outcomes from technical assistance. In 1992 interviews with UNDP officials, one made the comment: "The UNDP speaks with many voices" (Staudt, 1994a, p. 48).

Based on mid-1980s interviews, Nüket Kardam (1991) concluded that the World Bank had responded more fully on WID than the UNDP. Acknowledging the heavy economist orientation of the bank, she cited the UNDP's decentralization and philosophy about recipient-country sovereignty as sources for the UNDP's reluctance to support WID with concrete procedures. Herein lies again the question of who speaks for women, and on what terms. Kardam's analysis was done before the UNDP established a gender unit; she did not consider the likelihood that the UNDP did less damage than the World Bank's structural adjustment programs in the name of development.

Other large U. N. affiliates include the Food and Agriculture Organization (FAO) and the International Labor Organization (ILO). The FAO has produced a massive array of studies and internal procedures that call attention to women's agricultural work for its staff of agronomists, who somehow missed these details in their academic studies and visions of "modern" agriculture. Its WID unit came out of a post-war-style home economics, a philosophic orientation based on modernized domesticity. In the late 1980s, the WID unit has developed goals that fall under what Moser (1993) called "strategic interests" that reduce gender subordination: land reform, political empowerment, and legal rights.

The ILO is a tripartite organization that brings itself together with governments and workers' and employers' organizations. It is the oldest of U. N. affiliates (born in 1919, before the United Nations itself) and women workers were always one of its constituents. The ILO has been a leader in stretching the intellectual agenda of development to include what women do and to value all of their labor through studies on informal economies and unpaid work in the 1970s and 1980s.

Several U. N. agencies have a health and family planning focus. The World Health Organization (WHO) and the United Nations Fund for Population Activities (UNFPA) pursue missions compatible with, and central to, women: health, including reproductive health; people friendly health-delivery systems; and data collection, frequently broken down by gender. The challenge WID posed to these agencies was in broadening their conception of women from mothers to people who work, give birth, and do many other things. Simultaneously, they broadened their conception of health to include not only disease reduction but also socioeconomic factors influencing health, including tolerance for violence against women. The WHO and the UNFPA have supported income-generating and empowerment activities. With its gender focus, the UNFPA also targets males in terms of attitudes toward women and coresponsibility for contraception. Both groups emphasize partnerships with nongovernmental organizations (NGOs) more than most agencies.

UNICEF has long been a leader in work on maternal and child health and on women's practical interests in safe water and sanitation. Its previously mentioned influential study counteracted the structural-adjustment thrust of the World Bank and the WID-less International Monetary Fund with sound analyses that received wide play in the academic and practitioner worlds. UNICEF took the lead in focusing on girls—the incoming generation of women—and their educational needs.

Two agencies concentrate efforts on poverty alleviation. For the World Food Organization (WFO) and the International Fund for Agricultural Development (IFAD), attention to women is central to their effectiveness as agencies. They support food-for-work and innovative agricultural projects. While women were more quickly integrated into their mainstream, this may have occurred at increased cost to women's (unpaid) labor time. Until agencies address female farmers' control over the fruits of their labor and the means of production (land), agency effectiveness burdens women while meeting their practical needs.

The multilateral organizations use a variety of technical procedures to change their staff and staff work habits to mainstream women. Most have experimented with gender training. Rao et al. (1991) define this as "a way of looking at the world, a lens that brings into focus the roles, resources, and responsibilities of women and men within the system under analysis" (Rao et al., 1991, p. 7). The effects of training on staff performance have not been fully documented. Several days of training are unlikely to make big dents in gender ideology unless reinforced in personnel and redefined jobs with results-oriented evaluation.

Multilateral agencies use other techniques as well. Most have guidelines and checklists to call staff attention to relevant women's work. Most have plans of action and coordinating committees to build coalitions inside institutions. Outside of the UNFPA and the WHO, the primary political orientation is at international and regional conferences. The more decentralized organizations, however, have the potential to bring women's voices into national-level consultation. UNICEF, in particular, has experienced staff and links with NGOs at country-program levels.

Bilateral Agencies

Bilateral agencies have distinctive missions and bureaucratic operations. Some, such as Germany and the United States, focus on large numbers of countries, while others, such as the Nordic countries, focus on selected recipient countries. Only members of the latter, more selective group have long met the U. N. goal to provide 0.7 percent of their gross national product (GNP) in official development assistance. Germany represents the most centralized of agencies, with authority shared over several agencies at headquarters, while other aid programs sponsor country offices and expert staff (Arnold, 1982). Decades ago, Sweden was rare among bilateral countries to identify itself in "partnership" with recipient countries.

The pioneering countries to establish WID policies include Sweden (1968), the United States (1973), and Norway (1975). Canada, the Netherlands, and Germany later followed. Most bilaterals began with a welfare orientation to women, viewing them as recipients of maternal and child health, domestic training, and family planning. The latter was a risky undertaking in early years, before contraception and women's choice became legitimate aid activities.

The United States pioneered in steering WID away

from women as mothers and toward women as workers. In both the liberal and the socialist feminist academic approaches of the 1970s, productive work was central to analysis. In emphasizing productive work, academics laid the foundation for advocates to make the economic-efficiency case to often resistant agency staff, whether bilateral or multilateral. Attention was thereby diverted from the more politically charged cases of equity and empowerment.

The Netherlands comes closest to an empowerment approach, known as autonomy. Drawing on university action-oriented research projects, it has conceptualized what it calls women *and* development in terms of emancipation from multiple inequalities. By 1990, it had shifted to an emphasis on women's autonomy to enable women to control their lives in political, physical, economic, and sociocultural ways. Its agency takes cues from women in recipient countries, some of whom are part of a network trained at the world's largest advanced graduate program on women's development, at the Institute for Social Studies, The Hague.

Norway calls its assistance "women-oriented," integrated, and prioritized in a development strategy that questions traditional economic growth and bureaucratic models. It leads among bilateral agencies in the number of women professionals (43 percent), including senior managers (Jahan, 1995, p. 53). In 1984, it instituted the Women's Grant, providing flexibly administered seed money to support pilot activities that strategically address female subordination. Among those funded were NGOs such as the Ambedkar untouchables' social-justice movement in India and KALI for women, the first feminist publishing house in India.

Sweden says it is a gender-mainstreamed bilateral agency, a result of staff training and country programming with staff attentive to women (Himmelstrand, 1990). Its aid agency also pursues separate women's programming, especially support for what the United Nations calls women's machinery, such as government women's bureaus and ministries.

Canada took its cues from the limitations of separatist WID units, which were sidelined from the mainstream of agency operations in early years. It developed a corporatelike plan to spread responsibility to all staff. Gender training was mandated for male and female staff alike. By the late 1980s, Canada had the reputation as one of the most forward looking of bilaterals. Was mainstreaming complete? It cut back on its specialist staff, too soon perhaps. Such was the conclusion of a methodical reflection of 1992–1993, sponsored by its own evaluation unit.

Germany steers WID activity from a women, family, and youth unit. To preempt presumed staff resistance, it is careful to avoid women-specific approaches. It has an elaborate categorization scheme to differentiate projects according to expected impacts: women positive, women risky, women negative, and unspecified effects. The unit has authority to countersign and delay project approval, a labor-intensive task for its small staff. Germany emphasizes women's practical needs, couched in family discourse.

Bilateral agencies are set in political contexts that are more or less attentive to women's voices. Nordic countries not only have the world's highest representation of women in parliaments and cabinets, but they elevate the portfolio of equal rights to ministerial-level activity. Some analysts have referred to these countries as "state feminist" wherein governments have institutionalized equity concerns. Nordic countries see themselves as global leaders in U. N. equity activities.

WID efforts in several bilaterals were fostered by active women's movements and groups. Jahan (1995) analyzes the sources of WID effort in Norway and Canada as such. For these and other bilaterals, women's advisory committees help sustain public scrutiny over agency mainstreaming. In the Netherlands, the aid agency supports a women's lobby effort, university research, and internationally minded NGOs.

In the United States, a savvy women's network in Washington, D.C., prompted Congress to amend the Foreign Assistance Act to integrate women in 1973 (Staudt, 1985). Coalitions of women's groups continue to monitor WID's budgetary health inside the USAID and in Congress. Although female representation in the United States Congress is exceptionally low, below global averages, a strong bipartisan women's coalition within Congress cooperates in WID matters. Still, the mid-1990s antiequality, antiaid backlash makes defense difficult.

Bureaucratic practices regarding documentation of WID mainstreaming vary in degree of complexity. Among the most useful techniques, several bilaterals adopt targets and goals for women's participation in key development activities. For the Netherlands, women and development is a priority program, dramatically increasing funding. It adopted measurable objectives, against which performance can be judged. One goal states that 25 percent of all funding should result in increased autonomy for women, to rise to 50 percent by 1998. Norway develops accountability measures for goals within sectors. For example, it earmarks funds for credit, sets aside half of fellowship funds for women in education, and mandates a minimum of 20 percent assistance in agriculture. In contrast, without goals and earmarked funds, promises and voluntary goodwill fall through the cracks: Of the $5.8 billion that multilateral banks allocated for credit in 1990, only 5 percent reached rural women (UNDP, 1995, p. 39).

Other agencies take the next best step of making public information available about funding for women. The United States Agency for International Development (USAID) reports amounts regularly. The United States General Accounting Office (GAO), the investigatory arm of Congress, criticizes USAID for lack of agencywide accountability to WID. Further, when figures are set against total aid allocations, which include massive aid to Egypt and Israel on foreign-policy grounds, women's proportional share shrinks in significance.

Many agencies do not make public reports of their WID projects and integrated activities, outside of optimistic anecdotal evidence about innovative projects. Overall WID spending sometimes seems like the best-kept secret! But this, too, must be set against an overall agency picture in which internal accounting for gender is lacking and people-oriented evaluations are underdeveloped. Closed bureaucratic practice does little good, both for refining WID technical approaches and for supplying data to outside constituencies.

Aid's Impact

It is difficult to trace the impact of aid interventions on women. People-oriented evaluations are few and far between. Moreover, many factors affect people's lives, only one of which is international aid.

Jahan (1995) included connections with impacts on women in her comparative assessment of agencies. She focused on the World Bank, the UNDP, the NORAD, and the CIDA (the latter two, aid agencies in Norway and Canada, respectively). The NORAD and the CIDA are among the better bilateral agencies, with women integrated into almost 20 percent of aid efforts. Jahan could not point to any comparable movement of the multilaterals. Her interviews with women's-group leaders and government officials in two countries, Bangladesh and Tanzania, judged the impact of bilateral aid as indirectly positive, largely for the support it offered women's NGOs. Yet, bilateral aid agencies sometimes promote competing meanings of mainstreaming women; Kabeer's (1991) insightful analysis of Bangladesh describes the differences between Western countries and Arab donors promoting grassroots religious fundamentalism.

Goetz (1995a) compares women's machinery in six countries, based on 1992 interviews. On the whole, staff encounter some of the same dynamics as WID advisers in aid agencies with limited staff and resources and only occasional support from strategically placed men in bureaucracy and politics.

"Getting Institutions Right for Women in Development" is a challenge (Goetz, 1995b), not only for aid agencies and women's machinery, but also for government ministries in recipient countries. We have few WID studies that go deep into the bowels of bureaucracy to understand change that mainstreams women. One fine exception is an analysis of how a Zambian agricultural officer was able to build coalitions with strategically placed male officers to turn around a conservative Ministry of Agriculture provincial office, with opportunities provided through an FAO-instigated People's Participation Project and Netherlands-supported action research. She mobilized women's farmer groups, targeted 50 percent of trainees as women, and revised the formerly male-biased criteria for selecting "contact farmers" once the idea of women farmers became legitimate. She also maintained contact with women development committees in other ministries and with a scholar-activist WID network in the Zambian capital of Lusaka to promote change elsewhere (Jiggins et al., 1995).

How can both governments and aid agencies be transformed to make them accountable and responsive to women as well as men? This takes more than a female labor of love. As the UNDP so eloquently states: "Human development, if not engendered, is endangered" (UNDP, 1995, p. 1). Examining global data on a grander scale, one could point to improvements (and neglect) in which aid agencies played some part. As the UNDP's *Human Development Report* documents, gender gaps have narrowed in education and health; female life expectancy has increased; and fertility rates dropped by a third (UNDP, 1995, p. 3). Yet, it notes: "Poverty has a woman's face." Seventy percent of the 1.3 billion people in the world who live in poverty are women. Women's labor-force participation (paid work) has inched up merely 4 percent in two decades; male–female wage gaps are wide; and women's reproductive-health needs are neglected. Each year, a half-million women die from pregnancy and birth complications in developing countries. "Too often, the miracle of life becomes a nightmare of death" (UNDP, 1995, p. 4).

Women's representation in politics worldwide averages a mere 10 percent of seats on representative bodies, 7 percent of cabinet seats, and 4 percent of chief-executive positions (UNDP, 1995). As women's groups engage more actively in the political process, including the installation of quotas for female candidacies in political parties and gender-sensitive policy agendas, more genuine transitions to democracies (with women's greater participation) can occur. Over the last decades, a veritable explosion of women's groups has emerged in countries around the world.

Toward Strategies for Aid Questions: Technical and Political

Aid agencies share classic bureaucratic characteristics. These hierarchical grids divide and fragment labor, creating coordination problems. Although it is procedure laden, rule making is nontransparent. To protect themselves and mobilize future funds, bureaucracies operate somewhat secretly, defending their outcomes through public relations and symbolic action. Additionally, aid bureaucracies share some peculiar problems: They operate in highly uncertain contexts, and they reward staff for moving money rather than spending it well (Tendler, 1975; Wapenhans, in U. S. GAO, 1994). Such characteristics do not bode well for honest evaluations of program outcomes.

Strategies for the problems identified in this review fall into two broad categories: technical and political. Technical solutions are internal; political solutions are both internal and external. The agency profiles above illustrate these two approaches.

It should be noted that some criticize any solutions that perpetuate bureaucracies. Ferguson (1984) asks whether bureaucracy is compatible with broadly defined feminism that values and empowers women. Even bureaucratic feminists are thereby suspect, for the language and development discourse they use perpetuates the dominant male hierarchy. The response of most WID practitioners is that bureaucracies must be transformed, not abandoned for lack of hope (Staudt, 1990; Goetz, 1994).

The first approach is *technical.* It calls for changes in bureaucratic procedures to put the institutional house in order. This takes forms ranging from the adoption of procedural guidelines for bilateral organizations (OECD/DAC, 1992) to strategic diagnoses ("gender planning") of organizations to transform them (Moser, 1993).

In 1983, the Women in Development Committee of the Organization for Economic Development and Cooperation/Development Assistance Committee (OECD/DAC) adopted *Guiding Principles for Aid Agencies;* it revised and strengthened them in 1989. The guidelines call for bilateral agencies to adopt programs of action, checklists, evaluation procedures, and staff training. Periodically, this committee surveys members about their procedures and then makes comparative information available to member countries.

In the United Nations, no umbrella unit does for multilateral agencies quite what the OECD does for bilaterals. Each U.N.-affiliated development organization has its own distinctive mission, and some are more amenable to responding to women's "practical and strategic interests" (Moser, 1993) than others. Practical interests include those that meet everyday needs, such as water, sanitation, food, and housing. Strategic interests are those that undermine women's subordination to men, such as legal reform, curricular reform that overturns sex-stereotyped content, and political empowerment. For example, UNICEF historically fostered a mission friendly to mothers and children. In recent years, it has emphasized empowerment, even using such strong language as "gender apartheid" to criticize segregation once considered culturally compatible and commonplace.

However, the United Nations Division for the Advancement of Women (DAW) communicates with affiliated organizations. Besides serving the Committee on the Status of Women, the UN/DAW takes a lead role in organizing the U.N.-sponsored international conferences on women, the first of which was held in Mexico City (1975) and the latest, in Beijing (1995). International women's conferences supply agendas and programs of action around which multilateral agencies gear up and respond. The 1985 international women's conference in Nairobi is widely recognized as the meeting at which the mainstreaming idea was born.

The second approach to aid problems is *political,* responding to Vina Mazumdar's plea that we "not treat politics as a dirty word" (Mazumdar, 1989, p. 217). Political solutions treat institutional resistance to gender equity as a problem of powerlessness. Aid agencies change in response to pressure both from outside political constituencies and from inside coalitions for change (Staudt, 1985). To the extent that inside procedures can supply honest evaluations of bureaucratic outcomes, along with budgets that detail gender spending, fuller information will be available to promote change.

The international women's conferences demonstrate the power of women's and other nongovernmental organizations (NGOs) in pushing governments and multilateral agencies to change. Increasingly, aid agencies implement projects and programs through NGOs, but it is still a fraction of overall spending. Such strategies have the potential to widen political constituencies with a stake in representing women and responding to their needs. Of course, NGOs are a mix of people's organizations and private contractors, the latter of which are unlikely to sustain interests once funding ends (Korten, 1990). But women's NGOs are leaders in gender-sensitive development. Investments in these organizations should be a far higher priority than investments in bureaucracies.

Conclusion

This essay has analyzed the slow progress of most international aid agencies in mainstreaming women. Mainstreaming occurs on grounds that sometimes serve agencies more than women. Yet, agencies are a diverse lot. Some bilateral agencies, even though they are bureaucracies, demonstrate remarkable creativity, openness, and responsiveness to political constituencies. They should be an inspiration to other aid agencies and a model of how bureaucracies can, under certain conditions, be responsive to women. NORAD (Norway), Sida (Sweden), CIDA (Canada), DGIS (the Netherlands) are examples of such agencies.

This essay has also analyzed the technical and political motors to mainstreaming women in aid agencies. Political forces helped establish WID policies at the outset; they sustain, even rejuvenate, them periodically. The remarkable achievements of some bilateral agencies can be traced to progressive political forces that shared voices in agency advisory committees, political parties, and parliamentary/legislative bodies.

Yet, technical approaches cannot be discounted. They are crucial in those agencies that operate with little public scrutiny, such as multilateral agencies. With only occasional international and regional meetings to prompt change, bureaucratic techniques can sustain attention to women in the interim.

In both bilateral and multilateral institutions, technical approaches help legitimize gender analyses in ways that foster coalition building inside bureaucracies. Technical approaches can also institutionalize policy changes that empower women with skills, authority, capital, citizenship, and human rights. Finally, technical approaches can protect and sustain women's gains when prevailing political winds would undermine them. Such winds blew strongly in many Northern countries, beginning in the 1980s, and continue in the 1990s. In some Southern countries, political winds of reaction use cultural and religious rationales to control women, even as women defy or ignore those ideologies to support themselves and their families. In democratic contexts, bureaucracies cannot and should not be insulated from the popular will—that is, the political process—and that popular will must include women's voices in critical mass numbers.

Ultimately, bureaucratic responsiveness to women and people generally in the recipient countries counts most in this evaluative analysis. Who speaks for people in the South, the central question underlying this analysis? Agencies must create genuine partnerships in recipient countries in ways that empower women. When women occupy more space in government and politics, in all countries, they will have a greater voice in setting development agendas.

References

Anderson, Mary. *Focusing on Women: UNIFEM's Experience in Mainstreaming.* New York: UNIFEM, 1990.

Arnold, Steven H. *Implementing Development Assistance: European Approaches to Basic Needs.* Boulder: Westview, 1982.

Cornia, Giovanni, Richard Jolly, and Frances Stewart (eds.). *Adjustment with a Human Face.* 2 vols. Oxford: UNICEF/Clarendon, 1987.

Ferguson, Kathy. *The Feminist Case Against Bureaucracy.* Philadelphia: Temple University Press, 1984.

Goetz, Anne Marie. "From Feminist Knowledge to Data for Development: The Bureaucratic Management of Information on Women and Development," *IDS Bulletin,* vol. 25, no. 2, 1994, pp. 27–36.

———. *The Politics of Integrating Gender to State Development Processes: Trends, Opportunities, and Constraints in Bangladesh, Chile, Jamaica, Mali, Morocco, and Uganda.* UNRISD Geneva: United Nations Research Institute for Social Development, 1995a.

———. "Getting Institutions Right for Women in Development," Special Issue. *IDS Bulletin,* vol. 26, no. 3, 1995b.

Himmelstrand, Karen. "Can an Aid Bureaucracy Empower Women?" In Kathleen Staudt (ed.), *Women, International Development, and Politics: The Bureaucratic Mire.* Philadelphia: Temple University Press, 1990, pp. 101–113. (Second edition 1997.)

Jahan, Rounaq. *The Elusive Agenda: Mainstreaming Women in Development.* London: Zed, 1995.

Jiggins, Janice, Paul Maimbo, and Mary Masona. "Breaking New Ground: Reaching Out to Women Farmers in Western Zambia." In Ann Leonard (ed.), *Seeds 2: Supporting Women's Work Around the World.* New York: Feminist Press, 1995, pp. 17–40.

Kabeer, Naila. "The Quest for National Identity: Women, Islam, and the State in Bangladesh." In Deniz Kandiyoti (ed.), *Women, Islam, and the State.* Philadelphia: Temple University Press, 1991, pp. 115–143.

Kardam, Nüket. *Bringing Women In: Women's Issues in International Development Programs.* Boulder: Lynne Rienner, 1991.

Korten, David. *Getting to the 21st Century: Voluntary Action and the Global Agenda.* West Hartford: Kumarian, 1990.

Mazumdar, Vina. "Seeds for a New Model of Development: A Political Commentary." In Ann Leonard (ed.), *Seeds: Supporting Women's Work in the Third World.* New York: Feminist Press, 1989, pp. 213–217.

Mies, Maria. *Patriarchy and Accumulation on a World Scale: Women in the International Division of Labour.* London: Zed, 1986.

Moser, Caroline O.N. *Gender Planning and Development: Theory, Practice, and Training.* London: Routledge, 1993.

Organization for Economic Cooperation and Development (OECD)/Development Assistance Committee (DAC). *Third Monitoring Report on the Implementation of the DAC Revised Guiding Principles on Women in Development (1989).* Paris: OECD, 1992.

Rao, Aruna, Hillary Feldstein, Kathleen Cloud, and Kathleen Staudt. *Gender Training and Development Planning: Learning from Experience.* Bergen: Christen Michelsen Institute; New York: Population Council, 1991.

Sen, Gita, and Caren Grown. *Development, Crises, and Alternative Visions: Third World Women's Perspectives.* New York: Monthly Review Press, 1987.

Staudt, Kathleen. *Women, Foreign Assistance, and Advocacy Administration.* New York: Praeger, 1985.

———. *Getting Institutions Right: Crossing the Threshold to Mainstream Women.* Geneva: United Nations Research Institute for Social Development (UNRISD), 1994a.

———. *Technical Assistance and Women.* New York: United Nations Division for the Advancement of Women, 1994b.

———. *Women, International Development and Politics: The Bureaucratic Mire.* Philadelphia: Temple University Press, 1990. (Second edition fall 1997.)

Stromquist, Nelly. *Gender and Basic Education in International Development Cooperation.* New York: United Nations Children's Fund (UNICEF), 1994.

Tendler, J. *Inside Foreign Aid.* Baltimore: Johns Hopkins University Press, 1975.

United Nations Development Program (UNDP). *Women's Participation in Development: An Inter-Organizational Assessment.* New York: UNDP, 1985.

———. *Human Development Report 1994.* New York: Oxford University Press, 1994.

———. *Human Development Report 1995.* New York: Oxford University Press, 1995.

United States General Accounting Office (GAO). *Multilateral Development: Status of World Bank Reforms.* Washington, D.C.: GAO, 1994.

Girls and International Development

Neera Kuckreja Sohoni

Introduction

One of the significant lessons of the development decades sponsored by the United Nations has been that the development phenomenon is neither uniform nor egalitarian. Above all, it is sexist. Development strategies have been male centric—virtually of, for, and by men—with women, along with children, consigned to a subordinate or ancillary position. Inevitably, girls have received the least handouts from development planners and processes as a consequence of both their age and their gender. In that sense, they are twice denied or marginalized and in an even worse position than adult women.

The purpose of this essay is to establish that gender, when combined with age, provides a formidable nexus for disability. It accentuates the burden and consequences of sexism for girls in terms of their survival, health, and nutritional capability, as well as educational and employment preparation. More than women, girls bear the burden of discriminatory and inegalitarian treatment at all societal or personal socioeconomic levels.

> In situations of extreme poverty, the young female's unequal position is even more pronounced. It is the female child who gets sold or bonded first; it is the female child who first drops out of school, and when there is a food shortage, it is the woman and her daughters who eat last (Heyzer, 1986, p. 2).

Even when poverty and survival are not overriding concerns, the gender factor inhibits equal opportunities and aspirations for girls. Comparisons between countries reveal that the disadvantaged status of girls in relation to boys as well as women is not unique to developing countries, although it may be more rampant and visible there. Whereas for historical, cultural, and economic reasons there are expected differences among countries and cultures, in all societies girls face certain disadvantages compared to boys. It is therefore important to view international development in the context of a gendered social construct and to trace the experience of girls within that structure.

Demographic Status of Girls

International attention has focused singularly on the unequal demographic (specifically numerical) status of women in relation to men, neglecting the less favorable numerical status of girls. Widely quoted U. N. data estimate that of the world's 5.3 billion people in 1990, fewer than half (2.63 billion) were women (United Nations, 1991, p. 11). There were fewer girls than boys, too, in 1990 (1.09 billion girls compared to 1.14 billion boys). The sex ratio in 1990 for females of all ages was estimated at 987 for every 1,000 males in the world. Closer analysis (see Table 1), however, reveals that there were 954 girls (age 0–19) per 1,000 boys compared to 1,011 women (age 20 and older) per 1,000 men in 1990. Moreover, it is significant that the numerical status of girls does not appear to be linked to that of women or to the level of development: In both the developed and the developing regions of the world, girls are found to be proportionally fewer in number than boys. U.N. data also reveal a decline in the female/male sex ratio from 996 females per 1,000 males in 1970 to 987 females in 1990, with the decline largely attributed to the low socioeconomic and cultural status of the female in East, South, and West Asia. As the statistical data in Table 1 point out, girls in the 0–19 age group also experienced a sex-ratio decline from 957 in 1970 to 954 girls per 1,000 boys in 1990. However, unlike women, the negative sex ratio for girls is more universal, affecting both the developed and the developing regions. Surprisingly, in fact, the sex ratio in 1990 in Africa was 987 girls and in Latin America 976 girls per 1,000 boys compared to 952 girls in the developed regions. Among women (age 20 and older), adverse sex ratios are not universal and are confined to Asia. These statistics convincingly establish the weaker demographic status of girls vis-à-vis boys and women.

Table 1. Females per 1,000 Males, 1970 and 1990

	All Ages	0–19	20+
World	987	954	1,011
	(996)	(957)	(1,032)
More developed	1,061	952	1,108
	(1,075)	(959)	(1,144)
Less developed	966	954	976
	(966)	(956)	(977)
Africa	1,012	987	1,044
	(1,020)	(991)	(1,056)
Asia	954	942	963
	(957)	(947)	(967)
Latin America	1,002	976	1,025
	(996)	(977)	(1,019)

Percent Decline in Number of Females per 1,000 Males Between 1970 and 1990

	More-Developed Regions (%)	*Less-Developed Regions (%)*
0–19	0.73	0.21
20+	3.2	0.10
All ages	1.3	0.0

Source: United Nations, 1989, pp. 4–33.
Note: 1970 figures are expressed in parentheses.

Genderist Enculturation

Apart from their numerical presence, there are other parameters that serve as vital indices to the status of girls in development. The first of these is the process of acculturation that causes the girl to be perceived as the lesser child. One of the earliest notions a child picks up is the perception of gender or the idea that one is a girl or boy. Psychologists agree that gender is not a biologically but a culturally determined identity that is shaped by the child's nurturers and socializing agents. The evolution from being born a biological female to becoming a culturally defined girl is influenced by a variety of social institutions and forces, including the family, the school, the workplace, law, religion, tradition, the media, and politics. Parental and societal behavioral expectations, and patterns fostered in support thereof, reinforce a person's gender identity. Beginning with the choice of the child's name, the color of clothing, bed linen, room decor, games and toys, length and content of feeding and weaning—all set the mode for girlhood or boyhood. Girls' toys are geared to domesticity; boys', to exploration and adventure. The male child is socialized for aggression; the female, for docility. When a girl plays with dolls or kitchen utensils and a boy with guns or bows and arrows, they are considered as behaving in a manner appropriate to their gender. Crossing "toy" and "behavioral" lines confuses ascribed gender roles.

Sexual asymmetries thus are fundamentally social constructions. We raise, in effect, two different kinds of children: boys and girls. Girlhood becomes a preparation and an informal training ground for motherhood. But boyhood is not necessarily a preface of fatherhood. Rearing patterns consciously ensure that boys remain children while girls become little women. Enculturation leads also to different gender characteristics. Socialization of boys tends to be oriented toward achievement and self-reliance; that of girls, toward nurturance and responsibility. Girls are pressured to be involved with and connected to others; boys, to bypass, deny, or ignore this involvement and connection. Further, the sex role development of girls in modern society is complicated by dual, dissonant expectations. On the one hand, girls enter school to prepare for a technologically complex workplace. On the other, there is a sense that schooling is a "pseudo-training." It is not meant to interfere with the much more important training to be feminine and a wife and mother, which is embedded in the girl's unconscious development. A study (Lees, 1993) of a sample of adolescent girls in Great Britain notes that, by adolescence, the girls are already locked into a life centered on domesticity, subservience, subordination, and motherhood, with any career aspirations or personal ambition or freedom holding second place to the ideal of finding the man of their dreams. Worse, many accept the difference between girls and boys as biological (Aries, 1962; Chodorow, 1989; Lees, 1993; Lott, 1994; Richardson and Taylor,).

Genderism in the Family: Son Preference vs. Daughter Neglect

Within the family, genderism is both generic and chronic. A strong preference for the birth of sons rather than daughters, which is both a cause and an effect of the female's low status, is unmistakably present throughout most of the world. Even where there is no special preference for sons, very few cultures actively prefer daughters. A 1983 World Fertility Survey found that of approximately 40 developing countries, in only two, Venezuela and Jamaica, did daughter preference before birth prevail. The strongest son preference was expressed in Bangladesh, Jordan, Korea, Nepal, Pakistan, and Syria (Ware, 1981; Ravindran, 1986).

The desire for sons is so strong that parents in traditional cultures and elsewhere follow strict nutritional and intercourse guidelines to facilitate the conception of a male child. Astrologers are consulted in India to detect the most propitious time for conception of a male. Women in India also pray to certain gods and observe fasts on given days to maximize their probability of conceiving a male fetus. Use of baking powder, Epsom salts, and vinegar douches is also resorted to in Western countries to enhance the probability of a male fetus. In several cultures and countries, special treatment awaits the birth of the male child, while the female child is ignored, cast aside, abandoned, starved, or killed. Female infanticide is known to have been widely practiced in many cultures throughout history to eliminate female children. In ancient Greece and Rome, the advice to expectant mothers was: "If it is male, let it live; if it is female, expose it" (meaning let it die). In recent decades, technological tools have made it easier through genetic planning and sex screening to systematically prevent the female from being conceived and born. In an Indian study (Ramanamma and Bambawale, 1980, in Ware, 1981) of 700 pregnant women who were given genetic amniocentesis to determine the sex of the fetus, of the 450 who were told they would have a daughter, only 20 went through the pregnancy; of the 250 male infants predicted, even when a genetic disorder was likely, all were carried to full term. Recently, of the 8,000 abortions carried out in a clinic in Bombay, India, all but one of the aborted fetuses was female (UNICEF, 1994). The availability of sex-determining techniques in a Chicago clinic indicated an overwhelming preference for male babies in a sample of middle-class European American women. Press reports from China in the 1980s and 1990s suggest a resurgence of the ancient prejudice against girl babies and a rise in female infanticide believed to have been sparked by the government's one-child-per-family fertility policy. Evidence from other countries such as Korea and Singapore affirms parental willingness to prevent female births from occurring (Ravindran, 1986; Lott, 1994; UNICEF, 1994).

Anecdotes and proverbs in many countries refer to the pride with which the male child is welcomed and the shadow of gloom that is cast by the coming of a baby girl. Among orthodox Jews in Eastern Europe, the birth of a boy was time for rejoicing; the birth of a girl, for stoic acceptance. Siring a female child was a shameful act for which Hassidic Jews occasionally flogged a young father. More time and money are spent on the celebrations for the birth of a son than a daughter in Britain: a prince earns a 21-gun salute, a princess only 10. Popular sayings and rituals provide solid evidence of the girl's devalued status. Daughters are referred to as "water spilled on the ground" in Taiwan,

and as "only a prostitute" who will be exchanged for cattle at the time of marriage among the Iteso in Uganda. An old Chinese saying asserts that "girls are maggots in the rice. . . . When fishing for treasures in the flood, be careful not to pull in girls." Among Zulus, the girl is "only a weed." To be called the "father of daughters" is an insult in Arabic. An unexpected silence or conversational gap in an assembly invites the Arabic expression *"Khilqat bint"* or "Why the silence? Has a girl been born?" A phrase from the Korean language translates to: "A girl lets you down twice, once at birth and the second time when she marries." The girl thus represents a parental feeling of being let down, cheated, or burdened. Evidence of less enthusiasm for female babies is not confined to the developing countries. Although American parents do not selectively discard female infants or ridicule parents of girls, female babies continue to be seen as second best. Lesser desirability of a girl for a first or only child is well documented in the United States. Sadly, a study of 236 parents who had experienced the death of a child found that boys had been grieved far more than girls (Lott, 1994).

The reasons for son preference are well recognized. The son is a potential and permanent source of economic support; especially in societies without the protective cover of social security and pensionary provision, the son is the only protection parents have against privation. The male child also derives his favored status from being his father's genetic imprint and lineage carrier. In certain religions such as Hinduism, the son alone is vested with the right to perform postdeath rituals that assure peace and salvation to the departed soul. In those contexts, the son is perceived as a conduit of the parents' material as well as spiritual well-being. The daughter offers no such gains. Where she is desired, the underlying reasons are the expectation of her loyalty ("A son is a son till he gets a wife; a daughter is a daughter all her life"); her help within the home as a nurturer and caregiver; her value as an additional laboring hand in the family farm and business; her "doll-like cuteness" or dress-up potential; and, in rare cultures, her material worth in the marriage market. In such cultures, the girl commands a bride price that the bridegroom is expected to pay to the parents, a practice that makes her a commodity.

Controlling Girls' Sexuality and Reproductive Risks

Much of the girl's inferior value is traced to her sexuality and the reproductive risks it entails. In fact, it is ironic that although the girl has indispensable societal value as a potential bearer of the human race, that reproductive function has done little to enhance her status or worth in either the family or society. If anything, it places further constraints on her. In many cultures, menarche begins a process of emotional and physical trauma and nutritional and other types of denial that are totally demeaning of girlhood. In some African, Arab, and Asian countries, girls are barred from high-protein foods at menarche as a probable means of controlling and subduing sexuality in the adoles-

Table 2. Marital Status of 15–19-Year-Olds, by Sex, for Selected Countries

Country	Year	Female	Male
		(% Married)	
Ethiopia	1982	53.2	5.2
Senegal	1976	33.2	1.1
Sudan	1973	41.0	4.2
Brazil	1980	16.0	2.3
Indonesia	1980	27.3	3.3
Bangladesh	1974	71.8	7.4
Turkey	1980	21.4	8.0
United Arab Emirates	1975	55.0	8.4
United States	1980	8.2	2.7

Source: United Nations, 1986, pp. 88–110.

cent girl. Food taboos also prejudice the pregnant female's intake during childbearing years. In the more traditional cultures, moreover, menarche becomes the dividing line between freedom and incarceration, leading to the girl's isolation and even to her biological cauterization. The custom of foot-binding, which lasted in China from the tenth to the early part of the twentieth century, was a brutal way of controlling the mobility and sexuality of the young female causing the physical crippling and psychological mutilation of millions of girls. In other parts of the world, the girl's isolation has been enforced through the custom of purdah (seclusion), which bans any form of social intercourse with the male world and is widely practiced not only in the Islamic countries but also among non-Islamic populations in Bangladesh, India, and elsewhere. The girl's biological tampering and sealing occurs through infibulation, female circumcision, or other similar practices that are observed to prevail in at least 25 countries in Africa, Asia, and Arab regions and, in recent years as a result of migration, are surfacing also in Europe and North America. The number of girls affected by these practices is estimated at over 100 million (Sohoni, 1995).

Other cultures that do not practice such overt regulation seek to control the reproductive risk that girls' sexuality poses through the institution of early marriage. According to the United Nations (United Nations, 1991, p. 13), marriage occurs for girls, on the average, at age 20 in Africa, 21 in Asia, 22 in Latin America, and 23 in the developed countries. Averages, however, are not representative of countries and populations at either end of the spectrum. The average age at marriage of girls in Bangladesh, for instance, is estimated by the United Nations Fund for Population Activities (UNFPA) to be 11.6 years. In India, data from two villages showed the average marriage age for girls as 14.3 years. In Nigeria, one-fourth of all marriages of girls take place before age 13; more than one-third by age 14; one-half by age 15; and more than 80 percent before age 20. Boys, on the other hand, marry when they are considerably older. In Sierra Leone, for example, 60 percent of females but only 2 percent of males are married before age 19. Comparative marital-status data of eight

developing countries (see Table 2) reveal that the percent of married girls age 15–19 is from 3 to 10 times higher than the percent of married boys age 15–19 in all countries except Senegal, where the proportion of married girls is 30 times higher than boys. Even in the United States, the proportion of married girls age 15–19 is three times as high as the proportion of married boys.

Among traditional cultures, being married is highly prized as a prerequisite for attaining adult status. As individuals move from childhood directly into adulthood, adolescence as a stage in the life cycle is nonexistent and often has no special name. Where girls marry before or soon after the onset of menses, there is the paradox of wives not yet adults and of child mothers. The purpose of early marriage is to ensure that the girl does not lose her virginity or conceive, both of which would be a source of great shame in traditional cultures. Another advantage associated with early-age marriage is the greater malleability of the younger girl to adapt herself to the ways of her husband and his family. There is evidence that a rise in age at marriage of girls occurs when parents appreciate the earning capability of girls or when, as in India, parents are unable to afford a dowry and other costs of arranging a marriage. Where girls have access to life outside the home and can mix freely with members of the opposite sex, some amount of postponement of marriage is also likely.

Early marriage not only leads to the disturbing phenomenon of children bearing children, it also implies significantly greater reproductive and caregiving burdens on girls than boys at a very early age and for a greater length of time. There are sizable differences in global reproductive activity. Based on an average age at marriage of 17–19 years, the reproductive span for women in Africa, Asia, and Latin America has been estimated at 16–19 years, compared to 7 years for women in the developed countries (United Nations, 1991). Inevitably, longer reproductive spans account for maternal depletion and high levels of infant, child, and maternal mortality as well as morbidity. Although maternal mortality data by each year of mother's age are unavailable for most countries or globally, the burden of mortality is demonstrated to be borne unequally by young mothers or, more accurately, girl-mothers. In Bangladesh, for instance, the maternal mortality rate (MMR) among mothers age 10–14 is five times higher than among mothers aged 20–24; among those age 15–19, it is twice as high as among women age 20–24. In a wide variety of countries, including the Dominican Republic, El Salvador, Jamaica, Japan, Malaysia, Nigeria, Tanzania, and the United States, girls age 15 to 19 are twice as likely to die in childbirth as women age 20–24 (UNFPA, 1988; Sohoni, 1995). Early childbearing is also demonstrably related to low school enrollment and retention of girl-mothers, with attendant economic and social costs to the family and the economy arising from loss of educability and employability of the girl-mother.

The impact of raising marriage ages on female fertility is well documented. In Malaysia, two-thirds of the decline of fertility prior to 1970 was the direct result of a decline in the proportions of women who married early. More recently, Kerala in India provides another convincing example with the lowest infant and overall mortality rate, the highest expectation of life at birth, the lowest birth rate, and the highest average age at marriage of women in India (Government of India, 1995, in Sohoni, 1995).

Delayed marriage reduces fertility in marriage but not necessarily out of wedlock. Where virginity is neither prized nor enforced through religious and social control, the risk of early sexual activity is greater, causing abortion among schoolgirls to become a significant problem, as it has in coastal West Africa. In the developed countries, the contemporary social revolution has resulted in increasing numbers of teenage girls engaging in sexual activity and experiencing motherhood prematurely. In the United States, the Centers for Disease Control in Atlanta, Georgia, reports that 70 percent of U.S. girls now have sex by their high-school graduation day and that the percentage of girls who reported having had sex by age 15 rose from 4.6 percent in 1970 to 25.6 percent in 1988. The pregnancy rate for teenagers younger than 15 increased from 15.9 per 1,000 in 1980 to 18.6 per 1,000 in 1988, and teen births accounted for one-fifth of all births. Pregnancy among U.S. girls age 15–19 years is the highest among the 10 top industrial nations. More than one million girls age 12–19 become pregnant in the United States annually. Planned Parenthood estimates that 40 percent of U.S. girls 14 years old will become pregnant in their teens (*San Francisco Chronicle,* October 14, 1994, in Sohoni, 1995).

Almost two-thirds of all births to U.S. teenagers—about one million annually—are to unmarried girls, up from less than one-third in 1970. A study (UNFPA, 1988) notes that three-fourths of teenage births in Denmark and Sweden were out of wedlock and that in England more than 50 percent of all such births in 1982 were to teenage mothers. According to one estimate (Newland, 1979, in Sohoni, 1995), 40 percent to 50 percent of all children in Paraguay, Peru, and Venezuela and 32 percent in Ecuador were believed to be out-of-wedlock children.

A growing trend in the United States is for teen mothers to take their pregnancies to full term rather than undergo abortion. The findings of federal research show that more U.S. girls age 15 and younger are getting pregnant, and more are keeping their babies (Sohoni, 1995). Among that age group, there were 949 abortions for every 1,000 live births in 1988, compared to 1,408 abortions per 1,000 live births in 1980. Of the total legal abortions (about 1.4 million in 1988), according to the Centers for Disease Control, women below age 20 accounted for 25 percent in 1988, down from 29 percent in 1980.

Thus, although marriage rates may have declined in the United States and other developed countries, as well as in the modern sectors of some developing countries, sexuality and fertility are not necessarily declining among younger females. On the contrary, girls at a younger age are being trapped into similar burdens that early marriage and motherhood impose on younger girls compulsorily in traditional societies.

Largely as a result of the economic, educational, and health disadvantages that early sexuality and marriage pose for the young female, teen fertility has been accepted as a growing tragic phenomenon in women's health. Teen reproduction, out-of-wedlock especially, has grave consequences. In the United States, scholars commonly predict that a teen mother is likely to drop out of school; not find a steady job; not be able to earn enough to provide for herself and her child; be impelled to marry someone for the wrong reasons; have few life choices, and most of them bad; and experience an unstable environment for rearing her child. In the developing countries, all of those disabilities confronting the teen mother can only be further accentuated by poverty, social stigma, and the personal risk posed to her own survival through premature motherhood.

A critical issue related to the young female's sexuality is that of prostitution and illegal drug trafficking (UNICEF, 1997). Whereas prostitution has always prevailed, its highly organized promotion on an international scale as a lucrative adjunct to tourism has caused a phenomenal growth of young women engaged in commercial and recreational sex. AIDS and tourism jointly have increased the profitability and global spread of child prostitution. At least 32 countries in the world report sexual abuse of children by foreign visitors, businessmen or tourists (Hornblower, 1993). A 1991 conference of Southeast Asian women's organizations estimated that, since the mid-1970s, 30 million females had been sold worldwide. Such figures are, at best, guesses representing merely the tip of the iceberg. Prostitution is presented to young girls and their parents as a paying entrepreneurial venture. In Thailand and Nepal, girls as young as 11–14 years old are being lured into prostitution through deception and coercion, giving rise to sex slavery. Current estimates place the number of girl prostitutes under the age of 16 in Thailand alone at 800,000 (UNICEF, 1990). A study in Bombay by Chatterjee found 20 percent of an estimated 100,000 prostitutes to be minors (UNICEF, 1990). Commercialization of children for sex is a common feature of economic life in every region of the world. The young female's recruitment into prostitution and pornography is rampant also in the developed nations, including Bulgaria, Romania, Hungary, Czechoslovakia, the Netherlands, Germany, Belgium, Japan, Israel, Italy, Canada, the United States, and former Soviet Republics (UNICEF, 1994a). The attendant risk of AIDS to these young girls as well as to their offspring is a matter of growing concern. Countries that have expressed anxiety about teen prostitution and associated sexual health risks include Bangladesh, Belize, Djibouti, India, Jamaica, Nepal, Peru, and Rwanda. However, where sexual abuse and exploitation of girls are not openly noted, their absence can hardly be assumed (UNICEF, 1994b).

Health and Survival of Girls

Health and survival are telling measures of the relative values placed upon the female and the male in society. Discrimination in health care tends to be reflected through higher death and disease among girls than boys. In terms of sheer survival, females are presumed to be biologically superior to males. More male babies are conceived and born than female, but more male infants are presumed to die than female, allowing the female to overtake the male population. Traditionally, this pattern in favor of girls is presumed to exist in all populations except where the neglect of girls outweighs the substantial biological vulnerability that boys experience during the first year of life. Unfortunately, international data broken down by gender are available only in respect to infant (age 0–1 years) and not child (age 1–4) mortality, although it is at the latter stage that the relative neglect of girls can become more apparent in mortality statistics. The reliability of available data is also questionable, given that parents have a tendency to underreport the mortality of girls. Data drawn from 28 developing countries for the 1983 World Fertility Survey, however, do indicate higher female than male deaths among 1–2-year-olds in 13 countries and among 2–5-year-olds in 18 countries (UNFPA, 1988; Lott, 1994).

Life expectancy among girls is adversely affected by parental bias. For the Third World as a whole, although the average girl in 1985 could hope for an added 18 years of life compared to her mother's generation, female life expectancy continued to be unacceptably low in many developing countries. In contrast to the average of 77 years in developed countries, 14 of the poorer countries, representing a female population of 66 million, had not attained a life expectancy of 50 years by 1985; with one exception, these countries were in sub-Saharan Africa or South Asia. Although the difference in life expectancy between the countries with the highest and the lowest levels was reduced from 47 to 36 years during the years 1980–1985, female babies in poorer countries still had a shorter life ahead than those in richer countries had in 1950 (Sivard, 1985, p. 24).

The normal expectation is that, under equal-treatment conditions, the female should have a higher life expectancy than the male. This is not true of countries with strong male preference. In Bangladesh, for instance, expectation of life at each age group (0–1, 0–4, 1–4, 5–9, and 10–14) was found to be higher for male than female children. In India, age-specific death rates reveal that up to the age of 35 years, more females than males die at every age level (Sohoni, 1995). A 1986 survey in Jordan (Sohoni, 1995) revealed infant mortality to be up to 20 points lower for males in Amman: Son preference leading to the neglect of daughters was found to be responsible for the difference in mortality. Recent evidence from a variety of countries, including Egypt, Nigeria, and Pakistan, reiterates the continued gender bias in health care. More male children are immunized and treated by hospitals than female; mortality due to measles, diarrhea, and respiratory infections is higher among female children; girls are weaned more quickly; boys are breast-fed longer; and girls are usually taken to the hospital later and in worse condition than boys. This unequal treatment is estimated to have accounted for 100 million fewer females totally in the world's population than would be expected from general demographic trends (Sen, 1990).

Nutritional Intake of Girls

Whereas 40 percent or more of the world's children are born in poverty and face the consequences of poor nutrition, gender-directed nutritional research has established that the intrahousehold distribution of food in many developing countries is skewed in favor of the earning male and of the boy as potential earner, with women and girls getting the lesser share in terms of both quantity and quality. In most societies, it is customary for men to eat first, boys next, girls and women last. Where food, especially protein, is scarce, it goes primarily to the men. In many countries in East Africa, young women and children, especially girls—who can be regarded as most in need of optimal health for the task of child-bearing—tend too frequently to receive leftovers and to bear the brunt of traditional food taboos. Similar evidence is found elsewhere. In the state of Tlaxcala in Mexico, an examination of the intake of selected nutrients for a sample of families showed that the intake of males, relative to requirement, exceeded that of females for all nutrients examined. A Philippines study that looked into the intrafamily allocation of food in 97 households in rice-growing communities found that fathers had slightly better diets than mothers, and male children better than female children. Of all sex–age groupings among children, male preschoolers and female adolescents were, respectively, the best and the least adequately fed groups. A Bangladesh village study showed that at all ages under 5, girls received less food than boys. Differentials in the nutritional treatment and status of girls have been observed in a number of countries, including Bolivia, Colombia, Egypt, India, Iran, Jordan, Nepal, Pakistan, Saudi Arabia, and Syria. In Pakistan, a tendency for longer breast-feeding of boys and early introduction of solids for girls was apparent. Throughout the Arab world, a male baby is likely to be nursed until he is two-and-a-half, a year longer than a female baby (Ravindran, 1986, pp. 14–15).

Table 3. Difference Between Male and Female Gross Enrollment Ratios (Percentage points)

Region	First Level		Second Level		Third Level	
	1970	1986	1970	1986	1970	1986
World	15	15	9	10	4	3
Developed countries	1	1	1	−8	8	−1
Developing countries	19	18	11	14	3	4
Africa	22	18	8	17	2	4
Latin America/Caribbean	2	5	2	−4	4	3
North America	1	2	−2	−1	15	−9
Asia 20	20	12	14	3	4	
Europe/former USSR	1	1	2	−10	5	1
Oceania	4	4	3	−2	8	0
Arab States	32	22	16	17	5	7

Source: UNESCO, 1988, p. 6.

Note: Figures for different levels of education are not strictly comparable due to the different values of the enrollment ratios and the different age ranges used in the calculation of enrollment ratios at each level.

Discriminatory breast-feeding, weaning, and dietary practices begin early in the affected countries and continue through adolescence, accounting for lower heights and weights and making a lifelong difference. In Syria, it was found that at every age the median weight of girls was below the standard to a greater extent than the median weight of boys. Protein-energy malnutrition was marginally higher in a sample of preschool girls in Calendria, Colombia, and twice as common among girls as boys in Iran. In Bolivia, low weight-for-age was seven times as common for girls, and four times as common for boys, as in the United States. There is evidence of a differential and delay of at least one year in the age at menarche of Indian girls who come from the poorest families, compared to those from the highest socioeconomic groups. A similar age differential is observed between rural and urban girls. Body weights of rural and urban girls in India also record a difference of the order of 3–4 kg (7–9 lbs.) at each year between the ages of 12 and 16. The critical variables contributing to these age-weight and menarche differences were differences in income, purchasing power, distribution of household and other labor, food sufficiency, nutritional habits, and value systems.

Education is a key indicator and determinant of status. Despite the efforts of the last few decades, girls continue to enjoy a less favorable educational status than boys (King, 1990). Although more educated today than before, girls continue to be less educated than boys. According to UNESCO (1988), about 60 million girls in the world have no access to primary schooling, compared to 40 million boys. In the developing countries, fewer girls enroll in school and far more drop out than boys. Between 1970 and 1986, gross enrollment ratios at all three levels of education in the developing regions—primary, secondary, and tertiary—continued to show considerable disparities in favor of male children (see Table 3).

Parental reluctance to enroll and maintain girls in schools even when primary and secondary education may be free and compulsory is associated with their limited view of girls as liabilities rather than assets. As an Indian proverb crudely puts it, "Investing in a girl is like watering a plant in a neighbor's garden." Parents perceive disadvantages (opportunity costs in economic language) in taking girls away from their role as surrogate caregivers and as additional hands for the family-owned farm or other business. The risk of physical mobility and its inevitable impact on the girl's personal, sexual, intellectual, and financial autonomy are other factors that weigh against educating girls. Additional constraints include the absence of nearby schools and of women teachers; the paucity of school uniforms, books, and equipment; and the lack of toilets. Whereas poverty, parental illiteracy, and tradition have a lot to do with keeping girls out of schools, even where poverty is not an overriding factor various forms of discrimination inhibit equal opportunity for girls in education. A report issued by the American Association of University Women (1992) concludes that, on the basis of available research on the subject of girls in school, girls do not receive the same quality or even quantity of education as their brothers. They are denied equitable amounts of teacher attention and are less apt than boys to see themselves reflected in the materials they study and to pursue higher-level mathematics and science courses.

Both public (state) and private (family) decision making and allocation of resources tend to favor the access of boys more than girls to education. In Kenya and other East African nations, it was observed that the payment of school fees for secondary education serves as a deterrent for female attendance, especially if the family's resources are limited and there is a male around who is of school age. The tendency to invest more in college education for sons than daughters is not unknown among families in the developed countries or among the educated elite in developing countries. In several developing countries, the proportionally greater investment in higher levels of education also impacts negatively on girls compared to boys, causing fewer girls to continue to higher levels of learning.

A disturbing facet of discriminatory treatment of girls in education is that they tend to internalize parental perceptions and rationalize higher spending on their brothers by their parents. This has been documented by UNICEF in Bihar, India. In Malaysia, a study of Chinese girls revealed that they were not as concerned as males with education as an economic investment. They had internalized social attitudes about their sex, and their occupational and educational choices remained limited by internalized conceptions of what is feminine and what women are capable of doing. These beliefs determine their educational and job decisions to a much greater degree than does financial remuneration.

Gender stereotyping in school textbooks and curricula reinforces societal perceptions and parental beliefs, including those of the girls themselves, as to the limited potential of girls and the questionable value of educating and investing in them. In turn, the lack of education and formal or informal employable skills compels girls to be consigned as adults either to unemployment or to low-skill, low-pay, low-status occupations. As adult earners, since they are destined to, and do, earn less than men, the cycle of perceiving girls as being economically less worthy than boys continues unbroken.

Unenumerated and Unremunerated Work

Being threatened as a fetus and during infancy is not the only survival risk that confronts girls. Perhaps even more damaging is the callous flouting of their right to childhood. In that sense, it is hardly a boon for the surviving girls to be forced to experience prematurely the hardships associated with adult life. In several recent studies, there is considerable evidence that girls are not only neglected in terms of their nutritional and health-care needs but also are overworked, underpaid, or unpaid (UNDP, 1997; UNICEF, 1997).

The productive burden shouldered by the girl continues to be a gray area in development strategies and public policy. Current yardsticks to measure their unenumerated and unremunerated labor are not only inadequate but inequitable; they do not honor the real work situations of girls. Time-use studies in Africa and Asia show that the female child, in fact, bears a heavier burden of both household activities and care of siblings than the male child. Girls age 10–14 put in seven or more hours of labor per day in household production and domestic work. In a majority of the world, the greater share of work within the family seems to be reserved for girls rather than boys. In Java, for instance, most young girls spend at least 33 percent more hours per day working at home and in the market than boys of the same age. Malaysian girls age 5–6 work 75 percent more hours per week than boys of the same age. In Cote d'Ivoire, girls age 10–14 work three to five hours per day in household chores while their male counterparts work only two hours. In Nepal, girls spend more time in fetching water, collecting fuel, and growing and processing food than boys of the same age group. Even in the developed countries, disparities surface in the workloads and nature of work of girls and boys. Time-use studies of agricultural households in Italy reveal that girls put in 4.35 and

5.01 hours daily on remunerated and unremunerated work, respectively, compared to 7.20 and 2.11 hours put in by boys. In Australia, sons and daughters do different amounts and different types of unpaid work. Sons do only two-thirds of the unpaid work that daughters do, and most of it is done outdoors. For every hour of unpaid work done by a son, a daughter does one hour and 20 minutes. Wage differentials among children reflect the wage-discriminatory adult world. This disparity is also prevalent in developed countries. Evidence from a U.S. study indicates that parents are more likely to pay boys than girls for the work they do inside the home; this pattern also holds outside the house, so that girls earn less in general. Moreover, boys are encouraged and observed to work for "pay" and girls "out of love."

Regardless of the length and hardship of their work, the work that girls do continues to be invisible not only to the larger society and economy but also to their parents. Parental perceptions continue to underrate the economic value of the girl. A study of the value and cost of children among 600 rural wives and husbands in Nigeria revealed that the majority of parents believed that boys were more productive than girls of the same age. Both men and women also thought that parents should invest more in their male children, particularly on their feeding and education. Such parental attitudes explain why boys often receive greater opportunities to learn economically valuable skills than girls.

Those parental attitudes carry over into the state, the society, and the economy, enabling planners and policy framers to ignore the special vulnerability, as well as potential, of the girl. Worse, they impact negatively on girls' perception of themselves and their educational and occupational horizons. For instance, occupational aspirations of adolescent girls in the United States are found to be lower than those of comparable boys: Girls underplay future occupational and educational goals in relation to their academic ability and grades.

Psychologists such as Lott (1994) confirm that others' knowledge of a child's sex continues to influence perceptions beyond infancy of many of the child's characteristics and to affect their assumptions about the child's interests, abilities, and development. Picking up these perceptions from their environment, girls begin to show those differences in behavior and aptitude that the culture expects of them. Thus girls find themselves in a double bind occupationally speaking. On the one hand, parents and society make fewer investments in their employment preparation. On the other hand, social handling of girls compels them to self-limit their aspirations and capabilities.

Self-Esteem

Traditionally, as scholars such as Ware (1981) have pointed out, demographers and planners have confined their concern to the desire for sons and its impact on fertility levels. The impact of sex preference on the "unwanted" sex has not been given much attention as an area of study or public policy. Yet, the outcome for girls of the absence of an ethos of care is disturbing and lasting. The damage caused by their neglect is intergenerational and holistic. It affects the girl's entire growth process and her potential and performance as an adult. Sadly, a bruised and deprived girlhood is not unique to the developing regions of the world. Even in the developed countries, the emotional costs of girlhood are high. In the United States, for instance, there is a drop of 31 percentage points between elementary and high-school girls in self-esteem, and the gap between boys and girls on self-esteem increases with age (Dworkin, 1991). White girls, surprisingly, end up at the bottom of the self-image heap. Only 34 percent of White high-school girls say they are happy with the way they are, compared to 52 percent of Hispanic girls and 60 percent of Black girls. Girls learn early and continue to learn as they grow older that they are less valued, less important, less powerful, and less effective than boys. Lower self-esteem is also linked to higher attempted suicide rates. As scholars have noted, girls tend to self-destruct. Because they usually do not act out their troubles, as boys do, they often do not get the attention they need when they are hurting.

Personal and Familial Violence

As noted earlier, girls are prey to societal violence through such customs as foot binding, infibulation, and purdah to control their sexuality and to such systematic violence as infanticide and genetic selection. Additionally, there is need to acknowledge the rampant nature of personal violence perpetrated on girls by known and unknown males. Such violence entails demeaning, intrusive, insulting, violative, or oppressive behavior by males manifested through "eve-teasing" (verbal and physical harassment in streets, parks, buses, and other public places), sexual abuse, sexual harassment at work and in other controlling situations, battering, rape, and other forms of forced sex.

Male violence is imbibed culturally rather than biologically, as is female passivity. Although widespread, the magnitude and impact of male violence has not been systematically tracked by development analysts and social scientists. In the developing countries, much of the family-instituted violence against girls—with the exception of bride burning in India and self-immolation of widows through the custom of *sati*— is still not openly discussed, let alone tracked or tackled. But evidence from the developed countries is disturbing. In the United States, child sexual abuse is reported to be more than twice as prevalent among girls as boys. Only one-third of child molestations are believed to be committed by strangers, another one-third by acquaintances, and the remaining third by primary relatives. Date rape and date-related violence are also recurring themes in U.S. research. Regardless of their cultural or economic background, girls speak of the violence in their intimate relationships. As many as one in 10 teenage girls in the United States is estimated to experience date-related violence. One in four college women is a victim of actual or attempted rape while she is in college, mostly by someone who is known to her. Many more teen and col-

lege-age girls may be experiencing covert rather than overt violence by way of verbal abuse, put-downs, threats, and other forms of controlling and abusive behavior. Rape in marriage, like child abuse, is only just beginning to surface as an area of violative and violent conduct toward the female. Inevitably, the younger the age, the greater is the risk of such abuse. Street harassment, which prevails in most cultures and countries, is another facet of male complicity particularly linked to the denigration, and lack of safety, of girls. All of these forms of violence and control are ways of restricting the inner and outer space of the girl. They are also an infringement of her basic right to privacy and dignity.

Corrective Strategies

A chronic limitation of contemporary development has been to perceive the female mainly in her reproductive role. Childbearing and child care have been the female's nexus in most gender systems. Whereas this optic has evoked a certain amount of care for the female in her reproductive years, it has left females at either end of the spectrum untended. Yet, the need for protection and preparation is particularly crucial during girlhood when the physical, mental, and emotional circuits are being designed for human efficacy and fulfillment. The argument for integrating girls in development is familiar to those who in the 1960s advocated on behalf of children. It was then effectively argued and accepted that planning for tomorrow's adults must begin with caring for the children today and, further, that the costs of negligence of children are intergenerational, often irreversible. The girl offers the same irrefutable logic and link to the status of the adult woman and to the children she herself will bear. But quite independently of her link to the future as the prospective bearer and caretaker of the coming generation, there is her intrinsic right to her own childhood and to its equitable gender entitlements, which development strategists cannot any longer set aside.

A strategic shift is required also in the societal approach to girls. In human evolution, the principal basis for neglect of girls has been their suspected or feared sexuality. In most societies, in fact, it is the denial of sexuality that has caused girlhood—especially female adolescence—to be denied by default. It is evident that girls, and not their sexuality, are in need of the broadest and widest attention.

Fortunately, in the more traditional countries, laws are being formulated or tightened to help raise and enforce the minimum age at marriage of girls, to eliminate harmful health and social customs, and to vest girls with greater inheritable human rights within both the family and society. But laws alone are not enough to make a significant behavioral change, whether among sexually active teenagers in the freer context of developed countries or among parents determined to marry off their girls while still young, in conventional developing cultures that are committed to early marriage and motherhood. Better, more rigorous public education concerning the aggregate potential as well as the greater vulnerability of girls, along with more vigilant enforcement of corrective legislation and sociological reformation, will enable girls to overcome the risks posed to their bodies and life chances by all-round neglect and by such deleterious practices as early marriage or sexual cohabitation, purdah, infibulation, and dowry. Above all, the girl's economic, emotional, biological, and social autonomy has to be restored to her.

An overall ethos of care for the adolescent and younger girls needs to be created. Especially relevant in this context is the need for planners, strategists, and feminists to depart from a false complacency that gains resulting from "women's development" do or will automatically percolate to girls. It is the fatal combination of her age and her gender that causes the younger female's life chances to be further compressed compared to the adult female. It is imperative to recognize that childhood is politically mute, and girlhood is culturally so as well. It is this veil of silence that has to be lifted for girlhood to come into its own, and provide a strong foundation for future womanhood.

Affirmative Policies for Girls in International Development

This section, based on a review of the negative status of girls in national and international development, presents some action points for consideration of policymakers and development strategists.

The first step toward enhancing the status of girls is to recognize that the problem of their greater vulnerability exists. Clearly, girls have been the most sizable casualty of development processes. Fortunately, there is growing recognition in various regions of the world that the constraints and challenges girls face in exercising their full human potential are different from those faced by boys and women. Countries such as Bangladesh, India, Nepal, and Pakistan have created national and regional networks to expand the information and policymaking base so as to frame appropriate affirmative-action programs for enhancing the education, health, nutrition, and employment preparation of girls. Similar initiatives are being fostered in Arab regions and in a variety of countries in Africa, Latin America, and Oceania. In the developed countries, too, a movement has begun to examine the lesser participation of girls in mathematics and sciences and sports and their overrepresentation in other disciplines. Gender roles among children are being critically examined and recast through action research. Several countries have designated a recent year as the Year of the Girl, and member countries of the South Asian Association for Regional Cooperation have declared 1991–2000 as the Decade of the Girl Child. The United States House of Representatives passed a resolution designating the remaining years of the 1990s as the Years of the Girl Child. At the global level, the adoption by the United Nations General Assembly of the Convention on the Rights of the Child in 1989 and its ratification by 193 countries by 1997 (UNICEF, 1997) committed a majority of countries to the convention's explicit goal of equality and nondiscrimination among children. Article 2 of the convention sets out the principle that all of children's rights to survival, protection, and development apply equally to every child regardless of gender. It is imperative for signatory countries to enforce this key international mandate and for others to lend their explicit support to it. The global momentum generated by the convention's ratification provides a concrete basis for member nations to urge the United Nations to adopt the decade beginning with 2001 as the Decade for Girls.

Whereas declarations and signatures have considerable symbolic and catalytic value, what is required of countries is to take up concrete, integrated, and sustained action on behalf of girls. Such action includes, but is not limited to, the following:

- Countries, along with bilateral, multilateral, and private voluntary bodies, should commit to gender parity in all areas of public policy affecting children; in support of such policy, they should adopt legal and other instruments committing political and societal support to reduction of disparities in the short run and elimination of the basis for discrimination in the long run.

- Concrete, time-bound targets need to be adopted to equalize benefits for girls and boys in the key sectors of human development: survival, health, nutrition, education, employment preparation, marital and reproductive activity, and the like. Where, as in mortality and morbidity or school dropout rates, girls show up more adversely, those indicators need priority targeting and remedial action.

- All research, monitoring, and evaluation activity should produce gender-specific profiling (data developed separately for males and females at all ages and in all sectors of development) on a sustained basis and should seek to make up for the paucity of information disaggregated by gender in all sectors of human development. Furthermore, development research should promote careful documentation of sexism in development as it affects all areas of life, including attitudes, beliefs, behavior, and entitlements.

- All planning, programming, and budgeting must address gender-specific needs, balancing disparities where they exist and undertaking compensatory or affirmative action where necessary to bridge gaps in coverage. The futility of blanket policies and approaches to serve populations by age groups without reference to gender, ethnicity, or socioeconomic disability needs to be accepted, and barriers to resources based on age and gender eliminated.

- International development strategies and scholarship should end the misleading role they have played in generating and perpetuating the perception of girls as a Third World problem and accept that the burden of girlhood is universal.

- Greater understanding and treatment of the asymmetry in the status of girls vis-à-vis boys as well as women will require the combined attention of both the developed and the developing countries. Multilateral and bilateral cooperative arrangements that facilitate monitoring and exchange of relevant information and experience among countries will enhance the prospects for making development processes more fair and equitable to girls.

- Since gender, in contrast to sex, is socially constructed, no significant success in overcoming genderism as it affects girls can be achieved without the willingness of the international development machinery to acknowledge and confront the basic structuring of gender relations in childhood, as distinct from adulthood. Such an approach necessitates the grounding of girls independently of women in development and within a theoretical framework that is willing to identify and come to grips with the nature, causes, and sources of the subordination and marginalization of girls. The object of analysis can no longer be women alone but must be the relations among girls, boys, women, and men.

The ultimate goal and outcome of all of the above interventions can only be to end the patriarchal, sexist nature of contemporary life and its negative consequences for girls' development. Naturally, such a structural transformation requires synergistic cooperative action by the forces that condition human behavior, including intergovernmental and nongovernmental structures, governments, religion, culture, law, business, academia, the media, and parents themselves. All of them need to participate in a massive, well-orchestrated effort toward the attainment of a gender-just life for children.

In the final analysis, governments and people everywhere have to accept that even where or when women begin to enjoy a high status, girls do not automatically share that experience, but that the reverse is *not* true. Strategies for empowering girls are the beginning of women's empowerment.

References

American Association of University Women. *How Schools Shortchange Girls*. Washington, D.C.: American Association of University Women, 1992.

Aries, P. *Centuries of Childhood: A Social History of Family Life*. New York: Kanops, 1962.

Chodorow, Nancy. "Family Structure and Feminine Personality." In Laurel Richardson and Verta Taylor (eds.), *Feminist Frontiers II: Rethinking Sex, Gender, and Society*. 2nd ed. New York: McGraw-Hill, 1989, pp. 43–57.

Dworkin, Susan. "Can We Save the Girls?" *New Directions for Women*, September–October 1991, pp. 3–4.

Heyzer, Noeleen (ed.). *Working Women in South-East Asia: Development, Subordination, and Emancipation*. Philadelphia: Open University Press, 1986.

Hornblower, Margot. "The Skin Trade." *Time*, June 21, 1993, pp. 45–51.

King, Elizabeth M. *Educating Girls and Women: Investing in Development*. Summary report. Washington, D.C.: World Bank, 1990.

Lees, Sue. *Sugar and Spice: Sexuality and Adolescent Girls*. London: Penguin, 1993.

Lott, Bernice. *Women's Lives: Themes and Variations in Gender Learning*. 2nd ed. Pacific Grove: Brooks/Cole, 1994.

Ravindran, Sundari. *Health Implications of Sex Discrimination in Childhood*. Geneva: World Health Organization (WHO)/United Nations Childrens Fund (UNICEF), 1986.

Richardson, Laurel, and Verta Taylor (eds.). *Feminist Frontiers II: Rethinking Sex, Gender, and Society*. 2nd ed. New York: McGraw-Hill, 1989.

San Francisco Chronicle, October 14, 1994.

Sen, Amartya. "More Than 100 Million Women Are Missing." *New York Review of Books,* vol. 37, no. 20, December 20, 1990, pp. 61–67.

Sivard, Ruth Leger. *WOMEN . . . a World Survey*. Washington, D.C.: World Priorities, 1985.

Sohoni, Neera K. *The Burden of Girlhood—A Global Inquiry into the Status of Girls*. Oakland: Third Party, 1995.

United Nations. *Compendium of Statistics and Indicators on the Situation of Women*. New York: United Nations, 1986.

———. *Global Estimates and Projections of Population by Sex and Age*. New York: United Nations, 1989.

———. *The World's Women, 1970–1990: Trends and Statistics*. New York: United Nations, 1991.

United Nations Children's Fund (UNICEF). *First Call for Children*. New York: UNICEF, 1994a.

———. *The Girl Child: An Investment in the Future*. New York: UNICEF, 1994b.

———. *Status of the World's Children Report, 1997*. New York: UNICEF, 1997.

United Nations Development Program (UNDP). *Human Development Report 1997*. New York: Oxford University Press, 1997.

United Nations Educational, Scientific, and Cultural Organization (UNESCO). *A Review of Education in the World: A Statistical Analysis*. Paris: UNESCO, 1988.

United Nations Fund for Population Activities (UNFPA). *1988 Report*. New York: United Nations Population Fund, 1988.

Ware, Helen. *Women, Demography, and Development*. Development Studies Center Demography Teaching Notes no. 3. Canberra: Australian National University, 1981.

The United Nations Decade for Women and Beyond

Valentine M. Moghadam

The attainment of the goals and objectives of the [U.N.] Decade [for Women] requires a sharing of this responsibility by men and women and by society as a whole, and requires that women play a central role as intellectuals, policy-makers, decision-makers, planners, and contributors and beneficiaries of development (*Nairobi Forward-Looking Strategies,* paragraph 15).

Introduction

The United Nations Decade for Women (1976–1985) was a pathbreaking event in the evolution of the women's movement worldwide. In 1972, the United Nations General Assembly passed a resolution proclaiming 1975 International Women's Year, to be devoted to intensified action to promote equality between men and women, to ensure the full integration of women in the total development effort, and to increase women's contribution to the strengthening of world peace. The World Plan of Action for the Implementation of the Objectives of the International Women's Year, adopted by the World Conference of the International Women's Year (Mexico City, 1975) was endorsed by the General Assembly, which also proclaimed 1976–1985 the United Nations Decade for Women: Equality, Development, and Peace. A subsequent resolution decided that the subthemes of employment, health, and education would be the focus of a mid-Decade conference on women, to be held in Copenhagen in 1980. The Copenhagen World Conference adopted a program of action that further elaborated on the existing obstacles to women's advancement and on measures to be taken. At the end of the Decade, in July 1985, Nairobi was host to the United Nations World Conference to Review and Appraise the Achievements of the United Nations Decade for Women: Equality, Development and Peace. Also referred to as the Third World Conference on Women, the Nairobi meeting was a landmark conference, for it was there that *The Nairobi Forward-Looking Strategies for the Advancement of Women* (com-

monly called the *Forward-Looking Strategies*), were adopted unanimously. Given that the earlier conferences had not been able to reach unanimity in their deliberations, the adoption of the 1985 document by consensus signaled a new stage in the women's movement and in governments' approach to women-and-development issues.

This essay describes the process leading up to the Decade for Women, including the adoption of conventions, recommendations, and norms, and some of the subsequent developments, including new areas of research and forms of advocacy. Emphasis is placed on milestones in the process of U. N. efforts toward the advancement of women, feminist networking, and trends in gender analysis within the U. N. system.

Legal Instruments and World Conferences

Issues concerning women and of interest to them have gone through a process of varying treatment by the United Nations and its specialized agencies. During the 1950s and 1960s, women's issues were seen primarily within the context of human rights and were not related to the "larger" issues of development and peace (Pietilä and Vickers, 1994, p. 118). A change came about in the 1970s, partly as a result of the publication of Ester Boserup's now-classic study, *Woman's Role in Economic Development,* which examined the key role of women, especially in the fields of population and food, and partly as a result of the growing women's movement worldwide.

The human rights of women and the principle of the equality of men and women were recognized early on in the history of the United Nations. The United Nations Charter states: "We, the peoples of the United Nations, determined . . . to reaffirm faith in fundamental human rights, in the dignity and worth of the human person, in equal rights of men and women and of nations large and small . . . have resolved to combine our efforts to accomplish these aims." It also stipulates that one of the purposes of the United Nations is "to achieve international cooperation . . . in promoting and encouraging respect for human rights and for fundamental freedoms for all without distinction as to race, sex, language or religion." At the first session of the United Nations General Assembly, a commission, headed by Eleanor Roosevelt, was appointed to draft a Universal Declaration of Human Rights. When adopted in 1948, the declaration contained the words: "All human beings are born free and equal in dignity and rights." Article 2 of the declaration is even more specific: "Everyone is entitled to all rights and freedoms set forth in this Declaration, without distinction of any kind, such as race, color, sex, language. . . ."

In 1946, the Commission on the Status of Women was established with a mandate to study and prepare recommendations on human-rights issues of special concern to women. Its first task was to determine in which conditions and situations worldwide the most severe forms of discrimination against women occurred; four areas formed the point of departure for its work: (1) political rights and the possibility of exercising them; (2) legal rights of women, both as individuals and as family members; (3) access of girls and women to education and training, including vocational training; and (4) working life. Since then, recommendations and conventions have been prepared and adopted by the United Nations, the United Nations Educational, Scientific, and Cultural Organization (UNESCO), and the International Labor Organization (ILO) in the above fields. Table 1 lists the most important of those relating directly to women.

Of the various legal instruments providing for the equality of the sexes, the Convention on the Political Rights of Women, adopted in 1952, has been one of the more successful ones. When the United Nations Charter was signed in 1945, political rights of women were in force in only 30 of the 52 signatory states. By 1993, 104 countries had ratified this convention, and today few countries deny women such political rights as voting, taking part in the formal political process, and being elected to political office. Another successful legal instrument was the Convention on Consent to Marriage, Minimum Age of Marriage, and Registration of Marriages, adopted in 1962. Although many countries still have not signed the convention; registration of marriages is almost universal, and most countries have set a minimum age of marriage, usually 16 for girls. A crucial legal measure to improve the situation of women was the recognition of the right to family planning and access to the information and practical means necessary to exercise it. Mentioned for the first time in the 1968 Declaration of Teheran, and included in the General Assembly's 1969 Declaration on Social Progress and Development, it was subsequently inserted as an obligatory provision in the Convention on the Elimination of All Forms of Discrimination Against Women, adopted in 1979.

In the international development strategy for the U.N.'s Third Development Decade (the 1980s), a trend emerged toward seeing women as equals, "as agents and beneficiaries in all sectors and at all levels of the development process" (Pietilä and Vickers, 1994, p. ix). The year 1985 became a turning point in the history of women's issues in the U. N. system, with the world conference in Nairobi and its *Forward-Looking Strategies*. These specifically recognized women as "intellectuals, policy-makers, decision-makers, planners, and contributors and beneficiaries of development" (paragraph 15) and required implementation by both member governments and the U. N. system. They constituted the main U. N. accords for the advancement of women, and the

Table 1. Selected Conventions of Concern to Women

Adopted		In Force	Ratifications (as of Sept. 1993)
1949	Convention for the Suppression of Traffic in Persons and the Exploitation of the Prostitution of Others	1951	68
1951	Equal Remuneration for Men and Women Workers for Work of Equal Value (ILO No. 100)	1953	120
1952	Maternity Protection (Revised) (ILO No. 103)	1954	31
1952	Convention on the Political Rights of Women	1954	104
1958	Discrimination in Respect of Employment and Occupation (ILO No. 111)	1960	118
1960	International Convention Against Discrimination in Education (UNESCO)	1962	82
1962	Convention on Consent to Marriage, Minimum Age of Marriage, and Registration of Marriages	1964	41
1979	Convention on the Elimination of all Forms of Discrimination Against Women	1981	131
1981	Convention Concerning Equal Opportunities and Equal Treatment for Men and Women Workers: Workers with Family Responsibilities (ILO No. 156)	1982	20

Source: Pietilä and Vickers, 1994, p. 120; ILO, 1993.

U. N. system committed itself to implement them by adopting the System-Wide Medium-Term Plan for Women and Development in 1990–1995 and a corresponding plan for the years 1996–2001.

The Fourth World Conference on Women (Beijing, September 4–15, 1995) was a stocktaking of the progress made in the worldwide advancement of women: Among other things, government reports reviewed and assessed national implementation of the *Nairobi Forward-Looking Strategies*. The task of the Beijing conference was to adopt an action-oriented *Platform for Action* for the rest of the decade.

The milestones in the process of U. N. efforts toward the advancement of women may be delineated as follows:

1. The United Nations Convention on the Political Rights of Women (1952)
2. International Women's Year (1975)
3. The International Women's Year World Conference (Mexico City, 1975) and its Declaration and *World Plan of Action for Implementation of the Objectives of International Women's Year*
4. The United Nations Decade for Women: Equality, Development, and Peace, proclaimed by the General Assembly in Resolution 3520, 1975
5. The Convention on the Elimination of All Forms of Discrimination Against Women, adopted by the General Assembly in 1979
6. The World Conference on Women (Copenhagen, 1980) and its *Program of Action for the Second Half of the United Nations Decade for Women*
7. The World Conference to Review and Appraise the Achievements of the United Nations Decade for Women: Equality, Development, and Peace, (Nairobi, 1985) and its *Forward-Looking Strategies* for the period 1986–2000
8. The Fourth World Conference on Women: Action for Equality, Development, and Peace (Beijing, 1995), which appraised the implementation of the *Forward-Looking Strategies* and adopted a *Platform for Action*

The Women's Convention

By far the most important of the U. N. conventions on women's rights, the Convention on the Elimination of All Forms of Discrimination Against Women, was adopted in 1979 without any dissenting vote and entered into force in 1981 following ratification by the required 20 countries. Pietilä and Vickers (1994) describe it as "a concise and comprehensive conclusion to the long process which had taken place within the United Nations System during some 30 years to incorporate the principles of gender equality in the provisions of international law, covering all relevant provisions of previous, separate conventions and complementing them with regard to issues not yet covered" (Pietilä and Vickers, 1994, p. 126). It provides for the establishment of a Committee for the Elimination of Discrimination Against Women (CEDAW) to monitor its implementation—that is, implementation of human-rights conventions from the point of view of women.

CEDAW, composed of 23 elected experts nominated by the states that are parties to the convention, meets once a year to consider progress made in implementation of the convention and to review the periodic reports of governments. Each government must submit its initial report within one year after entry into force of the convention in the country concerned, then at least once every four years thereafter. CEDAW has the power to subject governments, one by one, to public scrutiny, and it can request additional or specific reporting whenever this appears necessary. Its published reports reveal a serious and often critical line of questioning of representatives of states' parties. However, CEDAW has no enforcing powers or monitoring capacity; nor is there an international court to which claims under the convention may be brought.

This convention, which is also known as the Women's Convention, gained ratifications more rapidly than any other international convention before it, and its impact has been significant. In Finland, for example, it speeded up the adoption of a general Equality Act and prompted the establishment of an Office of Equality Ombudsman in 1986 (Pietilä and Vickers, 1994, p. 128). In Turkey, the convention has inspired feminists to agitate around implementation and enforcement of its various provisions. In other Middle Eastern countries, activist lawyers have produced pamphlets that compare the status of women to the norms of the convention. For example, in Egypt, which signed the convention with reservations based on the presumed incompatibility of some provisions with Islamic law (Sharia), lawyer Mona Zulficar of the nongovernmental organization (NGO) New Civic Forum produced a document called *The Egyptian Woman in a Changing World*, which applies the provisions of the convention toward an analysis of de jure and de facto rights, discrimination, and inequalities faced by Egyptian women. One of the legal inequalities that Egyptian women have successfully protested is that which prevents an Egyptian woman married to a non-Egyptian to claim Egyptian nationality for her children. At the Arab regional preparatory meetings for the Beijing conference, held in Amman, Jordan, in November 1994, both the NGO Forum and the official intergovernmental meeting recommended in their final documents that all Arab governments sign, ratify, and implement the Women's Convention (Moghadam, 1997). And at the Beijing conference, a leaflet by the women's NGOs from the Islamic Republic of Iran announced that they were recommending that Iran sign, ratify, and implement the Women's Convention.

In a 1994 article entitled "International Standards of Equality and Religious Freedom: Implications for the Status of Women," the United Nations Division for the Advancement of Women argues that freedom of religion and equality between men and women are not incompatible. Moreover, because the convention, like the Universal Declaration on Human Rights and the *Nairobi Forward-Looking Strategies,* was framed by people from diverse cultures, religions, and nationalities, it makes no provision whatsoever for differential interpretation based on culture or religion. Instead, it states clearly in Article 2 that "States Parties . . . undertake . . . to take all appropriate measures, including legislation, to modify or abolish existing laws, regulations, *customs and practices* [emphasis added] which constitute discrimination against women. . . ." (United Nations, 1979).

Nevertheless, one problem with the convention is that states' parties are allowed to enter reservations or declarations. For example, the government of Maldives, upon accession, stated that:

> The Government of the Republic of Maldives will comply with the provisions of the Convention, except those which the Government may consider contradictory to the principles of the Islamic Shariah upon which the laws and traditions of the Maldives is founded. Furthermore, the Republic of Maldives does not see itself bound by any provisions of the Convention which obliges to change its constitution and laws in any manner (United Nations, 1989).

The governments of Finland and the Netherlands have raised objections to the reservations and declarations made by the government of Maldives, which they argue "create serious doubts about the commitment of the reserving State to fulfil its obligations under the Convention" and "are incompatible with the object and purpose of the Convention." Women Living Under Muslim Laws, a worldwide network of women's organizations in Muslim countries and Muslim women's organizations elsewhere, has called for the ratification of the convention without reservations.

The *Nairobi Forward-Looking Strategies*
An essential part of U. N. preparations for the Nairobi conference was the compilation of the *1989 World Survey on the Role of Women in Development* (United Nations, 1989), the first review and appraisal of global development undertaken from women's perspective, which gave baseline data on the situation of women worldwide upon which the *Nairobi Forward-Looking Strategies* could be based. The latter document, consisting of about 400 paragraphs organized in three chapters, covers issues of peace and war, development, human rights, natural resources and the environment, culture, participation in politics and the economy, education, food, water, agriculture, industry, trade and commerce services, housing and community development, energy, and social services. The document begins by defining the basic concepts of equality, development, and peace:

> Equality is both a goal and a means whereby individuals are accorded equal treatment under the law and equal opportunities to enjoy their rights and to develop their potential talents and skills so that they can participate in national political, economic, social, and cultural development, both as beneficiaries and as active agents.

For women in particular, equality means the realization of rights that have been denied as a result of cultural, institutional, behavioral, and attitudinal discrimination (*Nairobi Forward-Looking Strategies,* paragraph 11).

Development means total development, including development in the political, economic, social, cultural, and other dimensions of human life as well as the development of the economic and other material resources and the physical, moral, intellectual, and cultural growth of human beings.

More directly, the increasingly successful participation of each woman in societal activities as a legally independent agent will contribute to further recognition in practice of her right to equality.

Development also requires a moral dimension to ensure that it is just and responsive to the needs and rights of the individual and that science and technology are applied within a social and economic framework that ensures environmental safety for all life forms on our planet (*Nairobi Forward-Looking Strategies,* paragraph 12).

Peace includes not only the absence of war, violence, and hostilities at the national and international levels, but also the enjoyment of economic and social justice, equality, and the entire range of human rights and fundamental freedoms within society.

Peace cannot be realized under conditions of economic and sexual inequality, denial of basic human rights and fundamental freedoms, deliberate exploitation of large sectors of the population, unequal development of countries, and exploitative economic relations.

Without peace and stability there can be no development. Peace and development are interrelated and mutually reinforcing (*Nairobi Forward-Looking Strategies,* paragraph 13).

These formulations—especially the linkages among equality, development, and peace—were the product of many years of disagreement and debates among governmental representatives and female activists from what was then known as the First, Second, and Third Worlds. Earlier, the choice of the three major themes of the Decade for Women—equality, development, and peace—had signaled not only a comprehensive agenda but a compromise. Equality had been seen primarily as a feminist issue coming from Western industrialized countries; peace was included at the request of the Eastern socialist bloc; and development was perceived as key to the improvement of women's lives in the Third World (Bunch and Carrillo, 1990, p. 70). In the course of the Decade for Women, the necessary links among the three themes came to be appreciated. Hence, the *Forward-Looking Strategies* were able to articulate a new understanding—that the three objectives of equality, development, and peace are interrelated and mutually reinforcing, so that the advancement of one contributes to the advancement of the others. This integrative principle has informed many discussions since 1985, including preparations for the Fourth World Conference on Women held in Beijing in September 1995.

The Fourth World Conference on Women and the *Platform for Action*

The Beijing conference carried forward the themes of equality, development, and peace that began with the Mexico conference in 1975. Preparations included regional meetings and adoptions of regional plans of action, which subsequently fed into the draft *Platform for Action*. Groups in the Asia and the Pacific region met in Jakarta, Indonesia, in June 1994; Latin America and the Caribbean, in Mar del Plata, Argentina, in September 1994; Europe and North America, in Vienna, Austria, in October 1994; and the Arab region and the African region, in Amman, Jordan, and Dakar, Senegal, respectively, in November 1994. Countries were expected to prepare detailed reports that assessed the status of women since 1985 with reference to the *Nairobi Forward-Looking Strategies* in such areas as fertility, health, literacy and education, health, and employment, as well as in the new priority areas of violence against women, participation in decision making, and media images of women, all within the framework of the overarching themes of equality, development, and peace.

At its 38th session, held in New York in March 1995, the 45 governmental members of the Commission on the Status of Women, the sponsoring body for the Beijing conference, met to review the draft *Platform for Action*. This document identified 12 critical areas of concern: the burden of poverty on women, unequal access to education, inequalities in health care, violence against women, armed conflicts and their impact on women, unequal economic participation of women, the need for power sharing and women's participation in decision making, the underfinancing of national and international machineries for the advancement of women, the human rights of women, the problematical portrayal of women in the mass media, the need for environmental and development policies to take women into account, and the unequal status of the girl child. During the session, a wide range of issues was discussed and there was considerable disagreement over the draft document's paragraphs dealing with economic policies and recommendations, resource allocations and financial commitments, and women's sexual, reproductive, and human rights. Some countries argued that the *Platform for Action* neglected the importance of motherhood, family, and religious values; others insisted on inclusion of the right of sexual orientation; and there was considerable disagreement over prostitution, pornography, and abortion. Objections were also raised to the draft document's criticisms of structural-adjustment policies, economic growth, and unequal global patterns of production and consumption. As a result, the draft *Platform*

for Action presented at the Beijing conference itself was heavily bracketed.

In Beijing, the debates and discussions continued, but eventually all brackets were removed, and the *Platform for Action* was formally adopted, although 30 countries filed their reservations.

Feminist Networking

If the women's movement helped bring about the first women's conference in Mexico City in 1975, the Decade for Women helped spawn many women's organizations and networks, including information-and-exchange networks such as ISIS (based in Quezon City, the Philippines, and Santiago, Chile) and the International Women's Tribune Center (New York City), culminating in the 1985 Nairobi conference that established common ground among women worldwide. When the Decade for Women began, feminism was seen primarily as a Western phenomenon, and many argued that the demands and emphases of Western feminists were irrelevant to the problems of poor women in developing countries or to the overriding possibility of nuclear war caused by the contention between the United States and the Soviet Union. Gradually, however, the Decade for Women encouraged interaction among women activists from North and South, leading to better mutual understanding. As noted by Rounaq Jahan:

> Common ground was created not through a monolithic feminist vision but through a process of exchange and negotiation among women's groups from North and South. The many exchanges between North and South held during the Decade facilitated consensus-building. Gender equality and social transformation emerged as two common objectives . . ." (Jahan, 1995, p. 7).

In developing countries, women's movements emerged "that addressed the specific regional concerns of women's lives and that expanded the definition of what feminism means and can do in the future" (Bunch 1987, quoted in Bunch and Carrillo, 1990, p. 72).

The 1985 Nairobi conference encouraged contacts and better understanding among feminists worldwide, resulting in a proliferation of feminist networks throughout the world. Many writings have singled out the Nairobi conference as a turning point in national women's organizations and in the emergence of what is known as global feminism (Bunch and Carrillo, 1990). One of the best-known feminist networks to emerge during regional preparations for the Nairobi conference is DAWN (Development Alternatives with Women for a New Era), which is based in Barbados and comprises feminist groups and individuals from Latin America, the Caribbean, the Pacific, South Asia, Southeast Asia, Africa, and the Middle East. The "manifesto" of the DAWN group, *Development, Crises, and Alternative Visions: Third World Women's Perspectives* (Sen and Grown, 1987) is a classic in the field of women in development (WID) and in the feminist literature. Another feminist network that was launched during and after the Nairobi conference is WIDE, originally based in Dublin, Ireland, and now headquartered in Brussels, Belgium, with 12 national platforms in Europe dealing with issues of structural adjustment, alternative economic policies, and the impact of European development assistance on gender equality and empowerment.

When the United Nations declared the Decade for Women—a decision prompted by considerable pressure from feminist groups—it created high expectations not only that governments would be forced to move on issues pertaining to women's legal rights and women's participation in development, but also that international cooperation would be forthcoming to support the goal of the advancement of women. "The post-Nairobi decade witnessed a significant increase in donors' commitment to and resourcing of WID. Most agencies adopted WID policies and measures, and introduced various procedures to ensure agency compliance with WID mandates" (Jahan, 1995, p. 10). Among the objectives adopted by international donor agencies were: integration (making sure women are involved in the development process), mainstreaming (bringing women's issues into all policies, programs, and projects), and women's autonomy (emphasizing women's human rights), and over the years they have variously called their program or units Women in Development (WID), Women and Development (WAD), or, more recently, Gender and Development (GAD).

The women's agenda that emerged during the Decade for Women revolved around a number of key themes: legal rights and equality; entitlements to productive resources and services; investments in women to eliminate gender gaps in human development; voice for women in decision making and in the formulation of an alternative development agenda; policy interventions with respect to poverty and female-headed households; gender equality in reproductive labor and the social provisioning of services; security for women and freedom from violence; and the self-empowerment of women Jahan (1995, p. 5). Complementarity of the priority themes of women's organizations and researchers and of the United Nations is confirmed by the fact that these themes have found their way into numerous U.N. documents, including the *1989 World Survey on the Role of Women in Development* (United Nations, 1989), the *1994 World Survey on the Role of Women in Development* (United Nations, 1995a), publications of the regional commissions, and publications of the specialized agencies and departments.

Gender in Research and Policy in the U. N. System

The United Nations Decade for Women prompted the growth of the field of women-in-development, and the worldwide women's movement spurred academic research on women's issues. Many books and articles were published during this time, largely with a focus on women and economic development. One such book, *Women and National Development: The Complexities of Change* (Wellesley Editorial Committee, 1977), with contributions by Lourdes Arizpe, Elizabeth Jelin, Deniz Kandiyoti, Fatima Mernissi, Hanna Papanek, Helen Safa, Marie-Angelique Savané, and many others, and with a preface by Boserup, became very influential. This scholarship, in turn, contributed to increased interest in women-and-development issues across the U. N. system, in such agencies as the Food and Agriculture Organization (FAO), the International Labor Organization (ILO), the United Nations Industrial Development Organization (UNIDO), the United Nations Development Program (UNDP), and the World Bank. These and other agencies—such as the United Nations Research and Training Institute for the Advancement of Women (INSTRAW), the United Nations Women's Fund (UNIFEM), and the Division for the Advancement of Women (DAW)—also conduct research on women and development issues.

Many U. N. agencies shifted from an exclusively welfare approach to women—family-centered programs that assumed motherhood as the most important role for women in the development process—to a diversity of approaches emphasizing the productive role of women. The 1970s saw the first calls for greater equity between women and men, particularly in regard to education, employment, and other material benefits. WID specialists throughout the U. N. system have argued that investments in women's health, education, and training would increase women's effectiveness and efficiency at work, as well as reduce fertility, thus assisting both economic development and women's lives. They have called for more credit, greater access to land, legal reform, and greater female involvement in development planning. Income-generating projects and support for women's microenterprises have also been seen as means of alleviating poverty.

During the 1970s, equality for women and attention to the basic needs of the poor were major policy and advocacy themes. In the 1980s, the World Bank's structural-adjustment strategy forced WID proponents within the U.N. system and outside it to emphasize the ways in which returns on investments could be raised and balance of payments improved through investing in women. This has come to be known as the efficiency argument. Of course, even before structural adjustment, there were discussions within U. N. circles about how investing in people would solve socioeconomic problems. For example, the ILO's World Employment Conference in 1976 sought to link the basic-needs and poverty-alleviation strategy with increased employment opportunities, particularly in small enterprises and microenterprises in both rural and urban areas. One of the best-known and most successful WID projects that combines the efficiency and poverty-alleviation approaches for microentrepreneurship is the Grameen Bank in Bangladesh. When evidence began to mount that women borrowed and repaid loans from the Grameen Bank and similar institutions at rates far above male borrowers at any level of enterprise (Tinker 1990, p. 39), women's resourcefulness and usefulness in development terms became more widely appreciated. In the 1990s, the World Bank and the UNDP present economic analyses and policy recommendations linking investments in women's health and education ("human development" or "human capital") with desired economic outcomes.

Whereas the WID approach focused on making women visible and integrating them into the development process, the GAD approach also looks at women in relation to men, emphasizes empowerment for women, and calls for major changes to gender roles. Empowerment of women is especially prevalent in UNIFEM and the United Nations Fund for Population Activities (UNFPA). Across the U. N. system, therefore, policy approaches to women have encompassed welfare, equity, antipoverty, efficiency, and empowerment. Although different agencies have been identified with particular approaches at various points in time, these policy approaches are not chronological or mutually exclusive, and various agencies and policies have combined two or more.

A little more than a decade after the Nairobi conference, it is apparent that the U. N. system has adopted the WID or the GAD approach—at least in principle and in many cases in practice—and that the mainstreaming of women's issues and of the gender variable is the official objective. In the 1990s, U. N. documents have started to speak "gender-language" fluently, changing the emphasis from women as a homogeneous group to the relation between the sexes and recognizing that gender roles are not fixed but variable and in need of change. The WID focal points within the U. N. departments, specialized agencies, and affiliates have been instrumental in the efforts to mainstream gender in policies, programs, and projects. The Division for the Advancement of Women has played a pivotal role in this respect; it has facilitated the permeation of gender analysis throughout the U. N. system by hosting annual interagency consultations and expert-group meetings and by preparing reports to the secretary-general. It was also responsible for the *1989 World Survey on the Role of Women in Development.* Focusing on the global economic situation in the 1980s, the *1989 Survey* was one of several important studies published during the latter half of the 1980s that highlighted the gender dimension of structural adjustment and the adverse effects it was having on women. Gender-disaggregated data have been a priority since the Decade for Women, and U. N. statisticians have created a statistical database on women, WISTAT. This provided the baseline data for the valuable and popular *World's Women 1970–1990: Trends and Statistics,* published in 1991. A 1995 edition of *The World's Women* was prepared for the Beijing conference (United Nations, 1995b).

Women in a Changing Global Economy

The *1994 World Survey on the Role of Women in Development* (United Nations, 1995a), officially also known as the Second Regular Update, was one of the principal documents for the Fourth World Conference on Women in Beijing. It focuses on "the results of the restructuring process and the emergence of women as decisive elements in the global economy." Among other things, it calls for "public provision of the inputs required to perform human reproduction activities" as a way of alleviating the burden on women, especially in developing countries. Beyond its useful description of the process of global economic restructuring and its policy recommendations, the *1994 Survey* is a guiding tool for the evolution of principles and policies for planning and programming within the U. N. system and in the member states. The introduction states that its focus "goes beyond women, however, and examines development from a gender perspective" (United Nations, 1995a, p. 2). It elaborates:

> The complexity of gender relations must be taken into account when analyzing trends and policies. Certain changes may have a universal impact, as with policies to combat violence against women. Others may impact poor women more than middle-class women.
>
> [W]omen's issues cannot be resolved in isolation from the relationships between women and men or social and economic structures and trends. Instead, changing the status of women requires the entire society to rethink the type of development it pursues.
>
> Advancing women cannot be a marginal exercise in micro-level projects. Nor will gender-aware sectoral or macro-level plans be sufficient. The entire range of social and economic relations and policies needs to be reviewed from a gender perspective, and gender issues should help define development goals. Current trends, development strategies and State policies should be evaluated against a gender backdrop, particularly in the fields of education, employment, family law, population policy, and national development (United Nations, 1995a, p. 3).

The introduction includes a section on "Human Development and Sustainability" that endorses the objective, first put forward by the UNDP in 1990, of *human* development as the process of enlarging people's choices, and increasing their opportunities in the full range of human choices, from health and education to economic and political freedom. This new definition of development has special implications for women, in that it implies possibilities for women's autonomy and self-empowerment. Finally, the introduction contains a section on "Changes in the Enabling Environment," which examines two positive changes that have occurred since the *Nairobi Forward-Looking Strategies* were adopted at the 1985 Nairobi conference. The first is the improvement in women's legal status, as evidenced by the growing number of states' parties to the Convention on the Elimination of All Forms of Discrimination Against Women, although the document does note that "the situation of women in countries that are not yet party to the Convention remains a matter of concern" (United Nations, 1995a, p. 6). The second is the improvement around the world in women's educational attainment, which is nearing equity with men and in some cases is higher.

Women's Positions in the 1990s

A number of publications produced by various U. N. bodies have sought to describe and analyze women's positions at the end of the twentieth century, either through qualitative and comparative studies (Moghadam, 1996) or through quantitative assessments. Two examples of the latter are the above-mentioned *World's Women 1995: Trends and Statistics* (United Nations, 1995b), which provides a wealth of demographic, social, economic, and political indicators for the countries of the world, and the UNDP's *Human Development Report 1995,* which focuses on gender (UNDP, 1995). The consensus seems to be that although there has been considerable progress in the legal status of women; in women's access to fertility control, health, care, education, and paid employment; and in women's participation in political structures, there remain considerable gaps between men and women, and in some parts of the world these gaps are huge. Moreover, new problems have emerged that have affected women more adversely than men, including the transition to the market economy in the former socialist countries, while some old problems, such as domestic violence, rape in armed conflicts, and trafficking in women, continue and may be spreading.

A cross-regional survey drawn from the above sources of gains and setbacks to the advancement of women since the end of the Decade for Women shows a mixed picture. In Latin America and the Caribbean, women in urban areas have made some significant gains, according to indicators of health, childbearing, economic, social, and political participation, and especially education, where girls outnumber boys at both secondary and tertiary levels. But adolescent fertility and maternal mortality caused by unsafe abortions remain high. The serious macroeconomic deterioration of many Latin American and Caribbean countries in the 1980s may be a factor in the very high unemployment rates of women.

In most countries of the Middle East and North Africa, women have made gains in fertility control, health, education, and labor-force participation (paid work). But there are huge gaps with men in access to paid employment and participation in political and economic decision making. The rise of fundamentalist movements set back some of the progress women had made, in that family laws were revised in some countries to reflect more conservative values and women were pressured into veiling. Nevertheless, the rise of feminist and human-rights movements in the region is one of the most promising developments.

In sub-Saharan Africa, women's economic and social participation and contribution are high, but indicators in health and education levels are still far from even minimally acceptable in most countries. Literacy remains the lowest and fertility the highest in the world, and HIV infection rates continue to soar. Serious economic decline, armed conflicts, and rapid population growth have undermined the previous modest gains in health, education, and paid employment. On the other hand, the end of apartheid in South Africa and the establishment of a government popular at home and abroad, as well as the growth of women's organizations throughout the region, are positive signs.

In South Asia, life expectancy rose by 10 years between 1970 and 1990, but, at 58 years for both women and men, it is still lower than in any other region save sub-Saharan Africa. The average age of marriage for girls remains low, and although maternal mortality has decreased it remains high. There are large gender gaps in literacy and educational attainment. Fundamentalist and communalist violence has created difficulties for many women, but, as in the Middle East and North Africa, these movements are increasingly contending with more numerous and powerful feminist organizations.

In most of East and Southeast Asia, women's levels of living have improved steadily. Many of the inequalities between men and women in health, education, and employment have been reduced in both urban and rural areas, and fertility has declined considerably. Even so, considerable gender inequalities persist because women are confined to the lowest-paid and lowest-status jobs and sectors and because they are excluded from decision making.

Throughout the developed regions, the health of women is generally good and their fertility is low. Women's economic participation is high in Northern Europe, North America, and the former socialist bloc, and lower in Australia, Japan, New Zealand, and Southern and Western Europe. Almost everywhere, occupational segregation and discrimination in wages and training favor men, and women's unemployment rates are higher than men's. In Eastern and central Europe, women experienced setbacks in participation in formal politics and, in some countries, in their reproductive rights, following the collapse of Communism. The *Human Development Report 1995* ranked Sweden, Finland, Norway, and Denmark as highest in its gender-related development index, explaining this in terms of the countries' conscious implementation of national policies to effect gender equality and women's empowerment.

Conclusion

The United Nations Decade for Women has had a number of important impacts. It was a turning point for the global women's movement: It helped create common ground between women's activists from the North and from the South, spawned feminist networks and women's NGOs, and legitimized women's rights activities within countries. The Decade for Women contributed to the growth of the fields of women-in-development and gender-and-development and led to years of funding for women's projects by rich countries and international development agencies. It has led to the integration and mainstreaming of a gender perspective in the U. N. conferences and documents of the 1990s, such as *Agenda 21* and the *Rio Declaration* of the United Nations

Conference on Environment and Development (Rio, 1992), the *Vienna Declaration* and *Program of Action* of the Second World Conference on Human Rights (Vienna, 1993), the *Program of Action* of the International Conference on Population and Development (Cairo, 1994), and the *Program of Action* of the World Summit on Social Development (Copenhagen, 1995). Indeed, the last document includes a commitment to "encourage the ratification, removal of reservations, and implementation of all provisions of the Convention on the Elimination of All Forms of Discrimination Against Women and other relevant instruments, and implementation of the *Nairobi Forward-Looking Strategies for the Advancement of Women*." The Fourth World Conference on Women (Beijing, 1995) was not the culmination of the process—which is bound to continue, particularly in light of the disagreements over aspects of the *Platform for Action* and the implications of the use of the term "gender"—but there can be no doubt that it was another milestone in the history of U. N. efforts toward the advancement of women.

Finally, mention must be made of the U. N. efforts toward the advancement of women within its own house. At its 48th session, on December 20, 1993, the General Assembly adopted Resolution 48/106 on the improvement of the status of women in the Secretariat, and the secretary-general was asked to accord greater priority to the recruitment and promotion of women in posts subject to geographical distribution. Consequently, the United Nations established a policy of seeking to achieve 35 percent female representation, mainly in the higher professional and decision-making categories, by 1995. In mid-1994, the female shares in the higher professional categories were between 10 and 24 percent; especially underrepresented were women from Eastern Europe, the Middle East, and Africa. It is now acknowledged that a glass ceiling exists within the U. N. system and that more assertive affirmative action–type policies, to be coordinated by the Office of Human Resources Management, will be needed if the United Nations itself is to realize the objectives of gender equality and women's empowerment that are the legacy of the United Nations Decade for Women.

References

Boserup, Ester. *Woman's Role in Economic Development.* New York: St. Martin's Press, 1970 (1986, second edition).

Bunch, Charlotte, and Roxanna Carrillo. "Feminist Perspectives on Women in Development." In Irene Tinker (ed.), *Persistent Inequalities: Women and World Development.* New York: Oxford University Press, 1990, pp. 70–82.

Division for the Advancement of Women, United Nations, "International Standards of Equality and Religious Freedom: Implications for the Status of Women," in Valentine M. Moghadam, ed., *Identity Politics and Women: Cultural Reassertions and Feminisms in International Perspective.* Boulder, CO: Westview, 1994.

International Labor Organization (ILO). *The International Labor Organization and Women Workers' Rights.* Geneva: ILO, 1993.

Jahan, Rounaq. *The Elusive Agenda: Mainstreaming Women in Development.* London: Zed, 1995.

Moghadam, Valentine (ed.). *Patriarchy and Economic Development: Women's Positions at the End of the Twentieth Century.* Oxford: Clarendon, 1996.

———. "Women's NGOs in the Middle East and North Africa: Constraints, Opportunities, and Priorities." In Dawm Chatty and Anniko Rabo (eds.), *Organizing Women.* Oxford: Berg, 1997.

Pietilä, Hilkka, and Jeanne Vickers. *Making Women Matter: The Role of the United Nations.* 2nd ed. London: Zed, 1994.

Sen, Gita, and Caren Grown. *Development, Crises, and Alternative Visions: Third World Women's Perspectives.* New York: Monthly Review Press, 1987.

Tinker, Irene. "The Making of a Field: Advocates, Practitioners, and Scholars." In Irene Tinker (ed.), *Persistent Inequalities: Women and World Development.* New York: Oxford University Press, 1990, pp. 27–53.

United Nations. "Convention on the Elimination of All Forms of Discrimination Against Women." Adopted and opened for signature, ratification, and accession by General Assembly resolution 34/180, New York, 18 December 1979.

———. *The Nairobi Forward-Looking Strategies for the Advancement of Women.* Adopted by the 40th session of the General Assembly in resolution 40/180 on 13 December 1985. New York: Department of Public Information, 1986.

———. *1989 World Survey on the Role of Women in Development.* New York: United Nations, 1989.

———. *Women in a Changing Global Economy: 1994 World Survey on the Role of Women in Development.* New York: Department for Policy Coordination and Sustainable Development, United Nations, 1995a.

———. *The World's Women 1995: Trends and Statistics.* New York: United Nations, 1995b.

United Nations Development Program (UNDP). *Human Development Report 1995.* New York: UNDP, 1995.

Wellesley Editorial Committee. *Women and National Development: The Complexities of Change.* Chicago: University of Chicago Press, 1977.

See also Kathleen Staudt, "International and Bilateral Aid Agencies"

Women-Centered Nongovernmental and Grass-Roots Organizations

Patricia Ruiz Bravo
Karen Monkman

Introduction

Processes of development are dependent on the participation of all social actors, including, and particularly, women. "[T]he issue is not 'integrating women in development,'" said first lady of the Republic of Ghana Nana Agyeman Rawlings in 1990. "It is rather how to achieve integrated development through women, or, to put it another way, 'rooting development in women!'" (*Abuja Declaration,* 1989, p. 4). She went on to discuss the gap between policy and implementation so often experienced in development efforts. Nongovernmental organizations (NGOs) have emerged as crucial actors in more effectively bridging these gaps. In particular, women-centered NGOs have been the key force in framing development efforts around the actual conditions lived by women in the Third World and in shifting the attention to the necessity of empowering women and transforming gender-biased social structures. Women-centered NGOs, including grass-roots movements of locally based poor women and cross-class groups of women working together, have introduced innovative strategies and approaches to meeting basic needs and altering the conditions that give rise to those needs; they have had positive influences on the development scene.

While many mainstream NGOs offer a necessary alternative to state-oriented development efforts, their activities and structures have not been sufficient to address women's concerns. More mainstream NGOs do include such necessary characteristics as participatory and community-based orientations. They work with and strengthen local institutions and tend to achieve more positive results at less cost. They lack bureaucratic and historical constraints that hamper innovative, experimental, and flexible efforts, and they have the ability to undertake people-centered research. They tend to be better placed in positions to reach the poor and to articulate rural reality. Mainstream NGOs, however, can be just as blind to women as donor agencies are (Yudelman, 1987b). Women's experiences in life must be central to the determination of their needs, and

women must be central participants in determining those needs.

The first section of this essay discusses types and objectives of women-centered NGOs and grass-roots groups, their scope and range of activities and membership, and common characteristics. The second section focuses on the impact of women's NGOs on individuals, communities, the state, and broader social realms. The third section highlights unresolved issues, including the integration of women in development, funding, expansion, leadership and management, and interorganizational relations. Throughout, examples are presented of women-centered NGOs from various regions of the world. A comprehensive picture of women's NGOs in any one region, not to mention the world, is impossible due to the nature of available scholarship and circulating literature. Each region has its own peculiarities, just as each country does; this discussion is intended to give an overview of the important work that women's NGOs are promoting in the world.

Types and Objectives of Women-Centered NGOs and Grass-Roots Organizations

Many categorizations of women's NGOs have been put forth. Distinguishing between service-oriented NGOs and community-based grass-roots organizations is common. The first typically includes urban women from the middle and upper classes in positions of leadership in the organization; their position in society is often useful in channeling services and resources to poor women. The cross-class relations can sometimes present constraints to effective functioning of the group but can also be used constructively in building cross-class cooperative efforts and in raising the consciousness of all participants around class issues. The grass-roots groups typically arise out of local community concerns, often related to women's practical needs. What they lack in resources—strategic social connections and financial resources—they often make up for in integral and empowering participatory processes.

Antrobus (1987) discusses international NGOs, whose role is to support and strengthen indigenous NGOs. The latter group includes user organizations that address the direct and specific needs of the membership, as well as intermediary organizations, whose work is not for the benefit of its own membership. Intermediary organizations are instead multipurpose and multifunctional.

Sen and Grown (1987) identify six types of women's organizations in the Third World: (1) service-oriented women's organizations; (2) women's organizations affiliated with political parties; (3) worker-based organizations; (4) donor-funded development organizations; (5) grass-roots project-related women's organizations; and (6) research organizations. Many of these types of organizations have incorporated participatory processes both in the projects offered and in the organizational structures. Grass-roots project-related women's organizations have the greatest potential when they work with poor women. Research organizations and service-oriented women's organizations are best placed to make poor women's voices heard in more-mainstream, public arenas and to influence public policy. The experiences of political parties and worker organizations are often useful to women; attention to gender dynamics, however, between women's organizations and their (male) affiliates, or within mixed groups, is critical.

Perhaps the most critical gender-based issue distinguishing types of organizations is the underlying assumptions about gender inequality. There are two broad orientations of NGOs here. One addresses the *situation* of poverty of women. These groups can be women only or mixed groups of men and women. These projects do not consider discrimination and subordination of women on the basis of gender as a concern in identifying needs and determining objectives. The relations of power between men and women are not questioned. These NGOs, above all, carry out survival programs and in some cases work with women as a means to benefit children and the family, but they don't specifically work toward changing the social status of women. The second group comprises NGOs whose objectives are directed explicitly toward achieving change in the *position* of women without neglecting the satisfaction of material needs. In this group are the NGOs that have established a feminist perspective from the beginning and those that adopted a view of gender relatively recently. For these NGOs, the gender perspective goes together with a conception of development on a human scale that comes from an economics frame and places equity, justice, and human rights as basic foundations of development. These NGOs start from a realization that the material necessities of women are related to their condition of subordination. In theoretical terms, these NGOs are nourished by feminist perspectives, and, in some regions, particularly Latin America, they are sometimes influenced by or exhibit some charateristics also common to liberation theology, the popular education of Paulo Freire, and Marxist theory as an instrument of social analysis and interpretation. From Gramsci, they take the concept of the organic intellectual to refer to the agent in charge of bringing forward and implementing the projects.

With this point of departure, projects develop in the most diverse areas. NGOs sometimes focus on specific types of projects, and others are broad in their involvement. Many focus their efforts on projects that seek to enable women to expand their capacities in areas such as education, women's rights, reproductive health, nutrition, agricultural technology, entrepreneurial business, self-esteem, women's identity, and the like. Income-generation and savings-generation projects also are common; these can embody such forms as microenterprises, production workshops, and survival organizations. Organizational forms of support for women can include leadership training, promotion of women's organizations, occupational or social groups, unions, federations, and other groupings of women around common interests or activities. NGOs also facilitate the provision of services, two of which are legal assistance and health care. Research and dissemination of information as support for gender-sensitive social change are also represented in the activities of women's NGOs. In much of Latin America, NGOs have influenced public opinion through these efforts and have gained a presence and recognition as much in the academic arena as with the state. Professional women who work in the NGOs have been able to associate themselves with women from popular sectors through the projects and actions of training and networking, giving a space to a women's movement that is very important in Latin America.

The Scope and Range of Women's NGO Activities

While various broadly defined types of projects exist, comprehensive and detailed information on countrywide or regional bases is lacking. Some countries collect statistical data on NGOs operating within their borders, but many do not and fewer focus such survey efforts primarily on women's NGOs. Bangladesh is one country that reports nationwide data (see Table 1).

Women's NGOs in Bangladesh are a relatively recent phenomenon. As of 1992, there were 1,288 women's NGOs registered in the country (Chowdhury, et al., 1993). Still other NGOs have women's departments. Before independence (1971), it was mostly voluntary activity under government patronage that addressed women's issues. Since 1972, various types of charity-oriented NGOs have emerged, at first addressing relief issues and rehabilitation of people affected by the civil war and gradually becoming more diversified social-economic programs integrated into development efforts. The numbers of innovative NGOs increased during the 1970s, and after 1975 the government strengthened the development of NGOs. In 1984, the Department of Women's Affairs was formed; it includes an office that regulates and coordinates NGOs.

As is fairly common in many regions, there is a concentration of NGOs in urban areas in Bangladesh, with 42 percent in the Dacca Division and many others in the more-developed district towns (Chowdhury, et al., 1993).

Sixty-three of the 64 districts in the country have NGOs, and 66 percent of the NGOs operate in a district municipal area, as opposed to being national in scope or being limited to a local community. Most of the NGOs (56 percent) are multifunctional, more than one-third (37 percent) are bifunctional, and only 7 percent perform only one function. Most of the bifunctional NGOs concentrate on family planning and handicrafts programs. Women's NGOs have been categorized by Chowdhury, et al. (1993) into nine broad functional groups according to the types of programs they provide. Table 1 shows the number of NGOs in each area of focus. These categories are broad, so it is difficult to determine exactly what the NGOs do or how they go about doing it. It is of possible significance, however, that the emancipatory potential of these NGO programs is quite low, with less than 1 percent of the NGOs identifying legal rights and leadership training—areas closely linked to emancipation—as functions. This summary doesn't tell us the numbers of women who participate in the programs, but only the numbers of NGOs and the broad types of programs they reported. It is difficult to project the emancipatory relevance of the legal rights and leadership training projects. Similarly, the empowering potential of other types of projects is hidden within the broad categorizations.

Although most NGOs—and particularly women's NGOs—are small in size, three organizations in South Asia are well-known exceptions: the Bangladesh Rural Advancement Committee (BRAC) and the Grameen Bank in Bangladesh, and the Self-Employed Women's Association (SEWA) in India. SEWA began operation in 1972, and after more than 20 years in operation, its membership was about 40,000 women (Rose, 1992). Union members participated in more than 30 different trade groups. In addition, there were 40 cooperatives representing a diverse collection of activities, including dairy farming, crafts, trading and vending, land reclamation, and service cooperatives. By 1989, SEWA's bank had more than 12,000 shareholders and had made more than 7,000 loans. More than 25,000 self-employed women had opened savings accounts (Rose, 1992). The banking, unionization, and collective

organizational strategies of SEWA are but three of its strategies to address the multifaceted needs of poor women.

BRAC, similarly, engages in diverse activities. Its Credit Division was operating in more than 3,600 villages by 1990 (Ebdon, 1995). BRAC health workers have taught oral rehydration therapy to 13 million women in 68,000 villages, reaching virtually all poor, rural families in Bangladesh. In 1991, BRAC had 4,700 full-time employees and more than 6,000 part-time teachers. They had organized more than 550,000 men and women into more than 7,000 village organizations; 65 percent of their members are women. BRAC runs 30,000 schools or centers, serving nearly 900,000 children with a targeted goal of having a student body of 75 percent girls. By 1995 the Grameen Bank had developed a network in which over 1,000 bank branches and a staff of 14,000 were reaching more than 2 million members—96 percent of whom were women dispersed in nearly 40 percent of Bangladesh's villages (Kabeer, 1994; Ebdon, 1995; Grameen Bank, 1995). These are only a few of the numerous activities and programs of these large NGOs.

In preparation for the Fourth World Conference on Women (Beijing, 1995), many countries compiled directories of women's NGOs and organizations. China's list reflects its unique political context in that many of the so-called NGOs are affiliated in some way with the All-China's Women's Federation (ACWF), the mass-membership women's arm of the Communist Party, and thus integrally related to the state. The ACWF began in 1949 while most of the other NGOs listed in China's directory began between 1986 and 1994. There are two NGO network organizations—one focused on rural women and development, the other on health. Four media NGOs are listed, along with a journalists' organization, a feminist translation group, two magazines (one focused on rural women, the other on the international women's movement), a trade union, and one or two NGOs that focus on each of the following: health and family planning, social support, and rural and agricultural development. Although the listing does not include women's professional organizations or research or government organizations, several of the listed NGOs seem to serve professional

Table 1. Distribution of Women-Related NGOs by Program in Bangladesh

Categories	Number of NGOs	Percentage
Skill Development and Handicrafts Training	1203	93.40
Family Planning	842	65.47
Child, Adult, and Religious Education	462	35.87
Cottage Industry and Income Generation	207	16.07
Health, Nutrition, and Day-Care Center	60	4.66
Formalized Nonformal Education	35	2.72
Research and Publication	8	0.62
Legal Rights and Leadership Training	8	0.62
Other (Welfare and Rehabilitation)	19	1.47

Source: Adapted from Table 6 in Chowdhury, et al., 1993, p. 13. Numbers can exceed 100 since the activities are not mutually exclusive.

women, particularly research institutes and associations for promoting women in several professions (Gender and Development Group, 1995).

While such a broad vision of the activities conducted by women's NGOs is important, it does not provide a complete picture. There are some common themes in more-qualitative discussions and studies of women's NGOs that help flesh it out.

Local Experience, Knowledge, and Action. NGOs strive to work from local, traditional knowledge, and to focus on local concerns (Sen and Grown, 1987). By beginning with women's lived experiences, locally relevant needs are identified and strategies planned to address them. Strategies that place women as central participants are critical. Benton (1993) found that basic needs united groups of women in Bolivia who were then able to approach NGOs for help, having seen the benefits that other women had gained by working with the NGOs.

The Singamma Sreenivasan Foundation (SSF) in Bangalore, India, recognized three traditional survival strategies of women in rural India and used these familiar models to address newly identified concerns. The three traditional strategies of women are: (1) savings and thrift practices; (2) menstruation, pre- and postnatal practices, and other situations involving midwives; and (3) women's informal social networks. Two additional rural strategies were also recognized: rural markets and traditional agricultural practices. Through using informal savings and credit schemes for various purposes, SSF saw a strengthening of social ties that crossed traditional boundaries of class and caste. By using these strategies to address other concerns, SSF was able to validate and legitimize the women's knowledge and thus help facilitate the empowerment process.

Similarly, a self-help and soil conservation project in Kenya recognized the power of the local chief in mobilizing the voluntary self-help groups. By using this traditional manner of encouraging participation—the local chief asked people to participate—the projects were much more successful than if a person from outside the villages had been the primary mobilizer. In a study of four women's groups in Lesotho, their organizational and management approaches and successes, Morolong (1995) found that the one group that had outlasted all of the others—a funeral society—used traditional cooperative saving strategies to pay for the increasing costs of funerals. The group experienced more permanent relations among the members while the other three groups had failed to survive or had experienced internal turmoil to the point of its adversely affecting their efforts. The projects of these three unsuccessful groups originated from external consultants and were not based on the women's past experience or knowledge. Being personally invested in a project and organization makes possible the formation of relevant links between lived experience and culturally compatible strategies.

In successful women's NGOs, local women are positioned as central actors, not as clients or recipients (Kabeer, 1994). They actively participate in program planning, design, and implementation. The rationale behind this focus on participation is that it provides the women with the experiences within which they can develop skills and self-confidence, and it leads to the creation of projects that are more responsive to local women's situations and conditions. Unlike goals of traditional development organizations in which development—usually meaning economic development—is the intended outcome, the empowerment of women is central. Self-confidence is a crucial basis for women's empowerment processes, and participatory experiences are used as a tool in its development. Skill building, such as leadership and political action, is also made possible through participatory experiences. And the experience of participating instills a feeling of ownership, which translates into the taking of responsibility for the organization and project.

These participatory approaches specifically address and reflect women's lived experiences and gender interests (Yudelman, 1987b). The focus on participation reflects an acknowledgment by the NGOs of the important and necessary knowledge that the women bring to the organization and efforts (Stromquist, 1992). Their knowledge and experiences are valid and, indeed, vital for the operations of the NGOs. Active and mutually determined modes of engagement within the NGOs not only benefit the women participants but also make it possible for the organization to better understand the women, their lived experiences, and the communities in which they live. This deep level of understanding leads to more relevant projects, programs, and organizational structures in which women's needs are addressed. Women's NGOs do not restrict their activities and attention to local communities, however. Links with other segments of society are critical.

Local-Global Linkages. Women-centered NGOs are strategically placed in a position to link grass-roots groups with national and international entities and resources. Many of the paid staff and members of the boards of directors of the NGOs are university educated and have not only personal connections with those in other organizations but also the ability and the knowledge necessary to interact successfully with governmental organizations, international donor agencies, and other NGOs. These linkages are important for local women gaining access to specialized groups for specific purposes and also to enhance the development work that would otherwise be done by only one organization in a more isolated manner. These linkages facilitate networking and coordination of activities with other like-minded groups and access to resources held by national and international organizations. Yudelman (1987a) sees more of these linkages in service organizations, in which middle- and upper-class women are more involved, than in grass-roots movements. Innovative processes and organizational structures are critical for creating and maintaining these mutually beneficial cooperative relations and activities.

Other types of linkages are also important. Conceptually linking women's local, lived experiences with broader social phenomena offers a way to plan and direct action toward changing social conditions that constrain women from fulfilling their responsibilities and expanding the realms of possibility in their lives. Feminist influences are active in bridging this conceptual gap to raise consciousness and focus attention on social injustice and inequality. Popular-education methodologies and multifaceted programs facilitate these processes.

Impact of Women-Centered Organizations

The various organizational and procedural strategies used by women's NGOs have led to critical benefits on individual and community levels, as well as on institutional and societal levels. Shifting the center of focus from masculinist and economic perspectives to women's lived experiences and creating avenues of collaboration between poor and middle- and upper-class women, rural and urban women, grass-roots and international entities, and local women and the state are ways to make visible that which has been hidden.

Women's Voices: Impact on Individuals and the Community

Because of the women-centered strategies of these NGOs, space is created in which women's voices are heard. The NGOs try, often successfully, to bring the voices to regional, national, and international arenas. In these spaces in which women are heard, their experiences are validated and self-confidence develops. This process is reflected in a change in the direction of the flow of communication. Traditionally, community women hear much from regional, national, and international agencies but have little access to having their voices heard at those levels. Women's NGOs create this critical link. Hearing voices of the disempowered can lead to more innovative social relations and actions. The Chipko movement, a grass-roots effort by village women in northern India, began when women spontaneously clung to trees to protect them from being felled by unscrupulous contractors. Because of others hearing the village women's concerns and resonating with them, this local popular protest for environmental protection was later joined by women from other areas, as well as by urban, middle-class women.

Often, different sets of needs and priorities come to the surface as a result of the participatory process in which women's voices are heard. The Grameen Bank originally identified employment and wages as the needs of the landless (not exclusively women) in Bangladesh. By listening to poor women directly, however, the bank soon realized that credit and self-employment were more critical and that women were the "poorest of the poor" (Kabeer, 1994). They were able to move beyond coping strategies into more-dynamic activities—most important, providing credit to landless women. SEWA was also shaped by its ability to recognize the different sets of needs of women who work outside of formal employment situations (Kabeer, 1994). The initial sets of needs identified by the

Grameen Bank and SEWA were economic, but they were able to expand to address other needs that were not traditionally identified. Domestic violence and adequate sanitary facilities for women have also been identified in this way; they are critically linked to women's ability to work. SUTRA, also in India, originally focused on service delivery but, during an internal review, members realized that women were attracted to the organization because of the space it provided them that was not available elsewhere (Price, 1992, in Kabeer, 1994, p. 233). Consequently, development of women's organizations overtook development implementation as the group's highest priority. In this space provided for women, SUTRA enabled issues of domestic violence to come to the surface.

In addition to identifying different sets of needs, women-centered NGOs are particularly skilled at recognizing the interrelatedness of those needs and initiating multifaceted approaches to addressing them. SEWA and the Grameen Bank are both examples here also. SEWA recognizes that self-employed women struggle with exploitive contractors and money lenders, police harassment, and discriminatory laws. Through its four areas of operation—cooperatives, unions and trade groups, a bank, and service provision—and the many organized activities within each, SEWA addresses the complex needs of its membership. Recognizing that women's responsibilities in caring for the health of family members affects their income-generating activities, the Grameen Bank broadened the activities for which it provided credit to include health-related activities. In addition, the bank has formed borrower cooperatives to address various health and nutrition, sanitation, literacy, and family planning needs. Gonoshasthaya Kendra (People's Health Center), an NGO in Bangladesh that originally provided only health care, recognized the necessity and interrelatedness of income-generation for poor women and now provides both types of opportunities (Kabeer, 1994).

NGOs and the State: Societal Impact

Women's NGOs present an important alternative approach to traditional modes of political participation and critical force in the creation of civil society. Their impact has been most strongly felt in Latin America, and they present an important force in South Asia also (Kusterer, 1993). Since the mid-1970s, the increase of NGOs in Latin America has accelerated, and in the 1990s they represent an important sector of civil society. In Peru in the mid-1990s, there exist around 800 NGOs that work in rural and urban areas. In Chile, one finds a similar number. India has more than 12,000 voluntary action organizations and NGOs operating at various levels; they are actively supported by a vibrant women's movement. While numbers of organizations are important in creating a presence, strategies that impact the status quo, particularly political participation, are also critical. Women's NGOs—through their growing numbers and their strategies of interaction with, and challenge of, traditional sources of power—have been a critical force in altering political culture and building a civil society.

Latin American politics have traditionally fallen into two categories (Kusterer, 1993). In the patron-client machine political model, one joins the party or aligns oneself with and supports a political entity and receives favors in return. A programmatic, or totalistic, all-or-nothing progressivism, on the other hand, demands a complete, progressive platform of social change across all levels of the state and society. Even though these two models are quite different, they share a statist orientation: They both put state power at the center. They are both nonpluralist; that is, there are no temporary alliances of interest groups. They are also male-gendered cultures, much like the other major social forces in Latin America—the Catholic Church and the military. Women-centered NGOs have transformed the political culture in several ways. Kusterer (1993) outlines five:

> They (1) concretize and personalize previous ideologically abstract political discussions; (2) mobilize and activate people—many women, but men as well—who had not been motivated to participate in traditional Latin American win–lose power-seeking politics; (3) disengage politically active organizations from identification with and domination by a specific political party; (4) create competition among political parties for the support of independent, politically active constituencies; and (5) initiate autonomous community action to solve community problems, creating critical tension (whether through competition or through nondependent cooperation) with government programs to increase the programs' effectiveness (Kusterer, 1993, pp. 188–189).

In some countries, women's NGOs have emerged as an alternative to the exclusionary development policies and practices of the state and charity institutions. In Bolivia, women's NGOs operate freely because the state is unable to deal with many of society's problems. Brazil has seen a profusion of feminist businesses and endeavors. These include women's studies programs, conferences, and various types of subgroups, networks, and revitalized mobilization efforts. This "NGO-ization of the feminist movement," Alvarez (1994) notes, "attest[s] to the vitality of autonomous society-centered feminist thought and action" (Alvarez, 1994, pp. 49–50). New strands of feminism, as well as new feminist organizing modalities and strategies, have multiplied in Brazil. More important perhaps, feminism is being appropriated by women in the grass roots; a popular feminism has developed.

Women's NGOs in many countries have also been instrumental in promoting the transformation of political culture (Kusterer, 1993; Jaquette, 1994). The Madres de la Plaza de Mayo in Argentina and the communal kitchens of Peru have used strategies that have moved the democratization processes forward. The Madres of Argentina is a group of mothers, grandmothers, and wives who mobilized to demand answers about the disappearances of their family members. Through making personal issues public—by demanding to know where their children and spouses were—they bypassed the state political agenda. They asked "where" instead of challenging the state structure or seeking individual gains through political alliances. In Peru, women pooled resources to feed their families and demanded access to affordable food. They did not demand reform of Peru's economic policies; this wasn't their approach. Both of these groups used traditionally personal issues—information about family members and demands for food—as the basis for challenging the actions of the state. Because these were issues that would have been politically unpopular to ignore, the state was forced to listen to the voices of grass-roots women. Although the problems were not solved, women's presence in the public realm was recognized. More broadly, these strategies introduced modes of political involvement that deviated from the more traditional political ideologies common in Latin America.

Other examples include the Chilean situation in which intense repression under General and President Augusto Pinochet in the 1970s and 1980s unintentionally promoted women's resistance and the growth of women's organizations and movements. Many women's organizations arose, initially as social support to those affected by the repression. Gradually, many of the organizations realized the need and relevancy for deeper feminist perspectives and expanded their activities addressing basic needs to include such areas as education in political participation and analysis of gender oppression. Feminist and popular-sector women have worked together closely to build broad opposition coalitions against dictatorships and for democratic transition in many Latin American countries (for a variety of case studies, see Jaquette, 1994).

South Asia—India and Bangladesh in particular—is also noted for the dynamic role of the NGO sector (Viswanath, 1991). In India, as in other countries and regions, women's organizations in the early 1970s were often arms of unions, political parties, or other organizations. Many grass-roots efforts have emerged in India in the decades since then, targeting gender-biased cultural practices and social institutions for protest. NGOs have also been active, along with some governmental agencies, in sensitizing people to gender bias and in promoting consciousness-raising and social action. The National Literacy Mission, the Women's Development Program Mahila Samakhya (Education for Women's Equality), and several other programs have been active in mobilizing communities and using the media to raise issues. Women's studies centers and development institutes have researched and articulated gender issues and have created materials to be used in advocacy. In India governmental interventions for women's empowerment are generally implemented by NGOs, whose initiatives are rich and diverse. In addition, links are emerging in India between NGOs and mass movements (Stromquist, 1992).

Many countries in sub-Saharan Africa have also felt the impact of women's NGOs. Indeed, governments have called for an elevated role of such groups in African development (*Abuja Declaration,* 1989). On the attainment of independence from colonial rule, new African states witnessed a higher level of participation and involvement of women in development programs. In the preparatory phases of the elections of the new governments, eligible citizens were encouraged to cast their votes. The numerous political meetings and rallies that were held for the entire population regardless of social orientation, educational level, socioeconomic status, and political affiliation served as consciousness-raising exercises that prepared people, including women, for the new era.

While women's organizations do positively influence the state and development efforts, they can also be seen as limiting that influence by the ways they construct their needs and approaches. Lind (forthcoming) argues that the Ecuadorian neighborhood women's organizations she studied both proactively rearticulate the economic and political effects of development, and limit their influences by the ways they have constructed their political positions as consumers vis-à-vis the state and development apparatus. They make demands on the state, engage in direct protest, and organize themselves around issues of neoliberal state policies and structural adjustment policies. At the same time, they challenge gender roles in households and communities. With their primary focus on meeting the basic needs of families, within contexts of diminishing resources, they limit their efforts to pursuing these traditional gendered needs.

The newly independent countries attracted many international development agencies. The origins, agenda, and target groups of these agencies vary, and their contribution in helping governments implement microdevelopment programs is significant. These organizations have the advantage of being met by enthusiastic, motivated, and receptive populations who are eager to participate in the development processes of their countries to bring about change after oppressive colonial regimes and long periods of struggle and suffering for liberation. This state of affairs is true of Zimbabwe, Namibia, Eritrea, and South Africa, among others.

Because race in formerly apartheid South Africa was such a highly charged issue, gender has been less visible in defining the identity of NGOs. During this era, progressive NGOs and the African National Congress (ANC) ran nonformal education programs that promoted critical reflection on the oppressive social relations and conditions experienced by Black South Africans. Because of the transformative nature of some of these programs, the NGOs were often identified by the government as a threat and thus were more often victimized than officially endorsed. Many all-Black NGOs disappeared during the late 1970s as a result of the detainment of leaders of the community-based organizations. The groups that survived into the 1980s tended to be those who aligned themselves with more-established community organizations and trade unions, which could provide security as umbrella organizations against government harassment. In 1993, the ANC

called on all South African women to form a coalition for the purpose of influencing governmental policy in the form of legislative and constitutional protection for women and educational efforts aimed at the police and the courts around issues of domestic violence and rape. The Women's Charter, which was written by the coalition and is in the new constitution, is expected to be a vehicle for raising women's awareness of their rights and their newly acquired economic and political power in postapartheid South Africa.

While all religions influence lived experience, Islam is often discussed in terms of its impact on women. Women in various regions respond to Islamization forces using three main strategies (Hélie-Lucas, in Moghadam, 1993). Some women join fundamentalist groups, seeking to gain entry into spheres of influence. Among the typical goals they seek is access to education, which is encouraged by the Qur'an (Koran). Other women work for change from within the context of Islam at the levels of both religion and culture: SIS Forum, a Malaysian women's group, is working to reinterpret the Qur'an (Wadud-Mushin, 1995), and other groups are rewriting Islamic women's history. Still other women fight for secularism and laws more closely associated with international human-rights frameworks. Women Living Under Muslim Laws is a network operating in 18 countries seeking legal reforms (Moghadam, 1993).

Women's NGOs have evolved in relation to various types of state structures and political agendas, sometimes as allies and sometimes as adversaries. The NGOs in many Pacific Island nations and other countries, for example, have little alternative but to cooperate with the state, which acts as a gatekeeper for external funding (Antrobus, 1987). The inequitable power relations in such an arrangement can easily lead to increased government control in "monitor[ing] popular participation" (Snyder and Tadesse, 1995, p. 174). In many African countries where the private sector is underdeveloped, NGOs act as a source of pressure on the state for moving the agendas of women forward.

Impact of Women's NGOs Globally

In addition to acting as a significant force in deepening, broadening, and strengthening civil society at a national level (Kusterer, 1993), women's NGOs have been successful in influencing the activities of other organizations operating in the international development arena. Coordinated efforts among NGOs have had a growing impact on U.N. agencies and document production in relation to various world conferences, most notably the Fourth World Conference on Women (Beijing, 1995). The *Nairobi Forward-Looking Strategies,* prepared largely by the U.N. representatives at the Third World Conference on Women in Nairobi (1985), became a tool with which NGOs could pressure governments to more adequately address women's concerns. The NGO forums that take place alongside the U.N. world conferences present a situation in which women and NGO representatives can, in a deliberate and organized fashion, lobby the U.N. delegates to influence the content of the resulting U.N. documents. While the physical distance (about 35 miles) between the NGO Forum and the U.N. meeting in Beijing effectively minimized this activity in 1995, it was recognized that women's NGOs are important partners in development (Stromquist, 1996).

With increasing communication, collaboration, and coordinated efforts, women's NGOs are emerging as an important force in a global civil society. Through electronic communication—faxes and e-mail particularly—NGOs have been able to influence the U.N. conferences. Their potential for influencing worldwide agendas as well as cultural and social ideologies will be strongly felt in the years to come. The core of the global feminist movement is in the hands of the women's NGOs; it is through their collaboration and communication that the movement is carried forward.

This international level of influence is not universal, however, nor is it as deeply felt as it could be. Feminist organizations have not been as successful in shaping the priorities of some of the other international agencies, such as the World Bank. Learning from the greater successes of the environmental movement may be one strategy to increase the degree of influence. Siddharth (1995) suggests three strategies to better influence international policies: (1) women's NGOs should develop a closer alliance with the international women's movement so that their voice is more strongly supported; (2) the profile of Southern women's NGOs should be increased in the advocacy processes; and (3) gender advocacy should be focused on specific sectoral reforms—technical issues as well as processes—so that it is not so general that nothing concrete could be accomplished.

Despite varying degrees of action, the discourse of feminist NGOs has been appropriated by many actors in the international development arena. In the 1990s, women are involved in the design and implementation of programs in international agencies, bilateral donor agencies, and governmental organizations. The language discussing approaches to development in the *Abuja Declaration* (1989), for example, more closely approximates that which has been central in women-centered NGOs for some time. In the conference that produced this declaration, representatives from nearly all African states met to assess the implementation of the Arusha strategies and to reassess priorities. Not only did the language of the document reflect women as central actors in development, the governmental representatives also acknowledged the instrumental and necessary role played by NGOs in the process. The NGOs, "whatever their character," they stated, should (1) coordinate their efforts with governmental entities, (2) produce and disseminate materials to accelerate the impact of development actions for and with women, and (3) make an effort to establish and reinforce cooperation among the various organizations and agencies working toward the promotion of African women in development. While noting that the international community will help the states in planning processes and in acquiring funding, these rep-

resentatives acknowledged the expertise of NGOs in being better able to reach small and poor communities at grassroots levels and advised governmental organizations to learn from the experiences of NGOs. "[T]he experience and efficiency of . . . NGOs in the field is incontestable in terms of assistance and support for action direct to women as well as in the domain of techniques, technology and credit" (*Abuja Declaration,* 1989, pp. 28–29).

NGOs have served as an important training ground for women in positions of influence. Many women in positions of power in Zimbabwe and other areas had their origins in nongovernmental, community-based organizations. And as women's NGOs grow older, they acquire more public presence and legitimacy—and more power to influence various social arenas. SEWA, for example, founded in the early 1970s, had by the early 1990s spread to five states in India creating nine new brochures. It has achieved visibility and influence at political and policy levels and is used as a model for other organizations in other areas. SEWA has moved into new, more central arenas such as health insurance, and because of its influence trade unions are now listening to the self-employed when determining their own strategies (Rose, 1992).

Some Unresolved Issues

This section discusses several areas of tension in issues relating to women's NGOs, including development ideology, financial resources, organizational size, leadership issues, and interorganizational relations.

Integration

While integration of women into the processes of development is a stated goal of many NGOs, the road in that direction is not without its obstacles. Yudelman (1987a) studying groups in Latin America and the Caribbean, found that increased integration led to "loss of independence, diversion of funds for women's projects for other purposes, refusal to provide credit and technical assistance to women for other than traditional projects, inequality of relationships between male and female staff members, and lack of opportunity for professional advancement" (Yudelman, 1987a, p. 106). She recommends, instead of integration, a relationship of coordination with other organizations that aren't specifically and primarily woman centered. Organizational arrangements and relationships should ensure that women control their own resources and manage their own programs. Co-optation by others is not easily avoided, particularly in organizations that work closely with more powerful groups such as political parties, trade unions, and nationalist movements.

Funding

NGOs fulfill their work through projects that are financed by international and bilateral development agencies, and by private foundations and NGOs in Northern countries. Inadequate levels and duration of funding, cumbersome application processes, and steep competition for funds are realities of the financial climate for many NGOs in the South and women's NGOs in particular. Dependency on external funding, while permitting NGOs a certain autonomy in front of the state, makes them institutionally weak in front of the sources of external financing. Even so, governments often must approve projects to be funded, effectively limiting NGO activities to those in line with state agendas. Donor priorities are often not the same as those of the women's NGOs, and national NGOs have been known to divert funds allocated to women's projects (Yudelman, 1987b). It is often especially difficult to find funding for unconventional projects or organizations. This dependency climate limits the ability of NGOs to build their own capacity (Antrobus, 1987). They often remain small and isolated organizations and projects, staffed largely by volunteers (Yudelman, 1987b).

The United Nations Development Fund for Women (UNIFEM) is unique among U.N. agencies in that it specifically funds women's projects, many of which are run by small NGOs or community groups. Yet UNIFEM's funding has been very small relative to its mandate and is itself dependent on voluntary donor contributions. MATCH International, a Canadian NGO, is similar in its identification of preferred organizations. In addition, both of these organizations are women-oriented themselves.

Possible strategies to remedy these shortcomings include changes in resource-control and decision-making processes (Yudelman, 1987b). Those who engage in implementation should be in control of the budgets and the flow of resources allocated. This would encourage the resources to remain where they were intended. Boards of funding agencies should include representatives of the recipient community. Acceptance of alternative methodologies, needs identification, management systems, and the like could be promoted through the increased participation of women's NGOs in the decision-making process. Finally, a stronger financial commitment to women's NGOs is crucial. This commitment should include higher levels of funding, funding of more innovative projects and more experimental organizations, and support for longer periods of time. Sporadic funding patterns result in piecemeal and inconsistent development efforts.

Expansion

One issue that is on the minds of many NGOs is the size and scope of their efforts. While locally based, small-scale efforts are particularly effective at reaching the community with relevant content and processes, the need in much broader areas is great. Many small, community-based NGOs are "scaling up" into larger organizations (Goetz, 1995). This expansion process introduces new influences and conditions that affect—sometimes not positively—the empowerment of poor women.

The Honduras Federation of Peasant Women was begun as a program of rural homemakers' clubs established by the social-action arm of the Catholic Church. Through participation and mobilization experiences, inspired by

mística (vision) of some of the leaders, the clubs developed into a national peasant women's federation of 294 women's groups with more than 5,000 members in nearly all of the departments in Honduras (Yudelman, 1987a). The federation finds that some of their strengths in earlier times—informality, responsiveness to all project requests, decentralization—have become problematic. Transformation of the voluntary movement through the institutionalization of service provision to its membership has created different organizational needs and processes. Adaptation of management structures has been problematic, but Yudelman was optimistic given the federation's history of demonstrated resiliency and capacity to survive.

SEWA in India is an example of an organization that has expanded dramatically but still reaches poor women in meaningful ways while helping them address their own concerns. SEWA has been able to maintain that responsiveness by keeping the local women's groups as the basis of its organizational structure. With significant power remaining in the community-based groups, local women continue to be central participants in identifying issues, planning strategies, and participating in actions to solve their problems.

The Musasa Project in Zimbabwe has also "scaled up" in recent years (Stewart and Taylor, 1995). This project is an indigenous NGO that fosters change from within various social institutions to tackle the problems of rape and domestic violence. Working mostly in urban areas, project workers focus on public education, educational work with the police, and counseling. Increased bureaucratization and professionalization are common during expansion of NGOs (Clark, in Stewart and Taylor, 1995). Musasa, seeing a potential danger for women emanating from processes of professionalization, set up a triangular power relationship in which newly created membership-based regional committees balanced the involvement of the head office and the police. The regional committees are usually made up of senior women in the communities as opposed to younger professional women in order to retain some control that is locally based and indigenous in nature. Because they are not employees, but members, they enjoy some autonomy.

Expansion of organizations often creates new difficulties. With larger organizations come more-rigid bureaucratization and less responsiveness to local communities and participants. In addition, changes in the operations of one organization affect others working in the same area. Ebdon (1995) describes detrimental effects on small local NGOs in Bangladesh when larger NGOs rapidly expanded. Focusing on development of the organization, she suggests, diverts attention from empowerment of poor women to goals of organizational growth and compromised measures of success. In the Bangladesh situation, these included targeting limited-risk individuals (those with some assets for repayment) for credit instead of the poorest of the poor, and turning a blind eye to women obtaining credit that is then used by husbands.

Leadership and Management

Discussion of leadership and organizational management revolves around issues of strong leadership and participatory leadership. Many organizations find it difficult to promote both simultaneously. Strong leadership is necessary for the successful functioning of an organization, yet participatory leadership is integral to empowerment processes for participants. Yudelman (1987a) studied five women's development organizations: Centro de Orientación de la Mujer Obrera (COMO) in Mexico, Federación Hondureña de Mujeres Campesinas (FEHMUC) in Honduras, Federación de Organizaciones Voluntarias (FOV) in Costa Rica, Mujeres en Desarrollo Dominicano (MUDE) in the Dominican Republic, and the Women and Development Unit (WAND) in the Caribbean. Four of these were run by charismatic individuals who were instrumental in building the organizations and ensuring their continuing dynamic work. Charismatic leaders attract a staff that has a strong emotional commitment, but efforts often lead to internal conflict. Lack of a clear management structure and burnout due to heavy workloads and high expectations are two obstacles to continuing success in these organizations. Yudelman identifies two solutions: to train a second generation of leadership and to open up leadership, delegate responsibility and authority, and make sure that staff or members have access to leadership training. WAND took impressive steps in this second direction. Yudelman notes that the leader of WAND at the time of the study was "delegating authority and downplaying the powerful-leader syndrome" as well as seeking experienced and qualified staff (Yudelman, 1987a, p. 101). She downplayed her own image by focusing attention on the organization and had her consultancies paid to WAND for their benefit. Overall, a strong management system is necessary so that the structure of the organization and the role of the participants is clear (Yudelman, 1987a). Two other organizations studied by Yudelman—MUDE and FOV—made transitions away from charismatic leadership to institutionalized sharing of management, in which management procedures and opportunities for staff advancement exist.

Development of leadership abilities requires adequate time and space. The financial constraints and urgent material needs of poor women affect many women-centered NGOs, creating situations of overwork and tendencies toward trying to achieve results quickly. Leadership development is a longer-term endeavor and is often not central to NGO strategies. Provision of time and space must be incorporated into projects and organizational strategies if women are to develop the skills and experience necessary for sustained action.

The other main theme relating to leadership is ideologically based. While many women strive for an egalitarian model of shared leadership, the reality is sometimes different. Pohlmann (1995) found ambivalence about leadership and authority among women involved in women's organizations in Bangladesh. Social pressures to conform to hierarchical organizational models can reinforce traditional behavior and lead to competition. Lack of confidence can lead to reluctance to participate in leadership, while a desire for sharing experiences, including leadership, can be a motivating factor. As gender inequalities in society influence these ambivalent feelings, class inequity also impacts the ways in which the women in these organizations work together and share authority.

Organizational management approaches should include a gendered structuring of time and space that reflects the needs and experiences of women employees and participants. Flexible rules for service to organizations are integral to the operations of Saptagram, one of the few NGOs in Bangladesh with mainly female staff at all levels (Kabeer, 1994). Women employees, particularly those at the higher levels, are allowed flexible schedules to accommodate domestic responsibility and health and childbearing concerns. They can work in the office (as opposed to the field) during times of menstruation, bring their children into the office or headquarters, and have time off for family responsibilities. They can use conventional public transportation rather than ride bicycles because of the cultural taboos. The women work more intensively when required, in exchange for the flexibility. The results are a much lower rate of female staff attrition than in other Bangladeshi NGOs, and more women holding higher positions within the organization. On the other hand, Saptagram's rate of expansion is slow because it grows only as fast as it can find good female staff.

Interorganizational Relations

Examination of the relationships among women's NGOs and other entities and organizations—their funding sources, the state, peer organizations, and other actors in the development community—reveals tensions relating to ideological assumptions and balancing of cooperation and autonomy.

The *Nairobi Forward-Looking Strategies,* developed largely by governmental bodies as an expression of the goals on which they agreed, can be seen as the basis on which relations between governments and NGOs were built and as a jumping-off point from which NGOs can further define women's concerns and work to transform gender inequities. International cooperation among the various organizations, including women-oriented NGOs, has increased in recent years. One of the best-known cooperative ventures between a government and an NGO is BRAC, which has explicitly targeted women—who constitute between 70 percent and 80 percent of its beneficiaries—in most of its programs. The Bangladeshi government recognizes and supports BRAC's advantage in reaching people in poor, rural areas, particularly women, and government funding enables BRAC to broaden the scope of its services to reach more people.

The Women's Development Program (WDP) in Rajasthan is also a joint government-NGO effort but of another type: It builds alliances and networks so that local women have access to channels in which they can pressure the government when it is unresponsive or unfairly implementing its mandated programs (Kabeer, 1994).

Despite this increased cooperation, significant barriers exist. One set of impediments is related to ideological orientations of women-oriented NGOs and governments. These can reflect differing assumptions underlying various development theories or could relate more specifically to how gender is understood as a social phenomenon. Where women are perceived as requiring special treatment or care, cooperative development ventures would seek to provide services to address the perceived conditions. Where women are seen to be excluded from mainstream development efforts, inclusion would be the focus. Where transformation of societal and gender power relations is viewed as central, development efforts would attempt to analyze, critique, and change these dynamics. Much of the tension inherent in the working relationships of NGOs and governments emanates from the different perspectives to which each ascribes. These result in mutual suspicions and different focuses, goals, and approaches in the work that is done.

In addition to ideological differences, power relations impact working relations between different types of organizations. Changes in relations between Southern and Northern NGOs are crucial. The flow of money from the North to the South has for too long characterized communications and power relationships; a two-way exchange of information and experience is critical for a truly equitable relationship without undue influence being exerted from the North on projects and organizations in the South. Accountability should be bidirectional, also. NGOs often find themselves moved in one direction or the other on the basis of what kinds of funding are available. If Northern funders identify certain priorities, Southern NGOs may have to agree with them or go without funding.

One innovative collaborative arrangement is the South Asia Partnership (SAP), which links Canadian NGOs with NGOs in South Asian countries that receive Canadian NGO funding. The partnership was begun in 1981 with two primary objectives: to identify and support small locally based community groups and to share decision making over allocation of resources with people in the South Asian countries in which SAP operates. While SAP is not particularly women-centered in its membership, such an arrangement could be used with women-centered NGOs also. What is crucial are safeguards so that the local women and the small NGOs retain autonomy in their work. Collaboration should be mutually beneficial. A balance of autonomy and financial security is central.

Conclusion

This essay has highlighted trends and topics evident in the analysis of women-centered NGOs in the Third World. While neither complete nor representative of every country or every region, this discussion has revealed some common threads emanating from the work of many women's NGOs.

One can affirm that the more significant achievements are obtained in the NGOs that have a focus on gender and whose objectives are directed explicitly to the modification of the position of women. The women who have been part of the projects, not as beneficiaries but as participants, manifest important changes in their self-image and personal valuation. A second aspect is related to their organization and to the collective learning they have experienced: To go outside the home, to participate in an organization and be part of a respected collective in the community is an important change affecting the social recognition of the women in their communities. The training of more women as leaders and social and political directors has been one of the more significant changes in the NGOs. The participation of women from popular sectors in the public space is increasing in many parts of the world. Important advances in legislation and in the design of politics have been achieved through the work of the NGOs and the women's movement in various countries. The relationship between NGOs and the state is often cooperative, and in many cases it is the NGOs that carry out social programs themselves that the state, with reduced resources, has left undone. The increasing participation and resonance of the NGOs in the world summits on population, the summits on social development, and the women's conferences are indicators of the significance and importance of the NGOs and of women's demands put forth within processes of change in favor of equity, justice, and peace.

References

Abuja Declaration on Paticipatory Development: The Role of Women in Africa in the 1990s. United Nations Economic Commission for Africa (UNECA), Fourth Regional Conference on the Integration of Women in Development and on the Implementation of the Arusha Strategies for the Advancement of Women in Africa in 1990, Abuja, Nigeria, November 6–10, 1989.

Alvarez, Sonia E. "The (Trans)formation of Feminism(s) and Gender Politics in Democratizing Brazil." In Jane Jaquette (ed.), *The Women's Movement in Latin America: Participation and Democracy.* 2nd ed. Boulder: Westview, 1994, pp. 13–64.

Antrobus, Peggy. "Funding for NGOs: Issues and Options," *World Development,* vol. 15, supplement ("Development Alternatives: The Challenge for NGOs"), Autumn 1987, pp. 95–102.

Benton, Jane. "The Role of Women's Organisations and Groups in Community Development: A Case Study of Bolivia." In Janet H. Momsen and Vivian Kinnaird (eds), *Different Places, Different Voices: Gender and Development in Africa, Asia, and Latin America.* London: Routledge, 1993, pp. 230–242.

Chowdhury, Rofi Ahmed, Suraiya Hakim, Jowshan A. Rahman (eds.). *Inventory of Women's NGOs in Bangladesh.* 3rd ed. Dacca: Department of Women's Affairs, Government of the People's Republic of Bangladesh, 1993.

Ebdon, Rosamund. "NGO Expansion and the Fight to Reach the Poor: Gender Implications of NGO Scaling-Up in Bangladesh," *IDS Bulletin,* vol. 26, no. 3, July 1995, pp. 49–55. Special Issue: "Getting Institutions Right for Women in Development."

Freire, Paulo. *Pedagogy of the Oppressed.* 20th anniversary ed. New York: Continuum, 1993 [1973].

Gender and Development Group. *Interim Directory of Chinese Women's Organizations.* Beijing: Ford Foundation, 1995.

Goetz, Anne Marie. "Institutionalizing Women's Interests and Gender-Sensitive Accountability in Development," *IDS Bulletin,* vol. 26, no. 3, July 1995, pp. 1–10. Special Issue: "Getting Institutions Right for Women in Development."

Grameen Bank. *Grameen Dialogue,* vol. 24, October 1995.

Gramsci, Antonio. Edited by Frank Rosengarten. Translated by Raymond Rosenthal. *Letters from Prison,* vols. 1 and 2. New York: Columbia University Press, 1994.

Jaquette, Jane (ed.). *The Women's Movement in Latin America: Participation and Democracy.* 2nd ed. Boulder: Westview, 1994.

Kabeer, Naila. *Reversed Realities: Gender Hierarchies in Development Thought.* London: Verso, 1994.

Kusterer, Ken. "Women-Oriented NGOs in Latin America." In Gay Young, Vidyamali Samarasinghe, and Ken Kusterer (eds.), *Women at the Center: Development Issues and Practices for the 1990s.* West Hartford: Kumarian, 1993, pp. 182–192.

Lind, Amy. "Negotiating Boundaries: Women's Organizations and the Politics of Restructuring in Ecuador." In Anne Runyan and Marianne Marchand (eds.), *Gender and Global Restructuring: Sites and Sightings.* New York: Routledge, forthcoming.

Moghadam, Valentine M. *Modernizing Women: Gender and Social Change in the Middle East.* Boulder: Lynne Rienner, 1993.

Morolong, Bantu. "Sustainable Development in Lesotho: Women's Groups and Projects." Ph.D. diss., University of Alberta, Edmonton, 1995.

Pohlmann, Lisa. "Ambivalence About Leadership in Women's Organizations—a Look at Bangladesh," *IDS Bulletin,* vol. 26, no. 3, July 1995, pp. 117–124. Special Issue: "Getting Institutions Right for Women in Development."

Rose, Kalima. *Where Women Are Leaders: The SEWA Movement in India.* London: Zed, 1992.

Sen, Gita, and Caren Grown. *Development, Crises, and Alternative Visions: Third World Women's Perspectives.* New York: Monthly Review Press, 1987.

Siddharth, Veena. "Gendered Participation: NGOs and the World Bank," *IDS Bulletin,* vol. 26, no. 3, July 1995, pp. 31–38. Special Issue: "Getting Institutions Right for Women in Development."

Snyder, Margaret C., and Mary Tadesse. *African Women and Development: A History.* London: Zed, 1995.

Stewart, Sheelagh, and Jill Taylor. "Women Organizing Women—'Doing It Backwards and in High Heels,'" *IDS Bulletin,* vol. 26, no. 3, July 1995, pp. 79–85. Special Issue: "Getting Institutions Right for Women in Development."

Stromquist, Nelly P. "Empowering Women Through Knowledge: International Support for Nonformal Education." In Robin J. Burns and Anthony R. Welsh (eds.), *Contemporary Perspectives in Comparative Education.* New York: Garland, 1992, pp. 265–293.

———. "Beijing 1995: Euphoria, Unity, and the Morning After," *CIES Newsletter of the Comparative and International Education Society,* vol. 111, January 1996, pp. 1, 4, and 10.

Viswanath, Vanita. *NGOs and Women's Development in Rural South India: A Comparative Analysis.* Boulder: Westview, 1991.

Wadud-Mushin, Amina. " Sisters in Islam: Effective Against All Odds." In Doug A. Newsom and Bob J. Carrell (eds.), *Silent Voices.* Lanham: University Press of America, 1995, pp. 117–138.

Yudelman, Sally W. *Hopeful Openings: A Study of Five Women's Development Organizations in Latin America and the Caribbean.* West Hartford: Kumarian, 1987a.

———. "The Integration of Women into Development Projects: Observations on the NGO Experience in General and in Latin America in Particular," *World Development,* vol. 15, supplement ("Development Alternatives: The Challenge for NGOs"), Autumn 1987, pp. 179–187, 1987b.

See also Zelda Groener, "Women in South Africa"; Valentine M. Moghadam, "The United Nations Decade for Women and Beyond"; Karen Monkman, "Training Women for Change and Empowerment"; Moema Viezzer and Thais Corral, "Global Struggles for a Healthy Planet"; Shirley Walters, "Informal and Nonformal Education"

Training Women for Change and Empowerment

Karen Monkman

Introduction

Nonformal education and training can be used as strategies for maintenance of the status quo or for transformative purposes. Educational pursuits commonly aim to provide learners with skills, abilities, knowledge, and attitudes that enable them to enter into and participate in the dominant society. Individual change is the goal. In transformative or empowerment education, on the other hand, elements of the dominant social structure are seen as oppressive, and transformation of these elements through collective action is the goal (Sen and Grown, 1987).

Empowerment education is a strategy used with subordinated populations in numerous settings, including poor, marginalized women in the Third World. Various approaches within this type of strategy reflect the elements in society that are deemed significant by the program providers. Those with strong Marxist class orientations seek economic empowerment and assume that, for women, such empowerment will promote equality in other areas as well; class differences are the basis of inequality in such a view. Psychologically informed educational efforts seek to enhance self-concept and self-esteem; these qualities are thought to empower learners so that they are better able to be in charge of, and control, their own lives. Many educational and training programs seek cognitive empowerment: the acquisition of skills and knowledge that will make possible more options for the learner. Political empowerment seeks to promote participation in arenas in which power is central to relations among people. This can be within the family, at the community level, or nationally. When gender is central, and when all four components of empowerment are included, we see potential for the improvement of women's lives and for the dismantling of social barriers that promote and maintain gender inequality. Training and education for change and empowerment of women in the Third World result in individual empowerment and social transformation whereby women have more control over their lives and more of an effect in chang-ing society to eliminate or reduce gender bias and discrimination.

The first section of this essay discusses the terms "empowerment," "change," and "nonformal education and training" in relation to Third World women. The second section presents examples of nonformal training and educational efforts toward for women's empowerment and gender-equitable social change and discusses conceptual frameworks that can help shape these efforts. The last section discusses recommendations and challenges for the future.

Central Concepts

Empowerment

The term "empowerment" has been overused in the 1980s and 1990s; it is used as a synonym for participation, for speaking out, or for meeting some basic need. In its undiluted form, however, it is an important concept. Key to the concept is power, but rather than substituting one dominant power for another, notions of empowerment here are used to imply "power to" and not "power over." Women need more power to control their lives, to meet their practical and strategic needs, and to shape the worlds in which they live in ways that are not themselves oppressive. Power can be thought of as "a social relationship between groups that determines access to, use of, and control over the basic material and ideological resources in society" (Morgen and Bookman, 1988, p. 4). Those who are empowered are able to shape social relations so that resources are used for the benefit of everyone, especially those who are disadvantaged.

Various definitions of empowerment for women exist. Empowerment is a process of gaining understanding of, and control over, the political forces around one as a means of improving one's standing in society (Kindervatter, 1979). This requires awareness of one's situation, skill acquisition that enables change, and working jointly in effecting change (Kindervatter, 1979). It involves "claiming equality" instead of waiting for others to provide it (Hall, 1992). Empowerment can be used for social mobilization, changing women's state of mind, and gaining access to the bases of social power (Friedmann, 1992). Networking and organizing are central to Friedmann's goals of political, psychological, and social empowerment processes. Empowerment begins when women "change their ideas about the causes of their powerlessness, when they recognize the systemic forces that oppress them, and when they act to change the conditions of their lives" (Morgen and Bookman, 1988, p. 4). Morgen and Bookman use the term to "connote a spectrum of political activity ranging from acts of individual resistance to mass political mobilizations that challenge the basic power relations" in society (Morgen and Bookman, 1988, p. 4). They see empowerment as "a *process* aimed at consolidating, maintaining, or changing the nature and distribution of power in a particular cultural context" (Morgen and Bookman, 1988, p. 4). Stromquist further clarifies that this process of changing the distribution of power should focus on "interpersonal relations and in institutions throughout society" (Stromquist, 1995, p. 13).

Empowerment As a Collective Process. Process and collectivity are central to empowerment (Schuler and Kadirgamar-Rajasingham, 1992; Kabeer, 1994). Certain types of experiences will lead to feelings of self-confidence, and this self-confidence can generate more courage for women to venture into previously foreign arenas where they can exert pressure or challenge social situations that create difficulties for them. Achieving positive results from such an endeavor is reflective of an empowering process and a product or outcome of that process. But individual efforts to effect change in one's life or environment, though important, are limited in scope and results. Collective efforts not only increase the numbers of individuals involved in a social action, they also provide contexts in which empowerment is more actively and energetically pursued. Collective action has the potential to create a stronger voice and sustain a more powerul challenge to discriminatory structures, and to elicit a more adequate response. Experiences of a collective nature can be more dynamic and achieve greater results in empowering the participants and effecting positive social change.

Gender Basis of Empowerment. Gender analysis must also be central if women are to become empowered. Patriarchal social structures create barriers for women in the home and family, the workplace, the community, and the nation. Without direct analysis and understanding of these social dynamics and subsequent action to dismantle them, women (and men) will continue to be constrained by them. Analysis of issues relating to class, race, ethnicity, nationality, age, caste, and other dimensions is also important and, ideally, should include all pertinent intersections.

Dimensions of Empowerment. Empowerment has numerous dimensions. A study of manifestations of women's empowerment in the Grameen Bank and the Bangladesh Rural Advancement Committee (BRAC) revealed six general categories: (1) sense of security and vision of a future; (2) ability to earn a living; (3) ability to act effectively in the public sphere; (4) increased decision-making power in the household; (5) participation in nonfamily solidarity groups; and (6) mobility and visibility in the community (Schuler and Kadirgamar-Rajasingham, 1992). Stromquist (1995) groups the dimensions broadly as cognitive, psychological, economic, and political.

Cognitive empowerment refers to knowledge about, and understanding of, the conditions and causes of subordination. Sexuality and legal rights are two commonly necessary but highly volatile issues in this area. Although cognitive aspects of empowerment are not identified in the Grameen Bank and BRAC study, they can be considered a basis for the six categories used. Psychological empowerment, which is reflected in (1) above, relates to the development of self-esteem and self-confidence so that women are able to motivate themselves into action. Positive feelings and beliefs in one's ability to act are central to psychological empowerment. Positive experiences are the basis for the development of these feelings and beliefs. Economic empowerment, manifested in (2) above, is the ability to earn and control economic resources. Independence in controlling economic resources opens more options for addressing one's interests and often serves to improve one's status in social settings. Political empowerment, in (3–6) above, has to do with the ability to analyze one's world and to organize and mobilize for social change. Friedmann (1992) uses the terms "social empowerment" and "political empowerment" and distinguishes them in relation to involvement of the state. Friedmann's notion of social empowerment involves access to information, knowledge, and skills; participation in social organizations; and financial resources, all within the sphere of civil society. Political empowerment has to do with access to decision-making processes involving the state (and including local governmental entities), typically through voting, collective action, and other means of having one's voice heard. Other analyses of empowerment assume a broader notion of political (not linked to the state but referring to any type of power relations) that encompasses both of Friedman's types. Physical aspects of empowerment—control over one's body and sexuality and protecting oneself against sexual violence (Claessen and van Wesemael-Smit, 1992)—are also important.

These dimensions of empowerment are interrelated. Ignoring one aspect will limit the empowerment that results. Most education or training projects tend to focus on only one or two aspects. Empowerment can also be worked on at different levels, including intellectual (consciousness-raising), emotional (personal experiences), and instrumental (action plans) (Claessen and van Wesemael-Smit, 1992). Work in all levels is needed for empowerment to occur.

Change

Training and educational programs are intended to produce change. Change, however, exists in various forms; training and educational efforts should be explicit about the types of change they seek to promote, and methodologies should lead to the desired types of change. Educational programs are often weak in addressing change at societal levels; they tend to focus on psychological models—on change in the individual lives of the learners. To effect societal-level change, a program must look beyond transmission of knowledge to individuals and incorporate social action that challenges social assumptions and transforms institutions' social relations.

Incremental versus holistic change is also an issue that should be considered explicitly in educational programs. Where the intended transformation is designed to result in women being better able to care for their families, engage in productive activity, manage community affairs, and have time for themselves as individuals, a broad range of issues must be included in the parameters and efforts to promote change because of the need to address both reproductive and productive concerns. Many training programs focus on either productive issues *or* reproduction; the area that is not central is usually all but ignored. In addition, empowerment should be both an individual and a collective process. Individuals as well as social organizations can become empowered through educational or training endeavors.

Nonformal Education and Training

Nonformal education and training are identified as potential tools in promoting the empowerment of women in the Third World (Kindervatter, 1979; Antrobus, 1989). Nonformal modes of education or training, as opposed to formal modes, enjoy more potential flexibility in goal development, organization, and strategy. Nonformal education may be similar to formal schooling in terms of reproduction of the status quo, but it also has the potential for promoting empowerment, liberation, and transformative social change. Projects that promote these types of change are responsive to context, facilitate active participation, are flexible, and critically examine and reflect on life experiences. Projects must be based on the actual needs and perceptions of the learners and must develop appropriate and relevant strategies to address those needs. The active participation of the women learners in all processes of project conception, planning, implementation, and evaluation is important. Skill training should be both practical and broadly focused. Training in leadership skills, for example, can promote social change by altering power relations in mixed-sex groups and by enabling women's groups to have a greater impact on other segments of society through strong leadership. Consciousness-raising and critical reflection on one's life situation and social context increase chances of success. Gender is central to the issues examined. Making gender a central focus in educational or training projects is necessary if discriminatory and inequitable gender social relations are to be altered.

Training for Empowerment and Change

This section describes nonformal training and education programs that seek to empower women and transform gender and other social inequities. It has two main subsections: The first discusses programs in which poor women in the Third World are central participants; the second focuses on "gender training" programs that seek to improve the abilities of development workers to facilitate gender-based empowerment and social transformation. Within each subsection, various conceptual frameworks are discussed and illustrated using the examples. The first subsection discusses Molyneux's practical and strategic gender interests; Stromquist's productive, reproductive, and emancipatory training; elements of popular education; and Stromquist's four dimensions of empowerment of women. The second subsection discusses three main approaches to the training of development workers and the underlying assumptions about development and gender.

Training for Women in the Third World

Training approaches that seek to empower women and to promote individual and social change in gender relations within families, communities, organizations, and societies must operate from assumptions encompassed within transformative ideologies in which gender is central. Integral to the training process are the goals of promoting critical awareness of one's social world—including the gendered nature of that world—development of instrumental skills and acquisition of knowledge, and collective actions directed toward challenging and changing inequitable social structures and patterns of relating. Positive participatory experiences in these processes are the basis for constructing new knowledge that supports and generates individual and collective empowerment and social change. While these foundational characteristics are basic to transformative education, there are many varieties of program designs and implementation.

Boserup (1970) brought to our attention the exclusion of women in planned development projects, their integral contributions to locally generated development processes, and the lack of attention to women's lived experiences and needs in planned development. Since then, important efforts have been made to reverse those practices. Meaningful change for women doesn't occur unless women are actively involved in change processes and unless their positionality as women is central to defining needs, interests, and implementation strategies.

Gender Interests. Women's needs and interests are of two main types: practical and strategic (Molyneux, 1985). Practical gender interests reflect needs that are basic to life: food, shelter, clothing, and the like. Because of their traditional roles in caring for the family, women often identify these practical aspects as central concerns; they are also typically the focus of traditional development projects. Strategic interests are those that are directed at altering unjust social relations and structures that disadvantage women. These include such issues as domestic violence, the sexual division of labor, legal protection, and inequitable gender relations within the family. Practical and strategic needs can be thought of as a continuum, with basic life issues on one end and structures of social oppression on the other. Addressing one type of need without the other is not sufficient to eliminate unjust and oppressive conditions or to satisfy basic needs.

The Women and Development Unit (WAND) at the University of West Indies' Integration of Women in Rural Development Program offers nonformal education and training to women in rural areas. As with many other programs, rural women are targeted because of the greater inequities they experience, the degree of difficulty they face in meeting their basic needs, and their significance in facilitating development that has national implications. WAND's approach to training is participatory: Needs assessments, program planning, implementation, and evaluation are done by the women themselves with some organizational assistance. While they focus mostly on practical needs, the women are also exposed to new concepts, information, and skills that lead to a deeper understanding of how they as women participate in, and benefit from, development. Gender is central both in analyzing women's social experiences and development processes and in critiquing gender roles and relations and lived experiences. Ongoing educational experiences build self-confidence and changes in gender relations.

Literacy programs also have the potential to address both practical and strategic interests. Where literacy is defined as a set of discrete skills, however, training is unlikely to address needs beyond those related to deciphering and producing basic written language. Although this is an important skill, literacy in this framework leads to integration into the existing oppressive social structure and does not address strategic interests related to transformation. Literacy as an ability to engage in meaningful discourse in various settings and for various purposes includes using language—both written and oral—as a means of exploring one's world and engaging in efforts to transform it. Adult literacy programs that follow this second model use language skills to gain a deeper understanding of the learners' experiences of oppression in society and to plan and carry out strategies to change inequitable social structures. Experiences of raising one's consciousness and engaging in social-change efforts stimulate increased self-esteem and self-confidence in one's ability to operate effectively in new social situations. Literacy is then a means to transformation and empowerment.

Literacy programs of this type are often not gender specific, although, because women generally have lower literacy rates than men, they are often more numerous as participants in programs. "Women's issues" are sometimes targeted as the areas of interest or as the "generative themes" that are central to literacy development. The economic and domestic needs of the participants tend to be practical in nature and must be addressed. One literacy program in Kerala, India, addressed practical needs of income generation for women quarry workers by making them the subcontractors who distribute pay to themselves and other employees (Chunkath, 1994). The women learned literacy and numeracy skills with which to manage the books and subcontracting process. Because of previous problems of women receiving half of the amount that men were paid for similar work, and men not contributing adequate amounts of income to their families' support, the women were collectively organized and encouraged to determine salaries and pay employees, including themselves and their husbands, in a more gender-equitable manner. Their ability to control financial resources—a strategic gender interest—enhanced their ability to meet basic (practical) family needs. The program focused on a second strategic interest also—mobility. Women were taught to ride bicycles and thus were provided with a means of mobility previously denied them in a part of India where bicycle riding was considered a male activity. Control of resources and increased ability to expand one's physical arena are strategic interests that challenge patriarchal control of women and free them to realize their potential.

Another literacy project, in rural Nepal, used literacy as a vehicle with which to examine lived experiences. Using key words in the literacy classes, women learners held serious discussions about card playing and firewood through which they developed an awareness of inequality in gender relations and their local ecological crisis. By bringing their individual experiences into public discourse and recognizing that many of their problems related to the lack of a public space for women, they gained moral strength to demand land to build a women's center for meetings and training sessions and to seek a restructured temple committee that included women and low castes (Parajuli and Enslin, 1990).

Another approach to addressing practical and strategic interests occurs in many grass-roots organizations that are originally intended to meet basic needs but then expand their awareness and activities to challenge oppressive social barriers. Mothers' clubs are popular—that is, local, grass-roots organizations in much of Latin America in which women come together to meet practical needs related to motherhood. In some, the women organize communal kitchens and glass-of-milk programs so that children and families can eat even though individual mothers are unable to provide sufficiently for their own families. The collective space created by these endeavors provides an arena in which some groups attempt to address strategic needs as well.

In Brazil, for example, the NGO Rede Mulher used action research as a technique to train and create a gender awareness among the women already organized in 155 mothers' clubs in two zones of Sao Paulo (Stromquist, 1994). Through analyzing their experiences in the mothers' clubs, the women became more consciously aware of their experiences in a broader social context and of the nature of the processes and dynamics in their experiences. After learning about the action-research initiative, women in the clubs were trained to administer and analyze a questionnaire; they worked jointly with researchers. Presentation of the survey findings to the clubs was participatory, and extensive discussion was encouraged. The research and discussions addressed the questions: Who are we? How are we organized? What do we do? How do we work? Discussion of these questions became the basis for critiquing women's roles in the clubs and the clubs' organizational potential. Women in the clubs created a play called *For Being a Woman*, based on the knowledge they gained from the questionnaires and their newly raised consciousness; it was performed in the two zones. The women also created an audiovisual presentation, *And Now What, Maria?* It ends with these words:

> Our liberation is a long road. Prejudice is always present in our mass media. Our wages are lower. We have to fight against the education we receive; we must destroy the ghost of the weaker sex. Today I have more problems than before. But now I feel at ease with myself. My head has been changing; now I think differently. I have the courage to express many things, to fight to be heard, and [I] even learned to say no, which I do many times. In our society, women must be functional beings (in Stromquist, 1994, p. 277).

For nonformal education and training to be empowering, the knowledge and skills that structure the content of the program should address women's various roles in society and the family and have emancipatory potential. Moser (1993) identifies three primary roles of women: productive, reproductive, and community-management roles. Traditionally, development projects have recognized only reproductive roles; more recently, they have also acknowledged women's significant productive activities. Women's activities in relation to management of community relations are also important and can be understood when contrasted to men's activities in community leadership. Training and educational programs can acknowledge all three of these female roles and attempt to provide skills and knowledge that can enhance performance of the related activities. For empowerment to occur, however, emancipatory skills and knowledge must be central (Stromquist, 1994). Unfortunately, such skills and knowledge, which are determined by the contextual situation, are often peripheral or nonexistent in nonformal education and training programs. Legal rights and advocacy, leadership skills, and knowledge about sexuality and physical control over one's body are critical emancipatory areas.

Emancipatory Knowledge and Skills. Training programs that focus on legal rights and the development of advocacy skills with grass-roots women are an important strategic concern that was voiced at the Third World Conference on Women in Nairobi in 1985 (Schuler and Kadirgamar-Rajasingham, 1992). Learning about the law, the legal system, and how to make it work in one's favor is central in these transformative educational programs, which use feminist consciousness-raising and participatory-learning techniques to help women gain new understandings of their place in society as human beings with certain rights and help them develop new advocacy skills as well.

Leadership training is also essential for emancipation. The SWANIRVAR movement provides such training for landless, illiterate women in Bangladesh as part of a national program to promote village development. SWANIRVAR has a multifaceted program. It offers education and training in literacy, health, nutrition, child care, and family planning (all issues related to women's traditional reproductive role) and also engages in income-generation activities (productive roles) and community projects (community management role), including analyzing women's role and status in village development. Through such variety, women develop leadership skills and self-confidence to act in situations that previously excluded women's presence and/or active participation.

Another key for emanicpation is knowledge about sexuality. Sexual taboos, domestic violence, lack of access to information on family planning, and lack of reproductive rights are strong societal controls on women. The Grameen Bank, although focused mainly on providing credit to poor, landless women, requires that dowry not be received or given by members of the bank (Kabeer, 1994). By critiquing the practice of dowry, women deepen their understanding of its constraining effects on them and gain a basis on which to reject it. Emancipation from this practice offers women the potential to control their resources more directly and to change their position in society, individually and structurally.

Productive, reproductive, and emancipatory skills and knowledge can be put to use in pursuing different kinds of practical and strategic interests—sometimes in meeting basic needs, and sometimes in challenging societal barriers that inhibit the ability of women to meet those basic needs or to participate in certain domains in society. Emancipation from these oppressive structures is key. While training in areas of production and reproduction is more traditionally focused on maintaining the gender status quo, it also offers potential arenas for change.

Training in Domestic Activities. Traditionally, training projects for women have focused on skills and knowledge that extend domestic activities. Training in home arts, crafts, food production, family health care, and the like reflect an assumption that women can enhance the ways in which they perform their reproductive duties. Communal kitchens, for example, came into being because of the difficulty in acquiring adequate food for one's family and the burden of performing both productive and reproductive roles. With food communally prepared and distributed, women's reproductive burden is decreased and the food costs less, thereby also reducing productive demands. An example of training to meet health-care needs is BRAC's project on oral rehydration therapy (ORT), which trained local health workers to teach ORT to each and every rural woman in Bangladesh. It is unique in that it has reached a large number of women—virtually all poor, rural women in the country. While women can benefit from training in reproductive skills and knowledge, these programs can be enhanced if they critique and challenge the social structures that act as barriers to fulfilling basic needs and carrying out reproductive responsibilities—and that define reproduction as a female duty and production as primarily male. They must recognize the larger context in which women's reproductive responsibilities are carried out. This larger context includes productive activities that allow women to engage in the various activities that men engage in, and the social relations that are reflected in the division of labor and in other oppressive social structures. A common limitation in planning and implementing training in reproductive areas is the lack of recognition of the skills and knowledge that women already have acquired and constructed. Basic assumptions for successful adult learning include starting with what adults already know and valuing that knowledge.

Training for Productive Activity. Training women in productive skills and knowledge can lead to some forms of emancipation from oppressive structures based in a division of labor that relegates women to the home sphere. Since Boserup's landmark work in 1970 about women's role in economic development and exclusion from development programs that are strongly associated with economic goals, more effort has been made to include women in development processes and economic activity. Engaging in economic activity is not enough, however, to develop empowerment in women or to make development efforts more successful. Adding work responsibilities to women's already burdened activities in the home and community creates a triple burden (Moser, 1993). Additionally, some programs have neglected to recognize the productive activities that women already engage in and the social barriers that place limitations on their participation. While practical needs relating to income-generation and work are crucial, strategic needs relating to sharing domestic responsibilities, having access to child care, and the dismantling of a gender-biased division of labor are central if basic needs are to be met without increasing women's burden. Productive skills and knowledge that are the focus for training and educational efforts for women can be broadly categorized as those relating to women's position as workers and those intended to enhance women's ability to generate income, often in the informal sector. Training programs that focus on women as workers recognize class and gender as central forces that shape experiences. These training and education efforts reflect some balance of women's and workers' concerns. Training for income-generation assumes that skills and resources—knowledge as well as financial resources—are central to women's abilities to generate income. While both of these broad categories consider productive activity central, the particular projects within each that are discussed here have contextually relevant ways of balancing them with gender issues and of coordinating reproductive, productive, and emancipatory skills and knowledge.

Coordination between women's organizations and labor or worker organizations is central to the Self-Employed Women's Association (SEWA) in India, the Women's Work-

ers' Movement in Asia, an early 1990s effort in Brazil, and work within socialist revolutions in Cuba and Nicaragua (Cheng-Kooi, 1989; Kabeer, 1994). Cuba and Nicaragua have both, as part of their revolutionary goals, sought to link labor and gender more closely. Women's participation in worker movements has been encouraged through targeted mobilization and leadership training of women. Through participatory approaches to research and training, the Nicaraguan Association of Rural Workers (ATC) and the national women's association (AMNLAE) brought 8,000 grass-roots women to workshops intended to increase agricultural women's participation in the labor movement. This activity, as well as others within the context of Nicaragua's revolution, enabled women to raise their consciousness and develop a capacity for organizing, for critically analyzing society, and for union leadership (Chinchilla, 1994). However, both the Nicaraguan and the Cuban efforts are motivated mainly by a desire for nation building; where labor and women are seen to add to that effort, they are encouraged. Where goals diverge, however, support for gender equity has fallen short in both countries. Integrating gender into existing labor movements can lead to some positive influences in that movement. Because the primary concern is labor-oriented, however, gender interests can again be considered secondary. Only where efforts are informed by the women's movement can gender assume a central focus.

In Brazil in 1991 and 1992, a collaborative effort by a rural workers' group (Fifth National Congress of Rural Workers) and a feminist organization (SOS CORPO) brought discussion of gender issues into the union movement's analysis of class issues. Participants analyzed these issues in relation to their experiences as workers and as union members. Some women then took gender issues into their own unions and spoke out, challenging union policies that disadvantaged women.

The Women's Workers' Movement (WWM) is a loose coalition of women's workers' organizations in Asia that provides educational and training activities to bridge the issues of labor and gender. Similar to the Brazilian coalition and those efforts in Nicaragua and Cuba, WWM works to integrate gender into worker education, to raise the awareness of all workers and unions about the roles that women workers play in labor movements, and to introduce labor issues into women's groups. Leadership development is central to WWM's focus so that more women can participate in the running of labor organizations. Women workers are the trainers and facilitators in these efforts, and it is their experiences and struggles and those of the other women participants that are the focus in the training (Cheng-Kooi, 1989).

SEWA, perhaps the largest and longest-running program that seeks to link labor and gender issues, is unique in that it is a union for self-employed women in the informal sector. SEWA's activities are broad in scope and include providing legal assistance and credit, education and action for workers' rights, and representation for higher wages and better working conditions. SEWA members also address some reproductive concerns of women: they organize child-care and health cooperatives, as well as life-insurance and maternity-benefit programs. SEWA actively links its members with existing services, including governmental training programs (Kabeer, 1994). Although it targets women primarily in relation to economic activities, issues of reproduction and emancipation are central assumptions: Empowerment of self-employed women is the goal of SEWA (Bhatt, 1989). Its approach to empowerment is to combine collective struggle with cooperative development efforts. Training for vocational skills is linked with organizing cooperative worker units that work together and engage in collective change efforts. Through these deliberate training efforts (skill development) and indirect learning experiences (organizing and mobilization), women can use the cooperatives to break away from exploitive conditions (Chatterjee, 1993).

Productive activities for women also include access to credit and the ability to generate income outside of formal-employment situations as important areas of concern. Training for income generation broadens notions inherent in traditional job-skill training. For most women in poor countries, jobs are not available in the formal sector; income-generation approaches assume that they must create their own jobs. To successfully engage in entrepreneurial activities, one must recognize that skills are not enough; the training must link skill development to knowledge construction. The participants must recognize that enterprise is possible; that they can create their own opportunities for generating income and manage that activity. Income-generation training typically includes training in business skills, accounting, management, marketing, production and quality control, pricing, and the like. Ideally, training acknowledges the barriers women experience that limit their activities and seeks to overcome those barriers. Many of these projects fail, unfortunately. Time and a broad contextual view are necessary for entrepreneurial success; these are often lacking.

Through BRAC in Bangladesh, several thousand destitute women have been trained to rear and vaccinate poultry; the goals of the project are to increase the production of eggs and birds—a nutritional goal—and to provide income-generation skills (Lovell, 1992). Training is usually complemented by other necessary assistance, such as loans or free vaccines provided by the government to the women trained as poultry vaccinators. These additional components (credit and free products) reflect a recognition that skill development is not sufficient for successful generation of income. Other support is also instrumental.

Lack of money for start-up costs is a common limitation for women embarking on income-generation projects. Gender differences in access to credit have been well documented; men, with more collateral and experience, have greater access to credit, particularly from formal institutions such as banks. Programs that specifically target women often make important inroads in altering the access that women have to economic resources of various kinds. The

Grameen Bank, for example, provides credit to the poor; most of its members are women since it is they who tend to be the poorest of the poor and lack access to credit for entrepreneurial activities. Training in the Grameen Bank's program, which is required for members, takes several forms. The initial one-week training is to teach members how to sign their own names. Thereafter, training is linked to group formation and the internalization of the bank's rules and practices, which are called the Sixteen Decisions. Workshops and meetings are ritualized, with special salutes, exercises, the shouting of slogans, and the recitation of the Sixteen Decisions (Kabeer, 1994). In addition to teaching the operations of the bank, these training activities are designed to make more of a public presence for the women, to help them develop more self-confidence, and to build a collective identity.

Provision of credit to women is not the Grameen Bank's only focus, however. It recognizes that societal barriers exist that inhibit women's access to credit, and it seeks to change them. In conjunction with its priority of providing credit to women for durable housing, it requires that land on which the housing is built be registered in the woman's name (Kabeer, 1994). This requirement challenges societal preferences favoring men as landholders, making a space in which women are supported in their challenge to this practice. Challenging societal barriers and gender relations is critical in women's empowerment and social-change processes. Feminist popular education is one approach that links training with challenging gender-biased social structures and relations.

Popular Education: Participation, Consciousness-Raising, and Social Action. Popular education is built on three main components: participation, consciousness-raising, and collective social action. This approach acknowledges, at least indirectly, the existence of practical and strategic interests, and, although it can accommodate reproductive and productive types of training, it is particularly well suited for emphasizing emancipation. Participation is central to popular-education approaches: Learners are integrally involved in the processes of determining topics to discuss, and they are central in those discussions. Teachers are instead facilitators; the power relations between teachers and learners are thus restructured in a less hierarchical fashion. Through the altered relations and processes that privileges the knowledge and experiences of the learners, both practical and strategic interests can be pursued. Popular educators are particularly interested in facilitating discussions that reveal the strategic issues related to practical needs. For example, in the Nepal literacy project, women discussed the difficulties of access to firewood, a practical interest. Through exploration of this issue, they realized the impact of previous development projects in their community on limiting the availability of firewood and the resulting ecological consequences. This process reflects direct participation of the women in the process of exploring interests that are central to their lived experiences. They are not peripheral or passive learners of a detailed agenda set by others. This process becomes their experience, and through this experience they become more aware of the dynamics in their social realms; in other words, the women's subjugated knowledge is brought to a more conscious level. Putting this new awareness to use in transforming society is done through collective social action. Again, participation is central.

Collective mobilization and organization are central to the development of empowerment and to effecting change at a societal level. *Nigera Kora* in Bangladesh builds the organizational capacity of poor people so they can pressure public institutions to fulfill the government's obligations. *Nigera Kora* workers form groups, train people in human and skill development, hold meetings, have a legal aid program, and promote collective action and mobilization on social issues (Kabeer, 1994). While both women and men are participants in *Nigera Kora,* women are explicitly targeted by the Women's Development Program (WDP) in Rajasthan. They build alliances and networks so that village women can better put collective pressure on local institutions. SEWA also recognizes that collective support for individual struggles results in positive experiences for individuals and increases the pressure on the government and oppressive community factions. As one SEWA member who engages in agricultural labor said after describing a successful resolution to a land dispute in which SEWA mobilized and supported local women in court and against a powerful and violent local group, "Our greatest strength is our unity" (Jhabvala, 1994, p. 123).

WDP's training programs are intended to respond to local needs in each village, and they incorporate a process of moving from individual perceptions of problems to collective identification of priorities. These priorities become the main focus of training programs in which relevant information is disseminated and participants come up with possible solutions. (Kabeer, 1994). Active collective participation is central to these and other efforts to build women's confidence to act in public realms and to identify interests relevant to women and strategies to pursue them. WDP believes that:

> the struggle to learn, to describe, to understand, to educate is a central and necessary part of our humanity . . . [a] struggle [that] is not begun at second hand after reality has occurred [but that] is in itself a major way in which reality is continually formed and changed (in Kabeer, 1994, p. 250).

Components of Empowerment. The examples of training programs described above include reflections of at least four components of empowerment: cognitive, economic, political, and psychological. Cognitive empowerment is evident where the skills and knowledge that women learn are instrumental/integral in their processes of challenging gender bias. Emancipatory skills and knowledge are critical here. Learning of legal rights that benefit women and how they can be exercised opens windows of opportunity for effecting change in one's own life and in women's po-

sition in society. In many countries, laws that protect women against domestic violence are not well publicized. While there are often problems with enforcement and proof, where laws can be used to benefit women, knowing about them and exercising one's rights reflects enormous cognitive empowerment.

Economic empowerment is most easily seen in the programs that focus on training for production. Enhancing women's position as workers, learning to engage successfully in income-generation activities, and having access to credit and other economic resources all lead to increasing women's ability to successfully acquire and control economic resources. The literacy project in Kerala addressed this aspect of empowerment directly by putting women in subcontractor positions.

Political dimensions of empowerment are reflected in the primary goals of several programs in which collective social action is central. In addition, participating in women's organizations—experiencing group relations in a setting where gender dynamics are not as they are in the mainstream—can lead to political empowerment. It is here that leadership skills can more easily be developed and that success in effecting collective action and social change can be experienced. WDP uses training as a vehicle for linking individual experiences with collective action. Rede Mulher's action-research project led to women participating in the creation of media and theater productions in which their lives were made central and valued and in which gender inequities were challenged. Exercising leadership in new, broader social contexts, such as in labor unions (the WWM) or local politics (the SWANIRVAR movement's leadership training, or Latin America's mothers' clubs demanding infrastructural resources from municipalities), and engaging in social action that challenges gender inequities (the Kerala quarry subcontractors restructuring salary scales and controlling financial resources) are areas in which political empowerment is developed.

Successful experiences of putting pressure on social and governmental institutions to address local problems provides the new experiences through which self-confidence and self-esteem are developed. Psychological empowerment, then, is closely linked to experiencing other dimensions of empowerment. Newly developed self-confidence can then lead to further social action or economic activity that stimulates political and/or economic empowerment. While causal relationships are not well understood at this point, different dimensions of empowerment are often recognizable in the same settings, situations, and processes. At the same time, however, *all* dimensions are not usually present in one project, even though, as discussed previously, empowerment of women cannot occur unless all dimensions are developed.

Gender Training

The examples of training described above focus on local or grass-roots women as the target population. Training can also be directed toward development workers, educators, project planners, and others who work with local women and men.

Development workers are targeted for consciousness-raising and for learning strategies for use in development planning, implementation, and analysis so that their work will more directly facilitate women's empowerment and gender-equitable social change. Gender training occurs within development agencies and other organizations for their staff. The Center for Adult and Continuing Education (CACE) at the University of the Western Cape, South Africa, for example, held a series of "Talking Gender" workshops in 1990 and 1991 with development workers and educators in which gender awareness was raised. The aim of CACE was to develop educational methods so that people could better challenge gender bias in their organizations and educational programs. Popular-education approaches were used to promote consciousness-raising in the participants and to teach them how to design and run similar workshops in their organizations (Mackenzie, 1993). Similarly, the Thematic Committee for Women in PHILDHRRA, Mindanao, Philippines, developed a training program in 1992 to expose development workers to a gender-based framework to be used in project planning.

"Gender training" is a term that refers to training of development staff to increase their awareness of gender and development and to enable them to better integrate gender into development planning processes (Moser, 1993; Kabeer, 1994). Most gender training challenges gender biases in planning processes by encouraging planners to become aware of gender divisions in resources, practices, and responsibilities. Training is intended as a strategy for improving development processes and outcomes through addressing the needs of women and for decreasing women's subordination in families, organizations, and societies. The latter goal (to decrease subordination) is related to empowerment approaches to planning; the former, to efficiency models of development (Moser, 1993). Moser (1993) identifies three methodologies that are used in gender-training programs: gender analysis, gender planning, and gender dynamics. Each targets a different audience, for different purposes, and is related to different needs (or stages of gender awareness).

Gender-analysis training helps participants acquire a consciousness of gender issues, such as the gender-based division of labor in families, communities, organizations, and societies, and recognize who has access to and control over resources in these contexts. The World Bank and most North American donor agencies have participated in this type of training. Harvard University's case-method approach is typically used to help participants become aware of gender issues (Moser, 1993).

Gender-planning training is aimed at practitioners in development planning, including those in NGOs and donor agencies. Training seeks to provide tools for the diagnosis and translation into practice of gender concerns. It aims to effect change in the institutional structure and operational procedures within the organizations in which the planners work so that gender is an integral part of the planning process.

Gender-dynamics training has been developed in grass-roots groups and NGOs and focuses on consciousness-raising about issues of gender bias. Participants are trained to identify gender bias, analyze it through feminist lenses, and take actions that promote gender-fair values in their personal and professional work.

Kabeer (1994) sees the need for gender analysis within development organizations also. She identifies three common visions of gender equity, based on analysis of gender roles, triple roles, and social relations. A gender-roles approach attempts to distinguish men's and women's roles in societies and take these roles and responsibilities into account in planning development projects. This framework offers a way to integrate gender awareness into project design by looking at activities, who does them, and how they are balanced with or displace other activities. The triple-role framework looks at the various productive, reproductive, and community-management roles that women have—and must balance—in societies and relates development efforts to those, including attention to the overburdening of women because of these triple roles. Social-relations approaches look at the rules, resources, practices, and power structures within societies and the implications for development planning and practice. Kabeer's version of this framework also considers space, time, and resources to local women so that they can better articulate their own interests. All three frameworks are important in increasing gender awareness in project- and program-planning processes. Hearing the voices of the women themselves and making their participation central are requisites for empowerment.

Kabeer (1994) identifies three dimensions of gender analysis that are included in gender-training programs: professional, political, and personal training. Professional training focuses on developing the professional capacities of development workers. Political training focuses on analysis of gender in a society or an organization. Personal training challenges gender-biased attitudes and stereotypes of development workers themselves. Organizations that are more closely linked with feminism or grass-roots efforts tend to be more open to including personal change of the staff, while many Northern aid agencies and development banks resist this dimension (Kabeer, 1994).

Kabeer's analysis of internal dynamics and underlying gender assumptions within development agencies and the consequences for how gender training is formulated and implemented leads her to recommend more collaboration across these organizational boundaries. The Center for Women and Development Studies in India organized a program in which women in grass-roots organizations trained state officials, showing them that poor rural women were experienced and knowledgeable managers of their environments and could benefit from material assistance more than from top-down education projects (Kabeer, 1994). Bringing in participants from multiple levels or domains can stimulate increasingly beneficial mutual understanding and collective efforts that are more relevant to women in the Third World.

Conclusion

Collaborative efforts that cross traditional organizational boundaries could be instituted in various contexts. Offered here are some recommendations for improving programs and a discussion of some challenges. Consideration of these will help practitioners and planners develop innovative ways to stimulate women's empowerment and social change through training and education.

Recommendations

1. A broad contextual focus and recognition of the interconnectedness of various dynamics are necessary. Training and education programs should not be limited to narrowly defined educational goals, nor should work toward empowerment and change end at the classroom door or at the time a session ends. When meaningful and long-lasting social and individual change is the goal—as opposed to narrowly defined content-area goals—educators, trainers, and facilitators can restructure their activities to cross the many boundaries between education and training and other life processes. Training in literacy may also need to address income-generation needs of the learners. Child care and domestic duties often restrict the ability of women to participate in training; programs need to address the real-life constraints of women.

2. Gains that are made in training programs need to be sustained. Funding, of course, is primary. Promotion of autonomous women's structures is also necessary. Women can act collectively to exert more pressure to effect social change, which often enhances self-esteem and positively affects individual empowerment processes. These gains should not be isolated, but built upon and supported in ways determined by the participants. Decentralized decision making is often better able to promote locally relevant and sustainable change and empowerment processes.

3. Equitable social relations within training contexts are necessary. Participation by women from all levels within an educational effort is central to the discourse of many training programs, yet, in reality, action often does not reflect the ideal. It is difficult to move beyond traditional hierarchical relations between learners and trainers, educators, or facilitators. Few facilitators in training programs are able to handle participatory methods and gender issues (Claessen and van Wesemael-Smit, 1992), and they need to be trained to do so. Facilitators must also be expected to challenge themselves by reflecting on their own actions and situations and changing them where necessary. They are central figures in the educational processes of the learners, and their participation in that process should be a focus for examination also.

4. Rural women need to become more actively incorporated into training and education programs. Few gender-informed training programs are oriented toward rural

women. Although these women are the hardest to reach due to transportation difficulties, geographical isolation, and dispersed population patterns, they are often those who could benefit the most.

5. As a long-term goal, men must also participate in gender consciousness-raising and change processes. Men tend to be peripheral or nonexistent in gender programs. Changing women, as a goal and process in and of itself, won't have the desired impact in creating more equitable social contexts within which to live and work. Men's gender consciousness and social relations must also change. However, women first need to have their own space and experiences in which they can identify their problems from their perspective and develop confidence in addressing them. Eventually, after women are empowered, the participation of men becomes a more relevant issue.

6. Development agencies that fund the NGOs that do training for change and empowerment (women-oriented NGOs) need to understand what empowerment is, how it can be facilitated, and how they could best support that process. Priorities for funding, amounts of funding, and consistency of funding could be significantly affected should the agencies understand the importance of women's empowerment. Empowerment takes time; short-term funding strategies keep women-oriented NGOs focused too much on where their support will come from, and that takes their energies away from their work with women.

7. Evaluation of programs and of empowerment and change processes is crucial. Much of the literature that exists on training and education programs is descriptive or includes impressionistic views of successes and difficulties by those involved in the programs. While these have been important in informing the broader public of these efforts, more systematic evaluation of the programs is needed. Evaluations ideally should be long-term as well as short-term and should look at dynamics of social change and empowerment as well as skill development and participant satisfaction. They should also evaluate the process itself in addition to product outcomes. Evaluations should include qualitative approaches because of the complex nature of gender relations, social change, and empowerment. Many programs have begun to use action-research techniques in their efforts to evaluate their processes. In action research, the participants (grass-roots women) set the research agenda, do the research, and carry out change efforts indicated by the research. More activity in these areas is needed.

Challenges

1. Individual empowerment has to be accompanied by significant and related social change. Educational and training programs have been much better, overall, at facilitating individual change in the form of acquisition and construction of new knowledge and skills in the individual learners. These changed individuals need to be successful in using their new knowledge and skills. It has been difficult to sustain feelings of empowerment when little changes in the broader social environment. As Kabeer observes:

It is only when the participation of poorer women goes beyond participation at the project level to intervening in the broader policy-making agenda that their strategic interests can become an enduring influence on the course of development (Kabeer, 1994, p. 292).

Achieving multiple dimensions of empowerment ensures a combination of individual and social change. Political empowerment, for example, relies on individuals' participation in collective action that leads to effecting change at the structural level. Meeting strategic gender interests implies meaningful social change.

2. Microlevel successes and macrolevel needs should be linked. Transferability of aspects of successful small-scale programs into macrolevel efforts needs to be better understood. NGOs tend to be the more successful providers of training and education at microlevels. International, governmental, and bilateral agencies have the resources to support larger programs. With women in much of the world being less educated, less literate, less skilled, and more in need of training and education for change and empowerment, the needs are great.

Collaboration among organizations attempts to bridge this gap between microsuccesses of NGOs and large-scale programs that are only possible with national and international resources. BRAC has shown that collaboration between governmental entities and NGOs can bridge this gap in important ways. Where funding is provided with adequate freedom for local decision making, training programs can reach more people and be more finely tuned to their particular circumstances. Maintaining a broad scope in recognizing the context in which women live their lives and the related issues that arise is difficult when funding and priorities of more mainstream organizations attempt to shape projects in contradictory ways.

These challenges relate to balancing or interrelating one area of concern with another. Without tensions such as these, our work would undoubtedly be much easier. Many of the development and educational efforts inspired by feminist concerns are known for their creative ways of addressing similar conflicts. With more women involved, and more participation by those women, in training, education, and other change processes, we will have access to, and develop, many more innovative approaches to achieving a more gender-equitable social world.

References

Antrobus, Peggy. "The Empowerment of Women." In Rita S. Gallin, Marilyn Aronoff, and Anne Ferguson (eds.), *The Women and International Development Annual*, vol. 1. Boulder: Westview, 1989, pp. 189–207.

Bhatt, Ela. "Toward Empowerment," *World Development*, vol. 17, no. 7, 1989, pp. 1059–1065.

Boserup, Ester. *Woman's Role in Economic Development*. New York: St. Martin's Press, 1970.

Chatterjee, Mirai. "Struggle and Development: Changing the Reality of Self-Employed Workers." In Gay Young, Vidyamali Samarasinghe, and Ken Kusterer (eds.), *Women at the Center: Development Issues and Practices for the 1990s*. West Hartford: Kumarian, 1993, pp. 81–93.

Cheng-Kooi, Loh. "Women Workers' Education in Asia: Designing Their Own Program," *Convergence*, vol. 22, no. 2–3, 1989, pp. 12–19.

Chinchilla, Norma Stoltz. "Feminism, Revolution, and Democratic Transitions in Nicaragua." In Jane S. Jaquette (ed.), *The Women's Movement in Latin America: Participation and Democracy*. 2nd ed. Boulder: Westview, 1994, pp. 177–197.

Chunkath, Sheela Rani. "Mass Literacy Campaigns: Voluntarism and Women's Empowerment," lecture and video presentation sponsored by the Center for Feminist Research and the Center for Multiethnic and Transnational Studies, University of Southern California, February 17, 1994.

Claessen, Jeannette, and Lilian van Wesemael-Smit. *Reading the Word and the World: Literacy and Education from a Gender Perspective*. Oegstgeest: Vrouwenberaad Ontwikkelingssamenwerking, 1992.

Friedmann, John. *Empowerment: The Politics of Alternative Development*. Cambridge: Blackwell, 1992.

Hall, C. Margaret. *Women and Empowerment: Strategies for Increasing Autonomy*. Bristol: Hemisphere, 1992.

Jhabvala, Renana. "Self-Employed Women's Association: Organising Women by Struggle and Development." In Sheila Rowbotham and Swasti Mitter (eds.), *Dignity and Daily Bread: New Forms of Economic Organising Among Poor Women in the Third World and the First*. London: Routledge, 1994, pp. 114–138.

Kabeer, Naila. *Reversed Realities: Gender Hierarchies in Development Thought*. London: Verso, 1994.

Kindervatter, Suzanne. *Nonformal Education As an Empowering Process with Case Studies from Indonesia and Thailand*. Amherst: Center for International Education, University of Massachusetts, 1979.

Lovell, Catherine H. *Breaking the Cycle of Poverty: The BRAC Strategy*. West Hartford: Kumarian, 1992.

Mackenzie, Liz. "On Our Feet: Taking Steps to Challenge Women's Oppression: A Handbook on Gender and Popular Education Workshops." Institut für Internationale Zusammenarbeit/Des Deutschen Volkshochschul-Verbandes supplement to *Adult Education and Development*, no. 41, 1993.

Molyneux, Maxine. "Mobilization Without Emancipation? Women's Interests, State, and Revolution in Nicaragua," *Feminist Studies*, vol. 11, no. 2, 1985, pp. 227–254.

Morgen, Sandra, and Ann Bookman. "Rethinking Women and Politics: An Introductory Essay." In Ann Bookman and Sandra Morgen (eds.), *Women and the Politics of Empowerment*. Philadelphia:

Temple University Press, 1988, pp. 3–32.

Moser, Caroline O.N. *Gender Planning and Development: Theory, Practice, and Training.* London: Routledge, 1993.

Parajuli, Pramod, and Elizabeth Enslin. "From Learning Literacy to Regenerating Women's Space: A Story of Women's Empowerment in Nepal," *Convergence,* vol. 23, no. 1, 1990, p. 44–55.

Schuler, Margaret A., and Sakuntala Kadirgamar-Rajasingham (eds.). *Legal Literacy: A Tool for Women's Empowerment.* Washington, D.C.: OEF International, 1992.

Sen, Gita, and Caren Grown. *Development, Crises, and Alternative Visions: Third World Women's Perspectives.* New York: Monthly Review Press, 1987.

Stromquist, Nelly P. "Education for the Empowerment of Women: Two Latin American Experiences." In Vincent D'Oyley, Adrian Blunt, and Ray Barnhardt (eds.), *Education and Development: Lessons from the Third World.* Calgary: Detselig, 1994, pp. 263–282.

———. "The Theoretical and Practical Bases for Empowerment." In Carolyn Medel-Añonuevo (ed.) *Women's Education and Empowerment: Pathways Toward Autonomy.* Report of The International Seminar on Women's Empowerment, held at UNESCO Institute for Education (UIE), Hamburg, January 27–February 2, 1993, 1995, pp. 13–22.

See also Shelley Feldman, "Conceptualizing Change and Equality in the Third World Context"; Shirley Walters, "Informal and Nonformal Education"

Geographical Entries

Women in Some Liberal Modernizing Islamic Countries

Nagat El-Sanabary

Introduction

Islam plays an important role in the lives of all Muslims. The vast body of literature on Arab and Muslim women generally documents the diversity and complexity of women's existence among the various countries and within each. The relationship between the status of women and Islam is not a straightforward one. There is neither a monolithic Muslim woman nor a monolithic Muslim country. Traditionalism and modernism exist side by side in the same country and even in the same person. Some of the strongest feminist activities and feminist scholarship on women in recent years, such as the writings of Fatima Mernissi and her colleagues, have emanated from a relatively traditional country, Morocco, a country that has also produced two female Olympic winners. Some of the strongest voices of traditionalism and the most visible trends of Islamization of women appear in Egypt, a country with more than 100 years of activist feminist activities. Despite the tendency to explain everything in Islamic society in terms of Islam, a complex web of factors determines the status and roles of women in these and other countries. This is not to deny the key role of Islam in influencing women's status and roles. But, as Tucker (1993) asserts, "facile assumptions about the monolithic role 'Islam' or Arab culture plays in the seclusion, disempowerment, and oppression of women no longer pass as the accepted academic discourse on the topic" (Tucker, 1993, p. viii). The literature also documents the vital role that Muslim women have played throughout Islamic history, a role that is rarely documented or acknowledged by Western writers, in history books in Islamic countries, or by predominantly male government bureaucracies.

It is relatively easy to identify traditional countries, such as Saudi Arabia, postrevolutionary Iran, Afghanistan, and Yemen, by the visible symbols of tradition: mandatory veiling, segregation of the sexes, and a high degree of female seclusion. But it is difficult to identify the liberal modernizing ones. Turkey and Tunisia are usually identified as modernizing or modernized countries because of the radical fam-

ily-law reforms that have taken place, distinguishing them from all other Muslim countries. In the middle are countries like Egypt, Jordan, and Syria, which are relatively open and have relatively good development indicators.

This essay focuses on Turkey, Tunisia, and Egypt. Turkey and Tunisia were chosen because of their progressive family laws. Egypt is included because of its status as the largest Arab country, the cultural center of the Arab world for more than a century, and the birthplace of modernization and feminism in the Arab region, and because of the progress made by educated Egyptian women in many areas of public life. The three countries have generally high indicators of women in development, including generally high literacy rates (Egypt is an exception); open access to women of the waged labor force; generally low infant and maternal mortality rates; and significant women's participation in political and leadership positions. According to the World Bank (1993), Egypt is a low-income country, with an annual gross national product (GNP) per capita of U.S. $660.00 in 1992. Tunisia and Turkey are middle-income countries: In 1992, the GNP per capita of Tunisia was more than double that of Egypt, while Turkey's was more than four times as high as Egypt's. Turkey's population in 1992 was 58.4 million; Egypt's, 54.7 million; and Tunisia's, 8.6 million.

This essay begins with a history of Islam and women and some of the changes in women's situation over the centuries. Four major sections follow. The first deals with legal reforms, particularly in family laws; the second addresses women and education; the third deals with women and the economy; and the fourth addresses the political roles of women. There is then a brief discussion about the emerging fundamentalist movements and their impact on women. The essay concludes with a discussion of prospects for change in the future.

Historical Perspective on Islam and Women

At the dawn of Islam in the seventh century A.D., the new religion gave women many rights that were unavailable to women in other parts of the world: the right to own property in their own name, to choose their marriage partners, to maintain their name after marriage, to inherit and bequeath property, and to manage their affairs independently. Islam condemned the practice of female infanticide that was practiced at the time and urged Muslims to treat women with fairness and justice. Islam preached a philosophy that addressed all believers with no distinctions on the basis of such factors as sex and race. During the early eras of Islamic history, Muslim women were active in the religious and cultural affairs of the community. Some women accompanied men in their military battles and nursed the wounded in what are considered the first mobile hospitals in the world (El-Sanabary, 1993). With the decline of Islamic civilization, the situation of women deteriorated. At the turn of the nineteenth century, the majority of Muslim women and men were illiterate, and women were confined to the home and charged with family responsibilities.

Islamic governments, with the exception of Turkey, in their modernization efforts during the mid-nineteenth century, focused on men, the main participants in and beneficiaries of, education and industrialization. Women continued to shoulder the responsibility of caring for their families and supporting them through their unpaid productive activities in and outside the home. In addition to their participation in agriculture, some women were drawn into the new industries in Egypt, Turkey, and elsewhere in the Muslim world. However, seclusion and segregation of the sexes were the rule, especially among the upper classes.

Contact with the West came through trade relations, beginning in the seventeenth century; and Western colonialism in the nineteenth century; and educational missions to Europe in the late nineteenth and early twentieth centuries. These activities ushered in significant changes in education and society in general. Foreign missionary educational activities in many Muslim countries and the education of upper-class women by European governesses introduced new ideas about women and their role in society, including traditional ideas about gender-role divisions that prevailed in Europe in the late nineteenth and early twentieth centuries. At the same time, the spread of modern education opened up new opportunities for women outside their traditional family roles.

Feminism and nationalism played an important part in the modernization of gender roles in Turkey, Tunisia, and Egypt. In the mid-nineteenth and early twentieth centuries, Western-educated men and women in Egypt and Turkey called for changes in the situation of women. They considered women's low status the cause of the backwardness of these societies, and advocated new opportunities and roles. Nationalism and feminism were twin goals of the nationalists. Those who fought for the liberation of their countries from Western colonialism sought also the liberation of women from oppressive traditions and unjust legal and other practices, many of which predated Islam.

During the early twentieth century, Tunisian feminist Tahar Al-Haddad called for the freeing of women from outmoded traditional practices. In a book titled *Our Women in the Sharia and Society* (1930), Al-Haddad called for the school education of women and for moral education that "would make them aware of their duties in life and the legitimate advantages they could expect" (Tunisian External Communication Agency and National Union of Tunisian Women, 1993, pp. 12–13). He supported a woman's right not to be forced into marriage against her will, the right of access to legal divorce, and a woman's right to work and to dispose of her possessions freely. He argued that Islam is innocent of the oft-made accusation that it is an obstacle in the way of progress. Rather, it is the religion of progress par excellence. Thus Al-Haddad, like other Muslim reformers before him, based his call for the emancipation of women on true Islamic values. Affirming that Islam is a religion of equity and justice, he quoted the Qur'an (Koran) to affirm the equality of women and men: "It is he who created you from a single being, making it a complementary pair, so that the two might find plenitude each in the other" (Qur'an, Surat Al-Aaraf, Ayat 189).

As equal partners with men in the nationalist struggle for liberation, women also called for their own liberation and full exercise of their rights in society. Egyptian women actively participated in the nationalist revolution in 1919 against the British occupation. They organized street demonstrations defying British soldiers, and suffered arrest and death. Tunisian women participated in demonstrations in Tunis against French colonialism in April 1938. When the leaders of the Neo-Destour Party (New Constitution) were arrested, the women unfolded the party's banner in front of the French president to assert Tunisian rights to independence. As a result, some women were arrested and sentenced to 15 days in prison. As more men were arrested, women held their own secret meetings in hospitals, Turkish baths, *zaouias,* and shrines to sacred saints. Tunisian women accelerated their participation in the nationalist struggle in the early 1950s after forming the women's chapter of the Neo-Destour Party. In 1952 women demonstrators were arrested, and some died either under fire or by torture in prison. One woman died in childbirth in prison. Nationalist movements acted as catalysts for feminist movements. In a speech on August 13, 1956, on the occasion of Tunisia's independence President Habib Bourguiba stated that Tunisian women had won the respect of others through the competence they had shown in the national movement.

The pace of change accelerated in the mid-twentieth century and during the first and second United Nations Decades for Women (1976–1995). These changes were prompted and supported by the international conferences held during these decades and by the U.N. conventions on human rights, the rights of children, and women's rights.

For over a century, two competing ideologies existed to foster change in the status of women and modernization of gender roles: liberal religious reform and secular ideology. Some modernists believed that women's status could be changed within a liberal interpretation of Islam; while others, mainly the nationalists, had a secular orientation. Both sought in Islamic texts—the Qur'an and the Hadith (sayings of the prophet Muhammad)—support for more-positive and active roles for women in Islamic society. These two ideologies were found among both men and women in the early eras of reform, and they continued to compete for allegiance among the people in the 1990s.

Legal Reforms: A Key to Modernization of Gender Roles

Most Islamic countries have a dual legal system, a civil code and a personal-status or family law (also referred to as family-status law), often called Sharia law. Sharia law, which is based on the interpretation of the Qur'an and the Sunnah (the sayings and actions of the Prophet), regulates marriages, divorces, inheritance rights, and related matters. Because of the difficulties and contradictions arising from this duality, reformers, historically and in the present, have focused on the need to eliminate the duality through family law reforms.

As of the mid-1990s, Turkey and Tunisia were the only Islamic countries that had abolished this duality and established a unified legal system. The reforms in both countries were enacted by revolutionary Western-oriented leaders at periods of major political and social changes— the Turkish revolution and establishment of the Turkish republic in 1924, and the founding of independent Tunisia in 1956. Kemal Ataturk of Turkey and Bourguiba of Tunisia were able to institute changes that many other leaders of Islamic countries have not been able to enact in the face of strong religious opposition. These family law reforms were part of major legal reforms granting women extensive rights in the educational, employment and political arenas, thus making it possible for women to achieve a level of equality with men, in law and in practice, unparalleled elsewhere in the Muslim world.

In Turkey, legal reforms began in the nineteenth century, during a period of intellectual ferment when ideas about Ottoman society, the family, and the position of women started to be debated in a variety of forms and media, from newspaper column to novel. But radical changes did not occur until the 1920s after the establishment of the Turkish Republic. Kemal Ataturk, the father of modern Turkey and a secularist, "strongly believed that the modernization of Turkish women could only be realized by the reform of two major institutions: education and law" (Abadan-Unat, 1986, p. 45). With the adoption of the Swiss Civil Code in the new Turkish Civil Code in 1926 and the creation of Western-type courts, Orthodox Islamic codes and their application were discarded. A series of reforms affecting women's status were adopted, polygamy was abolished, and the Civil Service laws were amended to secure women's employment rights as civil servants.

In Egypt, by contrast, the struggle for family law reform has been underway for more than 100 years. Its advocates were male reformers in the mid-nineteenth century with strong feminist ideas, such as Rifaa Tahtawi,

Mohammad Abdu, and Qasim Amin, most of whom attempted to provide a "Muslim rationale for change." In the early twentieth century, women feminists advocated women's education and family law reforms as a means to elevate the situation of women and give them an active role in the affairs of the country. While it has been relatively easy to introduce and implement reforms in political and economic law granting women significant rights in both areas, family law lags far behind. New educational and employment opportunities for women resulted in significant changes in the status and roles of women with little impact on family status law (also known as personal status law). Consequently, major contradictions exist: Women have achieved considerable power in the public sphere but have sharp restrictions imposed upon them in the private sphere of the family.

The most significant family law reform was enacted by President Anwar Sadat in July 1979, when he approved Law 44 amending the 1929 family status law as an emergency measure during a recess of the Egyptian Parliament. The new law gave a married woman several new rights, including the right to seek and secure a divorce, claiming emotional harm, if her husband takes on a second wife. This law was intended to restrict polygamy and give rights to the first wife. Additionally, a divorced woman who had custody of minor children was entitled to remain in the marital residence until the children reached legal age, at which time the father assumed custody of the children and the family domicile. That law was challenged by Egypt's Supreme Constitutional Court on procedural grounds. It was repealed, thus restoring the family law of 1929. Some minor revisions were later introduced, but the struggle continues. On the occasion of the International Conference on Population and Development (Cairo, 1994) and the Fourth World Conference on Women (Beijing, 1995), feminists brought renewed pressure on the government for family law reform. In 1996, some feminist lawyers proposed a revision of the marriage contract, claiming that Sharia law gives women many rights that they themselves are unaware of. So revising the marriage contract to stipulate specific rights of married women, such as the right to work, will increase women's legal awareness and effectiveness in pursuing their legal rights.

In Tunisia, after independence from the French colonialism in 1956, President Bourguiba embarked on a massive program for the modernization of the country. He enacted the most progressive family code of all Arab countries. The new family law outlawed polygamy, set a minimum age of marriage of 17 for women, and required their consent to marriage; it outlawed a husband's unilateral right to divorce his wife by giving both spouses the right to seek legal divorce; and it granted the mother the guardianship of minor children in the case of the father's death. In August 1992, more reforms were introduced. Official Tunisian policy toward women aims, first, at providing the legal framework and adequate mechanisms to guarantee women's presence within governmental and nongovernmental structures, and, second, at ensuring women's active participation in the country's development.

Despite their progressive nature, both Turkish and Tunisian laws maintain a patriarchal view of the family by stipulating that the husband is the head of the family and is financially responsible for its support. The sensitivity of family law issues in all Islamic countries is revealed in Egypt's reservations concerning the United Nations Convention on the Elimination of All Forms of Discrimination Against Women, which are based on "respect for the sacrosanct nature of firm religious beliefs which govern marital relations in Egypt and which may not be called into question" (Mayer, 1995, p. 106).

Women's Education: The Best Investment

Turkey, Tunisia, and Egypt have made significant progress in girls' education since the 1950s (see Table 1). The data show the vast growth in enrollment and the reduction of the education gender gap at all levels; they also reveal some interesting differences among the three countries. The fe-male illiteracy rate in Egypt is more than double that of Turkey. Both Turkey and Tunisia seem to have achieved universal access of girls to primary education, while Egypt has not. However, Egypt, with a smaller population than Turkey, has had more students, and higher percentages of girls to total enrollment, in secondary education than Turkey since the 1970s. Before the 1990s, Egypt had more female students in higher education than Turkey, but the situation has been reversed in recent years as a result of an overall declining higher-education enrollment in Egypt and an attempt to give greater attention to primary and secondary education. The following brief discussion of the history of girls' education reveals the link of girls' education and the modernization of Islamic society.

Muslim reformers in the nineteenth and early twentieth centuries attributed the backwardness of Muslim society to women's lack of education. Hence, they advocated girls' education as a means to elevating women's status and upgrading living standards in the family and society. Women's education became a top priority on their agenda

Table 1. Educational Indicators by Educational Level, 1950–1992

Year	Turkey		Egypt		Tunisia	
			Illiteracy Rate, 15+ (Census Dates in Parentheses)			
Female	31.5 (1990)		67.2 (1986)		54.8 (1989)	
Male	10.1		41.6		30.8	
			Female Primary Gross Enrollment Ratios (%)[a]			
1975	97		60		N/A	
1980	90		65		88	
1985	110		82		106	
1989	108		85		109 (1990)	
1992	107		93		115 (1993)	
			Number of Female Students and Their Percentage of Total Students			
			Primary Education			
1950[b]	598,152	37	471,661	36	N/A	N/A
1970	2,120,332	42	1,433,270	39	N/A	N/A
1980	N/A	N/A	N/A	N/A	438,252	42
1989	3,225,259	47	2,751,022	45	643,910	46 (1990)
1991	3,238,599	47	2,942,755	45	688,292	47 (1993)
			Secondary Education			
1950	22,403	25	83,720	25	N/A	N/A
1970	373,167	29	465,901	32	N/A	N/A
1980	1,030,948	35[c]	1,081,504	37	107,074	37
1992	1,514,423	38	2,355,772	45	287,456	45
			Higher Education			
1980	N/A	N/A	225,562	32	9,437	30
1985	152,047	32	254,528	30	14,824	36
1992	316,690	35	224,700	35[d]	36,121	41

Source: Education data from UNESCO, 1994.

[a]Gross ratios are inflated (over 100%) because of the inclusion of over- and underage children.

[b]Education data for 1950 are estimates.

[c]Data are for 1985.

[d]Data are for 1990.

for social reform and modernization. They argued that Islam encourages the education of all Muslims, men and women, and that modern scientific developments in the world require educated and enlightened citizens, males and females.

Turkey pioneered the opening of girls' schools and the passage of compulsory education laws back in 1824 (Szyliowicz, 1973, p. 142). Initiatives to enhance female education were vigorous during the late nineteenth century when the Ottoman constitution reaffirmed the right of all children to primary education. Turkey's early emphasis on universal primary education provided the foundation for further education and opened new opportunities for women for employment and participation in public life. Disparities still exist, however, in the educational opportunities for girls and women in rural and urban areas (Szyliowicz, 1973; Abadan-Unat, 1986). Since the 1970s, Turkey has made significant progress in reducing the gender gap in education and literacy. As a result, the proportion of illiterate women in the labor force declined from 69 percent in 1970 to 29 percent in 1990 as compared to 29 and 8 percent, respectively, for males (World Bank, 1993, p. xv).

Egypt started its modern girls' education efforts in the nineteenth century, but, in contrast to Turkey, it continues to experience serious problems in universalizing primary education, especially among girls. The Egyptian government opened two girls' schools in Cairo in 1873. Early progress in female education was slow, however, because of a generally conservative social climate, hesitancy on the part of the government, and the stringent fiscal policies of the British colonial government. The colonial administration (1882–1922) made no effort to increase female enrollment—on the pretext of not offending local traditions, and it imposed tuition in primary education, making it less affordable to the masses. This precipitated a decline in female enrollment.

Egypt's first constitution, at the time of independence in 1923, stipulated that education be compulsory for all children age 6–12, male and female. The law was never enforced, however, for lack of educational facilities to accommodate the growing school-age population. The 1952 revolution and the founding of the Egyptian republic resulted in major educational progress, especially among females. This was a result of the institution of free education at all levels, and revolutionary affirmation of gender equality and the importance of women's role in national development, as enunciated in the Egyptian national charter, and the government's declared commitment to hire all university and high-school graduates. Despite egalitarian educational policies, the government never affirmed its commitment to basic primary education by providing the needed educational facilities and enforcing its compulsory education laws. Instead, it placed greater emphasis on secondary and higher education as part of a policy that emphasized high-level manpower development.

Consequently, after more than 120 years of girls' public education, universal primary education has not been achieved, and Egypt has one of the highest female illiteracy rates in the developing world. In 1994, the government estimated that one million children, the majority of whom are girls in the rural southern governorates, had no access to primary education. Providing education for all children of school age remains a major challenge for the Egyptian government in the 1990s. Egypt has experienced major economic difficulties that have had a negative impact on female education. Egyptian economist Nader Fergany argues that structural adjustment (measures to open the economy and reduce state expenditures) has reinforced the shortfall of access to education, especially for girls. The institution of incremental cost-recovery measures by the Ministry of Education in the 1980s has made it difficult for poor parents to educate their daughters. For poor people, the cost of fees, clothes, children's allowances, and private tutoring skyrocketed, making it difficult to send their children to schools. Girls are losing in the process as parents give greater priority to educating their sons (Nader, 1995).

By comparison, Tunisia was able to make major progress toward universal primary education, especially among girls, in a relatively short time. Female enrollment rates in primary education increased from 19 percent in 1950 to 83 percent in 1968 and close to 100 percent by 1980. This progress was achieved because of Tunisia's relatively small population and the aggressive modernization policies pursued by the government after independence in 1956. Nonetheless, as with the other two countries, adult illiteracy rates are still a problem in rural areas. The Tunisian government's commitments to girls' education and the integration of women in national development have been impressive. Tunisian women have become a strong force of development in their country.

Cursory analysis of data on female education and income levels in the three countries suggests that the greater investment in girls' education in Turkey and Tunisia may account in part for their higher development indicators, including their higher income levels. In both countries, the progress in girls' education, particularly basic education, coupled with progressive legal codes and women's relatively high participation rates in the labor force, may have contributed to, and in some ways been affected by, their substantially higher GNP per capita than Egypt's. The association between girls' education and a country's GNP per capita, documented in research by the World Bank for various countries, seems to apply in the case of Egypt, Turkey, and Tunisia. Research on advanced developing countries in Asia and the Near East indicates that those countries that have achieved sustainable social and economic development are the ones with both high levels of educational attainment among girls and women and a substantial share of women in the waged labor force, often above 40 percent (Joekes, 1991).

The progress may also be linked to the distribution of students among the various fields of education at the university level, particularly in medicine, commerce, and engineering. Egypt and Turkey have a relatively high female enrollment in engineering. In the natural and medical sciences, however, the enrollments are much higher in Turkey than in Egypt. In addition, Turkey has much higher enrollment rates in math and computer science, architecture, and town planning. Since development is generally linked with higher levels of enrollment in scientific and technical fields, the discrepancy between these two countries in enrollments in these fields may be linked to their different economic-development levels. The highest female enrollment in both countries continues to be in fields with low employment prospects, such as the social and behavioral sciences in Turkey and commerce and business administration in Egypt. Although trade and commerce seem to be market-related fields, their employment prospects in Egypt have been relatively limited. A major problem in Egypt is the mismatch between the educational output and labor-market requirements. A recent government report for Egypt acknowledged the problem when it declared that the type of education provided in Egyptian schools and universities has not allowed citizens to contribute to economic progress as much as in other countries.

The advances in girls' and women's education in all three countries, notwithstanding the problems, have had significant impact on women. Available data indicate that education improves women's standing in the family and opens new options for employment. El-Sanabary's analyses of data from the Demographic Health Survey for Egypt (1992) show that increased women's education is associated with declining infant and maternal mortality and fertility rates and an increase in life expectancy at birth. The same data indicate the increased power of women in the household in the urban governorates, where women's educational levels are higher than in the rural, more conservative southern governorates of Upper Egypt, where educational attainments among women are very low. The data suggest that patriarchal attitudes among both sexes decrease with education. An educated Egyptian woman has a greater say in fertility and household budgetary decisions, greater mobility outside the household, a desire for fewer children, more freedom to express a different point of view than her husband's in his presence, and is more likely to hold a salaried job outside the home (El-Sanabary, 1995).

In Turkey, education has been found to increase the participation in the paid labor force for both sexes, but particularly women. "A university diploma increases by 50 percent the probability of wage employment for women" (Tansel, 1992, pp. 19–20).

Women and the Economy: Necessity or Empowerment?

In Turkey, Egypt, and Tunisia, women have a legal right to equal access to employment and promotion, equal pay for equal work, and all related employment benefits, in addition to benefits designed to help women integrate their multiple reproductive and productive roles. In Turkey, for instance, Article 49 of the 1982 constitution guarantees all citizens the right and duty to work, provides for improving working conditions of all workers, and creates an environment conducive to full employment. Article 50, which addresses working conditions and the right to rest and leisure, however, allows women, together with minors and persons with physical or mental disabilities, "to enjoy special protection with regard to working conditions" (World Bank, 1993, p. 87). The law does not specify what these special protections are. Women are simply not required to do work deemed unsuited to their mental or physical capacities.

In Egypt, the first law regulating women's employment was issued in 1933. It forbade the employment of women in night work and in arduous jobs such as mining, quarrying, or similar work requiring strenuous physical activity. The law also provided for a maternity leave for working women. It did not apply to agriculture work (Abdel-Kader, n.d.). When the labor laws were drastically changed after the revolution of 1952, the new laws gave women extensive work rights based on the revolutionary ideology that socialism and speedy economic development required the participation of women and men in the building of the new society. The Egyptian national charter issued in 1960 (the first Egyptian constitution after 1952) viewed women as key participants in the country's development. It stated, "Woman must be regarded as equal to man and must, therefore, remove the remaining shackles that impede her free movement so that she might take a constructive and profound part in shaping life" (author's translation). Subsequently, a law was passed guaranteeing government employment for all university and high-school graduates, male and female. In the 1980s, however, the government could no longer guarantee employment for all graduates because of economic retrenchment and a shrinking public-sector employment base. This resulted in a serious unemployment problem among the educated, especially females.

The Tunisian Labor Code stipulates that men and women have the same right, without distinction, to access to employment, titularization, and pay. The Tunisian labor law issued in 1966 did not contain any articles that spelled out the principle of nondiscrimination against women. The government commitment in support of women's work, however, was declared by Tunisian president Ben Ali Bourguiba, who proclaimed in 1989 that "there would be no sense in talking about women's rights and liberty if women did not have the right to work. Without this right, all other rights are hollow slogans" (Tunisian External Communication Agency and National Union of Tunisian Women, 1993, p. 26–27). Accordingly, in 1993, a section was added to the labor law specifying that "there shall be no discrimination between men and women in the implementation of the provision of this law or any interpretive text thereof."

Working women in the three countries enjoy substantial maternity and childbearing benefits, as well as protective benefits. Egypt probably has the most liberal maternity benefits in all Arab and Islamic countries. In addition to a mandatory five-week paid maternity leave, a working woman can take up to six years of unpaid childrearing leave and still maintain her seniority and workers' benefits. Furthermore, all employees hiring 100 women or more are required to provide on-site child-care facilities. These benefits have proven to be a double-edged sword. Private-sector employers use them as an excuse not to hire women, invoking high costs of female employment and low productivity of married women.

These generally egalitarian labor laws have provided substantial opportunities for women, especially the educated, to assume larger roles in the modern economic sector. Hence, women have made major strides and are playing an active role in shaping the affairs of their countries. Hundreds of thousands of women in these countries occupy positions in the modern economy, and some have reached the highest-ranking positions in government, in the private sector, and in politics. Nonetheless, major discrepancies remain between law and practice. Some problems are common to all developing countries, such as the concentration of working women in the informal, unregulated economic sector; the gender division of labor; salary differentials; unequal access of the two sexes to policymaking positions; the difficulties women face in juggling their multiple roles; and the negative impact of structural adjustment and economic reforms upon women.

Has employment been the liberating and empowering force it is supposed to be? At the risk of oversimplication, the following sections provide a brief discussion of the level and pattern of female labor-force participation in these three countries and the opportunities and constraints involved. These sections avoid extensive use of comparative statistics on women's labor-force participation rates because "statistics on women's labor force behavior in developing countries, in general, are plagued by conceptual problems and inadequate data collection methods" (World Bank, 1993, p. 13). The limited reliance on statistical data makes it difficult to compare and analyze conditions in any one country, to say nothing of attempts to compare data among countries. Instead, the discussion focuses on the concerns mentioned above, showing some of the main patterns of female labor-force participation.

Predominance of Women in the Informal Sector

As with other developing countries, the majority of working women in Turkey, Tunisia, and Egypt are concentrated in the informal sector, either in agriculture or in unregulated or unpaid family work. According to a 1990 Labor Force Sample Survey (LFSS) in Turkey, about 75 percent of all female workers (4.8 million) were engaged in agriculture (50 percent of all agriculture workers). According to an Egyptian LFSS in 1988, 67 percent of the economically active females were in agriculture (CAPMAS, 1994). In Tunisia, during the same year, 33 percent of working women were in the agricultural sector (National Council on Childhood and Motherhood, 1995, p. 35).

The position of women in agriculture in these countries is precarious. Much of the work is unregulated and unpaid. Women working in agriculture and the informal economy generally enjoy none of the benefits or protections enjoyed by women in the formal economy. For instance, in Turkey, despite the importance of women to the agriculture sector, "women do not enjoy ready access to agriculture resources and support services such as extension and training, information, credit, or appropriate technology" (World Bank, 1993, p. 51).

No precise quantitative data are available on women's work in the informal urban sector, especially in micro and small enterprises. Numerous problems are faced by women in that sector, including lack of credit and training opportunities, as well as bureaucratic constraints.

Throughout the Middle East and North Africa, the shift from agriculture to industry has been underway since the nineteenth century, resulting in a reduction of the numbers of women and men in the agriculture sector and an increase in the industrial and service sectors. In the nineteenth century, Egyptian women were employed in light industries, including the tobacco, sugar, and garment industries. More than 90 percent of them were illiterate (Hammam, 1986). In Tunisia, women have increased their participation in the industrial sector as a result of the dramatic growth of export industries. In 1990, 55 percent of the female labor force in Tunisia was in that sector, providing an example of the importance of export industries in utilizing women's labor in the Middle East. In Turkey, 7 percent of women were in industry in 1990 (World Bank, 1993).

The service sector claims the highest proportion of working women outside agriculture: 12 percent in Turkey, 12 percent in Egypt, and 22 percent in Tunisia (UNDP, 1995). Women are relatively well represented among professional and technical workers: 31.9 percent in Turkey, 28.3 percent in Egypt, and 17.6 percent in Tunisia (UNDP, 1995), though the majority of these are in occupations traditionally considered suitable for women, such as teaching, nursing, and social work. There are also thousands of women in nontraditional occupations, such as medicine and engineering. The medical profession is considered highly suitable for women in Islamic countries because of its high status and because of a traditional preference for the treatment of women patients by women doctors (Abadan-Unat, 1986; El-Sanabary, 1993). Egyptian and Turkish women are relatively highly represented in engineering, and thousands graduate from engineering colleges every year. Although women are found in all engineering specialties and many of them work in factories, female engineers in Egypt are less likely to do field work than male engineers. They also come from higher socioeconomic backgrounds than male engineers. The legal profession attracts many women, especially in Turkey and Tunisia; the former is said to have the largest proportion of women in the legal profession of any Islamic country. Egypt does not allow women to be judges, claiming it would contradict

Islamic teachings, although women are judges in Turkey, Tunisia, Lebanon, and elsewhere. Female lawyers and feminists continue working to get the Egyptian government to appoint female judges, but, as of the mid-1990s, they had not been successful.

Some Key Gender Issues

For the majority of working women, however, lack of education and training keeps them locked in low-status, low-paying jobs. A 1980s LFSS for Turkey indicates that 37 percent of all working women were illiterate or literate with no diploma and 48 percent had only primary-school education. Only 4 percent had university degrees and 5.5 percent had high school diplomas. Even when working in the same occupation as men, women were found to be less educated, have fewer opportunities for advancement, and earn less money (Chamie, 1985).

Wage Differential. In Turkey, women in the salaried sector receive lower wages than men, although they work more hours. Fewer than 30 percent of working women receive monetary payment, and employers can fire a pregnant woman after giving her a proper notice. In Egypt, a woman earns 79.5 percent of a man's wage; in Turkey, 84.5 percent (UNDP, 1995, p. 36). In all three countries, women's gains from employment are not commensurate with men's, as suggested by U.N. data showing women's low share of the national income: 30.2 percent versus 69.8 percent for males in Turkey; 19.5 percent versus 80.5 percent in Tunisia; and only 8.2 percent versus 91.8 percent in Egypt (UNDP, 1995, p. 76). These figures seem to indicate a higher level of economic empowerment of women in Turkey than in either Tunisia or Egypt.

Despite the problems discussed above, women in all three countries have reached high-ranking positions in government, the public sector, and some flourishing businesses. Yet, as with their sisters in other parts of the world, a glass ceiling keeps more women from reaching key policymaking positions. Women are still a minority in high-level administrative and managerial positions: 10.3 percent in Egypt and 4.2 percent in Turkey (UNDP, 1995, p. 70). (Comparable data for Tunisia are not available.)

Impact of Waged Employment upon Women. Women's entry in the labor force has not been the liberating force it was supposed to be. "The idea that economic 'modernization' gradually brought women out of the 'traditional' confines of a *harem* or a peasant family into a modern labor force is no longer an accepted generalization: new research suggests far greater complexity" (Tucker, 1993, p. xi). A study in Egypt (Hatem, 1992) indicates that the majority of women who work women do so because of economic necessity. Hatem argues that "[e]conomic and political liberalization did not enhance the equality or the liberty of Egyptian women" (Hatem, 1992, p. 248). This is due not only to the persistence of patriarchal values but also to the generally low-status occupations, low salaries, and numerous difficulties and hassles faced by the majority of working women. Most say they would quit their jobs if they could afford to.

Women's position in the labor force is vulnerable because of the perception among government officials and the public alike that women are a reserve army of labor that can be mobilized at times of need and dispensed of when their labor is no longer needed for economic or other societal necessities. Hence, economic retrenchment and restructuring often lead to loss of jobs for working women. They also reinforce the traditional ideology that the place of woman is in the home. The case of Egypt is illustrative. During the economic recession of the 1980s precipitated by the fall of oil prices, opportunities for public-sector employment declined drastically. The government was no longer able to honor its commitment to employ all graduates. Graduate unemployment became a serious problem, especially for women: Twice as many females were unemployed as males. During that period, conservative writers began calling upon women to return to the home or take part-time employment to make room for men, the primary wage earners. Egyptian newspapers and magazines were filled with arguments against women's work and rebuttals by feminists. Recent economic reforms and privatization measures that started in the early 1990s have had a more negative impact on women than on men, as documented by research in Egypt and other countries experiencing economic restructuring. A study of data (unpublished report prepared for USAID, 1995) for the Principal Bank for Development and Agricultural Credit indicates that female employees were a majority among those taking advantage of early retirement incentives offered by the bank as a result of restructuring. Furthermore, many private-sector employers advertise jobs for males only, and some government officials have declared that men should be given preference in employment because of their presumed responsibilities in support of their families. The private sector is finding it convenient to discriminate against female job applicants, citing women's family responsibilities and the governments' generous maternity benefits as an added cost of female employment.

These modernizing countries show the complexities surrounding women's labor-force participation. While they still have a long way to go, they have made considerable progress in acknowledging and facilitating the economic contributions of women and in providing them opportunities for employment and advancement. Women activists will continue to pressure their governments to address gender issues in the labor force and in the society at large.

Women's Political Participation

Women's active participation in the formal political system and in civil society is another characteristic of these modernizing Islamic countries. For centuries, women have played an active role in the informal political system, and they have increased their participation in the formal system since the 1960s. Progress, however, has been uneven.

Turkish women gained their political rights much

earlier than women in many European countries. They were enfranchised for municipal elections in 1930 and for national elections in 1934 (Abadan-Unat, 1986). Egyptian and Tunisian women began their political activism during the movements for national liberation from Western colonialism in their countries and gained the right to vote and run for elections to public office in 1956. In all three countries, women occupy key cabinet positions. Turkey is one of five countries in the world that have had women prime ministers (two other Islamic countries, Pakistan and Bangladesh, have had women prime ministers also). Egypt currently has two women ministers, the minister of security and social affairs (a post occupied by women since 1956) and the minister of scientific research. Egypt also has several women ambassadors to foreign countries. Since the mid-1980s, Tunisia has had a woman minister of women and family affairs, and in 1994 a woman was elected as deputy of the national parliament. Tunisian women are active in the country's political life as members of parliament and elected representatives to national and municipal bodies.

Yet, obtaining the vote and having the right to be elected to public office are necessary, but not sufficient, conditions for women's political empowerment. Research indicates that women's participation in elections in these three countries has been limited, as has their representation in elected bodies, despite a relatively long history of enfranchisement and participation in civil society. In 1994, the percentage of women in the national parliaments was 6.8 in Tunisia, 2.2 in Egypt, and 1.8 in Turkey (UNDP, 1995, p. 85); all are below the developing-country average of 10 percent.

Furthermore, in Egypt, there has been a marked diminution of women's participation in political life. "Their participation . . . whether as registered voters or as candidates for public office, is not commensurate with their numerical weight in society, where they constitute 50 percent of the population" (Communications Group, 1992, p. 27). Research indicates that the majority of Egyptian women (92.6 percent) are not registered in the electoral polls. In addition, the percentage of those who are registered and who actually vote does not exceed 27.9 percent (National NGO Committee, 1994). These trends reflect the conservative forces that have prevailed in Egypt since the 1970s. Furthermore, Egyptian political parties "have not been active in promoting participation of women in the political activities they sponsor" (Zulficar, 1994, p. 9).

The presence of women in the Egyptian parliament (People's Assembly and Shoura Advisory Council) has diminished considerably—from 35 members (out of 357) in the People's Assembly in 1979 to 10 (out of 444) in 1992. To compensate, President Mubarak appointed 12 women to the Shoura Advisory Council (out of 258 members, or 4.7 percent). Not a single woman was elected to the council in 1992.

Egyptian women's participation in the municipal councils is also low. Not a single Egyptian woman has ever occupied the position of governor or mayor, although some have become members of the local councils in the governorates, districts, cities, and villages. The annulment of Law 43 of 1979, which reserved 10–20 percent of the seats in the parliamentary, local, and district councils for women, resulted in a fall in women's participation in the local assemblies from 11.2 percent in 1975 to 1.2 percent in 1992. The decline in representation was sharp in the village councils, from 6.2 percent in 1979 to 0.5 percent in 1992. In 1992, 437 women (out of a total of 37,632 members) were members in the local councils in all governorates. These were distributed as follows: 102 (out of 2,508) in the governorate councils, 130 (out of 9,834) in the administrative districts councils, 52 (out of 4,112) in the city councils, and 115 (out of 20,160) in the village councils.

The Egyptian experience illustrates several factors that limit women's participation in the political system in the three modernizing countries. First, adult female illiteracy limits the formal political involvement of women. Second, political alienation keeps women and men from active political participation in Egypt, as does the weakness of the institutions of civil society and their inability to attract citizens and to encourage a feeling of belonging. Third, the rising conservative trends in the country have fostered traditional notions about women's roles in society. Finally, Egyptian lawyer and activist Mona Zulficar cites the lack of an organization to "increase women's consciousness of their legal rights, assist them in asserting and enforcing such rights or promote the nomination of female candidates for public positions or for seats in the parliament" (Zulficar, 1994, p. 9).

Gender and the Emerging Conservative "Fundamentalist" Discourse

In the 1980s and 1990s, all three countries have experienced a rise in religious conservatism or fundamentalism, with calls for a return to veiling and traditional family values. The debate has been more visible in Egypt than in Tunisia or Turkey. Tunisia, for instance, has managed to control the traditionalist voices by discouraging any outward display of religious sentiment. In the mid-1990s, veiled women are a rarity in Tunisia, where Western attire is the norm in all strata of society.

The situation is different in Egypt, where the Islamist movement is so strong that liberals argue that there is a strong backlash against women. By 1990, the majority of Egyptian women in government offices, in Cairo streets, and in towns and villages across the country had donned Islamic *hijab* or *al zei Islami* attire, which is used as a symbol against Westernism and excessive materialism. Unveiled women may feel uncomfortable walking the streets in Cairo, and pressures have been exerted on women in schools and government offices to veil.

Strong conservative forces are heard in Turkey as well. Kandiyoti (1989) comments on changes in Turkey:

In 1987, Turkey offered the perplexing spectacle of a

sit-in and a hunger strike by ultra religious students demanding the right of veiled students to go to classes (a right officially denied) and a small group of feminists marched through the streets to demonstrate against violence against women, virtually in the same week. While for an outside observer, this may seem as a healthy manifestation of political pluralism, the roots of the contemporary situations have to be sought in the specificities of the women's question in Turkey and of its evolution through time (Kandiyoti, 1989, p. 126).

There has been an increase in conservative Islamic trends in Turkey in the 1990s under eased government controls and the rising number of television and radio channels. On Istanbul's Channel 7 in April 1995, Serpil Ocalan became Turkey's first TV anchorperson to appear with her head wrapped in an Islamic silken shawl. In the streets, across the airwaves, in parliament and the bazaars, the Turkish people have engaged in noisy argument over Islam, secularism, democracy, and modernism—an ongoing discourse that is changing the way Turks look at themselves and their nation.

This trend is a move away from the debate that went on in the nineteenth century when the calls for modernization began. In the 1990s, it is a call for a "return to Islam," albeit a narrow version of Islam. Dozens of books and monographs are being written on the subject of women and Islam, and the voices of conservative Muslim women, both moderate and leftist, are vehement. The ongoing debate about the United Nations Convention on the Elimination of All Forms of Discrimination Against Women and the *Platform for Action* of the Fourth World Conference on Women (Beijing, 1995) illustrates the sensitivity of, and difficulties surrounding, the subject of women in Islamic countries, modern and traditional alike.

Muslim women feminists are fighting back, rewriting Muslim history from a feminist perspective. A project of North African women, focused on modern interpretations of Islamic texts to promote a more active role for women in society, resulted in a book (yet unpublished), *le Dictionnaire des Versets et Hadiths que Sacralise les Droits des Femmes en Islam* (Dictionary of Islamic Verses and Hadiths That Sanction Women's Rights), by Farida Benani and Zeinab Maiadi. Benani, a university professor of jurisprudence, and Maiadi, a professor of Sharia (Islamic jurisprudence), are pioneering work that could have far-reaching implications throughout the Muslim world.

This work illustrates the efforts of Muslim women to legitimate women's rights in the family, workplace, and political life within the Islamic discourse. Writing in French and Arabic, presenting the Quranic verses and Hadiths in both languages, the authors argue that justifications customarily given to restrict women's roles and autonomy in the family and society have no basis in Islamic texts. They dispel the arguments that are used by conservative and radical Muslims to limit women's rights. They cite texts fully supporting women's rights in the family and society. Although this work was done in Morocco, the authors are promoting their ideas and work in other Islamic countries and in international forums.

Conclusion

The preceding discussion shows that women in these Islamic countries have made significant gains in the socioeconomic, political, and legal arenas but still face major impediments to achieving gender equality. It shows that legal reforms and policy reforms are necessary, but not sufficient, conditions for women's full participation in society on an equal footing with men. Laws guaranteeing women an equal right to employment have not guaranteed women's full participation in the salaried economy or their advancement to high-level policymaking and leadership positions. Women's political rights make it possible for women to participate in the political life of their country but do not ensure that the masses of women actually practice their rights of political participation. In all three countries, class differences and economic restructuring continue to place poor women at a major disadvantage.

The experiences of the three countries prove that girls' and women's education and training are essential for women's advancement, as are employment opportunities and political participation. Education has increased women's options for economic and political participation, improved the quality of life for women and their children, and increased the life options of future generations.

In all three countries, educated women have been able to reach high-ranking positions in the economic and political systems. The women professionals in these countries—doctors, university professors, parliament members, engineers, businesswomen, journalists, ambassadors, and even a prime minister—are shaping the affairs in their countries. But the masses of the poor, illiterate women continue to lag behind, deprived of education, adequate housing and health services, and access to paid employment. Grass-roots women's organizations are helping poor women meet their basic needs and those of their families.

Although not discussed in this essay, women's health and population issues are vitally linked to their empowerment. The ability of Tunisia to put a check on population growth has enabled it to provide basic primary education for the majority of children of school age, both females and males. It has also enabled Tunisian women to assume a more active role in the economy than women in other Islamic countries.

There are promising indications of change as these countries engage in development planning and create institutional structures to address gender issues, with women's organizations working with their governments as partners in development. For instance, Tunisia has a Ministry of Women and the Family, a Center for Research and Documentation on Women, a special adviser to the president on women's affairs, and several active women's organizations, especially the National Union of Tunisian Women. These organizations are able to work hand in hand with men for the advancement of Tunisian society.

Egypt has also created women's units in the ministries of Social Affairs, Agriculture, and Planning, and the Social Fund for Development. The National Committee on Women has created specialized committees to address gender issues and plans to set up a committee to develop a strategy for women to be included in its national development plan. Egypt also has strong women's organizations and other organizations of civil society working together on gender issues. But, as of the mid-1990s, it had yet to set up research institutes on women or women's studies departments or centers at academic institutions to address gender issues.

The Turkish government has established agencies to address gender issues and to speed up progress toward implementation of the mandate of the United Nations Convention on the Elimination of All Forms of Discrimination Against Women. These include the Directorate General for the Status and Problems of Women (DGWSP), established in 1990; the Ministry of Women, Family, and Social Services, founded in 1991; and the Family Research Institute, founded in 1989, under the prime ministry. The DGWSP was first placed under the prime ministry and then moved in 1991 to the Ministry of Women, Family, and Social Affairs.

Women in these countries are marching forward. Though there will be setbacks and possibly a continuing backlash, the important thing for these countries will be to avoid polarization between competing ideologies and seek the middle ground. That will help women, men, and their societies find agreement so as to create a better future for all. Islam and cultural traditions will continue to play important parts in the debate over women's roles in these and other Islamic societies. These are influences that cannot be ignored but have to be moderated to reflect the true and equitable spirit of Islam.

References

Abadan-Unat, N. *Women in the Developing World: Evidence from Turkey.* Denver: Graduate School of International Studies, University of Denver, 1986.

Abdel-Kader, Suha. *The Situation of Women in Egypt.* Cairo: CAPMAS and United Nations Children's Fund (UNICEF), n.d.

CAPMAS. *Women's Employment in Egypt.* Cairo: CAPMAS, 1994.

Chamie, Mary. *Women of the World: Near East and North Africa.* Washington, D.C.: Bureau of the Census, United States Department of Commerce, 1985.

Communications Group for the Enhancement of the Status of Women in Egypt. *The Legal Rights of Egyptian Women in Theory and Practice.* Cairo: Communications Group for the Enhancement of the Status of Women in Egypt, 1992.

El-Sanabary, Nagat. "The Education and Contribution of Women Health-Care Professionals in Saudi Arabia," *Social Science and Medicine,* vol. 37, no. 11, 1993, pp. 1331–1343.

———. "Education and Demographic Correlates in Egypt." Unpublished paper, 1995.

Hammam, Mona. "Capitalist Development, Family Division of Labor, and Migration in the Middle East." In Eleanor Leacock and Helen Safa (eds.), *Women's Work: Development and Division of Labor by Gender.* Boston: Bergin and Garvey, 1986, pp. 158–173.

Hatem, Mervat. "Economic and Political Liberation in Egypt and the Demise of State Feminism," *International Journal of Middle Eastern Studies,* vol. 24, 1992, pp. 231–251.

Joekes, Susan P. *Lessons Learned from Advanced Developing Countries.* Report for the Office of Women in Development, United States Agency for International Development (USAID). GENESYS Special Studies Series no. 3. Washington, D.C.: USAID, 1991.

Kandiyoti, Deniz. "Women and the Turkish State: Political Actors or Symbolic Pawns?" In Nira Yuval-Davis et al. (eds.), *Woman, Nation, State.* New York: St. Martin's, 1989, pp. 126–149.

Mayer, Ann. "Rhetorical Strategies and Official Politics on Women's Rights." In Mahnaz Afghani (ed.), *Women's Human Rights in the Muslim World.* Syracuse: Syracuse University Press, 1995.

Nader, Ferghany. Interview with author, Cairo, 1995.

National Council on Childhood and Motherhood. *Women in Egypt.* Cairo: National Council on Childhood and Motherhood, 1995.

National NGO Committee. "Egyptian NGO Report," platform document to the International Conference on Population and Development, Cairo, 1994.

Szyliowicz, J.S. *Education and Modernization in the Middle East.* Ithaca: Cornell University Press, 1973.

Tansel, Aysit. "Wage Employment, Earnings, and Returns to Schooling for Men and Women in Turkey," discussion paper no. 661. Economic Growth Center, Yale University, New Haven, April 1992.

Tucker, Judith (ed.). *Arab Women: Old Boundaries, New Frontiers.* Bloomington: Indiana University Press, 1993.

Tunisian External Communication Agency and National Union of Tunisian Women. *Women of Tunisia.* Tunisia: Tunisian External Communication Agency, Government of Tunisia, 1993.

United Nations Development Program (UNDP). *Human Development Report, 1995.* New York: UNDP, 1995.

United Nations Educational, Scientific, and Cultural Organization (UNESCO). *UNESCO Statistical Yearbook.* Paris: UNESCO, 1986–1994.

World Bank. *Turkey: Women in Development.* A World Bank Country Study. Washington, D.C.: World Bank, 1993.

Zulficar, Mona. *The Egyptian Woman in a Changing World.* Cairo: New Civic Forum, 1994.

Jewish and Palestinian Women in Israeli Society

Barbara Swirski
Manar Hasan

Introduction

By most of the accepted econometric standards, Israel is what is termed a "developed" country. It can boast a multiparty political system, judicial review, and a civil society. The state social welfare system is well developed, and civil law grants women formal equality with men. At the same time, when it comes to the status of women, Israel has many characteristics of Third World countries. This stems from the strong position of religion in the state, especially that of the patriarchal Jewish religious establishment, and an unholy alliance between the government and the patriarchal leadership of the Palestinian citizens of the Muslim, Christian, and Druze faiths.

There is no separation between state and religion in Israel. Moreover, the basic definition of the Israeli nation-state as a Jewish state has far-reaching consequences for all female citizens. It allows the ultimate status of Jewish women to be defined by patriarchal religious laws, and it accords non-Jewish citizens of Israel second-class citizenship—reflecting strongly on the even lower status of the non-Jewish women.

This essay first highlights the unique situation of Jewish and Palestinian female citizens of Israel, then takes a critical look at how the following affect them: marriage and divorce, family life, education, employment, politics, violence against women, and the feminist movement.

Jewish Women

Since the establishment of the state of Israel in 1948, governments have been formed by coalitions that have always included Jewish religious parties. These have been able to impose a certain degree of adherence to Jewish law on all Jews living in Israel. The most damaging stipulations of an arrangement in force since 1948, referred to as the "status quo," grants Rabbinic (Jewish religious) courts jurisdiction over personal-status matters. This means that, in domestic litigation, women's status is inferior at the very outset, women cannot be judged by their peers (other females), and they have to plead their cause in the framework of laws prejudicial to their sex. This arrangement was a political compromise between socialist and fundamentalist men, at the expense of women.

Another impediment to women's equality in Israel is the prolonged conflict between Israel and the Palestinians and neighboring Arab states. One direct result of this state of siege is that defense matters receive first priority, and, as wars are men's business, only men are perceived as qualified to make decisions concerning them; this situation has helped perpetuate women's lack of political power in Israel. Another result is the development of a large defense industry, also reserved for men only. The military spills over into civilian life: Women's power is also limited by the fact that a large number of career officers, most of them Jewish men, retire at the age of 40 to begin a second career in politics or administration, where they are invariably preferred over female aspirants.

The conflict contributes to women's lower status in yet another way: Military service is compulsory for Jewish women and men alike, and 70 percent of 18-year-old Jewish females are mobilized. Despite the trend of increasing job opportunities for women within the military, for most women the two years spent in army service contribute very little to personal or career development. Ninety percent of the combat positions—and all of the most prestigious ones—are closed to women. Within the military, male chauvinism is the norm, and female conscripts are far more likely to tote coffee cups than M-16 rifles. The old adage of guns or butter means that 25 percent of the state budget goes to defense, limiting social-service expenditures on which the lot of women depends for improvement.

The initiation of a peace process in 1993 gave many social observers cause for optimism. Indeed, the result may be the eventual decline of the importance of military matters and the conversion of the people's army into a volunteer one (Swirski, 1995)—developments that may augur well for women. On the other hand, there is no reason to believe that military expenditures will be drastically reduced in the near future. And if and when they are, the monies freed will not necessarily be rechanneled into women's development. The domestic agenda at the end of the 1990s—decreased government responsibility for social services and increased privatization—will benefit the privileged rather than the disadvantaged. Women are prominent among the latter, as they constitute the majority of the aged, a significant portion of the low-wage earners, and the majority of single parents.

The Gendered Workplace

Although Jewish women in Palestine fought for, and won, the right to work long before the state of Israel was created, the workplace continues to constitute a major focus of inequality. Women and men are employed in what are essentially separate labor markets, with different starting salaries and ceilings. Even when the basic salaries are similar, men receive much higher fringe benefits, like car and travel allowances and overtime pay. Women and men tend to work in different occupations; when they do work in the same profession, they tend to specialize in different areas or work for different types of employers. As Izraeli points out, workplaces are gendered institutions; her findings concerning female managers (1993) can be applied to other jobs as well: Women receive lower entry-level positions and salaries, and these result in accumulated disadvantage; they fail to receive promotions in accordance with their abilities and experience; they have limited access to information (which is usually passed on after normal work hours in social settings); if a woman does get promoted to a job formerly held by a man, she gets fewer benefits; and if women encounter discrimination on the job, they are unlikely to find support.

Palestinian Women

As Tzartzur notes:

> due to his [sic] very ascription, the Israeli Arab citizen cannot accept the ideology of the Zionist movement. . . . Israeli Arabs have low social status, because in Israel the social status of the citizen is determined not only by education and professional achievements, but by the degree of identification with the central goals of Israel as a Jewish Zionist state (Tzartzur, 1985, p. 479).

It should be added that those Arabs who do openly identify with the Jewish state, like some of the Druze and Bedouins who volunteer for military service, remain second-class citizens: National origin is more important than support of Zionism in the determination of social status.

When Israel was established in 1948, the Palestinian minority was perceived as hostile, or at least as opposed to the aims of Zionism, and, therefore, as a group that had to be controlled. Control was achieved by various means, from the military government, in effect until 1966,

through the Secret Service and various governmental advisers of Arab affairs, to the allocation of public-service jobs, the supervision of school curricula, and the limitations placed on the freedom of expression of Arab citizens, which included censorship of the Arabic press, Arabic literature, and even Arabic poetry (Benziman and Mansour, 1992). Until July 1995, Palestinians who wished to convert to Judaism were subject to Secret Service investigations. The correspondence of Arab citizens of Israel is opened by the mail censor, for the purpose of reporting to government agencies on the "general atmosphere" among Arab citizens of Israel. The latter is especially damaging to women, because of the tight social control to which they are subject in Palestinian society. In such a situation, letters become a means of expression, communication, and unburdening for women. Censorship turns their hopes and dreams into intelligence material that reinforces Israeli-Jewish stereotypes of Palestinian society.

Economically, Palestinian citizens of Israel are at the bottom of the ladder: The average monthly income of Arab urban households is only 50 percent of that of Jewish households (CBS, 1994a), and 44 percent of Arab children live in poverty, compared to 22 percent of Jewish children (Achdut, 1994). These disparities are the outcome of the segregation and separate development of the Jewish and Arab sectors, of unequal access to resources, and of unequal definition and provision of needs.

Palestinian society in Israel is predominately rural in character: Although agriculture is no longer the mainstay and the villages have grown, they have no urban infrastructure. The primary framework is still the *hamula* (clan). The process of urbanization slowed down after 1948, when most of the urban Arab population of Palestine, 280,000 out of 300,000 persons, were either expelled or left the country (Lustik, 1980). Their numbers included the political, cultural, and national leadership of Palestine. The majority of those who stayed behind were peasants. Whereas prior to 1948 they had constituted but one part of an entire society made up of different classes, after 1948 the peasants were practically the only Arab people left. They came to be known as the "Israeli Arabs"—Palestinians living within the 1948 borders of Israel. Subsequently, most of their lands were confiscated (Kamen, 1984), and they underwent proletarization. The urban exodus was very significant for women. Prior to 1948, women's organizations had proliferated in the cities. These disappeared after the establishment of Israel, and with a few notable exceptions, it is only in the 1990s that Arab women have begun to set up their own frameworks.

Religion As a Double-Control Mechanism

The lack of separation between religion and state in Israel affects Palestinian women in Israel—Muslims, Christians, and Druze—no less than Jewish women. The fact that civil courts cannot adjudicate in domestic matters relegates the fate of Palestinian women to the religious courts (Hasan, 1993a), which discriminate against them. While the legal system gives Jewish women the option of applying to civil courts for child custody and support (Jahshan, 1993), it gives the church and *Shara'i* (traditional Muslim laws) courts exclusive jurisdiction over Palestinian women not only in marriage and divorce but also in custody and child-support litigation. Thus, the control of the various religious authorities over Palestinian women cannot be ascribed to tradition alone but is also the result of government policy, which gives religious tradition its teeth.

To outward appearances, Israel takes a multicultural approach to the Palestinian minority, permitting minority ethnic groups to live according to their own traditions. The problem is that, in this case, the minority is defined not as part of a Palestinian nation but as diverse religious and ethnic groups: Muslims, Druze, Bedouins, and Christians of different sects. The buttressing of tradition, through the strengthening of religious leaders and patriarchal *hamula* heads in Palestinian society, was accompanied by the suppression of organization on a national basis, as well as by the encouragement of ethnic divisiveness. In 1957, the Druze were officially transformed into a separate entity. Subsequently, a separate school system was set up for them, and their religious courts became autonomous.

The government policy toward the Palestinian population, which emphasizes and enhances its division into religious and ethnic groups, affects women as well as men. In 1995, the office of the prime minister created four distinct "representative" bodies for Palestinian women, one each for Muslims, Christians, Druze, and Bedouins. While encouraging ethnic separation, the government also encourages the traditional *hamula* leadership, which opposes change in Palestinian society, through co-optation, which

> has contributed to the continued dependence of the Arab minority on the Jewish sector as well as to the internal fragmentation of the Arab population. Having sought out and supported traditionalist elements in Arab villages, the government has in effect helped give power to those in the Arab community who are often least desirous of rapid social and economic development (Lustik, 1980, p. 236).

Thus, the direct connection of the Palestinian citizen to the state is blocked by government policy and by traditional, patriarchal mechanisms in which women are considered inferior to men. This situation is responsible for the continuation of phenomena such as "family honor" murders, early marriage, endogamous marriage, and female circumcision.

The Jewishness of the state and the nonseparation of religion and state affect every aspect of life in Israel. This is clearly illustrated by the Basic Law of Human Freedom and Dignity, passed in 1994. While this law guarantees basic freedoms such as the right to life, property, and privacy, it stops short of abolishing the 1945 State of Emergency provisions—a legacy of the British mandate—that permit discrimination against Palestinian citizens, and it

does not supersede the authority of the religious courts that discriminate against all women in Israel.

The following sections examine a number of women's issues with a critical eye, pointing out recent changes as well as possible directions of future development.

Marriage and Divorce

Marriage is the norm in Israel, and divorce is much less common than it is in most of the developed world: four per 1,000 adults for Jews and 2.9 for Muslims (CBS, 1994a).

Jewish Law

One reason for the much-acclaimed stability of marriage is the difficulty of obtaining a divorce. A Jewish woman cannot divorce without the consent of her husband, even if she is battered or her husband is missing or insane. While a man cannot obtain a divorce without consent either, there are special conditions under which he may take a second wife. The religious courts discriminate against women in other ways as well: A man can commit adultery and eventually marry his lover, while a married woman is forbidden to marry hers, and any children born from such a union are considered bastards. A bastard cannot marry another Jew unless he or she is also a bastard; neither can a bastard marry outside the faith, as there is no provision for civil or intermarriage in Israel.

Another anomaly of Jewish religious law, still enforced in Rabbinic courts, is the levirate marriage. A woman whose husband dies leaving her childless must be released from her deceased husband's brother in a humiliating ceremony carried out in Rabbinic court. Often, extortion payments are involved before she gets her release. Women's organizations—Na'amat, the Women's International Zionist Organization (WIZO), the shelters for battered women, the Israel Women's Network, and the Agunot (anchored women) Association—have expended considerable energy in attempts to find religious or civil solutions to these problems, without much success.

The influx of 600,000 immigrants from the former Soviet Union between 1989 and 1995, 20 percent of them non-Jews, and of 18,000 immigrants from Ethiopia, may, in the long run, lead to a liberalization of Jewish law. (Israel has two main types of Jewish immigrants: the Mizrahi, or Jews from Asian and North African countries; and the Ashkenazi, or Jews from Europe and the Americas; the latter are the dominant group.) As the Jewish religious establishment in Israel does not consider the Ethiopian immigrants bona fide Jews, it requires couples to undergo a conversion ceremony before marriage. Many young couples refuse to comply. The non-Jews among the Soviet immigrants will not be able to marry Jews. If an alternative, like civil marriage, is not found, many young people will either forgo formal marriage or invent their own alternatives. Both possibilities will erode the authority of the Jewish religious establishment, which will also have to find a solution to the thorny problem of the status of the progeny born of these unions.

In Israel, as elsewhere, heterosexuality is the norm. Sexual preference became a public issue for Jews after a prominent female member of the Knesset (parliament) sponsored a hearing on homosexuality at the Knesset in 1993. Israel has lesbian and homosexual associations, composed almost exclusively of upper-middle-class Jews, that lobby for recognition and equal rights and engage in cultural and communal activities. Israeli liberals and fundamentalists line up on different sides of the homosexuality debate, with the latter negating gay rights. Although homosexual marriage is certainly not in the offing, the future will probably see greater respect for sexual preference.

In the case of divorce, Muslim law is no kinder to women than Jewish law. A woman is not defined as the natural custodian of her own children; *Shara'i* (Muslim religious) courts generally grant divorced women custody of young children (as they are viewed as in need of a woman's care), but when males reach age 7 and females age 9, custody reverts to the father. In such a situation, women may relinquish custody in the present so as to avoid a painful separation in the future. Moreover, pressure is often brought to bear on a woman by the husband's family to give up the children altogether in return for a divorce, and the religious courts accept extortion of this kind. Under Muslim law, the man is the master of the house; when a couple divorces, the woman leaves and the house and furniture remain with the man, unless the marriage contract states otherwise.

It is possible for Muslims in Israel to write into the marriage contract a provision that, in the event of divorce, the matter of property division is to be decided in civil rather than religious court (under Israeli civil law, a woman is entitled to half of the property accumulated in the course of the marriage). The question is how common such a provision actually is; obviously, it depends on the power of the bride's family and their willingness to invoke Israeli civil law rather than Muslim religious law. Some efforts have been made—by the shelters for battered women and by the Palestinian Feminist Organization—to move domestic litigation to the civil courts, thus far on a small scale and often to no avail.

Endogamous Marriage

Among Palestinians in Israel, endogamous marriage is common: A 1993 study found 43 percent of unions to be with members of the same *hamula*—that is, marriage with relatives (Jabar, 1993). Often, the wishes of the prospective bride are not taken into account. The arrangement reflects patriarchal *hamula* preferences, foremost among them the preservation of *hamula* unity and the prevention of fragmentation into many households. It serves the interests of both the *hamula* heads and the government. For the latter, an increase in the number of clan leaders resulting from fragmentation would make co-optation difficult (Lustik, 1980; Hasan, 1993b). It should be noted that although marriage with relatives is common throughout the Middle East, it is less common in Lebanon (25 percent), Algeria (21 percent), and Egypt (28 percent). It is only in Jordan,

Saudi Arabia, and Kuwait that the proportion of endogamous marriages exceeds 50 percent.

Despite the fact that endogamous marriages result in a high proportion of hereditary diseases, congenital birth defects, retardation, and rare diseases (Jabar, 1993), public-health officials do nothing to prevent them, not even elementary health education. When questioned about the phenomenon, the Israeli Ministry of Health spokesperson replied, "The policy of the Ministry of Health is not to intervene, and it is also impossible to change the Arab mentality" (*Hadashot*, 1993). It should be noted that the policy of nonintervention is applied selectively: While the authorities do not intervene in cultural matters that strengthen patriarchal tradition, like marriage with relatives, female circumcision, early marriage, and girls dropping out of school, they do intervene when the issue is, for example, the publication of nationalist poetry, the school curricula, or organization on a national basis. Thus, the politics of nonintervention, or, as it is otherwise referred to, "cultural relativism," functions to strengthen patriarchal elements, the outcome of which is to reinforce the inferior status of women and weaken progressive trends struggling against oppression.

Early Marriage and Exchange Marriage

In Israel, the minimum legal marriage age for girls is 17, except under special circumstances. Early marriage is much more common among Palestinians than among Jews: 15.5 percent of Palestinian girls, compared to fewer than 1 percent of Jewish girls, are under 17 years of age when they marry (Ben-Arieh, 1991). Among Palestinians, the proportion of minors marrying does not appear to have decreased in the course of the last 50 years: A 1943 survey of five Palestinian villages in the Lod area found that 12 percent of the women married between the ages of 13 and 17 (Dajany, 1943, in Washitz, 1947). Since the establishment of the state of Israel, no progress has been made with regard to early marriage by Palestinian women, which takes place when they are in high school, violating their right to education and a good family life. In fact, early marriage robs them of their childhood and turns them into "children bearing children." The custom is also a violation of Israeli law.

Another marriage custom that disadvantages Palestinian women is the exchange marriage (*badal*), under which a man avoids paying the bride price by giving his sister or daughter to the father or brother of his bride (Washitz, 1947; Rosenfeld, 1964b). Such marriages also serve to forge alliances between clans. The problem is that the fates of the two women are intertwined: If one marriage breaks up, the other does as well, regardless of the sentiments of the persons involved. The one who has the least to say is the woman who has been traded off. The problematics of this arrangement were highlighted by the 1995 murder of a woman from the village of Mashad, near Nazareth. Although the family was aware that her husband beat her regularly, they ignored it because the marriage had been part of a package deal they wished to keep intact.

Family Life

In Israeli and Palestinian societies, family still comes first, and a "real" family has at least two children. Social policy encourages childbearing, especially for Jewish women. Fertility rates are higher than those in many urban, industrialized countries: in 1993, 2.92, compared with 1.77—the average in OECD countries (calculated from OECD, 1996); 2.62 for Jews and 4.67 for Muslims in the years 1990–1994, compared to 3.28 and 8.47, respectively, in the years 1970–1974) (CBS, 1996, pp. 119–120).

The idea behind social policy was and still is to increase the Jewish population of Israel, in the face of what is referred to as the "demographic threat"—the higher birthrate of the Arab minority inside Israel and that of the Arab majority outside. Pregnant women cannot be fired from their jobs, and either new parent may take a three-month, fully paid leave of absence. Mothers may take another nine months' unpaid leave; they also receive birth allowances. Birthing is in hospital and is completely covered by national health insurance.

Children are highly valued among both Arabs and Jews. Inexpensive pre- and postnatal care is provided through a nationwide network of mother-and-child clinics. Birth control is widely practiced in the Jewish sector. Doctors routinely prescribe the pill for women who have not given birth and insert IUDs for those who have. Family planning clinics are not an open, institutionalized part of the health-care system, whereas in-vitro-fertilization clinics, which offer a wide range of reproductive technologies for women who have not succeeded in becoming pregnant after a year of trying, are. In fact, Israel boasted 19 such clinics in 1992, the highest clinic–population ratio in the world (Swirski et al., 1992). Legislation legalizing surrogacy was enacted in 1996.

Early childhood education is highly developed in the Jewish sector. In 1993–1994, 99 percent of 4-year-olds, 95 percent of 3-year-olds, and 69 percent of 2-year-olds attended day care or preschools; the corresponding figures for Arab children were 71 percent for 4-year-olds and 44 percent for 3-year-olds. There are no figures on Arab 2-year-olds, probably because the number in day care is negligible. The coming years will probably see the rapid expansion of early childhood education in Palestinian Arab communities in Israel, due in part to the pressure being brought to bear on state education officials and local government authorities by voluntary organizations, many of them feminist.

In both Jewish and Arab families, women are responsible for housework, child care, and care of elderly parents. Jewish men often help out at home, but all of the research points to women spending much more time on "homework" than men, regardless of their employment status (Katz and Peres, 1986; Izraeli, 1988). Jewish women do not believe that they have to choose between having a career or having a family; they usually opt for both.

About 6 percent of Israeli households with children are single-parent families headed by women, mainly widows and divorcees (CBS, 1994b). For both Arabs and Jews, singleness involves a loss of social and economic status. The immigration of large numbers of single-parent families to Israel from the former Soviet Union and from Ethiopia between 1989 and 1995 increased their visibility in Israeli society and resulted in the legislation of increased social benefits, including child-support payments, home-purchase loans, and special municipal-tax discounts.

Education

In Israel, education is nearly universal at the elementary level. Nearly all Jewish youths (93 percent) attend high school, but most (about 80 percent) do not attend university or college as degree students; 66 percent graduate without a matriculation certificate—a prerequisite for further study, and the number of places for students working toward college degrees in institutions of higher learning is limited to about 22 percent of the age group (Sprinzak et al., 1995; and calculations based on CBS, 1995a).

The education of all women in Israel has risen steadily over the years, but a large gap remains between Jewish and Arab women. In 1992, the median years of schooling was 12.0 for Jewish women (12.1 for Jewish men) and 9.0 for Arab women (10.2 for men). In 1993, 28 percent of Jewish women and 9.2 percent of Arab women cited an institution of higher learning as the last school they had attended (CBS, 1994b). Although Israel does not collect figures on illiteracy, the 1994 *Statistical Abstract* (CBS, 1994a) indicated that, in 1993 16.6 percent of Arab women had no education whatsoever and 6.6 percent had one to four years of schooling. No equivalent figures were presented for Jewish women.

In 1994, 70 percent of Palestinian 14–17-year-olds attended high school (official figures show no gender breakdowns), compared with 97 percent of Jewish girls and 89 percent of Jewish boys (Sprinzak et al., 1995). In 1992–1993, women constituted more than half of the recipients of university degrees (54 percent of bachelor's degrees and 44 percent of doctoral degrees). Jewish women accounted for 55 percent of Jewish degree recipients. For Arab women, opportunities for university study are much more constrained: in 1992–1993, only 5.3 percent of university-degree recipients were Arab women and men, while they constituted a far larger proportion (21.8 percent) of the age cohort. Forty-one percent of these students were women (CBS, 1995b, 1995c). The trend of increased educational achievements for women, Arab and Jewish alike, is expected to continue, as are the gaps between them.

Employment

Whether on the farm or in the factory or office, women's work has always been defined as different and less valuable than men's work. Israeli labor legislation, enacted in the 1950s (protective laws) and again in the 1980s (antidiscrimination laws) is progressive, but women and men continue to work in what are essentially different labor markets.

The trend is for increased workforce participation for Jewish women. It began with the expansion of public services in the 1970s and continued with the growth of the business and financial sectors of the economy in the 1980s and 1990s. Another factor that encouraged women's employment was the rising standard of living, which the average family could not maintain without two salaries.

At first, women entered occupations that were already defined as female, like teaching and nursing. They also filled new positions as bank tellers and as clerical workers in insurance companies and other businesses. Women also increased their presence in a number of traditionally male professional and managerial occupations, becoming lawyers and jurists, architects, managers in the public service, and school principals. Women constitute about 16 percent of managers, in specializations where competition with men is less intense (Izraeli, 1988).

Between 1970 and 1993, the proportion of adult women in the civilian labor force increased from 29.3 to 43.4 percent; the latest figure for Jewish women is 48 percent and that for Arab women 16 percent (CBS, 1994b). The figure for Arab women does not include the many rural women who work on family farms. The workforce participation of Arab women has been limited by the underdevelopment of Palestinian communities in Israel, which results in relatively few employment opportunities, as well as by lower educational opportunities and the expectation that women stay home after marriage—unless they are highly educated and work in acceptable fields like teaching or social work.

Women in Israel tend to concentrate in "female" occupations: They constitute 62 percent of public-service employees, 74 percent of education employees, and 69 percent of health workers. While they represent only 27 percent of industrial workers, women account for nearly 50 percent of those employed in the financial- and business-services sector (CBS, 1994b).

Women receive less remuneration than men for the same or equivalent work. The gap is 30 percent in the public services and 50 percent in the financial sector (Atzmon and Izraeli, 1993). In the public services, this represents a widening of the gap. Women tend to compare their salaries and working conditions with those of other women rather than with those of men, and they rarely initiate discrimination suits. Palestinian women earn less, on the average, than Jewish women.

For all Israeli women, workforce participation increases with educational level. While women's labor-force participation will probably continue to increase, that of Palestinian women will grow more slowly than that of Jewish women, as the factors inhibiting their participation are not about to disappear. There is no indication that the remuneration gap between women and men will decrease in the foreseeable future.

Politics and Power

It was the new feminist movement in Israel that called political equality between women and men the bluff it had always been. For women as well as men, politics is a realm reserved for the elite. But for women, it is even more so: Holding office requires time, money, and a power base— three elements largely absent from the lives of most Israeli women, even those of the elite.

Since the establishment of the state in 1948, women's representation in the Knesset has remained low, from 7 to 11 members of the 120-person body. Representation in local governing bodies has doubled, from about 4 percent to 8 percent, but is still very low. Women enter politics later than men. They train in political-party frameworks, including Na'amat, the largest women's organization, actually the women's branch of the National Federation of Labor Unions (Histadrut), in which top positions are distributed on a proportional, political-party basis. A study of women in local politics found a consistent picture over the years: The local female politico is middle-aged, well educated, and employed in a "female" occupation with above-average earnings. At home, she fulfills traditional female roles; she is a "superwoman" (Herzog, 1993).

In 1995, no Arab women were serving or had ever served on national bodies; two were serving on local government councils. Interviews with the three Arab women elected to local councils in 1989 revealed that they had much in common: All were Christians living in large urban communities. They came from politically active families; they were highly educated; and their aspirations for equality and political power originated in their activities in the framework of the Communist Party (Herzog, 1994).

Jewish women have occasionally served as ministers or deputy ministers. In early 1996, Israel boasted a record of two female ministers and one deputy minister. Most other positions of power are held by men, including the key positions in the Histadrut, the directorships of the state companies, and the higher managerial posts in the public-service sector. On the other hand, high positions in the state banks have opened for women, and the judiciary was never closed to them. Formerly, women sat on the benches of the lower courts; now they sit on the Supreme Court as well.

Jewish women in Israel have always been active in women's organizations and social movements, and they have been especially active in the peace movement, first in 1982 in the wake of the Lebanon War, and on a much larger scale following the Palestinian Intifada in 1987. (The Intifada was the popular uprising of the Palestinians living in the territories occupied by Israel in 1967 against the occupation, which set off a conflict that ended in attempts to make peace and effect a settlement between Israel and the Palestinian Authority.) Palestinian women have been active mainly in the framework of Democratic Women, the women's auxiliary of the Communist Party. The introduction of primaries into national elections is likely to result in decreased women's representation, since the method requires more financial resources than the previous system. Women's participation in organizations of their own making is expected to increase as civil society in Israel develops.

Violence Against Women

In Israel, as elsewhere, in developed as well as Third World societies, violence against women, especially marital violence, is common. Whereas Jewish religious law generally puts women at a disadvantage, the case of marital rape is an exception. In 1981, an Israel Supreme Court judgment defined marital rape as a crime and inconsistent with Jewish religious law. Civil legislation made wife battering a crime in 1991, enabling women to obtain a six-month restraining order to keep abusive husbands out of the house. In 1996, Israel had eleven shelters for battered women, one of which was set up by and for Arab women. The other shelters admit women of all religions and backgrounds. The creation of shelters and the raising of public consciousness to wife abuse is one of the major accomplishments of "new wave" feminism in Israel begun in the 1970s.

"Family Honor" Murders

The phenomenon of "family honor" murders has not been eradicated in contemporary Israel: Each year, an estimated 20–40 female Palestinian citizens of Israel are murdered for violating traditional codes of proper conduct (Regev, 1988). Before the 1990s, such murders were not openly covered by the press, and the existence of the phenomenon was denied. There are cases in which even raped women were murdered in order to purify their families of the shame of rape; other women were murdered on the strength of rumors or gossip. These murders are carried out by the women's own blood relations (fathers, brothers, uncles, sons) and not by the "injured" husband. Often the crimes are disguised as accidents or suicides (Kressel, 1981; Nejidat, 1992; Hasan, 1994). Until the 1990s, the police and the courts demonstrated leniency and consideration toward the murderers, an approach that was reinforced by the guardians of tradition, and, at the same time, reinforced their own position. For example, for many years the police, out of what they called "respect for tradition," would return girls who had run away from home (because they feared reprisal) directly to their families or to village notables, who would then turn them over to their families, where some were subsequently murdered.

Despite the efforts described above to preserve the power of the *hamula,* it has weakened over the years. One of the outcomes is the increasing number of wives murdered by their husbands, a crime that in the past was extremely rare in Palestinian society, because the laws of blood redemption required a husband to compensate her family for their loss. With the weakening of the *hamula,* these unwritten laws have lost much of their force.

Female Circumcision

Another form of violence against women practiced in Israel is the circumcision of nearly all teenage girls (age 12–17) among some of the Bedouin tribes in the Negev (Asli et al., 1992). Bedouin citizens of Israel number about 90,000; they constitute about 10 percent of the non-Jewish population of Israel and about 25 percent of the total population of the Negev, or southern part of Israel. Formerly nomads, about half of the Bedouin were induced to settle in seven government-planned townships; the other half reside in small, scattered, "unrecognized" (by the central government) settlements.

Among the Bedouins, clitoridectomy is performed by an older woman, while other women hold the girl down and gag her. It is done without anesthetic and without sterilization. The function of the practice is to control the woman's sexuality. Besides the physical damage, clitoridectomy prevents a woman from enjoying sex, turning her into a vessel/baby machine for her husband. The state health authorities do nothing to abolish the practice, which affects thousands of women, for fear of angering the Bedouin sheiks (appointed by a government commission), who might view such intervention as interference with their authority. The sheiks themselves contend that this is a women's matter that men have nothing to do with.

For the sake of comparison, it should be noted that clitoridectomy was also found to be practiced among some of the Ethiopian Jews. However, as immigration to Israel was accompanied by the destruction of the patriarchal family, the practice disappeared entirely.

Feminism

In the Jewish community in Palestine before 1948, women struggled for the right to work and the right to vote and won both. In the 1970s, a new movement developed, inspired by Western feminist activity. Started by university teachers, it trickled down to students and up to government bureaucracies. Its first successes were feminist projects: a publishing company, shelters for battered women, rape crisis intervention centers, and cultural institutions. The mid-1980s saw the development of liberal feminism, notably the jumping-on-the-bandwagon of the traditional women's organizations—Na'amat and WIZO—and the creation of the Israel Women's Network (the largest "new wave" women's organization, comparable to the United States National Organization for Women (NOW) or the French Choisir, which engages in a variety of activities but focuses primarily on political participation, legislation, and litigation); women's studies units at universities; and Status of Women committees within government ministries and local governments. The late 1980s saw the expansion of the women's peace movement and the proliferation of women's peace organizations. Foremost among them was Women in Black, loosely organized groups of predominately Jewish women who, from January 1988, following the outbreak of the Palestinian Intifada, until the signing of the Oslo peace accords, held silent vigils at as many as 25 different locations to protest the Israeli occupation of the West Bank and the Gaza Strip.

Developments in the 1990s may give feminism in Israel renewed relevance and power. The female legislators in the Knesset, many of whom are declared and committed feminists, work together in the framework of their own creation, a Knesset Status of Women Committee. Traditional women's organizations and grass-roots feminist organizations join forces around burning issues, like violence against women or proposed legislation perceived as harmful to women. The movement, consisting of prominently Jewish, Ashkenazi, and middle-class women, has diversified. Two Mizrahi feminist organizations have emerged, one middle-class and the other working class, as well as the Palestinian Feminist Organization, which organizes consciousness-raising groups and focuses on issues such as "family-honor" murders. Finally, since 1990, feminists from diverse backgrounds have gotten together on a regular basis for three-day national feminist conventions in which considerable efforts are made to assure equal representation of Palestinian, Mizrahi, Ashkenazi, and lesbian women.

Despite its small size, and despite the fact that most women who appear to be liberated would rather not identify themselves as feminists, the feminist movement in Israel has been successful in raising the general public consciousness to the idea of gender equality and to specific issues such as violence against women and equal opportunity in the workplace. A generation after the start of the modern movement, a change can definitely be detected in the direction of higher educational and career ambitions of young women, and of families that appear to be more egalitarian. A whole new body of legislation was enacted to bolster equality between the sexes. On the other hand, much of this progressive legislation is still honored more in the breach than in the observance. The Israeli feminist movement has never succeeded in mobilizing masses of women, it cannot boast a gender gap in voting patterns, nor has it boosted women into positions of economic and political power. The major impediments to women's development—the ongoing conflict, the control exercised over the Palestinian minority in Israel, the nonseparation of religion and state, and the gendered construction of the workplace—have not yet been abolished or significantly altered by the Israeli women's movement.

References

Achdut, Lea (ed.). *Annual Survey, 1993–1994.* Jerusalem: National Insurance Institute, 1994.

Asli, A., N. Hamaisi, Y. Abu-Rabiya, and H. Balamkar. "Public Health: Female Circumcision," *Harefuah,* no. 122, 1992, p. 7 (Hebrew).

Atzmon, Yael, and Dafna Izraeli. "Introduction: Women in Israel: A Sociological Overview." In Yael Atzmon and Dafna Izraeli (eds.), *Women in Israel.* New Brunswick: Transaction, 1993, pp. 1–13.

Ben-Arieh, Asher. *Equality of Opportunity of Children of Israel: A Comparative Study of the Rights of the Child in the Arab Sector.* Jerusalem: National Council for the Welfare of the Child and Sikkuy, 1991 (Hebrew).

Benziman, Uzi, and Atallah Mansour. *Subtenant.* Jerusalem: Keter, 1992 (Hebrew).

Central Bureau of Statistics (CBS). *Statistical Abstract of Israel 1994.* Jerusalem: Office of the Prime Minister, 1994a.

———. *Women in Israel 1993: Selected Figures.* Jerusalem: Office of the Prime Minister, 1994b (Hebrew).

———. *Statistical Abstract of Israel 1996.* Jerusalem: Office of the Prime Minister, 1996.

———. *Supplement to the Monthly Bulletin of Statistics,* no. 4. Jerusalem: Office of the Prime Minister, April 1995b.

———. *Supplement to the Monthly Bulletin of Statistics,* no. 5. Jerusalem: Office of the Prime Minister, May, 1995c.

———. *Statistical Abstract of Israel 1995.* Jerusalem: Office of the Prime Minister, 1995a.

Hadashot, November 28, 1993.

Hasan, Manar. "Growing Up Female and Palestinian in Israel." In Barbara Swirski and Marilyn Safir (eds.), *Calling the Equality Bluff: Women in Israel.* New York: College Teachers Press, 1993a, pp. 66–74.

———. "Marriage Between Relatives in Palestinian Society in Israel: Tradition or Product of Government Policy?" Jerusalem: Unpublished essay, 1993b.

———. "Murder of Women for 'Family Honor' in Palestinian Society and the Factors Promoting Its Continuation." Master's thesis, Greenwich, England, 1994.

Herzog, Hannah. "Profile of the Female Candidate for Local Office." In Barbara Swirski and Marilyn Safir (eds.), *Calling the Equality Bluff: Women in Israel.* New York: College Teachers Press, 1993, pp. 234–236.

———. *Realistic Women: Women in Local Politics.* Jerusalem: Jerusalem Institute for Israel Studies, 1994.

Izraeli, Dafna. "Women Managers in Israel." In Nancy Adler and Dafna Izraeli (eds.), *Women in Management Worldwide.* New York: M.E. Sharpe, 1988, pp. 186–212.

———. "Outsiders in the Promised Land: Women Managers in Israel." In Nancy Adler and Dafna Izraeli (eds.), *Competitive Frontiers: Women Managers in a Global Economy.* Cambridge: Blackwell, 1993, pp. 301–324.

Jabar, Loutfi. "Marriage of Relatives: Demographic and Medical Aspects," *Refuah,* vol. 123, no. 3, 1993, pp. 235–236 (Hebrew).

Jahshan. "For Civil Legislation," *Al-Fanar Newsletter,* September 1993.

Kamen, Charles. "After the Disaster: The Arabs in the State of Israel 1948–1950," *Mahkbarot L'mekhkar u'L'bikoret* (Haifa), December 1984 (Hebrew).

Katz, Ruth, and Yochanan Peres. "The Sociology of the Family in Israel: An Outline of Its Development from the 1950s to the 1980s," *European Sociological Review,* vol. 2, no. 2, 1986, pp. 148–159.

Kressel, Gideon. "Sorocide/Filiacide: Homicide for Family Honor," *Current Anthropology,* vol. 22, no. 2, 1981, pp. 141–158.

Lustik, Ian. *Arabs in the Jewish State: Israel's Control of a National Minority.* Austin: University of Texas Press,

1980.

Nejidat, Salah. "Sex Crimes in Bedouin Law." Master's thesis, Hebrew University, Jerusalem, 1992 (Hebrew).

Organization for Economic Cooperation and Development (OECD). *OECD Health Data 1996: A Software for the Comparative Analysis of 27 Health Systems.* Paris: OECD, 1996.

Regev, Matti. *Monitin,* January, 1988 (Hebrew).

Rosenfeld, Henry. "From Peasantry to Wage Labor and Residual Peasantry: The Transformation of an Arab Village." In R.A. Manners (ed.), *Process and Pattern in Culture.* Chicago: Aldine, 1964a, pp. 211–237.

———. *They Were Peasants.* Jerusalem: Hakibbutz Hameuchad, 1964b (Hebrew).

Sprinzak, Dalia, Ehud Bar, and Daniel Levy-Mazlum. *The Educational System As Reflected by the Figures.* Jerusalem: Department of Finance and Statistics, Israel Ministry of Education and Sports, 1995 (Hebrew).

Swirski, Shlomo. "The School and Army in the Construction of Israeliness." In *Seeds of Inequality.* Tel Aviv: Breirot, 1995, pp. 71–117 (Hebrew).

Swirski, Barbara, Hatem Kanaaneh, Avgar Hatem, and Michal Schonbrun. "Health Care in Israel," *Israel Equality Monitor,* no. 2. Tel Aviv: Adva Center, 1992.

Tzartzur, Sa'ad. "The Problem of the Education of a Minority Who Are Foreigners in Their Own Country." In Walter Ackerman, Arik Carmon, and David Zucker (eds.), *Education in an Evolving Society.* Tel Aviv: Hakibbutz Hameuchad and Van Leer Institute, 1985, pp. 473–526 (Hebrew).

Washitz, Yosef. *The Arabs in the Land of Israel.* Jerusalem: Merhavia, 1947.

Women in West Africa

N'Dri Assié-Lumumba

Introduction

The foundations of women's location in West African social systems have been laid down since ancient times by various ethnic and subcultural groups in the region within the broad framework of African civilization. Even before the large-scale and long-lasting devastating effects of the Atlantic slave trade that was triggered by the Europeans coming to the Americas at the end of the fifteenth century, Africa had constant and dynamic relations with other regions of the world.

The dynamics of the relations between preslavery West Africa and external influences must be taken into consideration when analyzing contemporary social facts. For example, in the eighth century, the penetration of Islam started, eventually settling in West Africa; it was still spreading toward the southern coastal regions in the nineteenth century at the time of European colonial expansion. This has created a specific cultural blend that has led to what has been labeled "Black Islam." The colonization of nearly the entire African continent by European nations had the most global, extended, and far-reaching impact of external forces on the African social systems. Some of the issues related to women in West Africa are the same as elsewhere on the continent, and some are specific to the region.

With regard to women's position in society, there are variations within the region. In some ethnic groups, women have had historically important political, economic, religious, and cultural power. In other groups, women were fighting two forces: the colonial rule and the African relative degree of patriarchal rule (Urdang, 1979).

Women and Education

Indigenous education in West Africa, as in other African regions, has provided knowledge about the society's mores, values, and technical skills. It has been gender specific, both philosophically and practically, and rarely has been coeducational, especially after early childhood education. In the contemporary rural communities that still rely on the Af-rican system of education to provide general knowledge, norms, and specific skills for agriculture, trade, art, handicraft, and political leadership, single-sex education is still prevalent. Scientific knowledge in the area of agriculture has been in great part the domain of the female population.

In some African societies, the educational system and its processes were predominantly informal; in other states, they were formal, elaborate, and complex. In West Africa, the traditional type of formal education has been organized, since the precolonial era, by different ethnic groups. For example, in Sierra Leone, northern Côte d'Ivoire, and Mali, an important stage in formal education has been organized within the *Bundu/Sande* for women/girls' society versus *Poro* for men/boys' society. Many technical skills, moral values, philosophies, and rights and obligations of individuals within the group as well as the communal ethos of the African worldview have been during this stage to culminate in what has been referred to generally as *rites de passage*. The emphasis of this education is preparation for adulthood. In this case of Sierra Leone, for example, the rites performed included circumcision for boys and girls, although the term is technically incorrect for girls, given the type of operation that is performed on them.

Within the philosophy of parallel male and female spheres, the female population has historically been in charge of its own sphere, including education. In Sierra Leone, this formal education used to be quite sophisticated, reaching what would be classified as higher education (Forde, 1976). Whatever the content and level of education, women have been in charge of their educational space: The learners and the teachers, as well as those who design this education, are women. The technical skills learned for economic (agriculture and trade) and cultural production, health care, and political participation are provided by women. Indeed, the socialization of girls into the idea that women can be, and are, in charge of specific key spheres in society has been an important element of the African system of education that builds girls' capacity to function with confidence later as women in both private and public life. This aspect has been more striking among, but not limited to, matrilineal societies, in which matriliny is not simply a means for tracing descent but is also associated with matriarchy, which refers to power. The Akan in Ghana and Côte d'Ivoire illustrate this case of matriliny and matriarchy. Even in patrilineal societies, a close analysis reveals that female space, in terms of power and influence, has been indeed extended in the indigenous socioeconomic, political, and religious context (Amadiume, 1987). Socialization into this philosophy of gender-specific social space with a fair distribution of power has been provided by education.

Besides formal training for general knowledge, specific groups, social classes, or individuals have been trained in apprentice settings to become specialists in specific fields such as medicine, religion, and entertainment; cottage industries for the production of tools and utensils; ironwork, pottery, jewelry, clothing, trade (for example, learning the elaborate system of the Akan gold weights that constitutes a special alphabet), and drumming (not only for entertainment but also as a means of communication that only people with special knowledge in drumming language can comprehend). In most societies in West Africa, the production of pottery used to be entirely controlled by women. Among the Akan, including the Baule in Côte d'Ivoire, women controlled the production of cotton and its final products.

With the centuries-long progression of Islam in Africa, another form of formal education has been fully integrated in many African societies. At the end of the twentieth century, more than two-thirds of the population in Mali, Mauritania, Niger, and Senegal are Muslims. Significant proportions of the population in many other coastal countries, such as Nigeria, Benin, Côte d'Ivoire, Ghana, Guinea, and Togo, are Muslims. In these countries, Quranic schools (known as *medersa* or *madrasa*) have been key institutions in the educational process, and, in some cases, Muslim institutions of higher education were also created.

Although Arabic, the language of Quranic instruction, is not an African language, some aspects of the Quranic philosophy of education, at least in terms of the relationship between the school and the community, are close to the African conception of education, especially in that the pupils participate in all social activities in the community, including economic production and social ceremonies such as weddings, baptisms, and funerals. Although there has been no gender discrimination, at the lower level, at least in some communities, access to the *medersas* has been unequal after the elementary level, which is where the majority of students ended their formal education. Post-elementary formal education includes the study of theology and law, and at this level the student population has been male. Throughout the centuries, the interpretation of the Qur'an (Koran) by some male leaders in search of a religious (nonchallengeable) base to legitimize their power gave more importance to men. The family code, the general code of conduct, and liberty of movement were set to favor the male. As the professions for which higher education has been provided have been occupied by men, formal education at the highest levels held little relevance for women. Thus male students who graduated from the local Quarnic schools in West Africa were the ones who attended the universities in West Africa and beyond.

Although the Muslim population challenged European education on religious and other grounds, in terms of its patriarchal foundations and gender imbalance Quranic education presented, at least at the higher levels, some structural similarities with European, Christian-based formal education. With a more diversified and secular use of some of the learning in the *medersas'* education and with the broader usefulness of literacy in Arabic, these schools have expanded in recent years in many West African countries.

Colonial education in West Africa, whether organized by the state or by the churches, shared the same Christian values regarding gender roles and what was considered at the time of colonial rule an appropriate education for females, both in Europe and in the colonies. This common Christian tradition stipulates that the place of the woman is in the private sphere while the man's domain is in the public sphere. Moreover, each sphere corresponds to a specific type of education. The education for the female population was designed and organized for the needs of a life that revolves around the domestic sphere. The transfer of this conception into African social conditions was one of the most eloquent testimonies of the irrelevance of European education to African societies, for this type of education is the result of European, not African, historical processes.

The colonial administrations, French as well as British, created schools for chiefs' sons in order to train Africans who would protect the Europeans' interests and would transmit orders to the people. In many cases, when the chiefs were forced to send their sons to these schools, they fooled the administration by sending children from lower social strata (Foster, 1965). It is important to note that, while there has been some gender-based discrimination and inequality in African societies, if girls and women had been consistently and systematically considered of lower social value and with fewer rights, the African male leaders would have forced them to go to European schools at a time when such education was not desired by the Africans because it was associated with negative outcomes for the community and for the individual who received it. It is true that, because the Europeans wanted to reproduce their system of a social division of labor in which women had little education and lived a domestic life, they would have rejected any attempts by African men to enroll girls and women in the schools, which were designed to train future leaders and clerical workers. However, even the home economics type of education that was provided specifically to girls and women was rejected by African parents. The resistance of African families to enrolling their female offspring must be analyzed in its correct historical context, for Africans did not anticipate the change in the social value of schooling. Unfortunately, some of the initial internal and external causes of low female attendance have had enduring impacts.

Educational statistics indicate that none of the countries in West Africa, and very few on the whole continent, have achieved universal primary enrollment. Those who do not have access to a European type of education have been, and will continue to be, receiving their education in the African systems. Thus, when dealing with education in general, or specifically education for the female population, it is important to present the relevant aspects of African indigenous education.

As indicated in Table 1, the rate of literacy is low in the general population, and it is particularly low among women. Illiteracy is the result of specific policies and social factors that prevent higher enrollment rates of girls, and of early dropout and dismissal of many among those who do enroll. Illiteracy has been defined almost exclusively as the inability to read and write the language of the former colonial master. Yet, in many countries, especially in Muslim regions and communities, the actual number of literate people is higher, as education is offered in other languages, especially Arabic. Mastery of European languages is a sine qua non for access to political office and economic positions in the highest economic echelons, particularly for those positions in the modern sectors. Schooling carries relative (although not automatic) economic and political power at the individual and societal levels. Illiteracy, as it has been officially defined, does not imply ignorance. Indeed, these illiterate people, especially women, not only contribute greatly to the economy and other dimensions of social life, they can clearly articulate their views and express their needs. Thus, they are literate in terms of the actual social knowledge that makes their lives possible despite many obstacles.

Analysis of educational reform in several West African countries indicates that gender, which is the most widespread indicator of unequal educational opportunity, is not specifically acknowledged as requiring special efforts and commitment, despite much rhetoric. Table 2 provides some illustrations of the persistence of gender imbalance in access to education.

From preschool on, other things being equal, an African boy is more likely to be enrolled. Although a boy has intellectual capacities similar to those of a girl, he is still more likely than she to have higher academic achievement scores, a factor that significantly contributes to ensuring his progression throughout primary and secondary school. He is also less likely to repeat his grade and to leave school. However, it is not the simple fact of being born a girl that will determine one's educational and social chances. Other key variables that have an impact on the educational opportunity of the female population include state policy, area of residence, social class, religion, availability of school facilities (including, at the post-primary level, boarding services and the distance between home and the school), the type of economy, the gender division of labor, and other divisions of social and religious tasks. Encouragement from relatives, teachers, and peers is another important variable, as are political will and actual commitment by the state to education in general and gender issues in particular. The coastal regions, which experienced the first and more intense contact with the Europeans, have higher enrollment rates for both boys and girls than do the hinterland regions and Sahelian countries. While, due to a lack of later data, some figures in the tables refer to periods in the early to mid-1980s, it is important to note that the severe economic crises and the negative impact of structural-adjustment programs have not led to improvement in the 1990s. As a matter of fact, some countries, such as Togo and Mali, experienced severe setbacks and declines in enrollments in the 1980s, and, up to 1997, there had not been any economic recovery that might lead toward an educational recovery.

Given the more systematic and considerable contribution of girls in high-yield economic activities for low-income families in rural and popular urban neighborhoods,

Table 1. Estimated Adult Literacy (in French) in Selected West African Countries (%)

Country	1980	1985	1990	1995
Benin				
Female	9.7	13.9	19.3	25.8
Male	28.0	34.6	41.6	48.7
Burkina Faso				
Female	4.3	5.6	7.2	9.2
Male	18.8	22.1	25.6	29.5
Côte d'Ivoire				
Female	13.7	18.2	23.6	30.0
Male	34.3	39.0	44.2	49.9
Nigeria				
Female	23.0	30.3	38.6	47.3
Male	46.7	53.8	60.8	67.3

Source: UNESCO, 1995, p. 81.

Table 2. Women Students Enrolled by Level of Education in Some West African Countries (%)

Country Year	Preprimary	Primary	Secondary	Tertiary
Benin				
1985	45	34	29	16
Burkina Faso				
1980	44	37	33	22
1985	N/A	37	34	23
Guinea				
1980	N/A	33	28	19
1989	N/A	31	N/A	12
Guinea-Bissau				
1988	49	36	32	06
1985	47	44	31	23
Niger				
1980	50	35	29	20
1988	48	36	30	16
Nigeria				
1985	N/A	44	43	27
Togo				
1980	49	38	24	15

Source: Adapted from UNESCO/BREDA, 1991.

the opportunity cost for girls is much higher than that of boys. In many pastoral societies in southern Africa, the opportunity cost for boys is very high, as they are needed to care for the cattle; moreover, many young men enter the migrant-labor cycle of the region dominated by the South African economy. However, West Africa is characterized by the centrality of women in trade, in which young girls offer valuable labor in addition to providing agricultural and domestic work. Those who enroll in school often have to use valuable homework time for their participation in trade, leading to poor attendance and achievement scores. In Togo, for example, the famous Nana Benz (women traders who constitute a small wealthy minority and are identifiable by their Mercedes Benz cars) usually have girls and young women, among other people, who assist them in their work. Some of the traders use a portion of their financial resources to pay for the education of their offspring, including their daughters and other female relatives. However, exposure of the young girls to the "illiterate" but financially successful women traders may constitute an incentive for nonenrollment, especially in the context of the rising number of unemployed school graduates.

As in other parts of Africa, early pregnancy constitutes an important factor in poor school performance, grade repetition, and dropout rates for female students. This is particularly a problem for rural students who leave their community to attend schools in urban areas, especially when boarding facilities are not available. Low initial enrollment rates and high attrition rates at the lower levels of the system lead to a persistently lower proportion of female students in higher-education institutions.

At the postsecondary level, the tendency for female students to enroll in fields that are already crowded in the labor market, such as the humanities, does not provide enough incentive to pursue education further. There are also socially hindering factors, such as the importance of marriage, which comes with its constraints, especially childbearing.

Women and Economic Production
African women play a vital role in the economy, especially in agricultural production, which historically has been the main economic endeavor in Africa and is the most important source of the revenues of African states. Food crops have been a basic domain of women's economic production. The introduction of cash crops during the colonial era constitutes one of the most important aspects of a profound change in the African social fabric, for it affects the family as a unit of production and consumption, as well as the production relations among family members, and between women and men, between the younger and older members.

One of the most dramatic changes export/cash crops have brought about is in the relations of men and women to land and agricultural products. From their position as managers of food crops that they used to, and continue to, process and commercialize, African women were projected during colonialism into capitalist cash-crop production as dependents of men. Women were put in the situation of offering free labor, and they became "invisible" yet key producers. While land used to be collective property accessible to women, newer laws have confirmed men's new relative economic power base through individual property ownership.

"Law has played a major role in enabling the invisibility of women in economic development," as "laws (or lack of laws) affect gender-equity in access, control and exploitation of economic resources" (Martin and Hashi, 1992a, p. 3). Whether the reference is to common law or enacted laws, if a practice that may not have had any significant negative or positive impact on one particular social group is transferred to a different socioeconomic context (especially one with a profit-making drive in economic production), it may result in direct equity problems. This has been the legacy of the transfer of European laws and practices of land tenure onto African societies. As summarized by Martin and Hashi:

> Given their key role in tradition agricultural production systems, women had significant use rights in land. In some matrilineal societies they also had management rights in land. In many societies these rights were diminished or lost during the colonial period by the establishment of arrangements which responded to the economic interests of colonists and African power brokers (Martin and Hashi, 1992b, p. 27).

Nevertheless, women in West Africa continue to produce and nearly control food production, which constitutes the ground for their further participation in trade. West African markets of the late twentieth century are one of the most distinct and dynamic social sectors in the economies of the countries of all of Africa. One of the most striking features of these markets is the presence of women, not simply in terms of sheer numbers but also, and more important, in terms of the level of power and authority they exercise. These markets and the role played by women in their creation and functioning date back centuries.

Before colonial domination, women in most of the West African subregions were visible in exchanging (bartering) or selling commodities at local markets. The first type of commodity to which women historically had access for their transactions was agricultural. They have sold the surplus of their production in raw, cooked, and processed forms. An important female political figure (Iyalode) among the Yoruba in Nigeria, for example, was directly linked to the market and its development. In Islamic societies such as in northern Nigeria, where purdah (seclusion) is practiced, even the secluded women are involved in trade. Part of the handicap created by the absence of freedom of movement of married women is alleviated by girls—daughters, nieces, cousins—who act as extensions of these women. The latter send the girls to the market with a variety of commodities processed within the family compound.

Women's market space has expanded from rural to urban areas. Three categories of women work together in a complementary manner: (1) rural women who are farmers, who sell their commodities directly at the market in rural or small urban centers; (2) women who buy large quantities of these commodities from local markets and sell them at wholesale to retailers in towns and cities; and (3) retailers who sell them at the urban markets. Moreover, in terms of their market participation, women have been responding to new demands. They have expanded their sphere of operations from the local rural markets to the cosmopolitan markets in the cities of Abidjan, Accra, Cotonou, Freetown, Lagos, Lomé, Monrovia, and so forth. They have even stretched beyond African state boundaries; a growing number of West African women operate around the world, in Western Europe, the United States, Brazil, the Middle East, and increasingly in Asia.

Women traders constitute a heterogeneous group in terms of the nature of their activities, the types of commodities they sell, their level of formal education, and their actual economic power. Most are involved in the commercialization of agricultural commodities, meeting the needs of growing urban centers. Those in international trade deal mainly with clothes, jewelry, cosmetics, utensils, and the like. The Nana Benz in Togo, Mammy in Ghana, and Tantie in Côte d'Ivoire, for example, control a substantial portion of the fabric trade between Europe and different African countries. Women involved in this long-distance trade likely have impressive (by any standard) financial power. Because of growing unemployment, there are more formally educated women with advanced academic degrees among such traders. Some even combine their regular waged jobs with trade, which has some specific constraints. However, many traders are still illiterate. This inability to read or write any of the European languages that dominate the new global village in which they operate constitutes a serious handicap when they are dealing with the international market and traveling in industrial countries, where literacy is assumed in commercial transactions. There also is a small but growing number of female entrepreneurs in construction and in the garment, cloth, and food production and processing industries. However, the extraordinary cultural heritage that set the stage for women's participation in these aspects of the economy, as well as the tremendous creativity, risk-taking capacity, and courageous actions of the women involved are hindered by many legal restrictions and other difficulties, including unequal access to credit and a lower educational level, that make it necessary for them to utilize more energy to pursue their activities.

Women in West Africa have been responsible for the preservation, protection, development, and extraordinary innovation of African arts. Arts, handicraft, and practical instruments for daily and ceremonial use have been interwoven. Although part of their creation has been classified by some Western experts as of nonvalue in arts, they have continued to be creative, always in search of new frontiers in design, clothing, and other areas. They have been less successful in contemporary areas of audiovisual art such as cinema, but they are present in the music, usually as singers or dancers.

The introduction of cash crops took a big chunk out of women's economic sphere, as they have had no power in the production or marketing of those crops or in the use of the profits realized from them. Their sphere of agricultural trade has been confined to food crops. Despite the cultural heritage of women's role in trade in West Africa, many in the 1990s enter trade for lack of better opportunities in wage labor, having little or no formal education to secure other jobs in the context of shrinking economies and the growing unemployment of degreed graduates.

Indeed, the unequal and progressively decreasing representation of the female population in the educational system is reflected in the lower number of women in the modern economic sector, especially in high-ranking administrative, political, policymaking, and technical positions. The proportion of women among administrators and managers is generally below 10 percent (UNDP, 1990). The distribution of teaching staff by educational-system level is a good illustration of women's underrepresentation in the top and most rewarding positions. Data for Guinea, Benin, Togo, and Niger (UNESCO/BREDA, 1991) indicate that women's representation among primary-school teachers ranges from 16 percent to 25 percent; at the university level, it ranges from 3 percent to 12 percent.

In general, the proportion of women in the modern sector is low. Data from 1984 concerning the occupational distribution of economically active women in Ghana showed that 56 percent were in agriculture, 24 percent in sales, 14 percent in production, 3 percent in clerical positions, and 3 percent in professional jobs. In the Sahelian countries, nearly 90 percent of the active population is in agriculture and a significant proportion are women. There are considerable differences between and within countries that can be explained by a whole range of factors in addition to educational attainment, such as state policies, area of residence, religion, local culture, the structure and type of economy, and women's age and marital status.

> Participation in the labor force varies considerably from one town to another, due to local attitudes about women's economic roles and the nature of the economic activities which provide opportunities for employment. Disapproval of the principle of employment is seldom as strong as in East Africa, where some people still assume that women who earn money must be prostitutes, but Islamic societies often discourage women's aspirations for economic independence. On the other hand, some societies strongly encourage women to develop careers of their own in order to support themselves and their children (Peil, 1979, p. 485).

Cultural and Health Issues

One of the African social institutions that has been the least altered by external influence is the family. In a traditional European sense, a family is a nuclear unit composed of a husband, a wife, and their child or children. In most of Africa, the family is an extended group composed of a larger number of people sharing various aspects of family life, including living space, along with rights and responsibilities in raising children, and daily and major decision-making processes. Cousins, aunts, uncles, nieces, nephews, and grandparents all constitute parts of the African household. One of the major characteristics of the African family is the legal practice of polygyny (having more than one wife).

There are some aspects of African culture that were not fundamentally changed by the European Christian influence. For example, although polygyny was not universal in Africa, it was legally accepted throughout the continent. Very few ethnic groups, such as the Bidjogo (or Bijagó) in Guinea-Bissau, systematically practiced monogamy. Polygyny also influenced the nature of the participation of women in economic production, in patrilineal as well as in matrilineal societies.

It has been argued that Islam spread in Africa because some of the Quranic prescriptions or laws, such as the one related to polygyny, were compatible with the African social context. While the Qur'an states that a man may marry up to four wives, it also stipulates that all the women must be treated equally. It appears that this condition has not been respected. Several West African countries, such as Mali, Senegal, Mauritania, and Niger, that have large Islamic populations, have adopted laws in their postcolonial constitutions and family codes that confirm the legality of polygyny. By contrast, Europeans introduced, along with their Christian values, monogamy, and the European laws inherited at the time of independence have led some countries to adopt monogamy laws. According to these laws, the man is declared the head of the family and has major rights that are denied the woman, who is viewed as his dependent. This is the case in Côte d'Ivoire. Under these conditions, monogamy does not necessarily lead to more equality for women. In some countries, such as Nigeria, different marriage laws coexist: the British-inherited Christian law of monogamy, the Islamic code, and the African code. In Senegal, although polygyny is legal, couples have the option of choosing between the monogamous and polygynous codes.

There are divergent arguments regarding the problems and benefits related to polygynous marriages. One observation that all seem to agree upon is that the polygynous marriage in the rural agrarian/foodstuff production societies of past and present offers life experiences to all of those involved that are different from urban polygyny. In contemporary African urban centers, different occupations and living arrangements have created a different form of polygyny, in which the women and their children do not necessarily have opportunities to have contact and share their experiences, to support each other, and to be, in practice, part of the same family. Each wife has increasing responsibility in fending for herself and her children, forming a separate living unit. Furthermore, women's low educational level deprives them of power and authority, and many men in polygynous relations have become the main, if not the sole, decision makers, in contrast to the traditional African context in which many members of the extended family, including the wives, took part in the decisions concerning the family, including decisions relating to subsequent marriages.

It is not simply polygyny per se, but the corruption of the institution, the unequal level of educational and economic power to the advantage of men, that has led to many of the socioeconomic problems women face. Furthermore, there are some new challenges in the area of health that are directly related to polygyny. For example, the husband and the wives in a polygynous marriage almost inevitably run the risk of contracting sexually transmittable diseases if one of them, through whatever means, contracts the disease. In the case of curable diseases, the issue may not be serious; however, due to lack of health facilities and limited resources, the majority of the population does not have access to well-equipped and affordable health services. The threat presented by the AIDS epidemic offers a new challenge. In the absence of economic resources and proper information that might allow use of protective means such as condoms, if one of the women or the husband contracts the HIV virus, it means a death sentence for them all. To address these and emerging health issues related to polygyny, the legal adoption of monogamy will not be sufficient. A major task remains for social education to effect a change in behavior that is compatible with the new social reality and challenges.

Based on the cultural gender division of education and occupation, women historically have held a central position in health care provided to the family. In colonial education, women were excluded from medical studies, except as assistants to the men who later became medical doctors; most women who entered the field became nurses. Even areas related to women's reproductive health—areas that, in the past, were entirely controlled by women who were specialists—were appropriated by male doctors in obstetrics and gynecology.

African women face many health problems, including cardiac failure, obstetric hemorrhage, occupational health hazards, genital malignancies such as cervical cancer, malaria in pregnancy, wife battering, the social effects of childlessness, unsanitary food, physical disability, and maternal morbidity and mortality (Kisekka, 1992). They frequently suffer from general and maternal malnutrition, pregnancy-related deaths (including abortion and delivery complications), unmet postmenopause and aging needs, and permanent exposure to unsafe water, among other problems. Some of the issues are exacerbated in West Africa due to the large number of extremely poor Sahelian countries, many of which are landlocked, such as Burkina Faso, Mali, and Niger.

An issue that elicits more attention, especially in the Western/industrial countries, is what has been labeled "female genital mutilation" (Toubia, 1993). West African societies that practice it use different forms of operation that affect different parts of the female sex organ. Although the most severe, the "pharaonic cut," is not prevalent in this region, there are a large number of medical problems that are associated with it. Besides the obvious physiological ones, there are also reported psychological problems.

It is important to note that these operations do not constitute the major source of health problems for women in Africa in general or West Africa specifically, although other health problems may be exacerbated by the practice. In the West African countries, genital-mutilation figures range from 20 percent of girls and young women in Niger and Senegal, and 25 percent in Mauritania, to 70 percent in Burkina Faso, 75 percent in Mali, and 90 percent in Sierra Leone (Toubia, 1993, p. 25). However, the general health of the population and development indicators such as life expectancy at birth do not show any particular pattern of correlation with the proportion of girls and women who have been through the operation. In all of the Sahelian countries, life expectancy at birth in 1990 was low: 48 years in Burkina Faso, Mali, and Senegal and 47 in Mauritania (UNDP/World Bank, 1992). Furthermore, the figures for maternal mortality per 100,000 live births in 1985 indicate poor and dangerous experiences of women in their reproductive years regardless of whether they had undergone this operation: 250 deaths in Mali, 420 in Niger, 450 in Sierra Leone, 600 in Senegal, 800 in Nigeria, 810 in Burkina Faso, and up to 2,000 in The Gambia (UNDP/World Bank, 1992).

Structural-adjustment programs (SAPs) that require payment of fees even before a patient can be examined in a health center have negatively affected the already generally precarious health conditions of the poorest sections of the populations, especially women and children in rural areas and urban peripheries, who do not have salaried jobs and thus no money or insurance to pay for the service. Under these conditions, health issues must be addressed with a sense of urgency, not for genital mutilation in isolation, but as part of global issues.

Women in the Political Process

African women, particularly those in West Africa, have played key roles in politics throughout the history of their respective societies. Legends and histories of female founding ancestors are abundant. Even the Hausa, who, as Muslims, practice purdah, still refer proudly to their Queen Amina (Sweetman, 1984) while the Baule of Côte d'Ivoire still honor Queen Abla Pokou, the eighteenth-century founding ancestor of the "Queendom" (Assié-Lumumba, 1996). Many other key figures are less known. What is even less known and understood are the social structure that produced such leadership and the process of change that led to the contemporary situation of political alienation and marginalization of women.

The often-cited Yaa Asantewa of Ghana (1840–1921), who died in detention, organized and led the struggle for freedom of the Ashanti against the British colonial power. She was eventually captured and was sent into exile by the British in the Seychelles, where she died after 20 years.

Indeed, during the entire second half of the nineteenth century, when the British were struggling to colonize the area, women were practically in charge of the Ashanti political scene.

The existence of dual or parallel systems of governance based on gender made it possible for women to participate in the political process. Through these political channels, women in the entire region took an active part, on equal footing with men, also being executed and being sent into exile during the anticolonial and decolonization struggle from the end of the nineteenth century to independence in the 1950s, 1960s, and 1970s. Besides the famous leaders, women at the grass-roots level, using various forms of organizations and networks, launched actions such as the Aba Riots where Igbo women protested against some colonial policies in Nigeria in 1929. They marched against the choices of some chiefs and preempted the administration's policy of levying taxes on women.

Although these resistance movements and actions are of great importance, the roles that African women played in them are not the most important testimony to their place in African politics. The ability of African women to organize, to resist colonial rule, and to take an active part in decolonization movements was made possible by the cultural and historical setting of the societies to which they belonged. Women in matrilineal societies had a more predictable role in this type of political involvement. However, analysis of gender in the distribution and exercise of power, even in some patrilineal societies, reveals that Africans had developed their own social arrangements in the area of gender. Amadiume (1987), who studied the structure of power in various Igbo societies, some of which are patrilineal and others matrilineal, found:

> Mother-focus seems to be a more appropriate term than matrilineality as it embraces other indispensable roles played by women in traditional Igbo societies other than their importance in succession and inheritance. Mother-focus/matrifocality covers the importance of women kinship terminology, domestic arrangement, and their central role in the economy as producers and providers (Amadiume, 1987, p. 17).

The roles that African women in general and West African women in particular have played in politics throughout history have been defined within various multifaceted organizations with cultural, religious, economic, and political components. Such organizations constitute a mechanism for women to voice their concerns, defend their interests, organize their activities, support each other, and play a structured and systematic role in the different spheres of society. They can be based on lineage, as is the case among the Igbo, where the women of each patrilineage organize in associations formed by lineage daughters (*umu okpu*), lineage wives *(inyom di),* and all of the lineage women *(inyom nnobi).*

While Christianity and Islam secured the religious justification of patriarchy, African religions and the dual conception of the ethos made room for men and women to exercise full and strong religious power. Among the Baule of Côte d'Ivoire, Adjanu is a religious dance and political association for women at the level of each smallest unit (village, town, city, or neighborhood in the larger urban centers). While its membership is strictly reserved for women, it does not serve the interests of women alone. Rather, it is a political forum for women to take part in major political decisions and activities. Another organization, called Bomanpi, which is even more strict in asserting women's religious and political power, is also used as a means for women to play a significant role in the political process.

A large variety of women's organizations exist in postcolonial countries. Among the most visible categories of political organizations have been the female wings of the major or only ruling parties, particularly from the 1960s to the 1980s. Such organizations often have been headed by first ladies; given their nature, they and their members cannot be critical of government actions. There are also nongovernmental organizations (NGOs), some of which are officially headed by first ladies and rely heavily on government support. As a result of their vested interests in the government, these organizations cannot be critical of public policies either, although they may concretely foster activities that aim at improving living conditions of women and the general population. This is the case of grass-roots movements such as the December 31st movement in Ghana and the Better Life Program (BLP) for rural women in Nigeria.

At the same time, given the inability of the modern state to deliver services, more critical and independent women's NGOs have been created. These have increased in number and in kind, particularly since the Third World Conference on Women (Nairobi, 1985) at the end of the United Nations Decade for Women. West African women are members of, and active in, many continental and regional organizations, including: AAWORD (Association of African Women for Research and Development), based in Dakar but with national chapters; the Society for Women Against Aids and the Association of African Women Professionals in the Media, also based in Dakar; CIFAD (Comité International des Femmes Africaines pour le Developpement), based in Abidjan; and REFAD (Le Réseau Sous-Régional des Femmes Africaines et Droits Humains), based in Burkina Faso.

Created by 15 West African countries that signed the Treaty of Lagos on May 28, 1975, ECOWAS (Economic Organization of West African States) was joined later by Cape Verde and is composed of 16 countries: Benin, Burkina Faso, Cape Verde, Côte d'Ivoire, The Gambia, Ghana, Guinea, Guinea-Bissau, Liberia, Mali, Mauritania, Niger, Nigeria, Senegal, Sierra Leone, and Togo. It was created to help minimize the negative impact of the colonial boundaries of modern African states. It is supposed to be the beginning of economic integration, and there are already some laws aimed at integration, such as the nonrequirement of entrance visas for citizens of member states. However, the law of free residence, which was supposed to be implemented in 1985, has had several setbacks. Free movement of goods has not been achieved either. Thus, for women traders, who contribute greatly to the regional economy and who need to travel to and from the countries in the region to buy and sell commodities, there are still major problems. Implementation of some of the laws would make it less difficult for the women in the region to expand, develop, and deal more effectively with the competition of multinational corporations. But real progress can occur only if women become key decision makers.

White (1988) notes: "In general, colonial rulers came from societies that did not recognize women's political power as legitimate. Once in control of West and Central Africa, they simply ignored the institutionalized power that women in many pre-colonial societies held" (White, 1988, p. 87). The contemporary African states that emerged from colonial rule presented barriers that previously were unknown to African women, often new barriers that reinforced some that were latent in the old context, and against which, due to their unequal access to formal education, women are not adequately prepared to fight effectively. Differences in educational attainment or even simply in the ability to speak a particular European language have become factors in modern, Westernized African politics. One of the de facto, if not de jure, conditions for running for public office (the presidency, the parliament, or the national assembly) is the ability to read and write European languages, which, in all the states in West Africa, are the actual official languages, even in countries such as Mali, where one African language is the lingua franca, or in countries that have adopted policies of African national languages. Thus, women who qualify and have support to run for office and succeed are very few.

Curiously, while in other regions of Africa there has been some effort, at least symbolically or through stated affirmative action, to decrease the gender imbalance and increase women's visibility, in West Africa, where women historically have had valuable political space, they have not achieved a new valued status in politics. Liberia has produced in the process of war and negotiation, the only West African case of a woman head of State. However, Ruth Perry is in office only during a transitional period. The forthcoming elections are not likely to produce a woman president. As of 1996, on the whole continent, only two small Central African countries (Burundi and Rwanda) had had, symbolically and for a very short period in both cases, the political appointment of a woman as prime minister. Only Uganda had had a woman occupy the post of vice president, the highest position held there by a woman since the colonial era. Women cabinet ministers were first appointed in the 1970s, following the U.N. declaration of the Decade for Women, the First World Conference on Women (Mexico City, 1975), and the creation of Ministries of Women's Condition or Advancement headed by these women ministers. Numerical representation of

women in parliaments has been very low, ranging from 0 percent in Mauritania and 2 percent in Mali and Nigeria, to 5 percent in Côte d'Ivoire, and 13 percent in Guinea-Bissau (UNDP, 1994).

There are cases of famous West African women at the international level. The government of the late President Sekou Toure of Guinea sent a woman, Madame Jeanne Martin Cisse, to the United Nations Security Council, which she chaired in 1974, at a time when this organ of the United Nations, in the context of the Cold War, played a major role in world politics. Angie Brook, the first woman to be seated in the International Court of Justice at The Hague in the 1970s, is Liberian. Yet, at home, none of the West African countries had had a woman prime minister or president in normal conditions—as of 1996, at least.

The postcolonial reality continues to be characterized by predominantly illiterate women who become politically relevant to politicians mostly as voters supporting a man who wants to keep or acquire a political position as an elected official. In many countries, with Togo as the extreme case in West Africa inspired by the disastrous Zairean experience of *animation politique,* women, who constitute the majority of the impoverished rural dwellers and inhabitants of urban peripheries, have been reduced to cheerleaders singing the praises of opulent male leaders whose policies and alliances with international financial institutions and multinational corporations create more miseries for them. The multiparty systems of the 1990s have not fundamentally changed the participation of the majority of women in the political process.

The low representation of women in legislative bodies has led to the adoption of many laws that take away women's previously held rights and deny women new equal status in the various spheres of society, including the family, the economy, and the legal system. Many laws passed in male-dominated parliaments have adopted a patriarchal Christian/Western or Islamic legal heritage and disregarded the female-empowering base of some of the African legal practices, leading to a lowered status for women.

The few women who have been elected to office have not received the same chances their male counterparts have to assume full political power and responsibility. In countries like Nigeria, which regressed since the 1993 presidential elections (when a woman, Sarah Djibril ran for the presidency) to an arbitrary military takeover, women have even fewer chances of being included. There have been a few exceptional cases of progressive military regimes, such as Burkina Faso from 1983 to 1987, where then president Sankara became the advocate of women's rights and of the promotion of women to key ministerial positions. However, as a general rule, the military and its power structure, with regard to gender, is one of the most eloquent symbols of women's marginalization. In Nigeria, during the military regime in 1984–1985, the military rulers, "influenced in part by Western education and the colonial experience, and also by growing, conservative Islam in Northern Nigeria, . . . felt that women's proper place was in the home, taking responsibility for the morality of their children and husbands" (White, 1988, p. 101).

Even women who have a relatively high rank in the army, air force, and navy are likely to be assigned to technical and administrative duties. In the context of democratic rhetoric, women in West Africa in the last decade of the twentieth century have not yet emerged as a source of challenge to past corrupt systems or as leaders of progressive and opposition parties.

Conclusion

Societies in West Africa present an exceptionally rich well of cultural arrangements that guaranteed a valuable space for women in the past. While women's continued role in the economy, despite their unequal access to education, constitutes a testimony to the space they occupied in the past, their dynamic and empowering cultural legacies have not been used to create a new space. Rather, some aspects of the culture that had the potential of weakening women when placed in a different sociopolitical and socioeconomic environment have been preserved in laws passed by predominantly male parliaments.

The complexity of the determining factors that operate at the national level makes it difficult to articulate the pace of change in the whole region. Yet, policies that take into consideration women's needs and the improvement of women's conditions must be strong at all levels—the local, national, and regional. However, the interests of male leaders in each country have been a stumbling block. Women can achieve highly influential political positions only if they achieve higher levels of education and create well-educated constituencies who understand the need for gender-focused policies at the local, national, regional, and continental levels.

References

Amadiume, Ife. *African Matriarchal Foundations: The Igbo Case.* London, Karnak House, 1987.

Assié-Lumumba, N'Dri. *Les Africaines dans la Politique: Femmes Baoulé de Côte d'Ivoire.* Paris: L'Harmattan, 1996.

Forde, T.J.L. "Indigenous Education in Sierra Leone." In Godfrey N. Brown and Mervyn Hiskett (eds.), *Conflict and Harmony in Education in Tropical Africa.*

Cranbury: Associated University Presses, 1976, pp. 65–75.

Foster, Philip. *Education and Social Change in Ghana.* London: Routledge and Kegan Paul, 1965.

Kisekka, Mere (ed.). *Women's Health Issues in Nigeria.* Zaria: Tamaza, 1992.

Martin, Doris, and Fatuma Hashi. "Gender, the Evolution of Legal Institutions, and Economic Development in Sub-Saharan Africa," working paper no. 3, Women in Development Unit, Washington, D.C.: World Bank, 1992a.

———. "Law As an Institutional Barrier to the Economic Empowerment of Women," working paper no. 2, Women in Development Unit, Washington, D.C.: World Bank, 1992b.

Peil, Margaret. "Urban Women in the Labor Force," *Sociology of Work and Occupations,* vol. 6, no. 4, 1979, pp. 482–501.

Sweetman, David. *Women Leaders in African History.* Oxford: Heinemann, 1984.

Toubia, Nahid. *Female Genital Mutilation: A Call for Global Action.* New York: Rainbo, 1993.

United Nations Development Program (UNDP). *Human Development Report 1990.* New York: UNDP, 1990.

———. *Human Development Report 1994.* New York: UNDP, 1994.

United Nations Development Program/World Bank. *African Development Indicators.* Washington, D.C.: World Bank, 1992.

United Nations Educational, Scientific and Cultural Organization (UNESCO). *1995 Statistical Yearbook.* Paris: UNESCO, 1995.

UNESCO/BREDA. *Sixth Conference of Ministers of Education and Those Responsible for Economic Planning in African Member States.* Dakar: UNESCO/BREDA, 1991.

Urdang, Stephanie. *Fighting Two Colonialisms: Women in Guinea-Bissau.* New York: Monthly Review Press, 1979.

White, Frances. "Women of Western and Western Central Africa." In Organization of American Historians (eds.), *Restoring Women to History.* Bloomington: Organization of American Historians, 1988, pp. 57–114.

Women in East Africa

Maria Nzomo

Introduction

The struggle for women's empowerment and rights in Africa, as elsewhere in the Third World, has been an uphill battle against patriarchy, poverty, and autocracy. In Africa, however, women are slowly moving forward, especially in the 1990s, following political liberalization and the movement toward multiparty democracy in many African countries. But numerous obstacles for the advancement of the status of women are still in place, some of which are deeply rooted in Africa's history and patriarchal cultures and have tended to promote politics of exclusion and discrimination against women in both private and public life. In Africa, then, as elsewhere, women's tremendous tenacity and capacity to organize, lead, innovate, produce, and reproduce human and material resources for society have not necessarily translated into gender equity and female empowerment. In this regard, it is worth nothing that even in those few and scattered cases of precolonial matrilineal societies in Africa in which women wielded enormous powers as queen mothers, tribal chiefs, and owners of landed property, most of this formal power for women was eroded during the colonial and postcolonial periods (Parpart, 1988). More important, perhaps, throughout history, in matrilineal as well as patrilineal societies, most African women had been able to exercise at best only indirect power and authority in their societies: "Direct authoritative power held through elected or appointed offices, with its concomitant control over resource allocation, has been less available to women" (Parpart, 1988, p. 209). Indeed, such power seems to have eluded African women throughout history, thus depriving them of the ability to allocate resources, roles, and statuses in society. The patriarchal ideology embedded in African customary laws not only legitimated women's exclusion from formal power and decision making but also sanctioned an unfair gender division of labor that allocated most of unpaid work to women. Colonial and international capitalism, before and since independence, and undemocratic postcolonial states have further reinforced the subordinate status of women.

In the postcolonial era, it is impossible to fully contextualize and analyze the politics of gender exclusion and discrimination anywhere in Africa without bringing in the role of the state. The African state, even in countries where it has been described as weak, soft, or collapsed (Rothchild and Chazan, 1988; Harbeson et al., 1994; Zartman, 1995), remains a force to reckon with, albeit in a highly contested political terrain. For women, the reality of their exclusion from formal politics and power in general reflects the gendered nature of the postcolonial state (Peterson, 1992). State power in Africa remains conspicuously male power, ingrained with predominantly male values, ideology, and vision of the world. This manmade political machinery (Nzomo and Staudt, 1994) codifies, institutionalizes, and legitimizes patriarchy (Parpart and Staudt, 1989), a system that manifests itself not only in social and economic life but also in the low and biased political and legal statuses of women:

> The state is male in the feminist sense. The law sees and treats women the way men see and treat women. The state coercively and authoritatively constitutes the social order in the interest of men as a gender, through its legitimizing forms, relation to society, and substantive policies. It achieves this through embodying and ensuring male control over women's sexuality at every level. . . . Substantively, the way the male point of view frames an experience is the way it is framed by the state policy (McKinnon, in Parpart and Standt, 1989, p. 55).

The postcolonial policy and legal frameworks in most African countries have tended to institutionalize discrimination against women by failing to legislate laws and policies to protect women's rights and to subject customary and religious laws to the universal standards of human rights.

To the extent that the male-dominated state in Africa has been the prime instrument for acquisition and distribution of power and status, entry into the ruling class has been virtually blocked for the majority of women. In this connection, it has been noted that women's past quest for status and wealth heavily depended on aligning themselves with powerful men. In so doing, they accept the male vision of the world and the patterns and processes of their own subordination. In the absence of such alignments, women have tended to withdraw from the public arena to build their own parallel and independent spheres of survival (Parpart and Staudt, 1989; and Nzomo, 1994b). There are also cases in which women's political action has created the impression that they sometimes undermine rather than promote their own advancement and autonomy (Obbo, 1980). Women's contribution to their own subordination and oppression can be attributed in part to the patriarchal sociocultural conditioning that pervades gender relations in African societies. As Fatton (1989) points out:

> In Africa, where patriarchal traditions are so ingrained in the fabric of society, women's struggle for emancipation is replete with contradictions, ambivalence, and silence. This is not to say that women fail to resist and protest, but that their resistance and protest are easily coopted or suppressed by the structural, political, and ideological powers of male supremacy (Fatton, 1989, p. 54).

The structural-adjustment programs (SAPS) implemented in most of Africa beginning in the 1980s have also significantly contributed to the deterioration of women's social, economic, and political condition, primarily due to the government's cutbacks on social expenditures in health, education, and employment. Because of the enormous social and economic responsibilities borne by women in Africa, the female gender has in general experienced greater negative impact of SAPS than the male gender. Consequently, women's response to the crisis of the 1980s and 1990s, has, in part, been to multiply their economic-survival activities to ensure family survival, but at great cost to their health and status, including political participation (Meena, 1989; Nzomo, 1994c) African women, on average, rank lowest in global terms of human development. More than 70 percent of rural women are illiterate and live below the poverty line. Women earn only 10 percent of the total continental income from formal employment and own only 1 percent of the continent's assets (UNDP, 1995).

Against this background of manmade barriers of gender discrimination and inequalities in the allocation of resources and opportunities, women in Africa continue to intensify their struggles for basic rights and freedoms. In particular, the new wave of democratization in the 1990s has opened up political spaces for broad-based political participation. Women in Africa appear to be seizing this political opportunity to ensure that gender concerns are on, and at the center of, the new democratic agendas being worked out in their respective countries. This essay examines trends, opportunities, and prospects for women's struggles in one African subregion—East Africa.

Women in East Africa: An Overview

The East African region, comprising Kenya, Uganda, and Tanzania, provides interesting contrasts as well as similarities. These three East African countries, apart from sharing common territorial boundaries, also share a common colonial history under British rule in a federation in which they shared commonly administered services in the areas of communication, transport, education, and the like.

The three East African countries attained independence in 1961 (Tanzania), 1962 (Uganda), and 1963 (Kenya). They all adopted national constitutional forms based on the British Westminster model, purporting to promote parliamentary democracy, the rule of law, equality for all, and popular participation in the processes of governance. Subsequently, however, the original independent constitutional provisions were not adhered to and, in many cases, were amended in a manner that gave more powers to the executive and enhanced the authoritarianism in governance. The three countries have, therefore, been ruled by a largely authoritarian postcolonial state, despite differences in regimes and leadership styles. All three have experimented with multipartyism, single partyism, and variations of capitalism conveniently labeled as "African socialism."

Kenya and Tanzania have had relatively smooth regime and leadership changes and, at least through 1996, have been spared the bloody military coups and protracted civil wars, as witnessed in neighboring Uganda until 1986, when President Yoweri Museveni's National Resistance Movement (NRM) took over the reigns of political power. Uganda has, in the process, experimented with not only party politics but also monarchism, military dictatorship, and the no-party system.

After independence, an East African Community was created to promote economic cooperation among the three East African countries. But leadership rivalries, ideological differences, and external interference in the affairs of the community led to its death in 1977. Fresh efforts at regional integration are under way and showing promising signs for their realization in the near future.

The authoritarian state, combined with the complete lack of gender sensitivity in the legal framework, greatly contributed to the political marginalization of women in East Africa; the national economies remain predominantly agrarian and highly dependent on women's labor, especially in the area of food production. In that regard, the female experience in East African politics bears a striking resemblance to that of women in other African countries, characterized largely by exclusion, inequality, neglect, and subordination of women to men, with women playing a minimal role in the creation of the modern state systems in their respective countries (Parpart and Staudt, 1989). Furthermore, the legal structures of the East African countries have buttressed and legitimated the political exclusion of, and discrimination against, women in social and economic aspects of private and public life.

While recognizing that women's struggles in East Africa are taking place at many different levels and on many different issues, this essay focuses on the political and legal issues. It reviews progress made and obstacles encountered and analyzes some of the strategic ways in which the women's movement in East Africa was moving forward during the 1990s toward the twenty-first century.

Legal Frameworks

Women and men in all three East African countries are governed by at least three systems of often contradictory legal regimes that operate concurrently: statutory/constitutional law, customary law, and religious law. Although statutory law constitutes the supreme national law in theory, in practice customary and religious laws tend to prevail, especially on issues deemed to be of a private rather than public nature, such as marriage, divorce, custody of children, inheritance, property ownership, and reproduction. It is precisely in the context of these issues that women's rights are grossly violated. This position is sanctioned by male-dominated African governments that prefer to treat issues of gender relations as falling within the realm of "African culture" and hence beyond the jurisdiction of government. Furthermore, due to low levels of functional literacy and of legal and gender awareness among women, women are, even in those situations in which their rights are provided for in law, held back from exercising such rights not only by tradition but also by ignorance of the existence of those rights. Further still, although the statutory legal frameworks of all three countries provide for fundamental and equal rights before the law, these constitutional provisions do not explicitly outlaw discrimination on the basis of sex—an omission that has created ambiguity in the status of women's rights and adequate grounds for gender discrimination in legal and social practice. These major legal issues are addressed below, with specific examples drawn from each of the three East African countries.

Uganda

As of 1996, Uganda continued to operate under its 1967 constitution, awaiting the coming into force of the new constitution, which has been in preparation since 1990. Chapter 2, Article 20 of the 1967 Ugandan constitution provided for equality of all. However, while it outlawed most forms of discrimination, it did not outlaw discrimination on the basis of sex (Tibatemwa, 1995), thus allowing the enactment of laws that foster inequality between women and men, especially in family matters, where customary law rather than statutory law takes precedence in many aspects of women's lives (Tibatemwa, 1995). Furthermore, even where positive legal-rights provisions exist, many women in Uganda have been unable to exercise such rights due to their ignorance about them and harassment by male family members who, under the guise of tradition

and culture, for instance, disinherit widowed women and render them destitute (WiLDAF, 1993, p. 15). Tibatemwa (1995) underscores this point, noting that the high incidence of illiteracy among women prevents them from taking advantage of the existing positive elements of the law and that women, even when they are aware of their rights, often cannot afford to hire the services of lawyers." The Uganda Association of Women Lawyers (FIDA-U) does provide some free legal service to needy women.

Commenting on the issue of gender discrimination in a 1993 interview, the permanent secretary in Uganda's Ministry of Women Development said, "Although the law of our nation does not discriminate against women, it also does not make specific reference to women's concerns in terms of inheritance or ownership" (Snyder and Tadesse, 1995, p. 137). Tibatemwa points out that there are several sections of the Ugandan law that constitute gender discrimination in regard to marriage and property ownership. They include: Section 3 of the Succession Act, which gives preference to male relatives in cases of inheritance; Rule 8 of Schedule 2 to the Succession Act, which gives different rights of occupancy to widows and widowers; and Section 27, which provides for the punishment of an adulterous wife but not an adulterous husband (Tibatemwa, 1995).

The legal-rights situation of women in Uganda is likely to improve when the new constitution becomes fully operational. Women participated in its preparation; it is said to be gender-sensitive and to provide women with more rights than before in such matters as marriage and divorce, inheritance and property ownership (Klemp, 1994). Klemp, a female Ugandan lawyer and activist, concludes: "If all of these propositions are maintained in the new constitution of Uganda, women will be clearly empowered to participate in the process of democratization in Uganda" (Klemp, 1994, p. 52).

Female activists and politicians have worked together in securing legal rights for Ugandan women. Uganda's female parliamentarians were instrumental in passing amendments to the Penal Code in 1990 that made rape a capital offense and punished hotel owners for allowing prostitution on their premises, while women's groups helped draw up the domestic-law bill that gives women more right's under the new constitution (Tripp, 1994).

Tanzania

Tanzania is among the African countries that have ratified most of the international conventions on human rights and incorporated them into national law, but it had not, as of 1996, effectively implemented them. Furthermore, as in Kenya and Uganda, the existence of the three legal systems (statutory, religious, and customary) has generally had a negative impact on women's human rights. "In Tanzania, ambiguity and conflict is facilitated by the existence of more than one code of law and what appears to be an inordinate desire to maintain the customary law even when it seems to be unfair and unjust" (Mascarenhas and Mbilinyi, 1983, p. 170). For example, whereas the marriage act under the statutory law provides for more monogamous unions, this provision comes into conflict with customary and Islamic/religious laws that uphold polygamy. In addition, whereas statutory laws confer on women the same rights as men in regard to property acquired during marriage, under customary law women in Tanzania cannot inherit property after the death of their spouses. Indeed, among the ethnic groups that are patrilineal, descent is almost exclusively through males. Even in matrilineal communities, the principal heirs are uterine brothers (same mother and father) and sister's sons. And even where changes have occurred, women usually get usufruct rights rather than outright ownership with full rights to dispose of the property by grant, will, or sale. On the issue of divorce, a man has the right to divorce his wife by virtue of repudiation, and such repudiation of his wife can be used as bona fide evidence of the irreparable disintegration of the marriage.

Section 13 of the Tanzanian constitution guarantees the right to equality, but, as in Uganda and Kenya, it does not explicitly outlaw discrimination based on sex. Other areas of women's rights and concerns that existing laws do not adequately address include reproductive rights and the question of violence against women. Gender violence in Tanzania is as rampant as in the other two countries, and conviction rates are low. For example, out of an estimated 228 reported cases of domestic violence in 1993, 112 were withdrawn, 60 were cited for nonappearance, and only 27 cases were determined (WiLDAF, 1993). In 1994, the Tanzanian Law Reform Commission (LRC) documented numerous cases of violence against Tanzanian women and noted:

> Although under our penal code the offenses of rape, attempted rape, and defilement of a girl under 14 years carry a maximum penalty of life imprisonment . . . in the majority of cases actual sentences have only been a fine or imprisonment for up to 5 years (LRC, 1994, in Nzomo and Halfani, 1995, p. 54)

The legal status of women in Tanzania is, however, changing in response to political and social pressures, especially since the beginning of the 1990s. Economic and political liberalization in the 1990s has facilitated gender and human rights activism and pressure for fundamental legal reforms. Women lawyers and university researchers continue to provide leadership in this regard, on the heels of work at the women's Research and Documentation Center at the University of Dar es Salaam contributing to a commission reviewing laws on women even prior to the emergence of political pluralism (Meena, 1989). These initiatives will likely benefit from coordination and networking among women's organizations, especially since the formation of the Tanzanian Gender Networking Program (TGNP) in 1992 and Baraza Ya Wanawake wa Tanzania (BAWATA) in 1995, both of which are discussed more fully later.

Kenya

The situation of women's legal rights in Kenya is similar in many ways to that prevailing in other African countries; the differences, where they occur, are of degree rather than kind.

At the international level, Kenya is a signatory to all of the major instruments on human rights, including the United Nations Convention on the Elimination of All Forms of Discrimination Against Women (CEDAW). However, as of 1996 at least, it had not acted to promote, implement, or observe the provisions of these instruments or to incorporate them into national law.

The Kenyan constitution presents a major bottleneck in the realization of women's human rights. While it dedicates a full chapter to the protection of fundamental human rights and freedoms of the individual, irrespective of race, tribe, place of origin or residence, political opinion, color, creed or sex, the succeeding Section 82 against discrimination does not specifically outlaw discrimination on the basis of sex. As a result, differential treatment on the basis of sex is not considered to be discrimination by law, and, indeed, no existing law can challenge it. Furthermore, Section 82 contains other provisions that virtually eliminate freedom from such discrimination.

As in the other East African countries, customary law in Kenya remains valid as long as the custom in question is not repugnant to natural justice equity and good conscience. But the question Tibatemwa (1995) rhetorically posed in the case of Uganda is also pertinent to Kenya: "With a judicial system dominated by men, can we expect courts to declare repugnant [those] customary Laws which deny women property rights in favor of men?" The answer, for Kenya, would appear to be negative. A good case in point is that of a Kenyan woman who in 1987 sought the right to bury her spouse on the strength of common law under which she and her spouse were married. But a male judge ruled against her, invoking customary law in the case (Stamp, 1991).

In most customary and religious marriages, a woman is not a consenting party. The woman is basically the "property" of her husband when she is married, and of her father and male relatives if she is unmarried. For many ethnic groups in Kenya, customary law permits wife beating as a form of discipline. As in Uganda and Tanzania, a husband can sue his wife for adultery, but a wife cannot similarly sue her husband. Circumcision of women is still practiced in some communities as a form of controlling women's sexuality. In certain Kenyan communities, widows are still inherited by the brother or a close male member of the dead husband's clan, thus denying such widows the right to select a partner and make decisions regarding the estate of the deceased spouse. Furthermore, some view rape as a form of sexual control of a man over a woman rather than as a gross violation of woman's human rights, and "when women are raped, whether in marriage or outside it, the law treats them not as victims but as prosecution witnesses on whom the burden of proof lies" (Muli,

1995, p. 79). The most celebrated rape case in Kenya in the 1990s was the 1991 St. Kizito tragedy that left 70 schoolgirls raped, 19 dead, and scores of others injured (Nzomo, 1994b). Government reaction was to treat this case as one of *undiscipline* in schools rather than a case of violence against women; some in the local community saw no harm in the mass rapes and were quick to defend the male killers and rapists, arguing that they meant no harm to the girls—they merely wanted to rape.

Women's freedom of movement, association, and assembly is highly constrained by sociocultural beliefs and vagrancy laws. In urban areas, a woman not accompanied by a man may be denied access to or service in, public utilities. If a woman is spotted walking alone at night, the police may arrest her under the Vagrancy Act and charge her with loitering with the intention of engaging in prostitution.

In sum, provisions or lack of them in the Kenyan constitution and other legislation that have the effect of discriminating against women and violating their human rights include: (1) Section 82 on discrimination; (2) the 1975 Employment Act, which contains built-in discrimination in terms, conditions, and type of work for women, including housing, medical benefits, maternity leave, and working hours; (3) the 1981 Law of Succession Act, which fails to provide for inheritance rights for pastoral women; (4) the Family Law, which is inadequate and creates ambiguities and the possibility for manipulation in interpretation of laws concerning women's rights in a manner that may infringe women's human rights; (5) the Penal Code (on violence against women), which provides for life imprisonment as a maximum sentence for rape but makes no provision for minimum sentence for rapists, thus facilitating extremely lenient punishment, while statutory law is silent on other forms of gender violence, including female circumcision, wife battery, sexual harassment, rape within marriage, widow inheritance, and forced childhood marriages; (6) the Citizenship Act, which under Section 91 allows a Kenyan man who marries a foreigner to pass on his citizenship to his foreign spouse but does not give a Kenyan woman who marries a foreigner the same rights; and (7) (the lack of) affirmative action (or positive discrimination) laws. There is simply no legislation that takes into account the historical disadvantage of women and minority groups and ensures that women are adequately represented in all key decision-making bodies, in politics, bureaucracy, and other public and private institutions—an omission that contributes to the marginalization of women in politics and other public decision-making positions.

Political Participation and Status of Women in East Africa

Uganda

At independence in 1962, the Milton Obote government appointed two women to ministerial posts to demonstrate

to the world that "it was prepared to meet the western standards of nominal female participation in mainstream institutional politics. The reality, however, was that parliamentary democracy was a disguise; actual power remained firmly in the hands of foreign nations and their Ugandan allies in high public offices . . . the main beneficiaries were men" (Byanyima, 1992, p. 132).

The situation of women did not improve in the subsequent years, as the country endured the worst forms of political crises in East Africa, at least until President Museveni's NRM came to power in 1986. In the intervening war years, the women's movement was subsumed in the political turmoil. Byanyima (1992) recalls women's oppression under Idi Amin's dictatorial regime (1971–1979).

> In the name of morality, Amin, through a series of decrees, banned women from wearing certain clothes, using cosmetics, and from having abortions. Amin's list of prohibited female appearance and behavior was so long that a woman walking on the street could never be certain she might not be arrested at any time for breaking the law. Amin's decrees served to reduce the social status of all women and to make them even more vulnerable in an already oppressive situation. . . . Women's organizations ceased to function (Byanyima, 1992, p. 133).

Co-optation and control of women's personal and associational life persisted through Obote's second regime (1980–present), which played a kind of divide-and-conquer game with the women's movement. The state-controlled women's wing of the ruling Uganda People's Congress was used to manipulate the national women's nongovernmental organization (NGO), the National Council of Women, thus further dividing an already fragmented women's movement.

Significantly, even under Museveni's NRM regime (1986–1995), which claimed to be more benevolent and gender sensitive, the state engaged in a similar pattern of controlling the women's movement and driving a wedge between women's groups and organizations. The strained relations between the Directorate of Women's Affairs, controlled by the NRM, and the Ministry of Women and the National Council of Women is a case in point. The directorate, much like the women's wing in Obote's second government, attempted, apparently without success, to manipulate the National Council of Women and, through it, to weaken independent women's NGOs. The state-controlled directorate was instrumental in the NGO's name change to the National Association of Women's Organizations of Uganda. Furthermore, as Tripp (1994) has noted, the NRM's positive response to pressure from the women's groups to establish a gender policy framework and institutional structures to advance the status of women in Uganda can be traced largely to the fact that the state has been significantly weakened by years of conflict with Tanzania and by serious economic decline. Thus it was in no position to restrict private social and economic initiatives when it could no longer provide comparable services or ensure economic well-being. At a broader socioeconomic level:

> The economic crises enhanced the position of many Ugandan women . . . and weakened the basis for men's domination. . . . This period prepared women for future political action (Byanyima, 1992, p. 134).

The response of the Ugandan state under Museveni to the gender question was inspired not by a burning desire to promote democracy and human rights but by what some observers describe as the paternalism of the Museveni regime toward all Ugandans, from whom the state first and foremost demanded obedience in exchange for services. Museveni could not afford to ignore women's demands, as most of the support for his NRM derived from women. Thus, Museveni's regime, with pressure from the women's movement, provided a significant platform for women's political empowerment, largely through an affirmative-action policy that applies at all levels of governance, including the NRM's village councils. This has enabled women to increase their numbers in parliament from 1of 142 members in 1980 to 40 of 278 in 1989. Women also have served in key cabinet posts, including minister of agriculture and minister of industry and technology (Byanyima, 1992) and as vice president. This is in addition to the creation of a Ministry of Women, albeit lumped together with Youth and Cultural Affairs, and the designation of a Women's National Day—all of which can be viewed as important gains for Ugandan women in their struggles for empowerment.

Despite these formal achievements, Ugandan women continue to experience the same problems that confront their counterparts in Kenya and Tanzania: discrimination in law, gender violence, the feminization of poverty, and an inability to influence public policy and decision making, despite official representation through affirmative action. Low levels of rights awareness among women, compounded by negative sociocultural conditioning, have slowed progress in the advancement of women's status. Lack of rights awareness and gender sensitization remains a major obstacle to women's participation in electoral politics. Some female voters, for example, join men in harassing and refusing to support female candidates, either because they believe that women like themselves cannot make good leaders or because of intimidation and threats by male relatives. Married women face additional constraints in Uganda when they seek electoral office. Miria Matembe, one of Uganda's female members of parliament (MP), put it succinctly:

> If a woman wants to run in the constituency in which she was born, she is told, "You left long ago. Go run somewhere else." And if she wants to run in her husband's constituency, she is told, "You didn't come here to rule, you came here to marry" (USAID, 1995, p. 3).

But the sociocultural barriers to women's political participation are likely to decrease as the women's movement intensifies gender-sensitization and rights-awareness programs. In this regard, since the late 1980s new women's associations have mushroomed and joined existing ones at the local and national levels and are involved in numerous activities ranging from income-generating projects to legal- and gender-rights-awareness programs for women. These include the Young Women's Christian Association, the Ugandan Women's Credit and Finance Trust, Action for Development Association, and the Women Lawyer's Association. The last two have spearheaded lobbying efforts for enhanced female representation in public and political decision-making positions. This new wave of the women's movement of the 1990s consists of a new breed of female activists who lobbied the state and actively sought to influence the process and content of the new democratic constitutional development and the first multiparty elections in Uganda.

Kenya

For the first two decades of its independence, Kenya was viewed as the most stable country in East Africa. Except for an army mutiny in 1964 and an attempted military coup in 1982, Kenya had escaped the worst forms of political crises manifested elsewhere in Africa in the forms of bloody military coups (as in Uganda), civil wars (neighboring Somalia and Sudan), and genocide (Rwanda and Burundi). But underneath a veneer of assumed political tranquility, Kenya's political economy has endured a highly repressive and autocratic political system that has tended to mask underlying political tensions, including politically instigated ethnic cleansing, assassinations and detention without trial, worker and student unrest, unbridled corruption, breakdown of the justice system, and deterioration in social services and delivery systems. Women have not only borne the greatest social costs for bad governance, they have largely been excluded from social politics and centers of decision making.

The late president Daniel Arap Moi managed to control political dissent and to repress social, economic, and political demands for popular participation until 1990. The return to a multiparty system in Kenya in December 1991 created some political space for civil-society groups, including women's groups, to participate actively in the multiparty democratic struggles of the 1990s. In the process, the women's movement demonstrated its potential as a formidable political force capable of seeking and influencing change in the oppressive state autocracy and patriarchy. As Nzomo (1995–1996) observes, 1992 was the year that political empowerment was the number one priority for women, notwithstanding the equally great concern for redressing economic and social injustices that underlie female subordination to men. Women activists argued that if women attained key political decision-making positions in large enough numbers—at least 30 percent of the total—they could ensure the removal or repeal of laws that discriminate against women at the social and economic levels. They would also participate in designing policies that would bring women into the mainstream.

The women's movement in Kenya in the 1990s, as in Uganda and Tanzania, has been seeking to provide leadership in a country divided along ethnic, religious, racial, and gender lines. In their own movement, women seek to practice unity in diversity and urge the rest of society, especially in pro-democracy movements, to do likewise. The push for engendered democratization and political empowerment of Kenyan women in the early 1990s greatly benefited from the emergence, beginning in 1991, of a number of new feminist lobby groups, including the National Commission on the Status of Women, the League of Kenya Women Voters, the Anti-Rape Organization, and the short-lived Mothers in Action. Unlike the 23,000-plus groups already in existence, these new lobby groups were much more political in their orientation and more assertive, innovative, and willing to take political risks in the pursuit of the women's agenda. At the same time, many of the existing groups and organizations, such as FIDA-K, the National Council of Women of Kenya (NCWK), the Young Women's Christian Association (YWCA), which had never before articulated a political agenda, also became vocal and critical of the undemocratic status quo. Jointly with the new lobby groups, they vigorously lobbied all political parties to integrate gender issues within the context of their democratic agendas and programs. This unity of purpose in the women's movement provided leadership that greatly facilitated gender activism in Kenya's first multiparty elections, in December 1992.

Women in Kenya have made significant advancements, in terms of setting a women's agenda for democratization and working out strategies for political and other forms of empowerment, but they have yet to implement fully the bulk of this agenda. This agenda, which was drawn up at a national women's convention in February 1992, identified key issues of concern to women and strategies for achieving them. The agenda first committed Kenyan women to a nonsectarian, nonpartisan, but political ideology, guided by the motto: "Unity in Diversity for Women's Empowerment." This entailed creating a coalition of numerous and diverse women's groups and organizations and then crafting a consensus and common approach to issues deemed fundamental to women's human rights and democratic entitlements. The issues agreed upon that the women's movement has since sought to implement include: (1) setting up gender-sensitization, legal-awareness, and voter-education programs for women and men; (2) building the capacity of female candidates across political parties through training on such issues as public speaking, fundraising, electoral laws and procedures, and the packaging of campaign issues; (3) lobbying for public and government support of the women's agenda through voting power, poster campaigns, the media, and seminars and workshops, and organizing peaceful demonstrations, such as the March–December 1992 hunger strike by moth-

ers of political prisoners; and (4) participating in all stages of the electoral process, from registration of voters to election monitoring, focusing on female candidates and ensuring that the democratic principles of freedom and fairness are observed during balloting.

The December 1992 elections did not result in a critical mass of women elected to decision-making bodies. But it did reflect the enormous efforts of the women's movement to empower female voters and candidates. Women increased their numbers from 2 to 6 in the 200-member parliament, and from about 20 to 50 in local government councils. One of those 50 was later elected mayor of the city of Embu—at the time, she was only the fourth female mayor since independence. In May 1995, the first female minister since independence was appointed to head the gender-stereotyped Ministry of Culture and Social Services. However, the male-dominated government has not demonstrated any commitment to gender equity in public decision making.

Although the issue of political empowerment took center stage in 1992, other issues related to women's empowerment remained important long after the elections. The women's movement continues to lobby the state for legal and institutional reform, to remove gender discrimination, and to restore rights and freedoms. Violence against women as a human-rights issue has sustained a united response among women. Although, by 1996, the women's movement had failed to take any concrete action beyond public statements and an occasional peaceful demonstration condemning gender violence and demanding stiff punishment for rapists, at least two crisis centers had been set up by women's NGOs to provide temporary shelters and counseling for women and girls in distress, and the Anti-Rape Organization, founded in 1992, lobbies and acts as a crisis center for rape victims. Women's lobby groups continue to conduct legal and gender-awareness education to sensitize the public on gender violence and other human-rights abuses against women, while the Greenbelt movement provides leadership in response to environmental degradation in Kenya. In the meantime, concern has also been voiced about the negative impact of economic liberalization on women. In particular in the 1990s, activist researchers have documented and publicized the negative impact of SAPs and have recommended gender-sensitive alternatives (NCSW, 1994).

Tanzania

Tanzania at independence was bequeathed, as Kenya and Uganda, a pluralistic political system and space for some autonomous civil associations to operate. But by the mid-1960s, the state had disbanded or taken control of most civic organizations, barred opposition political parties from 1965, and confined women's associational life to the government-controlled women's body, Umoja wa Wanawake wa Tanzania (UWT).

Between 1964 and 1977, the Tanzanian state asserted its supremacy and consolidated its hegemony over civil society and political institutions. This period, which started with the usurping of the autonomy of the trade-union movement and its subordination to the authority of the state, ended with the merger of the ruling parties of Tanganyika and Zanzibar, and the formation of the party Chama cha Mapinduzi (CCM) in 1977. In the period between 1977 and 1985, the ruling party and the state became fused, with the party becoming supreme and completely subordinating to itself the authority of government organs, including parliament and the judiciary. Furthermore, political dissent was not tolerated, and government response to it was either to co-opt or to repress dissenters. Tanzania, like Kenya, though spared the worst forms of political crises and despite its socialist ideology under President Julius Nyerere, masked a highly centralized and repressive political system.

The change of national political leadership in 1985, from Nyerere to Ali Hassan Mwinyi, provided an opportunity to reconstitute the state–society relationship. By early 1991, the basic principles of socialism and self-reliance, as articulated in the 1967 *Arusha Declaration,* were revoked, thus facilitating economic liberalization and paving the way for the introduction of political pluralism. Nyerere's departure from the political helm in the mid-1960s facilitated the implementation of economic liberalization and the reemergence of associational life.

With economic liberalization and the intensified economic crisis brought about by SAPs, women's income-earning associations increased (Tripp, 1994). However, many of these groups, though vital for the provision of welfare to their communities, had by 1994 not yet shifted their focus from practical and welfare needs to strategic needs that are empowering and transformative. Shayo and Koda (1994) note, with regard to women's informal associations in Tanzania:

> Much as these very useful associations do exist, most of them have tended to promote welfarism and are hence nontrasformative or else pose very little challenge to the status quo. There is for instance no grassroots-oriented initiative focusing on advocacy and lobbying for political and legal reforms and transformation at local and higher levels. Even during this era of political pluralism and democratic processes in Tanzania, not much effort is being taken to exploit the space provided for more transformative politically challenges. Most women therefore are politically conscious but largely inactive (Shayo and Koda, 1994).

Within the context of the institutional and policy framework for gender action, Tanzania, like its African neighbors, has demonstrated more rhetoric than action, and women and gender questions have remained marginalized, patronized, manipulated, and wooed—as befits political expedience at any given period. The government-controlled women's national machinery, UWT, like Kenya's Women's Bureau, constituted a small, highly underfunded department in one of the least important ministries—the Ministry of Women's Affairs, Youth, and Community Development.

Despite this, Tanzania has maintained a better record of women holding political decision-making positions than Kenya and Uganda before 1986. This has largely been the result of government exercising affirmative action in the political arena. Consequently, although women seeking political office in Tanzania through the normal electoral process have over the years been few, and even fewer have attained political office through the ballot, the number of women in parliament has been consistently higher than in Kenya and comparable to the figures for Uganda during Museveni's regime. Thus, unlike Kenya, where the average number of female MPs has been 2 and the highest 6, Tanzania's lowest number immediately after independence was 6 representing 8 percent of the total, while the highest after the 1990 elections was 28, or 11 percent of the total. However, if one separates elected from appointed parliamentary posts, it becomes clear that most female MPs are beneficiaries of the quota system rather than winners of regular competitive elections. Thus, in mainland Tanzania, only one woman won a parliamentary seat in the 1985 national elections and only two women marginally won parliamentary seats in the 1990 national elections. Outside the political arena in other decision-making positions in which the government has not applied affirmative action, women's marginalization is in some cases worse than that of Kenya, especially in the national bureaucracy, the judiciary, and the political-party hierarchy.

Since April 1992, when Tanzania moved to political pluralism, the political landscape in that country has allowed the reappearance and reinvigoration of civil-society groups. New forms of women's organizations are emerging as important actors in national politics. Unlike previous associations, these gender associations seek to participate in and transform structures of power and governance and to promote gender-sensitive democratic development and women's human rights through conferences, seminars, radio programs, training for institutional-building, publications, and general political education on democracy.

Among these new associations is the Tanzanian Gender Networking Program (TGNP), which was formed in 1992 to promote networking and solidarity among women by strengthening coalitions among groups and organizations that can lobby for polity reform and action at all levels. Another important development in the Tanzanian women's movement was the creation, in May 1995, of a broadly based organization, Baraza Ya Wanawake wa Tanzania (BAWATA), to coordinate and promote women's rights in the context of the new democratic climate. Although it is too early to predict its future, the national mandate and objectives that it has set for itself would indicate that BAWATA has the potential to bring some cohesion and vision to the women's movement and to galvanize Tanzanian women to become influential actors in advancing their status and democratic development. The political space created by pluralism has also invigorated a number of existing women's organizations formed since the mid-1980s, such as the Tanzania Media Women's Association, the Medical Women's Association of Tanzania, the Tanzanian Women Lawyers Association, and several university women's research groups. Most are now lobbying for women's empowerment and the inclusion of gender rights in the new democratic program.

The women's movement in Tanzania is still fragile and fragmented, reflecting its long history of complete demobilization under Nyerere's socialist regime. Consequently, although Tanzania has seen the growth of a vibrant women's movement since the early 1990s, it has not been possible to replicate the level of enthusiasm and political dynamism evident in Kenya during the 1992 election year. When Tanzanians went to the polls in October 1995, few women had declared their candidacy. As elsewhere, traditional constraints—lack of funds, multiple roles, political inexperience, and negative societal attitudes toward female candidates—combined to prevent women from running for political office. Furthermore, the male-dominated political parties have not been supportive of women's candidacies. In addition, there appeared to be no common women's agenda similar to the one prepared by Kenyan women in 1992.

For Tanzania, as for most of Africa, retrogressive sociocultural conditioning, illiteracy, and lack of rights awareness among most women have been the major obstacles to empowerment and effective participation in democratic change. In recognition of this, one of the most important responses made by the women's movements in Tanzania, Kenya, Uganda, and, indeed, most of Africa in the 1990s has been to establish programs for training, educating, and raising the consciousness of women and men concerning legal, gender, and civic rights and obligations.

Conclusions

This essay has analyzed the political and legal status of women in East Africa from a broader global context, in which women's rights as human rights are acknowledged but not yet implemented, and women's participation in key decision-making bodies and forums is still marginal—even within the U.N. system itself. This essay has noted that, whereas the lack of legal provisions to protect women's rights is a major obstacle to their empowerment, an equally important obstacle is the lack of legal- and gender-rights awareness among both women and men. Thus, even when legal statutes exist that could promote rights, women are unable to utilize them. In the East African region, the concurrent existence of three different systems of law—constitutional/statutory law, customary law, and religious laws—has generally tended to create special problems for women. Customary and religious laws tend to be more oppressive to women than national constitutional law since the former are based on patriarchal cultural values that define women in a status subordinate to men. On most critical issues of personal rights and human dignity, the customary and religious laws prevail. In this area, little progress has been made in all three East African countries,

although the women's movement and other human-rights groups have embarked on a lobbying campaign for comprehensive legal reform.

The undemocratic and male-dominated African state has also been a major stumbling block to women's advancement and has largely shaped the level and nature of women's political participation. The women's movement of the 1990s in all three countries is, however, more dynamic, aggressive, and politically inspired than earlier movements. The very fact that women have succeeded in getting gender issues on the national democratic agendas is itself a significant step forward. Where multiparty elections have been held in the 1990s, more women are getting elected into political office, as in Kenya. Women have managed to influence new democratic constitutions, as in Uganda, and to influence their governments to exercise political affirmative action, as in both Uganda and Tanzania. This has facilitated an increased presence of women in parliament, the appointment of a woman as vice president in Uganda, and the establishment of ministries of women in Uganda and Tanzania. Kenya, in May 1995, appointed the first female minister since it attained independence in 1963. She may not make a difference on her own, but she has opened doors for other women to lobby for greater and more effective political representation in the future.

Women in East Africa are still predominantly outsiders in the political machinery of both the state system and the opposition political parties. Even where multiparty elections have been held, the first elections have returned to power the same political parties (KANU in Kenya and CCM in Tanzania) that presided over political repression for more than 30 years. This is likely to derail the democratic transition, as already witnessed in Kenya. This possibility complicates the task of women's struggles for political inclusion and empowerment. However, these democratic setbacks also provide opportunities for women to develop alternative models of leadership and vision and, in so doing, influence and gender-sensitize the content of the democratic agenda.

References

Byanyima, W. Karagwa. "Women in Political Struggle in Uganda." In Jill M. Bystydzienski, *Women Transforming Politics: Worldwide Strategies for Empowerment.* Bloomington: Indiana University Press, 1992, pp. 129–142.

Fatton, Robert, Jr. "Gender, Class, and State in Africa." In Jane L. Parpart and Kathleen A. Staudt (eds.), *Women and the State in Africa.* Boulder: Lynne Rienner, 1989, pp. 47–66.

Harbeson, John W., Donald Rothchild, and Naomi Chazan (eds.). *Civil Society and the State in Africa.* Boulder: Lynne Rienner, 1994.

Klemp, L. (ed.). *Empowerment of Women in the Process of Democratization: Experiences of Kenya, Uganda, and Tanzania.* Dar-es Salaam: Friedrich Ebert Stiftung FES, 1994.

Mascarenhas, Ophelia, and Marjorie Mbilinyi. *Women in*

Tanzania: A Bibliography. Uppsala: Motala Grafiska, 1983.

Meena, Ruth. "Crisis and Structural Adjustment: Tanzanian Women's Politics," *Issue,* vol. 17, no. 2, Summer 1989, pp. 29–31.

Muli, Koki. "Help Me Balance the Load: Gender Discrimination in Kenya." In I. Peters and Andrea Wolper (eds.), *Women's Rights Human Rights: International Feminist Perspectives.* New York: Routledge, 1995, pp. 78–81.

National Commission on the Status of Women (NCSW). *The Impact of SAPS on the Female Gender in Kenya.* Nairobi: NCSW, 1994.

Nzomo, Maria. "The Status of Women's Human Rights in Kenya," *Issues,* vol. 22, no. 3, 1994a, pp. 17–20.

———. "Women in Politics and Public Decision Making." In Ulf Himmelstrand et al. (eds.), *In Search of a New Paradigm for the Study of African Development.* London: James Currey, 1994b.

———. "The Political Economy of the African Crisis: Gender Impacts and Responses," *International Journal,* vol. 51, 1995–1996, pp. 79–101.

Nzomo, Maria, and M. Halfani. *Toward a Reconstruction of State Society Relations: Democracy and Human Rights in Tanzania.* Montreal: International Center for Human Rights and Democratic Development (ICHRDD), 1995.

Nzomo, Maria, and Kathleen A. Staudt. "Man-Made Political Machinery in Kenya: Political Space for Women?" In Barbara J. Nelson and Najma Chowdhury (eds.), *Women and Politics Worldwide.* New Haven: Yale University Press, 1994, pp. 415–435.

Obbo, Christine. *African Women: Their Struggle for Economic Independence.* London: Zed, 1980.

Parpart, Jane L. "Women and the State in Africa." In Donald Rothchild and Naomi Chazan (eds.), *The Precarious Balance: State and Society in Africa.* Boulder: Westview, 1988, pp. 208–232.

Parpart, Jane L., and Kathleen A. Staudt (eds.). *Women and the State in Africa.* Boulder: Lynne Rienner, 1989.

Peterson, V. Spike (ed.). *Gendered States: Feminist (Re)visions of International Relations Theory.* Boulder: Lynne Rienner, 1992.

Rothchild, Donald, and Naomi Chazan (eds.). *The Precarious Balance: State and Society in Africa.* Boulder: Westview, 1988.

Shayo, R., and B. Koda. "Women and Politics in Tanzania." In L. Klemp (ed.), *Empowerment of Women in the Process of Democratization: Experiences of Kenya, Uganda and Tanzania.* Dar es Salaam: Friedrich Ebert Stiftung FES, 1994.

Snyder, Margaret C., and Mary Tadesse. *African Women and Development: A History.* London: Zed, 1995.

Stamp, Patricia. "Burying Otieno: The Politics of Gender and Ethnicity in Kenya," *SIGNS,* vol. 16, no. 4, Summer 1991, pp. 808–845.

Tibatemwa, Lilian Ekirikubinza. "Property Rights, Institutional Credit, and Gender in Uganda," *East African Journal of Peace and Human Rights,* vol. 2, no. 1, 1995, pp. 68–80.

Tripp, Aili Mari. "Gender, Political Participation, and the Transformation of Associational Life in Uganda and Tanzania," *African Studies Review,* vol. 37, no. 1, April 1994, pp. 107–131.

United Nations Development Program (UNDP). *Human Development Report 1995.* New York: UNDP, 1995.

United States Agency for International Development (USAID). "Gender and Democracy," background paper for USAID workshop, USAID, Washington, D.C., July, 1995.

Women in Law and Development in Africa (WiLDAF). *The World Conference on Human Rights: The WiLDAF Experience.* Harare: WiLDAF, 1993.

Zartman, William I. (ed.). *Collapsed States: The Disintegration and Restoration of Legitimate Authority.* Boulder: Lynne Rienner, 1995.

Women in Southern Africa, Excluding South Africa

Jean Davison

Introduction

Seismic shifts in southern Africa's political landscape provide the grist for reshaping gender relations in that part of the world. With the collapse of South Africa's pernicious apartheid system in 1994 and the blossoming of multiparty democracy in neighboring countries, a door has been pried open for the expansion of social justice. Critical to the broadening of social justice is the advancement of women.

Democracy, however, is not a sufficient condition to ensure that women will be included in a new government's vision of social justice. Nor are improved living conditions, often identified with democracy in Africa, a precursor of women's inclusion in key political forums. Women may experience better health conditions and increased education yet continue to be excluded from critical decision-making arenas in which policies are crafted that affect their lives. Botswana, with the highest rank of the Southern Africa Development Community (SADC) countries on the Human Development Index (UNDP, 1993), has the lowest proportion of women in its national legislative assembly (5 percent). In contrast, Mozambique, a former socialist state ranked among the lowest on the Human Development Index, has the highest proportion of women (16 percent) in its national assembly of all of the southern African states. Even though the winds of political change augur well for the expansion of human rights, the political shifts taking place in the region do not necessarily guarantee women the political, social, and economic rights they deserve.

This overview of women in southern Africa compares their status in the SADC states. These countries surround or are within the boundaries of the Republic of South Africa. They have significant economic ties and increasing trade relations with that key player in the region. The essay begins with a brief overview of differences and similarities among states in the region as a means of understanding the diversity of situations that characterize women's experience. Then, using the United Nations Development Program's (UNDP) Human Development Index as a measure, it compares the status of women cross-nationally on a number of internationally recognized indicators. The essay concludes with a discussion of maintenance law as an example of emerging legal issues that affect gender equity.

The Regional Context

SADC, formerly the Southern African Development Coordination Conference (SADCC), was created in 1980 as a countermeasure to economic dependence on apartheid South Africa and as a means of coordinating development efforts. In August 1992, the member countries that made up the organization met to change the name to the Southern Africa Development Community (SADC) to reflect the changing political situation with Namibia's independence in 1990. This regional organization includes the nine countries that are the focus of this essay: Botswana, Lesotho, Swaziland, Angola, Mozambique, Malawi, Zambia, Zimbabwe, and Namibia. The countries divide into subregional groupings based on their proximity to South Africa and their colonial past.

The "Captured" States: Botswana, Lesotho, and Swaziland

Botswana, Lesotho, and Swaziland, formerly known as BLS, were a target of South Africa's hegemony throughout the colonial period. However, by the mid-1960s the incorporation of BLS into South Africa became permanently stalled. Botswana has rich diamond deposits that have given this country a singularly affluent position compared to South Africa's other neighbors and, together with a relatively democratic state, have enabled Botswana to negotiate a very different relationship with South Africa than Lesotho or Swaziland. The mining of scarce mineral resources also means that Botswana has a more favorable balance-of-trade position with sufficient financial resources to spend on the development of communications infrastructure and social services such as health and education.

Embedded within the borders of South Africa, Lesotho and Swaziland became known as the "captive states of southern Africa" during the 1980s because they were forced into economic and military dependence on apartheid South Africa (Ajulu and Cammack, 1986). Lesotho and Swaziland and, to a lesser extent, Botswana continue to depend heavily on South Africa for employment opportunities and trade in the postapartheid era. This dependency has a gender bias in that those most directly affected are male migrant laborers and traders who reside in their own countries but work in South Africa. In Lesotho, for instance, out of a total population of nearly 1.7 million (1989), 90,000—overwhelmingly male—emigrated to work in South Africa (Government of Lesotho, 1989). The emigration of male labor, beginning as early as age 15, impinges on gender relations in each country.

The States of the Former British Raj: Malawi, Zambia, and Zimbabwe

Of the three landlocked states that were formerly under British colonial rule—Malawi, Zambia, and Zimbabwe—the avuncular state of Malawi is the most impoverished, having few mineral resources and being largely dependent on agriculture (tobacco, cotton, and tea) for foreign exchange. Ethnically, Malawi is diverse. The matrilineal, Chewa-speaking groups and Yao predominate, politically and numerically, over a minority of patrilineal groups concentrated largely in the northern region, although some have migrated to other regions to take up wage employment. Of the patrilineal groups, the Tumbuka, who were among the earliest to receive missionary schooling, were in a favorable position in the postindependence period to take up civil-service jobs. That the majority of Malawi's population (68 percent) comes from groups that ascribe to matrilineal ideology influences gender relations, especially in rural areas. It is not uncommon to find women acting as chiefs and village heads in the central and southern regions. Women have rights to inherit and control land and pass it down to their daughters and granddaughters. This control over land, for women who marry and remain in their natal villages after marriage, as many rural women in southern Malawi do, gives them relative economic security (Davison, 1992, 1994; Kaufulu, 1992). At the same time, older men as avunculates historically wielded local and regional political power. They account for the majority of those found within the state's bureaucracy and, in the late twentieth century, hold the vast majority of elected and appointed positions, an inequitable gender situation that is being addressed through Malawi's new constitution.

In contrast to Malawi, Zambia and Zimbabwe have significant mineral resources, whose development during the colonial period (from 1892) influenced the shape of gender relations; mining depended on African male labor, and it excluded women except as wives in Zambia from the 1930s onward (Chauncey, 1981; Heisler, 1984; Parpart, 1986). However, from that same time, despite colonial efforts to discourage single women from migrating, increasing numbers emigrated to urban and mining centers in search of income-generating activities (Schuster, 1979; Parpart, 1986). However, the majority of women, especially those with dependents to feed, remained rooted to agricultural production while their men migrated to the mines (Moore and Vaughan, 1994).

Zambia was among the world's three largest copper producers in the 1960s, along with the United States and Chile (Serpell, 1993). However, the African country's intense economic dependency on copper proved to be its Achilles' heel beginning in the 1970s, when externally controlled copper prices plummeted, precipitating an economic crisis. To counter the crisis and encourage unemployed urban dwellers to return to their indigenous villages, the state initiated a "back-to-the land" campaign, using hybrid maize as an incentive. Whereas 60 percent of Zambia's population was found in urban areas and mining towns in the late 1970s, by 1990 a reverse trend was well under way. Jobless men were the majority, but unemployed women also trickled back to rural areas to take up maize farming or other rural income-generating pursuits (Pottier, 1988).

Zimbabwe (formerly Southern Rhodesia) began as a settler colony with the northward migration of White South African farmers—called the Pioneer Column—in the late nineteenth century. Agriculture was the backbone of Southern Rhodesia's colonial economy, with mining also growing in importance. Zimbabwe in the 1990s has a robust agricultural sector, an active mining sector, and a burgeoning industrial sector that began well before independence in 1980. As such, Zimbabwe has had the advantage of a much more diversified economy than Zambia. Earlier experimentation with socialism has given way to a mixed economy increasingly dependent on capitalist forces and global markets. The two major political parties at independence were ZANU-PF (Zimbabwe African National Union-Patriotic Front), whose leadership was dominated by Shona speakers who made up roughly 85 percent of the population, and PF-ZAPU (Zimbabwe African Peoples Union) representing the minority Ndebele. After a bitter civil struggle for leadership in the first half of the 1980s, the two parties agreed to unify. Through the unity agreement of 1987, PF-ZAPU's leadership was integrated in the new ZANU-PF party. Although a smattering of minor opposition parties exists largely dependent on individual personalities, in the fourth general elections held in April 1995 all but two of the 65 contested seats in the parliament were won by ZANU-PF giving rise to criticism of an electoral process engineered by a single party (Laakso, 1995). Of the 63 ZANU-PF seats claimed, 18 were won by women.

Zimbabwe has always been patriarchal and largely patrilineal. Until relatively recently, women had limited rights to productive resources, except as wives. Women's legal status at the time of independence was one of the worst in the region: Women were considered legal minors who could not appear in a court of law without a male relative representing them. The Legal Age of Majority Act, signed in 1982, was a significant piece of legislation for women because it accorded all Zimbabweans over the age of 18 years, regardless of sex, full adult status. Still, a study cited by Gwaunza and Zana (1990) based on a sample of rural women found that more than 90 percent, when asked who benefited from this law, responded that young people most benefited because they used the law as a means of gaining their independence from their parents; in fact, some mothers declared that the law had caused older children to become rebellious. These women equated the law more with older children's rights than with their own improved rights and status, a situation attributed by Gwaunza and Zana (1990) to a lack of outreach to rural women providing them legal education following the law's passage.

The Former Socialist States: Mozambique and Angola

Angola and Mozambique form a separate subgroup, given their former occupation by the Portuguese and their socialist bent at independence. Both won independence in 1975 at great cost: The Portuguese destroyed much of the transport system and industry they had built. Moreover, these emerging socialist states had to contend with military and socioeconomic destabilization campaigns launched indirectly by South Africa through insurgent groups in each country during the 1980s. In Mozambique, in the mid-1980s, the brunt of the civil war between FRELIMO (Front for the Liberation of Mozambique) government forces and the South African–backed RENAMO (Mozambique National Resistance) forces was felt by the rural poor, especially women and children (Urdang, 1989; Sheldon, 1994). Of the total population of roughly 14 million in the mid-1980s, 2.9 million were forced off their land through acts of terror and crop destruction (Bowen, 1993, p. 351). It was only with the cessation of war and multiparty elections in October 1994 that Mozambicans began hesitantly to rebuild their lives. In Angola, under the auspices of a U.N.-brokered ceasefire and peace agreement, multiparty elections were held in 1993. The Marxist MPLA (Peoples Movement for Liberation of Angola) party in power won the elections, but the dissident party's leaders refused to recognize the MPLA's right to govern and instead returned to guerrilla warfare in the countryside. The situation in Angola remains precarious. For women in Angola and Mozambique, maintaining peace to rebuild their lives is the highest priority.

The Newest SADC Member: Namibia

South of Angola is Namibia, the last country to join SADC. Fronting the Atlantic coast, Namibia has rich mineral resources and a colonial past that includes both German and South African occupation. This country of stark deserts and abundant wildlife has a number of significant ethnic groups, from the cattle-herding Herero to the once-nomadic Basarwa (or San) of the Kalahari. The position of women in these groups varies, depending on the groups' gender ideologies. Among the patrilineal Herero, women in the nineteenth century adopted the European dress and Victorian customs of their patriarchal colonizers. Those without education are little changed. In contrast, Basarwa women, who formerly hunted and gathered with their men, have negotiated the changes invoked by the transition to a sedentary life—ignoring some, embracing others. Some women have taken up new crafts, such as painting, using age-old Basarwa designs. Others have turned to herding. As well as indigenous women are those of German and South African descent who have remained in independent Namibia. These women are located largely in the urban centers, and their needs differ from those of indigenous rural women. How women will fair, overall, in this newly liberated country may depend on the lessons they learn from other women in the SADC region.

Women and the State

The nine countries in the region share a gendered colonial legacy that privileged males regardless of women's status in precolonial societies. Among the changes that affected women were: (1) the state used African male labor to launch its mining and plantation enterprises, leaving women to sustain family-based agricultural production without benefit, often, of male labor; (2) women in matrilineal groups, who had been political leaders and who controlled land and labor, were increasingly marginalized; (3) economic development for Africans became gendered male; and (4) where educational institutions were set up, African males were singled out for opportunities. Disturbingly, the gendered nature of the state has continued to persist in the postcolonial era. Even in Angola and Mozambique, where Marxist states were committed to expanding social justice at the onset, except for cosmetic changes, women's status continued to trail men's (Organization of Angolan Women, 1984; Urdang, 1984; Davison, 1988; Sheldon, 1994). The patriarchy that began in some places prior to colonial occupation and that was initiated in others with colonial patriarchy, was extended through the independent state apparatus by educated, African male elites. These male bureaucrats often were, and in some cases still are, reluctant to share their newly won power, and they have used a number of strategies to thwart the aspirations of southern African women for empowerment (Longwe, 1988).

Human Development Indicators of Women's Status

One way of measuring women's status in the region is to analyze it in terms of social and economic development. The Human Development Index measures quantitatively women's status cross-nationally (UNDP, 1992, 1993). Sufficient data exist for a comparison of the nine SADC countries on six of the nine indicators (see Table 1).

Health Status

Key measures of women's health status are life expectancy, women's access to prenatal care, and maternal mortality. The country that has the highest life expectancy at birth for women is Botswana, where women live on average to

Table 1. Status of Southern African Women

Country	Life Expectancy at Birth in Years	Average Age at First Marriage	Adult literacy Rate (%), Age 15 +		Net School Enrollment Ratio (%)			Women in Labor Force (% of total)	Women in National Assembly (%)
			M	*F*	*Pr.*	*Sec.*	*Ter.*		
Botswana	62.8	26.4	84	65	93	47	3	35	5
Swaziland	58.6	N/A	N/A	N/A	84	49	3	40	N/A
Lesotho	61.8	19.6	N/A	N/A	76	31	6	44	N/A
Zimbabwe	61.4	20.4	74	60	N/A	46	2	35	12
Zambia	55.5	19.4	81	65	79	14	1	29	5
Namibia	58.8	N/A	N/A	N/A	N/A	38	N/A	24	7
Malawi	48.7	17.8	65	35	51	33	(.)	42	10
Mozambique	49.2	17.6	45	21	37	40	(.)	48	16
Angola	47.1	N/A	56	29	N/A	N/A	0	39	15

Source: Compiled from Table 16, UNDP, 1993.

be 62.8 years; Angola has the lowest life expectancy (47.1 years). (Although this is an intragender comparison, in all nine countries, women outlive men, in some cases by three to four years.) That a difference of 15.7 years exists between the country with the highest rate and the one with the lowest rate is not insignificant. It is indicative of the greater financial resources that Botswana, a country with rich mining resources and a stable political environment, has to invest in developing its health sector. Angola, although it does have oil resources in the north, has been plagued since independence by an economically destabilizing civil war that destroyed most of its infrastructure. It also is notable that the three most impoverished countries in the region—Angola, Mozambique, and Malawi—have the lowest life expectancy for women.

Botswana also has the lowest maternal mortality rate, 300 maternal deaths per 100,000 live births; Angola has the highest, 900 per 100,000 live births. However, at 71 percent Botswana ranks sixth in the proportion of women who receive prenatal health care (see Table 2), behind Zimbabwe (83 percent), Namibia (82 percent), Zambia (80 percent), and Swaziland and Malawi (each with 76 percent). The availability of trained medical workers and clinics may account for the discrepancy; Botswana's clinics are few and far between in rural areas.

The two countries with the highest maternal mortality rates, Angola and Mozambique, have the lowest proportion of women receiving prenatal care—only 27 percent in Angola and 54 percent of women in Mozambique. These countries' clinics were destroyed during the postindependence civil wars. An anomaly exists in Lesotho, which has a relatively low maternal mortality rate (350 deaths per 100,000 live births) for the region and a low prenatal-care-participation rate (50 percent). Lesotho's mountainous terrain, where most people travel by horseback, may preclude some women in rural areas from attending a prenatal clinic.

Average age at first marriage is another factor influencing women's health. It is related to educational status and childbearing; the more education women have, the later

they are apt to marry, and, with delayed marriage, one would assume that women will bear fewer children. However, in most southern African countries, especially Botswana, a woman often has a child (or two) prior to marriage. The woman may or may not marry the child's father. Consequently, age at marriage is not always a useful indicator of fertility in Africa, although it is associated with educational achievement.

Botswana's women marry on average seven years later than their southern African counterparts. The average age for Batswana women is 26.4 years, but their childbearing years often begin earlier. In Zimbabwe, age at first marriage is 20.4 years whereas for Mozambican and Malawian women it is younger—17.6 and 17.8 years, respectively. The educational pattern of females in these countries influences age at first marriage. Of girls (age 6–14), 93 percent are enrolled in primary school in Botswana, compared to 51 in Malawi and 37 percent in Mozambique. Among the SADC countries, Botswana also has one of the highest gross enrollment rates for girls in secondary school (47 percent). Only Swaziland has a higher rate (49 percent), but unfortunately no data exist for Swaziland on average age at first marriage.

Educational Status

Educational status of women is often measured in two ways, through adult literacy rates and through school enrollment rates at various levels. For adult literacy rates (inclusive of the population age 15 years and older), we have included male participation rates as a basis of comparison. In only three of the nine countries—Botswana, Zimbabwe, and Zambia—are more than half the adult females literate. In the other three countries for which data are available—Malawi, Mozambique, and Angola—the percentage drops to 35 percent, 21 percent, and 29 percent, respectively.

The contrast between these six countries is striking. In the first three countries, the majority of women have some formal education, which enables them to transact business, read newspapers, and follow written directions. Armed

Table 2. Women's and Infants' Health in Southern Africa

Country	Maternal Mortality per 100,000 Live Births	Women with Access to Prenatal Care (% of Total)	Infant Mortality per 1,000 Live Births
Botswana	300	71	62
Swaziland	400	76	76
Lesotho	350	50	82
Zimbabwe	330	83	61
Zambia	600	80	85
Namibia	400	82	73
Malawi	500	76	144
Mozambique	800	54	149
Angola	900	27	128

Source: Compiled from UNDP, 1993.

with literacy skills, these women have the potential for influencing policy. In the other three countries, two-thirds or more of adult women are nonliterate, making them dependent on literate relatives for supplying information acquired through print media. These women are effectively shut out of elected office and other policymaking positions. In several SADC countries, a literacy test in the dominant national language or in English (or Portuguese) is required to stand for election.

Finally, it is significant that in all six countries for which adult literacy data exist, women trail men by upward of 14 or more percentage points, indicating a gender gap within each country and for the region as a whole. Until more women achieve literacy, their comparative disadvantage with men will persist, crippling their ability to participate in civil decision-making arenas at all levels.

School enrollment rates are another measure of women's status. The Human Development Index gives net enrollment rates—the percentage of girls age 6–14 enrolled out of an estimated total number of girls of the same age group at the primary level—and gross enrollment rates—all girls enrolled regardless of their age—as a percentage of the relevant age group for the secondary level. Botswana has the highest net primary enrollment ratio (93 percent) and nearly the highest gross secondary enrollment ratio for girls (47 percent). Swaziland has the next-highest net primary enrollment ratio for girls (84 percent) and the highest secondary gross enrollment ratio (49 percent). The comparable figures for Lesotho are 76 percent and 31 percent. UNDP statistics for girls' net enrollment at the primary level were not available for Zimbabwe, but the percentage of girls of the total number of children enrolled in the first grade in 1991–1992 was 50 percent (Government of Zimbabwe, 1992, p. 8). However, a disturbing trend in completion rates for girls has emerged that is partly attributable to falling education quality in the latter part of the 1980s. The percentage of girls who entered primary school in the first grade in 1985 who completed the last year (in 1991) was 78 percent; the comparable rates for those enrolling in 1986 and 1987 were 73 percent and 70 percent, respectively (UNICEF/Zimbabwe, 1994, p. 95). The gross enrollment rate for girls at the secondary level was 46 percent in 1990, an increase of more than 90 percent over 1980 (UNICEF/Zimbabwe, 1994).

Malawi, which had the lowest female primary enrollment rate of the southern African countries in 1980 (UNESCO, 1993) increased girls' enrollment considerably between 1987 and 1992 from 43.8 percent to 51 percent (USAID/Malawi, 1995, p. 15). The increase is due largely to the commitment that the government of Malawi has made to expanding opportunities for basic education, especially for girls, by curtailing school tuition fees and by providing scholarships for girls who do not repeat a grade (Robinson et al., 1994). However at the secondary level, Malawi has not done as well. Of the total enrolled at this level in the 1990–1991 academic year, only 34 percent were girls (Government of Malawi, 1991).

Participation rates for Mozambique indicate that 37 percent of girls are enrolled in primary school while the gross enrollment rate for the secondary level is just 40 percent. Data do not exist for Angola.

Net tertiary enrollment data show Lesotho leading the way with 6 percent of the relevant cohort (age 18–21) being enrolled. The net tertiary female enrollment rate in Zimbabwe was 2 percent in 1990 (see Table 1); the overall net enrollment rate, 7 percent (UNESCO, 1991). Women students accounted for 26 percent of the total university enrollment in 1992 (Government of Zimbabwe, 1994, p. 47). Although in Malawi women's net enrollment rate is less than 1 percent of the appropriate cohort, the percentage of female students enrolled in 1990–1991 was 24 percent (Government of Malawi, 1991). It is at the tertiary level that the gender gap in education is most obvious, in terms of both net and gross enrollments. Males gained access to education earlier than females in all independent states. Consequently, the pool of available female candidates eligible for tertiary education in southern Africa lags far behind the pool of male candidates. Until females achieve educational parity with males at lower levels, they will continue to trail far behind men at the highest levels. Educational achievement affects the type of employment open to women and men. As women in southern Africa trail men in educational achievement, they are less likely to find the high-paying jobs that many seek. Those without a primary education will have little chance of securing employment.

Labor-Force Participation and Women's Work

Labor-force-participation rates in southern Africa are deceiving. In Zambia, for instance, employment statistics account for only 25 percent of the total work performed, and this is high compared to other southern African nations (Serpell, 1993). Moreover, even though the labor-force-participation rate measures women's access to employment, in particular it is not a measure of women's total labor. Labor-force participation measures only exchange-value labor in the formal sector; it does not measure labor in the informal sector, where women often are found, nor does it measure women's uncompensated work (child care and household maintenance).

In 1990, the percentage of women in the total labor force was highest in Mozambique (48 percent) and lowest in Namibia (24 percent) and in Zambia (29 percent). Much of labor recruitment in the latter two countries has been connected with mining, a specialization that has been deemed inappropriate and unsafe for women. Mining continues to be an occupation dominated by men. Additionally, cultural biases against women working in the formal sector are likely to be a factor. In both countries, as well as other countries in the region, historically women who earned a living of their own, especially if they were single and living in urban areas, were equated with being "loose" or "prostitutes." In some areas, this perception still holds. It is a stereotype that is rooted in colonial patriarchal attitudes and is found in both Christian and Muslim beliefs. The notion of men as breadwinners dies hard.

Women's work in the informal sector, including income-generating projects, microenterprises, and their use-value labor connected with agricultural production and household maintenance, cries out to be measured. Until a broader method of measuring labor is devised, it is difficult to get a realistic picture of women's work in southern Africa and the role that it has in shaping women's status in each country.

Women's Political Participation

The only measure of women's political participation in the Human Development Index is women's participation at the national level in a state's legislative body. Other macrolevel measures that might be included, but are not, are the number of women who hold leadership positions in ministries and in national political parties. The number of women holding positions as permanent or deputy secretaries of ministries is limited in the southern African countries. Often, women have held positions as secretaries of community services (Malawi, Zimbabwe, and Lesotho) or as secretaries of education (Malawi, Mozambique, and Zimbabwe). In 1992, a woman became deputy secretary of labor in Malawi and remained in that position until her death in 1993. In Zambia, after the multiparty elections of 1991, one woman was appointed a cabinet minister and by 1994, two more had been appointed (Ferguson and Liatto-Katundu, 1994). There are six female permanent secretaries, two additional acting permanent secretaries, and one female High Court judge. In 1995 Malawi had one woman minister and two deputy ministers, as well as a woman High Court judge. In 1995 the Minister of Foreign Affairs in Mozambique was a woman. Few women hold key political party positions except through the "women's wing." However, an increasing number of women are running for legislative assembly posts.

For the SADC countries for which data are available on women's participation in national legislative assemblies, the former socialist states of Mozambique and Angola had the largest percentage of women in their national assemblies in 1990—16 percent and 15 percent, respectively. That women made inroads into legislative positions in the early postindependence period is partly due to political pressure from women's organizations to include gender equity as a goal of socialist transformation in these two countries.

In the multiparty elections held in Mozambique in October 1994, 65 of the 250 seats in the national assembly (26 percent) were won by women (European Parliamentarians for Southern Africa, 1995). However, women ran not as individual candidates but as part of a political party slate, with 51 of FRELIMO's candidates (40 percent) and 13 of RENAMO's candidates (11.6 percent) being women. One woman was elected from a minor opposition party. Because voters voted by party rather than electing individual candidates, Mozambican women look to their representation in appointed offices as being a more accu-

rate measure of their political advancement (Sheldon, 1994). Of the 20 ministers appointed by President Joaquim Chissano in 1995, only one Alcinda Abreu, was a woman; she was named minister of social welfare coordination. Four other women were appointed deputy ministers: of foreign affairs and cooperation, justice, planning and finance, and labor. Although the gender balance was better than it had been in the former administration, it disappointed Mozambican feminists because it fell short of Chissano's campaign promises to promote more women in government (AIM, 1995, p. 7).

Zimbabwe and Malawi follow the two former socialist countries in the percentage of women in the national assembly or parliament. In Zimbabwe, which gained independence in 1980, halfway through the United Nations Decade for Women, gender equity became a key goal of women. Foremost were political and legal rights. Women represented 12 percent of Zimbabwe's parliament a decade after independence, but an active campaign began in late 1994 under the auspices of the newly formed Women Voters Association in Zimbabwe (WOVAZ) to recruit more women to run for national office and to educate voters on gender issues and who the gender-sensitive candidates were who would initiate change. The elections in 1995 saw more female candidates challenging men than ever before, and in most cases, in both the primaries and the national elections, they won. Of 65 contested seats in the national elections held in April, women representing the ZANU-PF party won the 18 seats for which they stood as candidates. Although some women ran from the minor opposition parties, they were not elected. Subsequently, two more ZANU-PF women were appointed by President Robert Mugabe, giving women 13.3 percent of the total 150 seats in the ZANU-PF-dominated parliament. However, only one woman was appointed a cabinet minister.

In Malawi, in 1972 then president H.K. Banda instituted a policy designating that at least one-third of Malawi's parliamentary positions be reserved for women, mainly from his Malawi Congress Party's (MCP) women's wing. He ensured their representation through presidental appointment if they were not elected in adequate numbers. However, by the 1980s, the number of MCP women in parliament had dwindled to 10 out of 100 seats; as of August 1992, they accounted for 11 percent of the total. The May 1994 multiparty elections saw a shift, with the number of members of parliament increasing to 177, 10 of whom were women. Of this number, five were from the winning UDF (United Democratic Front) party, four were from the MCP, and one was from AFORD. Overall, the proportion of women in the parliament had shrunk to 5.6 percent. A new constitution approved in May 1994 contained provisions to address gender inequities in the national assembly. It was largely through women's groups, led by the National Commission on Women in Development (NCWID), that women participated in the formulation of the constitution. A constitutional conference in February 1995, at which the newly formed Society for the Advancement of Women made a case for increasing women's representation in the political process along with that of chiefs, reaffirmed a bicameral legislature that included a senate and proposed that the senate include two women and two chiefs from each district to better represent rural areas (Funk, 1995). Members of NCWID, nongovernmental organizations, church leaders, women's leaders, and chiefs signed a petition presented to parliament for the retention of the senate in the constitution. Parliament voted in March 1995 to include a senate that would be made up equally of women and men members. However, it put off elections for the senate until 1999. As of 1994, of the appointed positions, only one woman has been named a cabinet minister and one woman is a High Court judge.

Women represented 7 percent (in Namibia) and 5 percent (in Botswana and Zambia) of their national assemblies in 1990. In Zambia, in particular, women began to form political-consciousness-raising groups and to put women forward for election with limited success. A nonpartisan group, the National Women's Lobby Group (NWLG), largely made up of female lawyers, was formed in July 1991 to encourage women to run for national elections. It was given a negative reception by leaders in the then ruling UNIP party and the opposition MMD (Movement for Multiparty Democracy) party. To counter criticisms that it was an elite group out of touch with rural women, the NWLG began encouraging women to stand for positions in local government elections in November 1992 with some success (Liatto-Katundu, 1993). In the 1991 national elections, women won 7 seats in parliament; in 1994, 10 of the 150 MPs were women. Women have served in numerous cabinet posts, as both permanent and acting secretaries on the High Court, and in diplomatic posts. Wome also are represented on the 7-member Human Rights Commission and on the 24-member Constitution Review Commission.

Legal Rights and Priorities

In 1988, a group of women from six SADC countries (Malawi is not a member) formed the Women and the Law in Southern Africa group (WLSA) to identify, prioritize, and research critical legal issues related to women's social justice in the region with a view to improving women's status under the law. Funded initially by a grant from four Scandivanian development agencies, the group identified two key issues for investigation: maintenance law covering women and children in the region, and inheritance law. What distinguishes legal issues in the region is that dual systems of law exist: customary and statutory. Women are affected by and use these two systems differently to achieve their goals. This section briefly analyzes issues related to maintenance under dual systems of law and describes strategies women are using to expand their legal rights and, accordingly, their overall empowerment.

Family Maintenance, Marital Status, and the Law

Why is family-maintenance law critical to women in southern Africa? Partly, the concern comes from women's cumu-

lative experience of male emigration in the region—a phenomenon that has rendered impotent customary laws that formerly ensured the socioeconomic sustenance of a woman and her children. Also, women's combined experiences in nationalist liberation struggles and in struggles for democratization have fueled their demands for greater gender equity and effective legal rights. Women in the region cannot be treated as an aggregate category, however. One distinguishing factor is marital status. In some countries, such as Botswana, the majority of children (60 percent) are born to unmarried mothers (Molokomme, 1990), while in other countries mothers are more apt to be married. The divorce rate is higher in Malawi and Zimbabwe than it is in Mozambique, a predominantly Catholic country. Secondly, a woman's socioeconomic status and her embeddedness in a rural or urban context influence the way she understands and uses the two systems of law. As the vast majority of women in southern Africa are rural peasants, both WLSA and, in Malawi, NCWID are especially concerned that rural peasant women be educated about, and empowered in, the use of their legal systems for their own advancement.

Customary laws pertaining to marriage and maintenance, although they differ among ethnic groups, are based on the assumption that husband and wife have a reciprocal responsibility to maintain and support each other and their children. This assumption covers both patrilineal and matrilineal groups. However, that children belong to the father's lineage or kin in patrilineal groups through the transfer of bride wealth from the husband's family to the wife's family means that the onus of support, in theory, is on the husband. The wife theoretically has no rights to her children once they are weaned unless laws of custody are enacted that change this situation. At the same time, countries vary in the way customary laws pertaining to a mother's rights to her children are practiced. In some matrilineal groups, where sizable bride wealth is transferred from a husband's to a wife's family, a husband may earn the right to have custody of his children and he has rights over many aspect of their lives in consultation with the mother's brothers. In matrilineal groups where matrilocality whereby a man on marriage moves to the wife's mother's home village or uxorilocality whereby a man moves to his wife's village on marriage are practiced and no bride wealth is transferred, children belong to a mother's lineage. In cases of divorce or nonmarriage, custody of a woman's children and their maintenance usually are the responsibility of the matrikin. The father, in theory, has no rights or maintenance responsibilities to the children, although in practice this may vary depending upon the resources of the husband and his desire to exercise parental rights. The next section outlines some of the issues women face specifically in negotiating dual systems of law.

Dual Systems in Botswana, Lesotho, and Swaziland

Southern African countries that ascribe to patrilineal customary law include Botswana, Lesotho, and Swaziland. In Botswana, the husband/father is required to supply wives and their children with the means to support themselves, including fields for cultivation, seeds, and animal draft power. A wife, in turn, provides household maintenance and sustenance. This reciprocal arrangement breaks down, however, with male emigration. Where a wife is not adequately supported, she can report the matter to her parents and a meeting will be called of members of the couple's two patrilineages to work out a resolution. An absent husband may be asked to pay a certain amount in remittances if he is employed. But no mechanism within customary law exists for enforcement. A similar situation exists in Lesotho and Swaziland.

The situation for unmarried mothers under customary law is even more fragile. Whereas paternity is increasingly illusive in Botswana, in Swaziland the social pressure on a father to marry the mother of his child—once paternity is established—is fierce. Alternatively, a man may try to "buy" his children by an unmarried Swazi woman by giving cattle to the woman's parents, a threat that prevents unmarried mothers from seeking maintenance. For divorced women, children remain with their paternal relatives while their mother leaves the father's patrilineal homestead to return to her own home.

In all three countries, enforcement of customary rules related to family maintenance is weak, especially if the father is a migrant laborer. Statutory law makes provision for conditional enforcement. A married woman's right to maintenance in Botswana is protected by the Deserted Wives and Children Protection Act. Similar codes, drawing on Roman-Dutch law, are in place in Lesotho and Swaziland. However, whereas in Botswana only wives and children are included, the law in Lesotho covers a broader range of relatives. The law also extends to illegitimate as well as legitimate children where paternity can be proved. In Lesotho, if a wife's husband fails in his duty to support and she can prove desertion as well as destitution, she may be able to pledge her husband's credit without his consent or the court may attach his wages. Proving desertion is the linchpin.

In Swaziland, the law pertaining to maintenance requires that a wife who seeks support under the Maintenance Act be assigned a social-welfare officer to investigate her financial situation and to assist her in obtaining food and shelter. Swazi women are averse to using this law to seek maintenance, however, because they feel humiliated having to air their problems before a public officer (Maphalala, 1990).

One of the problems facing women is that their husbands may have emigrated to work or may have a new address. If a husband does not appear in court, a wife is frustrated in her attempt to gain support. Botswana's system allows for civil action against a recalcitrant or absent husband, usually in the form of garnishment from an employer or a lump sum to be paid. In Lesotho and Swaziland, a criminal action—imprisonment—can also be brought against the husband, but the courts discourage women from using it (Monyamane, 1990).

For divorced women and their children in Botswana, the Matrimonial Causes Act is designed to make sure that a husband's duty of support does not end when the marriage does. For women in Botswana and Lesotho, the problem lies in obtaining ongoing maintenance since customarily when a woman leaves her husband she takes half of the communal property as her share of the settlement. It is assumed, even by women who bring charges, that the division of communal property is sufficient.

For unmarried mothers in Swaziland and Lesotho, the procedure is the same as for married mothers. However, in Botswana, where children of unmarried mothers predominate, a separate law applies. (It also applies in other southern African countries such as Malawi, Zambia, and Zimbabwe.) Referred to as the Affiliations Proceedings Act, it states that the natural father has a duty to support his children once paternity can be confirmed. A mother must bring a paternity case before the courts within 12 months of the birth of the child or within 12 months of a father's return to the country if he is employed elsewhere. A problem for women in all southern African countries where this law exists is that many are unaware of the 12-month limitation; once it elapses, their only recourse is through the customary system. A monthly sum is set for maintenance. It varies by country, but in all cases it is far below what most women need to maintain their families, a situation that women in the region are addressing. Some middle-class women avoid maintenance courts (either statutory or customary) for fear of losing control over their children in an adverse decision. Another problem is that, although a man may be ordered to pay, he often fails to do so. A study in Botswana found that only 36 percent of men ordered to pay maintenance actually paid the amount for more than six months (Molokomme, 1990). Even in the statutory system, enforcement remains an issue. Finally, women's fears of appearing in magistrate courts, which seem on the surface to be more formal and bureaucratic than customary law courts, also works against them. Particularly nonliterate women have difficulty approaching the civil-law courts.

Dual Systems in Mixed/Matrilineal Societies: Zambia and Malawi

Zambia's Local Courts Act, which pertains to customary law, is silent on the subject of maintenance. Children belong to, and are affiliated with, either the mother's lineage (matrilineal groups) or the father's lineage (patrilineal groups) regardless of a parent's marital status. These kin groups are responsible for children's maintenance. In cases of divorce, in theory neither the in-marrying husband nor the wife may seek maintenance under customary law; it is assumed they will return to their home village. In practice, the situation is more ambiguous and the in-marrying spouse may remain and also seek maintenance.

Civil law pertaining to family maintenance in Zambia is outdated. It derives from British statutes that date back, in some cases, to the nineteenth century, such as the Bastardy Laws Amendment Act of 1872 and the Summary

Jurisdiction (Married Women) Act of 1895. The former makes it possible to bring a case in a magistrate's court in order to adjudge a man's relation as a putative father and order him to make weekly or monthly payments to the mother or child. Under the second law, a wife can bring a case of neglect by her husband to the court and seek maintenance. She does not need to show destitution. In addition, the Maintenance Orders Act enables a woman to enforce maintenance from her husband by garnisheeing his earnings. The Matrimonial Causes Act of 1950 makes it possible for a divorced woman to seek alimony, maintenance, and custody of children from a former husband. Zambian women's greatest problems lie in their lack of knowledge of how to use the civil legal system. The British statutes need to be updated, simplified, and made more relevant to women's needs.

Whereas Zambia has mixed patrilineal and matrilineal customary laws, Malawi is predominantly a matrilineal country. Its matrilineal heritage has influenced its dual system of customary and civil law. A married Malawian woman retains her right to own and control property independent of her husband (NCWID, 1993). She can also carry on a business in her own name. Under customary law, a man may take more than one wife, similar to other southern African countries. However, polygyny is rare because it requires a man to provide bride wealth, land, and a house for each wife and maintenance for each wife's children. The Marriage Act passed by the Malawian Parliament in 1970 is more explicit: It allows only for monogamous marriage. Cohabitation is not recognized as marriage.

To divorce under customary law, one of five conditions must be present: (1) adultery on the part of either spouse, (2) lack of financial support, including failure to pay children's school fees, (3) desertion, (4) cruelty, and (5) incompatibility. In the last case, the spouse choosing to leave may return to his or her natal home provided compensation has been paid to the other party for breach of faith.

In the civil system, the Divorce Act stipulates that a person may not bring a divorce petition before the High Court in the first three years of marriage unless the wife is a victim of hardship, or the husband is guilty of depravity (NCWID, 1993). After three years, causes for divorce are similar to those of customary law: desertion, cruelty, or adultery. Under cruelty, cause for divorce includes physical beatings, assault or threat of assault, accusations (as of witchcraft), unreasonable tracking or spying, and infection from venereal disease. Lack of financial support is covered by a Maintenance Act similar to those found in other southern Africa countries.

Rights to children in the case of unmarried mothers belong to a mother and her matrikin in matrilineal groups and to fathers and their patrikin in patrilineal groups. Damages for making her pregnant are paid to the woman's parents. Under civil law, unmarried mothers are covered by the Affiliation Act, which is similar to that found in other southern African countries: A father is summoned and, if he does not appear, a warrant for his arrest can be brought; if the man can be shown to be the putative father, the court will assess his income and order maintenance payments.

Malawi has a dual system of law that in some ways better serves the maintenance needs of women than the dual systems of Botswana, Lesotho, and Zambia. However, areas exist for reform. The National Commission on Women in Development wants legislative action taken to standardize the grounds for divorce so that it more closely resembles "no-fault" divorce (NCWID, 1993). The group also wants jurisdiction for divorce expanded to include magistrate's courts so that women will have the option of getting a divorce at this level rather than going to Malawi's High Court. Regarding unmarried mothers, the group has recommended that the civil-court system, similar to Zambia's, be allowed to garnishee a putative father's salary to ensure maintenance of his child.

Laws at Cross Purpose: Mozambique

The maintenance law in Mozambique differs in that it does not recognize a separate customary-law system, although customary practices are acknowledged and sometimes run contrary to civil law. Maintenance law fits under civil law. However, a separate law, the Draft Family Law, applies to cases of separation and divorce.

A patriarchal perception persists in Mozambique that women need protection, provided by either father or husband (Pinto and Chicalia, 1990). It stems from this country's historical encounter first with Islam and later with the Portuguese. Awards for maintenance depend on a woman's moral conduct rather than her need, and, for single mothers and their children, maintenance is provided only through the first year of a child's life.

If a maintenance order is requested on its own, it is handled by a children's court. If it is connected to a divorce, it is handled by a city or provincial civil court. The latter disadvantages the majority of women who live some distance from cities or provincial centers. Local courts, which are more informal, exist, but they do not handle maintenance hearings. Rural semiliterate women have to go armed with formal documents that they may hardly be able to read themselves; they are usually ignorant of the law. Another constraint is that women who are economically dependent on men in urban areas and among patrilineal groups are hesitant to seek divorce even though they may have legitimate grounds. Divorce law is not applied uniformly. Under the civil code a court decides guilt, while under the Draft Family Law consideration of fault in a divorce is waived. This is confusing to women (and men) and does not help to clarify Mozambique's system to make it more accessible. The Women and the Law group in Mozambique wants maintenance law to be aligned with the Draft Family Law so that it will encourage women to use the legal system (Pinto and Chicalia, 1990).

Progress in Zimbabwe

Zimbabwe, prodded often by its women, has enacted substantive reforms in its legal systems. The Customary Law and Primary Courts Act of 1981 took the quantum step of codifying Zimbabwe's customary law. Its civil law is derived partly from Roman-Dutch law and partly from British law. Maintenance, in theory, is viewed as support given by spouses to one another and to their children. However, not uncommon for the region, in practice, 90 percent of all maintenance claims are brought by women against men (Gwaunza and Zana, 1990).

Under customary law, both parents have the responsibility for maintenance according to their means. Under the African Marriages Act of 1981, reciprocal maintenance of spouses is understood. When a couple divorces under this law, the communal property is divided so that the person with less means gets more property to make the division more equitable. This act also enables an unmarried woman to claim maintenance and provides that a child will be supported up to age 18. Men have resisted complying with this part of the law, believing that they are exempt from duty of support under what was previously uncodified customary law. By and large, women have custody of children on divorce.

The Matrimonial Causes Act of 1985 is unique in southern Africa; it empowers a court to award a maintenance order in case of divorce in favor of children—for their best interest. Fault is not an issue. The court has the right to refuse a divorce petition until it is satisfied that suitable arrangements have been made regarding the children's custody and maintenance. This includes an assessment of the family's income, members' financial needs and obligations, and the family's standard of living. It also includes an assessment of the physical and mental conditions of the parents and children.

Under civil law, the Maintenance Act of 1986 provides that a person who has custody and care of a dependent child can make a claim for maintenance. In terms of enforcement, the maintenance court not only may garnishee the wages of a parent but also may bring a criminal case for imprisonment up to one year if the parent proves delinquent. Even though bringing a case for maintenance to court is time consuming and expensive for women, they do use the system. In a three-year period between 1987 and 1989, 6,175 cases were brought to community courts and 4,538 were brought to magistrate's courts in Harare (Gwaunza and Zana, 1990). As women become more educated in the civil law, they are more likely to use it.

By coming together across national boundaries to pursue issues related to women's legal status, Women and the Law in Southern Africa has drawn regional attention to women's needs for greater justice. Its efforts have paved the way for women's collaboration on other pertinent issues, including women and the environment and women's political action, that affect the lives of southern African women.

Conclusion

This overview of southern African women illuminates both their diversity and similarities in the region. It also demonstrates that women's status in various countries does not reduce to a single indicator. Botswana is a country blessed with mineral resources that enables it to invest in social services. As such, girls and women have access to education, with girls outstripping boys at the primary level. Swaziland is not far behind. Botswana also has the best health record for women. Yet, labor-force-participation rates for women are low in both countries.

In patriarchal Botswana, married men generally have custody over their children, regardless of whether a wife is in residence or not. Divorced and single-status women must either depend on the good offices of a husband's patrikin through the customary legal system or petition a civil court for maintenance. As the proportion of single mothers is high in Botswana, the potential for maintenance proceedings is substantial. Being better educated, Batswana women are better able to use the system than women in Mozambique, for instance, most of whom have little education. Yet, the laws with which Batswana women have to cope—in terms of proving paternity, desertion, and/or destitution—are formidable. To push for reform, women need greater political power. One way of increasing their power is to regionalize so that their voices are multiplied with other women's voices. The Women and the Law in Southern Africa project has provided a model for such a forum.

At the same time, the analysis of maintenance law illuminates the critical role that education plays in liberating women. Without some schooling, a woman cannot read or write, and without these literacy tools, she must depend on others to read and sign legal documents. Particularly to rural peasant women, who lack knowledge of how to negotiate the formal bureaucracies that characterize government, the civil-court system appears insurmountable. Civil law needs to be humanized so that it comes closer to customary law in feeling and flexibility, and customary law needs to be codified so that it has more clout. Bringing the two into alignment, as Zimbabwe has begun to do, will help women feel less reluctant to use the system. At the same time, laws are not immutable; they must appropriately reflect changing social conditions. This calls for constant reevaluation of existing laws to ensure that women's human rights are protected and for legal education so that women of all walks of life know how to use laws to prevent their victimization and increase their opportunities for empowerment.

Negotiating the kaleidoscopic changes that have characterized southern Africa in the last decades of the twentieth century has been a major task of women in the region. Women for Change, a nongovernmental organization in Zambia, helps men and women collaborate in identifying, analyzing, and resolving problems such as domestic violence and women's triple workload. Similar NGOs are emerging throughout southern Africa and are proving effective. In Maseru, Lesotho, a group of women representing most of the southern African states came together in 1991 to tackle environmental problems that women in the region face as procurers of fuelwood, water, and other natural resources. Reversing the trend toward environmental degradation was part of their agenda. Other regional efforts are surfacing. An increased sense of empowerment means that women in southern Africa will not stand back; they will be at the forefront of the changes taking place in their individual countries and in the region as a whole.

References

AIM (Mozambique Information Agency). *Mozambiquefile,* no. 222, January 1995, pp. 4–7.

Ajulu, Rok, and Diana Cammack. "Lesotho, Botswana, Swaziland: Captive States." In P. Johnson and D. Martin (eds.), *Destructive Engagement: Southern Africa at War.* Harare: Zimbabwe Publishing House, 1986, pp. 139–169.

Bowen, Merle. "Socialist Transitions: Policy Reforms and Peasant Producers in Mozambique." In T.J. Bassett and D.E. Crummey (eds.), *Land in African Agrarian Systems.* Madison: University of Wisconsin Press, 1993, pp. 326–353.

Chauncey, G. "The Locus of Reproduction: Women's Labour in the Zambian Copperbelt," *Journal of Southern African Studies,* vol. 7, no. 2, 1981, p. 135–164.

Davison, Jean. "Land Redistribution in Mozambique and Its Effects on Women's Collective Production: Case Studies from Sofala Province." In J. Davison (ed.), *Agriculture, Women, and Land: The African Experience.* Boulder: Westview, 1988, pp. 228–249.

———. "Changing Relations of Production in Southern Malawi: Implications for Involving Rural Women in Development," *Journal of Contemporary African Studies,* vol. 11, no. 1, November 1992, pp. 72–84.

———. "Matriliny and the Durability of 'Banja' Household Production in Southern Malawi," paper presented at the annual meeting of the African Studies Association, Toronto, November 3–6, 1994.

European Parliamentarians for Southern Africa. *Mozambican Peace Process Bulletin,* no. 15, April 1995.

Ferguson, A., and B. Liatto-Katundu. "Women in Politics in Zambia: What Difference Has Democracy Made?" *African Rural and Urban Studies,* vol. 1, no. 2, 1994, pp. 11–30.

Funk, Stephanie. "Women and the Constitution: An Agenda for Fair Representation and Equal Protection." Paper presented at the National Constitutional Conference, Lilongwe, February 20–23, 1995.

Government of Lesotho, Bureau of Statistics. *Basuto Women and Their Men.* Maseru: Government Printing Office, 1989.

Government of Malawi. *Education Statistics, 1991.* Lilongwe: Ministry of Education and Culture, 1991.

Government of Zimbabwe. *Report of the Secretary for Education and Culture.* Harare: Ministry of Education and Culture, 1992.

———. *An Analysis of Developments in the Education Sector, 1990–1994.* Harare: Ministry of Education and Culture, 1994.

Gwaunza, Elizabeth, and Esther Zana. "Maintenance in Zimbabwe." In *Working Papers on Maintenance Law in Southern Africa.* Harare: WILSA, 1990.

Heisler, H. *Urbanization and the Government of Migration: The InterRelations of Urban and Rural Life in Zambia.* New York: St. Martin's, 1984.

Kaufulu, Febbie. "A Comparison of Women's Use-Value and Exchange-Value Labour in Three Areas in Zomba District." Master's thesis, University of Malawi, Zomba, 1992.

Laakso, Liisa. *Relationship Between the State and Civil Society in the Zimbabwe Election 1995.* Helsinki: University of Helsinki, 1995.

Liatto-Katundu, B. "The Women's Lobby and Gender Relations in Zambia," *Review of African Political Economy,* no. 56, 1993, pp. 79–102.

Longwe, Sarah H. "From Welfare to Empowerment," paper presented at the NGO African Women's Task Force meeting, Nairobi, April 11–15, 1988.

Maphalala, Lucia. "Child Maintenance Law in Swaziland." In *Working Papers on Maintenance Law in Southern Africa.* Harare: WLSA, 1990.

Molokomme, Athaliah. "Maintenance in Botswana." In *Working Papers on Maintenance Law in Southern Africa.* Harare: WLSA, 1990.

Monyamane, S.N. "The Law of Maintenance in Lesotho." In *Working Papers on Maintenance Law in Southern Africa.* Harare: WLSA, 1990.

Moore, Henrietta, and Megan Vaughan. *Cutting Down Trees: Gender, Nutrition, and Agricultural Change in the Northern Province of Zambia, 1890–1990.* Portsmouth: Heinemann; London: James Curry, 1994.

National Commission on Women in Development (NCWID). *Women and the Law in Malawi.* Lilongwe: NCWID, 1993.

Organization of Angolan Women. *Angolan Women Building the Future.* Translated by M. Holness. London: Zed, 1984.

Parpart, Jane L. "Class and Gender on the Copperbelt: Women in Northern Rhodesian Copper Mining Communities, 1926–1964." In Claire Robertson and Iris Berger (eds.), *Women and Class in Africa.* New York: Holmes & Meier, 1986, pp. 141–160.

Pinto, Ana P., and Isabel Chicalia. "Maintenance Law in Mozambique." In *Working Papers on Maintenance Law in Southern Africa.* Harare: WLSA, 1990.

Pottier, J. *Migrants No More: Settlement and Survival in Mambwe Villages, Zambia.* Manchester: University of Manchester Press, 1988.

Robinson, B., J. Davison, and J. Williams. *Malawi Education Sector Analysis.* Lilongwe: USAID/Malawi, 1994.

Schuster, I. *New Women of Lusaka.* Palo Alto: Mayfield, 1979.

Serpell, Robert. *The Significance of Schooling: Life Journeys in an African Society.* Cambridge: Cambridge University Press, 1993.

Sheldon, Kathleen. "Women and Revolution in Mozambique: A Luta Continua." In Mary A. Tetreault (ed.), *Women and Revolution in Africa, Asia, and the New World.* Columbia: University of South Carolina Press, 1994, pp. 33–61.

United Nations Childrens Fund (UNICEF)/Zimbabwe. *Children and Women in Zimbabwe: A Situation Analysis Update, 1994.* Harare: UNICEF, 1994.

United Nations Development Program (UNDP). *Human Development Report 1992.* New York: UNDP, 1992.

———. *Human Development Report 1993.* New York: UNDP, 1993.

United Nations Educational, Scientific, and Cultural Organization (UNESCO). *World Education Report.* Paris: UNESCO, 1991.

———. *Statistical Yearbook 1993.* Paris: UNESCO, 1993.

United States Agency for International Development (USAID)/Malawi. *Girls Attainment of Basic Literacy and Education Program.* PAAD document. Lilongwe: USAID/Malawi, 1995.

Urdang, Stephanie. "The Last Transition? Women and Development in Mozambique," *Review of African Political Economy,* no. 27–28, 1984, pp. 8–32.

———. *And Still They Dance: Women, War, and the Struggle for Change in Mozambique.* New York: Monthly Review Press, 1989.

Women in South Africa

Zelda Groener

Introduction

The history of women's oppression in South Africa predates the apartheid era. Gender inequality accompanied colonialism. The institutionalization of apartheid brought with it a racial division among people, and class and racial divisions among women. Apartheid created a complex of laws and customs that organized the labor force to permit a particular type of industrialization, to protect a particular category of workers (Whites), and to make it difficult for the majority of the labor force (Blacks) to organize to change its conditions; it also created a coherent cluster of values and cultural practices that reinforced solidarity among one part of the population and justified the discriminatory legislation. Apartheid linked racial discrimination with the natural order so that enforcing apartheid became a moral responsibility of citizenship. With the changing political economy and the legal, moral, and cultural context it created in South Africa, apartheid affected women's roles.

Under apartheid, legislation prevented Black women from living with their migrant spouses. Limited access to schooling and tertiary education, coupled with high attrition rates, relegated girls to the lowest echelons of the labor force. This oppression and exploitation of women prompted the formation of women's organizations. Among the first major Black women's organizations was the African National Congress Women's League (ANCWL), established in 1945, followed by the Federation of South African Women (FEDSAW) in 1956. In decades since, both engaged in mass protests, pickets, and boycotts. During this time, women directed their struggle not only against macro political and economic forces reproducing male domination, but also against male domination in antiapartheid organizations such as civic groups, trade unions, and political parties. During the transitional period toward democracy, 1990–1994, negotiations created new challenges for women's organizations, and women mobilized on several fronts to ensure that women's issues were placed on the political agenda of the negotiating forums. A victory was scored when the interim constitution of the Government of National Unity (GNU) in April 1994 stipulated that 30 percent of the positions in parliament be filled by women. Today, the more visible position of women in leadership positions provides an impetus for women challenging male domination and gender oppression in political institutions, educational institutions, and the home. As the constitutional nonracial, nonsexist democracy unfolds, women continue to challenge male domination and gender oppression and to redress gender inequalities in schools, universities, the workplace, and the military.

This essay is organized into four sections: (1) apartheid and the exploitation and oppression of women it created; (2) women's organizations and the struggle against exploitation and oppression; (3) women, politics, and gender-policy strategies during the period of transition toward democracy; and (4) gender-sensitive policies in a nonracial, nonsexist democracy. It argues that the composition of the labor market in South Africa reflects the class, race, and gender relations in educational institutions and suggests that changes in gender relations in educational institutions could affect changes in policy, the economy, and the home. Educational institutions, as intellectual institutions, play an instrumental role in generating and disseminating new ideas. Educational institutions in South Africa have been at the forefront of the struggle against apartheid and in the conceptualization of a nonracial, nonsexist democracy. Women's organizations, in particular, have been pivotal, not only as organs that have addressed women's issues but also as groups through which women have participated in the political struggle as a whole.

Feminist thinker bell hooks has observed that:

Feminism, as liberation struggle, must exist apart from and as part of a larger struggle to eradicate domination in all its forms. We must understand that patriarchal domination shares an ideological foundation with racism and other forms of group oppression, that there is no hope that it can be eradicated while these systems remain intact (hooks, 1989, p. 22).

The class, race, and gender inequalities in South Africa are often referred to as the "triple oppression," regarded as rhetoric as well as the basis for theoretical analysis. Charman et al. (1990) quote Walker:

That black women in South Africa suffer a triple oppression, of gender, race and class has become a rhetorical commonplace. White women, too, it is generally recognized, are discriminated against as women, although their membership of a privileged racial group softens the impact of gender discrimination and works against their identification with black women as women, with shared problems (Charman et al., 1990, p. 1).

Charman et al. argue that the theoretical analysis of race, class, and gender has obscured the issue of the subordination of women to men.

With regard to racial terminology, it should be noted that the apartheid government classified people into four racial groups: African, White, Colored, and Indian. Under the present nonracial, nonsexist democratic dispensation, the usage of the apartheid classification still continues. In this presentation, the term "Black" is used as a political construct to include African, Colored, and Indian. The terms "African," "Colored," and "Indian" will be used only to illustrate a particular phenomenon.

Class, Race, and Gender Inequalities Under Apartheid

Apartheid is a political philosophy that is based on beliefs that White people are superior to Black people. Embraced by the National Party, apartheid became the official ideology when it came to power in 1948. By the 1960s, this ideology had pervaded all forms of legislation in every sphere of life. Draconian apartheid laws not only stripped Black women of political rights, it also undermined and violated their fundamental human rights. Laws denied Black women and men the right to vote for a national government and to choose where to live, as well as access to schools, universities, and medical facilities of their choice and, in some instances, to any at all. Laws also denied African women the right to live with their migrant husbands. Several examples illustrate these conditions: Hospitals had separate entrances, separate wards, and different ambulances for people from different racial groups. Black students were required to apply for a special permit from the government when they wished to study at White universities; the number of permits issued to Black students was limited. African women were allowed to live in cities with their husbands only if they were granted a work permit. These permits were denied to most African women, compelling them to live in "homelands," residential areas designated for African people (similar to "Indian reservations"

in North America). While much of the apartheid legislation affected Black women and men alike, an analysis of class, race, and gender inequalities in education and the economy indicates that African women were most exploited and oppressed. Madlala (1994) describes the plight of Black women in South Africa:

The system of racial, class, and gender domination has effectively marginalized the majority, excluding them from major decisions affecting their lives. In the work-place they are abused and exploited by their bosses. In the community they suffer political and social repression. In their homes they experience domination and violence from their male relatives— their husbands, partners, brothers, and sons (Madlala, 1994, p. 7).

Inequalities in the Labor Force

Women constituted just over 50 percent of South Africa's total population in 1990 and about 36 percent of the labor force, excluding women in the former homelands. Of the women in the labor force, 60 percent were African, 23 percent White, 15 percent Colored, and 3 percent Indian (Budlender, 1991, p. 6; Commonwealth Secretariat, 1991).

Race and gender have determined the nature of Black and White women's employment in various economic sectors. African women made up 82 percent of the service sector and 85 percent of the farming sector. White women constituted 50 percent of all women workers in the sectors defined in the national plan. Colored women constituted 35 percent of women in production and 31 percent of those in unskilled work. Indian women made up 8 percent of the women in manufacturing (Budlender, 1991, p. 9).

The actual jobs done by women of the different racial groups within each sector and industry differ widely. White women accounted for half of the professional and trade-related women, two-thirds of the clerical, and almost all (88 percent) of the managerial. Within the service sector, 23 percent of the White women employed were classified as supervisory housekeepers and another 24 percent as beauticians. Also within the service sector, 49 percent of Indian women, 83 percent of Colored women, and 88 percent of African women were classified as domestic workers, cleaners, or chars. Therefore, there was a discernible occupational hierarchy: White women predominated in the more skilled, more pleasant, and better-paid jobs, Colored and Indian women filled the semiskilled occupations, and African women performed the unskilled jobs, particularly domestic service (Budlender, 1991).

The National Union of Metalworkers of South Africa (NUMSA) conducted a survey among its women members in 1988.

At one factory employing 45 women and 25 men, over 40 of the women were classified as operators and earned R144.00 per week. . . . The men classified as operators earned R171.45 per week. . . . The NUMSA survey also showed that those factories employing proportionately more women had lower minimum wages than those with more men. In factories where no women were employed the average minimum wage was R136.35 per week; for factories where less than half the work-force were women, it was R134.55; and for those where more than half the workers were women the average was R91.35 (Budlender, 1991, pp. 20–21).

According to Berger (1992), a large number of Black women entered industrial jobs, primarily in clothing and textiles, in the 1980s, and some female labor leaders in these industries have presented feminist perspectives. It remains to be seen how these women may influence postapartheid society, particularly in showing the connections between women's oppression at work, at home, and in the wider community.

Although the postapartheid economy is only in the infant stage, and few studies have occurred thus far, some questions can be posed: Are there likely to be changes, for example, in patterns of migrant labor and thus in the situations of women who are left behind to manage rural households or who try to follow their men to urban areas? To what extent is housing policy linked with thinking about changes (desired or undesired) in family relationships and employment? Are there to be special policies for women engaged in sidewalk or factory-gate sales and other sectors of what is called the informal economy?

Inequalities in Education

Historically, there have been high attrition rates among African girls, but this has changed dramatically. Statistics show that for every 100 White girls who started Standard 6 (grade 6) in 1970, 58 reached Standard 10 (grade 10) by 1974, compared to 2 of every 100 African girls. By 1990, the comparable figures for White and Black girls were 83 and 69, respectively (Truscott, 1994, p. 16). A 1990 head count at South Africa's 18 universities showed that Black women constituted 21 percent of the universities' total population compared to White men, who constituted 33 percent (Truscott, 1994, p. 92). With regard to the employment of professional women, in 1989 there were no Black women engineers (compared to 311 White women engineers); 234 Black women natural scientists (8,179 White women natural scientists); and 26 African women lawyers (1,348 White women lawyers) (Unterhalter et al., 1992).

Illiteracy rates reveal major differences by ethnic group. Official figures show that, in 1985, 67.3 percent of African women were illiterate, compared to 0.82 percent of White women (Unterhalter et al., 1992, p. 80). Bantu education under the apartheid system provided separate public education for Black Africans, an education characterized by substantially inferior programs.

Women's Organizations and the Struggle Against Exploitation

Throught the history of the oppressed people's struggle against apartheid, women's organizations have been pivotal, not only as organizations that have addressed women's issues but as entities through which women have participated in the political struggle at large. Over the years, these groups have ranged from mass-based organizations involved in national protests to small community-based groups responding to local needs.

During the period of apartheid, some urban women became involved in brewing illegal liquor and managing environments where it could be sold and drunk. These shebeens (speakeasies), generally located in homes, became a major feature of African urban communities, and their managers, the shebeen queens, became important—often political as well as economic and social—figures in the African community. Beyond their obvious social and escapist roles, the shebeens provided venues for young musicians (Miyriam Makeba, among others), gangsters, and others. Some of the most dramatic political mobilization among women during the apartheid period was focused on government efforts to stamp out the illegal liquor and the shebeens. Women's opposition to the requirement that they carry passes (men had been required to have passes for some time) reflected in part women's concerns about how the requirements would affect their economic roles and opportunities. When White military students burned the urban beer halls in the late 1970s (the students saw them as part of the strategy for maintaining White rule), they were less clear on what to do about the shebeens. In general, the students condemned escape into alcohol. At the same time, they recognized the important community roles of the shebeens.

The experiences of exploitation and oppression have provoked women, particularly Black women, to mobilize and organize on several fronts. Key figures involved in women's organizations and campaigns include: Nontsikelelo Albertina Sisulu, a veteran activist in the struggle against apartheid, founding member of the Federation of South African Women (FEDSAW), and former president of the United Democratic Front; Winnie Mandela, a prominent figure in women's struggles and former chair of the ANC's Women's League; Ray Alexander, a labor union leader; Ruth Mompati, a founding member of FEDSAW and member of the ANC's National Executive Committee; and Lilia Ngoyi, leading member of the ANC's Women's League and founding member of FEDSAW.

Traditionally, mass-based women's organizations have been led and dominated by Black women. Hence, many of the issues these organizations have addressed characteristically affect the lives of Black women. Leading women's organizations include: the African National Congress Women's League (ANCWL), which, during apartheid, contested policies that relegated Black women to the lowest echelons of society; FEDSAW, which struggles against apartheid, racism, sexism, and capitalist exploitation; the

United Women's Organization (UWO); the United Women's Congress (UWCO); the Natal Organization of Women (NOW); the Federation of Transvaal Women (FEDTRAW); the Federation of South African Women–Western Cape Region; the National Council for Women; the Women's Brigade; and the National Women's Coalition (NWC), which seeks to mobilize women across the entire political spectrum. At particular moments in history when women have organized national campaigns, there was a strong sense of a women's movement or, as some people may term it, a feminist movement. Yet, the majority of women activists in the country do not refer to themselves as feminists or to the women's movement as a feminist movement. This arises in part from the resistance of Black women to the concept of feminism and its historical association with White Western feminism.

Protests and Pickets: 1950–1990

Since the 1950s, women's organizations have conducted numerous campaigns, protests, pickets, boycotts, and marches related to a host of women's issues, such as maternity benefits, oppressive legislation, unequal pay, and sexual harassment. When the apartheid government promulgated pass laws during the 1950s requiring that African women carry passes, it met with resistance. Throughout this decade, Black women organized antipass campaigns, culminating on August 9, 1956, when approximately 20,000 African women marched to the Union Buildings in Pretoria. Under the leadership of FEDSAW, these women delivered a petition to the prime minister in an attempt to halt the oppressive pass laws. Despite continued protests, laws were instituted in 1961. With the banning of the ANC in 1961, the ANCWL and other leading women's organizations suspended their operations during the decade, which became known as the "silent 60s," and for much of the 1970s.

At the beginning of the 1980s, social movements, including women's organizations, proliferated throughout the country (Matiwana et al., 1989). Concepts such as sexism, male domination, women's subordination, gender, gender oppression, and feminism became more popular. It is significant to note that women's organizations, particularly during the early 1980s, took up issues that were not exclusively women's issues. These included rent increases, evictions, bus-fare increases, and education crises. A reason for this development was the banning of political organizations such as the ANC, the Pan Africanist Congress, and the South African Communist Party, which left a political vacuum—in their absence, women's organizations addressed the issues in question.

During the 1980s, the struggle against apartheid took on another form as the struggle for a nonracial and non-sexist democracy. The struggle for nonsexism thus became an integral part of the democratic movement's struggle for democracy. This particular formulation created a number of problems of its own. The struggle against exploitation and women's oppression was reduced to the struggle against sexism. In the way that nonracial, nonsexist democracy was

presented by the democratic movement as the alternative to apartheid, notions of nonracialism, nonsexism, and democracy appeared to have been conflated. During the 1980s, women began to turn their attention to forms of male domination and women's oppression that prevailed in progressive organizations that constituted the democratic movement. In these organizations, Women's Desks, Gender Commissions, and Women's Subcommittees were established to address these issues (Matiwana et al., 1989).

Social Transition and Gender-Policy Strategies

The release of Nelson Mandela in February 1990, the lifting of the ban on liberation movements, and the commencement of negotiations between the old and the new regime during the early 1990s created the space for women, albeit precarious, to place women's issues more firmly on the political agenda. These events signified a shift in the strategies women adopted in asserting their power in public policymaking. Women's organizations turned their attention once more to the sphere of public policy as an interim constitution and postapartheid policies were emerging. During the transitional period in South Africa, liberation movements, trade unions, social movements, and political organizations played leading roles in policymaking, yet women mere marginalized in these efforts.

The "social movement" space should have afforded women a greater opportunity to participate in policymaking, yet this did not occur because the organizations reflected the dominant gender relations. This represents an example of how women were denied a place of prominence in leadership (Cock, 1993). Given their marginalization in these efforts, women in these organizations devised various strategies to ensure that gender issues were addressed. Through these strategies, gendered policy options were generated in education policy initiatives of the Congress of South African Trade Unions (COSATU), the National Education Coordinating Committee (NECC), the African National Congress (ANC), and the Women's National Coalition (WNC).

Congress of South African Trade Unions (COSATU)

Reporting on the marginal status of women in COSATU, Cock found that, of the 83 officeholders at a national level in COSATU's affiliates in 1988, only eight were women and noted: "Women are prominent in the activities of COSATU yet they are visibly absent from positions of leadership" (Cock, 1991, p. 9). Cock argued that the labor movement is dominated by males and that few women have occupied leadership positions in trade unions. Since COSATU was formed in 1985, the position of women in it has been subject to debate. The inaugural conference passed resolutions pertaining to child care, sexual harassment, equal pay, and maternity and paternity benefits and proposed the creation of a Women's Subcommittee under the National Education Committee (Shefer, 1991, pp. 2–5). The marginalization of women in society prompted several initiatives within COSATU and a debate about the position of women. The year 1988 saw women in COSATU organize a special conference on women, called the COSATU Women's Congress, which "was a milestone for the struggle to get women and gender issues onto the federation's agenda and that of its affiliates" (Shefer, 1991, p. 2). Subsequent to the Women's Congress, COSATU established a Gender Commission in 1990. In the debate about the position of women in COSATU, two positions surfaced. First, that, as in the case of COSATU, women's subcommittees should be formed as substructures in the education committees of trade unions. Second, that a separate COSATU Women's Forum should be established with women's forums at trade-union level. Both positions were implemented; by 1991, the existence of both women's committees and women's forums in trade unions had become a source of conflict.

COSATU's vision for a future education and training system, expressed in a resolution adopted at its Fourth National Congress, reaffirms its commitment to work for a single nonracial and nonsexist educational system geared to meet the needs and aspirations of society as a whole. Like its counterparts, the NECC and the ANC, COSATU recommended that the constitution enshrine the right of all citizens to education and training, including women. This was accomplished. However, COSATU's recommendations failed to elaborate how its educational policies, as outlined in its proposals, would eliminate inequalities in the workplace in particular and in society in general.

The National Education Coordinating Committee (NECC)

When the NECC was established, its National Education Policy Investigation (NEPI) outlined its basic principles as "nonracism, nonsexism, democracy, a unity education system, and the redressing of historical imbalances" (Truscott, 1994, p. 1). In addition to its 12 working groups, the NEPI established a Gender in Education subgroup to develop policy options and processes for redressing gender inequalities in education. Noting the lack of empirical evidence and the paucity of research in the area of gender in education, the subgroup nevertheless made a number of recommendations to be considered in formulating gender-sensitive policy options (Truscott, 1994). It proposed that the national constitution embrace women's equality and that legislation on education and training comply with gender sensitivity by expressing a commitment to eliminate gender bias and by incorporating gender sensitivity into the structure of schooling, the curriculum, and the pedagogical approach. The subgroup also suggested that a Gender Curriculum Group be formed to review the gender content of all proposed curricula and that a Gender in Education Network of teachers, students, and researchers be formed "to act as a loose coordination or communications network to inform and support ongoing research or campaigns on gender in education" (Truscott, 1994, p. 70). Finally, it suggested that a Gender in Education Code be adopted that would

contain the main points (agreed after a process of democratic debate) underpinning the equal treatment of women in education. (For example, all girls have the right to go to school, not to be harassed, take subjects of their own choice, share cleaning tasks with boys, women teachers should have the same rights as men teachers, etc.) (Truscott, 1994, p. 53).

African National Congress Center for Education Policy Development

Proposals outlined in the "Policy Framework on Education and Training" (ANC, 1994) and its "Implementation Plan for Education and Training" (IPET) (ANC, 1994) emphasize the need for nonsexist education policies. The entitlement to education and training is a cornerstone of the ANC's policy framework:

The right to education and training should be enshrined in a Bill of Rights which should establish principles and mechanisms to ensure that there is an enforceable and expanding minimum floor of entitlements for all. All individuals should have access to lifelong education and training irrespective of race, class, gender, creed, or age (ANC, 1994, p. 3).

The ANC requested the Center for Education Policy Development to draft the IPET for the new Minister of Education (ANC, 1994, p. 1). Among the 20 task teams that were constituted to draft these implementation plans, a Gender Task Team was established to develop a plan for gender equity in education and training. The team used as its point of departure

the recognition, first, that gender inequalities in South Africa show marked dissimilarities from other Third World countries. Girls' entry into and participation in both schooling and higher education is in fact comparable with that of advanced industrial countries. As in these countries, their subject choices in education and training reflect expectations of a role in society which confirms their place in the home and family. Even as girls enter schools in large numbers, their career directions and participation in the labor market are gender specific. They are concentrated in low-paid, low skilled and low status work. The exceptions are teaching and nursing, which have hitherto provided black women in particular an entree into relatively high status jobs. Disproportionately large numbers of African women in rural areas remain functionally illiterate. Patriarchal relations and gender stereotyping continue to plague social relations at all levels (ANC, 1994, p. 1).

Education is seen by the most important democratic institutions in South Africa as the main mechanism with which to achieve gender equity. Legislation is also being used as a tool to institute change. It remains to be seen whether the policies and legislation are implemented.

The Women's National Coalition (WNC)

On the political front, approximately 70 women's organizations representing the entire political spectrum formed the WNC in 1992. By 1994, the WNC represented 80 organizations and two million women. The main objective of this coalition is to ensure that women play a central role in the process of policymaking for a nonracial, nonsexist democracy. Madlala describes the emergence of the WNC:

> On the eve of our first nonracial election, South African women have broken their silence. The fear of being marginalized even further helped spur women into united action. Through the Women's National Coalition (WNC), women for all walks of life—black and white, rich and poor, urban and rural—have joined in a campaign to draw up their demands (Madlala, 1994, p. 7).

In 1994, the members of the coalition adopted the "Women's Charter for Effective Equality," outlining their demands. This document emphasizes the significance of legal recognition and access to political power as the starting points in women's struggle for emancipation. In its preamble, the charter locates the position of South African women in a patriarchal society:

> At the heart of women's marginalization is the patriarchal order that confines women to the domestic arena and reserves for men the arena where political power and authority reside. Conventionally, democracy and human rights have been defined and interpreted in terms of men's experiences. Society has been organized and its institutions structured for the primary benefit of men (Women's National Coalition, 1994, p. 2).

With regard to the role of women in education policymaking, the charter stresses that "women shall be represented at all levels of policymaking, management, and administration of education and training" (Women's National Coalition, 1994, p. 6).

Women's Position in a Nonracial, Nonsexist Democracy

The election of an interim Government of National Unity (GNU) in April 1994 marked the beginning of a nonracial, nonsexist democracy. The abolition of oppressive apartheid legislation that denied women their fundamental rights and the adoption of an interim constitution that embraces women's rights as human rights have created the space for gender equality. Some of the moves toward gender equality in the political sphere, the education arena, and the economy are elaborated briefly.

The parliament under the apartheid regime had eight women members of which one served as a cabinet minister and another as a deputy cabinet minister. Women's organizations scored a victory when the interim constitution of the GNU, accepted in December 1993, stipulated that 30 percent of the positions in parliament be filled by women. After the election in April 1994, the number of women in parliament increased dramatically: 101 in the 400-member assembly and 16 in the 90-member senate; in addition, two of 27 cabinet ministers and three deputy cabinet ministers were women.

In industry, there has been a trend toward employing more Black people in management positions. At this stage, however, it is too early to assess whether democracy has effected significant changes in the economy and in economic institutions.

The education policies proposed by the NECC, the ANC, and COSATU have been incorporated into the education policies of the new Ministry of Education. Among other things, these policies, outlined in the White Paper on Education, propose that education be written into the Bill of Rights as a "basic human right" of all South African citizens. They also call for the appointment of a Gender Equity Task Team led by a full-time gender equity commissioner who would report to the director-general. While this team was appointed in late 1996, none of the gender-sensitive education policies had been implemented by early 1998.

Conclusions

Developments in South Africa indicate that women, despite their marginalization by the state in the public policymaking arena, effectively participate in policymaking though organizations in civil society. A historical overview shows that the strategies for effecting changes in public policies in the earlier period included protests, pickets, and petitions. More recently, women have been participating in gender-oriented teams within democratic organizations.

Considerable challenges remain. South Africa is described in its constitution as a nonracial and nonsexist democracy. In a setting in which racial discrimination has been the most visible and most widely discussed form of exploitation and oppression, all efforts to come to grips with other forms of exploitation and oppression must also deal with racism, both historical and persisting. In addition, South Africans are just beginning to implement a constitution, policies, and processes that will facilitate the development of gender equality, and much work remains to be done.

Rural women, in particular, should be supported in their struggle to survive poverty. Women need to work together to develop strategies to redress gender inequalities in parliament, schools, factories, and the home and to ensure that affirmative-action programs are implemented in every major institution in the society. Finally, women's organizations should play a central role in the country's reconstruction and development program to ensure that all

South African women enjoy a gender equality that allows them to develop their human potential to the fullest.

References

African National Congress (ANC). *A Plan For Gender Equity in Education and Training.* Johannesburg: Center for Education Policy Development and the African National Congress, 1994.

Berger, Iris. *Threads of Solidarity: Women in South African Industry, 1900–1980.* Bloomington: Indiana University Press, 1992.

Bhavnani, Kum-Kum, and Margaret Coulson. "Transforming Socialist Feminism: The Challenge of Racism," *Feminist Review,* no. 23, 1986, pp. 81–92.

Budlender, Debbie. "Women and the Economy," paper presented at a conference on "Women and Gender in Southern Africa," Durban, January 30–February 2, 1991.

Cock, Jacklyn. *Women and War in South Africa.* Cleveland: Pilgrim, 1993.

Commonwealth Secretariat. *Beyond Apartheid: Human Resources in a New South Africa.* London: Commonwealth Secretariat, 1991.

hooks, bell. *Talking Back: Thinking Feminist, Thinking Black.* Boston: South End, 1989.

Madlala, Nozize. "Building a Women's Movement." Johannesburg: 1994. Mimeographed.

Matiwana, Mizana, Shirley Walters, and Zelda Groener. *The Struggle for Democracy: A Study of Community Organizations in Greater Cape Town from the 1960s to 1988.* Bellville: Center for Adult and Community Education, University of the Western Cape, 1989.

Shefer, Tammy. "The Gender Agenda: Women's Struggle in the Trade Union Movement," paper presented at a conference on Women and Gender in Southern Africa, Durban, January 30–February 2, 1991.

Truscott, Kate. *Gender and Education.* Johannesburg: University of the Witwatersrand, 1994.

Unterhalter, Elaine, Harold Wolpe, and Thozamile Botha (eds.). *Education in a Future South Africa.* New Jersey: Africa World Press, 1992.

Women's National Coalition (WNC). "Women's Charter for Effective Equality," working document adopted at the National Convention organized by the WNC, February 25–27, 1994.

Women in India

Neera Kuckreja Sohoni

Introduction

The status of Indian women is best contexualized in terms of India's ancient history and mythology going back about 4,000 years and, in modern times, its subjugation as a colony and regeneration since 1947 as a free, democratic, secular country experiencing the paradox of an egalitarian constitution with an inegalitarian society. Different perspectives have emerged in the literature assessing Indian women's status. The antitraditionalist perspective traces Indian women's subjugation to patriarchal and feudal forces inherent in the Indian culture that denigrate the female's status from infancy through adulthood, leading to such abhorrent customs as female infanticide, child marriage, enforced widowhood, and *sati* (self-immolation of a widow on her husband's pyre). The traditionalist perspective eulogizes ancient Indian canonic tradition and scriptures and seeks to prevent Indian women and society from the debilitating consequences of modernism in the limiting sense of Westernism. The anticolonialist sees Westernism as an arbitrarily imposed choice on the colonized by the colonizer. Tracing the deterioration in women's position—particularly economic power—to colonial economic, civil, political, land revenue, and annexation policies, it contends that Western male-centered capitalism reinforced patriarchy; increased women's dependency on fathers, husbands, and sons; and relegated women's traditional roles to social and domestic spheres.

The Marxist-socialist perspective attributes the female's submergence, like that of other oppressed classes, to private property and individual control of the means of production. This view seeks to end private control in order to reintegrate women, along with the poor, into production processes and to socialize basic human services. The most recent perspective is the postcolonialist post-Marxist one that views "development" as a calculated process of subordination of India and other Third World countries to Western financial and survival interests. It perceives development as a euphemism for Western cultural and economic imperialism that reinforces the inferiority of not only women to men but also of Indian women and men to Western counterparts. The unifying theme of all of these perspectives is an acknowledgment of the inferior status of women compared to men and, irrespective of the length and causes of women's subordination, the need to change existing patriarchal structures to bolster the impact of legal and social reforms on the status of women.

The Religious Context

India is predominantly Hindu: nearly 83 percent of Indians follow Hinduism, 11 percent are Muslims, 2 percent each are Christians and Sikhs, and the rest are Buddhists, Jains, and others. Hindu scriptures and mythology have dominated Indian society and psyche since the very beginning. Early Hinduism perceived woman as a complement to man: All major male gods worshipped by Hindus carry a female consort endowed with autonomous qualities and powers. Society in the Vedic and Epic ages (1500–300 B.C.) considered the presence of the wife at religious and state rituals to be indispensable. Women were known to participate in religious, cultural, educational, and other assemblies and were seers, poets, and philosophers. In the post-Epic period known as the Pauranic and Smriti age (300 B.C.–A.D. 500), the status of Indian women declined. They began to be equated with the pariah and were denied the right to individual freedom. Eternally consigning the woman to the protection of her father, husband, and son during childhood, youth, and old age, respectively, *Manu Smriti* (an authoritative compendium of laws written by lawmaker Manu about A.D. 100) simultaneously acknowledged her indispensable value as a copartner with the male householder, asserting further that where women are honored, gods reside, and where they are not, all religious ceremonies are in vain. Elsewhere, the same canonical work hailed the woman as a mother, stating that a spiritual preceptor equals 10 teachers, a father a hundred teachers, but a mother has greater glory and grandeur than even a thousand fathers. These contradictory assertions marked the beginning of the dichotomous status of the Indian woman—on the one hand, indispensable for her domesticity but wholly dependent on the male, with no independent rights or entity, and, on the other hand, overpowering and venerable as a mother. Both perceptions held their ground throughout the ages as valid indicators of the Indian woman's status.

With the coming of Muslim rule, medieval India witnessed enhanced dependency of the female. Some Islamic practices, such as polygamy and divorce, favored men. The Islamic custom of purdah (veiling of women) forced the public world to be separated from the private world, with women confined to the latter. Following its subjugation by the Muslims, and fearing adverse outcomes for its women, a large part of Hindu India accepted the practice of veiling. Through this privatization, Indian women were forced to trade their mobility for safety. The challenge of Islamic aggression also made Hindu India defensive and introverted, causing a desperate return to orthodox Hindu beliefs and practices and further constraining the status of Indian women.

The British Colonial Context

British occupation of India began in 1772, and the British crown assumed charge of administering the colony from the British East India Company in 1858. Despite its stated official policy of laissez faire, or noninterference, in the personal law of India, the British colonial government did intervene selectively—not always out of altruism and often in response to the pressure of activist Indian reformers—to eliminate such brutal practices against women as female infanticide, child marriage, enforced widowhood, and *sati*. The choice of reform sometimes reflected British patriarchal biases, as indicated by the imposition of legal restrictions on matriliny in the south Indian Nayar community. At other times, their reluctance to push aggressively for reforms reflected their desire not to alienate the Hindu or Muslim orthodoxy. Serving the colonial interests, the judicial administration formally endorsed many of the customs and practices of Hindu Law inimical to women without reconsidering them in the light of natural justice (Liddle and Joshi, 1986).

Piecemeal and halfhearted as it was, British legal reform could make no significant difference culturally to the status of women. On the other hand, colonial politics and policies did undermine the institutional solidarity of the Indian family and of the woman's position in it. The destruction of agricultural self-sufficiency and Indian textile and other industries led rural men to migrate to cities, causing rural women to experience their first taste of female-headed households and the feminization of poverty. Moreover, since Indian men had to be educated in order to be hired, Western education—particularly knowledge of English—became (and still is) a dividing line between status or lack of it and between genders. The lack of mass education, caused largely by British apathy to educate In-

dians other than for the limited purpose of serving the colonial government's bureaucratic needs, had a negative impact on the enforceability of laws, dimming the prospects of social change and improvement in the female's status.

However limited in scale, Western education opened Indian minds to Western concepts of democracy and enlightened governance. At the same time, beginning in the nineteenth century, the work of Western and Indian Indologists caused a period of vigorous cultural renaissance. Indians began to rediscover their roots and to critically articulate and oppose the unacceptable components of their cultural legacy and tradition. India's fight for freedom thus coincided with a call for social reform. As Bipin Chandra Pal, an eminent freedom fighter, noted, "Our youthful intellectuals were not only anxious to acquire political freedom, but also equally, if not more, anxious to break through every shackle that interfered with their freedom of thought and action. Social reform was even more popular than political reform, and, in those early days, understanding of sacerdotal and social bondage was far keener than the understanding of political bondage."

The construction of gender under Western imperialism positioned men in the vanguard of the national struggle for self-emancipation. Initially, the women's question was raised by men under the aegis of broader sociopolitical reform. Reformist organizations such as the Brahmo Samaj, founded by Ram Mohan Roy in 1828; the Arya Samaj, set up by Dayanand Saraswati in 1875; and Ramakrishna Mission, created by Vivekanada in 1897; and individual male reformers like Ishwarchandra Vidyasagar, Ramakrishna Paramahansa, Keshab Chandra Sen, Maharishi Karve, Mahadev Ranade, and Gopal Gokhale led the fight against women's oppression by condemning such practices as polygamy, early marriage, and enforced and disinvested widowhood and by advocating female literacy. Although coming from different perspectives—the progressive reformist and the revivalist traditionalist—these institutions and individuals played a significant complementary role in redefining the value of women in Indian society and beginning the process of investing women with human rights. The Muslim Reform movement similarly advocated a reform agenda for Muslim women.

The unsatisfactory pace of social reform fanned the demand by Indians to have greater say in their governance. In 1885, with the support of enlightened British associates such as Allen Octavian Hume, the Indian National Congress was set up to crystallize India's political aspirations and spearhead the campaign for sociopolitical reform. Advocating strongly on behalf of women, Sir Hume noted that "political reformers of all shades of opinion should never forget that unless the elevation of female elements of the nation proceeds *pari passu* (with equal pace) . . . all their labor for the political enfranchisement will prove vain" (Hume, in Kaur, 1968, p. 84). Two years later, in 1887, the National Social Conference was formed specifically to lead the social reform campaign. To further support one of its major objectives of women's emancipation, it created a separate entity known as the Indian Women's Conference.

Simultaneously, beginning in the latter half of the nineteenth century, limited educational opportunities were becoming available to girls, although it took more than 50 years for the first women's university to be established in 1916. Education enabled reformist women from various regions and religious persuasions, of Indian and British ethnicity, like Pandita Ramabai, Margaret Cousins, Annie Besant, Dorothy Jinarjadasa, Sharadaben Mehta, Muthulakshmi Reddy, Cornelia Sorabji, Sarala Devi Choudharani, Abala Bose, Sarojini Naidu, Begum Hamid Ali, Abru Begum, Kamaladevi Chattopadhyaya, Vilasinidevi Shenai, Vidyagauri Nilkanth, Hilla Rustomji Faridoonji, and a host of others to emerge as key players in the Indian reform drive aimed at women. Organizations such as Poona Seva Sadan, the Servants of India Society, the Home Rule League, the Women's Indian Association, and the All-India Women's Conference, founded and managed by women, came into being to help reform the condition of, educate, rehabilitate, and mobilize women.

Beginning with the twentieth century, the emergence of Mohandes Gandhi on the political scene catapulted women to far greater visibility more quickly than any other leader or event. Gandhi openly held that sacred texts and customs which rationalized injustice to women or advocated female subservience deserved neither respect nor compliance. He supported women's induction into public life while asking also that their domestic role be fully honored and valued. His tool of mass agitational politics served him well in getting women out of the private into the public sphere. Led by his simple common-sense approach to indigenizing politics and economics, Indian women extended their full participation to the *swadeshi* (import-substitution) movement, which boycotted British goods and encouraged indigenous cottage and other industries. But the thing that did more than any other single factor to speed the process of women's rights was the *satyagraha* (nonviolent resistance) movement devised by Gandhi. Women already in public life "took to organizing pickets for liquor and cloth shops, processions and demonstrations . . . while many thousands came out of conditions of privacy and semi-seclusion to support the cause. Some of these were so ardent that at times, as in Delhi, they directed the whole Congress movement in an area until arrested" (Percival Spear, in Liddle and Joshi, 1986, p. 33). Women's open activist participation in the freedom movement convinced both men and women of the need to vest women with equal rights. The principle of gender equality was accepted by the Indian National Congress in 1931 and subsequently enshrined in free India's constitution. However, the dispute between public, or political, and personal, or civil, rights remained unresolved. Thus, whereas women's right to suffrage and political representation was widely supported, the same could not be said of their rights in marriage or inheritance. The dichotomy arose from the dissociation of the national (public) from the domestic (private) arena. Equality between the sexes was not welcome within the family. Thus nationalism and freedom remained gendered constructs.

Women's Status in Free India's Planned Development

Although formal global calls for the inclusion of women in national and international development began in the early 1960s, women were integrated as a special concern in the Indian development-planning process even before that—from the formulation of India's first development plan (1951–1956) and in its constitution, which guarantees justice, liberty, and equality to all citizens. Article 14 provides that the state shall not deny to any person equality before, or equal protection of, the law. Article 15 prohibits any discrimination. Article 16(1) guarantees equality of opportunity for all citizens in matters relating to employment. Article 16(2) prohibits discrimination in employment on the basis of religion, race, caste, and sex. Recognizing the greater vulnerability of women, the constitution makes special provisions for them: While prohibiting discrimination based on sex, Article 15(3) provides that nothing in the article shall prevent the state from making any special provision for women and children. There is thus a constitutional basis for discriminatory or affirmative action by the state in favor of women.

The socioeconomic goals of India as welfare state are embodied in Part IV of the constitution. Known as "Directive Principles of State Policy," these nonjusticiable policy pronouncements provide for the right to an adequate means of livelihood for men and women equally, stated in Article 39(a); clauses (d) and (e) of the same article provide for equal pay for equal work and protection of health and strength of workers. Article 42 ensures maternity relief and just and humane conditions of work; Article 45 provides for free and compulsory education for all children up to age 14; Article 47, for raising the level of nutrition and public health; and Article 44, for a uniform civil code.

The constitutional mandate regarding women was reflected in the five-year development-planning process on which the country embarked soon after independence in 1949. The First Five-Year Plan (1951–1956) provided welfare measures for women, and a national body known as the Central Social Welfare Board was set up in 1953 to implement welfare programs for women. Simultaneously, the community-development movement was launched, which sought to mobilize women through a network of community-level women's groups known as Mahila Mandals. Unfortunately, these groups turned out to be elitist in composition and agenda and had limited value for the large mass of rural and urban poor women. Conceptually, community development did not take into account the impact of caste, race, gender, and other variables in stratifying rural communities, power politics, and structures. Hence, its impact was minimal on those who were most in need of political representation and development assistance.

The Second Five-Year Plan (1956–1961) was geared toward intensive agricultural development, but, once again, without a defined approach to the gendered and caste construct of the agriculture sector and rural society, it failed to make an impact on rural women and the poor in general. The pivotal and extensive role that women play in agriculture was also ignored. On the positive side, the plan asked that women be protected from injurious work, receive maternity benefits, and have access to creches. It also asked for speedy implementation of the principle of equal pay for equal work and of training provisions to enable women to compete for higher jobs.

The Third Five-Year Plan (1961–1966) promoted women's education as a key strategy for women's welfare. Another focal area was women's reproductive health, with emphasis on providing services for maternal and child welfare, health education, family planning, and nutrition. After a brief disruption of the five-year planning cycle, the Fourth Five-Year Plan (1969–1974) came into force. The main areas of emphasis were family planning and mass education of women to reduce the birthrate from 40 to 25 per 1,000. Under the new strategy, child survival became a correlate of the birthrate-reduction policy. Immunization and the supplementary feeding of poor children and expectant and nursing mothers received priority as a means of reducing high infant and maternal mortality rates.

The approaches to women in each of the plans thus far were more welfare oriented and did not seek to address basic structural and cultural issues constraining women's roles, opportunities, and entitlements. In 1971, a major step was taken by the government of India in appointing a Committee on the Status of Women to undertake a comprehensive examination of all issues relating to the rights and status of women in the context of changing socioeconomic conditions in the country and new problems relating to the advancement of women. The report, entitled *Towards Equality* (Government of India, 1974), stressed that the dynamics of social change and development had adversely affected a large section of women, particularly poor women. It demonstrated a visible link between female poverty and the depressing demographic trends of the declining female–male sex ratio, lower life expectancy, higher infant and maternal mortality, declining work participation, and rising rates of illiteracy and immigration among women. The report's crucial conclusion was that disabilities and inequalities imposed on women have to be seen in the larger context of an exploitive society, and that any policy aimed at the emancipation and development of women must form a part of a total movement for the removal of social inequalities and oppressive social institutions.

The report's release more or less coincided with the designation of 1975 as International Women's Year and of 1976–1985 as the United Nations Decade for Women. Following the guidelines laid down in the United Nation's *World Plan of Action for Women,* a National Plan of Action for Women was adopted in 1976 that called for planned interventions in the areas of health, family planning, nutrition, education, employment, legislation, and social welfare to improve the conditions of women. The Women's Welfare and Development Bureau was created in the Ministry of Social Welfare to act as a nodal point within the government to coordinate policies and programs and initiate measures for women's development.

A significant outcome of the preceding developments was a shift in perspective: From being passive recipients of welfare policies, women began to be perceived as a critical mass for development. The Fifth Five-Year Plan (1974–1979) especially emphasized the training of women through income generation and functional literacy programs with a view to making women more productive on the public as well as the domestic front. The enhanced visibility of women's development potential influenced the Sixth Five-Year Plan (1980–1985), which contained for the first time in India's planning history a separate chapter on women and development. Overall, the sixth plan sought to promote women's development through economic independence, educational advance, and better access to health care and family planning. Repeated pregnancies, physical overload, lack of education, absence of independent ownership of economic assets, and social beliefs and practices that stigmatized women became the targeted areas of reform. For the first time, the government agreed to issue joint titles of ownership of key development assets such as land, housing, animals, and other units of economic production to husbands and wives.

Nevertheless, a major limitation of the sixth plan was that the family rather than the woman remained the basic unit of development programming. The Seventh Five-Year Plan (1986–1991) targeted women more centrally and qualitatively. Operationalizing the concepts of equity and empowerment propagated globally by the United Nations Decade for Women, it emphasized interventions to empower women by generating awareness of their rights and privileges and training women for more productive economic activity. The latter entailed wider continued support for improved access, control, and use of economic assets, services, and emerging technologies by women. The plan also asked for developing support services to reduce the excessive daily burden of domestic work on women. In addition, efforts were begun to enhance women's access to science and technology and to promote their participation in the creation of a sustainable environment.

The mid-term evaluation of the women's decade globally threw the door open to critical assessments of planned approaches to women's development in India. The difficulty of reaching poorer women adequately with existing channels was highlighted through the evaluations—as a result of which, sufficient momentum was generated to grant greater institutional visibility in government to the cause of women. In 1985, a separate Department of Women and Child Development was created in the Human Resource Development Ministry to oversee the implementation of more than 27 different programs for women covering a range of services, including employment and income generation, education and training, legal support, general awareness, and support services. In 1988, the Department of Women and Child Development formulated a *National Perspective Plan for Women 1988–2000 A.D.,* which presents a long-term comprehensive policy for Indian women and a framework for integrating women in the country's development process. A National Advisory Committee on Women chaired by the prime minister was also set up to ensure treatment of women's concerns at the highest level. In 1992, the National Commission on Women was created to monitor the enforcement of constitutional and legal safeguards, review existing legislative gaps, suggest amendments, and take *suo moto* notice of cases involving the infringement of women's rights (Goverment of India, 1994, p. 246).

Outcomes of Development Planning

Contrary to what was intended, the persistent gap between development theory and practice in India has accentuated the burden of poverty and underdevelopment for women as for the less privileged. Indian development's contradictions, especially in relation to the poor and women, started to become evident during the 1960s. By the 1970s, the realization came that the status of women had actually regressed. The 1971 census revealed that there were fewer women per 1,000 men in that year than at the turn of the century (930 in 1971, compared to 972 in 1901). Literacy had grown among girls but so had the school dropout rate. School enrollment ratios were consistently higher for boys than girls at all three stages of education—primary, secondary, and tertiary. Adult literacy was dramatically lower among females than males (around 19 and 40 percent, respectively, in 1971). The female work-force-participation rate in 1971 was about 12 percent, compared to almost 53 percent for males. The 1981 census showed marginal improvements over 1971 in sex ratio (933 females per 1,000 males), female literacy (25 percent), and employment (14 percent) (Government of India, 1985, 1988).

Health

Since 1981, the performance has been mixed. Sadly, the sex ratio has declined to below 1981 levels: The 1991 census placed it at 927 females per 1,000 males. The total female population in 1991 was estimated at 407 million, compared to 439 million men. As their negative sex ratio reveals, the overwhelming health issue for Indian women is that of survival. The Indian cultural norms that adversely affect female health and survival are the attitudes toward early marriage, fertility, and son preference and the self-abnegation traditionally practiced by women and girls. Food, nutrition, and medical care are proven to favor the male. The high Indian fertility rate (4.6 live births per woman, compared to 1.8 in the United States) is deleterious to women's health. Anemia, hemorrhage, toxemia, sepsis, and abortion are the major causes of maternal deaths, 70 percent of which are preventable through better health and nutritional care. Compared to less than 1

percent in the United States, 71 percent of deliveries in rural areas of India and 29 percent in urban areas are conducted by untrained personnel. The approach to family planning is gendered, with female sterilizations accounting for 90 percent of all sterilizations (Government of India, 1985). Fertility regulation is not used to empower women or to enhance their control over their bodies, but more to fit national and global population-control objectives. Nor does it adequately address the cultural and economic bases of high fertility and its symbiotic link with the female's low status. The overriding focus of health interventions on reproductive years leaves females at both ends of the spectrum untended, causing, in particular, irreversible damage to the young female's growth process and, through her, to the future generation.

Education

Educationally, the lot of Indian women has been improving, albeit slowly. At the turn of the century in 1901, female literacy was only 0.8 percent. There were 12 girls enrolled at the primary level and four at the secondary level for every 100 boys (Jain, 1975, p. 130). By 1984–1985, the female enrollment ratio at the primary level was almost 77 percent, compared to 94 percent for boys. (The 1991 census noted a further increase in girls' enrollment at the primary level to 86 percent.) However, the sharp decline at the secondary level—to 36 percent enrollment for girls, compared to nearly 51 percent for boys—points to the greater difficulty in retaining girls in school. The growth in adult literacy has seemingly favored women: Between 1951 and 1991, female literacy increased five times, from just below 8 percent to 39.3 percent. However, the gender literacy gap has not only persisted but widened, from 17 percentage points in 1951 to 22 in 1981, and 25 in 1991. Women accounted for 57 percent of the illiterate population in 1981, and girls formed 70 percent of the population not enrolled in school (Government of India, 1985). Furthermore, the likelihood of a rural girl being out of school is much higher than that of an urban girl. High opportunity costs of educating girls because of forgone earnings in poorer families, social beliefs regarding early marriage, the advisability of keeping girls uncontaminated from the social and psychological autonomy associated with education, the absence of single-sex schools and female teachers in adequate numbers (only 28 percent of all schoolteachers are women), discriminatory processes at work within the school system itself, low returns on educating girls, and subsidiary status of the girl within the family—all constitute major barriers to the greater induction and retention of girls in education (Desai and Krishnaraj, 1990).

Like health, literacy is a gendered and class construct. College-going is overwhelmingly an urban middle-class phenomenon, more so for women than men. Among those age 17–23, only 1.5 percent of females are estimated to be enrolled. The sex ratio in higher education translates to 38 women for every 100 men. Interestingly, it improves in favor of women at the graduate level compared to the undergraduate level (Desai and Krishnaraj, 1990). Lower literacy and education levels are regionally clustered, with states like Andhra Pradesh, Bihar, Madhya Pradesh, Rajasthan, and Uttar Pradesh accounting for the lowest and Kerala for the highest rates. Aware of these regional disparities, in 1983 the Indian government introduced a 90 percent–10 percent financial assistance plan for the educationally backward states with a view to stimulate the school enrollment of girls age 6–14 through nonformal education. Simultaneously, a major campaign was launched to bring women age 15–35 into the literacy stream through adult literacy projects. Although more attractive on account of greater structural and substantive flexibility, these schemes are unlikely to unilaterally solve the problems experienced by girls and women in accessing education and literacy in the absence of widespread adequate child-care arrangements and the reorientation of family perceptions and gender politics toward educating girls and women.

Work

Work in the narrow sense of paid employment, and the status ascribed to paid work, have eluded Indian women. As in other agricultural economies, Indian women work all of the time but their work remains invisible to national accounting systems that view only cash-worthy or valorized activity as work. Until recently, Indian censuses have taken a limited view of work. The 1971 census, for instance, defined a worker as a person whose main activity is economically productive work. Women's work, which is largely perceived as secondary activity, thus remained unenumerated. Owing partly to changing census definitions and the continued fuzziness in defining productive or gainful work, but mainly to socioeconomic "push" factors including structural and technological changes in the economy that eliminate female occupations and substitute female with male skills, the female work-participation rate has declined dramatically from 34 percent in 1911 to 20 percent in 1961, 12 percent in 1971, and 14 percent in 1981. The female worker's displacement by the male, like her status in education and health, shows strong regional variation. It was most pronounced in the states of Haryana and Punjab, where the decline between 1961 and 1981 was an estimated 90 percent (Desai and Krishnaraj, 1990).

Women workers accounted for just 17 percent of the total labor force in 1971 compared to 34 percent in 1911 and 32 percent in 1961 (Government of India, 1988). Revised, more credible definition and measurement of women's productivity in the 1991 census enabled women's work in the informal sector to be captured for the first time. The provisional census estimates show higher female participation rates in organized as well as unorganized sectors of employment (Government of India, 1994).

The World Bank (1991) estimates that Indian women make up one-third of the labor force. An estimated four-fifths (80 percent) of all economically active women are employed in agriculture, which has seen a dramatic increase in women's presence. This is due to the feminization of subsistence agriculture, which occurs in response to commercialization of smallholder agriculture by the imposition of cash crops; the introduction of new agricultural technologies that permit greater use of female labor; and male migration or the movement of men to nonfarm employment (Government of India, 1988).

A systematic gender bias in industrial and agricultural sectors of employment invariably surfaces in micro studies. Operations like ploughing that fetch higher wages are male preserves, whereas transplanting, which commands lower wages, is a female preserve. In the construction industry, men do bricklaying while women carry bricks and mortar—and get paid less for their labor. Female professional and skilled workers in the organized industrial sector, too, report lower remuneration for equal work, as well as occupational segmentation by gender, with some of the lowest-paid occupations filled with disproportionately high numbers of women (Government of India, 1985, p. 30). There is a higher degree of job segregation among women than men, with 12 occupations accounting for more than 80 percent of female employees. The medical and teaching professions account, respectively, for 21 and 18 percent of female professionals; civil service for 9 percent; and management and industry executive jobs for 2 percent. The marginalization resulting from gender bias is reinforced by class and caste biases. Professional women are predominantly from the upper castes and urban areas, and micro studies indicate that the representation of each caste declines with decreasing caste status (Liddle and Joshi, 1986, pp. 124–125). Tribal women have special work-related problems linked to their topography. They are generally engaged in forestry-based occupations, and deforestation by industrial users and land encroachment by vested interests add to their economic distress. Surveys in the 1980s revealed that 30 percent–35 percent of the rural households were headed by women due to male migration, neglect, or abandonment (Government of India, 1988).

Occupational immobility among women is an integral part of the social construction of gender embedded in the interlocking religious, economic, and kinship structures that define the social domains of females and males. These domains are associated with an inside/outside dichotomy (World Bank, 1991) whereby women's intrinsic association with reproduction and the family puts them in the private, or "inside," sphere, while men interact with the markets, government, courts, and other institutions in the public, or "outside," sphere. Women's links with the outside are thus mediated by male relatives. The strength of the dichotomous phenomenon varies considerably by class and region. On the whole, the barriers to women's transition from the private to the public sphere are greater in Northern than in Southern India; stronger among caste Hindus than among the scheduled castes and tribes (disadvantaged groups protected by the constitution); and greater among landowning cultivators than among landless laborers or marginal farm owners. In the final analysis, the overwhelming determinant of the female's occupational activity is poverty. The fact that families with no or poor resources put their women and girls to work, whatever the type of work and at whatever cost, makes poverty and, therefore, class and caste the proxy indicators of female employment and economic exploitation.

Women's occupational burden outside the home is matched by the work they have to do in the home. As carriers of water and fuel; preparers, processors, and preservers of food; caregivers; childrearers; and in-home multipurpose workers, they carry a heavy dual or even triple burden. Some of that burden is passed on early by surrogacy to girls, whose demanding labor remains invisible and unvalorized not only to the development planner but also to the parents. It is important to change this valuation of female labor within the household before a difference can be made at the societal level. Governmental intervention in the private domain, where gender relations are rooted, is problematic.

> The most effective—and perhaps the only legitimate—means by which public policy can affect household processes and reduce women's dependency is to alter the economic environment. In a sense, this means that market forces should be allowed to influence the boundaries of culturally acceptable women's activity (World Bank, 1991, p. xvi).

It is widely recognized that to change women's productivity and, through that, their overall role and status, women must be assured unmediated access to human-capital resources, including education, skills training, and extension advice, as well as to credit, ownership of land, raw materials, and technology. There is also recognition that making women more educated, productive, and, therefore, better income earners and spenders will reduce their dependency and enhance their status. As in other countries, it has been demonstrated in India, too, that improved female educability and employability help reduce fertility and slow population growth, improve child survival, enhance the share of family income allocated to food consumption and health care of children, raise the level of life through added household income, and speed national economic growth by increased aggregate labor productivity.

Political Participation

Equal political participation as a woman's right in modern times was first articulated in India in 1917 by a delegation of distinguished Indian women led by Sarojini Naidu to the British Parliament. By 1937, women had achieved limited franchise rights, as well as the right to contest elections to the legislatures. Although elected women were few and drawn mainly from the educated urban elite, inspired by Gandhi, many more women—and from all socioeconomic levels—were active in agitational politics for freeing India. However, contemporary feminists argue that the female's public (that is, political) role was perceived and argued in gendered terms even by Gandhi, who celebrated women's

"natural" greater resilience, self-sacrifice, and moral as opposed to brute force as being particularly suited to his brand of peaceful agitational politics.

Post-independence political participation by women tended to fall along the elite–subaltern binary. The class divide between those who could enter national political decision making (mainly on account of their education, family connections, or public recognition through their role in the freedom movement) and those (average anonymous women) who could hope to influence it at all persisted through the early decades of freedom. There was then also a false complacency that the gains from self-determination and independence would somehow filter automatically to all women. The 1970s showed the fallacy in that belief as a major mid-decade report (Government of India, 1974) demonstrated that women's status had regressed in many ways from what it had been before independence. While the government moved vigorously but bureaucratically to try to stem that regressive trend, women themselves took to heart the significance of their vulnerability. Beginning in the 1980s, a different, more fiery brand of female participative and coalitional politics came into being to assure women a more authentic voice in, and fairer share of, the country's development. Across classes and regions—occasionally even across religious ethnicity—women began to link, drawing upon community-centered strengths and structures to broaden the base and nature of political discourse. Inevitably, this new wave of women's activism was a composite of Marxist, feminist, secularist, and nationalist streams. Women were now seen as one of many victims of patriarchy, casteism, and global capitalism.

Disenchanted with the privileging formal modes of political participation, keeping in view the limited potential of female representation in elected and appointed as well as administrative positions, women devised a variety of parallel nonformal participatory mechanisms and activities to enable political engagement by the "common woman" and around her concerns. In addition to exercising their voting power, women resorted to organizing rallies and street processions to publicly articulate their concerns and convey their support to specific political issues and groups. Other strategies women used effectively included making representations to legislators; disseminating political and social views and opinions among the illiterate and semiliterate underprivileged people; engaging in labor-union activity; undertaking *gheraos* (encirclement) and other means of nonviolent protest, as well as tactics to combat and change the politics of decision making within the family and community, nationally, and, when necessary, even globally. In the 1990s, this type of active political behavior has led Indian women to mobilize and agitate against a variety of pressing issues, ranging from inequality in the workplace to dowry, rape, alcohol and drug addiction, domestic and custodial violence, eve-teasing (that is, verbal or physical abuse of the female in public spaces), price rises, corruption, inefficiency of subsidized food distribution, food adulteration, unwanted and risky contraception, lack of cooking fuel, deforestation, and other environmental safety and sustainability issues, including those arising from economic structural adjustment. Linked to external pressure by monetary lending agencies, this latter devious device to restructure economies and societies so that they can better honor their debts to lender countries has been proven through micro studies in India and other countries to further marginalize women and worsen the preexisting gender and class inequalities.

To offset the many persistent inequities experienced by women, successful organizations have emerged, some of which have attained national or international eminence as paradigms. Among these are SEWA (Self-Employed Women's Association), Working Women's Forum, Annapuram Mahila Mandal, and Samakhya, all of which provide self-employment, credit, and other services to poor, assetless women and women-headed households living below the poverty line; and the women's sustainable-environment movement known as Chipko (hugging), which began with women hugging or strapping themselves to trees to prevent them from being felled. Examples of the efficacy of issue-centered activist and proactive organizations are the Forum Against Oppression of Women; the Dahej Virodh Chetana Manch, a forum dedicated to consciousness-raising mobilizing women against dowry; the Janwadi Mahila Samiti and Bhopal Gas Peedit Mahila Udyog Sangathan, which spearheaded a massive mobilization of female victims against the disastrous 1984 toxic gas leak at the chemical pesticide factory of American chemical company Union Carbide at Bhopal, Madhya Pradesh, India; Stree Shakti Sangathan, Nari Raksha Samiti, and Chhatra Yuva Sangharsha Vahini, which seek to unite women against violence inflicted on women and girls; and Vimochana, which serves as a catalyst on political issues relevant to women, elected candidates' accountability, and informed political engagement by women. Other nationally reputed organizations of historical standing, some with origins in the preindependence era, are the All-India Women's Conference, the Guild of Service, the Kasturba Gandhi Memorial Trust, the Harijan Sevak Sangh, the Bharatiya Adimjati Sevak Sangh, and the Rama Krishna Mission, which are dedicated to offering welfare, education, and support services mainly to socioeconomically marginalized women.

As the political sophistication of a broad cross-section of women is growing, there is increasing frustration with the glass ceiling in representative structures manifested through women's modest share of elected and executive positions. Globally, as of January 1, 1997, women accounted for a mere 12 percent of the Lower House, and 9.8 percent of the Upper House or Senate. Indian women's parliamentary strength at 7.2 percent of the Lower and 7.8 percent of the Upper House was even less impressive ("Democracy Still in the Making," 1997). This is a decline from their proportional strength in earlier years. The voting record of Indian women, on the other hand, is less dismal. Forty percent–50 percent of women exercise their franchise (Government of India, 1994). Although fewer women

than men have voted in each election since the first in 1952, the gender gap (male minus female) has narrowed from a 17 percentage point difference in 1957 to 11 percentage points in 1980, with male voter turnout at 62 percent and female at 51 percent (Desai and Krishnaraj, 1990, p. 274).

Major political parties have traditionally set up women's wings or units, such as the National Federation of Women aligned to the CPI (Communist Party of India), and the All-India Democratic Women's Association, which acts as the women's wing of the CPI-M (Communist Party of India-Marxist). While political parties historically have capitalized on women's vote and dangled impressive visions of higher representation to women in the party and government structures, they are hesitant to field female candidates. Combined with their reluctance, the high cost of election campaigns and the general feeling that politics is becoming meaner and dirtier have tended to keep women away. However, once elected, Indian women have gone well beyond the political horizons accessed by women in most of the world's countries. India has had a female prime minister, as well as female cabinet ministers, state governors, chief ministers, and ambassadors. At the administrative level, a 1988 report notes that there are 339 women out of a total of 4,548 officers in the federal Indian Administrative Service (Government of India, 1988). Women account for much lower proportions in the police, forestry, and other federal services. The gender divide also pervades trade unions. The latter had kept women and their issues fairly marginalized until recently, but women miners, farmworkers, and domestic workers, and women in other occupations are increasingly forming solidarities either as part of the larger male unions or on their own.

Amendment 73 of the Indian constitution (1992) mandates that 30 percent of local *Panchayat* (village council) seats be reserved for women. A historic measure, it should add significantly to women's visibility at the grassroots level in politics. But utmost vigilance will be required to enforce it. As in national and state-level politics, reserved grass-roots representation for women risks being co-opted by caste and class forces. In the past, women of upper castes, wives of big landowners, and politically influential persons have tended to dominate the seats reserved for women in local bodies (Government of India, 1988, p. 155).

The level of women's membership in political parties has been guestimated at 10 percent–12 percent. The rise in fundamentalist, revivalist, and secessionist politics is believed by most observers to have caused a dramatic upsurge in women's political engagement. To women, no differently than to men, religion and ethnocentrism are addictive as political opiates. Whatever the ethics of such politicization, women's en masse engagement in regional, ethnic, and even terrorist politics challenges the earlier stereotype of the passive benign Indian female. Micro studies of women's voting behavior further indicate that even illiterate rural women are politically alert and aware of the key issues, are able to discuss and analyze them, and can vote according to what they want—with the secret ballot enabling them to exercise their individual preference freely (Rao, 1983; Government of India, 1988).

Violence Against Women and Girls

Like every culture, India has components of institutionalized violence against women and girls. Such violence affects all stages of the female's life cycle, beginning with preconception and the overwhelming value given to males, especially sons. Indian parents have relied on religious and other rituals to conceive a male fetus. Despite such precautionary measures, if a girl is born they have been known to resort to infanticide. Son preference does not stop with expulsion of the unborn female or infanticide of the female after birth. It is sustained through preferential feeding, care, and other discriminatory treatment of the male. The girl is treated as a transit passenger on her way to her marital household: Investing in her survival, safety, and development is considered nonproductive, like "watering a tree in a neighbor's courtyard" (quoted in Sohoni, 1995).

Estimates of the female's physical and sexual abuse within the family are impossible, given the paucity of data, but lack of data cannot be assumed to mean the absence of abuse. Visible acts of violence directed at the female consist of spousal battering and forced sex, eve-teasing or intimidation of the female in public spaces, sexual harassment on the job, rape, kidnapping, and a generally unsafe environment that restricts female mobility to daytime. Marriage accentuates the risk of violence, causing the female to face psychological or physiological maltreatment and at times even a violent death at the hands of the spouse or the in-laws. Dowry, which de facto amounts to placing a price on the bride's "eligibility" and "asset value" by her inlaws and spouse, figures largely in this contentious context. Absence or insufficiency of dowry becomes a source of the bride's maltreatment, victimization, and even "accidental death." The latter amounts to clandestine murder often made to look like a suicide. Female sexuality is feared as posing a risk of promiscuity, which is the basis for stringent behavioral control of the female, first by her natal and later by her marital family. Rape, prostitution, drugs, smuggling, and terrorism are increasingly affecting women in India and are symptomatic of the attitude of criminality and violence toward the female. The catchment area of these traffickers is steadily expanding to include urban, rural, and even tribal women.

Legislation Regulating Women's Position

The earliest legal initiatives on behalf of women's rights occurred in colonial India. Led by the determined efforts of social reformers and British sympathizers, laws were passed to ban *sati* (1829); legalize widow remarriage (1856); reform marriage (1872) by prohibiting polygamy, setting an age limit for marriage at 14 years for girls and 18 for boys, providing for a legal allowance for divorced women, and making no reference to the caste of the marriage partners; and ban infant marriage (1891 and 1929). One of the first legal reform efforts in free India was aimed at the Hindu institution of marriage. Customarily, Hindu marriage was not viewed as a contract between two consenting individuals. This position is similar to Christian law in India. The Hindu Marriage Act of 1955 legalized marriage between different castes and within the same *gotra* (an eponymous group reputed to descend in its entirety from a common ancestor), abolished exogamy, specified the prohibitive degree of relationship for the purpose of marriage, prescribed the minimum age at marriage as 15 for girls and 18 for boys, and conferred on the parties the right to dissolution of marriage by divorce. The Marriage Law Amendment Act, passed in 1976, facilitates divorce by mutual consent or on grounds of cruelty and desertion. It also recognizes a girl's right to repudiate a child marriage (occurring before the age of maturity) regardless of whether the marriage was consummated. The unfortunate aspect of these laws is that they exclude Muslim and other minority (Christian, Jewish, and Parsee) women, whose position in regard to marriage, divorce, or polygamy is governed by their personal (religious) law.

One significant amendment concerns the concept of rape and reflects a somewhat unique approach. The Criminal Law Amendment Act of 1983 protects the victim from the glare of publicity during investigation and trial; it changed the definition of rape to remove the element of consent and to add the crime of custodial rape, enhanced punishment for the crime, and shifted the onus of proof onto the accused. In a similar vein, the Criminal Law (Second Amendment) Act of 1983 gave, for the first time, legal recognition to domestic violence by making cruelty inflicted by the husband or his relatives an offense. Recognizing dowry to be at the root of the female's abuse, a 1984 amendment to the 1961 Dowry Prohibition Act makes women's subjection to cruelty a cognizable offense. A second amendment to the act, in 1986, allows for the punishment of the husband or in-laws if a woman commits suicide within seven years of her marriage and cruelty has been proved. Further, a criminal offense of "dowry deaths" has been incorporated in the Indian Penal Code. The Law of Evidence has also been amended to provide that, if a married woman commits suicide within seven years of her marriage, the presumption will be that her husband or his relatives abetted the suicide.

Unfortunately, the amended laws do not cover violence inflicted on girls and women in the natal home. A survey in Greater Bombay showed that 61 percent of women who died of burns were 15–19 and never married (Government of India, 1988, p. 136). Violence against the female is so generic to Indian society that a drastic attitudinal change is required for crimes of violence against females to be acknowledged, reported, and investigated, and the perpetrators apprehended and punished.

Vesting women with the right to ownership of property and assets is crucial to their economic autonomy and cultural empowerment. Traditionally, neither the personal laws based on religion nor the secular laws have given Indian women equal rights to property. The Hindu Succession Act of 1956 for the first time provided women an equal share in their fathers' property. However, certain restrictions continue: A woman cannot ask for the property to be partitioned, nor can she be a member of the coparcenary. Other laws address the economic rights of women as workers and range from the 1976 Equal Remuneration Act, to laws that protect employment conditions of women in mines, plantations, factories and contractual labor, to those that regulate maternity and other benefits, including abortion.

Despite the impressive array of laws, knowledge among women of their rights under various pieces of legislation and the enforcement machinery are still too sparse to permit a radically reformed social position of all Indian women on all fronts.

There are solutions. One is to force the pace not only of adequate legislation but also its enforcement. Given its magnitude, the task of enforcement cannot be left to the government alone. Voluntary agencies need to step in to play an activist, catalytic, connecting, even vigilante role to help bridge the gap between women and legal entitlements and public amenities. Legal and functional literacy and overall education and empowerment of socioeconomically marginalized women are crucial in this context.

Conclusion

Despite extraordinary efforts and achievements, for all but a handful of Indian women equality within the home and in the workplace remains an unrealized human right. There are formidable institutional and material hurdles in India in enabling women to play an equal role or have equal status vis-à-vis men. Equalities remain to be realized in key socioeconomic respects, including equal opportunities for education and development, equal wages for equal work, equal status in the family, valorization of the female's reproductive and sustenance role, and economic, cultural, and personal autonomy. These are the main areas on which the contemporary and prospective women's movements need to focus constantly. Moreover, the basic question of whether cumulative gender inequality embedded in a pluralistic but patriarchal, stratified society can be rectified merely by assertion of the principle of equality in the constitution needs to be addressed. How effectively and comprehensively women are able to build on the political and constitutional foundation of equality in the context of gender and class hegemony is the crux of the woman question in India.

The task ahead is formidable. The Indian woman is caught in a double bind. Despite stated intentions to the contrary, the government and society remain mired in patriarchy and elitism. Admitting women as equals necessitates diluting the first and devolving the second. But the tide for changing women's status is surging and neither class nor patriarchy can stem it.

References

"Democracy Still in the Making." Poster on Men and Women in Politics. Geneva: Inter-Parliamentary Union, 1997.

Desai, Neera, and Maithreyi Krishnaraj. *Women and Society in India.* Rev. ed. Delhi: Ajanta, 1990.

Government of India. *Towards Equality.* Report of the Committee on the Status of Women in India. New Delhi: Department of Social Welfare, Government of India, 1974.

———. *Women in India Country Paper.* New Delhi: Ministry of Social and Women's Welfare, Government of India, 1985.

———. *National Perspective Plan for Women 1988–2000 A.D.* New Delhi: Department of Women and Child Development, Government of India, 1988.

———. *India 1993.* New Delhi: Research and Reference Division, Government of India, 1994.

Jain, Devaki (ed.). *Indian Women.* New Delhi: Publications Division, Ministry of Information and Broadcasting, Government of India, 1975.

Kaur, Manmohan. *Role of Women in the Freedom Movement (1857–1947).* New Delhi: Sterling, 1968.

Liddle, Joanna, and Rama Joshi. *Daughters of Independence: Gender, Caste, and Class in India.* New Brunswick: Rutgers University Press, 1986.

Rao, Usha N.J. *Women in a Developing Society.* New Delhi: Ashish, 1983.

Sohoni, Neera K. *The Burden of Girlhood—a Global Inquiry into the Status of Girls.* Oakland: Third Party, 1995.

World Bank. *Gender and Poverty in India.* A World Bank Country Study. Washington, D.C.: World Bank, 1991.

Women in South Asia:
Pakistan, Bangladesh, and Nepal

Madhuri Mathema

Introduction

South Asia, an extremely diverse region with a range of religious faiths including Islam, Hindu, Buddhist, Jain, and Sikh, is also recognized as a "patriarch belt" (Caldwell, 1982) where women are subordinated to men in kin-ordered social structure. These societies are ordered by a powerful ideology of female subordination, composed mainly of patrilineal-patrilocal families, with control of land, capital, and the female labor process firmly in male hands (Bardhan, 1986).

Both Pakistan and Bangladesh, secular states at the time of independence, have declared Islam the state religion. Nepal is by constitution a Hindu state. Both Islam and Hinduism hold strict views about how women should lead their lives. A great value is attached to virginity (purity) and chastity, which necessitates early marriage, segregation of the workplace, and restriction of women's mobility. Marriage is almost universal in these countries and women exercise little choice in whom they marry. The match is usually organized by male guardians, although a network of female relatives and friends may operate at an informal level to ensure suitable matches.

In addition to religion and culture, economic and political forces impact the lives of women in South Asia, who share high levels of poverty, high infant mortality rates, and the unique situation of lower life expectancy rates than men. Rapidly growing populations and low rates of urbanization are also common characteristics.

Gender Ideologies in South Asia

Economic situations within families and at the national level, as well as social, cultural, and religious norms in South Asia, shape gender ideologies that affect women's and girls' lives and potential. Practices such as dowry, early marriage, and purdah, along with cultural notions of women's roles in society, are strong forces in creating gender inequities in health, education, economic activity, and other social institutions, including the legal and political systems.

Female Seclusion (Purdah) in Pakistan and Bangladesh

Women in Pakistan and Bangladesh are not only segregated from men but are also subject to seclusion, or purdah (literally, "veil" or "curtain"). Institutionalized in the practice of purdah, the ideal is expressed in avoidance rules designed to regulate contact between the sexes and to segregate most male and female activities socially and spatially. Purdah is believed to have originated as a means of controlling women of the dominant feudal or tribal groups. This is clear from the uneven distribution of the practice of purdah in Pakistan, where the strongholds are the feudal families of Punjab and Sind, tribal Baluchistan, and the North West Frontier Province. Thus one of the underlying reasons for purdah may have been the desire to guard against the possibility of a woman having contact with males who are not related, thereby protecting the virginity and fidelity of the woman— the latter necessary to confirm the paternity of a son who is the sole heir of paternal inheritance. Thus it ensures that property remains within the family of patrilineage. Moreover, in Muslim society, male honor is said to be dependent on the honor of women, and purdah is one effective means of preventing the dissolution of male honor.

Though there are numerous references to gender equality in both the Qur'an (Koran) and the Hadith (interpretive moral codes based on the sayings of the prophet Muhammad) women are rarely allowed to attend religious gatherings, enter a mosque, or assist at public meetings and festivities. Besides the economic factor, some scholars argue that the fundamental rationale behind female seclusion is that it gives men needed protection from women. Muslim women are regarded as extremely powerful, capable of making men lose their reason through *fitna* (disorder or chaos provoked by sexual attraction), and threatening in terms of their potential to divert men's devotion from Allah. The Muslim social order, then, can be seen as an attempt to control women's power and neutralize its disruptive effects through the seclusion of women.

Among Muslims, purdah restrictions do not apply within the immediate kin unit but only outside it, a rule that stresses unity of kindred vis-à-vis the outside. The degree to which purdah is practiced and the particular form of its observance vary considerably in both Pakistan and Bangladesh, depending on class and regional background. Purdah is practiced more strictly by upper-class women as a sign of affluence. Women from poor households cannot maintain purdah in the strict sense, since they are forced to work outside of their homes for survival. While women in much of Bangladesh contribute to tasks outside of the home such as fetching water and fuel, the activities of poor women in Baluchistan, Pakistan, tend to be restricted to activities within the household compound, namely cooking, cleaning, and childrearing (Stromquist and Murphy, 1995). Most women believe that they observe purdah whether they work or not and wear a veil or not (Weiss, 1992). For many women, purdah refers to a decent way of behaving. How "decent" is defined is a reflection of a woman's particular life circumstances. In both countries, as families improve their economic standing they tend to seclude women as a symbol of social status.

The constraints imposed on women through purdah extend well beyond the parameters of family life to encompass all other spheres. The division of society into public and private sectors impedes a woman's access to information, education, employment, and, therefore, to independence and other forms of power. Purdah is not the only form of control in women's lives, however. Accessibility to public facilities is also a strong factor in South Asia. Where facilities outside the home exist (schools, jobs, political organizations, and the like), women do become involved. Where these institutions do not exist, options for women are more limited. Nevertheless, purdah has been changing, and, paradoxically, the restrictions in the 1990s appear to be less severe for upper-class women, especially in relation to socioeconomic activities.

Purity of Caste and Control of Women's Sexuality in Nepal

In Nepal, as in Pakistan and Bangladesh, interpretation of cultural norms is influenced by social status. Among high-caste Hindus, purity of caste is considered very important. To maintain purity, the caste system restricts interaction and assimilation through sanctions enforcing caste-specific behavior and attributes. High-caste people assert that caste membership is determined by birth. They also claim that castes are ranked on the basis of ritual cleanliness. At the outer layer, to maintain caste purity, the higher castes (Brahmin or Kshatriya) are not supposed to mingle with lower castes or eat or drink from the hands of lower castes. But at the inner layer, it comes down to restricting women's sexuality and reproductive powers so the paternity of a child, especially of a son, is not in doubt. Only through such control can purity of lineage be maintained, a purity that is vital to maintaining a caste system. In Nepal, the political component of caste has been definitive through the medium of legal codes on caste identity and interaction. The national code (Mulki Ain of 1962) allows a man to bring a second wife home if the first wife fails to bear a son, thus keeping the Hindu patrilineal ideology intact. The high-caste Hindus in Nepal are also governed by Hindu patrilineal ideology, which holds that lineage can be maintained only by sons. It is only the sons who can perform funeral ceremonies that ensure the spiritual peace of the departed soul.

The Hindus are so concerned about the purity of their women that they require a bride to be a virgin and forbid widow marriage (widows are regarded as polluted). One way of ensuring virginity is to have marriage arranged before the onset of puberty, a practice commonly known as *kanyadan* (gift of a virgin daughter). *Kanyadan* is also considered to be one of the greatest religious duties that parents can perform; it secures them a place in heaven. Though the national code forbids child marriage, in rural areas it is still practiced by the high-caste Hindus. In rural Nepal, most of the girls are married by the time they are 14 or 15 years old. Many parents fear that their grown-up daughters might go astray and bring shame to the family (Mathema, 1992). Parents usually do not allow their unmarried grown-up daughters to frequent the marketplace for fear that they may develop a relationship with men whose caste is not known.

Hindus have contradictory views about women. In their own natal homes before puberty, daughters are considered pure and virgin; the idea of impurity creeps in as they reach puberty, which is associated with their sexuality. The nascent unattached sexuality is considered an anomaly that endangers a woman and her consanguineous male relatives.

The groom's family views the bride as an outsider. It is only the birth of a son that gives a woman common interest with the patriline. Hence, the birth of a son is as important to a woman as it is to a man. It is the son who gives a woman a daughter-in-law who works for her and over whom she, as mother-in-law, exercises supreme authority.

There is no purdah in Nepal, but a similar notion is expressed in "respect–avoidance" idioms—a women should respect all of the elderly males but, at the same time, avoid closeness with them. It is a distancing technique used in "situations of ambivalence, ambiguity and imminence of role conflict" (Bennett, 1983). It also serves to eliminate suspicion of sexual relations between the daughter-in-law and her father-in-law or elder brother-in-law by replacing ambiguity with strict prescribed mutual behavior patterns.

Fear of, and suspicion about, a woman in Hindu society also deprive her of owning any immovable property. The concept that a woman does not need property because she herself is a part of her father's or husband's movable property is found in many Vedic and epic writings. Hindu caste purity associated with female purity suggests underlying economic motives in retaining property within the patriarchal family.

Besides these ideas of purity and pollution, and respect and avoidance, there is also a long list of codes of conduct (virtues) prescribed for women, and violation of these codes may have serious consequences in women's lives. A woman should avoid any sexual contact with a man outside of marriage, should worship her husband like A god (even though he may be devoid of any moral virtues) and be faithful to him, should always serve men first, should not laugh, and should not frequent the "public" places. A widow should lead a celibate life (the same rule does not apply to widowers), should not view festive occasions, and should practice self-discipline and other ideologically identified austerities.

Though most of the high-caste Hindu women do take part in religious fasting to honor the longevity of their husband, keep a certain distance from their consanguineous males, and observe purity and pollution rules, not everybody observes the strict Hindu codes of conduct in everyday life. There are many instances in rural Nepal in which young unmarried girls become pregnant, widows remarry, married women run away with other men, lower-caste women marry higher-caste men or vise versa, and married women bear another man's child. These anomalous situations are known to village people but are not talked about in public. Nonetheless, such women are ostracized in overt or covert ways, in many instances left alone to fend for themselves (Mathema, 1992).

Though Nepal is a Hindu state constitutionally, not everybody there is Hindu. There are more than 200 non-Hindu ethnic groups that have their own cultural practices, traditions, and history. Many of them belong to Tibeto-Burman ethnic groups that practice Buddhism and have their own social organizations. Among some of these groups, polyandry (having more than one husband) is practiced such as those in the Mustang district near Tibet. The Tibeto-Burman women lead a less-restrictive life than their Hindu counterparts and are involved in many enterprising activities, such as running tea shops, brewing beer and selling it in the market, and other kinds of trade. Some of these groups have matrilineal succession systems, and husbands move to their wives' natal homes. In some areas, there is also bride price or bride service, in which men are required to offer services to the wife's family. Some groups practice cross-cousin marriage while others forbid men to marry women from their mother's natal clan. Many Tibeto-Burman ethnic groups do not practice child marriage. Virginity of the bride is not stressed. Because of the dominating influence of Hindu culture, many ethnic groups have incorporated some of the Hindu practices into their own social organization. However, in none of these groups do women inherit immovable property. These ethnic groups have received less attention and have remained peripheral. A few studies have been done of them by Western scholars, and some of these ethnic groups, especially the educated ones, have challenged the ways they were portrayed by native or foreign scholars. Political change in Nepal—a multiparty parliamentary democracy with the king as the constitutional head of state replaced a previous autocratic kingdom in 1990—has inspired many non-Hindu groups to speak up about their own history and culture, and they are demanding to be recognized. Although the national code states that customary practices and traditions are allowed to exist, if they violate the fundamental law of the land or other formal or state laws then the law of the land shall prevail.

Cultural practices and religious beliefs in Nepal, Bangladesh, and Pakistan, as in other parts of the world, are strong forces in shaping the conditions in which people live. While these influences do not produce unidimensional, unchanging social systems, there are some similarities. These three countries share high levels of poverty, for example, as well as gender ideologies that include many gender-based double standards and restrictions on women, and governmental and development policies that either do not address women's needs or do not adequately support implementation of efforts that do address their concerns.

The Condition of Women in South Asia

These three South Asian countries have a rapidly growing population with low urbanization rates. In 1981, population growth for Bangladesh, Nepal, and Pakistan was 2.31, 2.66, and 3.06 percent. In 1991, it was 2.16 for Bangladesh, 2.08 for Nepal, and 3.10 for Pakiston (UNICEF, 1992). The total populations of Bangladesh, Nepal, and Pakistan in 1994 are estimated at 117.8, 21.4, and 136.6 million (UNICEF, 1996). These growing populations are predominantly rural. Less than 10 percent of Nepal's population and about one-third of Pakistan's live in cities. Proportionately fewer are urban: 4–6 percent of Nepal's women are urban dwellers, 8–14 percent of Bangladeshi's, and 26–28 percent of Pakistan's (United Nations, 1991).

Despite the fact that approximately 80 percent of the people in Bangladesh are directly dependent on agriculture for their income, the ownership of land has become increasingly concentrated in the hands of a few. Eleven percent of rural households own more than 52 percent of all land, while more than 30 percent own no land at all. The landless and those owning less than half an acre constitute 48 percent of the rural population (Schendel, 1981). Studies have suggested that in the mid-1980s the richest 10 percent of the village population controlled between 25 and 50 percent of the land, while the bottom 60 percent of the population controlled less than 25 percent. This situation is accompanied by high rural underemployment and a decline in real agricultural wages. In addition, Bangladesh is confronted with frequent natural disasters. It has been estimated that 62 percent of rural people receive enough income to satisfy 90 percent or less of the necessary intake of calories (Schendel, 1981). Poverty is overwhelming, and women account for the largest share of the poor. As in other areas, because women tend to be responsible for meeting basic needs of the family, they are especially hard hit by the difficult situation. As men move out in search of employment (both internal and international migration), women are left alone to fend for themselves and their children. They, too, are seeking employment in a labor market in which access has been restricted on the basis of gender.

Similar conditions prevail in Nepal. Although 90 percent of the people depend on farming, 51 percent of the households are landless and near landless (Shrestha, 1990). Poverty, together with cultural restrictions on women's mobility, make for severe conditions in women's lives.

Although both Bangladesh and Pakistan have elected women as national leaders (Khaleda Zia in Bangladesh and Benazir Bhutto in Pakistan), the lives of the majority of women have not changed dramatically. Women in South Asia lag behind men in the areas of health, education, labor-force participation, and legal status. In Pakistan, the lives of women in the middle and upper classes contrast starkly with those of the majority of Pakistani women. The more privileged women hold jobs in professions that have opened up to them relatively recently: In 1990, they were one-third of the teachers and medical doctors. The cloistered women of the lower middle class are perhaps the most restricted of all Pakistani women (Khan, 1993; Shaheed and Mumtaz, 1993). Change is perhaps more likely in the higher classes, yet even these women are constricted by social norms and attitudes that devalue women and identify them in relationship to male family members. The government-sponsored Commission on the Status of Women report summarized the conditions of Pakistani women thus:

> The actual status of women . . . is today at the lowest ebb. Women in general are dehumanized and exercise little control over either themselves or affairs affecting their well-being. They are treated as possessions rather than as self-reliant self-regulating humans. They are bought and sold, beaten and mutilated, even killed with impunity and social approval. They are dispossessed and disinherited in spite of legal safeguards. The vast majority are made to work 16 to 18 hours a day, without any payment. . . . The participation . . . of women in national life is marginal and most of them are still mute spectators of the changes taking place around them (PCSW, 1985, p. 3).

Discrimination against women manifests itself in their poor health and survival rates, limited levels of education and labor-force participation, and even in inequitable legal systems. To compound this situation, the availability of schools, health care, and employment opportunities that agree with cultural and religious restrictions are scarce for girls and women and further reduce options in these areas.

Women and Health

Women's subordinate position in society and within the family manifests itself in their poor health and survival rates. Women are less than half of the population of all three countries. For Bangladesh, the World Bank (1989b) documented a significant decline in the proportion of fe-

males in the 0–4 age group from 98 females per 100 males in 1965 to 94 per 100 in 1985. Pakistan shows the largest discrepancy, however, at 92.1 women per 100 men (United Nations, 1991). South Asia is the only region of the world where women do not have a higher life expectancy than men (Khan, 1993; United Nations, 1991). Both female children's and mothers' mortality rates are high, which leads to men outnumbering women in the population (World Bank, 1990).

Maternal mortality is estimated at 570 per 100,000 live births in Bangladesh, 600 per 100,000 in Pakistan, and a staggering 1,500–2,000 per 100,000 births in Nepal (World Bank, 1990). Each year, 30,000 women in Bangladesh die from causes related to childbirth and pregnancy (World Bank, 1990). Females have a higher mortality rate than males in Nepal and Bangladesh at all ages; there is indication that this gap is widening. In Pakistan, the mortality rate for children age 2–5 shows the widest gap: 54.4 deaths per year per 1,000 population for girls and 36.9 for boys (United Nations, 1991).

Several factors have contributed to this dismal picture: inadequate food intake throughout the lifetime because of preferential intrahousehold food allocation to males, the young age of mothers and short intervals between pregnancies, traditional beliefs and practices around childbearing, and lack of prenatal and postnatal care. Cultural norms related to purdah often prevent consultation with male doctors and health staff when female staff are not available. Not only are there limited health facilities, especially in rural areas, but there is a dearth of female health professionals.

Chronic malnutrition among pregnant and lactating mothers is reported in Bangladesh and Pakistan. In Pakistan, 45 percent of pregnant and lactating women suffer from iron-deficiency anemia, and vitamin A deficiency is also found to be highest within this group. In Nepal, most pregnant women are affected by iron-deficiency anemia (UNICEF, 1992). Almost all low-income pregnant women in Bangladesh weigh less than 50 km (about 110 pounds) (World Bank, 1990). Female children have almost three times the rate of malnutrition as male children in Bangladesh, and, among severely malnourished children, the mortality rate of females is 45 percent higher than that of males (World Bank, 1990). In South Asia, it is common for the men to eat the most and the best, leaving the women and children to eat last, and then mothers feed their sons the best of what is left (UNICEF, 1996). Because the boys are valued as economic assets and for old-age security more attention is given to boys' health.

Early marriage and childbearing practices also adversely affect women's health. Starting at age 15, most women in Nepal are either pregnant or lactating through the majority of their childbearing years. Although, by law, marriage is not permitted under age 16, 40 percent of girls are married by the age of 14, with as many as 7 percent by the time they are 10 years old. An estimated 40 percent of girls bear at least one child when they are between the ages of 15 and 19 (World Bank, 1989a). In Bangladesh, 73 percent of women who have ever been married were married by the age of 15, and 21 percent of these women had at least one child by that same age (United Nations, 1991). The total fertility rate is about 5.9 in Nepal, 6.5 in Pakistan, and 5.5 in Bangladesh (United Nations, 1991). A heavy work load for women—both for income and in the domestic realm—and a lower worth attached to women's lives compound their health condition.

Women and Education

Cultural norms and beliefs that relegate women to a lower position in the family and society, parental attitudes toward female education, a lack of available educational facilities for females, and the effects of poverty adversely affect women's educational participation and achievement.

Economic costs in educating girls are high for poor families. Girls are frequently not sent to school beyond primary level. The forgone labor of girls who are needed for work in helping with domestic chores, looking after siblings, and, in Nepal, grazing cattle and collecting fodder and firewood is too high a cost for many families to bear (World Bank, 1989a). Additionally, the notion that "a daughter belongs to somebody else's house" also discourages families from investing in a daughter's education (Mathema, 1992). Education is often seen as interfering with the controllability (obedience) of women and creating the possibility that an educated woman will bring shame to a family. Men often show their fear by saying that educated women are too "independent and hard to control" (Khan, 1993, p. 227). Whereas men's increasing education has led to a shift from bride price to dowry demand, women's education has lowered their chance of marriage (Rozaria, 1992). Where dowry is customary, the higher dowry needed for more highly educated daughters discourages otherwise willing parents to continue their daughter's education (Khan, 1993). Additionally, fees are charged beyond primary-school level, and it is at this stage that girls are pulled out of school while boys are allowed to continue.

Lower educational attainment by girls is also attributed to lack of female teachers, distance of school from home, lack of all-girls schools (especially from secondary school onward), and negative attitudes of society toward female education. At most, girls attend up to primary school, but as they reach puberty they slowly drop out and do not continue schooling. Moreover, there are few secondary schools exclusively for girls in communities where co-education is not culturally acceptable.

Educational expenditure and policy implementation directly affect the availability of educational facilities for girls. Pakistan (in 1991), Bangladesh (in 1992), and Nepal (in 1995) spent, respectively, 2.3 percent, 2.0 percent, and 2.7 percent of their gross national product (GNP)—and Bangladesh and Nepal spent 8.7 percent and 7.8 of their total government expenditure—on education (United Nations, 1995). These low levels of spending result in limited access for girls to formal education. There are one-third as many women's schools as there are men's schools in Pakistan, for example (Shaheed and Mumtaz, 1993).

In Pakistan, the average national literacy rate in 1993 was 34 percent, while women's literacy was 21 percent. In rural areas, literacy was lower: 9.3 percent for women and 26.6 percent for men. In Nepal, the national literacy rate was 25.5 percent, 38.7 percent for men and 11.9 for women (Asian Development Bank, 1995).

Enrollment rates in all three countries are low overall, with lower rates for girls. In Bangladesh, despite compulsory education, only 60 percent of the eligible primary-school-age children are enrolled. In Pakistan in 1990 about 30 percent of boys and 3 percent of girls in rural areas were enrolled were in school and fewer than one-sixth of them completed five years of education (AED, 1994). Although the number of girls enrolled in school has been increasing, the gaps between male and female enrollment levels remain considerable. In Nepal, female enrollment is half that of male enrollment. About half of the boys and one-third of the girls age 6–10 are actually enrolled in Nepal; of those, 35 percent drop out without completing the first grade, and more than half drop out without completing primary schooling (World Bank, 1989a). Dropout rates among Pakistani girls are even more disturbing. Only 3 percent of rural girls were still in school by age 12, compared with 13 percent of boys; by age 14, fewer than 1 percent of girls remained, compared to 7 percent of boys. At higher levels of education the picture is somewhat better, though young women are still underrepresented. They are 15 percent–30 percent of all students, averaging 26 percent at levels beyond the intermediate grades (grades 11–12) (World Bank, 1989a). In Bangladesh by the early 1990s, primary completion rates had risen to about 43 percent; girls represented 45 percent of those enrolled in the last year of primary school but only 37 percent in the first year of secondary school (Stromquist and Murphy, 1995).

The prevailing culture, which values women's reproductive capacities much more than their productive ones, inhibits investment in education. Purdah observance also restricts girls' mobility outside the home and makes it necessary for separate schools to be provided for boys and girls. Provision of separate facilities and the low numbers of trained teachers who are female (and available to teach) reflects a gender bias. Only about one-third of the primary-school teachers in Pakistan are women. Parents are reluctant to send their girls to distant schools even if all-girls schools are available, without supervision, implying potentially higher costs to educate girls. Parents perceive the cost of educating girls to be too high and the benefits to accrue not to themselves but to others.

In some areas, however, there is evidence of changing priorities and practices. With increasing poverty and an eroding material base, people in Bangladesh are looking at education as the only alternative with which to improve their lives (Rozaria, 1992). Slowly, cultural values are changing. Most parents want education for their children—both boys and girls (Weiss, 1992).

Education could free women from their social shackles, or (gender-specific) education could further consolidate the constraints on women (Shaheed and Mumtaz, 1993). Despite this tension in Pakistan, and the dismal picture throughout the region, the gender gaps in schooling have not been ignored. Since the mid-1980s, Bangladeshi nongovernmental organizations (NGOs), in conjunction with the government, have made tremendous strides toward providing education to most disadvantaged children, especially girl children. The most prominent of these efforts is the Bangladesh Rural Advancement Committee (BRAC), which has programs for girls not in the formal system. BRAC's nonformal primary-education program offers a three-year curriculum that does not alienate rural children and a schedule that is adapted to the needs of the local community. The teachers are paraprofessionals, recruited from the villages and trained, and are mostly women. In the early 1990s BRAC was running nearly 30,000 schools with 900,000 students; it aims for enrollment in which girls make up 75 percent of all student enrollment. Dropout rates are lower in the BRAC schools than in government schools: There is a 95 percent completion rate, and nearly 80 percent of the students transfer into grade 4 of the public-school system (Stromquist and Murphy, 1995)

In its Sixth Five-Year Plan (1983–1988) Pakistan initiated the Mosque Schools Program to extend schooling to rural children, particularly girls. Unfortunately, the enrollment of girls in these schools has been very low. Several other governmental initiatives, while optimistic and impressive in the planning stages, have been failures (Shaheed and Mumtaz, 1993). Some nongovernmental initiatives have been more promising. One such program is a system of neighborhood-based home schools in low-income Karachi neighborhoods. Literate women in the neighborhoods run classes in their homes for dropouts and girls not in school. What began as an effort to provide an income for the women and literacy for girls has also become a forum for social action and change. The women monitor child growth, for example, and the group has established a women's training and production center (Shaheed and Mumtaz, 1993). Community-based schools for girls are emerging in Baluchistan (Stromquist and Murphy, 1995). The Society for Community Support for Primary Education in Baluchistan, popularly known as "the Society," established in 1993, has been very active in mobilizing parents and communities for girls' education. By 1994, the Society had established 198 girls' primary schools (Stromquist and Murphy, 1995).

Other indications signal change in cultural norms. A random survey of lower- and middle-class urban parents in Pakistan in the early 1990s revealed that the average man considers education to be a necessity for all of his children, girls and boys. Most of the interviewees were stretching limited resources to send their children, including their daughters, to school. While this changing trend is optimistic, daughters tended to be sent to cheaper and less prestigious schools than sons. To the surprise of some govern-

ment officials, the Pakistan census revealed that about 28,000 girls (21 percent of all girls enrolled in primary school) in Baluchistan, the poorest province, were attending boys' schools. This signaled to the government that families wanted more educational opportunity for daughters. Perhaps this is indicative of a positive trend (Stromquist and Murphy, 1995).

In Nepal, the government has attempted to improve girls' education by distributing free textbooks in grades 4 and 5, providing scholarships and uniforms for 5 percent of all girls, increasing the numbers of female teachers, and organizing nonformal classes for young girls who cannot attend the regular schools (Shrestha, 1993). Many foreign agencies are involved in this endeavor. In one project, low-caste girls attend classes in the early morning before household chores begin. Teachers are locally recruited and trained, and learning is based on the girls' own experiences and not on irrelevant curricula and content (Khan, 1993).

Women and Labor-Force Participation

Lower educational levels are one factor in women being relegated to particular areas of work and restricted from participating in others. In Bangladesh, the Labor Force Survey of 1984–1985 (World Bank, 1990) showed that, as a proportion of all sectors of employment, females predominated in only one occupational classification in the household sector, where they constitute 79 percent of the total. The other sectors in which females constituted a significant portion were manufacturing (24 percent) and community and personal service (15 percent). In all other sectors, females were 5 percent or less of the total employed (World Bank, 1990). Official labor-force statistics have yet to recognize the vital role women play in national agricultural production.

However, a survey commissioned by the United Nations Development Program (UNDP) and the United Nations Development Fund for Women (UNIFEM) (World Bank, 1990) indicates gross underestimation of female participation in the Labor Force Survey. It showed that more than 54 percent of rural females who listed agriculture as their primary occupation after housework were in the labor force, as were 21 percent of the women who listed agriculture as their second occupation. Some 60 percent of the landless and virtually 100 percent of the female-headed households reported female income-earning activities (waged or self-employed).

Due to methodological and definitional problems of what constitutes work, conventional classifications of work in measuring female labor-force participation are grossly inadequate. Women's home-based work, which constitutes a vital portion of total family income, is ignored. The World Bank indicated that 28 percent–47 percent of total family income is from homestead production, which includes vegetables, fruit, spices, and the like. About 50 percent of the labor opportunities for rural women come from rice processing, with many others engaged in processing other crops. Forty-one percent are hired for domestic work, including cooking (World Bank, 1990).

When they are hired, women are usually hired at lower wages and levels and as part-time workers in order to avoid government regulations. On the other hand, female participation in nontraditional jobs, such as export-oriented industries and construction, is increasing. Changes in Bangladesh's economy, including export-led growth and privatization and the New Industrial Policy (NIP) of the 1980s, and the related inability for many households in the rural middle class to meet subsistence costs through a family wage, have been reflected in the creation of new employment opportunities and demand for female workers (Feldman, 1992); 80 percent–90 percent of the 200,000 employees in the new export-manufacturing firms in Dacca are educated, unmarried young women between the ages of 12 and 21. This is the segment of the population that is usually more restricted by purdah than the poor, who, by necessity, must engage in work that takes them outside of the household.

The process of creating jobs and the desire for them shows that ideologies such as purdah can be manipulated and reinterpreted. Recruiters tend to be from the girls' village of origin and act as paternalistic protectors. Collective living arrangements have been created. While these accommodations were intended to preserve the women's honor and to respect religious and cultural norms through supervision and segregation, they have also enabled women to come together in ways previously unavailable to them. New patterns of urban migration, collective living arrangements, unaccompanied use of public transportation, increased market activity, new consumption patterns, and new forms of political expression have resulted (Feldman, 1992).

Given a cultural milieu in which women's work for income generation is still the source of social disapproval, and given the absence, in many cases, of market orientation and the home-based nature of work, it is difficult to distinguish women's economic activities from domestic work. Moreover, inappropriate data collection methods, inappropriate definitions of activities, stress on recording only one activity, and the cultural reluctance to admit to women's working have led to a gross underestimation of female work participation in Pakistan (Kazi and Raza, 1990).

In Pakistan, only one-third of workers are wage earners, and most live in cities. While others, mainly rural, tend to be self-employed or unpaid family helpers. Some categories of work even in cities are not classified as employment: these workers are mostly women and include housemaids, cooks, cleaners, and pieceworkers at small factories. This gap in data collection, along with the failure to record female work in family enterprises, contributes to the chronic underrepresentation of female employment in the statistics (Asian Development Bank, 1995).

Women constitute a small but rising portion of professionals and related workers. Between 1984 and 1988, their numbers rose from 15.5 percent to 18.3 percent of the total. The major increase was in teaching and medicine:

In 1988, nearly one-third of all teachers and one-fifth of all doctors were women. Among other white-collar jobs, such as managerial and clerical positions, the proportion of women is still extremely low, though rising: Although only 2.9 percent of clerical workers were female, this rate nearly doubled between 1984 and 1988 (Kazi and Raza, 1990).

In Nepal, the 1989 Multipurpose Household Budget Survey (Nepal Rastra Bank, 1989) estimated that 71.7 percent of the population was economically active (those age 10 or older who worked with or without pay during the year in one of the occupational categories used). Of the female population age 10 and older, 67.9 percent worked for pay or profits or as unpaid family workers. In rural areas, 73 percent of those 10 and older were found to be economically active, of whom 69.9 percent were females and 76.1 percent males. The percentage of females working in government was found to be very small—0.3 percent, compared to 4 percent of the males at the national level. A majority of economically active females worked as unpaid family workers, with 68 percent of all rural and 48 percent of all urban females in this category, compared to 24 percent of rural and 13 percent of urban males. The survey also indicated that 70 percent of the economically active population—86.1 percent of the women and 72.9 percent of the men—were engaged in agriculture.

Women and Law

The legal status of women, and legal protections for women, in South Asia are directly related to religious and cultural notions of gender. In Bangladesh and Pakistan, the legal system is influenced by Islam, with fundamentalism exerting increasing direct pressure in Pakistan. In Nepal, as in the other two countries, a double-standard based on gender is in place. In areas of family law, which typically covers marriage, divorce, child custody, and the like, goals of reproduction and clearly identifiable paternity are paramount. In addition, while some legal protections exist, enforcement is less than adequate. This section describes some of the prominent features of the legal systems that affect women in Nepal, Bangladesh, and Pakistan.

Nepal. The constitution of the Kingdom of Nepal, promulgated in 1990, includes protection against discrimination on grounds of religion, race, sex, caste, tribe, and ideology. Nepalese women, however, do not enjoy equal property rights with men. Women do not inherit paternal property, especially land. But as long as they remain faithful, married women have a right to their husband's property. An unchaste wife is punishable by law and disinherited from property, especially land, but the same is not true of men. While a son has an inalienable right to his father's property, a daughter has such right only after she reaches age 35 and remains unmarried. A woman loses her right to her husband's property upon divorce. Similarly, a widow loses her right to property should she enter another union or indulge in sexual activity. Thus, a married woman's security is firmly tied to her chastity. The traditional Hindu ideal of absolute female purity is reinforced by economic sanctions that do not apply to men. Since men have rights to property by virtue of the biological fact of birth, their economic security is not tied to their sexual morality.

Marital status and age determine the few property rights a woman has. A woman is entitled to paternal property after she reaches age 35 and if she remains unmarried. But she does not have any legal recourse if she is neglected by her father or brothers. A married woman has, at least, a right to claim her share of property if she is neglected by her husband; a widow can do the same if she is neglected by her husband's family. However, young widows have been blamed for adultery so as to bar them from claiming their husband's property. A widow who is younger than 30 cannot claim her share of property as long as she is provided with food and clothing and expenses for religious observance. In all of these cases, obtaining proof of neglect is very difficult for a woman, whose position within the family and the society is so subordinate.

Related to the problem of divorce and separation is that of child custody. Nepalese law considers child custody and maintenance of the child the right and responsibility of the father. Though an amendment to the National Code in 1975 has made some changes regarding custody and gives the mother important new custody rights, the provisions of the law are not clear. However, even in the new law the mother loses her right to custody if she remarries.

The legal code accepts bigamy if the wife fails to bear any children, especially a son: It assumes that it is a woman's fault if she fails to conceive a child. But the law does not recognize as valid the tradition of polyandry, which is customary in northern regions of the country. In effect, the laws concerning marriage, divorce, adultery, and bigamy mandate a double standard: Higher standards of behavior are enforced and harsher punishments are meted out for women than for men.

Bangladesh. The Bangladesh constitution of 1977 grants equal rights to women, but some discriminatory provisions are found in legislation concerning marriage and divorce, among other things. Unequal treatment of women within the family, workplace, and educational institutions is rampant. Despite a commitment to equal access to educational institutions, women are denied employment in, and admission to, the Islamic University and the Islamic Center for Vocational Training and Research (World Bank, 1990).

Most of the family laws in Bangladesh relating to marriage, divorce, custody of children, inheritance, and maintenance are governed by religious laws. In the areas of marriage and divorce, significant changes occurred as a result of the Muslim Family Law Ordinance of 1961 (as amended) and the Muslim Family Law Ordinance of 1982 (an amendment). Still, a Muslim wife, unlike her husband, does not have a unilateral right to divorce. A husband does not need to go to court but a wife must litigate in this forum, which calls for an unnecessarily long procedure.

Polygamy is restricted somewhat. Under the 1961 ordinance, husbands who wish to take additional wives are required to notify the chairman of the arbitration council. Although violators are subject to punishment, few men obey the law. Wives who lodge complaints rarely receive justice because of legal and socioeconomic barriers (Bhuiyan, 1986).

The Dowry Prohibition Act of 1980 made giving and taking a dowry an offense punishable by fine and imprisonment, but the law has not been enforced, and violence related to dowry has increased (Bhuiyan, 1986). It was a campaign launched by Mahila Parishad, a women's organization, against dowry practice and the attendant violence against women that sparked government passage of some of the laws, (such as the 1980 Dowry Prohibition Act and the 1983 Cruelty to Women [Deterrent Punishment] Act), to initially address these issues (Kabeer, 1991).

With regard to labor laws, women enjoy many benefits, including maternity leave (six weeks before and six weeks after giving birth), child-care facilities, and exemption from night work in factories. However, they are often forced to work overtime in appalling conditions in garment factories, and, because many women work as casual or temporary workers, they do not benefit from these legal provisions. Labor law, in reality, acts against women: Many industries and factories do not want to hire them full time and so have to give them the benefits prescribed by law (World Bank, 1990).

Although a signatory to the World Plan of Action of the United Nations Decade for Women, the Bangladesh government refused to ratify a number of clauses relating to inheritance, marriage, child custody, and divorce in the United Nations Convention on the Elimination of All Forms of Discrimination Against Women (CEDAW) on the grounds that they conflicted with Sharia law (the socioreligious law of Islam based on the Qur'an and Hadith).

Pakistan. Islam has also had a profound influence on Pakistan's legal system. One result of the Islamization program during the regime of Prime Minister Muhammad Zia ul-Haq (1977–1988) was deterioration in the condition of women in the arena of law. Since 1979, fundamentalist groups have labored mightily to redefine the proper demeanor for women and to push women back into their "*chador and chardevari*" (veil and four walls). The rights hard won by women during prime ministers Ayb Khan and Zulfiqar Bhutto's time came under serious attack during Zia's Islamization process. Mullahs (religious leaders) became instrumental in extending the scope of Sharia law. New rules of evidence and retribution have been promulgated that have adversely affected women and threaten to take them back to medieval ages. All venues affecting women are under question, including dress, choice of marriage partner, divorce, inheritance, and education.

Of prime concern to women are the Enforcement of Hudood Ordinances of 1979, and the Law of Evidence of 1984: The 1979 law excludes women's evidence as proof for awarding maximum punishment while the 1984 law accords their evidence only half the status of male testimony. Another law affecting women is Offense of *Zina* (adultery) and *Zina-bil-jabir* (rape) Act. It states that a man and a woman commit *zina* if they willfully have sexual intercourse without being validly married to each other. Although severe punishment could be prescribed for such an act, rules of evidence make it less likely for men to receive the maximum penalty. If the accused woman is a virgin or is married, the maximum penalty for the woman is death by stoning; if she is widowed, divorced, or a prostitute, the penalty is 100 lashes. Self-confession or the testimony of four Muslim males of known moral repute is sufficient for establishing the guilt of the suspects. Without such evidence, the penalty is at the court's discretion. The ordinance does not address the problem of an unmarried woman subjected to rape whose pregnancy could incriminate her in adultery. Because the law says that women's evidence alone is insufficient for maximum punishment, there have been innumerable cases in which victims of rape have been convicted of adultery while the accused rapist has been released for lack of evidence (Mumtaz and Shaheed, 1987; Kandiyoti, 1991).

With the enactment of the *Qisas* (retaliation) and *Diyat* (blood money) Ordinance in 1990, the compensation for a woman, if bodily harmed or murdered, is to be half that for a man. Yet if found guilty of murder, a woman is to receive the same sentence as a man. Such discriminatory law stipulates in black and white that the life of a woman (and non-Muslims, for they are categorized similarly to women) is worth half that of a Muslim man. The law also allows offenders to absolve themselves of the crime by paying compensation to the victim or heir. This means that wealthier persons can, literally, get away with murder.

Not all Pakistanis accept these inequities, however, and women have fought back. The Women's Action Forum (WAF), a group formed by educated urban middle- and upper-class women in 1981 in Karachi after a military court sentenced a 15-year-old girl to flogging because she married a man of a lower-class background contrary to her parents' wishes, has, since its inception, organized protests against discriminatory laws and the Islamization process (Mumtaz and Shaheed, 1987). A number of women's organizations, most notably the All Pakistan Women's Association, have given support to WAF in advancing women's rights (Jalal, 1991).

Conclusion

Challenges to gender inequities occur in all parts of the world, be they the action of one woman resisting her husband or a mobilized group of women directly protesting government actions and demanding change. The path that those challenges take reflects the particular situation as well as the larger cultural and social contexts in which they arise. Just as cultural and social contexts are not static or unchanging, neither are gender ideologies. The particular relations between the state and religion in the three countries discussed shape different notions of how gender is used and viewed in society. Both forces have used gender for their own purposes. Creating spaces in highly restrictive societies in which women can more actively analyze and challenge inequitable social relations and practices can be difficult. Still, even in hostile environments, women are finding ways of shaping and challenging their own lives, households, communities, and nations.

References

AED. *The Primary Education Development Program, Pakistan.* Washington, D.C.: Academy for Educational Development, 1994.

Asian Development Bank. *Key Indicators of Developing Asian and Pacific Countries.* Manila: ADB, 1995.

Bardhan, Kalbana. "Women: Work, Welfare, and Status: Forces of Tradition and Change in India," *South Asia Bulletin,* vol. 6, no. 1, Spring 1986, pp. 3–16.

Bennett, L. *Dangerous Wives and Sacred Sisters: Social and Symbolic Roles of High-Caste Women in Nepal.* New York: Columbia University Press, 1983.

Bhuiyan, Rabia. "Bangladesh: Personal Law and Violence Against Women." In Margaret A. Schuler (ed.), *Empowerment and the Law: Strategies of Third World Women.* Washington, D.C.: OEF International, 1986, pp. 48–51.

Butler, L. "Basic Education in Pakistan: Policies, Practice, and Research Directives," *Journal of South Asian and Middle Eastern Studies,* vol. 11, no. 4, Summer 1988, pp. 85–104.

Caldwell, John. *Theory of Fertility Decline.* London: Academic Press, 1982.

Feldman, Shelley. "Crisis, Islam, and Gender in Bangladesh: The Social Construction of a Female Labor Force." In Lourdes Benería and Shelley Feldman (eds.), *Unequal Burden: Economic Crises, Persistent Poverty, and Women's Work.* Boulder: Westview, 1992, pp. 105–130.

Jalal, Ayesha. "The Convenience of Subservience: Women and the State of Pakistan." In Deniz Kandiyoti (ed.), *Women, Islam, and the State.* London: Macmillan, 1991.

Kabeer, Naila. "The Quest for National Identity: Women, Islam, and the State in Bangladesh," *Feminist Review,* no. 37, Spring 1991, pp. 38–58.

Kandiyoti, D. (ed.). *Women, Islam, and the State.* London: Macmillan, 1991.

Kazi, S., and B. Raza. "The Duality in Female Employment in Pakistan," *South Asia Bulletin,* vol. 10, no. 2, 1990, pp. 1–8.

Khan, Shahrukh R. "South Asia." In Elizabeth M. King and M. Anne Hill (eds.), *Women's Education in Developing Countries: Barriers, Benefits, and Policies.* Baltimore: Johns Hopkins University Press for the World Bank, 1993, pp. 211–246.

Maskiell, Michelle. "The Impact of Islamization Policies on Pakistani Women's Lives," working paper, no. 69, Michigan State University, East Lansing, 1984.

Mathema, Madhuri. "Improving Rural Women's Lives: A Case Study in Nepal." Ph.D. diss., Stanford University, Stanford, California, 1992.

Mumtaz, K., and F. Shaheed (eds.). *Women of Pakistan: Two Steps Forward, One Step Backward?* London: Zed Books, 1987.

Nepal Rastra Bank. *Multipurpose Household Budget Survey: Study of Income Distribution, Employment, and Consumption Patterns in Nepal.* Kathmandu: Nepal Rastra Bank, 1988.

Pakistan Commission on the Status of Women (PCSW). *Report on the Pakistan Commission on the Status of Women.* Islamabad: PCSW, 1985.

Rozaria, S. *Purity and Communal Boundaries.* London: Zed, 1992.

Schendel, W.V. "After the Limelight: Long-Term Effects of Rural Development in a Bangladesh Village," *Bulletin of Concerned Asian Scholars,* vol. 43, 1981, pp. 28–34.

Shaheed, Farida, and Khawar Mumtaz. "Women's Education in Pakistan." In Jill Ker Conway and Susan C. Bourque (eds.), *The Politics of Women's Education: Perspectives from Asia, Africa, and Latin America.* Ann Arbor: University of Michigan Press, 1993, pp. 59–75.

Shrestha, C.K. *A Study on Supply and Demand for Nonformal Education in Nepal.* Kathmandu: World Education, 1993.

Shrestha,, Nanda. *Landless and Migration in Nepal.* Boulder: Westview, 1990.

Stromquist, Nelly P., and Paud Murphy. *Leveling the Playing Field: Giving Girls an Equal Chance for Basic Education: Three Countries' Efforts.* Washington, D.C.: Economic Development Institute, World Bank, 1995.

United Nations. *The World's Women, 1970–1990: Trends and Statistics.* New York: United Nations, 1991.

United Nations Children's Fund. *Perspectives on Children in South Asia: A Statistical Profile.* Kathmandu: UNICEF, 1992.

———. *The Progress of Nations.* New York: UNICEF, 1996.

United Nations Educational, Scientific, and Cultural Organization (UNESCO). *Compendium of Statistics on Literacy.* Paris: UNESCO, 1988.

———. *Statistical Yearbook 1990.* Paris: UNESCO, 1990.

Weiss, A.M. *Walls Within Walls: Life Histories of Working Women in the Old City of Lahore.* Boulder: Westview, 1992.

World Bank. *Nepal Policies for Improving Growth and Alleviating Poverty.* Washington, D.C.: World Bank, 1988.

———. *Nepal Social Sector Strategy Review,* vol. 1. April 19, 1989. Washington, D.C.: World Bank, 1989a.

———. *Social Sector Strategy Review,* vol. 111. Population and Health, Asia Region. Washington, D.C.: World Bank, 1989b.

———. *Women in Pakistan: An Economic and Social Strategy.* Washington, D.C.: World Bank, 1989c.

———. *Bangladesh: Strategies for Enhancing the Role of Women in Economic Development.* Washington, D.C.: World Bank, 1990.

Women in China

Li Xiaojiang
Liang Jun

Introduction

If we want to make an analysis of the condition of women in China in the late twentieth century, it is necessary to trace the history of this country; its vast expanse of land, deep-seated cultural tradition, modern revolutionary war, political movements, and ongoing social reform—all have exerted a great influence on the lives and development of Chinese women.

During modern times, Chinese women have experienced three historical periods that can be divided roughly into three stages:

Before 1949. Under the patriarchal-feudal system that lasted for thousands of years, the "three obediences [to father, husband, and son] and the four virtues [morality, proper speech, modest manner, and diligent work]" were behavioral principles and moral norms by which women had to abide. Generally, women did not take part in formal employment; they were engaged chiefly in auxiliary farm work and house work. Foot binding was one of the most corrupt customs that tied women down in old China. Women did not have any freedom in their marriage, they had to rely on males economically, and they had no right to receive an education or to participate in social activities. But women's position in the family was relatively stable, and mothers enjoyed some liberty and authority in family affairs.

1949–1978. The founding of the People's Republic of China in 1949 had a great impact on Chinese women's lives. In 1950, the Chinese government issued its first law—the Marriage Law of the People's Republic of China—which established a strictly monogamous system, abolished the child-daughter-in-law system by which little girls were sold to other families, and opposed parent-arranged engagements and marriages. The majority of women were freed from the fetters of oppression of the feudalistic marriage system.

In 1953, the Electoral Law of the People's Republic of China was published, which further defined equality of the sexes in respect to civil and political rights in society. To raise women's economic position, during the land-reform movement the government adopted the principle of allotting land according to the number of members in a family so as to ensure that rural women had the same opportunity as men to get land. In urban areas, women were encouraged to take part in various forms of work, initiating an upsurge in the number of women who stepped out of their homes and joined productive labor. In the mid-1950s, many working women learned to read and write during the illiteracy-elimination campaign. (The Chinese constitution guarantees that girls and boys have an equal right to education.)

During this 30-year period, Chinese women benefited from path-breaking gender reforms. In terms of occupation, for example, they enjoy equal rights with men to jobs, they receive equal pay for equal work, and their jobs are stable.

In old China, employed women constituted a mere 0.2 percent of the total female population, and female workers made up 7.5 percent of the total number of workers in the country. After the establishment of the People's Republic of China, the number of women in occupations increased, and the fields available for employment grew to include industry, agriculture, construction, traffic and communications, business, health care, education, the Communist Party and administrative positions, social organizations, and more. The number of women working in technical fields, especially in technologically advanced fields, also increased greatly.

In 1955, when China was still at an early stage of its socialist construction, the country's leader, Mao Zedong, said: "All the women who are capable of working should participate in work under the principle of equal pay for equal work. This goal is to be reached within a shortest possible time." The program was expected to be completed in less than 10 years.

During this period, women's jobs were stable. They were not affected by marriage and reproduction. Enterprises were not allowed to dismiss married and pregnant women, and women could have a 56-day leave after delivery. Beginning in the late 1970s, women who received a one-child certificate enjoyed a six-month leave to take care of their babies. Promotions, pay raises, and the welfare of these women were not to be affected by the leave.

The rapid economic and social liberation of women notwithstanding, their level of education was low. Moreover, influenced by frequent political movements and the Cultural Revolution (1966–1976), the "10-year upheaval," the enhancement of women's schooling could not keep pace with the development of society, and the lack of education checked their actual progress.

1978–. The reform in China that started at the end of the 1970s has offered many opportunities for the advancement of Chinese women. But because of their double burdens (i.e., working at home and in the workplace) and low levels of education—particularly among rural women, who make up 80 percent of the total female population—many women are feeling pressured.

The reform of the economic system of rural China in effect since 1978 has relaxed the relationship between peasants and the land. It also has changed the past unitary structure of the rural economy. Rural women have actively responded to the reform in two ways: Many have flooded into the cities, and many who remain in their villages are engaged in the processing industry, turning agricultural products directly into commodities. Those who stay in the village are middle-age or young, married women. Tied by household chores and their children, they become the major agricultural production force. To ameliorate their condition, they take up various jobs: They farm, participate in commodities production, and keep their houses at the same time. The important role they play in rural areas helps raise their status in society and in the family.

At the same time, large numbers of young girls leave their hometowns to seek opportunities in the big cities. The earliest group worked first as children's nurses in others' homes. It is these girls who "have trodden a country road in the city." Many did not want to return to their native villages: Some found jobs with the help of their former employers; some have started private businesses by means of resource-pooling; and a few settled down in the city and got married.

Shortly after that influx, another group of rural girls poured into the cities and took jobs as temporary workers, often in factories that needed workers with little technical knowledge. But these female workers' pay was low, and they did not have adequate job benefits. Like those who came before them and those who followed, those who leave their village are typically young and unmarried, cherishing the hope of changing their way of life rather than just making money.

By 1992, the number of female workers in cities and towns had reached 56 million, or 38 percent of all workers in China. They are the backbone of the city women and the direct beneficiaries and embodiment of the "socialist women's liberation." The majority of these women work in the service sector and in posts that do not require special skills; their lives basically consist of job and family. This dual role, and its attendant responsibilities, contributes to the dire experiences of many of these women, who are the main sufferers in the upheavals brought about by the economic reform in the city. Many of these women lack occupational and technical skills, and they have much heavier burdens of housework and childrearing than do males. Quite a few become "extra" (i.e., redundant) workers in the factories' "optimum classification" (the personnel adjustment conducted by factories to meet production needs in the economic reform cities). Some breast-feeding mothers are compelled into a "job-waiting" position.

There is another group of women in the city—the professional women—including scientists, technicians, teachers, doctors, administrative workers and clerks, leading party members, writers, and actresses. Their educational level is higher than that of ordinary women, even higher than that of many males. They are armed with professional skills and enjoy many privileges in society. Regarding their profession as their life cause, they identify themselves with their jobs. Some of them are highly respected in society. On their path of growth and in their work, they encounter almost no social obstacles such as sex discrimination so that they fully possess the same social rights as males. However, compared to male professionals, many of them have to work twice as hard to achieve recognition and success. In this way, some become fine examples of female self-reliance.

Social reform in China has not exerted a great pressure on intellectual women. Nevertheless, they have their own problems, such as lack of feminist self-awareness, conflict and adjustment between work and family, marriage difficulties, imperfectness in emotional life (i.e., less strong than men), and the "masculinization" (i.e., the strong imitation of men) of some successful women. The above-mentioned problems, to varying degrees, reflect profound historical contexts. Under the circumstances of relatively full equality between men and women, Chinese women intellectuals have engaged in theoretical explorations in the realm of society, culture, life, and spirit.

The Road to Women's Liberation

The road of Chinese women's liberation can be divided into three segments, each of which is discussed briefly here.

The Women's National Liberation Movement (late 1800s–1949). The first call for women's liberation was heard from constitutional reformers Kang Youwei, Liang Qichao, and others at the end of the nineteenth century who included a space for women's liberation in their blueprint to change China. They started a "foot-unbound" movement, set up schools for women, and initiated women's newspapers. However, they did not act only for the sake of women but also for the benefit of the whole Chinese nation, believing that the disintegration of the feudal society, the improvement of marriage conditions, and even men's self-realization all depended on women's liberation.

In the early twentieth century, quite a few upper-class women went to Japan to study. They were deeply impressed by the feminist trend they found and began to consider Chinese women's liberation. Qiu Jin is an outstanding example of these women. She devoted herself to the women's cause in the struggle against the corrupt Qing government. Her achievements and her *Letter to 200 Million Chinese Women* pioneered the way for the Chinese women's self-reliance movement.

Around the time of the Revolution of 1911 against the Qing dynasty, women from other social classes joined the revolutionary ranks, and the question of women's participation in politics was put forward publicly. But the idea of women's liberation resonated only among upper- and upper-middle-class women and had almost no impact on the majority of women. The May Fourth Movement in 1919, regarded as the prelude to the Chinese national revolution, was the true starting point of the Chinese women's revolution. At this time, women's liberation became an important constituent of the national revolution. From the May Fourth Movement onward to the founding of the People's Republic of China in 1949, the liberation of Chinese women was blended with the liberation of the Chinese nation from feudal and imperial control (in the anti-Japanese War [1937–1945] and with the whole society in the Civil War, which ended in 1949). Many proletarian women revolutionary heroines and soldiers emerged as fighters in the Chinese revolution and vanguards of the women's liberation movement. However, there seemed to be no clear banner for women's liberation throughout China. Women's major concern at that time was to combat feudalism and foreign imperialism as women were placed between clan and country. Whether it was in the Women's New Life Movement or the Women's Japanese Aggression Resistance Movement, in the Kuo Mingdang Party–controlled areas or land reform and war support in the Communist Party–controlled areas, women were all on the track of a counterfeudalist and counterimperialist national revolution. But the majority of laboring women were still living in backwardness, ignorance, and prejudice: The feudal society had not changed for them.

The Women's Movement in the Socialist Transformation Period (1949 to 1978). "Freeing Women" was the first policy objective for women sought by the Communist Party after the establishment of the People's Republic of China. The purpose was to free women from the husband's authority in the family, the idea being that women should be components not only of a family but also citizens of a society and of a country. Thus began a new stage of the women's liberation movement. The government, under the leadership of the Communist Party, called on women to join labor activities outside the home. It ensured that women got equal pay for equal work, and it issued the law for equal sex rights. It also implemented a series of policies concerning women's welfare, supported women's political participation, and helped set up women's organizations. The government's transformation of the conditions of women's existence and development made possible the realization of women's equal rights and extensive social participation.

Chinese women's revolutionary enthusiasm ran high, as they had been cruelly oppressed in old China. Through the Women's Federation, they actively answered the party's call and earnestly explored the way to liberation. In every period of China's advancement, such as land reform, the Assistant War to Korea, the Industry and Commerce Transformation, the upsurge of the socialist construction, and the "Cultural Revolution," which was a major effort to revitalize socialism in China, Chinese women were closely united around the government and the party as one of the main forces in society, and they have worked diligently for the country. Supported by the government, Chinese women have become liberated. However, this "liberation" has been obtained at the expense of women's consciousness as individuals and as a group with common interests, and it lays on women an extremely heavy burden of the dual role they have to play in society and in the family.

Women's Studies in the New Period (1978–). In the course of the social reform that began in the last years of the 1970s, there appeared a series of new problems that caused women to face unprecedented difficulties and, at the same time, stimulated women's enthusiasm for research about women.

The end of the 1970s and the beginning of the 1980s was a period of constant problems for women. The concept of the superiority of males was gaining ground in the family planning campaign, as families showed preference for sons over daughters. Intellectual women began to look closely at women's destiny in political movements after the establishment of the People's Republic of China, and they denounced the distortion by government of women's roles toward traditional practices.

During the "theoretical exploration period" (1980–1985), women's problems arising in the reform became the focus of concern. Some female (and a few male) scholars explored these problems in their own fields, and their work has served to guide subsequent actions in the women's movements.

The years 1986–1988 were the heyday of the women's movement in the New Period, the time of China's reform during which the possibility of building up the social market economy was explored. This period witnessed the establishment of many nongovernmental women's organizations and women's study institutes. The alliance of the Women's Federation with female intellectuals of all circles generated a momentum for cooperative ventures and strategies. Women's magazines and newspapers played an active role in promoting the movement. The field of women's studies came into being and moved beyond the confines of politics and academia.

The years 1990–1993 were a period of far-reaching changes for the women's movement and women themselves, as much was done to improve social conditions and to consolidate women's social status. Rural women's fields of work were enlarged. Urban women who had lost their jobs in the early stages of the economic reform were able to find new ones. A number of social organizations and service groups, founded by women of different social status and with different interests, emerged. Universities and other academic institutions established women's studies institutes, and thus education about women began to be systematic. The founding in 1990 of the Women and Children Work Committee of the State Council, discussed later in this essay, is one of the achievements in the women's-rights struggle.

The activities and strategies of the women's movement in the New Period have not been determined by the party and the government but by women themselves. The main force in the movement is a new generation of working women and scholars who have been brought up under new China's policies on women. It is not only a women's movement but also a social movement that is distinctive in its mutual promotion and encouragement of social development and women's progress.

One of the prominent features of the women's movement during the New Period is separatism: Theoretically, the movement is separated from the traditional theory of women's liberation; academically, it is separated from traditional humanistic studies. The movement is separated also from the traditional image of women molded by Chinese national policy and keeps its distance as well from the image of women set forth by feminism in the West. An obvious deficiency of the women's movement in the New Period is that Chinese women pay too much attention to their own national problems and are thus unable to share experiences with the international women's movement.

Conditions in Contemporary China

Reproduction and Health Care

Under the great pressure of China's dense population, the majority of women, rural as well as urban, can understand the family planning policy of the government that has brought about a change in people's reproductive concepts and behaviors. City women who were born in the 1960s or later tend to have one child, regardless of the sex of the child. The number of women who do not want to have children has increased over the years. Rural women and women of minority nationalities tend to have two children, ideally one boy and one girl. In some areas where the natural environment is harsh, people wish to have boys who are able to support the family when they grow up. From 1970 to 1992, the birthrate in China fell from 5.81 births per woman to just over two.

The government pays considerable attention to the health of women and children. Since 1992, a health-care network has been formed throughout the country. Ninety-eight percent of urban pregnant women and 70 percent of rural pregnant women are able to get adequate health examinations before delivery. The rate of a new, modern method of delivery has reached 84 percent. The death rate of women in pregnancy and delivery has been reduced from 1,500 per 100,000 women to 95. Some of the common diseases that once threatened the lives of women are now controlled and preventable. Aided by the Ford Foundation, researchers have studied women's reproduction and health in many parts of China since 1991, and their projects have spurred efforts to improve women's health. The conditions of women's reproduction, health, and sanitation and the negative influences caused by certain birth control methods and policies have caught the attention of researchers and the government departments concerned.

Education

Since 1949, the Chinese government has made a great effort to eliminate the inequality between boys and girls in education. The government legally ensures that boys and girls have an equal opportunity to receive an education. The rate of girls' enrollment has been raised from 15 percent before the founding of the People's Republic of China to 96 percent in 1992. Accordingly, the rate of illiterates has been lowered from 90 percent to 32 percent in the same

Table 1. Key Demographic Features of the Chinese Population (in Millions)

Category	Number	Percent of Total
Number of females	550.00	48.6
Rural female workers	240.00	46.0
Urban female workers	56.00	38.0
Education—Women, Age 16+		Percent of Woman
University graduates	4.80	1.2
Senior high-school graduates		
(including polytechnics)	34.96	8.8
Junior high-school graduates	96.71	24.2
Primary-school graduates*	134.96	33.8
Illiterates	127.25	32.0

Source: Fourth Census of China, 1990.

*Includes those who learned to read and write in the illiteracy-elimination classes.

period. Census data from 1990 provide pertinent educational information about women in China (see Table 1).

A nine-year compulsory education period is being popularized throughout China. However, the number of female students in senior high schools and universities shows a tendency to decrease primarily because of the difficulty in finding jobs after graduation. Since the beginning of the reform, job-training education has grown, helping satisfy the urgent needs of the industrial and service sectors and, to some extent, eased women's difficulty in finding jobs. Some rural children who have moved into cities with their parents are left out in the drive for universal education in cities, a problem the appropriate government departments need to address.

The primary challenge for women's education is in the rural areas, where the vast majority (80 percent) of women reside and where most of China's illiterates and semi-literates are found. The quality of education in these areas is poor, and raising the educational levels of rural women will be a historic task.

During the rural economic reform, governments at various levels achieved great success in providing practical-skills education to rural women by arousing their interest in getting rid of poverty. The grass-roots chapters of the Women's Federation have played an important role in this effort.

Another development in women's education in China has been the growing emphasis on education for self-realization. Seen as a means of helping women strengthen their consciousness as masters of their lives and of society and empowering themselves to create change in both, women's consciousness education is provided mainly by women's organizations, women's studies institutes, and the mass media. It needs a scientific and systematic development. But the educational policymaking bodies in the government have not accepted women's consciousness education as part of their educational plan.

Marriage and Family

Chinese women have considerable independence in their family affairs; many, both urban and rural, are in charge of the family income. This economic self-reliance lays a solid foundation for women's position in the family. They have rights to marriage, divorce, and remarriage. They can keep their names after marriage. They, together with their husbands, manage their families, take up family duties, look after their parents and grandparents, and bring up their children. China has thousands of nurseries and kindergartens that relieve the burden of childrearing for employed women. In cities and towns, husbands usually contribute one-third of the housework, though in rural areas women do most of it. Still, the burden is not what it once was, thanks to the proliferation of domestic electrical appliances, even in rural areas, and the growth of the service trades.

A typical Chinese family is made up of parents and unmarried children. But women may choose to remain single, to be heads of single-parent families, or not to have children. In most families, the husband's schooling, occupational rank, and salary are higher than the wife's. The dependence of wives on husbands and emotional distance between husbands and wives are the main obstacles to enhancing the quality of family life. In 1992, the divorce rate in China was 2 percent, but in recent years the number of divorces has risen substantially, as has the number of marriages without legal certificate. At least to the mid-1990s, family violence was not a major problem in Chinese families.

Legal Position and Political Participation

The legal rights of Chinese women are equal to those of men in many documents: the constitution of the People's Republic of China, the Marriage Law (1950), the Electoral Law (1953), the Inheritance Law (1985), the Civil Law (1986), the Compulsory Education Law (1986), the Standards and Requirements of the Health Care for Women During Pregnancy (1985), Principles of the Health Care for Women and Children (1986), Labor Protection for Women Staff and Workers (1988), and the Notice of Several Problems on the Treatment of Women's Reproduction (1988). In 1992, the Women's Interests Protection Law was formally issued and put into practice. It contains more detailed and more comprehensive regulations to protect women in the areas of politics, culture, education, labor, property, body health, marriage, and family.

Unfortunately, gender inequality does exist in practice—in departments and factories that discriminate against women and girls in employment and student enrollments, for instance, and in enterprises that fail to implement legal protections of female workers. Violations of women's rights can also be found in marriage and family. All of these indicate that there is a long way to go before sexual equality as legally required is fully realized.

The Electoral Law improved the condition of Chinese women's political participation considerably by permitting them to run for public office. But compared to other fields, women's participation in the political arena is low. Women's opportunities for political engagement (both voting and representation) are less than men's in cities and in rural areas. Women's consciousness about participation is faint especially in regard to the highest levels of political leadership. Among the 51 million members of the Chinese Communist Party in 1990, 7 million, or 14 percent, were women; women constituted 6 percent of the Party Central Committee; but there were no women in the Party Political Bureau. There were 17 female provincial governors and vice governors in the country, 12 percent of the total, and 17 female ministers and vice ministers on the State Council, 7 percent of the total. There was one female committee member on the State Council, and all of the vice premiers were men.

Thanks to a series of party policies mandating a given proportion of men and women in leading posts, women are able to work at various administrative levels. But the situ-

Table 2. Selected Occupations of Women and Men in
China (%)

Occupation	Women	Men
Administrators	11.5	88.5
Technicians and experts	45.2	54.8
Office workers	25.7	74.3
Commercial personnel	46.6	53.4
Factory workers	35.7	64.3
Service personnel	51.6	48.4
Farmers	47.9	52.1

Source: Fourth Census in China, 1990

ation of women's political participation can be improved substantially only if the mass of women's awareness of political democracy is awakened and China's political system is changed—neither of which is likely anytime soon.

Occupations and Social Security

The status of women's occupations is comparatively inferior to that of men, as shown in Table 2 below. Few women are found in leading administrative positions and as technicians and experts.

In the seven job categories listed there, men predominate in all but one, the service trades. By the end of 1992, more than 8.5 million women owned private businesses—one-third of all private business holders. Of the unemployed population in cities, 57 percent were women.

Owing to the disparate structure and location of occupations, urban women, who are burdened with family and childraising responsibilities and bound by their low levels of education, are in an unfavorable position in the competition for jobs. As China is in a transitional period toward a market economy, the problem of women's occupations will continue to be a major issue for Chinese women.

Social security for working women is not as good as it was before the reform. There are scarcely any labor-protection measures taken in many village factories and foreign investment enterprises, and pregnant women and young mothers are often overworked and unfairly treated.

Women of Minority Nationalities

According to the 1990 census, there are 55 minority nationalities in China distributed over a wide area. The minority population comprises 91.3 million people, among whom 44.5 million are females, representing 8.1 percent of all women in China. Eighteen minority nationalities each have populations that exceed one million.

The rate of female illiteracy among the five main nationalities (Korean, Manchu, Mongolian, Uigur, and Kazakh) is lower than the country's average rate. The vote is lowest among Koreans, at 11 percent; Tibetans have the highest rate, 82 percent.

Family planning policies came into force rather late in the areas of minority nationalities compared with the Han

(Chinese) people, and the policy is not as strict with them as with the Han people. The birthrates of the Kazakhs, the Uigurs, and the Tibetans are the highest among minority nationalities, above 4 percent.

Most minority peoples live in bordering and backward areas. Although there has been a great improvement in their condition as a result of a series of government policies, their living standards are low and health-care facilities are scarce, which is disadvantageous to the rise of women's status as women continue to be burdened by domestic and family responsibilities.

Women's Organizations and Women's Studies

Governmental Women's Organizations

Several important governmental women's organizations exist in China. The Women and Children Work Committee of the State Council was established in February, 1990. The committee consists of representatives from 16 departments of the State Council and from four social organizations (the National Women's Federation, the National Trades Union, the National Youth Federation, and the National Academic Association). Provincial governments have their own corresponding organizations. The function of these organizations is to coordinate work on women, secure the rights of women and children, and offer assistance to further the cause of women and children.

The National Women's Federation, also called the All-China Women's Federation, was established in April, 1949. This is a social organization of women under the leadership of the Communist Party of China to provide a bridge to join the party and the masses. It aims at safeguarding women's rights and ensuring equality between men and women. The federation receives funds from the government. Through a combination of local organizations and federation members, the federation has a nationwide women's network. By 1992, there were more than 68,000 federation branches at the town level and 810,000 subunits at the village level.

The Special Group for Women and Children of the Internal Affairs Judicial Committee under the National People's Congress was established in April 1989. The group consists of people who represent women and child workers, field workers, experts, and scholars. The chief task of the group is to handle, discuss, and propose legal cases concerning women's and children's legal rights and to advocate and supervise the implementation of the law.

The Youth and Women Committee of the National Political Consultative Conference was established in April, 1988. Its main function is to see that all of the laws and regulations with regard to women, youth, and children are put into effect.

Nongovernmental Organizations (NGOs)

In the late 1980s, NGOs came into being. The social structure that made the All-China Women's Federation the only legal women's organization in China at all levels has been broken. There are two types of women's NGOs. The first, which are supported by the Women's Federation, are groups of women from all walks of life. The unions comprise various women's associations, such as those of scientists and technicians, entrepreneurs, engineers, judges, lawyers, calligraphers, and the like. These associations are usually members of the Women's Federation, which gives policy directions and helps obtain governments funds for the associations.

The second type of women's NGO is truly nongovernmental, established by women who share a common goal. The leaders of such organizations, in most cases, are influential women in academia; funding is private.

Representative of such NGOs is the Future Women Studies Association of Henan Province; set up in 1985, it was the first NGO for women's studies in mainland China. The leader of the association is Li Xiaojiang, a professor at Zhengzhou University. Another NGO, the Women's Research Group of Beijing Foreign Languages Institutes, was established in 1985 by Chinese and foreign teachers at the institute. It was the first nongovernmental women's studies salon in Beijing. The group sponsors discussion classes that explore Chinese women's problems and the experiences of women's liberation movements abroad.

Women's Studies

The legal equality between men and women in China was a major factor blocking the advancement of women's studies as it was not seen as a necessary field. This situation did not improve until the recent reform in the 1990s. The reemphasis on women's problems by feminist scholars resulted in the rebirth of women's studies and the awakening of women's consciousness on the part of other women intellectuals. Books on feminism that flooded into China following the Cultural Revolution have served as a kind of catalyst, as well as references, for China's women's studies, which, from the beginning, have emphasized activism and practicality.

Since 1986, the All-Women's Federation at all levels has set up women's studies associations to discuss theoretical as well as practical problems of women. These activities, which were most intense from 1986 to 1989, reflect upon the lives of women and serve to alert society to women's issues. Until the end of the 1980s, few of these women's studies associations had any connection with the educational world or other institutions except the Women's Research Institute and some research offices of the Women's Federation.

Within academia itself, in the early 1980s female scholars on their own began exploring issues and constructing theories on women in their own fields of work. Two academic papers, "The Starting Point and the Essential Point of Marxist Theory on Women" (1983) and "The Characteristics of Chinese Women's Liberation" (1984), are the earliest attempts to construct a Marxist theory on women in the field of philosophy. Research on the history of women is also an important component in women's studies, and women's literary creation and literary criticism are becoming increasingly influential in society. Some notable books are *Charm of Love and Coquettishness,* a study of classical poems by Chinese women (Kang Zhengguo, 1988); *On the Horizon of History,* a study of modern women's literature (Meng Yue, Dai Jinghua, 1989), and *The Late-Coming Tide,* women's literary creation in the New Period (Yue Shuo, 1989). Chinese women's studies and movements are gradually developing open, mutually referenced exchanges with Western feminism, and centering on the creation of a new cultural research system.

Remaining Problems

Although progress is taking place, particularly within women's studies, several important problems in society await resolution:

Illiteracy. In 1990, 180 million people age 16 and older were illiterate in China, of whom 70 percent were women and more than 90 percent resided in rural areas. Fewer rural girls than boys were enrolled in school, and the proportion of female dropouts outnumbered male dropouts—creating a vicious circle that accounts for the low educational level of girls. Among the causes are the scarcity of schools in rural areas, poverty, and the feudal sex discrimination that still exists.

In 1994, the government of China set out to make nine-year compulsory education universal throughout the country. The rate of school-age girls who are kept out of school is not allowed by government to surpass 2 percent. It has set a target of reducing the dropout rate to less than 2 percent. At the same time, the government is working to eliminate adult illiteracy; with a goal of achieving, by the end of the twentieth century, 85 percent literacy among women age 15–40.

Poverty. As late as 1992, 87 million people in China (60 percent of them women and children) did not have enough to eat and wear. The government has implemented a project called the Helping 87 Million Poor People Plan aimed at moving all of these people out of poverty by the year 2000. One of the components of the plan is to help rural women by offering them cultural- and practical-skills training and setting up enterprises to provide them with jobs, and, in some cases, by helping them leave the poverty-stricken areas in which they live to work in other places.

Abduction and Selling of Women. Abductions and sales of women were widespread before 1949. After the establishment of the People's Republic of China, the practice died out with the adoption of centralized collective production and a strict management of census registration. As China started to move toward a market economy at the end of the 1970s, and as the management of population registration eased, the abductions and sales of women reappeared, becoming a serious problem in the late 1980s. Some people even engaged in these activities as if they were a regular job. The practice is conducted mainly in rural areas: two factors probably account for it: not enough women of marriageable age in a village or surrounding area, or not enough jobs or marriage prospects for poor young women themselves. To rid the country of abductions and sales of women, the government needs to severely punish those who have engaged in such practices, rescue the women who have been abducted and sold, and implement social-service programs to provide women the assistance they need to satisfy their basic needs.

Prostitution. All but eradicated in the late 1940s and early 1950s, prostitution reappeared at the end of the 1970s and expanded rapidly. In 1984, the city of Shanghai established its first prostitutes' house to rehabilitate prostitutes through vocational education. By 1992, China had 111 such houses, with a capacity for 28,000 prostitutes. Coercive measures and education failed to check the rapid expansion of prostitution. As a result, venereal diseases, which had been eradicated for many years in China, began to spread. Given the imperfections of the marriage system in China, the growing disparity between the rich and the poor, and the government's inability to control rural–urban population shifts, it is likely that prostitution and the complex problems that accompany it will continue to defy solution.

In all, China transformed itself substantially during the socialist period of its history, and, in that process, the status and condition of women improved. It remains to be seen whether women will benefit from the current changes in the country's political and economic culture.

References

Li Xiaojiang. *Eve's Quest.* Zhengzhou: Henan People's Publishing House, 1988.

———. *The Way Out for Women.* Shenyang: Liaoning People's Publishing House, 1989.

———. *Farewell to Yesterday.* Zhengzhou: Henan People's Publishing House, 1995.

Li Xiaojiang and Tan Sheng. Women's Studies in China. Zhengzhou: Henan People's Publishing House, 1991.

Li Xiaojiang, Zhu Hong, and Dong Xiuyu. *Sex and China.* Beijing: Life Books and Knowledge Store, 1994.

Li Xiu. "Population Situation of China's Minority Nationality Women," *Encyclopaedia,* no. 10, 1994.

People's Republic of China. *The Report of People's Republic of China on the Implementation of the Nairobi Forward-Looking Strategies for the Advancement of Women.* Beijing: People's Republic of China, 1994.

State Council of China. *Conditions of Chinese Women.* Beijing: News Agency of the State Council of China, 1994.

Tan Sheng. "Analysis and Anticipation of the Conditions for Chinese Women," *Sociological Studies* (Institute of Sociology, Beijing), no. 3, 1994.

Women in South Korea

Young Ok Kim
Young Hee Kim
Wha Soon Byun
Jae In Kim
Yanghee Kim

Introduction

South Korea has witnessed many cultural and economic changes in the last three decades of the twentieth century. So much has changed, in fact, that it can be asserted that this country is no longer a developing country and that its emergence as a newly industrialized nation is irreversible. A fast-growing economy and legislation to improve the condition of women are crucial factors affecting the condition of Korean women. In the context of a traditional Confucianist society, the shifts in the economic participation of women and in the definition of the family and the entrance of women into political arenas provide illuminating information about the course of change in one Asian country.

Economic Activities

The increase in female labor-force participation is one of the important characteristics of the Korean labor market. In the export-led industrialization, the female labor-force-participation rate increased from 36 percent in 1963 to 48 percent in 1994, while the rate for the male population held constant at 76 percent (National Statistical Office, 1964, 1995).

There have been profound structural changes in the female labor supply behind this growth in female participation in the labor force, including migration into the urban area and an increase of married women participating in the labor force. The concentration of the labor-force population in the urban labor market has been remarkable for both men and women: While 59 percent of male workers and 64 percent of female workers were in rural areas in 1963, in 1994 only 13 percent of male workers and 17 percent of female workers remained there (National Statistical Office, 1964, 1995).

Age is a factor in women's labor-force participation. Figure 1 shows that, in contrast to the inverted "U" distribution of male workers, the distribution of the female workers by age groups is M-shaped, indicating that women participate most actively in their early 20s and 40s. This,

in turn, indicates that most employed women discontinue work due to the burden of childbearing, and childrearing, and other domestic responsibilities. This interruption crucially influences their position as workers: Because of severe age discrimination in firms, women find it hard to get reemployed after their childrearing period, and even when they succeed they suffer from various personnel disadvantages because their careers are not counted under the seniority-based personnel practices. Employers justify their discrimination against women employees by referring to women's discontinuity of working as an excuse.

Though the average unemployment rate of Korean women was as low as 2 percent (men's was 3 percent) in 1994 (National Statistical Office, 1995), there are vast numbers of women who are the "discouraged unemployed," defined as those who have given up job seeking but are willing to work if suitable work is available. In 1992, those hidden unemployed women numbered 1.2 million (National Statistical Office, 1993).

In terms of employment status, the period 1963–1994 saw declines in the rate of unpaid family workers from 56 percent of female workers to 23 percent and in own-account or self-employed women from 22 percent to 19 percent; on the other hand, the proportion of women workers in paid employment increased from 22 percent to 59 percent during the same period (National Statistical Office, 1964, 1995). Although the rate of paid employees is on the increase, the rate of nonwage workers such as self-employed and unpaid family workers is still high. In the case of the self-employed, many are street-shop owners or peddlers. Even among women wage workers, 17 percent are casual or daily employees, and large numbers work for small firms of five or fewer employees. These figures suggest that many women are engaged in unstable employment or in the informal sector of the economy.

Distribution of Women Workers by Industry

The proportion of women workers in the primary sector

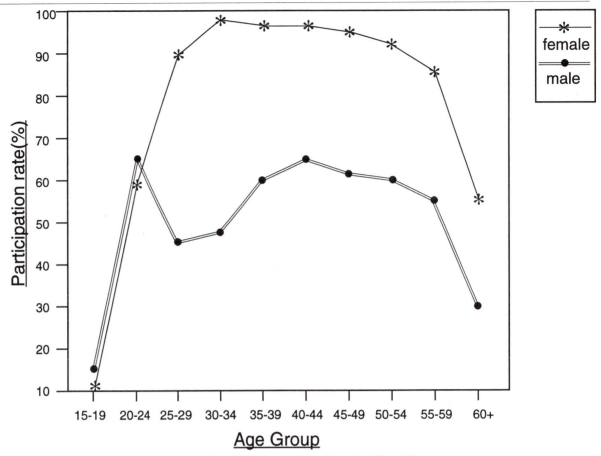

Figure 1. Labor Force Participation Rates by Sex and Age Groups. Source: National Statistical Office, 1995.

(i.e., the agricultural and fishery activities) has decreased rapidly while the proportions in manufacturing and the tertiary sectors (i.e., manufacturing and services) have increased. In 1963, 69 percent of female workers were employed in the agricultural and fishery sector, followed by 24 percent in the service and sales sector, and 7 percent in the manufacturing sector. In 1994, 62 percent were in service and sales industries, followed by 22 percent in manufacturing, and the remaining 16 percent in agriculture and fishery (National Statistical Office, 1964, 1995).

Occupationally, in 1994 the largest number of female workers (31 percent) were engaged in sales and service work, followed by production work (29 percent), clerical work (15 percent), farming (15 percent), and professional and managerial jobs (11 percent) (National Statistical Office, 1995).

Men and women are concentrated in different occupations. According to Ministry of Labor (1989) figures for 1988, for example, 65 percent of female workers were concentrated in just 20 occupations, while only 10 percent of male workers were engaged in those same jobs. In the professional classification, which listed 77 occupations, nursing accounted for 45 percent of all female professional workers. Within occupational categories, women and men are often found disproportionately at different levels, with men concentrated in the more skilled, responsible, and better-paid positions—a fact the 1988 figures illustrate,

showing that while 11 percent of male employees were classified as section chiefs or above only 0.6 percent of female workers held that rank (Ministry of Labor, 1989).

Looking at the age distribution of female workers by occupation, in 1994, 63 percent of clerical workers were age 15–24. This implies the existence of a voluntary or involuntary retirement-with-marriage practice in this occupation, in which working women are regarded by Korean culture as "flowers in the office" rather than as coworkers. Among women service and sales workers, 78 percent were 30 or older, while 84 percent of agricultural workers were older than 40. It is notable that the proportion of women production workers age 15–24 remained above 60 percent until 1985 but declined to 15 percent in 1994 (National Statistical Office, 1995). Two possible factors causing this phenomenon are that the general educational level of young girls has risen continuously (and thus the girls are at school) and that since the mid-1980s, labor-intensive manufacturing industries have been declining and have provided comparatively inferior working conditions. Manufacturing industries no longer attract young girls, who are substituted by middle-aged women workers.

Wages

The average monthly wage of female workers in 1994 was only 57 percent that of male workers (Ministry of Labor, 1995). As recent as 1989, the two countries with the lowest women's wages were Korea and Japan, with women earning only 55 percent and 51 percent, respectively, of the men workers's wages (ILO, 1990).

An analysis of wage differentials between the sexes in 1988 showed that only 38 percent stemmed from the differences in human capital (educational level, experience, skills, etc.), and the remaining 62 percent was due to discrimination against women (KWDI, 1990).

The institutional factors accounting for this wage differential seem to be the heavy concentration of female workers in low-wage firms (discrimination at the point of entry), occupational segregation, and internal labor-market discrimination—all of which allow women limited opportunities for promotion and on-the-job training (Ministry of Labor, 1989, 1995).

Economic Prospects

Korean economic success owes much to working women, and women's economic participation has increased in terms of quantity and quality. Yet, the majority of women workers remain in low-paid, unskilled manual labor and service-sector jobs.

Women's groups, labor unions, and concerned scholars across the political spectrum have strongly advocated the introduction of institutional measures to improve the employment structure and to correct sex discrimination practices. One such measure, the Equal Employment Opportunity Act of 1987, enshrines the principles of equal pay for equal work, maternity protection, and the right to continue employment after marriage or childbirth. However, it needs to be revised to include provisions for prohibiting sexual harassment at work and indirect discrimination.

Women's groups successfully forced the government to undertake regular inspections of companies for observance of the law and to announce correction orders for the abolition of discriminatory practices against female bank employees, who used to be paid far less than men and were restricted in promotions. These groups have also pressed the government and private enterprises to take the initiative in abolishing discriminatory personnel practices; a quota system was implemented in 1996 in the recruitment of high-level government officials as an institutional solution for women's low representation in the decision-making processes.

In South Korea, individual firms are solely responsible for the coverage of maternity-protection expenses, which presents a major obstacle to the hiring of women. The questions of how to transfer some of this cost to medical insurance or unemployment insurance programs, how to increase proper child-care facilities, and how to protect growing numbers of irregular (part-time, temporary, seasonal) women workers are contemporary issues related to women and employment in Korea.

Legal Aspects

First promulgated in 1948, the constitution of the Repub-

lic of Korea states that no citizen shall be discriminated against in political, economic, social, or cultural activities for reasons of sex, religion, or social status. The current Constitution, amended in 1987, makes clear that women shall not be discriminated against in recruitment, wage, and other working conditions (Article 32) and that the state shall make efforts to guarantee maternity protection (Article 36).

The Basic Act for Women's Development. The Basic Act for Women's Development, enacted in December 1995, and effective in June 1996, regulates and provides for affirmative action in areas in which women have been significantly underrepresented (Article 6), a basic plan for women's policies (Articles 7 and 9), a women's development fund (Article 29), governmental support for women's organizations (Article 32), and the establishment of committees on women's policies (Articles 10 and 12).

This Act seeks the promotion of equality between men and women, the protection of motherhood, the liquidation of sexually discriminatory attitudes, and the development of women's capabilities. Through achievement of these goals, it aims to provide a basis for equal participation, and equal allotment of responsibilities between men and women, in developing the state, society, and healthy families.

Family Law. After more than 30 years of persistent lobbying by women's organizations, the Family Law was revised in December 1989. The new law, which went into effect in 1991, was a significant breakthrough in equalizing the status of women and men in the family, especially in the area of property rights.

While the revised law continues to recognize the husband as the head of the family, provisions on the succession of family headship have been deleted, thereby falling back on traditional practice. The custody of children is no longer automatically granted to the father upon divorce but is to be determined either by mutual agreement of the parents or by the family court; likewise, the domicile of the children is to be decided by both parents and not by the husband only. The wife and the husband are equally responsible for the children's living expenses. The revised law expands women's rights of inheriting property by eliminating discrimination against daughters. In the absence of a will, the property is distributed evenly among the children, regardless of sex. Also, a childless widow is now entitled to receive half of her husband's inheritance, with the remaining half going to the parents of the deceased. Both wife and husband have an equal right to properties gained after marriage, as well as the right to ask for an equal division of properties. The revised law thereby entitles a woman to seek a share of the couple's property proportional to her contribution to its accumulation in case of divorce. According to the revised Family Law, relatives are no longer defined as those who are third cousins or closer on the father's side and first cousins on the mother's side; instead they are defined as those who are third cousins or closer on both the mother's and the father's side.

The Labor Standards Act. The Labor Standards Act was enacted in May 1953 and, as of 1996, had been revised eight times. The act sets standards for employment conditions to conform with the constitution and thereby secure and improve employees' standards of living and achieve a well-balanced development of the nation's economy.

This act upholds the constitutional principle of equality by stating that the employer shall not discriminate against employees by sex. It guarantees basic equality between men and women in employment and special protection for pregnancy and maternity, with provisions for one year of child-care leave as well as 60 days of paid maternity leave. It protects women and minors against employment in any immoral or dangerous work, prohibits nightshifts and work on holidays, and limits overtime work for women. The act also requires employers to provide child-care facilities in the workplace, though this latter provision has been implemented only partly.

The Equal Employment Opportunity Act. This act, established in 1987 to guarantee equality between men and women in employment and revised in 1989 and 1995, makes discrimination against women in recruitment, employment, placement, promotion, and retirement punishable. Among other things, it prohibits employers from requiring certain standards of appearance, height, weight, or marital status that are unessential for the performance of the job (Article 6). It also guarantees child-care leave for both the working woman and her male partner who assumes her role in child care (Article 11). Article 12 states that "the employer should provide necessary facilities for child care such as places for breastfeeding and day care to support the continued employment of working women."

The Mother-Child Welfare Act. The Mother-Child Welfare Act was enacted in April 1989 to ensure the security and welfare of fatherless families. It establishes basic principles, various kinds of welfare services, and welfare facilities. The act applies to women who are left behind, divorced, or abandoned by their spouses; women whose spouses are unable to find steady employment due to mental or physical handicaps; unmarried women who are heads of the household; and unwed mothers with children under 18 years of age.

The act stipulates that the government provide mother-headed families with welfare allowances to cover the costs of bringing up the children, including education through the high-school level and vocational training. The government also provides welfare loans to strengthen the security and self-sufficiency of mother-headed families, and it can give preference to members of such families in employment, in the use of public facilities, in contracting for shops and other small businesses in public establishments, and in entering permanent rental apartments.

The Child Care Act. The Child Care Act was promulgated in January 1991 to contribute to the promotion of home welfare by bringing up infants and preschoolers whose guardians have difficulty protecting them due to work loss, illness, or other circumstances as healthy members of society through mental and physical protection and education. This act mandates that the central and local governments, as well as all citizens, are responsible for the healthful nurturing of these children.

Act Relating to Punishment of Sexual Violence and Protection of Victims. This act, enacted in December 1993 and effective from April 1994, was the result of active campaigning by women's organizations in Korea to eradicate sexual violence. It designates as crimes rape committed by relatives, unresisted rape of handicapped persons, disgraceful sexual conduct using one's official position or authority, disgraceful sexual acts in public places, and obscenity expressed through communications media. Recognizing that such crimes often are not prosecuted because victims are afraid to come forward, the law provides that, in cases of rape by a relative and crimes against the handicapped, a third party can report the violence to the authorities for punishment. It also extends from six months to one year the period for filing a legal suit in criminal cases requiring the victim's complaint and allows summary court procedures and closed-door trials to protect victims. This act was the first legislation of its kind in Asia. Its effectiveness has yet to be evaluated.

The Family

In 1990, Korea had a population of approximately 43.4 million, of whom 21.6 million, almost 50 percent, were women. Two-generation households accounted for 66.3 percent of the total; households with three or more generations, 12.5 percent; one-generation households, 10.7 percent; and single-person households, 9 percent. Changes since 1970 indicate a trend away from extended, and toward nuclear, families and a growth in the number of single-person households (see Table 1).

An average household in 1990 numbered 3.7 persons. The fertility rate stood at 1.6 births, down drastically from six births in 1960, under a strong family planning policy. By 1990, the average age at first marriage had increased to 25.5 for women and 28.6 for men; and younger generations were expected to marry even later. Never-married single women accounted for 28 percent of the female population over age 15; married women, 59 percent; widowed women, 12 percent; and divorced women, 0.9 percent. Life expectancy was 68 years for men, and 76 years for women.

In 1993, there were 308,000 marriages and 46,832 divorces recorded, for a divorce rate of 15 percent (almost one of seven marriages). Since 1980, the number of marriages has been decreasing and the number of divorces increasing—changes attributed to changes in attitudes about marriage and family (KWDI/KFAW, 1991), some of which are discussed below.

Longer life expectancy, later marriage age, and fewer births spell drastic long-term changes in women's life cycles. Looking at three generations of women in a family (daughter, mother, and grandmother), one can see this illustrated (KIHS, 1987).

Since the daughter's generation has less than two chil-

Table 1. Transformation of Household Types (by Percent of Households)

Year	A	B	C	D	E	Total Households (Thousands)
1970	6.8	70.0	23.2	0.0	0.0	N/A
1980	8.3	68.5	17.0	4.8	1.5	7,969
1990	10.7	66.3	12.5	9.0	1.5	11,355

Source: National Statistical Office, 1979, 1982, 1992.

A	=	One-generation households
B	=	Two-generation households
C	=	Three-or-more-generation households
D	=	One-person households
E	=	Households of unrelated persons

dren, the birth-giving period for her will last just 2 to 3 years. Once her last child enters elementary school, the period of intensive child care can be supposed to be over, and she typically will have 40 years or more of life remaining. Furthermore, when her youngest child gets married, she will be in her middle to late 50s and can expect to spend almost 20 more years with her husband and survive about 8 years alone. From grandmother, to mother, to daughter, the periods after intensive child care and their children's marriages are becoming longer, and these changes in circumstances are bringing changes in women's attitudes and lifestyles.

The economic activity rate of women in general (15 years of age and above, either full- or part-time) was 48 percent in 1994. The rate for married women was 47 percent; for unmarried women, 51 percent. As the number of married working women increases, the issue of child care becomes more important for making work and family responsibilities compatible with each other.

In sum, the increasing equality between women and men in higher education, the trend toward smaller families, the decrease in the birthrate, the increase in life expectancy, the decrease in the number of marriages and the higher rates of divorce, and the increase in the economic participation of women are bringing about changes in the lives of Korean women.

Attitudes and Practices Related to Family

The patriarchal familism that originated with Confucianism has deeply influenced the attitudes of Korean people regarding the family; which is considered to include not only those members who live in the same house but also their ancestors and even their unborn descendants. This is a rather abstract concept, but it affects many aspects of family life in Korea. Family headship can only be transmitted through paternal lineage.

To Koreans, filial piety, which requires responsibility and devotion to one's parents and ancestors, is considered the foundation of human relationships. In this context, the relationship between parents and children often takes precedence over the conjugal relationship.

The imbalance in the sex ratio between boys and girls (the sex ratio at birth in 1993 was 115.6 boys to 100 girls) clearly reveals a "son preference," which can be attributed to sex discrimination in the family and in society in general.

Rapid industrialization and urbanization since the 1960s have fostered major changes in the structure and functions of the Korean family. While in some ways similar to changes accompanying industrialization in the Euro-American countries, those in Korea were much more abrupt since they occurred in a relatively shorter period of time.

In the traditional view of the family, the strong patriarchal father is the breadwinner, and the wife devotes herself to taking care of the family; the modern viewpoint holds that the wife should join the social and economic activities and maintain an equal partnership with her husband.

In the 1960s and 1970s, people were absorbed in enriching the family economically. Integration of the family took priority over the life of the person and the individuality of family members was easily neglected. But since the 1990s, the importance of individuality and conjugal relationships has gained recognition, and younger generations now consider that individuality should take precedence over collectivity in the family.

As to the conjugal power relationships, on the surface women seem to exercise more power than men since in most families it is the wife who makes decisions regarding the children's education, the purchase of household commodities, and family events and occasions. Nevertheless, a woman's power is limited—the wife has control only over things that are of "little importance" (or for limited purposes)—as long as she is economically dependent upon her husband.

The imbalance of power between marital partners is reflected in the increase of domestic violence reported in the mass media. The family is no longer a shelter from violence. This new phenomenon demands the establishment of protective social support mechanisms and the enactment of laws that will address the problem directly.

Education

Since 1948, three years after national liberation, the constitution of the Republic of Korea has provided that all citizens shall be equal before the law and that there shall be no discrimination on account of sex. This provided a momentum for women to participate in social activities and to make their social status firm. Since then, the educational opportunity of women (i.e., access and completion) has been increasing remarkably, but there remain many questions as to whether education is provided equally for women and men. Some of these are addressed briefly here.

Educational Access and Attainment. Figures for 1990 show that Korean women averaged 8.6 years of schooling and Korean men 10.6 years, an increase of about 1 year for both women and men since 1985, though the 2-year gap remained steady (KWDI, 1995).

Girls' gross enrollment rate was 101 percent for both elementary and middle schools in 1995, indicating mass participation of girls in schooling at those levels. As for high-school enrollment, there is no apparent gender gap: In 1995, 89 percent of girls and 90 percent of boys were enrolled in high schools. However, in regard to high-school enrollment by type of school, significant differences exist. For example, men students are more likely than women students to be enrolled in academic schools—62 percent compared to 55 percent, respectively, in 1994. In the same year, 62 percent of women students in vocational high schools were enrolled in commercial fields, while only 16 percent of their male counterparts were (Kim, 1995). The proportion of girls enrolled in technical specializations is gradually increasing.

Enrollment in higher educational institutions also shows marked gender disparity. In 1995, 39 percent of university-age women and 70 percent of men were enrolled in higher educational institutions. Women represented 31 percent of those enrolled in universities and colleges, 75 percent of those in teachers' colleges, and 37 percent of students in the less prestigious junior colleges.

Coeducation is widespread in Korea: By 1994, all elementary schools were coeducational, and 55 percent of middle schools and 42 percent of high schools were so. The Korean government, by policy, encourages all newly established middle and high schools to be coeducational.

Curriculum and Textbooks. As part of the Sixth Five-Year Economic and Social Development Plan (1987–1991), the government made it its policy to remove gender bias from the curriculum structure, textbooks and other materials, and from career guidance processes. One of the specific features of the revisions is the integration of technical and home-management subjects at the middle-school level and the requirement that both sexes participate in the integrated course. Another measure has been to remove from the textbooks published since 1987 any content prejudicial to women, such as discriminatory descriptions of gender roles and women's status (Kim, 1993), and to balance male and female representations in the illustrations.

Higher Education and Professional Preparation. In 1994, 56 percent of the undergraduate students in linguistics were women, 64 percent of students in the arts, and 40 percent of students in the humanities. In contrast, women accounted for 6 percent of the students in engineering, 15 percent in fisheries and marine studies, and 29 percent in agriculture—a graphic indication that women

are not adequately preparing for the future society, which will be even more highly technological than the present one (Kim 1995).

Access to Science and Technology. To develop female representation in the technological field and, thus prepare women for high-tech industries, the government encourages the establishment of technical high schools for girls and the admission of higher percentages of girls into all existing technical high schools. The number of female students enrolled in the girls-only technical schools reached 3,752 in 1994. Training courses in multiskilled engineering are provided at all technical colleges for women and men; this training includes detailed measurement, machine design, electronics, fashion design, precious metals, and office automation. In addition, Ehwa Women's University established a school of engineering and began admitting students to it in 1966.

Nonformal Education

Under the Educational Law, various types of nonformal education (termed "quasi-schools" in Korea) can be established and managed with financial support from the government or from industries. Vocational and technical education or special skill training is provided for adults who have missed opportunities for schooling, as well as youths working in industries. Nonformal education includes part-time classes, night classes, and seasonal classes and is open to anybody regardless of age. There is a great degree of flexibility in entrance requirements and pace of progress; emphasis is put on practice, and licenses are granted at the end of the program.

As more students are entering regular schools, the number of quasi-schools at the high-school level is decreasing. At the same time, the number of quasi-schools at the university level is on the rise. While men outnumber women at regular schools, the reverse is true in quasi-schools. However, women's enrollment is still low in technology-related schools such as technical high schools.

Vocational Training. As of 1994, there were 74 public vocational training centers in Korea, 40 of which had programs in 12 occupations for women. In addition, in-plant vocational training is provided by individual companies for the placement of needed personnel. There were 220 such training centers in 1994, 170 of which provided vocational training for women while the other 50 centers offered courses only for men. Women represented 12 percent of all trainees (14,654 students) and were found in 67 occupations. Authorized vocational training is also provided by nonprofit organizations and individuals in occupations that are not covered by the public or the in-plant training. There were 106 such training centers in 1994, 84 of which provided training for a total of 9,151 women (30 percent of the total trainees) in 30 occupations (Ministry of Labor, 1995b).

Among these, the Ansung Technical College for Women located in Koung-ki Province is the only one geared specifically toward fostering women's increased economic participation and the diversification of occupations, preparing women as an industrial resource with professional capabilities, and promoting the social participation of women and their status in society.

Leadership Training. Training is provided for women in various types of work to raise their consciousness about women's issues and to help them improve their leadership skills. The programs include basic training courses, educational tours, and consignment education [organized and implemented by the Korean Women's Development Institute (KWDI) at the request of other institutes, industrial companies, and local provinces]. The curriculum covers administration, education, economics, health, welfare, counseling, and local community leadership. There are two-night/three-day courses for leaders in each of these topics; a one-day course (an outreach program for those living in the countryside) is provided for local community women leaders in cities and provinces. There is also a special training course for women leaders in the suburbs of Seoul. The Lecturers' Bank consists of 200 lecturers (mostly women) in Seoul and local areas.

Leisure Education. An important part of nonformal education for women is leisure-liberal education. Institutions that have programs of this type are women's associations, public-welfare institutions (women's welfare and youth-welfare centers), educational institutions attached to colleges, cultural centers attached to mass-media companies, and private educational institutes.

Educational Policies

To promote gender equity in education, the Ministry of Education, which is in charge of female schooling and nonformal education, participates in the National Committee to Review Women's Policies. The minister is a member of the committee, and a high-ranking ministry officer is a member of the committee's working group. Table 2 lists policies concerning women's education in Korea.

For women's policies to be effectively implemented, government officials should clarify the reasons behind gen-

---

Table 2. Formal and Nonformal Education Policies in Korea

Formal Education Policies
 Gender Equity in Schooling
  Realization of gender equity in educational environments
  Elimination of sexual discrimination in textbooks
 Expansion of Educational Opportunities for Girl Students
  Expansion of free middle-school education
  Expansion of opportunities to enter colleges and universities
 Expansion of Opportunities to Major in Traditionally Nonfeminine Fields
  Adoption of measures to raise the enrollment rates of female students in science high schools and
   engineering high schools up to half the number of male students by 1996
  Increase in the number of girls in engineering colleges up to half that of boys
 Career Guidance for Girl Students
  Integration of home economics and technology in middle schools that have been combined into one to
   adjust to social change for teaching boys and girls
 Open-Door Policy of Special Educational Institutes
  Enrollment of women in the three military academies and ROTC
  Expansion of opportunities for women to enter special-purpose universities
Nonformal Education Policies
 Expansion of Opportunities for Women's Nonformal Education
  Increase in the establishment of adult-education centers or nonformal education centers in colleges
  Increase in the percentage of women in open colleges
 Gender Equity in the Content of Women's Nonformal Education
  Education and program development for consciousness-raising of gender equity
 Increase in Vocational Training Opportunities for Women
  Establishment of various women-only classes to produce skilled human resources
  Establishment of additional women's vocational training centers
  Education of women for special occupations
 Establishment of High-Tech Training Classes
  Establishment of high-tech and area-specialty classes
  Assistance for women entering high-tech departments
 Training of Female Technicians at Vocational Training Centers
  Encouragement of women to enter technical colleges
  Provision of assistance for women's entrance into industrial technical colleges
 Reinforcement of Support Systems for Vocational Training
  Provision of public relations and scholarships, facilities for women trainees, adjustment of training periods,
   establishment of day-care centers within the training centers

der discrimination through a comprehensive monitoring of the status of women; this action should be followed by the elimination of discriminatory laws and by education to eradicate discriminatory attitudes, behavior, and customs.

The premise underlying women's education is that women must be enlightened about their own problems first in order to assume a full partnership with men in handling social problems. Such issues as Korean unification, the environment, and the condition of youth should be discussed by men and women in formal and nonformal education systems. Change is possible only through education.

Culture

Despite developments such as improved educational levels and increased participation in society that have led to substantial progress in the status of Korean women in the formal sense, this progress is not always reflected in women's daily lives. This glaring cultural lag has led to widespread recognition that the development of women's cultural identity and cultural empowerment is crucial to solving women's problems in Korea, a society where discrimination in favor of the male has traditionally dominated all aspects of women's lives.

The essence of Korean patriarchy, based on feudal Confucianism, is "a male monopoly over culture" (Kim, 1992). Within this framework, women are not only restricted in their access to cultural resources, but their lives and worldviews are considered merely personal and too insignificant to be worthy of public attention. The scope of cultural experience for women is limited within the roles, responsibilities, and lifestyles imposed on them by men.

However, as women have become better educated and have entered more fully into society, their awareness of their cultural deprivation and their urge for self-expression have increased. Recognition of women's situation in Korea has grown among those who have been influenced by women's studies, which were introduced in Korea in the 1970s, as well as among female workers as they became aware of their role as the providers of cheap labor in the national economic development.

Korean women have pursued their cultural identity through organizational activities and their own cultural movement.

Women's Organizations

Since the first women's organization, Chan Yang Hoe, was formed in 1898 by 400 Yangban (upper-class) women, women's groups have taken an active part in mass education, the national independence movement, and civic enlightenment. After the establishment of the Republic of Korea and a democratic constitution in 1948, women's social participation began in earnest. Scores of new women's organizations were formed, and previously existing ones were invigorated. The Korean War (1950–1953) proved how strong and self-reliant Korean women could be under the most adverse conditions. Women started working as breadwinners for the family and engaged in such organizational activities as medi-

cal services, education, and social welfare programs for the peaceful reconstruction of Korea.

As Korean society has undergone rapid industrialization and modernization since the 1960s, women's groups have expanded in number and scope, working to improve the laws and institutions concerning women, expand women's economic and political participation, promote environmental protection and consumer rights, and eradicate violence against women. With the revival of local administrative autonomy in 1995, women's organizations are pooling their energy in joint strategies to enhance women's political profile and participation in policymaking processes. As of 1994, there were 4,050 women's organizations nationwide, including two major umbrella organizations (the Korean National Council of Women and the Korean Women's Associations United).

Some of these organizations focus on promoting the status of working women, concentrating on the effective implementation of the Equal Employment Opportunity Act by publicizing instances of discrimination in the workplace and providing counseling and other services.

Others work to encourage women's participation in politics by mobilizing women at the grass roots, providing educational programs to enhance the consciousness of female voters, and seeking out female candidates and training and supporting them. Women's groups have also united to demand quotas for women in party nominations.

Many women's organizations have joined to speed up the enactment of relevant legislation, to educate women and children toward the prevention of sexual violence, to provide counseling services, and to run shelters for victims. Another advancement in the women's movement concerned justice for the "comfort women" used by the Japanese army during its occupation of Korea (1910–1945). Thirty-six women's NGOs formed a coalition in 1990 demanding that Japan make a full disclosure of the truth and pay reparation for the victims. Their international campaign activities around the United Nations Commission on Human Rights succeeded in inducing an official investigation by the United Nations and in raising this issue as a form of extreme violence against women during wartime.

On another front, women's organizations have been promoting consumers' rights since the 1960s. They have also pushed forward the environmental movement to save and recycle resources and to reduce wastes and disposables, to provide educational programs for women in local communities about the environment, and to boycott goods produced by pollution-generating companies.

The Women's Cultural Movement

Beginning in the 1980s, the women's movement in Korea emerged as a quest for full expression and cultural identity of women. Within the patriarchy that prejudiced any kind of "women's culture" and relegated women to objectified roles, the movement's main objective became the discovery and development of a culture in which women could act as "subject" and not as "the other." Women pursued cultural means such as literature, drama, movies, and art to cope with and address problems stemming from a male-dominated culture.

Feminism, advancing in the field of literature, emphasized its distance from so-called women's literature, which views the essence of womanhood as an ambiguous sensitivity and tenderness. The movement in academia was represented by such publications as *Alternative Culture,* which started in 1985, and *Women,* which started in 1988 and has played an important role in publicizing theories concerning the oppressive realities of women.

Feminism in the fine arts has changed its emphasis from the aggressive depiction of sexual discrimination to the expression of women and women's freedom from oppression. Exhibitions depicting women's issues started in Seoul in 1986, and since 1987 the women's committee of a progressive art association (the National Artists Association) has undertaken an annual exhibition titled "Women and Reality," covering all forms of art.

In the field of dance, new themes emerged to feature the realities of women's lives, the male-dominated social systems (including marriage), and other social problems such as environmental degradation.

In the film movement, women's screen organizations have contributed to the production of films on female clerical workers, day-care centers, maternal protection, sexual slavery by the Japanese army, and other issues relevant to women.

Women's groups began paying attention to the issue of women and media. As of 1995, 14 percent of the workers in the field of mass communications were women (Korean Press Institute, 1996). But mostly they are kept away from political, economic, and social reporting. Television programs concerning women's affairs are popular, but images of women represented in the mass media stick to the traditional notion of separate gender roles, and female sexuality is often commercially abused. Women's groups have been involved in monitoring and campaigning against sex and violence in the mass media so as to strengthen the foundation for a culture based on gender equality.

Conclusion

The women's cultural movement in Korea reveals that in order to achieve a fundamental solution to women's problems, there must be changes in attitudes and in the consciousness of women themselves and the society in general. Among the forces that account for the success of the women's movement in South Korea are the increased level of women's educational attainment (Kim, 1995), an increased level of participation in economic activities, and the growing social democratic movement (led by workers and college students). At the dawn of the twenty-first century, Korean people, both women and men, should recognize the contradictions and irrationality of the dominant patriarchal culture. It is their task to develop a new set of values regarding women and women's experiences and also to develop new modes of gender relationship. Activists in the women's movement, in particular, should place emphasis on the establishment of cultural identity for women. Amassing the objectives regarding women's cultural identity is not an easy task. An effective point of departure will be the development of means and opportunities for expressing women's everyday experiences through various cultural activities. An additional source of support for changes in the condition of women will derive from the implementation of the existing gender-focused legislation.

References

International Labor Organization (ILO). *Yearbook of Labor Statistics.* Geneva: International Labor Organization, 1992.

Kim, J.A. Strategies on Women's Non-formal Education for the Empowerment of Women. Informal meeting of experts on "Social Education for Women in the Context of Human Rights," 13–16 June, 1995b, Doe-ku, Korea.

Kim, J.I. "A Study on Gender Roles in the Elementary and Secondary School Textbooks," *Women's Studies Forum,* vol. 9, 1993, pp. 63–83.

———. "The Impact of Women's Education on the Status of Women in Korea," *Women's Studies,* vol. 48, 1995, pp. 201–206.

Kim, J.M. "Patriarchal Discourse and Power in Ritual and Daily Life: A Study of Samni Village in Korea's Honam Region." Ph.D. diss., Seoul National University, Seoul, 1992.

Korean Institute for Health and Society (KIHS). *The Structural Change of the Korean Family.* Seoul: Korean Institute for Health and Society, 1987.

Korean Press Institute. *Korean Newspaper and Broadcasting Yearbook.* Seoul: Korean Press Institute, 1996.

Korean Women's Development Institute (KWDI). *A Study on Male–Female Wage Differentials.* Seoul: KWDI, 1990.

———. *Statistical Yearbook on Women.* Seoul: KWDI, 1995.

Korean Women's Development Institute (KWDI)/Kitakyushu Forum on Asian Women (KFAW). *A Comparative Study on the Family Consciousness Between Korea and Japan.* Seoul: KWDI and KFAW, 1991.

Ministry of Labor. *Report on Occupational Wage Survey.* Seoul: Ministry of Labor, 1989.

———. *Survey Report on Wage Structure.* Seoul: Ministry of Labor, 1995a.

———. *Women and Employment.* Seoul: Ministry of Labor, 1995b.

National Statistical Office. *Annual Report on the Economically Active Population Survey.* Seoul: National Statistical Office, 1964.

———. *Population and Housing Census Report.* Seoul: National Statistical Office, 1979, 1982, 1992.

———. *Report on the Employment Structure Survey.* Seoul: National Statistical Office, 1993.

———. *Annual Report on the Economically Active Population Survey.* Seoul: National Statistical Office, 1995.

Women in Central America

Claudia María Vargas

Introduction

Women in Central America, with the exception of Costa Rica, seem to have progressed at a slower pace than women in other parts of Latin America, in terms of social development. Yet, after generations of repression and violence, they have, in fact, moved toward a future very different from their social reality of the past through political, social, and even revolutionary action. The decades of the 1970s and 1980s, a period of pervasive war and civil strife, particularly in Guatemala, El Salvador, and Nicaragua, posed numerous challenges to women in their day-to-day lives, as well as in their efforts to enter a new age. They faced difficult choices: marginalization or defiance, disfranchisement or organization, subjugation or action, and death or life.

The war and the pervasive violation of human rights in the region propelled women into social and political activism. It forced them to organize and, more important, to take their movement into the open. The population of refugees and displaced persons rose to more than 40 million (Marsella et al., 1994). Many women found a sense of purpose in organizing to protest military abuses, torture, and the deaths of their sons, husbands, and brothers, as well as attacks on their own *compañeras* (comrades-in-action). Others organized for the purpose of fostering new cultural values about the role of women in society to overcome gender discrimination and traditional roles casting women as mothers, in reproductive roles, and men as breadwinners, in productive roles. Still others struggled to fight on both fronts.

These same forces that compelled them to action also victimized them. Thousands were forced to migrate, joining a worldwide community of refugees numbering in the millions. Many have remained "refugees in orbit"—unable to stay in their homeland for fear of reprisals and unable to find refuge elsewhere (Gallagher et al., 1989, p. 341). Their refugee status brought with it complex problems inherent in that condition, from displacement, to arduous and dangerous journeys, to poverty and unsettled status in the country of asylum, not to mention physical and psychological trauma (Farias, 1994; Marsella et al., 1994). The least fortunate women fell victim to the violence in their home countries and the abuses of military regimes in the region, particularly in Guatemala and El Salvador. In Nicaragua, female health workers, teachers, students, and mothers suffered the continuous attack by the Contra rebels operating from Honduras.

In the midst of such a hostile environment, education, a key to meaningful opportunity for Third World women, was relegated to a low priority, if not eliminated, as women attempted to survive and to protect and care for their families. However, even in these oppressive political conditions, Central American women emerged from historical and cultural oppression to assert their rights to education and participation in the labor market on a status equal with men. They also demanded their legal rights, including protection against sexual harassment, family violence, rape, access to legalized family planning services and abortion, and more equitable treatment in family law.

In sum, although women have been adversely affected by social change, they have played a central role as active participants in the process of change. In addressing this dual reality for Central American women, this essay focuses specifically on the impact of the upheavals on women and the impact of women on life in Central America. The first part explores the effect of war and displacement on women in their new roles as heads of household in the family, on their representation in the labor force, and on access to education. The second part considers the impact women have had on social development through participation in social movements and particularly in women's movements.

Home, Work, and Education for New Lives: The Impact of Social Upheaval on Women

The pace and nature of change since the 1970s in Central America have reshaped several major aspects of women's lives. First, military and economic turmoil have turned many families into refugees, and even more into persons displaced within their own countries. In the midst of this instability and uncertainty, women have faced difficulties obtaining the education needed to prepare them for lifestyles different from the submissive and underdeveloped profile so familiar in much of Central American history. Even when they could establish a stable base for their lives and obtain an education, the nature of the labor market prevented many from obtaining good jobs at reasonable wages due to discriminatory labor practices, particularly accentuated in the private sector. Finally, women have faced new obligations and new roles that have sometimes brought not greater opportunities but, rather, increased burdens.

Refugees in Central America

Among all of these important forces, the nature and consequences of refugee status have had some of the greatest impacts on women. These consequences have included the socioeconomic pressures that refugees face, the physical and psychological trauma that come to persons displaced from their homes, and the cultural and linguistic difficulties associated with attempts to build lives in a new place while retaining connections with home and extended family.

Socioeconomic Effect. Women and children make up 80 percent of the refugee population worldwide (Martin, 1995). Central American refugees share much in common with other refugees but face additional barriers to resettlement. Many have made their way to the United States only to be categorized as economic immigrants instead of political refugees. These barriers prevented large numbers from attaining asylum. Beyond that, based upon foreign-policy preferences, the United States treated persons fleeing turmoil in Nicaragua as political refugees but refused to acknowledge as refugees people from El Salvador and Guatemala. The government ultimately conceded that discrimination as part of a settlement brought by the American Baptist Church. Furthermore, international recognition in the 1970s and 1980s of repressive regimes in Guatemala, El Salvador, and Nicaragua belied claims by the ruling military forces that they were protecting the people, building a democratic government, and developing a stable nation. This situation left refugees unable to return to their home country and unwelcome in what they had hoped would be a country of asylum.

The double bind of being unable to stay and unable to find asylum has taken a toll on Central American women, who have suffered from "chronic political violence, extreme poverty, unrelenting trauma and loss" (Hunter-Jenkins, 1991, p. 141). Although the majority of those who suffered from physical violence during those years were men, women were victims as well. Many who lost family members became victims, as they were forced to find new homes and new lives. The traumatization of women who have consequently been displaced or forced to migrate extends to children, since they, too, carry the fear, the loss, and the frequent exposure to violent scenes with them into exile.

Those women who have managed to get their children and themselves to a country of asylum and start a new life face new difficulties in the host country, including displacement, family breakups, and domestic violence; cultural and linguistic impediments; lack of affordable housing; unemployment or underemployment; and difficulties in accessing medical care and psychological services for the trauma and loss that come with flight. Many of the Salvadorans and Guatemalans unable to establish refugee status have also been unable to access education and health services (Hunter-Jenkins, 1991).

While many governments, including the United States, have sought to distinguish between economic immigrants and political refugees, the reality is that the two are often mixed. Economic instability in a war-stricken country burdened women particularly as they have faced the stark reality of being the sole supporters of their families. Frequently, the only viable option has been migration to neighboring countries, most often Mexico and the United States. Unfortunately, the socioeconomic landscape has been grim. In Mexico City, for example, Central American refugees have depended for their survival on being able to establish new familial relationships, a traditional cultural pattern. When they are unable to form those relationships, women feel socially, psychologically, and economically isolated. Sadly, some of them have resorted to what appears to them to be the only viable source of survival, prostitution (Rapold, 1985). If, however, a woman is able to find a refugee family with whom to bond, her situation improves significantly.

In many cases, women who are refugees from countries where they had been economically marginalized (Nunez, 1985, p. 13) have been forced to take sole economic responsibility for themselves and their children, a task made more difficult by flight and resettlement. Those who expect to return to their home countries in the future must build a life while also contemplating the need to start all over again in years to come. When wives go into refuge with their husbands, the situation may still be difficult. Even if both obtain employment in the host country, they are usually dramatically underemployed at minimum wages or less. Also, in part because of economic stress, there is a high rate of divorce and abandonment among refugees, which can put the woman back into the difficulties discussed earlier.

Psychological Trauma. In the midst of war and violence, the mental health of women and children, for whom mothers may be the sole support, often has gone unnoticed. Yet, the state-sponsored terror and the creation of a "political ethos" that fosters collective fear, anxiety, and a "culture of terror" (Hunter-Jenkins, 1991) have had a very real impact. The experience of terror and violence can lead to an intergenerational perpetuation of illness. Moreover, women, in particular, along with their children, have had to endure these emotional injuries compounded by extreme poverty, which accentuates the sense of loss, increases hopelessness, and intensifies feelings of inferiority and inadequacy.

The mental-health issues of refugees must be understood both as a problem for individual counseling or psychotherapy and as collective psychosocial trauma (Hunter-Jenkins, 1991). The siege atmosphere in El Salvador, Guatemala, and Nicaragua created a pervasive collective psychology of terror but with a concurrent spirit of denial. For example, *nervios* (nerves or pervasive anxiety leading to both illness and sickness) has commonly resulted from internalized experiences of political and social violence by a community. The link between *nervios* and state politics is real, systemically affecting and threatening the family. The intensity of this situation has been particularly felt by women in terms of personal injury, family violence, and desperation over poverty.

The physical and psychological trauma experienced by many women often reinforced their feelings of powerlessness and hopelessness. Even though there may be psychological consequences from these experiences that require treatment, culture-specific features may lead to misunderstanding and misdiagnosis by clinicians working from a modern Western medical paradigm. Moreover women may, in turn, transmit their illness to their children (Guarnaccia and Farias, 1988, p. 1223). Daughters are often the most vulnerable, since they frequently play the role of surrogate mothers to other children as mothers scramble to earn a living, sometimes working several jobs.

Linguistic and Cultural Issues. For those in flight or under stress, maintaining indigenous languages and cultures is extremely important. At the same time, the constraints of language have presented difficulties for refugee women. In both Mexico and the United States, Mayan refugees have suffered marginalization due to cultural and linguistic differences. For the Mayan refugee population, linguistic restrictions have prevented them from accessing badly needed health and social services (Farias, 1994). Cultural differences mean that mere linguistic translation is not sufficient to ensure effective communication. Women's lack of education and inability to communicate in English, or sometimes even in Spanish, have translated into malnutrition, high incidence of infant mortality, and heightened emotional distress. Furthermore, cultural lenses for viewing health and disease—traditions that do not recognize mental or medical health treatment as appropriate or viable—prevent some refugee women from seeking health care for themselves and their children.

The issues of language and culture are not merely barriers to the acquisition of services. They do more than stand in the way of efforts by mothers to provide for their children. The lack of ability to communicate can exacerbate a loss of a sense of efficacy and intensify feelings of inadequacy already fueled by the mere fact of life as a refugee. Even where women develop language capability, the lack of cultural understanding in asylum communities adds to a sense of rootlessness felt by refugees everywhere.

Women must not only find a way to get their children into the educational process, but they must also find a way to obtain for themselves the education and training needed to build a new life in a strange, and sometimes hostile, country or community.

Education and Women in Central America

Although it seems obvious, it has taken years for many societies to learn that educating women has definite positive impacts. The more obvious benefits include higher productivity, lower infant mortality rate, fewer low-birth weight infants, and reduced family size. The children of educated women are likely to experience better health and be better educated. Equally important, educated women, especially those who are heads of households, are clearly better prepared to enter the workforce and command better wages.

Educational Opportunities. Even though women in Central America have made some strides in literacy, their educational levels are among the lowest in Latin America (Costa Rica, whose rates resemble those of the more developed Southern Cone countries, is an exception). At the higher-education level, a reduced minority ever attends college, and those who do focus on traditionally segregated fields of study.

Schooling at the primary and secondary levels has increased for women, but they still lag behind men, especially in the rural areas. Only about 15 percent of females pursue secondary education. At the university level, only 3 percent of women complete their education (Rivera-Cira, 1993). On the other hand, Costa Rica's female enrollment figures are on par with those of males. In some cases, enrollment rates are even higher than those of men—for example, at the secondary level and at the university level, specifically in medicine and law. Honduras presents a unique case of a dual-society phenomenon, with the highest representation of women with university degrees in traditional male careers in Latin America and, simultaneously, large numbers of women who are illiterate (García and Gomariz, 1989a).

Table 1. Illiteracy Rate by Gender and Country (Age 15+, 1985)

	Costa Rica	El Salvador	Guatemala	Honduras	Nicaragua	Panama*
Total	7.4	27.9	44.0	40.5	24.8	12.0
Men	7.3	25.0	37.0	39.3	23.4	12.0
Women	7.4	30.7	50.0	41.6	26.0	12.0

Source: García and Gomariz, 1989a, p. 448.
*Figures on Panama come from World Resources Institute, 1992, p. 254. According to the United Nations (United Nations, 1991), for the 15–24 age cohort, the illiteracy rates are 13.2 percent for females and 12.1 percent for males.

In Central America, excepting Costa Rica and Panama, women's literacy rates are abysmal. Gaps exist between men and women (see Table 1), and are more marked in the rural and indigenous areas. The problem is most dramatic for rural indigenous and ladino (those of predominately European origin) females in Guatemala. Costa Rica, with the lowest illiteracy rate (7.4 percent), has a higher-than-average gross national product (GNP) for the region; Honduras and Guatemala, with the highest illiteracy rates (40 percent and 44 percent, respectively) are among the countries in the region with the lowest GNP. Nicaragua, during the Sandinista regime, was successful in reducing the illiteracy rate from more than 50 percent to well below 26 percent by focusing on preschool and basic education for both men and women and conducting an exemplary nationwide literacy campaign in 1980. Unfortunately, the illiteracy numbers rose in the early and mid-1990s following the 1990 election of Violeta Chamorro, whose government shifted the emphasis from social development to development of a market economy, implementing policies that have had a clearly negative impact on health and education indicators. President Arnoldo Alemán, elected in 1996, will continue Chamorro's market-based policies.

It is important to note the increased marginalization of the urban female population, given its continued state of economic crisis due to the marked polarization between the rich and the poor. Often, the illiteracy rates of poor urban women, as well as their socioeconomic standards, are similar to those of the rural population, transcending the traditional rural–urban dichotomy.

Even assuming that literacy rates can be improved by ensuring wider access for women to elementary and secondary schools, significant changes in social mobility and improvements in socioeconomic status will require a substantially enhanced opportunity for women to attend universities over what has been true in the past. Of course, that means not merely the opportunity to attend the university but also the opening of various traditionally male-dominated professional programs within the university to women.

In spite of progress, gender discrimination, complicated by social-class distinction, is present even for women with higher-education degrees. Women in Costa Rica, for example, face new barriers in the labor market even upon completion of a higher education depending on the type of university they attended and their socioeconomic status (Mendiola, 1992). Discrimination is particularly accentuated in the labor market, especially in the private sector. The public sector in Costa Rica has been more open to women in administrative positions, including the judicial system, since the mid-1980s. One explanation may be the comparable number of women to men entering law schools, about 50 percent each, and the dramatic completion rate of females in the field of law, 40 percent, which is higher than in other fields, except in medicine (Rivera-Cira, 1993).

The increased participation of women in the judicial system may make the legal community more sensitive to the need for legal remedies and fairer treatment, compared to the traditional pattern of discrimination against women, especially in family law, the labor code, and the penal code. Thus, the benefit of change appears to be twofold. First, professional women are now entering a traditionally male-dominated field, including taking seats in the Supreme Court in El Salvador. Second, there is an expectation that the presence of professional women in the judicial system may help alleviate some of the past unfair treatment of women. Even so, there is a long way to go before it can be said that full equal access to higher education is a reality.

Education and Access to the Labor Market. The perpetual economic crisis in Central America has forced women as well as girls to engage in productive activity mainly through the informal sector or through home-based activities to generate resources for survival rather than personal development. Although the informal sector has traditionally been the most viable alternative to the paid labor market, especially for poor women who have worked as domestic workers, seamstresses, ambulant vendors, or waitresses, it poses serious disadvantages, such as a lack of job stability and of social benefits protected by law. In addition, survival has demanded alternative familial arrangements in which individuals who are not related live together in order to join production forces. Some women have had to limit family size, even resorting to abortions; other women, and often adolescents, have had to engage in prostitution, delinquency, and beggary; still others resort to migration in the hope of self-survival and with the goal of sending money to those left behind.

The dominant model of development has segregated domestic work from productive work, and in Central America the distinction between reproductive roles and productive roles has not been made, a situation common to poor and rural women. Saakes (1990, p. 12) makes an important distinction between paid and unpaid house work. Historically, women who have been productive but uncompensated in domestic work have been viewed as dependent on their husbands. For example, women in Nicaragua have been traditionally categorized as housewives, yet they have actively participated in the productive process by generating resources through work done at home, such as making tortillas and doing laundry, that has been instrumental to family survival. In fact, in poor countries, ". . . a great majority of women are not financially supported by a man on a regular basis or protected by a partner" (Saakes, 1990, p. 13).

García and Gomariz (1989a) report a much higher level of female parents without spouse who are heads of households: approximately 95 percent for Costa Rica, El Salvador, and Honduras and 85 percent for Nicaragua (see Table 2).

Thus, for women from the lower classes, "housework (money-generating activities conducted at home) remains the only stable job while that [the work] of men and their presence seem to be unstable" (Saakes, 1990, p. 13). Poor women face a double burden: financial responsibility and motherhood. This is particularly significant when considering the high birthrate among teenage mothers in Latin

Table 2. Heads of Housholds (%) and With or Without Partners (%) by Gender and Country, 1984–1987

| | Head of Household | | | |
| | With Spouse | | Without Spouse | |
Country	Male	Female	Male	Female
Costa Rica	82.5	17.5	9.5	95.5
El Salvador	73.4	26.6	13.0	94.5
Guatemala	85.0	15.0	N/A	N/A
Honduras	79.6	20.4	8.8	96.6
Nicaragua	75.7	24.3	9.7	85.3

Source: García and Gomariz, 1989a, p. 440.

America; four Central American countries, Guatemala, Honduras, El Salvador, and Nicaragua, present the highest rate, more than 130 births per 1,000 teenagers. Costa Rica and Panama have lower rates, 96 and 97 per 1,000, respectively. The social protection and benefits from employment, such as health insurance, social security, labor legislation, and job protection through unions, are not available to these women (United Nations, 1995).

Education is an important determinant of the level and type of labor participation by women. The lack of education locks into poverty those who are most marginalized yet most in need of inclusion in the market. In particular, rural women and poor urban women suffer an inability to participate because they lack adequate access to education. Women's experience in the labor market, stability of employment, the type of sector available, and earnings vary considerably according to their educational background and their socioeconomic status. Women in rural areas carry a heavier burden of domestic and economic labor, affecting educational attainment and subsequent labor participation. Unemployment rates for women in the region are twice those of men and are aggravated by early marriage, high fertility rates, and scarce access to social services, except in Costa Rica.

Even though women have made progress in some traditionally male-dominated fields and in the service sector, women's participation in the agricultural and industrial sectors has diminished due to increases in population and male competition for the same jobs. In the service sector, where women hold 70 percent of the jobs in the region, a marked increase occurred, but for the wrong reasons: First, women's access to agricultural and industrial sectors is limited or blocked; second, women are hired for the lowest-paid jobs; third, women are paid less than men for the same jobs.

The employment crisis of the 1980s, the result of global recession, foreign debt, and structural adjustment measures, hit women especially hard, with unemployment rates twice as high as those of males (United Nations, 1991a, 1991b). However, for those entering the labor market for the first time, women were two or three times more likely to find jobs than their male counterparts (García and Gomariz, 1989a). Unfortunately, these figures are overcast by salary discrimination: Across the Central American region, women are paid 25 percent less than men for comparable jobs.

Women in Central America make up 30 percent–40 percent of professional workers but only 15 percent–20 percent of managerial and administrative positions. Most professional women are teachers or health workers.

Several suggestions have been made to address some of these gender differences in the various sectors. Catanzarite (1992, p. 80) recommends improving wages and stability in the informal sector, since women entering it are doing so out of need because they are excluded from the other two sectors—agriculture and industry. Thus, women enter the least stable and most vulnerable sector as a last resort for family survival. The impact is particularly significant as their roles as heads of households have increased dramatically. Other alternatives include legal protection of job stability and legal guarantees of employment benefits.

In the labor market, women face two major problems. First, they need to have access to education, vocational, technical, and professional, because education increases their income-earning power (Catanzarite, 1992). Second, they need mechanisms or policies to remedy historical inequities. Costa Rica has established a precedent in 1986 with the *Ley de Igualdad Social de la Mujer* (Women's Social Equality Law), considered one of the most progressive in the world. Adopted to remedy past inequities and discrimination against women, it contains provisions to increase women's participation in the electoral and political decision-making process and to protect women against violence.

Other suggestions proposed in Nicaragua and Costa Rica, for example, include one that the media play a more active role in raising social consciousness about the role of women in society and the inherent problems they face, and another that gender-specific content and materials be eliminated from school curricula. With new learning may come a new awareness, one that transcends class and gender inequities. Still, for Central American women, the combination of factors such as high birthrate, an average of 5.5 children per woman, lack of education, and difficulty in entering the labor force accentuates the complexity of their situation.

Social Movements in Central America

Women's emancipation in Central America has come at a high price, in part at least as a response to the war. Still, despite threats and political and military controls, women have dared to assert their own rights, as well as those of their families and their communities. Thus, although the development of organizations and gender awareness is still weak in Central America, the crisis of the 1980s promoted the creation of small groups of women focused on survival and labor projects. The second half of the 1980s witnessed the blossoming of women's organizations as political and military strife intensified (García and Gomariz, 1989b).

Women's participation can be analyzed from three perspectives: (1) women as researchers of women's issues; (2) women as a research topic; and (3) women's participation in popular movements. Costa Rica provides leadership in all three, owing to the political stability that allows women to expand their demands following those made by women in industrialized nations. In other countries, particularly El Salvador and Nicaragua, women have focused on social and political liberation. In Guatemala and Honduras, no distinct feminist movements have emerged with a clear purpose of demanding gender equality, although in Guatemala women have been active in peasant, union, and guerrilla movements. Panama presents a less dynamic picture compared to its neighbors. Clearly, women's movements in Central America present different ideological and practical approaches.

Consequences of the Civil War on Women: Women's Movements

Rural and urban poor women launched their activism independently of, and for reasons other than, gender concerns or Western-style feminist movements. Their urgency was based on reactions to the war waged around them. The Asociación Centroamericana de Familiares de Detenidos y Desaparecidos (ACAFADE) was one of the first organizations that fought for social justice and for accountability for disappeared or detained political prisoners, family members. Women's movements emerged with an agenda that appeared divorced from the goals of the feminist movement. However, the decade of the 1990s has witnessed a coming together of gender issues and the struggle for peace and justice, and there are more than 60 women's organizations in Central America, including Belize (Gargallo, 1993).

Historically, women's movements have not always organized around a social agenda. Some represented by upper-class women—for example, those who demanded the right to suffrage—were divorced from the struggles of poor women or indigenous women, as was the case in Guatemala. In fact, in some cases women have organized to protest government measures intended to remedy inequities, supporting the right-wing politics of the time. In Chile in the early 1970s, women drumming on pots and pans rallied against plans of the Marxist Allende government for food rationing. In Mexico in the 1930s, they came together to protest the dissolution of large landholdings as part of the land-reform distributions to the poor and landless peasants. In Guatemala in the early 1950s, upper-class women, wives of the wealthy landowners, joined forces to protest the Arbenz government's democratic policies, such as land reform.

Yet, women have made major strides in the region. Although they are still undergoing a process of consciousness-raising, women have organized in an attempt to transcend class origin, race, and ethnicity and instead fight for social justice with a sense of the need to integrate gender equality and class equity. There has emerged a growing thrust for peace, social justice, and respect for human rights. Guatemalan Rigoberta Menchú, the 1992 Nobel Peace Prize winner, presented the interconnection well when she said that she felt the need simultaneously to fight three types of discrimination: "sexual for being a woman, racial for being indigenous, and economic for being poor" (Charles, 1992, p. 8).

Exile has, perhaps surprisingly, also mobilized women. Central American refugees have found meaning in their refugee experience through social and political activism launched from abroad, particularly those from El Salvador and Guatemala. Exile catalyzed Mayan women, and men also, to organize politically in their host country and, upon their return, in their homeland. In spite of repressive military measures against the autochthonous, or indigenous, population in the region, especially in Guatemala, these Mayans have launched a serious and adamant campaign to preserve the cultures and, contrary to tradition, make them known to those in the outside world sensitive to, and caring of, them. The number of Mayans who have acquired higher education, including doctoral degrees, committed to ethnic pluralism through the preservation and teaching of Mayan languages has increased. The Academy of Mayan Languages was established in 1990 with a "founding mission of introducing the new standard alphabet in all twenty-one Mayan linguistic communities" (Perera, 1993).

Many of these Mayans had been refugees in the United States or Europe and, upon returning to their homeland, committed themselves to the preservation of their ancestral cultures and languages. Transitional bilingual programs as advocated by the Summer Institute of Linguistics, a Protestant evangelical organization based in the United States, with an agenda to convert Mayans and to wean them from their native languages and culture into Spanish are no longer acceptable. Instead, Mayan organizations, including the National Program for Bilingual Education (PRONEBI), have been created to capture the identity and languages so long denied by the ladino culture.

Women's presence in such endeavors has been significant as has their involvement in organizing to demand long-promised but still unfulfilled land redistributions. Menchú's work in Guatemala has been significant. Calixta Canek, a Cakchiquel schoolteacher, who was an undocumented refugee in California, joined forces with other Mayans connected with peasant labor movement in that state. Since returning to her homeland in 1986, she has focused on political organizing both nationally and internationally, even establishing ties with the Scandinavian autochthonous people and other world native people connected with the Mayans.

Some Central American women, specifically Nicaraguans, have viewed feminist positions as a counterweight to machismo, "The name 'women's liberation movement' does not imply that it pretends only to liberate women, or that women must oppose themselves to men, but that they must start with their own interests, uniting with all other oppressed sectors which are also seeking a revolutionary change for all" (Malta Lamas, in Flynn, 1983, p. 415). Notwithstanding this position by some women, others have focused their efforts on general liberation movements. Women in Nicaragua, El Salvador, and Guatemala from all socioeconomic strata organized and participated in the war in the 1980s against the military-backed regimes. Military movements in the region included a large component of women, some of whom held leadership positions. In fact, 30 percent of the political and military movements have been integrated by armed women (Reif, 1986). What is important here is not that women have taken up arms, with all of the destruction and sadness associated with armed conflict. Instead, these figures show that women are involved in transformative efforts in all aspects of life of their communities.

Conclusion

Central American women have made difficult journeys as victims of social and political turmoil and as dynamic actors in the process of social change. In both capacities, they have confronted cultural and historical asymmetries in various contexts, at home, in school, in the labor market, in asylum or displacement, and in the political arena. Certainly, the road has been arduous, particularly for poor women and women from rural areas, and equally harsh for members of indigenous groups and for those women who are displaced or in search of refuge. However, these obstacles have forced women to forge a new destiny for themselves and their families. They strive for a fair and just future, one sculpted of social justice, peace, and gender equality. The tensions inherent in building that new society are real, particularly given the serious economic pressures and the complexity of refugee status. However, Central American women are transcending past barriers as they face the future.

References

Bustillo, Inés. "Latin America and the Caribbean." In Elizabeth M. King and M. Anne Hill (eds.), *Women's Education in Developing Countries: Barriers, Benefits, and Policies.* Published for the World Bank. Baltimore: Johns Hopkins University Press, 1993, pp. 175–210.

Catanzarite, Lisa. "Gender, Education, and Employment in Central America: Whose Work Counts?" In Nelly P. Stromquist (ed.), *Women and Education in Latin America: Knowledge, Power, and Change.* Boulder: Lynne Rienner, 1992, pp. 67–84.

Charles, Mercedes C. "'Soy Indígena y, Además, Soy Mujer . . . ,'" *fem.,* vol. 16, no. 118, December 1992, pp. 8–12.

Farias, Pablo. "Central and South American Refugees: Some Mental Health Challenges." In Anthony Marsella, Thomas Bornemann, Solvig Ekblad, and John Orley (eds.), *Amidst Peril and Pain: The Mental Health and Well-Being of the World's Refugees.* Washington, D.C.: American Psychological Association, 1994, pp. 101–113.

Flynn, Patricia. "Women Challenge the Myth." In Stanford Central America Action Network (ed.), *Revolution in Central America.* Boulder: Westview, 1983, pp. 414–422.

Gallagher, Dennis, Susan Martin, and Patricia Weiss-Fagen. "Temporary Safe Haven." In Gil Loescher and Laila Monahan (eds.), *Refugees and International Relations.* Oxford: Clarendon, 1989, pp. 331–353.

García, Ana Isabel, and Enrique Gomariz. *Mujeres Centroamericanas ante la Crisis, la Guerra y el Proceso de Paz, Tendencias Estructurales,* vol 1. San Jose: FLACSO, CSUCA, and Universidad para la Paz, 1989a.

———. *Mujeres Centroamericanas: Efectos del Conflicto,* vol 2. San Jose: FLACSO, CSUCA, and Universidad para la Paz, 1989b.

Gargallo, Francesca. "Los feminismos centroamericanos: Sus surgimientos, sus negaciones, sus participaciones y sus perspectivas: Un acercamiento a la política femenina," *fem.,* vol. 17, no. 119, January 1993, pp. 13–21.

Guarnaccia, Peter J., and Pablo, Farias. "The Social Meanings of 'Nervios': A Case Study of a Central American Woman," *Social Science Medicine,* vol. 26, no. 12, 1988, pp. 1223–1231.

Hunter-Jenkins, Janis. "The State Construction of Affect: Political Ethos and Mental Health Among Salvadoran Refugees," *Culture, Medicine, and Psychiatry,* no. 15, 1991, pp. 139–165.

Loescher, Gil, and Laila Monahan (eds.). *Refugees and International Relations.* Oxford: Clarendon, 1990.

Marsella, Anthony, Thomas Bornemann, Solvig Ekblad, and John Orley (eds.). *Amidst Peril and Pain: The Mental Health and Well-Being of the World's Refugees.* Washington, D.C.: American Psychological Association, 1994.

Martin, Susan. *Refugee Women.* London: Zed, 1995.

Mendiola, Haydée. "Gender Inequalities and the Expansion of Higher Education in Costa Rica." In Nelly P. Stromquist (ed.), *Women and Education in Latin America: Knowledge, Power, and Change.* Boulder: Lynne Rienner, 1992, pp. 125–145.

Nunez, Kyra. "Las refugiadas están urgidas de solidaridad femenina," *fem.,* vol. 8, no. 38, February/March 1985, pp. 13–14.

Perera, Víctor. *The Unfinished Conquest: The Guatemalan Tragedy.* Berkeley: University of California Press, 1993.

Rapold, Dora. "Jovenes Refugiadas en la Ciudad de México," *fem.,* vol. 8, no. 40, June/July 1985, pp. 32–33.

Reif, L.L. "Women in Latin American Guerrilla Movements: A Comparative Perspective," *Comparative*

Politics, vol. 18, no. 2, 1986, pp. 147–169.

Rivera-Cira, Tirza. "Women Judges in Central America," *Hemisphere,* vol. 5, no. 3, Summer/Fall 1993, pp. 8–9.

Saakes, Sylvia. "Las estadísticas y el trabajo de las Mujeres," *Documentos sobre la mujer,* no. 12, July/ August/September 1990, pp. 12–13.

United Nations. *The World's Women 1970-1990: Trends and Statistics.* New York: United Nations, 1991.

———. *Women in a Changing Global Economy: 1994 World Survey on the Role of Women in Development.* New York: United Nations, 1995.

World Resources Institute. *World Resources, 1992–1993.* New York: Oxford University Press, 1992.

Women in Contemporary Cuba

Berta Esperanza Hernández Truyol

Introduction

This being an essay in an encyclopedia of Third World women, it is pertinent to ask whether Cuba is a Third World country. The social and human-development indicators reveal that Cuba does not match the definition of a Third World state neatly. Cuba's health, education, and welfare figures rival those of industrial states. The economic-development figures, on the other hand, paint a radically different landscape: Economic considerations certainly permit labeling the island a "developing country," particularly since the onset of the 1986 recession, which was exacerbated by the demise of Communism and Cuba's consequent loss of Soviet-bloc aid and subsidies.

To be certain, the status of Cuba was debated even before Fidel Castro's revolution; some Cuban historians cite 1959 statistics to show that Cuba was not, even then, a Third World state. These figures place Cuba second or third in the Americas in terms of numbers of radios, television sets, cars, telephones, refrigerators, daily newspapers and magazines, and doctors and dentists per capita and rank it high in industrial salaries, fiscal stability, meat consumption, and literacy rates. In addition, these historians note that Cuba had increasingly high percentages of domestic ownership of important industries, and it was widely believed that Cuba's post–World War II economic picture was promising (Márquez Sterling and Márquez Sterling, 1975; Clark, 1992).

On the other hand, for the same period—the time at the outset of his revolution—Castro labeled Cuba an underdeveloped country. In support of this evaluation, in his 1960 speech to the United Nations he cited the following figures: 1960 unemployment rates rivaling those existing in the United States during the Great Depression, lack of electricity and housing with sanitary facilities by almost 50 percent of the population, a 37.5 percent illiteracy rate, high infant mortality, low life expectancy, and large foreign ownership of public services and industries (Castro, 1992). These statistics, contrasted to those of the historians, simply cannot be reconciled.

More than 30 years later, the debate is ongoing, but interest has peaked concerning the status of Cuba, now a sole socialist survivor. Depending on one's political leanings, the Cuban situation in the mid-1990s is characterized either as a magnificently successful or a wholly failed experiment. Pictures of Cuba range from a portrait of a thriving tourist industry and cutting-edge medical center, to a landscape of epidemics sustained by lack of food and ox-driven plowing due to a lack of fuel

A review of the status of women in Cuba reinforces these general dichotomies. For women, Cuba is in many ways a different country today than it was before the socialist revolution of 1959. Nevertheless, pre- and postrevolutionary Cuban society has been consistent with respect to women in one critical aspect: The laws and constitutional provisions affecting women and protecting women's rights have always been among the most progressive in the world. Cuban women have been integral to every struggle for independence in the island—from the 1898 wars of independence from Spain to the socialist revolution of 1959. However, regardless of women's participation in these struggles and of a progressive legal structure, women's roles in society have failed to change from the traditional models inherited initially from Spain.

In looking at women's movements and progress in Cuba, one will not find goals identical to those of Western feminists. Although the Cuban feminist movements were for egalitarianism and sought protections for, and achievement of, certain rights, feminism's goal never was for formal equality. Rather, Cuban feminism operated within "traditional" social roles and mores, always revering motherhood—which is seen as giving women social status and as a stepping-stone to power—and seeking to preserve femininity (Stoner, 1991). These roles have survived the revolution, resulting in women's second-class citizenship with respect to representation in high levels of government and the double burden of holding down jobs outside the home while still being primarily responsible for the family, including child care and house work.

To help readers understand the role of women in Cuba today, this essay traces the struggles and achievements of Cuban women and Cuban feminism from the turn of the twentieth century to the eve of the twenty-first. It evaluates the changes in the role and status of women in Castro's Cuba, according to existing rules and laws, and reviews various factors that perpetuate gender differences, including three laws that, while aimed at alleviating the burdens on working women, effectively entrench gender stereotypes. It concludes that, although Cuba is perhaps one of the most progressive countries in the world with respect to its treatment of women under the law, Cuban women have not reached gender equality.

The Early Feminist Movement: An Overview

The first Cuban feminists were the *mambisas* (woman warriors). These women, wives and mothers, redefined womanhood by leaving the safety of their homes to join in the fight for Cuban independence from Spain (the Ten Year's War [1868–1878] and the War of 1895). The *mambisas* performed traditional male and female tasks, nursing the wounded, growing and foraging for food, fighting and spying, and providing emotional support. In fulfilling these varied responsibilities, they left their own communities and moved into *maniguas,* communities of a few families and peasant houses up in the mountains. The *mambisas* were the first women in Cuba to own their own property, challenging the whole notion of male authority over family property and establishing a woman's right to control her own property without a prenuptial agreement (Stoner, 1991).

Property and Family Rights

Cuba won independence from Spain in 1898. The property rights for which the *mambisas* fought became law in 1917, when a bill was enacted granting married women the right to administer their dowries and their property. Ironically, the passage of this bill depended on the arguments of two senators who maintained that giving women this small right would prevent a revolution and dissolution of the family. They further noted that allowing women the right to control their own property would protect a man's fortune and prevent it from passing to another family line when a family had only daughters. This bill, in addition to granting married women control over their dowries and family properties, granted women the right to sue and be sued and gave women authority over their children from previous marriages.

A decade later, in 1928—even before obtaining the right to vote—women obtained further protection by virtue of a measure that repealed the adultery law, one of the most deeply ingrained and most dangerous aspects of male domination. The law that was repealed, Article 437 of the Spanish Penal Code, allowed a husband who encountered his wife in the act of adultery to kill her and the offending lover without being tried for homicide. The maximum penalty for such acts was exile. Indeed, the husband suffered no penalty at all if he only wounded his wife or her lover. This same conduct was deemed appropriate for a father who encountered his minor daughter with a lover.

In effect, the adultery law gave men complete authority over their wives and daughters and basically sanctioned violence against women. Women's groups had made it a priority to change this law and to turn adultery from a tool of violence against women into a ground for divorce for either party. Feminists achieved this goal when the law was repealed. Notwithstanding the legal change effected by the 1928 bill that abolished the adultery law, societal underpinnings failed to change. Interestingly, the law passed without presidential signature because President Machado did not want to be associated with "the loosening of sexual standards" that presumptively would take place if the men could not put/keep "their" women in their place.

Also before obtaining the right to vote, women received protection of the 1918 divorce law, which separated church and state and lessened the church's influence over Cuban women. This law passed without women's advocacy or even their expression of the law's desirability. The divorce law, revised in 1930 and again in 1934, was one of the most progressive in the world. This law treated women and men equally and provided more protections for women than any other laws. Grounds for divorce were expanded to include separation of spouses for five years or more, corruption of one marriage partner by the other, use of drugs, bigamy, and abandonment of six months. In addition, under the law, husbands and wives were charged with equal responsibility for the stability of the family and the support of the children. Custody was awarded on the basis of ability to support, rather than gender of the parent. Unfortunately, this signified that women, who were less financially solvent than men, more often lost custody of their children. The new divorce law also entitled women either to liquidate or to assume control of their husband's estate in the event that the husband could not meet his alimony payments. However, alimony was not need based, and it was only awarded to innocent (i.e., nonadulterous) wives (Stoner, 1991).

In 1934, Cuba's legislature passed the first national maternity law; it included a grant of 12 weeks' leave for maternity (6 weeks before and 6 weeks after giving birth), payment of a subsistence pension to mothers while they were absent from work, and a requirement that all public and private industrial factories that employed more than 50 women maintain a nursery for children under age 2. Three years later, the maternity law was expanded to include provisions that gave nursing mothers half an hour off in the morning and half an hour off in the afternoon to breast-feed the children until they reached one year of age and that made it illegal for employers to fire female employees when they married (Stoner, 1991). Although these maternity laws were a great legal advance for many women, their tragic flaw was the exclusion of coverage of domestic servants, who made up 32 percent of the population.

Moreover, while the paper rights existed, in reality many women failed to receive the services guaranteed.

Significantly, women obtained these broad rights without any acknowledgment by anyone of a notion or goal of achieving women's equality. Moreover, the Cuban feminists who fought for these rights were far from representative of Cuban society as a whole. They were White, middle-to-upper-class and educated women about 40 percent of whom were mothers, 42 percent of whom were married, 60 percent of whom at some time had been employed, 75 percent of whom had graduated from a university, 33 percent of whom had postgraduate degrees, and every single one of whom had at least one servant (Stoner, 1991). These women were the product of a Cuban culture that reflected many years of Spanish influence and dictated that women were not to work but were supposed to stay at home and prepare for marriage. Even if a woman did work, it was understood that it was temporary and that once she married, she would give up her job.

However, the cultural aspirations and expectations regarding equality were different for Black, mulatto, and poor White women who had no option but to leave home and work. So, when only a small number of women were employed during the early years of Cuban independence, approximately three-quarters of the employed women were Black, and most women were engaged as domestics, laundresses, and tobacco plantation workers (Leahy, 1986). These women, in contrast to the educated feminists, faced poverty, ill health, sexual exploitation by the men with whom they worked, and disrespect for their race and/or class. Thus, notwithstanding the broad legal rights pertaining to property and family, women, for the most part, were far from achieving equality.

The 1940 Constitution: The Rights to Vote and to Equality

Another one of the main goals of the early feminist movement was to secure the vote for women. Although President Ramón Grau San Martín granted the vote to women in the provisional constitution of 1934, that right, together with the right to be elected to public office, became a confirmed reality in the constitution of 1940, one of the most progressive for women. Like the right to own property, however, neither the right to vote nor the express provisions regarding gender equality found in the constitution effectively signified an acknowledgment of women's equality nor did they result in equality.

Grau San Martín's 1934 decree, Law 589, was the precursor of some of the 1940 constitution's equality provisions. This law mandated that salary should correspond to the nature of the work, not to the sex of the worker. It also assured women equal access to work, except for protective legislation preventing them from working in dangerous environments, which effectively restricted women from some 400 jobs. In addition, Article 8 of Law 589 sought to protect domestic servants by requiring them to register their employers' names and their salaries with the secretary of labor so that officials could detect underpayment and so that domestics would have access to national insurance (Stoner, 1991).

Notwithstanding the reality of the inequality of women, the 1940 constitution has several noteworthy provisions regarding technical legal rights to equality. For example, Article 23 stated: "All Cubans are equal before the law. The state does not recognize special privileges or status. It is illegal and punishable by law to discriminate based on sex, race, color, class, and any other prejudice against human dignity." In addition, Article 62 provided equal pay for equal work, regardless of sex, race, or nationality; Article 68 made it illegal to distinguish between married and single women in the workplace; and Article 77 mandated an 8.5-hour workday, restricting night work for women. Once these extensive gender-based legal rights were achieved, the feminist movement in Cuba basically died.

Unfortunately, although Cuba's constitution was one of the most progressive in the world, especially in respect to its treatment of women, this level of paper rights did not translate to real changes in women's lives. In 1953, while Cuba ranked third highest in Latin America for its literacy rate, with 78.8 percent of the population literate, women fared worse than men. In higher education, only 1.6 percent of women had received some secondary education, compared to 2.4 percent of men; and while 1.6 percent of men had received some university education, only 0.8 percent of women had (Purcell, 1975).

Similarly, women did not fare well in employment. Women made up only 17 percent of the work force (United Nations, 1991). While one-third of the labor force was unemployed or partly employed between 1956 and 1958 (Bodrova and Anker, 1985), women had less opportunity than men, with the economy allowing few options for support other than marriage in a culture that, despite broad legal pronouncements, reinforced the view of a woman's place being in the home (Leahy, 1986). Even when women entered the workforce, they were concentrated in the "pink ghetto" fields that were considered appropriate and acceptable for women. Traditionally, these were low-paying, low-status jobs. Before the revolution, for example, between one-quarter and one-third of the Cuban women who worked were employed as domestics, and more than 60 percent of the female professionals were schoolteachers in the lower grades while most of the teachers at the secondary level and above were men.

Finally, women had constitutionally dictated equal rights to being elected to public office. However, they rarely ran for office, and few were elected. Thus, although before the revolution women had very strong legal rights to equality on paper, these were aspirational; the reality did not reflect the existence of an egalitarian society.

The Revolution and Beyond

The Cuban revolution was founded on the concept of egalitarianism. In fact, in his first address to the nation Castro noted the need to end discrimination against women's participation in the labor force. Shortly thereafter, the Labor Ministry started to enforce labor legislation regarding women more strictly and enacted new regulations regarding the rights of pregnant women to their jobs (Pérez-Stable, 1993). In fact, in his address to the General Assembly of the United Nations on September 26, 1960, Castro clearly stated that one of the aims of the revolution was to eradicate inequality and discrimination, specifically noting that one of the principles of the Cuban revolution was the condemnation of discrimination against Blacks and Indians and of inequality and exploitation of women. Castro designated this move toward equality for women as a revolution within a revolution (Castro, 1992). The 1976 constitution encoded this right to equality into law, with chapter 5 expressly providing that "all citizens enjoy equal rights and are subject to equal obligations." The discussion following details the role and status of women during and after the revolution.

Just as in the war to free Cuba from Spanish domination, in the revolution of 1959 women fought alongside men. Women were involved in every aspect of the revolution, including producing rebel uniforms, participating in action in sabotage units, transporting arms, and hiding in the mountains. In fact, by 1958 about 5 percent of the rebel army's 3,000 soldiers were women. Although some men opposed female platoons, Castro fully supported them, noting that "there are men who ran, M-1 and all, but we do not have a single example of one of the *Mariana Grajales* women who ran from combat" (Riveira, 1989).

To institutionalize women's place in the revolution, in August 1960 Castro created the Federación de Mujeres Cubanas (Federation of Cuban Women, FMC). Although initial membership was less than 100,000, by 1968 membership had swelled to almost 1 million women, and by 1970, to 1.3 million, or 54 percent of the Cuban women over the age of 14 (Purcell, 1975). By the 1990s, 80 percent of Cuban women over the age of 14 are dues-paying members of the FMC. The goals of the FMC have been to prepare women educationally, politically, and socially to participate in the revolution, as well as to incorporate women into the workforce and raise their educational consciousness. Although the women workers who first organized women were especially militant, the composition, if not the leadership, of the FMC is a representative cross-section of Cuban women. Significantly, the head of the FMC was Vilma Espín, Castro's sister-in-law and the only woman ever to serve in the ruling Politburo, albeit for a very short time.

The United Nations *Human Development Report* (UNHDR) labels women as a "nonparticipating majority" because while they are "a majority of the world's population [they] receive only a small share of developmental opportunities. They are often excluded from education or from the better jobs, from the political systems or from adequate health care." As discussed below, health, education, and employment figures for Cuban women show that what the UNHDR reports as the condition of women around the world is simply not the case in Cuba, largely because programs instituted during Castro's rule and in which the FMC has played a salient role have benefited this segment of the population. The FMC's work to improve the condition of women has enjoyed an impressive success rate.

Women's Advancement: Retraining and Education Programs

One of Cuba's first revolutionary campaigns was to abolish illiteracy, and it succeeded: Illiteracy dropped from 23.6 percent at the time of the revolution to 3.9 percent of the population by the 1980s, making approximately 750,000 persons, 56 percent of whom were women, literate. The campaign was carried out mainly by volunteers among Cuban youth who went to the isolated rural areas to teach reading and writing while living in the homes of peasants and sharing in the work of planting, harvesting, cattle raising, and so forth. The FMC figured prominently in this campaign, mobilizing women for the volunteer work; in fact, 55 percent of the volunteers (*brigadistas*) were young women.

Another major undertaking of the revolution in which the FMC played a major role was the reeducation or retraining of domestic servants and prostitutes—a goal consistent with the revolutionary aim of ending gender-, class-, and race-based distinctions in society. Under specially designed programs run by the FMC, domestics, who were mostly women, were allowed to leave their employment and study full-time or stay at their posts and take courses at night. These women were taught literacy, history, geography, mathematics, typing, accounting, and revolutionary studies, allowing many former domestics to take jobs in banking or in state agencies.

The reeducation of the prostitutes took place on farms, where they stayed for one to two years. During their stay, they learned either the traditional primary-school curriculum or took more practical classes such as dressmaking, hairstyling, and typing. If the women were on the farm at the end of the program, and many were not because adaptation to rural conditions was so difficult, they were then relocated. Many of the estimated 100,000 prostitutes at the end of ousted dictator Fulgencio Batista's regime benefited.

Cuban women also have fared well in formal education. The initial significant step for progress in traditional education was to make education free at all levels. This resulted in an increase of women attending and completing school. For example, for every woman receiving a primary education in 1956–1957, 39 women were registered in the 1965–1966 school year (Purcell, 1975). By 1967, women represented 49 percent of students in elementary schools, 55 percent of students in junior high schools and high schools, and 40 percent of university students. Not only were women well represented in enrollment, they also were well represented in most fields—by 1975, women were 30 percent of all engineering students and 35 percent of all agricultural science students (Leahy, 1986).

In the mid-1990s, the literacy rate for Cuban women age 15–24 was 99 percent, compared to 65 percent for females in all developing countries, 82 percent for medium human-development states, and 93 percent for high human-development countries. The primary-school enrollment ratio for females in Cuba was 95, compared to 86 for females in all developing countries; comparable figures for the secondary level were 96 and 36; for the tertiary level, 25 and 5 (UNDP, 1993).

Despite such incontrovertible progress, an event relating to medical-school enrollment suggests that women may still fail to enjoy full equality in educational opportunities, particularly when state foreign policy conflicts with gender equality. At an FMC Congress in 1974, Castro recognized that women's equality was not yet a reality. A decade later, however, the government imposed quotas limiting the number of women who would be allowed to attend medical school. The government's explanation for the quotas was twofold and ostensibly based on the critical role played by civilian medical assistance in Cuba's foreign policy. Both rationales proffered by the government revealed the status of women as one of *in*equality. First, the government explained that because of women's greater family responsibilities it would be difficult for them to travel overseas for extended periods to deliver medical assistance. Second, the government rationalized the quota by indicating that male doctors were needed to deliver medical assistance in countries that had not recognized (as Cuba had) the changing role of women in society. By instituting the medical-school quota, the government did not allow women to decide for themselves whether to travel and contend with sexist attitudes. Thus the notion of equality ceded to the governmental policy to deliver aid. Moreover, the first rationale blindly accepts women's disparate roles and disproportionately higher responsibilities and concomitant time commitment within the family structure. Significantly, the FMC, which was charged with ensuring and promoting gender equality, did not challenge the discriminatory medical-school quota.

Women's Health

Women's health figures also are impressive. As with education, all health care is free in Cuba, including contraception and abortions, which are available on request and performed in government hospitals. As with other social-concern programs, the FMC played an active role in promoting women's health and developed an excellent sex-education program. For example, as early as 1962 the FMC was giving women first-aid training and promoting personal hygiene as well as pre- and postnatal care, particularly in rural areas.

The health-initiative successes are plainly evident from the life expectancy, maternal and infant mortality, and birth figures. Whereas the life expectancy of women at birth in all developing countries is 64.2 years, the figure for Cuba is 77.3 years, virtually identical to the industrialized states' figure of 77.4 years and only slightly lower

than the U.S. figure of 79.5 years. The maternal mortality rate per 100,000 live births in Cuba is 54, compared to 26 for the industrial countries and 420 for developing countries. The infant mortality rate per 1,000 live births is 14, far below the average of 71 for all developing countries and identical to the figure for industrial countries. Nevertheless, perhaps reflecting the state of the world economy and the impact on the Cuban economy of the end of Soviet subsidies, this 1992 figure shows an increase in infant mortality from 1990, when the rate in Cuba was 11, a figure that was better than the rate of 13 for industrialized states (UNDP, 1993).

Employment and Representation

Progress for women also is evident in employment, another area in which the FMC has made great efforts. In fact, the labor situation of women is telling. In 1964, only 282,000 women were gainfully employed; by 1970, the number of women in the labor force had reached 600,000 (Purcell, 1975). By the end of the 1970s, nearly 31 percent of the workforce was female. In 1990, women were 32 percent of the labor force, about average for all developing countries but below the 42 percent figure for industrial states (UNDP, 1993).

Aside from increasing their numbers, women also have broadened their role. Nevertheless, while women hold many jobs that once were held exclusively by men—they are cane cutters, citrus-fruit packers, auto mechanics, dentists, doctors, engineers, and traffic police—female participation in the labor force still follows traditional patterns. For example, the most common occupational group for women is "other intellectual activities," a classification that includes clerical and secretarial jobs. Also, percentages of women are high in the garment and service industries as well as in teaching and research, reflecting the fact that females continue to be clustered in traditional women's occupations, such as primary-school teaching and nursing (Diaz-Briquets, 1989).

Notwithstanding their concentration in traditionally female jobs, however, women make up the majority in such "nontraditional female" fields as medicine and, increasingly, in law. In fact, women constitute 56 percent of all working professionals, and there are more female than male doctors and judges. However, retirement age is sooner for women (55) than men (60) (Diaz-Briquets, 1989).

Cuba, as a socialist state, guarantees employment to every citizen who wants to work. With this guarantee, the state influences the size of the labor force by policy proclamations. Some of these proclamations, however, are strong evidence that true equality is but an aspirational goal. One of the ways that Cuba has manipulated the size of the labor force is by granting or removing economic and other incentives for female labor-force participation and by modifying the regulations under which women are allowed to work (Diaz-Briquets, 1989). As in other cultures, women are considered a supply of labor when there are shortages. However, when the state found it necessary, it took measures to discourage female employment, including classifying some jobs as male only. In fact, government regulations have limited women's access to certain jobs. These restrictions are explained as aimed at protecting women's health. However, they are, in reality, pretextual and aimed at resolving employment problems.

Through the years, the FMC has objected to the government-imposed work restrictions as violative of principles of equality, but there are still some job categories that are off limits to women. However, the list in the mid-1990s is down to approximately 25 categories, compared to 300 in the 1970s.

Still, as with the quotas imposed on women entering medical school, the rights women enjoy are largely only a manifestation of the state-supported reinforcement and perpetuation of traditional gender roles. This stereotyping includes the notion that work outside the home is more important for men, a notion that finds support in, and is facilitated by, state policies. In its 1990 Congress, the FMC noted the obstacles to achieving gender equality of the double burden of work and home. It also recognized the continued discrimination in job promotions, the lack of women's progress regarding promotions to leadership positions, and the significantly lower earnings of women compared to men even when women were better educated. So while women have achieved progress in the workforce, true equality remains an elusive ideal.

Women in Cuba have made great strides in parliamentary representation. In 1991, women held 34 percent of parliamentary seats in Cuba, a percentage exceeded worldwide by only Norway, Romania, and the Soviet Union. This figure was three times the average of 12 percent female representation for all developing countries, close to four times the 9 percent figure for all industrial states, and almost nine times the 6 percent figure for the United States. However, it was only in 1986 that a woman—Vilma Espín, Castro's sister-in-law, became a full Politburo member, and she was excluded in 1991 when that body, again, became all male. Overall, women lag far behind in the "more powerful and prestigious occupational levels in revolutionary Cuba. Although one out of every three workers is a woman, less than one in five directors of state, political and economic organizations is female. This female underrepresentation in the pinnacles of political and economic power . . . remains puzzling if one is to accept the official rhetoric calling for equality between the sexes" (Diaz-Briquets, 1989, p. 109). Thus, as with employment, women enjoy representative positions, but they are at the lower levels of government, rendering this achievement less than "equal."

The Myth of Equality

The revolutionary notions of equality, mostly encoded in law, were noble and partly realized insofar as one considers the improvements in women's health, education, and welfare discussed above.

However, these noble sentiments have often been inconsistently pursued. The revolution reinforced traditional

domestic arrangements by paying for honeymoons while at the same time insisting that women work outside the home in order to provide the state with "productive" labor. It considered paid employment a key to women's emancipation and encouraged women to join the labor force on the theory that housewives were "unintegrated" and would raise unintegrated children—that is, nonproductive members of the revolutionary society. While domestic tasks were seen as unproductive and unpleasant, the state failed to provide the means to relieve women of their double burden: working outside of the home all day and inside the home caring for their children and husbands at night. An in-depth analysis of workers' schedules revealed that working women had an average of only 2 hours and 59 minutes of free time a day; they spent 6 hours and 29 minutes at their job and 4 hours and 4 minutes in domestic chores. In contrast, men spent 7 hours and 48 minutes on the job and only 32 minutes on house work (Riveira, 1989). As these schedules plainly show, women continued to be seen primarily as mothers—a view that persists today and effectively results in their exclusion from representation in many areas of the labor force.

Legal Changes Reinforce Traditional Gender Roles

As a matter of theoretical rights in Cuba, the state emphasizes the need for men's equal participation in house work and child care. In reality, though, men are viewed as inherently unreliable regarding family responsibilities (Smith and Padula, 1990), and formal government policies reinforce traditional roles that place primary family-care burdens on women. During the 1960s and 1970s, the government promulgated three major legal changes designed to lift the double burden of work and homemaking on women. While they accomplished that goal to a degree, the traditional views behind the perpetuation of the burden were not transformed: Those family-related laws and policies, and subsequent ones based on them, continue to presume that women's place is in the home.

The Family Code

The Family Code was enacted in 1975 to stipulate a new equality between men and women within marriage. Articles 24 and 26 provide that marriage partners have equal rights and duties and that they both must care for the family they have created. Article 27 states that both spouses must contribute to the needs of the family by sharing household duties and child care, regardless of whether only one works outside the home or both do. Article 28 states that both partners have the right to practice their profession or skill and that each has a duty to help the other to make this possible.

The Family Code has been interpreted to require the equal division of house work and upbringing of children, thus allowing women more free time in which to work or study. It has been incorporated into the Cuban marriage ceremony and is read by judges performing all civil marriages. However, the problem lies in the enforcement of the code: It is dependent on women taking their husbands to court for violations, something women are loathe to do. The result is that, notwithstanding legal changes, in Cuba women still disproportionately bear the burden of family and household obligations.

A study conducted three years after the adoption of the code found that working women's share of domestic chores was eight times that of working men. Indeed, government regulations provide support for such inequitable domestic-chore distribution by giving women workers special shopping privileges to speed up purchase of groceries; allowing only women to stay in hospitals with relatives to give aid and provide company; and calling mothers, not fathers, when children become ill at state child-care facilities. Interestingly, the justification for the imbalance is that men are irresponsible, untrustworthy, and would abuse privileges (Smith and Padula, 1990).

The Maternity Law
The second major legal change was the Maternity Law of 1974, which gave women the right to an 18-week paid maternity leave (6 weeks prior to delivery and 12 weeks after delivery), paid days off for doctors' visits, and an optional nine-months' unpaid leave for new mothers and six-months' unpaid leave for women with children under age 16 to attend to family matters. In addition, if a woman's child had special needs and required more care, the mother was guaranteed an additional nonpaid leave of absence of up to one year, with her same position and pay waiting for her upon her return to work. Working mothers were also granted one hour per day with pay to nurse and care for their infants. There is guaranteed medical care during pregnancy, childbirth, and the postnatal period for the mother and the newborn child.

Beneficial as this law might be for women and children, it helps entrench gender roles; for one thing, it provides for no paternity leave. In fact, fathers are required to work in Cuba, where work is not only a right but a duty: Every citizen who is of age to work and able to do so must do so. Exempting women with small children, but not similarly situated men, from this obligation does nothing to achieve equality and does much to bolster the cultural and social perceptions of, and expectations from, parents based on sex.

Children's Circles
The third major legal reform aimed at facilitating women's participation, and one for which the FMC provided major impetus, was the establishment of child-care facilities called "children's circles." These cooperatives take children in from 45 days of age, an age roughly corresponding to the end of the paid maternity leave, until age 6, when the children are ready to begin school. The child can stay for the duration of the average workday—from approximately 6:30 A.M. to approximately 6:30 P.M.—or for the entire week until Saturday afternoon. The children are given three

meals a day, two snacks, a bath, a nap, and both learning and play time. Originally, there was a sliding-scale fee in accordance with the family income. In 1967 all child-care facilities became free, but in 1977 the sliding scale was reinstated.

Like the Family Code and the Maternity Law provisions, these children's circles do little to achieve gender equality or to break the sex-role stereotyping molds. Women's "productive" employment was facilitated by the availability of such day-care services, workers' cafeterias, and low-cost fast-food restaurants and laundromats, but these accommodations are based upon the acceptance of traditional views that these tasks are principally women's responsibilities. Thus, the laws essentially provide a state-substitute for mothers that "allows" women to work outside the home.

Other Sources Perpetuating Gender Differences
The entrenchment of gender stereotypes is effected even by programs that have been exceptionally helpful to women in other contexts. For instance, the FMC's very focus is assisting in health services by keeping track of pregnant women and newborns, organizing cervical-cancer screenings and vaccination programs for homemakers, and holding sex-education and reproductive-freedom education sessions for women—all traditional gender-role-related activities. Even family-related economic policies in Cuba perpetuate the prevailing traditional gender roles. For example, after divorce, jointly owned assets are sold and evenly distributed (Smith and Padula, 1990). However, by law, primary custody of children is awarded to mothers, and child-support payments are based on need and the father's income.

Women are also treated differently under other laws, including criminal laws. For example, charges against women and men for violations of Articles 132–136 (acts against the security of the state), Article 249 (acts that affect the right of extraterritoriality), and Article 119 (acts against heads and diplomatic representatives of foreign states) of the 1979 Penal Code resulted in different sentences based on the sex of the offender. Women were sentenced to 20 years while men were sentenced to 30.

In a society founded on a revolutionary philosophy of egalitarianism, it is noteworthy that some groups in Cuba enjoy privileged status—with women more often than not filling in the not-privileged category. One privileged class comprises high-level government functionaries, a class from which women are largely absent. Ironically, this creates a two-tiered society within a government that had as one of its primary goals the elimination of class-based hierarchies. In this respect, Vilma Espín is an embarrassing example in particular because she is the only woman who has reached a high level of government. The leader of the FMC, while waging a national campaign to retrain and reeducate domestic workers, enjoyed domestic help paid for by the FMC. Also, while engaged in a revolution aimed at obliterating class distinctions, she enjoyed the haute couture of Christian Dior, who regularly dressed her (Oppenheimer, 1992).

Another privileged group are the tourists, welcome to Cuba in the 1990s because of the economic downturn. A consequence of the move to attract tourism that has had a particularly offensive impact on women is the resurgence of prostitution, one of the practices the revolutionary government claimed to have terminated.

As with the ability to chart Cuba's progress in health, education, and welfare by looking at the status of women, so, too, the deterioration of the status of citizens can be scrutinized by looking at the position of women. It is accepted that neither women nor Blacks have met the stated revolutionary goals of equality. Although their representation has increased substantially from pre-Castro days, they have both been largely excluded from power (i.e., high-level government positions) (UNDP, 1993).

Foreign observers living in Cuba in the 1980s commented, with respect to the status of women, that in the '80s "many theoretical aspects seem to have been left behind in the transformations brought about by the construction of [Cuban] Socialism" (Smith and Padula, 1990, p. 177). They noted, among other things, that the collectivization of domestic labor had not been realized and that women were still raised with "feminine" traits (Smith and Padula, 1990). Thus, more than 30 years after the start of a revolution aimed at producing an egalitarian society, Cuba is still far from achieving that goal. In many respects, the rule for women remains inequality.

Conclusion

The postrevolutionary era has brought significant advances in some aspects of women's lives. Second- and third-generation human rights, such as access to health, education, and welfare, are universally available and are indeed praiseworthy achievements of Castro's government. Women have benefited from these social programs.

However, as far as social and family roles and expectations, women have not achieved much progress in spite of the broad paper rights that have been enacted. Certainly, first-generation human rights to individual rights and freedoms, such as freedom of speech and association, are sorely lacking for everyone. In hindsight, it seems that the way Castro addressed the issue of discrimination in his 1960 speech at the United Nations offered a hint of things to come: He seemed to view racial discrimination—against Blacks and Indians, probably men—as different from gender discrimination. He implied as much when he said that one of the principles of the Cuban revolution was the condemnation of discrimination against Blacks and Indians and inequality and exploitation of women. Castro's declaration that "the right of blacks and indians to the 'full dignity of *man*,' [and] the right of women to civil, social and political equality . . ." (Castro, 1992, p. 92) were centerpieces of his revolution was evidence that he viewed racial discrimination as different from, and more important than, discrimination based on sex. While the "dignity of man" is a goal of the government and Castro has increased his gender sensitivity, as testified by numerous feminists who

have visited the island, women still experience in Cuba a sense of lesser equality.

This sense of a less-equal "equality" for women challenges the egalitarianism upon which the revolution was founded. To be sure, advances in health, education, and welfare have taken place and women have been among the beneficiaries of these advancements, but the benefits have not included the achievement of the egalitarian society the revolution promised. Instead, the advances often have been pursued by firmly imprinting traditional sex roles on society and its institutions. Moreover, the advances themselves are being threatened in the 1990s by the severe economic crunch; as the island suffers as the sole socialist survivor in its part of the world, women's progress is consistently placed at risk.

References

Bodrova, Valentina, and Richard Anker (eds.). *Working Women in Socialist Countries: The Fertility Connection*. Geneva: International Labor Organization, 1985.

Castro, Fidel. "The Case of Cuba Is the Case of All Underdeveloped Countries: Address to United Nations General Assembly, September 26, 1960." In Mary Alice Waters (ed.), *To Speak the Truth: Why Washington's 'Cold War' Against Cuba Doesn't End*. New York: Pathfinder, 1992, pp. 27–92.

Clark, Juan. *Cuba: Mito y Realidad Testimonios de un Pueblo*. Miami and Caracas: Saeta Ediciones, 1992.

Diaz-Briquets, Sergio. "The Cuban Labor Force in 1981 and Beyond." In Antonio Jorge and Jaime Suchlicki (eds.), *The Cuban Economy: Dependency and Development*. Miami: Research Institute for Cuban Studies, University of Miami, 1989, pp. 99–120.

Leahy, Margaret E. *Development Strategies and the Status of Women*. Boulder: Lynne Reinner, 1986.

Márquez Sterling, C., and M. Márquez Sterling. *Historia de la Isla de Cuba*. Regents, 1975.

Oppenheimer, Andres. *Castro's Final Hour*. New York: Simon & Schuster, 1992.

Pérez-Stable, Marifel. *The Cuban Revolution*. New York: Oxford University Press, 1993.

Purcell, Susan Kaufman. "Modernizing Women for a Modern Society: The Cuban Case." In Ann Pescatello (ed.), *Female and Male in Latin America*. Pittsburgh: University of Pittsburgh Press, 1975.

Riveira, Diana M. "Women's Legal Advancements in Cuba." Master's thesis, University of Miami, 1989.

Smith, Lois M., and Alfred Padula. "The Cuban Family in the 1980s." In S. Halebsky and J.M. Kirk (eds.), *Transformation and Struggle: Cuba Faces the 1990s*. New York: Praeger, 1990.

Stoner, K. Lynn. *From the House to the Streets: The Cuban Woman's Movement for Legal Reform, 1898–1940*. Durham: Duke University Press, 1991.

United Nations. *Women: Challenges to the Year 2000*. New York: United Nations, 1991.

United Nations Development Program (UNDP). *Human Development Report 1993*. New York: UNDP, 1993.

The Women's Movements in the Southern Cone and Brazil

Beatriz Schmukler
María Elena Valenzuela
Sandra Mara Garcia
Graciela Sapriza
Graciela di Marco

Introduction

During the 1980s, Latin America faced a severe economic crisis, considered the last manifestation of a deferred social crisis. Its implications and social consequences demonstrate a crisis of social systems, in the sense that it has been impossible to restore the social rationale of development that prevailed from the post–World War II period up to the late 1970s. The crisis provoked in Latin America a reduction of 5 percent of the per-capita domestic product between 1980 and 1987, falling back to 1978 levels. The shrinkage in economic activity took place at the same time as the growth rate of the labor force increased, thus creating high rates of unemployment, a rise in underemployment, and a considerable deterioration in real remunerations. At the same time, social services provided by the governments deteriorated due to adjustment policies.

While the crisis affected all sectors of the population, the impact has probably been strongest among women and young people. The structural-adjustment policies designed to cope with the debt crisis made heavier demands on women. The deteriorating economic situation led large contingents of poor urban women to initiate collective survival strategies, designed to satisfy the basic needs of their families. As a result, women's work has increased significantly within, as well as outside, the home, in what has been called the invisible face of adjustment policies, meaning the unseen adjustment that has played a compensatory role in relation to the disruptions that have occurred in the labor market and in Latin American society at large.

The Southern Cone (Argentina, Chile, and Uruguay) has been no exception. Women almost doubled their economic participation in Brazil and Uruguay from the 1950s to the 1990s, the level of female illiteracy is lower than that of men in urban Uruguay, the rate of secondary education among women in the area is similar to or greater than that of men, the number of female university graduates increased relative to male graduates between 1975 and 1988, and fertility rates are among the lowest in Latin America; nevertheless, the indicators of poverty are similar to some of the other countries of the region. Women work primarily in the informal sector and in the lowest-paying jobs of the service sector; in 1990, approximately 80 percent of women in the paid labor force in Argentina, Chile, and Uruguay were in the service sector, while in Brazil 66 percent worked in the service sector and 15 percent in the agricultural sector (Bonilla, 1990). In Montevideo, for instance, the number of self-employed women grew by 71 percent between 1979 and 1986, compared to a 31 percent increase in the number of self-employed men during the same period.

Although the new economic organizations created by women addressed problems of survival as their main issue, most women active in these organizations went a step further by developing a gender identity. This had important effects on the development of a women's social movement. The organizations called attention to areas of conflict that had previously been ignored and questioned whether class contradictions should be the only focus of social conflict.

In spite of the fact that these women's organizations did not propose an end to gender discrimination—as had the feminist movement based in the middle class—changes in women's lives caused by the economic crisis (such as new tasks and responsibilities) led to changing attitudes and a greater sense of personal worth. Women's experience of leaving their homes, making contact with other women who were suffering the same problems, and discovering their own unsuspected capacities and abilities had an important impact on their lives and even led them to challenge power relations within their families.

It is ironic that the very features of the crisis in the Southern Cone created important preconditions for women's collective action. And it was because of the response of women to the economic crisis that the state had to focus increased attention on women's multifaceted roles.

The organization of women into their own autonomous organizations can be considered the focal point for the process of social transformation that occurred during this period (for a detailed analysis of women's economic organizations, see Serrano, 1988). The crisis obliged women to strengthen their ties with informal mutual support networks in the neighborhood and with the extended family, and it gave rise to a variety of community-based solutions, which attempted to create cooperatives to satisfy the demands involved in maintaining the household. These experiences provided an instrument for socializing domestic problems, which, in the context of poverty and crisis, transformed these problems into collective and political issues.

Poor urban women had often joined in protest movements to keep down the prices of basic foodstuffs and urban services that were regulated by the government, but the crisis of the 1980s gave rise to new organizations, such as health committees, and a whole set of popular economic organizations. Through these, women organized small cooperatives or workshops, and they knitted together, made tapestry, collected and sold old clothes, and tended collective gardens. They also operated communal kitchens (*ollas comunes*) and communal dining halls (*comedores populares*), cooking together to enable themselves to feed their families and the children of the neighborhood. They also formed shopping collectives, buying at wholesale and then selling the goods at lower-than-normal retail prices.

While it might be tempting to view these as housewives' organizations having purely economic goals, that would be incorrect. These organizations provided these women with a focus for political organizing and self-education, as well as a means for empowerment.

Empowerment can be considered a process by which oppressed persons gain some control over their lives by participating, with others, in the development of activities and structures that allow deeper involvement in matters that affect them directly. While one might think that the decade of the 1980s was a lost one in Latin America in many ways, one should note that women in those years began to shape the content and structure of their daily existence, redefine their gender identity, and participate in the process of social change. This was not a lost decade for women.

This essay contains four presentations, one for Brazil and each country in the Southern Cone, emphasizing developments since the mid-1970s, particularly those in the political arena, that represent the most notable Latin American contribution to the feminist movement. The women's role in these countries' democratic transition (the return to democratic rule after many years of military rule) has been substantial.

Among the issues in the Southern Cone, five are most salient. First is the rise of women's visibility in the military years when the political parties were silent, linking the women's movement with the struggle against authoritarianism and military repression. Second, between 1975 and 1985, women defined their own program: The struggle for democracy and political amnesty was linked to the struggle against the increasing cost of living and the struggle against

injustices of marriage and life in the private (home) sphere (Álvarez, 1994). This movement influenced changes in family law—such as the Divorce Law and antipatriarchal changes in the Patria Potestad Law in Argentina.

A third issue or theme is the increased relevance of mothers, not only in neighborhood or service organizations but also in the human-rights struggle. In the 1990s, feminists have discussed the impact that these movements had in culturally redefining the concept of motherhood. Some see them as having politicized motherhood, changing it from a private and fragmented structure based on individual altruism into a collective and political activity (Ackelsberg and Shanley, 1992; Schmukler, 1994). Others fear that, on the contrary, mothers' movements strengthened the traditional virtues of motherhood and reproduced the traditional gender division of labor (Dietz, 1987; Feijóo, 1989; Pollit, 1992). This essay argues that despite a lack of verbal redefinition of many of the traditional elements of motherhood, within both communal-kitchen organizations and human-rights movements, motherhood served to break down the boundaries between public and private grief and also served as a means for raising gender consciousness. Furthermore, it helped women recognize that individual rights are linked to collective action and to the consolidation of democracy (Schmukler, 1994).

Fourth, as the section on Brazil shows, there were efforts in the 1980s to fashion a women's movement that linked women from the poor communities with women from political parties and feminists. One example is the CNDM (National Council on Women's Rights). Other examples in the Southern Cone are the Encuentro Nacional de Mujeres (National Women's Meeting) in Argentina, the group Mujer (Woman) in Uruguay and the CNMD (National Union of Women for Democracy) in Chile. However, the articulation between the feminist movement and the wider women's movement remains weak.

Finally, women broadened the concept of politics when they started to consider the authority structures in the family and violence in the private sphere as political issues. This has not only created new demands on political parties to redefine their programs but has also motivated changes in the civil codes of these countries. Furthermore, this redefinition has helped society to begin to accept the expansion of the concept of citizenship to include women's negotiations and struggles within the private sphere.

Chile

In analyzing the situation of Latin America, one must take into consideration the role of the social movements, since these have cooperated to produce a transformation in political practice. In this field, women's organizations have played a particularly important role.

In Chile, women mobilized in the first half of the twentieth century during the suffrage movement, struggling to be incorporated into the political system. This movement was composed mostly of middle-class women who had increased their educational levels as access to education improved in the late nineteenth century. External factors also had an undeniable impact, as women in most countries mobilized to gain the vote. Following a period of intense activity in 1949 supporting women's right to vote, there began the "feminist silence" years. This period, from 1950 to 1973, witnessed important democratic changes, including broader access to education and health, higher standards of living, moderate but sustained economic development, and increased political participation of new social groups. However, the problem of gender inequality was not explicitly addressed.

Paradoxically, the end of the long feminist silence came during the military dictatorship under General Augusto Pinochet. The new organization of women under the military regime was both a response to economic and political crises and a manifestation of opposition to authoritarianism. The military coup of 1973 and the 17-year-long dictatorship that followed produced hardship, repression, and misery. Women were particularly active in the struggle for survival, defending human rights and organizing for their own and their families' subsistence. Chilean women organized by creating diverse associations, not necessarily interrelated, around issues such as human rights, economic survival, political participation, and gender. Women mobilized throughout the Southern Cone for similar issues.

The impetus for this second wave of activism among Latin American women is often attributed to external factors, mainly the United Nations Decade for Women, inaugurated at the International Women's Year conference in Mexico City in 1975. The international women's movement clearly influenced the generation of gender identity in the struggle for democracy. As in the case of the suffragist movement, a combination of internal and external factors was behind the reappearance of a women's social movement (Chinchilla, 1992).

The Struggle Against the Dictatorship

Under military rule, the organizations of civil society became a substitute political arena that contributed to the politicization of the private and social spheres, facilitating the emergence of specific demands over and above ideological alignments. Thus, these organizations acquired greater freedom in the promotion of demands that had previously been subsumed by other national priorities.

As a result, the concept of politics expanded to include the daily universe that had been invaded by the dictatorship. The dividing line between the public and the private became increasingly hazy as the repressive policies of the regime affected domestic unity and as its economic policies pushed women into the workforce. This breakdown of public space as male space and private space (or the home) as female space allowed gender demands to become more visible and political—to assume an antiauthoritarian character that was both antimilitarist and antipatriarchal.

Given the diversity of women's organizations and their shared confrontational nature, an umbrella group was created, the Chilean Movement for the Emancipation of Women 1983 (MEMCH-83), which took its name from the suffragist movement that led the struggle between 1935 and 1953 for women's right to vote. MEMCH-83 originally embraced 24 women's groups and organized several demonstrations repudiating the regime. The objective was to promote opposition activities among the women's groups and to facilitate their coordination. Even though these organizations did not mobilize a majority of women, they gave important visibility to women's demands.

Middle-class women organized, raising gender issues in connection with the struggle for democracy. They expressed the growing contradiction between the traditional policy of the military government toward women and the new roles women had already assumed. While by the mid-1980s most Latin American countries had started to enforce legal changes in favor of equal rights, the Chilean regime refused even to ratify the United Nations Convention on the Elimination of All Forms of Discrimination Against Women until December 1989, a few weeks before the presidential and congressional elections.

Policies developed by the military regime toward women were based on a traditional conception of women's social roles. The government promoted women's return to family life and discouraged their participation in the workforce and in government positions, focusing instead on their roles as mothers. In the political arena, the government assigned women the role of educating children for the fatherland, thus assuring the ideological continuity of the regime. In addition, the government assigned women a leading role in maintaining social order but excluded them from the exercise of power (Valenzuela, 1987; Arteaga, 1989). During the 17 years of military government, only two women occupied the office of minister of state, and there were never more than two women in undersecretarial posts at the same time. Moreover, women could not be members of the legislative body, which was reserved for the commanders in chief of the four branches of the armed forces.

Changing roles led women to demand a change in the opportunities open for them in the political, economic, and social spheres. Their rapid incorporation in the labor force, from a 25 percent participation in the mid-1970s to 33 percent by 1992—breaking the trend of a stable and low participation rate—was preceded by an increase in educational levels. The percentage of women in the workforce with a university education increased from 3 percent in 1960 to 15 percent in 1982; the participation of men with the same educational level increased from 2 percent to 8 percent during the same period. Another change was a decline in the fertility rate, from 3.6 children per woman in the 1970–1975 period to 2.9 children per woman in the 1980–1985 period (INE, 1992).

The women's movement was composed of groups with diverse goals, organizational systems, and social compositions. Women organized themselves in defense of human rights and developed ingenious survival strategies to endure the economic crisis and the effects of regime policies on the poor. By the end of the dictatorship when President Patricio Aylwin took over in 1990, women had mobilized specifically as women and had begun to redefine their relationship to politics. This led them to question authoritarian relations in all areas of society, which later resulted in a reconceptualization of democracy. While not all women's groups assumed gender demands, their actions played an important role in the reappraisal of women's contribution to politics.

A virtual explosion of women's organizations came about in the context of the progressive decomposition and atomization of the previous social fabric. This allowed greater autonomy for women. Both government and opposition women's groups were determined to construct their own spaces, directed and composed by women, outside the traditional tutelage of political parties and other organizations with historically male leadership. This led to serious tensions between the two types of organizations. Since political parties had no channels of expression during the dictatorship, they tended to function through social organizations. Once the parties began to reconstruct their spaces for action after 1990, they tried to control and co-opt the social organizations—including women's groups—that had developed autonomously.

Feminism had also an undeniable impact on the political agenda. During the parliamentary and presidential campaigns for the December 1989 elections, the starting point of the redemocratization process, "gender" was recognized by the political parties as an issue for the first time in Chilean political history. In this context, the women's movement was able to define and include its demands within the coalition that opposed General Augusto Pinochet, the Chilean dictator.

The large and active Chilean women's movement saw the return to democracy as an important opportunity to reach the goals of powersharing sought during the military dictatorship. It marked the beginning of the incorporation of gender issues into the public agenda through parties and the state. But, at the same time, it caused a demobilization in all women's organizations. The coalition that was created by these organizations to put an end to the dictatorship left them without a common reference, thus allowing the specific interests of other groups to emerge. With respect to their long-term proposal to change power relations between genders, the feminist sectors established alliances with the parties. While this allowed the incorporation of gender demands into the political agenda of transition, the women's movement lost its autonomy. The long-term proposal was incorporated by the parties as part of a wider agenda of change and lost some potential. Leaders of the women's movement lost their capacity to mobilize by incorporating themselves through the political parties into the political system. They began to participate from inside, competing for limited space.

The Women's Movement in a Democratic Context
Among the first experiences that brought feminism into the realm of the state in Chile was the creation of the National Women's Service (Servicio Nacional de la Mujer, SERNAM), whose goal was to grant women a voice within the state government, giving them influence over several positions inside the state administration, rather than isolating them in a separate department.

Among the issues already outlined in the agenda of the CNMD and assumed by the government were proposals for an equal-rights amendment to the constitution, elimination of discriminatory policies regarding employment and education, prevention and punishment of domestic violence, and special programs for women heads of households.

Women and the CNMD succeeded in influencing the policy process by incorporating gender issues into the transitional political agenda. The return to democracy, however, did not directly impact the empowerment of women in political professions. The political representation of women, an important measure of power, has not improved substantially, although there has been some increase in the number of high-level administrative appointments. President Aylwin's cabinet had only one woman.

In 1989, with the return to democracy, only 2 women were elected as senators in Congress and only 6 as representatives, bringing to 6 percent the total number of women legislators. In the parliament inaugurated on March 1994, there were 9 women representatives (7.5 percent) and the same number of women senators. This was fewer than in 1973, before the military coup, when there were 14 women representatives. Women accounted for only 10 percent of the individuals elected to local councils in the municipal elections in 1992.

Brazil
Between 1964 and 1974, the Brazilian civil society organized in resistance to the authoritarian regime. There was little room for the individualization of actors. On the contrary, the political organizations that opposed the military emphasized the unity of the Brazilian people, with no differentiation by either gender or race.

In the late 1970s, after a long period of demobilization imposed by the repressive military government, intense political activism arose. Labor unions, community movements, and grass-roots movements emerged on the Brazilian scene, breaking the opposition consensus into various political forces and bringing new elements to the struggle for democracy. The women's movement was one of the most widespread of these movements and comprised a broad array of goals and forms (Garcia, 1991).

In the large cities, thousands of small groups of low-income people have fought for water, sanitation, day-care centers, and other urgent neighborhood needs. This process is well documented by Schmink (1981) and Álvarez (1994). Working-class women mobilized to defend their rights as wives and mothers, rights that were affected by regressive wage policies, a rise in the cost of living, and the low priority given the social sectors in state policies. The basis for such a politicization of motherhood was provided by a large network of women's organizations, generally known as mothers' clubs, promoted by the Catholic Church. Since their action was confined to motherhood, it was difficult for the regime to be openly repressive. The women acted as mothers exerting their legitimate rights to defend their children. Therefore, they did not confront the ruling patriarchal structure and ideology, according to which the primary role of women is as mothers and wives catering to the needs of their children and husbands.

Meanwhile, a process of social change was transforming the status of women within and outside the family and setting the scene for new ideas. This change comprised the increase of female employment in the modern sectors of the economy (although confined to "feminine tasks" within industries), higher levels of university enrollment by women, and the wider dissemination of feminist ideas coming from the North. The legitimacy afforded by the United Nations International Women's Year, in 1975, and the disenchantment with the political position of women in left-wing parties were among the factors that helped to create the environment in which women's groups started to grow. Initially, this occurred among the educated middle-class women in large cities. When this movement emerged in 1975, it joined forces with the newly revived general struggle for democracy. The women's movement was defined as a struggle for civil liberties and against the high cost of living (Álvarez, 1994).

Between 1975 and 1979, there was resistance to issues raised by feminism in Brazil. During these years, debates in the universities, seminars, and meetings, and public demonstrations against sexual violence and discrimination brought to visibility the questions raised by feminism, which found reasonable dissemination in the media and contributed to a new political culture, extending the concept of democracy and bringing new elements to the political debate.

The Women's Movement and the Debate over Participation in State Politics

In the 1980s, the Brazilian economic situation underwent a radical change. Adjustment policies provoked a severe recession and cutbacks in government social spending, particularly for health. Increased inflation and unemployment aggravated the social inequality inherited from the period of the "economic miracle." This situation, along with changes in health, education, and agriculture policies, had severe consequences for the living conditions of the low-income masses. Women from the lower strata experienced a drop in their real income, new difficulties in the labor market, overwork, and tension (Garcia, 1991).

The growing crisis of legitimacy and the deepening economic collapse brought the military regime to a gradual end. This process started with the election of opposition state governors in 1982. The coalition that came into power in state governments, mainly allied within the Brazilian Democratic Movement—the major opposition party—was a blend of disparate political forces, including progressive elements. The transition to democracy presented new opportunities and new risks to Brazilian feminists. Some women had entered party politics and lobbied from within the state for gender-specific issues. Others were reluctant to do so, either because they had joined the Workers' Party (a progressive party), which remained in opposition, or because their view of feminism was not related to state policies.

In 1983, a group of women in the state of Sao Paulo, who were members of the MDB party (the Democratic Brazilian Movement), created the State Council on the Feminine Condition within the state-level bureaucracy, and in the state of Minas Gerais a similar council was created. In 1985, the first police station specializing in fighting violence against women was opened. In 1983 and 1984, the Brazilian society was mobilized by an unprecedented mass movement for direct presidential elections (*movimento das diretas-já*). At that moment, it was not just the Brazilian people who went to the streets, but a representation of diverse social movements.

In September 1985, the National Council on Women's Rights (CNDM) was inaugurated at the federal level. An intense collaboration between women's groups who were outside the government and the women elected in 1982 was the result of the CNDM political articulation. When the debate on the new constitution started, the CNDM acted as a lobby group for the women's movement. The new constitution, voted in 1988, incorporated many positive aspects related to the rights of rural women and domestic women workers. Moreover, it established equal rights and responsibilities for wives and husbands, made racial and gender discrimination and domestic violence crimes, increased maternity leave from 89 days to 120 days, and introduced a 5-day paternity leave.

The CNDM also campaigned for public education, the legal rights of married women, and health and reproductive rights, and it unsuccessfully sought the legalization of abortion. Strong political pressure inside the state, especially opposition to the abortion issue since the beginning of 1988, led to the resignation of all members of the CNDM. A new council was formed, composed mostly of lawyers who did not have strong links with the women's movement. Nevertheless, the CNDM was an important experiment in democratization of the state apparatus itself.

In the gap left by the dismantling of the CNDM, the National Forum of Presidents of Feminine Condition State Councils was created as a political and technical link between state councils, responsible for policies concerning women. In 1996, this forum is composed by the 10 state councils and three municipal councils.

NGOs and the Women's Movement: New Questions for the 1990s

Since the 1980s, nongovernmental organizations (NGOs) have emerged as new actors in the Brazilian scene. They have been financed by multilateral agencies, such as the United Nations Children's Fund (UNICEF) and the United Nations Development Fund for Women (UNIFEM), and the World Bank, and philanthropic organizations, such as the Ford

Foundation and the MacArthur Foundation, dedicated to developing countries. The multilateral agencies, which previously had financed only governments, redirected part of their resources to NGOs following assessments of governmental programs. A good example is a report by the United Nations Fund for Population Activities (UNFPA) presented to the II Prepcom for Population and Development in May 1993. This report recommended that 20 percent of international agency resources should go to NGOs; it cited inefficiency in resource allocation and turnover in the government technical staff as reasons for this move (Schumaher, 1993). The civil society entities were considered a good alternative for investment because they presented clearer plans of action and the ability to work in partnership with the popular movements.

The Brazilian NGOs were formed by professionals from the social sciences and by former political militants, who were keen on working autonomously without the state's interference. This previous militancy experience helped to strengthen the relationship between the larger popular movement and the women's movement. Apparently, this new relationship has been many times mixed up with the representation of those movements themselves. This is the case of some NGOs that deal with women's issues (Schumaher and Vargas, 1993).

Whether to join the state or be politically active outside it was the dilemma of the 1980s. The postdictatorship period opened up some possibilities to participants in the women's movement, including continued activism in social movements as well as service in legislative and executive bodies of the government. This duality has marked the women's movement. In the end, individual responses were influenced by party affiliation.

The women's movement struggled for its recognition by the state and its representation in the larger social movement. Nevertheless, it is undeniable that "the state question" has not been resolved by the substitution of NGOs for the state since they are both institutions with different characteristics, objectives, and visions.

The experience of the women's movement as a political force in the democratization of Brazilian political life has been established by the women's physical presence in public institutions and by their role in bringing to the national debate new themes and questions. From this perspective, the women's movement brought visibility to women's issues, which reached legitimacy and influenced initiatives not only in the legal field but also in the social-policy field. Such issues as contraception, reproductive rights, and racial discrimination entered into the political scene and others, such as abortion and homosexuality, became more legitimate. Some of the women's demands, such as the curbing of violence against women, were translated into concrete measures, for example, the establishment of women's police stations. Other demands were met by legal prescriptions, such as the setting up of day-care centers (Schumaher and Vargas, 1993).

The women's movement has been, and continues to be, influential because of its ability to weather the storms and adapt to shifts in the Brazilian political scene.

Uruguay

Women's movements in Uruguay have questioned society's organization, demanded an extension of citizenship, and redefined the concepts of work, family, and personal relationships. Women's opinions have been heard on each of these points.

Models of citizen participation are based on the constitution of 1830 and on the Civil Code of 1868. The constitution excluded women (along with illiterate persons, laborers, and dependents) from voting or being elected. Because of the Civil Code, which was based on the Napoleonic Code, women were considered—and treated—as unable to administer their own property or to choose their place of residence; in case of adultery, their lives were not protected.

This legal situation was contradictory, since the country was in the midst of an accelerated process of modernization, and caused women to organize to modify their place in society. They came from diverse areas and social backgrounds: bourgeois and workers, liberals and anarchists. They approached the problem from different angles and defended different proposals, but all had the same objective: the emancipation of women. They primarily fought to win citizenship rights, criticizing an educational system that discriminated against women and programmed them to "annihilate their personality," as feminist Paulina Luisi has put it.

In 1911, teacher María Abella founded the Uruguayan section of the Panamerican Women's Federation in Montevideo's Ateneo. In 1916, she and Luisi, Uruguay's first female doctor, founded the National Women's Council. A long struggle and numerous disappointments awaited women. The first was the constitution of 1917, which gave women the right to vote only in municipal matters. They did not give up and were eventually offered support from the most progressive sectors of the country's political parties. This process produced its first milestone in 1932, with passage of the vote for women.

On March 27 of that year, women were allowed to vote in national elections for the first time. Obtaining the vote did not give rise to significant formal political representation, since this never reached more than 3 percent in parliament, but the legislative work of individual women had significant results, including the Women's Civil Rights Law, which passed in 1946.

During the 1950s and 1960s, citizens of Uruguay, women as well as men, grew accustomed to seeing their prosperous country as an exception in the Latin American context. This state of well-being was put into question by the economic crisis that began to manifest itself in the late 1960s and that submerged the country into growing social conflict. In June 1973, a coup d'état destroyed one of the myths of Uruguayan society: that of an immutable political stability. The events provoked by the coup also came to destroy another established myth: the supposed equality between men and women.

The Breakdown of the Dictatorship

To institutional rupture and repression was added the operational transformation of the economic and social spheres, the most concrete expression of which was the drop in wages, which fell 50 percent between 1968 and 1984. Reduction in government social spending meant that the provision of social services, mainly in the areas of health and education, was curtailed.

Women played a central role as families developed survival strategies to confront the crisis. A phenomenon of those years was the massive entry of women into the workforce (while in 1973 they represented 32 percent of the economically active population, by 1986 they had become 42 percent), taking on the double role of being both wage-earners and the ones responsible for domestic work. In addition, a change occurred in the composition of the female workforce because of the increased participation of women who were married, divorced, and heads of household. In this way, the exploitation became twofold: patriarchal and capitalistic.

This provoked a more or less explicit questioning of intrafamilial relationships with regard to the division of labor within the household, the administration of resources, and the exercise of power—a context that led to the rebirth of the women's movement.

The Transition to Democracy and the Rebirth of Feminism

In Uruguay, the military regime began in 1973. Beginning with the 1980 plebiscite, in which Uruguayans surprised the military government by rejecting its proposal for a new constitution to suit its purposes, women's groups organized in defense of basic needs and particularly in defense of human rights. All of this took place in an international context in which, once again, women's rights were being discussed. During the preelection period in 1984, women's groups networked among themselves and created an umbrella group called *Mujer* (Woman), which participated in the National Program Agreement (the equivalent of a long-term national plan). Five issues were discussed: education, work, health, juridical order, and women's political participation. The military was forced to offer democratic elections in 1984, and President Julio Sanguinetti started his government in 1985. Women were not excluded from parliament, but they had little representation in it.

The 1986 Statute of Limitation Law of All Criminal Acts, created during the dictatorship and approved by the transitional parliament, provoked a prompt response from civil society, which organized the National Pro-Referendum Movement to revoke the law. This law permitted military officials who had committed criminal acts to go unpunished. Women were the main protagonists of this movement, providing its leadership and basic organizational structure, and managed to mobilize the country for over two years.

The Pro-Referendum Movement became the social movement of greatest importance during this period because it submitted proposals for laws for consideration by society and made the terms of democratic transition explicit. The movement continued to challenge the military regime and to question the existence of a "controlled democracy" (or limited democratic rule); it also demonstrated its ability to organize an important sector of civil society (40 percent of the electorate). It questioned the top-down rule of the regime and the incidence of partisan loyalty found in nonpartisan activities and trade-union organizations. What is more, "As if it were a heresy, it [the Pro-Referendum Movement] was born from a group of women and was directed by them, in this kingdom of machismo, where women are one point left of zero" (Galeano, 1989).

The plebiscite on the Statute of Limitation Law took place in April 1989. The law was passed, giving amnesty to the military. In the same year, the second elections of the new democracy were carried out, with significant victories for women: six were elected representatives in the new parliament and a considerable number of women won mayoral posts. The November 1994 elections sent even more women to parliament: three senators and seven representatives.

At this point it may be appropriate to reflect on future prospects for the women's movements in the context of economic crisis as a consequence of structural adjustment policies. The crisis also affects the collective identities on which modern Uruguay was constructed—social relations as a whole and gender relations in particular. The process of change, which will either open or close doors for women, should be considered from several perspectives.

The rise of women's NGOs, dedicated to specific activities such as the creation of women's programs and women's departments in a variety of organizations, has provided a new issue for public discussion: the need to sensitize government organizations about gender issues. These NGOs have succeeded in incorporating this issue on the national level and have developed actions to improve the condition of specific groups of women. As a result, attention to gender inequalities has increased in the civil society and in national politics.

Practically all services dedicated to women came under the charge of these civic organizations, groups, and networks, in large part directed and composed by women. The work is financed nearly 100 percent by external and international aid. In 1991, there were 96 women's organizations in Uruguay, most in the capital of Montevideo and about one-third in the interior. More than half were for-profit organizations; the rest included feminist, self-help, consciousness-raising, and church and volunteer groups and organizations, as well as trade unions, coordinating organizations, federations, and networks. The rise of groups specific to Black women, female heads of household, rural women, and so forth, each with their respective demands, reflects the diversity of situations faced by women and the variety of interests and needs within the women's movement.

Mechanisms for women generated by the government such as women's offices, even though they operate at low levels in the existing hierarchy and have minimal resources at their disposal, show better prospects than those at the beginning of the decade. Among these are the Women's Institute and the Woman's Hall, both created in 1987, and the Women's Defense Commission, created in 1988 and reconstituted in 1992 as the Women and Family Institute. The latter has prepared public officers and established women's- and family-rights information centers at the state administration level. Other agencies created by governmental action include the Women's Commissions in the Chamber of Representatives and in the State Council; the Women's Management Commission Office of Canelone, which was initiated by female mayors and representatives; and the Women's Management Commission of Montevideo, which, among other things, operates a telephone line for battered women. Programs and activities for women also exist within individual government ministries. All of the above represent official mechanisms to which, and through which, the women's movement can direct its demands.

Argentina

In Argentina the military regime started in 1976. In 1983 there were democratic elections and President Raúl Alfonsín, from the Radical Party won the elections. The Justicialist and the Radical Party have been the main parties in Argentina and they have alternated in the government. Since the reestablishment of democracy in Argentina, two main issues have dominated the political landscape: institutional and economic stability. In cases where decisions did not achieve consensus, the executive took increasingly to governing through decrees. While between 1983 (the time the new constitution was passed) and 1989 there were 23 emergency decrees; in contrast, between 1989 and 1992, 100 were (Ferreyra Rubio and Goretti, 1992).

Argentine society, under the powerful impact of the crisis provoked by the terrorism of the state and by the economic terrorism of hyperinflation, is fragmented in the 1990s. The social system comprises heterogeneous groupings, including neighborhood clubs, neighbors' cooperative associations, development organizations, human-rights organizations, NGOs, NGO networks, women's organizations, youth organizations, and so forth. These groups may achieve very specific short-term objectives, working especially in those areas left vacant by political parties that are considered insignificant in terms of capturing political power and government resources.

Madres de Plaza de Mayo

The most important of these organizations is, undoubtedly, the *Madres de Plaza de Mayo.* (The term here includes the organization that bears the name as well as the *Línea Fundadora,* a group of mothers who separated from the original group in January 1986. Though the groups are different, they have the same criteria regarding essential issues.) The other types include women's organizations in

poor neighborhoods that seek to develop, and organize women around, various strategies for surviving the economic crisis.

The *Madres de Plaza de Mayo* are the paradigm of political activity stemming from individual sorrow, in open rebellion against the military government as well as against the politics of the parties that assumed power during the 1980s. These women, who comprised mothers and grandmothers of missing persons, demonstrated publicly and regularly in one of the main squares of Buenos Aires (Plaza de Mayo) to demand government response regarding whether their relatives were dead or alive—persons who had been arrested or questioned by the military at one point and then had disappeared. They captured the respect of people by openly challenging military rule by displaying the pictures of missing relatives and parading in silence. The women represent an expression of social motherhood: individual motherhood turned public when they decided to look for their own children and thus for all of the children. Schmukler (1992) makes a distinction between the two types of motherhood: traditional, taking place specifically within private walls, accepting of women in a subordinate position with respect to male authority; and motherhood related to public activity, demanding justice, and fighting against authoritarianism. The *Madres* represented the new concept of motherhood: "Motherhood, after this moment, was no longer a symbol of privacy and isolation within segmented and disarticulated family groups. The altruism of motherhood was no longer tied to political passiveness and resignation" (Schmukler, 1992).

Into the mid-1990s, the *Madres* remain one of the clearest voices defending human rights, not only with their motto *Aparición con vida* (We want our children back alive) but also with their criticism of all forms of oppression, of the adjustment plan, of unemployment, of the division of society into progressively differentiated sectors. With the motto *Nuestros hijos tenían razón* (Our children were right), they assume the legitimacy of their children's political struggle, reminding people of the genesis of this structural-adjustment plan, which began during the "proceso" (trial) and which has gathered strength since.

The permanence of the organization, along with its isolation from political parties until the march of December 1994 (which gathered about 10,000 people), is related to the firmness of the claim *Aparición con vida,* and to the Madres' decision to make the political struggle of their abducted, tortured, and missing children their own struggle.

The mothers have been criticized for their non-negotiating attitude toward the government and their non-compromising attitude with political parties. They are politically active in a new way, and this has irritated many politicians. This type of deeply ethical and noncompromising movement arises as a countermodel to the party system and to a political class that waves the banner of political instability and hyperinflation in order to strengthen the neoliberal model.

Women's Neighborhood Organizations

In most Latin American countries, women in the poorer sectors organized to defend their children and families' right to survival, to improve their neighborhoods, and to have their own land and house (Feijóo, 1982, 1984; Sara-Lafosse, 1984; Raczynski and Serrano, 1981; Barrig, 1986, 1988; Campero, 1987; Blondet, 1991). Where the mothers themselves ran the organizations, they tended to create associational practices that allowed them to enjoy more symmetrical relationships among their members and, at the same time, to produce democratic arrangements in their families, even more so than the women from other organizations.

We must not think that because women do not explain their actions in terms of gender, they are not defending the interests of their gender. Two types of consciousness can be distinguished, reflected in the fragmented speeches and practices of some of the women. The conceptualization that dichotomizes two types of interests—strategic and practical (Molyneux, 1985)—obscures the understanding of the process through which women acquire authority in their families as well as within organizations when their effort stems from a collective interest in the defense of a right. The most striking difference between the women's neighborhood organizations and the *Madres* is that the latter has included all sociopolitical issues—foreign debt, economy, employment, education, health, prisons, and street children, to name a few—with the intention of contributing to a change in the sociopolitical system of the country. Neighborhood organizations, on the other hand, are the expression of social demands for the satisfaction of basic needs; they question authority around specific issues within the existing sociopolitical system. The first group might be best termed a "social movement" while the second group exemplifies "civic participation in politics" (Pizzorno, 1976; Pásara and del Pino, 1991).

Middle-Class Women and Social Participation

Middle-class women participate in their traditional associations, such as neighborhood organizations, the school, and hospital cooperative associations. When they participate, they never occupy top positions, which are the realm of men.

However, in addition to these traditional forms of participation, a new area is opening up for middle-class women in the form of health and education committees developed to defend public institutions threatened by adjustment and privatization. These institutions, especially the school, are dear to the middle class because they represent a channel for climbing the social ladder in the Argentine society. This is the reason their defense may mean an opportunity for middle-class women to become collective subjects (having common goals and a public agenda) much as the poor women did: united in the defense of a very strong interest deeply tied to their own class identity.

Women in the middle classes, as well as their husbands, parents, and children, fight not to slip lower on the social ladder. They give their support to the government's economic plan (the stability plan) and lay the blame for the crisis on economic "interests" but do not establish a relationship with the present economic model, whose privatization of social services and hyperinflation benefits mostly the upper classes. In this sense, the participation of middle-class women is worth observing, since it does not take the form of criticism of the socioeconomic situation, but rather of defense of what the women consider to be their vested rights—which could eventually disappear as a consequence of the adjustment plan.

The Women's Movement and the Feminist Movement

In the fragmented social scene that is characteristic of the 1990s in Argentina, the women's movement represents the continuity of the fight for women's rights and against discrimination. Diverse associations, organized in various forums and networks, have constituted the base from which women take up the fight and the task of building people's awareness. The issues addressed including divorce and shared parental rights (acts passed during the Radical Party administration between 1983 and 1989), the acknowledgment of domestic violence and sexual harassment, reproductive rights, women's health, and protection of the environment, all show a movement rich in programs and activism.

One of the more remarkable events is the *Encuentro Nacional de Mujeres* (Women's National Meeting), which takes place once a year in one of the Argentine provinces. An adhoc committee is in charge of its organization. Women from lower and middle social classes, feminists and nonfeminists, attend this meeting. Women share their achievements and their demands concerning living conditions, health, education, job opportunities, and the unemployment problems of their husbands. Those meetings have been characterized as autonomous, heterogeneous, multiclass, pluralistic, progressively massive, self-expanding (by word of mouth), not institutionalized, ignored by the media, and challenging to the government, especially as they raise issues related to foreign debt, structural adjustment, and state corruption.

Within this framework, women, rather than organizations, are the leading characters. Feminists are a minority in these meetings and have modified their strategy in the course of time: At first they would gather in their own workshops, but now they try to join with the rest. This trend has been accompanied by a progressive reduction in the number of feminists, while the *Encuentro*'s overall turnout has increased from 2,000 women in the first year to 5,000 in the most recent year.

The confluence of the women's movement with the feminist movement, as well as the articulation between different women's organizations, remains to be achieved. It is a difficult task, as relationships have to be built without using the traditional methods of the party system or those of the existing organizations. The horizontal approach and internal democracy and debate within this confluence still need to create new ways to make politics, ways that reach beyond those built under the old patriarchal order.

Political Parties, the State, and Women
The political parties have progressively adapted to the presence of women's organizations and frequently take into account the struggle for women's rights when they design their political programs. The movement's influence on the passing of the previously mentioned legislation has been felt with varying degrees of intensity. The administrations of both presidents Alfonsín (1983–1989) and Carlos Ménem (1990–present) established government councils and deparments to deal with women and women's issues, and women were appointed to various government departments and councils.

The Law of the Minimum Quota (Act 24.0112, which sets aside for women 30 percent of the nominations within the political parties), enacted in 1995 to regulate the political parties' lists of candidates, will be very important for the women's movement. Its first consequence has been a social trend toward follow the same guidelines in choosing board members of other organizations, which, in turn, has allowed for debate about association characteristics and the process to make them more democratic. The law has spurred women to strengthen their capacity for political negotiation so as to compete with their male counterparts in government elections (Di Marco, 1993).

A final question to be posed is whether the women's movement and the women in political parties will be able to follow the path led by the *Madres* and become *Madres* of a new political practice—a path marked by solidarity, ethical coherence, and the defense of human rights in the male arena of politics.

References
Ackelsberg, Marta, and Mary Lyndon Shanley. "From Resistance to Reconstruction? Madres de Plaza de Mayo, Maternalism and the Transition to Democracy in Argentina," paper presented at the 17th Congress of the Latin American Studies Association, Los Angeles, 1992.

Álvarez, S. "Women's Movements and gender politics in the Brazilian Transition." In Jane S. Jaquette (ed.), *The Women's Movement in Latin America: Feminism and the Transition for Democracy.* Boston: Unwyn Hyman, 1989, pp. 18–72.

———. "La (trans)formación de (los) feminismo(s) y la política de género en la democratización del Brasil." In Magdalena León (ed.), *Mujeres y participación política: Avances y desafíos en América Latina.* Bogota: Tercer Mundo Editores, 1994.

Arteaga, Ana María. "Politización de lo privado y la subversión de lo cotidiano." In Centro de Estudios de la Mujer (CEM) (ed.), *Mundo de mujer.* Santiago: Centro de Estudios de la Mujer, 1989.

Barrig, Maruja. *Democracia emergente y movimiento de mujeres.* Lima: DESCO, 1986.

———. *De Vecinas a Ciudadanas: La Mujer en el Desarrollo Urbano.* Lima: SUMBI, 1988.

Blondet, Cecilia. *Las mujeres y el poder: Una historia de Villa El Salvador.* Lima: Instituto de Estudios Peruanos, 1991.

Bonilla, Elsy. "Working Women in Latin America." In *Economic and Social Progress in Latin America, 1990 Report.* Washington, D.C.: Johns Hopkins University Press, 1990.

Campero, Guillermo. *Entre la sobrevivencia y la acción política: Las organizaciones de Pobladoras en Santiago.* Santiago: Ediciones ILET, 1987.

Chinchilla, Norma S. "Marxism, Feminism, and the Struggle for Democracy in Latin America." In Arturo Escobar and Sonia Álvarez (eds.), *The Making of Social Movements in Latin America: Identity, Strategy, and Democracy.* Boulder: Westview, 1992, pp. 37–51.

Dietz, Mary G. "Citizenship with a Feminist Face: The Problem with Maternal Thinking," *Political Theory,* vol. 13, no. 1, February 1987.

Di Marco, Graciela. "Experiencias de Participación sectorial en la ciudad de Buenos Aires." In *Biblioteca Política Argentina.* Buenos Aires: Centro Editor de America Latina, 1993.

Feijóo, María del Carmen. *Las luchas de un barrio y la memoria colectiva.* Buenos Aires: Estudios CEDES, 1982.

———. "The Challenge of Constructing Civilian Peace: Women and Democracy in Argentina." In Jane S. Jaquette (ed.), *The Women's Movement in Latin America: Feminism and the Transition to Democracy.* Boston: Unwin Hyman, 1984.

Facultad Latinoamericana de Ciencias Sociales (FLACSO), 1993.

Ferreyra Rubio and Goretti. "Gobernar por Decreto," *Poder Ciudadano,* no. 15, 1992.

Galeano, Eduardo. "Las mujeres en el movimiento por el referendum," *Brecha,* April 21, 1989.

Garcia, Sandra. "The Politics of Population Control: Who Sets the Agenda? The Case of Brazil." Master's thesis, Institute of Development Studies, University of Sussex, England, 1991.

Instituto Nacional de Estadística (INE). *Encuesta Nacional de Empleo.* Santiago: INE, 1992.

Molyneux, M. "Mobilization Without Emancipation? Women's Interests, State, and Revolution." In R. Fagen, C.D. Deere and J.L. Coraggio (eds.), *Transition and Development: Problems of Third World Socialism.* New York: New York Review Press, 1985, pp. 280–302.

Pásara, Luis, and Nina del Pino. *La otra cara de la luna: Nuevos actores sociales en el Perú.* Lima: CEDYS, 1991.

Pizzorno, Alessandro. *Participación y cambio social en la problemática contemporánea.* Buenos Aires: Editorial SIAP, 1976.

Pollit, Katha. "Are Women Morally Superior to Men?" *Nation,* December 28, 1992, pp. 709–807.

Raczynski, Dagmar, and Claudia Serrano. *Vivir la pobreza: Testimonios de mujeres.* Santiago: CIEPLN/PISPAL, 1985.

Sara-Lafosse, Violeta. *Comedores comunales: La mujer frente a la crisis.* Lima: SUMBI, 1984.

Schmink, M. "Women in Brazilian *Abertura* Politics," *Signs,* vol. 7, no. 1, 1981, pp. 118–134.

Schmukler, Beatriz. "Second Thoughts About Maternalist Politics," paper presented at the 17th Congress of the Latin American Studies Association, Los Angeles, 1992.

———. "Maternidad y Ciudadanía Femenina." In Cecilia Talamante, Fanny Salinas, and María de Lourdes Valenzuela (eds.), *Repensar y Politizar la Maternidad.* Mexico: Grupo de Educacion Popular con Mujeres, 1994, pp. 51–59.

Schumaher, Maria Aparecida, and Elizabeth Vargas. "Lugar no governo: Álibi ou conquista?" *Revista de Estudios Feministas,* vol. 1, no. 2, 1993, pp. 348–364

Serrano, Claudia. "Pobladoras en Santiago: Algo más que la crisis." In *Mujeres, crisis y movimiento.* Santiago: CIEPLAN, 1988.

Valdez, Teresa, and Enrique Gomariz. *Mujeres Latinoamericas en Cifras.* Madrid: Instituto de la Mujer, 1993.

Valenzuela, María Elena. *La Mujer en el Chile Militar.* Santiago: CESOC, 1987.

Women in Oceania

Sharon W. Tiffany

Introduction

The islands of Oceania encompass four major ethnographic regions scattered across 65 million square miles of the Pacific and four time zones. These regions are: (1) Melanesia in the southwest Pacific, dominated by New Guinea, the second-largest island in the world with a population of nearly four million, greater than that of the other three regions combined; (2) the huge central Pacific triangle of Polynesia, with its corners at Hawaii, Easter Island, and New Zealand; (3) the numerous small islands of Micronesia, most of which lie north of the equator and west of the International Date Line, consisting largely of four principal island groups (the Marianas, Carolines, Marshalls, and Gilberts); and (4) the continent of Australia.

The diversity of cultures and colonial histories of a region that covers one-third of the earth's surface does not permit generalizations about the lives and experiences of Pacific Islander women. Furthermore, some areas of Oceania have been intensively researched, notably Melanesia, while other regions have received comparatively less anthropological attention.

In contrast to the anthropological record, there is relatively little accessible, published material by Pacific Islander women. Computers and satellites will undoubtedly change this situation as increased numbers of Islander women gain access to these and other communication technologies. Important venues for indigenous women's views include materials published by the University of the South Pacific in Suva, the Australian National University in Canberra, and the University of Papua New Guinea in Port Moresby. For now, the voices of indigenous women are heard primarily within the context of published ethnographies and, in a few instances, as authors of their own texts that have received distribution to a broader readership than the university-based publications noted above.

Gender Issues in Oceania: An Overview

Influenced by the feminist movement of the 1960s and 1970s, women anthropologists entered the field with new concepts and research questions. This ethnographic literature provided new perspectives about gender relations in Oceanic societies and enhanced anthropological and Western understanding of indigenous women's lives. Landmark publications during this formative period of gendered perspectives in Pacific anthropology include ethnographies of Tiwi women of Aboriginal Australia (Goodale, 1971 [1994]), of Hagen women of the Western Highlands of New Guinea and the gendered politics of domestic and public space (Strathern, 1972 [1995]), of Trobriand Islander women and exchange (Weiner, 1976), and the ethnohistory of Polynesian gender relations and state formation in Tonga (Gailey, 1987).

Additional anthropological sources on the Pacific include Sinclair's (1986) useful bibliographic review and an introduction to gendered perspectives on Melanesian rituals of female initiation by Lutkehaus (1995). Another work of interest is the collection of ethnographic case studies on reproductive decision making in rural Papua New Guinea (McDowell, 1988). Finally, several chapters in *Family and Gender in the Pacific: Domestic Contradictions and the Colonial Impact* (Jolly and Macintyre, 1989) consider gender and domestic relations in the context of colonial institutions, with an emphasis on missions.

The task of choosing specific works for this essay proved difficult. I wanted this work, directed toward a nonspecialist audience, to include Oceania's four ethnographic regions as well as examples of Islander women's voices. The research and writing of this review took place during two momentous international events: Thousands of women convened in Beijing for the 1995 NGO Forum on Women and the U.N.-sponsored Fourth World Conference on Women. And in September 1995, France, despite worldwide protests, began its series of nuclear tests at Mururoa and Fangataufa atolls in French Polynesia. Accordingly, I have selected topics from the corpus of Pacific ethnography on gender and feminism that reflect concerns

raised in Beijing, in addition to issues articulated by Islander women active in political movements and local-level
organizations.

The first two sections of this essay consider the challenges posed by a gendered perspective to the androcentric
legacy of Pacific ethnography and its representations of
indigenous women. Sections three and four examine gender in the political economies of contemporary Pacific
societies. Women's political activism is the focus of section
five. The conclusion briefly considers issues of appropriation and resistance as Pacific Islander women contest images imposed upon them by others.

Gender and Ethnographic Representation

> Hawai'i—the word, the vision, the sound in the
> mind—is the fragrance and feel of soft kindness.
> Above all, Hawai'i is "she," the Western image of the
> Native "female" in her magical allure. And if luck
> prevails, some of "her" will rub off on you, the visi-
> tor.—Haunani-Kay Trask, Native Hawaiian activist
> (Trask, 1993, p. 180).

A significant issue associated with this far-flung island
world concerns the history of anthropological and Western ideas about gender and the South Seas, the latter occupying a geographic and cultural space as well as representing a state of mind. Historically characterized by
androcentrism, ethnographic research in Oceania focused
on issues such as rank, politics, warfare, cannibalism, male
rituals, and misogynist ideologies of sex and gender.

This male-focused tradition of anthropological literature represented indigenous women as muted and able to
articulate little of significance about society and culture
(Tiffany, 1987). Defined by their procreative and domestic functions, indigenous women were relegated to the
social margins in ethnography. Micronesian women were
represented as male-dominated despite the widespread
principle of matrilineal descent, which provided women
with access to economic and political resources through kin
ties. Polynesian women, regardless of rank and status, were
dismissed as politically and economically peripheral. Gender relations in Melanesian societies reflected phallic power
and institutionalized hostility between the sexes; Australian Aboriginal women were brutalized drudges. In these
representations, women were of primary significance when
they posed problems for men (Tiffany, 1987).

These earlier ethnographic images have assumed a life
of their own beyond the discipline of anthropology. Indeed, a legacy of this androcentric anthropological tradition, combined with a Western romance of the South Seas,
has served to reinforce popular notions of Oceania as paradise, an exotic place where quiescent, sexually alluring Islander women are willing participants in realizing the
dream-worlds that outsiders bring to tropical shores. The
realities for Pacific Islanders of colonialism, militarism,
nuclear testing, and tourist development, among other

problems of the New World Order, are overlooked in this
Western mythmaking.

Margaret Mead: An Anthropology of Gender Comes of Age

Modern anthropology came of age with fieldwork conducted
on the sexual lives of Pacific Islanders and the situating of
indigenous women on the periphery of male-defined, societal concerns. Bronislaw Malinowski, the "forefather" of
modern ethnographic field research, wrote many books on
the Trobriand Islanders of Melanesia. His texts, including the
provocatively titled *The Sexual Life of Savages in North-Western Melanesia* (Malinowski, 1929), are notable for the invisibility of women. The first major study about the lives of
Trobriand Islanders from a gendered perspective was not
published until nearly 60 years after Malinowski's ethnographic expeditions (see Weiner, 1976).

Published a year earlier than Malinowski's *Sexual Life
of Savages*, Margaret Mead's (1928 [1961]) *Coming of Age
in Samoa* has sold millions of copies worldwide. Translated
into 16 languages, it remains the best-selling anthropological text in history. Mead's research in 1925 on the island
of Ta'ū in the Territory of American Samoa centered on the
lives of 25 Samoan girls and young women, who ranged
in age from 14 to 20 years. A half-century later, her work
would be the subject of fierce contention.

Anthropologist Freeman (1983) criticized Mead's conclusions in his controversial text, *Margaret Mead and Samoa: The Making and Unmaking of an Anthropological
Myth,* published five years after Mead's death. This text
generated unprecedented media coverage and a considerable academic industry of anthropological debate. Whereas
Mead found ease, cooperation, and guilt-free sex, Freeman
found a pathology of conflict, violence, and rape. Rejecting Mead's view of nurturing and sharing, Freeman found
jealousy, competition, and explosive sexual rivalry. What
are we to make of these contrary representations of bliss-
ful sexuality and savage lust?

Review articles by Nardi (1984) and McDowell (1984)
provide a gendered perspective for understanding the issues
raised in the Mead–Freeman debate. Nardi conducted research on Western Samoan women's work and reproductive decision making; McDowell has extensively researched
the published and unpublished Oceanic material of Mead.
Both authors provide a cultural-historical context for examining Mead's Samoan research and Freeman's criticisms,
many of which are embroidered with a rhetorical tone of
dismissive contempt. What has often been overlooked in
the heat of debate is that Mead's pioneering fieldwork in
Samoa advanced a new avenue of inquiry by focusing on
the lives of girls and young women at a time when academic anthropology considered men the significant social
actors.

Little has changed in the anthropological canon on
Samoa since Mead's work nearly three-quarters of a century ago. More recently published ethnographies of Samoa,
written by men and strongly male focused, represent Sa-

moan women—at best—as accessories to men's lives. Meanwhile, Samoan women's perspectives on the controversy have yet to be heard.

Australian Aboriginal Women and a Feminist Ethnography

Daughters of the Dreaming (Bell, 1993) is a significant text of woman-centered fieldwork and writing. Bell integrates description, method, reflexivity, and interpretation throughout her ethnography of the women of Warlpiri, a central desert Aboriginal community. The author invites the reader to share the ethnographer's experiences as a student of anthropology and as a student mentored by senior Aboriginal women. Bell had to reassess a male-oriented framework of Aboriginal social organization—the legacy of her formal academic training. This process of relearning enabled her to shift her focus and to acquire the skills of listening to, following, and eventually understanding Warlpiri women's secret world of ritual knowledge and sacred purpose.

Daughters of the Dreaming demonstrates the ritual and political importance of Aboriginal women in Warlpiri social life, thus challenging the "man equals culture" approach in anthropology and its historical representations of Aboriginal gender relations premised on a Western model of hierarchy and power asymmetry. The equality of gendered domains of responsibility and ritual knowledge in Aboriginal society has, however, broken down in the colonial frontier. Warlpiri men gain power from differential access to the institutions and ideologies of White Australian society, which privilege Aboriginal men and denigrate Aboriginal women as domestics and sex objects. Resisting the dual burdens of race and gender, Aboriginal women have responded by creating their own organizations to combat problems such as domestic violence and to reassert their participation in community affairs.

The extensive epilogue added to the second edition of *Daughters of the Dreaming* discusses the problematic relationship of the mainstream academy to woman-centered knowledge and writing. Bell also tackles the difficulties of speaking out as a White woman activist-anthropologist. Having publicized the politically sensitive issue of Aboriginal men's violence against girls and women in the Northern Territory, Bell cites increasing incidence of Aboriginal intraracial rape as violations of human rights.

Gendered Lives in a Massim Motherland

Fruit of the Motherland (Lepowsky, 1993) reflects recent currents in feminist ethnography by moving beyond questions of women's status and dualisms of nature/culture and domestic/private—themes that preoccupied the academic literature of feminist theory and women-focused ethnography during the 1970s and 1980s. Reflecting concern with narrative, gender, and colonial history, Lepowsky's ethnography presents a multitextured approach to understanding the lives of Islander women as they pass through the life course.

Research on big men, large-scale ceremonial exchanges, and male bonding has dominated anthropological inquiry in Melanesia for generations, producing models of social life premised on relations of sexual antagonism and gender inequality. As a graduate student influenced by the feminist movement of the early 1970s, Lepowsky was not interested in conducting mainstream Melanesian ethnography: "I had already lived with sexual inequalities in the United States. I did not want to spend my time in the Pacific trying to cajole my way into the men's cult house in order to see interesting rituals or hear esoteric ancestral lore" (Lepowsky, 1993, p. x). The author eventually conducted fieldwork on the remote island of Vanatinai (Sudest) in the Massim region of Papua New Guinea, an area long associated with matrilineal descent, women's high status, and the mystique of Bronislaw Malinowski.

Presenting a critical analysis of anthropological gender theory, *Fruit of the Motherland* contests the notion that relations between women and men are necessarily structured by asymmetry and contest. Vanatinai gender ideology promotes women as life givers: Women nurture yams (the staple of life) and children, as well as the relatives of deceased kin through participation in mortuary rituals. Vanatinai women and men share equal access to productive resources and sacred knowledge. Like big men, big women may also participate in public ceremonial transactions and gain renown for accumulating and giving away valuables.

Indigenous Women and the Political Economy in Postcolonial Societies

> When we started buying food, our people did not think what that would mean. Now we are dependent on imported foods. Big business is getting rich because we have forgotten that we can grow our own food.— Isabella Sumang, Belauan activist (in Ishtar, 1994, p. 59).

The comparatively small literature on Pacific Islander women and development has focused on women's organizations, church groups, and issues of women's unequal access to education, employment, and technical support. Development agencies and planners may acknowledge women's economic contributions but nonetheless define women's needs as secondary concerns. As a result, Pacific Islander women continue to shoulder "a disproportionately onerous role in the maintenance and improvement of living standards" in their societies and receive few benefits from development agendas (Hughes, 1985, pp. 3–4).

Part of the problem rests with Western presumptions—held by expatriates and Islanders—that women's work is not "real" work and that social problems defined as women's issues, such as reproductive health and employment discrimination, are concerns of low priority. Poorly represented in the public and private sectors, Islander women are not socially positioned to affect policies at the national or regional level of their societies.

Mission Groups and Gender Politics

Missions, associated with colonial imperium, were perceived by Pacific Islanders as powerful, authoritative organizations that articulated new realities and provided unique social and economic opportunities. Expatriate and indigenous women—as wives, missionaries, members of religious orders, educators, deacons, and organizers of church groups—assumed many roles. Some promoted Victorian models of women as self-sacrificing helpmates to the pastoral endeavors of men. Others created new organizations or modified established groups to meet their own needs and aspirations (Forman, 1984; Jolly and Macintyre, 1989).

Silent Voices Speak (Marshall and Marshall, 1990), a case study of Trukese gender and social change, focuses on women's political activism against male drinking and alcohol abuse. In 1978, an ordinance prohibiting alcohol was inaugurated on Moen Island, Truk (of the Federated States of Micronesia). Trukese women mobilized a modern temperance campaign as public perceptions of crime and social problems, particularly domestic violence and family dissolution, came to be linked with alcohol.

Utilizing the organizational structure of women's church groups, Trukese women skillfully enlisted the support of male church leaders and local politicians—who recognized advantages to be gained by supporting prohibition, despite the potential loss of significant revenues from alcohol taxes and licenses. Underrepresented in formal political life, Trukese women mounted a successful campaign against alcohol, supported by a gender ideology compatible with women's roles as moral standard-bearers of family and domestic life.

Women's demonstrations and public sit-ins, unique in Truk's political history, overrode subsequent attempts to overturn prohibition, which remains in effect but is rarely enforced. Anecdotal evidence suggests that Trukese women continue to perceive their temperance campaign as a success. Alcohol, expensive but available, is not considered a major social problem, as incidents of public drunkenness and domestic conflict have declined (MacMarshall, personal communication, October 1995).

The experiences of Trukese women activists—as devout Christians and supporters of well-organized church groups—suggest important structural parallels with nineteenth-century Protestant evangelism and the Women's Temperance Crusade in the United States. Both Trukese and American women perceived the negative social costs of male drinking and its consequences for women and children. In these instances, "alcohol use became the focus of struggle between women and men, and a symbolic arena in which women fought for and gained greater participation in the political process" (Marshall and Marshall, 1990, p. 119).

Women Farmers in French Polynesia

Tahitian Transformation (Lockwood, 1993) is an ethnography of gender and development centered in the political economy of welfare-state colonialism. Lockwood's study of potato farming on the rural island of Tubuai is situated within the household economy and, at a broader level, within the context of a Polynesian society transformed by its colonial history. Most foreign visitors remain happily unaware that the tourist image of paradise is an artifact set in a militarized zone of nuclear testing.

In 1994, France supported the local economy of its Polynesian nuclear colony at a cost of US $1.5 billion. This artificially high standard of living is sustained at considerable costs of social disruption. These include rural depopulation, inflation, unemployment and crowding in expensive port-town areas, prostitution, and increasing rates of domestic violence and suicide, in addition to the largely unreported health and ecological consequences of nuclear testing in the region. Islanders pay these costs to sustain France's geopolitical interests in the South Pacific (see also Ishtar, 1994, pp. 185–203).

For most Tubuai Islander women, potato farming is a small-scale enterprise to raise supplemental income for household consumption in this cash-oriented, rapidly Westernizing society. Women farmers, including widows or wives in contentious marriages who grow potatoes out of economic necessity, do not view themselves as entrepreneurial role models. Government agricultural workers reinforce this view by defining women farmers in general as "nonserious planters." Thus, they are the most likely candidates to be cut from subsidized programs if commercial necessity warrants. Tubuai Islander women's access to land through kin ties may well prove to be a critical factor for their continued economic survival, given the context of changing international policies and the cyclical, boom-and-bust nature of Pacific Island economies.

Wok Meri: Women's Organizations in Papua New Guinea

Indigenous women in the Eastern region of the New Guinea Highlands have developed a savings-and-exchange system called *Wok Meri,* which means "women's work" in the Pidgin (Neo-Melanesian) lingua franca (Sexton, 1986). Small groups of women meet regularly to save money earned from selling coffee, garden produce, or, on occasion, their labor. Local *Wok Meri* groups may be integrated into larger, regional organizations in one of two ways. First, a local group initiates a relationship of exchange, articulated in a ceremonial framework of traditional marriage payments, with a more experienced group. Second, a regional network of *Wok Meri* organizations enables members of local groups to lend small amounts of money among themselves. A *Wok Meri* group that receives such loans during a public ceremony called "washing hands" may invest the proceeds in a joint business venture, such as passenger transport. Alternatively, individual members may use their accumulated savings for family obligations.

Much of the ceremony associated with *Wok Meri* involves women's rituals of marriage and childbirth. Members "give birth" to bags of money as symbolic daughters, thereby cementing "marriage" between mentor and protegé *Wok Meri* groups. Such rituals serve to emphasize women's ability to manage and control money, the newest form of wealth in these rural societies associated with subsistence horticulture and cash cropping.

The Lutheran mission, which provided indigenous women with opportunities to organize and acquire financial experience, was instrumental in the development of *Wok Meri* in the New Guinea Highlands during the 1930s. The development and elaboration of *Wok Meri* groups accelerated in the 1960s as coffee production expanded. Indigenous women asserted their rights to economic property and to participate in public displays and exchanges of wealth—focal domains of men's power and status-striving.

An important element of *Wok Meri* ideology cites men's mismanagement of money. Men's drinking and gambling are major sources of marital discord. Alcohol abuse and domestic violence are on the rise, and women have no effective recourse in government or the courts at the local, regional, and national levels. *Wok Meri* savings associations thus enable women to cope with the spendthrift ways of men, thereby safeguarding savings from the importunate demands of husbands or male relatives.

Women, Violence, and the State

Pacific Islander women are aware of their increasing economic and political marginalization in the postcolonial era as indigenous men monopolize government, technology, and employment. Male-dominant discourse, framed as "culture" or "custom" is a powerful ideological weapon "used as an instrument of oppression to deny women equality which is theirs as a moral right" (Ishtar, 1994, p. 219). The military presence in French Polynesia, Micronesia, and Hawaii, along with tourism and its attendant economic and sexual exploitation, are contributing factors in violence directed against indigenous women. The rising incidence of rape and domestic violence further constrains women's mobility and access to educational and economic opportunities.

Zimmer-Tamakoshi (1993) discusses the rhetoric of Papua New Guinea nationalism and the failure of postcolonial elite Islander males to meet the expectations of indigenous women in the process of postindependence nation building. Women's disappointment in the pace of change is understandable. Only a handful of women have ever served in the national parliament of Papua New Guinea; few women hold senior positions in government, the private sector, or church organizations. Women's issues are frequently derided, as exemplified by an incident in which members of the Law Reform Commission, while presenting their interim report on domestic violence, were booed off the floor by an all-male parliament.

Elite men, primarily from the patrilineal Highlands, utilize a rhetoric of conservative sexual politics to deny the aspirations of women. An ideology of male dominance and of women's traditional role as selfless supporters of men's

interests serves to divide women of Papua New Guinea. Elite women, who come primarily from the matrilineal coastal regions of the country, are associated with Western consumption patterns and lifestyles by rural women and elite men. Rural women shoulder most of the onerous demands of cash cropping and subsistence agriculture as a result of male migration. These women are more likely to look to local-level groups like *Wok Meri,* rather than to national organizations, for support of their concerns.

Belau: Domestic Violence in the Context of Political Violence

Nero's (1990) study of domestic violence in the Micronesian islands of the Republic of Belau (Palau) is situated in three contexts: (1) Belauan political turmoil and American strategic interests (see also Wilson, 1995, discussed below); (2) the transformation of gender roles among elites; and (3) urbanization (two-thirds of Belauan Islanders reside on the densely populated island of Koror).

Wife beating, considered shameful by both Islander men and women, occurs overwhelmingly in nuclear-family households of young, elite, and educated couples who hold salaried jobs. Both spouses are increasingly pressured to contribute economically to their kin groups but are often geographically isolated from the network of extended family members that would normally intervene in domestic conflicts.

Belauan cases of domestic violence do not easily lend themselves to Western feminist explanations of women's comparative inequality and powerlessness, since elite Islander women exercise economic and political power within the public and domestic domains. Nero suggests that the geographic isolation of these violence-prone households and the disinhibiting effects of alcohol consumption enable incidents of wife abuse to occur with greater frequency.

Fiji: Militarization and Violence Against Women

One of the least documented areas of anthropological research is the impact on women of the Fiji military coups in 1987 (Lateef, 1990). The two coups and resulting political instability divided the Indo-Fijian population (descendants of South Asian workers imported under British colonial auspices) and the indigenous Fijian population, which in the mid-1990s continued to hold 83 percent of the country's land under traditional clan ownership. The political climate of ethnic hostility in postcoup Fiji has many implications for women. This climate is reinforced by government promotion of a conservative, traditionalist ideology of women as mothers and as supporters of men's political causes.

The military coups created an economic and social crisis for women. As the lowest-paid salaried workers, many women lost their jobs or experienced wage reductions in service-sector and tourism-related employment as a result of the general economic decline in the postcoup period. The Women's Crisis Center in Suva, a nongovernmental organization (NGO), has been deluged with divorced and abandoned mothers desperate for economic assistance, in addition to victims of rape and domestic violence. The breakdown of judicial institutions and the lack of control over military and police activities have resulted in an escalation of human-rights violations. Incidents of violence directed at women cross all ethnic and social sectors of the population. Despite harassment and threats, many courageous women have publicly demonstrated against this rising tide of sexual violence.

Women and Political Activism

The islands of the Pacific are located at the crossroads of pleasure and danger. Swaying palm trees and "lovely hula hands" promise the romance of paradise:

> Beautiful islands, calm, clear water. Beautiful women—very cheap. You can get anything you want. Lies. Lies after lies. You don't hear anything about the war, or about nuclear testing, or about people who are dying slowly. No! You will just see beautiful pictures. Paradise in the Pacific!—Chailang Palacios, Chamorro activist, Northern Marianas (quoted in Ishtar, 1994, p. 84).

Islander women bear the sexual, health, and economic consequences of this imposed burden as they occupy the nuclear front lines and confront the militarization of their lands and societies.

"Women ARE the Government"

Speaking to Power (Wilson, 1995), based on research in the Western Caroline Islands of Belau (Palau) in Micronesia, illustrates the trend toward applied research in feminist anthropology. Wilson first learned about the Republic of Belau while conducting research on the international women's peace camp at Greenham Common, England, organized to protest NATO deployment of U.S. nuclear missiles in Europe. The author's interest in learning how communities organized opposition to nuclear militarism eventually brought her to Micronesia, a region where American military interests have dominated island politics since the end of World War II. Belau is of special interest in this context because elder women have been in the forefront of protesting U.S. proposals to maintain permanent military access to the islands' lands, reefs, and waters, as formulated in the Compact of Free Association, commonly referred to as "the Compact."

The protests of these elite, mid-life women have frequently put them at odds with their families and clan relatives (many of whom hold government positions), and with U.S. administrative authorities, who have pushed several referendums to modify the nuclear-free provision of the Belauan constitution. U.S. pressure has helped create schisms within and between Islander families, deepened the sense of economic insecurity, and contributed to a climate of political violence and instability. Despite fears for their personal safety, Belauan women have persisted in their campaigns of opposition, thus propelling the activists of this Pacific microstate into the international (and, by definition, masculine) arena of geopolitics.

The key to understanding the organizational and negotiating skills of Belauan women lies in their social positioning within matrilineal clans. According to Gabriela Ngirmang, a major participant in the anti-Compact movement, Belauan women have always had power and a sense of responsibility:

> When they [the women] got together to oppose the Compact, at first, there were men with them. But most of the men there would not see it through to the end. . . . No woman would be afraid to say something, because women are the stronghold of the clan. And so it's the women who got together around the Compact because they are stronger (Wilson, 1995, p. 188).

In Ngirmang's analysis, Belauan women have the strength to "pick up the pieces" when men lose their resolve. Belauan women's relationships are reinforced by participation within a matrix of kin ties. These are continuously affirmed through gift exchanges and wealth transfers in a society in which gender, age, and rank remain significant principles of social structure.

Wilson (1995) was also determined to write women back into Belauan ethnography, as well as to "write about gender and politics . . . in ways that create alternatives to the objectification of 'informants' and 'cultures' that so many anthropological representations contain" (Wilson, 1995, p. 1). The result of the ethnographer's collaboration with Gabriela Ngirmang and her daughter, Cita Morei, is a narrative reflecting the interests and concerns of Islander women who speak about their lives and histories during a tumultuous period of change.

Maori Women: Feminism and Sovereignty

Influenced by feminism and other indigenous nationalist movements, women activists for Maori self-determination have reinvented a Maori tradition that supports participation of indigenous women in tribal and national affairs, notably through various women's organizations. Traditionalist activists focus organizational abilities on their Maori identities and roles as wives and mothers. Separatists struggle for gender equality within Maori society and serve as leaders in confrontations with the state over land, mining, language, and other issues (Dominy, 1990; see also Awatere, 1984).

Challenging their double oppression as women and as Maoris in New Zealand society, activist Awatere also seeks to change the colonial burden of Maori women "as the largest alienated group in *New Zealand*. Too often rejected by our men as mates. Too often used in the family as dogsbodies" (Awatere, 1984, p. 86, emphasis in original). This latter theme is forcefully presented in the 1994 film *Once Were Warriors* based on the novel of the same title by

Maori writer Alan Duff (University of Hawai'i Press, 1990).

Testimonies of Resistance and Survival in "Paradise"

Daughters of the Pacific presents the eloquent, often poignant stories by women about issues omitted from the tourist brochures (Ishtar, 1994). The author, an Australian-Irish feminist, is cofounder of the British network of Women Working for a Nuclear Free and Independent Pacific. Ishtar lived at Greenham Common Women's Peace Camp and traveled extensively in Oceania in 1986–1987 to listen to and record the words of Islander women.

These testimonies have special significance because they are told by those whose voices are often ignored. Australian Aboriginal women, for instance, speak of their pain as members of the "Stolen Generation" who were raised in orphanages or adopted by White Australians. A government-sanctioned policy, thousands of Aboriginal babies and children were forcibly removed from their families of birth by missionaries and social workers from 1918 through the late 1960s (Hamilton, 1989).

Consider the words of Betty Edmond, a Rongelapese (Marshall Islander) who was 7 years old when she witnessed the hydrogen bomb explosion (code-named Bravo) of March 1954 at Bikini atoll:

> I was so frightened. I kept running to everybody. There was a bright light all over Rongelap. I saw the coconut trees bending down. I went into the coffee shop, the food was covered with white powder, fallout. It was everywhere. I was playing in the fallout with the other kids, throwing it everywhere. When the Americans came to the island they were wearing things like astronauts. They didn't say anything and left as quickly as they can [could] (in Ishtar, 1994, pp. 22–23, insert added).

Nor are readers likely to forget heartbreaking stories about "jelly-fish babies" born to Micronesian women radiated by nuclear fallout.

Much of *Daughters of the Pacific* deals with indigenous women's activism against militarism and environmental degradation. Several chapters consider Islander women's perspectives on nuclear testing in French Polynesia, the struggle of Australian Aborigines to end uranium mining and military testing on traditional lands, and the efforts of Native Hawaiians to take back their sacred island of Kaho'olawe, used for decades as a bombing and shelling target.

Land is another important issue for Pacific Islander women. In Hawaii, the Wai'anae Women's Association has been successful in acquiring affordable government housing on Oahu Island for displaced indigenous residents (in Ishtar, 1994, pp. 101–109). Belauan women speak movingly of their land stewardship and "taro-patch politics." In the gardens, Cita Morei says:

> [y]ou're thinking about the land. You are thinking,

"This is what I value." You are not thinking of politics or of money. You are thinking about what it is to be Belauan. . . . If we want to keep coming to the taro patches then we have to look after Belau. We got to keep on going. Taro-patch politics. Men, they think about politics, they think about money. But women have been strong, because of the taro (in Ishtar, 1994, p. 57).

Gardening work enables women to communicate, to build a consensus on issues, and to preserve their land.

Activists in Guam and the Northern Marianas Commonwealth have challenged military interests as well as tourist development. Members of the Northern Marianas Women's Association are speaking out against environmental pollution and the "wholesale conversion of Saipan [Island] into a tourist colony" (in Ishtar, 1994, p. 84). Tumon Bay in Guam, a "ghetto of Japanese tourism," includes a prostitution zone that provides Filipina sex workers for foreign (mainly Asian) visitors. Developers have destroyed prehistoric Chamorro sites to build hotels in Guam, while U.S. military forces continue to control one-third of the island. "They made their wars here," writes Chamorro activist Chailang Palacios, "and now they're getting rich, the Japanese and the Americans" (in Ishtar, 1994, p. 75).

Colonialism and Appropriation in the Aloha State

Trask (1993), a Native Hawaiian leader in the self-determination movement, writes passionately about Hawaiian Islanders' oppression under American colonial rule:

> "No matter what Americans believe, most of us in the colonies do not feel grateful that our country was stolen, along with our citizenship, our lands and our independent place among the family of nations. We are not happy Natives" (Trask, 1993, p. 2).

Trask prefers the term "Native Hawaiian," adopted by members of various sovereignty movements, to clarify that indigenous Hawaiians are neither Americans, nor Asians, nor immigrants.

Using the metaphor of prostitution, Trask criticizes the commoditization of Hawaiian lands and culture for tourist profit, as well as the psychological and material colonization of Native Hawaiian minds and bodies. She cites the ecological degradation of land and reefs by resort developments and evictions of Hawaiians from land and housing in a state where the median price of a house in the early 1990s was $450,000. Many Islanders are forced to accept their political and economic co-optation by working in tourist-related occupations, which dominate the state's economy. There are few alternatives for Islanders other than the military, unemployment, or emigration to the mainland. "[W]e can't understand our own cultural degradation," Trask (1993, p. 195) writes, "because we are living it."

Conclusion: De-Centering the Canon

Belauan Lorenza Pedro has observed: "First know that we

exist: we are not on your maps of the world. Then tell other people" (in Ishtar, 1994, p. 251). Recognizing women and writing women into ethnography is a critical but preliminary step in the process of feminist anthropological reassessments of gender in Pacific Island societies. One of the goals of feminist anthropology is to engage in dialogue and collaboration with indigenous women and to address their priorities. The trend toward applied research in feminist anthropology represents a positive step in recognizing issues that indigenous activists have raised about the legacies of the colonial past and the continuing violence against their cultures, lands, reefs, and waters. We can expect that Islander women's activism will increase as issues of nuclear testing (French Polynesia), armed conflict (Bougainville and New Caledonia), and indigenous sovereignty movements (Hawaii and New Zealand) increase.

Anthropology, a discipline shaped by its colonial history and androcentric study of Exotic Others, is often criticized for researching and publishing social practices and institutions distant from the contemporary concerns of indigenous women and men. Despite cosmetic changes in language and chapter titles, readers of undergraduate anthropology texts (and viewers of ethnographic films) continue to receive a consistent message about women—a message consistent with a Western paradigm of androcentrism and gender hierarchy.

Not surprisingly, Pacific Islanders are critical of anthropological inquiry, citing the asymmetry of anthropologist–Exotic Other relationships and the ideological and material consequences for indigenous peoples burdened with labels imposed by science. Nahau Rooney, a prominent activist in Papua New Guinea, considers the legacy of Margaret Mead's ethnographic research on Rooney's native island of Manus:

> [W]e gave more to Margaret Mead as an anthropologist and to her profession than she gave to us in return. The only value that her work produced, if anything at all, is that she put our little island [Manus] on the map of the world. . . . By using our culture and writing about us, we felt that she was given the fame, the economic status, and the popularity that she held in her community (in Gilliam, 1992, p. 41).

Activists from other Pacific Islands have also taken issue with anthropological research, challenging the problematic connection between anthropologists and their "subjects." "There should be a moratorium," Trask (1993, p. 172) asserts, "on studying, unearthing, slicing, crushing, and analyzing us."

This essay reflects a tension between mainstream anthropological representations of indigenous societies and the challenges to those representations by feminist anthropologists and by Pacific Islander women. The diversity of indigenous women's voices suggests a continuing de-centering of the anthropological canon. Pacific Islander women have their own narratives, grounded in history and society, to tell.

References

Awatere, Donna. *Maori Sovereignty.* Auckland, New Zealand: Broadsheet, 1984.

Bell, Diane. *Daughters of the Dreaming.* 2nd ed. Minneapolis: University of Minnesota Press, 1993 [1st ed., 1983].

Dominy, Michele D. "Maori Sovereignty: A Feminist Invention of Tradition." In Jocelyn Linnekin and Lin Poyer (eds.), *Cultural Identity and Ethnicity in the Pacific.* Honolulu: University of Hawai'i Press, 1990, pp. 237–257.

Forman, Charles W. "'Sing to the Lord a New Song': Women in the Churches of Oceania." In Denise O'Brien and Sharon W. Tiffany (eds.), *Rethinking Women's Roles: Perspectives from the Pacific.* Berkeley: University of California Press, 1984, pp. 153–172.

Freeman, Derek. *Margaret Mead and Samoa: The Making and Unmaking of an Anthropological Myth.* Cambridge: Harvard University Press, 1983.

Gailey, Christine Ward. *Kinship to Kingship: Gender Hierarchy and State Formation in the Tongan Islands.* Austin: University of Texas Press, 1987.

Gilliam, Angela. "Leaving a Record for Others: An Interview with Nahau Rooney." In Lenora Foerstel and Angela Gilliam (eds.), *Confronting the Margaret Mead Legacy: Scholarship, Empire, and the South Pacific.* Philadelphia: University of Pennsylvania Press, 1992, pp. 31–53.

Goodale, Jane C. *Tiwi Wives: A Study of the Women of Melville Island, North Australia.* Seattle: University of Washington Press, 1971; reprint, Prospect Heights: Waveland, 1994.

Hughes, Helen. "Women in the Development of the South Pacific." In Rodney V. Cole (ed.), *Women in Development in the South Pacific: Barriers and Opportunities.* Canberra: Development Studies Center, Australian National University, 1985, pp. 3–10.

Ishtar, Zohl Dé. *Daughters of the Pacific.* Melbourne: Spinifex, 1994.

Jolly, Margaret, and Martha Macintyre. "Introduction." In Margaret Jolly and Martha Macintyre (eds.), *Family and Gender in the Pacific: Domestic Contradictions and the Colonial Impact.* Cambridge: Cambridge University Press, 1989, pp. 1–18.

Lateef, Shireen. "Current and Future Implications of the Coups for Women in Fiji." *Contemporary Pacific,* vol. 2, no. 1, 1990, pp. 113–130.

Lepowsky, Maria. *Fruit of the Motherland: Gender in an Egalitarian Society.* New York: Columbia University Press, 1993.

Lockwood, Victoria S. *Tahitian Transformation: Gender and Capitalist Development in a Rural Society.* Boulder: Lynne Rienner, 1993.

Lutkehaus, Nancy C. "Feminist Anthropology and Female Initiation in Melanesia." In Nancy C. Lutkehaus and Paul B. Roscoe (eds.), *Gender Rituals: Female Initiation in Melanesia.* New York:

Routledge, 1995, pp. 3–29.

Malinowski, Bronislaw. *The Sexual Life of Savages in North-Western Melanesia.* New York: Harcourt, Brace, & World, 1929.

Marshall, Mac, and Leslie B. Marshall. *Silent Voices Speak: Women and Prohibition in Truk.* Belmont: Wadsworth, 1990.

McDowell, Nancy. "Book Review Forum: Margaret Mead and Samoa, by Derek Freeman," *Pacific Studies,* vol. 7, no. 2, 1984, pp. 99–140.

———, (ed.). *Reproductive Decision Making and the Value of Children in Rural Papua New Guinea.* Monograph no. 27. Boroko, Papua New Guinea: Papua New Guinea Institute of Applied Social and Economic Research, 1988.

Mead, Margaret. *Coming of Age in Samoa: A Psychological Study of Primitive Youth for Western Civilization.* New York: William Morrow, 1928; reprint, New York: William Morrow, 1961.

Nardi, Bonnie A. "The Height of Her Powers: Margaret Mead's Samoa," *Feminist Studies,* vol. 10, no. 2, 1984, pp. 323–337.

Nero, Karen L. "The Hidden Pain: Drunkenness and Domestic Violence in Palau," *Pacific Studies,* vol. 13, July 1990, pp. 63–92. Special Issue: "Domestic Violence in Oceania."

Sexton, Lorraine. *Mothers of Money, Daughters of Coffee: The Wok Meri Movement.* Ann Arbor: UMI Research Press, 1986.

Sinclair, Karen. "Women in Oceania." In Margot I. Duley and Mary I. Edwards (eds.), *The Cross-Cultural Study of Women: A Comprehensive Guide.* New York: Feminist Press, 1986, pp. 271–289.

Strathern, Marilyn. *Women in Between: Female Roles in a Male World: Mount Hagen, New Guinea.* London: Seminar Press, 1972; reprint, Lanham: Rowman & Littlefield, 1995.

Tiffany, Sharon W. "Politics and Gender in Pacific Island Societies: A Feminist Critique of the Anthropology of Power," *Women's Studies,* vol. 13, no. 4, 1987, pp. 333–355.

Trask, Haunani-Kay. *From a Native Daughter: Colonialism and Sovereignty in Hawaii.* Monroe: Common Courage, 1993.

Weiner, Annette B. *Women of Value, Men of Renown: New Perspectives in Trobriand Exchange.* Austin: University of Texas Press, 1976.

Wilson, Lynn B. *Speaking to Power: Gender and Politics in the Western Pacific.* New York: Routledge, 1995.

Zimmer-Tamakoshi, Laura. "Nationalism and Sexuality in Papua New Guinea," *Pacific Studies,* vol. 16, no. 4, 1993, pp. 61–97.

Annotated Bibliography

Annotated Bibliography

Karen Monkman

Ahmed, Leila. *Women and Gender in Islam: Historical Roots of a Modern Debate*. New Haven, Connecticut: Yale University Press, 1992. (ISBN 0-300-05583-8 pbk)

This historical examination of the discourses on women and gender in different periods of Middle Eastern Arab history focuses on changes in, and varieties of, such discourses from ancient (pre-Islamic times in Mesopotamia and the Mediterranean Middle East) to more modern periods. The later chapters discuss social, cultural, political, and intellectual change since the turn of the nineteenth century, with primary focus on Egypt as a mirror of the modern Arab world. Topics in this section include the historical discourse of the veil and the "return of the veil" as a symbol of resistance, socioeconomic changes and their impact on women in the twentieth century, and the changing feminist discourses.

Ballara, Marcela. *Women and Literacy*. Women and World Development Series. London and Atlantic Highlands, New Jersey: Zed Books, 1992. (ISBN 0-86232-981-7 pbk)

This short book examines the impact of literacy on women in the Third World and its relationship to development. It shows how literacy projects aimed at women can contribute to their improved status, better health care, greater environmental protection, and productive economic activity. Pertinent issues for program development are discussed, including choice of language, selection of personnel, and coordination with women's other responsibilities. It includes a "how-to" guide on preparing literacy activities, a guide to education and action, a bibliography, a list of organizations, and information on preparing a needs assessment and evaluation. Descriptions of literacy programs in many countries are included. (The book is free to groups in developing countries from the United Nations Nongovernmental Liaison Service, Palais des Nations, 1211 Geneva 10, Switzerland.)

Benería, Lourdes, and Shelley Feldman (eds.). *Unequal Burden: Economic Crises, Persistent Poverty, and Women's Work*. Boulder, Colorado: Westview Press, 1992. (ISBN 0-8133-8230-0 pbk)

This collection of essays is offered as a comparative-historical approach to analyzing economic change and responses. The burden of responding to change and crises is largely carried by women and women's community groups. The eight country or regional studies reported in this work focus on the effects of structural adjustment and other economic-change policies on vulnerable sectors of the populations in Jamaica and the Dominican Republic, Mexico, Bangladesh, Bolivia, Tanzania, South Asia, Italy, and Nicaragua. The studies recognize that restructuring creates space for transformation and that women play key roles in the transformation of everyday life during economic crises. Transformation affects the divisions of labor within the household and at work, where normative behavior and ideology that legitimate female behavior are re-created.

Benería, Lourdes, and Martha Roldán. *The Crossroads of Class and Gender: Industrial Homework, Subcontracting, and Household Dynamics in Mexico City*. Chicago, Illinois: University of Chicago Press, 1987. (ISBN 0-226-04232-4 pbk)

Based on fieldwork in Mexico City in 1981 and 1982, this book explores the interaction between economic processes and social relations, specifically, the effect of industrial piecework on gender and family dynamics. The work and lives of the women who do piecework at home are connected through subcontracting to the national systems of production. It also shows how productive and labor-market structures are closely connected to household dynamics. Material and ideological elements in these realms converge to shape gender and class inequalities.

Boserup, Ester. *Woman's Role in Economic Development*. Washington, D.C.: Island Press; London: Earthscan Publications, 1989. (ISBN 1-85383-040-2 pbk) (Originally published in 1970.)

This landmark comparative analysis by an economist (re-analyzing available statistical data and pulling together findings from several studies) emphasizes gender as a basic factor in the division of labor in relation to economic development and social change in developing countries. Boserup explains a variety of factors (gendered farming systems, population density, land holding, polygamy, land reform, subsistence activities, domestic work and family labor, migration, education, and the introduction of development, colonialism, and capitalism) that reduce women's productive functions and, therefore, negatively affect the development of the country or region.

Buijs, Gina (ed.). *Migrant Women: Crossing Boundaries and Changing Identities.* New York: Berg Publishers, 1993. (ISBN 0-85496-869-5 pbk)

These case studies look at women's varied responses to migration worldwide, whether related to poverty or to political circumstances. Despite hoping to retain some aspects of their original cultures and lifestyles, migrant women often lose their self-identities and are forced to remake themselves in traumatic ways, significantly changing their relationships with men. For some women, the changes achieved through migration were ones they actively sought and would have been denied at home—most notably, social and physical mobility and education. These case studies focus on women from Chile in the United States; from Palestine in West Berlin; from Bangladesh and East Asia in Great Britain; and from Vietnam in Hong Kong; British domestic servants in South Africa around the turn of the twentieth century; migrants within South Africa; international and internal migrants, in and from, Gao, India; and Quechuans within Peru. Most authors are anthropologists; among the others are a sociologist, a legal adviser, and an ethnic-health researcher.

Carrillo, Roxanna. *Battered Dreams: Violence Against Women as an Obstacle to Development.* New York: UNIFEM, 1992.

This 37–page book links domestic violence to development. Carrillo's thesis is that development cannot succeed without addressing the limitations that gender-based violence presents to development processes. In the first section of the book, gender violence is discussed in relation to psychological, social, cultural, and economic dependency of women on men, child development, health concerns, and economic and social costs to society. The second section summarizes recent (1990–1992) U.N. initiatives about gender violence. Organizations that address gender violence in 54 countries in the Third World, as well as many industrialized countries, are identified.

Chaney, Elsa M., and Mary Garcia Castro (eds.). *Muchachas No More: Household Workers in Latin America and the Caribbean.* Philadelphia, Pennsylvania: Temple University Press, 1989. (ISBN 0-87722-571-0 pbk)

This eclectic collection of 22 essays and articles from academics, activists, and domestic workers focuses on domestic work in Latin America and the Caribbean, where at least 20 percent of all women in the paid labor force are household workers. Articles address historical trends; relationships with ideology, feminism, and the state; unionization, pay, and education of household workers; and representation in the media.

Personal testimonies and specific case studies, as well as theoretical contributions, examine the politics of household labor. The bibliographic section describes three computerized databases that attempt to remedy the fragmentary nature of data in this field. The first database covers researchers who focus on the study of domestic service; the second is a bibliography of materials about domestic service around the world, including conventional bibliographic citations, disciplinary specialty of author(s), subject matter, location of study, time period, and other information that can be sorted and used in computer searches; and, the third contains an annotated bibliography. The approximately 425 bibliographic citations from the second database make up the bibliography in Chaney and Castro's book.

Charlton, Sue Ellen M., Jana Everett and Kathleen Staudt (eds.). *Women, the State, and Development.* Albany, New York: State University of New York Press, 1989. (ISBN 0-7914-0065-4 pbk)

This volume explores intersections of state, development, and feminism. The editors present a theoretical model to analyze the state and gender in various contexts of development. This model focuses on influences by state officials, by state policies and institutions, and by state ideologies, or definitions of policy. Seven chapters present regional or country studies that explore these levels of state influence and their relationship to gender and development. Various historical periods are represented.

Conway, Jill Ker and Susan C. Bourque (eds.). *The Politics of Women's Education. Perspectives from Asia, Africa, and Latin America.* Ann Arbor, Michigan: University of Michigan Press, 1993. (ISBN 0-472-08328-7 pbk)

These 17 chapters demonstrate the importance of education for women and girls and point to the need to examine underlying assumptions relating to the roles, and goals, of education that lead to misguided, gender-biased policy. The book is organized geographically, with several chapters for each of the three regions of Asia, Africa, and Latin America. Formal and nonformal education are included, as well as intended and unintended consequences. Implicit is the need to find a new concept of "development" that is not tied to a Western, or colonial, conception of industrial growth but is more closely related to local contexts and characteristics and has at its core the full potential of all citizens, both male and female. The last two chapters offer critiques of economic models (their market orientation and lack of consideration of family dynamics) and historical perspectives (in which outcomes should be central) that are commonly used to assess educational impact and shape public policy in relation to women's and girls' education.

Dagenais, Huguette, and Denise Piché (eds.). *Women, Feminism and Development—Femmes, Féminisme, et Développement.* Montreal, Quebec, and Kingston, Ontario: McGill-Queen's University Press for the Canadian Research Institute for the Advancement of Women—L'Institut canadien de recherches sur les femmes, 1994 (ISBN 0-7735-1185-7 pbk).

This Canadian bilingual (French and English) collection consists of 19 articles by feminists working in academia, government, or nongovernmental organizations (NGOs) in fields related to development. The book is divided into four sections: (1) chapters dealing with conceptual and methodological issues; (2) case studies of consequences of development policies (the Green Revolution in Malaysia; rural income-generation activities, household production units, and the nursing profession in China; commercial sexual exploitation in Thailand; interests at stake in procreation in Mexico); (3) studies of empowerment through mobilization and social action in Uganda, Mexico, and northern Canada and with the NGO MATCH International; and (4) testimonials of struggles that voice the need for legal recognition and rights of native women; changing behavior and beliefs in communities and individuals; and grounding experience in actual conditions. Most articles are in English, the introductory chapter is bilingual, and three chapters are in French.

Dankelman, Irene and Joan Davidson. *Women and Environment in the Third World: Alliance for the Future.* London: Earthscan Publications, in association with the International Union for Conservation of Nature and Natural Resources, 1988. (ISBN 1-85383-003-8 pbk)

This book examines the relationships between women and their natural surroundings, shows how women deal with environmental crises, and looks at the response of international agencies. The first six chapters look at women's involvement in the use and management of natural resources: agriculture, water, forestry and wood, and other energy sources. The second part includes four chapters that examine women's positions in environmental conservation, in activities such as education and training and family planning, and in local organizations. Other chapters discuss activities and policies of international agencies and present a strategy for action based on sustainability. Ten of the chapters include case studies from numerous settings.

Dixon-Mueller, Ruth. *Population Policy and Women's Rights: Transforming Reproductive Choice.* Westport, Connecticut, and London: Praeger Publishers, 1993. (ISBN 0-275-94611-8 pbk)

The history and complexity of the population-control debate is the focus of this book, in which the demographer-author outlines the diverse realities of Third World women's experiences. Part One addresses several broad issues related to women's rights as human rights and, in particular, sexual and reproductive rights as human rights. Part Two includes chapters about the politics of feminism and population control as social movements; these chapters illustrate the ideological differences and divergent agendas in this area. Part Three looks at women's lives and their sexual and reproductive choices in developing countries based on surveys and ethnographic research. Part Four lays out a woman-centered reproductive policy and program that includes essential feminist components yet is minimal enough for women in diverse contexts to define and interpret their own needs. It advocates a policy approach that places women's entitlement to high-quality, comprehensive re-

productive health services at the center of a focused program for promoting women's economic, social, and political rights.

Dwyer, Daisy, and Judith Bruce (eds.). *A Home Divided: Women and Income in the Third World.* Stanford, California: Stanford University Press, 1988. (ISBN 0-8047-2213-7 pbk)

This collection of 12 essays explores the role of income (and, less so, other less negotiable currencies: the bearing of children, education, training, social networking, and household-based production) in strengthening or altering household arrangements. Eleven of the essays focus on various regions or countries (Africa, particularly Cameroon; Zambia; Nigeria; Egypt; Bangladesh; India; Taiwan; Indonesia; Honduras; Mexico; and migrants from the Dominican Republic in the United States. The final essay critiques economic theories of household dynamics, arguing that both Marxist and neoclassical views ignore the issue of inequality within the home and that altruism and cooperation need further exploration in theories of the household. The authors include economists, sociologists, demographers, and anthropologists. The essays are intended to inform development policy in that they contribute to a theoretical link between macrolevel policy and microlevel impacts (at the household and subhousehold levels).

Genovese, Michael A. (ed.). *Women as National Leaders.* Newbury Park, California: Sage Publications, 1993. (ISBN 0-8039-4338-5 pbk)

This is a collection of case studies of women who have been world leaders, mostly in developing countries. Each case explores the context in which these women became leaders and the circumstances surrounding their success or failure. In all cases, the role that gender played in their leadership and political career is explored. In particular, the case studies review how each of these women used power—as a tool to influence and persuade others, to set their agenda for action, and to promote consciousness among their followers.

Heyzer, Noeleen, Geertje Lycklama à Nijeholt, and Nedra Weerakoon (eds.). *Trade in Domestic Workers: Causes, Mechanisms and Consequences of International Migration.* Published for the Asian and Pacific Development Center. London and Atlantic Highlands, New Jersey: Zed Books, 1995. (ISBN 1-85649-286-9 pbk)

This collection of papers resulted from the Regional Policy Dialogue on Foreign Domestic Workers held in 1992. The chapters address the socioeconomic impact of international migration; labor policies, mechanisms of recruitment and reintegration; worker protection and labor legislation for foreign domestic workers; and interventions and action by the International Labor Organization (ILO) and nongovernment organizations (NGOs). The last chapter presents recommendations that grew out of the regional policy dialogue. One of the three appendices in the book is a model employment contract intended to alleviate some of the injustices. The authors (and the meeting participants) explore ways to form policies and programs that affect the international migration and overseas employment of women domestic workers. The essays analyze interrelated international and national structural forces, particularly the state, as they affect women migrant workers within the family, household, and kinship and community networks.

Hondagneu-Sotelo, Pierrette. *Gendered Transitions: Mexican Experiences of Immigration.* Berkeley and Los Angeles: University of California Press, 1994. (ISBN 0-520-07514-5 pbk)

This qualitative sociological study of a Mexican immigrant community in California focuses on the intersection of gender and the processes of immigration and settlement. It explores how gender relations facilitate and constrain immigration and settlement and how immigration and settlement processes reconstruct gender. Commonly used analytical frameworks of immigration and the household are critiqued. The gendered nature of three approaches to immigration (migration of single, independent people; migration of family units; and "family stage" migration in which, typically, men migrate first and the rest of the family follows at a later date) are examined in detail. Hondagneu-Sotelo offers a model of three concentric circles of influence on immigration and settlement, with microlevel concerns surrounded by family-level and social-network influences, surrounded by broader economic and political situations.

Jabbra, Joseph G., and Nancy W. Jabbra (eds.). *Women and Development in the Middle East and North Africa.* Leiden, the Netherlands, and New York: E. J. Brill, 1992. (ISBN 90-04-09529-2 pbk)

This interdisciplinary book explores women and development in various Middle Eastern and North African countries. A number of theoretical models are used in the articles, with one by Eva M. Rathgeber discussing theoretical approaches to women and Third World development. Other chapters focus on a particular country (Afghanistan, Palestine, Iran, Algeria, Iraq, Saudi Arabia, Morocco, and Egypt) in conjunction with issues such as political ideology and activities, employment, empowerment, and brain drain.

Jaquette, Jane S. (ed.). *The Women's Movement in Latin America: Participation and Democracy.* (2nd ed.) Boulder, Colorado: Westview Press, 1994. (ISBN 0-8133-8488-5 pbk)

The nine authors in this work discuss women's political participation and the women's movements in Argentina, Brazil, Chile, Mexico, Nicaragua, Peru, and Uruguay. Women mobilize as mothers in human-rights groups and in urban neighborhoods to address practical needs. The essays discuss the roles of women's movements in transitions from military rule or dictatorship to democracy and, in the case of Peru, also in the context of increasing terrorism (by Sendero Luminoso) that has targeted women leaders of nongovernmental (NGOs) and the government. Women's movements often operate as an oppositional politics—opposition to military rule or to structural adjustment and other severe economic constraints—but have difficulty making the transition to a more participatory involvement in cooperation with state interests. Common concerns include the autonomy of women's groups and efforts and the focus on social needs as opposed to economic change.

Jayawardena, Kumari. *Feminism and Nationalism in the Third World.* London and Atlantic Highlands, New Jersey: Zed Books, 1986. (ISBN 0-86232-265-0 pbk)

Jayawardena analyzes internal and external forces affecting the changing nature of feminism in the context of nationalist struggles during the nineteenth and early twentieth centuries in 12 Eastern countries: Afghanistan, China, Egypt, India, Indonesia, Iran, Japan, Korea, the Philippines, Sri Lanka, Turkey, and Vietnam. Country studies discuss factors relating to feminism such as colonialism, imperialism, capitalist penetration, nationalist and social-change movements, religion, women's labor-force participation, women's education, and women's organizations. Also addressed are ideological and material changes affecting women that have occurred as a result of certain historical circumstances, including, but not limited to, imperialism and Western thought.

Kabeer, Naila. *Reversed Realities: Gender Hierarchies in Development Thought.* London: Verso, 1994. (ISBN 0-86091-584-0 pbk)

This ten-chapter book focuses on the development process and how it has affected women. The author first presents theoretical underpinnings of women and development, then analyzes the relationship between development theory and practice. The book follows the development process from the earlier Women in Development (WID) models, through a Marxist approach, and then a gender-relations approach, calling for "reversed realities" in development-planning approaches by offering ways in which the prevailing development practice should be reversed to enable the voices of women to be heard and to allow women to make decisions that affect them. The author advocates the empowerment of women through the provision of time, resources, and space in ways that address both practical and strategic gender needs and interests.

Kardam, Nüket. *Bringing Women In: Women's Issues in International Development Programs.* Boulder, Colorado: Lynne Rienner, 1991. (ISBN 1-55587-205-0 out of print)

Using analytical frameworks from regime and organization theories, Kardam explores how the international women's movement has tried to alter norms within the development-assistance regime and how the United Nations Development Program (UNDP), the World Bank, and the Ford Foundation have responded. Each agency's level of response to Women in Development (WID) issues reflects the degree of independence the agency has within the development-assistance regime. The UNDP offered the least response to incorporating gender into their development work; the World Bank made some progress; and the Ford Foundation went the furthest in developing procedures and providing resources for the inclusion of gender in all programs. These responses are affected by agency goals, core values, and procedures, including resource allocation and staffing patterns.

Kerr, Joanna (ed.). *Ours by Right: Women's Rights as Human Rights.* London and Atlantic Highlands, New Jersey: Zed Books, and Ottawa: North-South Institute, 1993. (ISBN 1-85649-228-1 pbk)

This book is based on 24 lectures from an international conference in 1992 ("Linking Hands for Changing Laws: Women's Rights as Human Rights Around the World") in which scholars and activists discussed the present state of women's rights as human rights, and strategic actions to further this effort. Development agencies and donor institutions are seen as lagging in the area of women's rights as human rights. The book is divided into four parts, convering (1) women's rights in relation to human rights and the need to eliminate male bias in addressing this issue; (2) the women's-rights experiences of eight regions or countries; (3) mechanisms for change; and (4) feminist discussions of strategic areas for action. The five priority areas of action addressed are institutions (including pressure to acknowledge women's rights as human-rights); global conferences (U.N. conferences on human rights, population, and development, and Women, plus linkages among grass-roots women and at various organizational levels); national laws (legal reform, alternative methods of conflict resolution, more involvement by women in the legal system); religion (critique of fundamentalism, promotion and acknowledgment of religious interpretations that respect rights of women); and the community (changing sexist attitudes, including more grass-roots involvement in change processes and education). The book discusses the difficult and unbalanced relations between donors and recipients, and it advocates greater understanding between them and alteration of the relations of power. Women in developing countries need to have their voices heard, while feminists in industrialized countries must recognize how their activities and lifestyles affect women in the Third World.

King, Elizabeth M. and M. Anne Hill (eds.). *Women's Education in Developing Countries. Barriers, Benefits, and Policies.* Published for the World Bank. Baltimore: The Johns Hopkins University Press, 1993. (ISBN 0-8018-4534-3)

This book is a synthesis of literature on women's education in five developing regions: sub-Saharan Africa, the Middle East and North Africa, Latin America and the Caribbean, South Asia, and East Asia. The regional chapters examine factors that affect girls' access to education (mostly primary and secondary schooling) and are preceded by an overview of the conditions of women in developing countries and an essay discussing the private and social, monetary and nonmonetary returns on women's education. Crucial benefits that accrue from the education of women, such as reduction in infant and maternal mortality, decreased family size, and increased male life expectancy, are identified. The gender policies considered are limited to efforts that have occurred through educational projects funded by international development agencies.

Koblinsky, Marge, Judith Timyan, and Jill Gay (eds.). *The Health of Women: A Global Perspective.* Boulder, Colorado: Westview Press, 1993. (ISBN 0-8133-1608-1 pbk)

The main argument in these essays is that definitions of women's health must go beyond reproductive roles to include considerations of socioeconomic factors and gender politics.

Issues addressed include poverty, nutrition, abortion, mortality, infection, access to and quality of health care, violence against women, and mental health—in various locations and across generations. Overall, the book urges the development of new initiatives to understand and improve women's health, taking into account biological elements, cultural constraints, and socioeconomic realities. This book is a product of the 1991 National Council for International Health's conference on "Women's Health: The Action Agenda," which brought together people from 74 countries.

Mencher, Joan P. and Anne Okongwu (eds.). *Where Did All the Men Go? Female-Headed/Female-Supported Households in Cross-Cultural Perspective.* Boulder, Colorado: Westview Press, 1993. (ISBN 0-8133-85409-7)

Derived from two symposia, these 14 chapters in this work look at ways that female-headed/female-supported households survive and function in the face of often hostile economic, social, political, and military policies. One chapter discusses theoretical issues of the political economy of mother-child families. Issues common in many of the essays include factors that may lead to female headship (widowhood, political unrest, desire to escape oppressive relationships, state eligibility criteria that separates unemployed men from families, economic climates, migration, and polygamy), cultural factors that impact female headship (religion, sex-roles and other ideologies), and the importance of social networks to women heading households. Policy concerns and recommendations are included.

Mies, Maria. *Patriarchy and Accumulation on a World Scale: Women in the International Division of Labour.* London and Atlantic Highlands, New Jersey: Zed Books, 1986. (ISBN 0-86232-342-8 pbk)

The author explores the state of the women's movement worldwide, the history of colonization processes, the witch hunt and housewifization, women's work in the international division of labor, violence against women, and the relationship between women's liberation and national liberation struggles. Using the U.S.S.R., China, and Vietnam as examples, Mies shows that socialist societies do not escape patriarchal coercion. Suggestions for change include a consumer liberation movement by women in overdeveloped regions so that consumerism is more responsive to the conditions of the Third World and the environment, and a production liberation movement in Third World countries where women can be freed from the oppressive production systems. Linking these efforts through a worldwide women's movement would be the first step toward creation of an alternative economy and a feminist conception of labor that would alter the relationship between patriarchy and capitalism. Much of the analysis relies on the author's extensive experience in India.

Mies, Maria, and Vandana Shiva. *Ecofeminism.* London and Atlantic Highlands, New Jersey: Zed Books, 1993. (ISBN 1-85649-156-0 pbk)

In this collaborative work, these two women, one from the

First World and one from the Third, argue that ecological destruction and industrial catastrophes are a direct threat to everyday life. Because women have primary responsibility for everyday life, it is women who are most affected by wars, ethnic chauvinism, and economic malfunctioning. The authors critique economic theories, conventional concepts of women's emancipation, and myths of "catch-up" development; they offer a vision of subsistence that is based not on transcending nature but on living within its confines. Their message is that globalism has local victims. Spiritual dimensions should be recognized, but not to the exclusion of political dimensions. Through recognizing and understanding diversity as well as interconnectedness and commonalities, the authors see potential for a global movement that recognizes local diversity.

Moghadam, Valentine M. (ed.). *Identity Politics and Women: Cultural Reassertions and Feminisms in International Perspective.* Boulder, Colorado: Westview Press, 1994. (ISBN 0-8133-8692-6 pbk)

This book explores identity politics—how gender influences, and is used by, cultural and political movements and discourses in the formation, and political use, of religious, ethnic, and national identities. Five initial chapters present theoretical, comparative, and historical approaches to the study of the politization of identity. Thirteen case studies focus mostly on Muslim countries or communities but also include those that are Christian, Jewish, and Hindu, in Africa, Asia, the Middle East, Europe, and North America. Three concluding essays discuss dilemmas and strategies of women living under Muslim personal laws in addressing ethnicity in feminist politics and working toward sexual equality without compromising religious freedom. The essays suggest that we deconstruct woman as symbol and reconstruct her as human being.

Mohanty, Chandra, Ann Russo, and Lourdes Torres (eds.). *Third World Women and the Politics of Feminism.* Bloomington and Indianapolis, Indiana: Indiana University Press, 1991. (ISBN 0-253-20632-4 pbk)

A collection of 15 essays exploring feminism in the Third World, and particularly women as active subjects, this work is divided into the four sections: (1) knowledge production about and by Third World women; (2) the intersection of race, gender, and the state; (3) nationalism and sexuality; and (4) identity and feminist practice. The editors' introduction and conclusion explore in depth definitions of "feminism" and the "Third World." The "Third World" includes not only countries commonly thought to be in the Third World, but also certain populations in First World countries that have similar socioeconomic and ideological relationships to the state. To understand women involved in feminist movement in the Third World, the editors look to contexts of political and historical junctures (decolonialization and national liberation movements; the consolidation of White, liberal, capitalist patriarchies in industrialized countries; and multinational capital in a global economy), as well as discursive contexts (anthropology and storytelling or autobiography).

Morgan, Robin (ed.). *Sisterhood Is Global: The International Women's Movement Anthology.* New York: Anchor Books, 1984. (ISBN 0-385-17797-6 pbk)

Entries for 70 countries and the United Nations include a statistical preface that gives basic information concerning location, area, population, capital and the like; sections on demography, government, economy, gynography (marriage, family, contraception, and other pertinent information), herstory (a women's history), and mythography; and articles by a wide range of contributors. This anthology was intended as an "opening statement" to further international dialogue about women, women's movements, and global feminism. Each entry also offers suggested further reading. Entries vary in style from first-person experiences, to journalistic-style reporting, to theoretical analyses.

Moser, Caroline O. N. *Gender Planning and Development: Theory, Practice, and Training.* New York: Routledge, 1993. (ISBN 0-415-05621-7 pbk)

This book examines feminist theories and Women in Development/Gender and Development debates on development planning. Assumptions about family structure and households are examined. Practical and strategic gender interests are distinguished and discussed in terms of the related gender needs and the state's control over strategic needs. Different macroeconomic development models and policy approaches (modernization, redistribution, structural adjustment) mirror some of the gender approaches (welfare, equity, antipoverty, efficiency, and empowerment approaches) that have been used. The author discusses performance indicators; pros and cons of approaching gender planning as an independent realm of planning, an add-on to other planning, or incorporated into the mainstream planning processes; and ways to implement these approaches. The different strategies are illustrated with examples from development agencies. Gender planning is set within the political context of women-generated movements. Moser identifies entry points for women's nongovernmental organizations for change to work.

Mosse, Julia Cleves. *Half the World, Half a Chance.* Oxford, England: Oxfam, 1994. (ISBN 0-85598-186-5 pbk)

This is an introduction to gender and development. Mosse explores ways in which gender inequalities are socially constructed and how this affects participation in the development process. She first presents the theoretical background to gender and development and then provides case studies from various Third World countries. Some of the pertinent topics include social construction of women's work, gender subordination, and the empowerment approach to development.

Nieuwenhuys, Olga. *Children's Lifeworlds: Gender, Welfare, and Labour in the Developing World.* New York: Routledge, 1994. (ISBN 0-415-09751-7 pbk)

This ethnography looks at the lives of child workers in Kerala, India, the poorest state in India but the one where children remain longest in school. The author challenges contemporary thinking about children's work and reveals its inadequacy for understanding children working in rural areas. These children of the rural poor are generally not paid for their work and are often exploited. Their work activity is often complementary to attending school, and some of her informants worked in order to pay for their schooling, which their parents couldn't afford. Work is a response to economic processes and is shaped by gender and seniority in the household and by kinship power and prestige in the extended family and broader community. The work of children in this community centers on the fishing industry for boys and yarn making for the girls.

Ogundipe-Leslie, Molara. *Re-Creating Ourselves: African Women and Critical Transformations.* Trenton, New Jersey: Africa World Press, 1994. (ISBN 0-86543-412-3 pbk)

This collection of essays explores how African women have transformed their lives socially, politically, and spiritually. These experiences are organized into two sections: theory and practice. The theory section includes essays that reflect six different "mountains," or constraints, African women have to deal with in the development process: foreign intrusion, the heritage of tradition, backwardness, men, race, and women themselves (in terms of negative self-images and having taken in patriarchal ideologies). The practice section is represented by speeches, essays, and poems that address the need for social change and ways in which women have organized for "critical transformations" in the home and families, in the workplace, in contexts concerned with spirituality, and in arenas in which material concerns are central.

Ong, Aihwa. *Spirits of Resistance and Capitalist Discipline: Factory Women in Malaysia.* Albany, New York: State University of New York Press, 1987. (ISBN 0-88706-381-0 pbk)

This ethnographic study of an agricultural district in Malaysia that is undergoing rapid change focuses on women workers in Japanese factories. Indigenous and borrowed cultural values and practices are reworked and reconstituted in the industrial hierarchical setting. The author interprets the periodic hysterical episodes, violent incidents, and other disruptive activities as a reflection of the women's and their families' difficulties and ambivalence in incorporating corporate culture and capitalist discipline into their lives. These forms of indirect resistance are culturally consistent with their subordinate female status and reflect experiences of discontinuity in cultural change.

Pankhurst, Helen. *Gender, Development, and Identity: An Ethiopian Study.* London and Atlantice Highlands, New Jersey: Zed Books, 1992. (ISBN 1-85649-158-7)

This ethnographic study of Menz, Ethiopia, explores the relationships between different spheres of the state and the peasantry with particular focus on the lives of women. The study was done in the years 1988–1989 and 1992, when Ethiopia's revolutionary government attempted ambitious social change, including a villagization campaign (an effort to settle people in villages so they aren't so dispersed). Some state strategies had no effect on the peasantry, while others had unintended influences. Pankhurst found that the peasantry used a strategy of continuous adjustment between resistance of, and compliance with, the state. While the state largely ignored women in their social-change efforts, women emerge as significant actors who are directly involved in the cash economy to support household subsistence. The appendices present methodological reflections, a personal note about how the research was undertaken, and a summary of the quantitative data used.

Pietilä, Hilkka, and Jeanne Vickers. *Making Women Matter: The Role of the United Nations.* 2nd ed. London and Atlantic Highlands, New Jersey: Zed Books, 1994. (ISBN 1-85649-270-2 pbk)

This book, which is based on a survey of the material the United Nations has collected on women, presents a picture of how women have begun to shape the ways in which development occurs, of how much more needs to be done, and of how the United Nations is playing a part in making women matter. The book, written from a nongovernmental perspective, highlights what the United Nations does to further the advancement of women and urges women to make sure that governments keep the promises they make when they make decisions at the United Nations. The 10 chapters discuss events such as the Nairobi conference and preparation for the Beijing conference, the World Survey on the Role of Women in Development, the *Forward-looking Strategies,* the United Nations Decade for Women, (1976–1985) other world conferences, and various institutional structures and practices within the United Nations.

Radcliffe, Sarah A., and Sallie Westwood (eds.). *"VIVA": Women and Popular Protest in Latin America.* New York: Routledge, 1993. (ISBN 0-415-07313-8 pbk)

This work is mostly a collection of case studies focusing on women's responses to specific oppressive situations. The first chapter presents theoretical issues related to gender, racism, nationalism, popular culture, and the politics of identity. The eight chapters that follow illustrate the diversity and specific ways in which racism, gender, and class relations are articulated in different locales. These include studies of the development of a gender consciousness in the mothers and widows of the disappeared in El Salvador and Guatemala, gendered patterns of participation in environmental movements in Venezuela, the influences of the church and the feminist movement in the emergence of a health movement in Brazil, political participation in poor, urban neighborhoods in Mexico, consciousness raising in community-based struggles in Brazil, political manipulation of images of women in Chile, women's time and their triple role in low-income urban areas in Ecuador, and state constructions and representations of femininities and resistance to these by women in Peru.

Robertson, Claire, and Iris Berger (eds.). *Women and Class in Africa.* New York: Africana Publishing Company, 1986. (ISBN 0-8419-0979-2 pbk)

The essays in this interdisciplinary book discuss three aspects of the relationship of class and gender: access to resources, such as land, livestock, and education; autonomy and dependence as related to social and economic organization; and female solidarity as linked to class-related political action. Three chapters focus on these issues theoretically; other essays present more contextualized analyses. The essays are informed by Marxist and feminist thought to varying degrees.

Schuler, Margaret (ed.). *Empowerment and the Law: Strategies of Third World Women.* Washington, D.C.: OEF International, 1986. (ISBN 0-912917-11-3 pbk)

Fifty-three case studies (in 30 countries) represent the starting point of the ongoing dialogue of how women are gaining the skills needed to enforce the law or to challenge it in order to assert rights, to redress injustices or gain access to economic and political resources, and to empower themselves to understand their own oppression and the social forces that shape and maintain it. Land-tenure systems, labor laws and practices, and family law are discussed within the context of the state, the law, and development. The intersection and contradictions that exist among custom, customary law, religion, state law, and women's rights are explored. A section on "Violence and Exploitation" precedes the "Strategies" section, which focuses on education and organization, law reform, and advocacy. This book is also available in Spanish as *Poder y Derecho: Estrategias de las Mujeres del Tercer Mundo* (ISBN 0-912917-17-2)

Schuler, Margaret (ed.). *Freedom from Violence: Women's Strategies from Around the World.* Washington, D.C.: OEF International/UNIFEM, 1992. (ISBN 0-912917-24-5 pbk)

The case studies from 12 countries (Bolivia, Brazil, Chile, India, Malaysia, Mexico, Pakistan, Sri Lanka, Sudan, Thailand, the United States, and Zimbabwe) describe how women challenge physical and psychological violence at home, at work, and in the street. They present a practical framework to develop objectives and strategies, integrating concerns in education, legal services, and judicial reform. Resources include a section on publications and organizations. The book is directed primarily at women activists and advocates.

Sen, Gita, Adrienne Germain, and Lincoln C. Chen (eds.). *Population Policies Reconsidered: Health, Empowerment, and Rights.* Boston, Massachusetts: Harvard University Press, 1994. (ISBN 0-674-69003-6 pbk)

This collection of essays raises critical questions about population policies and proposes alternatives. Three themes are featured. The first proposes that new ways for formulating population policies should consider human rights, including civil, political, social, and economic rights. The second underscores the empowerment of women as central to any population-policy consideration and argues that education for women should aim to promote social change through empowerment. The last proposes that current services for reproductive and sexual health should be reconsidered in terms of objectives, expanding coverage to various population groups, and improving resource allocation.

Sen, Gita, and Caren Grown (DAWN). *Development, Crises, and Alternative Visions: Third World Women's Perspectives.* New York: Monthly Review Press, 1987. (ISBN 0-85345-718-2 pbk)

This is a collective effort of Development Alternatives with Women for a New Era (DAWN), an organization of Third World activists and researchers committed to creating alternative development processes to attain social and economic justice, peace and development free of oppression. This work, first presented at the end (1985) of the United Nations Decade for Women, links the roots of women's oppression to current economic and political crises—debt, famine, militarization, and fundamentalism—and places the existing body of microlevel case studies, projects, and organizing attempts in a broader context. Chapter One looks at how women's experiences with economic growth, commercialization and market expansion are determined by class and gender. Chapter Two links past development policies and strategies with current crises and shows women's potential for mitigating the effects of these crises. The last chapter presents a tentative vision and strategies, both long- and short-term, and a critique of different types of organizations and their potential in shaping a new agenda.

SIGNS: Journal of Women in Culture and Society. (Special Issue: "Women, Family, State, and Economy in Africa"), vol. 16, no. 4, Summer 1991.

This special issue of the SIGNS journal has seven articles about women, and four reports from women's groups, in sub-Saharan Africa. Many of the articles focus on the gendered construction of the state, including not only the attempts to change discriminatory laws, but also the potential influences of peasant women on state action should they be integrated into the political system. Struggles for empowerment within the family and household, for access to, and control of, economic resources, and in the political realm are common themes. Family, the economy, and the state are contexts that are interwoven in the articles. Women are recognized not as an undifferentiated class but as various communities that relate to others (both men and women) in ways that are gendered. Women are shown as active agents of history during precolonial periods, early and later periods of colonialism, and postcolonial periods. The four reports from women's groups discuss initiatives to promote women's studies and research in Botswana, Nigeria, Tanzania, and Uganda. The Botswana group was also engaged in legal reform efforts.

Sontheimer, Sally (ed.). *Women and the Environment: A Reader: Crisis and Development in the Third World.* New York: Monthly Review Press, 1991. (ISBN 0-85345-835-9 pbk)

This collection of eight essays, prepared for the Italian Association of Women in Development, addresses issues relating to land, forests, and water and includes three chapters describing women's initiatives. One essay on land discusses legal rights and control over land resources and includes brief case studies for Togo, Mexico, Brazil, Iran, and Ethiopia. Another examines the ecological and sociocultural impacts of desertification, based on studies in six Sahelian countries (Burkina Faso, Cape Verde, Mali, Mauritania, Niger, and Senegal). The forestry chapters discuss forest use and management and cooking fuel in South Asia. The chapter on water addresses issues relating to water supply and sanitation. The three case studies of women's initiatives describe work done in urban settlements in Ecuador and Costa Rica, and, in India, the Chipko Movement to preserve forests and the Bankura project for the rehabilitation of wastelands.

Stromquist, Nelly P. *Literacy for Citizenship: Gender and Grassroots Dynamics in Brazil.* Albany, New York: State University of New York Press, 1997. (ISBN 0-7914-3166-5 pbk)

A case study of Movimento de Alfabetizacão de Jovens e Adultos (MOVA) in São Paulo, Brazil, this book looks at sociocultural conditions that shape the acquisition, uses, and outcomes of literacy skills for the women participants in this emancipatory literacy program administered by Paulo Freire. Through examining the women's changing lives, the grassroots groups and the political party implementing the program, the study reveals contradictions, ambiguities, and antagonisms that operate and influence the women's development of literacy skills, how they feel about themselves and their lives, and how they use literacy in their lives. Literacy acquisition was minimal, while psychological gains were much more impressive. Literacy use in daily lives is hampered by women's family maintenance responsibilities and the lack of available materials to read. MOVA's intention of using literacy to build citizenship was weak with the women participants due in part to its failure to address the lived experiences of women.

Stromquist, Nelly P. (ed.). *Women and Education in Latin America: Knowledge, Power, and Change.* Boulder, Colorado: Lynne Rienner, 1992. (ISBN 1-55587-286-7)

The 13 essays in this work use a variety of disciplinary perspectives and sociogeographical contexts in examining the role of education in shaping women's experiences in Latin America. Chapter themes include gender perspectives of relations between education and work, higher education, leftist university politics, coeducation, beliefs and choices about educational and social opportunities, sexual stereotypes in the curriculum and teaching practices, participation in teachers' unions, parental participation in schools, and popular education. While reproduction of inequality is seen clearly in many of the educational experiences, also clear are efforts to transform ideological understanding and material conditions.

Stromquist, Nelly P. and Paud Murphy. *Leveling the Playing Field: Giving Girls an Equal Chance for Basic Education—Three Countries' Efforts.* Washington, DC: Economic Development Institute of the World Bank, 1995.

This profile of three countries' innovative steps to develop equality in education for girls was prepared for the Fourth U.N. Conference on Women (Beijing, 1995). It features efforts in Pakistan's Balochistan Province, in Malawi, and by the Bangladesh Rural Advancement Committee (BRAC) that seek to improve the condition of education for girls in

quantitative and qualitative terms. Among the characteristics all three share are active government initiative and support, including partnerships with others—especially nongovernmental organizations (NGOs); quality-control mechanisms; and community involvement. Lessons learned include the need to consider the interaction of supply and demand forces when planning; to provide a package of measures, not one isolated intervention; and to specifically target girls. Implementation suggestions are also made. The study is based on short field studies, particularly observation and interviews.

Subbarao, K., and Laura Raney. *"Social Gains from Female Education: A Cross-National Study,"* discussion paper no. 194, World Bank, Washington, D.C.:, 1993. (ISBN 0-8213-2387-3 pbk)

This discussion paper examines the role of secondary female education relative to, and/or in combination with, health and family-planning programs and policies that reduce fertility and infant mortality. Using cross-country data from 72 developing countries covering the period 1970–1985, this study statistically estimates the social gains—improved maternal and child health, decreased infant mortality, and reduced fertility—from simulating increased enrollment of female students at the secondary level. Secondary education for girls is found to be the best single policy lever for achieving these gains, with benefits far greater than those provided by family-planning services, more physicians, or health services alone. When combined with other services, secondary education is found to promote even greater gains. These results are particularly promising for countries where enrollment of girls in secondary education is low.

Tetreault, Mary Ann (ed.). *Women and Revolution in Africa, Asia, and the New World.* Columbia, South Carolina: University of South Carolina Press, 1994. (ISBN 1-57003-016-2)

This book explores the participation of women in revolutionary movements in the Third World and how such participation impacts their postrevolutionary status. Case studies from Africa, Asia, and the Americas are discussed under three themes: (1) analysis of the status of women and men in pre-revolutionary society; (2) relationship of family, society, and the state under prerevolutionary conditions; and, (3) postrevolutionary state commitment to women's liberation. The cases show that, in almost all cases, women have participated actively in revolution in many ways and were always promised a change in terms of their status in development. However, except for Cuba, all gains women made during war have not lasted but have been eschewed by the renewed postwar patriarchal structures. In Cuba, socialism supports the maintenance of gender gains in employment and affirmative policies, but even there patriarchal influences have not disappeared.

Tiano, Susan. *Patriarchy on the Line: Labor, Gender, and Ideology in the Mexican Maquila Industry.* Philadelphia, Pennsylvania: Temple University Press, 1994. (ISBN 1-56639-196-2 pbk)

This comprehensive study of women in electronic- and apparel-assembly jobs in U.S. export-processing companies and in service jobs in Mexicali, Mexico, explores women's reasons for entering these types of jobs and how their lives are consequently changed. The author analyzes material and ideological transformations, including changing cultural ideas about appropriate roles for women, and how these changes are shaping the female labor market and creating a pool of labor for the *maquila* industry (assembly factories that export components). In contrast to assumptions related to viewing women as passive, docile workers, women take these jobs for a variety of reasons, and they are conscious of their productive and reproductive influence on the developing international division of labor.

Tucker, Judith E. (ed.). *Arab Women: Old Boundaries, New Frontiers.* Bloomington and Indianapolis, Indiana: University of Indiana Press, 1993. (ISBN 0-253-20776-2 pbk)

Overall, the essays in this book debate Western feminist evaluations of the position of Arab women and challenge assumptions about the monolithic character of Islam and of Arab culture. The 12 chapters reveal the diversity within the Arab world. The present as a time of change and tension for women in the region is central, and the book reveals the emergence of an Arab women's movement. The authors, Western-educated Arab women, are from the disciplines of history, anthropology, Arabic studies, comparative literature, political science, and sociology.

United Nations. *The World's Women 1995: Trends and Statistics.* New York: United Nations, 1995. (ISBN 92-1-161372-8 pbk)

This resource book compiles U.N. statistics on women in 51 developed countries and 159 countries in developing regions over a 20–25-year period (through 1990 or 1995), with some population projections to the year 2010. Numerous topics are analyzed in light of statistical patterns, including population, families and households, domestic violence, political participation. women in the media, peace, literacy and education, health risks and causes of death, and life expectancy, AIDS, migration, the environment, work and the informal sector, child labor, and structural adjustment.

United Nations, Department for Policy Coordination and Sustainable Development. *Women in a Changing Global Economy: 1994 World Survey on the Role of Women in Development.* New York: United Nations, 1995. (ISBN 91-1-130163-7 pbk)

This Third World survey is a statistical compilation that analyzes issues of poverty, productive employment, and economic decision making, using a gender perspective that recognizes male and female roles in society. This U.N. General Assembly document, prepared for the Fourth World Conference on Women (Beijing, 1995), is based on statistical data from 1970, 1980, and 1990, recovered from the Women's Indicators and Statistical Data Base (WISTAT), as well as on an exhaustive review of specialized and microstudies. Sections cover (1) global restructuring and the impact on women, as well as changes in the enabling environment, specifically gains in women's legal status in the direction of equality and achievement of equal access to education and training; (2) trends in gender dimensions of poverty; (3) gender analysis of productive employment; (4) women and economic decision making;

and (5) policies and strategies for women's effective participation in development, including policymakers considering gender as a key variable, addressing gender dimensions of poverty and incorporating women actors in poverty alleviation, promoting gender equality in the working world, and supporting women's participation in economic decision making, particularly as managers and entrepreneurs.

Vickers, Jeanne. *Women and War.* London and Atlantic Highlands, New Jersey: Zed Books, 1993. (ISBN 1-85649-230-3 pbk)

This is an introduction to the interrelationship between the condition of women and the occurrence of forms of aggression. The author investigates the impact on women of war generally and, particularly, of conflicts within the last decade. She looks at the ways women have worked and can contribute toward peaceful settlement of confrontations. An agenda for action includes specific suggestions on how to work for peace. One appendix is an education guide for use in leading discussions or conducting workshops. Another appendix includes a list of U.N. declarations, conventions, and other documents, as well as the U.N. agencies and nongovernmental organizations (NGOs) concerned with peace, security, and women in development.

Ward, Kathryn (ed.). *Women Workers and Global Restructuring.* Ithaca, New York: ILR Press, Cornell University, 1990. (ISBN 0-87546-162-X pbk)

The nine studies in this edited book explore women's work in relation to the emerging global assembly line. Part One includes three chapters that focus on the impact of transnational corporations and factory work in the informal sector in Java, Colombia, and Greece. The two chapters in Part Two discuss how the Irish and the Japanese governments simultaneously promote development and reinforce patriarchal ideologies that define women primarily as wives and mothers. The three chapters in Part Three look at gender- and race-specific tactics that employers use to control female workers in factories in the United States, Taiwan, and Mexico, and how women resist such strategies. The studies suggest the need to better understand women's work—in the informal sector as well as formal employment—and resistance to gender- and race-based oppression.

White, Sarah C. *Arguing with the Crocodile: Gender and Class in Bangladesh.* London and Atlantic Highlands, New Jersey: Zed Books, 1992. (ISBN 1-85649-086-6 pbk)

A study of the gender and class aspects of life in a rural Bangladesh area, this book critiques the discourse of international aid to show that, as discourse frames the ways we look at issues, it obscures other perspectives that should be explored. This ethnographic study of how Bangladeshi women see their own lives looks at the relationships within which work is performed and power is exercised, particularly within the household. Through critical reflection, the last chapter raises some alternative issues (political participation, religious and cultural practices, sexuality) that could add other dimensions to understanding gender and class in Bangladesh. The field research was done between 1985 and 1990.

Women's Feature Service. *The Power to Change: Women in the Third World Redefine Their Environment.* London and Atlantic Highlands, New Jersey: Zed Books, 1993. (ISBN 1-85649-226-5 pbk)

This journalistic collection, compiled from article submissions to the Women's Feature Service (WFS), focuses on news about women and international development from a progressive perspective. The WFS is the only wire service that seeks to insert women's voices into the mainstream media in relation to development. Stories from 24 countries illustrate the necessity of locating woman at the center of development.

World Bank. *Toward Gender Equality: The Role of Public Policy.* Development in Practice series. Washington, D.C.: World Bank, 1995. (ISBN 0-8213-3337-2 pbk)

This report, written for the Fourth World Conference on Women (Beijing, 1995), brings together issues relating to private and social returns to education for women. The first chapter presents and discusses gender inequalities in education, health, employment and work activity, mostly by broad world regions. Chapter two discusses how gender inequality within the household (through decision making) and social structural inequalities, such as access to resources and services and participation in the labor force, restrict economic growth and efficiency. The last chapter argues that laws, regulations, and public policy be changed so that access to opportunities is equalized across genders; that women's lives be supported so that they are better able to fully participate in productive activity; and women be targeted as beneficiaries in programs. Two types of policy changes are recommended: macroeconomic stability and removal of price distortions, and promotion of labor-demanding growth in agriculture and industry, as well as more accessible basic social services and infrastructures that yield high social returns, especially education, health care, and water supply.

Young, Gay, Vidyamali Samarasinghe, and Ken Kusterer (eds.). *Women at the Center: Development Issues and Practices for the 1990s.* West Hartford, Connecticut: Kumarian Press, 1993. (ISBN 1-56549-029-0 pbk)

This is a collection of essays emanating from the Fifth International Forum of the Association for Women in Development. The core theme is that women are central to all development activities, and, as such, they should be the focus of the development process. The book is organized into three parts. Part One outlines some of the areas in which women participate to meet basic human needs, including the economic sector. The second part reviews some of the strategies women use for transformation and social change. Part Three moves into current challenges—the environment, AIDS, and war—and how women cope with them.

Karen Monkman

Appendices

Appendix A

Convention on the Elimination of All Forms of Discrimination Against Women

Part I

Article 1

For the purposes of the present Convention, the term "discrimination against women" shall mean any distinction, exclusion or restriction made on the basis of sex which has the effect or purpose of impairing or nullifying the recognition, enjoyment or exercise by women, irrespective of their marital status, on a basis of equality of men and women, of human rights and fundamental freedoms in the political, economic, social, cultural, civil or any other field.

Article 2

States Parties condemn discrimination against women in all its forms, agree to pursue by all appropriate means and without delay a policy of eliminating discrimination against women and, to this end, undertake:

(a) To embody the principle of the equality of men and women in their national constitutions or other appropriate legislation if not yet incorporated therein and to ensure, through law and other appropriate means, the practical realization of this principle;

(b) To adopt appropriate legislative and other measures, including sanctions where appropriate, prohibiting all discrimination against women;

(c) To establish legal protection of the rights of women on an equal basis with men and to ensure through competent national tribunals and other public institutions the effective protection of women against any act of discrimination;

(d) To refrain from engaging in any act or practice of discrimination against women and to ensure that public authorities and institutions shall act in conformity with this obligation;

(e) To take all appropriate measures to eliminate discrimination against women by any person, organization or enterprise;

(f) To take all appropriate measures, including legislation, to modify or abolish existing laws, regulations, customs and practices which constitute discrimination against women;

(g) To repeal all national penal provisions which constitute discrimination against women.

Article 3

States Parties shall take in all fields, in particular in the political, social, economic and cultural fields, all appropriate measures, including legislation, to ensure the full development and advancement of women, for the purpose of guaranteeing them the exercise and enjoyment of human rights and fundamental freedoms on a basis of equality with men.

Article 4

1. Adoption by States Parties of temporary special measures aimed at accelerating de facto equality between men and women shall not be considered discrimination as defined in the present Convention, but shall in no way entail as a consequence the maintenance of unequal or separate standards; these measures shall be discontinued when the objectives of equality of opportunity and treatment have been achieved.

2. Adoption by States Parties of special measures, including those measures contained in the present Convention, aimed at protecting maternity shall not be considered discriminatory.

Article 5

States Parties shall take all appropriate measures:

(a) To modify the social and cultural patterns of conduct of men and women, with a view to achieving the elimination of prejudices and customary and all other practices which are based on the idea of the inferiority or the superiority of either of the sexes or on stereotyped roles for men and women;

(b) To ensure that family education includes a proper understanding of maternity as a social function and the recognition of the common responsibility of men and women in the up-bringing and development of their children, it being understood that the interest of the children is the primordial consideration in all cases.

Article 6

States Parties shall take all appropriate measures, including legislation, to suppress all forms of traffic in women and exploitation of prostitution of women.

Part II

Article 7

States Parties shall take all appropriate measures to eliminate discrimination against women in the political and public life of the country and, in particular, shall ensure to women, on equal terms with men, the right:

(a) To vote in all elections and public referenda and to be eligible for selection to all publicly elected bodies;

(b) To participate in the formulation of government policy and the implementation thereof and to hold public office and perform all public functions at all levels of government;

(c) To participate in non-governmental organizations and associations concerned with the public and political life of the country.

Article 8

States Parties shall take all appropriate measures to ensure to women, on equal terms with men and without any discrimination, the opportunity to represent their governments at the international level and to participate in the work of international organizations.

Article 9

1. States Parties shall grant women equal rights with men to acquire, change or retain their nationality. They shall ensure in particular that neither marriage to an alien nor change of nationality by the husband during marriage shall automatically change the nationality of the wife, render her stateless or force upon her the nationality of the husband.
2. States Parties shall grant women equal rights with men with respect to the nationality of their children.

Part III

Article 10

States Parties shall take all appropriate measures to elimi-

nate discrimination against women in order to ensure to them equal rights with men in the field of education and in particular to ensure, on a basis of equality of men and women:

(a) The same conditions for career and vocational guidance, for access to studies and for the achievement of diplomas in educational establishments of all categories in rural as well as in urban areas; this equality shall be ensured in pre-school, general, technical, professional and higher technical education, as well as in all types of vocational training;

(b) Access to the same curricula, the same examinations, teaching staff with qualifications of the same standard and school premises and equipment of the same quality;

(c) The elimination of any stereotyped concept of the roles of men and women at all levels and in all forms of education by encouraging coeducation and other types of education which will help to achieve this aim and, in particular, by the revision of textbooks and school programs and the adaptation of teaching methods;

(d) The same opportunities to benefit from scholarships and other study grants;

(e) The same opportunities for access to programs of continuing education, including adult and functional literacy programs, particularly those aimed at reducing, at the earliest possible time, any gap in education existing between men and women;

(f) The reduction of female student drop-out rates and the organization of programs for girls and women who have left school prematurely;

(g) The same opportunities to participate actively in sports and physical education;

(h) Access to specific educational information to help to ensure the health and wellbeing of families, including information and advice on family planning.

Article 11

1. States Parties shall take all appropriate measures to eliminate discrimination against women in the field of employment in order to ensure, on a basis of equality of men and women, the same rights, in particular:

(a) The right to work as an inalienable right of all human beings;

(b) The right to the same employment opportunities, including the application of the same criteria for selection in matters of employment;

(c) The right to free choice of profession and employment, the right to promotion, job security and all benefits and conditions of service and the right to receive vocational training and retraining, including apprenticeships, advanced vocational training and recurrent training;

(d) The right to equal renumeration, including benefits, and to equal treatment in respect of work of equal value, as well as equality of treatment in the evaluation of the quality of work;

(e) The right to social security, particularly in cases of retirement, unemployment, sickness, invalidity and old age and other incapacity to work, as well as the right to paid leave;

(f) The right to protection of health and to safety in working conditions, including the safeguarding of the function of reproduction.

2. In order to present discrimination against women on the grounds of marriage or maternity and to ensure their effective right to work, States Parties shall take appropriate measures:

(a) To prohibit, subject to the imposition of sanctions, dismissal on the grounds of pregnancy or of maternity leave and discrimination in dismissals on the basis of marital status;

(b) To introduce maternity leave with pay or with comparable social benefits without loss of former employment, seniority or social allowances;

(c) To encourage the provision of the necessary supporting social services to enable parents to combine family obligations with work responsibilities and participation in public life, in particular through promoting the establishment and development of a network of child-care facilities;

(d) To provide special protection to women during pregnancy in types of work proved to be harmful to them.

3. Protective legislation relating to matters covered in this article shall be reviewed periodically in the light of scientific and technological knowledge and shall be revised, repealed or extended as necessary.

Article 12

1. States Parties shall take all appropriate measures to eliminate discrimination against women in the field of health care in order to ensure, on a basis of equality of men and women, access to health care services, including those related to family planning.

2. Notwithstanding the provisions of paragraph 1 of this article, States Parties shall ensure to women appropriate services in connection with pregnancy, confinement and the post-natal period, granting free services where necessary, as well as adequate nutrition during pregnancy and lactation.

Article 13

1. States Parties shall take all appropriate measures to eliminate discrimination against women in other areas of economic and social life in order to ensure, on a basis of equality of men and women, the same rights, in particular:

(a) The right to family benefits;

(b) The right to bank loans, mortgages and other forms of financial credit;

(c) The right to participate in recreational activities, sports and all aspects of cultural life.

Article 14

1. States Parties shall take into account the particular problems faced by rural women and the significant roles which rural women play in the economic survival of their families, including their work in the non-monetized sectors of the economy, and shall take all appropriate measures to ensure the application of the provisions of this Convention to women in rural areas.

2. States Parties shall take all appropriate measures to eliminate discrimination against women in rural areas in order to ensure, on a basis of equality of men and women, that they participate in and benefit from rural development and, in particular, shall ensure to such women the right:

(a) To participate in the elaboration and implementation of development planning at all levels;

(b) To have access to adequate health care facilities, including information, counseling and services in family planning;

(c) To benefit directly from social security programs;

(d) To obtain all types of training and education, formal and non-formal, including that relating to functional literacy, as well as, *inter alia,* the benefit of all community and extension services, in order to increase their technical proficiency;

(e) To organize self-help groups and co-operatives in order to obtain equal access to economic opportunities through employment or self-employment;

(f) To participate in all community activities;

(g) To have access to agricultural credit and loans, marketing facilities, appropriate technology and equal treatment in land and agrarian reform as well as in land resettlement schemes;

(h) To enjoy adequate living conditions, particularly in relation to housing, sanitation, electricity and water supply, transport and communications.

Part IV

Article 15

1. States Parties shall accord to women equality with men before the law.

2. States Parties shall accord to women, in civil matters,

a legal capacity identical to that of men and the same opportunities to exercise that capacity. In particular, they shall give women equal rights to conclude contracts and to administer property and shall treat them equally in all stages of procedure in courts and tribunals.

3. States Parties agree that all contracts and all other private instruments of any kind with a legal effect which is directed at restricting the legal capacity of women shall be deemed null and void.

4. States Parties shall accord to men and women the same rights with regard to the law relating to the movement of persons and the freedom to choose their residence and domicile.

Article 16

1. States Parties shall take all appropriate measures to eliminate discrimination against women in all matters relating to marriage and family relations and in particular shall ensure, on a basis of equality of men and women:

   (a) The same right to enter into marriage;
   (b) The same right freely to choose a spouse and to enter into marriage only with their free and full consent;
   (c) The same rights and responsibilities during marriage and at its dissolution;
   (d) The same rights and responsibilities as parents, irrespective of their marital status, in matters relating to their children; in all cases the interests of the children shall be paramount;
   (e) The same rights to decide freely and responsibly on the number and spacing of their children and to have access to the information, education and means to enable them to exercise these rights;
   (f) The same rights and responsibilities with regard to guardianship, wardship, trusteeship and adoption of children, or similar institutions where these concepts exist in national legislation; in all cases the interests of the children shall be paramount;
   (g) The same personal rights as husband and wife, including the right to choose a family name, a profession and an occupation;
   (h) The same rights for both spouses in respect of the ownership, acquisition, management, administration, enjoyment and disposition of property, whether free of charge or for a valuable consideration.

2. The betrothal and the marriage of a child shall have no legal effect, and all necessary action, including legislation, shall be taken to specify a minimum age for marriage and to make the registration of marriages in an official registry compulsory.

New York, United Nations General Assembly, 1979.

Appendix B

Beyond the Debt Crisis: Structural Transformation

1. Because women play a central role in economic and social life, we must look directly to women's experiences to reveal the real depth of the current global economic crisis. Women in their vast majority are concentrated in the most impoverished and oppressed sectors of our societies. Yet it is women who are being made to bear the brunt of what have been called "structural adjustment" strategies for managing international debt.

   It is within this context that women are struggling for viable long-term alternatives: transformative approaches with a gender perspective. The burden of adjustment programmes must be lifted from the shoulders of the poor and shifted to those who profited most from the economic policies that led to the current crisis.

2. Under the present system, aid from the World Bank and the International Monetary Fund (IMF) is conditional on the adoption of "adjustment" packages that are imposed regardless of the cultural and economic diversities of women around the world, as well as internal political conflicts. These rigid prescriptions have added to women's responsibilities for family survival and social reproduction, reduced the resources at our disposal, and restricted options for women's participation and leadership in public life.

3. The use of an unchecked and unregulated market model for economic restructuring has aggravated existing inequities in both "developed" and "developing" countries. Women have found our traditional usage of land undermined, our access to other resources circumscribed, and the return on our labor reduced. Under the pressure of prevailing market forces, poor women are compelled to submit our labor to exploitative relations in both the formal and informal sector of the economy.

   East and West, North and South, women are pitted in competition against one another for investment, jobs, and aid, although in fact we share many common experiences of oppression. Transnational corporations, with a power surpassing that of many national governments, are a central institution in promoting such competitions.

   Governments remain unaccountable to women. Instead, the logic of the "free market" has led many governments to turn to militarism and ideological oppression to control women's labor and our bodies. Our resistance to such oppression is inextricably linked to sexual politics, because government and market control has extended even to the realms of sexual relations and biological reproductions.

4. Women are demanding an equal voice in redefining development priorities for our societies. All too often, development agencies—many times working under the misleading label of "women in development"—make exploitative appeals of women's needs to solicit funds for the perpetuation of unjust and unequal economic and social arrangements.

5. We are concerned about how the sweeping changes in Eastern Europe are being interpreted. While they expose the false promises of authoritarian governments, they do not discredit effort to create cooperative alternatives to conventional, market-driven development strategies, nor do they justify the imposition of such strategies by economic intimidation or overt military intervention. Many Eastern Europeans see their responsibility in avoiding any rivalry for resources. Already women from Eastern Europe are seeking new forms of cooperation with Third World women's movements.

6. We as women recognize our differences as well as our commonalities, and the need to analyze our situation in terms of gender as it related to class, caste, nationality, race, ethnicity, and religion. This in no way condones emerging currents of national chauvinism and religion fundamentalism. Different values, needs, and perceptions of different groups of women must be

acknowledged and engaged within the process of struggling to make a better and more just world.

7. Women are by no means passive victims, who are unaware of own situation. Women have developed multiple, complex, and innovative responses to the global economic crisis, both individually and collectively. Researchers and activists have joined forces with grassroots women at the local, national, and international level to organize for survival and change.

Donors and governments are called upon to recognize and support efforts by women to act on our own behalf. Such actions express the true meaning of "empowerment."

8. We propose an alternative approach to give women equal participation in decision-making and control over resources: the alternative of economic democracy. Economic democracy means that all women must have access to all resources; that women must be active shapers in decisions at every level of society—within the family, in our communities, and in the political process, nationally and internationally. Poor women in particular must be included at the negotiating table in setting funding priorities and development policies.

To make this vision a full-fledged reality for ourselves, we as women will need to organize collectively—in our communities as well as nationally and internationally. Key to the success of such collective efforts is joining forces in coalition with other sectors.

9. Development and strengthening of a critical population economics education is one of the most important strategies for realizing the vision outlined in this declaration. In addition, women's organizations worldwide need to increase their capacity for international cooperation—in research efforts, policy proposals, and action campaigns. There is a pressing need for more direct interchanges among grassroots women from different nations and regions, and for more research into the effect of "structural adjustment" on women worldwide.

Likewise, more dialogue is needed among women at the grassroots, researchers and scholars, and policy makers. We see a crucial role for nongovernment organizations as facilitators of a two-way communications flow between grassroots women's organizations and national and international institutions.

Final declaration, produced by the participants in the all-women seminar on global economic issues, in preparation for the Women's Alternative Economic Summit.

New York, April 1990

Appendix C

A Women's Creed: The Declaration of the Women's Global Strategies Meeting

We are female human beings poised on the edge of the new millennium. We are the majority of our species, yet we have dwelt in the shadows. We are the invisible, the illiterate, the laborers, the refugees, the poor.

And we vow: No more.

We are the women who hunger—for rice, home, freedom, each other, ourselves.

We are the women who thirst—for clean water and laughter, literacy, love.

We have existed at all times, in every society. We have survived femicide. We have rebelled—and left clues.

We are continuity, weaving future from past, logic with lyric.

We are the women who stand in our sense, and shout Yes.

We are the women who wear broken bones, voices, minds, hearts—but we are the women who dare whisper No.

We are the women whose souls no fundamentalist cage can contain.

We are the women who refuse to permit the sowing of death in our gardens, air, rivers, seas.

We are precious, unique, necessary. We are strengthened and blessed and relieved at not having to be all the same. We are the daughters of longing. We are the mothers in labor to birth the politics of the 21st Century.

We are the women men warned us about.

We are the women who know that all issues are ours, who will reclaim our wisdom, reinvent our tomorrow, question and redefine everything, including power.

We have worked now for decades to name the details of our need, rage, hope, vision. We have broken our silence, exhausted our patience. We are weary of listing refrains on our suffering—to entertain or be simply ignored. We are done with vague words and real waiting; famishing for action, dignity, joy. We intend to do more than merely endure and survive.

They have tried to deny us, define us, denounce us; to jail, enslave, exile, gas, rape, beta, burn, bury—and bore us. Yet nothing, not even the offer to save their failed system, can grasp us.

For thousands of years, women have had responsibility without power—while men have had power without responsibility. We offer those men who risk being brothers a balance, a future, a hand. But with or without them, we will go on.

For we are the Old Ones, the New Breed, the Natives who came first but lasted, indigenous to an utterly different dimension. We are the girlchild in Zambia, the grandmother in Burma, the woman in El Salvador and Afghanistan, Finland, and Fiji. We are whale-song and rainforest; the depth-wave rising hinge to shatter glass power on the shore; the lost and despised, who weeping, stagger into the light.

All this we are. We are intensity, energy, the people speaking—who no longer will wait and who cannot be stopped.

We are poised on the edge of the millennium—ruin behind us, no map before us, the taste of fear sharp on our tongues.

Yet we will leap.

The exercise of imagining is an act of creation.

The act of creation is an exercise of will.

ALL THIS IS POLITICAL. AND POSSIBLE.

Bread. A clean sky. Active Peace. A woman's voice singing somewhere, melody drifting like smoke from the cookfires. The army disbanded, the harvest abundant. The wound healed, the child wanted, the prisoner freed, the body's integrity honored, the lover returned. The magical skill that reads marks into meaning. The labor equal, fair, and valued. Delight in the challenge for consensus to solve problems. No hand raised in any gesture but greeting. Secure interiors—of heart, home, land—so firm as to make secure borders irrelevant at last. And everywhere laughter, care, celebration, dancing, contentment. A humble, early paradise, in the now.

We will make it real, make it our own, make policy, history, peace, make it available, make mischief, a difference, love, the connections, the miracle, ready.

BELIEVE IT.
WE ARE THE WOMEN
WHO WILL TRANSFORM THE WORLD.

Written by Robin Morgan, in collaboration with Perdita Huston, Sunetra Puri, Mahnaz Afkhami, Diane Faulkner, Corrine Kumar, Sima Wali, and Paola Melchiori at the WEDO Global Strategies Meeting, New York, United States, 1994 .

Appendix D

Commitments for Social Development

Commitment 1

We commit ourselves to create an enabling economic, political, social, cultural and legal environment that will enable people to achieve social development.

Commitment 2

We commit ourselves to the goal of eradicating poverty in the world, through decisive national actions and international cooperation, as an ethical, social, political and economic imperative of humankind.

Commitment 3

We commit ourselves to promoting the goal of full employment as a basic priority of our economic and social policies, and to enabling all men and women to attain secure and sustainable livelihoods through freely chosen productive employment and work.

Commitment 4

We commit ourselves to promoting social integration by fostering societies that are stable, safe and just . . . based on the promotion and protection of all human rights, and on non-discrimination, tolerance, respect for diversity, equality of opportunity, solidarity, security and participation of all people including disadvantaged and vulnerable groups and persons.

Commitment 5

We commit ourselves to promoting full respect for human dignity and to achieving equality and equity between women and men, and to recognizing and enhancing the participation and leadership roles of women in political, civil, economic, social and cultural life and in development.

Commitment 6

We commit ourselves to promote the goals of universal and equitable access to quality education, the highest attainable standard of physical and mental health, and the access of all to primary health care.

Commitment 7

We commit ourselves to accelerating the economic, social, and human resource development of Africa and the least developed countries.

Commitment 8

We commit ourselves to ensuring that when structural adjustment programmes are agreed to they should include social development goals, in particular, of eradicating poverty, promoting full and productive employment and enhancing social integration.

Commitment 9

We commit ourselves to increase significantly and/or utilize more efficiently the resources allocated to social development in order to achieve the goals of the Summit through national action, and regional and international cooperation.

Commitment 10

We commit ourselves to an improved and strengthened framework for international, regional and subregional cooperation for social development, in a spirit of partnership, through the United Nations and other multilateral institutions.

Approved by representatives from 185 U.N. member states at the World Summit for Social Development in Copenhagen, Denmark, 1995.

Appendix E

The Beijing Declaration

1. We, the Governments, participating in the Fourth World Conference on Women,

2. Gathered here in Beijing, in September 1995, the year of the fiftieth anniversary of the founding of the United Nations,

3. Determined to advance the goals of equality, development and peace for all women everywhere in the interest of all humanity,

4. Acknowledging the voices of all women everywhere and taking note of the diversity of women and their roles and circumstances, honoring the women who paved the way and inspired by the hope present in the world's youth,

5. Recognize that the status of women has advanced in some important respects in the past decade but that progress has been uneven, inequalities between women and men have persisted and major obstacles remain, with serious consequences for the well-being of all people,

6. Also recognize that this situation is exacerbated by the increasing poverty that is affecting the lives of the majority of the world's people, in particular women and children, with origins in both the national and international domains,

7. Dedicate ourselves unreversably to addressing these constraints and obstacles and thus enhancing further the advancement and empowerment of women all over the world, and agree that this requires urgent action in the spirit of determination, hope, cooperation and solidarity, now and to carry us forward into the next century.

We reaffirm our commitment to:

8. The equal rights and inherent human dignity of women and men and other purposes and principles enshrined in the Charter of the United Nations, to the Universal Declaration of Human Rights and other international human rights instruments, in particular the Convention on the Elimination of All Forms of Discrimination Against Women and the Convention on the Rights of the Child, as well as the Declaration on the Elimination of Violence against Women and the Declaration on the Right to Development;

9. Ensure the full implementation of the human rights of women and of the girl child as an inalienable, integral and indivisible part of all human rights and fundamental freedoms;

10. Build on consensus and progress made at previous United Nations conferences and summits—on women in Nairobi in 1985, on children in New York in 1990, on environment and development in Rio de Janeiro in 1992, on human rights in Vienna in 1993, on population and development in Cairo in 1994 and on social development in Copenhagen in 1995 with the objectives of achieving equality, development and peace;

11. Achieve the full and effective implementation of the Nairobi Forward-looking Strategies for the Advancement of Women;

12. The empowerment and advancement of women, including the right to freedom of thought, conscience, religion and belief, thus contributing to the moral, ethical, spiritual and intellectual needs of women and men, individually or in community with others and thereby guaranteeing them the possibility of realizing their full potential in society and shaping their lives in accordance with their own aspirations.

We are convinced that:

13. Women's empowerment and their full participation on the basis of equality in all spheres of society, including participation in the decision-making process and access to power, are fundamental for the achievement of equality, development and peace;

14. Women's rights are human rights;

15. Equal rights opportunities and access to resources, equal sharing of responsibilities for the family by men and women, and a harmonious partnership between them are critical to their well-being and that of their families as well as to the consolidation of democracy;

16. Eradication of poverty based on sustained economic growth, social development, environmental protection and social justice requires the involvement of women in economic and social development, the equal opportunities and the full and equal participation of women and men as agents and beneficiaries of people-centered sustainable development;

17. The explicit recognition and reaffirmation of the right of all women to control all aspects of their health, in particular their own fertility, is basic to their empowerment;

18. Local, national, regional and global peace is attainable and is inextricably linked with the advancement of women, who are a fundamental force for leadership, conflict resolution and the promotion of lasting peace at all levels;

19. It is essential to design, implement and monitor, with the full participation of women, effective, efficient and mutually reinforcing gender-sensitive policies and programs, including development policies and programs, at all levels that will foster the empowerment and advancement of women;

20. The participation and contribution of all actors of civil society, particularly women's groups and networks and other nongovernmental organization and community based organization, with full respect for their autonomy, in cooperation with Governments, are important to the effective implementation and follow-up of the Platform for Action;

21. The implementation of the Platform for Action requires commitment from Governments and the international community. By making national and international commitments for action, including those made at the Conference, Governments and the international community recognize the need to take priority action for the empowerment and advancement of women.

We are determined to:

22. Intensify efforts and actions to achieve the goals of the Nairobi Forward-looking Strategies for the Advancement of Women by the end of this century;

23. Ensure the full enjoyment by women and the girl child of all human rights and fundamental freedoms, and take effective action against violations of these rights and freedoms;

24. Take all necessary measures to eliminate all forms of discrimination against women and the girl child and remove all obstacles to gender equality and the advancement and empowerment of women;

25. Encourage men to participate fully in all actions towards equality;

26. Promote women's economic independence, including employment, and eradicate the persistent and increasing burden of poverty on women by addressing the structural causes of poverty through changes in economic structures, ensuring equal access for all women, including those in rural areas, as vital development agents, to productive resources, opportunities and public services;

27. Promote people-centered sustainable development, including sustained economic growth through the provision of basic education, life-long education, literacy and training, and primary health care for girls and women;

28. Take positive steps to ensure peace for the advancement of women, and recognizing the leading role that women have played in the peace movement, work actively towards general and complete disarmament under strict and effective international control, and support negotiations on the conclusion, without delay, of a universal and multilaterally and effectively verifiable comprehensive nuclear-test-ban treaty which contributes to nuclear disarmament and the prevention of the proliferation of nuclear weapons in all its aspects;

29. Prevent and eliminate all forms of violence against women and girls;

30. Ensure equal access to and equal treatment of women and men in education and health care and enhance women's sexual and reproductive health as well as education;

31. Promote and protect all human rights of women and girls;

32. Intensify efforts to ensure equal enjoyment of all human rights and fundamental freedoms for all women and girls who face multiple barriers to their empowerment and advancement because of such factors as their race, age, language, ethnicity, culture, religion, or disability, or because they are indigenous people;

33. Ensure respect for international law, including humanitarian law, in order to protect women and girls in particular;

34. Develop the fullest potential of girls and women of all ages, ensure their full and equal participation in building a better world for all and enhance their role in the development process.

We are determined to:

35. Ensure women's equal access to economic resources, including land, credit, science and technology, vocational training, information, communication and markets, as a means to further the advancement and empowerment of women and girls, including through the enhancement of their capacities to enjoy the benefits of equal access to these resources, *inter alia,* by means of international cooperation;

36. Ensure the success of the Platform for Action which will require a strong commitment on the part of Governments, international organizations and institutions at all levels. We are deeply convinced that economic development, social development and environmental protection are interdependent and mutually reinforcing components of sustainable development, which is the framework for our efforts to achieve a higher quality of life for all people. Equitable social development that recognizes empowering the poor, particularly women living in poverty, to utilize environmental resources sustainably is a necessary foundation for sustainable development. We also recognize that broad-based and

sustained economic growth in the context of sustainable development is necessary to sustain social development and social justice. The success of the Platform for Action will also require adequate mobilization of resources at the national and international levels as well as new and additional resources to the developing countries from all available funding mechanisms, including multilateral, bilateral and private sources for the advancement of women; financial resources to strengthen the capacity of national, subregional, regional and international institutions; a commitment to equal rights, equal responsibilities and equal opportunities and to the equal participation of women and men in all national, regional and international bodies and policymaking processes; the establishment of strengthening of mechanisms at all levels for accountability to the world's women;

37. Ensure also the success of the Platform for Action in countries with economies in transition, which will require continued international cooperation and assistance;

38. We hereby adopt and commit ourselves as Governments to implement the following Platform for Action, ensuring that a gender perspective is reflected in all out policies and programs. We urge the United Nations system, regional and international financial institutions, other relevant regional and international institutions and all women and men, as well as nongovernmental organizations, with full respect for their autonomy, and all sectors of civil society, in cooperation with Governments, to fully commit themselves and contribute to the implementation of this Platform for Action.

Beijing, China, 1995

Index